SOLUTIONS MANUAL

Jan William Simek
California Polytechnic State University

ORGANIC CHEMISTRY

SIXTH EDITION

L. G. Wade, Jr.

PEARSON
Prentice
Hall

Upper Saddle River, NJ 07458

Assistant Editor: Carole Snyder
Project Manager: Kristen Kaiser
Executive Editor: Nicole Folchetti
Executive Managing Editor: Kathleen Schiaparelli
Assistant Managing Editor: Becca Richter
Production Editor: Kathryn O'Neill
Supplement Cover Manager: Paul Gourhan
Supplement Cover Designer: Joanne Alexandris
Manufacturing Buyer: Ilene Kahn
Cover Image Credit: Joseph Galluccio (2004)

© 2006 Pearson Education, Inc.
Pearson Prentice Hall
Pearson Education, Inc.
Upper Saddle River, NJ 07458

The author and publisher of this book have used their best efforts in preparing this book. These efforts include the development, research, and testing of the theories and programs to determine their effectiveness. The author and publisher make no warranty of any kind, expressed or implied, with regard to these programs or the documentation contained in this book. The author and publisher shall not be liable in any event for incidental or consequential damages in connection with, or arising out of, the furnishing, performance, or use of these programs.

Printed in the United States of America

10   9   8   7   6

ISBN   0-13-147882-6

Pearson Education Ltd., *London*
Pearson Education Australia Pty. Ltd., *Sydney*
Pearson Education Singapore, Pte. Ltd.
Pearson Education North Asia Ltd., *Hong Kong*
Pearson Education Canada, Inc., *Toronto*
Pearson Educación de Mexico, S.A. de C.V.
Pearson Education—Japan, *Tokyo*
Pearson Education Malaysia, Pte. Ltd.

TABLE OF CONTENTS

PREFACE

Hints for Passing Organic Chemistry

Do you want to pass your course in organic chemistry? Here is my best advice, based on over thirty years of observing students learning organic chemistry:

Hint #1: *Do the problems.* It seems straightforward, but humans, including students, try to take the easy way out until they discover there is no short-cut. Unless you have a measured IQ above 200 and comfortably cruise in the top 1% of your class, *do the problems.* Usually your teacher (professor or teaching assistant) will recommend certain ones; try to do all those recommended. If you do half of them, you will be half-prepared at test time. (Do you want your surgeon coming to your appendectomy having practiced only *half* the procedure?) And when you do the problems, keep this Solutions Manual CLOSED. Avoid looking at *my* answer before you write *your* answer—your trying and struggling with the problem is the most valuable part of the problem. Discovery is a major part of learning. Remember that the primary goal of doing these problems is *not* just getting the right answer, but understanding the material well enough to get right answers to the questions you haven't seen yet.

Hint #2: *Keep up.* Getting behind in your work in a course that moves as quickly as this one is the Kiss of Death. For most students, organic chemistry is the most rigorous intellectual challenge they have faced so far in their studies. Some are taken by surprise at the diligence it requires. Don't think that you can study all of the material the couple of days before the exam—well, you can, but you won't pass. Study organic chemistry like a foreign language: try to do some every day so that the freshly-trained neurons stay sharp.

Hint #3: *Get help when you need it.* Use your teacher's office hours when you have difficulty. Many schools have tutoring centers (in which organic chemistry is a popular offering). Here's a secret: absolutely the best way to cement this material in your brain is to get together with a few of your fellow students and make up problems for each other, then correct and discuss them. When *you* write the problems, you will gain great insight into what this is all about.

Purpose of this Solutions Manual

So what is the point of this Solutions Manual? First, I can't do your studying for you. Second, since I am not leaning over your shoulder as you write your answers, I can't give you direct feedback on what you write and think—the print medium is limited in its usefulness. What I *can* do for you is: 1) provide correct answers; the publishers, Professor Wade, Professor Kantorowski (my reviewer), and I have gone to great lengths to assure that what I have written is correct, for we all understand how it can shake a student's confidence to discover that the answer book flubbed up; 2) provide a considerable degree of rigor; beyond the fundamental requirement of correctness, I have tried to flesh out these answers, being complete but succinct; 3) provide insight into how to solve a problem and into where the sticky intellectual points are. Insight is the toughest to accomplish, but over the years, I have come to understand where students have trouble, so I have tried to anticipate your questions and to add enough detail so that the concept, as well as the answer, is clear.

It is difficult for students to understand or acknowledge that their teachers are human (some are more human than others). Since I am human (despite what my students might report), I can and do make mistakes. If there are mistakes in this book, they are my sole responsibility, and I am sorry. If you find one, PLEASE let me know so that it can be corrected in future printings. Nip it in the bud.

What's New in this edition?

Better answers! Part of my goal in this edition has been to add more explanatory material to clarify how to arrive at the answer. The possibility of more than one answer to a problem has been noted. The IUPAC Nomenclature appendix has been expanded to include bicyclics, heteroatom replacements, and the Cahn-Ingold-Prelog system of stereochemical designation.

Better graphics! The print medium is very limited in its ability to convey three-dimensional structural information, a problem that has plagued organic chemists for over a century. I have added some graphics created in the software, Chem3D®, to try to show atoms in space where that information is a key part of the solution. In drawing NMR spectra, representational line drawings have replaced rudimentary attempts at drawing peaks from previous editions.

Better jokes? Too much to hope for.

Some Web Stuff

Prentice-Hall maintains a web site dedicated to the Wade text: try **www.prenhall.com/wade**. Two essential web sites providing spectra are listed on the bottom of p. 270.

Acknowledgments

No project of this scope is ever done alone. These are team efforts, and there are several people who have assisted and facilitated in one fashion or another who deserve my thanks.

Professor L. G. Wade, Jr., your textbook author, is a remarkable person. He has gone to extraordinary lengths to make the textbook as clear, organized, informative and insightful as possible. He has solicited and followed my suggestions on his text, and his comments on my solutions have been perceptive and valuable. We agreed early on that our primary goal is to help the students learn a fascinating and challenging subject, and all of our efforts have been directed toward that goal. I have appreciated our collaboration.

My new colleague, Dr. Eric Kantorowski, has reviewed the entire manuscript for accuracy and style. His diligence, attention to detail and chemical wisdom have made this a better manual. Eric stands on the shoulders of previous reviewers who scoured earlier editions for errors: Jessica Gilman, Dr. Kristen Meisenheimer, and Dr. Dan Mattern. Mr. Richard King has offered numerous suggestions on how to clarify murky explanations. I am grateful to them all.

The people at Prentice-Hall have made this project possible. Good books would not exist without their dedication, professionalism, and experience. Among the many people who contributed are: Lee Englander, who connected me with this project; Nicole Folchetti, Advanced Chemistry Editor; and Kristen Kaiser and Carole Snyder, Project Managers.

The entire manuscript was produced using *ChemDraw®*, the remarkable software for drawing chemical structures developed by CambridgeSoft Corp., Cambridge, MA. We, the users of sophisticated software like ChemDraw, are the beneficiaries of the intelligence and creativity of the people in the computer industry. We are fortunate that they are so smart.

Finally, I appreciate my friends who supported me throughout this project, most notably my good friend of almost forty years, Judy Lang. The students are too numerous to list, but it is for them that all this happens.

Jan William Simek
Department of Chemistry and Biochemistry
Cal Poly State University
San Luis Obispo, CA 93407
Email: jsimek@calpoly.edu

DEDICATION

To my inspirational chemistry teachers:

Joe Plaskas, who made the batter;

Kurt Kaufman, who baked the cake;

Carl Djerassi, who put on the icing;

and to my parents:

Ervin J. and Imilda B. Simek,

who had the original concept.

SYMBOLS AND ABBREVIATIONS

Below is a list of symbols and abbreviations used in this Solutions Manual, consistent with those used in the textbook by Wade. (Do not expect all of these to make sense to you now. You will learn them throughout your study of organic chemistry.)

BONDS

—————— a single bond

══════ a double bond

≡≡≡≡≡ a triple bond

━━━▬ a bond in three dimensions, coming out of the paper toward the reader

∥∥∥∥∥∥ a bond in three dimensions, going behind the paper away from the reader

- - - - - - a stretched bond, in the process of forming or breaking

ARROWS

———→ in a reaction, shows direction from reactants to products

⇌ signifies equilibrium (not to be confused with resonance)

◄——► signifies resonance (not to be confused with equilibrium)

⌒⌒ shows direction of electron movement:
the arrowhead with one barb shows movement of one electron;
the arrowhead with two barbs shows movement of a pair of electrons

┼——► shows polarity of a bond or molecule, the arrowhead signifying the more negative end of the dipole

SUBSTITUENT GROUPS

Me	a methyl group, CH_3
Et	an ethyl group, CH_2CH_3
Pr	a propyl group, a three carbon group (two possible arrangements)
Bu	a butyl group, a four carbon group (four possible arrangements)
R	the general abbreviation for an alkyl group (or any substituent group not under scrutiny)
Ph	a phenyl group, the name of a benzene ring as a substituent, represented:

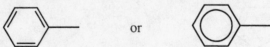

or

Ar   the general name for an aromatic group

continued on next page

SUBSTITUENT GROUPS, continued

Ac   an acetyl group: $CH_3-\overset{\displaystyle O}{\overset{\|}{C}}-$

Cy   a cyclohexyl group:

Ts   tosyl, or *p*-toluenesulfonyl group: CH_3-

Boc   a *t*-butoxycarbonyl group (amino acid and peptide chemistry): $(CH_3)_3C-O-\overset{\displaystyle O}{\overset{\|}{C}}-$

Z, or   a carbobenzoxy (benzyloxycarbonyl) group (amino acid and peptide chemistry):
Cbz

$-CH_2-O-\overset{\displaystyle O}{\overset{\|}{C}}-$

REAGENTS AND SOLVENTS

DCC   **di**cyclohexyl**c**arbodiimide $-N=C=N-$

DMSO   **d**i**m**ethylsul**fo**xide

$H_3C-\underset{\displaystyle CH_3}{\overset{\displaystyle O}{\overset{\|}{S}}}$

ether   diethyl ether, $CH_3CH_2OCH_2CH_3$

MCPBA   *meta*-**c**hloro**p**eroxy**b**enzoic **a**cid $\overset{\displaystyle O}{\overset{\|}{C}}-O-OH$

MVK   **m**ethyl **v**inyl **k**etone

$H_3C-\overset{\displaystyle O}{\overset{\|}{C}}-$

NBS   *N*-**b**romo**s**uccinimide

continued on next page

REAGENTS AND SOLVENTS, continued

PCC   pyridinium chlorochromate, $CrO_3 \cdot HCl \cdot N$

Sia_2BH disiamylborane

$$\begin{array}{ccccccccc} & CH_3 & H & & H & & H & CH_3 & \\ & | & | & & | & & | & | & \\ H- & C- & C- & B- & C- & & C- & H \\ & | & | & & | & & | & \\ & CH_3 & CH_3 & & CH_3 & CH_3 & \end{array}$$

THF   tetrahydrofuran

SPECTROSCOPY

IR	infrared spectroscopy
NMR	nuclear magnetic resonance spectroscopy
MS	mass spectrometry
UV	ultraviolet spectroscopy
ppm	parts per million, a unit used in NMR
Hz	hertz, cycles per second, a unit of frequency
MHz	megahertz, millions of cycles per second
TMS	tetramethylsilane, $(CH_3)_4Si$, the reference compound in NMR
s, d, t, q	singlet, doublet, triplet, quartet, referring to the number of peaks an NMR absorption gives
nm	nanometers, 10^{-9} meters (usually used as a unit of wavelength)
m/z	mass-to-charge ratio, in mass spectrometry
δ	in NMR, chemical shift value, measured in ppm
λ	wavelength
ν	frequency

OTHER

a, ax	axial (in chair forms of cyclohexane)
e, eq	equatorial (in chair forms of cyclohexane)
HOMO	highest occupied molecular orbital
LUMO	lowest unoccupied molecular orbital
NR	no reaction
o, m, p	*ortho, meta, para* (positions on an aromatic ring)
Δ	when written over an arrow: "heat"; when written before a letter: "change in"
δ^+, δ^-	partial positive charge, partial negative charge
$h\nu$	energy from electromagnetic radiation (light)
$[\alpha]_D$	specific rotation at the D line of sodium (589 nm)

CHAPTER 1—INTRODUCTION AND REVIEW

1-1 Na $1s^22s^22p^63s^1$ P $1s^22s^22p^63s^23p_x^13p_y^13p_z^1$

Mg $1s^22s^22p^63s^2$ S $1s^22s^22p^63s^23p_x^23p_y^13p_z^1$

Al $1s^22s^22p^63s^23p_x^1$ Cl $1s^22s^22p^63s^23p_x^23p_y^23p_z^1$

Si $1s^22s^22p^63s^23p_x^13p_y^1$ Ar $1s^22s^22p^63s^23p_x^23p_y^23p_z^2$

1-2 In this book, lines between atom symbols represent covalent bonds between those atoms. Nonbonding electrons are indicated with dots.

(a) H—N̈—H (with H below)
(b) H—Ö—H
(c) H—Ö—H (with H below, + charge)
(d) H—C—C—C—H (three H on top, three H on bottom)

(e) H—C—C—N̈—H (H—C—C—N—H with H substituents)
(f) H—C—Ö—C—H
(g) H—C—C—F̈:

(h) H—C—C—C—H (with O above middle C, H substituents)

(i) H—B—H (with H below)
(j) :F̈—B—F̈: (with :F̈: below)

The compounds in (i) and (j) are unusual in that boron does not have an octet of electrons—normal for boron because it has only three valence electrons.

1-3

(a) :N≡N:
(b) H—C≡N:
(c) H—Ö—N̈=Ö
(d) Ö=C=Ö

(e) H—C̈=N̈—H (with H below)
(f) H—C—Ö—H (with O above C)
(g) H—C=C—C̈l: (with H, H below)
(h) H—N̈=N̈—H

(i) H—C=C—C—H (with H substituents)
(j) H—C=C=C—H (with H substituents)
(k) H—C≡C—C—H (with H substituents)

1-4

(a) :N≡N:
(b) H—C≡N:
(c) H—O—N=O
(d) O=C=O

(e) H—C=N—H (with H below)
(f) H—C—O—H (with O above)
(g) H—C=C—Cl (with H, H below)
(h) H—N=N—H

There are no unshared electron pairs in parts (i), (j), and (k).

1

1-5 The symbols "δ⁺" and "δ⁻" indicate bond polarity by showing partial charge. (In the arrow symbolism, the arrow should point to the partial negative charge.)

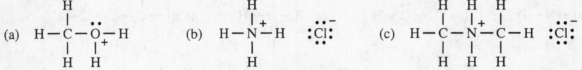

(a) $\overset{\delta^+}{C}-\overset{\delta^-}{Cl}$ (b) $\overset{\delta^+}{C}-\overset{\delta^-}{O}$ (c) $\overset{\delta^+}{C}-\overset{\delta^-}{N}$ (d) $\overset{\delta^+}{C}-\overset{\delta^-}{S}$ (e) $\overset{\delta^-}{C}-\overset{\delta^+}{B}$

(f) $\overset{\delta^+}{N}-\overset{\delta^-}{Cl}$ (g) $\overset{\delta^+}{N}-\overset{\delta^-}{O}$ (h) $\overset{\delta^-}{N}-\overset{\delta^+}{S}$ (i) $\overset{\delta^-}{N}-\overset{\delta^+}{B}$ (j) $\overset{\delta^+}{B}-\overset{\delta^-}{Cl}$

1-6 Non-zero formal charges are shown beside the atoms.

(a) [H–C–O–H structure with +O, H, H, H]   (b) [H–N⁺–H structure with Cl⁻]   (c) [H–C–N⁺–C–H structure with Cl⁻]

In (b) and (c), the chlorine is present as chloride ion. There is no covalent bond between chlorine and other atoms in the formula.

(d) Na⁺ [⁻O–C–H structure]   (e) [H–C⁺–H structure]   (f) [H–C⁻–H structure]   (g) Na⁺ [H–B⁻–H structure]

(h) Na⁺ [H–B⁻–C≡N: structure]   (i) [H–C–O⁺–C–H / F–B⁻–F / F structure]   (j) [H–O–N⁺–H structure]

(k) K⁺ [⁻O–C–C–H structure with CH₃ groups]   (l) [H–C=O⁺–H structure]

As shown in (d), (g), (h), and (k), alkali metals like sodium and potassium form only ionic bonds, never covalent bonds.

1-7 Resonance forms in which all atoms have full octets are the most significant contributors. In resonance forms, ALL ATOMS KEEP THEIR POSITIONS—ONLY ELECTRONS ARE SHOWN IN DIFFERENT POSITIONS.

(a) [carbonate resonance structures] ⁻:O–C(=O):O⁻ ⟷ ⁻:O–C=O with O⁻ ⟷ O=C–O⁻

(b) [nitrate resonance structures] ⁻:O–N⁺(=O):O⁻ ⟷ ⁻:O–N⁺=O with O⁻ ⟷ O=N⁺–O⁻

1-7 continued

(c) :Ö—N=Ö⁻ ⟷ Ö=N—Ö:⁻

(d)

$$H-C=C-\overset{+}{C}-H \quad \longleftrightarrow \quad H-\overset{+}{C}-C=C-H$$

with H below each carbon

(e)

$$H-C=C-\overset{\cdot\cdot}{\underset{\cdot\cdot}{C}}{}^{-}-H \quad \longleftrightarrow \quad H-\overset{\cdot\cdot}{\underset{\cdot\cdot}{C}}{}^{-}-C=C-H$$

with H below each carbon

(f) Sulfur can have up to 12 electrons around it because it has d orbitals accessible.

:Ö—S—Ö:⁻ ⟷ :Ö—S=Ö ⟷ Ö=S—Ö:⁻ ⟷ Ö=S—Ö:⁻ (with O above and below S)

:Ö—S=Ö ⟷ Ö=S=Ö (with O above and below S)

(g)

[resonance structures of a carbocation with three forms: carbocation, then C=O⁺ double bond forms]

1-8 Major resonance contributors would have the lowest energy. The most important factors are: maximize full octets; maximize bonds; put negative charge on electronegative atoms; minimize charge separation.

(a)

$$H-\overset{-}{\underset{\underset{H}{|}}{C}}-\overset{+}{N}=\overset{\cdot\cdot}{O} \quad \longleftrightarrow \quad H-\overset{-}{\underset{\underset{:O:}{|}}{C}}-\overset{+}{N}-\overset{\cdot\cdot}{O}: {}^{-} \quad \longleftrightarrow \quad H-C=\overset{+}{N}-\overset{\cdot\cdot}{O}:{}^{-}$$

(negative charge on electronegative atoms)

minor   minor   major

3

1-8 continued

(b)

$$\underset{H}{\overset{H}{\diagdown}}C=C-\overset{+}{N}=\overset{..}{\underset{..}{O}} \longleftrightarrow \underset{H}{\overset{H}{\diagdown}}C=C-\overset{+}{N}-\overset{..}{\underset{..}{O}}\overset{-}{:} \longleftrightarrow \underset{H}{\overset{H}{\diagdown}}\overset{+}{C}-C=\overset{+}{N}-\overset{..}{\underset{..}{O}}\overset{-}{:}$$

major   major   minor

These two forms have equivalent energy and are major because they have full octets, more bonds, and less charge separation than the minor contributor.

(c)

$$H-C=\overset{..}{\underset{+}{O}}-H \longleftrightarrow H-\overset{..}{\underset{H}{C}}-\overset{+}{\underset{..}{O}}-H$$

major   minor
(octets, more bonds)

(d)

$$H-C=\overset{+}{N}=\overset{..}{\underset{..}{N}}: \longleftrightarrow \overset{-}{H}-\overset{..}{\underset{H}{C}}-\overset{+}{N}\equiv N:$$

major (negative charge   minor
on electronegative atom)

(e)

$$H-\overset{-}{\overset{..}{\underset{H}{C}}}-C\equiv N: \longleftrightarrow H-C=C=\overset{-}{\underset{..}{N}}:$$

minor   major (negative charge
on electronegative atom)

(f)

$$H-\overset{..}{N}-\overset{+}{C}-C=C-\overset{..}{N}-H \longleftrightarrow H-\overset{+}{N}=C-C=C-\overset{..}{N}-H$$

minor   major

these two forms are major contributors
because all atoms have full octets

$$H-\overset{..}{N}-C=C-\overset{+}{C}-\overset{..}{N}-H \longleftrightarrow H-\overset{..}{N}-C=C-C=\overset{+}{N}-H$$

minor   major

(g)

$$H-\overset{\overset{\textstyle :O:}{\|}}{C}-\overset{..}{\underset{H}{C}}-\overset{\overset{\textstyle :O:}{\|}}{C}-H \longleftrightarrow H-\overset{\overset{\textstyle :\overset{-}{O}:}{|}}{C}=\overset{..}{\underset{H}{C}}-\overset{\overset{\textstyle :O:}{\|}}{C}-H \longleftrightarrow H-\overset{\overset{\textstyle :O:}{\|}}{C}-\overset{..}{\underset{H}{C}}=\overset{\overset{\textstyle :\overset{-}{O}:}{|}}{C}-H$$

minor   major   major

these two have equivalent energy and are major because the
negative charge is on the more electronegative oxygen atom

4

1-8 continued

(h)

```
      :O:                        :O:⁻
      ‖         ..               |      +
  H—C—N—H    ⟷     H—C=N—H
      |    ..                     |
major H                   minor  H
(no charge separation)
```

1-9 Your Lewis structures may *appear* different from these. As long as the atoms are connected in the same order and by the same type of bond, they are equivalent structures. For now, the exact placement of the atoms on the page is not significant. A Lewis structure is "complete" with unshared electron pairs shown.

(a)
```
    H   H   H   H   H   H
    |   |   |   |   |   |
H — C — C — C — C — C — C — H
    |   |   |   |   |   |
    H   H   H   H   C   H
                  H/|\H
                    H
```

(b)
```
    H   H   H
    |   |   |    ..
H — C — C — C — Cl:
    |   |   |    ..
    H   C   H
      H/|\H
        H
```

(c)
```
    H   H   :O:        H
    |   |   ‖         /
H — C — C — C — C = C
    |   |        |    \
    H   H        H     H
```

(d)
```
    H   H   :O:
    |   |   ‖
H — C — C — C — H
    |   |
    H   H
```

Always be alert for the implied double or triple bond. Remember that C has to have four bonds, nitrogen has three bonds, oxygen has two bonds, and hydrogen has one bond. The only exceptions to these valence rules are structures with formal charges.

(e)
```
    H   :O:
    |   ‖
H — C — C — C ≡ N:
    |
    H
```

(f)
```
          H
        H | H
         \|/
          C   :O:
      H   |   ‖
      |   |   ‖
H — C — C — C — O — H
      |   |        ..
      H   |
          C
        H/|\H
          H
```

(g)
```
    H   H   :O:  H   H
    |   |   ‖    |   |
H — C — C — C — C — C — H
    |   |        |   |
    H   H        H   H
```

1-10 Complete Lewis structures show all atoms, bonds, and unshared electron pairs.

(a)
```
              H
        H   H | H
        |   \ |/
   H     \    C — H
    \     C
H — C        C — H
    |     |
H — C     C — H
    |   .. |
    H   N   H
        |
        H
```

(b)
```
         H       H
     H   |       | H
      \  |      /
   H — C       C — H
           \  C /
            \ |
        H    H
        |    |
   H — C —— C
    |        C — H
    H        |
   \         H
H — C      C — H
     \    /
      C  ..
    H/| O |\H
      H   H
```

(c)
```
    H   H
    |   |
    C — C
    ‖    ‖
    C .. C
   / \ N / \
  H   |   H
      H
```

5

1-10 continued

(d)

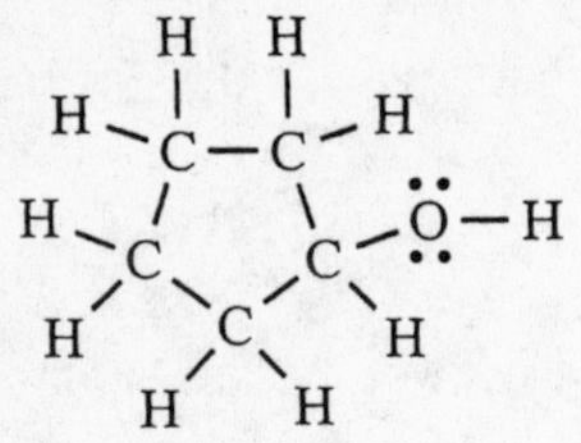

```
        H   H
        |   |
   H — C — C — H
        |   |
  H — C     C — Ö — H
       |   |    ··
       H   C   H
           |
         H   H
```

(e)

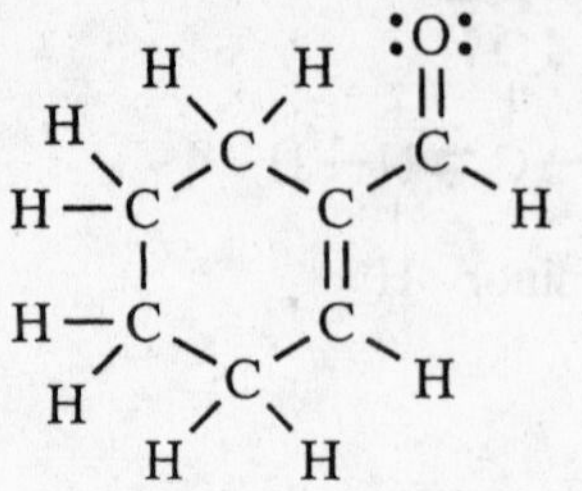

```
              :O:
               ||
     H    H    C
      \   |   / \
   H — C   C     C — H
      / \ / \ /
   H — C     C
      / \   / \
     H   C     H
         |
       H   H
```

(f)

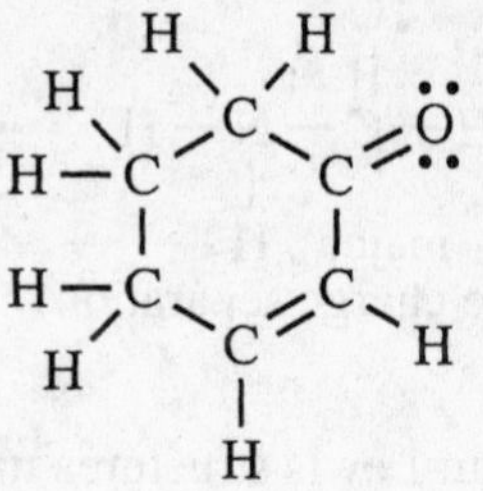

```
            H   H
            |   |
       H    C   C === Ö
        \  / \ /      ··
   H — C     C
      / \   / \
     H   C     C — H
         |     ||
         H     C
               |
               H
```

(g)

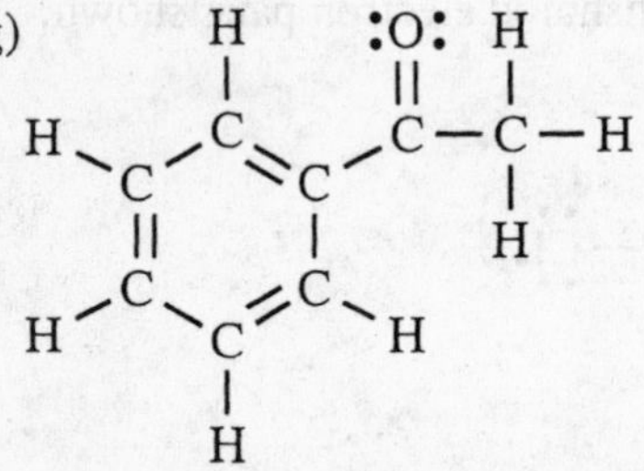

```
         H   :O:  H
         |    ||   |
    H — C    C — C — H
       /  \ /     |
      C    C      H
      ||   |
      C    C — H
     / \  /
    H   C
        |
        H
```

(h)

```
   H  :O:  H   H   H
   |   ||   |   |   |
   H — C — C — C — C — C — H
   |        |   |   |
   H        H   H   H
```

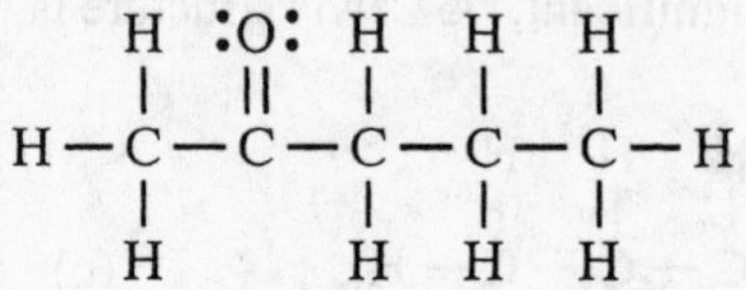

1-11 There is often more than one correct way to write condensed structural formulas. You must often make inferences about what a condensed formula means according to valence rules, especially in structures with C=O as shown in parts (a) and (d).

(a) $CH_3COCH_2CH_2CH_3$
(the O has a double bond to the carbon preceding it)

(b) $(CH_3)_2CHCH_2CH_2OH$

or $CH_3CH(CH_3)CH_2CH_2OH$

(c) $(CH_3)_2CHCH_2CH(OH)CH_3$

or $CH_3CH(CH_3)CH_2CH(OH)CH_3$

(d) $CH_3CH_2CH(CH_3)CH_2CHO$
(the O has a double bond to the carbon preceding it)

1-12 If the percent values do not sum to 100%, the remainder must be oxygen. Assume 100 g of sample; percents then translate directly to grams of each element.

There are usually many possible structures for a molecular formula. Yours may be different from the examples shown here.

(a) $\dfrac{40.0 \text{ g C}}{12.0 \text{ g/mole}} = 3.33$ moles C ÷ 3.33 moles = 1 C

$\dfrac{6.67 \text{ g H}}{1.01 \text{ g/mole}} = 6.60$ moles H ÷ 3.33 moles = 1.98 ≈ 2 H

$\dfrac{53.33 \text{ g O}}{16.0 \text{ g/mole}} = 3.33$ moles O ÷ 3.33 moles = 1 O

empirical formula = $\boxed{CH_2O}$ ⟹ empirical weight = 30

molecular weight = 90, three times the empirical weight ⟹

three times the empirical formula = molecular formula = $\boxed{C_3H_6O_3}$

some possible structures:

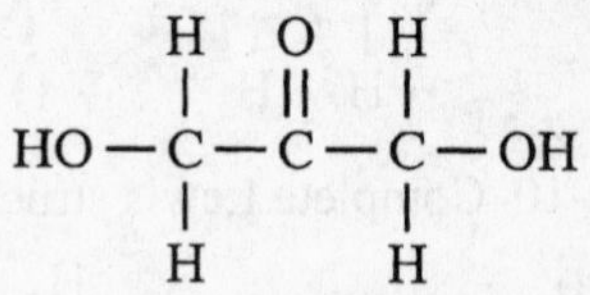

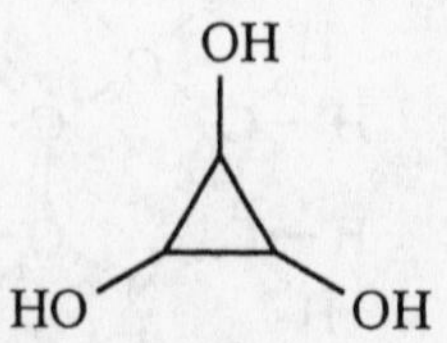

```
      H   O   H
      |   ||  |
 HO — C — C — C — OH
      |       |
      H       H
```

```
         OH
         |
        / \
      /     \
    HO       OH
```

MANY other structures possible.

6

1-12 continued

(b) $\dfrac{32.0 \text{ g C}}{12.0 \text{ g/mole}}$ = 2.67 moles C ÷ 1.34 moles = 1.99 ≈ 2 C

$\dfrac{6.67 \text{ g H}}{1.01 \text{ g/mole}}$ = 6.60 moles H ÷ 1.34 moles = 4.93 ≈ 5 H

$\dfrac{18.7 \text{ g N}}{14.0 \text{ g/mole}}$ = 1.34 moles N ÷ 1.34 moles = 1 N

$\dfrac{42.6 \text{ g O}}{16.0 \text{ g/mole}}$ = 2.66 moles O ÷ 1.34 moles = 1.99 ≈ 2 O

empirical formula = $\boxed{C_2H_5NO_2}$ ⟹ empirical weight = 75

molecular weight = 75, same as the empirical weight ⟹

empirical formula = molecular formula = $\boxed{C_2H_5NO_2}$

some possible structures:

MANY other
structures possible.

(c) $\dfrac{37.2 \text{ g C}}{12.0 \text{ g/mole}}$ = 3.10 moles C ÷ 1.55 moles = 2 C

$\dfrac{7.75 \text{ g H}}{1.01 \text{ g/mole}}$ = 7.67 moles H ÷ 1.55 moles = 4.95 ≈ 5 H

$\dfrac{55.0 \text{ g Cl}}{35.45 \text{ g/mole}}$ = 1.55 moles Cl ÷ 1.55 moles = 1 Cl

empirical formula = $\boxed{C_2H_5Cl}$ ⟹ empirical weight = 64.46

molecular weight = 64, same as the empirical weight ⟹

empirical formula = molecular formula = $\boxed{C_2H_5Cl}$

There is only one structure
possible with this molecular
formula:

(d) $\dfrac{38.4 \text{ g C}}{12.0 \text{ g/mole}}$ = 3.20 moles C ÷ 1.60 moles = 2 C

$\dfrac{4.80 \text{ g H}}{1.01 \text{ g/mole}}$ = 4.75 moles H ÷ 1.60 moles = 2.97 ≈ 3 H

$\dfrac{56.8 \text{ g Cl}}{35.45 \text{ g/mole}}$ = 1.60 moles Cl ÷ 1.60 moles = 1 Cl

empirical formula = $\boxed{C_2H_3Cl}$ ⟹ empirical weight = 62.45

molecular weight = 125, twice the empirical weight ⟹

twice the empirical formula = molecular formula = $\boxed{C_4H_6Cl_2}$

some possible structures:

MANY other
structures possible.

7

1-13

(a) 5.00 g HBr x $\dfrac{1 \text{ mole HBr}}{80.9 \text{ g HBr}}$ = 0.0618 moles HBr

0.0618 moles HBr $\Longrightarrow$ 0.0618 moles H_3O^+ (100% dissociated)

$\dfrac{0.0618 \text{ moles } H_3O^+}{100 \text{ mL}}$ x $\dfrac{1000 \text{ mL}}{1 \text{ L}}$ = $\dfrac{0.618 \text{ moles } H_3O^+}{1 \text{ L solution}}$

pH = $-\log_{10}[\,H_3O^+\,]$ = $-\log_{10}(0.618)$ = $\boxed{0.209}$

(b) 1.50 g NaOH x $\dfrac{1 \text{ mole NaOH}}{40.0 \text{ g NaOH}}$ = 0.0375 moles NaOH

0.0375 moles NaOH $\Longrightarrow$ 0.0375 moles ^-OH (100% dissociated)

$\dfrac{0.0375 \text{ moles } ^-OH}{50. \text{ mL}}$ x $\dfrac{1000 \text{ mL}}{1 \text{ L}}$ = $\dfrac{0.75 \text{ moles } ^-OH}{1 \text{ L solution}}$ = 0.75 M

$[\,H_3O^+\,] = \dfrac{1 \times 10^{-14}}{[\,^-OH\,]} = \dfrac{1 \times 10^{-14}}{0.75} = 1.33 \times 10^{-14}$

pH = $-\log_{10}[\,H_3O^+\,]$ = $-\log_{10}(1.33 \times 10^{-14})$ = $\boxed{13.88}$ (the number of decimal places in a pH value is the number of significant figures)

1-14

(a) By definition, an acid is any species that can donate a proton. Ammonia has a proton bonded to nitrogen, so ammonia can be an acid (although a very weak one). A base is a proton acceptor, that is, it must have a pair of electrons to share with a proton; in theory, any atom with an unshared electron pair can be a base. The nitrogen in ammonia has an unshared electron pair so ammonia is basic. In water, ammonia is too weak an acid to give up its proton; instead, it acts as a base and pulls a proton from water to a small extent.

(b) water as an acid: $H_2O + NH_3 \rightleftharpoons {}^-OH + NH_4^+$

water as a base: $H_2O + HCl \rightleftharpoons H_3O^+ + Cl^-$

(c) methanol as an acid: $CH_3OH + NH_3 \rightleftharpoons CH_3O^- + NH_4^+$

methanol as a base: $CH_3OH + H_2SO_4 \rightleftharpoons CH_3OH_2^+ + HSO_4^-$

1-15

(a) HCOOH   + $^-$CN   ⇌   HCOO$^-$ +   HCN   **FAVORS**
   stronger   stronger   weaker   weaker   **PRODUCTS**
   acid   base   base   acid
   pK_a 3.76 pK_a 9.22

(b) CH$_3$COO$^-$ +   CH$_3$OH ⇌ CH$_3$COOH   +   CH$_3$O$^-$ **FAVORS**
   weaker   weaker   stronger   stronger   **REACTANTS**
   base   acid   acid   base
               pK_a 15.5 pK_a 4.74

(c) CH$_3$OH   +   NaNH$_2$ ⇌   CH$_3$O$^-$ Na$^+$ +   NH$_3$ **FAVORS**
   stronger   stronger   weaker   weaker   **PRODUCTS**
   acid   base   base   acid
   pK_a 15.5 pK_a 33

(d) Na$^+$ $^-$OCH$_3$ +   HCN   ⇌   HOCH$_3$ +   NaCN   **FAVORS**
   stronger   stronger   weaker   weaker   **PRODUCTS**
   base   acid   acid   base
               pK_a 9.22 pK_a 15.5

(e) HCl   +   H$_2$O   ⇌   H$_3$O$^+$ +   Cl$^-$ **FAVORS**
   stronger   stronger   weaker   weaker   **PRODUCTS**
   acid   base   acid   base

The first reaction in Table 1-5 shows the K_{eq} for this reaction is 160, favoring products.

(f) H$_3$O$^+$ +   CH$_3$O$^-$ ⇌   H$_2$O   +   CH$_3$OH   **FAVORS**
   stronger   stronger   weaker   weaker   **PRODUCTS**
   acid   base   base   acid
   pK_a –1.7 pK_a 15.5

The seventh reaction in Table 1-5 shows the K_{eq} for the *reverse* of this reaction is 3.2 x 10^{-16}. Therefore, K_{eq} for this reaction as written must be the inverse, or 3.1 x 10^{15}, strongly favoring products.

1-16

Protonation of the double-bonded oxygen gives three resonance forms (as shown in Solved Problem 1-5(c)); protonation of the single-bonded oxygen gives only one. In general, the more resonance forms a species has, the more stable it is, so the proton would bond to the oxygen that gives a more stable species, that is, the double-bonded oxygen.

1-17 In Solved Problem 1-4, the structures of ethanol and methylamine are shown to be similar to methanol and ammonia, respectively. We must infer that their acid-base properties are also similar.

(a) This problem can be viewed in two ways. 1) Quantitatively, the pK_a values determine the order of acidity. 2) Qualitatively, the stabilities of the conjugate bases determine the order of acidity (see Solved Problem 1-4 for structures): the conjugate base of acetic acid, acetate ion, is resonance-stabilized, so acetic acid is the most acidic; the conjugate base of ethanol has a negative charge on a very electronegative oxygen atom; the conjugate base of methylamine has a negative charge on a mildly electronegative nitrogen atom and is therefore the least stabilized, so methylamine is the least acidic.

$$\text{acetic acid} \quad > \quad \text{ethanol} \quad > \quad \text{methylamine}$$

$$pK_a \ 4.74 \qquad pK_a \approx 15.5 \qquad pK_a \approx 33$$

strongest acid   weakest acid

(b) Ethoxide ion is the conjugate base of ethanol, so it must be a stronger base than ethanol; Solved Problem 1-4 indicates ethoxide is analogous to hydroxide in base strength. Methylamine has pK_b 3.36. The basicity of methylamine is between the basicity of ethoxide ion and ethanol.

$$\text{ethoxide ion} \quad > \quad \text{methylamine} \quad > \quad \text{ethanol}$$

strongest base   weakest base

1-18 Curved arrows show electron movement, as described in text section 1-14.

(a) $CH_3CH_2 - \overset{..}{\underset{..}{O}} - H \ + \ CH_3 - \overset{-}{\underset{..}{N}} - H \ \rightleftharpoons \ CH_3CH_2 - \overset{..}{\underset{..}{O}}{:}^- \ + \ CH_3 - \overset{H}{\underset{H}{N}} - H$

  stronger acid   stronger base   conjugate base   conjugate acid
                                                       weaker base   weaker acid

equilibrium favors PRODUCTS

(b) $CH_3CH_2 - \overset{\overset{:O:}{\|}}{C} - \overset{..}{\underset{..}{O}} - H \ + \ CH_3 - \overset{}{\underset{H}{N}} - CH_3 \ \rightleftharpoons \ CH_3 - \overset{H}{\underset{H}{\overset{+}{N}}} - CH_3 \ + \ \left\{ CH_3CH_2 - \overset{\overset{:O:}{\|}}{C} - \overset{..}{\underset{..}{O}}{:}^- \longleftrightarrow CH_3CH_2 - \overset{\overset{:O:^-}{|}}{C} = \overset{..}{\underset{..}{O}} \right\}$

  stronger acid   stronger base   conjugate acid   conjugate base
                                                      weaker acid   weaker base

equilibrium favors PRODUCTS

(c) $CH_3 - \overset{..}{\underset{..}{O}} - H \ + \ H - \overset{..}{\underset{..}{O}} - \overset{\overset{:O:}{\|}}{\underset{\underset{:O:}{\|}}{S}} - \overset{..}{\underset{..}{O}} - H \ \rightleftharpoons \ CH_3 - \overset{H}{\underset{}{\overset{+}{O}}} - H \ + $

  stronger base   stronger acid   conjugate acid
                                                                 weaker acid

equilibrium favors PRODUCTS $\left\{ {}^-\overset{..}{\underset{..}{O}} - \overset{\overset{:O:}{\|}}{\underset{\underset{:O:}{\|}}{S}} - \overset{..}{\underset{..}{O}} - H \longleftrightarrow \overset{..}{\underset{..}{O}} = \overset{\overset{:O:}{\|}}{\underset{\underset{:O:_-}{|}}{S}} - \overset{..}{\underset{..}{O}} - H \longleftrightarrow \overset{..}{\underset{..}{O}} = \overset{\overset{:O:^-}{|}}{\underset{\underset{:O:}{\|}}{S}} - \overset{..}{\underset{..}{O}} - H \right\}$

conjugate base, weaker base

1-18 continued

(d)

Na^+ $:\overset{..}{\underset{..}{O}}-H$ + $H-\overset{..}{\underset{..}{S}}-H$ $\rightleftharpoons$ $H-\overset{..}{\underset{..}{O}}-H$ + Na^+ $:\overset{..}{\underset{..}{S}}-H$

stronger base   stronger acid   conjugate acid   conjugate base
                                       weaker acid   weaker base

equilibrium favors PRODUCTS

(e)

$CH_3-\overset{\overset{\displaystyle H}{|}}{\underset{\underset{\displaystyle H}{|}}{\overset{+}{N}}}-H$ + $CH_3-\overset{..}{\underset{..}{O}}^-$ $\rightleftharpoons$ $CH_3-\overset{..}{\underset{\underset{\displaystyle H}{|}}{N}}-H$ + $CH_3-\overset{..}{\underset{..}{O}}-H$

                            stronger base   conjugate acid
                                                         weaker acid

stronger acid   conjugate base
equilibrium favors PRODUCTS   weaker base

(f)

$CH_3-\overset{\overset{\displaystyle :O:}{\|}}{C}-\overset{..}{\underset{..}{O}}-H$ + $CH_3-\overset{..}{\underset{..}{O}}^-$ $\rightleftharpoons$ $CH_3-\overset{..}{\underset{..}{O}}-H$ + $\left\{ CH_3-\overset{\overset{\displaystyle :O:}{\|}}{C}-\overset{..}{\underset{..}{O}}:^- \longleftrightarrow CH_3-C\overset{\overset{..}{}}{=}\overset{..}{\underset{..}{O}}:^- \right\}$

stronger acid   stronger base   conjugate acid   conjugate base
                                      weaker acid   weaker base

equilibrium favors PRODUCTS

(g)

$CH_3-\overset{\overset{\displaystyle :O:}{\|}}{C}-\overset{..}{\underset{..}{O}}-H$ + $\left\{ {}^-:\overset{..}{\underset{..}{O}}-\overset{\overset{\displaystyle :O:}{|}}{\underset{\underset{\displaystyle :O:}{\|}}{S}}-CH_3 \longleftrightarrow O\overset{\overset{\displaystyle :O:}{\|}}{=}\overset{}{\underset{\underset{\displaystyle :O:}{|}}{S}}-CH_3 \longleftrightarrow O\overset{\overset{\displaystyle :O:}{|}}{=}\overset{}{\underset{\underset{\displaystyle :O:}{\|}}{S}}-CH_3 \right\}$

weaker acid   weaker base

equilibrium favors REACTANTS

$\rightleftharpoons$ $\left\{ CH_3-\overset{\overset{\displaystyle :O:}{\|}}{C}-\overset{..}{\underset{..}{O}}:^- \longleftrightarrow CH_3-C\overset{\overset{\displaystyle :\overset{..}{O}:^-}{|}}{=}\overset{..}{\underset{..}{O}} \right\}$ + $H-\overset{..}{\underset{..}{O}}-\overset{\overset{\displaystyle :O:}{\|}}{\underset{\underset{\displaystyle :O:}{\|}}{S}}-CH_3$

                                 conjugate base   conjugate acid
                                 stronger base   stronger acid

1-19 Solutions for (a) and (b) are presented in the Solved Problem in the text. Here, the newly formed bonds are shown in bold.

(c)

$H-\overset{\underset{\displaystyle H}{|}}{B}-H$ + $CH_3-\overset{..}{\underset{..}{O}}-CH_3$ $\rightleftharpoons$ $H-\overset{\overset{\displaystyle H}{|}}{\underset{\underset{\displaystyle \mathbf{|}}{\mathbf{|}}}{B}}-H$

acid   H   base $CH_3-\overset{+}{\underset{..}{O}}-CH_3$

(d)

$CH_3-\overset{\overset{\displaystyle :O:}{\|}}{C}-H$ + ${}^-:\overset{..}{O}-H$ $\rightleftharpoons$ $CH_3-\overset{\overset{\displaystyle :\overset{..}{O}:^-}{|}}{\underset{\underset{\displaystyle :O-H}{|}}{C}}-H$

      acid   base

11

1-19 continued

(e) Bronsted-Lowry—proton transfer

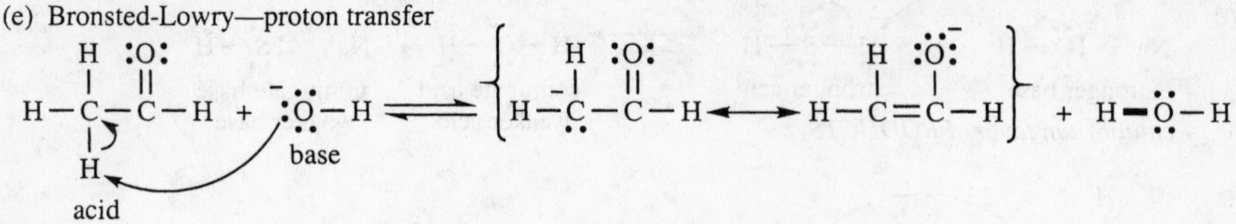

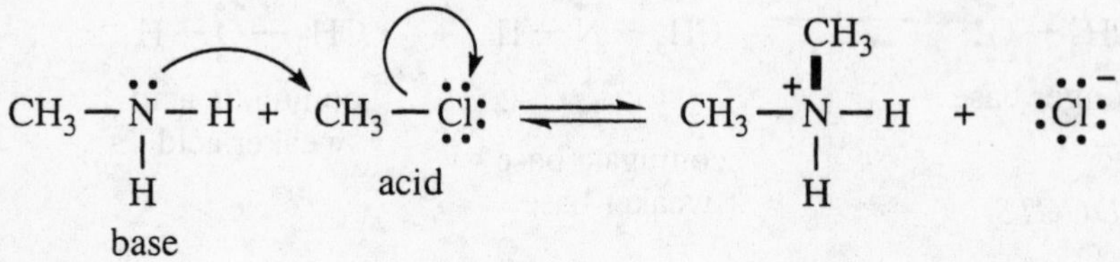

(f)

1-20

Learning organic chemistry is similar to learning a foreign language: new vocabulary, new grammar (the reactions), some new concepts, and even a new alphabet (the symbolism of chemistry). This type of definition question is intended to help you review the vocabulary and concepts in each chapter. All of the definitions and examples are presented in the Glossary and in the chapter, so this Solutions Manual will not repeat them. Use these questions to evaluate your comprehension and to guide your review of the important concepts in the chapter.

1-21 (a) CARBON!   (b) oxygen   (c) phosphorus   (d) chlorine

1-22

valence e⁻ →	1	2	3	4	5	6	7	8
	H							He (2e⁻)
	Li	Be	B	C	N	O	F	Ne
					P	S	Cl	
							Br	
							I	

1-23

(a) ionic only   (b) covalent (H—O⁻) and ionic (Na⁺ ⁻OH)

(c) covalent (H—C and C—Li), but the C—Li bond is strongly polarized

(d) covalent only   (e) covalent (H—C and C—O⁻) and ionic (Na⁺ ⁻OCH₃)

(f) covalent (H—C and C=O and C—O⁻) and ionic (HCO₂⁻ Na⁺)   (g) covalent only

(a) [Lewis structures of PCl₃ and PCl₅]

(b) [Lewis structures of NCl₃ and NCl₅]

CANNOT EXIST

NCl_5 violates the octet rule; nitrogen can have no more than eight electrons (or four atoms) around it. Phosphorus, a third-row element, can have more than eight electrons because phosphorus can use d orbitals in bonding, so PCl_5 is a stable, isolable compound.

1-25 Your Lewis structures may look different from these. As long as the atoms are connected in the same order and by the same type of bond, they are equivalent structures. For now, the exact placement of the atoms on the page is not significant.

(a) [Lewis structure H-N-N-H with lone pairs and H substituents]

(b) [Lewis structure H-N=N-H with lone pairs]

(c) [Lewis structure of tetramethylammonium with Cl⁻]

(d) [Lewis structure H-C-C≡N with H substituents]

(e) [Lewis structure H-C-C-H with =O]

(f) [Lewis structure H-C-S-C-H with =O]

(g) [Lewis structure H-O-S-O-H with two =O]

(h) [Lewis structure H-C-N=C=O with H substituents]

(i) [Lewis structure H-C-O-S-O-C-H with =O groups]

(j) [Lewis structure H-C-C-C-H with =N-H]

(k) [Lewis structure with C-C-N=O and methyl groups]

1-26

(a) [Lewis structure of long chain carboxylic acid H-C-C=C-C-C=C-C-O-H]

(b) [Lewis structure :N≡C-C-C-C-C-H with =O groups]

13

1-26 continued

(c)
```
        H—O:   H  :O:
         |     |   ||
  H—C=C—C—C—C—Ö—H
    |  | | |  ..
    H  H H H
```

(d)
```
              H   H
               \ | /
                C
              H | H
               \|/
                C
      H   H     |    :O:
       \ | \    |     ||
  H—C—C—C—C—C—H
       |   |   |
       H   C   H   C
          /|\     /|\
         H H H   H H H   H
                  |
                  C
                 /|\
                H | H
                  H
```

1-27 In each set below, the second structure is a more correct line formula. Since chemists are human (surprise!), they will take shortcuts where possible; the first structure in each pair uses a common abbreviation, either COOH or CHO. Make sure you understand that COOH does not stand for C—O—O—H. Likewise for CHO.

(a) [line structure] COOH

OR [line structure with OH and =O]

(b) N≡C—[structure with O]—CHO

OR N≡C—[structure with two =O and terminal H]

(c) [line structure] COOH with OH

OR [line structure] OH with OH and =O

(d) [line structure] CHO   OR   [line structure with H and =O]

1-28

(a)
```
     H  H  H  H
     |  |  |  |
  H—C—C—C—C—H    and
     |  |  |  |
     H  H  H  H
```
```
     H  H  H
     |  |  |
  H—C—C—C—H
     |  |  |
     H  C  H
       /|\
      H | H
        H
```
these are the only two possibilities, but your structures may appear different—making models will help you visualize these structures

(b)
```
     H  H  H
     |  |  |
  H—C—C—N:    and
     |  |  |
     H  H  H
```
```
     H  H  H
     |  |  |
  H—C—N—C—H
     |  ..  |
     H      H
```
these are the only two possibilities, but your structures may appear different—making models will help you visualize these structures

1-28 continued

(c) There are several other possibilities as well. Your answer may be correct even if it does not appear here. Check with others in your study group.

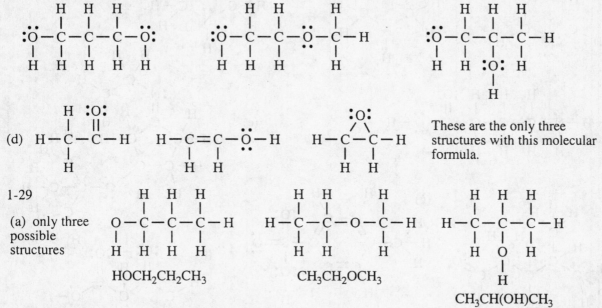

(d) These are the only three structures with this molecular formula.

1-29

(a) only three possible structures

HOCH₂CH₂CH₃   CH₃CH₂OCH₃   CH₃CH(OH)CH₃

(b) There are several other possibilities as well.

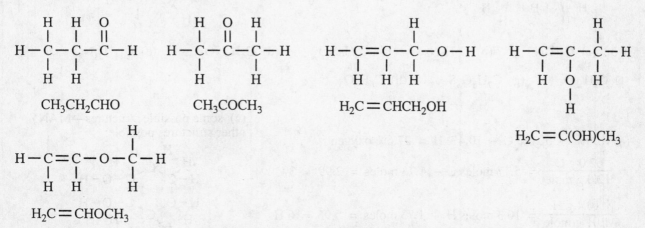

CH₃CH₂CHO   CH₃COCH₃   H₂C═CHCH₂OH   H₂C═C(OH)CH₃

H₂C═CHOCH₃

1-30 General rule: *molecular formulas of stable hydrocarbons must have an even number of hydrogens.* The formula CH₂ does not have enough atoms to bond with the four orbitals of carbon.

one carbon:

H—C—H   CH₄

two carbons: H—C≡C—H   C₂H₂   H—C═C—H   C₂H₄   H—C—C—H   C₂H₆

three carbons: H—C≡C—C—H   C₃H₄   H—C═C—C—H   C₃H₆   H—C—C—C—H   C₃H₈

1-31

(a)

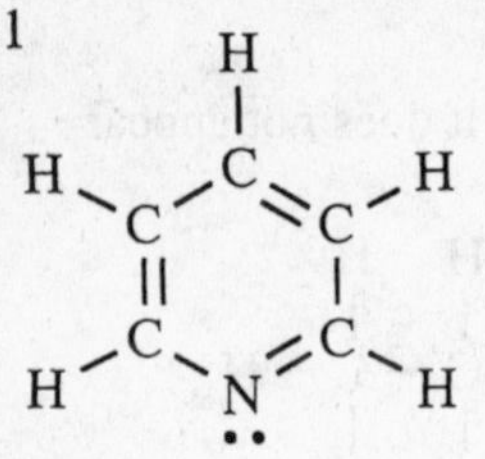

(b)

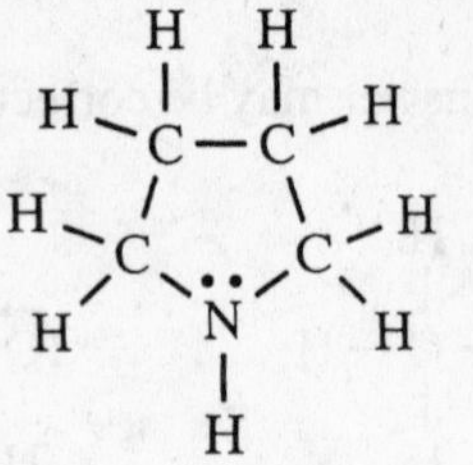

(c)

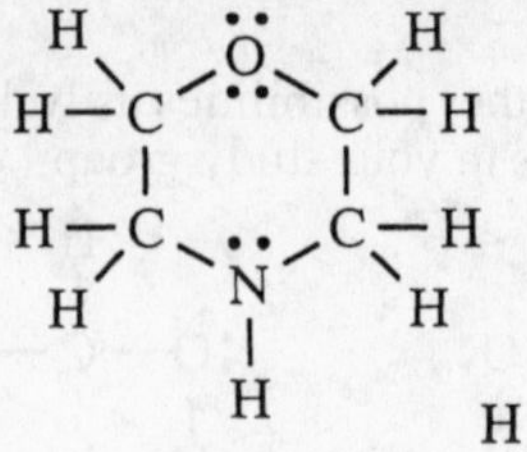

(d)

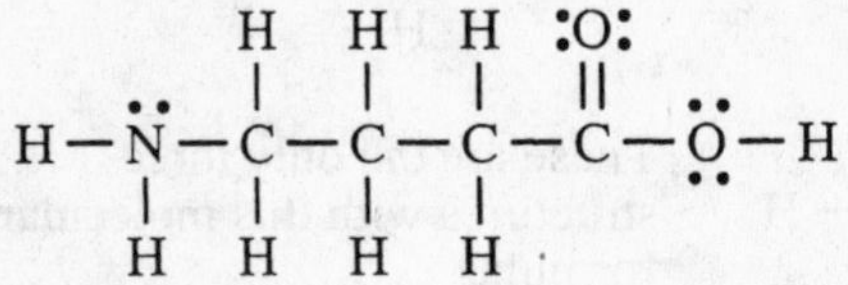

(e)

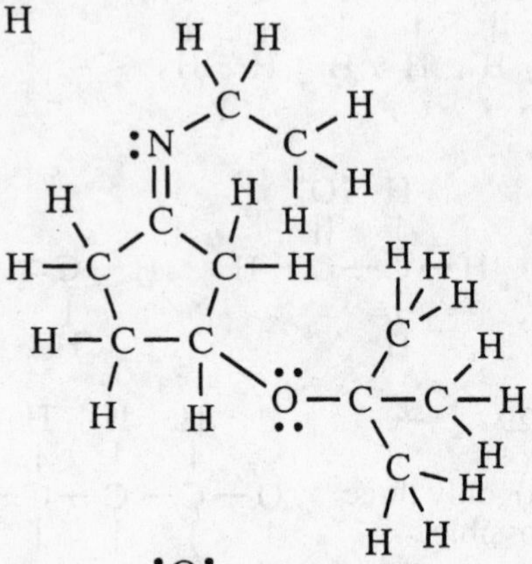

(f)

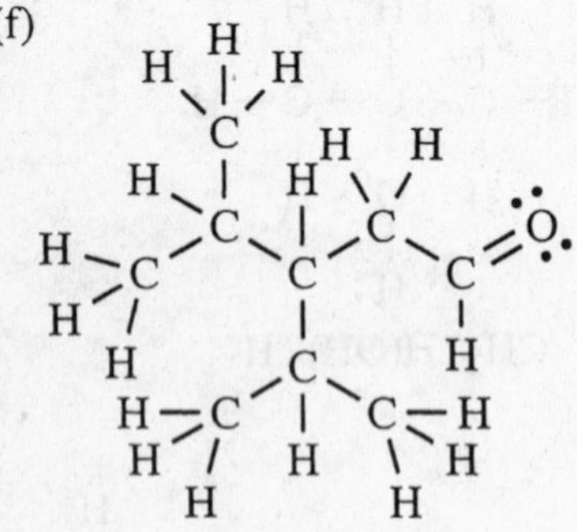

(g)

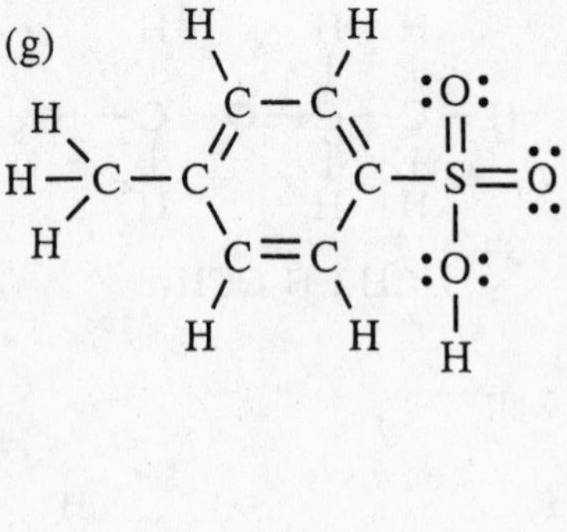

(h)

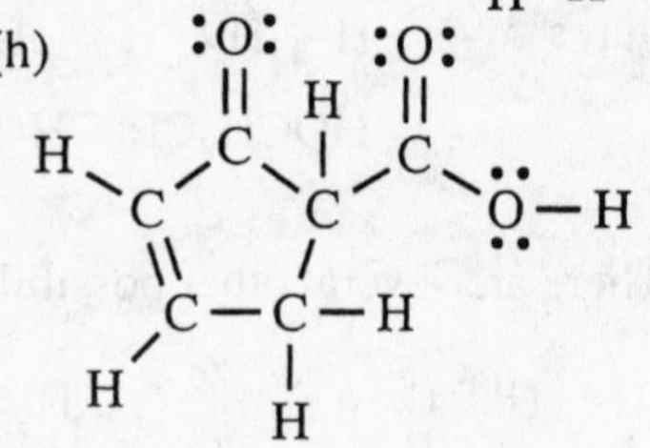

1-32 (a) C_5H_5N (b) C_4H_9N (c) C_4H_9NO (d) $C_4H_9NO_2$ (e) $C_{11}H_{21}NO$

(f) $C_9H_{18}O$ (g) $C_7H_8O_3S$ (h) $C_6H_6O_3$

1-33

(a) 100% − 62.0% C − 10.4% H = 27.6% oxygen

$$\frac{62.0 \text{ g C}}{12.0 \text{ g/mole}} = 5.17 \text{ moles C} \div 1.73 \text{ moles} = 2.99 \approx 3 \text{ C}$$

$$\frac{10.4 \text{ g H}}{1.01 \text{ g/mole}} = 10.3 \text{ moles H} \div 1.73 \text{ moles} = 5.95 \approx 6 \text{ H}$$

$$\frac{27.6 \text{ g O}}{16.0 \text{ g/mole}} = 1.73 \text{ moles O} \div 1.73 \text{ moles} = 1 \text{ O}$$

(b) empirical formula = $\boxed{C_3H_6O}$ $\Rightarrow$ empirical weight = 58

molecular weight = 117, about double the empirical weight

$\Rightarrow$ double the empirical formula = molecular formula =

$$\boxed{C_6H_{12}O_2}$$

(c) some possible structures—MANY other structures possible:

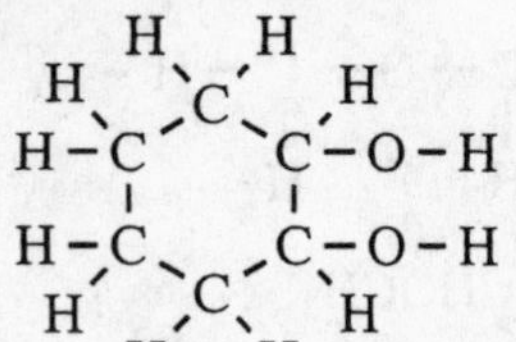

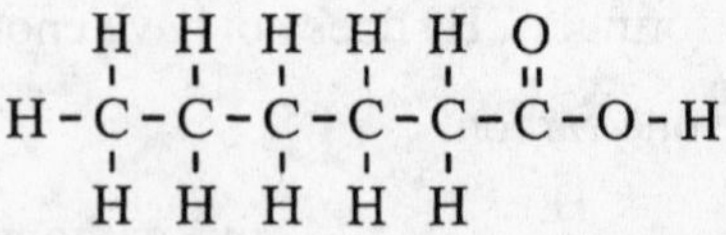

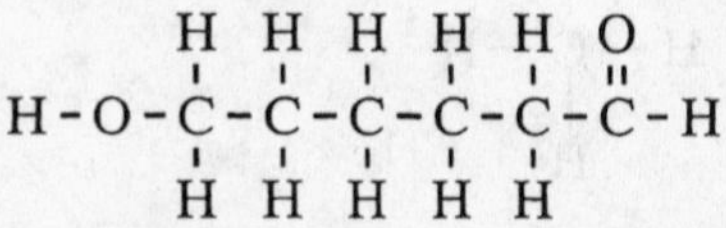

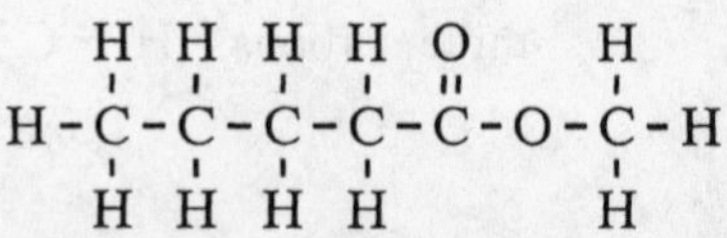

16

1-34 Non-zero formal charges are shown by the atoms.

(a)
$$H-\overset{\underset{|}{H}}{C}=\overset{+}{N}=\overset{-}{\underset{\cdot\cdot}{N}}\colon \quad\longleftrightarrow\quad H-\overset{\underset{|}{H}}{\overset{\cdot\cdot}{C}}{}^{-}-\overset{+}{N}\equiv N\colon$$

(b)
$$H_3C-\overset{+}{\underset{\overset{|}{CH_3}}{\underset{|}{N}}}\!\!\!\overset{CH_3}{\overset{|}{}}-\overset{-}{\underset{\cdot\cdot}{O}}\colon$$

(c)
$$H-\overset{\underset{|}{H}}{C}\!\equiv\!\overset{\underset{|}{H}}{C}-\overset{+}{\underset{\underset{|}{H}}{C}}-H$$

(d)
$$H-\overset{\underset{|}{H}}{C}-\overset{+}{\underset{\underset{|}{\underset{\cdot\cdot}{O}}{}^{-}}{N}}=\overset{\cdot\cdot}{O}$$

(e)
$$H-\overset{\underset{|}{H}}{C}-\overset{+}{\underset{\underset{|}{CH_3}}{\overset{\cdot\cdot}{O}}}-\overset{\underset{|}{H}}{C}-H$$

1-35 The symbols "δ⁺" and "δ⁻" indicate bond polarity by showing partial charge. Electronegativity differences greater than or equal to 0.5 are considered large.

(a) $\overset{\delta^+\ \ \delta^-}{C-Cl}$ large

(b) $\overset{\delta^-\ \ \delta^+}{C-H}$ small

(c) $\overset{\delta^-\ \ \delta^+}{C-Li}$ large

(d) $\overset{\delta^+\ \ \delta^-}{C-N}$ small

(e) $\overset{\delta^+\ \ \delta^-}{C-O}$ large

(f) $\overset{\delta^-\ \ \delta^+}{C-B}$ large

(g) $\overset{\delta^-\ \ \delta^+}{C-Mg}$ large

(h) $\overset{\delta^-\ \ \delta^+}{N-H}$ large

(i) $\overset{\delta^-\ \ \delta^+}{O-H}$ large

(j) $\overset{\delta^+\ \ \delta^-}{C-Br}$ small

1-36 Resonance forms must have atoms in identical positions. If any atom moves position, it is a different structure.

(a) different compounds—a hydrogen atom has changed position
(b) resonance forms—only the position of electrons is different
(c) resonance forms—only the position of electrons is different
(d) resonance forms—only the position of electrons is different
(e) different compounds—a hydrogen atom has changed position
(f) resonance forms—only the position of electrons is different
(g) resonance forms—only the position of electrons is different
(h) different compounds—a hydrogen atom has changed position
(i) resonance forms—only the position of electrons is different
(j) resonance forms—only the position of electrons is different

1-37

(a)
$$H-\overset{\underset{|}{H}}{C}-\overset{\overset{\textstyle :O:}{\|}}{C}-\overset{-}{\underset{\underset{|}{H}}{\overset{\cdot\cdot}{C}}}-H \quad\longleftrightarrow\quad H-\overset{\underset{|}{H}}{C}-\overset{\overset{\textstyle :\!\overset{\cdot\cdot}{O}\!:^{-}}{|}}{C}=\overset{\underset{|}{H}}{C}-H$$

(b)
$$H-\overset{\overset{\textstyle :O:}{\|}}{C}-\overset{\underset{|}{H}}{\overset{\cdot\cdot}{C}}{}^{-}\!=\overset{\underset{|}{H}}{C}-\overset{\underset{|}{H}}{C}-H \quad\longleftrightarrow\quad H-\overset{\overset{\textstyle :O:}{\|}}{C}-\overset{\underset{|}{H}}{\overset{\cdot\cdot}{C}}{}^{-}-\overset{\underset{|}{H}}{C}=\overset{\underset{|}{H}}{C}-H \quad\longleftrightarrow\quad H-\overset{\overset{\textstyle :\!\overset{\cdot\cdot}{O}\!:^{-}}{|}}{C}=\overset{\underset{|}{H}}{C}-\overset{\underset{|}{H}}{C}=\overset{\underset{|}{H}}{C}-H$$

17

1-37 continued

(c)

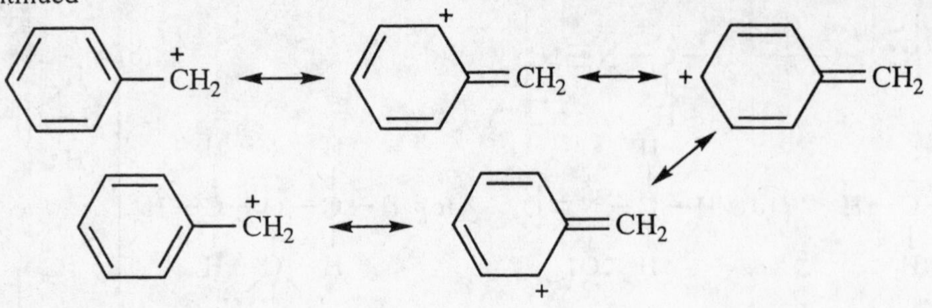

(d)

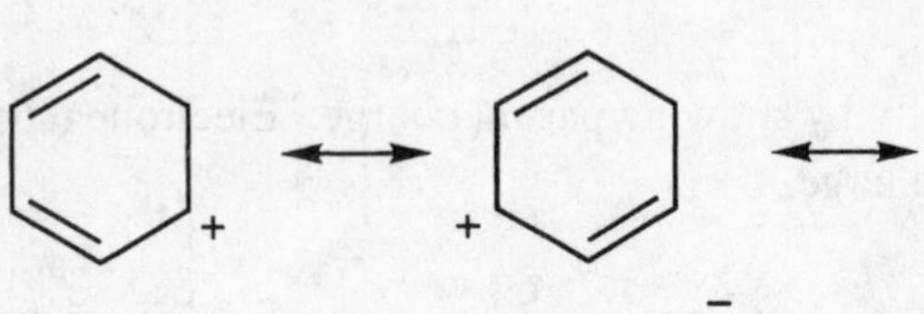

(e)

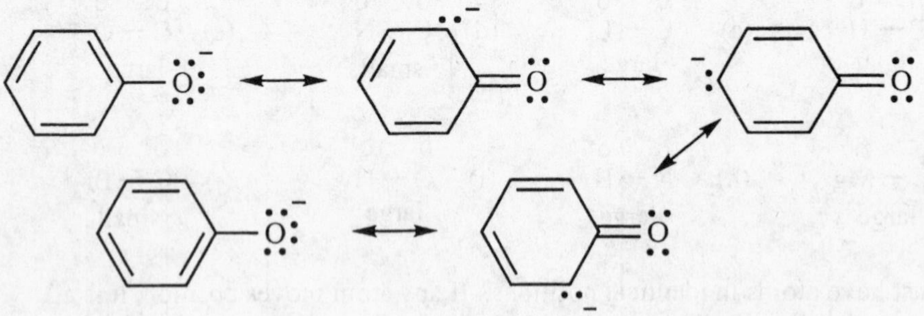

(f)

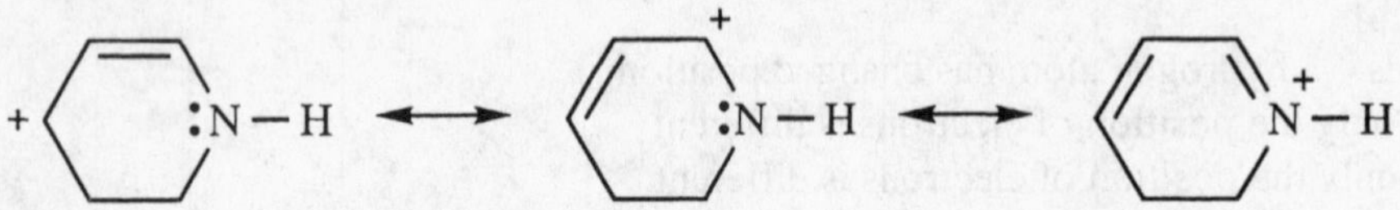

(g)

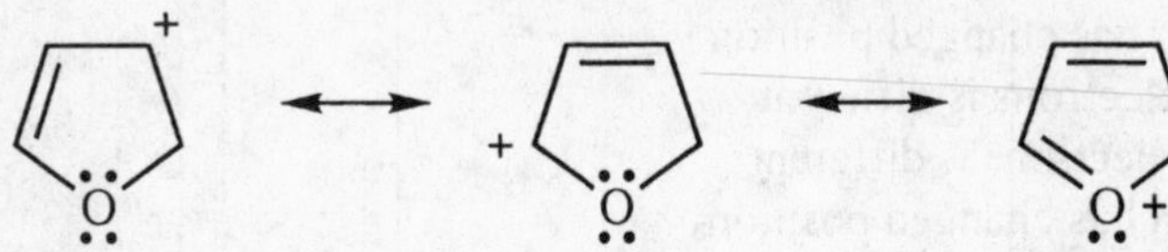

(h)

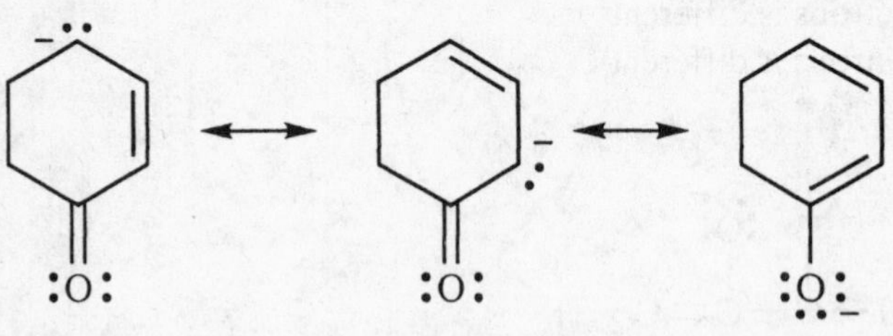

(i) $CH_3-C=C-C=C-\overset{+}{C}-CH_3 \longleftrightarrow CH_3-C=C-\overset{+}{C}-C=C-CH_3$

$CH_3-\overset{+}{C}-C=C-C=C-CH_3$

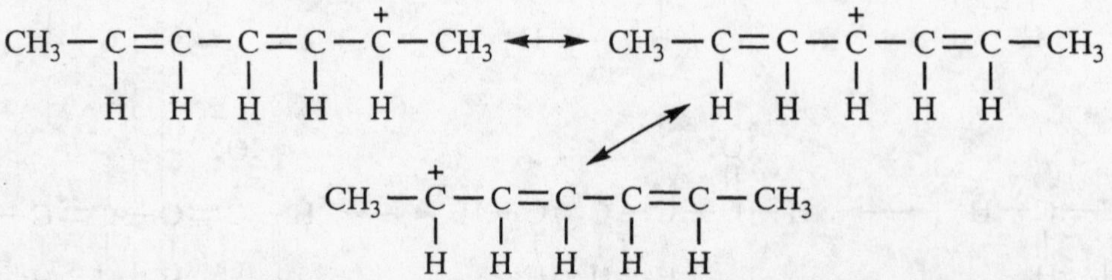

(j) no resonance forms—the charge must be on an atom next to a double or triple bond, or next to a non-bonded pair of electrons, in order for resonance to delocalize the charge

18

1-38

(a) $\overset{..}{\underset{..}{O}}=\overset{..}{\underset{+}{S}}-\overset{..}{\underset{..}{O}}:^{-}$ ⟷ $^{-}:\overset{..}{\underset{..}{O}}-\overset{..}{\underset{+}{S}}=\overset{..}{\underset{..}{O}}$ ⟷ $\overset{..}{\underset{..}{O}}=\overset{..}{S}=\overset{..}{\underset{..}{O}}$

(b) $\overset{..}{\underset{..}{O}}=\overset{..}{\underset{+}{O}}-\overset{..}{\underset{..}{O}}:^{-}$ ⟷ $^{-}:\overset{..}{\underset{..}{O}}-\overset{..}{\underset{+}{O}}=\overset{..}{\underset{..}{O}}$

(c) The last resonance form of SO_2 has no equivalent form in O_3. Sulfur, a third row element, can have more than eight electrons around it because of d orbitals, whereas oxygen, a second row element, must adhere strictly to the octet rule.

1-39

(a)

$CH_3 - N - C - NH_2$ with NH group (#3) and substituents #1, #2 labeled; H below N.

H⁺ to #1 →

$CH_3 - \overset{+}{N} - C - NH_2$ (with H, NH, H groups)
no other resonance forms

H⁺ to #3 →

$CH_3 - N - C - \overset{+}{NH_2}$ (with NH₂, H groups)

H⁺ to #2 →

$CH_3 - N - C - \overset{+}{NH_3}$ (with NH, H groups)
no other resonance forms

$CH_3 - \overset{+}{N}=C - \overset{..}{N}H_2$ ⟷ $CH_3 - \overset{..}{N} - \overset{+}{C} - \overset{..}{N}H_2$ ⟷ $CH_3 - \overset{..}{N} - C = \overset{+}{N}H_2$
(each with NH₂ and H groups)

(b) Protonation at nitrogen #3 gives four resonance forms that delocalize the positive charge over all three nitrogens and a carbon—a very stable condition. Nitrogen #3 will be protonated preferentially, which we interpret as being more basic.

1-40

(a) $CH_3 - \overset{-}{\underset{..}{C}} - C \equiv N:$ ⟷ $CH_3 - C = C = \overset{-}{\underset{..}{N}}:$ (with H below C)

minor   major (negative charge on electronegative atom)

(b) $CH_3 - \overset{:\overset{..}{O}:^{-}}{C} = \overset{+}{C} - C - CH_3$ ⟷ $CH_3 - \overset{:\overset{..}{O}:^{-}}{C} - \overset{+}{C} = C - CH_3$ ⟷ $CH_3 - \overset{:\overset{..}{O}:}{C} - C = C - CH_3$ (with H, H below carbons)

minor   minor   major—full octets, no charge separation

19

1-40 continued

(c)

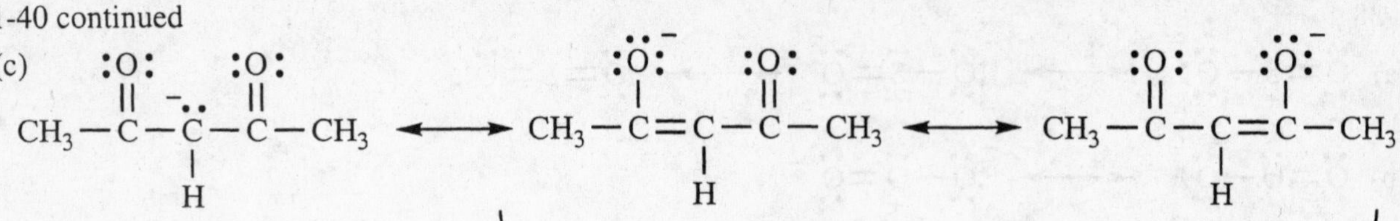

$$CH_3-\overset{\overset{\displaystyle :\ddot{O}:}{\|}}{C}-\overset{\cdot\cdot}{\underset{\underset{\displaystyle H}{|}}{C}}-\overset{\overset{\displaystyle :\ddot{O}:}{\|}}{C}-CH_3 \longleftrightarrow CH_3-\overset{\overset{\displaystyle :\ddot{O}:^{-}}{|}}{C}=\underset{\underset{\displaystyle H}{|}}{C}-\overset{\overset{\displaystyle :\ddot{O}:}{\|}}{C}-CH_3 \longleftrightarrow CH_3-\overset{\overset{\displaystyle :\ddot{O}:}{\|}}{C}-\underset{\underset{\displaystyle H}{|}}{C}=\overset{\overset{\displaystyle :\ddot{\ddot{O}}:^{-}}{|}}{C}-CH_3$$

minor   major   major

negative charge on electronegative atoms—equal energy

(d)

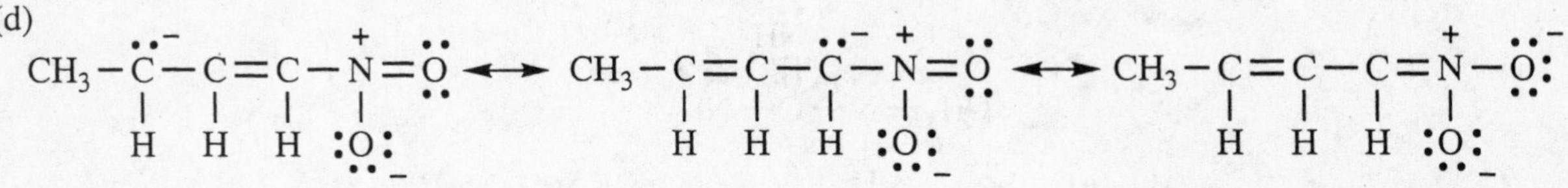

$$CH_3-\overset{\cdot\cdot}{\underset{\underset{\displaystyle H}{|}}{C}}^{-}-\underset{\underset{\displaystyle H}{|}}{C}=\underset{\underset{\displaystyle H}{|}}{C}-\overset{+}{\underset{\underset{\displaystyle :\underset{\cdot\cdot}{\ddot{O}}:^{-}}{|}}{N}}=\overset{\cdot\cdot}{\underset{\cdot\cdot}{O}} \longleftrightarrow CH_3-\underset{\underset{\displaystyle H}{|}}{C}=\underset{\underset{\displaystyle H}{|}}{C}-\overset{\cdot\cdot}{\underset{\underset{\displaystyle H}{|}}{C}}^{-}-\overset{+}{\underset{\underset{\displaystyle :\underset{\cdot\cdot}{\ddot{O}}:^{-}}{|}}{N}}=\overset{\cdot\cdot}{\underset{\cdot\cdot}{O}} \longleftrightarrow CH_3-\underset{\underset{\displaystyle H}{|}}{C}=\underset{\underset{\displaystyle H}{|}}{C}-\underset{\underset{\displaystyle H}{|}}{C}=\overset{+}{\underset{\underset{\displaystyle :\underset{\cdot\cdot}{\ddot{O}}:^{-}}{|}}{N}}-\overset{\cdot\cdot}{\underset{\cdot\cdot}{\ddot{O}}}{:}^{-}$$

minor   minor   major—negative charge on electronegative atoms

NOTE: The two structures below are resonance forms, varying from the first two structures in part (d) by the different positions of the double bonds in the NO_2. Usually, chemists omit drawing the second form of the NO_2 group although we all understand that its presence is implied. It is good idea to draw all the resonance forms until they become second nature. The importance of understanding resonance forms cannot be overemphasized.

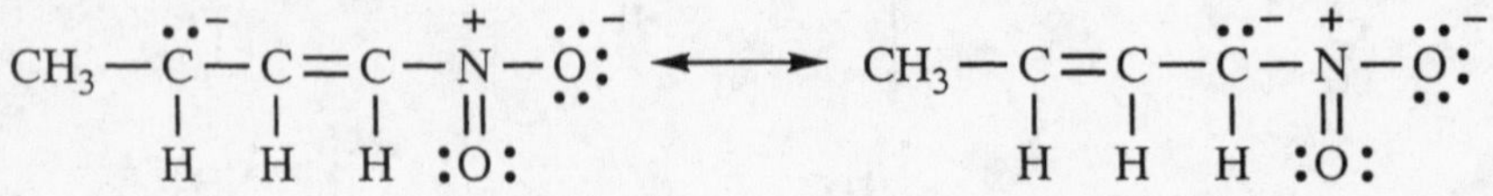

$$CH_3-\overset{\cdot\cdot}{\underset{\underset{\displaystyle H}{|}}{C}}^{-}-\underset{\underset{\displaystyle H}{|}}{C}=\underset{\underset{\displaystyle H}{|}}{C}-\overset{+}{\underset{\underset{\displaystyle :\ddot{O}:}{\|}}{N}}-\overset{\cdot\cdot}{\underset{\cdot\cdot}{\ddot{O}}}{:}^{-} \longleftrightarrow CH_3-\underset{\underset{\displaystyle H}{|}}{C}=\underset{\underset{\displaystyle H}{|}}{C}-\overset{\cdot\cdot}{\underset{\underset{\displaystyle H}{|}}{C}}^{-}-\overset{+}{\underset{\underset{\displaystyle :\ddot{O}:}{\|}}{N}}-\overset{\cdot\cdot}{\underset{\cdot\cdot}{\ddot{O}}}{:}^{-}$$

(e)

$$CH_3CH_2-\overset{\overset{\displaystyle \ddot{N}H_2}{|}}{\underset{\underset{\displaystyle +}{}}{C}}-\overset{\cdot\cdot}{\ddot{N}}H_2 \longleftrightarrow CH_3CH_2-\overset{\overset{\displaystyle \overset{+}{N}H_2}{\|}}{C}-\overset{\cdot\cdot}{\ddot{N}}H_2 \longleftrightarrow CH_3CH_2-\overset{\overset{\displaystyle \ddot{N}H_2}{|}}{C}=\overset{+}{N}H_2$$

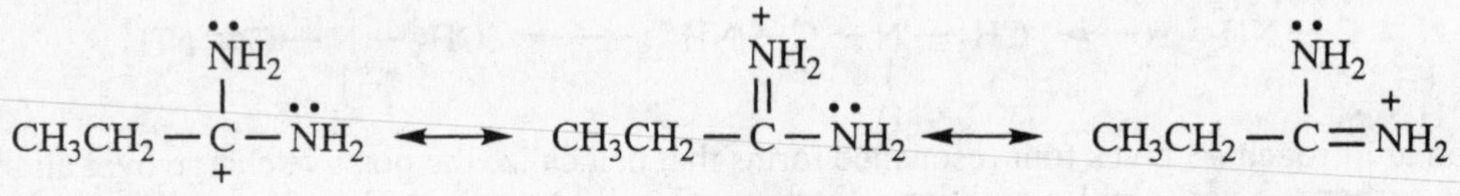

minor   major—full octets   major—full octets

equal energy

1-41

(a)

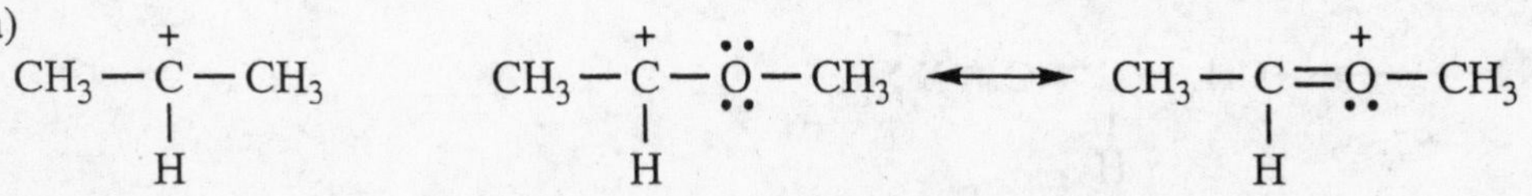

$$CH_3-\overset{+}{\underset{\underset{\displaystyle H}{|}}{C}}-CH_3 \qquad CH_3-\overset{+}{\underset{\underset{\displaystyle H}{|}}{C}}-\overset{\cdot\cdot}{\ddot{O}}-CH_3 \longleftrightarrow CH_3-\underset{\underset{\displaystyle H}{|}}{C}=\overset{+}{\underset{\cdot\cdot}{O}}-CH_3$$

no resonance stabilization   more stable—resonance stabilized

(b)

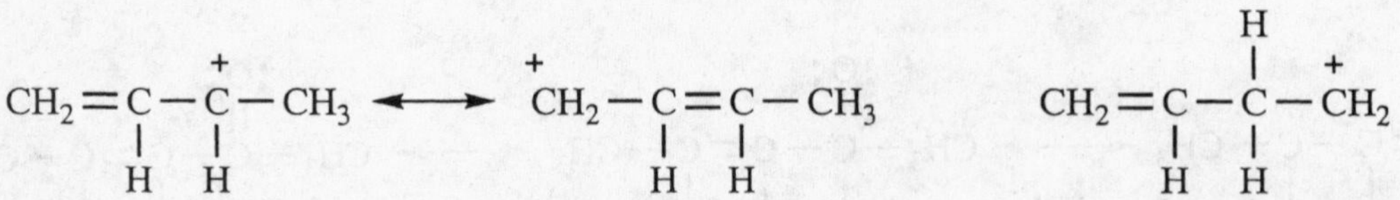

$$CH_2=\overset{+}{\underset{\underset{\displaystyle H}{|}}{C}}-\underset{\underset{\displaystyle H}{|}}{C}-CH_3 \longleftrightarrow \overset{+}{CH_2}-\underset{\underset{\displaystyle H}{|}}{C}=\underset{\underset{\displaystyle H}{|}}{C}-CH_3 \qquad CH_2=\underset{\underset{\displaystyle H}{|}}{C}-\overset{\overset{\displaystyle H}{|}}{\underset{\underset{\displaystyle H}{|}}{C}}-\overset{+}{CH_2}$$

more stable—resonance stabilized   no resonance stabilization

20

1-41 continued

(c) H—C̈⁻—CH₃   H—C̈⁻—C≡N: ⟷ H—C=C=N:⁻

          | | |
          H   H   H

no resonance stabilization   more stable—resonance stabilized

(d)

more stable—resonance stabilized   no resonance stabilization

(e)

CH₃—N̈—CH₃   ⟷   CH₃—N⁺—CH₃   CH₃—C—CH₃

   | || |
CH₃—C—CH₃   CH₃—C—CH₃   CH₃—C—CH₃
     +   +

more stable—resonance stabilized   no resonance stabilization

1-42   These pK_a values from the text, Table 1-5, and Appendix 5 provide the answers. The lower the pK_a, the stronger the acid.

least acidic			most acidic
NH_3	$< \quad CH_3OH$	$< \quad CH_3COOH$	$< \quad H_2SO_4$
33	15.5	4.74	–5

1-43 Conjugate bases of the weakest acids will be the strongest bases. The pK_a values of the conjugate acids are listed here. (The relative order of the first entry was determined from the pK_a value of sulfuric acid in Appendix 5 of the textbook.) The pK_a of protonated HF was not listed, so it cannot be correctly placed in this list. One reference suggests H_2F^+ is stronger than sulfuric acid meaning that HF is the least basic of all.

least basic					most basic
HSO_4^-	$< \quad H_2O$	$< \quad CH_3COO^-$	$< \quad CH_3O^-$	$< \quad NaOH$	$< \quad ^-NH_2$
from –6.1	from –1.7	from 4.74	from 15.5	from 15.7	from 33

1-44
(a) $pK_a = -\log_{10} K_a = -\log_{10} (5.2 \times 10^{-5}) = \mathbf{4.3}$ for phenylacetic acid

for propionic acid, pK_a 4.87: $K_a = 10^{-4.87} = \mathbf{1.35 \times 10^{-5}}$

(b)   phenylacetic acid is 3.9 times stronger than propionic acid

$$\frac{5.2 \times 10^{-5}}{1.35 \times 10^{-5}} = 3.9$$

(c)

⬡—CH₂COO⁻ + CH₃CH₂COOH ⇌ ⬡—CH₂COOH + CH₃CH₂COO⁻

          weaker acid   stronger acid

Equilibrium favors the weaker acid and base. In this reaction, **reactants** are favored.

21

1-45 The newly formed bond is shown in bold.

(a) CH₃—O:⁻ + CH₃—Cl: ⟶ CH₃—O—CH₃ + :Cl:⁻

nucleophile   electrophile
Lewis base   Lewis acid

(b) CH₃—O⁺—CH₃ + H—O—H ⟶ CH₃—O—CH₃ + H—O⁺—H
         | |
        H₃C   nucleophile   CH₃
                   Lewis base
electrophile
Lewis acid

(c)
:O: H   :O:⁻
 ‖ | |
H—C—H + :N—H ⟶ H—C—H
                   | |
                   H   H—N⁺—H
                                          |
electrophile   nucleophile   H
Lewis acid   Lewis base

(d) CH₃—NH₂   CH₃CH₂—Cl: ⟶ CH₃—N⁺—CH₂CH₃ + :Cl:⁻
                                        |
nucleophile   electrophile   H
Lewis base   Lewis acid

(e)
    :O: :O: :O—H⁺   :O:
     ‖   ‖   ‖   ‖
CH₃—C—CH₃ + H—O—S—OH ⟶ CH₃—C—CH₃ + ⁻:O—S—OH
                       ‖   ‖
nucleophile   :O: :O:
Lewis base
                  electrophile
                  Lewis acid

(f) (CH₃)₃C—Cl: + AlCl₃ ⟶
                                        CH₃   Cl
                                         | |
                                   H₃C—C⁺ + :Cl—Al—Cl⁻
                                         | |
                                        CH₃   Cl
nucleophile   electrophile
Lewis base   Lewis acid

This may also be written in two steps: association of the Cl with Al, and a second step where
the C—Cl bond breaks.

(g)
    :O: :O:⁻
     ‖ |
CH₃—C—CH₂ + ⁻:O—H ⟶ CH₃—C=CH₂ + H—O—H
         |
         H   nucleophile
electrophile   Lewis base
Lewis acid

22

1-45 continued

(h)

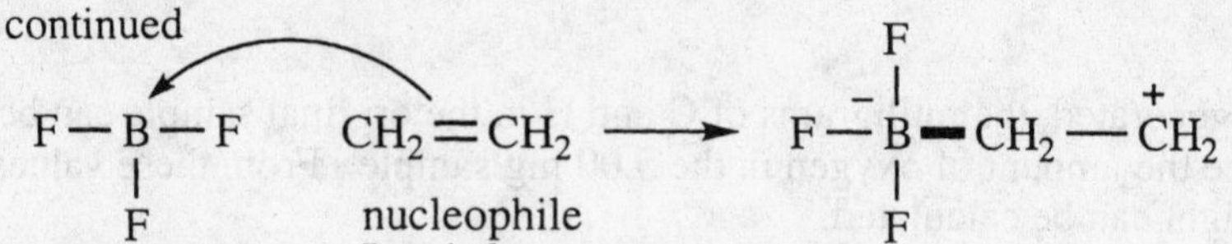

$$F-\overset{\underset{\displaystyle |}{F}}{\underset{\displaystyle |}{B}}-F \quad CH_2=CH_2 \longrightarrow F-\overset{\underset{\displaystyle |}{F}}{\overset{-}{B}}-CH_2-\overset{+}{CH_2}$$

electrophile   nucleophile
Lewis acid   Lewis base

(i) $^-BF_3-CH_2-\overset{+}{CH_2} + CH_2=CH_2 \longrightarrow {}^-BF_3-CH_2-CH_2-CH_2-\overset{+}{CH_2}$

electrophile   nucleophile
Lewis acid   Lewis base

1-46

(a) $H_2SO_4 + CH_3COO^- \rightleftharpoons HSO_4^- + CH_3COOH$

(b) $CH_3COOH + (CH_3)_3N\colon \rightleftharpoons CH_3COO^- + (CH_3)_3\overset{+}{N}-H$

(c)

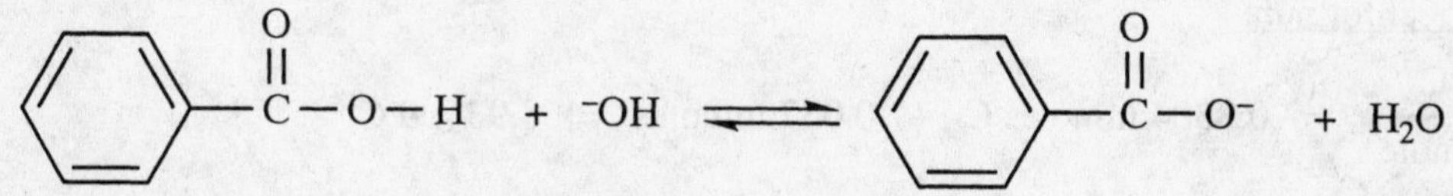

(d) $(CH_3)_3\overset{+}{N}-H + {}^-OH \rightleftharpoons (CH_3)_3N\colon + H_2O$

(e)

$$HO-\overset{\overset{\displaystyle O}{\parallel}}{C}-OH + 2\,{}^-OH \rightleftharpoons {}^-O-\overset{\overset{\displaystyle O}{\parallel}}{C}-O^- + 2\,H_2O$$

(f) $H_2O + NH_3 \rightleftharpoons HO^- + {}^+NH_4$

(g) $HCOOH + CH_3O^- \rightleftharpoons HCOO^- + CH_3OH$

1-47

(a) $CH_3CH_2-O-H + CH_3-Li \longrightarrow CH_3CH_2-O^-\ Li^+ + CH_4$

(b) The conjugate acid of CH_3Li is CH_4. Table 1-5 gives the pK_a of CH_4 as > 40, one of the weakest acids known. The conjugate base of one of the weakest acids known must be one of the strongest bases known.

1-48

From the amounts of CO_2 and H_2O generated, the milligrams of C and H in the original sample can be determined, thus giving by difference the amount of oxygen in the 5.00 mg sample. From these values, the empirical formula and empirical weight can be calculated.

(a) how much carbon in 14.54 mg CO_2

$$14.54 \text{ mg CO}_2 \times \frac{1 \text{ mmole CO}_2}{44.01 \text{ mg CO}_2} \times \frac{1 \text{ mmole C}}{1 \text{ mmole CO}_2} \times \frac{12.01 \text{ mg C}}{1 \text{ mmole C}} = 3.968 \text{ mg C}$$

how much hydrogen in 3.97 mg H_2O

$$3.97 \text{ mg H}_2O \times \frac{1 \text{ mmole H}_2O}{18.016 \text{ mg H}_2O} \times \frac{2 \text{ mmoles H}}{1 \text{ mmole H}_2O} \times \frac{1.008 \text{ mg H}}{1 \text{ mmole H}} = 0.444 \text{ mg H}$$

how much oxygen in 5.00 mg estradiol

5.00 mg estradiol - 3.968 mg C - 0.444 mg H = 0.59 mg O

calculate empirical formula

$$\frac{3.968 \text{ mg C}}{12.01 \text{ mg/mole}} = 0.3304 \text{ mmoles C} \div 0.037 \text{ mmoles} = 8.93 \approx 9 \text{ C}$$

$$\frac{0.444 \text{ mg H}}{1.008 \text{ mg/mole}} = 0.440 \text{ mmoles H} \div 0.037 \text{ mmoles} = 11.9 \approx 12 \text{ H}$$

$$\frac{0.59 \text{ mg O}}{16.00 \text{ mg/mole}} = 0.037 \text{ mmoles O} \div 0.037 \text{ mmoles} = 1 \text{ O}$$

empirical formula = $\boxed{C_9H_{12}O}$ $\Longrightarrow$ empirical weight = 136

(b) molecular weight = 272, exactly twice the empirical weight

twice the empirical formula = molecular formula = $\boxed{C_{18}H_{24}O_2}$

CHAPTER 2—STRUCTURE AND PROPERTIES OF ORGANIC MOLECULES

2-1 The fundamental principle of organic chemistry is that a molecule's chemical and physical properties depend on the molecule's structure: the structure-function or structure-reactivity correlation. It is essential that you understand the three-dimensional nature of organic molecules, and there is no better device to assist you than a molecular model set. You are strongly encouraged to use models regularly when reading the text and working the problems.

(a) requires use of models

(b)

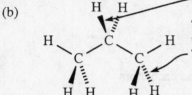

The wedge bonds represent bonds coming out of the plane of the paper toward you.
The dashed bonds represent bonds going behind the plane of the paper.

2-2 (a) The hybridization of oxygen is sp^3 since it has two sigma bonds and two pairs of nonbonding electrons. The reason that the bond angle of 104.5° is less than the perfect tetrahedral angle of 109.5° is that the lone pairs in the two sp^3 orbitals are repelling each other more strongly than the electron pairs in the sigma bonds, thereby compressing the bond angle.

(b) The electrostatic potential map for water shows that the hydrogens have low electron potential (blue), and the area of the unshared electron pairs in sp^3 orbitals has high electron potential (red).

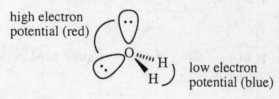

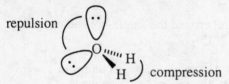

2-3 Each double-bonded atom is sp^2 hybridized with bond angles about 120°; geometry around sp^2 atoms is trigonal planar. In (a), all four carbons and the two hydrogens on the sp^2 carbons are all in one plane. Each carbon on the end is sp^3 hybridized with tetrahedral geometry and bond angles about 109°. In (b), the two carbons, the nitrogen, and the two hydrogens on the sp^2 carbon are all in one plane. The CH_3 carbon is sp^3 hybridized with tetrahedral geometry and bond angles about 109°.

(a)

(b)

2-4 The hybridization of the nitrogen and the triple-bonded carbon are sp, giving linear geometry (C—C—N are linear) and a bond angle around the triple-bonded carbon of 180°. The CH_3 carbon is sp^3 hybridized, tetrahedral, with bond angles about 109°.

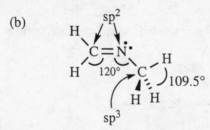

25

2-5

(a) linear, bond angle 180°

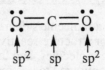

(b) all atoms are sp³; tetrahedral geometry and bond angles of 109° around each atom

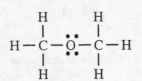

not a bond—shows lone pair coming out of paper

not a bond—shows lone pair going behind paper

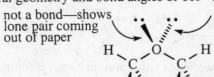

(c) all atoms are sp³; tetrahedral geometry and bond angles of 109° around each atom

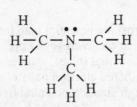

not a bond—shows lone pair going behind paper

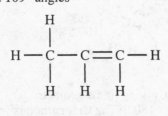

(d) trigonal planar around the carbon, bond angles 120°; tetrahedral around the single-bonded oxygen, bond angle 109°

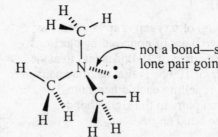

all atoms in one plane

(e) carbon and nitrogen both sp, linear, bond angle 180°

$$H-C\equiv N:$$ all three atoms in a line

(f) trigonal planar around the sp² carbons, bond angles 120°; around the sp³ carbon, tetrahedral geometry and 109° angles

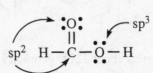

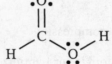

(g) trigonal planar, bond angle about 120°

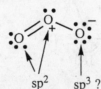

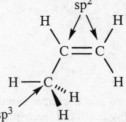

(the other resonance form of ozone shows that BOTH end oxygens must be sp²— see Solved Problem 2-8)

2-6 Carbon-2 is sp hybridized. If the p orbitals making the pi bond between C-1 and C-2 are in the plane of the paper (putting the hydrogens in front of and behind the paper), then the other p orbital on C-2 must be perpendicular to the plane of the paper, making the pi bond between C-2 and C-3 perpendicular to the paper. This necessarily places the hydrogens on C-3 in the plane of the paper. (Models will surely help.)

model of *perpendicular* π bonds

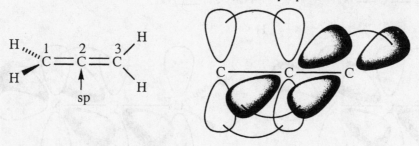

2-7 For clarity, electrons in sigma bonds are not shown.

(a) carbon and oxygen are both sp² hybridized

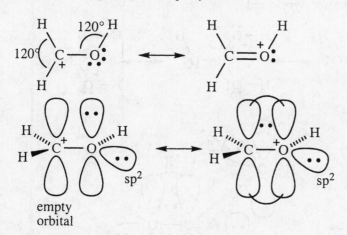

One pair of electrons on oxygen is always in an sp² orbital. The other pair of electrons is shown in a p orbital in the first resonance form, and in a pi bond in the second resonance form.

(b) oxygen and both carbons are sp² hybridized

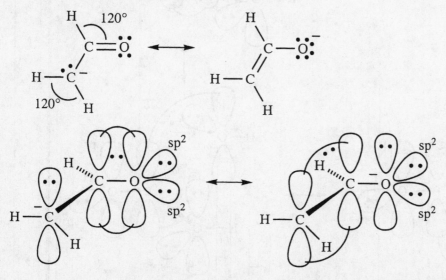

27

(c) the nitrogen and the carbon bonded to it are sp hybridized; the left carbon is sp²

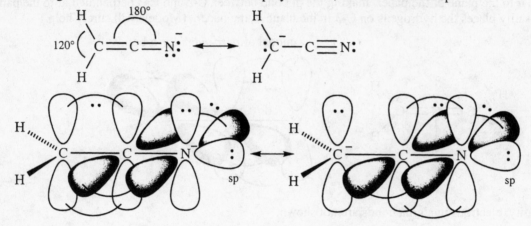

(d) the boron and the oxygens bonded to it are sp² hybridized

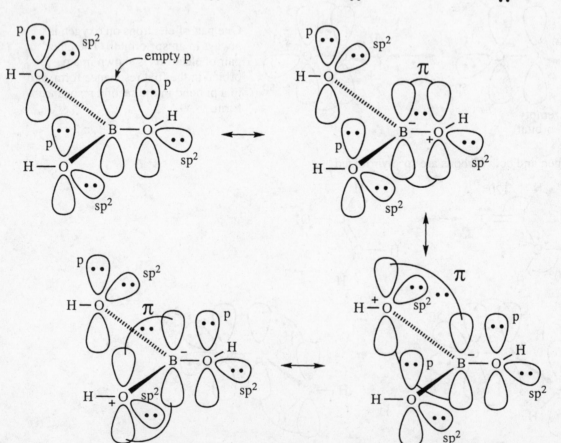

28

2-8 Very commonly in organic chemistry, we have to determine whether two structures are the same or different, and if they are different, what structural features are different. In order for two structures to be the same, all bonding connections have to be identical, and in the case of double bonds, the groups must be on the same side of the double bond in both structures. (A good exercise to do with your study group is to draw two structures and ask if they are the same; or draw one structure and ask how to draw a different compound.)

(a) different compounds; H and CH_3 on one carbon of the double bond, and CH_3 and CH_2CH_3 on the other carbon—same in both structures; drawing a plane through the p orbitals shows the H and CH_3 are on the same side of the double bond in the first structure, and the H and the CH_2CH_3 are on the same side in the second structure, so they are DIFFERENT compounds

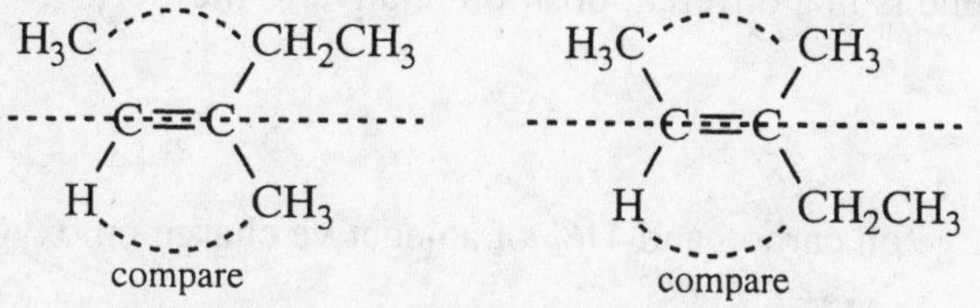

these are DIFFERENT

(b) same compound; in the structure on the right, the right carbon has been rotated, but the bonding is identical between the two structures

(c) different compounds; H and Br on one carbon, F and Cl on the other carbon in both structures; H and Cl on the same side of the plane through the C=C in the first structure, and H and F on the same side of the plane through the C=C in the second structure, so they are DIFFERENT compounds

(d) same compound: in the structure on the right, the right carbon has been rotated 120°

2-9

(a)

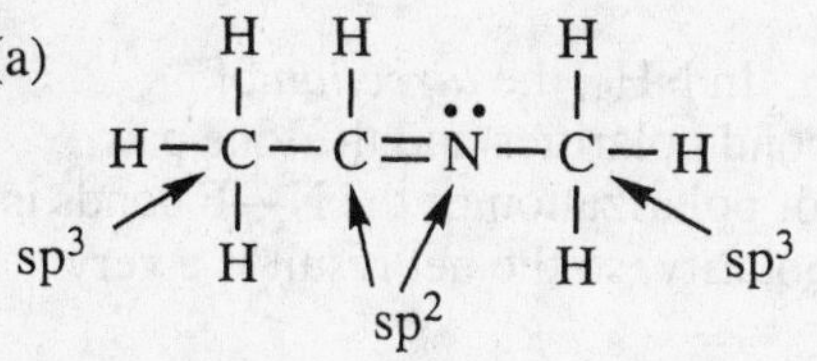

(b)

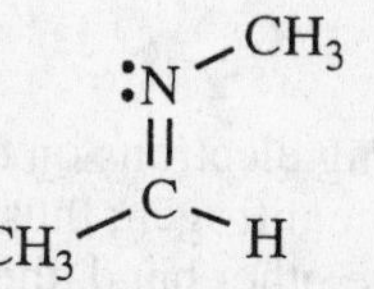

NOT INTER-CONVERTIBLE

two CH_3's on opposite sides of the C=N

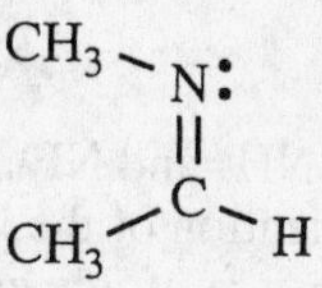

two CH_3's on the same side of the C=N

(c) the CH_3 on the N is on the same side as another CH_3 no matter how it is drawn—only one possible structure

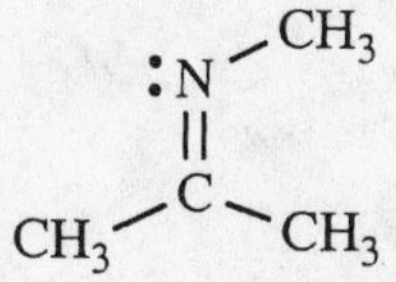

2-10

(a)

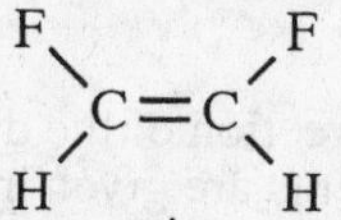

and

cis

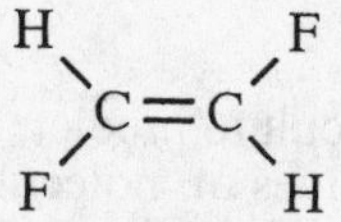

trans

(b) no *cis-trans* isomerism
(c) no *cis-trans* isomerism   two identical groups on one
(d) no *cis-trans* isomerism   carbon of the double bond

(e)

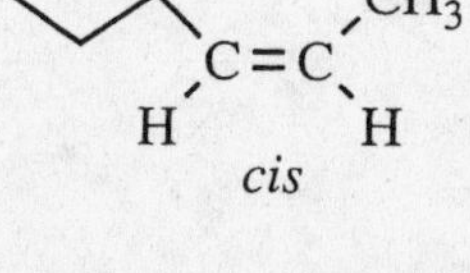

and

cis

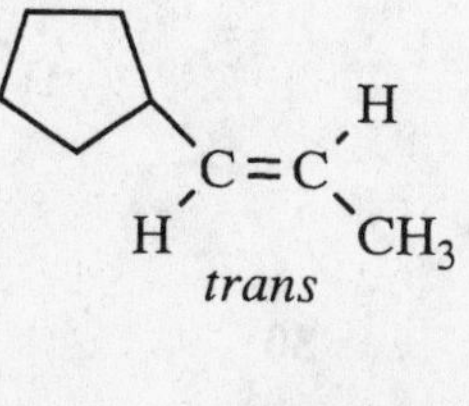

trans

(f)

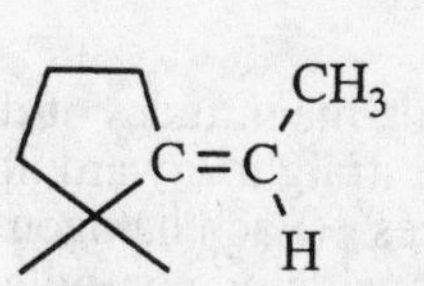

and

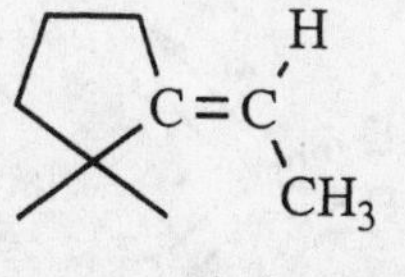

"cis" and *"trans"* not defined for this example

2-11 Models will be helpful here.
(a) constitutional isomers—the carbon skeleton is different
(b) cis-trans isomers—the first is trans, the second is cis
(c) constitutional isomers—the bromines are on different carbons in the first structure, on the same carbon in the second structure
(d) same compound—just flipped over
(e) same compound—just rotated
(f) same compound—just rotated
(g) not isomers—different molecular formulas
(h) constitutional isomers—the double bond has changed position
(i) same compound—just reversed
(j) constitutional isomers—the CH_3 groups are in different relative positions
(k) constitutional isomers—the double bond is in a different position relative to the CH_3

2-12

(a) $2.4 D = 4.8 \times \delta \times 1.21 Å$

$\delta = 0.41$, or 41% of a positive charge on carbon and 41% of a negative charge on oxygen

(b)

Resonance form A must be the major contributor. If B were the major contributor, the value of the charge separation would be between 0.5 and 1.0. Even though B is "minor", it is quite significant, explaining in part the high polarity of the C=O.

2-13

Both NH_3 and NF_3 have a pair of nonbonding electrons on the nitrogen. In NH_3, the *direction* of polarization of the N—H bonds is *toward* the nitrogen; thus, all three bond polarities and the lone pair polarity reinforce each other. In NF_3, on the other hand, the direction of polarization of the N—F bonds is *away* from the nitrogen; the three bond polarities cancel the lone pair polarity, so the net result is a very small *molecular* dipole moment.

polarities reinforce;
large dipole moment

polarities oppose;
small dipole moment

2-14 Some magnitudes of dipole moments are difficult to predict; however, the direction of the dipole should be straightforward, in most cases. Actual values of molecular dipole moments are given in parentheses. (Each halogen atom has three nonbonded electron pairs, not shown below.) The C—H is usually considered non-polar.

(a)

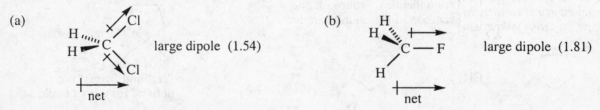

large dipole (1.54)

(b)

large dipole (1.81)

net

net

30

2-14 continued

(c)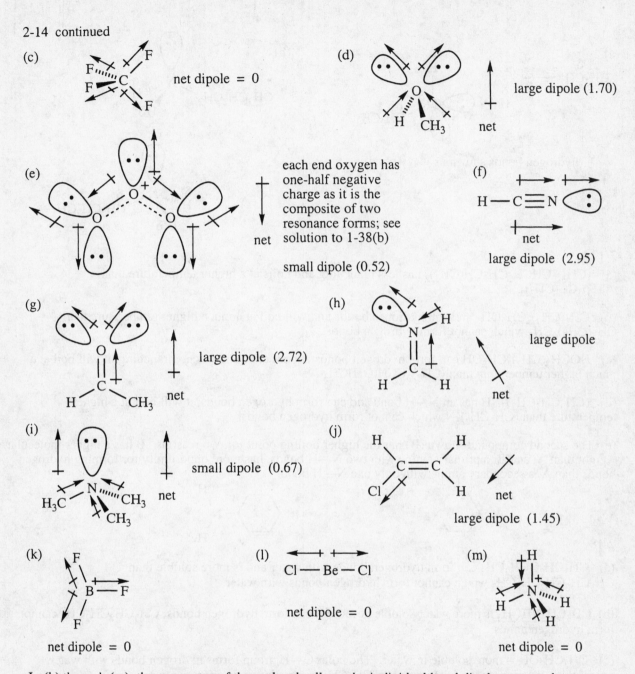
net dipole = 0

(d)
large dipole (1.70)
net

(e)
each end oxygen has
one-half negative
charge as it is the
composite of two
resonance forms; see
net solution to 1-38(b)

small dipole (0.52)

(f)
H—C≡N
net
large dipole (2.95)

(g)
large dipole (2.72)
net

(h)
large dipole
net

(i)
small dipole (0.67)
net

(j)
net
large dipole (1.45)

(k) F
net dipole = 0

(l)
Cl—Be—Cl
net dipole = 0

(m)
net dipole = 0

In (k) through (m), the symmetry of the molecule allows the individual bond dipoles to cancel.

2-15 With chlorines on the same side of the double bond, the bond dipole moments reinforce each other, resulting in a large net dipole. With chlorines on opposite sides of the double bond, the bond dipole moments exactly cancel each other, resulting in a zero net dipole.

large net dipole

net dipole = 0

2-16

(a)

$\overset{\delta^+}{CH_3CH_2}\!-\!\overset{\delta^-}{\ddot{O}\!:}$

(hydrogen bonds shown as wavy bond)

(b)

2-17

(a) $(CH_3)_2CHCH_2CH_2CH(CH_3)_2$ has less branching and boils at a higher temperature than $(CH_3)_3CC(CH_3)_3$.

(b) $CH_3(CH_2)_5CH_2OH$ can form hydrogen bonds and will boil at a much higher temperature than $CH_3(CH_2)_6CH_3$ which cannot form hydrogen bonds.

(c) $HOCH_2(CH_2)_4CH_2OH$ can form hydrogen bonds at both ends and has no branching; it will boil at a much higher temperature than $(CH_3)_3CCH(OH)CH_3$.

(d) $(CH_3CH_2CH_2)_2NH$ has an N—H bond and can form hydrogen bonds; it will boil at a higher temperature than $(CH_3CH_2)_3N$ which cannot form hydrogen bonds.

(e) The second compound shown (**B**) has the higher boiling point for two reasons: **B** has a higher molecular weight than **A**; and **B**, a primary amine with two N—H bonds, has more opportunity for forming hydrogen bonds than **A**, a secondary amine with only one N—H bond.

A   N - H   **B**   N

2-18

(a) $CH_3CH_2OCH_2CH_3$ can form hydrogen bonds with water and is more soluble than $CH_3CH_2CH_2CH_2CH_3$ which cannot form hydrogen bonds with water.

(b) $CH_3CH_2NHCH_3$ is more water soluble because it can form hydrogen bonds; $CH_3CH_2CH_2CH_3$ cannot form hydrogen bonds.

(c) CH_3CH_2OH is more soluble in water. The polar O—H group forms hydrogen bonds with water, overcoming the resistance of the non-polar CH_3CH_2 group toward entering the water. In $CH_3CH_2CH_2CH_2OH$, however, the hydrogen bonding from only one OH group cannot carry a four-carbon chain into the water; this substance is only slightly soluble in water.

(d) Both compounds form hydrogen bonds with water at the double-bonded oxygen, but only the smaller molecule (CH_3COCH_3) dissolves. The cyclic compound has too many non-polar CH_2 groups to dissolve.

2-19

(a) 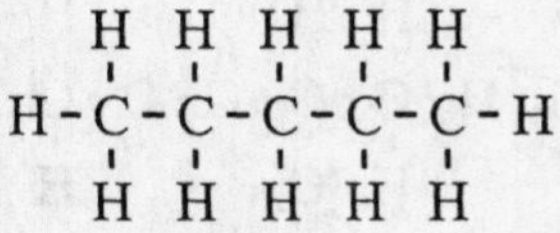 alkane

(b) 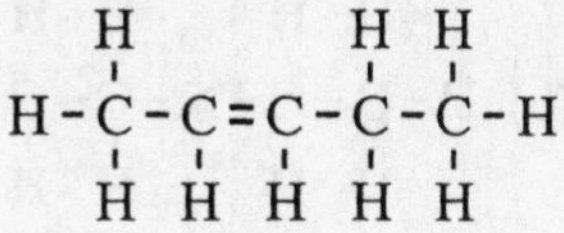 alkene

(c) 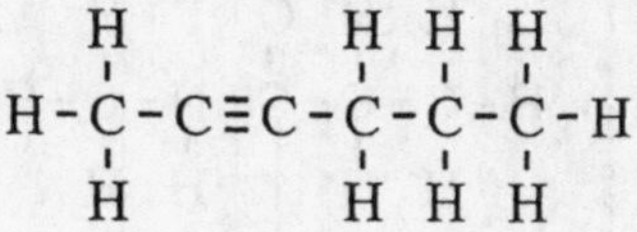 alkyne

(Usually, we use the term "alkane" only when no other groups are present.)

(d) 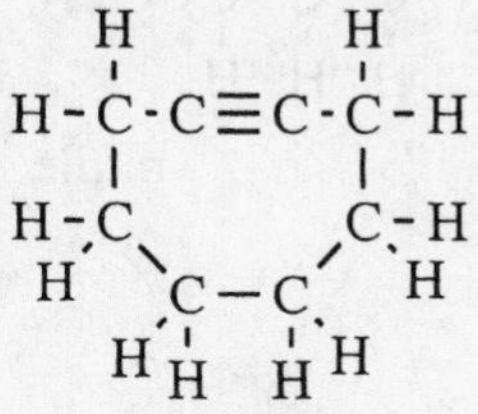 cycloalkyne

(e) 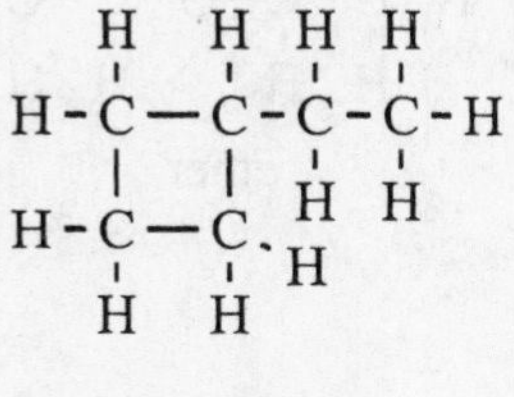 cycloalkane

(f) 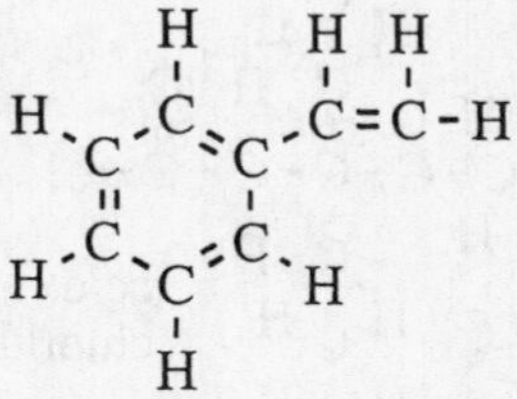 aromatic hydrocarbon and alkene

(g) 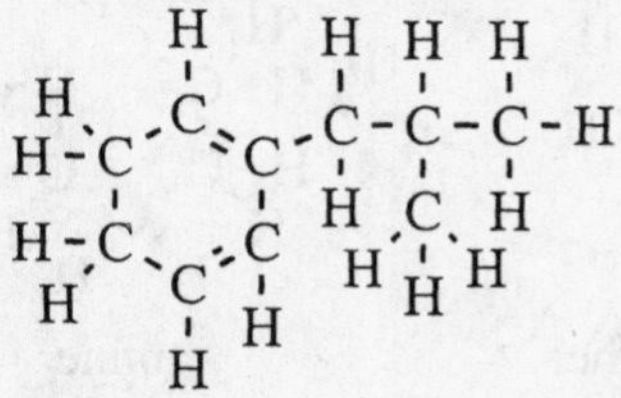 cycloalkene

(h) 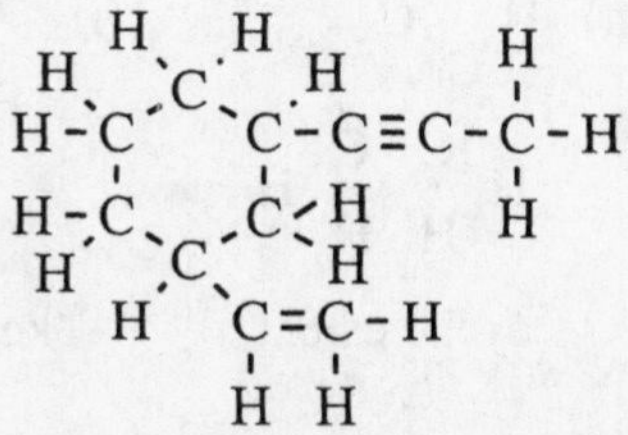 alkyne, alkene, cycloalkane

(i) 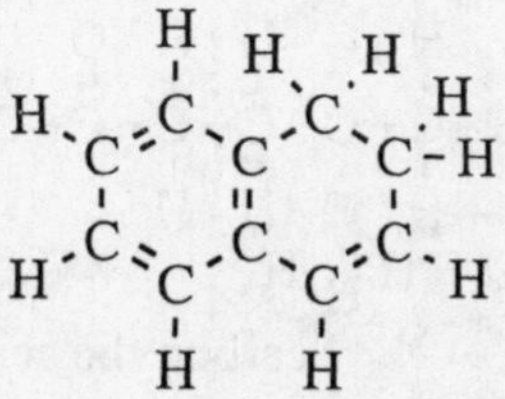 aromatic hydrocarbon and cycloalkene

2-20

(a) 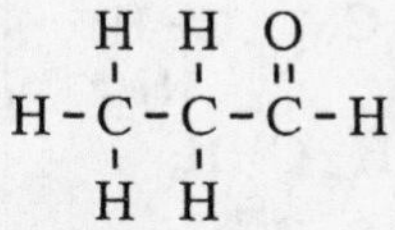 aldehyde

(b) 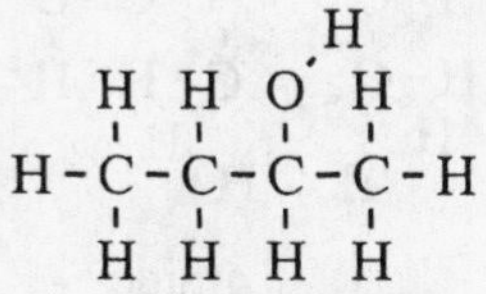 alcohol

(c) 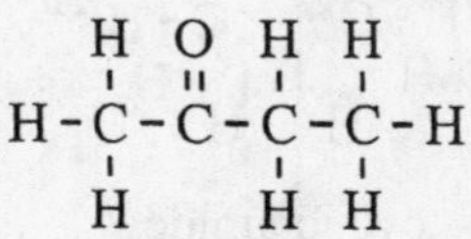 ketone

(d) 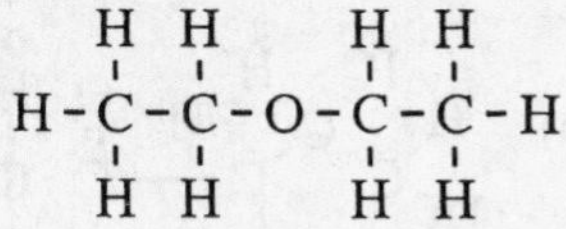 ether

(e) 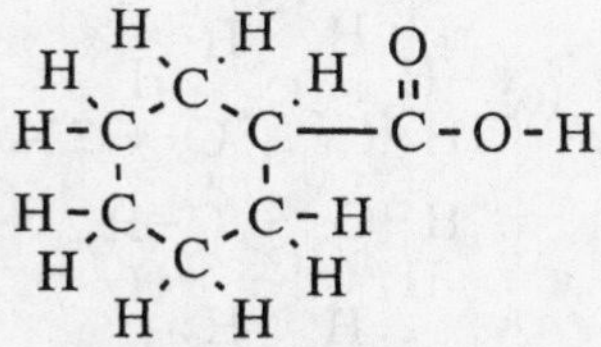 carboxylic acid

(f) 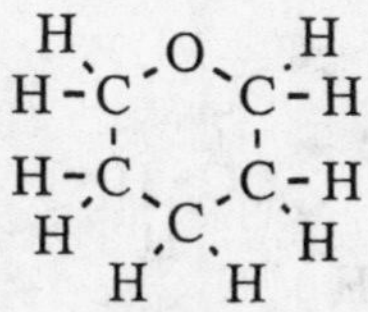 ether

(g) 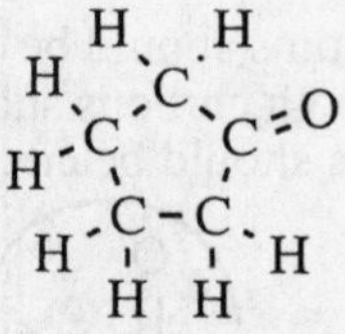 ketone

(h) 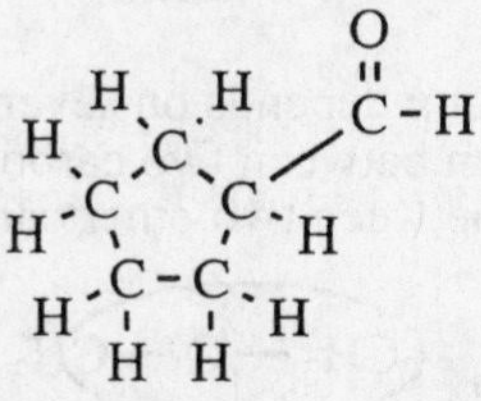 aldehyde

(i) 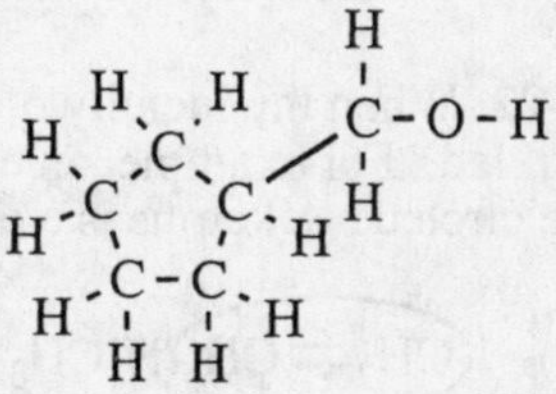 alcohol

2-21

(a) 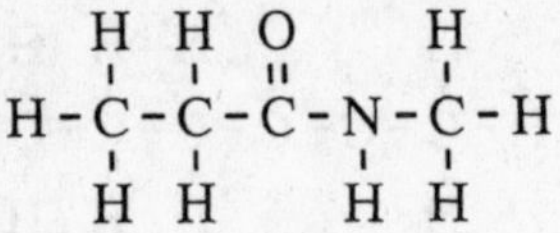 amide

(b) 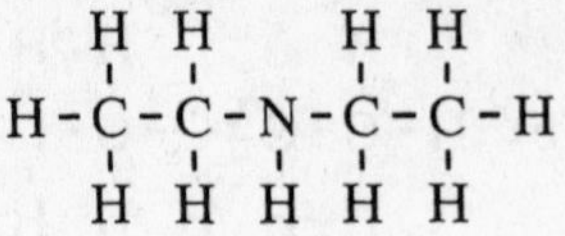 amine

(c) 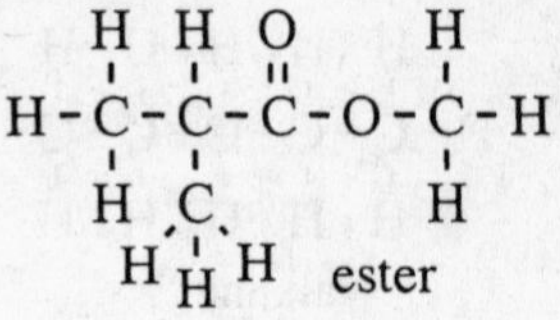 ester

(d) 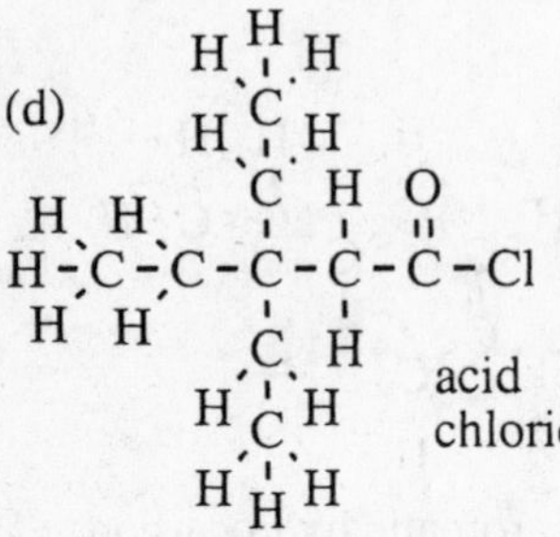 acid chloride

(e) 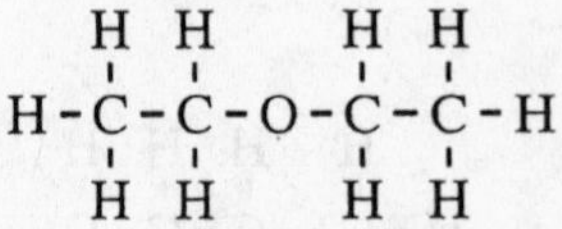 ether

(f) 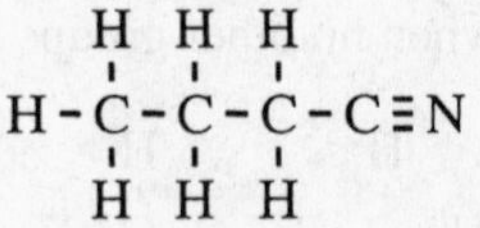 nitrile

(g) 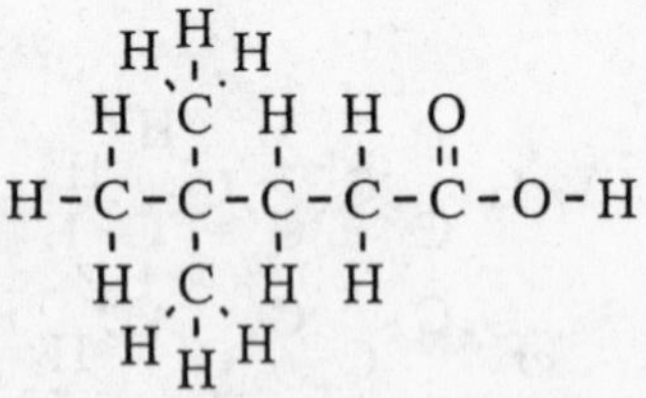 carboxylic acid

(h) 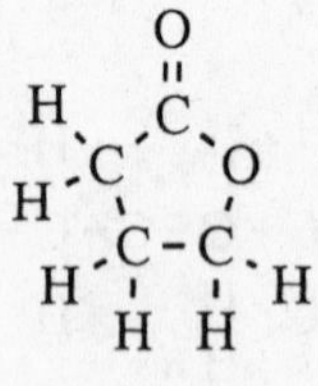 cyclic ester

(i) 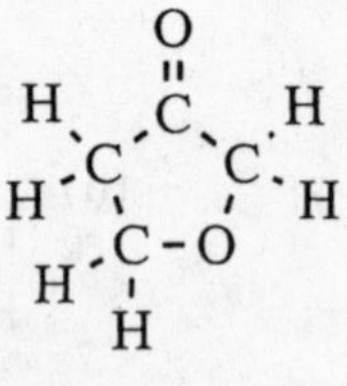 ketone and ether

(j) 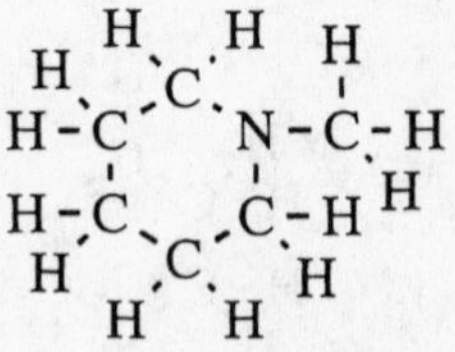 amine

(k) 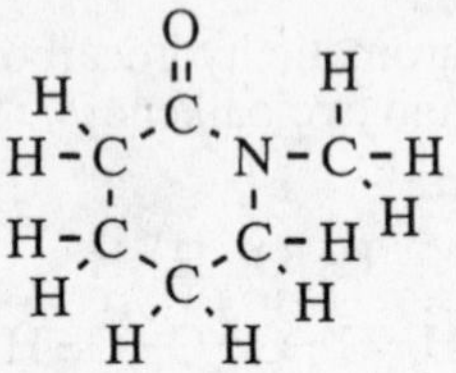 cyclic amide

(l) 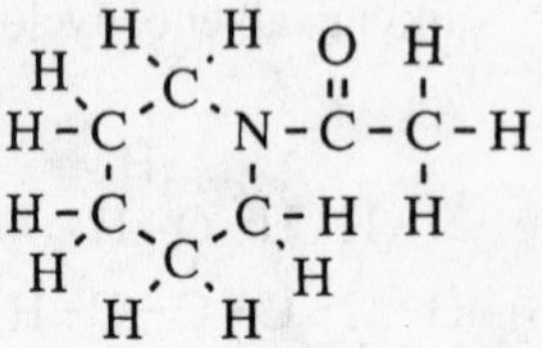 amide

(m) 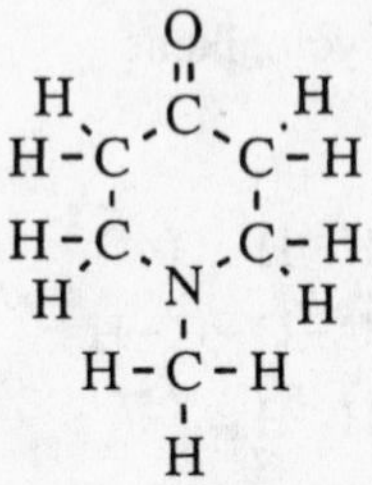 ketone and amine

(n) 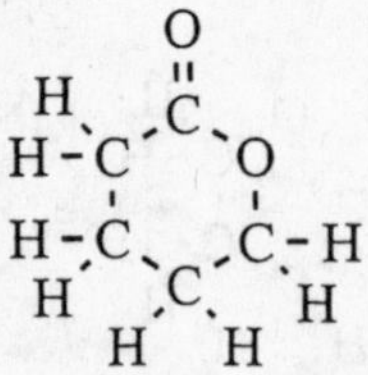 cyclic ester

(o) 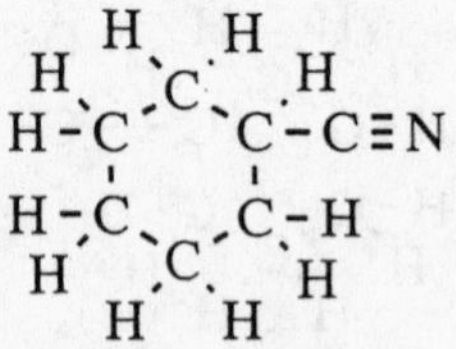 nitrile

(p) 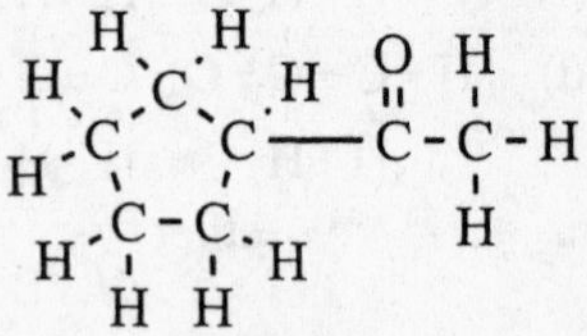 ketone

2-22 When the identity of a functional group depends on several atoms, all of those atoms should be circled. For example, an ether is an oxygen between two carbons, so the oxygen and both carbons should be circled. A ketone is a carbonyl group between two other carbons, so all those atoms should be circled.

(a) 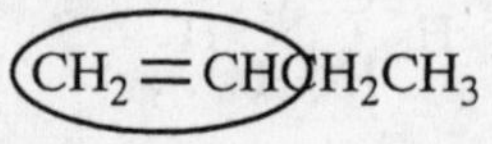 $CH_2 = CHCH_2CH_3$

alkene

(b) 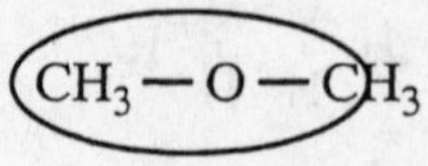 $CH_3 - O - CH_3$

ether

(c) 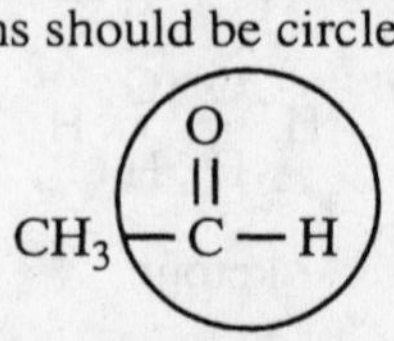 $CH_3 - C - H$ with O double bonded to C

aldehyde

2-22 continued

(d)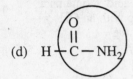

amide

(this also looks like an aldehyde, but an amide has higher "priority" as you will see later)

(e)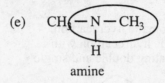

amine

(f)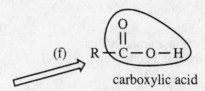

carboxylic acid

Suggested by student Richard King: R is the symbol that organic chemists use to represent alkyl and aryl groups. As you will see in the course of your study, there are quite a few ways that carbon and hydrogen atoms can go together to form alkyl and aryl groups. So when you see this symbol, you should know that it represents ONLY some combination of carbon and hydrogen atoms—except when it includes other atoms.

(g)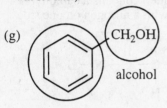

alcohol

aromatic

(h)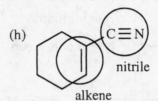

nitrile

alkene

(i)

ketone

(j)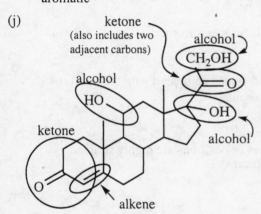

ketone (also includes two adjacent carbons)

alcohol

alcohol

ketone

alkene

(k)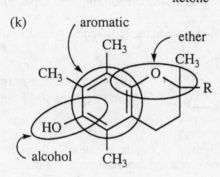

aromatic

ether

alcohol

$R = (CH_2CH_2CH_2CH(CH_3))_3CH_3$

2-23 Please refer to solution 1-20, page 12 of this Solutions Manual.

2-24 The examples here are representative. Your examples may be different and still correct. What is important in this problem is to have the same functional group.

(a) alkane: hydrocarbon with all single bonds; can be acyclic (no ring) or cyclic

(b) alkene: contains a carbon-carbon double bond

(c) alkyne: contains a carbon-carbon triple bond

(d) alcohol: contains an OH group on a carbon

(e) ether: contains an oxygen between two carbons

(f) ketone: conatins a carbonyl group between two carbons

2-24 continued

(g) aldehyde: contains
a carbonyl group with a
hydrogen on one side

$$\begin{array}{ccc} H & H & O \\ | & | & \| \\ H-C-C-C-H \\ | & | \\ H & H \end{array}$$

(h) aromatic hydrocarbon:
a cyclic hydrocarbon with
alternating double and single
bonds

(i) carboxylic acid: contains a
carbonyl group with an OH
group on one side

$$\begin{array}{cc} H & O \\ | & \| \\ H-C-C-O-H \\ | \\ H \end{array}$$

(j) ester: contains a carbonyl
group with an O—C on one side

$$\begin{array}{ccc} H & O & H \\ | & \| & | \\ H-C-C-O-C-H \\ | & | \\ H & H \end{array}$$

(k) amine: contains a nitrogen
bonded to one, two, or three
carbons

$$\begin{array}{cc} H & H \text{ or R group} \\ | & | \\ H-C-N-H \text{ or R group} \\ | \\ H \end{array}$$

(l) amide: contains a carbonyl
group with a nitrogen on one side

$$\begin{array}{ccc} H & O & H \\ | & \| & | \\ H-C-C-N-H \\ | \\ H \end{array}$$

(m) nitrile: contains the carbon-nitrogen triple bond: $H_3C-C\equiv N$

2-25 Models show that the tetrahedral geometry of CH_2Cl_2 precludes stereoisomers.

2-26

(a)

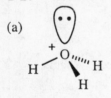

(b) Cyclopropane must have 60° bond angles compared with the usual sp^3
bond angle of 109.5° in an acyclic molecule.

(c) Like a bent spring, bonds that deviate from their normal angles or positions are highly strained.
Cyclopropane is reactive because breaking the ring relieves the strain.

2-27

(a)

$$H-\overset{+}{O}\cdots H$$
$$\quad H$$

sp^3, ≈ 109°

(b)

$$H\cdots\overset{-}{O}$$

sp^3, no bond angle because
oxygen is bonded to only one atom

(c)

both sp^3, all ≈ 109°

(d)

behind the plane
of the paper

all sp^3, all ≈ 109°

(e)

all sp^3, all ≈ 109°

(f)

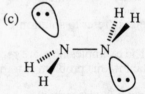

angles around sp^3 atoms ≈ 109°
angles around sp^2 carbon ≈ 120°

36

2-27 continued

(g)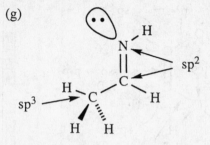

angles around sp³ atom ≈ 109°
angles around sp² atoms ≈ 120°

(h)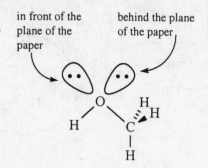

in front of the plane of the paper

behind the plane of the paper

both sp³, all ≈ 109°

(i)

both sp², all ≈ 120°

2-28 For clarity in these pictures, bonds between hydrogen and an sp³ atom are not labeled; these bonds are s-sp³ overlap.

(a)

sp³-sp³

H—C—N—C—H

109°

(b)

sp³-sp³

H—C—C—O—H

109°

(c)

109°   sp³-sp²   sp²-sp³   109°

H—C—C═N—C—H

sp²-s   sp²-sp²   120°
        + p-p

(d)

109°   sp³-sp²   120°   109°

H—C—C═C

sp²-s   sp²-sp²   sp²-sp³   109°
        + p-p

(e)

sp-sp + two p-p   120°

109°

H—C—C≡C—C—H

sp³-sp   180°   sp-sp²   sp²-s

sp²-sp² + p-p

(f)

109°   180°

H—N—C—C≡N:

sp³-sp³   sp³-sp   sp-sp + two p-p

(g)

120°   sp²-sp² + p-p

H—C—C—C—H

109°   sp³-sp²   sp²-sp³   109°

37

2-28 continued

(h)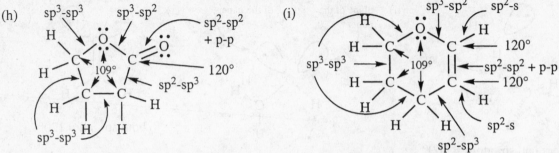

(i)

2-29 The second resonance form of formamide is a minor but significant resonance contributor. It shows that the nitrogen-carbon bond has some double bond character, requiring that the nitrogen be sp² hybridized with bond angles approaching 120°.

$$:O: \qquad\qquad :O:^-$$
$$\parallel \qquad\qquad\qquad | \quad sp^2$$
$$H-C-\overset{..}{N}-H \quad\longleftrightarrow\quad H-C=\overset{+}{N}-H$$
$$\qquad | \qquad\qquad\qquad\qquad |$$
$$\qquad H \qquad\qquad\qquad\qquad H$$

2-30

(a) The major resonance contributor shows a carbon-carbon double bond, suggesting that both carbons are sp² hybridized with trigonal planar geometry. The CH₃ carbon is sp³ hybridized with tetrahedral geometry.

$$H \quad :O: \qquad\qquad\qquad H \quad :\overset{..}{O}:^-$$

H—C—C—C—H ⟷ H—C—C=C—H

minor   major

(b) The major resonance contributor shows a carbon-nitrogen double bond, suggesting that all three carbons and the nitrogen are sp² hybridized with trigonal planar geometry.

H—N—C=C—C⁺—H ⟷ H—N—C⁺—C=C—H ⟷ H—N⁺=C—C=C—H

minor   minor   major

2-31 In (c) and (d), the unshadowed p orbitals are vertical and parallel. The shadowed p orbitals are perpendicular and horizontal.

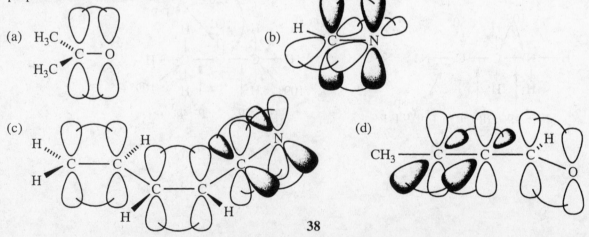

(a) H₃C⋯C—O

(b) H—C—N

(c) H, H—C—C—C—C—N

(d) CH₃—C—C—C—O

38

2-32

(a)

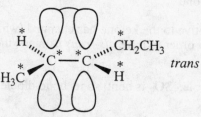

cis

(b) The coplanar atoms in the structures to the left and below are marked with asterisks.

(c)

trans

There are still six coplanar atoms.

(d)

2-33 Collinear atoms are marked with asterisks.

2-34

(a)

and

cis   *trans*

(b) no *cis-trans* isomerism around a triple bond

(c) no *cis-trans* isomerism; two groups on each carbon are the same

(d) Theoretically, cyclopentene could show cis-trans isomerism. In reality, the *trans* form is too unstable to exist because of the necessity of stretched bonds and deformed bond angles. *trans*-Cyclopentene has never been detected.

cis   "*trans*"—not possible because of ring strain

(e)

and

these are *cis-trans* isomers, but the designation of *cis* and *trans* to specific structures is not defined because of four different groups on the double bond

2-35

(a) constitutional isomers—the carbon skeletons are different
(b) constitutional isomers—the position of the chlorine atom has changed
(c) cis-trans isomers—the first is *cis,* the second is *trans*
(d) constitutional isomers—the carbon skeletons are different
(e) cis-trans isomers—the first is *trans,* the second is *cis*
(f) same compound—rotation of the first structure gives the second
(g) cis-trans isomers—the first is *cis,* the second is *trans*
(h) constitutional isomers—the position of the double bond relative to the ketone has changed (while it is true that the first double bond is *cis* and the second is *trans,* in order to have *cis-trans* isomers, the rest of the structure must be identical)

2-36 CO_2 is linear; its bond dipoles cancel, so it has no net dipole. SO_2 is bent, so its bond dipoles do not cancel.

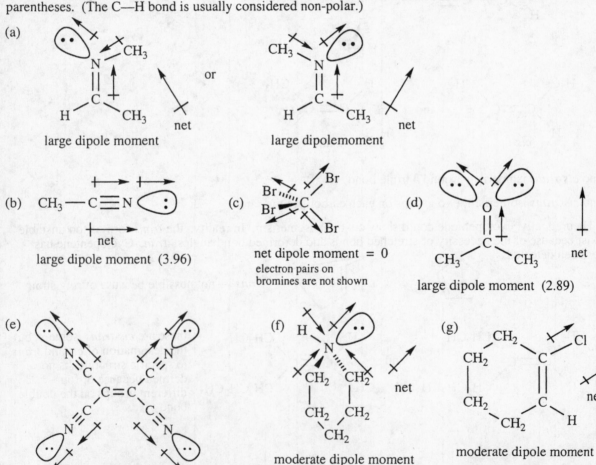

2-37 Some magnitudes of dipole moments are difficult to predict; however, the direction of the dipole should be straightforward in most cases. Actual values of molecular dipole moments are given in parentheses. (The C—H bond is usually considered non-polar.)

40

2-38   Diethyl ether and 1-butanol each have one oxygen, so each can form hydrogen bonds with water (water supplies the H for hydrogen bonding with diethyl ether); their water solubilities should be similar. The boiling point of 1-butanol is much higher because these molecules can hydrogen bond with each other, thus requiring more energy to separate one molecule from another. Diethyl ether molecules cannot hydrogen bond with each other, so it is relatively easy to separate them.

$$CH_3CH_2 \!-\! O \!-\! CH_2CH_3 \qquad\qquad CH_3CH_2CH_2CH_2 \!-\! OH$$

diethyl ether   1-butanol
can hydrogen bond with water   can hydrogen bond with water
cannot hydrogen bond with itself   can hydrogen bond with itself

2-39

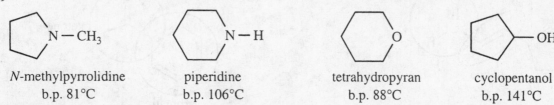

N-methylpyrrolidine   piperidine   tetrahydropyran   cyclopentanol
b.p. 81°C   b.p. 106°C   b.p. 88°C   b.p. 141°C

(a) Piperidine has an N—H bond, so it can hydrogen bond with other molecules of itself. *N*-Methylpyrrolidine has no N—H, so it cannot hydrogen bond and will require less energy (lower boiling point) to separate one molecule from another.

(b) Two effects need to be explained: 1) Why does cyclopentanol have a higher boiling point than tetrahydropyran? and 2) Why do the oxygen compounds have a greater difference in boiling points than the analogous nitrogen compounds?

The answer to the first question is the same as in (a): cyclopentanol can hydrogen bond with its neighbors while tetrahydropyran cannot.

The answer to the second question lies in the text, Table 2-1, that shows the bond dipole moments for C—O and H—O are much greater than C—N and H—N; bonds to oxygen are more polarized, with greater charge separation than bonds to nitrogen.

How is this reflected in the data? The boiling points of tetrahydropyran (88°C) and *N*-methylpyrrolidine (81°C) are close; tetrahydropyran molecules would have a slightly stronger dipole-dipole attraction, and tetrahydropyran is a little less "branched" than *N*-methylpyrrolidine, so it is reasonable that tetrahydropyran boils at a slightly higher temperature. The large difference comes when comparing the boiling points of cyclopentanol (141°C) and piperidine (106°C). The greater polarity of O—H versus N—H is reflected in a more negative oxygen (more electronegative than nitrogen) and a more positive hydrogen, resulting in a much stronger intermolecular attraction. The conclusion is that hydrogen bonding due to O—H is much stronger than that due to N—H.

2-40

(a) can hydrogen bond with itself and with water
(b) can hydrogen bond only with water
(c) can hydrogen bond with itself and with water
(d) can hydrogen bond only with water
(e) cannot hydrogen bond
(f) cannot hydrogen bond

(g) can hydrogen bond only with water
(h) can hydrogen bond with itself and with water
(i) can hydrogen bond only with water
(j) can hydrogen bond only with water
(k) can hydrogen bond only with water
(l) can hydrogen bond with itself and with water

2-41 Higher-boiling compounds are listed.

(a) $CH_3CH(OH)CH_3$ can form hydrogen bonds with other identical molecules
(b) $CH_3CH_2CH_2CH_2CH_3$ has a higher molecular weight than $CH_3CH_2CH_2CH_3$
(c) $CH_3CH_2CH_2CH_2CH_3$ has less branching than $(CH_3)_2CHCH_2CH_3$
(d) $CH_3CH_2CH_2CH_2CH_2Cl$ has a higher molecular weight AND dipole-dipole interaction compared with $CH_3CH_2CH_2CH_2CH_3$

2-42

(a) 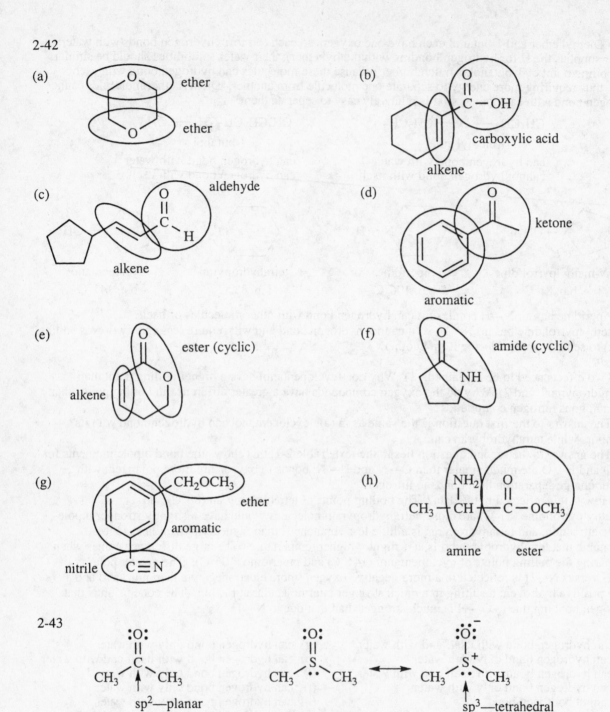 ether ether

(b) carboxylic acid alkene

(c) aldehyde alkene

(d) ketone aromatic

(e) ester (cyclic) alkene

(f) amide (cyclic)

(g) CH₂OCH₃ ether aromatic nitrile C≡N

(h) amine ester

2-43

sp²—planar

sp³—tetrahedral

The key to this problem is understanding that sulfur has a *lone pair of electrons*. The second resonance form shows four pairs of electrons around the sulfur atom, an electronic configuration requiring sp³ hybridization. Sulfur in DMSO cannot be sp² like carbon in acetone, so we would expect sulfur's geometry to be pyramidal (the four electron pairs around sulfur require tetrahedral geometry, but the three atoms around sulfur define its shape as pyramidal).

2-44

(a) penicillin G

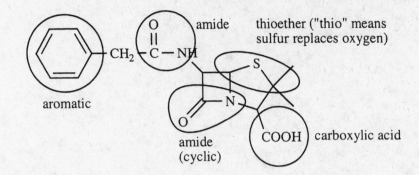

aromatic   amide   thioether ("thio" means sulfur replaces oxygen)

amide (cyclic)

carboxylic acid

(b) dopamine

alcohol

alcohol   aromatic   amine

(In later chapters, you will learn that the OH group on a benzene ring is a special functional group called a "phenol". For now, it fits the broad definition of an alcohol.)

(c) thyroxine

alkyl iodide (four of these)   aromatic   amine

alcohol

ether

aromatic

carboxylic acid

(d) testosterone

OH alcohol

ketone

alkene

43

CHAPTER 3—STRUCTURE AND STEREOCHEMISTRY OF ALKANES

3-1

(a) C_nH_{2n+2} where n = 25 gives $C_{25}H_{52}$

(b) C_nH_{2n+2} where n = 44 gives $C_{44}H_{90}$

> Note to the student: The IUPAC system of nomenclature has a well defined set of rules determining how structures are named. You will find a summary of these rules as Appendix 1 in this Solutions Manual.

3-2 Use hyphens to separate letters from numbers. Longest chains may not always be written left to right.

(a) 3-methylpentane (always find the longest chain; it may not be written in a straight line)
(b) 2-bromo-3-methylpentane (always find the longest chain)
(c) 5-ethyl-2-methyl-4-propylheptane ("When there are two longest chains of equal length, use the chain with the greater number of substituents.")
(d) 4-isopropyl-2-methyldecane

3-3 This Solutions Manual will present line formulas where a question asks for an answer including a structure. If you use condensed structural formulas instead, be sure that you are able to "translate" one structure type into the other.

(a) 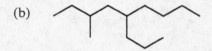 (b) (c)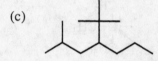

(d)

3-4 Separate numbers from numbers with commas.

(a) 2-methylbutane
(b) 2,2-dimethylpropane
(c) 3-ethyl-2-methylhexane
(d) 2,4-dimethylhexane
(e) 3-ethyl-2,2,4,5-tetramethylhexane
(f) 4-t-butyl-3-methylheptane

3-5 (Hints: *systematize* your approach to these problems. For the isomers of a six carbon formula, for example, start with the isomer containing all six carbons in a straight chain, then the isomers containing a five-carbon chain, then a four-carbon chain, *etc.* Carefully check your answers to AVOID DUPLICATE STRUCTURES.)

(a)

n-hexane   2-methylpentane   3-methylpentane

2,2-dimethylbutane   2,3-dimethylbutane

45

3-5 continued

(b)

n-heptane   2-methylhexane   3-methylhexane   2,2-dimethylpentane

3,3-dimethylpentane   2,3-dimethylpentane   2,4-dimethylpentane   3-ethylpentane   2,2,3-trimethylbutane

3-6 For this problem, the carbon numbers in the substituents are indicated in italics.

(a)
$$CH_3$$
$$—CHCH_3$$
$$\quad 1 \quad 2$$
1-methylethyl
common name = isopropyl

(b)
$$CH_3$$
$$—CH_2CHCH_3$$
$$\quad 1 \quad 2 \quad 3$$
2-methylpropyl
common name = isobutyl

(c)
$$CH_3$$
$$—CHCH_2CH_3$$
$$2° \quad 1 \quad 2 \quad 3$$
1-methylpropyl
common name = sec-butyl

(d)
$$CH_3$$
$$—C—CH_3$$
$$\quad 1 \quad 2$$
$$\quad CH_3$$
$$3°$$
1,1-dimethylethyl
common name = t-butyl or tert-butyl

3-7

(a)   (b)

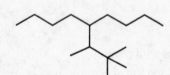

3-8 Once the number of carbons is determined, C_nH_{2n+2} gives the formula.
(a) $C_{10}H_{22}$ (b) $C_{15}H_{32}$

3-9

(a) (lowest b.p.) hexane < octane < decane (highest b.p.) —molecular weight

(b) $(CH_3)_3C—C(CH_3)_3$ < $CH_3CH_2C(CH_3)_2CH_2CH_2CH_3$ < octane —branching
    (lowest b.p.)   (highest b.p.)

3-10

(a) (lowest m.p.) hexane < octane < decane (highest m.p.) —molecular weight

(b) octane < $CH_3CH_2C(CH_3)_2CH_2CH_2CH_3$ < $(CH_3)_3C—C(CH_3)_3$ —branching
  (lowest m.p.)   (highest m.p.)

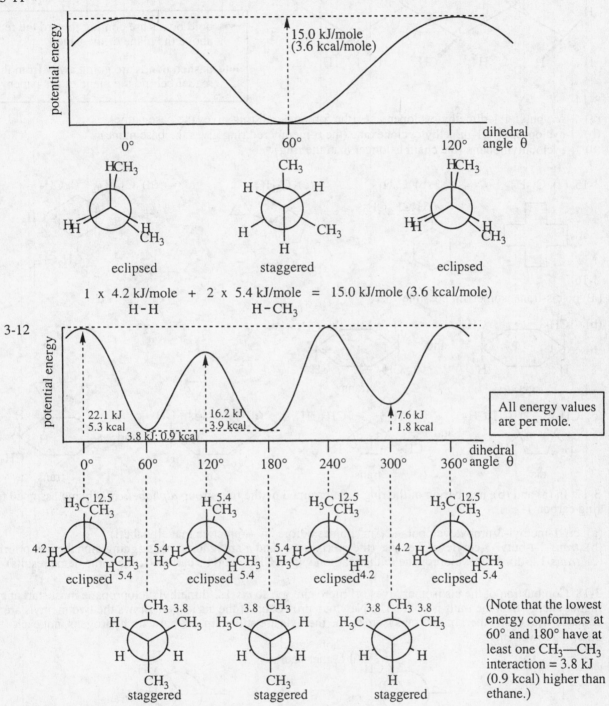

3-11

$$1 \times 4.2 \text{ kJ/mole} + 2 \times 5.4 \text{ kJ/mole} = 15.0 \text{ kJ/mole (3.6 kcal/mole)}$$
H–H   H–CH$_3$

3-12

All energy values are per mole.

(Note that the lowest energy conformers at 60° and 180° have at least one CH$_3$—CH$_3$ interaction = 3.8 kJ (0.9 kcal) higher than ethane.)

Relative energies on the graph above were calculated using these values from the text:
3.8 kJ/mole (0.9 kcal/mole) for a CH$_3$–CH$_3$ gauche (staggered) interaction; 4.2 kJ/mole (1.0 kcal/mole) for a H–H eclipsed interaction; 5.4 kJ/mole (1.3 kcal/mole) for a H–CH$_3$ eclipsed interaction; 12.5 kJ/mole (3.0 kcal/mole) for a CH$_3$–CH$_3$ eclipsed interaction. These values in kJ/mole are noted on each structure and are summed to give the energy value on the graph. Slight variation between this graph and the one in the text are due to rounding.

47

3-13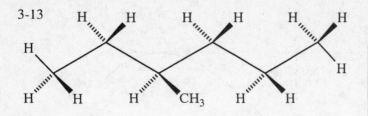

all bonds are staggered

> ━━ bold bonds are coming toward the reader above the plane of the paper
>
> ⋯⋯ dashed bonds are going away from the reader behind the plane of the paper

3-14

(a) 3-*sec*-butyl-1,1-dimethylcyclopentane (the 5-membered ring gives the base name)
(b) 3-cyclopropyl-1,1-dimethylcyclohexane (the 6-membered ring gives the base name)
(c) 4-cyclobutylnonane (the chain is longer than the ring)

3-15 (a) $C_{12}H_{24}$　　　(b) C_9H_{18}　　　(c) C_8H_{14}　　　(d) $C_{10}H_{20}$

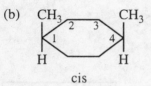
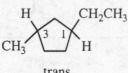
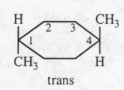

3-16

(a) no cis-trans isomerism possible

(b)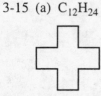

3-17 In (a) and (b), numbering of the ring is determined by the first group *alphabetically* being assigned to ring carbon 1.

(a) *cis*-1-methyl-3-propylcyclobutane ("m" comes before "p"—practice that alphabet!)
(b) *trans*-1-*t*-butyl-3-ethylcyclohexane (the prefixes *t*, *s*, and *n* are ignored in assigning alphabetical priority)
(c) *trans*-1,2-dimethylcyclopropane (either carbon with a CH_3 could be carbon-1; the same name results)

3-18 Combustion of the cis isomer gives off more energy, so *cis*-1,2-dimethylcyclopropane must start at a higher energy than the trans isomer. The Newman projection of the cis isomer shows the two methyls are eclipsed with each other; in the trans isomer, the methyls are still eclipsed, but with hydrogens, not each other—a lower energy.

more strain = higher energy

cis

trans

3-19 *trans*-1,2-Dimethylcyclobutane is more stable than *cis* because the two methyls can be farther apart when trans, as shown in the Newman projections.

CH₃) more strain =
CH₃) higher energy

cis

H
CH₃

trans

H CH₃

H H

In the 1,3-dimethylcyclobutanes, however, the *cis* allows the methyls to be farther from other atoms and therefore more stable than the *trans*.

CH₃ ⟨ ⟩ CH₃   cis
H   H

H ⟨ ⟩ CH₃   trans
CH₃ ⟷ H

3-20

equatorial only

axial only

showing both axial
and equatorial

3-21 The abbreviation for a methyl group, CH₃, is "Me". Ethyl is "Et", propyl is "Pr", and butyl is "Bu".

(a)

Me   Me
   Me H
H   H
H   H
   H   Me
Me   Me

all methyls axial
(all H's equatorial)

(b)

H   H
   H Me
Me   Me
Me   Me
   Me   H
H   H

all methyls equatorial
(all H's axial)

Note that axial groups alternate
up and down around the ring.

3-22 Carbons 4 and 6 are behind the circles.

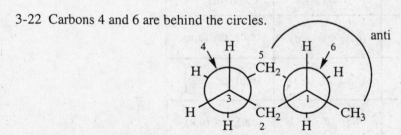

anti

3-23 The isopropyl group can rotate so that its hydrogen is near the axial hydrogens on carbons 3 and 5, similar to a methyl group's hydrogen, and therefore similar to a methyl group in energy. The *t*-butyl group, however, must point a methyl group toward the hydrogens on carbons 3 and 5, giving severe diaxial interactions, causing the energy of this conformer to jump dramatically.

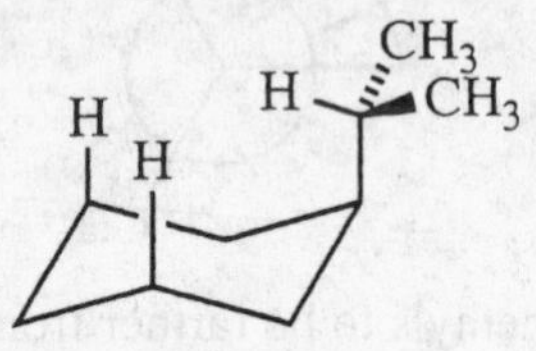

isopropylcyclohexane

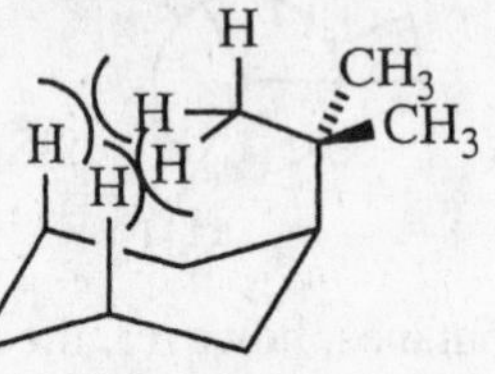

t-butylcyclohexane

3-24 The most stable conformers have substituents equatorial.

(a)   (b) 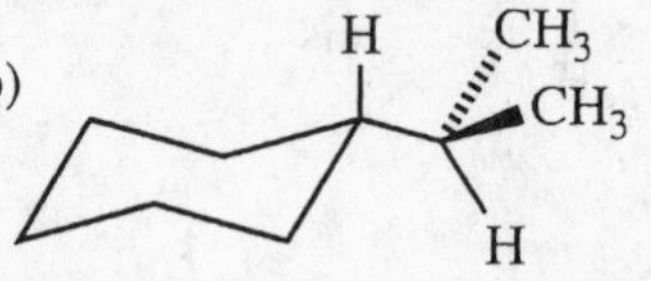   (c)

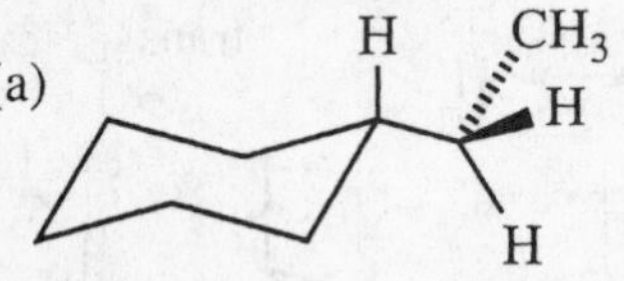

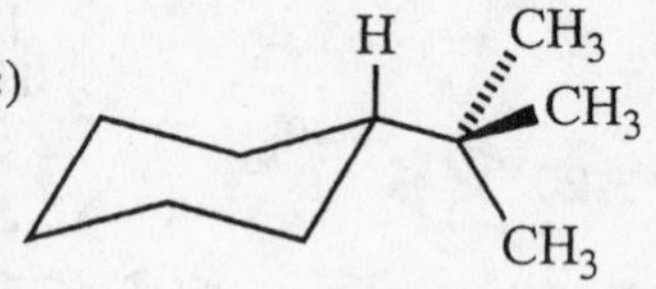

3-25

(a) cis

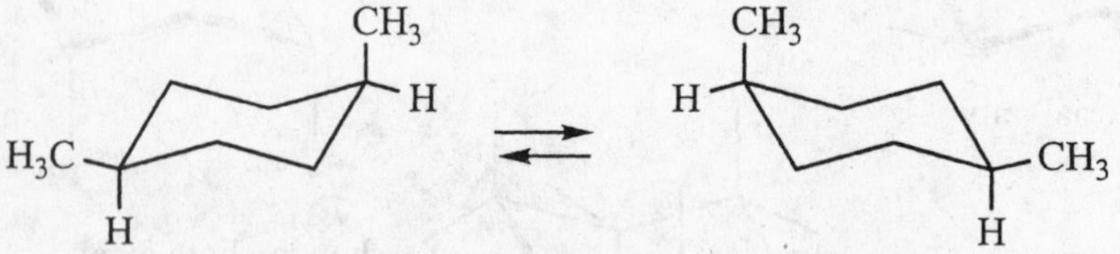

equatorial, axial   axial, equatorial

EQUAL ENERGY

(b) trans

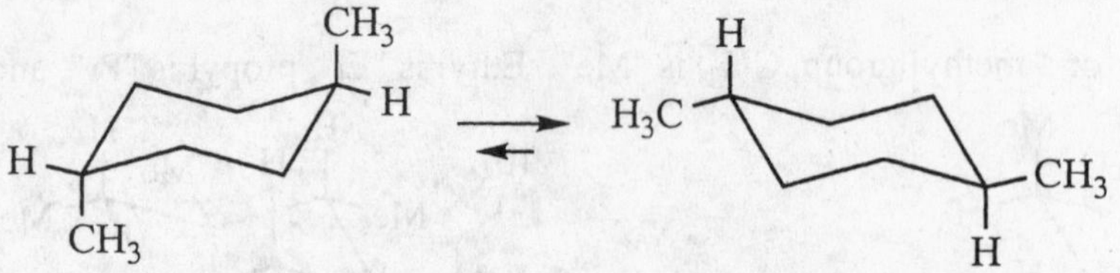

axial, axial
higher energy

equatorial, equatorial
lower energy

(c) The trans isomer is more stable because BOTH substituents can be in the preferred equatorial positions.

3-26

Positions	cis	trans
1,2	(e,a) or (a,e)	(e,e) or (a,a)
1,3	(e,e) or (a,a)	(e,a) or (a,e)
1,4	(e,a) or (a,e)	(e,e) or (a,a)

50

3-27 The more stable conformer places the larger group equatorial.

(a)

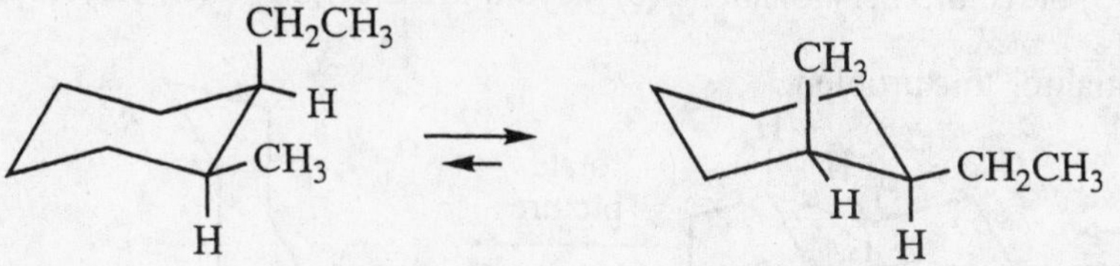

more stable

(b)

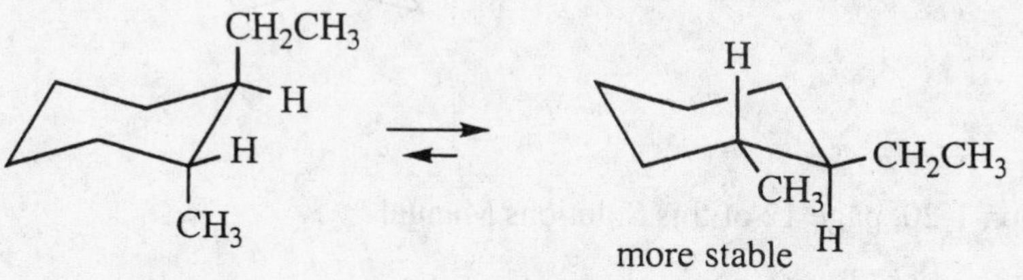

more stable

(c)

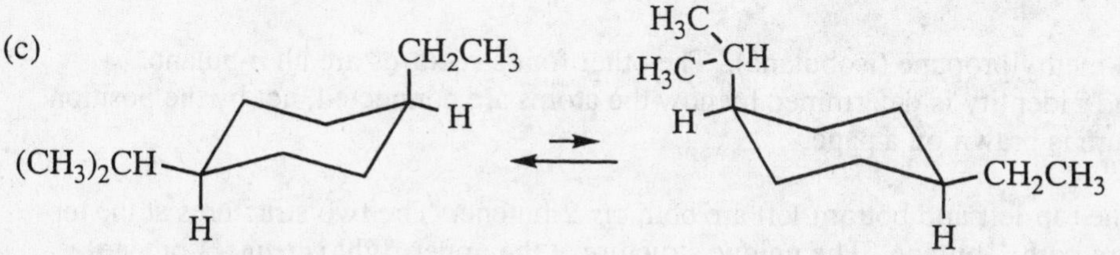

more stable

(d)

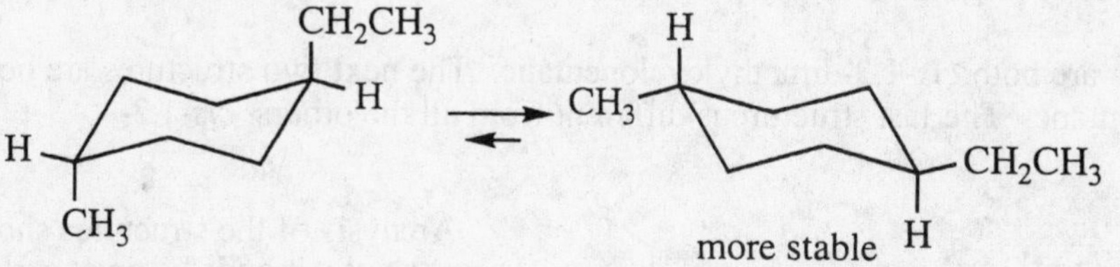

more stable

3-28 The key to determining cis and trans around a cyclohexane ring is to see whether a substituent group is "up" or "down" *relative to the H at the same carbon*. Two "up" groups or two "down" groups will be cis; one "up" and one "down" will be trans. This works independent of the conformation the molecule is in!

(a) *cis*-1,3-dimethylcyclohexane
(b) *cis*-1,4-dimethylcyclohexane
(c) *trans*-1,2-dimethylcyclohexane

(d) *cis*-1,3-dimethylcyclohexane
(e) *cis*-1,3-dimethylcyclohexane
(f) *trans*-1,4-dimethylcyclohexane

3-29

(a)

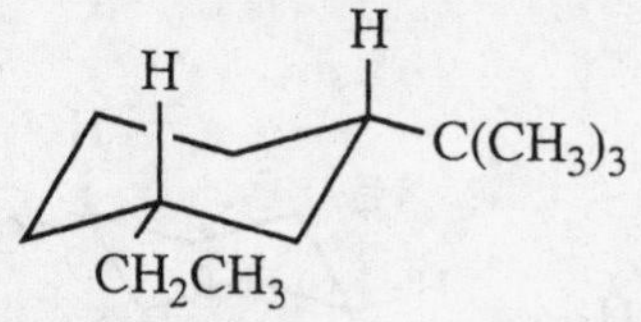

(b)

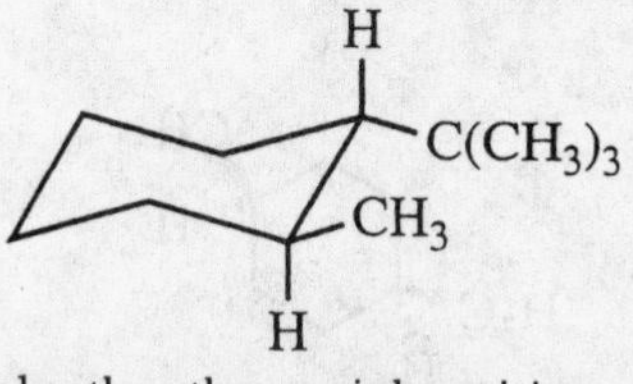

(c) Bulky substituents like *t*-butyl adopt equatorial rather than axial positions, even if that means altering the conformation of the ring. The twist boat conformation allows both bulky substituents to be "equatorial".

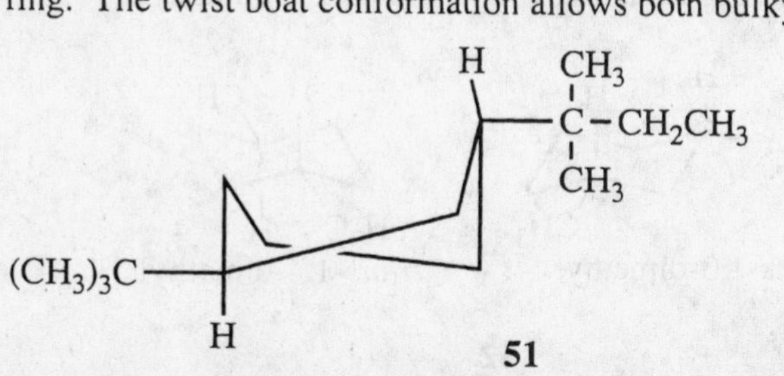

51

3-30 The nomenclature of bicyclic alkanes is summarized in Appendix 1.
(a) bicyclo[3.1.0]hexane (b) bicyclo[3.3.1]nonane (c) bicyclo[2.2.2]octane (d) bicyclo[3.1.1]heptane

3-31 Using models is essential for this problem.

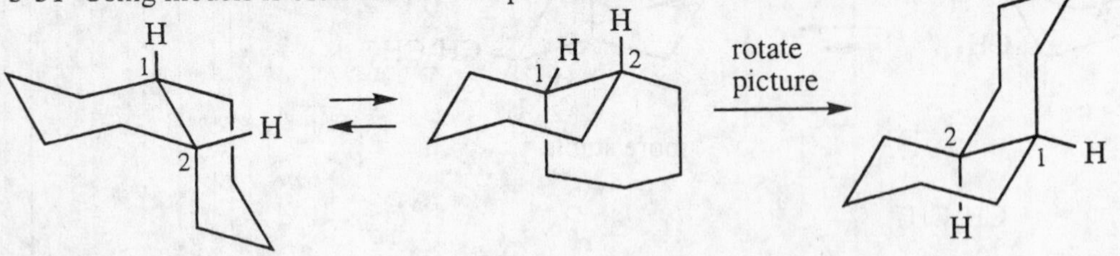

from text Figure 3-29

3-32 Please refer to solution 1-20, page 12 of this Solutions Manual.

3-33

(a) The third structure is 2-methylpropane (isobutane). The other four structures are all *n*-butane. Remember that a compound's identity is determined by how the atoms are connected, not by the position of the atoms when a structure is drawn on a page.

(b) The two structures at the top-left and bottom-left are both *cis*-2-butene. The two structures at the top-center and bottom-center are both 1-butene. The unique structure at the upper right is *trans*-2-butene. The unique structure at the lower right is 2-methylpropene.

(c) The first two structures are both *cis*-1,2-dimethylcyclopentane. The next two structures are both *trans*-1,2-dimethylcyclopentane. The last structure is different from all the others, *cis*-1,3-dimethylcyclopentane.

(d)

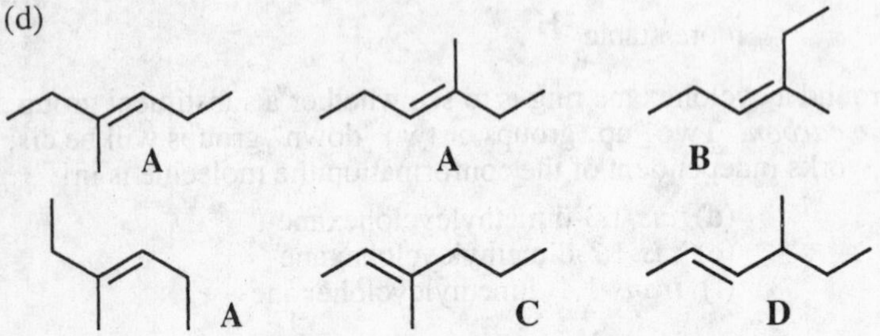

Analysis of the structures shows that some double bonds begin at carbon-2 and some at carbon-3 of the longest chain. The three structures labeled **A** are the same, with the double bond *trans*; **B** is a geometric isomer (*cis*) of **A**. **C** and **D** are constitutional isomers of the others.

(e) Naming the structures shows that three of the structures are *trans*-1,4-dimethylcyclohexane, two are the *cis* isomer, and one is *cis*-1,3-dimethylcyclohexane. Although a structure may be shown in two different conformations, it still represents only one compound.

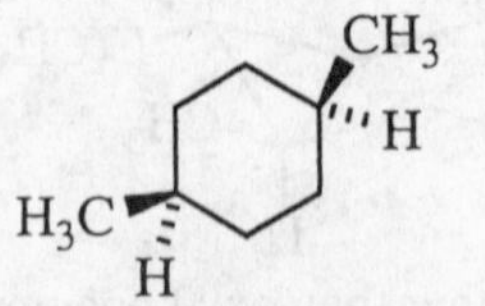

cis-1,4-dimethyl

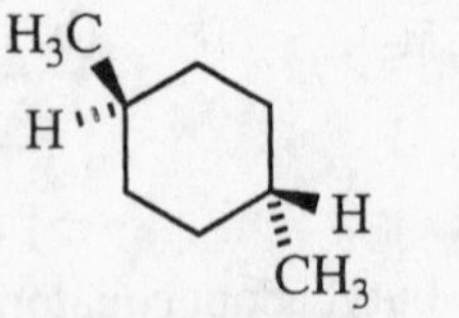

trans-1,4-dimethyl

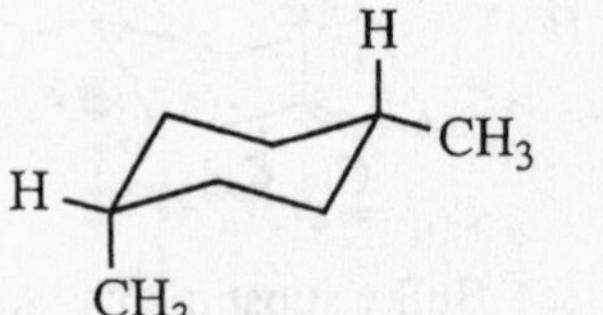

cis-1,4-dimethyl

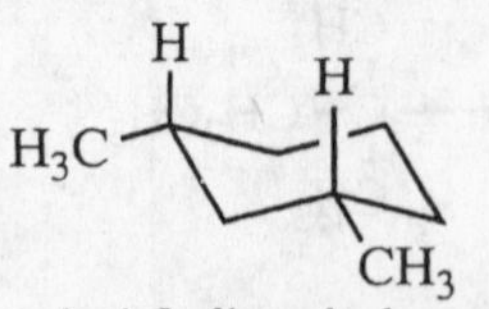

cis-1,3-dimethyl

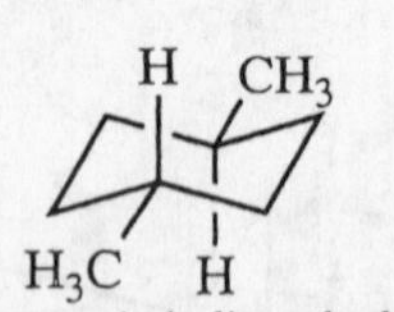

trans-1,4-dimethyl

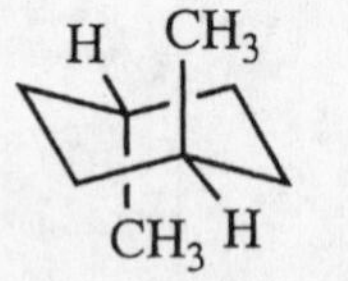

trans-1,4-dimethyl

3-34 Line formulas are shown.

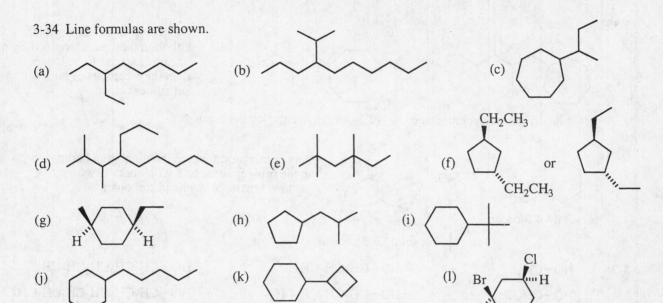

(a)

(b)

(c)

(d)

(e)

(f)   or

(g)   H   H

(h)

(i)

(j)

(k)

(l)   Cl
         Br   H
         H

3-35 There are many possible answers to each of these problems. The ones shown here are examples of correct answers. Your answers may be different AND correct. Check your answers in your study group.

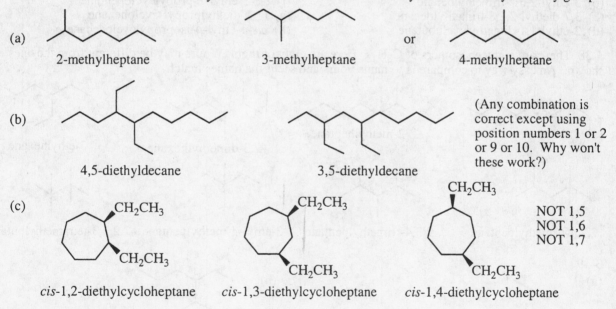

(a)

2-methylheptane   3-methylheptane   or   4-methylheptane

(b)

4,5-diethyldecane   3,5-diethyldecane

(Any combination is correct except using position numbers 1 or 2 or 9 or 10. Why won't these work?)

(c)   CH₂CH₃   CH₂CH₃   CH₂CH₃

         CH₂CH₃   CH₂CH₃   CH₂CH₃

NOT 1,5
NOT 1,6
NOT 1,7

cis-1,2-diethylcycloheptane   cis-1,3-diethylcycloheptane   cis-1,4-diethylcycloheptane

(d) only two possible answers

         CH₃   CH₃
              CH₃

                                    CH₃

trans-1,2-dimethylcyclopentane   trans-1,3-dimethylcyclopentane

3-35 continued

(e)

(2,3-dimethylpentyl)cycloheptane

(2,3-dimethylpentyl)cyclooctane

Other ring sizes are possible, although they must have 6 or more carbons to be longer than the 5 carbons of the substituent chain.

(f)

bicyclo[4.4.0]decane

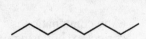

bicyclo[3.3.2]decane

Any combination where the number of carbons in the bridges sums to 8 will work. (Two carbons are the bridgehead carbons.)

3-36   HO–CH₃   HO–CH₂CH₂CH₃   HO–CH₂CH₂CH₂CH₂CH₃

HO–CH₂CH₃   HO–CH₂CH₂CH₂CH₃   HO–CH₂CH₂CH₂CH₂CH₂CH₃

3-37

(a) 3-ethyl-2,2,6-trimethylheptane
(b) 3-ethyl-2,6,7-trimethyloctane
(c) 3,7-diethyl-2,2,8-trimethyldecane
(d) 2-ethyl-1,1-dimethylcyclobutane

(e) bicyclo[4.1.0]heptane
(f) *cis*-1-ethyl-3-propylcyclopentane
(g) (1,1-diethylpropyl)cyclohexane
(h) *cis*-1-ethyl-4-isopropylcyclodecane

3-38 There are eighteen isomers of C₈H₁₈. Here are eight of them. Yours may be different from the ones shown. An easy way to compare is to name yours and see if the names match.

n-octane

2-methylheptane

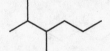

2,3-dimethylhexane

3-ethylhexane

2,2,4-trimethylpentane

2,3,4-trimethylpentane

3-ethyl-3-methylpentane   2,2,3,3-tetramethylbutane

3-39

(a)

correct name: 3-methylhexane
(longer chain)

(b)

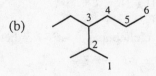

correct name: 3-ethyl-2-methylhexane
(more branching with this numbering)

(c)

Cl

correct name: 2-chloro-3-methylhexane
(begin numbering at end closest to substituent)

(d)

correct name: 2,2-dimethylbutane (include a position number for each substituent, regardless of redundancies) (e)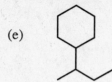

correct name: *sec*-butylcyclohexane or
(1-methylpropyl)cyclohexane
(the longer chain or ring is the base name)

(f)
CH$_2$CH$_3$
CH$_2$CH$_3$

correct name: 1,2-diethylcyclopentane
(position numbers are the lowest possible)

3-40

(a) *n*-Octane has a higher boiling point than 2,2,3-trimethylpentane because linear molecules boil higher than branched molecules of the same molecular weight (increased van der Waals interaction).

(b) 2-Methylnonane has a higher boiling point than *n*-heptane because it has a significantly higher molecular weight than *n*-heptane.

(c) *n*-Nonane boils higher than 2,2,5-trimethylhexane for the same reason as in (a).

3-41   The point of attachment is shown by the bold bond at the left of each structure.

—CH$_2$CH$_2$CH$_2$CH$_2$CH$_3$
1°   *n*-pentyl

—CHCH$_2$CH$_2$CH$_3$
2° CH$_3$ 1-methylbutyl

—CH$_2$CHCH$_2$CH$_3$
1°   CH$_3$ 2-methylbutyl

—CH$_2$CH$_2$CHCH$_3$
1°   CH$_3$ 3-methylbutyl
                          (isopentyl)

—CHCH$_2$CH$_3$
2° CH$_2$CH$_3$ 1-ethylpropyl

CH$_3$
—C—CH$_2$CH$_3$
3° CH$_3$ 1,1-dimethylpropyl
                (*t*-pentyl)

CH$_3$
—CHCHCH$_3$
2° CH$_3$ 1,2-dimethylpropyl

CH$_3$
—CH$_2$—C—CH$_3$
1°   CH$_3$ 2,2-dimethylpropyl
                    (*neo*-pentyl)

3-42 In each case, put the largest groups on adjacent carbons in anti positions to make the most stable conformations.

(a) 3-methylpentane

2
3

C-2 is the front carbon with H, H, and CH$_3$
C-3 is the back carbon with H, CH$_3$, and CH$_2$CH$_3$

4   5
CH$_2$CH$_3$
H   H
2
H   CH$_3$
1 CH$_3$

carbon-3 cannot be seen;
it is behind carbon-2

Newman projections were introduced in text section 3-7. They are used extensively to show the three-dimensional arrangement of atoms in a molecule.

3-42 continued
(b) 3,3-dimethylhexane

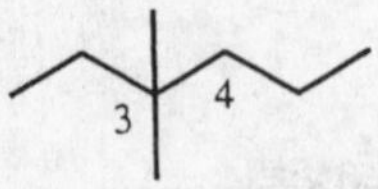

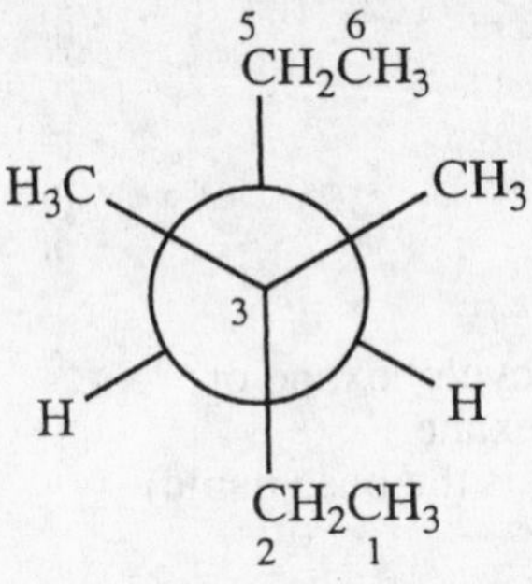

C-3 is the front carbon with CH_3, CH_3, and CH_2CH_3
C-4 is the back carbon with H, H, and CH_2CH_3

carbon-4 cannot be seen;
it is behind carbon-3

3-43
(a)

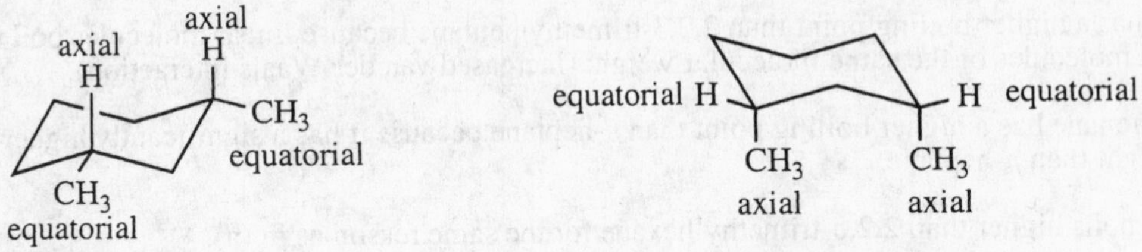

(b)   more stable   less stable
           (lower energy)   (higher energy)

(c) From Section 3-14 of the text, each gauche interaction raises the energy 3.8 kJ/mole (0.9 kcal/mole), and each axial methyl has two gauche interactions, so the energy is:
2 methyls x 2 interactions per methyl x 3.8 kJ/mole per interaction = 15.2 kJ/mole (3.6 kcal/mole)

(d) The steric strain from the 1,3-diaxial interaction of the methyls must be the difference between the total energy and the energy due to gauche interactions:
        23 kJ/mole − 15.2 kJ/mole = 7.8 kJ/mole
        (5.4 kcal/mole − 3.6 kcal/mole = 1.8 kcal/mole)

3-44 The more stable conformer places the larger group equatorial.

(a)

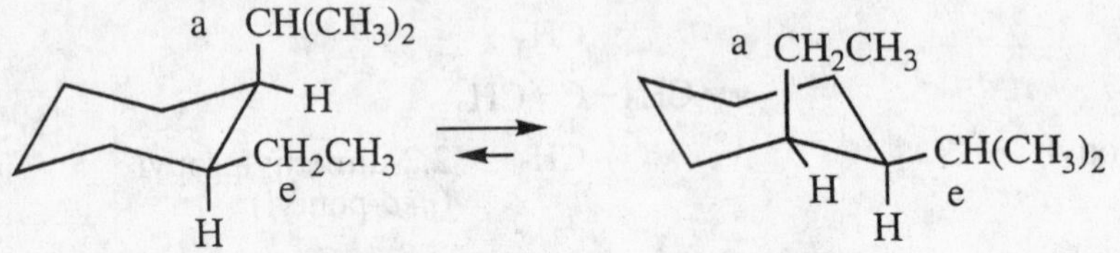

more stable

(b)

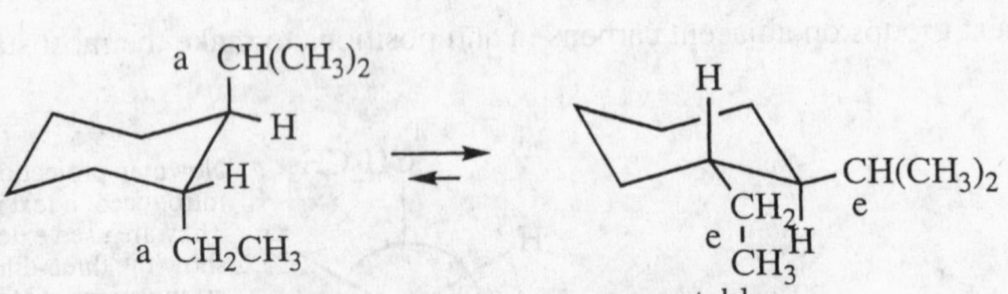

more stable

(c)

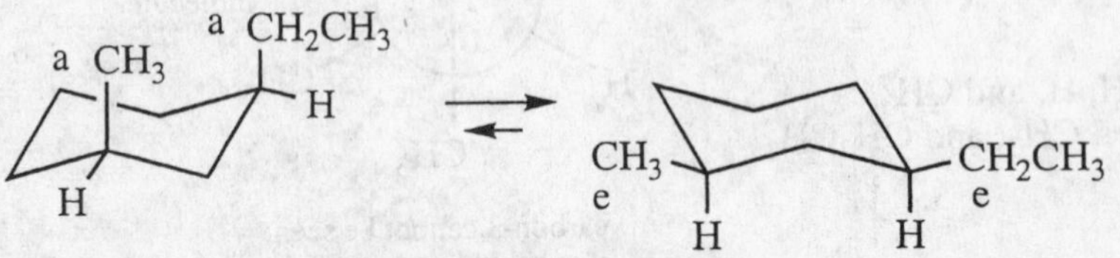

more stable

56

3-44 continued

(d)

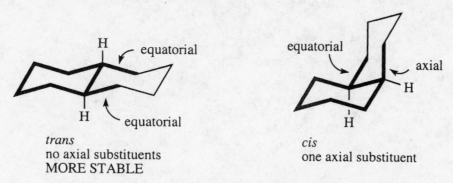

a CH₂CH₃

H

H

CH₃
e

⇌

H

CH₂CH₃
e

a CH₃
H

more stable

(e)

a CH₂CH₃

H

CH₃
e

H

⇌

a CH₃

H

CH₂CH₃
e

more stable H

3-45 (Using models is essential to this problem.)

In both *cis-* and *trans*-decalin, the cyclohexane rings can be in chair conformations. The relative energies
will depend on the number of axial substituents.

H ← equatorial

H ← equatorial

trans
no axial substituents
MORE STABLE

equatorial →

← axial
H

H

cis
one axial substituent

3-46 chair form of glucose—all substituents equatorial

OH

H

CH₂

H

HO

HO

H

H

OH

H

O

OH

OH

H

OH

CH₂

HO

HO

OH

O

OH

OH

(without ring H's shown)

4-1

(a)

H H H
| | |
H—C—C—C•
| | |
H H H

(b)

H   H
| • |
H—C—C—C—H
| | |
H   C   H
   / | \
  H H H

(c)

H   •   H
| | |
H—C—C—C—H
| | |
H   H   H

(d)

$\cdot \ddot{I} \cdot$

4-2

(a)

(b) Free-radical halogenation substitutes a halogen atom for a hydrogen. Even if a molecule has only one type of hydrogen, substitution of the first of these hydrogens forms a new compound. Any remaining hydrogens in this product can compete with the initial reactant for the available halogen. Thus, chlorination of methane, CH_4, produces all possible substitution products: CH_3Cl, CH_2Cl_2, $CHCl_3$, and CCl_4.

    If a molecule has different types of hydrogens, the reaction can generate a mixture of the possible substitution products.

(c) Production of CCl_4 or CH_3Cl can be controlled by altering the ratio of CH_4 to Cl_2. To produce CCl_4, use an excess of Cl_2 and let the reaction proceed until all C—H bonds have been replaced with C—Cl bonds. Producing CH_3Cl is more challenging because the reaction tends to proceed past the first substitution. By using a very large excess of CH_4 to Cl_2, perhaps 100 to 1 or even more, a chlorine atom is more likely to find a CH_4 molecule than it is to find a CH_3Cl, so only a small amount of CH_4 is transformed to CH_3Cl by the time the Cl_2 runs out, with almost no CH_2Cl_2 being produced.

4-3

(a) This mechanism requires that one photon of light be added for each CH_3Cl generated, a quantum yield of 1. The actual quantum yield is several hundred or thousand. The high quantum yield suggests a chain reaction, but this mechanism is not a chain; it has no propagation steps.

(b) This mechanism conflicts with at least two experimental observations. First, the energy of light required to break a H–CH_3 bond is 435 kJ/mole (104 kcal/mole, from Table 4-2); the energy of light determined by experiment to initiate the reaction is only 251 kJ/mole of photons (60 kcal/mole of photons), much less than the energy needed to break this H–C bond. Second, as in (a), each CH_3Cl produced would require one photon of light, a quantum yield of 1, instead of the actual number of several hundred or thousand. As in (a), there is no provision for a chain process, since all the radicals generated are also consumed in the mechanism.

4-4

(a) The twelve hydrogens of cyclohexane are all on equivalent 2° carbons. Replacement of any one of the twelve will lead to the same product, chlorocyclohexane. *n*-Hexane, however, has hydrogens in three different positions: on carbon-1 (equivalent to carbon-6), carbon-2 (equivalent to carbon-5), and carbon-3 (equivalent to carbon-4). Monochlorination of *n*-hexane will produce a mixture of all three possible isomers: 1-, 2-, and 3-chlorohexane.

4-4 continued

(b) The best conversion of cyclohexane to chlorocyclohexane would require the ratio of cyclohexane/chlorine to be a large number. If the ratio were small, as the concentration of chlorocyclohexane increased during the reaction, chlorine would begin to substitute for a second hydrogen of chlorocyclohexane, generating unwanted products. The goal is to have chlorine attack a molecule of cyclohexane before it ever encounters a molecule of chlorocyclohexane, so the concentration of cyclohexane should be kept high.

4-5

(a) $K_{eq} = e^{-\Delta G°/RT}$ $\quad^{-2.1\ kJ/mole\ =\ -2100\ J/mole}$ $\qquad\qquad$ K = temperature in the kelvin scale

$\qquad = e^{-(-2100\ J/mole)/((8.314\ J/K\text{-}mole)\cdot(298\ K))}$

$\qquad = e^{2100\ /\ 2478} = e^{0.847} = \boxed{2.3}$

(b) $K_{eq} = 2.3 = \dfrac{[CH_3SH]\ [HBr]}{[CH_3Br]\ [H_2S]}$

$\qquad\qquad\qquad$ [CH$_3$Br] $\quad$ [H$_2$S] $\quad$ [CH$_3$SH] $\quad$ [HBr]

initial concentrations: $\qquad$ 1 $\qquad$ 1 $\qquad$ 0 $\qquad$ 0
final concentrations $\qquad$ 1 – x $\quad$ 1 – x $\quad$ x $\quad$ x

$K_{eq} = 2.3 = \dfrac{x\cdot x}{(1-x)(1-x)} = \dfrac{x^2}{1-2x+x^2} \implies x^2 = 2.3\,x^2 - 4.6\,x + 2.3$

$0 = 1.3\,x^2 - 4.6\,x + 2.3 \implies x = 0.60,\ 1-x = 0.40$
$\qquad\qquad\qquad\qquad\qquad\qquad\qquad$ (using quadratic equation)

$\boxed{[CH_3SH] = [HBr] = 0.60\ M;\ [CH_3Br] = [H_2S] = 0.40\ M}$

4-6 $\quad$ 2 acetone $\rightleftharpoons$ diacetone

Assume that the initial concentration of acetone is 1 molar, and 5% of the acetone is converted to diacetone. NOTE THE MOLE RATIO.

	[acetone]	[diacetone]
initial concentrations:	1 M	0
final concentrations:	0.95 M	0.025 M

$K_{eq} = \dfrac{[diacetone]}{[acetone]^2} = \dfrac{0.025}{(0.95)^2} = \boxed{0.028}$

$\Delta G° = -2.303\ RT\ \log_{10} K_{eq} = -2.303\ ((8.314\ J/K\text{-}mole)\cdot(298K))\cdot\log(0.028)$

$\qquad = \boxed{+8.9\ kJ/mole \quad (+2.1\ kcal/mole)}$

4-7 $\Delta S°$ will be negative since two molecules are combined into one, a loss of freedom of motion. Since $\Delta S°$ is negative, $-T\Delta S°$ is positive; but $\Delta G°$ is a large negative number since the reaction goes to completion. Therefore, $\Delta H°$ must also be a large negative number, necessarily larger in absolute value than $\Delta G°$. We can explain this by formation of two strong C–H bonds (410 kJ/mole each) after breaking a strong H–H bond (435 kJ/ mole) and a WEAKER C=C pi bond.

4-8

(a) $\Delta S°$ is positive—one molecule became two smaller molecules with greater freedom of motion

(b) $\Delta S°$ is negative—two smaller molecules combined into one larger molecule with less freedom of motion

(c) $\Delta S°$ cannot be predicted since the number of molecules in reactants and products is the same

4-9 (a) initiation   (1) $Cl-Cl \xrightarrow{h\nu} 2\,Cl\cdot$

propagation
- (2) $Cl\cdot + H-CH_2CH_3 \longrightarrow H-Cl + \cdot CH_2CH_3$
- (3) $Cl-Cl + \cdot CH_2CH_3 \longrightarrow Cl-CH_2CH_3 + Cl\cdot$

termination
- (4) $Cl\cdot + \cdot Cl \longrightarrow Cl-Cl$
- (5) $Cl\cdot + \cdot CH_2CH_3 \longrightarrow Cl-CH_2CH_3$
- (6) $CH_3CH_2\cdot + \cdot CH_2CH_3 \longrightarrow CH_3CH_2CH_2CH_3$

(b) step (1)   break Cl–Cl $\Delta H° = +242$ kJ/mole (+ 58 kcal/mole)

step (2)   break H–CH$_2$CH$_3$ $\Delta H° = +410$ kJ/mole (+ 98 kcal/mole)
         make H–Cl $\underline{\Delta H° = -431\text{ kJ/mole}\ \ (-103\text{ kcal/mole})}$
              step (2) $\Delta H° = -21$ kJ/mole (–5 kcal/mole)

step (3)   break Cl–Cl $\Delta H° = +242$ kJ/mole (+ 58 kcal/mole)
         make Cl–CH$_2$CH$_3$ $\underline{\Delta H° = -339\text{ kJ/mole}\ \ (-81\text{ kcal/mole})}$
              step (3) $\Delta H° = -97$ kJ/mole (–23 kcal/mole)

(c) $\Delta H°$ for the reaction is the sum of the $\Delta H°$ values of the individual propagation steps:

-21 kJ/mole $+$ -97 kJ/mole $=$ **–118 kJ/mole**
$(-5$ kcal/mole $+$ -23 kcal/mole $=$ **–28 kcal/mole**)

4-10 (a) initiation   (1) $Br-Br \xrightarrow{h\nu} 2\,Br\cdot$

propagation
- (2) $Br\cdot + H-CH_3 \longrightarrow H-Br + \cdot CH_3$
- (3) $Br-Br + \cdot CH_3 \longrightarrow Br-CH_3 + Br\cdot$

step (1)   break Br–Br $\Delta H° = +192$ kJ/mole (+ 46 kcal/mole)

step (2)   break H–CH$_3$ $\Delta H° = +435$ kJ/mole (+ 104 kcal/mole)
         make H–Br $\underline{\Delta H° = -368\text{ kJ/mole}\ \ (-88\text{ kcal/mole})}$
              step (2) $\Delta H° = +67$ kJ/mole (+16 kcal/mole)

step (3)   break Br–Br $\Delta H° = +192$ kJ/mole (+ 46 kcal/mole)
         make Br–CH$_3$ $\underline{\Delta H° = -293\text{ kJ/mole}\ \ (-70\text{ kcal/mole})}$
              step (3) $\Delta H° = -101$ kJ/mole (–24 kcal/mole)

(b) $\Delta H°$ for the reaction is the sum of the $\Delta H°$ values of the individual propagation steps:

$+67$ kJ/mole $+$ -101 kJ/mole $=$ **–34 kJ/mole**
$(+16$ kcal/mole $+$ -24 kcal/mole $=$ **–8 kcal/mole**)

4-11

(a) first order: the exponent of [(CH$_3$)$_3$CCl] in the rate law = 1
(b) zeroth order: [CH$_3$OH] does not appear in the rate law (its exponent is zero)
(c) first order: the sum of the exponents in the rate law = 1 + 0 = **1**

4-12

(a) first order: the exponent of [cyclohexene] in the rate law = 1
(b) second order: the exponent of [Br$_2$] in the rate law = 2
(c) third order: the sum of the exponents in the rate law = 1 + 2 = **3**

4-13

(a) the reaction rate depends on neither [ethylene] nor [hydrogen], so it is zeroth order in both species. The overall reaction must be zeroth order.
(b) rate = k$_r$
(c) The rate law does not depend on the concentration of the reactants. It must depend, therefore, on the only other chemical present, the catalyst. Apparently, whatever is happening on the surface of the catalyst determines the rate, regardless of the concentrations of the two gases. Increasing the surface area of the catalyst, or simply adding more catalyst, would accelerate the reaction.

4-14

(a)

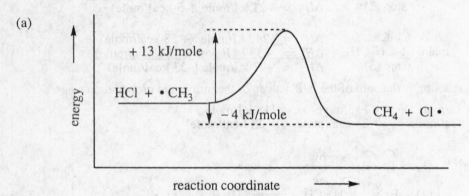

(b) E_a = + 13 kJ/mole   (+ 3 kcal/mole)
(c) $\Delta H°$ = – 4 kJ/mole (– 1 kcal/mole)

4-15

(a)

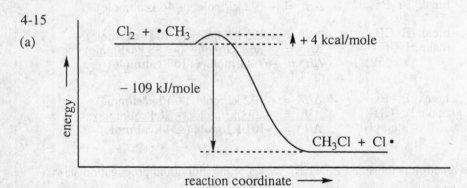

(b) reverse: CH$_3$Cl + Cl • $\longrightarrow$ Cl$_2$ + • CH$_3$

(c) reverse: E_a = + 109 kJ/mole + + 4 kJ/mole = **+ 113 kJ/mole**
        (+ 26 kcal/mole + + 1 kcal/mole = **+ 27 kcal/mole**)

4-16

(a)

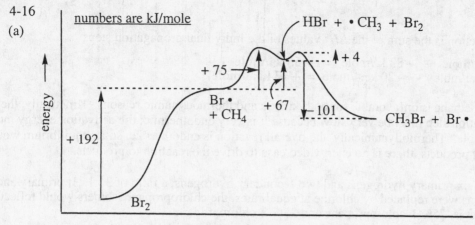

numbers are kJ/mole

HBr + •CH₃ + Br₂

+ 75

+ 4

Br•
+ CH₄

+ 67

− 101

CH₃Br + Br •

+ 192

Br₂

energy →

reaction coordinate ⟶

(b) The step leading to the highest energy transition state is rate-limiting. In this mechanism, the first propagation step is rate-limiting:

$$ Br• + CH_4 \longrightarrow HBr + •CH_3 $$

(c)   (1) $\left[\begin{array}{cc} \delta• & \delta• \\ Br\text{-------}Br \end{array}\right]^{\ddagger}$ ("δ•" means partial radical character on the atom)

(2) $\left[\begin{array}{c} H \quad \delta• \\ | \\ H-C\text{--------}H\text{------}Br \\ | \\ H \end{array}\right]^{\ddagger}$

(3) $\left[\begin{array}{c} H \quad \delta• \\ | \\ H-C\text{--------}Br\text{------}Br \\ | \\ H \end{array}\right]^{\ddagger}$

(d) $\Delta H°$ for the reaction is the sum of the $\Delta H°$ values of the individual propagation steps (refer to the solution to 4-10 (a) and (b)):

$$ + 67 \text{ kJ/mole} \quad + \quad - 101 \text{ kJ/mole} \quad = \quad -\textbf{34 kJ/mole} $$
$$ (+ 16 \text{ kcal/mole} \quad + \quad - 24 \text{ kcal/mole} \quad = \quad - \textbf{8 kcal/mole}) $$

4-17

(a)   initiation   (1) $I-I \xrightarrow{h\nu} 2\ I•$

propagation ⎰ (2) $I• \ + H-CH_3 \longrightarrow H-I \ + \ •CH_3$

⎱ (3) $I-I \ + \ •CH_3 \longrightarrow I-CH_3 + I•$

step (1)	break I–I	$\Delta H° = + 151$ kJ/mole (+ 36 kcal/mole)

step (2)	break H–CH₃	$\Delta H° = + 435$ kJ/mole (+ 104 kcal/mole)
	make H–I	$\Delta H° = - 297$ kJ/mole (− 71 kcal/mole)
	step (2)	$\Delta H° = + \textbf{138 kJ/mole} \ (+ \textbf{33 kcal/mole})$

step (3)	break I–I	$\Delta H° = + 151$ kJ/mole (+ 36 kcal/mole)
	make I–CH₃	$\Delta H° = - 234$ kJ/mole (− 56 kcal/mole)
	step (3)	$\Delta H° = -\textbf{83 kJ/mole} \ (-\textbf{20 kcal/mole})$

4-17 continued

(b) $\Delta H°$ for the reaction is the sum of the $\Delta H°$ values of the individual propagation steps:

+ 138 kJ/mole + – 83 kJ/mole = **+ 55 kJ/mole**
(+ 33 kcal/mole + – 20 kcal/mole = **+ 13 kcal/mole**)

(c) Iodination of methane is unfavorable for both kinetic and thermodynamic reasons. Kinetically, the rate of the first propagation step must be very slow because it is very endothermic; the activation energy must be at least + 138 kJ/mole. Thermodynamically, the overall reaction is endothermic, so an equilibrium would favor reactants, not products; there is no energy decrease to drive the reaction to products.

4-18 Propane has six primary hydrogens and two secondary hydrogens, a ratio of 3 : 1. If primary and secondary hydrogens were replaced by chlorine at equal rates, the chloropropane isomers would reflect the same 3 : 1 ratio, that is, 75% 1-chloropropane and 25% 2-chloropropane.

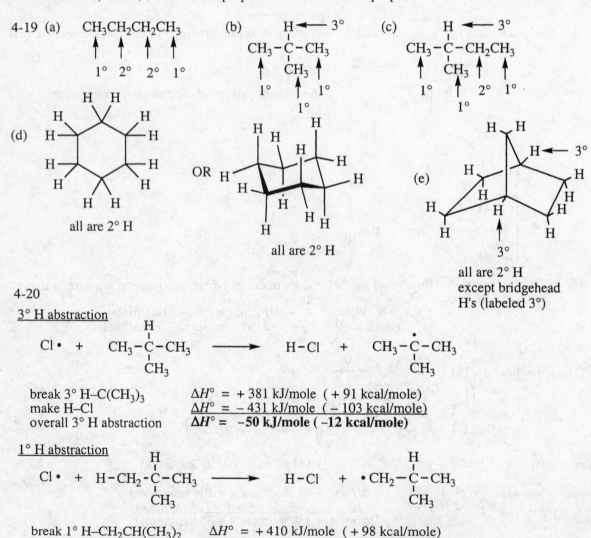

4-20
3° H abstraction

break 3° H–C(CH$_3$)$_3$	$\Delta H° = + 381$ kJ/mole (+ 91 kcal/mole)
make H–Cl	$\Delta H° = - 431$ kJ/mole (– 103 kcal/mole)
overall 3° H abstraction	$\Delta H° = $ **–50 kJ/mole (–12 kcal/mole)**

1° H abstraction

break 1° H–CH$_2$CH(CH$_3$)$_2$	$\Delta H° = + 410$ kJ/mole (+ 98 kcal/mole)
make H-Cl	$\Delta H° = - 431$ kJ/mole (– 103 kcal/mole)
overall 1° H abstraction	$\Delta H° = $ **–21 kJ/mole (–5 kcal/mole)**

(Note: $\Delta H°$ for abstraction of a 1° H from both ethane and propane are + 410 kJ/mole (+ 98 kcal/mole). It is reasonable to use this same value for abstraction of the 1° H in isobutane.)

4-20 continued

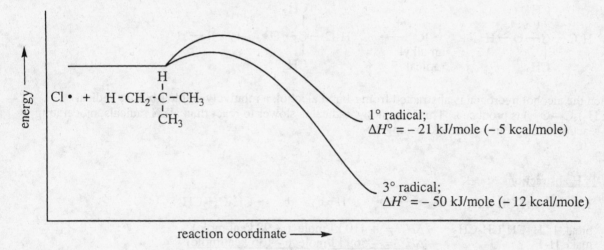

Since $\Delta H°$ for forming the 3° radical is more negative than $\Delta H°$ for forming the 1° radical, it is reasonable to infer that the activation energy leading to the 3° radical is lower than the activation energy leading to the 1° radical.

4-21
2-Methylbutane can produce four mono-chloro isomers. To calculate the relative amount of each in the product mixture, multiply the numbers of hydrogens which could lead to that product times the reactivity for that type of hydrogen. Each relative amount divided by the sum of all the amounts will provide the percent of each in the product mixture.

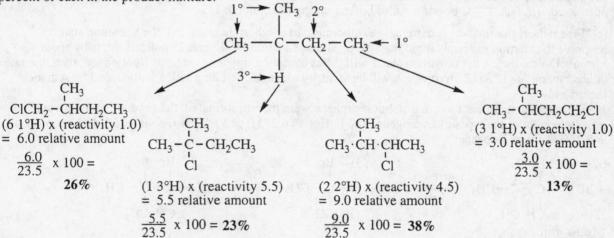

4-22
(a) When *n*-heptane is burned, only 1° and 2° radicals can be formed (from either C—H or C—C bond cleavage). These are high energy, unstable radicals which rapidly form other products. When isooctane (2,2,4-trimethylpentane, below) is burned, 3° radicals can be formed from either C—H or C—C bond cleavage. The 3° radicals are lower in energy than 1° or 2°, relatively stable, with lowered reactivity. Slower combustion translates to less "knocking."

isooctane (2,2,4-trimethylpentane)

Any indicated bond cleavage will produce a 3° radical.

4-22 continued

(b)

$$H_3C-\underset{\underset{CH_3}{|}}{\overset{\overset{CH_3}{|}}{C}}-O-H \; + \; \cdot R \; \longrightarrow \; H_3C-\underset{\underset{CH_3}{|}}{\overset{\overset{CH_3}{|}}{C}}-O\cdot \; + \; R-H$$

$\qquad\qquad\qquad\qquad$ an alkyl
$\qquad\qquad\qquad\qquad$ radical

When the alcohol hydrogen is abstracted from *t*-butyl alcohol, a relatively stable *t*-butoxy radical
($(CH_3)_3C-O\cdot$) is produced. This low energy radical is slower to react than alkyl radicals, moderating
the reaction and producing less "knocking."

4-23

(a) 1° H abstraction

$$F\cdot \; + \; CH_3CH_2CH_3 \; \longrightarrow \; H-F \; + \; \cdot CH_2CH_2CH_3$$

break 1° H–CH$_2$CH$_2$CH$_3$	$\Delta H° = +410$ kJ/mole ($+98$ kcal/mole)
make H–F	$\Delta H° = -569$ kJ/mole (-136 kcal/mole)
overall 1° H abstraction	$\Delta H° = -159$ **kJ/mole** (-38 **kcal/mole**)

2° H abstraction

$$F\cdot \; + \; CH_3CH_2CH_3 \; \longrightarrow \; H-F \; + \; CH_3\overset{\bullet}{C}HCH_3$$

break 2° H–CH(CH$_3$)$_2$	$\Delta H° = +397$ kJ/mole ($+95$ kcal/mole)
make H–F	$\Delta H° = -569$ kJ/mole (-136 kcal/mole)
overall 2° H abstraction	$\Delta H° = -172$ **kJ/mole** (-41 **kcal/mole**)

(b) Fluorination is extremely exothermic and is likely to be indiscriminate in which hydrogens are
abstracted. (In fact, C—C bonds are also broken during fluorination.)

(c) Free radical fluorination is extremely exothermic. In exothermic reactions, the transition states
resemble the starting materials more than the products, so while the 1° and 2° radicals differ by about 13
kJ/mole (3 kcal/mole), the transition states will differ by only a tiny amount. For fluorination, then, the rate
of abstraction for 1° and 2° hydrogens will be virtually identical. Product ratios will depend on statistical
factors only.

$\qquad$ Fluorination is difficult to control, but if propane were monofluorinated, the product mixture would
reflect the ratio of the types of hydrogens: six 1° H to two 2° H, or 3 : 1 ratio, giving 75% 1-fluoropropane
and 25% 2-fluoropropane.

4-24

$$CH_3-\underset{\underset{CH_3}{|}}{\overset{\overset{H}{|}}{C}}-\underset{\underset{CH_3}{|}}{\overset{\overset{H}{|}}{C}}-CH_3 \; \xrightarrow[h\nu]{Br_2} \; CH_3-\underset{\underset{CH_3}{|}}{\overset{\overset{H}{|}}{C}}-\underset{\underset{CH_3}{|}}{\overset{\overset{Br}{|}}{C}}-CH_3 \; \xrightarrow[h\nu]{Br_2} \; CH_3-\underset{\underset{CH_3}{|}}{\overset{\overset{Br}{|}}{C}}-\underset{\underset{CH_3}{|}}{\overset{\overset{Br}{|}}{C}}-CH_3$$

Mechanism

initiation $\qquad Br-Br \; \xrightarrow{h\nu} \; 2 \; Br\cdot$

propagation $\qquad CH_3-\underset{\underset{CH_3CH_3}{|\;\;|}}{\overset{\overset{H\;\;H}{|\;\;|}}{C-C}}-CH_3 \; + \; Br\cdot \longrightarrow \; HBr \; + \; CH_3-\underset{\underset{CH_3CH_3}{|\;\;|}}{\overset{\overset{H\;\;\bullet}{|\;\;|}}{C-C}}-CH_3$

$\qquad\qquad CH_3-\underset{\underset{CH_3CH_3}{|\;\;|}}{\overset{\overset{H\;\;\bullet}{|\;\;|}}{C-C}}-CH_3 \; + \; Br-Br \; \longrightarrow \; CH_3-\underset{\underset{CH_3CH_3}{|\;\;|}}{\overset{\overset{H\;\;Br}{|\;\;|}}{C-C}}-CH_3 \; + \; Br\cdot$

4-25 (a) <u>Mechanism</u>

initiation   Br—Br $\xrightarrow{h\nu}$ 2 Br •

propagation   + Br • $\longrightarrow$ + HBr

+ Br—Br $\longrightarrow$ + Br •

(b) Energy calculation uses the value for the allylic C—H bond from Table 4-2.

First
propagation
step
{
break allylic H–CH[ring] $\Delta H° = +364$ kJ/mole (+ 87 kcal/mole)
make H–Br $\underline{\Delta H° = -368 \text{ kJ/mole } (-88 \text{ kcal/mole})}$
overall allylic H abstraction $\Delta H° = -4$ **kJ/mole (−1 kcal/mole)**

Second
propagation
step
{
break Br–Br $\Delta H° = +192$ kJ/mole (+ 46 kcal/mole)
make 2° C–Br $\underline{\Delta H° = -285 \text{ kJ/mole } (-68 \text{ kcal/mole})}$
overall C–Br formation $\Delta H° = -93$ **kJ/mole (−22 kcal/mole)**

The first propagation step is rate-limiting.

structure of the transition state:

(c) The Hammond postulate tells us that, in an exothermic reaction, the transition state is closer to the reactants in energy and in structure. Since the first propagation step is exothermic (although not by much), the transition state is closer to cyclohexene + bromine radical. This is indicated in the transition state structure by showing the H closer to the C than to the Br.

(d) A bromine radical will abstract the hydrogen with the lowest bond dissociation energy at the fastest rate. The allylic hydrogen of cyclohexene is more easily abstracted than a hydrogen of cyclohexane because the radical produced is stabilized by resonance. (Energy values below are per mole.)

aliphatic: 397 kJ (95 kcal) $\longrightarrow$ $\longleftarrow$ 364 kJ (87 kcal): allylic
SLOWER   *FASTER*

4-26
BHA radical

67

4-27

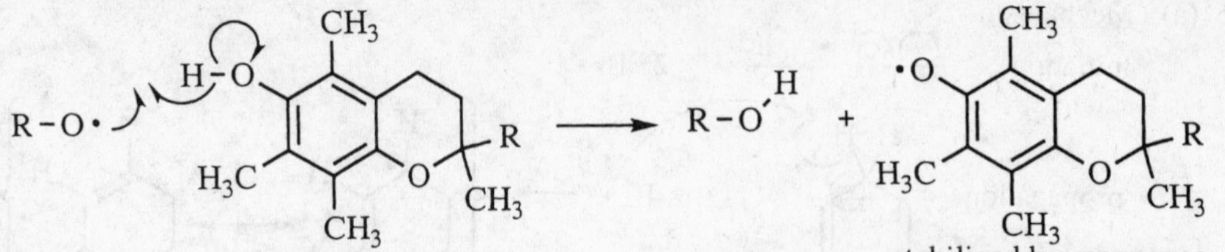

Vitamin E

stabilized by resonance—
less reactive than RO•

4-28 The triphenylmethyl cation is so stable because of the delocalization of the charge. The more resonance forms a species has, the more stable it will be.

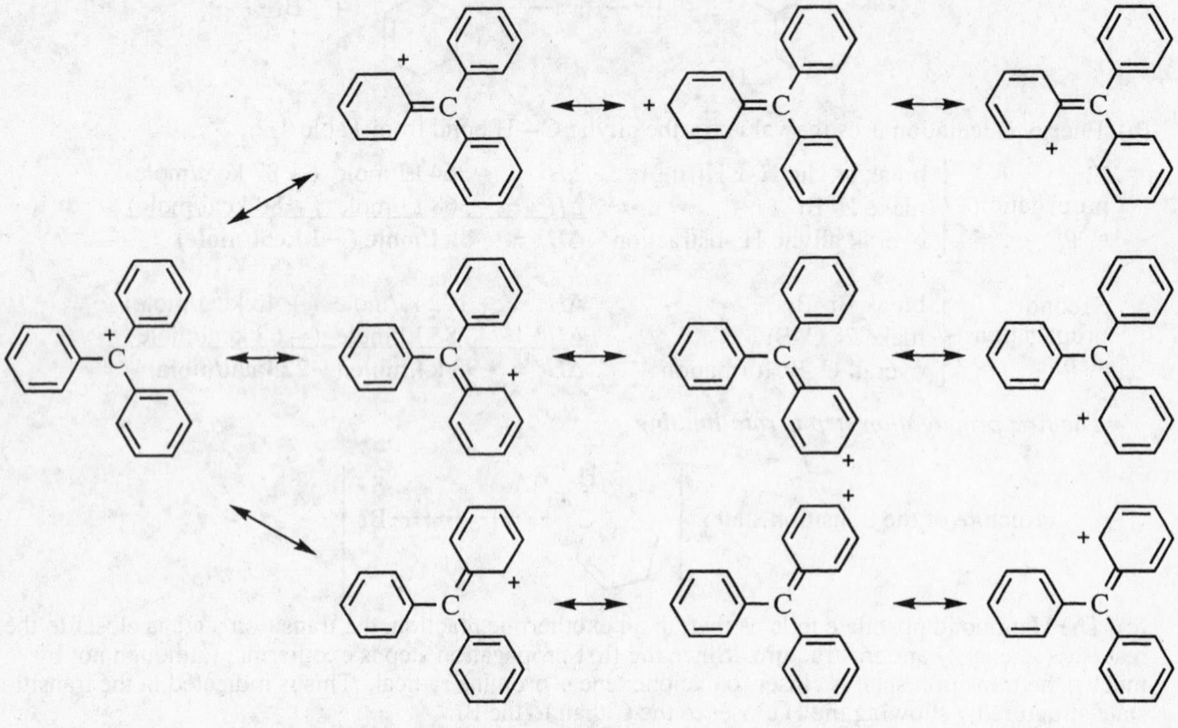

(Note: these resonance forms do not include the simple benzene resonance forms as shown below; they are significant, but repetitive, so for simplicity, they are not drawn here.)

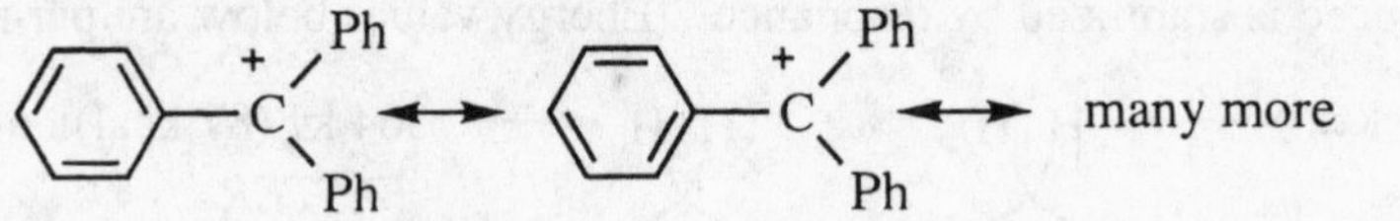

many more

4-29   most stable (c) > (b) > (a) least stable

(c) $CH_3-\overset{+}{\underset{CH_3}{C}}CH_2CH_3$ (b) $CH_3-\overset{+}{\underset{CH_3}{C}}HCHCH_3$ (a) $CH_3-\underset{CH_3}{C}HCH_2\overset{+}{C}H_2$

      3°   2°   1°

4-30   most stable (c) > (b) > (a) least stable

(c) $CH_3-\overset{\bullet}{\underset{CH_3}{C}}CH_2CH_3$ (b) $CH_3-\overset{\bullet}{\underset{CH_3}{C}}HCHCH_3$ (a) $CH_3-\underset{CH_3}{C}HCH_2\overset{\bullet}{C}H_2$

      3°   2°   1°

68

4-31

$$H_3C-\overset{\overset{\displaystyle :O:}{\|}}{C}-\overset{\overset{\displaystyle H}{|}}{\underset{\underset{\displaystyle H}{|}}{C}}-\overset{\overset{\displaystyle :O:}{\|}}{C}-CH_3 \quad + \quad {}^-\!:\!\overset{..}{\underset{..}{O}}\!-H \quad \longrightarrow \quad H_2O \quad +$$

base

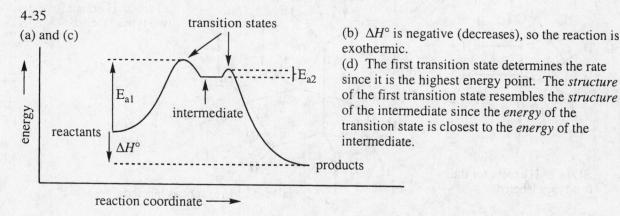

4-32

$$H-\overset{\overset{\displaystyle H}{|}}{\underset{\underset{\displaystyle H}{|}}{C}}-C\equiv N: \quad + \quad :\!\overset{-}{Base} \quad \longrightarrow \quad H-Base \quad + \quad \left\{ H-\overset{\overset{\displaystyle ..}{\overset{-}{C}}}{\underset{\underset{\displaystyle H}{|}}{}}-C\equiv N: \quad \longleftrightarrow \quad H-\underset{\underset{\displaystyle H}{|}}{C}=C=\overset{..}{\underset{..}{N}}:^- \right\}$$

4-33

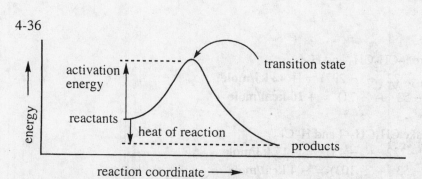

4-34 Please refer to solution 1-20, page 12 of this Solutions Manual.

4-35
(a) and (c)

(b) $\Delta H°$ is negative (decreases), so the reaction is exothermic.

(d) The first transition state determines the rate since it is the highest energy point. The *structure* of the first transition state resembles the *structure* of the intermediate since the *energy* of the transition state is closest to the *energy* of the intermediate.

4-36

4-37

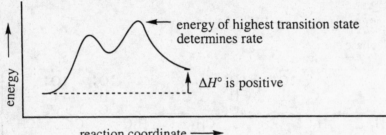

4-38

The rate law is first order with respect to the concentrations of hydrogen ion and of *t*-butyl alcohol, zeroth order with respect to the concentration of chloride ion, second order overall.

rate $= k_r\,[(CH_3)_3COH][H^+]$

4-39 (a)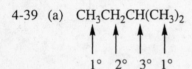

(b)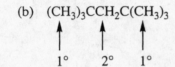

(c)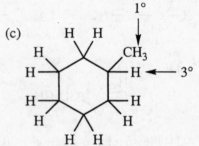

all are 2° H except for the two types labeled

(d)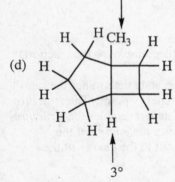

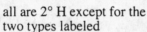

all are 2° H except for the two types labeled

(e)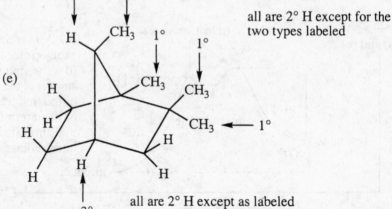

all are 2° H except as labeled

4-40

(a) break H–CH$_2$CH$_3$ and I–I, make I–CH$_2$CH$_3$ and H–I

kJ/mole: $(+410 + +151) + (-222 + -297) = $ **+42 kJ/mole**

kcal/mole: $(+98 + +36) + (-53 + -71) = $ **+10 kcal/mole**

(b) break CH$_3$CH$_2$–Cl and H–I, make CH$_3$CH$_2$–I and H–Cl

kJ/mole: $(+339 + +297) + (-222 + -431) = $ **−17 kJ/mole**

kcal/mole: $(+81 + +71) + (-53 + -103) = $ **−4 kcal/mole**

4-40 continued

(c) break $(CH_3)_3C–OH$ and H–Cl, make $(CH_3)_3C–Cl$ and H–OH

kJ/mole: $(+381 + +431) + (-331 + -498) = -17 \text{ kJ/mole}$

kcal/mole: $(+91 + +103) + (-79 + -119) = -4 \text{ kcal/mole}$

(d) break $CH_3CH_2–CH_3$ and H–H, make $CH_3CH_2–H$ and $H–CH_3$

kJ/mole: $(+356 + +435) + (-410 + -435) = -54 \text{ kJ/mole}$

kcal/mole: $(+85 + +104) + (-98 + -104) = -13 \text{ kcal/mole}$

(e) break $CH_3CH_2–OH$ and H–Br, make $CH_3CH_2–Br$ and H–OH

kJ/mole: $(+381 + +368) + (-285 + -498) = -34 \text{ kJ/mole}$

kcal/mole: $(+91 + +88) + (-68 + -119) = -8 \text{ kcal/mole}$

4-41 Numbers are bond dissociation energies in kcal/mole in the top line and kJ/mole in the bottom line.

$$\phi–\overset{\bullet}{C}H_2 \ > \ CH_2=CH\overset{\bullet}{C}H_2 \ > \ (CH_3)_3\overset{\bullet}{C} \ > \ (CH_3)_2\overset{\bullet}{C}H \ > \ CH_3\overset{\bullet}{C}H_2 \ > \ \overset{\bullet}{C}H_3$$

| | 85 | 87 | 91 | 95 | 98 | 104 |
| | 356 | 364 | 381 | 397 | 410 | 435 |

most stable   least stable

4-42

(a)   Only one product; chlorination would work. Bromination on a 2° carbon would not be predicted to be a high-yielding process.

(b)

Chlorination would produce four constitutional isomers and would not be a good method to make only one of these. Monobromination at the 3° carbon would give a reasonable yield.

(c) $CH_3–\overset{\overset{\displaystyle H}{|}}{\underset{\underset{\displaystyle CH_3}{|}}{C}}—\overset{\overset{\displaystyle H}{|}}{\underset{\underset{\displaystyle CH_3}{|}}{C}}–CH_3 \longrightarrow CH_3–\overset{\overset{\displaystyle H}{|}}{\underset{\underset{\displaystyle CH_3}{|}}{C}}—\overset{\overset{\displaystyle Cl}{|}}{\underset{\underset{\displaystyle CH_3}{|}}{C}}–CH_3 + CH_3–\overset{\overset{\displaystyle H}{|}}{\underset{\underset{\displaystyle CH_3}{|}}{C}}—\overset{\overset{\displaystyle H}{|}}{\underset{\underset{\displaystyle CH_3}{|}}{C}}–CH_2Cl$

Chlorination would produce two constitutional isomers and would not be a good method to make only one of these. Monobromination would be selective for the 3° carbon and would give an excellent yield.

(d) $CH_3–\overset{\overset{\displaystyle CH_3}{|}}{\underset{\underset{\displaystyle CH_3}{|}}{C}}—\overset{\overset{\displaystyle CH_3}{|}}{\underset{\underset{\displaystyle CH_3}{|}}{C}}–CH_3 \longrightarrow CH_3–\overset{\overset{\displaystyle CH_3}{|}}{\underset{\underset{\displaystyle CH_3}{|}}{C}}—\overset{\overset{\displaystyle CH_3}{|}}{\underset{\underset{\displaystyle CH_3}{|}}{C}}–CH_2Cl$ Only one product; chlorination would give a high yield. Monobromination would be very difficult since all hydrogens are on 1° carbons.

4-43

initiation   (1) $Cl–Cl \xrightarrow{h\nu} 2\ Cl\bullet$

propagation

(2)

(3)

71

Termination steps are any two radicals combining.

4-44

(a) $CH_2=CH-\overset{\bullet}{C}H_2 \longleftrightarrow \overset{\bullet}{C}H_2-CH=CH_2$

(b)

(c)
$$CH_3-\overset{:\overset{..}{O}:}{\underset{}{C}}-\overset{..}{\underset{..}{O}}\cdot \longleftrightarrow CH_3-\overset{:\overset{..}{O}\cdot}{\underset{}{C}}=\overset{..}{\underset{..}{O}}$$

(d)

(e)

4-45

(a) <u>Mechanism</u>

initiation $\quad Br-Br \xrightarrow{h\nu} 2\ Br\cdot$

propagation

The 3° allylic H is abstracted selectively (faster than any other type in the molecule), forming an intermediate represented by two *non-equivalent* resonance forms. Partial radical character on two different carbons of the allylic radical leads to two different products.

(b) There are two reasons why the H shown is the one that is abstracted by bromine radical: the H is 3° and it is allylic, that is, neighboring a double bond. Both of these factors stabilize the radical that is created by removing the H atom.

4-46 Where mixtures are possible, only the major product is shown.

(a)
only one product possible

(b)
3° hydrogen abstracted selectively, faster than 2° or 1°

(c)
$$\underset{\underset{CH_3}{\overset{CH_3}{|}}}{\overset{\overset{H}{|}}{CH_3-C-C-CH_3}} \longrightarrow \underset{\underset{CH_3}{\overset{CH_3}{|}}}{\overset{\overset{Br}{|}}{CH_3-C-C-CH_3}}$$
3° hydrogen abstracted selectively, faster than 1°

72

4-46 continued

(d)

decalin

3° hydrogen abstracted selectively, faster than 2°

(e)

3° hydrogen abstracted selectively, faster than 2° or 1°

(f)

both 2°—formed in equal amounts

(g)

from resonance-stabilized benzylic radical

(h)

All hydrogens at the starred positions are equivalent and *allylic*. The H from the lower right carbon has been removed to make an intermediate with two non-equivalent resonance forms, giving the two products shown. *Drawing the resonance forms is the key to answering this question correctly.*

4-47

(a) As CH_3Cl is produced, it can compete with CH_4 for available $Cl \bullet$, generating CH_2Cl_2. This can generate $CHCl_3$, *etc.*

propagation steps

$$CH_4 + Cl \bullet \longrightarrow HCl + \bullet CH_3$$
$$\bullet CH_3 + Cl_2 \longrightarrow ClCH_3 + Cl \bullet$$
$$ClCH_3 + Cl \bullet \longrightarrow HCl + \bullet CH_2Cl$$
$$\bullet CH_2Cl + Cl_2 \longrightarrow CH_2Cl_2 + Cl \bullet$$
$$CH_2Cl_2 + Cl \bullet \longrightarrow HCl + \bullet CHCl_2$$
$$\bullet CHCl_2 + Cl_2 \longrightarrow CHCl_3 + Cl \bullet$$
$$CHCl_3 + Cl \bullet \longrightarrow HCl + \bullet CCl_3$$
$$\bullet CCl_3 + Cl_2 \longrightarrow CCl_4 + Cl \bullet$$

(b) To maximize CH_3Cl and minimize formation of polychloromethanes, the ratio of methane to chlorine must be kept high (see problem 4-2).

To guarantee that all hydrogens are replaced with chlorine to produce CCl_4, the ratio of chlorine to methane must be kept high.

4-48

(a) *n*-Pentane can produce three monochloro isomers. To calculate the relative amount of each in the product mixture, multiply the numbers of hydrogens which could lead to that product times the reactivity for that type of hydrogen. Each relative amount divided by the sum of all the amounts will provide the percent of each in the product mixture.

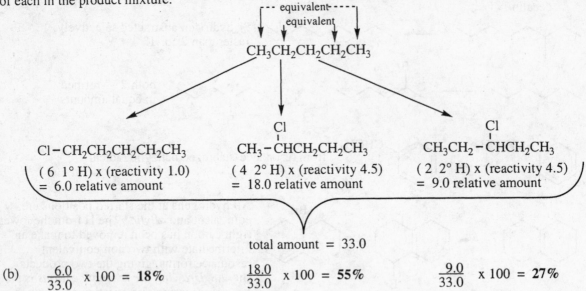

total amount $= 33.0$

(b) $\quad \dfrac{6.0}{33.0}$ x 100 = **18%** $\qquad \dfrac{18.0}{33.0}$ x 100 = **55%** $\qquad \dfrac{9.0}{33.0}$ x 100 = **27%**

4-49

(a) The second propagation step in the chlorination of methane is highly exothermic ($\Delta H° = -109$ kJ/mole (-26 kcal/mole)). The transition state resembles the reactants, that is, the Cl–Cl bond will be slightly stretched and the Cl–CH$_3$ bond will just be starting to form.

$$\overset{\delta\bullet}{Cl}\text{-----}Cl\text{---------------}\overset{\delta\bullet}{CH_3}$$
$$\text{stronger} \qquad \text{weaker}$$

(b) The second propagation step in the bromination of methane is highly exothermic ($\Delta H° = -101$ kJ/mole (-24 kcal/mole)). The transition state resembles the reactants, that is, the Br–Br bond will be slightly stretched and the Br–CH$_3$ bond will just be starting to form.

$$\overset{\delta\bullet}{Br}\text{-----}Br\text{------------}\overset{\delta\bullet}{CH_3}$$
$$\text{stronger} \qquad \text{weaker}$$

4-50 Two mechanisms are possible depending on whether HO • reacts with chlorine or with cyclopentane.

Mechanism 1 $\qquad$ all in kJ/mole

initiation
{
(1) $\quad$ HO–OH $\longrightarrow$ 2 HO • $\qquad$ $\Delta H° = +213$ kJ/mole

(2) $\quad$ HO • + Cl–Cl $\longrightarrow$ HO–Cl + Cl • $\qquad$ **$\Delta H° = + 242 – 210 = +32$**

propagation
{
(3) $\quad$ Cl • + $\longrightarrow$ H–Cl + $\qquad$ $\Delta H° = + 397 – 431 = –34$

(4) $\quad$ Cl–Cl + $\longrightarrow$ + Cl • $\quad$ $\Delta H° = + 242 – 335 = –93$

Mechanism 2

initiation
{
(1) $\quad$ HO–OH $\longrightarrow$ 2 HO • $\qquad$ $\Delta H° = +213$ kJ/mole

(2) $\quad$ HO • + $\longrightarrow$ H–OH + $\qquad$ **$\Delta H° = + 397 – 498 = –101$**

propagation
{
(3) $\quad$ Cl–Cl + $\longrightarrow$ + Cl • $\quad$ $\Delta H° = + 242 – 335 = –93$

(4) $\quad$ Cl • + $\longrightarrow$ H–Cl + $\qquad$ $\Delta H° = + 397 – 431 = –34$

In this case, the energies of initiation steps determine which mechanism is followed. The bond dissociation energy of HO—Cl is about 210 kJ/mole (about 50 kcal/mole), making initiation step (2) in mechanism 1 *endothermic* by about 30 kJ/mole (about 8 kcal/mole). In mechanism 2, initiation step (2) is *exothermic* by about 101 kJ/mole (24 kcal/mole); mechanism 2 is preferred. One strongly endothermic step can be enough to stop a mechanism.

4-51

a carbene

4-52 This critical equation is the key to this problem: $\Delta G = \Delta H - T \Delta S$

At 1400 K, the equilibrium constant is 1; therefore:

$K_{eq} = 1$ $\Longrightarrow$ $\Delta G = 0$ $\Longrightarrow$ $\Delta H = T \Delta S$

Assuming ΔH is about the same at 1400 K as it is at calorimeter temperature:

$$\Delta S = \frac{\Delta H}{T} = \frac{-137 \text{ kJ/mole}}{1400 \text{ K}} = \frac{-137,000}{1400} \text{ J/K-mole}$$

$$= -98 \text{ J/K-mole} \ (-23 \text{ cal/K-mole})$$

This is a large *decrease* in entropy, consistent with two molecules combining into one.

4-53

Assume that chlorine atoms (radicals) are still generated in the initiation reaction. Focus on the propagation steps. Bond dissociation energies are given below the bonds, in kJ/mole (kcal/mole).

$$Cl\cdot \; + \; H-CH_3 \longrightarrow H-Cl \; + \; \cdot CH_3 \qquad \Delta H = +4 \text{ kJ/mole } (+1 \text{ kcal/mole})$$
$$435\,(104) 431\,(103)$$

$$Cl-Cl \; + \; \cdot CH_3 \longrightarrow Cl-CH_3 \; + \; Cl\cdot \qquad \Delta H = -109 \text{ kJ/mole } (-26 \text{ kcal/mole})$$
$$242\,(58) 351\,(84)$$

What happens when the different radical species react with iodine?

$$Cl\cdot \; + \; I-I \longrightarrow I-Cl \; + \; \cdot I \qquad \Delta H = -60 \text{ kJ/mole } (-14 \text{ kcal/mole})$$
$$151\,(36) 211\,(50)$$

$$I-I \; + \; \cdot CH_3 \longrightarrow I-CH_3 \; + \; I\cdot \qquad \Delta H = -83 \text{ kJ/mole } (-20 \text{ kcal/mole})$$
$$151\,(36) 234\,(56)$$

Compare the second reaction in each pair: methyl radical reacting with chlorine is more exothermic than methyl radical reacting with iodine; this does not explain how iodine prevents the chlorination reaction. Compare the first reaction in each pair: chlorine atom reacting with iodine is very exothermic whereas chlorine atom reacting with methane is slightly endothermic. Here is the key: chlorine atoms will be scavenged by iodine before they have a chance to react with methane. Without chlorine atoms, the reaction comes to a dead stop.

4-54 (a)

initiation

(1) $Br-Br \xrightarrow{h\nu} 2\ Br\cdot$

(2) $Br\cdot + H-SnBu_3 \longrightarrow H-Br + \cdot SnBu_3$

propagation

(3) [cyclohexane with H and Br] $+ \cdot SnBu_3 \longrightarrow$ [cyclohexyl radical with H] $+ Br-SnBu_3$

(4) [cyclohexyl radical with H] $+ H-SnBu_3 \longrightarrow$ [cyclohexane with H and H] $+ \cdot SnBu_3$

(b) All energies are in kJ/mole. The abbreviation "cy" stands for the cyclohexane ring.

Step 2: break H—Sn, make H—Br: +310 + −368 = −58 kJ/mole

Step 3: break cy—Br, make Br—Sn: +285 + −552 = −267 kJ/mole WOW!

Step 4: break H—Sn, make cy—H: +310 + −397 = −87 kJ/mole

The sum of the two propagation steps are: −267 + −87 = −354 kJ/mole —a hugely exothermic reaction.

4-55

Mechanism 1:

(a) $Cl \bullet + O_3 \longrightarrow ClO \bullet + O_2$

(b) $2\ ClO \bullet \longrightarrow Cl-O-O-Cl$

(c) $Cl-O-O-Cl \xrightarrow{h\nu} O_2 + 2\ Cl \bullet$

The biggest problem in Mechanism 1 lies in step (b). The concentration of Cl atoms is very small, so at any given time, the concentration of ClO will be very small. The probability of two ClO radicals finding each other to form ClOOCl is virtually zero. Even though this mechanism shows a catalytic cycle with Cl• (starting the mechanism and being regenerated at the end), the middle step makes it highly unlikely.

Mechanism 2:

(d) $O_3 \xrightarrow{h\nu} O_2 + O$

Step (d) is the "light" reaction that occurs naturally in daylight. At night, the reaction reverses and regenerates ozone.

(e) $Cl \bullet + O_3 \longrightarrow ClO \bullet + O_2$

(f) $ClO \bullet + O \longrightarrow O_2 + Cl \bullet$

Step (f) is the crucial step. A low concentration of ClO *will* find a relatively high concentration of O atoms because the "light" reaction is producing O atoms in relative abundance. Cl• is regenerated and begins propagation step (e), continuing the catalytic cycle.

Mechanism 2 is believed to be the dominant mechanism in ozone depletion. Mechanism 1 can be discounted because of the low probability of step (b) occurring, because two species in very low, catalytic concentration are required to find each other in order for the step to occur.

4-56

(a)

$$\left[\begin{array}{c} H \\ | \\ H-C \cdots\cdots H\cdots\cdots Cl \\ | \\ H \end{array} \right]^{\ddagger} \qquad \left[\begin{array}{c} D \\ | \\ D-C \cdots\cdots D\cdots\cdots Cl \\ | \\ D \end{array} \right]^{\ddagger}$$

In each case, the bond from carbon to H (D) is breaking and the bond from H (D) to Cl is forming.

(b) $C_2H_5-D + Cl_2 \longrightarrow \underbrace{C_2H_5-Cl + DCl}_{7\%} + \underbrace{C_2H_4DCl + HCl}_{93\%}$

D replacement: $7\% \div 1\ D = 7$ (reactivity factor)
H replacement: $93\% \div 5\ H = 18.6$ (reactivity factor)
relative reactivity of H : D abstraction = $18.6 \div 7 = 2.7$

Each hydrogen is abstracted 2.7 times faster than deuterium.

(c) In both reactions of chlorine with either methane or ethane, the first propagation step is rate-limiting. The reaction of chlorine atom with methane is *endothermic* by 4 kJ/mole (1 kcal/mole), while for ethane this step is *exothermic* by 21 kJ/mole (5 kcal/mole). By the Hammond Postulate, differences in activation energy are most pronounced in *endothermic* reactions where the transition states most resemble the products. Therefore, a change in the methane molecule causes a greater change in its transition state energy than the same change in the ethane molecule causes in its transition state energy. Deuterium will be abstracted more slowly in both methane and ethane, but the rate effect will be more pronounced in methane than in ethane.

CHAPTER 5—STEREOCHEMISTRY

> Note to the student: Stereochemistry is the study of molecular structure and reactions in three dimensions. Molecular models will be especially helpful in this chapter.

5-1   The best test of whether a household object is chiral is whether it would be used equally well by a left- or right-handed person. The chiral objects are the corkscrew, the writing desk, the can opener, the screw-cap bottle (only for refilling, however; in use, it would not be chiral), the rifle and the knotted rope; the corkscrew, the bottle top, and the rope each have a twist in one direction, and the rifle, can opener, and desk are clearly made for right-handed users. All the other objects are achiral and would feel equivalent to right- or left-handed users.

5-2

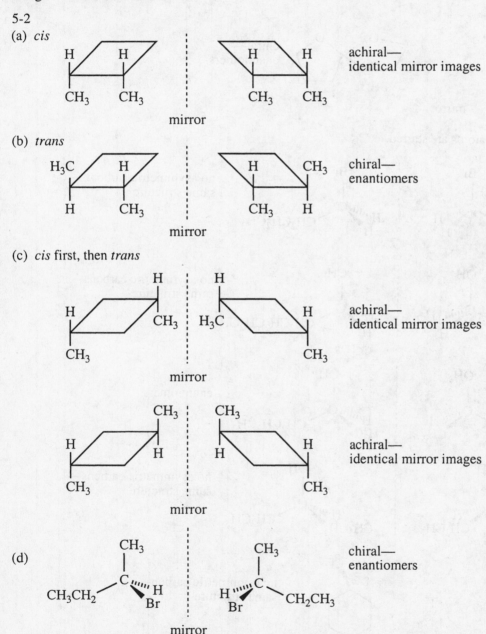

(a) *cis*

achiral—
identical mirror images

mirror

(b) *trans*

chiral—
enantiomers

mirror

(c) *cis* first, then *trans*

achiral—
identical mirror images

mirror

achiral—
identical mirror images

mirror

(d)

chiral—
enantiomers

mirror

79

5-2 continued

(e)

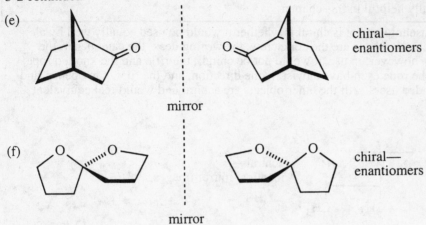

mirror

chiral—
enantiomers

(f)

mirror

chiral—
enantiomers

5-3 Asymmetric carbon atoms are starred.

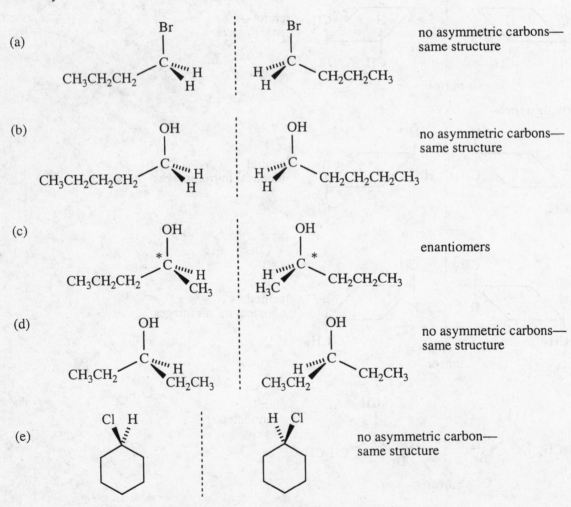

(a) no asymmetric carbons—
same structure

(b) no asymmetric carbons—
same structure

(c) enantiomers

(d) no asymmetric carbons—
same structure

(e) no asymmetric carbon—
same structure

5-3 continued

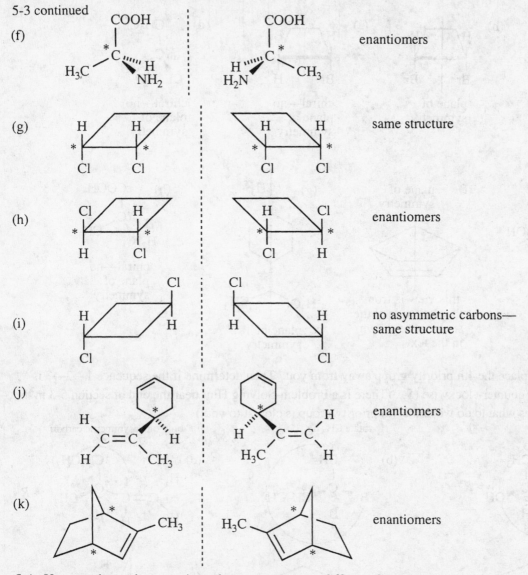

(f) COOH   COOH   enantiomers

(g) same structure

(h) enantiomers

(i) no asymmetric carbons—same structure

(j) enantiomers

(k) enantiomers

5-4 You may have chosen to interchange two groups different from the ones shown here. The type of isomer produced will still be the same as listed here.

Interchanging any two groups around a chirality center will create an enantiomer of the first structure.

CH₂CH₃   Interchanging the Br and the H creates an enantiomer of the structure in Figure 5-5.

H₃C C Br H

CH₂CH₂CH₃   Interchanging the ethyl and the isopropyl creates an enantiomer of the structure in Figure 5-5.

H₃C N CH(CH₃)₂ CH₂CH₃

On a double bond, interchanging the two groups on ONE of the stereocenters will create the other geometric (*cis-trans*) isomer. However, interchanging the two groups on BOTH of the stereocenters will give the original structure.

original structure is *cis*

H CH₃
C
||
C
H CH₃

⟹ interchange H and CH₃ on top stereocenter to produce *trans*

H₃C H
C
||
C
H CH₃

81

5-5

(a)

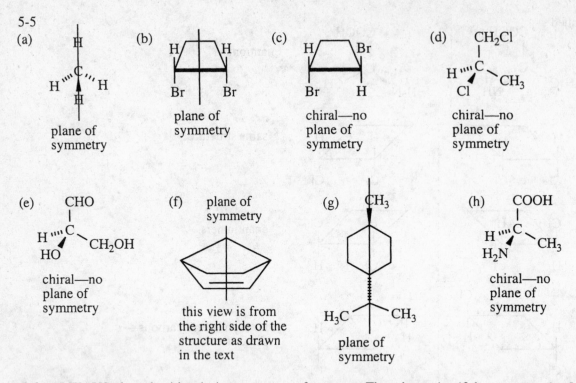

plane of
symmetry

(b)

plane of
symmetry

(c)

chiral—no
plane of
symmetry

(d) CH₂Cl / H⁗C—CH₃ / Cl

chiral—no
plane of
symmetry

(e) CHO / H⁗C—CH₂OH / HO

chiral—no
plane of
symmetry

(f) plane of
symmetry

this view is from
the right side of the
structure as drawn
in the text

(g) CH₃ / H₃C CH₃

plane of
symmetry

(h) COOH / H⁗C—CH₃ / H₂N

chiral—no
plane of
symmetry

5-6 ALWAYS place the 4th priority group away from you. Then determine if the sequence 1→2→3 is clockwise (*R*) or counter-clockwise (*S*). (There is a Problem-Solving Hint near the end of section 5-3 in the text that describes what to do when the 4th priority group is closest to you.)

rotate CH₃ up   only one asymmetric carbon

(a)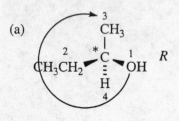

CH₃ (3) / CH₃CH₂ (2) *C 1 OH / H (4)   *R*

(b) CH₃ (3) / Br (1) *C (2) CH₂CH₃ / H (4)   *S*

(c) 2 CH(CH₃)₂ / H C=C H / *C 3 CH₃ / H (4)   *R*

(d) Cl H / *   * / H Cl   *S* → ← *S*

(e) H H / *   * / Cl Cl   *R* → ← *S*

(f) 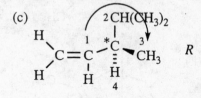 H *S* / * H *R*

(g) H *R* / * =O / H *R*

(h) 2 3 / *C H (4) / 1   *R*

(i) O=C—H (2) / (H₃CO)₂HC (1) *C CH₂ / C—H (3) / CH(CH₃)₂ (4)   *S*

see next page for an
explanation of part (i)

82

5-6 continued

Part (i) deserves some explanation. The difference between groups 1 and 2 hinge on what is on the "extra" oxygen.

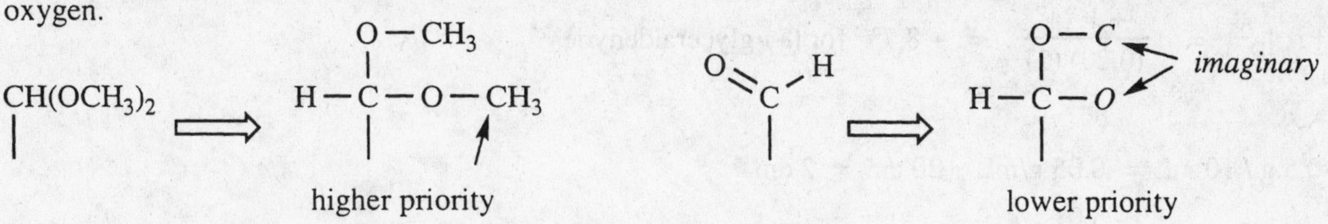

CH(OCH₃)₂ ⟹ H—C—O—CH₃ with O—CH₃ above, **higher priority** (arrow)

O=C—H (aldehyde) ⟹ H—C—O with O—C imaginary, **lower priority**

CH(OCH₃)₂ → $CH(OCH_3)_2$

5-7 There are no asymmetric carbons in 5-3 (a), (b), (d), (e), or (i).

(c)

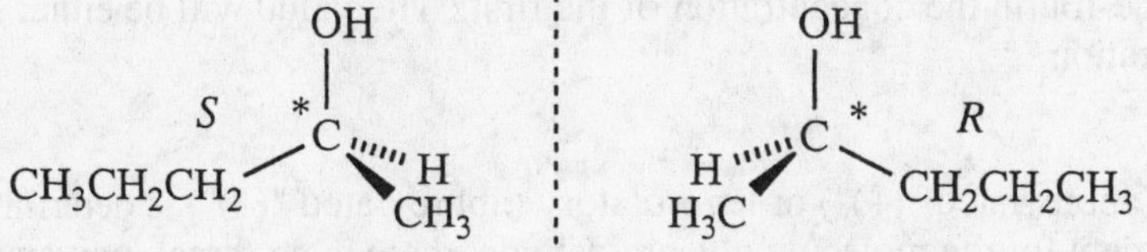

S *C with OH up, H and CH₃; CH₃CH₂CH₂ left | R *C with OH up, H₃C and CH₂CH₂CH₃

(f)

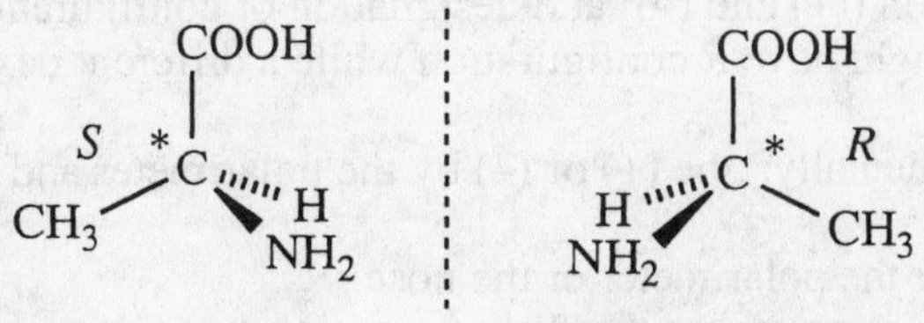

S *C with COOH up, H and NH₂; CH₃ left | R *C with COOH up, NH₂ and CH₃; H left

(g) cyclohexane chair: R * and S * with Cl, Cl and H, H | R * and S * with Cl, Cl

(h) S * and S * with Cl, H / H, Cl | R * and R * with H, Cl / Cl, H

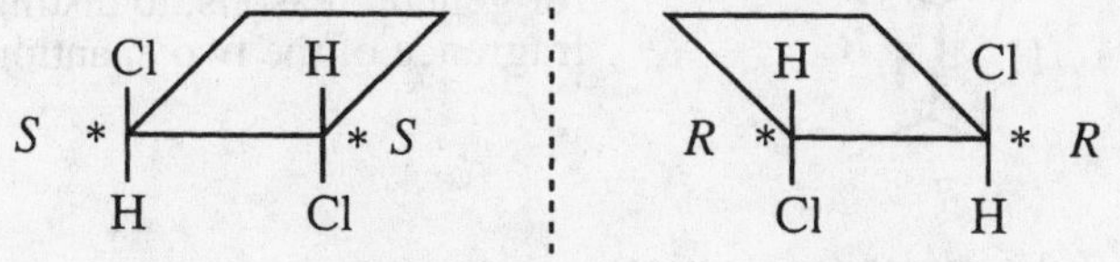

(j)

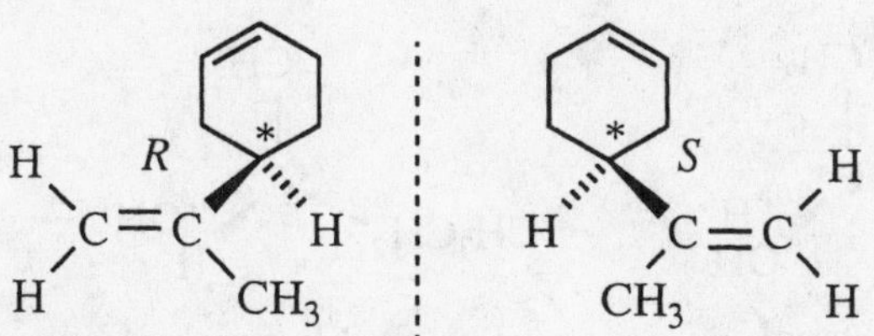

H₂C=C with R* cyclohexenyl, H, CH₃ | S* cyclohexenyl, C=CH₂ with CH₃, H

(k) bicyclic: S* R* with CH₃ | R* S* with CH₃

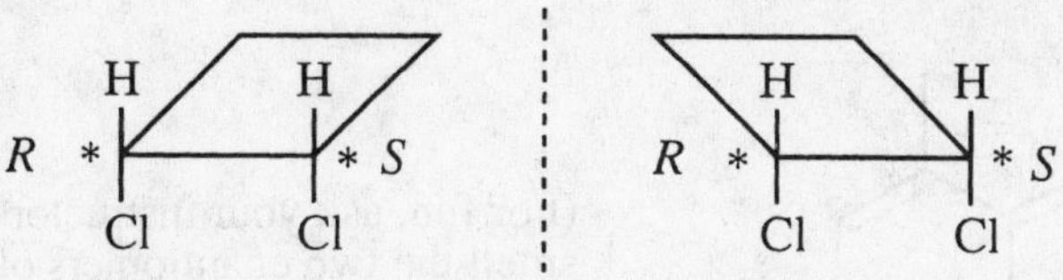

83

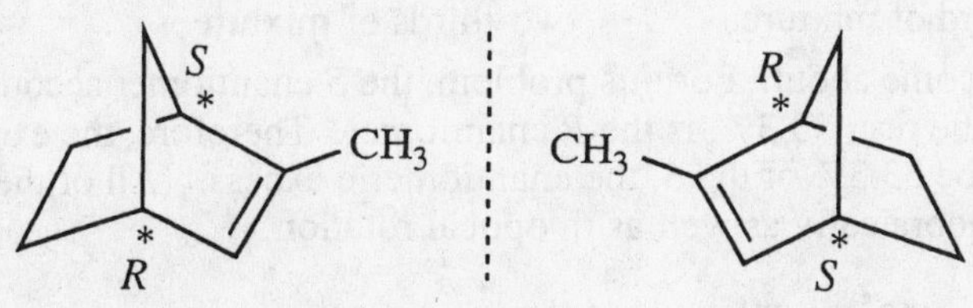

5-8

$2.0 \text{ g} / 10 \text{ mL} = 0.20 \text{ g/mL} ; \; 100 \text{ mm} = 1 \text{ dm}$

$$[\alpha]_D^{25} = \frac{+1.74°}{(0.20)\,(1)} = +8.7° \text{ for (+)-glyceraldehyde}$$

5-9

$0.5 \text{ g} / 10 \text{ mL} = 0.05 \text{ g/mL} ; \; 20 \text{ cm} = 2 \text{ dm}$

$$[\alpha]_D^{25} = \frac{-5.0°}{(0.05)\,(2)} = -50° \text{ for (–)-epinephrine}$$

5-10

Measure using a solution of about one-fourth the concentration of the first. The value will be either $+45°$ or $-45°$, which gives the sign of the rotation.

5-11

Whether a sample is dextrorotatory (abbreviated "(+)") or levorotatory (abbreviated "(–)") is determined experimentally by a polarimeter. Except for the molecule glyceraldehyde, there is no direct, universal correlation between direction of optical rotation ((+) and (–)) and designation of configuration (*R* and *S*). In other words, one dextrorotatory compound might have *R* configuration while a different dextrorotatory compound might have *S* configuration.

(a) Yes, both of these are determined experimentally: the (+) or (–) by the polarimeter and the smell by the nose.

(b) No, *R* or *S* cannot be determined by either the polarimeter or the nose.

(c) The drawings show that (+)-carvone from caraway has the *S* configuration and (–)-carvone from spearmint has the *R* configuration.

(+)-carvone (caraway seed)   (–)-carvone (spearmint)

(For fun, ask your instructor if you can smell the two enantiomers of carvone. Some people are unable, presumably for genetic reasons, to distinguish the fragrance of the two enantiomers.)

5-12

(*R*)-2-bromobutane   (*R*)-2-butanol   (*S*)-2-butanol
                           one-third of mixture   two-thirds of mixture

Chapter 6 will explain how these mixtures come about. For this problem, the *S* enantiomer accounts for 66.7% of the 2-butanol in the mixture and the rest, 33.3%, is the *R* enantiomer. Therefore, the excess of one enantiomer over the racemic mixture must be 33.3% of the *S*, the enantiomeric excess. (All of the *R* is "canceled" by an equal amount of the *S*, algebraically as well as in optical rotation.)

The optical rotation of pure (*S*)-2-butanol is $+13.5°$. The optical rotation of this mixture is:

$$33.3\% \; \times \; (+13.5°) = +4.5°$$

5-13   The rotation of pure (+)-2-butanol is $+13.5°$.

$$\frac{\text{observed rotation}}{\text{rotation of pure enantiomer}} = \frac{+0.45°}{+13.5°} \times 100\% = 3.3\% \text{ optical purity}$$
$$= 3.3\% \text{ e.e.} = \text{excess of (+) over (−)}$$

To calculate percentages of (+) and (−):   (two equations in two unknowns)

$$(+) + (−) = 100\% \implies (−) = 100\% − (+)$$

$$(+) − (−) = 3.3\% \implies (+) − (100\% − (+)) = 3.3\%$$

$$\mathbf{2}\,(+) = 103.3\%$$

> (+) = 51.6% (rounded)
> (−) = 48.4%

5-14 Drawing Newman projections is the clearest way to determine symmetry of conformations.

(a)

chiral—
optically active

$$H_3C - \overset{\overset{\displaystyle H}{|}}{\underset{\underset{\displaystyle Br}{|}}{C}}{}^{*} - Cl$$

(b)

plane includes
Br and Cl

plane of symmetry containing
Br—C—C—Cl; not optically active

$$Br - \overset{\overset{\displaystyle H}{|}}{\underset{\underset{\displaystyle H}{|}}{C}} - \overset{\overset{\displaystyle H}{|}}{\underset{\underset{\displaystyle H}{|}}{C}} - Cl$$

no asymmetric carbons

(c)

chiral—
optically active

$$ClCH_2 - \overset{\overset{\displaystyle H}{|}}{\underset{\underset{\displaystyle Br}{|}}{C}}{}^{*} - Cl$$

(d)

plane of symmetry—not optically
active despite the presence of two
asymmetric carbons

(e)

no plane of symmetry—
optically active
(other chair form is
equivalent—no plane of
symmetry)

(f)

plane of symmetry through
C-1 and C-4—not
optically active

no asymmetric carbons

Part (2) Predictions of optical activity based on asymmetric centers give the same answers as predictions
based on the most symmetric conformation.

5-15

(a)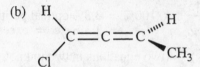

no asymmetric carbons, but the molecule is chiral (an allene); the drawing below is a three-dimensional picture of the allene in (a) showing there is no plane of symmetry because the substituents of an allene are in different planes

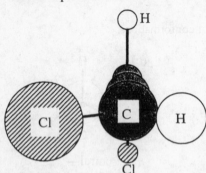

(b)

no asymmetric carbons, but the molecule is chiral (an allene)

(c)

no asymmetric carbons; this allene has a plane of symmetry between the two methyls (the plane of the paper), including all the other atoms because the two pi bonds of an allene are perpendicular, the Cl is in the plane of the paper and the plane of symmetry goes through it; not a chiral molecule

(d)

planar molecule—no asymmetric carbons; not a chiral molecule

(e)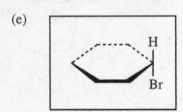

plane of symmetry bisecting the molecule; no asymmetric carbons; not a chiral molecule

(f)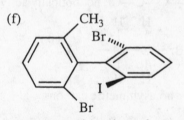

no asymmetric carbons, but the molecule is chiral due to restricted rotation; the drawing below is a three-dimensional picture showing that the rings are perpendicular (hydrogens are not shown)

(g)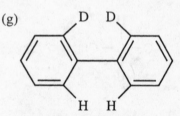

no asymmetric carbons, and the groups are not large enough to restrict rotation; not a chiral compound

(h)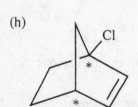

two asymmetric carbons; a chiral compound

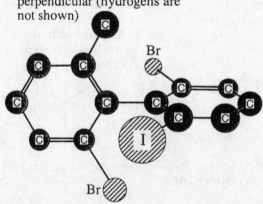

5-16

(a)

$$\begin{array}{c} COOH \\ H-\!\!\!-OH \\ CH_3 \end{array}$$

$$\begin{array}{c} COOH \\ HO-\!\!\!-H \\ CH_3 \end{array}$$
enantiomer

$$\begin{array}{c} H \\ CH_3-\!\!\!-COOH \\ OH \end{array}$$
enantiomer

$$\begin{array}{c} CH_3 \\ HO-\!\!\!-H \\ COOH \end{array}$$
same

(b)

$$\begin{array}{c} CH_2CH_3 \\ H-\!\!\!-Br \\ CH_3 \end{array}$$

$$\begin{array}{c} CH_3 \\ Br-\!\!\!-H \\ CH_2CH_3 \end{array}$$
same

$$\begin{array}{c} CH_2CH_3 \\ Br-\!\!\!-H \\ CH_3 \end{array}$$
enantiomer

$$\begin{array}{c} CH_3 \\ H-\!\!\!-Br \\ CH_2CH_3 \end{array}$$
enantiomer

(c)

$$\begin{array}{c} CH_3 \\ HO-\!\!\!-H \\ CH_2CH_3 \end{array}$$
(R)-2-butanol

$$\begin{array}{c} CH_3 \\ H-\!\!\!-OH \\ CH_2CH_3 \end{array}$$
enantiomer

$$\begin{array}{c} CH_3 \\ HO-\!\!\!-H \\ CH_2CH_3 \end{array}$$
same

$$\begin{array}{c} CH_2CH_3 \\ H-\!\!\!-OH \\ CH_3 \end{array}$$
same

Rules for Fischer projections:
1. Interchanging any two groups an odd number of times (once, three times, etc.) makes an enantiomer. Interchanging any two groups an even number of times (e.g. twice) returns to the original stereoisomer.
2. Rotating the structure by 90° makes the enantiomer. Rotating by 180° returns to the original stereoisomer.
(The second rule is an application of the first. Prove this to yourself.)

5-17

(a)
$$\begin{array}{c} CH_2OH \\ HO-\!\!\!-H \\ CH_3 \end{array}$$

(b)
$$\begin{array}{c} CH_2OH \\ H-\!\!\!-Br \\ CH_2CH_3 \end{array}$$

(c)
$$\begin{array}{c} CH_2Br \\ Br-\!\!\!-H \\ CH_2CH_3 \end{array}$$

(d)
$$\begin{array}{c} CH_3 \\ HO-\!\!\!-H \\ CH_2CH_3 \end{array}$$

(e)
$$\begin{array}{c} CHO \\ H-\!\!\!-OH \\ CH_2OH \end{array}$$

5-18   mirror

(a)
$$\begin{array}{c} CHO \\ H-\!\!\!-OH \\ CH_2OH \end{array}$$

$$\begin{array}{c} CHO \\ HO-\!\!\!-H \\ CH_2OH \end{array}$$

180° rotation of the right structure does not give left structure; no plane of symmetry: chiral—**enantiomers**

(b)
$$\begin{array}{c} CH_2OH \\ H-\!\!\!-Br \\ CH_2OH \end{array}$$

$$\begin{array}{c} CH_2OH \\ Br-\!\!\!-H \\ CH_2OH \end{array}$$

180° rotation of the right structure gives same structure as on the left; also has plane of symmetry: **same structure**

(c)
$$\begin{array}{c} CH_2Br \\ Br-\!\!\!-Br \\ CH_3 \end{array}$$

$$\begin{array}{c} CH_2Br \\ Br-\!\!\!-Br \\ CH_3 \end{array}$$

plane of symmetry: **same structure**

5-18 continued   mirror

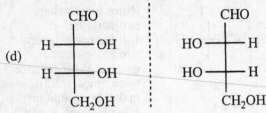

(d)   180 ° rotation of the right structure does not give left structure; no plane of symmetry: chiral—**enantiomers**

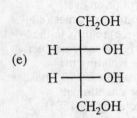

(e)   180° rotation of the right structure gives same structure as on the left; also has plane of symmetry: **same structure**

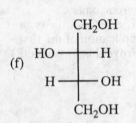

(f)   180 ° rotation of the right structure does not give left structure; no plane of symmetry: chiral—**enantiomers**

5-19 If the Fischer projection is drawn correctly, the most oxidized carbon will be at the top; this is the carbon with the greatest number of bonds to oxygen. Then the numbering goes from the top down.

(a) *R*
(b) no chiral center
(c) no chiral center

(d) *2R,3R*
(e) *2S,3R* (numbering down)
(f) *2R, 3R*

(g) *R*
(h) *S*
(i) *S*

5-20
(a) enantiomers—configurations at both asymmetric carbons inverted
(b) diastereomers—configuration at only one asymmetric carbon inverted
(c) diastereomers—configuration at only one asymmetric carbon inverted (the left carbon)
(d) constitutional isomers—C=C shifted position
(e) enantiomers—chiral, mirror images
(f) diastereomers—configuration at only one asymmetric carbon inverted (the top one)
(g) enantiomers—configuration at all asymmetric carbons inverted
(h) same compound—superimposable mirror images (hard question! use a model)
(i) diastereomers—configuration at only one chirality center (the nitrogen) inverted

5-21

(a)

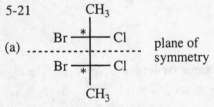

meso structure
not optically active

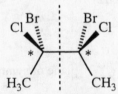

plane of symmetry

5-21 continued

(b)

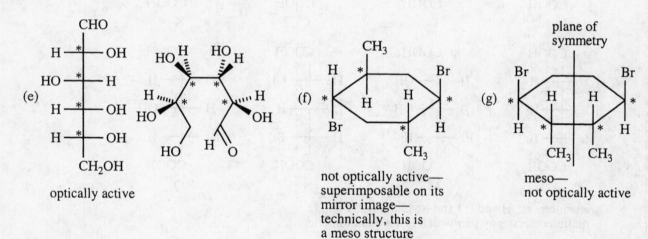

50:50

racemic mixture—not optically active, although each enantiomer by itself would be optically active

(c) planes of symmetry

not chiral
not optically active

(d) optically active

(e) optically active

(f) not optically active—
superimposable on its
mirror image—
technically, this is
a meso structure
(may require models!)

(g) plane of
symmetry

meso—
not optically active

5-22

(a)

A   B   C   D

enantiomers: A and B; C and D
diastereomers: A and C; A and D; B and C; B and D

89

5-22 continued

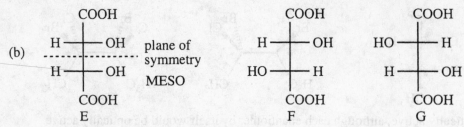

(b)

COOH
H — OH — plane of symmetry
H — OH MESO
COOH
E

COOH
H — OH
HO — H
COOH
F

COOH
HO — H
H — OH
COOH
G

enantiomers: F and G
diastereomers: E and F; E and G

(c)

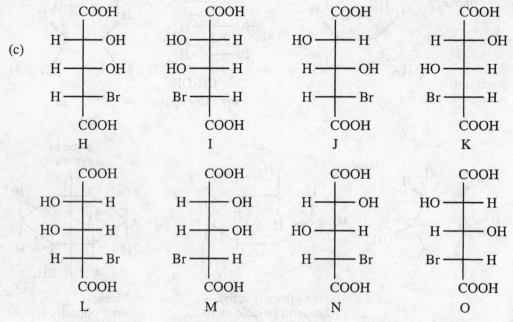

COOH
H — OH
H — OH
H — Br
COOH
H

COOH
HO — H
HO — H
Br — H
COOH
I

COOH
HO — H
H — OH
H — Br
COOH
J

COOH
H — OH
HO — H
Br — H
COOH
K

COOH
HO — H
HO — H
H — Br
COOH
L

COOH
H — OH
H — OH
Br — H
COOH
M

COOH
H — OH
HO — H
H — Br
COOH
N

COOH
HO — H
H — OH
Br — H
COOH
O

enantiomers: H and I; J and K; L and M; N and O
diastereomers: any pair which is not enantiomeric

(d)

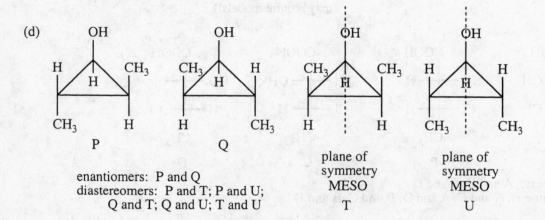

OH
H   CH₃
    H
CH₃   H
P

OH
CH₃   H
    H
H   CH₃
Q

OH
CH₃   CH₃
    H
H   H
plane of
symmetry
MESO
T

OH
H   H
    H
CH₃   CH₃
plane of
symmetry
MESO
U

enantiomers: P and Q
diastereomers: P and T; P and U;
        Q and T; Q and U; T and U

90

5-23 Any diastereomeric pair could be separated by a physical process like distillation or crystallization. Diastereomers are found in parts (a), (b), and (d). The structures in (c) are enantiomers; they could not be separated by normal physical means.

5-24

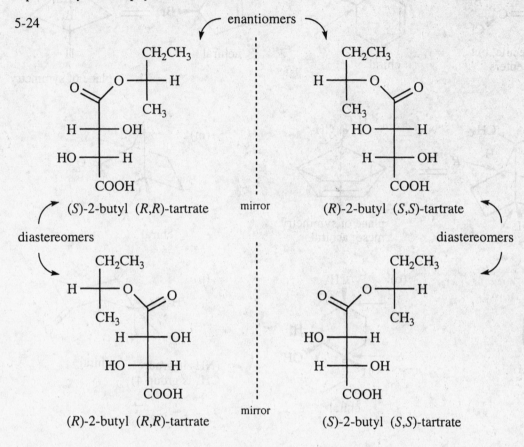

5-25 Please refer to solution 1-20, page 12 of this Solutions Manual.

5-26

(a)
chiral

(b)
chiral

(c)
chiral

(d)
chiral

(e)
meso; achiral

(f)
chiral

(g)
chiral

(h)
chiral

5-26 continued

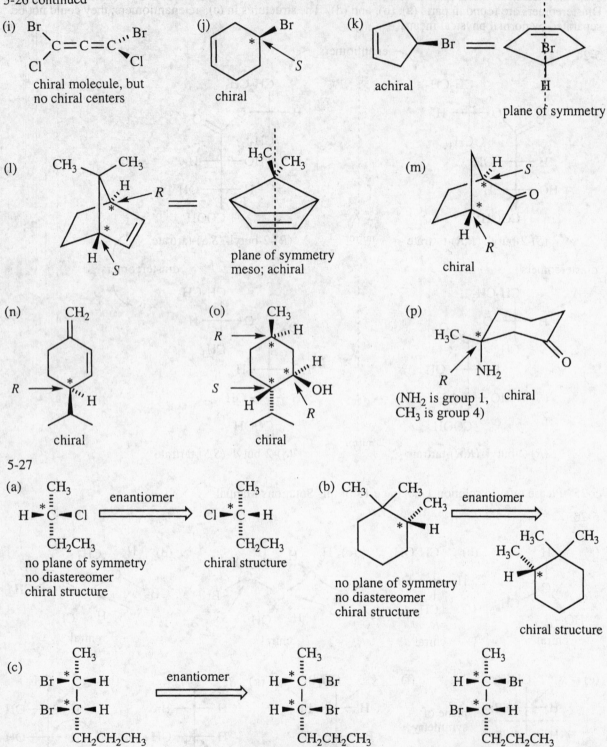

(i) chiral molecule, but
     no chiral centers

(j) chiral

(k) achiral   plane of symmetry

(l) plane of symmetry
     meso; achiral

(m) chiral

(n) chiral

(o) chiral

(p) (NH₂ is group 1,   chiral
      CH₃ is group 4)

5-27

(a) enantiomer
     no plane of symmetry   chiral structure
     no diastereomer
     chiral structure

(b) enantiomer
     no plane of symmetry   chiral structure
     no diastereomer
     chiral structure

(c) enantiomer
     no plane of symmetry   chiral structure   chiral structure
     chiral structure   (inverting two groups on the
                                                   bottom asymmetric carbon
                                                   instead of the top one would
                                                   also give a diastereomer)
     diastereomer

5-27 continued

(d)

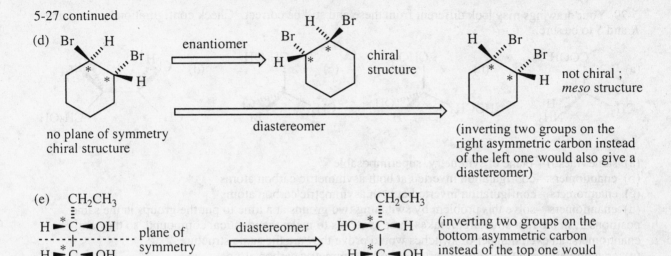

no plane of symmetry
chiral structure

diastereomer

not chiral ;
meso structure

(inverting two groups on the
right asymmetric carbon instead
of the left one would also give a
diastereomer)

(e)

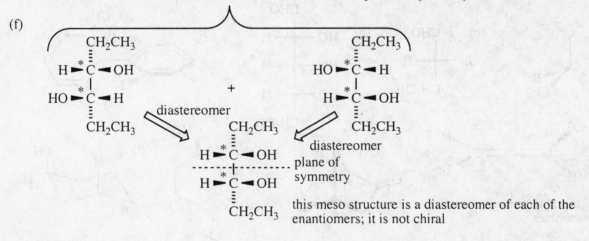

CH₂CH₃ | H—*C—OH | plane of symmetry (dashed)
H—*C—OH | CH₂CH₃

no enantiomer
a meso structure, not chiral

diastereomer

CH₂CH₃ | HO—*C—H | H—*C—OH | CH₂CH₃

(inverting two groups on the
bottom asymmetric carbon
instead of the top one would
also give a diastereomer)

chiral structure

racemic mixture of enantiomers; each is chiral with no plane of symmetry

(f)

CH₂CH₃
H—*C—OH
HO—*C—H
CH₂CH₃

+

CH₂CH₃
HO—*C—H
H—*C—OH
CH₂CH₃

diastereomer

diastereomer

CH₂CH₃
H—*C—OH
H—*C—OH
CH₂CH₃

plane of symmetry

this meso structure is a diastereomer of each of the
enantiomers; it is not chiral

5-28

(a)

CH₂OH
H———OH
CH₃

(b)

CHO
H———Br
CH₃

(c)

CH₂OH
H———Br
HO———H
CH₃

(d)

CH₂OH
HO———H
H———OH
CH₃

93

5-29 Your drawings may look different from these and still be correct. Check configuration by assigning *R* and *S* to be sure.

(a)

COOH
CH₃ — C —''''H
      NH₂

(b)

CHO
HOCH₂ — C —''''OH
           H

(c)

CH₂OH
CH₃ — C —''''Cl
           Br

(d)

Cl—H   H—Br
    \   /
     CH₃   CH₂OH

5-30

(a) same (meso)—plane of symmetry, superimposable
(b) enantiomers—configuration inverted at both asymmetric carbon atoms
(c) enantiomers—configuration inverted at both asymmetric carbon atoms
(d) enantiomers—solve this problem by switching two groups at a time to put the groups in the same positions as in the first structure; it takes three switches to make the identical compound, so they are enantiomers; an even number of switches would prove they are the same structure
(e) enantiomers—configuration inverted at both asymmetric carbon atoms
(f) diastereomers—configuration inverted at only one asymmetric carbon
(g) enantiomers—configuration inverted at both asymmetric carbon atoms
(h) same compound—rotate the right structure 180° around a horizontal axis and it becomes the left structure

5-31 Drawing the enantiomer of a chiral structure is as easy as drawing its mirror image.

(a)

CH₃
H — C —''''Br
        Cl

(b)

CHO
Br — + — H
  CH₂OH

(c)

CHO
HO — + — H
HO — + — H
HO — + — H
  CH₂OH

(d)

(e)

CH₃   H
   \   /
    C=C=C''''
   /   \
  H   Br

(f) plane of symmetry—
no enantiomer

(g)

(h)

(i)

5-32

(a)   1.00 g / 20.0 mL = 0.050 g/mL ;   20.0 cm = 2.00 dm

$$[\alpha]_D^{25} = \frac{-1.25°}{(0.0500)(2.00)} = -12.5°$$

(b)   0.050 g / 2.0 mL = 0.025 g/mL ;   2.0 cm = 0.20 dm

$$[\alpha]_D^{25} = \frac{+0.043°}{(0.025)(0.20)} = +8.6°$$

5-33 The 32% of the mixture that is (–)-tartaric acid will cancel the optical rotation of the 32% of the mixture that is (+)-tartaric acid, leaving only (68 – 32) = 36% of the mixture as excess (+)-tartaric acid to give measurable optical rotation. The specific rotation will therefore be only 36% of the rotation of pure (+)-tartaric acid: (+ 12.0°) x 36% = + 4.3°

5-34

(a)

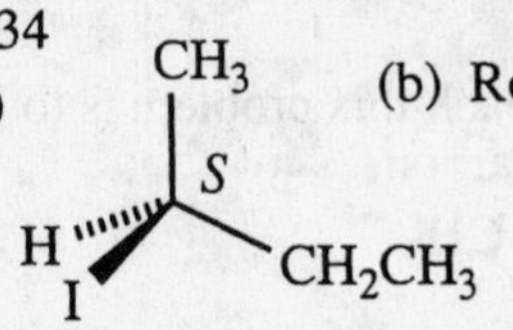

(b) Rotation of the enantiomer will be equal in magnitude, opposite in sign: – 15.90°.

(c) The rotation – 7.95° is what percent of – 15.90°?

$$\frac{-7.95°}{-15.90°} \times 100\% = 50\% \text{ e.e.}$$

There is 50% excess of (R)-2-iodobutane over the racemic mixture; that is, another 25% must be R and 25% must be S. The total composition is 75% (R)-(–)-2-iodobutane and 25% (S)-(+)-2-iodobutane.

5-35 All structures in this problem are chiral.

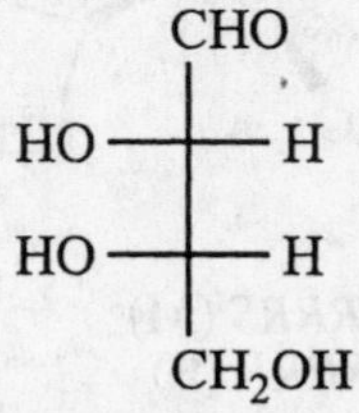

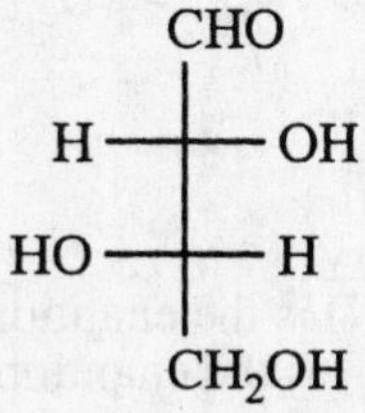

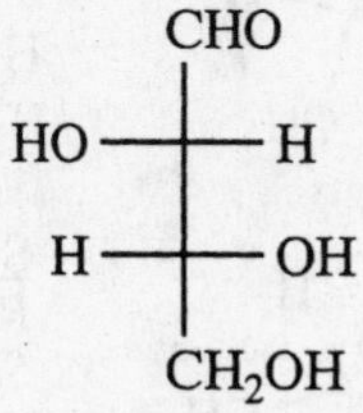

(a)

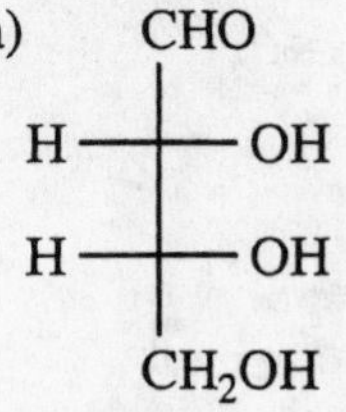

enantiomers: **A** and **B; C** and **D**
diastereomers: **A** and **C; A** and **D; B** and **C; B** and **D**

(b)

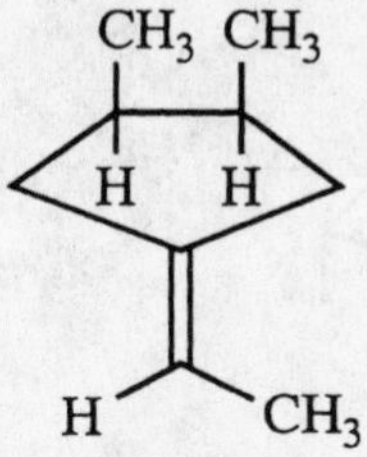

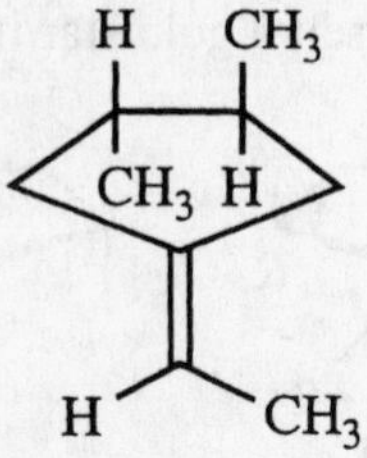

enantiomers: **E** and **F; G** and **H**
diastereomers: **E** and **G; E** and **H; F** and **G; F** and **H**

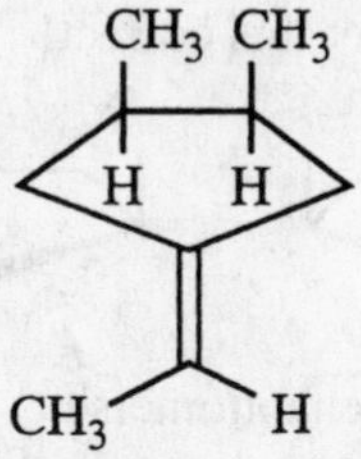

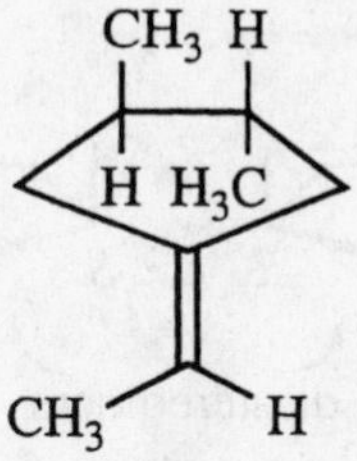

5-35 continued

(c) This structure is a challenge to visualize. A model helps. One way to approach this problem is to assign R and S configurations. Each arrow shows a change at one asymmetric carbon.

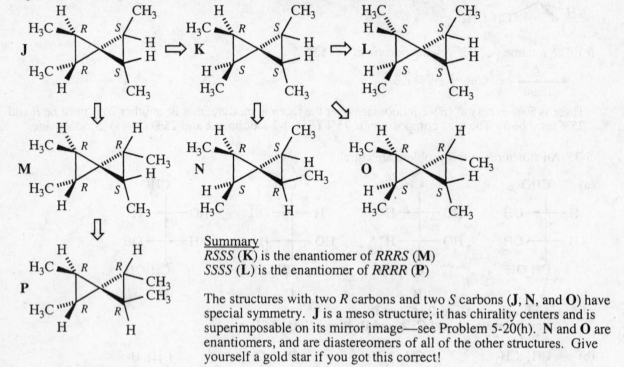

Summary
RSSS (**K**) is the enantiomer of RRRS (**M**)
SSSS (**L**) is the enantiomer of RRRR (**P**)

The structures with two R carbons and two S carbons (**J**, **N**, and **O**) have special symmetry. **J** is a meso structure; it has chirality centers and is superimposable on its mirror image—see Problem 5-20(h). **N** and **O** are enantiomers, and are diastereomers of all of the other structures. Give yourself a gold star if you got this correct!

5-36

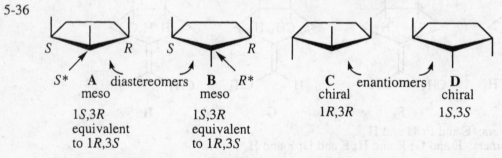

S*	**A**	diastereomers	**B**	R*
	meso		meso	
	1S,3R		1S,3R	
	equivalent		equivalent	
	to 1R,3S		to 1R,3S	

C	enantiomers	**D**
chiral		chiral
1R,3R		1S,3S

C and **D** are enantiomers. All other pairs are diastereomers.

*Structures **A** and **B** are both meso structures, but they are clearly different from each other. How can they be distinguished? One of the advanced rules of the Cahn-Ingold-Prelog system says: When two groups attached to an asymmetric carbon differ only in their absolute configuration, then the neighboring (R) stereocenter takes priority. Now the configuration of the central asymmetric carbon can be assigned: S for structure **A**, and R for structure **B**. This rule also applies to problem 5-37. (Thanks to Dr. Kantorowski for this explanation.)

5-37 This problem is similar to 5-36.

(a)

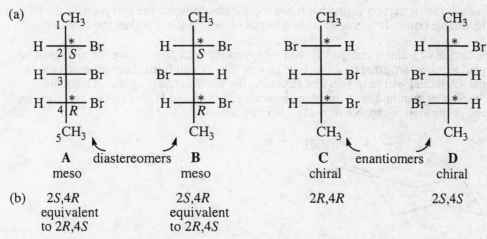

A diastereomers	**B**	**C** enantiomers	**D**
meso	meso	chiral	chiral

(b)   2S,4R   2S,4R   2R,4R   2S,4S
      equivalent   equivalent
      to 2R,4S   to 2R,4S

The configuration of carbon-3 in **A** and **B** can be assigned according the rule described above in the solution to problem 5-36: C-3 in **A** is *S*, and C-3 in **B** is *R*.

(c) According to the IUPAC designation described in text Section 5-2B, a chirality center is "any atom holding a set of ligands in a spatial arrangement which is not superimposable on its mirror image." An asymmetric carbon must have four different groups on it, but in **A** and **B**, C-3 has two groups that are identical (except for their stereochemistry). C-3 holds its groups in a spatial arrangement that is superimposable on its mirror image, so it is not a chirality center. But it is stereogenic: in structure **A**, interchanging the H and Br at C-3 gives structure **B**, a diastereomer of **A**; therefore, C-3 is stereogenic.

(d) In structure **C** or **D**, C-3 is not stereogenic. Inverting the H and the Br, then rotating the structure 180°, shows that the same structure is formed. Therefore, interchanging two atoms at C-3 does *not* give a stereoisomer, so C-3 does not fit the definition of a stereogenic center.

5-38 The Cahn-Ingold-Prelog priorities of the groups are the circled numbers in (a).

(a)

(R)-3,4-dimethyl-1-pentene   (S)-2,3-dimethylpentane

(b) The reaction did not occur at the asymmetric carbon atom, so the configuration has not changed—the reaction went with retention of configuration at the asymmetric carbon.

(c) The *name* changed because the *priority* of groups in the Cahn-Ingold-Prelog system of nomenclature changed. When the alkene became an ethyl group, its priority changed from the highest priority group to priority 2. (We will revisit this anomaly in problem 6-21(c).)

(d) There is no general correlation between *R* and *S* designation and the physical property of optical rotation. Professor Wade's poetic couplet makes an important point: do not confuse an object and its properties with the *name* for that object. (Scholars of Shakespeare have come to believe that this quote from Juliet is a veiled reference to designation of *R,S* configuration versus optical rotation of a chiral molecule. Shakespeare was *way* ahead of his time.)

5-39

(a) The product has no asymmetric carbon atoms but it has three stereocenters: the carbon with the OH, plus both carbons of the double bond. Interchange of two bonds on any of these makes the enantiomer.

(b) The product is an example of a chiral compound with no asymmetric carbons. Like the allenes, it is classified as an "extended tetrahedron"; that is, it has four groups that extend from the rigid molecule in four different directions. (A model will help.) In this structure, the plane containing the COOH and carbons of the double bond is perpendicular to the plane bisecting the OH and H and carbon that they are on. Since the compound is chiral, it is capable of being optically active.

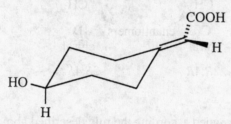

(c) As shown in text Figure 5-16, Section 5-6, catalytic hydrogenation that creates a new chirality center creates a racemic mixture (both enantiomers in a 1:1 ratio). A racemic mixture is not optically active. In contrast, by using a chiral enzyme to reduce the ketone to the alcohol (as in part (b)), an excess of one enantiomer was produced, so the product was optically active.

6-1 In problems like part (a), draw out the whole structure to detect double bonds.

(a) vinyl halide   (c) aryl halide   (e) vinyl halide
(b) alkyl halide   (d) alkyl halide   (f) aryl halide

6-2 (a)
$$I-\overset{\displaystyle H}{\underset{\displaystyle H}{C}}-I$$
(b)
$$Br-\overset{\displaystyle Br}{\underset{\displaystyle Br}{C}}-Br$$
(c) [structure]
(d)
$$I-\overset{\displaystyle H}{\underset{\displaystyle I}{C}}-H$$
(e) [structure with Br]

(f) [structure with Br] (g) [cyclohexane structure with H, CH$_2$F, F] (h) [structure] Cl or (CH$_3$)$_3$CCl

6-3 IUPAC name; common name; degree of halogen-bearing carbon

(a) 1-chloro-2-methylpropane; isobutyl chloride; 1°
(b) diiodomethane; methylene iodide; methyl
(c) 1,1-dichloroethane; no common name; 1°
(d) 2-bromo-1,1,1-trichloroethane; no common name; all 1°
(e) trichloromethane; chloroform; methyl
(f) 2-bromo-2-methylpropane; *t*-butyl bromide; 3°
(g) 2-bromobutane; *sec*-butyl bromide; 2°
(h) 1-chloro-2-methylbutane; no common name; 1°
(i) *cis*-1-bromo-2-chlorocyclobutane; no common name; both 2°
(j) 3-bromo-4-methylhexane; no common name; 2°
(k) 4-fluoro-1,1-dimethylcyclohexane; no common name; 2°
(l) *trans*-1,3-dichlorocyclopentane; no common name; all 2°

6-4

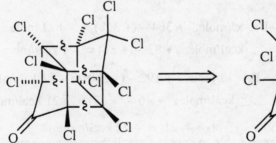

Kepone®

Chlordane®

99

6-5 From the text, Section 2-9B, the bond dipole moment depends not only on bond length but also on charge separation, which in turn depends on the difference in electronegativities of the two atoms connected by the bond. Because chlorine's electronegativity (3.2) is significantly higher than iodine's (2.7), the C—Cl bond dipole is greater than that of C—I, despite C—I being a longer bond.

6-6
(a) *n*-Butyl bromide has a higher molecular weight and less branching, and boils at a higher temperature than isopropyl bromide.
(b) *t*-Butyl bromide has a higher molecular weight and a larger halogen, and despite its greater branching, boils at a higher temperature than isopropyl chloride.
(c) *n*-Butyl bromide has a higher molecular weight and a larger halogen, and boils at a higher temperature than *n*-butyl chloride.

6-7 From Table 3-2, the density of hexane is 0.66; it will float on the water layer (d 1.00). From Table 6-2, the density of chloroform is 1.50; water will float on the chloroform. Water is immiscible with many organic compounds; whether water is the top layer or bottom layer depends on whether the other material is more dense or less dense than water. (This is an important consideration to remember in lab procedures.)

6-8
(a) Step (1) is initiation; steps (2) and (3) are propagation.

(1) $Br-Br \xrightarrow{h\nu} 2\ Br\cdot$

(2) $H_2C=C(H)-CH_2 + Br\cdot \longrightarrow HBr + \{ H_2C=C(H)-\dot{C}H_2 \longleftrightarrow H_2\dot{C}-C(H)=CH_2 \}$

(3) $H_2C=C(H)-\dot{C}H_2 + Br-Br \longrightarrow H_2C=C(H)-CH_2-Br + Br\cdot$

(b) Step (2): break allylic C—H, make H—Br: kJ/mole: $+364 - (+368) = -4$ kJ/mole
kcal/mole: $+87 - (+88) = -1$ kcal/mole

Step (3): break Br—Br, make allylic C—Br: kJ/mole: $+192 - (+280) = -88$ kJ/mole
kcal/mole: $+46 - (+67) = -21$ kcal/mole

ΔH overall $= -4 + -88 = -92$ kJ/mole $(-1 + -21 = -22$ kcal/mole$)$
This is a very exothermic reaction; it is reasonable to expect a small activation energy in step (1), so this reaction should be very rapid.

6-9 (a) propagation steps

repeats chain mechanism

100

The resonance-stabilized allylic radical intermediate has radical character on both the 1° and 3° carbons, so bromine can bond to either of these carbons producing two isomeric products.

(b) Allylic bromination of cyclohexene gives 3-bromocyclohex-1-ene regardless of whether there is an allylic shift. Either pathway leads to the same product. If one of the ring carbons were somehow marked or labeled, then the two products can be distinguished. (We will see in following chapters how labeling is done experimentally.)

the second structure from an allylic shift is *identical* to the first structure—only one compound is produced

6-10

(a)

This compound has only one type of hydrogen—only one monochlorine isomer can be produced.

(b)

Bromination has a strong preference for abstracting hydrogens (like 3°) that give stable radical intermediates.

(c)

Bromine atom will abstract the hydrogen giving the most stable radical; in this case, the radical intermediate will be stabilized by resonance with the benzene ring.

(d)

the second structure from an allylic shift is *identical* to the first structure—only one compound is produced

Bromine atom will abstract the hydrogen giving the most stable radical; in this case, the radical intermediate will be stabilized by resonance with the benzene ring.

6-11

(a) substitution—Br is replaced
(b) elimination—H and OH are lost
(c) elimination—both Br atoms are lost

6-12

(a) $CH_3(CH_2)_4CH_2 - OCH_2CH_3$ (b) $CH_3(CH_2)_4CH_2 - CN$ (c) $CH_3(CH_2)_4CH_2 - OH$

6-13 The rate law is first order in both 1-bromobutane, C_4H_9Br, and methoxide ion. If the concentration of C_4H_9Br is lowered to one-fifth the original value, the rate must decrease to one-fifth; if the concentration of methoxide is doubled, the rate must also double. Thus, the rate must decrease to 0.02 mole/L per second:

$$\text{rate} = (\,0.05 \text{ mole/L per second}\,) \times \frac{(\,0.1\text{ M}\,)}{(\,0.5\text{ M}\,)} \times \frac{(\,2.0\text{ M}\,)}{(\,1.0\text{ M}\,)} = 0.02 \text{ mole/L per second}$$

original rate   change in   change in   new rate

                                   C_4H_9Br $NaOCH_3$

A completely different way to answer this problem is to solve for the rate constant k, then put in new values for the concentrations.

rate $= k$ $[C_4H_9Br]$ $[NaOCH_3]$ $\implies$ 0.05 mole L^{-1} sec $^{-1}$ $= k$ $(0.5$ mol $L^{-1})$ $(1.0$ mol $L^{-1})$ $\implies$

rate constant k = 0.1 L mol^{-1} sec^{-1}

rate $= k$ $[CH_3I]$ $[NaOH]$ = $(0.1$ L mol^{-1} sec$^{-1})$ $(0.1$ mol $L^{-1})$ $(2.0$ mol $L^{-1})$ = 0.02 mole L^{-1} sec $^{-1}$

6-14 Organic and inorganic products are shown here for completeness.

(a) $(CH_3)_3C-O-CH_2CH_3$ + KBr

(b) $HC\equiv C-CH_2CH_2CH_2CH_3$ + NaCl

(c) $(CH_3)_2CHCH_2-\overset{+}{N}H_3$ Br^- $\xrightarrow{NH_3}$ $(CH_3)_2CHCH_2-NH_2$ + $\overset{+}{N}H_4$ Br^-

(d) CH_3CH_2-CN + NaI

(e)  $\diagup\!\!\diagdown\!\!\diagup\!\!\diagdown^{I}$ + NaCl

(f) $\diagup\!\!\diagdown\!\!\diagup\!\!\diagdown^{F}$ 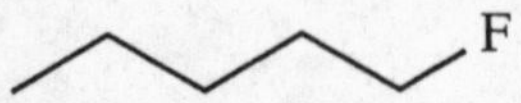   + KCl   (18-crown-6 is the catalyst and does not change; CH_3CN is the solvent)

6-15 All reactions in this problem follow the same pattern; the only difference is the nucleophile ($^-$:Nuc). Only the nucleophile is listed below. (Cations like Na$^+$ or K$^+$ accompany the nucleophile but are simply spectator ions and do not take part in the reaction; they are not shown here.)

       $\diagup\!\!\diagdown\!\!\diagup\!\!\diagdown$Cl   + $^-$:Nuc $\longrightarrow$ $\diagup\!\!\diagdown\!\!\diagup\!\!\diagdown$Nuc   +   Cl$^-$

   1-chlorobutane

(a) HO$^-$ (b) F$^-$ from KF/18-crown-6   (c) I$^-$ (d) $^-$CN   (e) HC$\equiv$C$^-$

(f) $^-OCH_2CH_3$ (g) excess NH$_3$ (or $^-$NH$_2$)

6-16
(a) $(CH_3CH_2)_2NH$ is a better nucleophile—less hindered
(b) $(CH_3)_2S$ is a better nucleophile—S is larger, more polarizable than O
(c) PH$_3$ is a better nucleophile—P is larger, more polarizable than N
(d) CH_3S^- is a better nucleophile—anions are better than neutral atoms of the same element
(e) $(CH_3)_3N$ is a better nucleophile—less electronegative than oxygen, better able to donate an electron pair
(f) CH_3S^- is a better nucleophile—anions are generally better than neutral atoms, and S is larger and more polarizable than O
(g) $CH_3CH_2CH_2O^-$ is a better nucleophile—less branching, less steric hindrance
(h) I$^-$ is a better nucleophile—larger, more polarizable

6-17 A mechanism must show *electron movement*.

$$CH_3-\overset{..}{\underset{..}{O}}-CH_3 \ + \ H^+ \ \rightleftharpoons \ CH_3-\overset{\overset{H}{|}}{\underset{..}{\overset{+}{O}}}-CH_3 \ + \ :\overset{..}{\underset{..}{Br}}:^- \ \longrightarrow \ CH_3-\overset{\overset{H}{|}}{\underset{..}{O}}: \ + \ CH_3-\overset{..}{\underset{..}{Br}}:$$

Protonation converts OCH_3 to a good leaving group.

6-18 The type of carbon with the halide, and relative leaving group ability of the halide, determine the reactivity.

methyl iodide > methyl chloride > ethyl chloride > isopropyl bromide >> neopentyl bromide, } *least*
most reactive   *t*-butyl iodide } *reactive*

Predicting the relative order of neopentyl bromide and *t*-butyl iodide would be difficult because both would be extremely slow.

6-19 In all cases, the less hindered structure is the better S_N2 substrate.

(a) 2-methyl-1-iodopropane (1° versus 3°)

(b) cyclohexyl bromide (2° versus 3°)

(c) isopropyl bromide (no substituent on neighboring carbon)

(d) 2-chlorobutane (even though this is a 2° halide, it is easier to attack than the 1° neopentyl type in 2,2-dimethyl-1-chlorobutane—see below)

(e) isopropyl iodide (same reason as in (d))

a neopentyl halide—
hindered to backside attack
by neighboring methyl groups

6-20 All S_N2 reactions occur with inversion of configuration at carbon.

(a)

trans   transition state   *cis*
                 inversion

(b)

6-20 continued

(c)

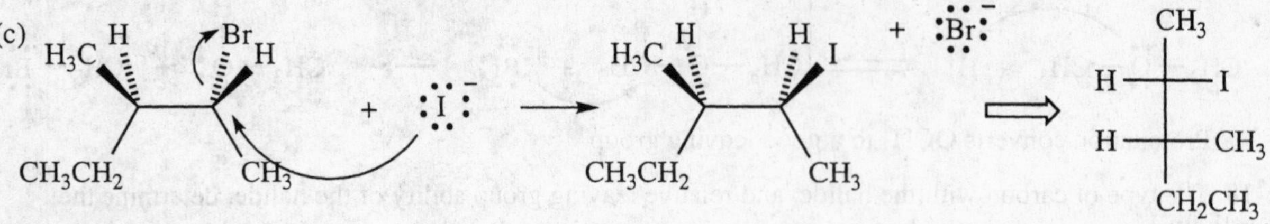

(d)

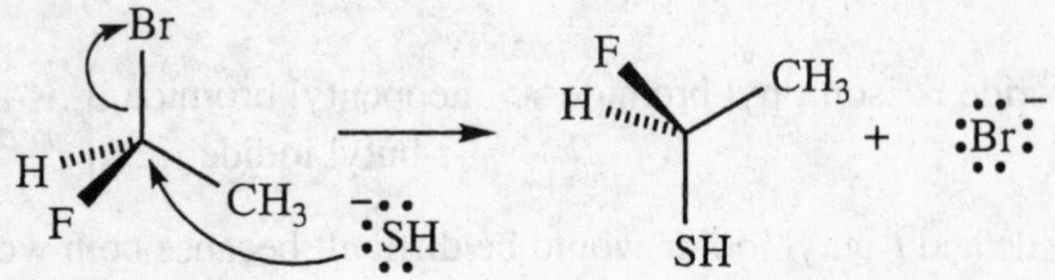

(e)

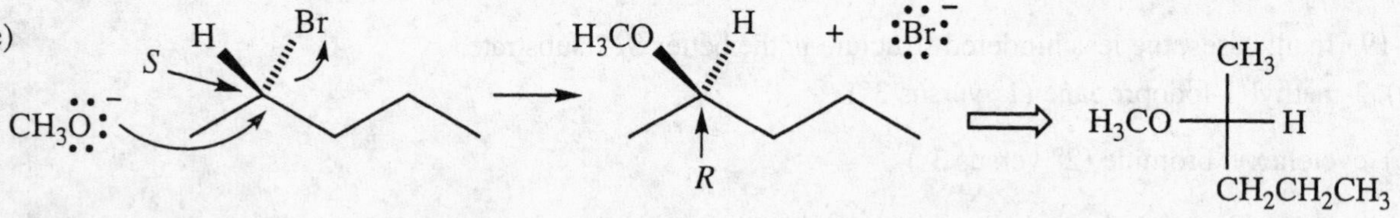

(f)

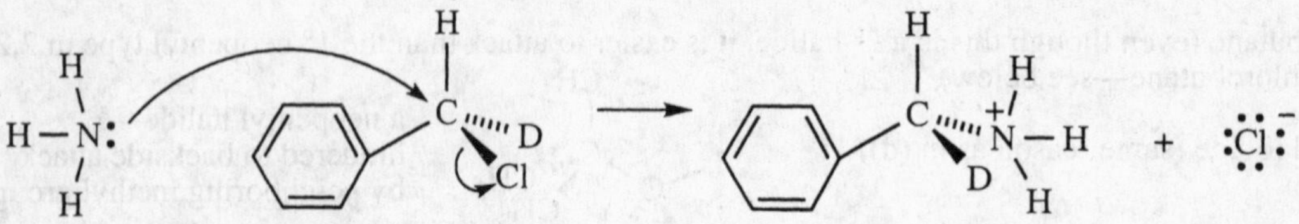

6-21

(a) The best leaving groups are the weakest bases. Bromide ion is so weak it is not considered at all basic; it is an excellent leaving group. Fluoride is moderately basic, by far the most basic of the halides. It is a terrible leaving group. Bromide is many orders of magnitude better than fluoride in leaving group ability.

(b)

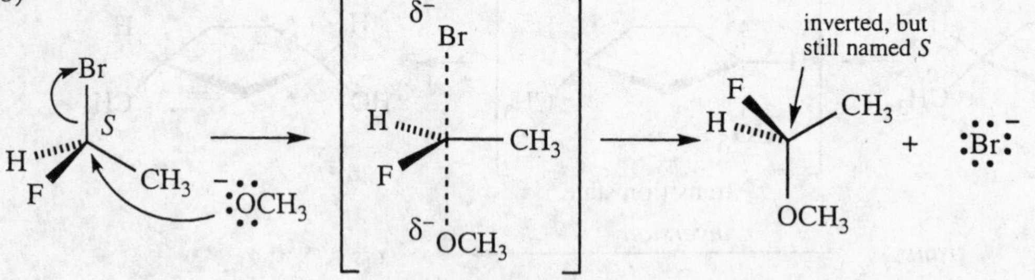

transition state

(c) As noted on the structure above, the configuration is inverted even though the designations of the configuration for both the starting material and the product are S; the oxygen of the product has a lower priority than the bromine it replaces. Refer to the solution to problem 5-38, p. 97, for the caution about confusing absolute configuration with the *designation* of configuration.

(d) The result is perfectly consistent with the S_N2 mechanism. Even though both the reactant and the product have the S designation, the configuration has been inverted: the nomenclature priority of fluorine changes from second (after bromine) in the reactant to first (before oxygen) in the product. While the designation may be misleading, the structure shows with certainty that an inversion has occurred.

104

6-22

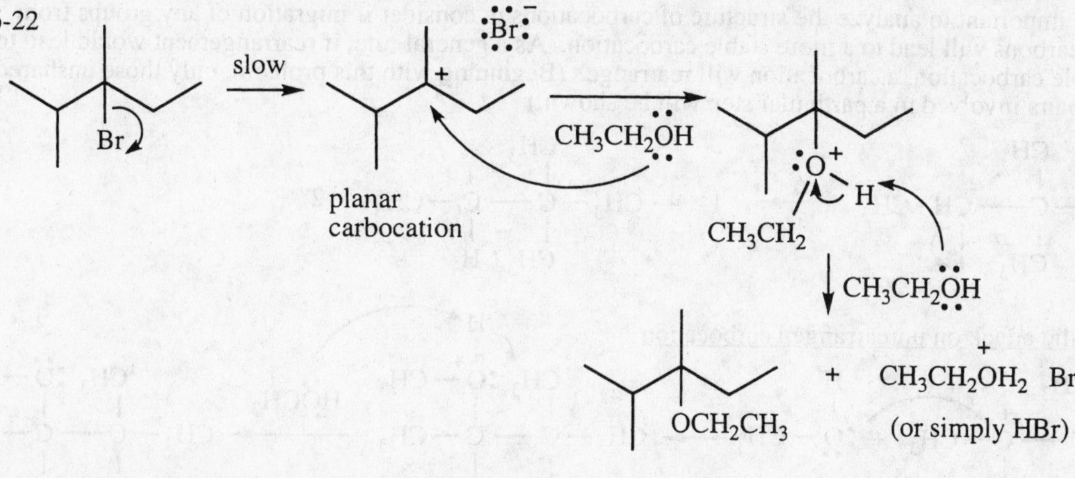

planar
carbocation

CH₃CH₂ÖH

CH₃CH₂

CH₃CH₂ÖH

OCH₂CH₃   +   CH₃CH₂OH₂ Br⁻

(or simply HBr)

6-23 The structure that can form the more stable carbocation will undergo S_N1 faster.

(a) 2-bromopropane: will form a 2° carbocation
(b) 2-bromo-2-methylbutane: will form a 3° carbocation
(c) allyl bromide is faster than *n*-propyl bromide: allyl bromide can form a resonance-stabilized intermediate.
(d) 2-bromopropane: will form a 2° carbocation
(e) 2-iodo-2-methylbutane is faster than *t*-butyl chloride (iodide is a better leaving group than chloride)
(f) 2-bromo-2-methylbutane (3°) is faster than ethyl iodide (1°); although iodide is a somewhat better leaving group, the difference between 3° and 1° carbocation stability dominates

6-24 Ionization is the rate-determining step in S_N1. Anything that stabilizes the intermediate will speed the reaction. Both of these compounds form resonance-stabilized intermediates.

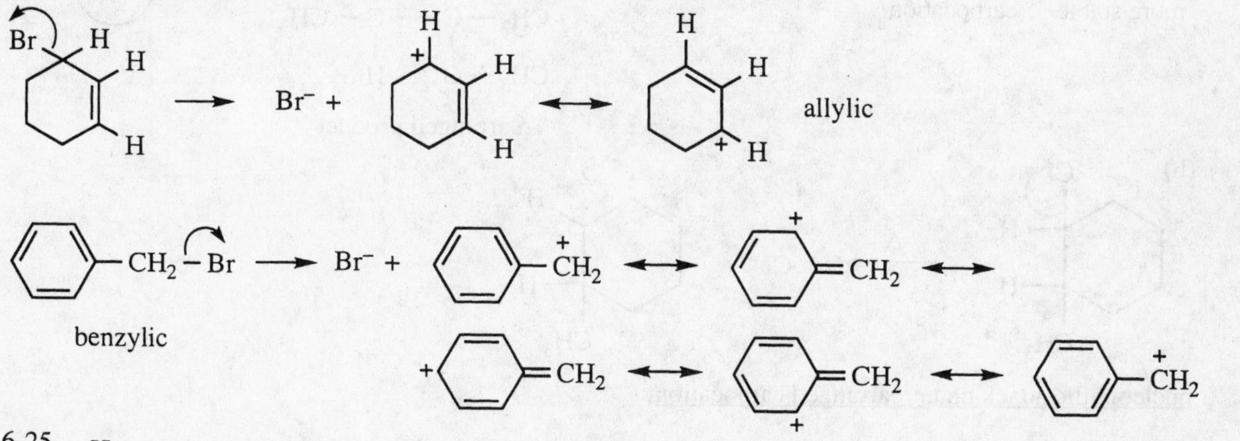

allylic

benzylic

6-25

105

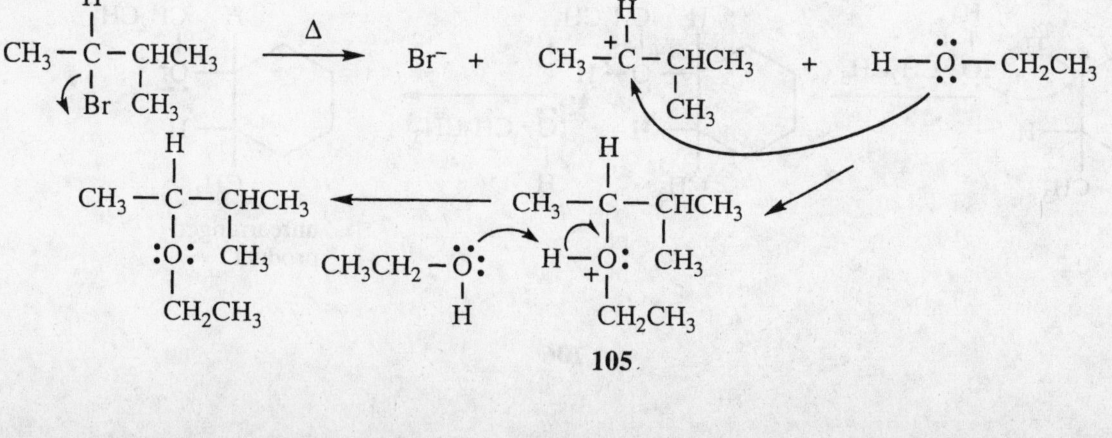

6-26 It is important to analyze the structure of carbocations to consider if migration of any groups from adjacent carbons will lead to a more stable carbocation. As a general rule, if rearrangement would lead to a more stable carbocation, a carbocation will rearrange. (Beginning with this problem, only those unshared electons pairs involved in a particular step will be shown.)

(a)

$$CH_3-\underset{\underset{\displaystyle CH_3}{|}}{\overset{\overset{\displaystyle CH_3}{|}}{C}}-\underset{\underset{\displaystyle I}{|}}{CH}-CH_3 \xrightarrow{\Delta} I^- + CH_3-\underset{\underset{\displaystyle CH_3}{|}}{\overset{\overset{\displaystyle CH_3}{|}}{C}}-\overset{+}{\underset{\underset{\displaystyle H}{|}}{C}}-CH_3 \quad 2°$$

nucleophilic attack on unrearranged carbocation

$$CH_3-\underset{CH_3}{\overset{CH_3}{C}}-\overset{+}{\underset{H}{C}}-CH_3 + \:\ddot{O}-CH_3 \longrightarrow CH_3-\underset{CH_3}{C}-\underset{H}{\overset{CH_3\;\overset{+}{\ddot{O}}-CH_3}{C}}-CH_3 \xrightarrow{H\ddot{O}CH_3} CH_3-\underset{CH_3}{C}-\underset{H}{\overset{CH_3\;\ddot{O}-CH_3}{C}}-CH_3$$

unrearranged product

nucleophilic attack after carbocation rearrangement

$$CH_3-\underset{CH_3}{\overset{CH_3}{C}}-\overset{+}{\underset{H}{C}}-CH_3 \longrightarrow CH_3-\overset{+}{\underset{CH_3}{C}}-\underset{H}{\overset{CH_3}{C}}-CH_3 + \:\ddot{O}-CH_3 \longrightarrow CH_3-\underset{\underset{H}{\overset{+}{\ddot{O}}}}{\overset{CH_3}{C}}-\underset{H}{\overset{CH_3}{C}}-CH_3$$

Methyl shift to the 2° carbocation forms a more stable 3° carbocation.

$$CH_3-\underset{CH_3-\ddot{O}}{\overset{CH_3}{C}}-\underset{H}{\overset{CH_3}{C}}-CH_3$$

rearranged product

(b)

$$\xrightarrow{\Delta} Cl^- +$$ 2° ring carbocation

nucleophilic attack on unrearranged carbocation

unrearranged product

106

6-26(b) continued

nucleophilic attack after carbocation rearrangement

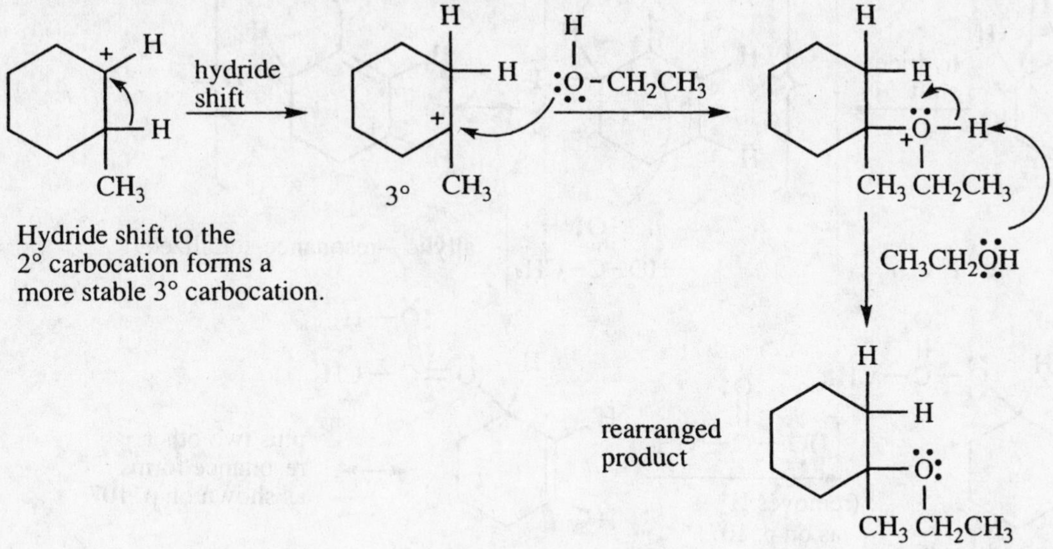
hydride shift
3°
Hydride shift to the 2° carbocation forms a more stable 3° carbocation.
rearranged product

(c)

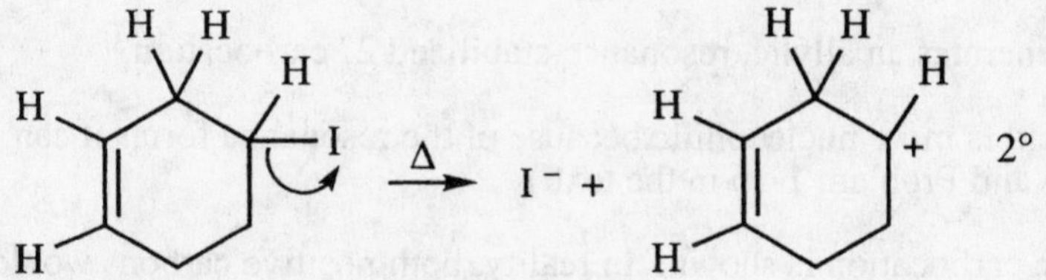
Δ
2°

Note: braces are used to indicate the ONE chemical species represented by multiple resonance forms.

nucleophilic attack on unrearranged carbocation

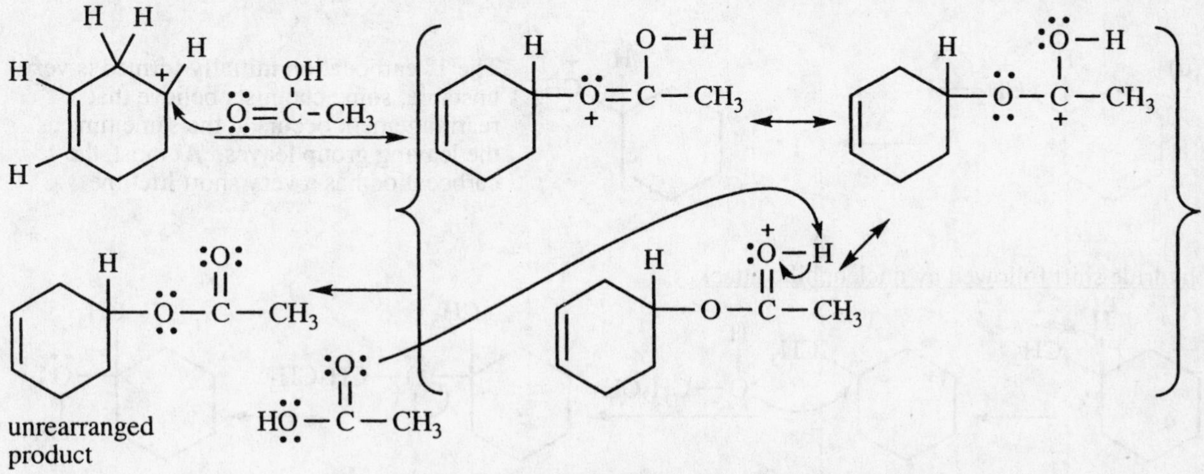
unrearranged product

The most basic species in a mixture is the most likely to remove a proton. In this reaction, acetic acid is more basic than iodide ion.

6-26(c) continued

nucleophilic attack after carbocation rearrangement

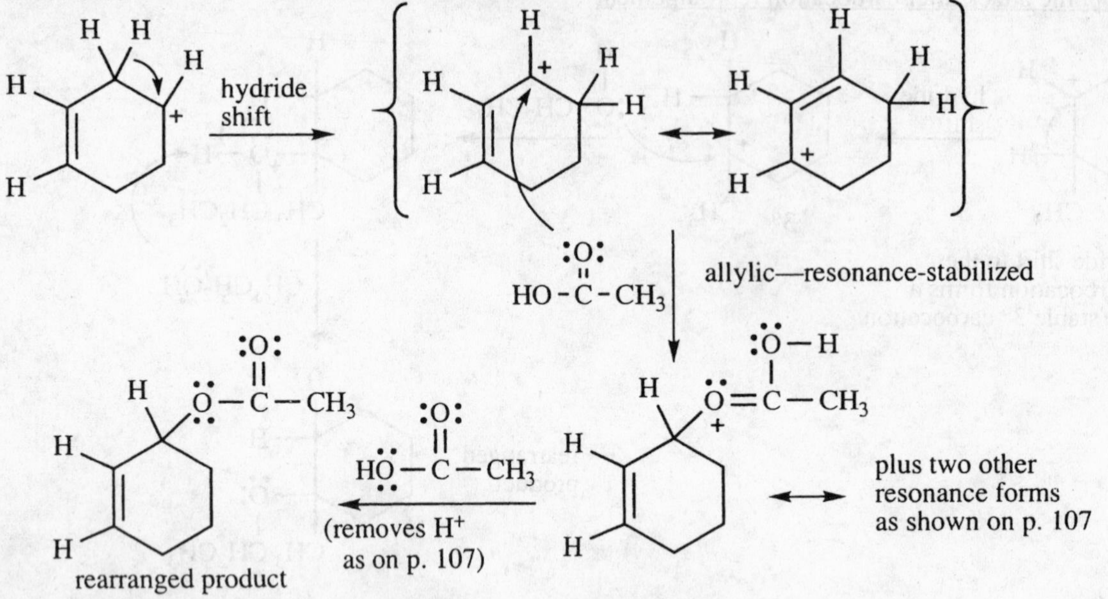

allylic—resonance-stabilized

plus two other
resonance forms
as shown on p. 107

(removes H⁺
as on p. 107)

rearranged product

Comments on 6-26(c)

(1) The hydride shift to a 2° carbocation generates an allylic, resonance-stabilized 2° carbocation.

(2) The double-bonded oxygen of acetic acid is more nucleophilic because of the resonance forms it can have after attack. (See Solved Problem 1-5 and Problem 1-16 in the text.)

(3) Attack on only one carbon of the allylic carbocation is shown. In reality, both positive carbons would be attacked in equal amounts, but they would give the identical product *in this case*. In other compounds, however, attack on the different carbons might give different products. ALWAYS CONSIDER ALL POSSIBILITIES.

(d)

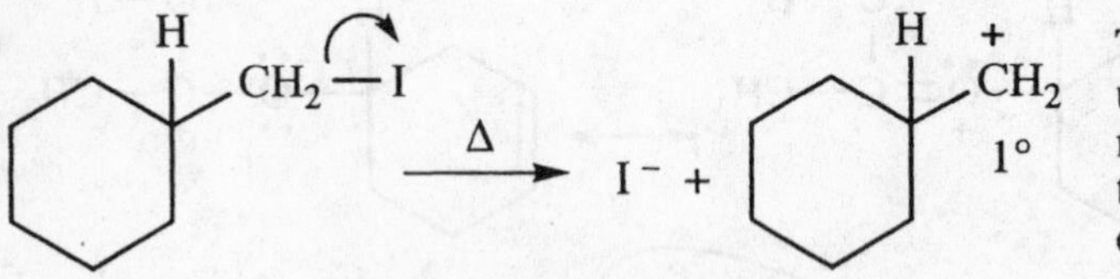

The 1° carbocation initially formed is very unstable; some chemists believe that rearrangement occurs at the same time as the leaving group leaves. At most, the 1° carbocation has a very short lifetime.

hydride shift followed by nucleophilic attack

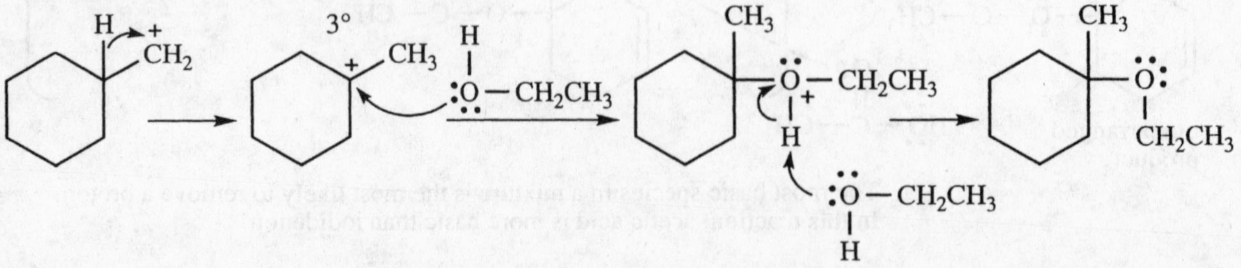

108

6-26 (d) continued

alkyl migration (ring expansion) followed by nucleophilic attack

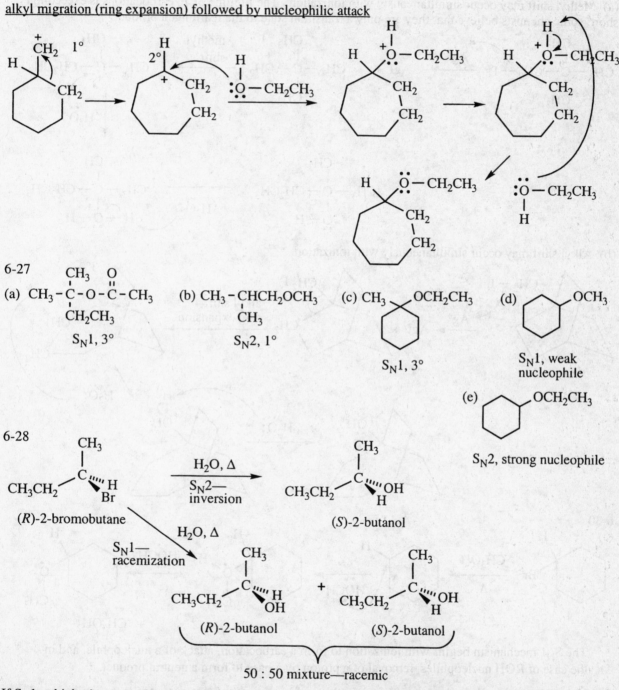

6-27

(a) $CH_3 - \underset{\underset{CH_2CH_3}{|}}{\overset{\overset{CH_3}{|}}{C}} - O - \overset{\overset{O}{||}}{C} - CH_3$

$S_N1, 3°$

(b) $CH_3 - \underset{\underset{CH_3}{|}}{CH}CH_2OCH_3$

$S_N2, 1°$

(c) $CH_3 \quad OCH_2CH_3$

$S_N1, 3°$

(d) OCH_3

$S_N1,$ weak nucleophile

(e) OCH_2CH_3

$S_N2,$ strong nucleophile

6-28

$$CH_3CH_2 - \underset{\underset{Br}{}}{\overset{\overset{CH_3}{|}}{C}}\!\!\!\!\!\text{'''}H$$

(R)-2-bromobutane

$\xrightarrow[\substack{S_N2\!-\!\\inversion}]{H_2O, \Delta}$

$CH_3CH_2 - \underset{\underset{H}{}}{\overset{\overset{CH_3}{|}}{C}}\!\!\!\!\!\text{'''}OH$

(S)-2-butanol

$\xrightarrow[\substack{S_N1\!-\!\\racemization}]{H_2O, \Delta}$

$CH_3CH_2 - \underset{\underset{OH}{}}{\overset{\overset{CH_3}{|}}{C}}\!\!\!\!\!\text{'''}H$

(R)-2-butanol

$+$

$CH_3CH_2 - \underset{\underset{H}{}}{\overset{\overset{CH_3}{|}}{C}}\!\!\!\!\!\text{'''}OH$

(S)-2-butanol

50 : 50 mixture—racemic

If S_N1, which gives racemization, occurs exactly twice as fast as S_N2, which gives inversion, then the racemic mixture (50 : 50 $R + S$) is 66.7% of the mixture and the rest, 33.3%, is the S enantiomer from S_N2. Therefore, the excess of one enantiomer over the racemic mixture must be 33.3%, the enantiomeric excess. (In the racemic mixture, the R and S "cancel" each other algebraically as well as in optical rotation.)

The optical rotation of pure (S)-2-butanol is + 13.5°. The optical rotation of this mixture is:

$$33.3\% \times +13.5° = +4.5°$$

109

6-29

(a) Methyl shift may occur simultaneously with ionization. The lifetime of 1° carbocations is exceedingly short; some chemists believe that they are only a transition state to the rearranged product.

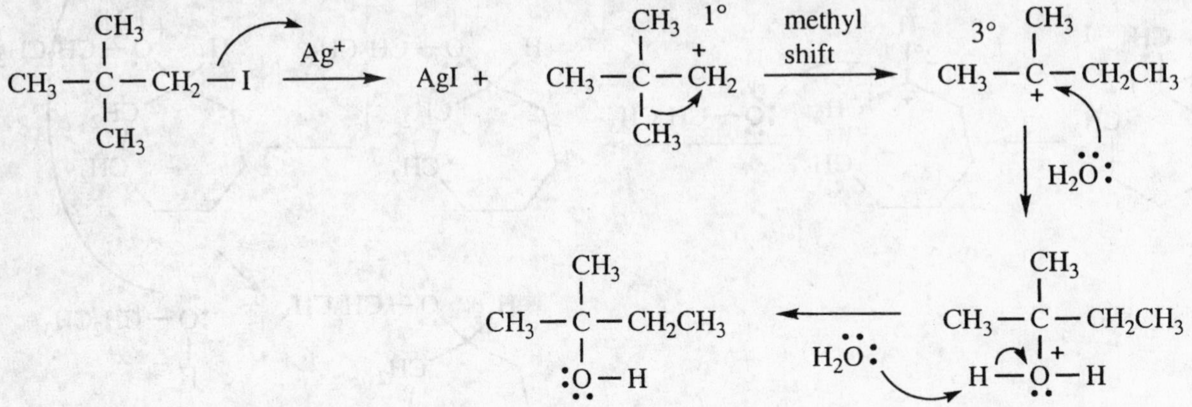

(b) Alkyl shift may occur simultaneously with ionization.

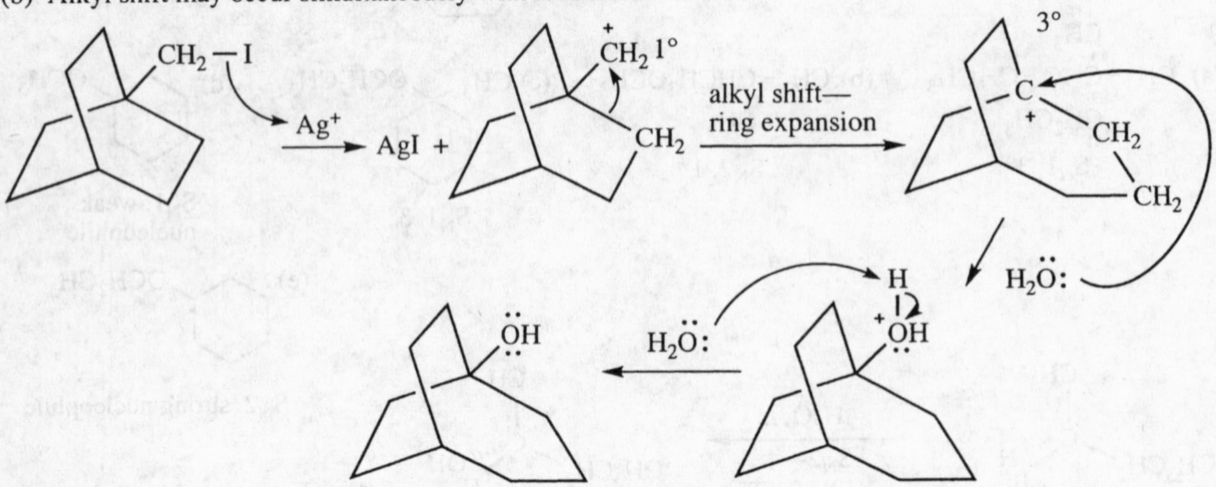

6-30
(a)

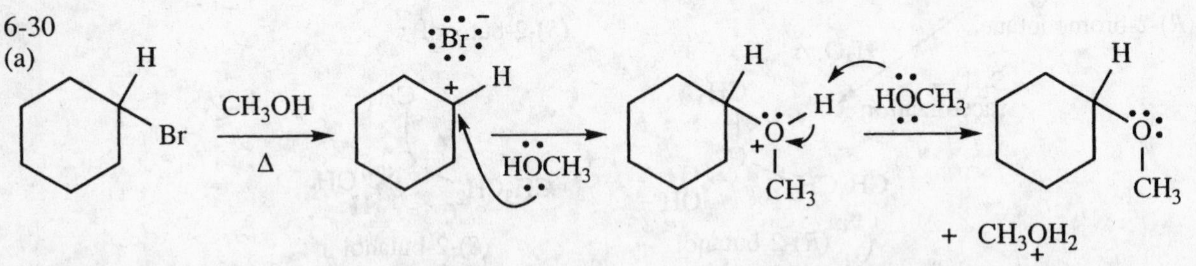

The S_N1 mechanism begins with ionization to form a carbocation, attack of a nucleophile, and in the case of ROH nucleophiles, removal of a proton by a base to form a neutral product.

(b) In the E1 reaction, the solvent (methanol, in this case) serves two functions: it aids the ionization process by solvating both the leaving group (bromide) and the carbocation; and second, it serves as a base to remove the proton from a carbon adjacent to the carbocation in order to form the carbon-carbon double bond. The S_N1 mechanism adds a third function to the solvent: the first step is the same as in E1, ionization to form the carbocation; the second step has the solvent acting as a nucleophile—this step is different from E1; third, the solvent acts like a base and removes a proton, although from an oxygen (S_N1) and not a carbon (E1). Solvents are versatile!

110

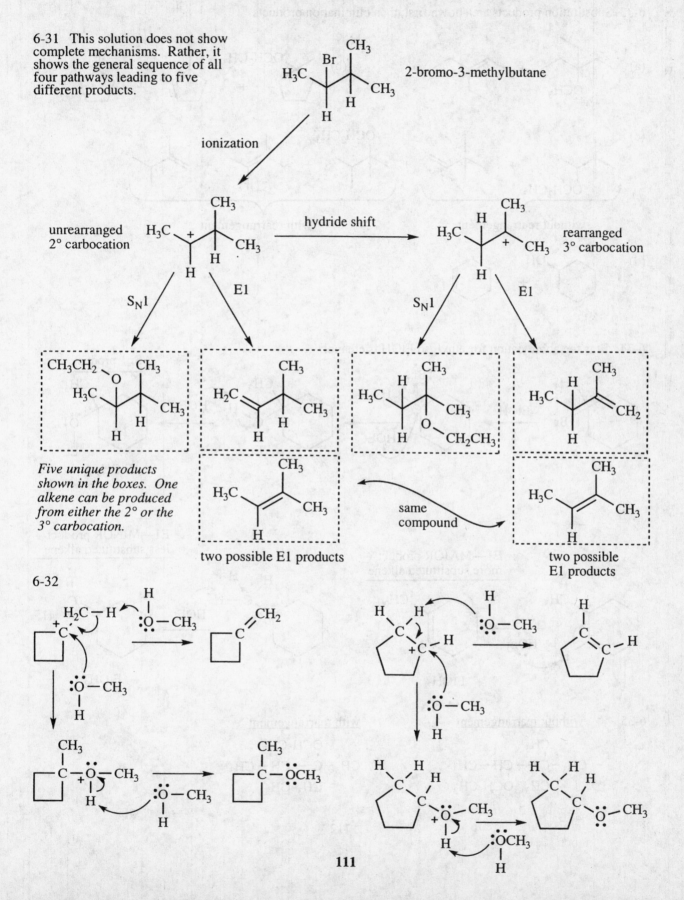

6-31 This solution does not show complete mechanisms. Rather, it shows the general sequence of all four pathways leading to five different products.

2-bromo-3-methylbutane

ionization

unrearranged 2° carbocation

hydride shift

rearranged 3° carbocation

S_N1

E1

S_N1

E1

Five unique products shown in the boxes. One alkene can be produced from either the 2° or the 3° carbocation.

two possible E1 products

same compound

two possible E1 products

6-32

111

6-33 Substitution products are shown first, then elimination products.

(a)

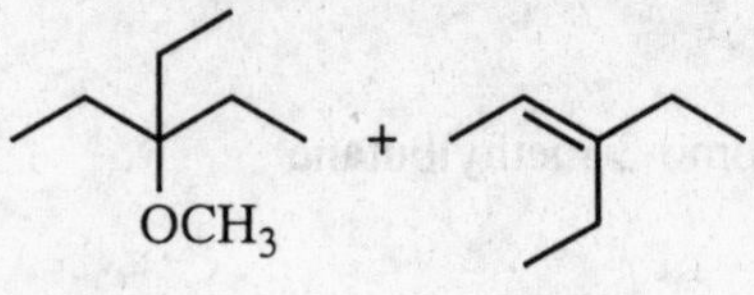

(b)

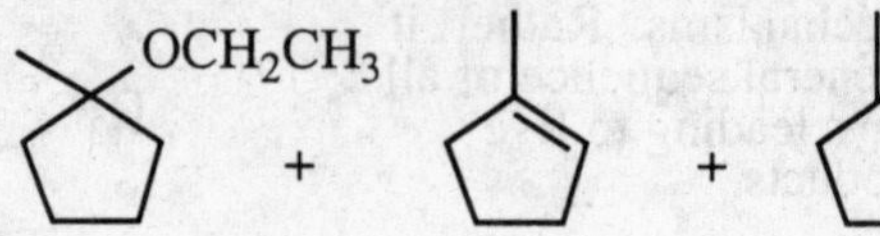

(c)

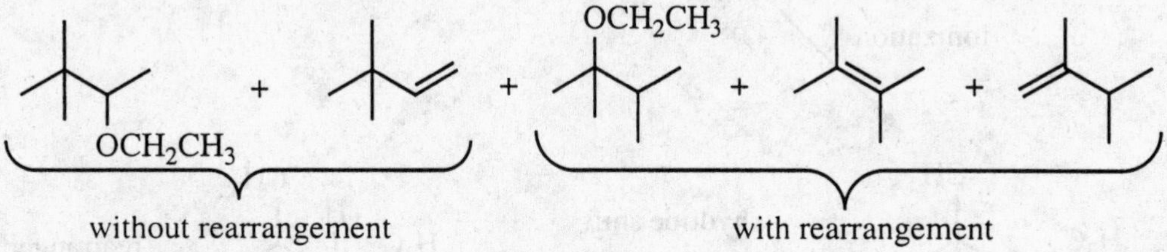

without rearrangement   with rearrangement

(d)

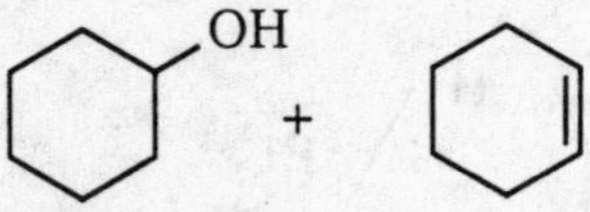

6-34 Et is the abbreviation for ethyl, so EtOH is ethanol.

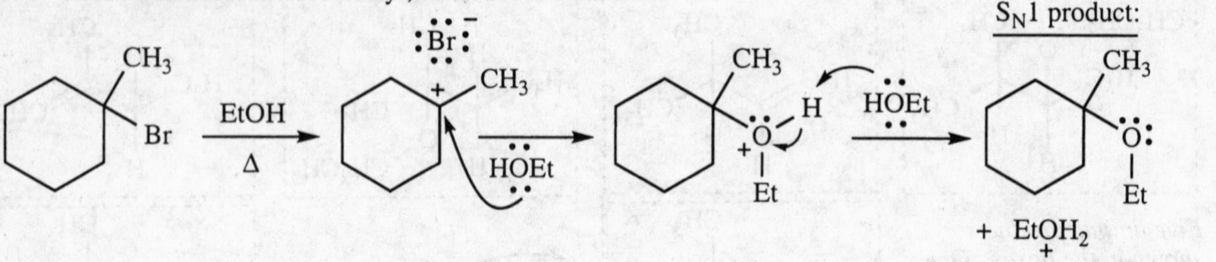

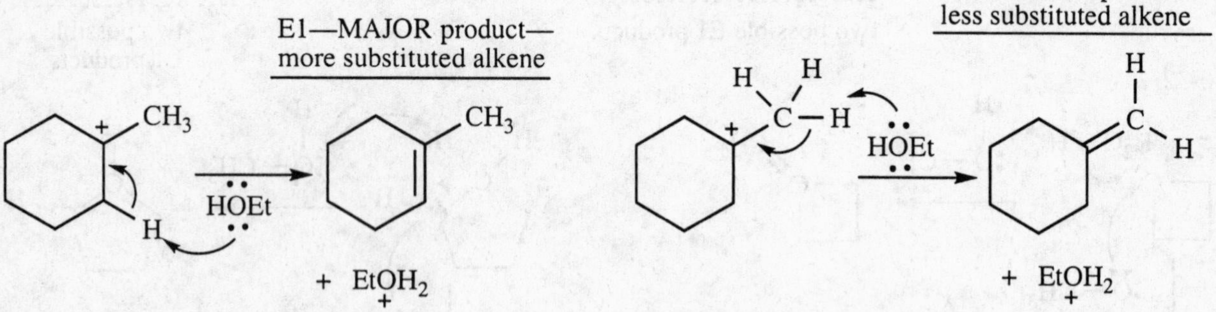

S_N1 product:

E1—MAJOR product—
more substituted alkene

E1—MINOR product—
less substituted alkene

6-35   without rearrangement   with rearrangement

$$CH_3-\underset{\underset{CH_3}{|}}{\overset{\overset{CH_3}{|}}{C}}-\underset{\underset{OCH_2CH_3}{|}}{CH}-CH_3$$

$$CH_3-\underset{\underset{CH_3}{|}}{\overset{\overset{OCH_2CH_3}{|}}{C}}-\underset{\underset{CH_3}{|}}{CH}-CH_3$$

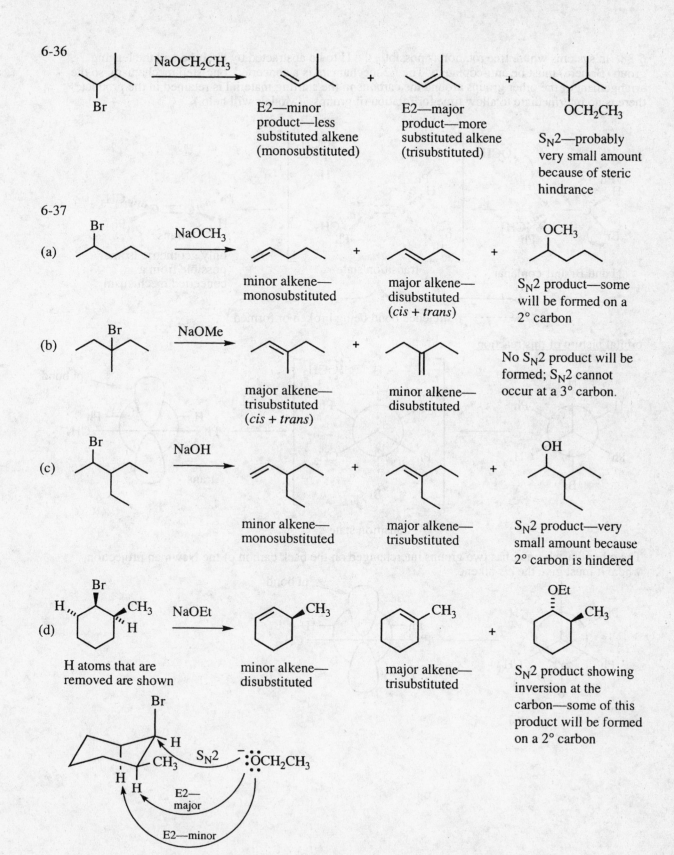

6-36

NaOCH₂CH₃

E2—minor product—less substituted alkene (monosubstituted)

+

E2—major product—more substituted alkene (trisubstituted)

+

OCH₂CH₃

S_N2—probably very small amount because of steric hindrance

6-37

(a) NaOCH₃

minor alkene—monosubstituted

+

major alkene—disubstituted (*cis + trans*)

+

OCH₃

S_N2 product—some will be formed on a 2° carbon

(b) NaOMe

major alkene—trisubstituted (*cis + trans*)

+

minor alkene—disubstituted

No S_N2 product will be formed; S_N2 cannot occur at a 3° carbon.

(c) NaOH

minor alkene—monosubstituted

+

major alkene—trisubstituted

+

OH

S_N2 product—very small amount because 2° carbon is hindered

(d) NaOEt

H atoms that are removed are shown

minor alkene—disubstituted

major alkene—trisubstituted

+

OEt

S_N2 product showing inversion at the carbon—some of this product will be formed on a 2° carbon

S_N2   :ÖCH₂CH₃

E2—major

E2—minor

113

6-38 In systems where free rotation is possible, the H to be abstracted by the base and the leaving group (Br here) must be anti-coplanar. The E2 mechanism is a concerted, one-step mechanism, so the arrangement of the other groups around the carbons in the starting material is retained in the product; there is no intermediate to allow time for rotation of groups. (Models will help.)

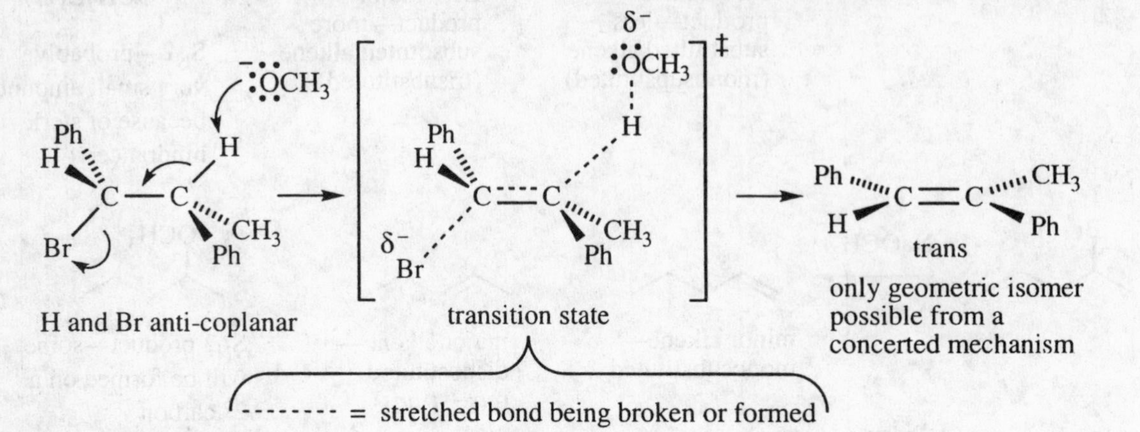

H and Br anti-coplanar

transition state

only geometric isomer possible from a concerted mechanism

- - - - - - - = stretched bond being broken or formed

orbital picture of this reaction

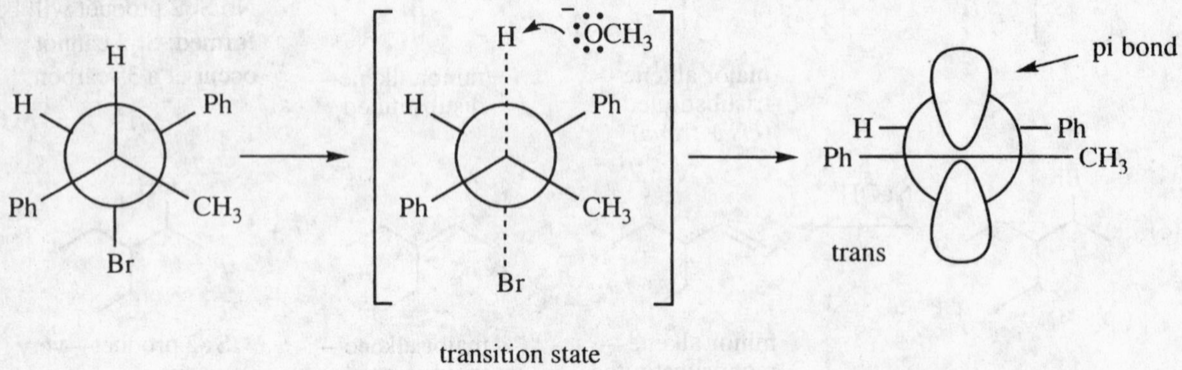

transition state

pi bond

trans

The other diastereomer has two groups interchanged on the back carbon of the Newman projection, where it must give the *cis*-alkene.

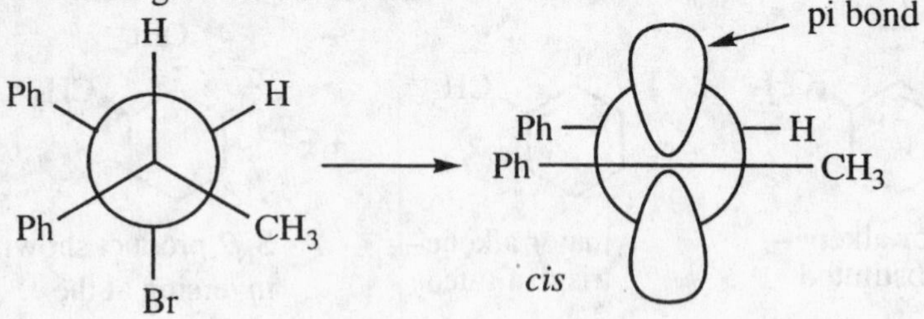

pi bond

cis

114

6-39 Iin this problem, all new internal alkenes form *cis* isomers as well as the *trans* isomers shown.

Section 1

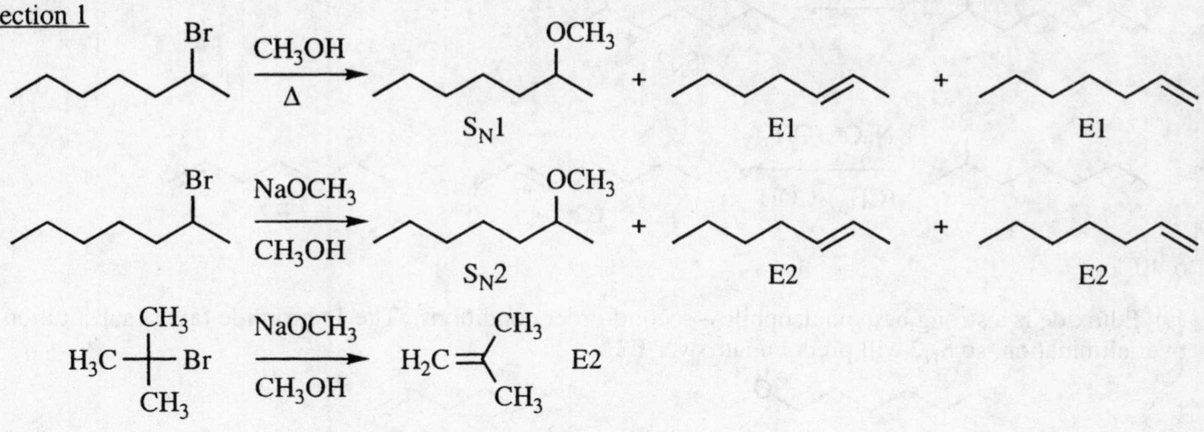

Br $\quad$ CH$_3$OH $\quad$ OCH$_3$

Δ

S$_N$1 $\qquad$ + $\qquad$ E1 $\qquad$ + $\qquad$ E1

Br $\quad$ NaOCH$_3$ $\quad$ OCH$_3$

CH$_3$OH

S$_N$2 $\qquad$ + $\qquad$ E2 $\qquad$ + $\qquad$ E2

H$_3$C—Br (with CH$_3$, CH$_3$) $\quad$ NaOCH$_3$ $\quad$ H$_2$C=C(CH$_3$)(CH$_3$) $\quad$ E2

CH$_3$OH

Section 2

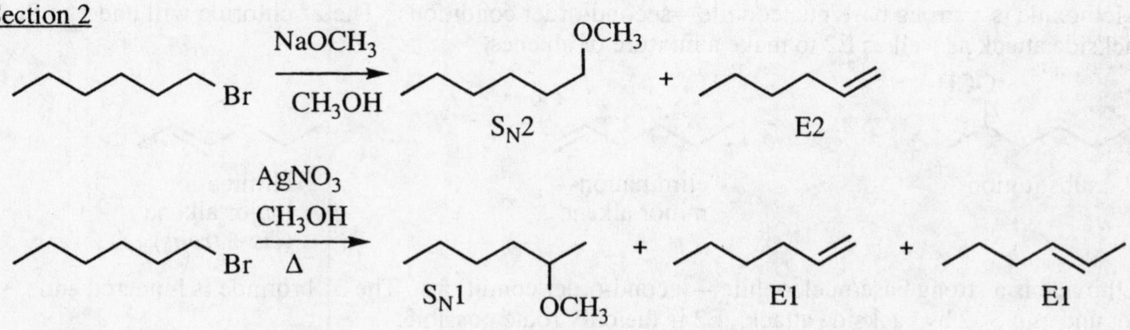

Br $\quad$ NaOCH$_3$ $\quad$ OCH$_3$

CH$_3$OH

S$_N$2 $\qquad$ + $\qquad$ E2

Br $\quad$ AgNO$_3$

CH$_3$OH

Δ $\qquad$ S$_N$1 OCH$_3$ $\qquad$ + $\qquad$ E1 $\qquad$ + $\qquad$ E1

Section 3

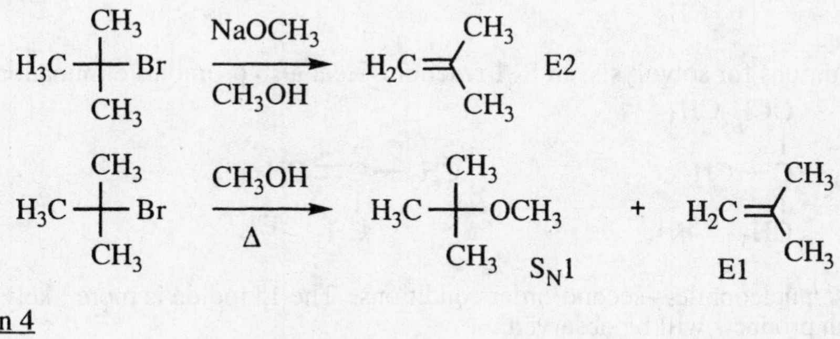

H$_3$C—Br (with CH$_3$, CH$_3$) $\quad$ NaOCH$_3$ $\quad$ H$_2$C=C(CH$_3$)(CH$_3$) $\quad$ E2

CH$_3$OH

H$_3$C—Br (with CH$_3$, CH$_3$) $\quad$ CH$_3$OH $\quad$ H$_3$C—OCH$_3$ (with CH$_3$, CH$_3$) S$_N$1 $\quad$ + $\quad$ H$_2$C=C(CH$_3$)(CH$_3$) E1

Δ

Section 4

Br $\quad$ NaOCH$_3$ $\quad$ OCH$_3$

CH$_3$OH

S$_N$2 $\qquad$ + $\qquad$ E2 $\qquad$ + $\qquad$ E2

Br $\quad$ CH$_3$OH $\quad$ OCH$_3$

Δ

S$_N$1 $\qquad$ + $\qquad$ E1 $\qquad$ + $\qquad$ E1

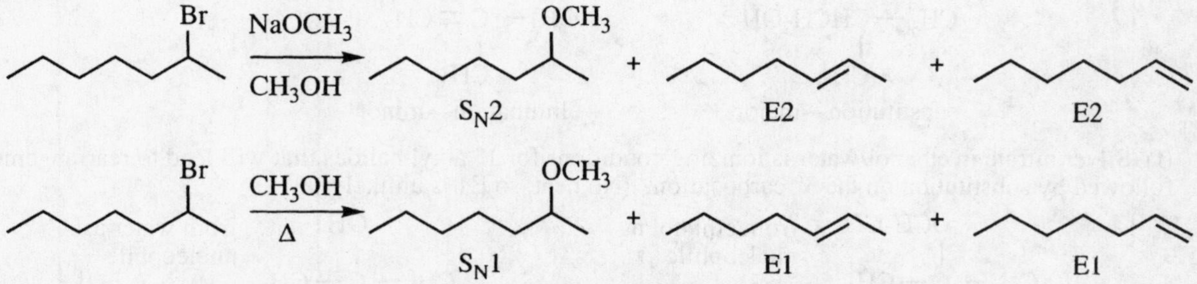

6-39 continued

Section 5

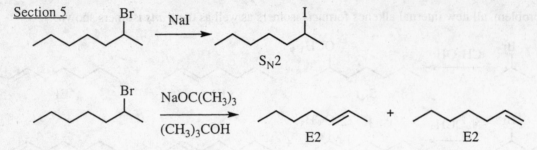

S_N2

E2   +   E2

6-40

(a) Ethoxide is a strong base/nucleophile—second-order conditions. The 1° bromide favors substitution over elimination, so S_N2 will predominate over E2.

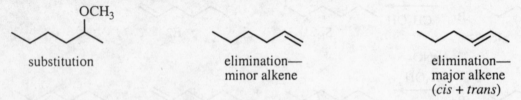

substitution—major   elimination—minor

(b) Methoxide is a strong base/nucleophile—second-order conditions. The 2° chloride will undergo S_N2 by backside attack as well as E2 to make a mixture of alkenes.

OCH₃

substitution   elimination—   elimination—
                      minor alkene   major alkene
                                            (*cis* + *trans*)

(c) Ethoxide is a strong base/nucleophile—second-order conditions. The 3° bromide is hindered and cannot undergo S_N2 by backside attack. E2 is the only route possible.

$$CH_3 - C = CH_2$$
$$\qquad\quad |$$
$$\qquad\; CH_3$$

(d) Heating in ethanol is conditions for solvolysis, an S_N1 reaction. Heat also promotes elimination.

OCH₂CH₃
|
CH₃ — C — CH₃
|
CH₃ S_N1

CH₃ — C = CH₂
|
CH₃   E1

(e) Hydroxide is a strong base/nucleophile—second-order conditions. The 1° iodide is more likely to undergo S_N2 than E2, but both products will be observed.

CH₃ — CHCH₂OH
|
CH₃

substitution—major

CH₃ — C = CH₂
|
CH₃

elimination—minor

(f) Silver nitrate in ethanol/water is ionizing conditions for 1° alkyl halides that will lead to rearrangement followed by substitution on the 3° carbocation. (No heat, so E1 is unlikely.)

OCH₂CH₃   from ethanol as
| nucleophile
CH₃ — C — CH₃
|
CH₃

OH   from water as
| nucleophile
CH₃ — C — CH₃
|
CH₃

116

6-40 continued

(g) Silver nitrate in ethanol/water is ionizing conditions for 1° alkyl halides that will lead to rearrangement followed by substitution on the 3° carbocation. (No heat, so E1 is unlikely.)

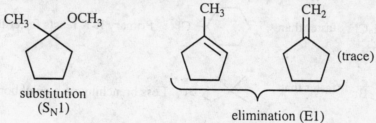

$$CH_3 - \underset{\underset{CH_3}{|}}{\overset{\overset{OCH_2CH_3}{|}}{C}} - CH_2CH_3 \qquad \text{from ethanol as nucleophile}$$

$$CH_3 - \underset{\underset{CH_3}{|}}{\overset{\overset{OH}{|}}{C}} - CH_2CH_3 \qquad \text{from water as nucleophile}$$

(h) Heating a 3° halide in methanol is quintessential first-order conditions, either E1 or S_N1 (solvolysis).

substitution
(S_N1)

elimination (E1)

(i) Ethoxide in ethanol on a 3° halide will lead to E2 elimination; there will be no substitution.

CH₃

major alkene
trisubstituted

CH₂

minor alkene
disubstituted

6-41 Please refer to solution 1-20, page 12, of this Solutions Manual.

6-42   (a) [structure] Cl   (b) [structure] Br   (c) Br / Br [structure]   (d) $Cl - \underset{\underset{Cl}{|}}{\overset{\overset{Cl}{|}}{C}} - CH_2OH$

(e) I⠇⠇⠇ [cyclohexane] I   (f) $H - \underset{\underset{Cl}{|}}{\overset{\overset{Cl}{|}}{C}} - H$ (g) $Cl - \underset{\underset{Cl}{|}}{\overset{\overset{Cl}{|}}{C}} - H$ (h) Cl [structure]   (i) $I - \underset{\underset{CH_3}{|}}{\overset{\overset{CH_2CH_3}{|}}{C}} - CH_3$

6-43

(a) 2-bromo-2-methylpentane
(b) 1-chloro-1-methylcyclohexane
(c) 1,1-dichloro-3-fluorocycloheptane

(d) 4-(2-bromoethyl)-3-(fluoromethyl)-2-methylheptane
(e) 4,4-dichloro-5-cyclopropyl-1-iodoheptane
(f) cis-1,2-dichloro-1-methylcyclohexane

6-44 Ease of backside attack (less steric hindrance) decides which undergoes S_N2 faster in all these examples except (b).

(a) [structure] Cl   faster than   [structure] Cl   Primary R-X reacts faster than 2° R-X.

(b) [structure] I   faster than   [structure] Cl   Iodide a better leaving group than chloride.

117

6-44 continued

(c)   faster than [structure] Less branching on a neighboring carbon.

(d) [structure] Br faster than [structure] Br Same neighboring branching, so 1° faster than 2°.

(e) [cyclohexane]—CH₂Cl faster than [cyclohexane]—Cl Primary R-X reacts faster than 2° R-X.

(f) [structure] Br faster than [structure] Br Less branching on a neighboring carbon.

6-45 Formation of the more stable carbocation decides which undergoes S_N1 faster in all these examples except (d).

(a) 3° [structure] Cl faster than [structure] Cl 2°

(b) [structure] faster than [structure] Cl 1°
     2° Cl

(c) [cyclohexane]—Br faster than [cyclohexane]—CH₂Br
     2°   1°

(d) [cyclohexane]—I faster than [cyclohexane]—Cl (leaving group ability)
                                                113

(e) [structure] Br faster than [structure]
     3°   Br   2°

(f) [cyclohexene] Br faster than [cyclohexane] Br
     2° **allylic**   2°

6-46 For S_N2, reactions should be designed such that the nucleophile attacks the least highly substituted alkyl halide. ("X" stands for a halide: Cl, Br, or I.)

(a) [cyclohexyl]—CH₂X + HO⁻ ⟶ [cyclohexyl]—CH₂OH

(b) [cyclohexyl]—S⁻ + X–CH₂CH₃ ⟶ [cyclohexyl]—SCH₂CH₃

(c) [chain with OH and X] $\xrightarrow{HO^-}$ [ring with O⁻ and X] ⟶ [O ring] (some [chain with OH and OH] also produced)

(d) [cyclohexyl]—CH₂X + NH₃ (excess) ⟶ [cyclohexyl]—CH₂NH₂

(e) $H_2C = CHCH_2X + {}^-CN \longrightarrow H_2C = CHCH_2CN$

(f) $(CH_3)_3C - O^- + X - CH_3 \longrightarrow (CH_3)_3C - O - CH_3$

(g) $HC \equiv C^- + X - CH_2CH_2CH_3 \longrightarrow HC \equiv C - CH_2CH_2CH_3$

6-47

(a) (1)
$$CH_3\text{-}CH\text{-}O^- \text{ (with CH}_3\text{)} + \overset{X}{CH_2CH_3} \text{ (1°)} \longrightarrow CH_3\text{-}CH\text{-}O\text{—}CH_2CH_3 \text{ (with CH}_3\text{)}$$

this bond formed

(2)
$$CH_3\text{-}CH\text{—}X \text{ (with CH}_3, 2°) + \overset{O^-}{CH_2CH_3} \longrightarrow CH_3\text{-}CH\text{-}O\text{-}CH_2CH_3 \text{ (with CH}_3\text{)}$$

this bond formed

Synthesis (1) would give a better yield of the desired ether product. (1) uses S_N2 attack of a nucleophile on a 1° carbon, while (2) requires attack on a more hindered 2° carbon. Reaction (2) would give a lower yield of substitution, with more elimination.

(b)   CANNOT DO S_N2 ON A 3° CARBON!

bumps into H before it can find C

$$CH_3 - \overset{\overset{\displaystyle CH_3}{|}}{\underset{\underset{\displaystyle CH_3}{|}}{C}} - Cl + {}^-OCH_3 \longrightarrow CH_3 - \overset{\overset{\displaystyle CH_2}{||}}{\underset{\underset{\displaystyle CH_3}{|}}{C}}$$ elimination (E2) competes

6-47 (b) continued on next page

119

6-47 (b) continued

Better to do S_N2 on a methyl carbon:

$$CH_3-\underset{\underset{CH_3}{|}}{\overset{\overset{CH_3}{|}}{C}}-O^- \quad + \quad CH_3-X \quad \longrightarrow \quad CH_3-\underset{\underset{CH_3}{|}}{\overset{\overset{CH_3}{|}}{C}}-O-CH_3$$

6-48

(a) S_N2—second order: reaction rate doubles
(b) S_N2—second order: reaction rate increases six times
(c) Virtually all reaction rates, including this one, increase with a temperature increase.

6-49 This is an S_N1 reaction; the rate law depends only on the substrate concentration, not on the nucleophile concentration.

(a) no change in rate
(b) the rate triples, dependent only on [*t*-butyl bromide]
(c) Virtually all reaction rates, including this one, increase with a temperature increase.

6-50 The key to this problem is that iodide ion is both an excellent nucleophile AND leaving group. Substitution on chlorocyclohexane is faster with iodide than with cyanide (see Table 6-3 for relative nucleophilicities). Once iodocyclohexane is formed, substitution by cyanide is much faster on iodocyclohexane than on chlorocyclohexane because iodide is a better leaving group than chloride. So two fast reactions involving iodide replace a slower single reaction, resulting in an overall rate increase.

6-51

(a)   (b)   (c)   (d)

rearrangement   rearrangement

6-52

(a)

120

6-52 continued

(b)

(i) equivalent resonance forms

(ii)

(iii) equivalent resonance forms

(iv) loss of bromide gives unstable 1° carbocation that quickly rearranges to a 2° allylic carbocation

equivalent resonance forms—same as from part (iii)

(c) Only one substitution product arises from equivalent resonance forms.

(i) OCH₂CH₃

(ii) CH₂OCH₂CH₃ ... H + CH₂ ... OCH₂CH₃

(iii) OCH₂CH₃

cis + trans

(iv) OCH₂CH₃

cis + trans   same as part (iii)

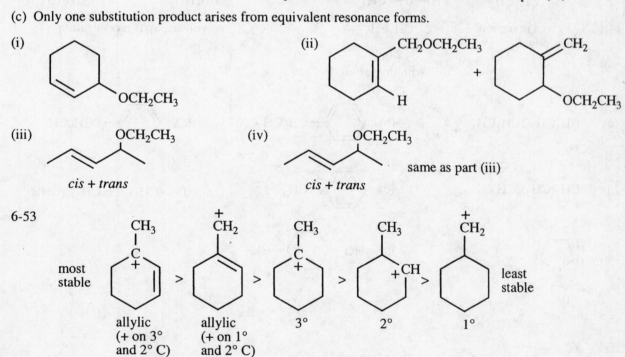

6-53

most stable

CH₃ / +C > +CH₂ > CH₃ / +C > CH₃ / +CH > +CH₂   least stable

allylic (+ on 3° and 2° C)   allylic (+ on 1° and 2° C)   3°   2°   1°

6-54

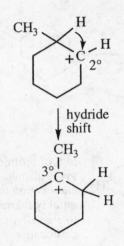

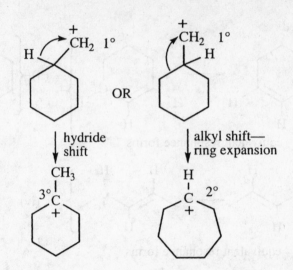

hydride shift

hydride shift

alkyl shift—ring expansion

OR

CH₃ |
3°C ⟨H / H⟩
+

CH₃ |
3°C
+

H |
C 2°
+

6-55 Reactions would also give some elimination products; only the substitution products are shown here.

(a)

```
        CH₃
        ┊
N≡C━━C◄H
        ┊
        CH₂CH₃
```

S_N2 gives inversion:
only product:

(b)

```
      CH₃
      ┊
H━C◄OH
      |
H►C◄CH₃
      ┊
      CH₂CH₃
```

S_N2 only
with inversion

(c)

```
           CH₂CH₃
           ┊
CH₃CH₂O►C◄CH₃       +
           ┊
           CH(CH₃)₂
```

```
         CH₂CH₃
         ┊
CH₃►C◄OCH₂CH₃
         ┊
         CH(CH₃)₂
```

solvolysis, S_N1, racemization

6-56

(a) $CH_3CH_2OCH_2CH_3$

(b) ⟨phenyl⟩—CH_2CH_2CN

(c) ⟨phenyl⟩—SCH_2CH_3

(d) $CH_3(CH_2)_8CH_2$—$C≡CH$

(e) ⟨pyridine⟩$\overset{+}{N}$—CH_3 I^-

(f) $(CH_3)_3C$—$CH_2CH_2NH_2$

(g) ⟨tetrahydrofuran ring with O⟩

(h) HO···⟨cyclohexane⟩····CH_3

6-57   <u>substitution</u>

2S,3S   2R,3R

racemic mixture

Regardless of which bromine is substituted on each molecule, the same mixture of products results.

Each of the substitution products has one chiral center inverted from the starting material. The mechanism that accounts for inversion is S_N2. If an S_N1 process were occurring, the product mixture would also contain 2R,3R and 2S,3S diastereomers. Their absence argues against an S_N1 process occurring here.

↓ KOH

2R,3S   2S,3R

racemic mixture

<u>elimination</u>

H and Br
anti-coplanar

trans

The other enantiomer gives the same product (you should prove this to yourself).

The absence of *cis* product is evidence that only the E2 elimination is occurring in one step through an anti-coplanar transition state with no chance of rotation. If E1 had been occuring, rotation around the carbocation intermediate would have been possible, leading to both the *cis* and *trans* products.

6-58

(a) $\dfrac{+15.58°}{+15.90°}$ x $100\% = 98\%$ of original optical activity $= 98\%$ e.e.

Thus, 98% of the *S* enantiomer and 2% racemic mixture gives an overall composition of 99% *S* and 1% *R*.

(b) The 1% of radioactive iodide has produced exactly 1% of the *R* enantiomer. Each substitution must occur with inversion, a classic S_N2 mechanism.

6-59

(a) An S_N2 mechanism with inversion will convert *R* to its enantiomer, *S*. An accumulation of excess *S* does not occur because it can also react with bromide, regenerating *R*. The system approaches a racemic mixture at equilibrium.

R   S

123

6-59 continued

(b) In order to undergo substitution and therefore inversion, HO⁻ would have to be the leaving group, but HO⁻ is never a leaving group in S$_N$2. No reaction can occur.

(c) Once the OH is protonated, it can leave as H$_2$O. Racemization occurs in the S$_N$1 mechanism because of the planar, achiral carbocation intermediate which "erases" all stereochemistry of the starting material. Racemization occurs in the S$_N$2 mechanism by establishing an equilibrium of *R* and *S* enantiomers, as explained in 6-59(a).

inverted

planar carbocation

racemic mixture

6-60

(a)

major   minor   trace amount

The carbocation produced in this E1 elimination is 3° and will not rearrange. The product ratio follows the Zaitsev rule.

(b)

trace—from unrearranged 2° carbocation

from either unrearranged 2° carbocation or rearranged 3° carbocation

from rearranged 3° carbocation

The carbocation produced in this E1 elimination is 2° and can either eliminate to give the first two alkenes, or can rearrange by a hydride shift to a 3° carbocation which would produce the last two products. The amounts of the last two products are not predictable as they are both trisubstituted, but the first product will certainly be the least.

(c)

trace—from unrearranged 2° carbocation

major
from rearranged 3° carbocation

minor
from rearranged 3° carbocation

The carbocation produced in this E1 elimination is 2° and can either eliminate to give the first alkene, or can rearrange by a methyl shift to a 3° carbocation which would produce the last two products. The middle product is major as it is tetrasubstituted versus disubstituted for the last structure and monosubstituted for the first structure.

6-61 The allylic carbocation has two resonance forms showing that two carbons share the positive charge. The ethanol nucleophile can attack either of these carbons, giving the S_N1 products; or loss of an adjacent H will give the E1 product.

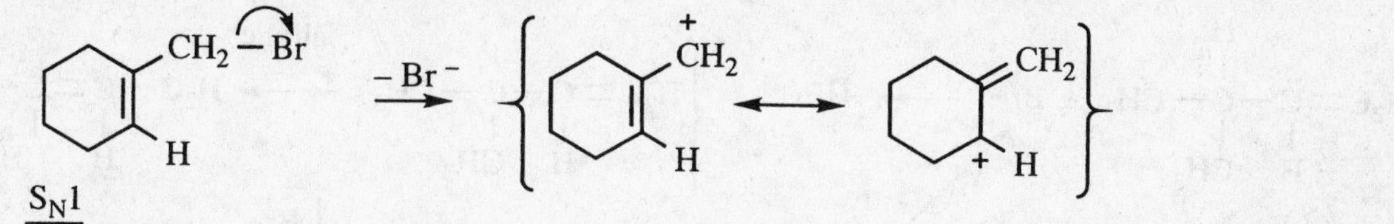

$\underline{S_N1}$

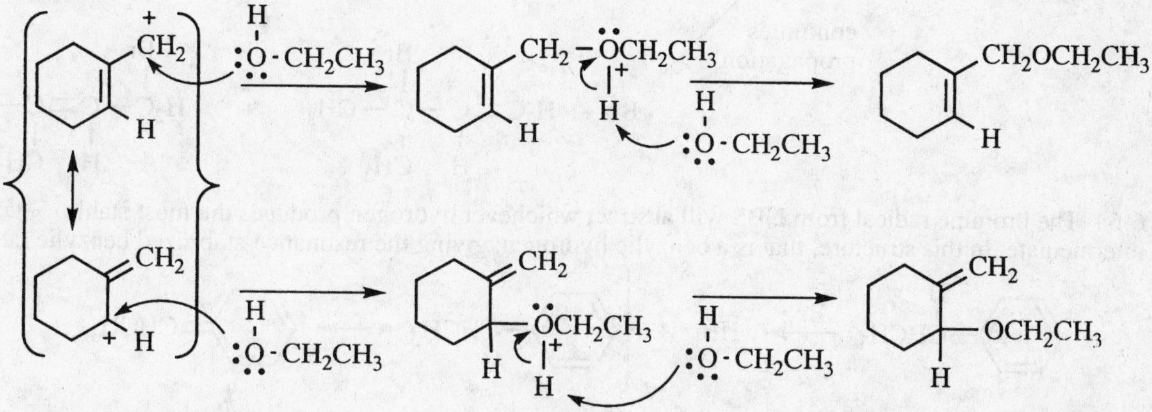

$\underline{E1}$

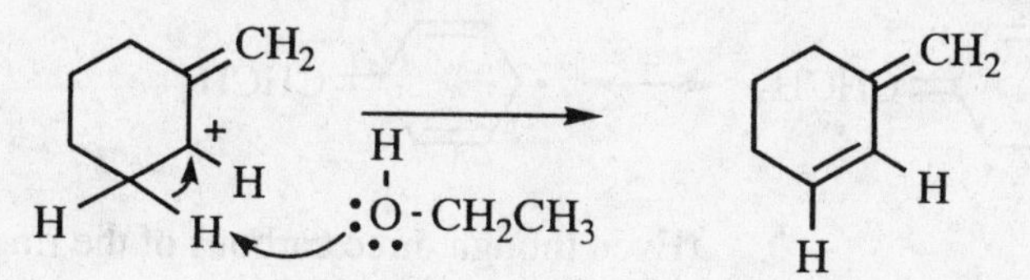

6-62

(a)

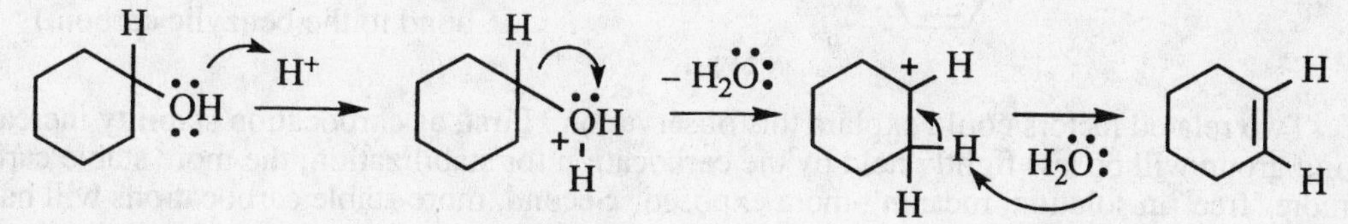

(b)

S_N1

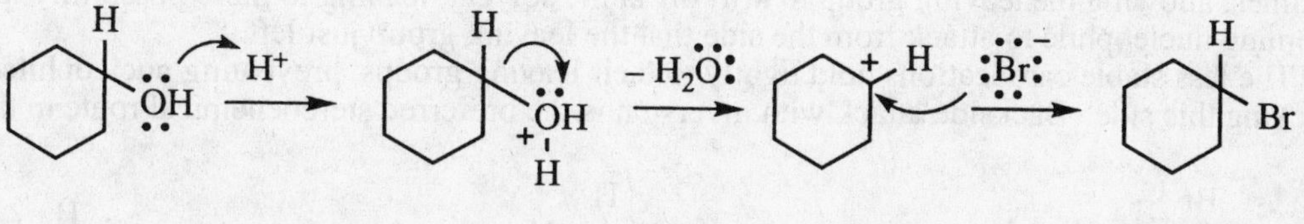

S_N2

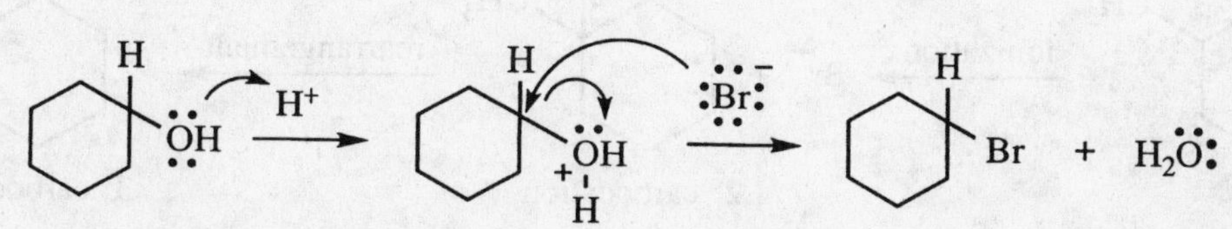

125

6-63 NBS generates bromine which produces bromine radical. Bromine radical abstracts an allylic hydrogen, resulting in a resonance-stabilized allylic radical. The allylic radical can bond to bromine at either of the two carbons with radical character.

continues
propagation

6-64 The bromine radical from NBS will abstract whichever hydrogen produces the most stable intermediate; in this structure, that is a benzylic hydrogen, giving the resonance-stabilized benzylic radical.

(Even though three carbons of the ring have some radical character, these are minor resonance contributors. The product is most stable when the ring has all three double bonds intact, necessitating that the bromine bond to the benzylic carbon.)

6-65 Two related factors could explain this observation. First, as carbocation stability increases, the leaving group will be less tightly held by the carbocation for stabilization; the more stable carbocations are more "free" in solution, meaning more exposed. Second, more stable carbocations will have longer lifetimes, allowing the leaving group to drift off in the solvent, leading to more possibility for the incoming nucleophile to attack from the side that the leaving group just left.

The less stable carbocations hold tightly to their leaving groups, preventing nucleophiles from attacking this side. Backside attack with inversion is the preferred stereochemical route in this case.

6-66

2° carbocation

3° carbocation

mechanisms continued
on next page

6-66 continued

substitution on 2° carbocation

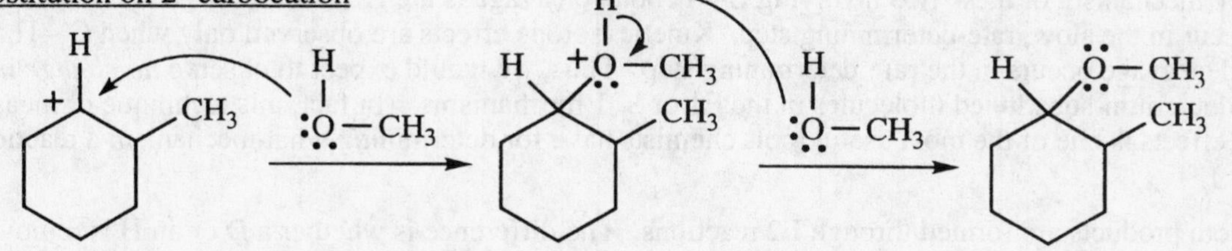

substitution on rearranged 3° carbocation

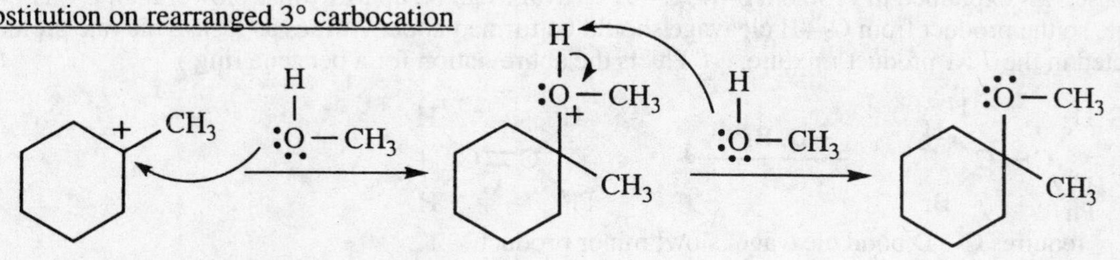

elimination from rearranged 3° carbocation

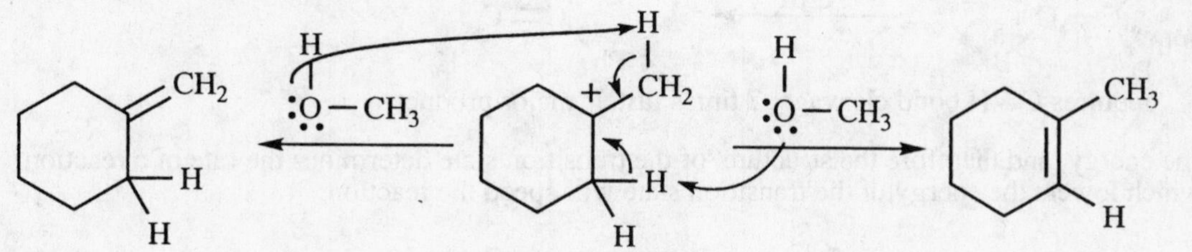

6-67

(a) E2—
one step

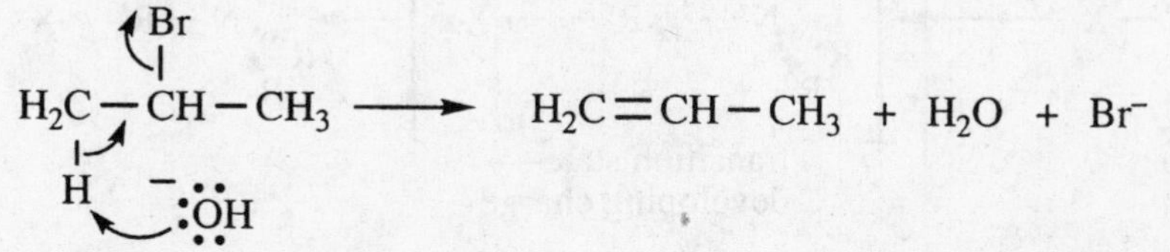

$$H_2C=CH-CH_3 + H_2O + Br^-$$

S_N2—
one step

$$CH_3-CH-CH_3 + Br^-$$

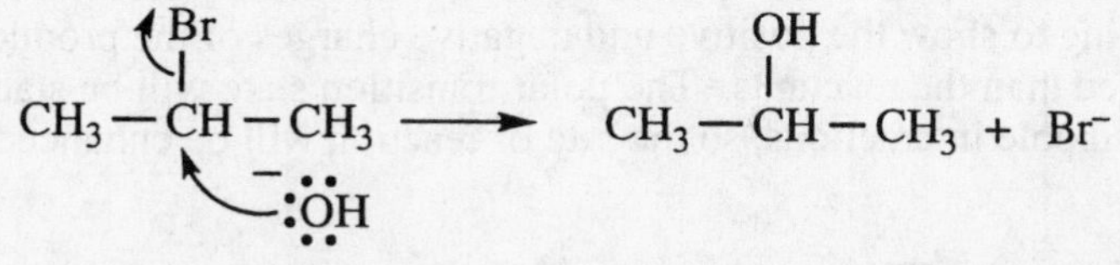

(b) In the E2 reaction, a C—H bond is broken. When D is substituted for H, a C—D bond is broken, slowing the reaction. In the S_N2 reaction, no C—H (C—D) bond is broken, so the rate is unchanged.

(c) These are first-order reactions. The slow, rate-determining step is the first step in each mechanism.

Explanation on next page.

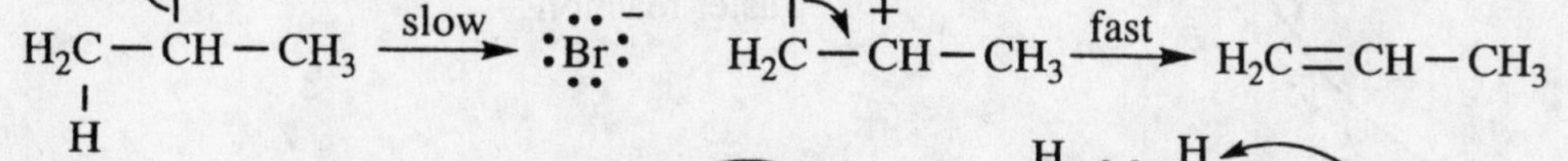

127

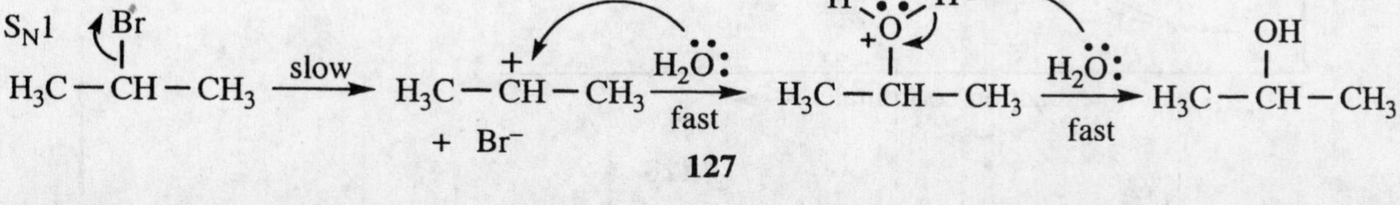

The only mechanism of these two involving C—H bond cleavage is the E1, but the C—H cleavage does NOT occur in the slow, rate-determining step. Kinetic isotope effects are observed only when C—H (C—D) bond cleavage occurs in the rate-determining step. Thus, we would expect to observe *no change in rate* for the deuterium-substituted molecules in the E1 or S_N1 mechanisms. (In fact, this technique of measuring isotope effects is one of the most useful tools chemists have for determining what mechanism a reaction follows.)

6-68 Both products are formed through E2 reactions. The difference is whether a D or an H is removed by the base. As explained in Problem 6-67, C—D cleavage can be up to 7 times slower than C—H cleavage, so the product from C—H cleavage should be formed about 7 times as fast. This rate preference is reflected in the 7 : 1 product mixture. ("Ph" is the abbreviation for a benzene ring.)

requires C—D bond cleavage; slow; minor product

requires C—H bond cleavage; 7 times faster; major product

6-69 The energy, and therefore the structure, of the transition state determines the rate of a reaction. Any factor which lowers the energy of the transition state will speed the reaction.

transition state—
developing charge

This example of S_N2 is unusual in that the nucleophile is a neutral molecule—it is not negatively charged. The transition state is beginning to show the positive and negative charges of the products (ions), so the transition state is more charged than the reactants. The polar transition state will be stabilized in a more polar solvent through dipole-dipole interactions, so the rate of reaction will be enhanced in a polar solvent.

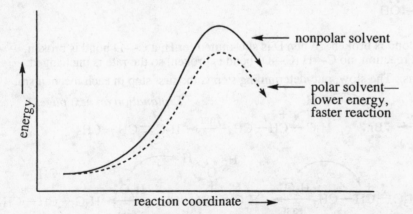

nonpolar solvent

polar solvent—
lower energy,
faster reaction

energy

reaction coordinate ⟶

6-70 The problem is how to explain this reaction:

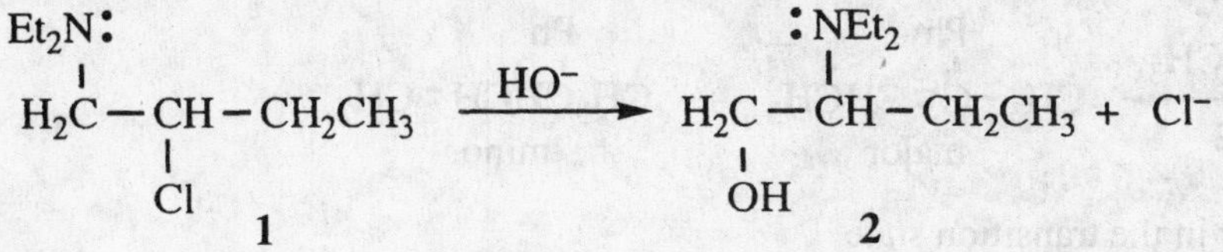

facts
facts
1) second order, but several thousand times faster than similar second order reactions without the NEt_2 group
2) NEt_2 group migrates

Solution

Clearly, the NEt_2 group is involved. The nitrogen is a nucleophile and can do an internal nucleophilic substitution (S_Ni), a very fast reaction for entropy reasons because two different molecules do not have to come together.

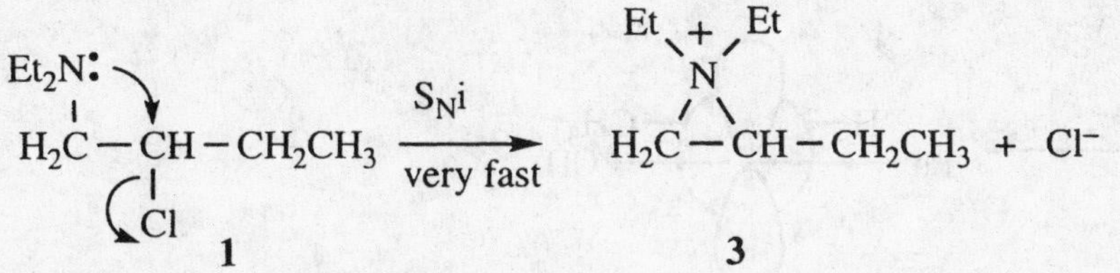

The slower step is attack of HO^- on intermediate **3**; the N is a good leaving group because it has a positive charge. Where will HO^- attack **3**? On the less substituted carbon, in typical S_N2 fashion.

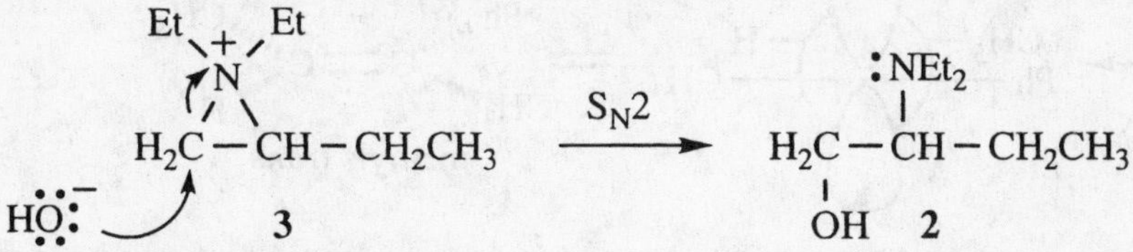

This overall reaction is fast because of the *neighboring group assistance* in forming **3**. It is second order because the HO^- group and **3** collide in the slow step (not the *only* step, however). And the NEt_2 group "migrates", although in two steps.

6-71 The symmetry of this molecule is crucial.

(a)

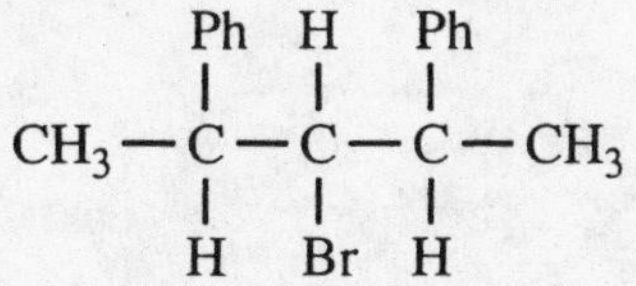

Regardless of which adjacent H is removed by *t*-butoxide, the product will be 2,4-diphenyl-2-pentene.

(b) Here are a Newman projection, a three-dimensional representation and a Fischer projection of the required diastereomer. On both carbons 2 and 4, the H has to be anti-coplanar with the bromine while leaving the other groups to give the same product. Not coincidentally, the correct diastereomer is a meso structure.

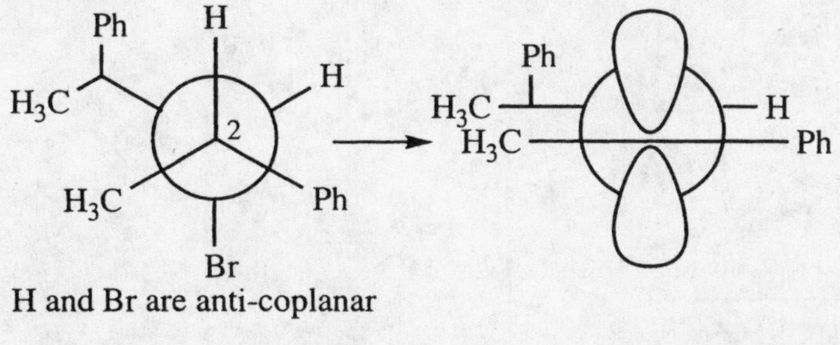

H and Br are anti-coplanar

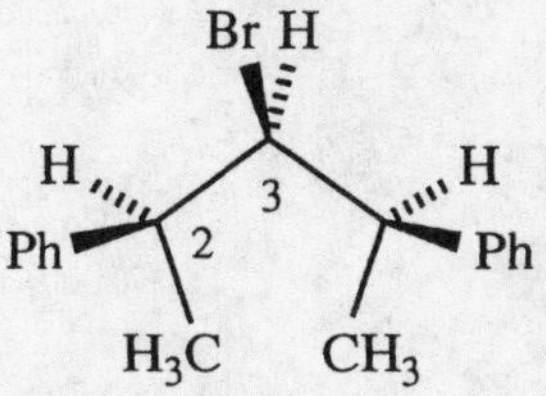

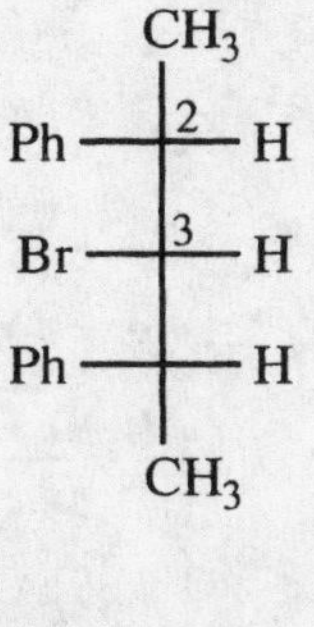

6-72 "Ph" = phenyl

(a)

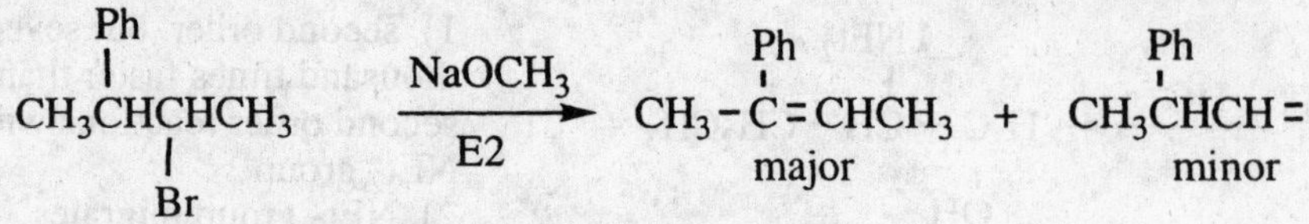

$$CH_3CHCHCH_3 \xrightarrow[E2]{NaOCH_3} CH_3-C=CHCH_3 + CH_3CHCH=CH_2$$

with Ph substituents; Br leaving group

major   minor

(b) H and Br must be anti-coplanar in the transition state

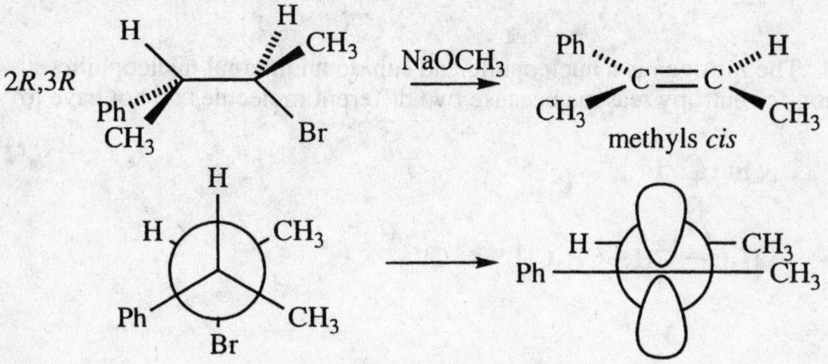

2R,3R

$$\xrightarrow{NaOCH_3}$$

methyls *cis*

(c)

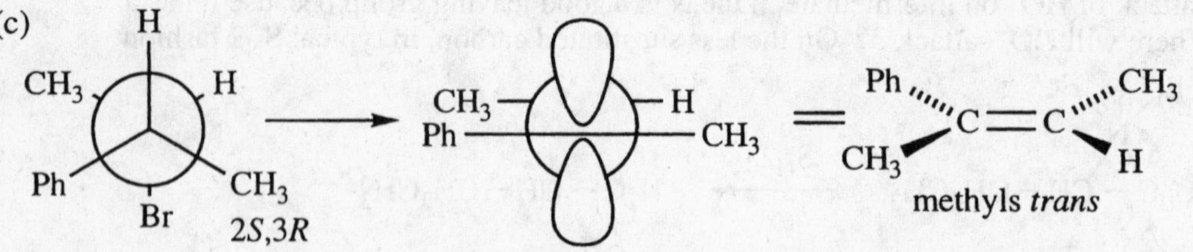

2S,3R

$$\longrightarrow$$

methyls *trans*

(d) The 2S,3S is the mirror image of 2R,3R; it would give the mirror image of the alkene that 2R,3R produced (with two methyl groups *cis*). The alkene product is planar, not chiral, so its mirror image is the same: the 2S,3S and the 2R,3R give the same alkene.

130

6-73 All five products (boxed) come from rearranged carbocations. Rearrangement, which may occur simultaneously with ionization, can occur by hydride shift to the 3° methylcyclopentyl cation, or by ring expansion to the cyclohexyl cation.

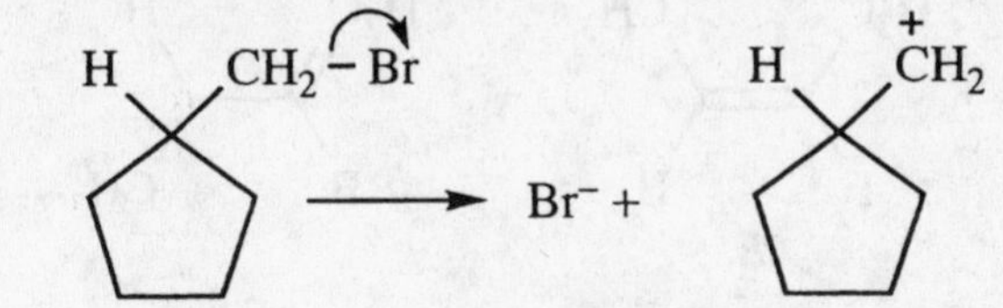

hydride shift

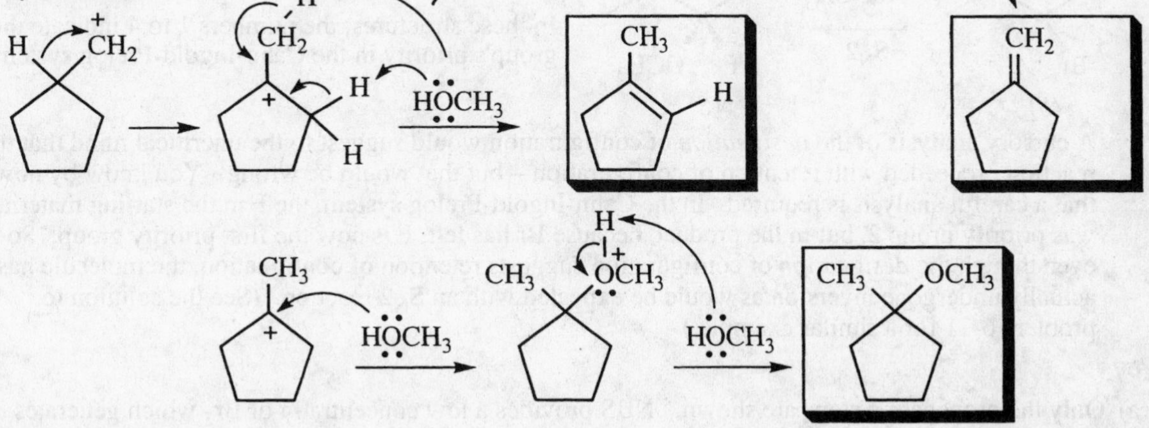

alkyl shift (ring expansion)

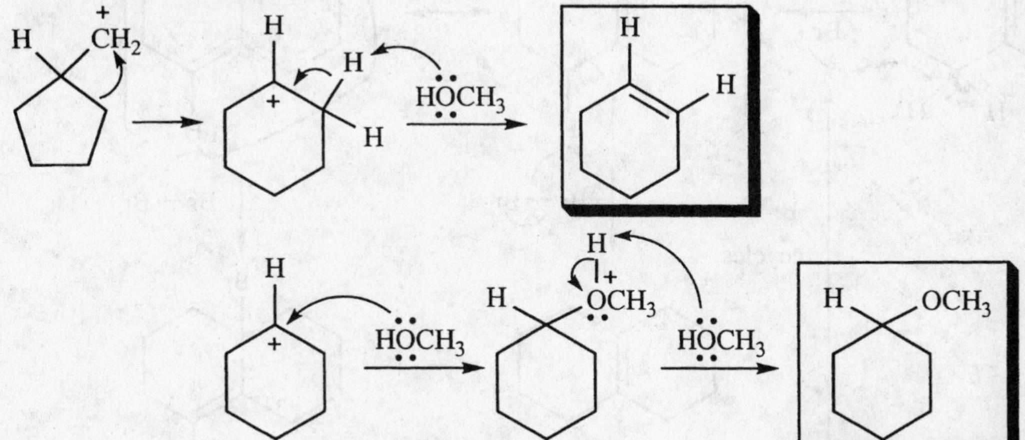

6-74 Begin with a structure of (S)-2-bromo-2-fluorobutane. Since there is no H on C-2, the lowest priority group must be the CH_3. The Br has highest priority, then F, then CH_2CH_3, and CH_3 is fourth. Sodium methoxide is a strong base and nucleophile, so the reaction must be second order, E2 or S_N2.

(a)

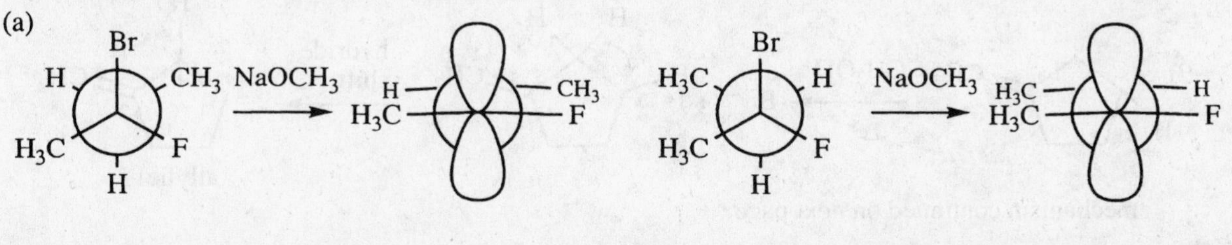

In regular structural formulas, the reaction would give three products including the stereoisomers shown above.

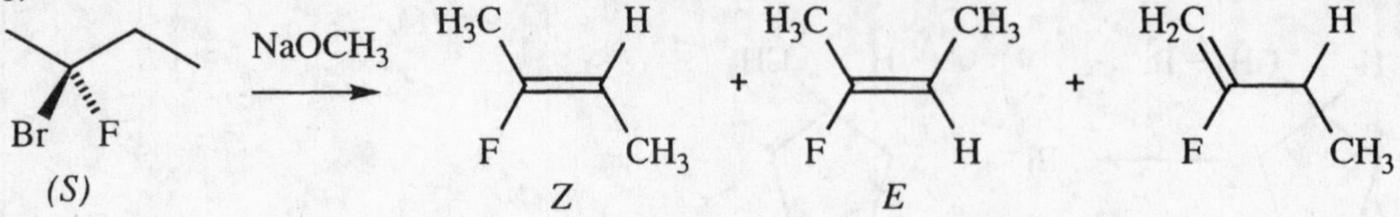

(b)

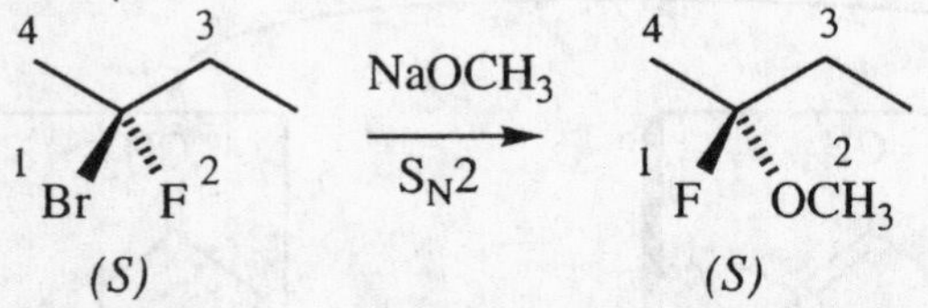

In these structures, the numbers 1 to 4 indicate the group's priority in the Cahn-Ingold-Prelog system.

A cursory analysis of the *designation* of configuration would suggest to the uncritical mind that this reaction proceeded with retention of configuration—but that would be wrong! You know by now that a careful analysis is required. In the Cahn-Ingold-Prelog system, the F in the starting material was priority group 2, but in the product, because Br has left, F is now the first priority group. So even though the *designation* of configuration suggests retention of configuation, the molecule has actually undergone inversion as would be expected with an S_N2 reaction. (See the solution to problem 6-21 for a similar example.)

6-75

(a) Only the propagation steps are shown. NBS provides a low concentratin of Br_2 which generates bromine radical in ultraviolet light. In the starting material, all 8 allylic hydrogens are equivalent.

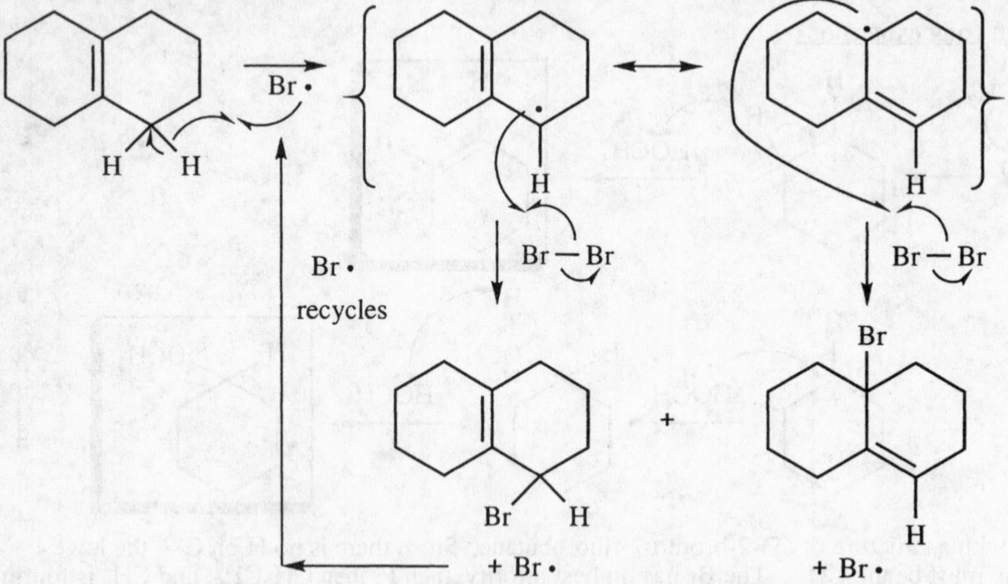

(b) The first step in this first-order solvolysis is ionization, followed by rearrangement.

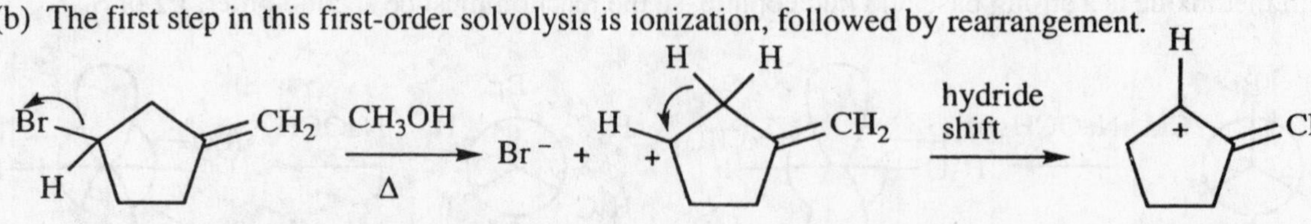

mechanism continued on next page

6-75 (b) continued

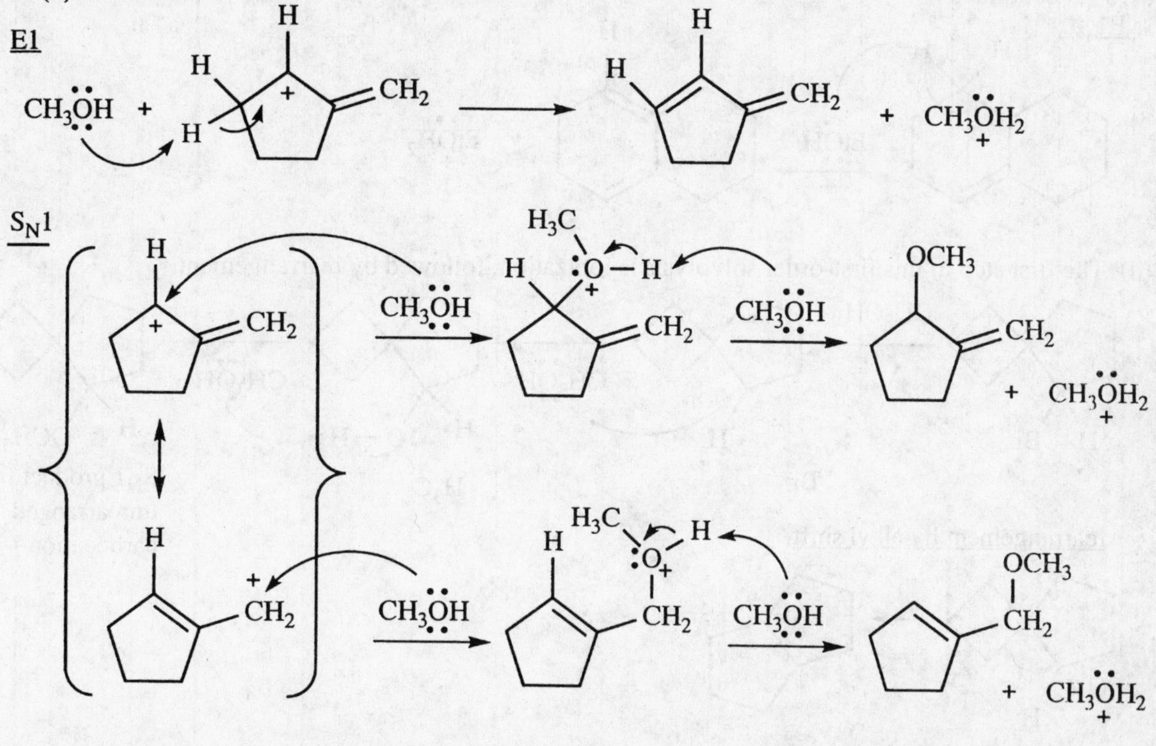

<u>E1</u>

$CH_3\overset{..}{\underset{..}{O}}H$ + [pentane ring structure with CH₂ and + charge] ⟶ [cyclopentene with CH₂] + $CH_3\overset{..}{\underset{+}{O}}H_2$

<u>S_N1</u>

(c) This first-order solvolysis generates an allylic carbocation in the ionization step.

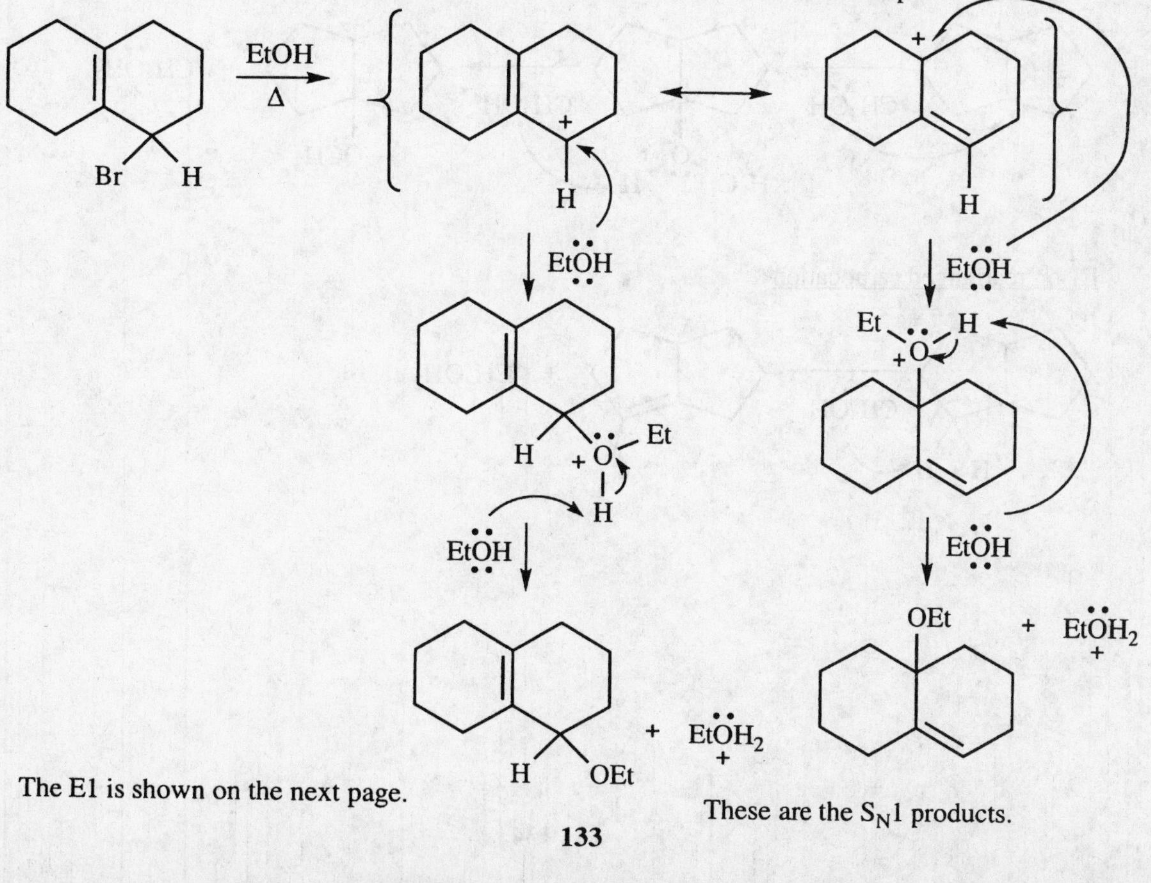

The E1 is shown on the next page.

These are the S_N1 products.

133

6-75 (c) continued

<u>E1</u>

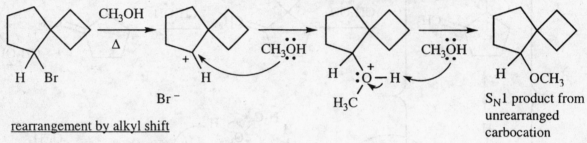

(d) The first step in this first-order solvolysis is ionization, followed by rearrangement.

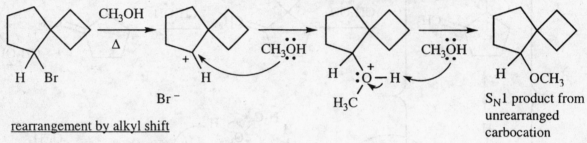

S$_N$1 product from
unrearranged
carbocation

<u>rearrangement by alkyl shift</u>

<u>S$_N$1 on rearranged carbocation</u>

<u>E1 on rearranged carbocation</u>

134

CHAPTER 7—STRUCTURE AND SYNTHESIS OF ALKENES

7-1 The number of elements of unsaturation in a hydrocarbon formula is given by:

$$\frac{2(\#C) + 2 - (\#H)}{2}$$

(a) $C_6H_{12} \Rightarrow \dfrac{2(6) + 2 - (12)}{2} = 1$ element of unsaturation

(b) Many examples are possible. Yours may not match these, but all must have either a double bond or a ring, that is, one element of unsaturation.

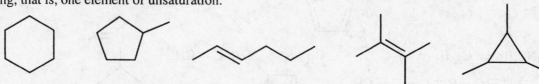

7-2 $C_4H_6 \Rightarrow \dfrac{2(4) + 2 - (6)}{2} = 2$ elements of unsaturation

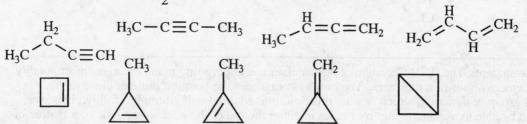

7-3 Hundreds of examples of C_4H_6NOCl are possible. Yours may not match these, but all must contain two elements of unsaturation.

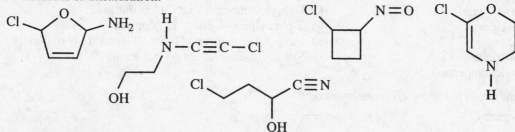

7-4 Many examples of these formulas are possible. Yours may not match these, but correct answers must have the same number of elements of unsaturation.

(a) $C_3H_4Cl_2 \Rightarrow C_3H_6 = 1$

(b) $C_4H_8O \Rightarrow C_4H_8 = 1$

7-4 continued

(c) $C_4H_4O_2 \Rightarrow C_4H_4 = 3$

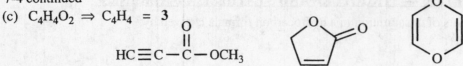

$HC\equiv C-\overset{\overset{\displaystyle O}{\|}}{C}-OCH_3$

(d) $C_5H_5NO_2 \Rightarrow C_{5.5}H_5 = 4$

$\underset{O}{\overset{O}{\bigcirc}}\!-C\equiv C-NH_2$

(e) $C_6H_3NClBr \Rightarrow C_{6.5}H_5 = 5$

$HC\equiv C-\underset{\underset{Cl}{\overset{|}{NH}}}{\overset{|}{C}}=\underset{Br}{\overset{|}{C}}-C\equiv CH$

Note to the student: The IUPAC system of nomenclature is undergoing many changes, most notably in the placement of position numbers. The new system places the position number close to the functional group designation, which is what this Solutions Manual will attempt to follow; however, you should be able to use and recognize names in either the old or the new style. Ask your instructor which system to use.

7-5

(a) 4-methylpent-1-ene
(b) 2-ethylhex-1-ene
(c) penta-1,4-diene
(d) penta-1,2,4-triene
(e) 2,5-dimethylcyclopenta-1,3-diene

(f) 4-vinylcyclohex-1-ene ("1" is optional)
(g) 3-phenylprop-1-ene ("1" is optional)
(h) *trans*-3,4-dimethylcyclopent-1-ene ("1" is optional)
(i) 7-methylenecyclohepta-1,3,5-triene

7-6 (b), (e), and (f) do not show *cis,trans* isomerism

(a)

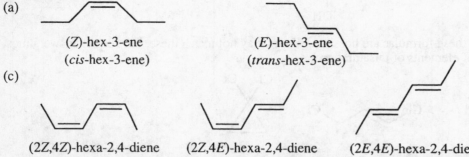

(Z)-hex-3-ene
(*cis*-hex-3-ene)

(E)-hex-3-ene
(*trans*-hex-3-ene)

(c)

(2Z,4Z)-hexa-2,4-diene
cis, cis-hexa-2,4-diene

(2Z,4E)-hexa-2,4-diene
cis, trans-hexa-2,4-diene

(2E,4E)-hexa-2,4-diene
trans, trans-hexa-2,4-diene

(d)

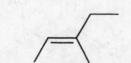

"*Cis*" and "*trans*" are not clear for this example; "*E*" and "*Z*" are unambiguous.

(Z)-3-methylpent-2-ene   (E)-3-methylpent-2-ene

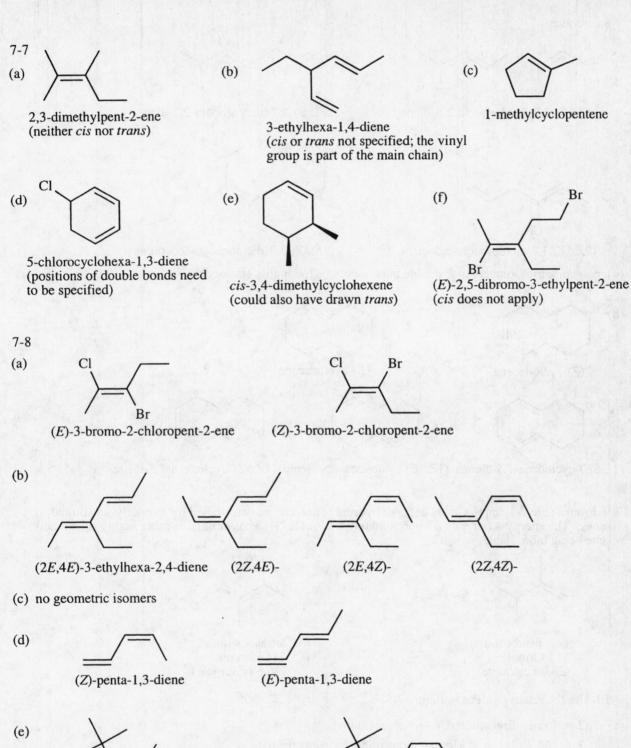

7-7

(a) 2,3-dimethylpent-2-ene
(neither *cis* nor *trans*)

(b) 3-ethylhexa-1,4-diene
(*cis* or *trans* not specified; the vinyl group is part of the main chain)

(c) 1-methylcyclopentene

(d) 5-chlorocyclohexa-1,3-diene
(positions of double bonds need to be specified)

(e) *cis*-3,4-dimethylcyclohexene
(could also have drawn *trans*)

(f) (*E*)-2,5-dibromo-3-ethylpent-2-ene
(*cis* does not apply)

7-8

(a) (*E*)-3-bromo-2-chloropent-2-ene   (*Z*)-3-bromo-2-chloropent-2-ene

(b) (2*E*,4*E*)-3-ethylhexa-2,4-diene   (2*Z*,4*E*)-   (2*E*,4*Z*)-   (2*Z*,4*Z*)-

(c) no geometric isomers

(d) (*Z*)-penta-1,3-diene   (*E*)-penta-1,3-diene

(e) (*E*)-4-*t*-butyl-5-methyloct-4-ene   (*Z*)-4-*t*-butyl-5-methyloct-4-ene

137

7-8 continued

(f)

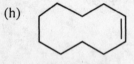

(2Z,5E)-3,7-dichloroocta-2,5-diene

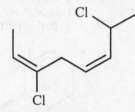

(2E,5E)-3,7-dichloroocta-2,5-diene

(2Z,5Z)-3,7-dichloroocta-2,5-diene

(2E,5Z)-3,7-dichloroocta-2,5-diene

(g) no geometric isomers (an *E* double bond would be too highly strained)

(h)

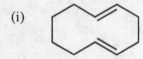

(*Z*)-cyclodecene

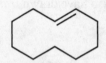

(*E*)-cyclodecene

(i)

(1*E*,5*E*)-cyclodeca-1,5-diene   (1*Z*,5*E*)-cyclodeca-1,5-diene   (1*Z*,5*Z*)-cyclodeca-1,5-diene

7-9 From Table 7-1, approximate heats of hydrogenation can be determined for similarly substituted alkenes. The energy difference is approximately 6 kJ/mole (1.4 kcal/mole), the more highly substituted alkene being more stable.

gem-disubstituted
117 kJ/mole
(28.0 kcal/mole)

tetrasubstituted
111 kJ/mole
(26.6 kcal/mole)

7-10 Use the relative values in Figure 7-8.

(a)   2 x (*trans*-disubstituted − *cis*-disubstituted) =
       2 x (22 − 18) = 8 kJ/mole more stable for *trans,trans*
       (2 x (5.2 − 4.2) = 2 kcal/mole)

(b)   *gem*-disubstituted − monosubstituted = 20 − 11 = 9 kJ/mole
                                    (4.8 − 2.7 = 2.1 kcal/mole)
       2-methylbut-1-ene is more stable

138

7-10 continued

(c)   trisubstituted − *gem*-disubstituted = 25 − 20 = 5 kJ/mole

$$(5.9 - 4.8 = 1.1 \text{ kcal/mole})$$

2-methylbut-2-ene is more stable

(d)   tetrasubstituted − *gem*-disubstituted = 26 − 20 = 6 kJ/mole

$$(6.2 - 4.8 = 1.4 \text{ kcal/mole})$$

2,3-dimethylbut-2-ene is more stable

7-11

(a) strained but stable

(b) could not exist—ring size must be 8 atoms or greater to include *trans* double bond

(c) Despite the ambiguity of this name, we must assume that the *trans* refers to the two methyls since this compound can exist, rather than the *trans* referring to the alkene, a molecule which could not exist.

(d) stable—*trans* in 10-membered ring

(e) unstable at room temperature—cannot have *trans* alkene in 7-membered ring (possibly isolable at very low temperature—this type of experiment is one of the challenges chemists attack with gusto)
(f) stable—alkene not at bridgehead
(g) unstable—violation of Bredt's Rule (alkene at bridgehead in 6-membered ring)
(h) stable—alkene at bridgehead in 8-membered ring
(i) unstable—violation of Bredt's Rule (alkene at bridgehead in 7-membered ring)

7-12
(a) The dibromo compound should boil at a higher temperature because of its much larger molecular weight.
(b) The *cis* should boil at a higher temperature than the *trans* as the *trans* has a zero dipole moment and therefore no dipole-dipole interactions.
(c) 1,2-Dichlorocyclohexene should boil at a higher temperature because of its much larger molecular weight and larger dipole moment than cyclohexene.

7-13

(a)

7-13 continued

(b)

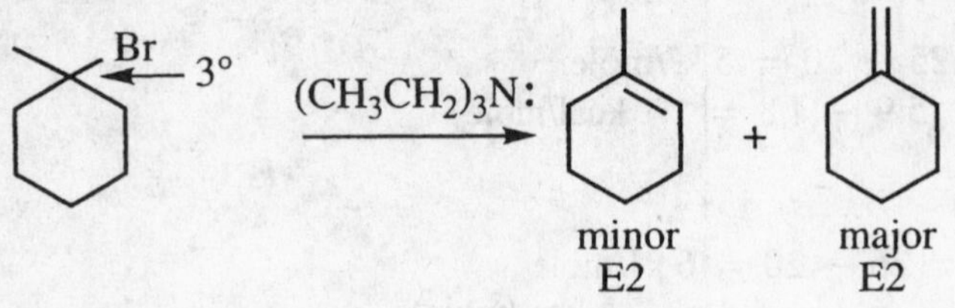

$(CH_3CH_2)_3N:$

hindered base gives Hofmann product as major isomer

minor
E2

major
E2

(c)

2°

$\dfrac{NaOCH_3}{CH_3OH}$

S_N2 +   E2

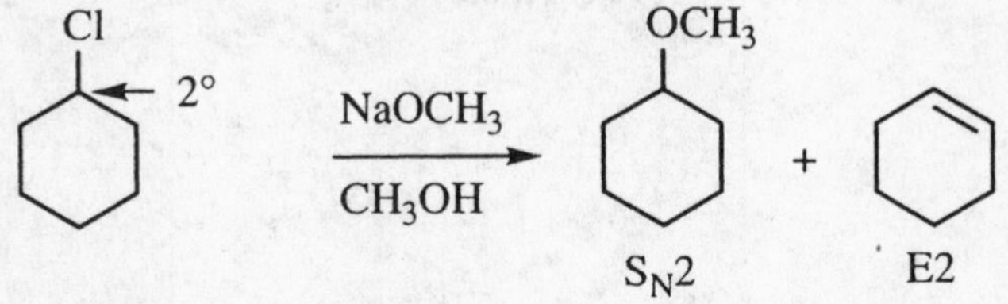

major product is difficult to predict for 2° halides

(d)

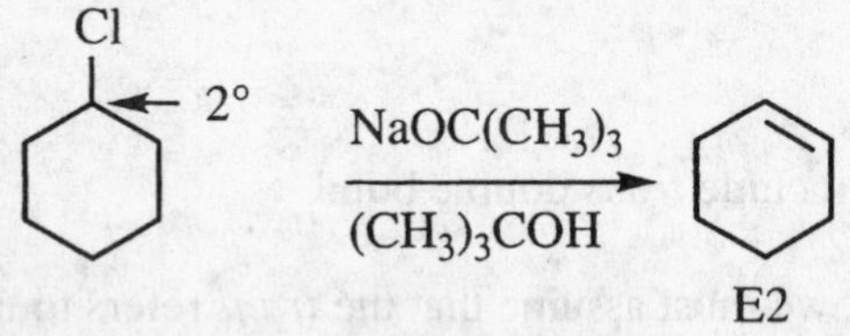

2°

$\dfrac{NaOC(CH_3)_3}{(CH_3)_3COH}$

E2

$NaOC(CH_3)_3$ is a bulky base and a poor nucleophile, minimizing S_N2

7-14

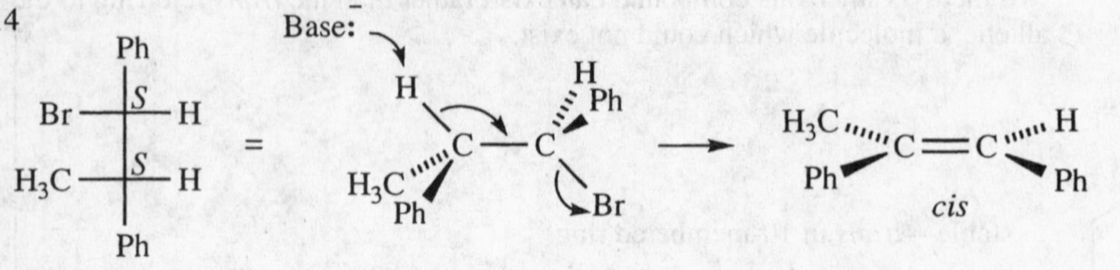

H_3C⸺C=C⸺H with Ph and Ph

cis

7-15 (a)

rotate

HO^-

HO^-

E2 elimination

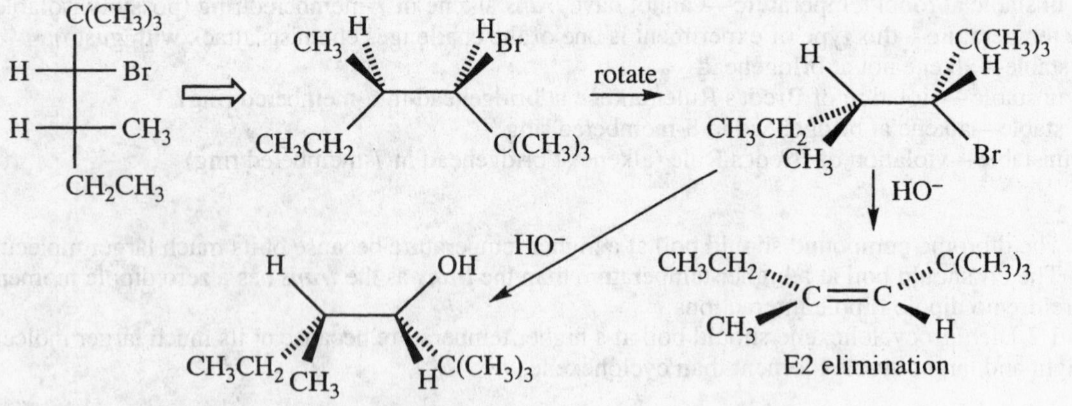

substitution product—since the substrate is a neopentyl halide and highly hindered, the S_N2 substitution is slow, and elimination is favored

(b)

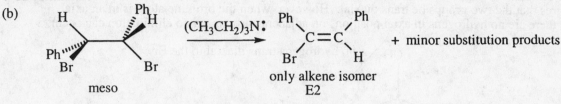

Ph
C=C
Br   H

only alkene isomer
E2

+ minor substitution products

(c)

H   Ph
        H
Br
Ph   Br

one enantiomer
of the *d,l* pair

$(CH_3CH_2)_3N\colon$

Br   Ph
C=C
Ph   H

only alkene isomer
E2

+ minor substitution products

(d)

H   Cl

$\xrightarrow[\text{acetone}]{\text{NaOH}}$

HO   H

S_N2 is the only mechanism possible because no H can get coplanar with Cl; an elimination product would violate Bredt's Rule

(e)

1
2   Cl
3   D
H
H

$\xrightarrow[(CH_3)_3COH]{(CH_3)_3CO^-}$

H

H

See Appendix 1 in this manual for numbering and naming bicyclic systems.

Models show that the H on C-3 cannot be anti-coplanar with the Cl on C-2. Thus, this E2 elimination must occur with a *syn*-coplanar orientation: the D must be removed as the Cl leaves.

7-16 As shown in Solved Problem 7-3, the H and the Br must have trans-diaxial orientation for the E2 reaction to occur. In part (a), the cis isomer has the methyl in equatorial position, the $NaOCH_3$ can remove a hydrogen from either C-2 or C-6, giving a mixture of alkenes where the most highly substituted isomer is the major product (Zaitsev). In part (b), the *trans* isomer has the methyl in the axial position at C-2, so no elimination can occur to C-2. The only possible elimination orientation is toward C-6.

(a)

H_3C Br
1
6
H
H

either H
can be removed

$\xrightarrow[\text{CH}_3\text{OH}]{\text{NaOCH}_3}$

CH_3

major

+

CH_3

minor

Zaitsev orientation

(b)

H   Br
1
6
H_3C
H

only this H
can be removed

$\xrightarrow[\text{CH}_3\text{OH}]{\text{NaOCH}_3}$

CH_3

only alkene formed; stereochemistry of E2 precludes other isomer from forming

141

7-17 E2 elimination requires that the H and the leaving group be anti-coplanar; in a chair cyclohexane, this requires that the two groups be trans diaxial. However, when the bromine atom is in an axial position, there are no hydrogens in axial positions on adjacent carbons, so no elimination can occur.

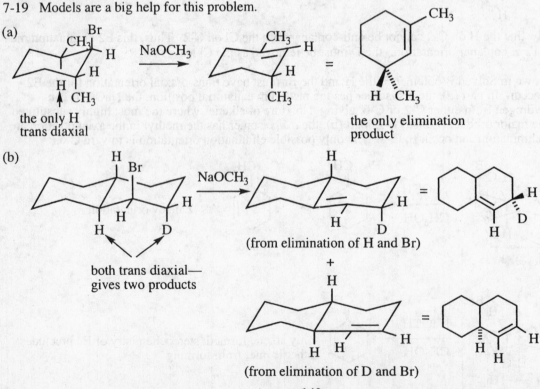

7-18

(a)

(b) Showing the chair form of the decalins makes the answer clear. The top isomer locks the H and the Br into a trans-diaxial conformation—optimum for E2 elimination. The bottom isomer has Br equatorial where it is exceedingly slow to eliminate.

7-19 Models are a big help for this problem.

(a)

the only H
trans diaxial

the only elimination
product

(b)

both trans diaxial—
gives two products

(from elimination of H and Br)

+

(from elimination of D and Br)

142

7-20

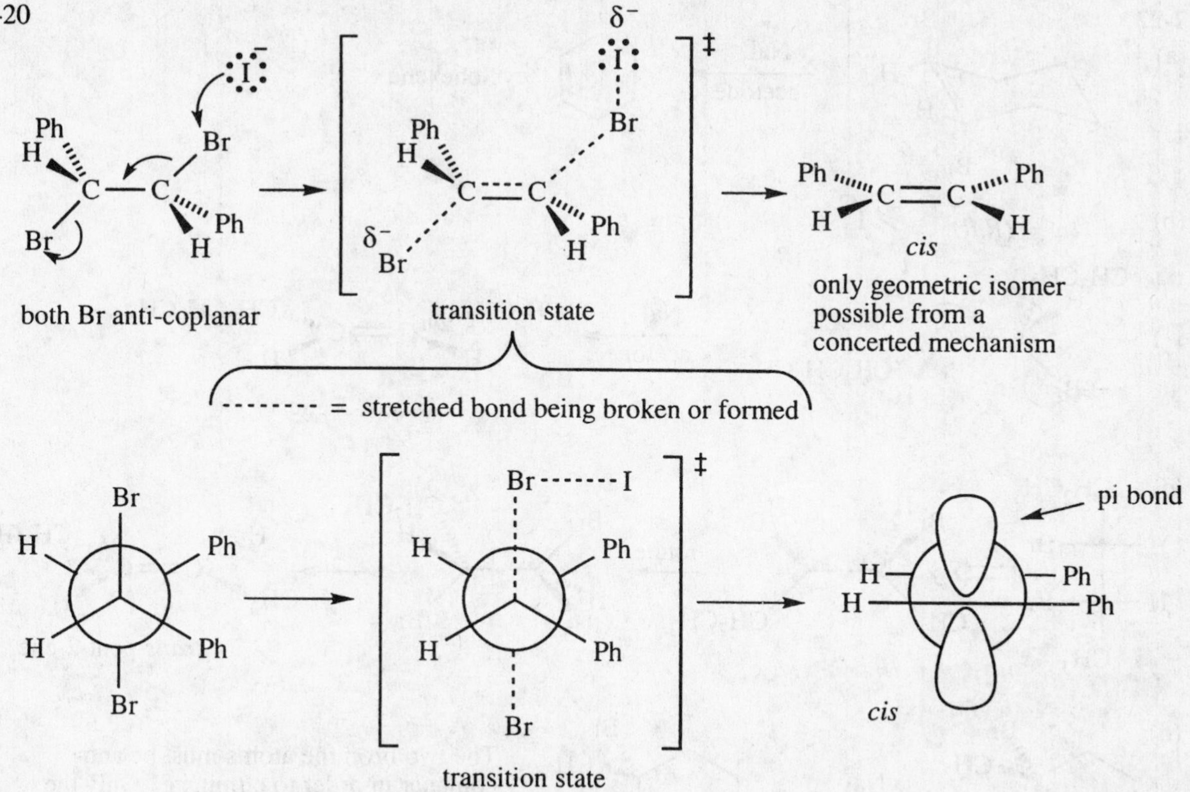

both Br anti-coplanar

transition state

------- = stretched bond being broken or formed

Ph ⁗C=C⁗ Ph
H   H
cis

only geometric isomer
possible from a
concerted mechanism

transition state

pi bond

cis

7-21

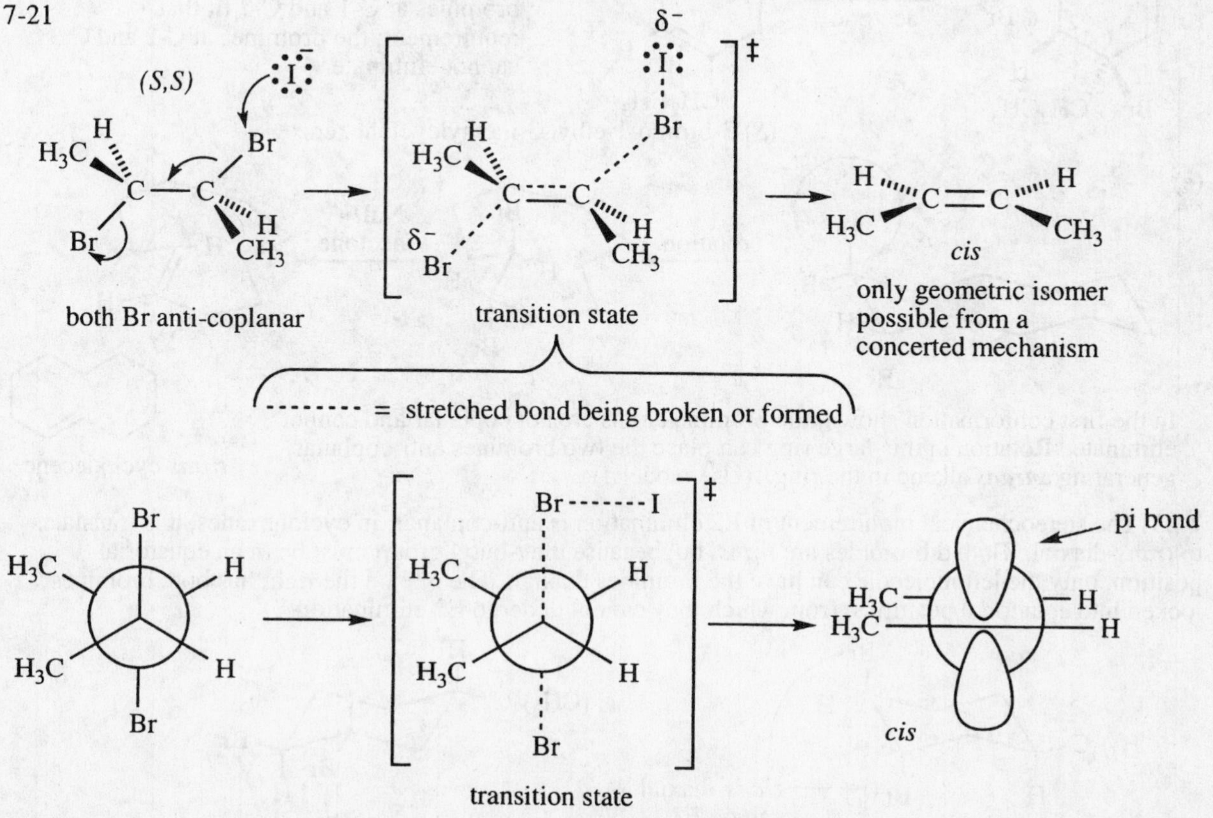

(S,S)

both Br anti-coplanar

transition state

------- = stretched bond being broken or formed

H ⁗C=C⁗ H
H₃C   CH₃
cis

only geometric isomer
possible from a
concerted mechanism

transition state

pi bond

cis

143

7-22

(a)

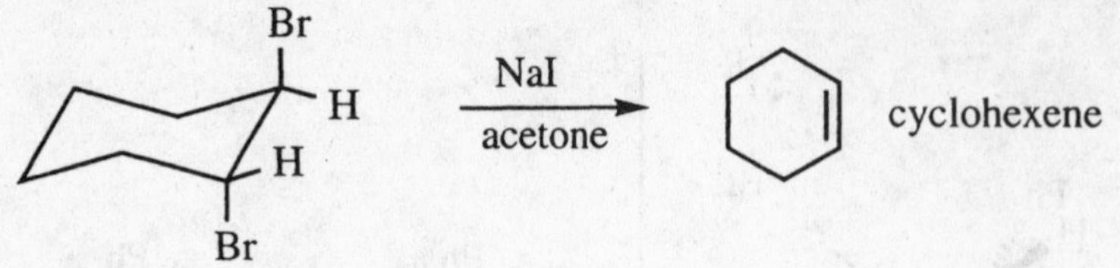

NaI / acetone → cyclohexene

(b) *(R,R)*

NaI / acetone →

cis-hept-3-ene

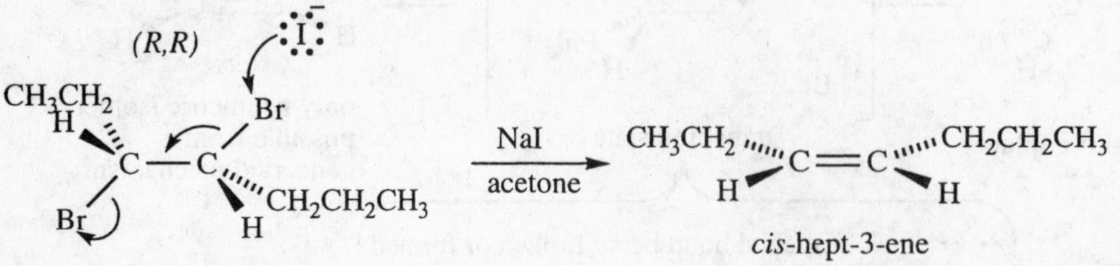

(c)

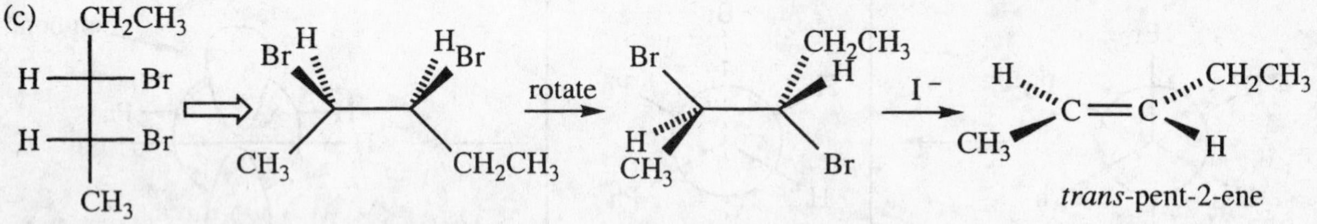

rotate → I⁻ →

trans-pent-2-ene

(d)

NaI / acetone →

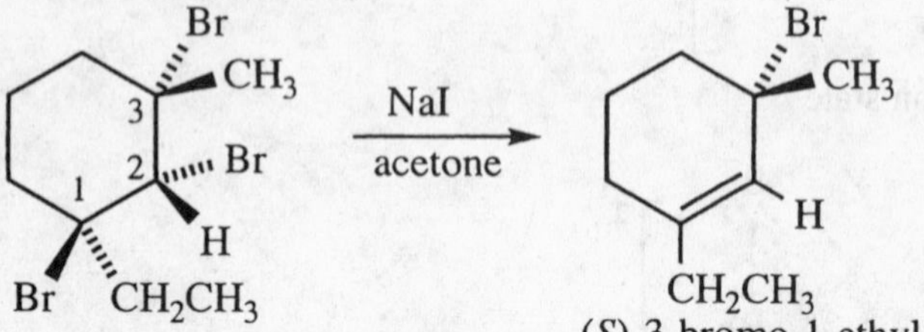

(S)-3-bromo-1-ethyl-3-methylcyclohexene

The two bromine atoms must be anti-coplanar in order to eliminate. Only the bromines at C-1 and C-2 fit that requirement; the bromines at C-2 and C-3 cannot eliminate.

(e)

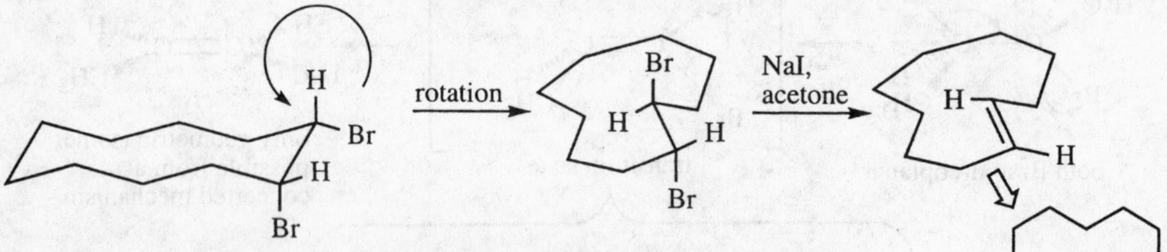

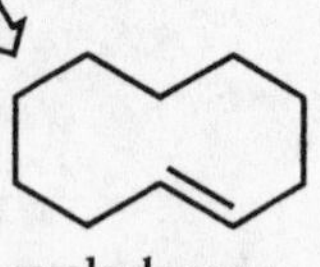

rotation → NaI, acetone →

trans-cyclodecene

In the first conformation shown, the bromine atoms are not coplanar and cannot eliminate. Rotation in this large ring can place the two bromines anti-coplanar, generating a *trans* alkene in the ring. (Use models!)

7-23 The stereochemical requirement of E2 elimination is anti-coplanar; in cyclohexanes, this translates to *trans*-diaxial. Both dibromides are *trans*, but because the *t*-butyl group must be in an equatorial position, only the left molecule can have the bromines diaxial. The one on the right has both bromines locked into equatorial positions, from which they cannot undergo E2 elimination.

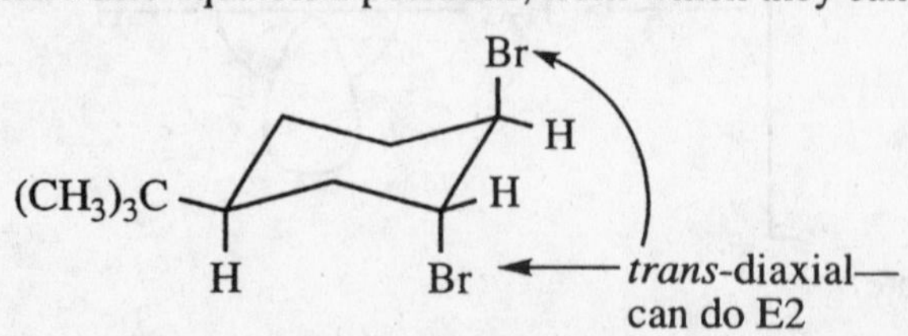

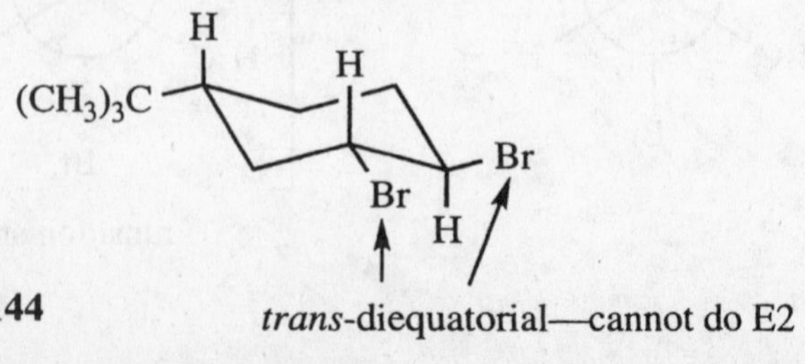

trans-diaxial— can do E2

trans-diequatorial—cannot do E2

7-24
(a)

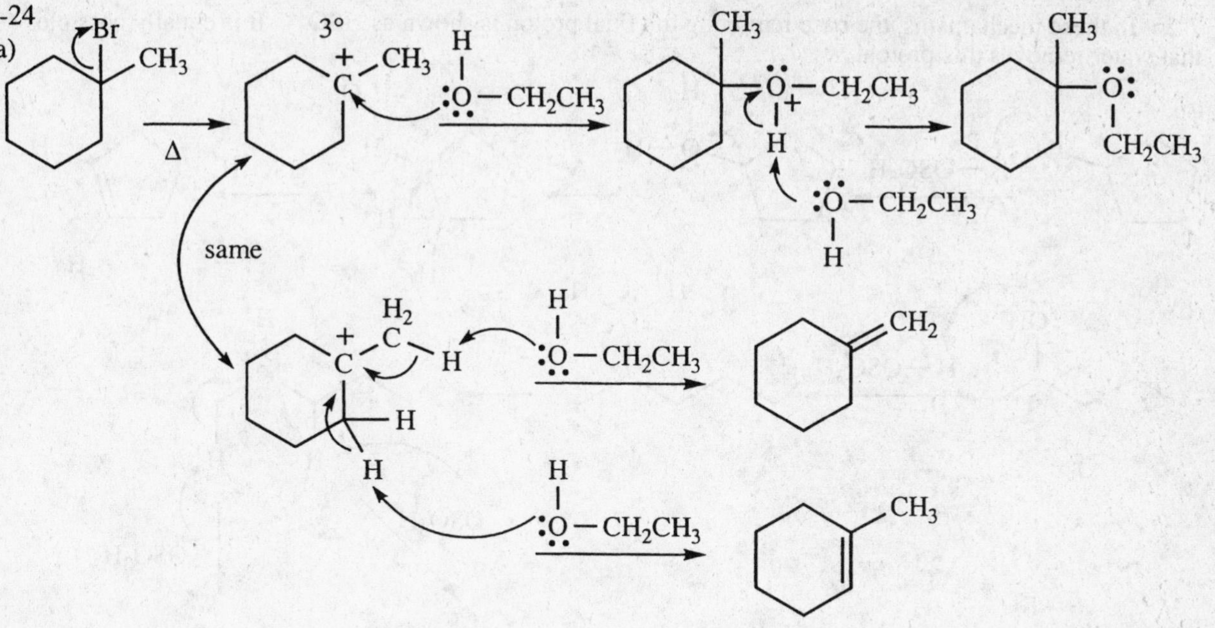

(b)

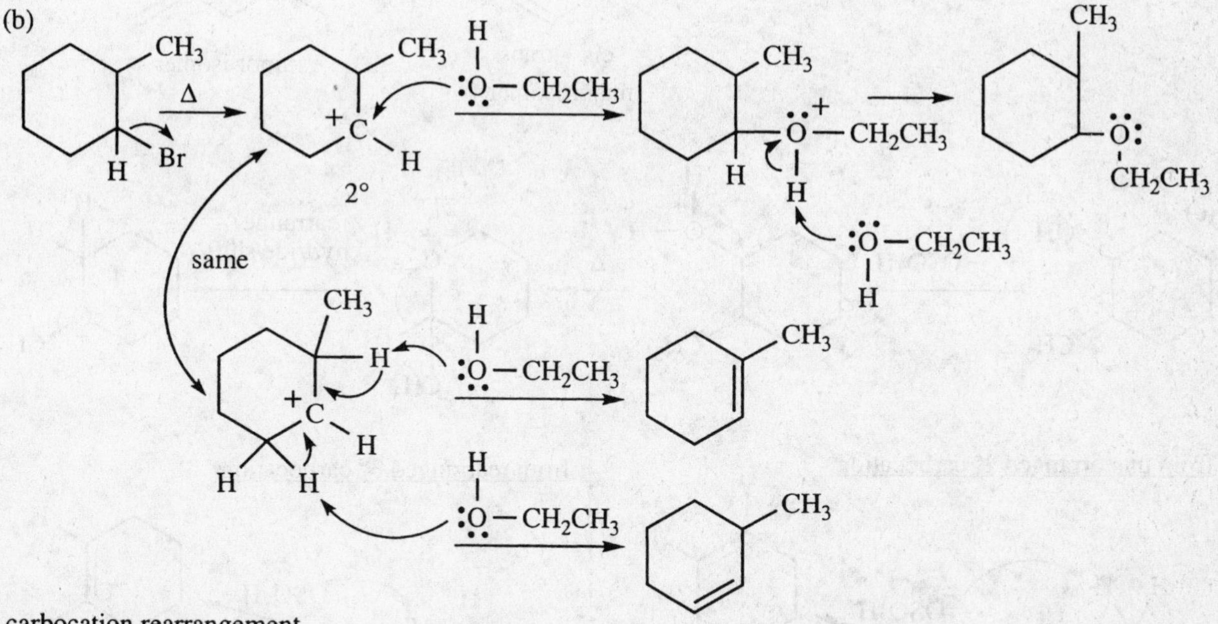

carbocation rearrangement

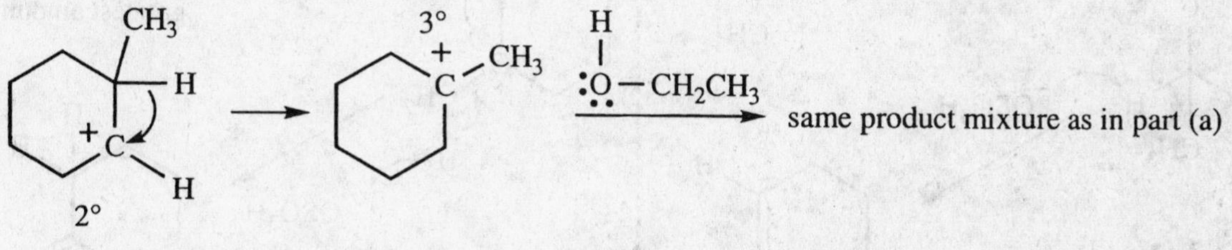

same product mixture as in part (a)

145

7-25 In these mechanisms, the base removing the final proton is shown as HSO_4^-. It is equally possible that water removes this proton.

(a)

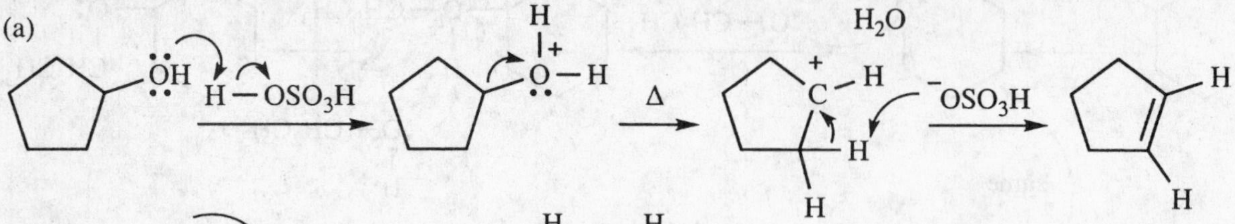

(b)

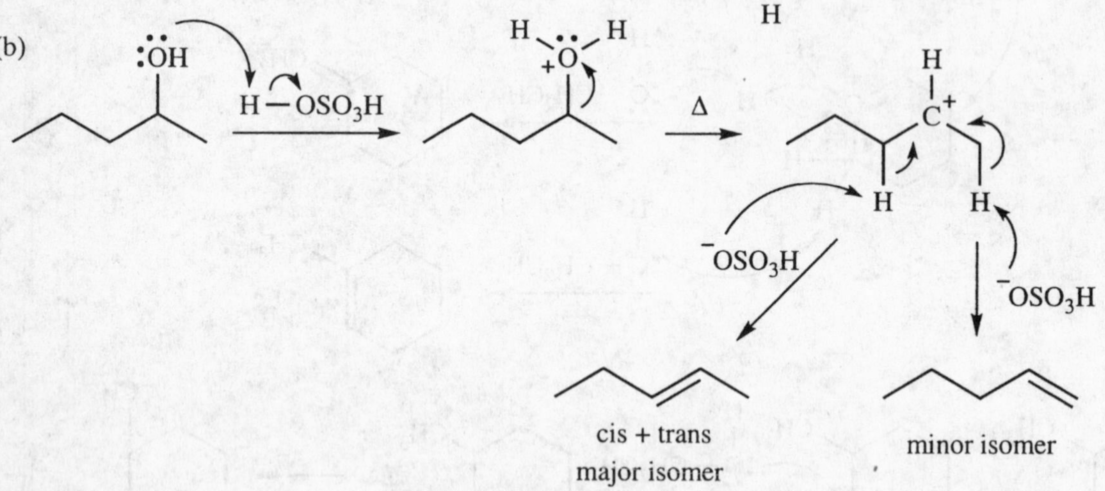

cis + trans
major isomer

minor isomer

(c)

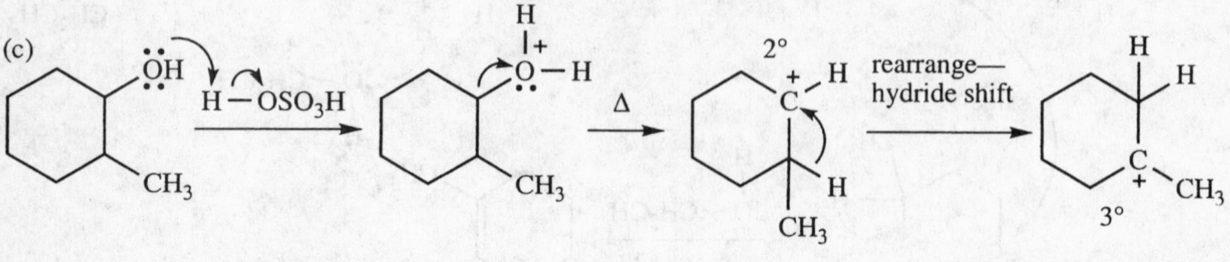

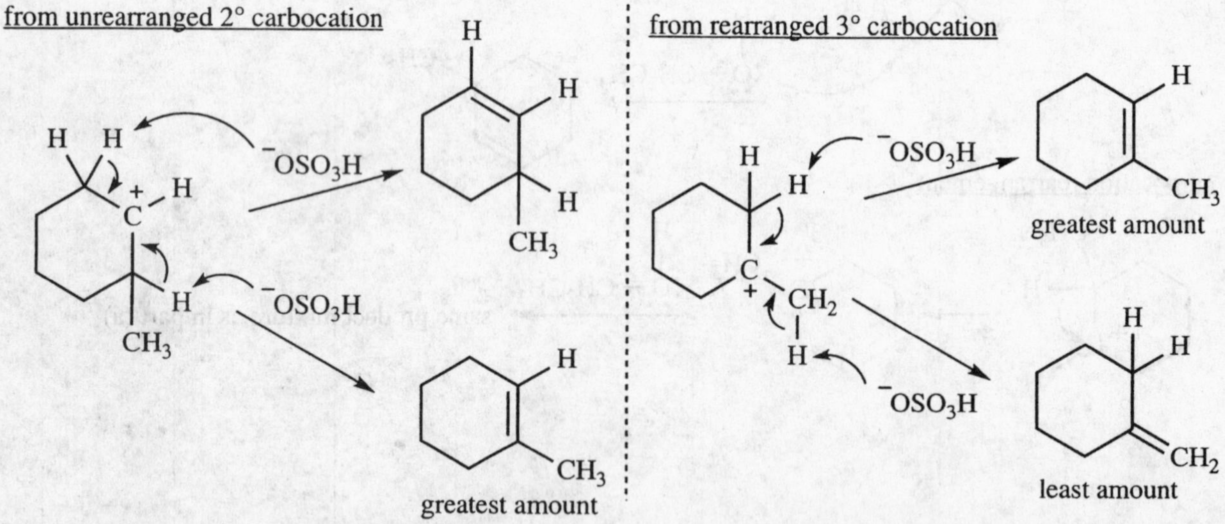

from unrearranged 2° carbocation

greatest amount

from rearranged 3° carbocation

greatest amount

least amount

146

7-26

(a) $\Delta G° = \Delta H° - T\Delta S°$

$= +116{,}000 \text{ J/mol} - 298 \text{ K } (117 \text{ J/K} \bullet \text{mol})$

$= +81{,}100 \text{ J/mol} = +81.1 \text{ kJ/mol}$ (+19.3 kcal/mole)

$\Delta G°$ is positive, the reaction is **disfavored** at 25°C

(b) $\Delta G_{1000} = +116{,}000 \text{ J/mol} - 1273 \text{ K } (117 \text{ J/K} \bullet \text{mol})$

$= -32{,}900 \text{ J/mol} = -32.9 \text{ kJ/mol}$ (- 8.0 kcal/mole)

ΔG is negative, the reaction is **favored** at 1000°C

7-27

(a) basic and nucleophilic mechanism: $Ba(OH)_2$ is a strong base
(b) acidic and electrophilic mechanism: the catalyst is H^+
(c) free radical chain reaction: the catalyst is a peroxide that initiates free radical reactions
(d) acidic and electrophilic mechanism: the catalyst BF_3 is a strong Lewis acid

7-28

(a)

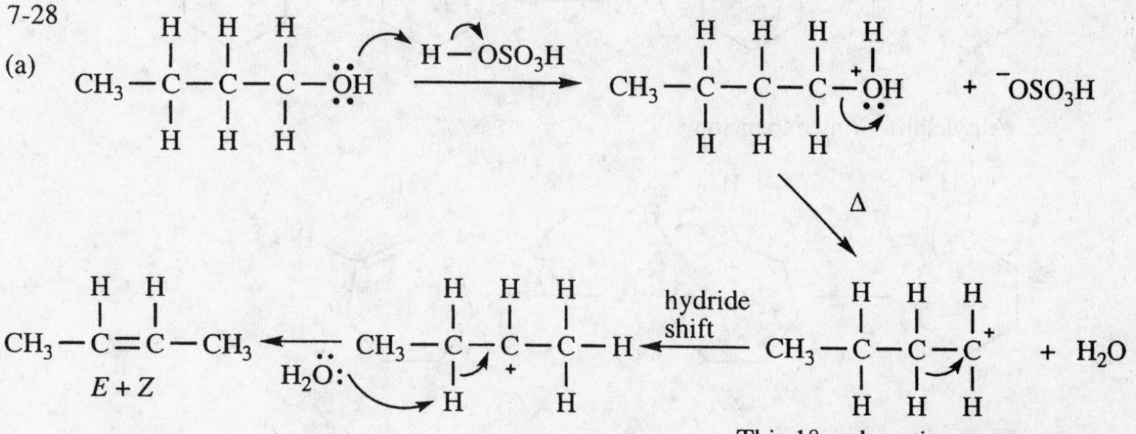

This 1° carbocation may or may not exist. It is shown for clarity.

(b)

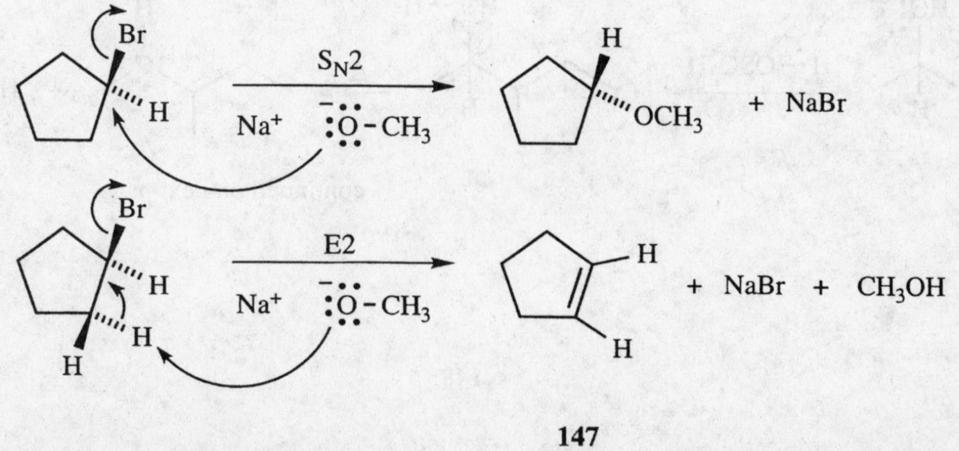

147

7-29

(a)

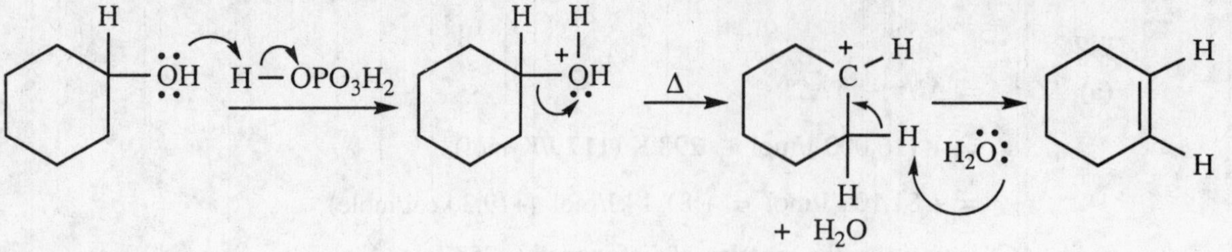

$+ H_2O$

(b)

$CH_2-\overset{..}{\underset{..}{O}}H$

$H-\overset{..}{\underset{..}{O}}SO_3H$

$CH_2-\overset{H}{\overset{|}{\underset{..}{\overset{+}{O}}}}H$

$\overset{+}{C}H_2$

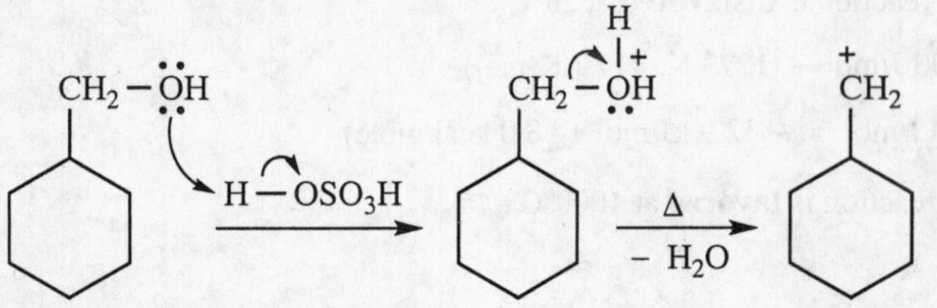

This 1° carbocation may or may not exist. It is shown for clarity.

$\xrightarrow{\underset{-H_2O}{\Delta}}$

Two possible rearrangements

1. Hydride shift

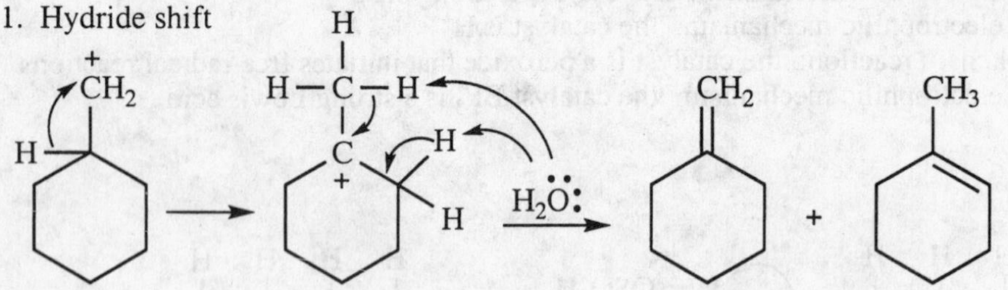

$H_2\overset{..}{\underset{..}{O}}$:

2. Alkyl shift—ring expansion

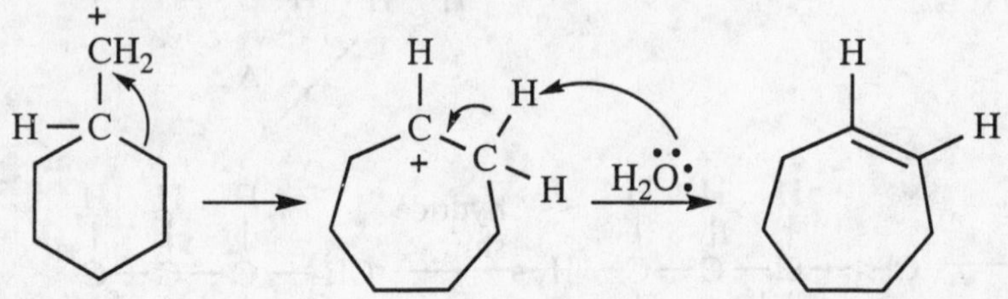

$H_2\overset{..}{\underset{..}{O}}$:

(c) carbocation formation

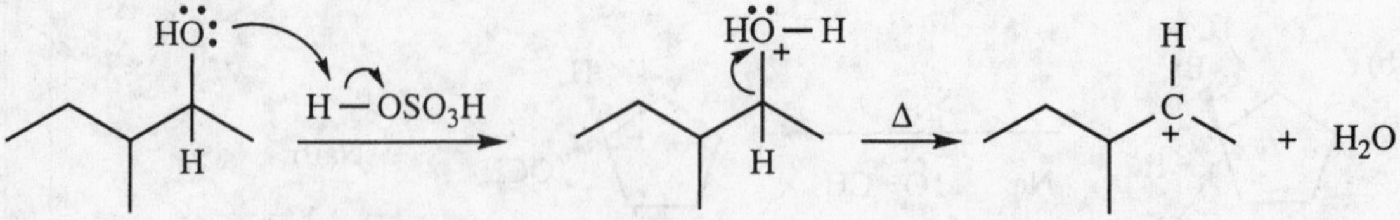

$+ H_2O$

continued on next page

148

7-29 (c) continued

<u>without rearrangement</u>

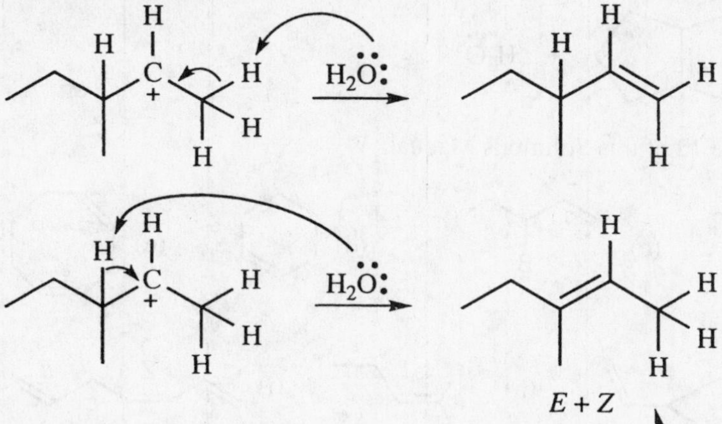

<u>with hydride shift</u>

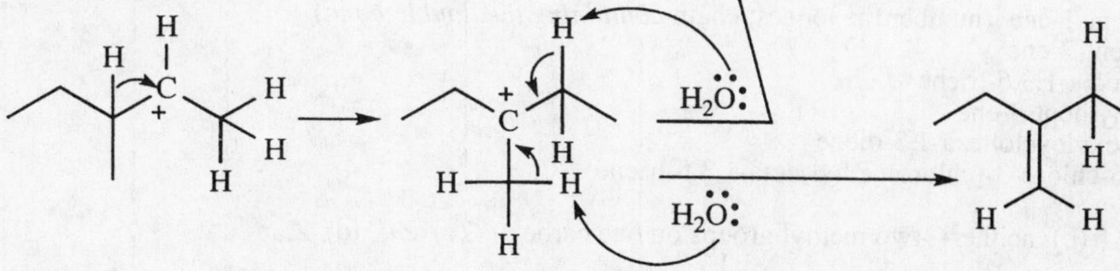

(d)

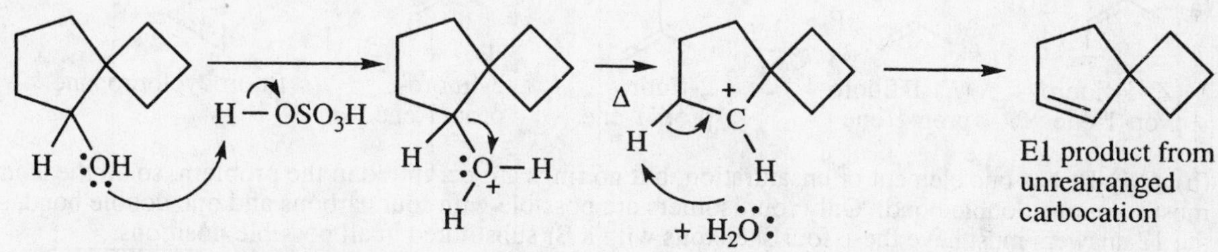

E1 product from unrearranged carbocation

<u>rearrangement by alkyl shift</u>

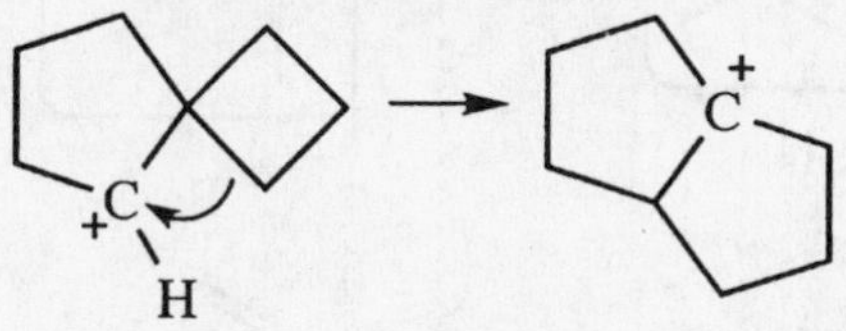

<u>E1 on rearranged carbocation</u>

$+ \ H_3\ddot{O}^+$

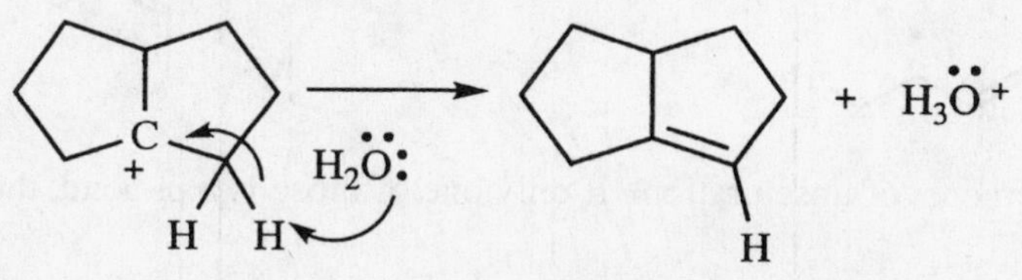

7-29 (d) continued

<u>E1 on rearranged carbocation</u>

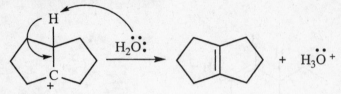

7-30   Please refer to solution 1-20, page 12 of this Solutions Manual.

7-31

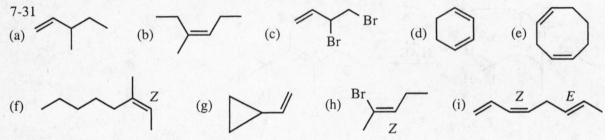

(a)   (b)   (c)   (d)   (e)

(f)   (g)   (h)   (i)

7-32
(a) 2-ethylpent-1-ene (number the longest chain *containing the double bond*)
(b) 3-ethylpent-2-ene
(c) (3*E*,6*E*)-octa-1,3,6-triene
(d) (*E*)-4-ethylhept-3-ene
(e) 1-cyclohexylcyclohexa-1,3-diene
(f) (3*Z*,5*E*)-6-chloro-3-(chloromethyl)octa-1,3,5-triene

7-33 (a) *E*   (b) neither—two methyl groups on one carbon   (c) *Z*   (d) *Z*

7-34
(a)

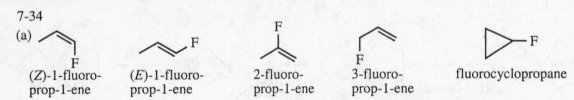

(*Z*)-1-fluoro-   (*E*)-1-fluoro-   2-fluoro-   3-fluoro-   fluorocyclopropane
prop-1-ene   prop-1-ene   prop-1-ene   prop-1-ene

(b) C_4H_7Br has one element of unsaturation, but no rings are permitted in the problem, so all the isomers must have one double bond. Only four isomers are possible with four carbons and one double bond, so the 12 answers must have these four skeletons with a Br substituted in all possible positions.

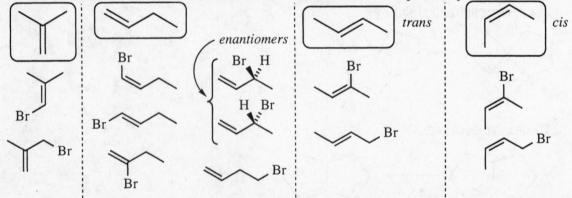

(c) Cholesterol, $C_{27}H_{46}O$, has five elements of unsaturation. If only one of those is a pi bond, the other four must be rings.

150

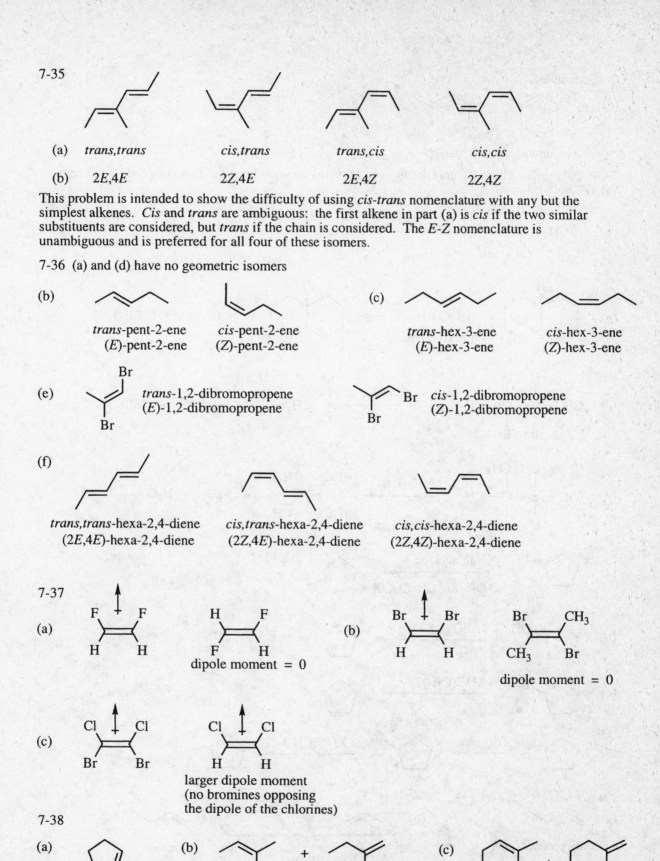

7-35

(a) *trans,trans*   *cis,trans*   *trans,cis*   *cis,cis*

(b)   2E,4E   2Z,4E   2E,4Z   2Z,4Z

This problem is intended to show the difficulty of using *cis-trans* nomenclature with any but the simplest alkenes. *Cis* and *trans* are ambiguous: the first alkene in part (a) is *cis* if the two similar substituents are considered, but *trans* if the chain is considered. The *E-Z* nomenclature is unambiguous and is preferred for all four of these isomers.

7-36 (a) and (d) have no geometric isomers

(b)   *trans*-pent-2-ene   *cis*-pent-2-ene
            (*E*)-pent-2-ene   (*Z*)-pent-2-ene

(c)   *trans*-hex-3-ene   *cis*-hex-3-ene
            (*E*)-hex-3-ene   (*Z*)-hex-3-ene

(e)   *trans*-1,2-dibromopropene   *cis*-1,2-dibromopropene
            (*E*)-1,2-dibromopropene   (*Z*)-1,2-dibromopropene

(f)   *trans,trans*-hexa-2,4-diene   *cis,trans*-hexa-2,4-diene   *cis,cis*-hexa-2,4-diene
        (2E,4E)-hexa-2,4-diene   (2Z,4E)-hexa-2,4-diene   (2Z,4Z)-hexa-2,4-diene

7-37

(a)   dipole moment = 0

(b)   dipole moment = 0

(c)   larger dipole moment
                   (no bromines opposing
                   the dipole of the chlorines)

7-38

(a)   (b)   major   +   minor   (c)   major   +   minor

7-38 continued

(d)

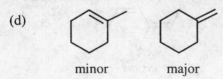

minor   major

7-39 Only major alkene isomers are shown. Minor alkene isomers would also be produced in parts (a), (b) and (d).

(a)
$$CH_3 - \overset{\overset{\displaystyle H}{|}}{\underset{\underset{\displaystyle H}{|}}{C}} - \overset{\overset{\displaystyle H}{|}}{\underset{\underset{\displaystyle OH}{|}}{C}} - CH_3 \xrightarrow[\Delta]{H_2SO_4} CH_3 - \overset{\overset{\displaystyle H}{|}}{C} = \overset{\overset{\displaystyle H}{|}}{C} - CH_3 + H_2O$$
$$E + Z$$

(b) [bicyclic structure with H and Br] + NaOC(CH$_3$)$_3$ → [bicyclic alkene with H] + NaBr + HOC(CH$_3$)$_3$
hindered base

(c)
$$CH_3 - \overset{\overset{\displaystyle H}{|}}{\underset{\underset{\displaystyle Br}{|}}{C}} - \overset{\overset{\displaystyle H}{|}}{\underset{\underset{\displaystyle Br}{|}}{C}} - CH_3 + Zn \xrightarrow{CH_3COOH} CH_3 - \overset{\overset{\displaystyle H}{|}}{C} = \overset{\overset{\displaystyle H}{|}}{C} - CH_3 + ZnBr_2$$
$$E + Z$$

(d)
$$CH_3 - \overset{\overset{\displaystyle CH_3}{|}}{\underset{\underset{\displaystyle H}{|}}{C}} - \overset{\overset{\displaystyle CH_3}{|}}{\underset{\underset{\displaystyle Br}{|}}{C}} - CH_3 \xrightarrow{NaOH, \Delta} \begin{matrix} H_3C \\ H_3C \end{matrix} C = C \begin{matrix} CH_3 \\ CH_3 \end{matrix} + NaBr + H_2O$$

7-40

(a) [cyclopentane with two Br] $\xrightarrow[\text{OR Zn, CH}_3\text{COOH}]{\text{NaI, acetone}}$ [cyclopentene]

(b) [cyclopentanol, OH] $\xrightarrow{H_2SO_4, \Delta}$ [cyclopentene]

(c) [cyclopentyl bromide, Br] $\xrightarrow{KOC(CH_3)_3}$ [cyclopentene]

(d) [cyclopentane] $\xrightarrow{Br_2, h\nu}$ [cyclopentyl bromide, Br] $\xrightarrow{KOC(CH_3)_3}$ [cyclopentene]

7-41

(a) [alkene structure]

only product

(b) [alkene] + [alkene]   +   [alkene]

major   minor

152

7-41 continued

(c) 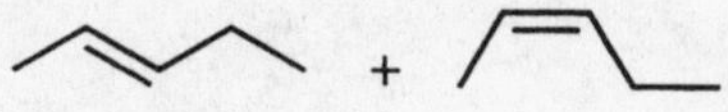 +   (d) 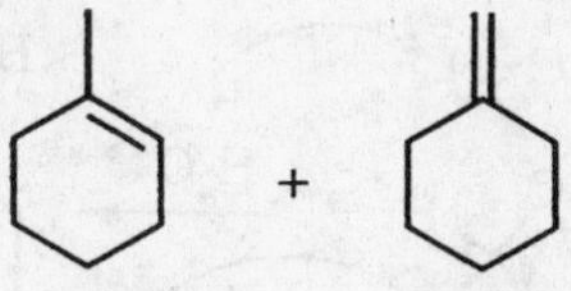 +   (e) 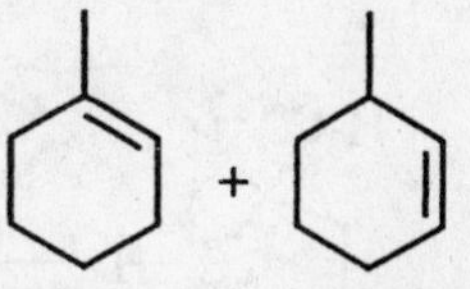 +

major   minor   major   minor

(f) 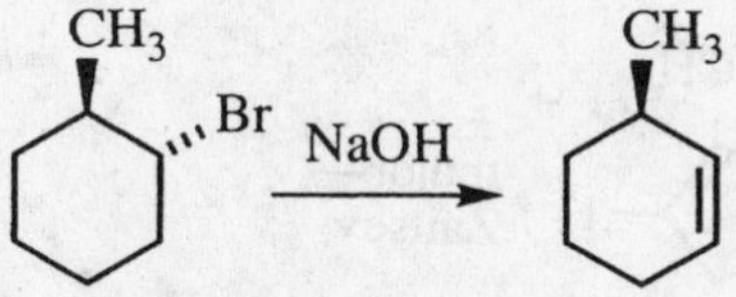 $\xrightarrow{\text{NaOH}}$ The E2 mechanism requires anti-coplanar orientation of H and Br.

7-42 The bromides are shown here. Chlorides or iodides would also work.

(a) 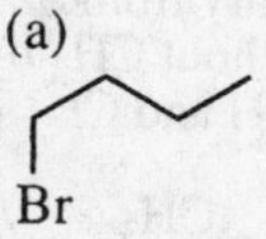   (b) 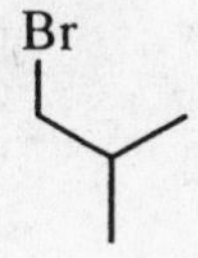 or 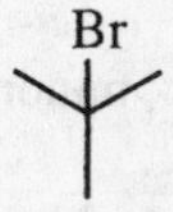   (c) 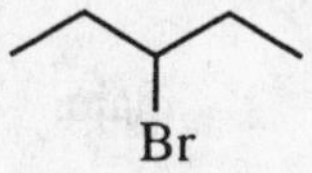   (d) 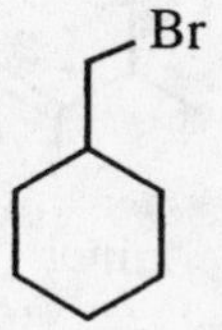   (e)

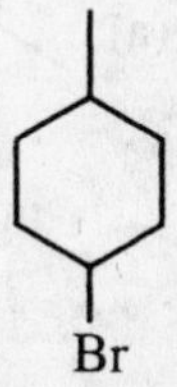

7-43

(a) There are two reasons why alcohols do not dehydrate with strong base. The potential leaving group, hydroxide, is itself a strong base and therefore a terrible leaving group. Second, the strong base deprotonates the —OH faster than any other reaction can occur, consuming the base and making the leaving group anionic and therefore even worse.

$$\text{ROH} + {}^-\text{OC(CH}_3)_3 \rightleftharpoons \text{RO}^- + \text{HOC(CH}_3)_3$$

(b) A halide is already a decent leaving group. Since halides are extremely weak bases, the halogen atom is not easily protonated, and even if it were, the leaving group ability is not significantly enhanced. The hard step is to remove the adjacent H, something only a strong base can do—and strong bases will not be present under strong acid conditions.

7-44

(a) 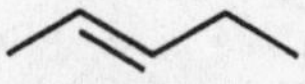   (b) 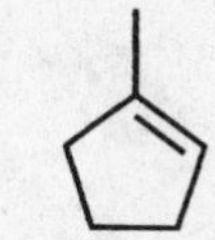   (c) 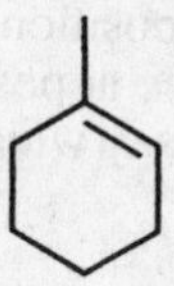   (d) 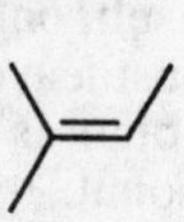 rearrangement

7-45

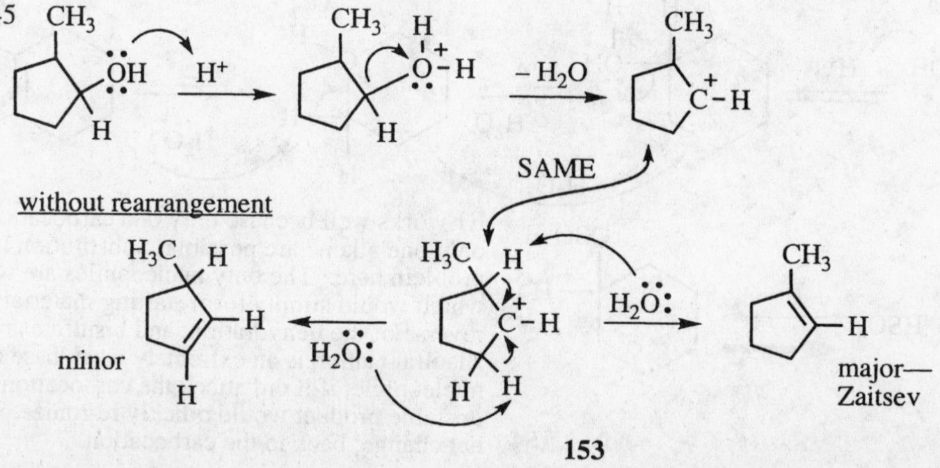

without rearrangement

minor   major—
Zaitsev

7-45 continued
with rearrangement

H₃C, H ... C–H (2°) → $\xrightarrow{\text{hydride shift}}$ → C (3°) with $H_2\ddot{O}$: → minor and major—Zaitsev

7-46

(a) major + minor

(b) major + minor

only product from E2—anti-coplanar is possible only from carbon without CH₃ by removing H and Cl

(c) major + minor

(d) CH_3 major + CH_3 minor

(e) CH_3 ... D

(f) $(CH_3)_3C$ — 1, 3, 4, Cl, Cl $\implies$ $(CH_3)_3C$ with $\overset{ax}{Cl}$ and $\underset{eq}{Cl}$ $\xrightarrow[\Delta]{KOH}$ $(CH_3)_3C$ ... Cl

+

$(CH_3)_3C$ ... Cl

The conformation must be considered because the leaving group must be in an axial position. The *t*-butyl group is so large that it must be equatorial, locking the conformation into the chair shown. As a result, only the Cl at position 4 is axial, so that is the one that must leave, not the one at position 3. Two isomers are possible but it is difficult to say which would be formed in greater amount.

7-47

$\xrightarrow{H^+}$ ... $\xrightarrow[-H_2O]{\Delta}$ ... $\xrightarrow{H_2\ddot{O}:}$ ...

... $+ HSO_4^-$ ⇌ OSO_3H ... H

E1 works well because only one carbocation and only one alkene are possible. Substitution is not a problem here. The only nucleophiles are water, which would simply form starting material by a reverse of the dehydration, and bisulfate anion. Bisulfate anion is an extremely weak base and poor nucleophile; if it did attack the carbocation, the unstable product would quickly re-ionize, with no net change, back to the carbocation.

7-48

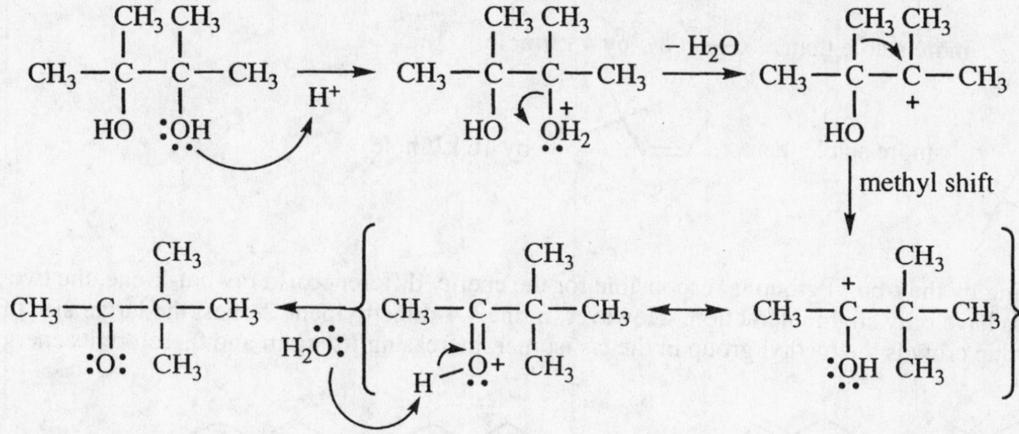

The driving force for this rearrangement is the great stability of the resonance-stabilized, protonated carbonyl group.

7-49 NBS generates bromine which produces bromine radical. Bromine radical abstracts an allylic hydrogen, resulting in a resonance-stabilized allylic radical. The allylic radical can bond to bromine at either of the two carbons with radical character. See the solution to problem 6-63.

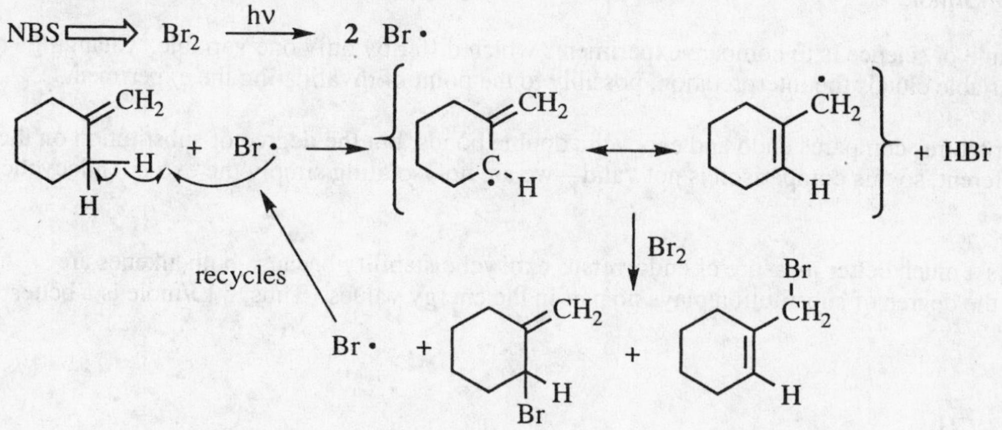

recycles

7-50

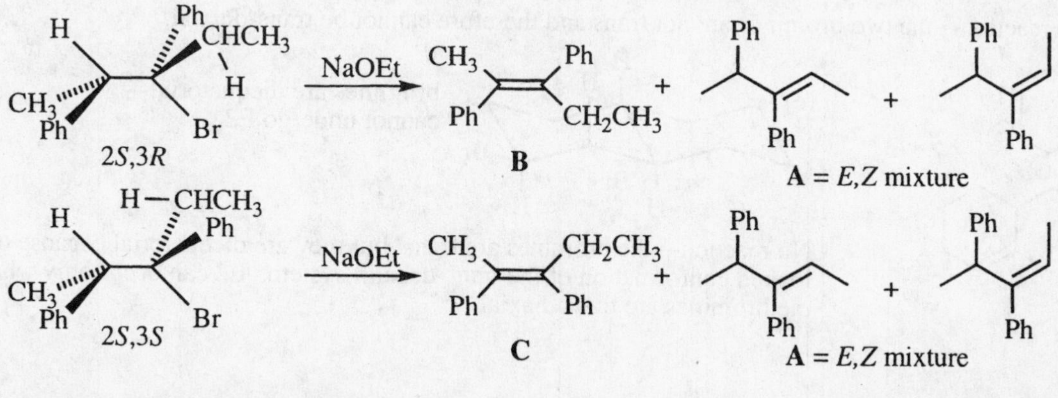

E2 dehydrohalogenation requires anti-coplanar arrangement of H and Br, so specific cis-trans isomers (**B** or **C**) are generated depending on the stereochemistry of the starting material. Removing a hydrogen from C-4 (achiral) will give about the same mixture of *cis* and *trans* (**A**) from either diastereomer.

7-51

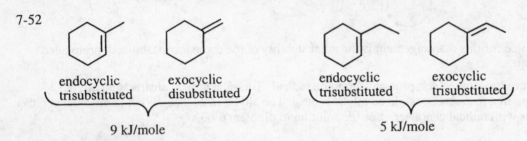

more stable than ⟍=∕ by 4 kJ/mole

more stable than ⟍= by 16 kJ/mole

Steric crowding by the *t*-butyl group is responsible for the energy difference. In *cis*-but-2-ene, the two methyl groups have only slight interaction. However, in the 4,4-dimethylpent-2-enes, the larger size of the *t*-butyl group crowds the methyl group in the *cis* isomer, increasing its strain and therefore its energy.

7-52

endocyclic   exocyclic   endocyclic   exocyclic
trisubstituted   disubstituted   trisubstituted   trisubstituted

9 kJ/mole   5 kJ/mole

A standard principle of science is to compare experiments which differ by only one variable. Changing more than one variable clouds the interpretation, possibly to the point of invalidating the experiment.

The first set of structures compares endo and exocyclic double bonds, but the degree of substitution on the alkene is also different, so this comparison is not valid—we are not isolating simply the exo or endocyclic effect.

The second pair is a much better measure of endo versus exocyclic stability because both alkenes are trisubstituted, so the degree of substitution plays no part in the energy values. Thus, 5 kJ/mole is a better value.

7-53

(a) $CH_3CH_2CH=CH_2$

(b) No reaction—the two bromines are not trans and therefore cannot be trans-diaxial.

(c)

(e) bromines are diequatorial— cannot undergo E2

(d)

No reaction—the bromines are trans, but they are diequatorial because of the locked conformation of the trans-decalin system. E2 can occur only when the bromines are trans diaxial.

7-54 In E2, the two groups to be eliminated must be coplanar. In conformationally mobile systems like acyclic molecules, or in cyclohexanes, anti-coplanar is the preferred orientation where the H and leaving group are 180° apart. In rigid systems like norbornanes, however, SYN-coplanar (angle 0°) is the only possible orientation and E2 will occur, although at a slower rate than anti-coplanar.

The structure having the H and the Cl syn-coplanar is the *trans*, which undergoes the E2 elimination. (It is possible that the *other* H and Cl eliminate from the *trans* isomer; the results from this reaction cannot distinguish between these two possibilities.)

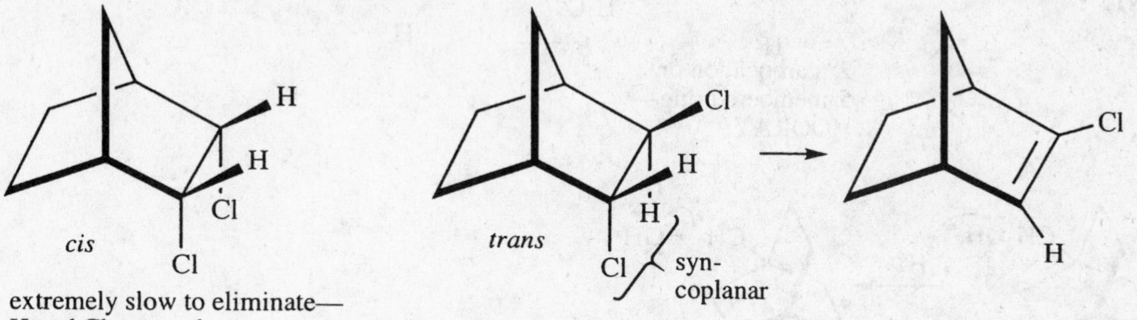

cis

extremely slow to eliminate—
H and Cl not coplanar

trans

syn-coplanar

7-55 It is interesting to note that even though three-membered rings are more strained than four-membered rings, three-membered rings are far more common in nature than four-membered rings. Rearrangement from a four-membered ring to something else, especially a larger ring, will happen quickly.

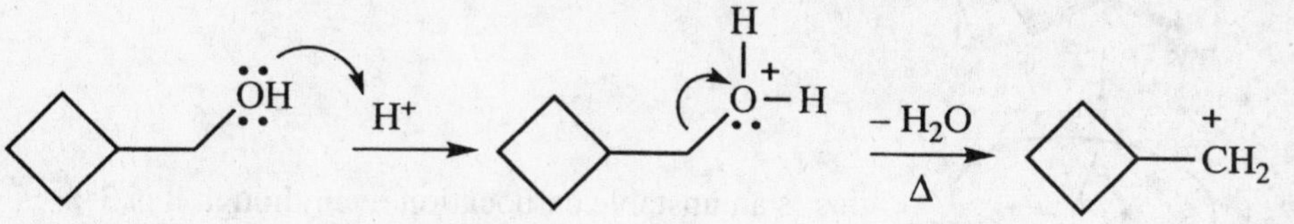

unstable 1° carbocation—
short lifetime if it exists at all

without rearrangement

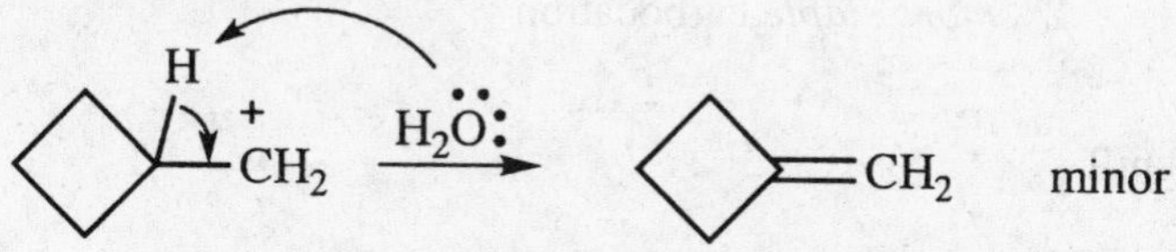

minor

with rearrangement—hydride shift

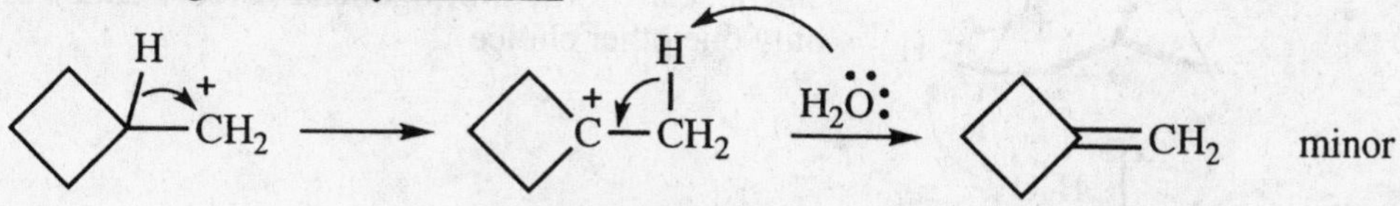

3° carbocation, but still in a strained 4-membered ring

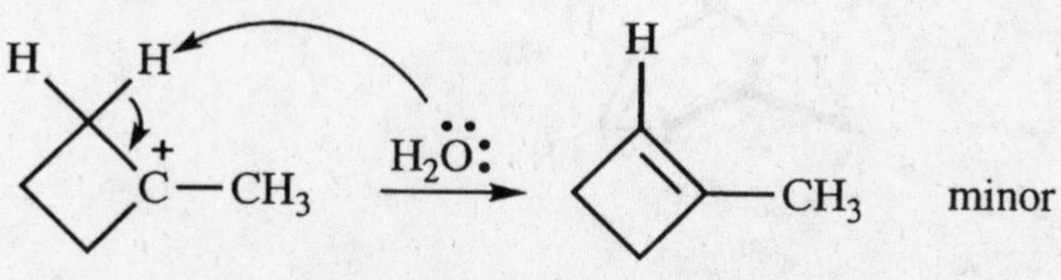

minor

minor

157

7-55 continued

<u>with rearrangement—alkyl shift—ring expansion</u>

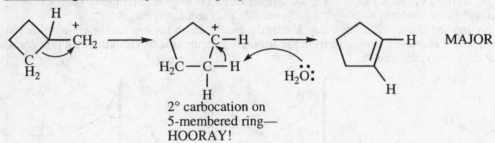

2° carbocation on
5-membered ring—
HOORAY!

7-56

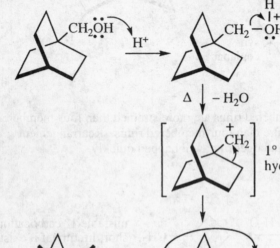

1° carbocation—terrible!—will rearrange; can't do
hydride shift, must do alkyl shift = ring expansion

this is an unstable carbocation even though it is 3°;
bridgehead carbons cannot be sp² (planar—try to make
a model), so this carbocation does a hydride shift to a
2°, *more stable* carbocation

abstraction of adjacent
H gives bridgehead
alkene—violates
Bredt's Rule

 hydride shift

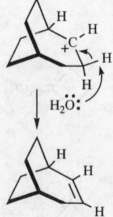

this 2° carbocation loses an adjacent H to form an
alkene; can't form at bridgehead (Bredt's Rule)—
only one other choice

CHAPTER 8—REACTIONS OF ALKENES

8-1 *Major* products are produced in greatest amount; they are not necessarily the *only* products produced.

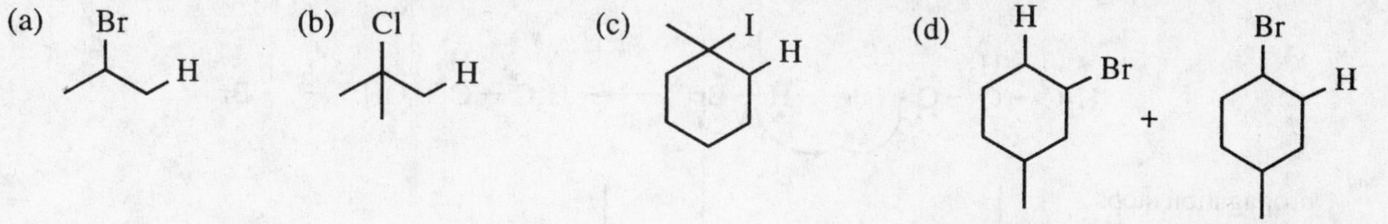

(a)

(b)

(c)

(d)

produced in about equal amounts
(mixture of *cis* and *trans*)

8-2

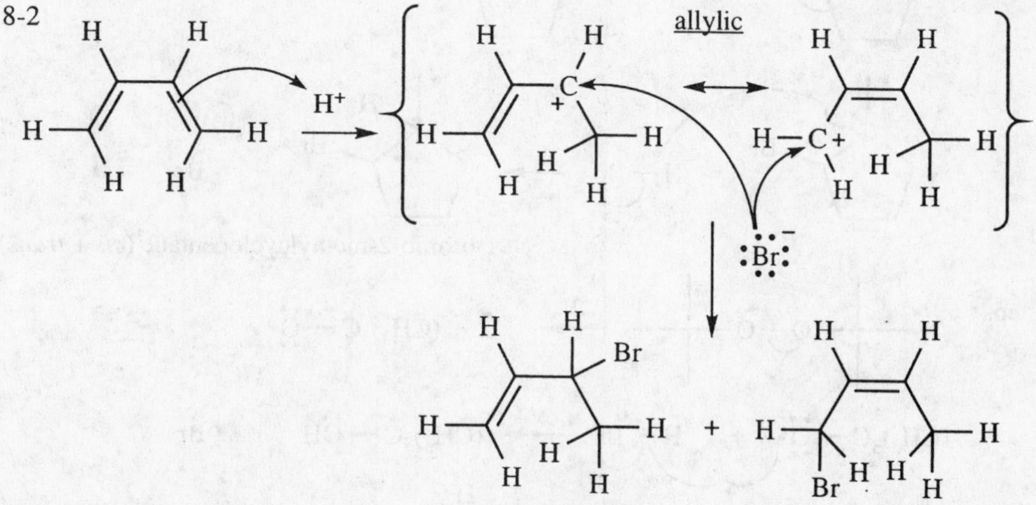

3-bromobut-1-ene   1-bromobut-2-ene

Because the allylic carbocation has partial positive charge at two carbons, the bromide nucleophile can bond at either electrophilic carbon, giving two products.

8-3

(a) initiation steps

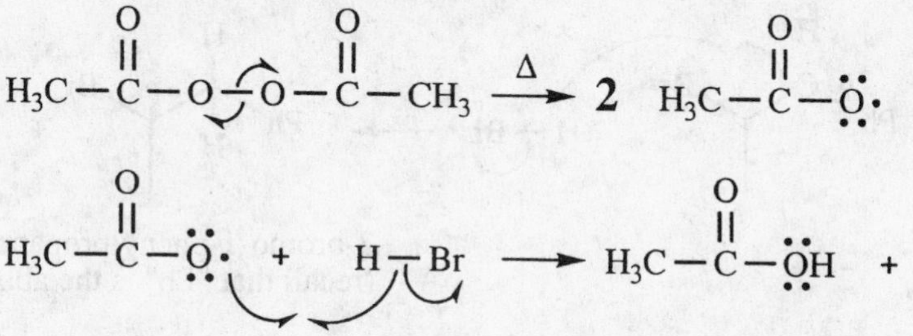

(this radical has another resonance form but it is not germane to this mechanism)

propagation steps

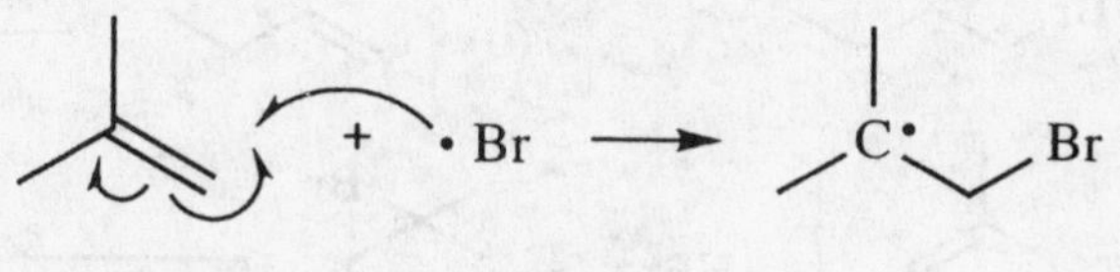

the 3° radical is more stable than the 1° radical

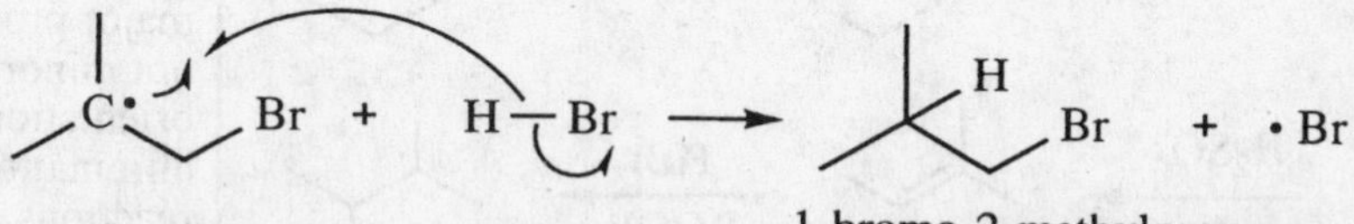

1-bromo-2-methylpropane

159

8-3 continued
(b) initiation steps

$$H_3C - \overset{H_2}{\underset{}{C}} - O \overset{\curvearrowright}{-} O - \overset{H_2}{\underset{}{C}} - CH_3 \xrightarrow{\Delta} 2 \ H_3C - \overset{H_2}{\underset{}{C}} - \ddot{\underset{..}{O}} \cdot$$

$$H_3C - \overset{H_2}{\underset{}{C}} - \ddot{\underset{..}{O}} \cdot \ + \ H - Br \longrightarrow H_3C - \overset{H_2}{\underset{}{C}} - \ddot{\underset{..}{O}}H \ + \ \cdot Br$$

propagation steps

the 3° radical is more stable than the 2° radical

1-bromo-2-methylcyclopentane (*cis + trans*)

(c) initiation steps

$$\underset{}{\overset{}{\diagup}} - O \overset{\curvearrowright}{-} O - \underset{}{\overset{}{\diagup}} \xrightarrow{\Delta} 2 \ (CH_3)_3C - \ddot{\underset{..}{O}} \cdot$$

$$(CH_3)_3C - \ddot{\underset{..}{O}} \cdot \ + \ H - Br \longrightarrow (CH_3)_3C - \ddot{\underset{..}{O}}H \ + \ \cdot Br$$

propagation steps

the benzylic radical is stabilized by resonance, and more stable than the aliphatic radical

2-bromo-1-phenylpropane
(recall that "Ph" is the abbreviation for "phenyl")

8-4
(a) $\xrightarrow[\text{ROOR}]{\text{HBr}}$

(b) $\xrightarrow{\text{HBr}}$

(c) $\xrightarrow[\Delta]{H_2SO_4}$ $\xrightarrow{\text{HBr}}$

(d) $\xrightarrow[\Delta]{H_2SO_4}$ $\xrightarrow[\text{ROOR}]{\text{HBr}}$

Note: A good synthesis uses major products as intermediates, not minor products. Knowing orientation of addition and elimination is critical to using reactions correctly.

8-5

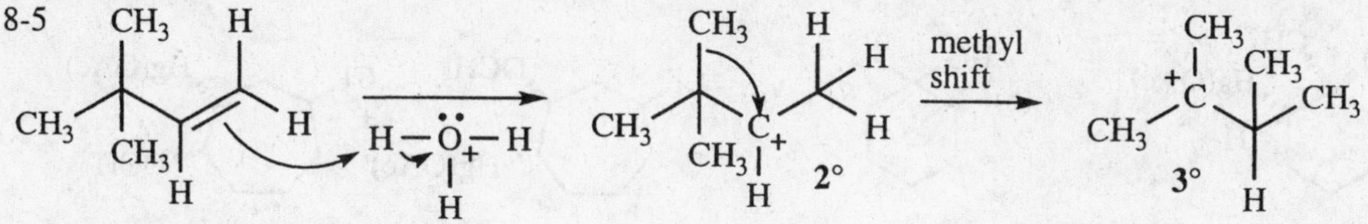

nucleophilic attack by water

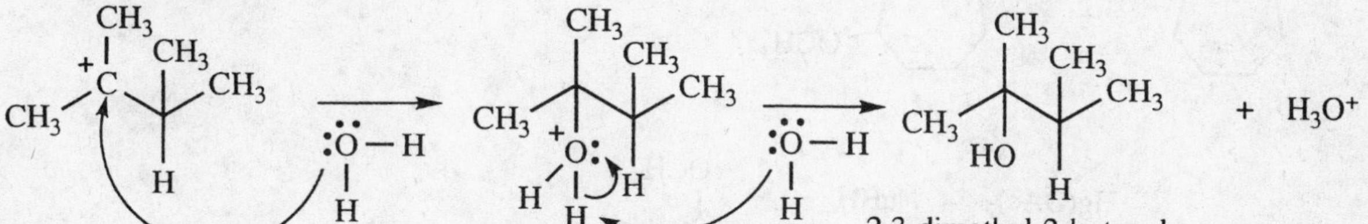

2,3-dimethyl-2-butanol

proton removal

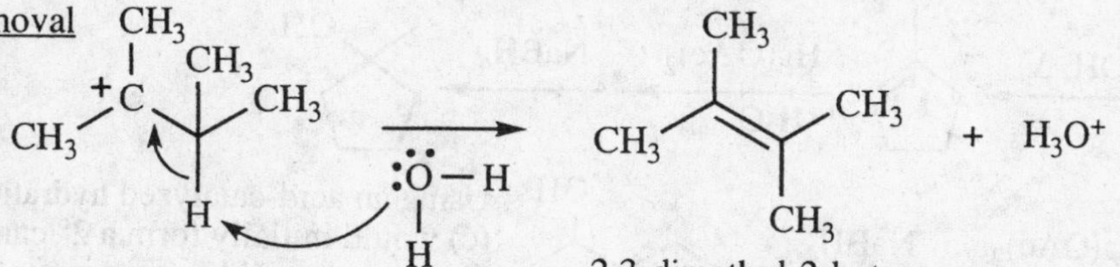

2,3-dimethyl-2-butene

8-6

(a)

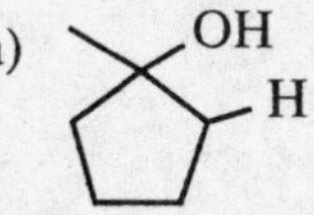

from 3° carbocation

(b)

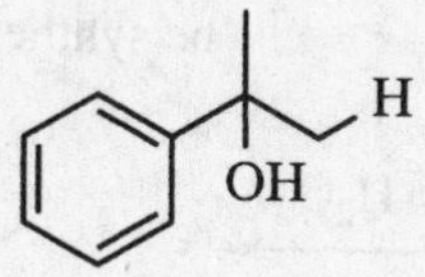

(c)

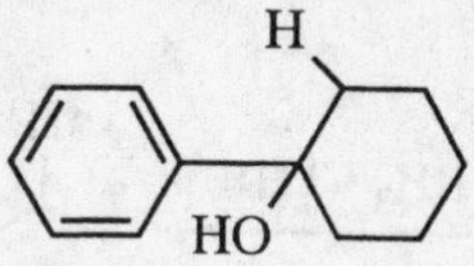

from 3° benzylic carbocation

8-7

(a)

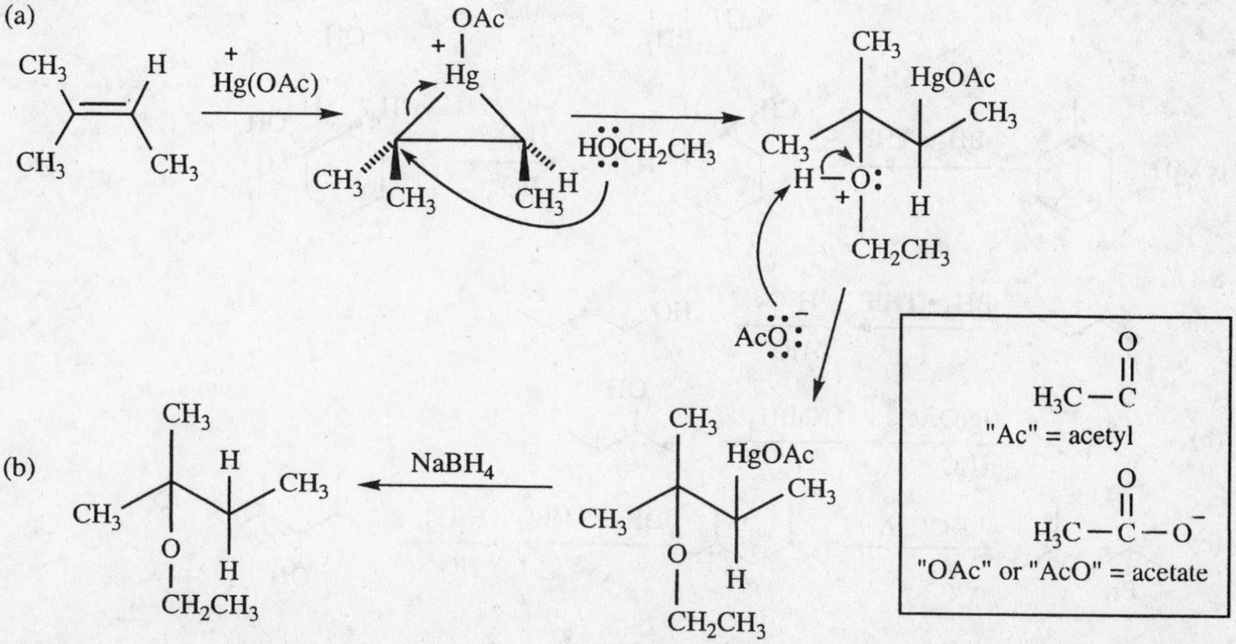

(b)

$$\begin{array}{c} O \\ \| \\ H_3C-C \end{array}$$

"Ac" = acetyl

$$\begin{array}{c} O \\ \| \\ H_3C-C-O^- \end{array}$$

"OAc" or "AcO" = acetate

161

8-8

(a) HO, CH₃, Hg(OAc), H (structure on cyclohexane)

(b) HO, CH₃ (structure on cyclohexane)

(c) Cl, OCH₃, Hg(OAc) (cycloheptane) + Cl, Hg(OAc), OCH₃ (cycloheptane)

(d) Cl, OCH₃ (cycloheptane) + Cl, OCH₃ (cycloheptane)

8-9

(a) $\xrightarrow[\text{CH}_3\text{OH}]{\text{Hg(OAc)}_2}$ $\xrightarrow{\text{NaBH}_4}$ (product with OCH₃)

(b) $\xrightarrow{\text{KOH, }\Delta}$ $\xrightarrow[\text{H}_2\text{O}]{\text{Hg(OAc)}_2}$ $\xrightarrow{\text{NaBH}_4}$ (product with OH)

(c) $\xrightarrow[\text{H}_2\text{O}]{\text{Hg(OAc)}_2}$ $\xrightarrow{\text{NaBH}_4}$ (product with OH)

Using an acid-catalyzed hydration in part (c) would initially form a 2° carbocation that would quickly rearrange to 3° on carbon-3. The desired product would not be synthesized.

8-10

(a), (b) $\xrightarrow{\text{BH}_3 \cdot \text{THF}}$ (H, BH₂) $\xrightarrow[\text{HO}^-]{\text{H}_2\text{O}_2}$ (OH)

(c), (d) $\xrightarrow{\text{BH}_3 \cdot \text{THF}}$ (H, BH₂) $\xrightarrow[\text{HO}^-]{\text{H}_2\text{O}_2}$ (OH)

(e), (f) $\xrightarrow{\text{BH}_3 \cdot \text{THF}}$ (CH₃, H, BH₂, H) $\xrightarrow[\text{HO}^-]{\text{H}_2\text{O}_2}$ (CH₃, H, OH, H)

8-11

(a) $\xrightarrow{\text{BH}_3 \cdot \text{THF}}$ $\xrightarrow[\text{HO}^-]{\text{H}_2\text{O}_2}$ HO (product)

(b) $\xrightarrow[\text{H}_2\text{O}]{\text{Hg(OAc)}_2}$ $\xrightarrow{\text{NaBH}_4}$ (product with OH)

(c) Br (structure) $\xrightarrow{\text{KOH, }\Delta}$ $\xrightarrow{\text{BH}_3 \cdot \text{THF}}$ $\xrightarrow[\text{HO}^-]{\text{H}_2\text{O}_2}$ (product with OH)

8-12 Instead of borane attacking the bottom face of 1-methylcyclopentene, it is equally likely to attack the top face, leading to the enantiomer.

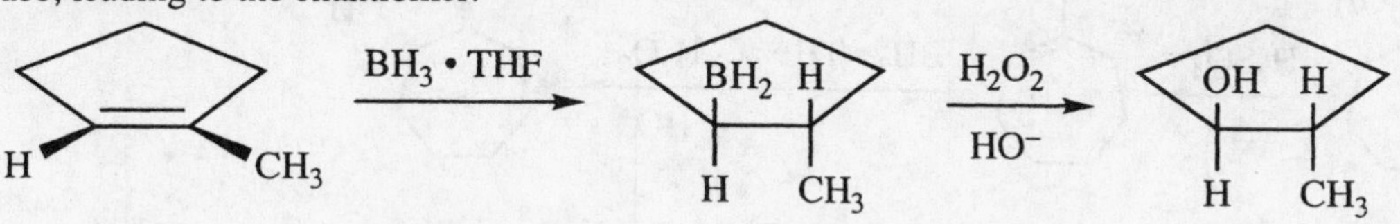

8-13

(a)

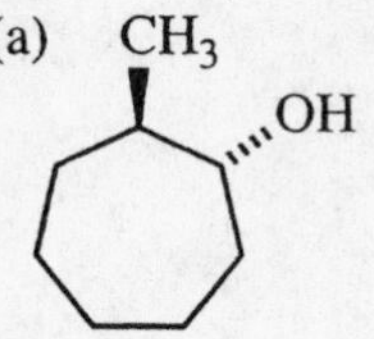

(b)

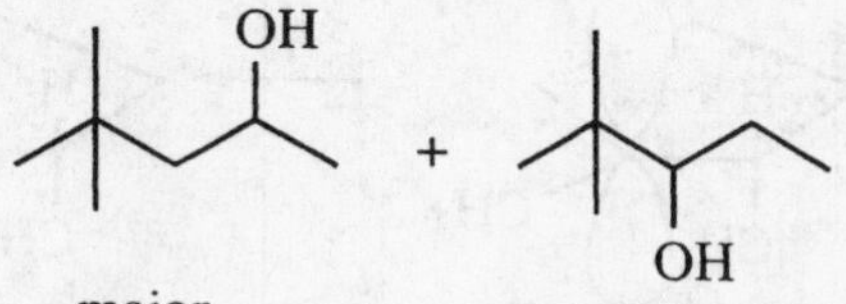

major   +   minor—steric
hindrance to
attack of BH_3

(c)

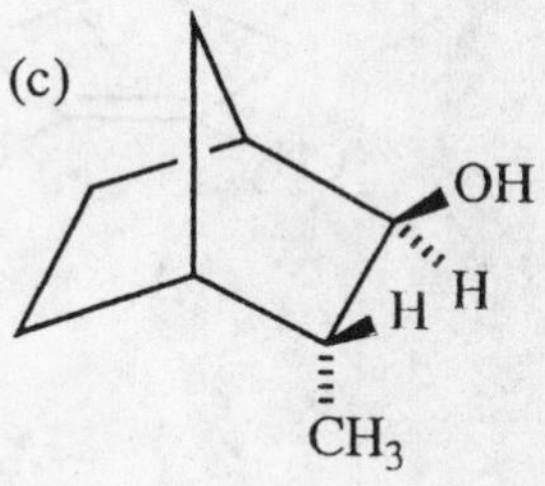

8-14

(a)

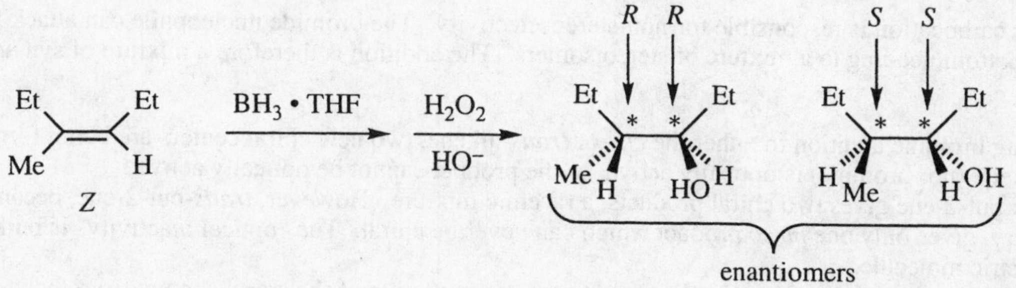

enantiomers

(b)

enantiomers

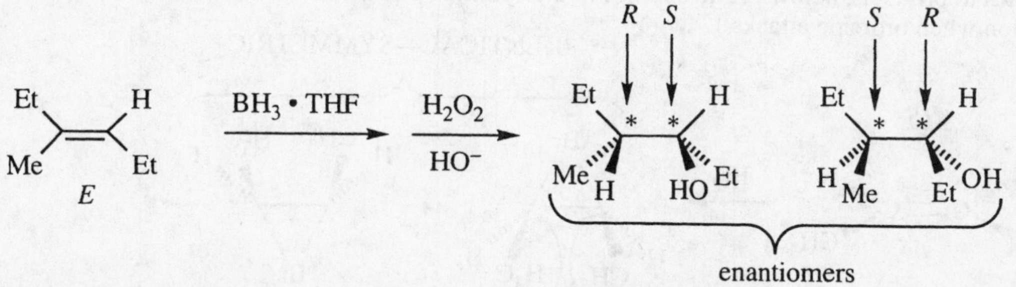

The enantiomeric pair produced from the Z-alkene is diastereomeric with the other enantiomeric pair produced from the E-alkene. Hydroboration-oxidation is stereospecific, that is, each alkene gives a specific set of stereoisomers, not a random mixture.

8-15

(a)

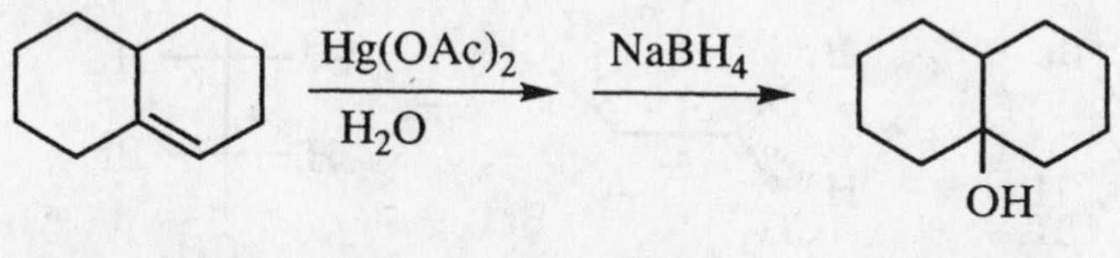

(b)

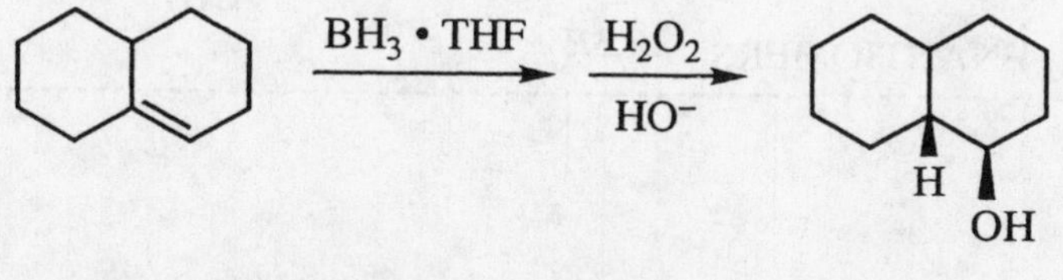

163

8-15 continued

(c)

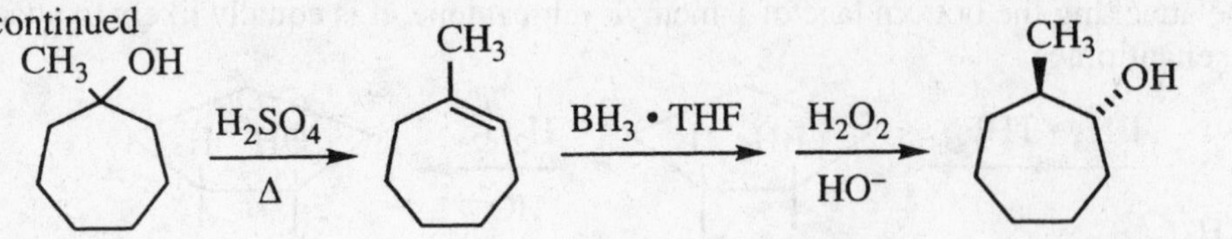

8-16

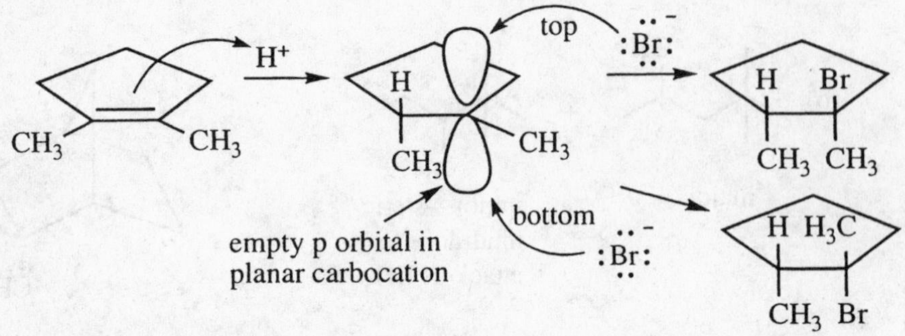

empty p orbital in
planar carbocation

The *planar* carbocation is responsible for non-stereoselectivity. The bromide nucleophile can attack from the top or bottom, leading to a mixture of stereoisomers. The addition is therefore a mixture of syn and anti addition.

8-17 During bromine addition to either the *cis-* or *trans*-alkene, two new chiral centers are being formed. Neither alkene (nor bromine) is optically active, so the product cannot be optically active.

The *cis*-but-2-ene gives two chiral products, a racemic mixture. However, *trans*-but-2-ene, because of its symmetry, gives only one *meso* product which can never be chiral. The "optical *in*activity" is built into this symmetric molecule.

This can be seen by following what happens to the configuration of the chiral centers from the intermediates to products, below. (The key lies in the symmetry of the intermediate and *inversion* of configuration when bromide attacks.)

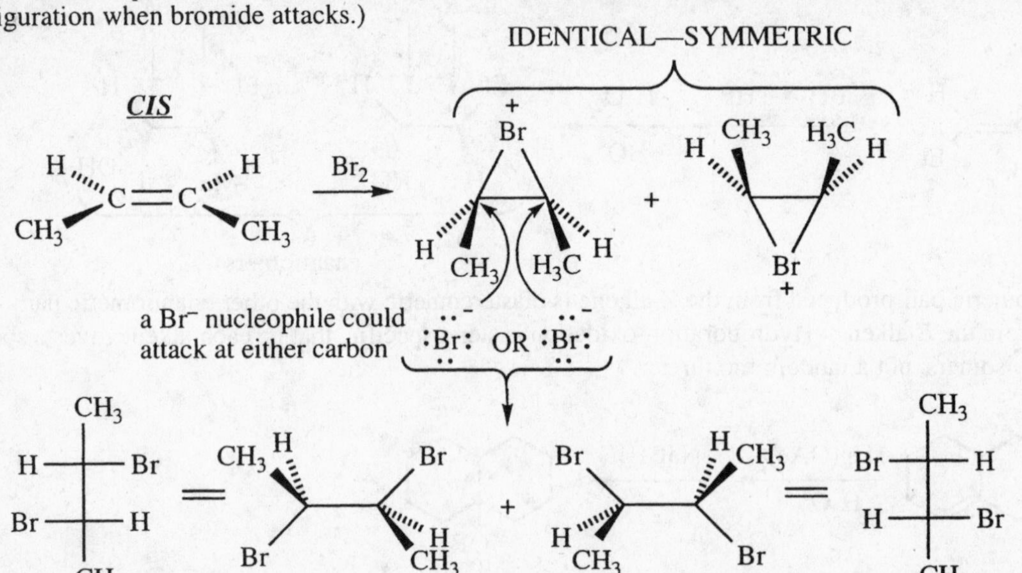

IDENTICAL—SYMMETRIC

continued on next page

164

TRANS

a Br⁻ nucleophile could attack at either carbon

ENANTIOMERIC

bromide attack on this bromonium ion gives the same results

:Br:⁻ OR :Br:⁻

CH₃
Br —|— H
Br —|— H
CH₃

2R,3S   IDENTICAL—MESO   2R,3S

CH₃
H —|— Br
H —|— Br
CH₃

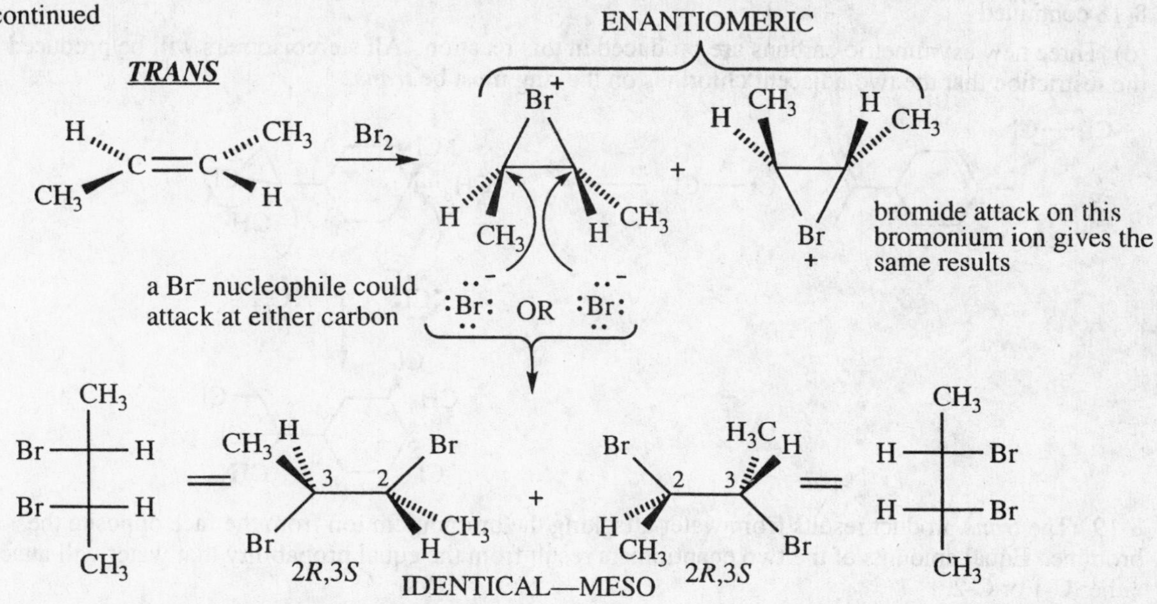

CONCLUSION: anti addition of a symmetric reagent to a symmetric *cis*-alkene gives racemic product, while anti addition to a *trans*-alkene gives meso product. (We will see shortly that syn addition to a *cis*-alkene gives meso product, and syn addition to a *trans*-alkene gives racemic product. Stay tuned.)

8-18 Enantiomers of chiral products are also produced but not shown.

(a)

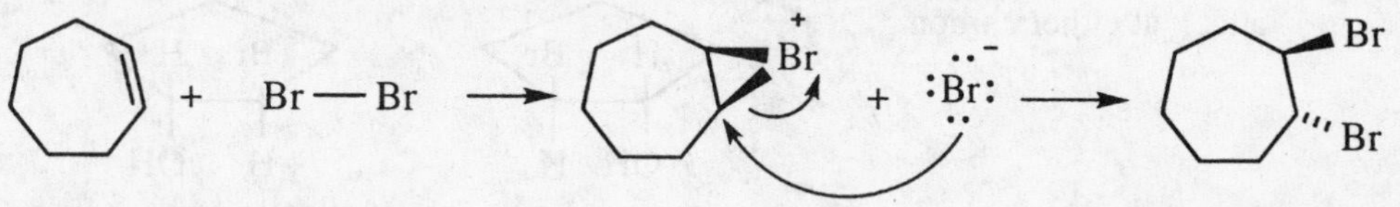

(b)

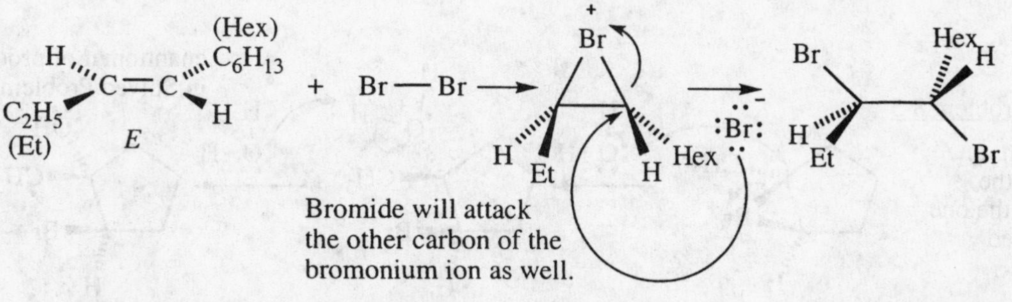

Bromide will attack the other carbon of the bromonium ion as well.

(c)

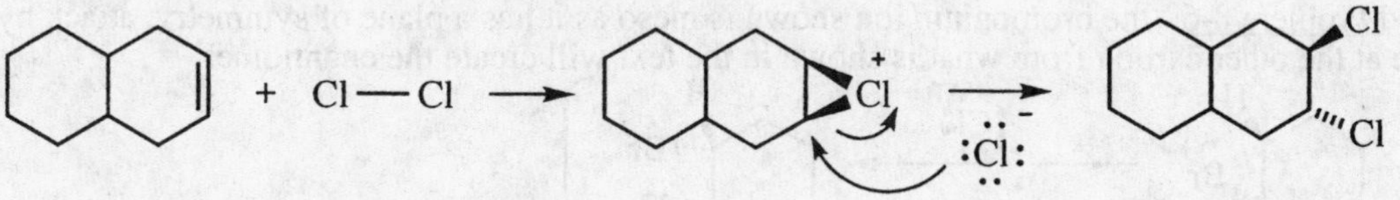

8-18 continued

(d) Three new asymmetric carbons are produced in this reaction. All stereoisomers will be produced with the restriction that the two adjacent chlorines on the ring must be *trans*.

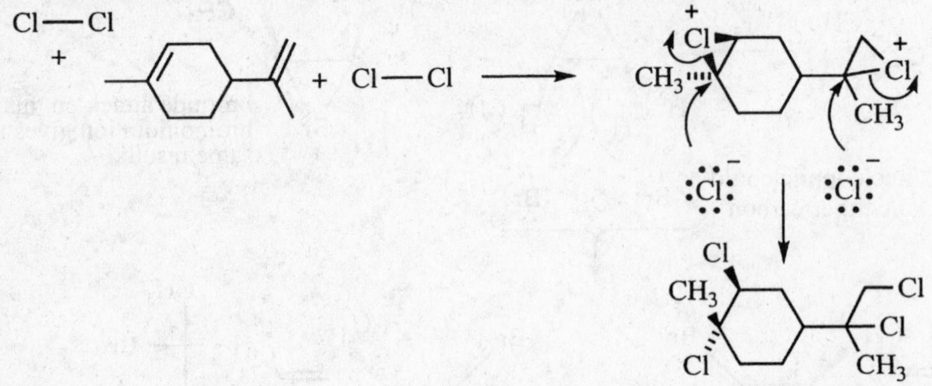

8-19 The *trans* product results from water attacking the bromonium ion from the face opposite the bromine. Equal amounts of the two enantiomers result from the equal probability that water will attack either C-1 or C-2.

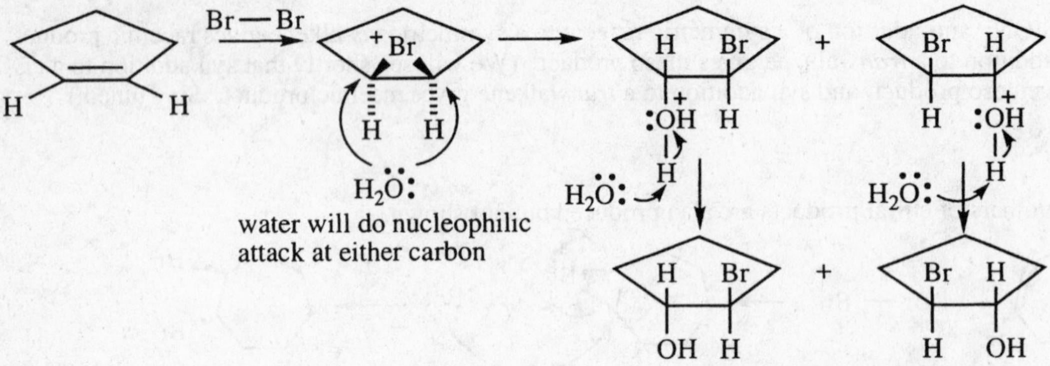

water will do nucleophilic attack at either carbon

equal amounts of enantiomers

8-20

from Solved Problem 8-5

the bromonium ion shown here is the enantiomer of the one shown in Solved Problem 8-5

enantiomer of product in Solved Problem 8-5

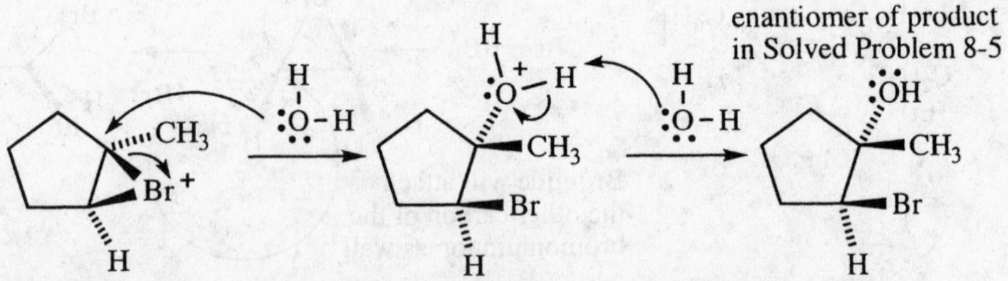

from Solved Problem 8-6: the bromonium ion shown is meso as it has a plane of symmetry; attack by the nucleophile at the other carbon from what is shown in the text will create the enantiomer

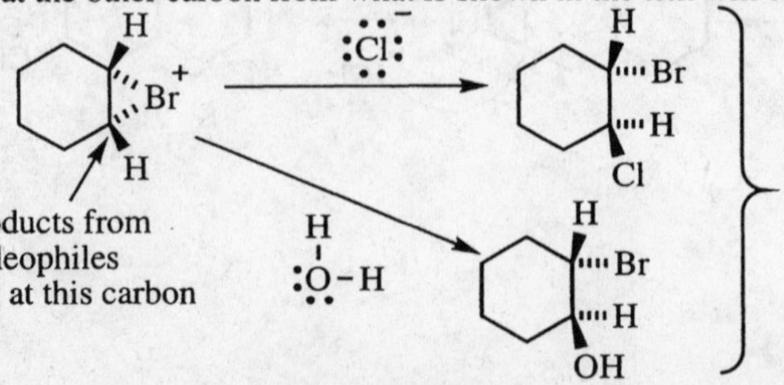

show products from both nucleophiles attacking at this carbon

these are the enantiomers of the structures shown in the text

166

8-21 The chiral products shown here will be racemic mixtures.

(a) CH₃ OH

Cl

plus enantiomer

(b)

HO Br

(c) H₃C H Cl

HO CH₃ H

plus enantiomer

(d) H₃C H Cl

HO H CH₃

plus enantiomer

(e) CH₃ OH

Br

+

CH₃ Cl

Br

plus enantiomers

8-22

(a)

$\xrightarrow[\text{H}_2\text{O}]{\text{Cl}_2}$

Cl OH

(b)

Cl

$\xrightarrow[\Delta]{\text{KOH}}$

$\xrightarrow[\text{H}_2\text{O}]{\text{Cl}_2}$

Cl OH

(c) CH₃ OH

$\xrightarrow[\Delta]{\text{H}_2\text{SO}_4}$

CH₃

$\xrightarrow[\text{H}_2\text{O}]{\text{Cl}_2}$

CH₃ OH Cl

8-23

(a)

(b)

(c)

(d)

8-24 Limonene, $C_{10}H_{16}$, has three elements of unsaturation. Upon catalytic hydrogenation, the product, $C_{10}H_{20}$, has one element of unsaturation. Two elements of unsaturation have been removed by hydrogenation—these must have been pi bonds, either two double bonds or one triple bond. The one remaining unsaturation must be a ring. Thus, limonene must have one ring and either two double bonds or one triple bond. (The structure of limonene is shown in the text in Problem 8-23(d), and the hydrogenation product is shown above in the solution to 8-23(d).)

8-25 The BINAP ligand is an example of a conformationally hindered biphenyl as described in text section 5-9A and Figure 5-17. The groups are too large to permit rotation around the single bond connecting the rings, so the molecules are locked into one chiral twist or its mirror image.

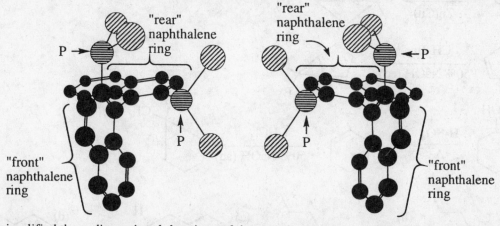

These simplified three-dimensional drawings of the enantiomers show that the two naphthalene rings are twisted almost perpendicular to each other, and the large —P(Ph)₂ substituents prevent interconversion of these mirror images.

167

8-26 Methylene inserts into the circled bonds.

(a)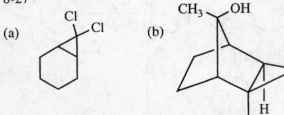

(b)

(c)

8-27

(a)

(b)

(c)

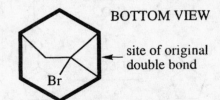

BOTTOM VIEW

site of original
double bond

Br

*the original 6-membered ring
is shown in bold bonds*

SIDE VIEW

Br

8-28

(a)

$$\xrightarrow{\substack{CH_2I_2 \\ Zn(Cu)}}$$

(b)

$$\xrightarrow{\substack{CH_2Br_2 \\ 50\% \; NaOH \; (aq)}}$$

H

Br

(c)

OH

$$\xrightarrow{\substack{H_2SO_4 \\ \Delta}}$$

$$\xrightarrow{\substack{CHCl_3 \\ 50\% \; NaOH \; (aq)}}$$

Cl
Cl

8-29

(a)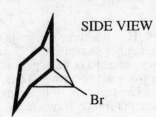

(b)

(c)

H

O

H

(d)

H

O
H

168

8-30

(a)

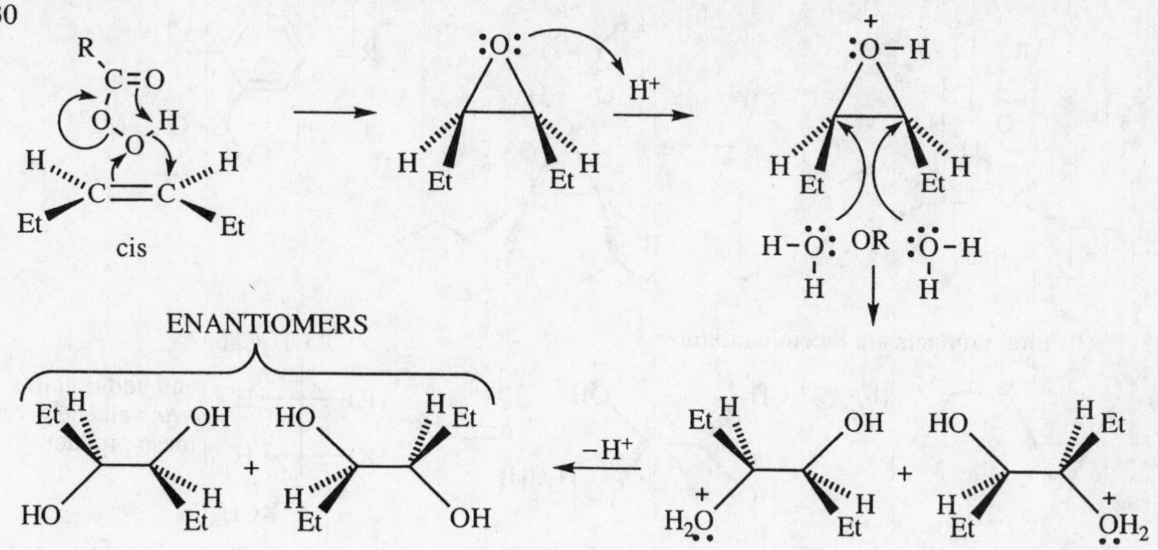

ENANTIOMERS

(b)

IDENTICAL—MESO

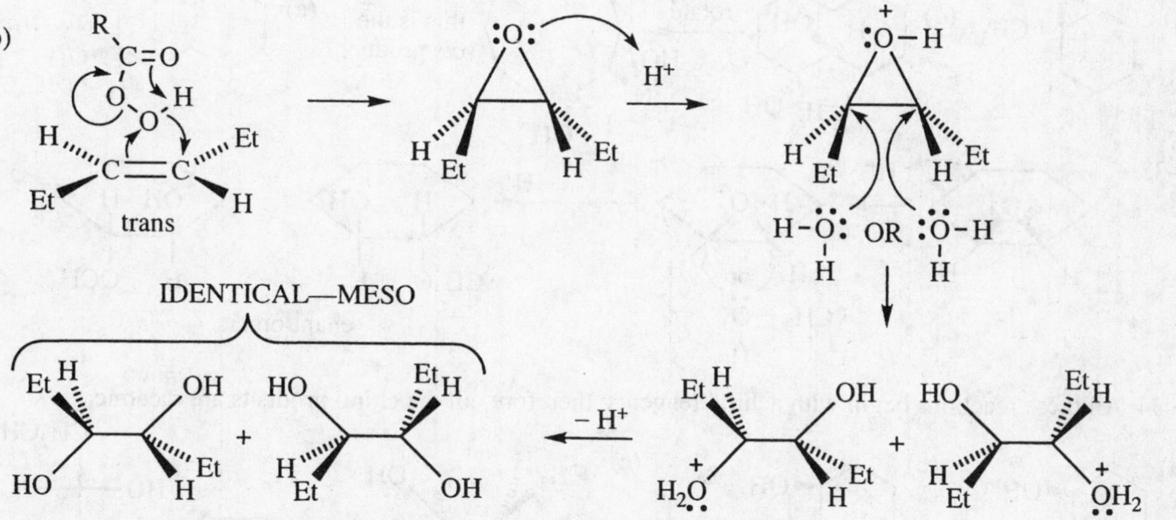

stereochemistry shown in Newman projections:

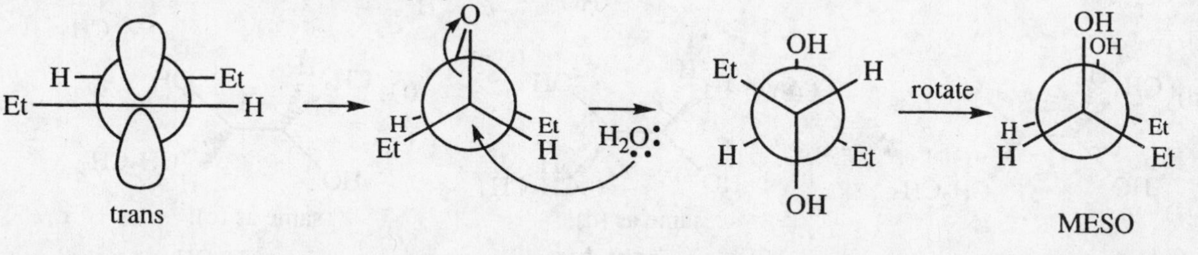

Remember the lesson from Problem 8-17: anti addition of a symmetric reagent to a symmetric *cis*-alkene gives racemic product, while anti addition to a *trans*-alkene gives meso product. This fits the definition of a *stereospecific* reaction, where different stereoisomers of the starting material (*cis* and *trans*) are converted into different stereoisomers of product (a *dl-pair* and *meso form*).

8-31

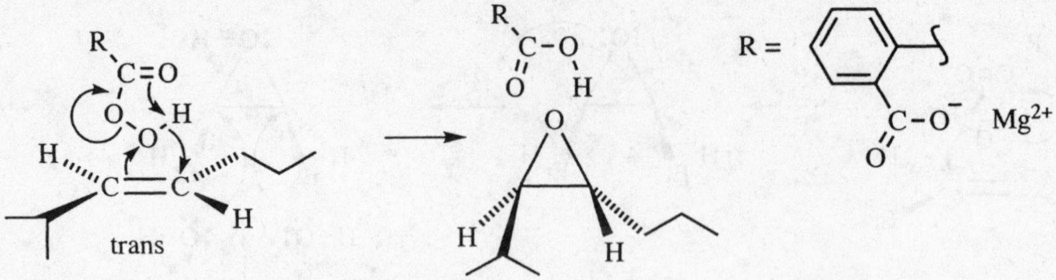

8-32 All chiral products are racemic mixtures.

(a)

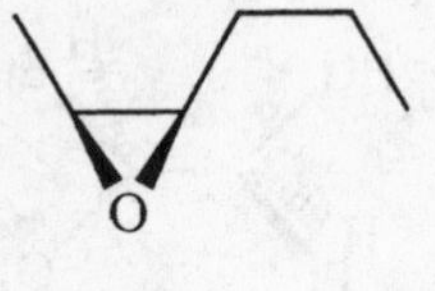

(b)

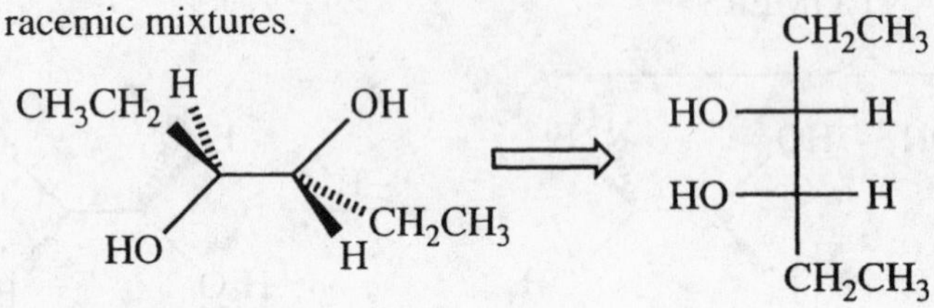

anti addition to a *trans*-alkene gives meso product

(c)

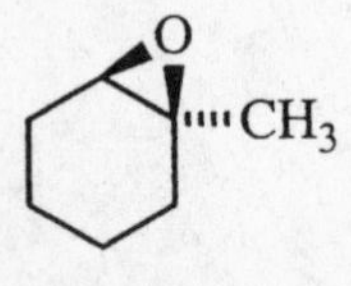

(d)

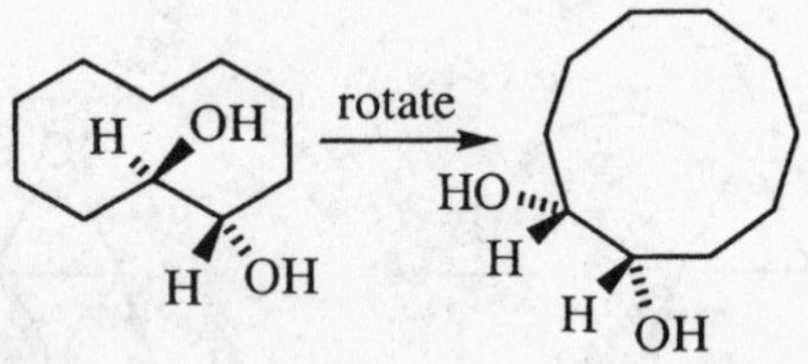

rotate

this is the *cis* product

(e)

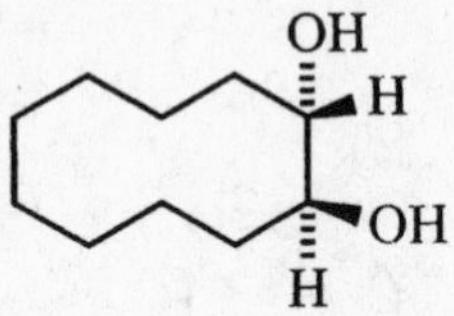

trans

8-33

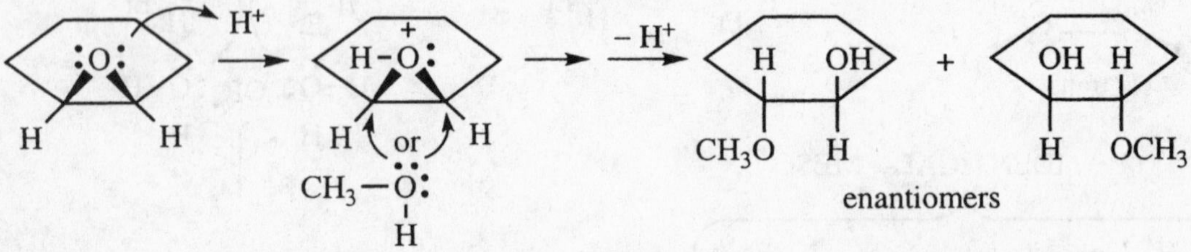

enantiomers

8-34 All these reactions begin with achiral reagents; therefore, all the chiral products are racemic.

(a)

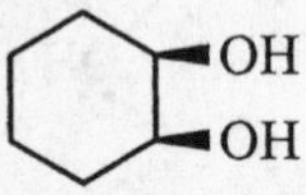

(b)

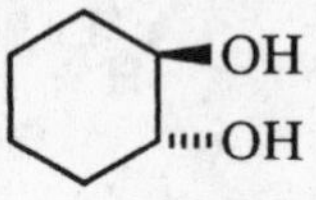

(c)

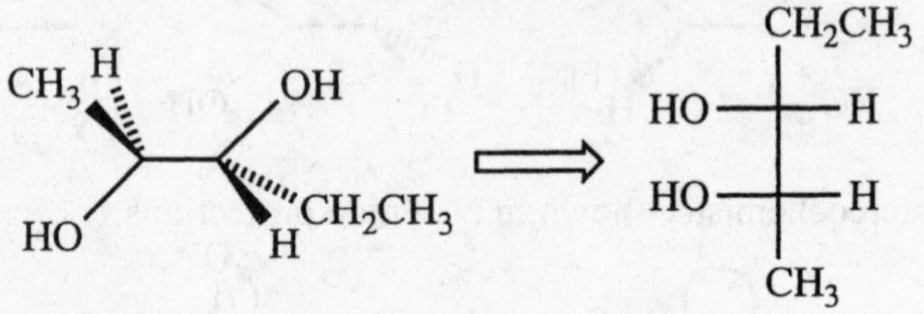

(d)

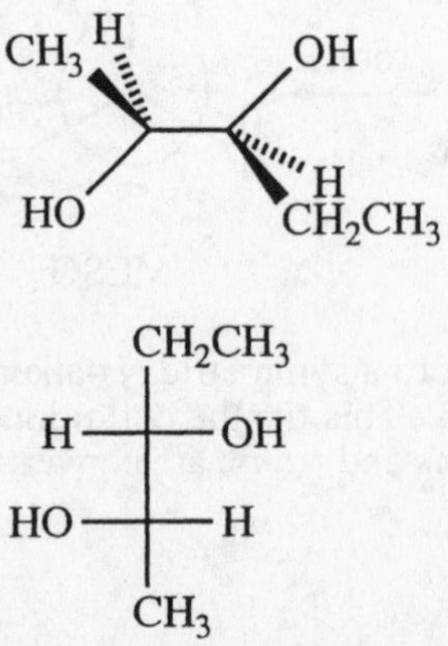

(e)

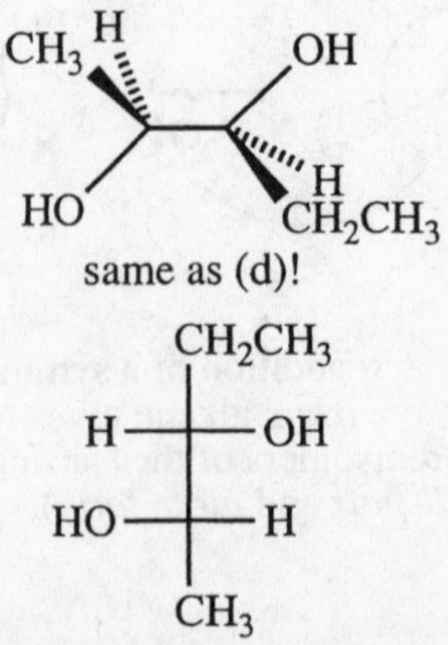

same as (d)!

(f)

same as (c)!

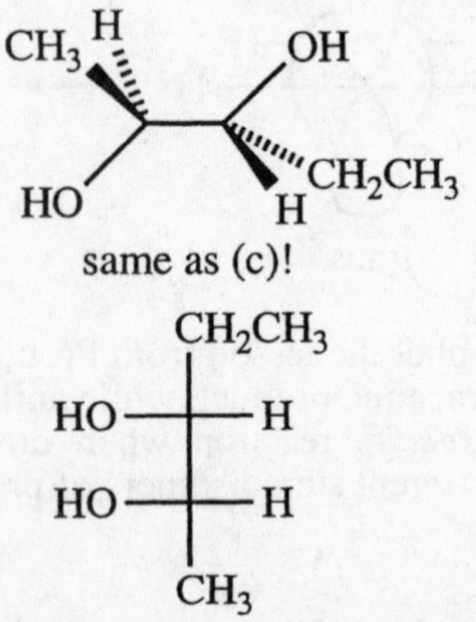

Refer to the observation in the solution to Problem 8-35 on the next page.

(a)

$$\xrightarrow[\text{H}_2\text{O}_2]{\text{OsO}_4}$$

HO   OH   meso

(b)

$$\xrightarrow[\text{H}_2\text{O}]{\text{CH}_3\text{CO}_3\text{H}}$$

HO   OH   racemic (*d*,*l*)

(c)

$$\xrightarrow[\text{H}_2\text{O}]{\text{CH}_3\text{CO}_3\text{H}} \xrightarrow{\text{rotate}}$$

HO   OH   meso

(d)

$$\xrightarrow[\text{H}_2\text{O}_2]{\text{OsO}_4} \xrightarrow{\text{rotate}}$$

HO   OH   racemic (*d*,*l*)

Have you noticed yet? For symmetric alkenes and symmetric reagents (addition of two identical X groups):

cis-alkene + **syn** addition → meso

cis-alkene + **anti** addition → racemic

trans-alkene + **syn** addition → racemic

trans-alkene + **anti** addition → meso

Assume that cis/syn/meso are "same", and trans/anti/racemic are "opposite". Then any combination can be predicted, just like math!

$$+1 \times +1 = +1$$
$$+1 \times -1 = -1$$
$$-1 \times +1 = -1$$
$$-1 \times -1 = +1$$

8-36 Solve these ozonolysis problems by working backwards, that is, by "reattaching" the two carbons of the new carbonyl groups into alkenes. Here's a hint. When you cut a circular piece of string, you still have only one piece. When you cut a linear piece of string, you have two pieces. Same with molecules. If ozonolysis forms only one product with two carbonyls, the alkene had to have been in a ring. If ozonolysis gives two molecules, the alkene had to have been in a chain.

(a) two carbonyls from ozonolysis are in a chain, so alkene had to have been in a ring

(b) two carbonyls from ozonolysis are in two different products, so alkene had to have been in a chain, not a ring

(c) two carbonyls from ozonolysis are in two different products, so alkene had to have been in a chain, not a ring

E or *Z* of alkene cannot be determined from products

8-37

(a)

$$\xrightarrow{\text{O}_3} \xrightarrow{\text{Me}_2\text{S}}$$

+

(b)

$$\xrightarrow[\Delta]{\text{KMnO}_4}$$

+

8-37 continued

(c)

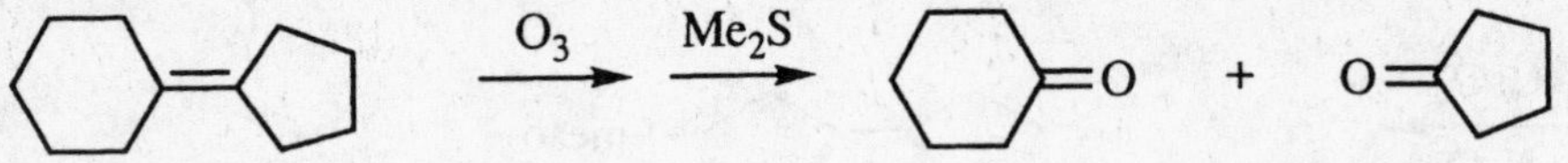

$$\xrightarrow{O_3} \quad \xrightarrow{Me_2S}$$

(d)

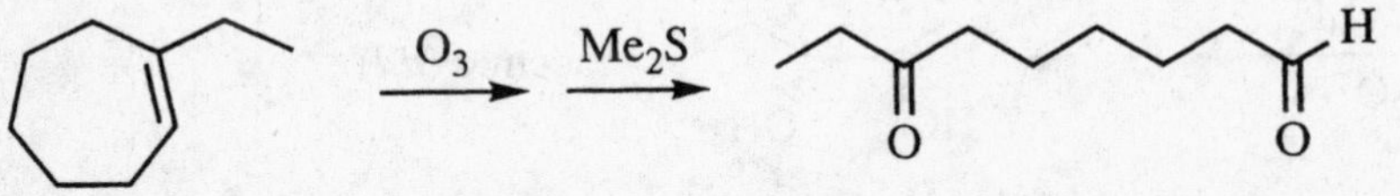

$$\xrightarrow{O_3} \quad \xrightarrow{Me_2S}$$

(e)

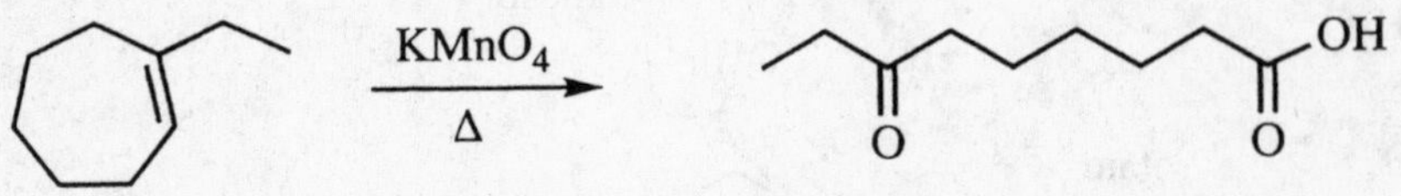

$$\xrightarrow[\Delta]{KMnO_4}$$

(f)

$$\xrightarrow[cold]{KMnO_4}$$

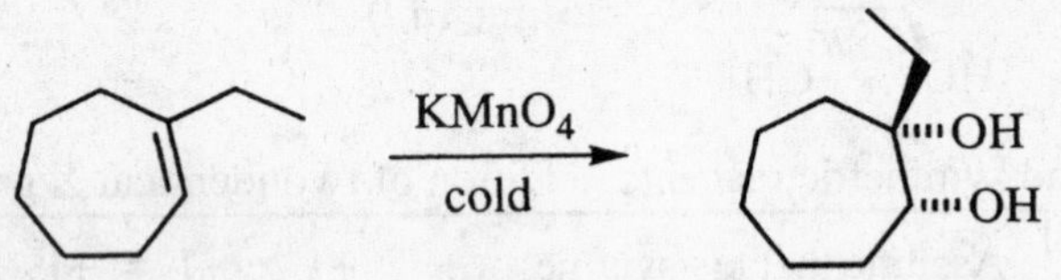

8-38 The representation for a generic acid will be H—B, where B is the conjugate base.

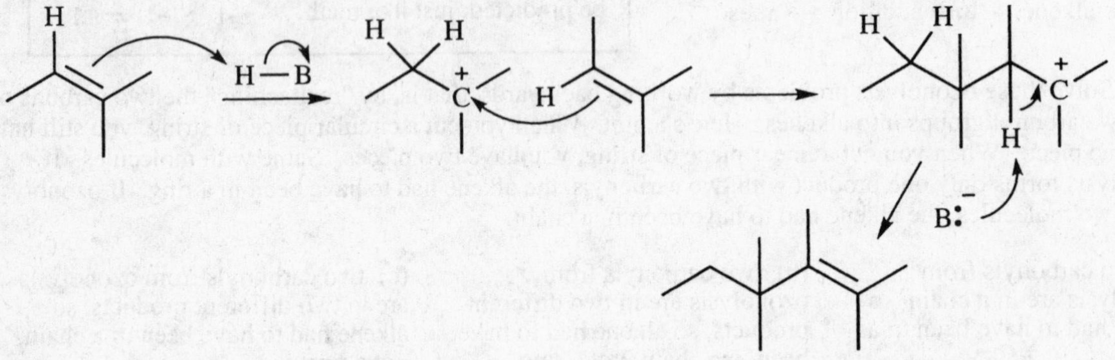

8-39 Catalytic BF_3 reacts with trace amounts of water to form the probable catalyst:

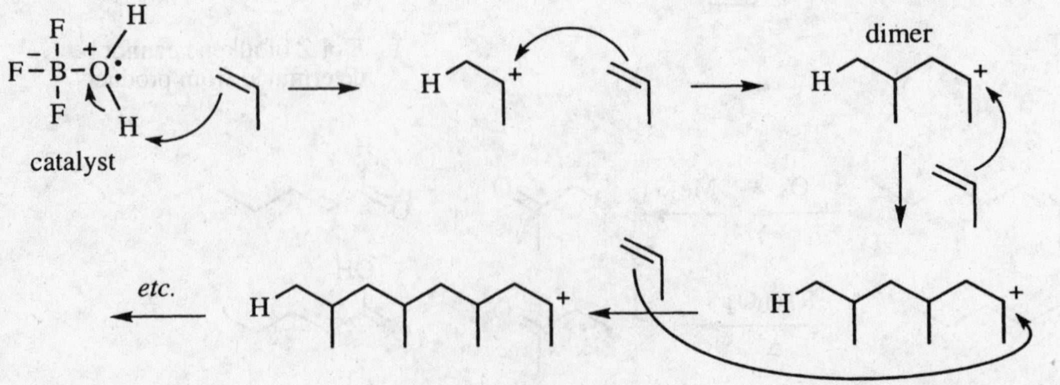

catalyst

dimer

tetramer

trimer

etc.

172

8-40 H—B is used here to symbolize a generic acid.

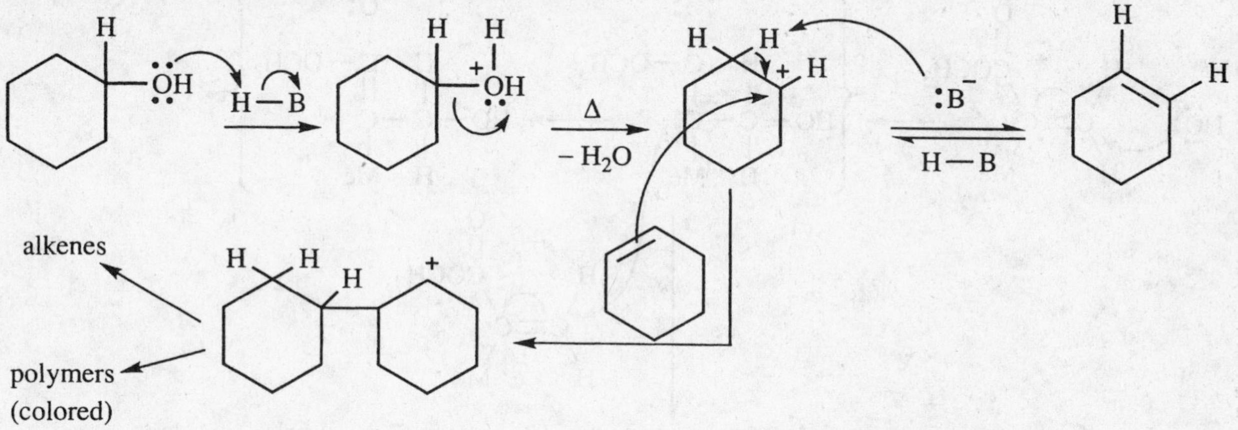

alkenes

polymers
(colored)

Note: the dashed bond symbol is used here to indicate the continuation of a polymer chain.

8-41

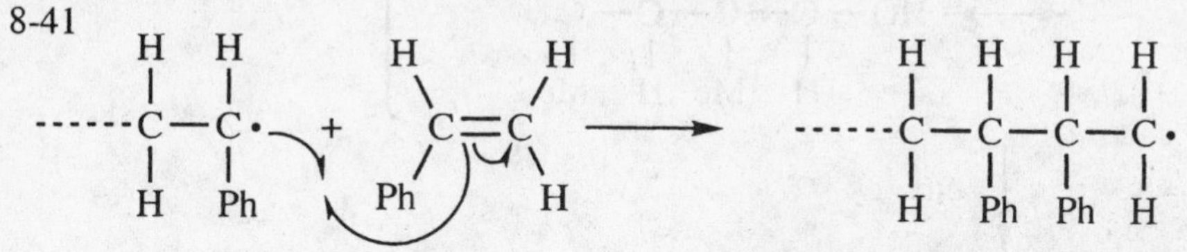

1° radical, and *not* resonance-stabilized—
this orientation is not observed

Orientation of addition always generates the more stable intermediate; the energy difference between a 1° radical (shown above) and a benzylic radical is huge. The phenyl substituents must necessarily be on alternating carbons because the orientation of attack is always the same—not a random process.

8-42 For clarity, the new bonds formed in this mechanism are shown in bold.

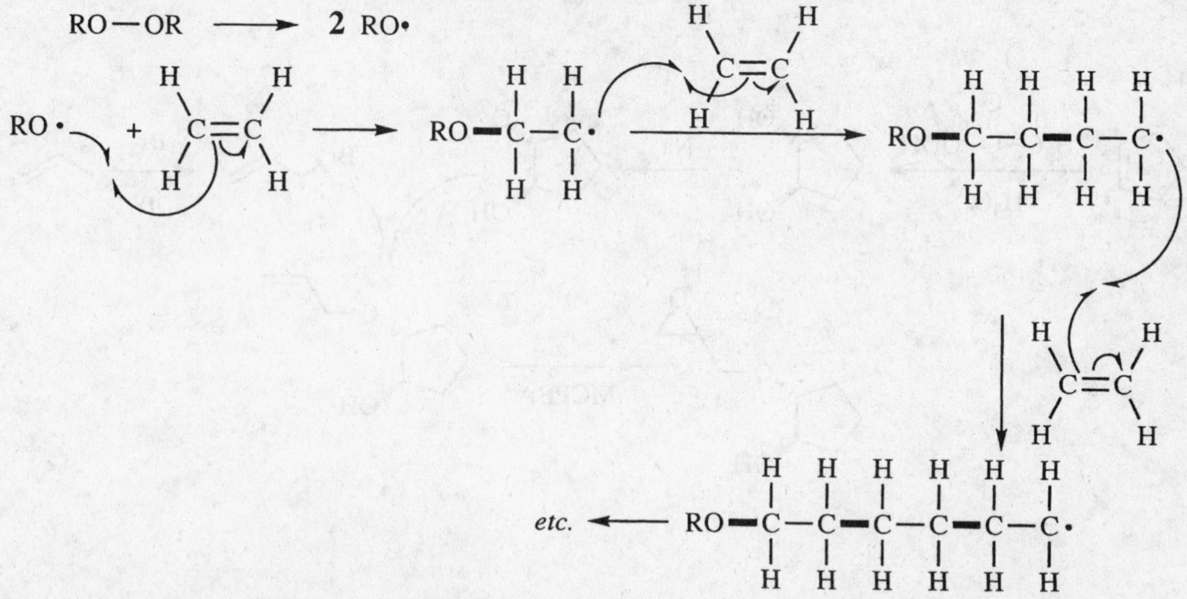

RO—OR $\longrightarrow$ **2** RO•

etc.

173

8-43

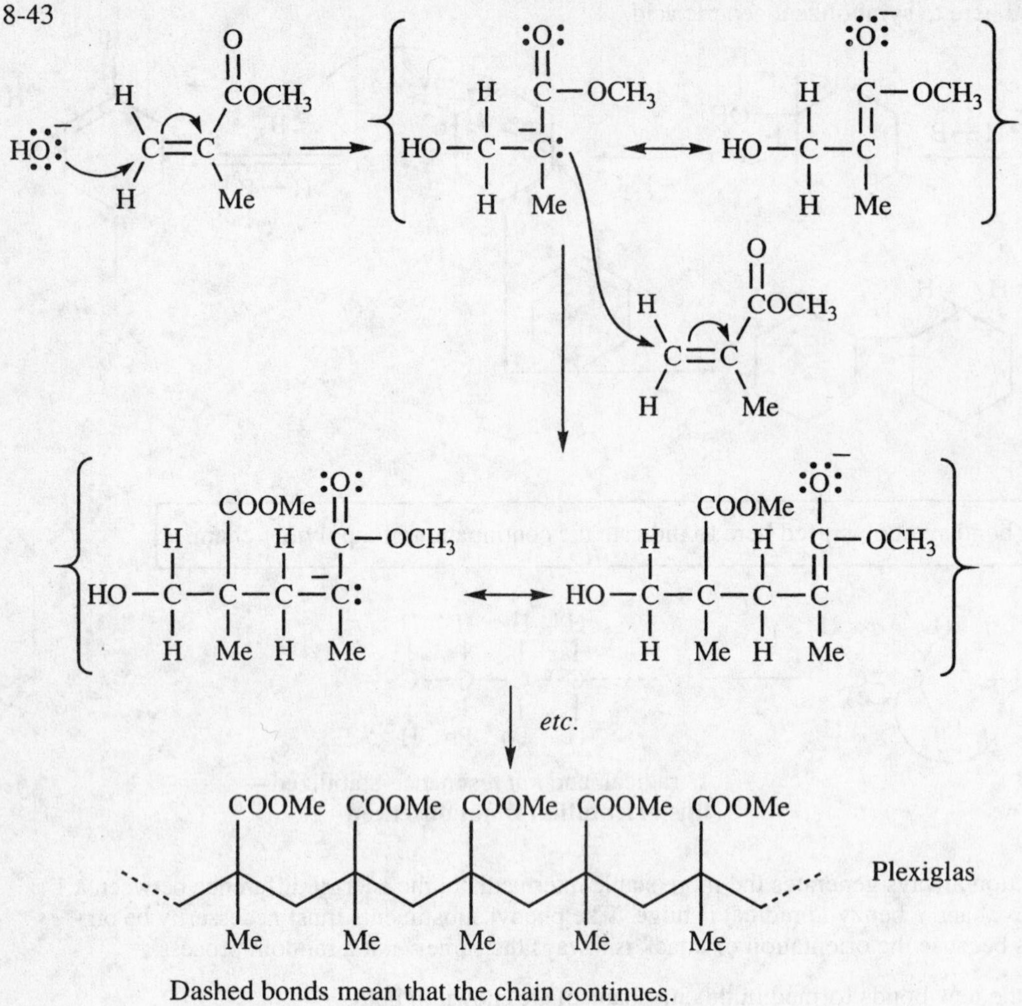

Dashed bonds mean that the chain continues.

8-44

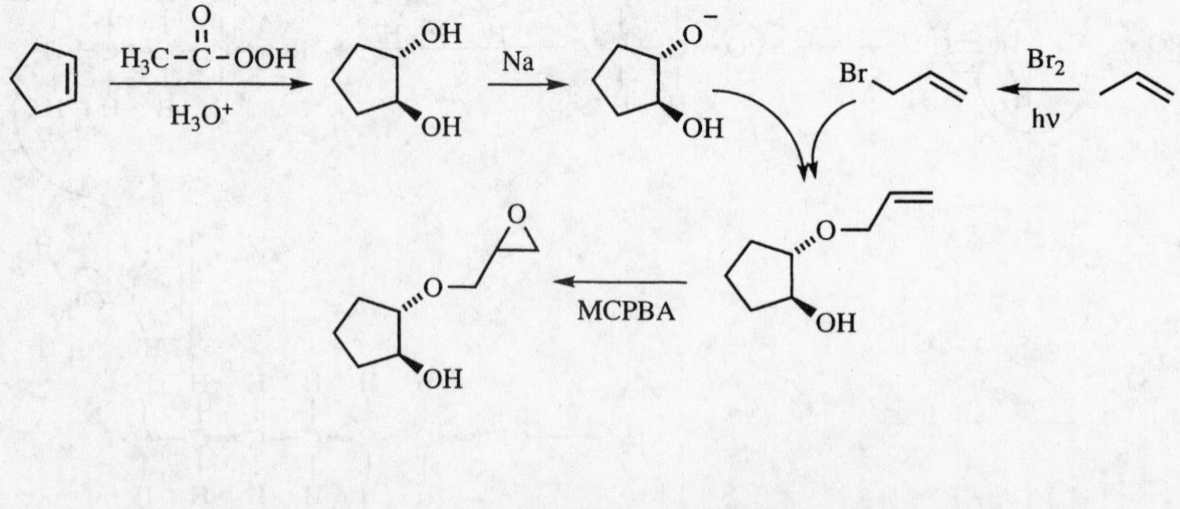

174

8-45 In the spirit of this problem, all starting materials will have six carbons or fewer and only one carbon-carbon double bond. Reagents may have other atoms.

(a)

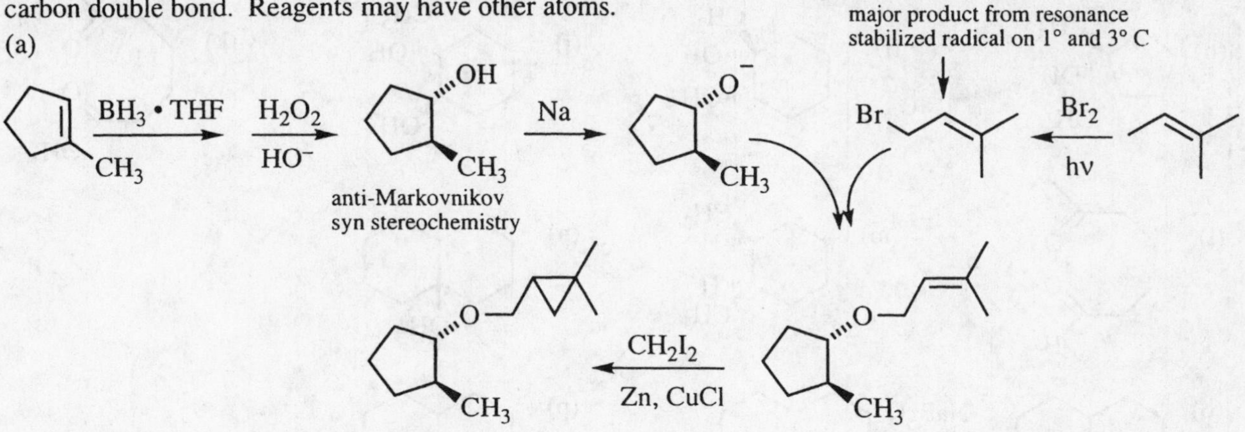
major product from resonance stabilized radical on 1° and 3° C
BH3 • THF
H2O2
HO−
OH
CH3
anti-Markovnikov
syn stereochemistry
Na
O−
CH3
Br
Br2
hν
O
CH3
CH2I2
Zn, CuCl
O
CH3

(b)

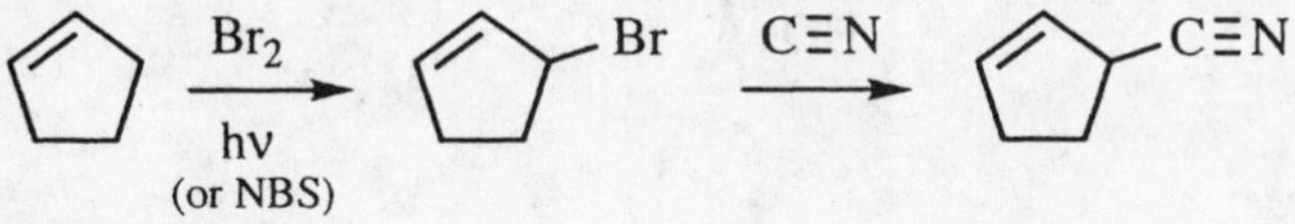
Br2
hν
(or NBS)
Br
−C≡N
C≡N

(c)

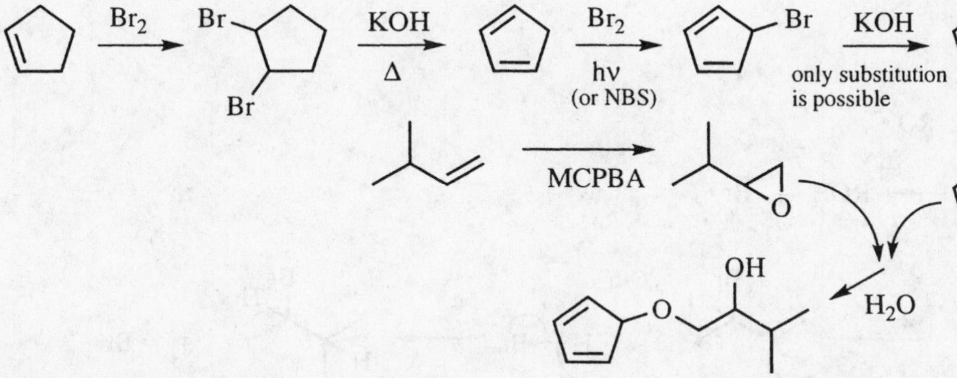
Br2
Br
Br
KOH
Δ
Br2
hν
(or NBS)
Br
KOH
only substitution
is possible
OH
Na
O−
MCPBA
O
H2O
OH
O

8-46 Please refer to solution 1-20, page 12 of this Solutions Manual.

8-47

(a) 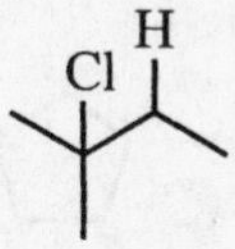
H
Cl

(b) 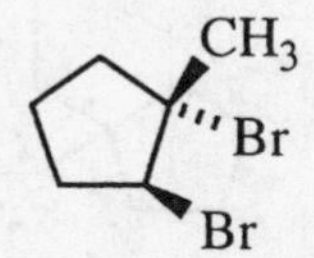
CH3
Br
Br

(c) 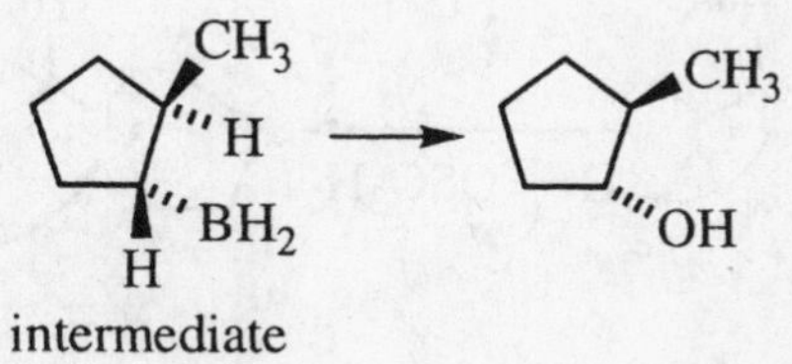
CH3
H
BH2
H
intermediate
CH3
OH

(d) 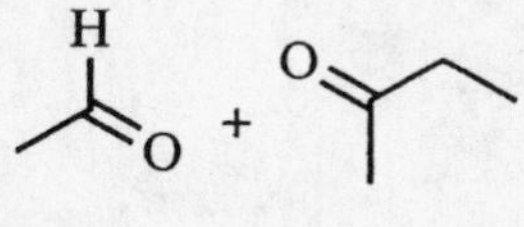
H
O
+
O

(e) 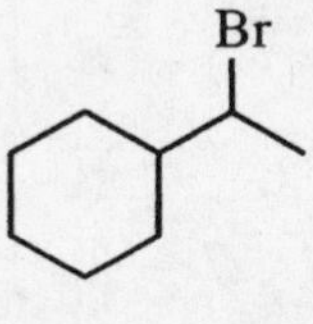
Br

(f) 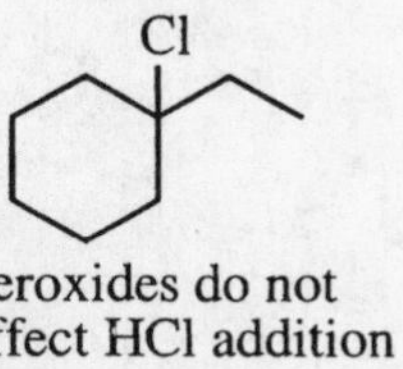
Cl
peroxides do not
affect HCl addition

(g) 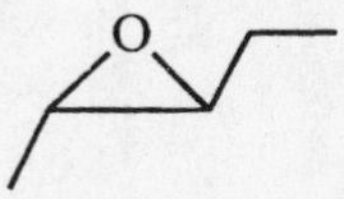
O

8-47 continued

(h) 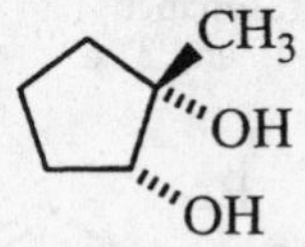

(i) 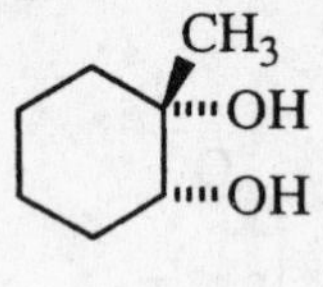

(j) 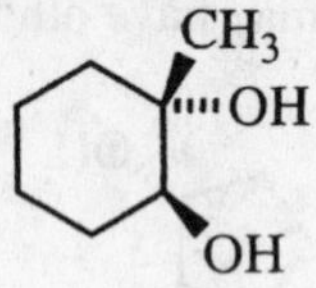

(k) 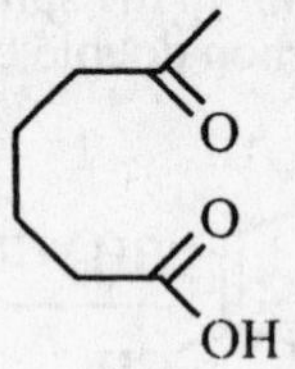

(l) 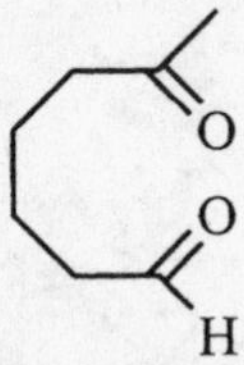

(m) 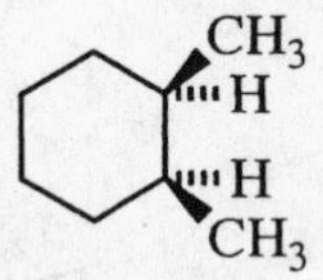

(n) 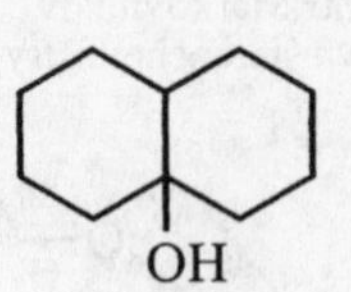

(o) 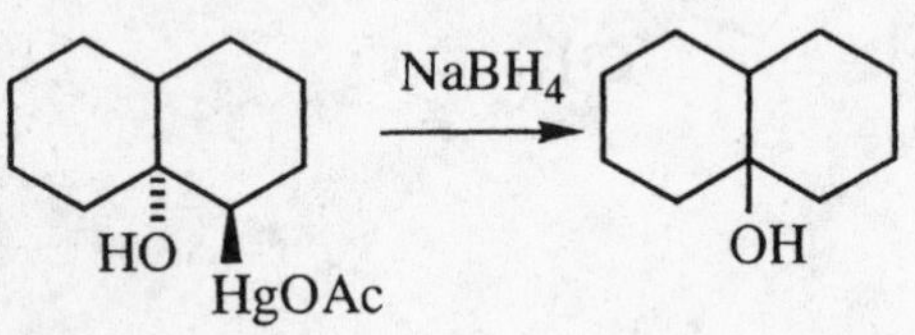

intermediate

(p) 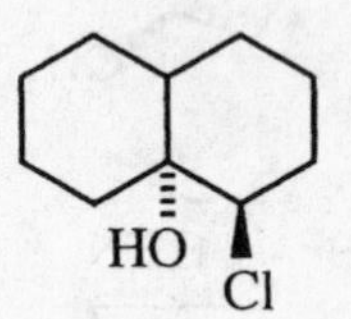

8-48

(a)

initiation

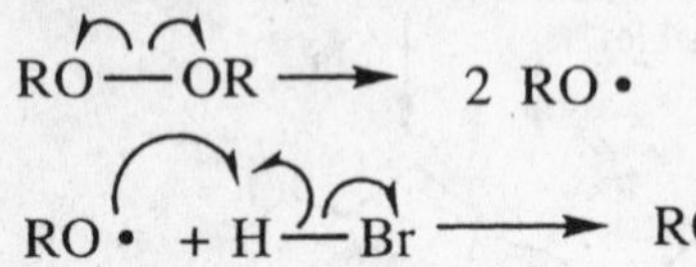

propagation

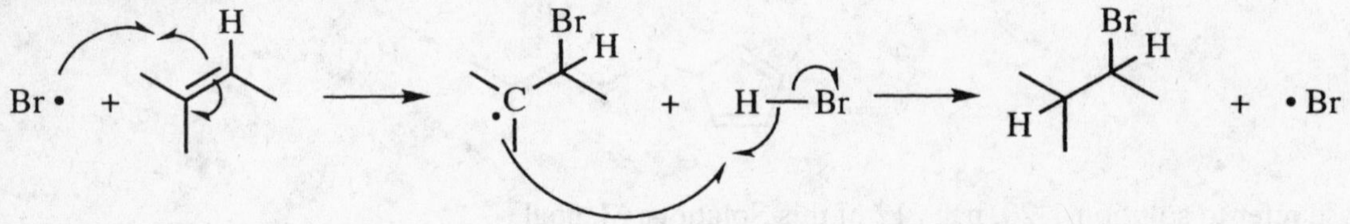

(b) 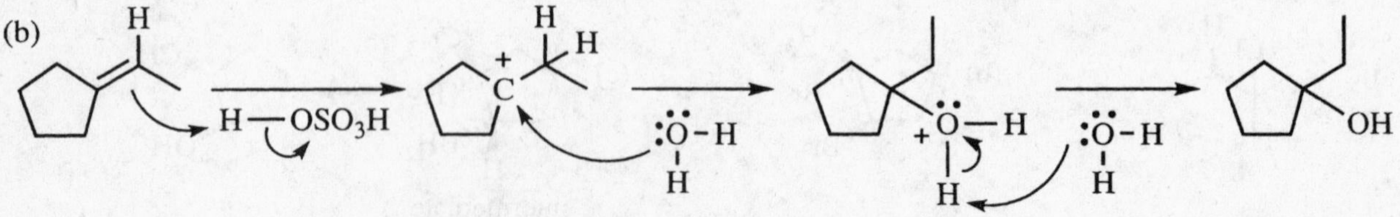

8-48 continued

(c)

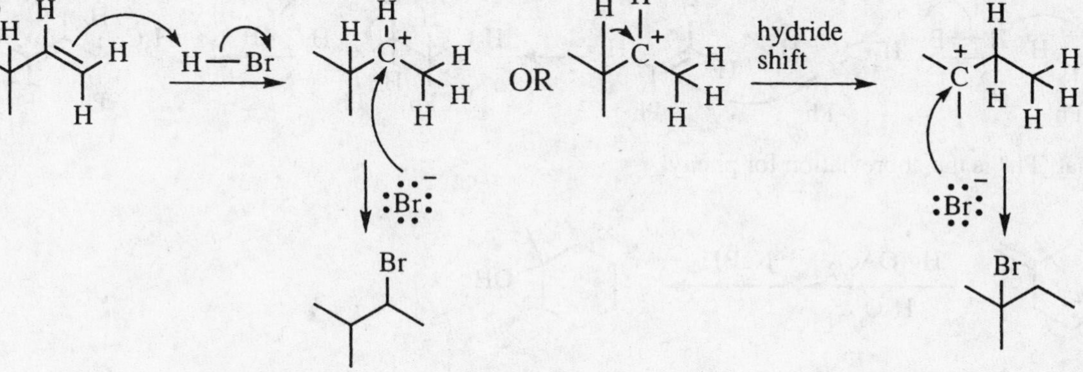

(d)

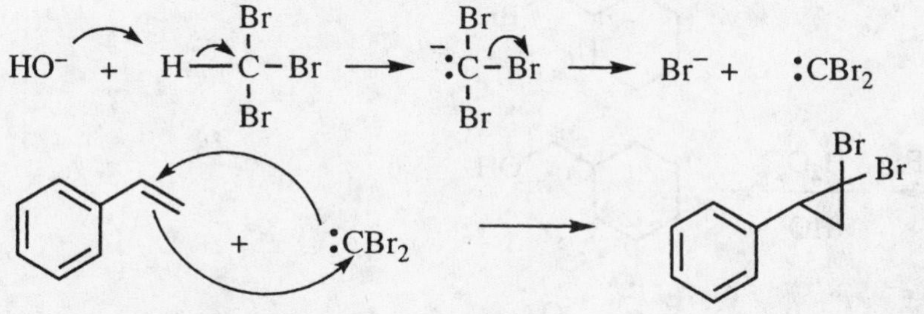

(e)

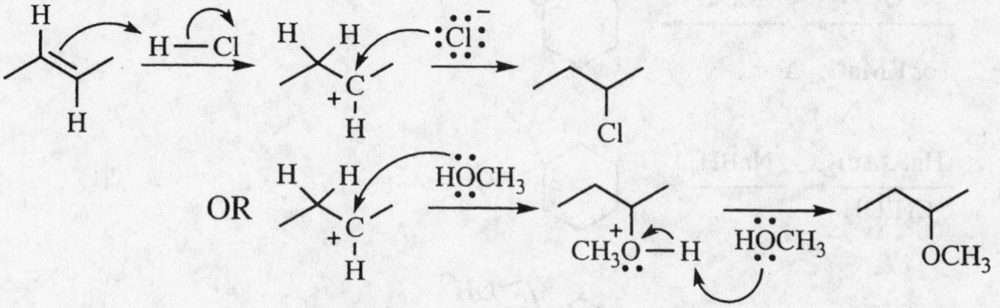

(f)

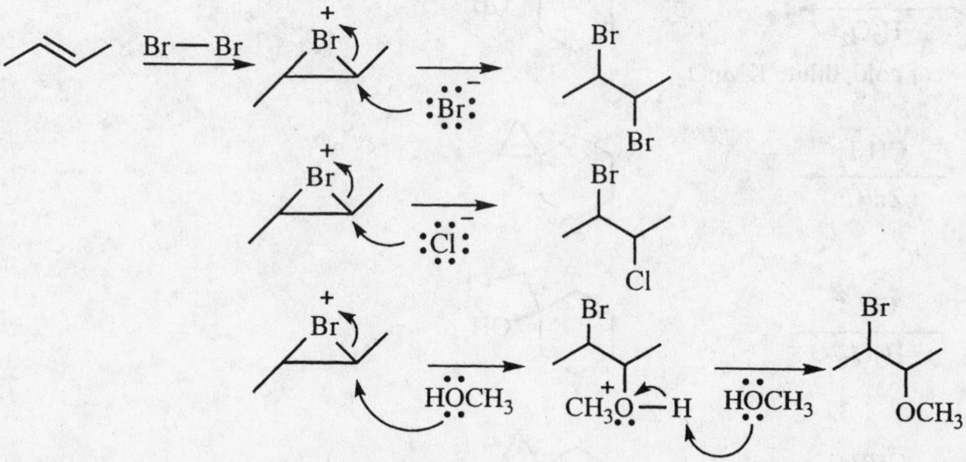

8-48 continued

(g)

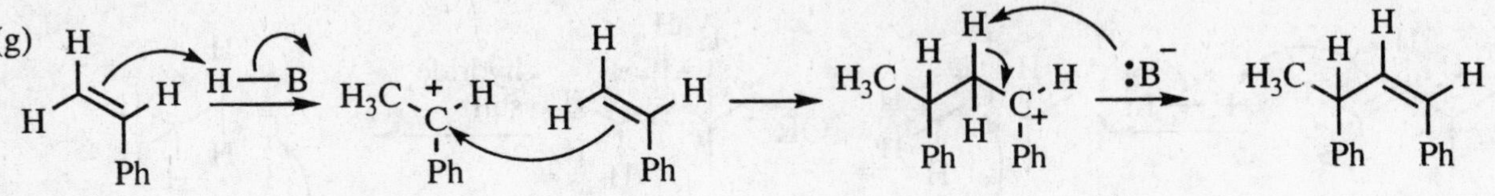

Recall that "Ph" is the abbreviation for phenyl.

8-49

(a)

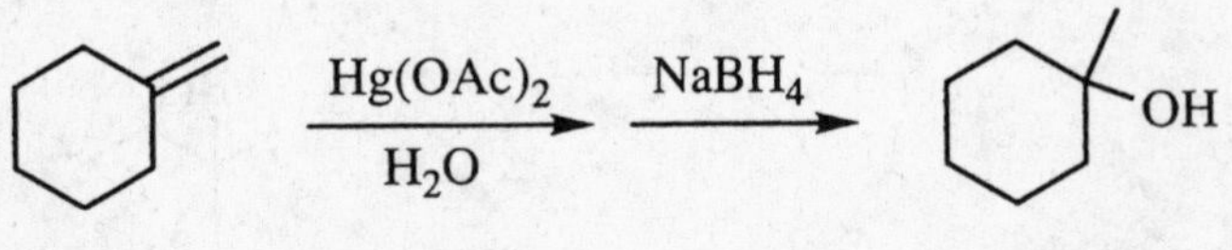

$$\text{(cyclohexane ring with =CH}_2\text{)} \xrightarrow[\text{H}_2\text{O}]{\text{Hg(OAc)}_2} \xrightarrow{\text{NaBH}_4} \text{(1-methylcyclohexanol)} \text{—OH}$$

(b)

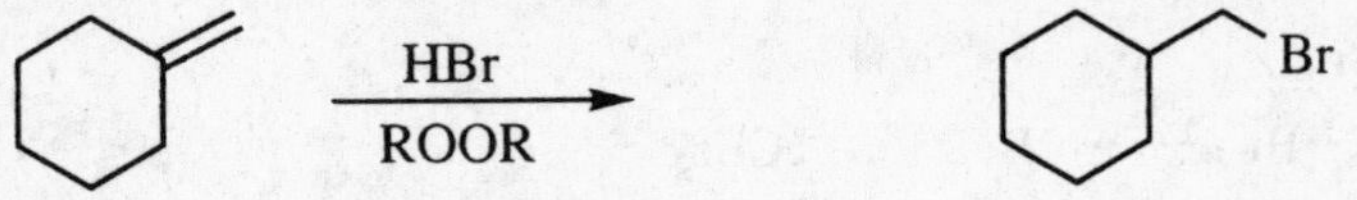

$$\xrightarrow[\text{ROOR}]{\text{HBr}} \text{(cyclohexylmethyl bromide)} \text{—Br}$$

(c)

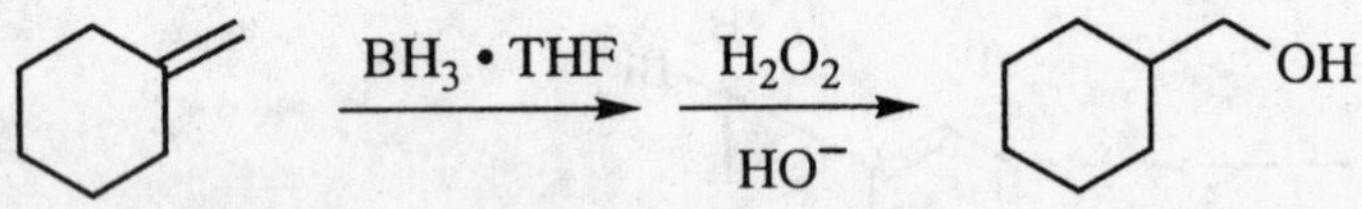

$$\xrightarrow{\text{BH}_3 \cdot \text{THF}} \xrightarrow[\text{HO}^-]{\text{H}_2\text{O}_2} \text{(cyclohexylmethanol)} \text{—OH}$$

(d)

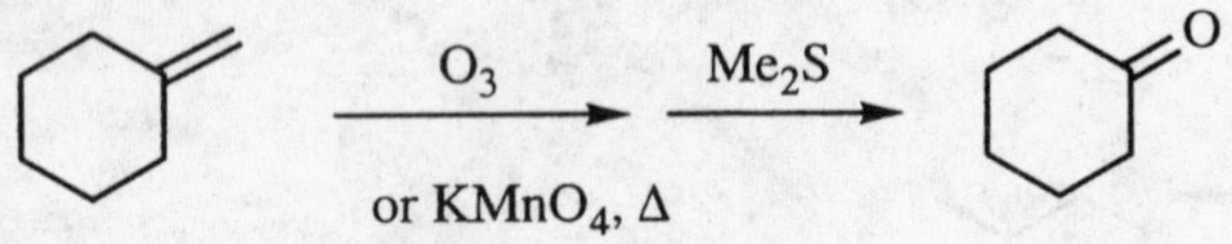

$$\xrightarrow{\text{O}_3} \xrightarrow{\text{Me}_2\text{S}} \text{(cyclohexanone)}$$

or KMnO$_4$, Δ

(e)

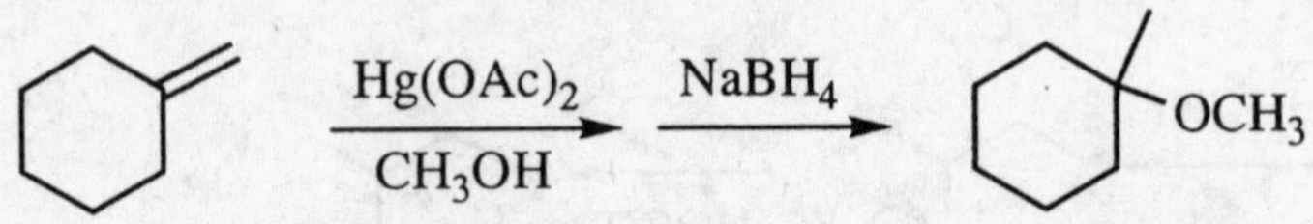

$$\xrightarrow[\text{CH}_3\text{OH}]{\text{Hg(OAc)}_2} \xrightarrow{\text{NaBH}_4} \text{—OCH}_3$$

(f)

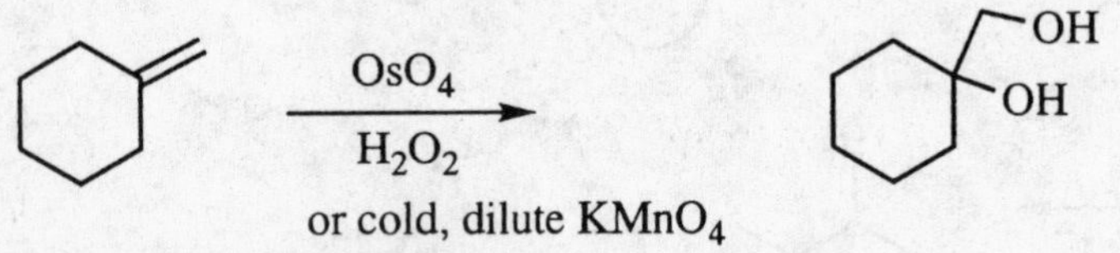

$$\xrightarrow[\text{H}_2\text{O}_2]{\text{OsO}_4} \begin{array}{l}\text{—OH}\\ \text{—OH}\end{array}$$

or cold, dilute KMnO$_4$

(g)

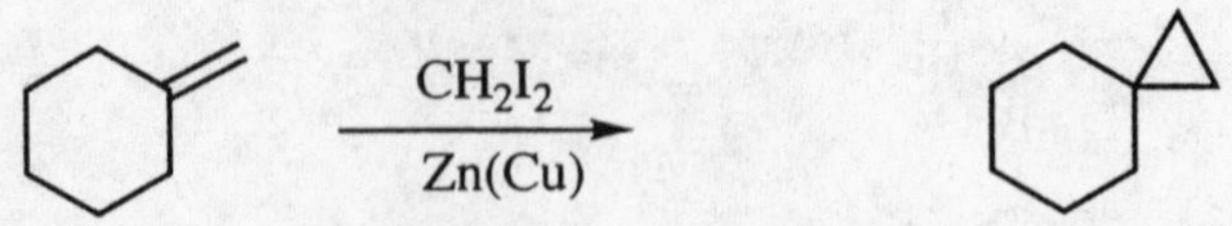

$$\xrightarrow[\text{Zn(Cu)}]{\text{CH}_2\text{I}_2} \text{(spiro cyclopropane)}$$

(h)

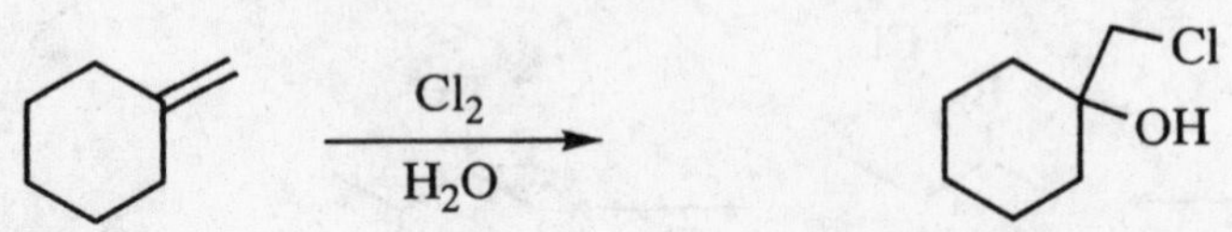

$$\xrightarrow[\text{H}_2\text{O}]{\text{Cl}_2} \begin{array}{l}\text{—Cl}\\ \text{—OH}\end{array}$$

(i)

$$\xrightarrow[\text{50\% NaOH (aq)}]{\text{CHBr}_3} \begin{array}{l}\text{—Br}\\ \text{Br}\end{array}$$

178

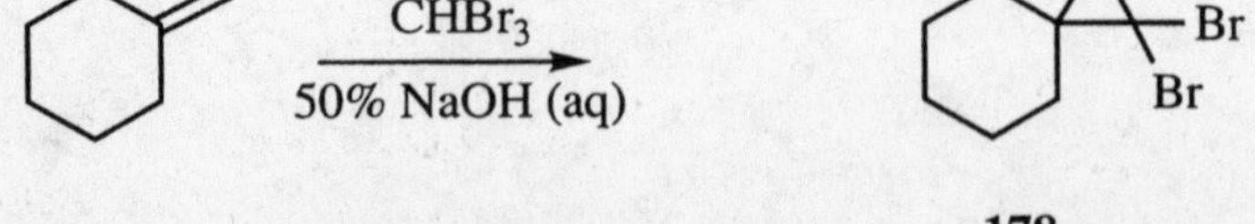

8-50

(a) [structure: cyclohexane with CH₃, HO-CH₂ and CH(CH₃)CH₂OH substituents]

(b) [structure: cyclohexane with CH₃, epoxide, and isopropyl epoxide]

(c) [structure: diketone aldehyde] + $CH_2=O$

(d) [structure: cyclohexane with HO, HO, CH₃, CH₂OH and C(CH₃)(OH)]

(e) [structure: keto acid] + CO_2

(f) [structure: cyclohexane with HO, HO, CH₃, CH₂OH and C(CH₃)OH]

(g) [structure: cyclohexane with isopropyl]

(h) [structure: cyclohexane with Br and C(CH₃)₂Br]

(i) [structure: cyclohexane with Br and CH(CH₃)CH₂Br]

(j) [structure: cyclohexane with Br, HO, CH₃ and C(CH₃)(OH)CH₂Br]

(k) [structure: cyclohexane with Cl, Cl, CH₃ and C(CH₃)(Cl)CH₂Cl]

(l) [structure: cyclohexane with CH₃O and C(CH₃)₂OCH₃]

(m) [structure: cyclohexane with CH₃ and 1-methylcyclopropyl]

8-51 Each monomer has two carbons in the backbone, so the substituents on the monomer will repeat every two carbons in the polymer. Dashed bonds indicate continuation of the polymer chain.

<u>line formula representation</u>

[structure: vinyl chloride $CH_2=CHCl$] → ---- $CH_2CHCH_2CHCH_2CHCH_2CH$ ---- (each CH bearing Cl)
polyvinyl chloride, PVC

[structure: tetrafluoroethylene $CF_2=CF_2$] → ---- backbone of $-C-C-C-C-C-C-$ each carbon bearing F, F ----
polytetrafluoroethylene, PTFE, Teflon

[structure: acrylonitrile $CH_2=CH-C≡N$] → ---- $CH_2CHCH_2CHCH_2CHCH_2CH$ ---- (each CH bearing C≡N)
polyacrylonitrile, Orlon

179

8-52 Without divinylbenzene, individual chains of polystyrene are able to slide past one another. Addition of divinylbenzene during polymerization forms bridges, or "crosslinks", between chains, adding strength and rigidity to the polymer. Divinylbenzene and similar molecules are called crosslinking agents.

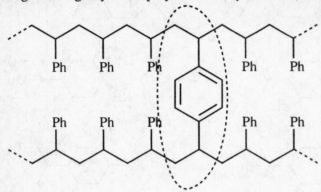

two polystyrene chains crosslinked by a divinylbenzene monomer shown in the dashed oval

8-53 A peroxy radical is shown as the initiator. Newly formed bonds are shown in bold.

8-54

$H_2C=CH$ — $\overset{O}{\underset{\|}{C}}$ — OCH_2CH_3 ethyl acrylate

8-55 In each case, the compound (boxed) that produces the more stable carbocation is more reactive.

(a)   3°

(b)   3°

(c)   allylic

8-56 Once the bromonium ion is formed, it can be attacked by either nucleophile, bromide or chloride, leading to the mixture of products.

8-57 Two possible orientations of attack of bromine radical are possible:

(A) anti-Markovnikov

$3°$ radical

(B) Markovnikov

$1°$ radical

The first step in the mechanism is endothermic and rate determining. The $3°$ radical produced in anti-Markovnikov attack (A) of bromine radical is several kJ/mole more stable than the $1°$ radical generated by Markovnikov attack (B). The Hammond Postulate tells us that it is reasonable to assume that the activation energy for anti-Markovnikov addition is lower than for Markovnikov addition. This defines the first half of the energy diagram.

The relative stabilities of the final products are somewhat difficult to predict. (Remember that stability of final products does not necessarily reflect relative stabilities of intermediates; this is why a thermodynamic product can be different from a kinetic product.) From bond dissociation energies (kJ/mole) in Table 4-2:

anti-Markovnikov		Markovnikov	
H to $3°$ C	381	H to $1°$ C	410
Br to $1°$ C	285	Br to $3°$ C	272
	666 kJ/mole		682 kJ/mole

If it takes more energy to break bonds in the Markovnikov product, it must be lower in energy, therefore, more stable—OPPOSITE OF STABILITY OF THE INTERMEDIATES!

Now we are ready to construct the energy diagram; see the next page.

8-57 continued

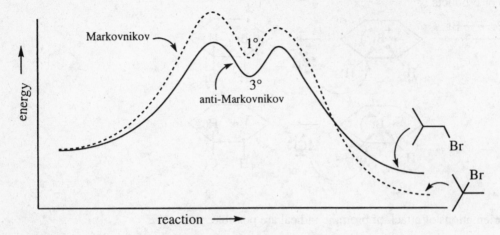

It is the anti-Markovnikov product that is the kinetic product, not the thermodynamic product; the anti-Markovnikov product is obtained since its rate-determining step has the lower activation energy.

8-58 Recall these facts about ozonolysis: each alkene cleaved by ozone produces two carbonyl groups; an alkene in a chain produces two separate products; an alkene in a ring produces one product in which the two carbonyls are connected.

(a) [structure with O, H, O, H, O] + $CH_2=O$

(b) [structure with H, O] + [structure with O]

(c) [bicyclic structure with O, H, H, O] = [cyclohexane with CHO, CHO]

(d) [structure with O, H, O]

8-59

(a) [cyclohexene with CH₃] $\xrightarrow[\text{H}_2\text{O}]{\text{CH}_3\text{CO}_3\text{H}}$ [cyclohexane with CH₃, OH, OH]

(b) [cyclooctene] $\xrightarrow[\text{H}_2\text{O}_2]{\text{OsO}_4}$ [cyclooctane with OH, OH] $\xleftarrow[\text{H}_2\text{O}]{\text{CH}_3\text{CO}_3\text{H}}$ [cyclooctene]

(or cold, dilute $KMnO_4$)

cis + **syn**
(*cis* double bond plus
syn stereochemistry of addition)

trans + **anti**
(*trans* double bond plus
anti stereochemistry of addition)

182

(c)

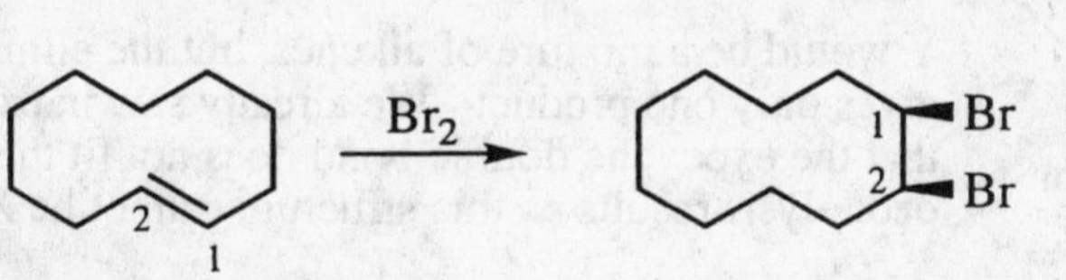

anti addition of Br_2 *requires trans* alkene to give meso product

This structure shows a *trans* alkene in a 10-membered ring, just the rotated view of the structure to the right.

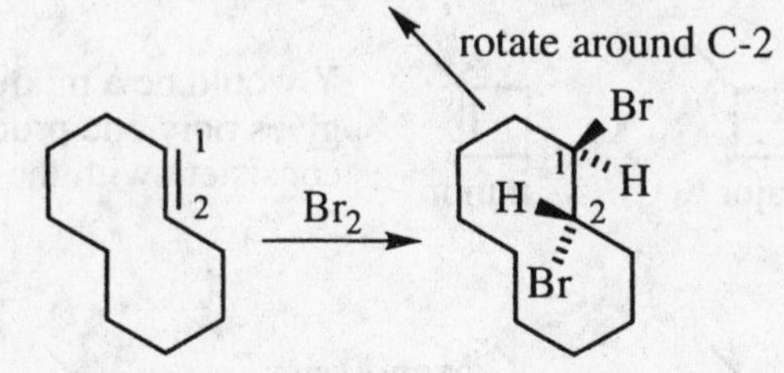

rotate around C-2

trans-cyclodecene

(d)

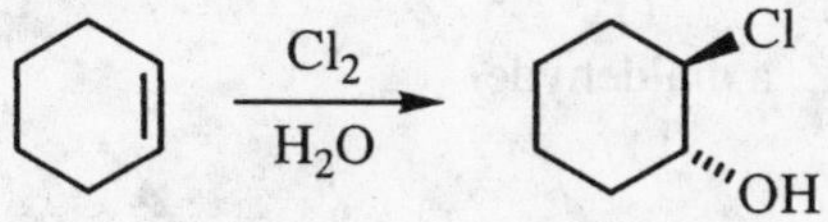

(e)

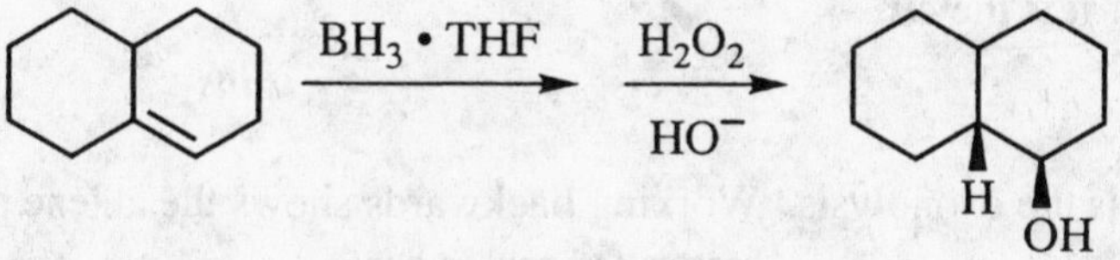

(f)

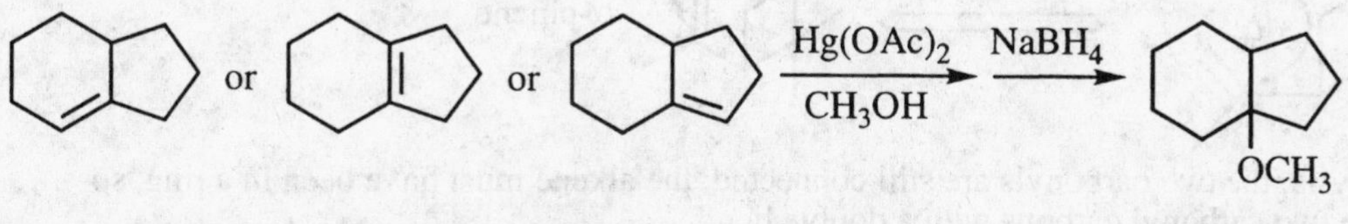

8-60

A) Unknown X, C_5H_9Br, has one element of unsaturation. X reacts with neither bromine nor $KMnO_4$, so the unsaturation in X cannot be an alkene; it must be a ring.

B) Upon treatment with strong base (*t*-butoxide), X loses H and Br to give Y, C_5H_8, which does react with bromine and $KMnO_4$; it must have an alkene and a ring. Only one isomer is formed.

C) Catalytic hydrogenation of Y gives methylcyclobutane. This is a BIG clue because it gives the carbon skeleton of the unknown. Y must have a double bond in the methylcyclobutane skeleton, and X must have a Br on the methylcyclobutane skeleton.

D) Ozonolysis of Y gives a dialdehyde Z, $C_5H_8O_2$, which contains all the original carbons, so the alkene cleaved in the ozonolysis had to be in the ring.

Let's consider the possible answers for X and see if each fits the information.

1)

if this is X

this must be Y; only one product

DOES NOT FIT OZONOLYSIS RESULTS so this cannot be X and Y

more possibilities on the next page

8-60 continued

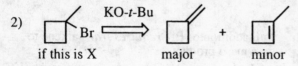

2) **if this is X** — KO-*t*-Bu ⟹ **major** + **minor**

Y would be a mixture of alkenes, but the elimination gives only one product. We already saw in example 1 that the exocyclic double bond does not fit the ozonolysis results so this structure cannot be X.

3) **if this is X** — KO-*t*-Bu ⟹ **major** + **minor**

Y would be a mixture of alkenes, but the elimination gives only one product, so this structure of X is not consistent with the information provided.

4) **if this is X** — KO-*t*-Bu ⟹ SAME COMPOUND; this must be Y; only one product — ozonolysis ⟹ **Z, a dialdehyde**

The correct structures for X, Y, and Z are given in the fourth possibility. The only structural feature of X that remains undetermined is whether it is the *cis* or *trans* isomer.

cis ? *trans* ?

8-61 The clue to the structure of α-pinene is the ozonolysis. Working backwards shows the alkene position.

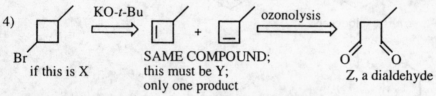

backwards — became carbonyl carbons — α-pinene

After ozonolysis, the two carbonyls are still connected; the alkene must have been in a ring, so reconnect the two carbonyl carbons with a double bond.

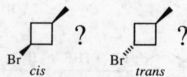

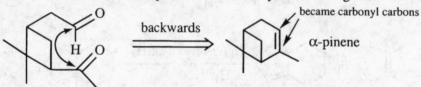

A — Br₂ ⟶ *A* or *A*

B — Br₂ / H₂O ⟶ *B* — H₂SO₄ / Δ ⟶ *C* **Zaitsev product**

— PhCO₃H ⟶ *D* — H₃O⁺ ⟶ *E*

8-62 The two products from permanganate oxidation must have been connected by a double bond at the carbonyl carbons. Whether the alkene was E or Z cannot be determined by this experiment.

$$CH_3(CH_2)_{12}CH=CH(CH_2)_7CH_3$$

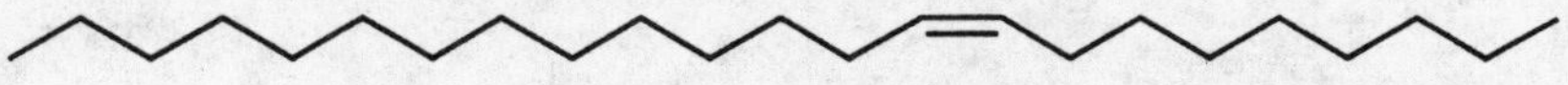

shown here as Z which is the naturally-occurring isomer

8-63

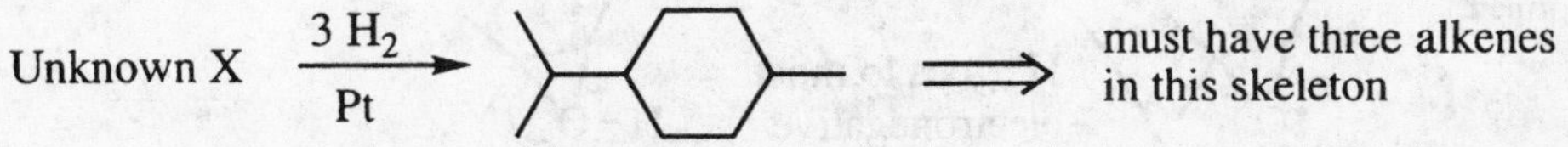

Unknown X $\xrightarrow[\text{Pt}]{3\ H_2}$ (structure) $\Longrightarrow$ must have three alkenes in this skeleton

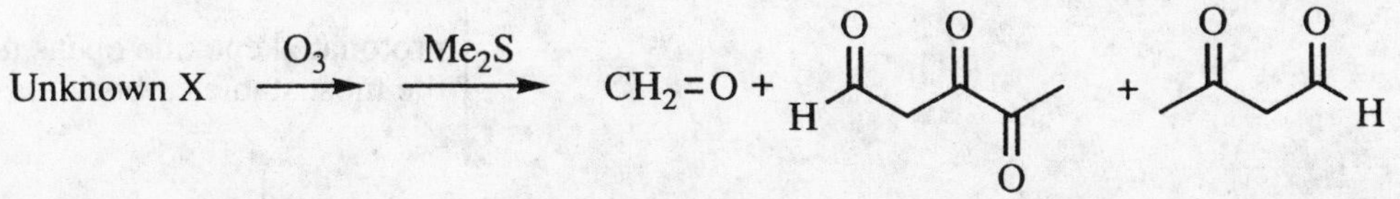

Unknown X $\xrightarrow{O_3} \xrightarrow{Me_2S}$ $CH_2=O$ + (structures)

There are several ways to attack a problem like this. One is the trial-and-error method, that is, put double bonds in all possible positions until the ozonolysis products match. There are times when the trial-and-error method is useful (as in simple problems where the number of possibilities is few), but this is not one of them.

Let's try logic. Analyze the ozonolysis products carefully—what do you see? There are only two methyl groups, so one of the three terminal carbons in the skeleton (C-8, C-9, or C-10) has to be a $=CH_2$. Do we know which terminal carbon has the double bond? Yes, we can deduce that. If C-10 were double-bonded to C-4, then after ozonolysis, C-8 and C-9 must still be attached to C-7. However, in the ozonolysis products, there is no branched chain, that is, no combination of C-8 + C-9 + C-7 + C-1. What if C-7 had a double bond to C-1? Then we would have acetone, CH_3COCH_3, as an ozonolysis product—we don't. Thus, we can't have a double bond from C-4 to C-10. One of the other terminal carbons (C-8) must have a double bond to C-7.

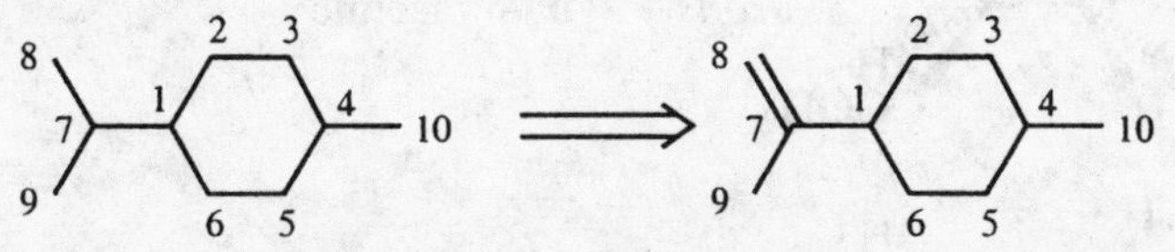

(skeletal structures with numbered carbons) $\Longrightarrow$ (skeletal structure)

The other two double bonds have to be in the ring, but where? The products do not have branched chains, so double bonds must appear at both C-1 and C-4. There are only two possibilities for this requirement.

I (structure) or (structure) **II**

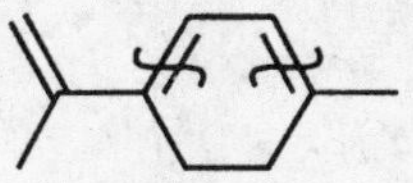

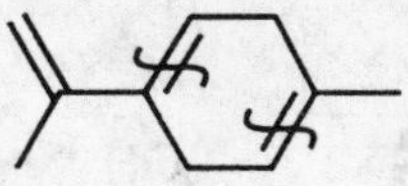

Ozonolysis of **I** would give fragments containing one carbon, two carbons, and seven carbons. Ozonolysis of **II** would give fragments containing one carbon, four carbons, and five carbons. Aha! Our mystery structure must be **II**.

(Editorial comment: Science is more than a collection of facts. The application of observation and logic to solve problems by *deduction* and *inference* are critical scientific skills, ones that distinguish humans from algae.)

8-64 In this type of problem, begin by determining which bonds are broken and which are formed. These will always give clues as to what is happening.

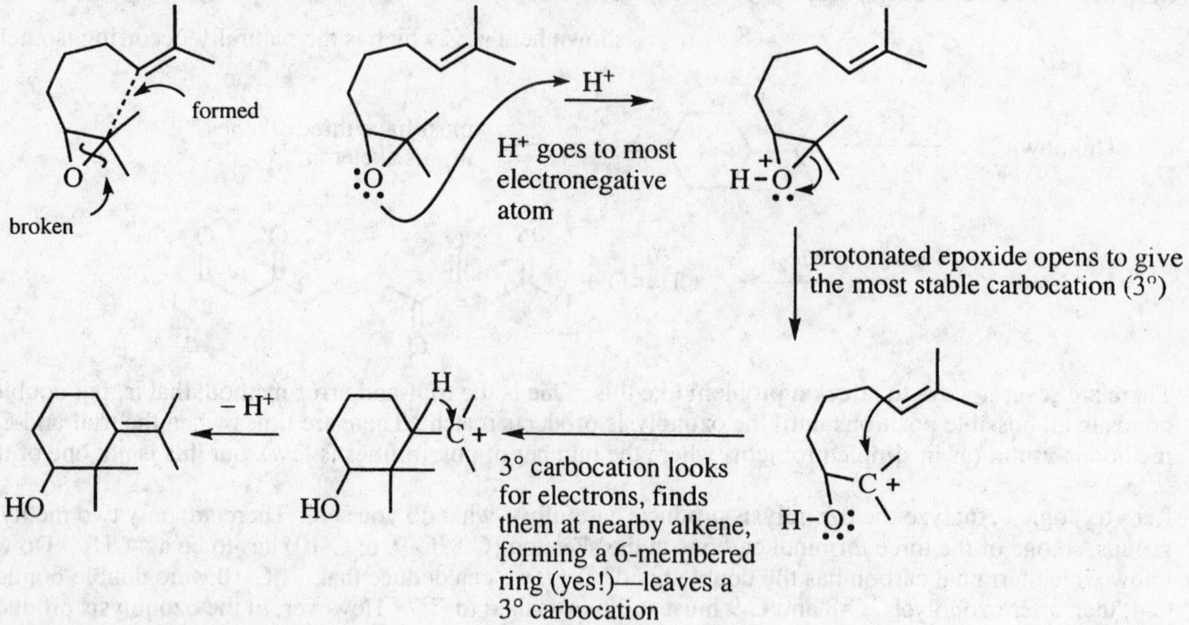

formed

broken

H⁺ goes to most electronegative atom

protonated epoxide opens to give the most stable carbocation (3°)

3° carbocation looks for electrons, finds them at nearby alkene, forming a 6-membered ring (yes!)—leaves a 3° carbocation

8-65 See the solution to Problem 8-35 for simplified examples of these reactions.

(a)

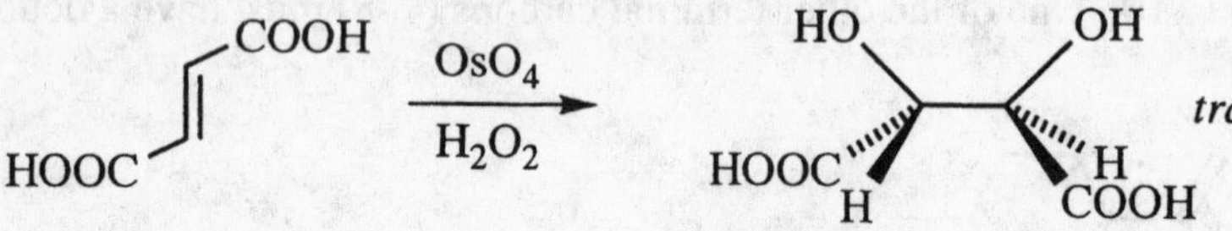

trans + **syn** → racemic

(b)

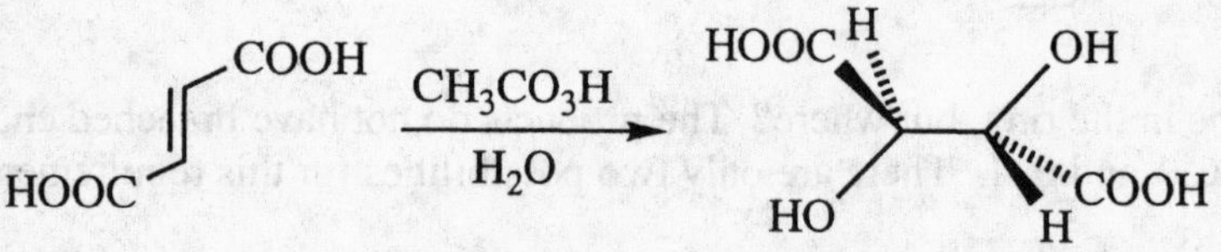

trans + **anti** → meso

(c)

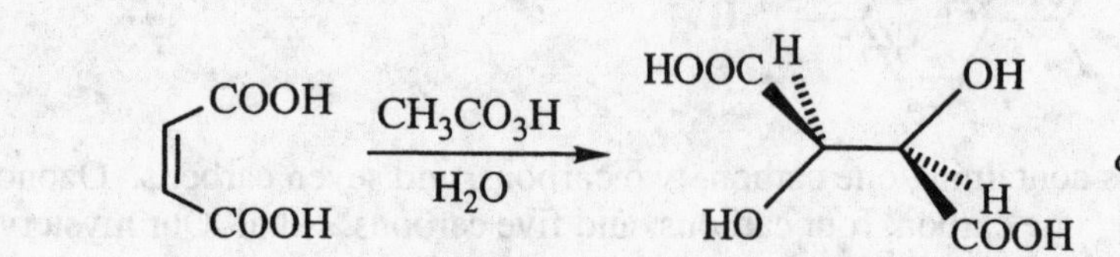

cis + **anti** → racemic

(d)

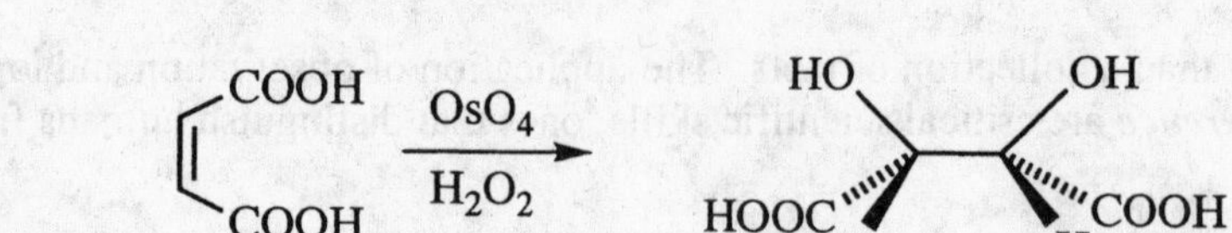

cis + **syn** → meso

8-66

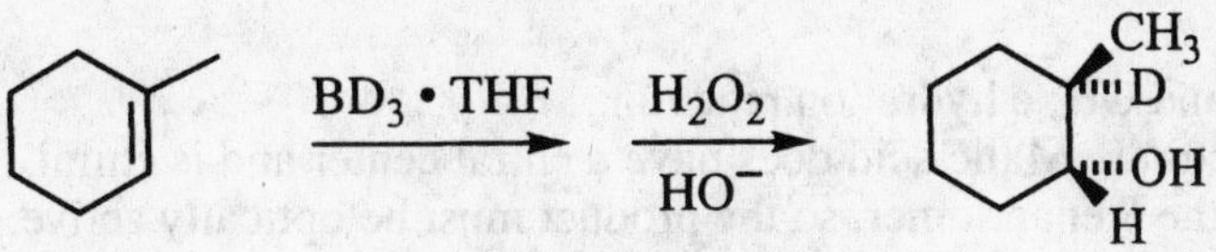

8-67 By now, these rearrangements should not be so "unexpected".

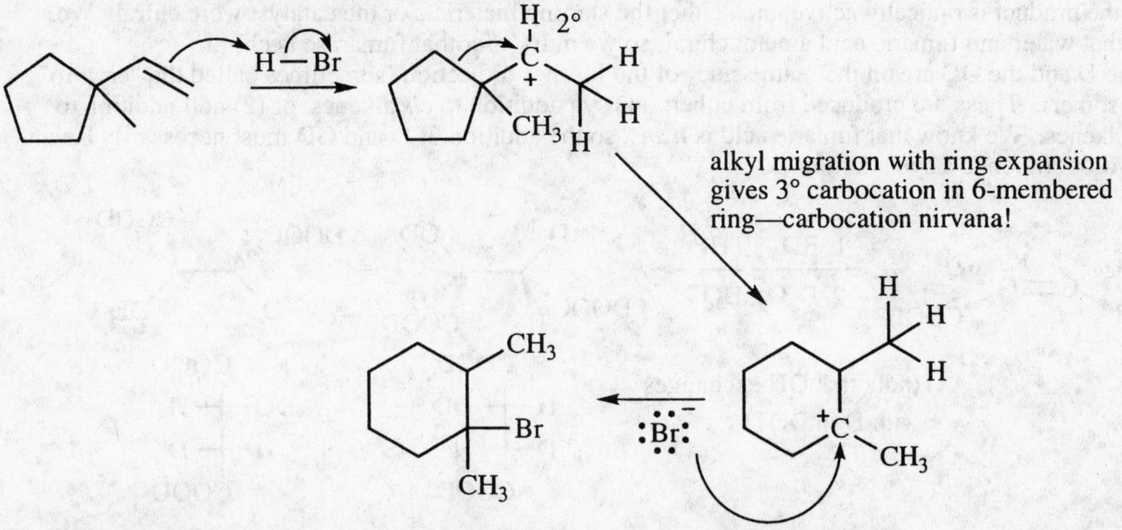

alkyl migration with ring expansion gives 3° carbocation in 6-membered ring—carbocation nirvana!

You must be asking yourself, "Why didn't the methyl group migrate?" To which you answered by drawing the carbocation that would have been formed:

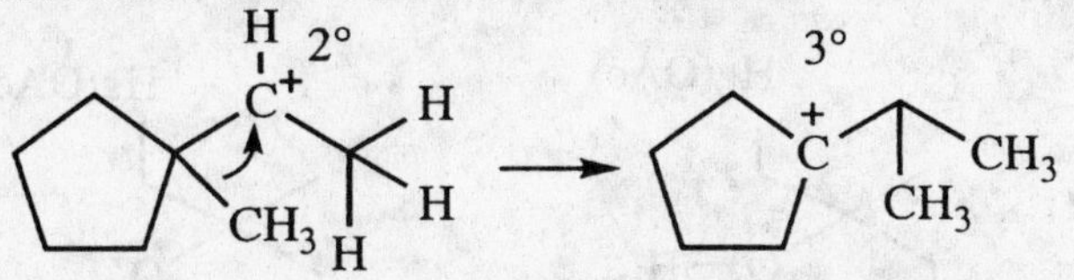

The new carbocation is indeed 3°, but it is only in a 5-membered ring, not quite as stable as in a 6-membered ring. In all probability, some of the product from methyl migration would be formed, but the 6-membered ring would be the major product.

8-68 Each alkene will produce two carbonyls upon ozonolysis or permanganate oxidation. Oxidation of the unknown generated four carbonyls, so the unknown must have had two alkenes. There is only one possibility for their positions.

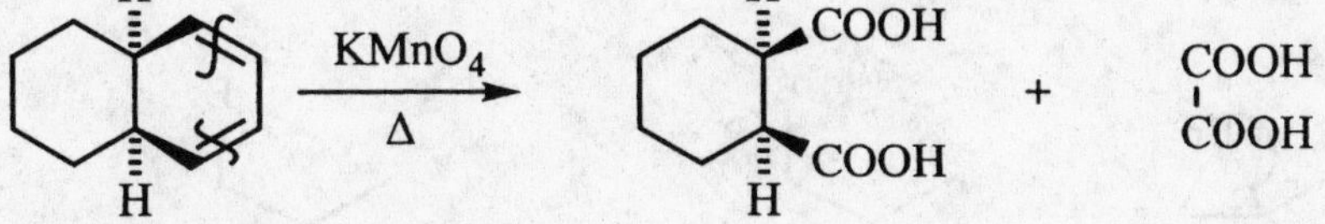

the unknown

187

8-69

(a) Fumarase catalyzes the addition of H and OH, a hydration reaction.

(b) Fumaric acid is planar and cannot be chiral. Malic acid does have a chiral center and is chiral. The enzyme-catalyzed reaction produces only the *S* enantiomer, so the product must be optically active.

(c) One of the fundamental rules of stereochemistry is that optically inactive starting materials produce optically inactive products. Sulfuric-acid-catalyzed hydration would produce a racemic mixture of malic acid, that is, equal amounts of *R* and *S*.

(d) If the product is optically active, then either the starting materials or the catalyst were chiral. We know that water and fumaric acid are not chiral, so we must infer that fumarase is chiral.

(e) The D and the OD are on the "same side" of the Fischer projection (sometimes called the "erythro" stereoisomer). These are produced from either: (1) syn addition to *cis* alkenes, or (2) anti addition to *trans* alkenes. We know that fumaric acid is *trans*, so the addition of D and OD must necessarily be anti.

(f) Hydroboration is a syn addition.

(note that OH exchanges
with D in DO⁻)

As expected, *trans* alkene plus syn addition puts the two groups on the "opposite" side of the Fischer projection (sometimes called "threo").

8-70

(a)

mercurinium ion

(b)

bromonium ion

$C_7H_{13}BrO$

8-71 The addition of BH$_3$ to an alkene is reversible. Given heat and time, the borane will eventually "walk" its way to the end of the chain through a series of addition-elimination cycles. The most stable alkylborane has the boron on the end carbon; eventually, the series of equilibria lead to that product which is oxidized to the primary alcohol.

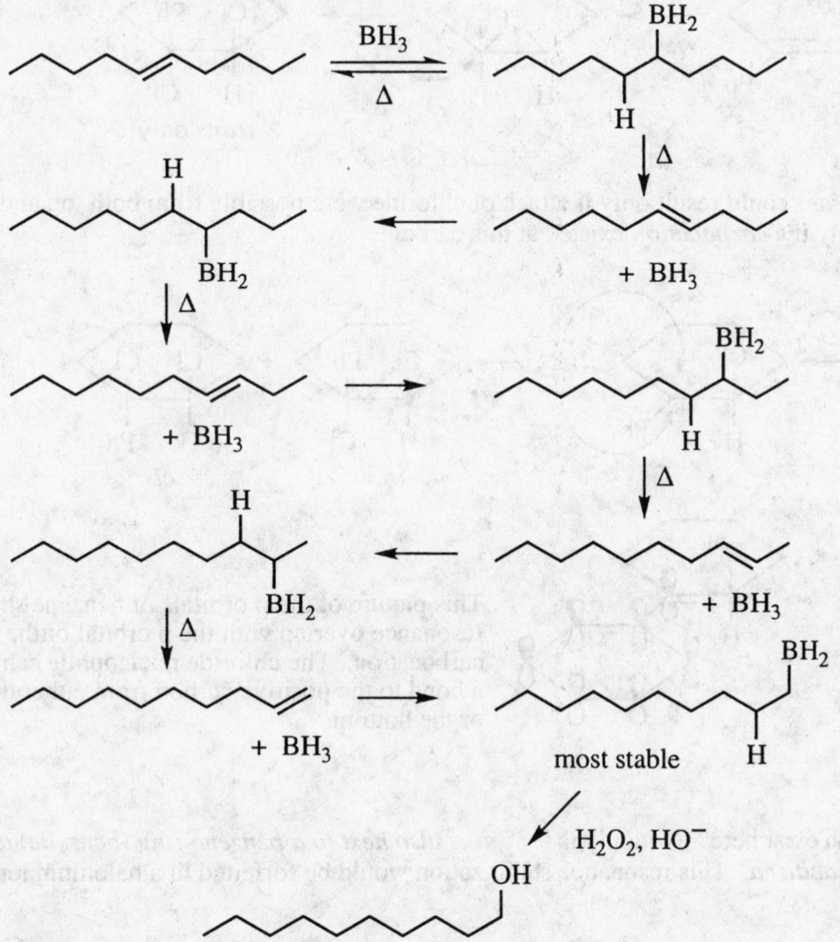

most stable

H_2O_2, HO^-

OH

8-72 First, we explain *how* the mixture of stereoisomers results, then *why*.

We have seen many times that the bridged halonium ion permits attack of the nucleophile only from the opposite side.

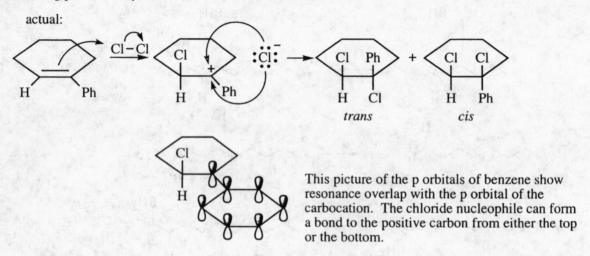

expected:

trans only

A mixture of *cis* and *trans* could result only if attack of chloride were possible from both top and bottom, something possible only if a *carbocation* existed at this carbon.

actual:

trans   *cis*

This picture of the p orbitals of benzene show resonance overlap with the p orbital of the carbocation. The chloride nucleophile can form a bond to the positive carbon from either the top or the bottom.

Why does a carbocation exist here? Not only is it 3°, *it is also next to a benzene ring (benzylic) and therefore resonance-stabilized.* This resonance stabilization would be forfeited in a halonium ion intermediate.

CHAPTER 9—ALKYNES

9-1 Other structures are possible in each case.

(a) C_6H_{10} $CH_3CH_2C\equiv CCH_2CH_3$ $CH_3CH_2CH_2CH_2C\equiv CH$

(b) C_8H_{12} ⬡—$C\equiv CH$ $HC\equiv C—CH_2CH=CHCH_2CH_2CH_3$

(c) C_7H_{10} ⬠—$C\equiv CH$ $CH_2=CH—C\equiv C—CH_2CH_2CH_3$

9-2 New IUPAC names are given. The asterisk (*) denotes acetylenic hydrogens of terminal alkynes.

(a) $CH_3CH_2—C\equiv C—\overset{*}{H}$ $CH_3—C\equiv C—CH_3$
       but-1-yne   but-2-yne

(b) $CH_3CH_2CH_2—C\equiv C—\overset{*}{H}$ $CH_3CH_2—C\equiv C—CH_3$ $CH_3\overset{\displaystyle CH_3}{\overset{|}{CH}}—C\equiv C—\overset{*}{H}$
       pent-1-yne   pent-2-yne   3-methylbut-1-yne

9-3 The decomposition reaction below is exothermic ($\Delta H° = -234$ kJ/mole) as well as having an increase in entropy. Thermodynamically, at 1500°C, an increase in entropy will have a large effect on ΔG (remember $\Delta G = \Delta H - T\Delta S$). Kinetically, almost any activation energy barrier will be overcome at 1500°C. Acetylene would likely decompose into its elements:

$$HC\equiv CH \xrightarrow{1500° C} 2\ C\ +\ H_2$$

9-4 Adding sodium amide to the mixture will produce the sodium salt of hex-1-yne, leaving hex-1-ene untouched. Distillation will remove the hex-1-ene, leaving the non-volatile salt behind.

$$\left.\begin{array}{l}CH_2=CHCH_2CH_2CH_2CH_3\\ H—C\equiv C—CH_2CH_2CH_2CH_3\end{array}\right\}\xrightarrow{NaNH_2}\left\{\begin{array}{l}CH_2=CHCH_2CH_2CH_2CH_3\\ Na^+\ {}^-{:}C\equiv C—CH_2CH_2CH_2CH_3 \quad\text{non-volatile salt}\end{array}\right.$$

9-5 The key to this problem is to understand that *a proton donor will react only with the conjugate base of a weaker acid*. See Appendix 2 at the end of this Solutions Manual for an in-depth discussion of acidity.

(a) $H—C\equiv C—H$ + $NaNH_2$ $\longrightarrow$ $H—C\equiv C{:}^-$ Na^+ + NH_3

(b) $H—C\equiv C—H$ + CH_3Li $\longrightarrow$ $H—C\equiv C{:}^-$ Li^+ + CH_4

(c) no reaction: $NaOCH_3$ is not a strong enough base

(d) no reaction: $NaOH$ is not a strong enough base

(e) $H—C\equiv C{:}^-$ Na^+ + CH_3OH $\longrightarrow$ $H—C\equiv C—H$ + $NaOCH_3$ (opposite of (c))

(f) $H—C\equiv C{:}^-$ Na^+ + H_2O $\longrightarrow$ $H—C\equiv C—H$ + $NaOH$ (opposite of (d))

(g) no reaction : $H—C\equiv C{:}^-$ Na^+ is not a strong enough base

(h) no reaction: $NaNH_2$ is not a strong enough base

(i) CH_3OH + $NaNH_2$ $\longrightarrow$ $NaOCH_3$ + NH_3

9-6

$$H-C\equiv C-H \xrightarrow{NaNH_2} H-C\equiv C:^- Na^+ \xrightarrow{CH_3CH_2Br} H-C\equiv C-CH_2CH_3$$

$$\downarrow NaNH_2$$

$$CH_3(CH_2)_5-C\equiv C-CH_2CH_3 \xleftarrow{CH_3(CH_2)_5Br} Na^+ \; ^-:C\equiv C-CH_2CH_3$$

9-7

(a) $H-C\equiv C-H \xrightarrow[\text{2) } CH_3CH_2CH_2CH_2Br]{\text{1) } NaNH_2} H-C\equiv C-CH_2CH_2CH_2CH_3$

(b) $H-C\equiv C-H \xrightarrow[\text{2) } CH_3CH_2CH_2Br]{\text{1) } NaNH_2} H-C\equiv C-CH_2CH_2CH_3 \xrightarrow[\text{2) } CH_3I]{\text{1) } NaNH_2} CH_3-C\equiv C-CH_2CH_2CH_3$

(c) $H-C\equiv C-H \xrightarrow[\text{2) } CH_3CH_2Br]{\text{1) } NaNH_2} H-C\equiv C-CH_2CH_3 \xrightarrow[\text{2) } CH_3CH_2Br]{\text{1) } NaNH_2} CH_3CH_2-C\equiv C-CH_2CH_3$

(d) cannot be synthesized by an S_N2 reaction—would require attack on a 2° alkyl halide

$$CH_3-C\equiv C:^- \; Na^+ \; + \; \underset{CH_3}{CHCH_2CH_3}\overset{Br \; 2°}{} \xrightarrow{\quad\times\quad} CH_3-C\equiv C-\underset{CH_3}{CHCH_2CH_3}$$

strong nucleophile;
must be second order

low yields; not practical

(e) $H-C\equiv C-H \xrightarrow[\text{2) } CH_3I]{\text{1) } NaNH_2} CH_3-C\equiv C-H \xrightarrow[\underset{CH_3}{\text{2) } BrCH_2CHCH_3}]{\text{1) } NaNH_2} CH_3-C\equiv C-CH_2\underset{CH_3}{CHCH_3}$

(f) $H-C\equiv C-H \xrightarrow[\text{2) } Br(CH_2)_8Br]{\text{1) } NaNH_2}$

$H-C\equiv C-CH_2$

Br⌒⌒⌒

$$\downarrow NaNH_2$$

$H_2C-C\equiv C-CH_2$ ⬡ $\longleftarrow$ $Na^+ \; ^-:C\equiv C-CH_2$

Br⌒⌒⌒

Intramolecular cyclization of large rings must be carried out in dilute solution so the last S_N2 displacement will be *intra*molecular and not *inter*molecular.

9-8

(a) $H-C\equiv C-H \xrightarrow[\text{2) } H_2C=O]{\text{1) } NaNH_2} \xrightarrow{H_2O} H-C\equiv C-CH_2OH$

9-8 continued

(b) 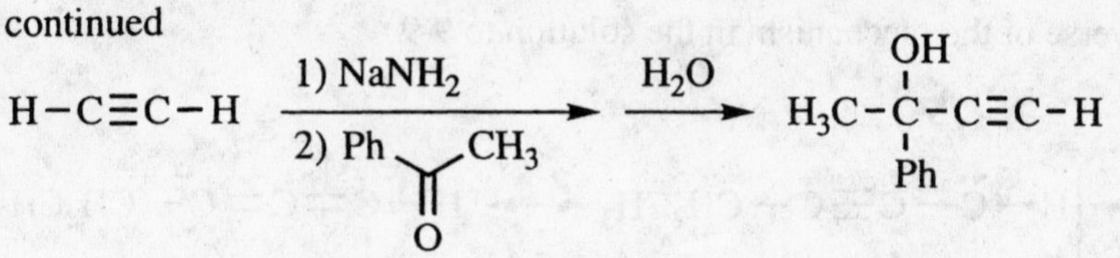 $H-C\equiv C-H$ $\xrightarrow[\substack{2)\ Ph \\ \ \ \ \ \ C \\ \ \ \ \ \ \parallel \\ \ \ \ \ \ O}]{1)\ NaNH_2}$ $\xrightarrow{H_2O}$ $H_3C-\underset{\underset{Ph}{|}}{\overset{\overset{OH}{|}}{C}}-C\equiv C-H$

(c) $H-C\equiv C-H$ $\xrightarrow[2)\ CH_3I]{1)\ NaNH_2}$ $CH_3-C\equiv C-H$ $\xrightarrow[\substack{2)\ HCCH_2CH_2CH_3 \\ \ \ \ \ \ \parallel \\ \ \ \ \ \ O \\ 3)\ H_2O}]{1)\ NaNH_2}$ $CH_3-C\equiv C-\overset{\overset{OH}{|}}{C}HCH_2CH_2CH_3$

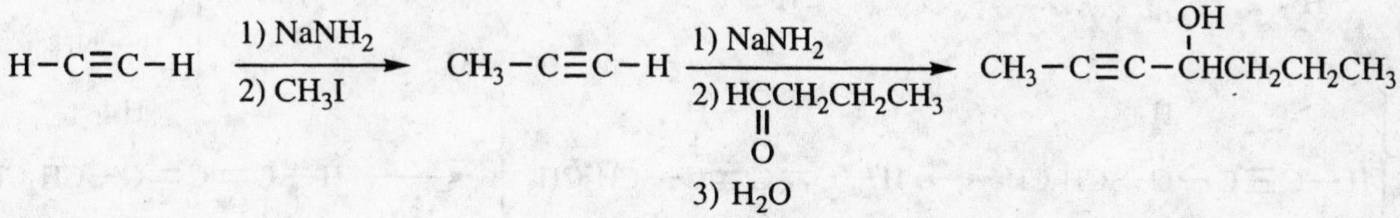

(d) 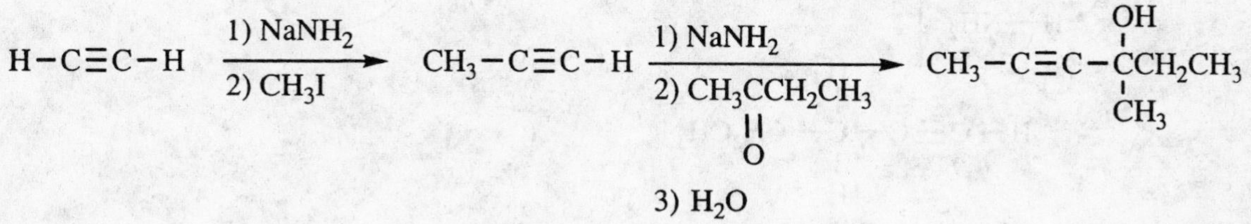 $H-C\equiv C-H$ $\xrightarrow[2)\ CH_3I]{1)\ NaNH_2}$ $CH_3-C\equiv C-H$ $\xrightarrow[\substack{2)\ CH_3CCH_2CH_3 \\ \ \ \ \ \ \parallel \\ \ \ \ \ \ O \\ 3)\ H_2O}]{1)\ NaNH_2}$ $CH_3-C\equiv C-\underset{\underset{CH_3}{|}}{\overset{\overset{OH}{|}}{C}}CH_2CH_3$

9-9

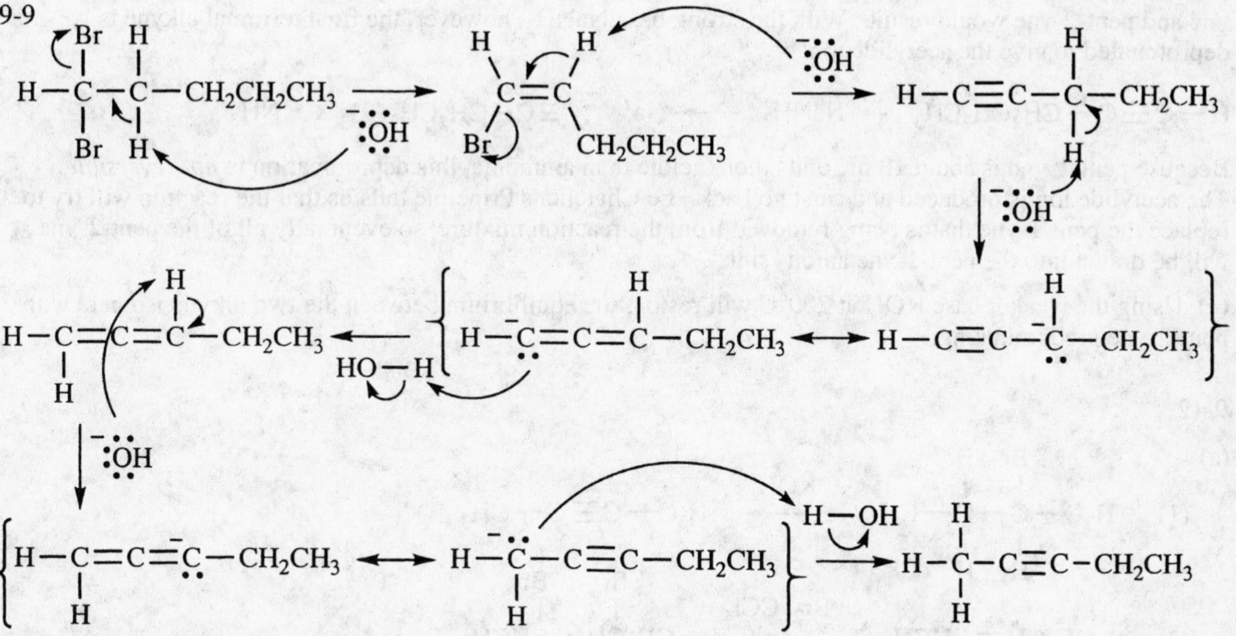

9-10 To determine the equilibrium constant in the reaction:

terminal alkyne $\rightleftharpoons$ internal alkyne

$\Delta G = -17.0$ kJ/mole

$\quad (-4.0$ kcal/mole)

$R = 8.314$ J/K•mole

$\Delta G = -RT \ln K_{eq}$

$K_{eq} = e^{\left(-\frac{\Delta G}{RT}\right)} = e^{\left(\frac{-(-17,000)}{(8.314)(473)}\right)} = e^{4.32} = 75$

$\dfrac{[\text{internal}]}{[\text{terminal}]} = \dfrac{75}{1} = \dfrac{98.7\%\ \text{internal}}{1.3\%\ \text{terminal}}$

9-11 (a) This isomerization is the reverse of the mechanism in the solution to 9-9.

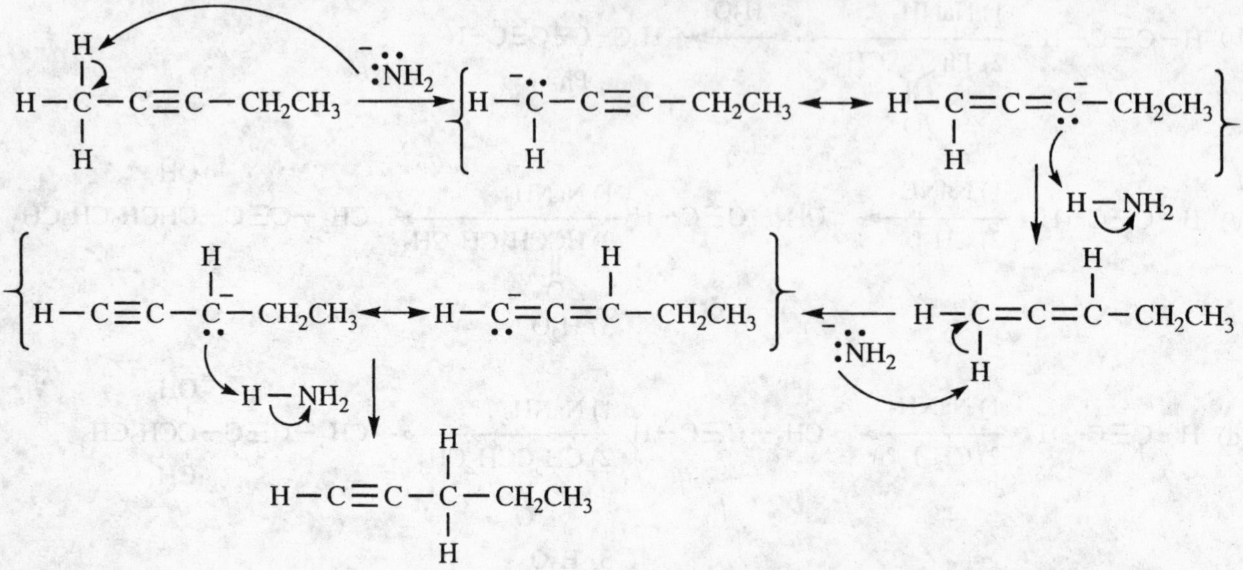

(b) All steps in part (a) are reversible. With a weaker base like KOH, an equilibrium mixture of pent-1-yne and pent-2-yne would result. With the strong base $NaNH_2$, however, the final terminal alkyne is deprotonated to give the acetylide ion:

$$H-C\equiv C-CH_2CH_2CH_3 \;+\; NaNH_2 \;\longrightarrow\; Na^+ \;\bar{:}C\equiv C-CH_2CH_2CH_3 \;+\; NH_3$$

Because pent-1-yne is about 10 pK units more acidic than ammonia, this deprotonation is *not reversible*. The acetylide ion is produced and can't go back. Le Châtelier's Principle tells us that the reaction will try to replace the pent-1-yne that is being removed from the reaction mixture, so eventually all of the pent-2-yne will be drawn into the pent-1-yne anion "sink".

(c) Using the weaker base KOH at 200°C will restore the equilibrium between the two alkyne isomers with pent-2-yne predominating.

9-12

(a)

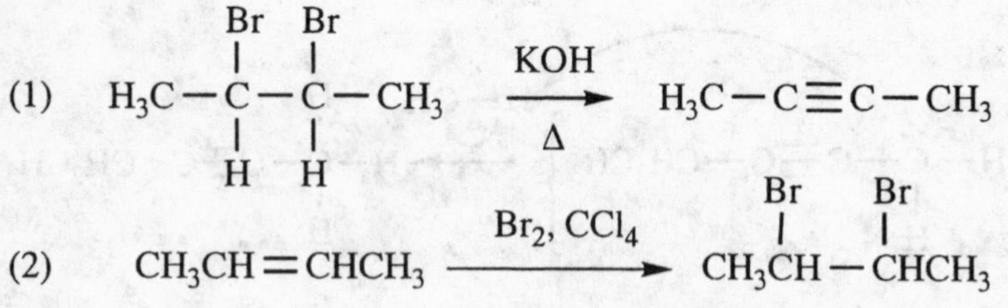

(1)

(2)

(b)

(1)

(2)

194

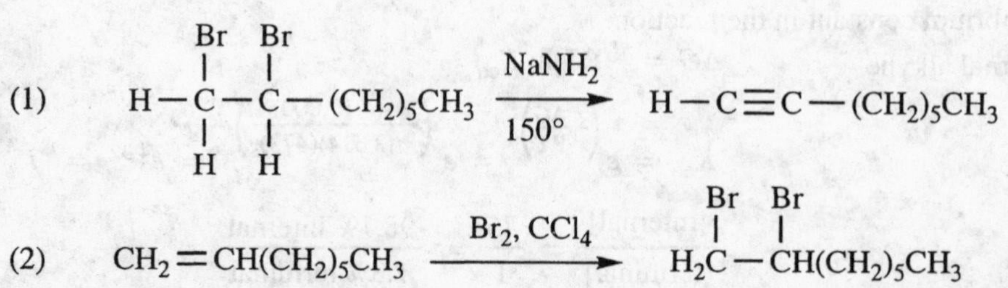

9-12 continued

(c)

(1) $\underset{\substack{H}}{\overset{\substack{Br}}{H_3C-\underset{\underset{H}{|}}{\overset{\overset{Br}{|}}{C}}-\underset{\underset{H}{|}}{\overset{\overset{Br}{|}}{C}}-(CH_2)_4CH_3}} \xrightarrow[\Delta]{KOH} H_3C-C\equiv C-(CH_2)_4CH_3$

(2) $CH_3CH=CH(CH_2)_4CH_3 \xrightarrow{Br_2,\ CCl_4} CH_3\underset{\overset{|}{Br}}{\overset{\overset{Br}{|}}{CH}}-\underset{}{CH(CH_2)_4CH_3}$

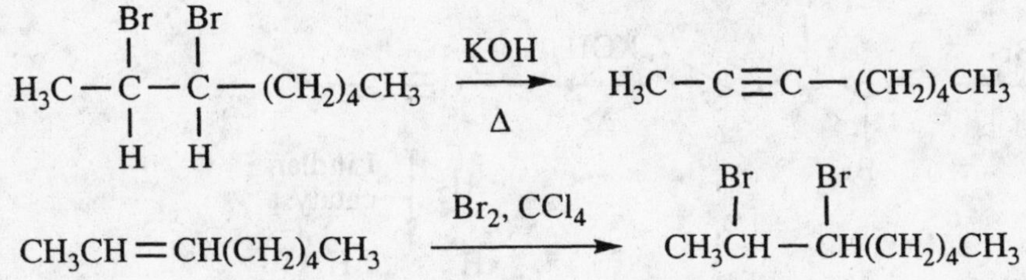

(d)

(1)

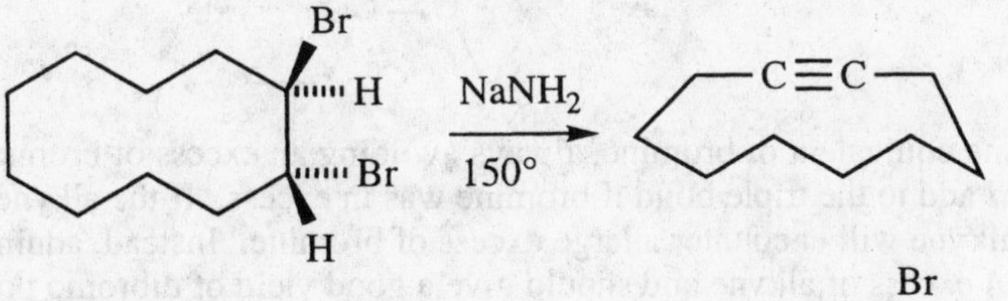

KOH at 200°C could have been used instead of NaNH$_2$

(2)

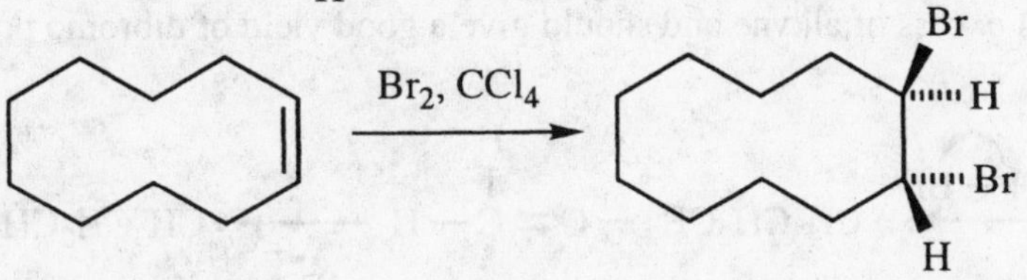

9-13

(a) $CH_3CH_2-C\equiv C-CH_2CH_2CH_2CH_3 \xrightarrow[\text{Lindlar catalyst}]{H_2}$

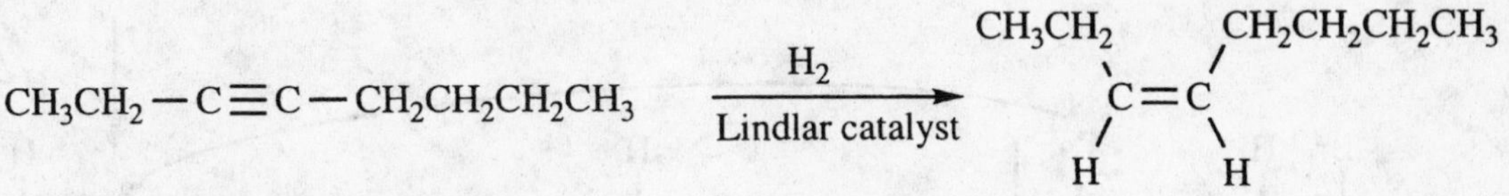

(b) $H_3C-C\equiv C-CH_2CH_3 \xrightarrow[NH_3]{Na}$

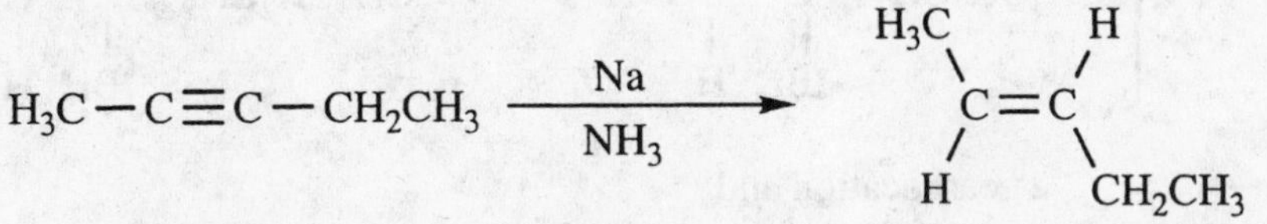

(c)

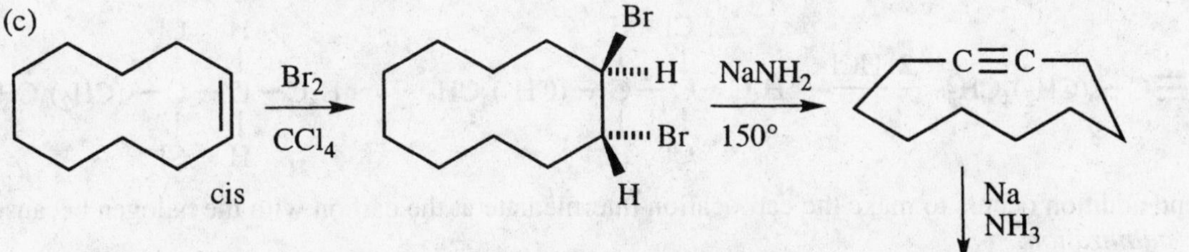

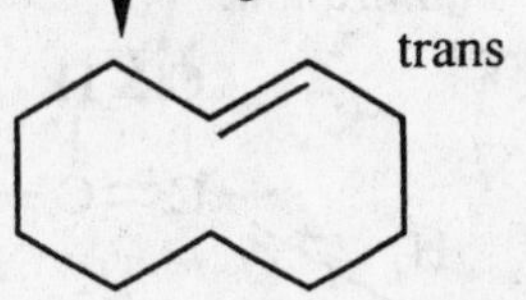

One of the useful "tricks" of organic chemistry is the ability to convert one stereoisomer into another. Having a pair of reactions like the two reductions shown in parts (a) and (b) that give opposite stereochemistry is very useful, because a common intermediate (the alkyne) can be transformed into either stereoisomer. This principle is used again in part (d).

9-13 continued

(d)

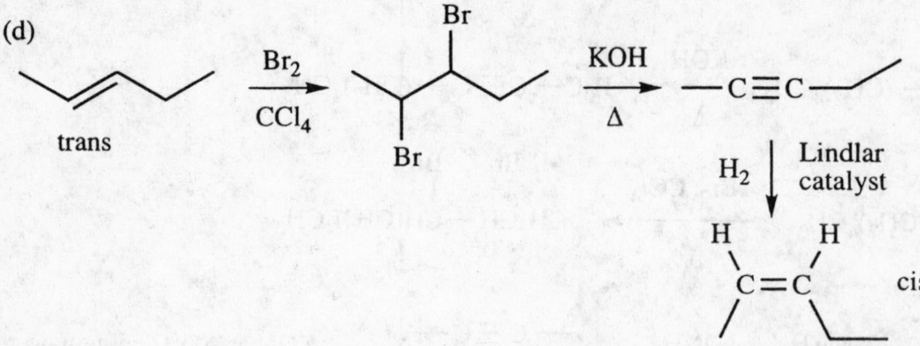

9-14 The goal is to add only one equivalent of bromine, always avoiding an excess of bromine, because two molecules of bromine could add to the triple bond if bromine was in excess. If the alkyne is added to the bromine, the first drops of alkyne will encounter a large excess of bromine. Instead, adding bromine to the alkyne will always ensure an excess of alkyne and should give a good yield of dibromo product.

9-15

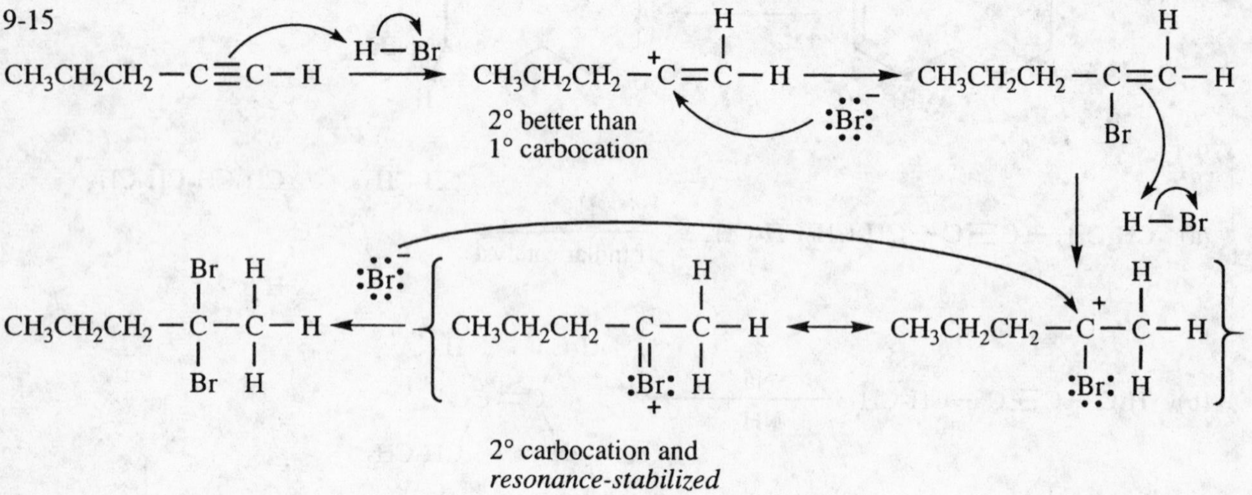

2° better than
1° carbocation

2° carbocation and
resonance-stabilized

9-16

(a)

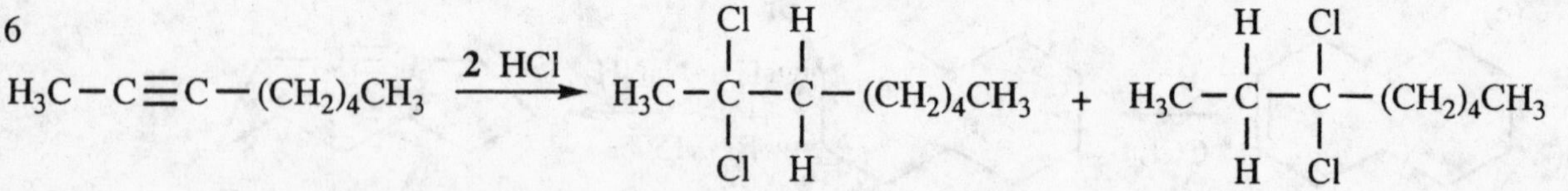

(b) The second addition occurs to make the carbocation intermediate at the carbon with the halogen because of *resonance stabilization*.

resonance stabilization

no stabilization

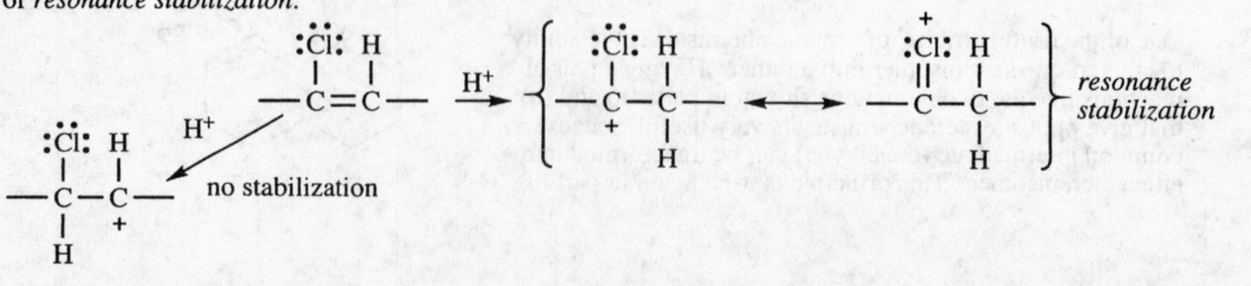

9-17

<u>initiation:</u>

$$RO-OR \xrightarrow{h\nu} 2\ RO\cdot$$

$$RO\cdot\ +\ H-Br \longrightarrow RO-H\ +\ Br\cdot$$

<u>propagation:</u>

$$Br\cdot\ +\ H-C\equiv C-CH_2CH_2CH_3 \longrightarrow \underset{\underset{Br}{|}}{H-C}=\overset{\cdot}{C}-CH_2CH_2CH_3$$

$$\underset{\underset{Br}{|}}{H-C}=\overset{\cdot}{C}-CH_2CH_2CH_3\ +\ H-Br \longrightarrow \underset{\underset{Br}{|}}{H-C}=\underset{\underset{H}{|}}{C}-CH_2CH_2CH_3\ +\ Br\cdot$$

The 2° radical is more stable than 1°. The anti-Markovnikov orientation occurs because the bromine radical attacks first to make the most stable radical, which is contrary to electrophilic addition where the H⁺ attacks first (see the solution to 9-15).

9-18

(a) $H-C\equiv C-(CH_2)_3CH_3 \xrightarrow{\text{1 eq. } Cl_2} \underset{\underset{Cl}{|}}{H-C}=\underset{\underset{Cl}{|}}{C}-(CH_2)_3CH_3 \quad E+Z$

(b) $H-C\equiv C-(CH_2)_3CH_3 \xrightarrow[\text{ROOR}]{\text{HBr}} \underset{\underset{Br}{|}}{H-C}=\underset{\underset{H}{|}}{C}-(CH_2)_3CH_3 \quad E+Z$

(c) $H-C\equiv C-(CH_2)_3CH_3 \xrightarrow{\text{HBr}} \underset{\underset{H}{|}}{H-C}=\underset{\underset{Br}{|}}{C}-(CH_2)_3CH_3$

(d) $H-C\equiv C-(CH_2)_3CH_3 \xrightarrow[\text{CCl}_4]{\text{2 Br}_2} \underset{\underset{Br}{|}}{\overset{\overset{Br}{|}}{H-C}}-\underset{\underset{Br}{|}}{\overset{\overset{Br}{|}}{C}}-(CH_2)_3CH_3$

(e) $H-C\equiv C-(CH_2)_3CH_3 \xrightarrow[\substack{\text{Lindlar}\\\text{catalyst}\\\text{(or Na, NH}_3)}]{\text{H}_2} \underset{\underset{H}{|}}{H-C}=\underset{\underset{H}{|}}{C}-(CH_2)_3CH_3 \xrightarrow{\text{HBr}} CH_3-\underset{\underset{Br}{|}}{CH}-(CH_2)_3CH_3$

(f) $H-C\equiv C-(CH_2)_3CH_3 \xrightarrow{\text{2 HBr}} \underset{\underset{H}{|}}{\overset{\overset{H}{|}}{H-C}}-\underset{\underset{Br}{|}}{\overset{\overset{Br}{|}}{C}}-(CH_2)_3CH_3$

$$CH_3 - C \equiv C - CH_2CH_3$$

H⁺   H⁺

$$CH_3 - \overset{+}{C} = C - CH_2CH_3 \qquad CH_3 - \overset{+}{C} = C - CH_2CH_3$$
| H   both 2° *vinyl*   H |

both 2° *vinyl* carbocations

H₂O:

$$CH_3 - C = C - CH_2CH_3 \qquad CH_3 - C = C - CH_2CH_3$$

H₂O:

$$CH_3 - C = C - CH_2CH_3 \qquad CH_3 - C = C - CH_2CH_3$$

H⁺   H⁺

$$CH_3 - \overset{H}{\underset{H}{C}} - \overset{+}{C} - CH_2CH_3 \qquad CH_3 - \overset{+}{C} - \overset{H}{\underset{H}{C}} - CH_2CH_3$$

resonance-stabilized carbocations   (resonance forms not shown)

H₂O:

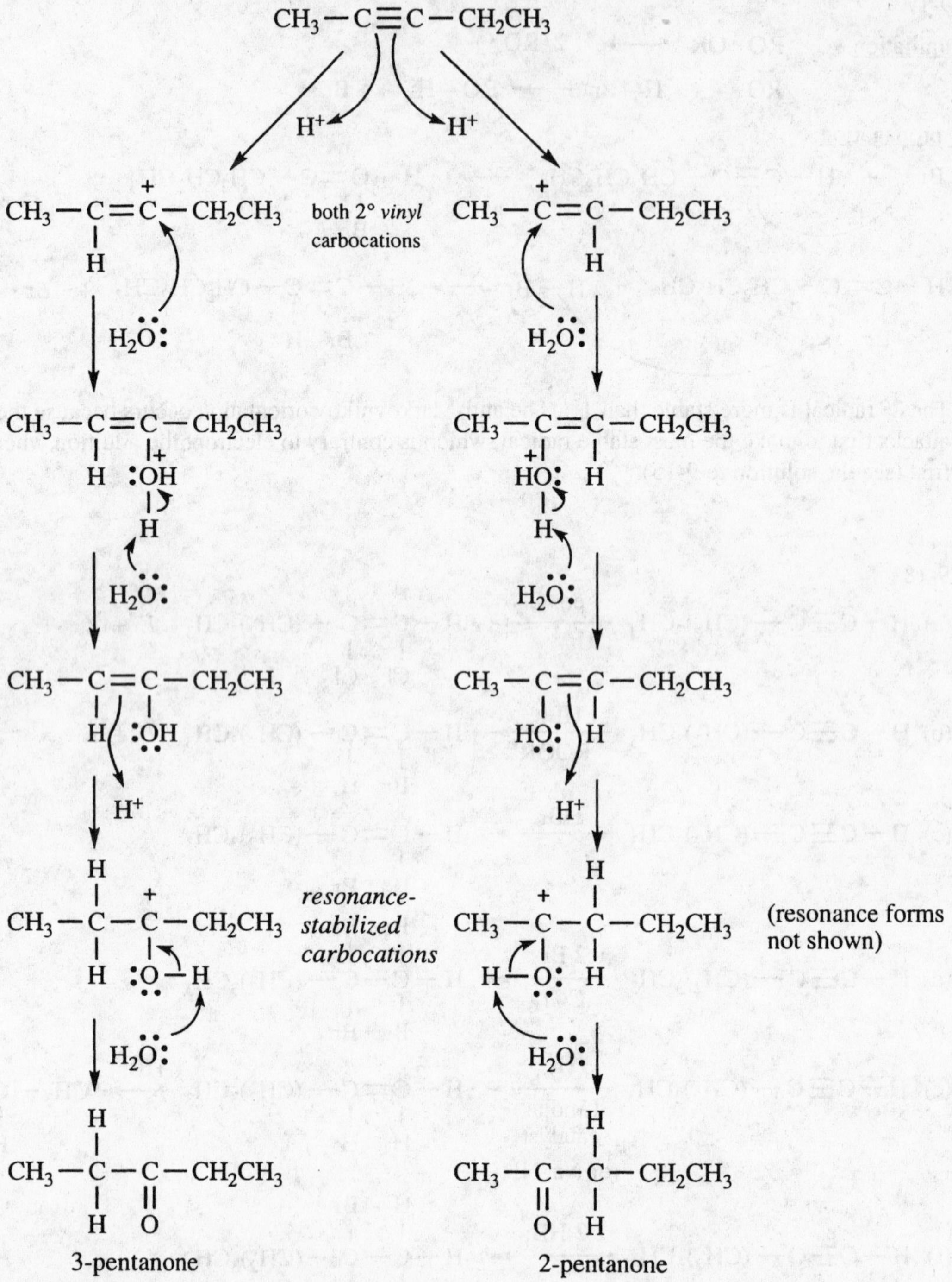

3-pentanone   **2-pentanone**

The role of the mercury catalyst is not shown in this mechanism. As a Lewis acid, it may act like the proton in the first step, helping to form vinyl cations; the mercury is replaced when acid is added.

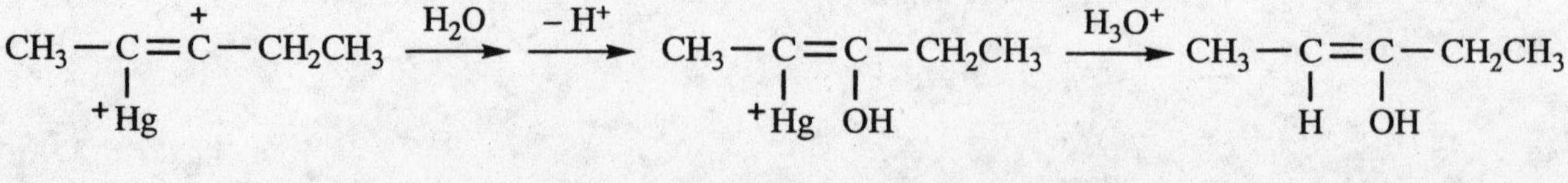

$$CH_3 - \overset{+}{C} = C - CH_2CH_3 \xrightarrow[\;]{H_2O \;\; -H^+} CH_3 - C = C - CH_2CH_3 \xrightarrow{H_3O^+} CH_3 - C = C - CH_2CH_3$$

9-20

(a) But-2-yne is symmetric. Either orientation produces the same product.

$$CH_3-C\equiv C-CH_3 \xrightarrow{Sia_2BH} \overset{\displaystyle CH_3\quad CH_3}{\underset{\displaystyle H\qquad BSia_2}{C=C}} \xrightarrow[HO^-]{H_2O_2} \overset{\displaystyle CH_3\quad CH_3}{\underset{\displaystyle H\qquad OH}{C=C}} \xrightarrow{HO^-} CH_3CH_2-\overset{\displaystyle O}{\overset{\|}{C}}-CH_3$$

(b) Pent-2-yne is not symmetric. Different orientations of attack will lead to different products on any unsymmetrical internal alkyne.

$$CH_3-C\equiv C-CH_2CH_3$$

Sia$_2$BH   Sia$_2$BH

$$\overset{\displaystyle CH_3\quad CH_2CH_3}{\underset{\displaystyle H\qquad BSia_2}{C=C}} \qquad\qquad \overset{\displaystyle CH_3\quad CH_2CH_3}{\underset{\displaystyle Sia_2B\qquad H}{C=C}}$$

↓ H$_2$O$_2$, HO$^-$ ↓ H$_2$O$_2$, HO$^-$

$$\overset{\displaystyle CH_3\quad CH_2CH_3}{\underset{\displaystyle H\qquad OH}{C=C}} \qquad\qquad \overset{\displaystyle CH_3\quad CH_2CH_3}{\underset{\displaystyle HO\qquad H}{C=C}}$$

↓ HO$^-$ ↓ HO$^-$

$$CH_3CH_2-\overset{\displaystyle O}{\overset{\|}{C}}-CH_2CH_3 \qquad\qquad CH_3-\overset{\displaystyle O}{\overset{\|}{C}}-CH_2CH_2CH_3$$

9-21

(a) (1) $CH_3\overset{\displaystyle O}{\overset{\|}{C}}CH_2CH_2CH_2CH_3$ (2) $H\overset{\displaystyle O}{\overset{\|}{C}}CH_2CH_2CH_2CH_2CH_3$

(b) (1) $CH_3\overset{\displaystyle O}{\overset{\|}{C}}CH_2CH_2CH_2CH_3$ + $CH_3CH_2\overset{\displaystyle O}{\overset{\|}{C}}CH_2CH_2CH_3$

(2) same mixture as in (b) (1)

(c) (1) $CH_3CH_2\overset{\displaystyle O}{\overset{\|}{C}}CH_2CH_2CH_3$ (2) $CH_3CH_2\overset{\displaystyle O}{\overset{\|}{C}}CH_2CH_2CH_3$

(d) (1)   (2)

199

9-22

(a)

$$H_3C-C(CH_3)=C(H)(CH_3) + BH_3 \longrightarrow H-\underset{CH_3}{\underset{|}{C}}-\underset{CH_3}{\underset{|}{C}}(CH_3)(H)-BH_2 \; + \; \underset{H_3C}{\overset{H}{C}}=\underset{CH_3}{\overset{CH_3}{C}}$$

$$\downarrow$$

$$H-\underset{CH_3}{\overset{CH_3}{\underset{|}{\overset{|}{C}}}}-\underset{CH_3}{\overset{H}{\underset{|}{\overset{|}{C}}}}-B-\underset{CH_3}{\overset{H}{\underset{|}{\overset{|}{C}}}}-\underset{CH_3}{\overset{CH_3}{\underset{|}{\overset{|}{C}}}}-H$$

disiamylborane, Sia_2BH

(b) There is too much steric hindrance in Sia_2BH for the third B—H to add across another alkene. The reagent can add to alkynes because alkynes are linear and attack is not hindered by bulky substituents.

9-23

(a) (1) $HO-\overset{O}{\overset{||}{C}}-\overset{O}{\overset{||}{C}}-(CH_2)_3CH_3$
oxidation of a terminal alkyne with neutral $KMnO_4$ produces the ketone and carboxylic acid without cleaving the carbon-carbon bond

(2) $CO_2 + HO-\overset{O}{\overset{||}{C}}-(CH_2)_3CH_3$
oxidation of a terminal alkyne with warm, basic $KMnO_4$ cleaves the carbon-carbon bond, producing the carboxylic acid and carbon dioxide

(b) (1) $CH_3-\overset{O}{\overset{||}{C}}-\overset{O}{\overset{||}{C}}-CH_2CH_2CH_3$

(2) $CH_3-\overset{O}{\overset{||}{C}}-OH + HO-\overset{O}{\overset{||}{C}}-CH_2CH_2CH_3$

(c) (1) $CH_3CH_2-\overset{O}{\overset{||}{C}}-\overset{O}{\overset{||}{C}}-CH_2CH_3$

(2) $CH_3CH_2-\overset{O}{\overset{||}{C}}-OH$

(d) (1) $\underset{CH_3}{\underset{|}{CH_3CH}}-\overset{O}{\overset{||}{C}}-\overset{O}{\overset{||}{C}}-CH_2CH_3$

(2) $\underset{CH_3}{\underset{|}{CH_3CH}}-\overset{O}{\overset{||}{C}}-OH + HO-\overset{O}{\overset{||}{C}}-CH_2CH_3$

(e) (1)

(2)

9-24

(a) $CH_3-C\equiv C-(CH_2)_4-C\equiv C-CH_3$

(b)

200

9-25 When proposing syntheses, begin by analyzing the target molecule, looking for smaller pieces that can be combined to make the desired compound. This is especially true for targets that have more carbons than the starting materials; immediately, you will know that a carbon-carbon bond forming reaction will be necessary.

People who succeed at synthesis *know the reactions*—there is no shortcut. Practice the reactions for each functional group until they become automatic.

(a) analysis of target

3° acetylenic alcohols made from acetylide plus ketones

from acetylene

from alkylation of acetylide

forward direction: put on less reactive group first

$H-C\equiv C-H$ + $NaNH_2$ $\longrightarrow$ $H-C\equiv C:^-$ Na^+ $\longrightarrow$ $H-C\equiv C-$

$\downarrow NaNH_2$

$\xleftarrow{H_2O}$ $\xleftarrow{}$ Na^+ $:C\equiv C-$

(b) analysis of target: cyclopropanes are made by carbene insertion into alkenes; to get *cis* substitution around cyclopropane, stereochemistry of alkene must be *cis; cis* alkene comes from catalytic hydrogenation of an alkyne

$H-C\equiv C-H$ $\xrightarrow{NaNH_2}$ $\xrightarrow{CH_3I}$ $H_3C-C\equiv C-H$ $\xrightarrow{NaNH_2}$ $\xrightarrow{CH_3CH_2Br}$ $H_3C-C\equiv C-CH_2CH_3$

H_2 $\downarrow$ Lindlar catalyst

$\xleftarrow[Zn(Cu)]{CH_2I_2}$

(c) analysis of target: epoxides are made by direct epoxidation of alkenes; to get *trans* substitution around epoxide, stereochemistry of alkene must be *trans* ; *trans* alkene comes from sodium/ammonia reduction of an alkyne

$H-C\equiv C-H$ $\xrightarrow{NaNH_2}$ $\xrightarrow{CH_3CH_2Br}$ $CH_3CH_2-C\equiv C-H$ $\xrightarrow{NaNH_2}$ $\xrightarrow{CH_3CH_2CH_2Br}$

$\xleftarrow{MCPBA}$ $\xleftarrow[NH_3]{Na}$ $CH_3CH_2-C\equiv C-CH_2CH_2CH_3$

9-26 Please refer to solution 1-20, page 12 of this Solutions Manual.

9-27

(a) $CH_3CH_2-C\equiv C-(CH_2)_4CH_3$ (b) $H_3C-C\equiv C-(CH_2)_4CH_3$ (c) (phenyl)$-C\equiv C-H$

(d) (cyclohexyl)$-C\equiv C-H$ (e) $CH_3CH_2-C\equiv C-\underset{\underset{CH_3}{|}}{CH}CH_2CH_2CH_3$

(f) (cyclic structure) $C\equiv C$ with Br (wedge) and Br   (g) $CH_3\underset{\underset{OH}{|}}{CH}-C\equiv C-(CH_2)_3CH_3$ (h) $\underset{\underset{H}{|}}{\overset{H_3C}{C}}=\underset{\underset{H}{|}}{C}\quad C\equiv C-\underset{\underset{CH_2CH_3}{|}}{CH}CH_2CH_3$

(i) $H-C\equiv C-CH_2-C\equiv C-CH_2CH_3$ (j) $H-C\equiv C-CH=CH_2$ (k) $H-C\equiv C-\overset{\overset{CH_3}{|}}{\underset{\underset{H}{|}}{C}}\overset{H}{\underset{}{}}\,C=CH_2$

9-28

(a) ethylmethylacetylene   (c) *sec*-butyl-*n*-propylacetylene
(b) phenylacetylene   (d) *sec*-butyl-*t*-butylacetylene

9-29

(a) 4-phenylpent-2-yne   (c) 2,6,6-trimethylhept-3-yne   (e) 3-methylhex-4-yn-3-ol
(b) 4,4-dibromopent-2-yne   (d) (*E*)-3-methylhept-2-en-4-yne   (f) cycloheptylprop-1-yne

9-30

(a)

internal alkynes	terminal alkynes	acetylide ions
$CH_3-C\equiv C-CH_2CH_2CH_3$ hex-2-yne	$H-C\equiv C-(CH_2)_3CH_3$ hex-1-yne	$^-C\equiv C-(CH_2)_3CH_3$
$CH_3CH_2-C\equiv C-CH_2CH_3$ hex-3-yne	$H-C\equiv C-\underset{\underset{CH_3}{|}}{CH}CH_2CH_3$ 3-methylpent-1-yne	$^-C\equiv C-\underset{\underset{CH_3}{|}}{CH}CH_2CH_3$
$CH_3-C\equiv C-\underset{\underset{CH_3}{|}}{CH}CH_3$ 4-methylpent-2-yne	$H-C\equiv C-CH_2\underset{\underset{CH_3}{|}}{CH}CH_3$ 4-methylpent-1-yne	$^-C\equiv C-CH_2\underset{\underset{CH_3}{|}}{CH}CH_3$
	$H-C\equiv C-\underset{\underset{CH_3}{|}}{\overset{\overset{CH_3}{|}}{C}}-CH_3$ 3,3-dimethylbut-1-yne	$^-C\equiv C-\underset{\underset{CH_3}{|}}{\overset{\overset{CH_3}{|}}{C}}-CH_3$

$\xrightarrow{\text{NaNH}_2}$

(b) All four terminal alkynes will be deprotonated with sodium amide.

9-31

$CH_3CH_2-\underset{\underset{\underset{CHCl}{\|}}{HC}}{\overset{\overset{O}{\|}}{C}} \quad ^-:C\equiv CH \xrightarrow{H_2O} CH_3CH_2-\underset{\underset{\underset{CHCl}{\|}}{HC}}{\overset{\overset{OH}{|}}{C}}-C\equiv CH \quad$ ethchlorvynol

$\underset{\text{NaNH}_2}{\overset{}{}} \quad HC\equiv CH$

9-32

$$H-C\equiv C-H \xrightarrow{NaNH_2} \xrightarrow{CH_3(CH_2)_7Br} CH_3(CH_2)_7-C\equiv C-H \xrightarrow{NaNH_2}$$

$$\downarrow CH_3(CH_2)_{12}Br$$

$$CH_3(CH_2)_7-C\equiv C-(CH_2)_{12}CH_3$$

$$\xleftarrow[\text{Lindlar catalyst}]{H_2}$$

$$\underset{\text{muscalure}}{\overset{CH_3(CH_2)_7 \quad (CH_2)_{12}CH_3}{\underset{H \qquad\qquad H}{C=C}}}$$

9-33

(a) $\overset{\quad Cl}{CH_2=CCH_2CH_2CH_3}$

(b) $H_3C-\overset{\overset{\displaystyle Cl}{|}}{\underset{\underset{\displaystyle Cl}{|}}{C}}-CH_2CH_2CH_3$

(c) $CH_3CH_2CH_2CH_2CH_3$

(d) $CH_2=CHCH_2CH_2CH_3$

(e) $\underset{Br \qquad CH_2CH_2CH_3}{\overset{H \qquad Br}{C=C}}$

(f) $H-\overset{\overset{\displaystyle Br}{|}}{\underset{\underset{\displaystyle Br}{|}}{C}}-\overset{\overset{\displaystyle Br}{|}}{\underset{\underset{\displaystyle Br}{|}}{C}}-CH_2CH_2CH_3$

(g) $HO-\overset{\overset{\displaystyle O}{||}}{C}-\overset{\overset{\displaystyle O}{||}}{C}-CH_2CH_2CH_3$

(h) $CO_2 + HO-\overset{\overset{\displaystyle O}{||}}{C}-CH_2CH_2CH_3$

(i) $H_2C=CHCH_2CH_2CH_3$

(j) $Na^+ \ {}^-{:}C\equiv C-CH_2CH_2CH_3$

(k) $H_3C-\overset{\overset{\displaystyle O}{||}}{C}-CH_2CH_2CH_3$

(l) $H-\overset{\overset{\displaystyle O}{||}}{C}-CH_2CH_2CH_2CH_3$

9-34

(a) $H_3C-\overset{\overset{\displaystyle Br}{|}}{\underset{\underset{\displaystyle Br}{|}}{C}}-CH_2CH_3 \xrightarrow[150°]{NaNH_2} \xrightarrow{H_2O} H-C\equiv C-CH_2CH_3$

(b) $H_3C-\overset{\overset{\displaystyle Br}{|}}{\underset{\underset{\displaystyle Br}{|}}{C}}-CH_2CH_3 \xrightarrow[200°]{KOH} H_3C-C\equiv C-CH_3$

(c) $CH_3CH_2-C\equiv C-H \xrightarrow{NaNH_2} CH_3CH_2-C\equiv C{:}^- \ Na^+$

$$\downarrow CH_3CH_2CH_2CH_2Br$$

$$CH_3CH_2-C\equiv C-CH_2CH_2CH_2CH_3$$

(d) $\underset{H \qquad CH_2CH_2CH_3}{\overset{H_3C \qquad H}{C=C}} \xrightarrow[CCl_4]{Br_2} \overset{\quad\;\; Br\; Br}{\underset{\quad\;\; H\;\; H}{CH_3C-CCH_2CH_2CH_3}} \xrightarrow[200°]{KOH} H_3C-C\equiv C-CH_2CH_2CH_3$

this product could contain minor amounts of 3-hexyne from rearrangement

(e) $\underset{H_3C \qquad CH_2CH_2CH_3}{\overset{H \qquad\quad H}{C=C}} \xrightarrow[CCl_4]{Br_2} \overset{\quad\;\; Br\; Br}{\underset{\quad\;\; H\;\; H}{CH_3C-CCH_2CH_2CH_3}} \xrightarrow[\text{2) }H_2O]{\text{1) }NaNH_2, 150°} H-C\equiv C-CH_2CH_2CH_2CH_3$

203

9-34 continued

(f)

$$CH \equiv C \xrightarrow{\begin{array}{c} H_2 \\ \text{Lindlar} \\ \text{catalyst} \end{array}} \quad \text{cis}$$

(g)

$$C \equiv C \xrightarrow{\begin{array}{c} Na \\ \hline NH_3 \end{array}} \quad \text{trans}$$

(h) $HC \equiv C - CH_2CH_2CH_2CH_3 \xrightarrow{\begin{array}{c} H_2O \\ H_2SO_4 \\ HgSO_4 \end{array}} \left[\begin{array}{c} OH \\ | \\ CH_2 = CCH_2CH_2CH_2CH_3 \\ \text{unstable enol} \end{array} \right] \longrightarrow H_3C - \overset{O}{\overset{||}{C}}CH_2CH_2CH_2CH_3$

(i) $HC \equiv C - CH_2CH_2CH_2CH_3 \xrightarrow{\begin{array}{c} 1) \ Sia_2BH \\ \hline 2) \ H_2O_2, \\ HO^- \end{array}} \left[\begin{array}{c} OH \\ | \\ CH = CHCH_2CH_2CH_2CH_3 \\ \text{unstable enol} \end{array} \right] \longrightarrow H - \overset{O}{\overset{||}{C}}CH_2CH_2CH_2CH_2CH_3$

(j)

this product could contain minor amounts of 3-hexyne from rearrangement

$$\underset{H}{\overset{H_3C}{>}} C = C \underset{CH_2CH_2CH_3}{\overset{H}{<}} \xrightarrow{\begin{array}{c} Br_2 \\ \hline CCl_4 \end{array}} \underset{H \ H}{\overset{Br \ Br}{CH_3\overset{|}{C} - \overset{|}{C}CH_2CH_2CH_3}} \xrightarrow{\begin{array}{c} KOH \\ \hline 200° \end{array}} \underset{\text{major}}{H_3C - C \equiv C - CH_2CH_2CH_3}$$

$$\xrightarrow[\text{Lindlar catalyst}]{H_2} \quad \underset{H \qquad H}{\overset{H_3C \qquad CH_2CH_2CH_3}{>C = C<}}$$

9-35

$$\underset{\overset{|}{Br}}{\overset{Br}{\diagdown}} \xrightarrow{\begin{array}{c} KOH \\ \hline 200° \end{array}} HC \equiv \quad + \quad \text{---} \left. \right\} \text{Mixture A}$$

from rearrangement—
could contain 3-hexyne

only the terminal alkyne reacts
with NaNH$_2$ and acetone

$$\xrightarrow{\text{1) NaNH}_2 \quad \text{2)} \quad \overset{O}{\overset{||}{\diagup}} \quad \text{3) H}_3O^+}$$

$$\overset{OH}{|} \diagdown \diagup C \equiv C \quad + \quad \text{---} C \equiv C \quad \left. \right\} \text{Mixture B}$$

distill

Compound **D**
b.p. 140-150°C
under vacuum

$$\overset{OH}{|} \diagdown \diagup C \equiv C$$

Compound **C**
b.p. 80-84°C

9-36

(a) $CH_3CH_2-C\equiv C-CH_2CH_3$

(b) $CH_3CH_2-C\equiv C-H$ + $CH_2=C{CH_3 \atop CH_3}$ elimination on 3° halide

(c) $CH_3CH_2-C\equiv C-CH_2OH$
(after H_2O workup)

(d) $CH_3CH_2-C\equiv C-\overset{OH}{\underset{}{C}}$ (cyclohexyl)
(after H_2O workup)

(e) $CH_3CH_2-C\equiv C-\overset{OH}{\underset{}{C}}HCH_2CH_2CH_3$
(after H_2O workup)

(f) $CH_3CH_2-C\equiv C-H$ + Na^+ ^-O—(cyclohexyl)

(g) $CH_3CH_2-C\equiv C-\overset{OH}{\underset{CH_3}{C}}-CH_2CH_3$
(after H_2O workup)

9-37

(a) $HC\equiv C-H$ $\xrightarrow{NaNH_2}$ $HC\equiv C:^-$ Na^+ $\xrightarrow{CH_3CH_2CH_2CH_2Br}$ $HC\equiv C-CH_2CH_2CH_2CH_3$

(b) $HC\equiv C-H$ $\xrightarrow{NaNH_2}$ $HC\equiv C:^-$ Na^+ $\xrightarrow{CH_3CH_2CH_2Br}$ $HC\equiv C-CH_2CH_2CH_3$

$H_3C-C\equiv C-CH_2CH_2CH_3$ $\xleftarrow[CH_3I]{NaNH_2}$

(c) $H_3C-C\equiv C-CH_2CH_2CH_3$ $\xrightarrow[\text{Lindlar catalyst}]{H_2}$ $\underset{H \quad\quad H}{\overset{H_3C \quad CH_2CH_2CH_3}{C=C}}$
synthesized in part (b)

(d) $H_3C-C\equiv C-CH_2CH_2CH_3$ $\xrightarrow[NH_3]{Na}$ $\underset{H \quad\quad CH_2CH_2CH_3}{\overset{H_3C \quad\quad H}{C=C}}$
synthesized in part (b)

(e) $H_3C-C\equiv C-CH_2CH_2CH_3$ $\xrightarrow[Pt]{\substack{\text{2 equiv.}\\ H_2}}$ $CH_3CH_2CH_2CH_2CH_2CH_3$
synthesized in part (b)

(f) $HC\equiv C-CH_2CH_2CH_2CH_3$ $\xrightarrow{\textbf{2 HBr}}$ $H_3C-\overset{Br}{\underset{Br}{C}}-CH_2CH_2CH_2CH_3$
synthesized in part (a)

9-37 continued

(g) $H-C\equiv C-CH_2CH_2CH_3$ $\xrightarrow{\text{1) Sia}_2\text{BH}}$ $\xrightarrow{\text{2) H}_2\text{O}_2,\ \text{HO}^-}$ $\underset{\displaystyle H-\overset{\displaystyle O}{\overset{\|}{C}}CH_2CH_2CH_2CH_3}{}$
from (b)

(h) $H-C\equiv C-CH_2CH_2CH_3$ $\xrightarrow[\substack{H_2SO_4 \\ HgSO_4}]{H_2O}$ $\underset{\displaystyle H_3C-\overset{\displaystyle O}{\overset{\|}{C}}CH_2CH_2CH_3}{}$
from (b)

(i) $HC\equiv C-H$ $\xrightarrow{\text{NaNH}_2}$ $HC\equiv C\colon^-\ Na^+$ $\xrightarrow{CH_3CH_2Br}$ $HC\equiv C-CH_2CH_3$ $\xrightarrow{\text{NaNH}_2}$ $\Big\downarrow CH_3CH_2Br$

alkene must be cis to produce the (±) product from anti addition

Review the stereochemistry in the solution to Problem 8-35, p. 171 of this Solutions Manual.

$\xleftarrow{Br_2}$ C=C (with CH_3CH_2 and CH_2CH_3 cis, H and H) $\xleftarrow[\substack{\text{Lindlar} \\ \text{catalyst}}]{H_2}$ $CH_3CH_2-C\equiv C-CH_2CH_3$

(j) $HC\equiv C-H$ $\xrightarrow{\text{NaNH}_2}$ $HC\equiv C\colon^-\ Na^+$ $\xrightarrow{CH_3I}$ $HC\equiv C-CH_3$ $\xrightarrow[\text{2) CH}_3\text{I}]{\text{1) NaNH}_2}$ $H_3C-C\equiv C-CH_3$ $\Big\downarrow^{H_2}$ Lindlar catalyst

Alternatively, *trans*-but-2-ene could be anti-hydroxylated with aqueous peracetic acid.

Review the stereochemistry in the solution to Problem 8-35, p. 171 of this Solutions Manual.

meso

$\xleftarrow[\text{H}_2\text{O}_2]{\text{OsO}_4}$ C=C (H_3C, CH_3 / H, H)

alkene must be cis to produce the meso product from syn addition

9-38

Compound **X** $\xrightarrow[\text{Pt}]{\textbf{5 } H_2}$ (cyclohexyl)$-CH_2CH_2CH_2CH_3$

$\Longrightarrow$ the fact that five equivalents of hydrogen are consumed says that **X** must have five pi bonds in the above carbon skeleton

$\Big\downarrow \substack{O_3 \\ Me_2S}$

$H-\overset{O}{\overset{\|}{C}}-CH_2CH_2-\overset{O}{\overset{\|}{C}}-\overset{O}{\overset{\|}{C}}-H$ $+$ $H-\overset{O}{\overset{\|}{C}}-\overset{O}{\overset{\|}{C}}-H$ $+$ $H-\overset{O}{\overset{\|}{C}}-\overset{O}{\overset{\|}{C}}-OH$ $+$ $H-\overset{O}{\overset{\|}{C}}-OH$

from $C\equiv C$

6 carbonyls $\Longrightarrow$ 3 alkenes   2 carboxylic acids $\Longrightarrow$ 1 alkyne

(phenyl)$-CH=CH-C\equiv CH$

Compound X

Whether the alkene is cis or trans cannot be determined from these results.

206

9-39 Compound **Z**

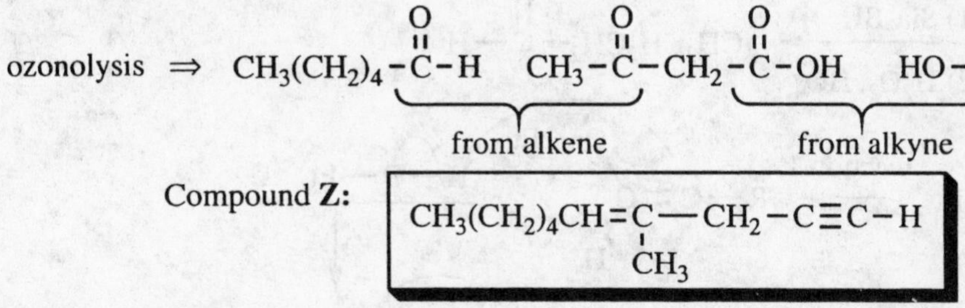

ozonolysis $\Rightarrow$ $CH_3(CH_2)_4-\overset{O}{\overset{\|}{C}}-H$ $CH_3-\overset{O}{\overset{\|}{C}}-CH_2-\overset{O}{\overset{\|}{C}}-OH$ $HO-\overset{O}{\overset{\|}{C}}-H$

$\underbrace{\hspace{4cm}}_{\text{from alkene}}$ $\underbrace{\hspace{4cm}}_{\text{from alkyne}}$

Compound **Z**:

$$CH_3(CH_2)_4CH=\underset{\underset{CH_3}{|}}{C}-CH_2-C\equiv C-H$$

Whether the alkene is *E* or *Z* cannot be determined from this information.

9-40 All four syntheses in this problem begin with the same reaction of benzyl bromide with acetylide ion:

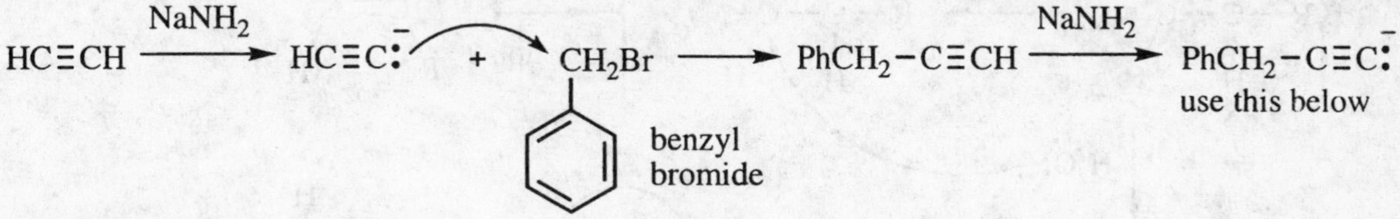

$HC\equiv CH \xrightarrow{NaNH_2} HC\equiv C:^- \; + \; CH_2Br \longrightarrow PhCH_2-C\equiv CH \xrightarrow{NaNH_2} PhCH_2-C\equiv C:^-$

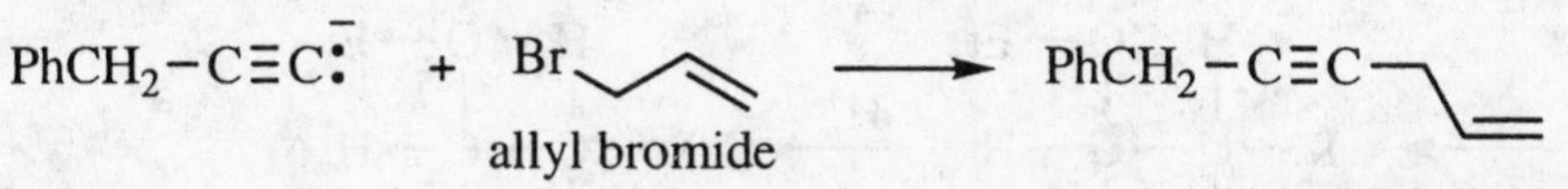

benzyl bromide

use this below

(a) $PhCH_2-C\equiv C:^- \; + \; Br\diagdown\diagup \longrightarrow PhCH_2-C\equiv C\diagup\diagdown$ 6-phenylhex-1-en-4-yne

allyl bromide

(b) $PhCH_2-C\equiv C:^- \; + \; Br\diagup\diagdown \longrightarrow PhCH_2-C\equiv C\diagdown\diagup \xrightarrow[\substack{Pd/BaSO_4 \\ quinoline}]{H_2}$

ethyl bromide

Lindlar catalyst

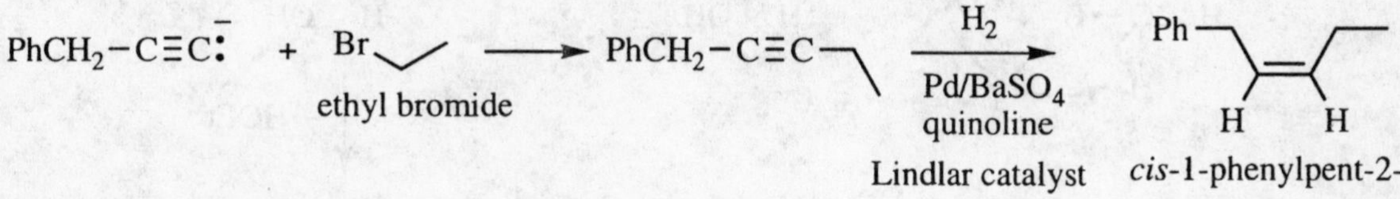

cis-1-phenylpent-2-ene

(c) $PhCH_2-C\equiv C:^- \; + \; Br\diagup\diagdown \longrightarrow PhCH_2-C\equiv C\diagdown\diagup \xrightarrow[NH_3]{Na}$

ethyl bromide

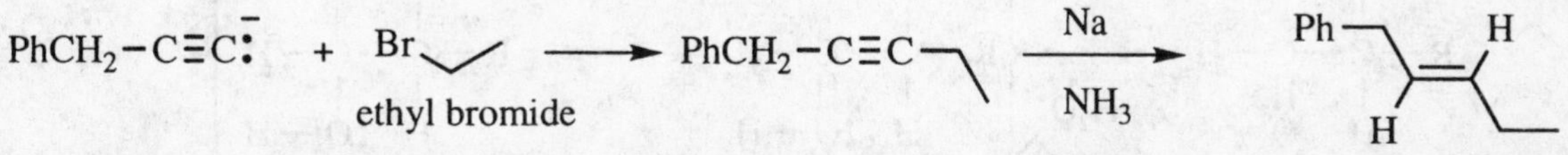

trans-1-phenylpent-2-ene

(d) The diol with the two OH groups on the same side in the Fischer projection is the equivalent of a meso structure, although this one is not meso because the top and bottom group are different. Still, it gives a clue as to its synthesis. The "meso" diol can be formed by either a syn addition to a cis double bond, or by an anti addition to a trans double bond. We saw the same thing in the solution to 9-37 (j).

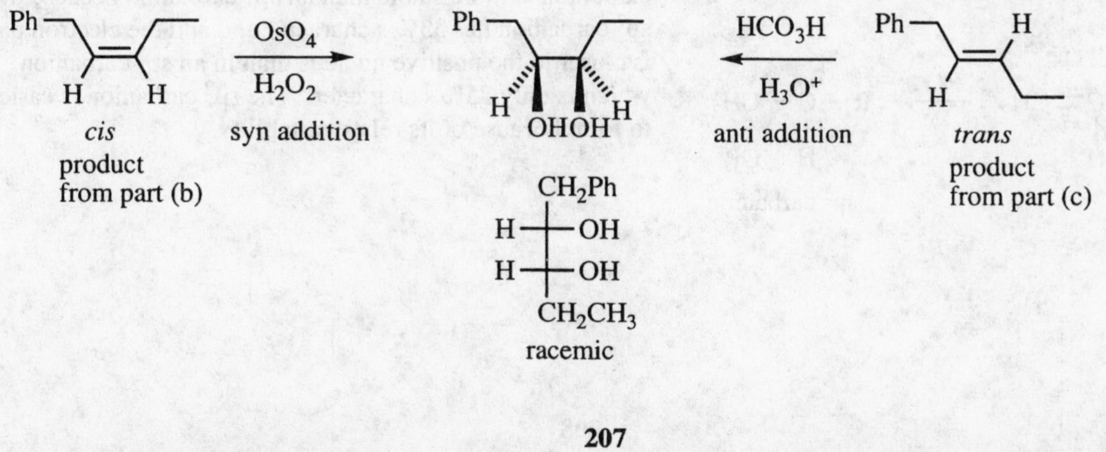

$\xrightarrow[H_2O_2]{OsO_4}$ syn addition

$\xleftarrow[H_3O^+]{HCO_3H}$ anti addition

cis product from part (b)

trans product from part (c)

$$\begin{array}{c} CH_2Ph \\ H-\!\!\!-OH \\ H-\!\!\!-OH \\ CH_2CH_3 \end{array}$$

racemic

207

9-41

(a) $CH_3CH_2-C\equiv C-H$ $\xrightarrow[\text{2) } H_2O_2, \text{ HO}^-]{\text{1) Sia}_2BH}$ $CH_3CH_2CH_2-\overset{\displaystyle O}{\overset{\displaystyle \|}{C}}-H$

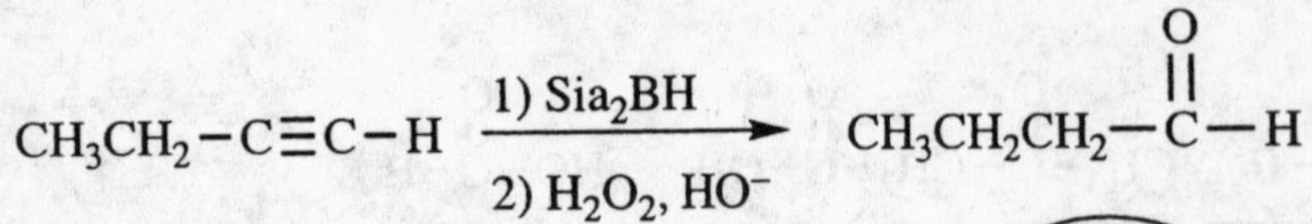

(b)

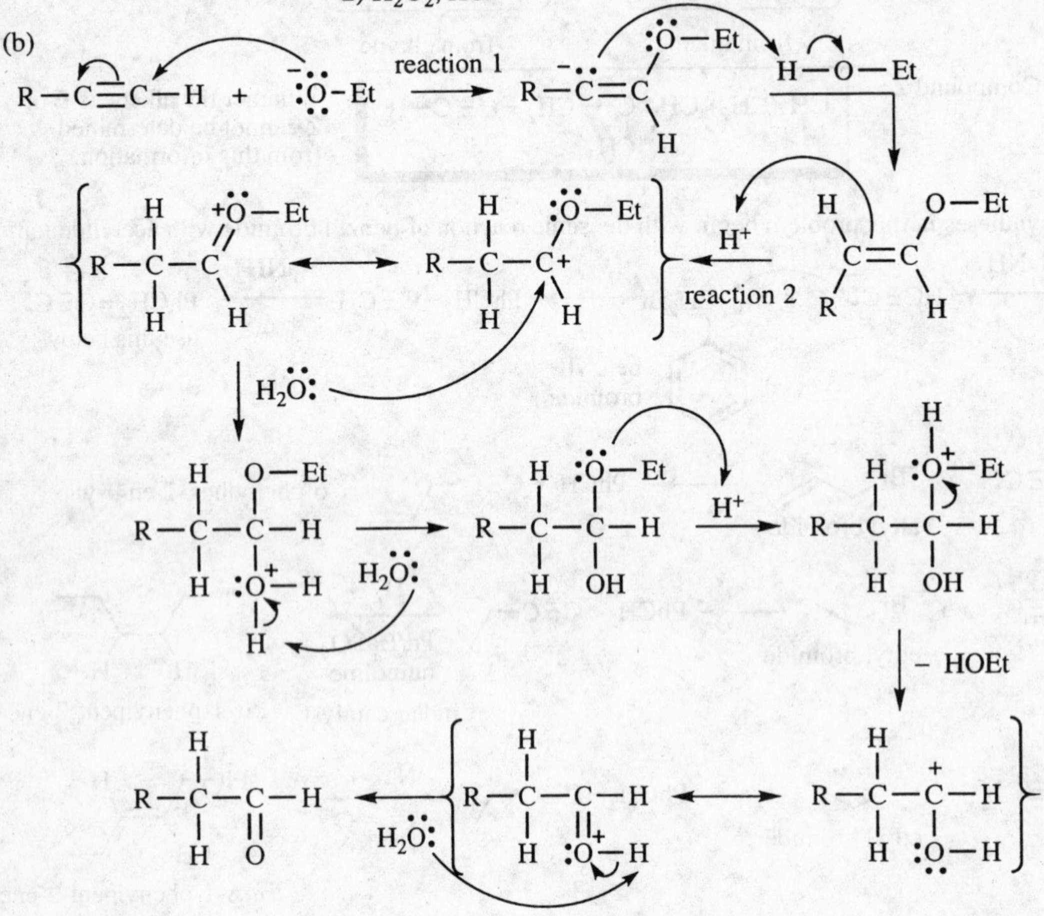

(c) <u>alkyne</u>

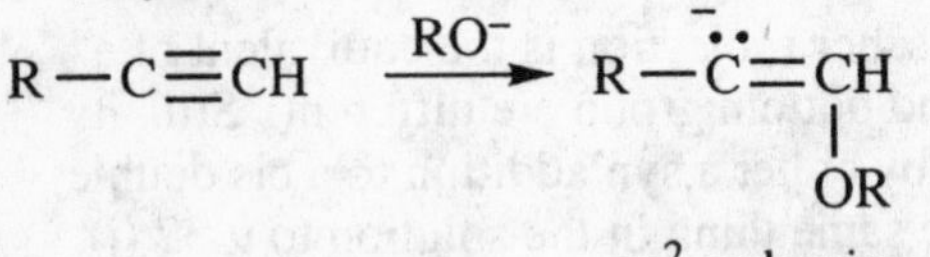

$R-C\equiv CH$ $\xrightarrow{RO^-}$ $R-\overset{..}{\overset{-}{C}}=CH$ | OR

sp^2 carbanion

<u>alkene</u>

$R-\underset{H}{C}=CH_2$ $\xrightarrow{RO^-}$ $R-\underset{H}{\overset{..}{\overset{-}{C}}}-\underset{OR}{CH_2}$

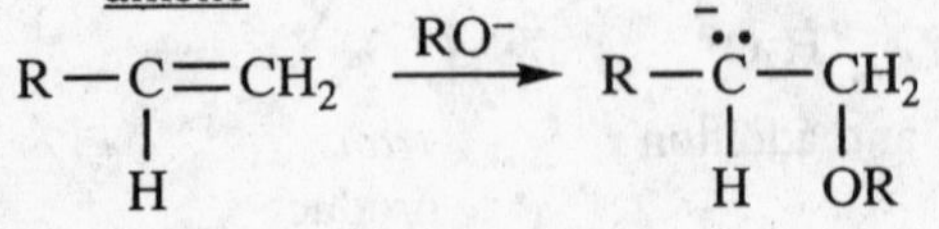

sp^3 carbanion

The closer that electrons are to the nucleus, the more stable. An s orbital is closer to a nucleus than a p orbital is, as p orbitals are elongated away from the nucleus. An sp^2 carbanion is more stable than an sp^3 carbanion because the sp^2 carbanion has 33% s character and and the electron pair is closer to the positive nucleus than in an sp^3 carbanion which is only 25% s character. The sp^2 carbanion is easier to form because of its relative stability.

9-42 Diols are made by two reactions from Chapter 8 and revisited in 9-41 (d): either syn-dihydroxylation with OsO_4, or anti-dihydroxylation via an epoxide using a peroxyacid and water. As this problem says to use inorganic reagents, the solution shown here will use OsO_4.

Recall the stereochemical requirements of syn addition as outlined in this Solutions Manual, p. 171, Problem 8-35:

cis-alkene + **syn** addition → meso

cis-alkene + **anti** addition → racemic (±)

trans-alkene + **syn** addition → racemic (±)

trans-alkene + **anti** addition → meso

Part (a) asks for the synthesis of the meso isomer, so syn addition will have to occur on the cis-alkene. Part (b) will require syn addition to the trans-alkene to give the (±) product.

(a)

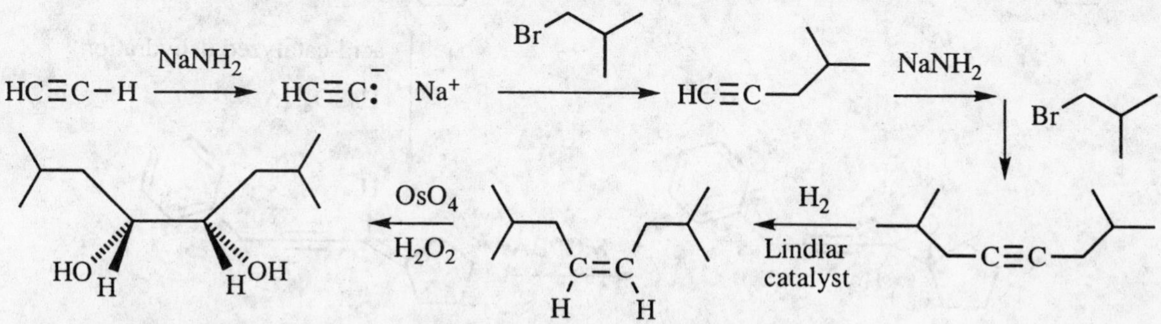

alkene must be cis to produce the meso product from syn addition, so reduction is done with Lindlar catalyst to produce the cis alkene

(b)

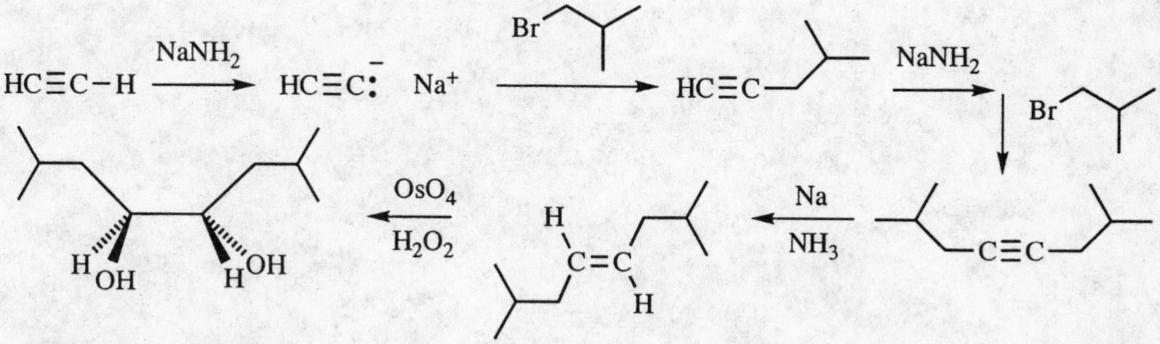

alkene must be trans to produce the (±) product from syn addition, so reduction is done with Na/NH₃

9-43 This synthesis begins the same as the solution to problem 9-40:

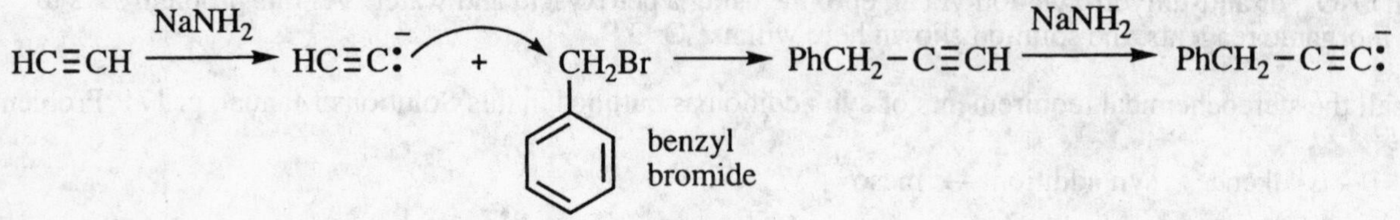
NaNH2
HC≡CH
HC≡C:⁻
+
CH2Br
PhCH2−C≡CH
NaNH2
PhCH2−C≡C:⁻
benzyl
bromide

The anion will add across the carbonyl group of the aldehyde:

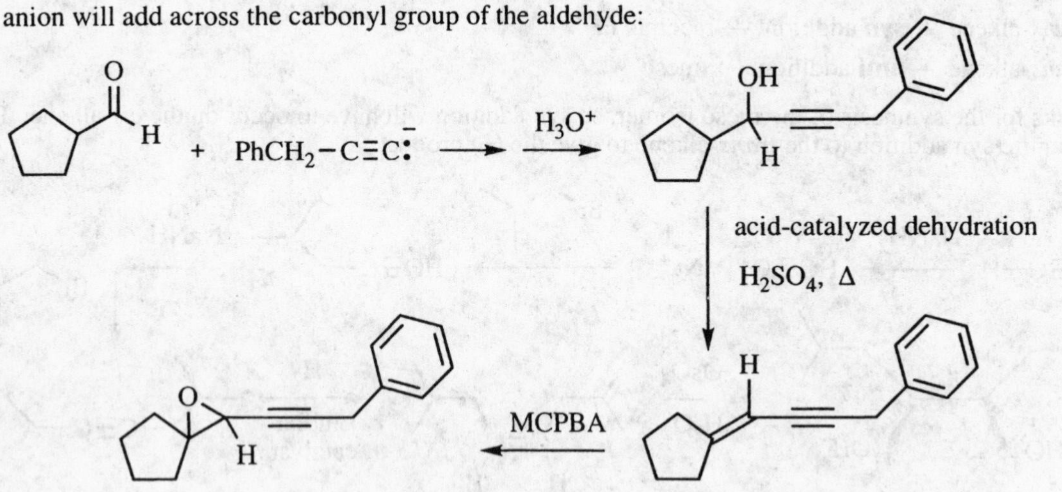
O
H
+ PhCH2−C≡C:⁻
H3O+
OH
H
acid-catalyzed dehydration
H2SO4, Δ
H
MCPBA
O
H

CHAPTER 10—STRUCTURE AND SYNTHESIS OF ALCOHOLS

10-1 Please see the note on p. 136 of this Solutions Manual regarding placement of position numbers.

(a) 2-phenylpropan-2-ol
(b) 5-bromoheptan-2-ol
(c) 4-methylcyclohex-3-en-1-ol ("1" is optional)
(d) *trans*-2-methylcyclohexan-1-ol ("1" is optional)
(e) (*E*)-2-chloro-3-methylpent-2-en-1-ol
(f) (2*R*,3*S*)-2-bromohexan-3-ol

10-2 IUPAC name first, then common name.

(a) cyclopropanol; cyclopropyl alcohol
(b) 2-methylpropan-2-ol; *t*-butyl alcohol
(c) 1-cyclobutylpropan-2-ol; no common name
(d) 3-methylbutan-1-ol; isopentyl alcohol
    (also isoamyl alcohol)

10-3 Only constitutional isomers are requested, not stereoisomers, and only structures with an alcohol group.

(a) C_3H_8O propan-1-ol   propan-2-ol

(b) $C_4H_{10}O$

    butan-1-ol   butan-2-ol   2-methylpropan-1-ol   2-methylpropan-2-ol

(c) C_4H_8O has one element of unsaturation, either a double bond or a ring.

cyclobutanol   cyclopropylmethanol   1-methylcyclopropanol   2-methylcyclopropanol
                                                             cis or *trans*

but-3-en-1-ol   but-3-en-2-ol   but-1-en-2-ol   but-1-en-1-ol (*E* or *Z*)

but-2-en-1-ol   but-2-en-2-ol   2-methyl-   2-methylprop-2-en-1-ol
(*E* or *Z*)   (*E* or *Z*)   prop-1-en-1-ol

(d) C_3H_4O has two elements of unsaturation, so each structure must have either a triple bond, or two double bonds, or a three-membered ring and a double bond. All structures must contain an OH. (In the name, the "e" is dropped from "yne" because it is followed by a vowel in "ol".)

HO–C≡C–CH₃ HC≡C–CH₂   H₂C=C=CH *   OH
  *
prop-1-yn-1-ol   OH   OH
       prop-2-yn-1-ol propa-1,2-dien-1-ol cycloprop-1-en-1-ol cycloprop-2-en-1-ol

*These compounds with the OH bonded directly to the carbon-carbon double bond are called "enols" or "vinyl alcohols." The structure with OH on a carbon-carbon triple bond is called an ynol. These are unstable, although the structures are legitimate.

10-4 (a) 8,8-dimethylnonane-2,7-diol
      (b) octane-1,8-diol
      (c) *cis*-cyclohex-2-ene-1,4-diol
      (d) 3-cyclopentylheptane-2,4-diol
      (e) *trans*-cyclobutane-1,3-diol

10-5 There are four structural features to consider when determining solubility in water: 1) molecules with fewer carbons will be more soluble in water (assuming other things being equal); 2) branched or otherwise compact structures are more soluble than linear structures; 3) more hydrogen-bonding groups will increase solubility; 4) an ionic form of a compound will be more soluble in water than the nonionic form.

(a) Cyclohexanol is more soluble because its alkyl group is more compact than in 1-hexanol.
(b) 4-Methylphenol is more soluble because its hydrocarbon portion is more compact than in 1-heptanol, and phenols form particularly strong hydrogen bonds with water.
(c) 3-Ethylhexan-3-ol is more soluble because its alkyl portion is more spherical than in octan-2-ol.
(d) Cyclooctane-1,4-diol is more soluble because it has two OH groups which can hydrogen bond with water, whereas hexan-2-ol has only one OH group. (The ratio of carbons to OH is 4 to 1 in the former compound and 6 to 1 in the latter; the smaller this ratio, the more soluble.)
(e) These are enantiomers and will have identical solubility.

10-6 Dimethylamine molecules can hydrogen bond among themselves so it takes more energy (higher temperature) to separate them from each other. Trimethylamine has no N-H and cannot hydrogen bond, so it takes less energy to separate these molecules from each other, despite its higher molecular weight.

10-7   See Appendix 2 at the back of this Solutions Manual for a review of acidity and basicity.
(a) Methanol is more acidic than *t*-butyl alcohol. The greater the substitution, the lower the acidity.
(b) 2-Chloropropan-1-ol is more acidic because the electron-withdrawing chlorine atom is closer to the OH group than in 3-chloropropan-1-ol.
(c) 2,2-Dichloroethanol is more acidic because two electron-withdrawing chlorine atoms increase acidity more than just the one chlorine in 2-chloroethanol.
(d) 2,2-Difluoropropan-1-ol is more acidic because fluorine is more electronegative than chlorine; the stronger the electron-withdrawing group, the more acidic the alcohol.

10-8

most   least
acidic   acidic
sulfuric acid >> 2-chloroethanol > water > ethanol > *t*-butyl alcohol > ammonia > hexane

H_2SO_4 $ClCH_2CH_2OH$ H_2O CH_3CH_2OH $(CH_3)_3COH$ NH_3 C_6H_{14}

Sulfuric acid is one of the strongest acids known. On the other extreme, alkanes like hexane are the least acidic compounds. The N-H bond in ammonia is less acidic than any O-H bond. Among the four compounds with O-H bonds, the tertiary alcohols are the least acidic. Water is more acidic than most alcohols including ethanol. However, if a strong electron-withdrawing substituent like chlorine is near the alcohol group, the acidity increases enough so that it is more acidic than water. (Determining exactly where water appears in this list is the most difficult part.)

10-9 Resonance forms of phenoxide anion show the negative charge delocalized onto the ring only at carbons 2, 4, and 6:

212

10-9 continued

Nitro group at position 2

Nitro at position 2
delocalizes
negative charge.

Nitro group at position 3

Nitro at position 3 cannot delocalize negative charge at position 2 or 4—
no resonance stabilization.

Nitro group at position 4

Nitro at position 4 delocalizes negative charge.

Only when the nitro group is at one of the negative carbons will the nitro have a stabilizing effect (via resonance). Thus, 2-nitrophenol and 4-nitrophenol are substantially more acidic than phenol itself, but 3-nitrophenol is only slightly more acidic than phenol (due to the inductive effect).

10-10

A

B

(a) Structure **A** is a phenol because the OH is bonded to a benzene ring. As a phenol, it will be acidic enough to react with sodium hydroxide to generate a phenoxide ion that will be fairly soluble in water. Structure **B** is a 2° benzylic alcohol, not a phenol, not acidic enough to react with NaOH.
(b) Both of these organic compounds will be soluble in an organic solvent like dichloromethane. Shaking this organic solution with aqueous sodium hydroxide will ionize the phenol **A**, making it more polar and water soluble; it will be extracted from the organic layer into the water layer, while the alcohol will remain in the organic solvent. Separating these immiscible solvents will separate the original compounds. The alcohol can be retrieved by evaporating the organic solvent. The phenol can be isolated by acidifying the basic aqueous solution and filtering if the phenol is a solid, or separating the layers if the phenol is a liquid.

10-11 The Grignard reaction needs a solvent containing an ether functional group: (b), (f), (g), and (h) are possible solvents. Dimethyl ether, (b), is a gas at room temperature, however, so it would have to be liquefied at low temperature for it to be a useful solvent.

10-12

(a) CH_3CH_2MgBr (b) ⟍⟋⟍Li + LiI   (c) F—⟨ ⟩—MgBr   (d) ⟍⟋⟍ + LiCl

10-13 Any of three halides—chloride, bromide, iodide, but not fluoride—can be used. Ether is the typical solvent for Grignard reactions.

(a) ⬡—MgCl + H₂C=O →(ether)→(H₃O⁺)→ ⬡—CH₂OH

(b) ⟍⟋⟍MgBr + H₂C=O →(ether)→(H₃O⁺)→ ⟍⟋⟍OH

(c) ⬠—MgI + H₂C=O →(ether)→(H₃O⁺)→ ⬠—CH₂OH

Note: the alternative arrow symbolism could also be used, where the two steps are numbered around one arrow:

1) ether
2) H_3O^+ OK

NO! Me BAD!
This means that water is present with ether during the Grignard reaction.

~~ether / H₃O⁺~~

10-14 Any of three halides—chloride, bromide, iodide, but not fluoride—can be used. Grignard reactions are always performed in ether solvent; ether is not shown here.

(a) two methods

Where two methods can be used to form the target compound, the newly formed bond is shown in bold.

(b) two methods

214

10-14 continued

(c)

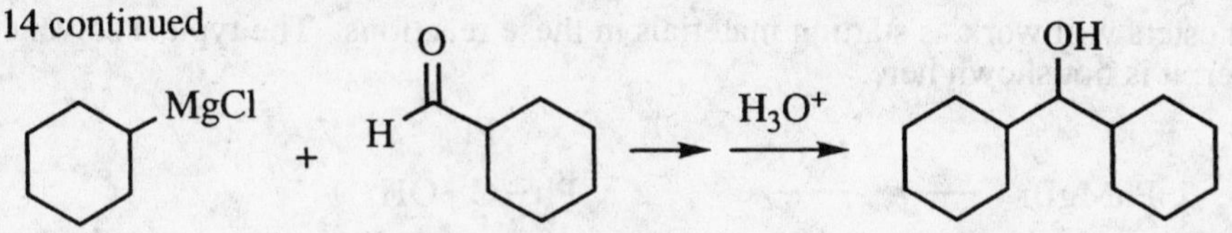

10-15 Grignard reactions are always performed in ether. Here, the ether is not shown.

(a) Any of the three bonds shown in bold can be formed by adding a Grignard reagent across a ketone.

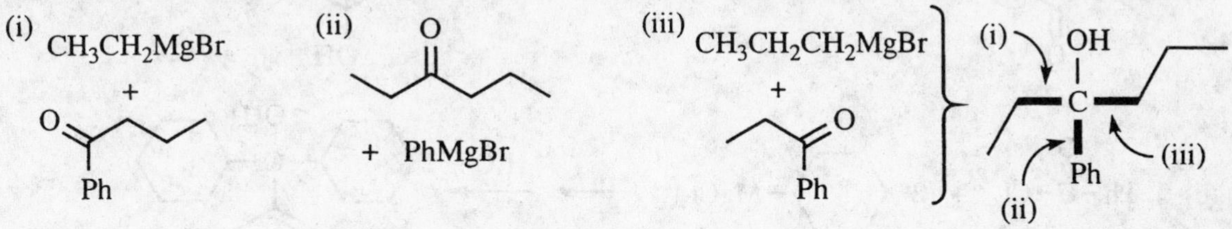

(i) CH_3CH_2MgBr
+

(ii)

(iii) $CH_3CH_2CH_2MgBr$
+

+ PhMgBr

(b)

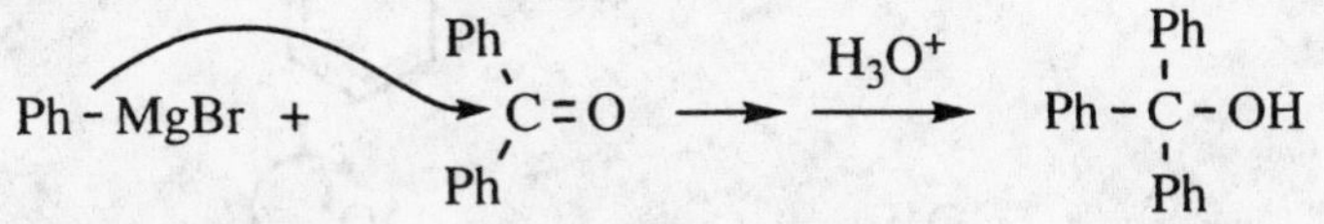

$Ph-MgBr$ +

$\xrightarrow{H_3O^+}$

(c) CH_3-MgI

$\xrightarrow{H_3O^+}$

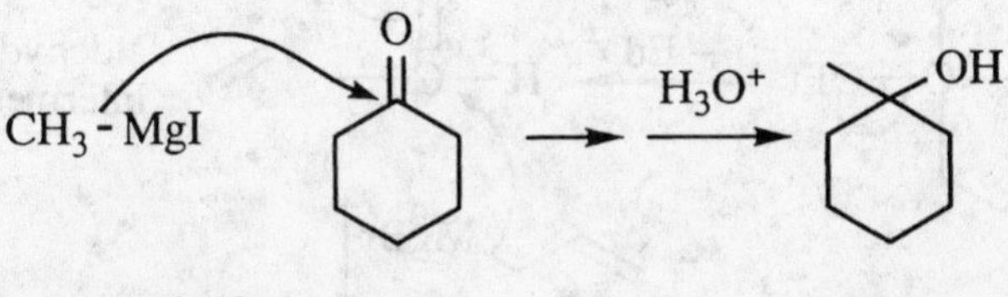

(d) two methods

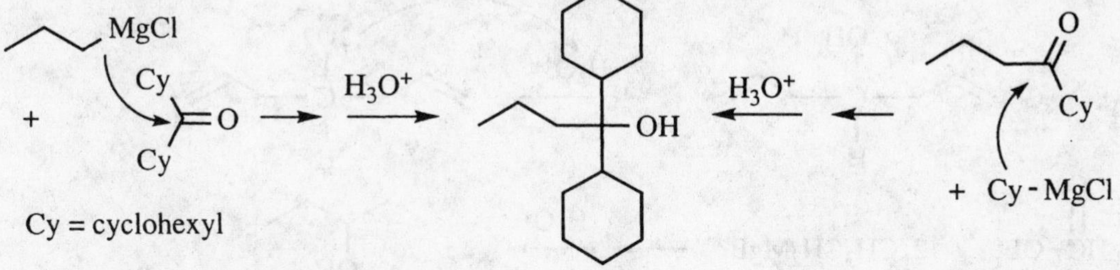

Cy = cyclohexyl

10-16

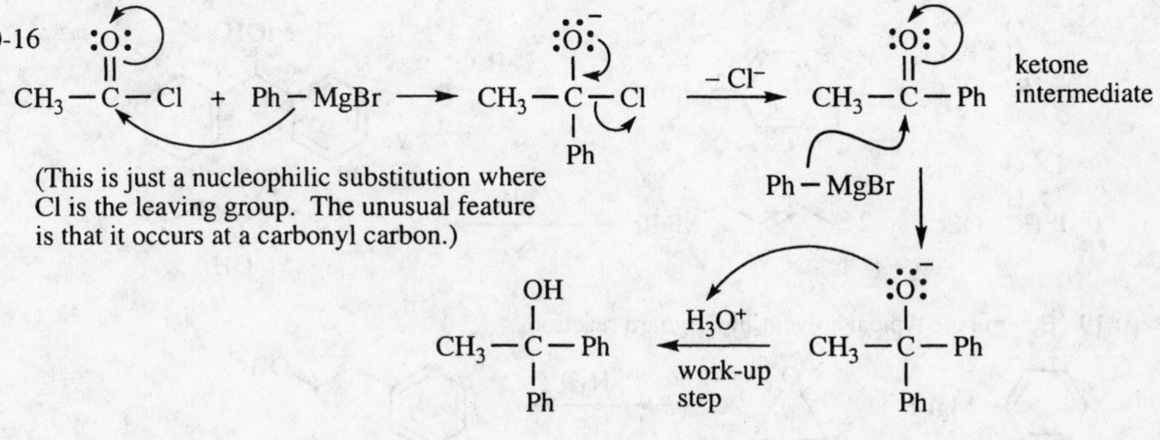

(This is just a nucleophilic substitution where Cl is the leaving group. The unusual feature is that it occurs at a carbonyl carbon.)

ketone intermediate

215

10-17 Acid chlorides or esters will work as starting materials in these reactions. The typical solvent for Grignard reactions is ether; it is not shown here.

(a) 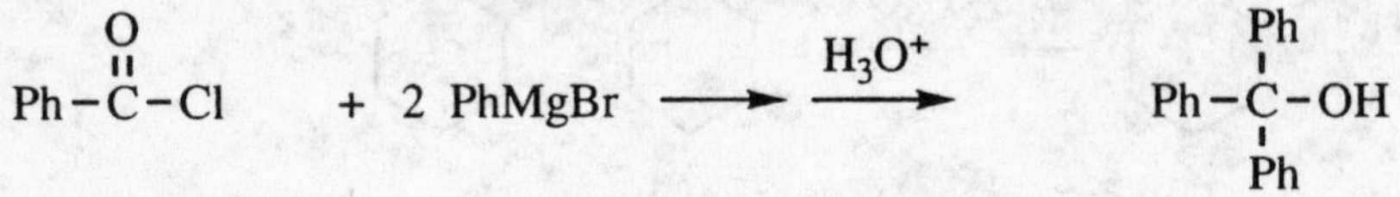 Ph—$\overset{\overset{\text{O}}{\|}}{\text{C}}$—Cl + 2 PhMgBr $\xrightarrow{}$ $\xrightarrow{\text{H}_3\text{O}^+}$ Ph—$\overset{\overset{\text{Ph}}{|}}{\underset{\text{Ph}}{\text{C}}}$—OH

(b) [isobutyrate ester with OCH₃] + 2 CH₃CH₂MgI $\xrightarrow{}$ $\xrightarrow{\text{H}_3\text{O}^+}$ [tertiary alcohol with OH]

(c) Ph—$\overset{\overset{\text{O}}{\|}}{\text{C}}$—Cl + 2 [cyclohexyl]—MgCl $\xrightarrow{}$ $\xrightarrow{\text{H}_3\text{O}^+}$ [dicyclohexyl phenyl carbinol with OH]

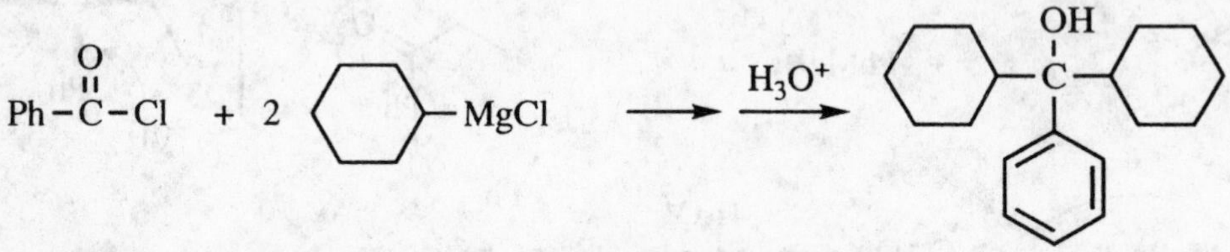

10-18

(a) 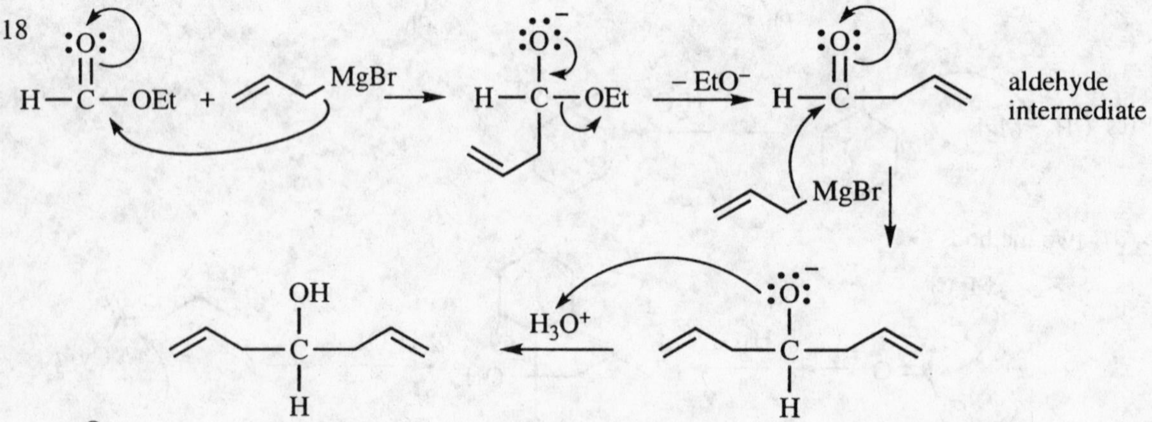 H—$\overset{\overset{:\text{O}:}{\|}}{\text{C}}$—OEt + [allyl]—MgBr $\longrightarrow$ H—$\overset{\overset{:\ddot{\text{O}}:^-}{|}}{\text{C}}$—OEt $\xrightarrow{-\text{EtO}^-}$ H—$\overset{\overset{:\text{O}:}{\|}}{\text{C}}$—[allyl]   aldehyde intermediate

MgBr $\downarrow$

[product with OH, two allyl groups] $\xleftarrow{\text{H}_3\text{O}^+}$ [alkoxide intermediate with :Ö:⁻, two allyl groups and H]

(b) (i) 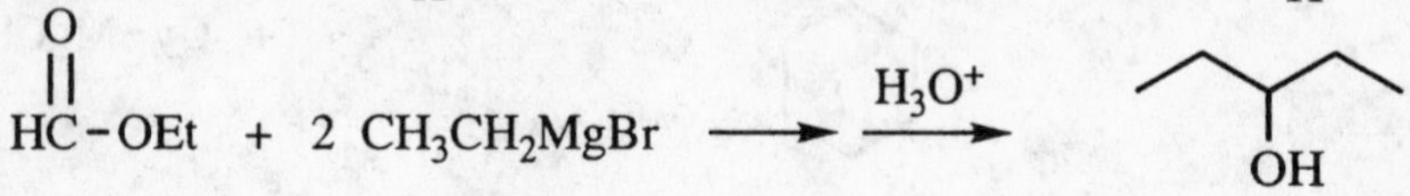 $\overset{\overset{\text{O}}{\|}}{\text{HC}}$—OEt + 2 CH₃CH₂MgBr $\xrightarrow{}$ $\xrightarrow{\text{H}_3\text{O}^+}$ [3-pentanol, OH]

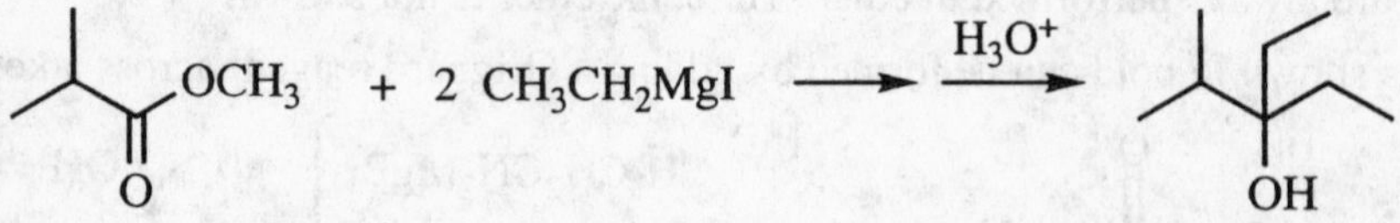

(ii) 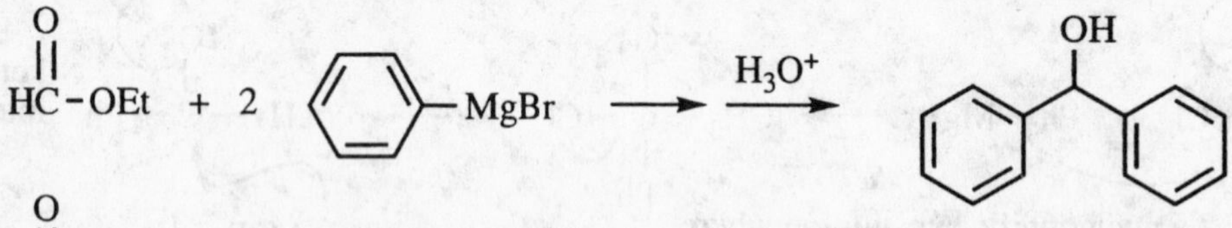 $\overset{\overset{\text{O}}{\|}}{\text{HC}}$—OEt + 2 [phenyl]—MgBr $\xrightarrow{}$ $\xrightarrow{\text{H}_3\text{O}^+}$ [diphenylmethanol, OH]

(iii) 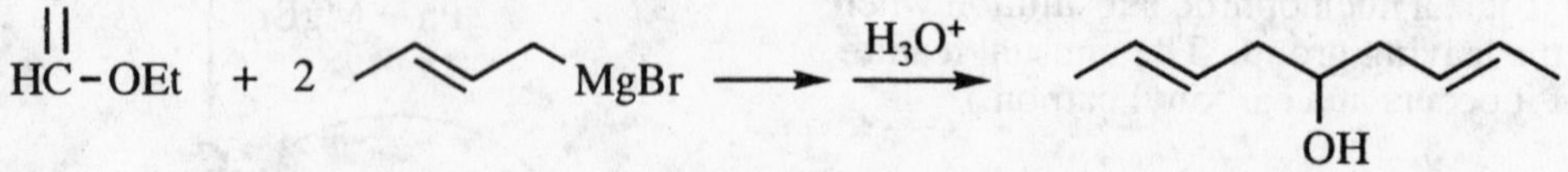 $\overset{\overset{\text{O}}{\|}}{\text{HC}}$—OEt + 2 [crotyl]—MgBr $\xrightarrow{}$ $\xrightarrow{\text{H}_3\text{O}^+}$ [diene alcohol, OH]

10-19 Ether is the typical solvent in Grignard reactions.

(a) 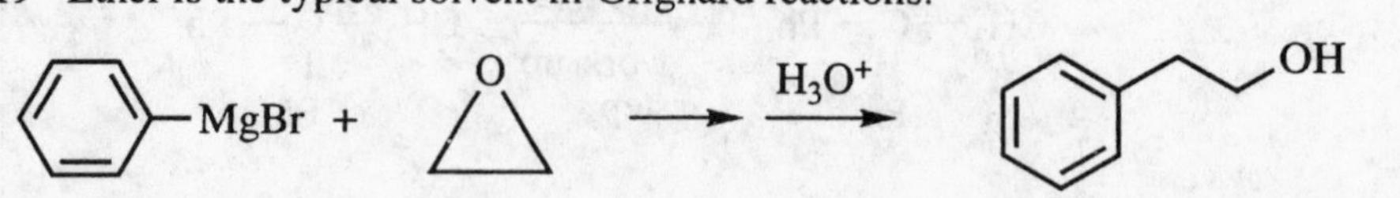 [phenyl]—MgBr + [epoxide] $\xrightarrow{}$ $\xrightarrow{\text{H}_3\text{O}^+}$ [phenethyl alcohol, OH]

10-19 continued

(b) 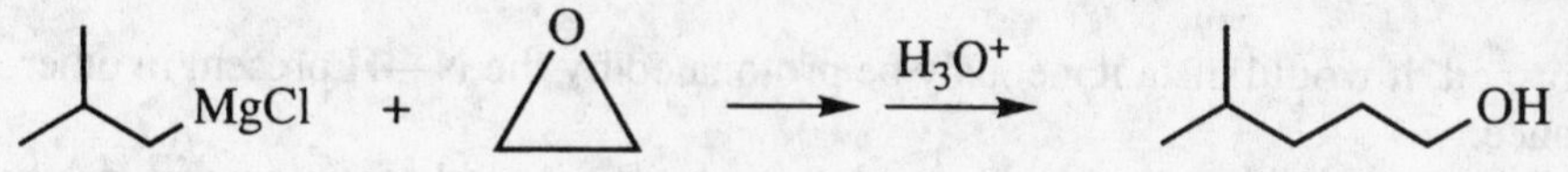 isobutyl-MgCl + epoxide → H₃O⁺ → product-OH

(c) 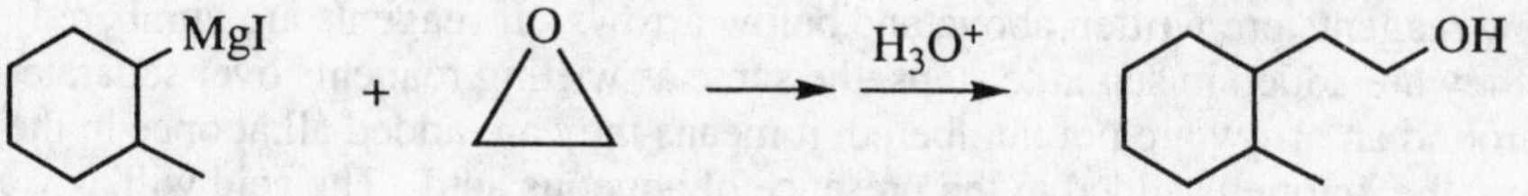 cyclohexyl-MgI + methyl substituent + epoxide → H₃O⁺ → product-OH

10-20

(a) 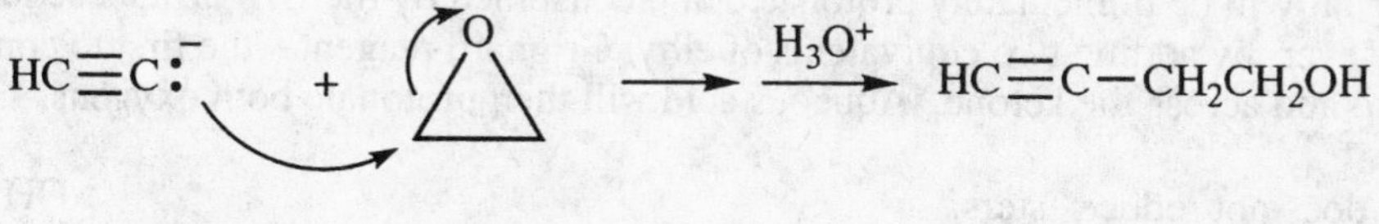 $HC{\equiv}C:^-$ + epoxide → H_3O^+ → $HC{\equiv}C-CH_2CH_2OH$

(b) 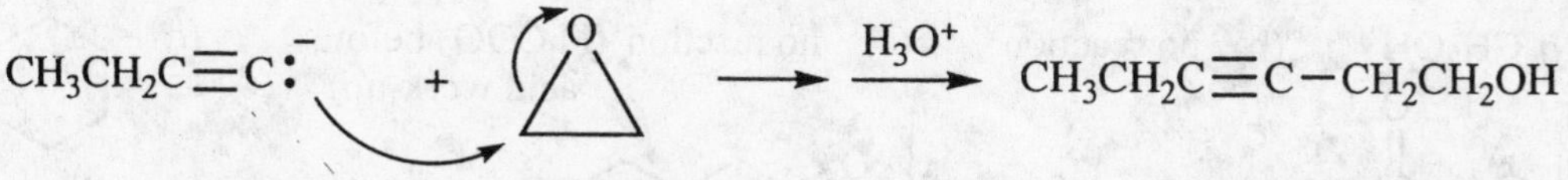 $CH_3CH_2C{\equiv}C:^-$ + epoxide → H_3O^+ → $CH_3CH_2C{\equiv}C-CH_2CH_2OH$

10-21 There are more than one synthetic route to each structure; the ones shown here are representative. The new bonds formed are shown here in bold. Your answers may be different and still be correct.

(a) 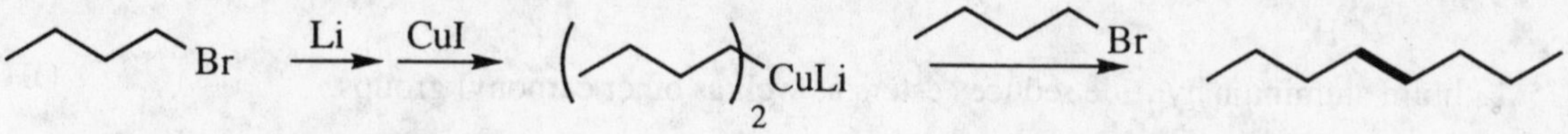 $\diagdown$Br $\xrightarrow{\text{Li}}$ $\xrightarrow{\text{CuI}}$ $\left(\diagdown\right)_2$CuLi $\xrightarrow{\diagup\diagdown\text{Br}}$ product

(b) 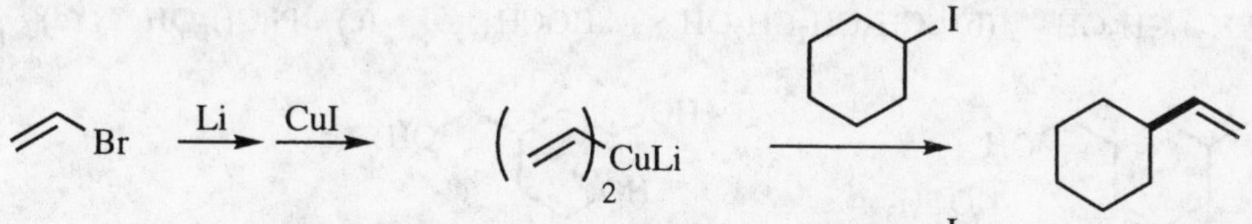 $\diagup$Br $\xrightarrow{\text{Li}}$ $\xrightarrow{\text{CuI}}$ $\left(\diagup\right)_2$CuLi $\xrightarrow{\text{cyclohexyl-I}}$ product

(c) 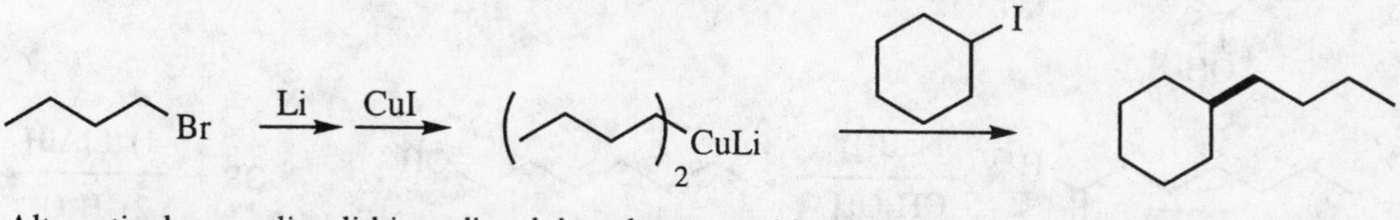 $\diagdown$Br $\xrightarrow{\text{Li}}$ $\xrightarrow{\text{CuI}}$ $\left(\diagdown\right)_2$CuLi $\xrightarrow{\text{cyclohexyl-I}}$ product

Alternatively, coupling lithium dicyclohexylcuprate with 1-bromobutane would also work. As this mechanism is not a typical S_N2, it is not as susceptible to steric hindrance like acetylide ion substitution or a similar S_N2 reaction.

(d) 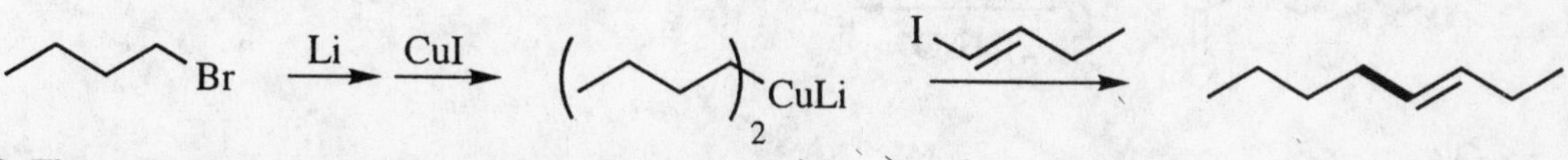 $\diagdown$Br $\xrightarrow{\text{Li}}$ $\xrightarrow{\text{CuI}}$ $\left(\diagdown\right)_2$CuLi $\xrightarrow{\text{I}\diagup\diagdown}$ product

10-22 These reactions are acid-base reactions in which an acidic proton (or deuteron) is transferred to a basic carbon in either a Grignard reagent or an alkyllithium.

(a) CH_3D + Mg(OD)I

(b) $CH_3CH_2CH_2CH_3$ + $LiOCH_2CH_3$

(c) 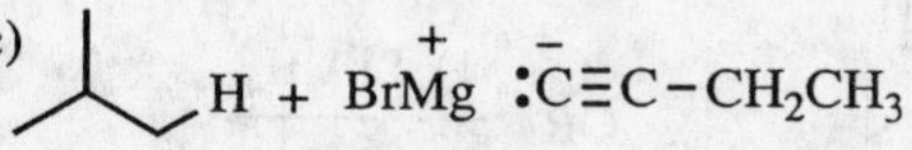 isobutyl-H + $BrMg$ $:C{\equiv}C-CH_2CH_3$

(d) 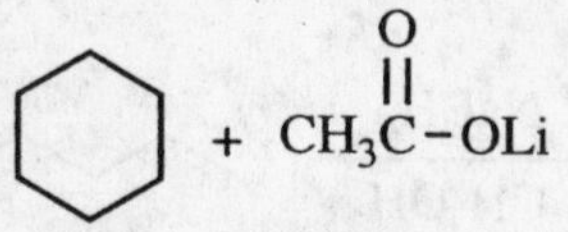 cyclohexane + $CH_3\overset{O}{\overset{\|}{C}}-OLi$

(e) 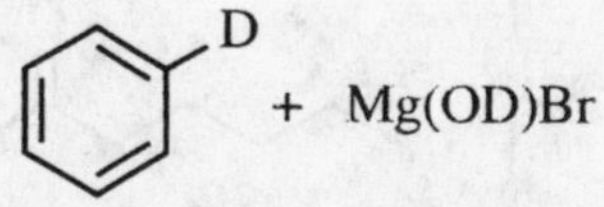 phenyl-D + Mg(OD)Br

217

10-23 Grignard reagents are incompatible with acidic hydrogens and with electrophilic, polarized multiple bonds like C=O, NO$_2$, etc.

(a) As Grignard reagent is formed, it would instantaneously be protonated by the N—H present in other molecules of the same substance.

(b) As Grignard reagent is formed, it would immediately attack the ester functional group present in other molecules of the same substance.

(c) Care must be taken in how reagents are written above and below arrows. If reagents are numbered "1. ... 2. ... *etc.*", it means they are added in separate steps, the same as writing reagents over separate arrows. If reagents written around an arrow are not numbered, it means they are added all at once in the same mixture. In this problem, the ketone is added in the presence of aqueous acid. The acid will immediately protonate and destroy the Grignard reagent before reaction with the ketone can occur.

(d) The ethyl Grignard reagent will be immediately protonated and consumed by the OH. This reaction *could* be made to work, however, by adding two equivalents of ethyl Grignard reagent—the first to consume the OH proton, the second to add across the ketone. Aqueous acid will then protonate both oxygens.

10-24 Sodium borohydride does not reduce esters.

(a) CH$_3$(CH$_2$)$_8$CH$_2$OH   (b) no reaction   (c) no reaction (PhCOO$^-$ before acid work-up)   (d)

(e) HO...OCH$_3$ (f) ester

10-25 Lithium aluminum hydride reduces esters as well as other carbonyl groups.

(a) CH$_3$(CH$_2$)$_8$CH$_2$OH   (b) CH$_3$CH$_2$CH$_2$OH + HOCH$_3$ (c) PhCH$_2$OH   (d)

(e) HO...OH + HOCH$_3$ (f) HO...OH

10-26

(a) NaBH$_4$ / CH$_3$OH   OR   1) LiAlH$_4$ 2) H$_3$O$^+$

OR

1) LiAlH$_4$ 2) H$_3$O$^+$

OR

1) LiAlH$_4$ 2) H$_3$O$^+$

(b) NaBH$_4$ / CH$_3$OH   OR   1) LiAlH$_4$ 2) H$_3$O$^+$

(c)

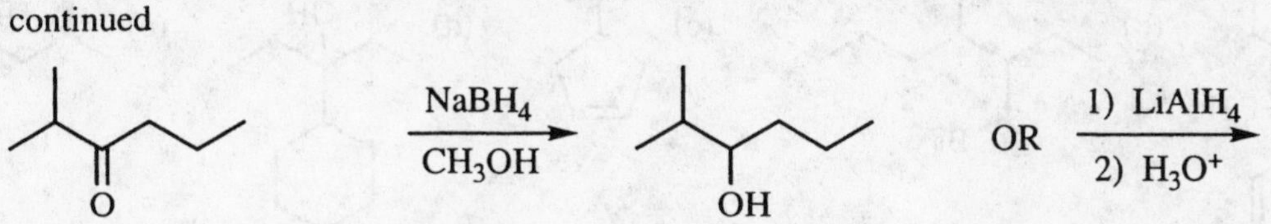

$$\xrightarrow[\text{CH}_3\text{OH}]{\text{NaBH}_4}$$

OR $$\xrightarrow[\text{2) H}_3\text{O}^+]{\text{1) LiAlH}_4}$$

(d)

$$\xrightarrow[\text{CH}_3\text{OH}]{\text{NaBH}_4}$$

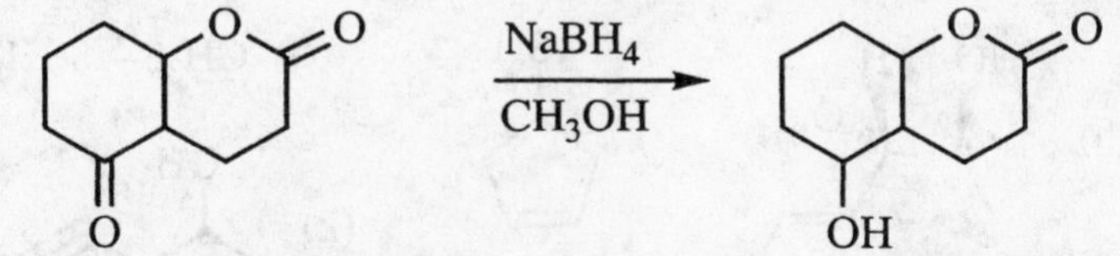

LiAlH$_4$ will NOT give the desired product. LiAlH$_4$ will also reduce the ester in addition to the ketone.

10-27 Approximate pKa values are shown below each compound. Refer to text Tables 1-5, 9-2, and 10-3, and Appendix 5 at the back of the text.

$$\text{CH}_3\text{SO}_3\text{H} \; > \; \text{CH}_3\text{COOH} \; > \; \text{CH}_3\text{SH} \; > \; \text{CH}_3\text{OH} \; > \; \text{CH}_3\text{C}\equiv\text{CH} \; > \; \text{CH}_3\text{NH}_2 \; > \; \text{CH}_3\text{CH}_3$$

| < 0 | 4.74 | ≈ 10.5 | 15.5 | 25 | ≈ 35 | 50 |

most acidic   least acidic

10-28
(a) 4-methylpentane-2-thiol
(b) (Z)-2,3-dimethylpent-2-ene-1-thiol ("1" is optional)
(c) cyclohex-2-ene-1-thiol ("1" is optional)

10-29

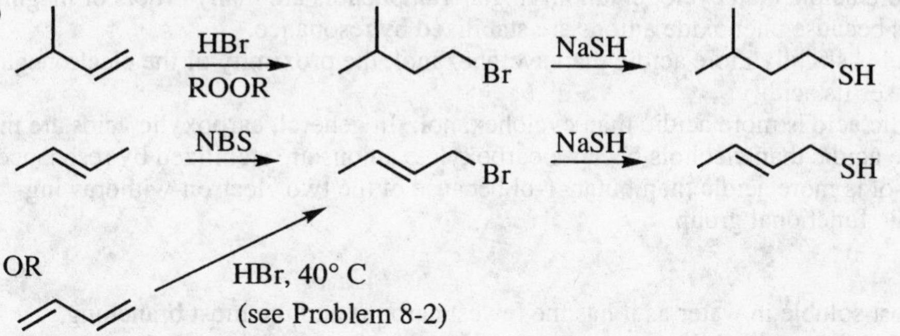

$$\xrightarrow[\text{ROOR}]{\text{HBr}}$$

Br

$$\xrightarrow{\text{NaSH}}$$

SH   3-methylbutane-1-thiol

$$\xrightarrow{\text{NBS}}$$

Br

$$\xrightarrow{\text{NaSH}}$$

SH   2-butene-1-thiol
(but-2-ene-1-thiol)

OR

HBr, 40° C
(see Problem 8-2)

10-30 Please refer to solution 1-20, page 12 of this Solutions Manual.

10-31
(a) 5-methyl-4-*n*-propylheptan-2-ol; 2°
(b) 4-(1-bromoethyl)heptan-3-ol; 2°
(c) (*E*)-4,5-dimethylhex-3-en-1-ol; 1°
(d) 3-bromocyclohex-3-en-1-ol; 2° ("1" is optional)
(e) *cis*-4-chlorocyclohex-2-en-1-ol; 2° ("1" is optional)
(f) 6-chloro-3-phenyloctan-3-ol; 3°
(g) (1-cyclopentenyl)methanol; 1°

10-32

(a) 4-chloro-1-phenylhexane-1,5-diol
(b) *trans*-cyclohexane-1,2-diol
(c) 3-nitrophenol
(d) 4-bromo-2-chlorophenol

10-33

(a) triphenyl C-OH

(b) OH, Br substituted octanol

(c) OH cyclopentenol

(d) OH

(e) OH OH

(f) OH, OH cyclopentane diol

(g) OH, I

(h) HO H, H OH

(i) SH

(j) $CH_3S—SCH_3$

(k) OH

10-34

(a) Hexan-1-ol will boil at a higher temperature as it is less branched than 3,3-dimethylbutan-1-ol.

(b) Hexan-2-ol will boil at a higher temperature because its molecules hydrogen bond with each other, whereas molecules of hexan-2-one have no intermolecular hydrogen bonding.

(c) Hexane-1,5-diol will boil at a higher temperature as it has two OH groups for hydrogen bonding. Hexan-2-ol has only one group for hydrogen bonding.

(d) Hexan-2-ol will boil at a higher temperature because it has a higher molecular weight than pentan-2-ol. All other structural features of the two molecules are the same, so they should have the same intermolecular forces.

10-35 Refer to Table 10-4 to compare acidities of different functional groups.

(a) 3-Chlorophenol is more acidic than cyclopentanol. In general, phenols are many orders of magnitude more acidic than alcohols because phenoxide anions are stabilized by resonance.

(b) 2-Chlorocyclohexanol is slightly more acidic than cyclohexanol; the proximity of the electronegative chlorine to the OH increases its acidity.

(c) Cyclohexanecarboxylic acid is more acidic than cyclohexanol. In general, carboxylic acids are many orders of magnitude more acidic than alcohols because carboxylate anions are stabilized by resonance.

(d) 2,2-Dichlorobutan-1-ol is more acidic than butan-1-ol because of the two electron-withdrawing substituents near the acidic functional group.

10-36

(a) Propan-2-ol is the most soluble in water as it has the fewest carbons and the most branching.

(b) Cyclohexane-1,2-diol is the most soluble as it has two OH groups for hydrogen bonding. Cyclohexanol has only one OH group; chlorocyclohexane cannot hydrogen bond and is the least soluble.

(c) Cyclohexanol is the most soluble as it can hydrogen bond. Chlorocyclohexane cannot hydrogen bond, and 4-methylcyclohexanol has the added hydrophobic methyl group, decreasing its water solubility.

10-37

(a)

$$\xrightarrow[\text{H}_2\text{O}]{\text{Hg(OAc)}_2} \xrightarrow{\text{NaBH}_4}$$ OH

(b)

$$\xrightarrow{\text{BH}_3 \cdot \text{THF}} \xrightarrow[\text{HO}^-]{\text{H}_2\text{O}_2}$$ CH₃, H, OH

10-37 continued

(c)

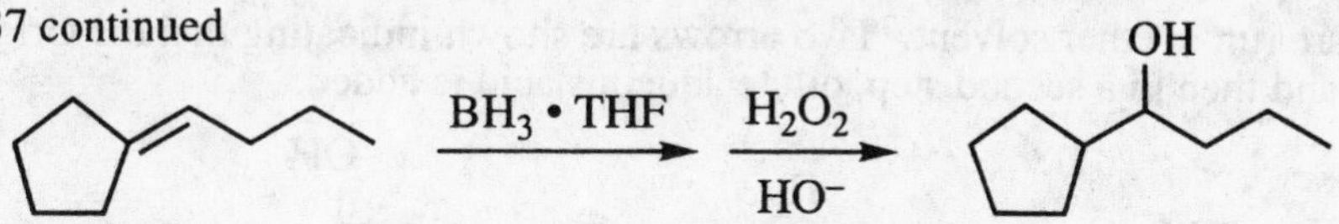

$$\xrightarrow[\text{HO}^-]{\text{BH}_3 \cdot \text{THF} \quad \text{H}_2\text{O}_2}$$

(d)

$$\xrightarrow[\text{H}_2\text{O}]{\text{Hg(OAc)}_2 \quad \text{NaBH}_4}$$

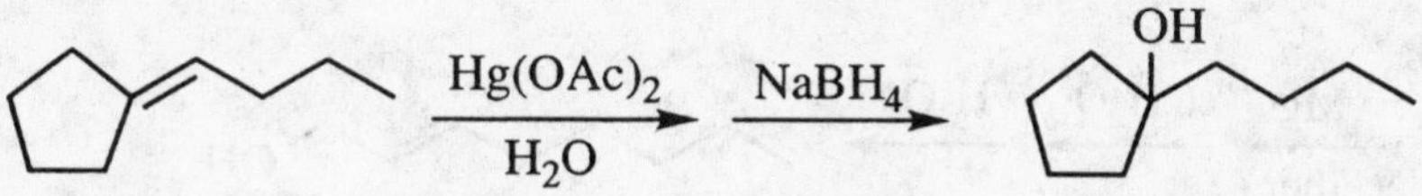

10-38

(a)

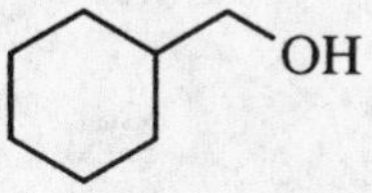

(b)

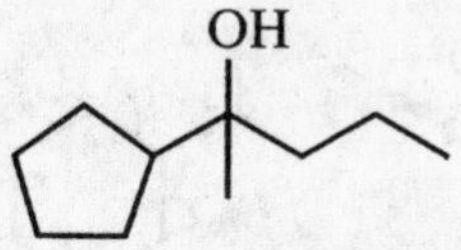

(c)

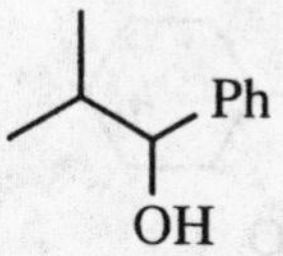

(d) This problem confuses a lot of people. When a Grignard reagent is added to a compound that has an OH group, the first thing that happens is that the Grignard reacts by removing the H$^+$ from the O$^-$.

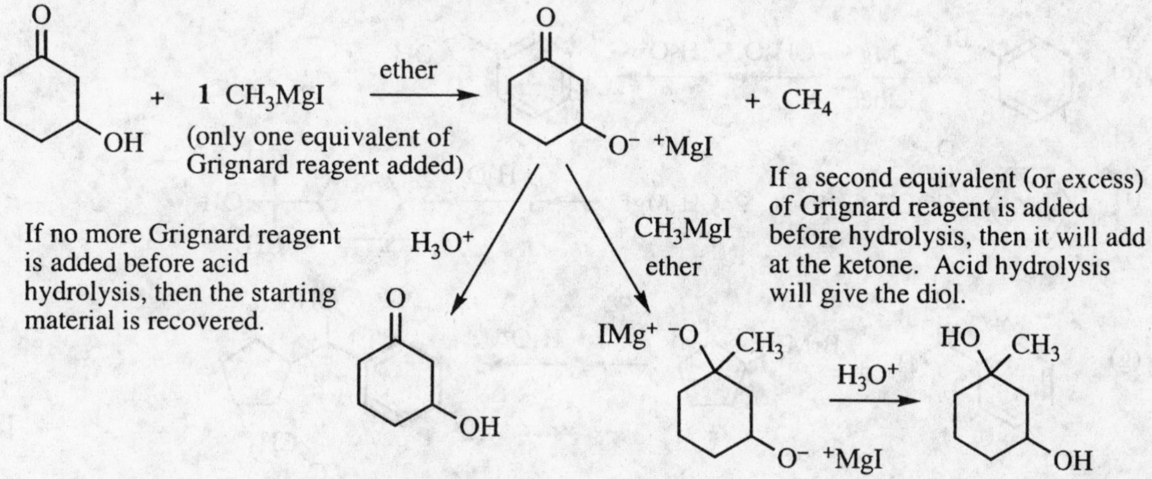

+ **1** CH$_3$MgI $\xrightarrow{\text{ether}}$

(only one equivalent of Grignard reagent added)

+ CH$_4$

If no more Grignard reagent is added before acid hydrolysis, then the starting material is recovered.

H_3O^+

$\xrightarrow[\text{ether}]{\text{CH}_3\text{MgI}}$

If a second equivalent (or excess) of Grignard reagent is added before hydrolysis, then it will add at the ketone. Acid hydrolysis will give the diol.

$\xrightarrow{\text{H}_3\text{O}^+}$

(e)

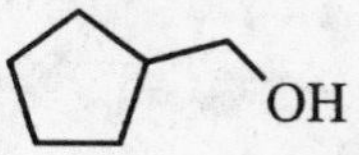

(f) 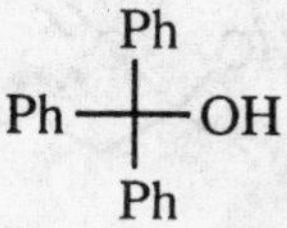 Ph—OH with Ph above and Ph below

(g) 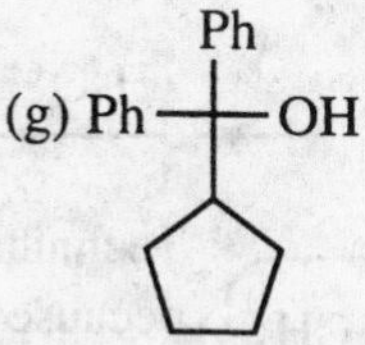 Ph—OH

(h)

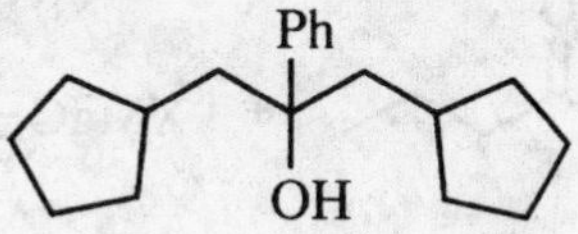

(i)

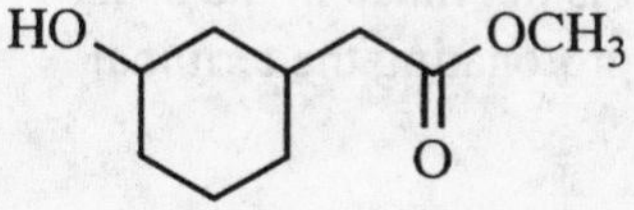

(j)

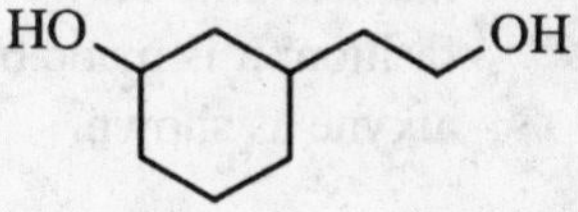

(k)

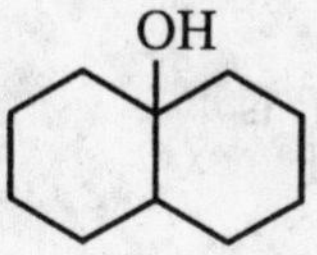

(l)

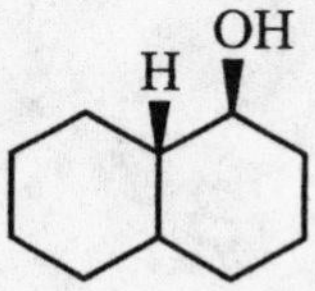

(m)

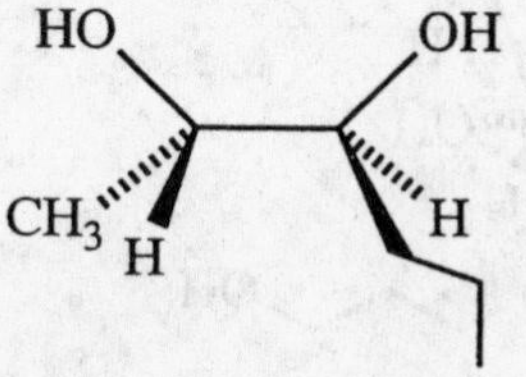

(n)

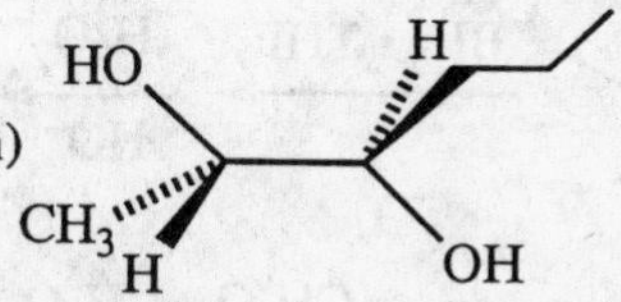

(o)

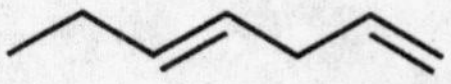

10-39 All Grignard reactions are run in ether solvent. Two arrows are shown indicating that the Grignard reaction is allowed to proceed, and then in a second step, dilute aqueous acid is added.

(a) (pentanal) + BrMg—CH₂CH₃ →(H₃O⁺)→ (heptanol with OH)

(b) (heptyl bromide) →(Mg, ether)→(CH₂O)→(H₃O⁺)→ (primary alcohol chain with OH)

(c) (acetaldehyde) + BrMg—(cyclohexyl) →(H₃O⁺)→ (secondary alcohol with OH)

(d) (cyclohexyl bromide) →(Mg, ether)→(epoxide)→(H₃O⁺)→ (cyclohexyl ethanol with OH)

(e) (bromobenzene) →(Mg, ether)→(CH₂O)→(H₃O⁺)→ (benzyl alcohol with OH)

(f) (cyclohexyl)—CO-CH₂CH₃ + 2 CH₃MgI →(H₃O⁺)→ (cyclohexyl C(CH₃)₂OH)

(g) (benzaldehyde, Ph-CHO) + BrMg—(cyclopentyl) →(H₃O⁺)→ (Ph-CH(OH)-cyclopentyl)

(h) (2-pentanone) + XMgC≡C—CH₃ →(H₃O⁺)→ (HO-C≡C-CH₃ addition product)

$$\underbrace{RMgX + H-C≡C-CH_3}$$

any Grignard reagent where R is alkyl or alkenyl

Technically, XMgC≡CCH₃ is a Grignard reagent because it is an organometallic compound of magnesium. However, it is not made in the usual fashion; it is made by deprotonating the terminal alkyne as shown.

10-40

(a) (propylcyclopentene) →(BH₃ • THF)→(H₂O₂, HO⁻)→ (cyclopentane with H, OH, propyl)

(b) Ph—CH₂CH₂—Cl →(Mg, ether)→(CH₂O)→(H₃O⁺)→ Ph—CH₂CH₂CH₂—OH

10-40 continued

(c)

2-cyclohexenone $\xrightarrow[\text{CH}_3\text{OH}]{\text{NaBH}_4}$ 2-cyclohexenol

(d)

2-cyclohexenone $\xrightarrow[\text{Pt}]{\text{1 eq. H}_2}$ cyclohexanone

(e)

ethyl 4-oxopentanoate $\xrightarrow[\text{CH}_3\text{OH}]{\text{NaBH}_4}$ ethyl 4-hydroxypentanoate

(f)

ethyl 4-oxopentanoate $\xrightarrow{\text{LiAlH}_4}$ $\xrightarrow{\text{H}_3\text{O}^+}$ pentane-1,4-diol

10-41

(a) benzyl MgBr + CH_2O $\xrightarrow{\text{ether}}$ $\xrightarrow{\text{H}_3\text{O}^+}$ 2-phenylethanol

(b) styrene $\xrightarrow{\text{BH}_3 \cdot \text{THF}}$ $\xrightarrow[\text{HO}^-]{\text{H}_2\text{O}_2}$ 2-phenylethanol

(c) cyclohexylmethyl bromide $\xrightarrow{\text{NaOH}}$ cyclohexylmethanol

(d) cyclohexylmethyl bromide $\xrightarrow[\text{ether}]{\text{Mg}}$ $\xrightarrow{\text{epoxide}}$ $\xrightarrow{\text{H}_3\text{O}^+}$ 3-cyclohexylpropanol

(e) allyl-type bromide + NaSH $\longrightarrow$ thiol

(f) isobutyl bromide $\xrightarrow{\text{Li}}$ $\xrightarrow{\text{CuI}}$ $\left(\text{isobutyl}\right)_2\text{CuLi}$ + isobutyl bromide $\longrightarrow$ 2,5-dimethylhexane

10-42 The position of the equilibrium can be determined by the strength of the acids or the bases. The stronger acid and stronger base will always react to give the weaker acid and base, so the side of the equation with the weaker acid and base will be favored at equilibrium. See Appendix 2 in this Solutions Manual for a review of acidity.

(a) $CH_3CH_2O^-$ + phenol-OH $\rightleftharpoons$ CH_3CH_2OH + phenol-O$^-$

     stronger   stronger   weaker   weaker

     base   acid   acid   base

products favored

10-42 continued

(b) KOH + Cl—⬡—OH ⇌ H₂O + Cl—⬡—O⁻ K⁺

with Cl substituent below each ring

stronger   stronger   weaker   weaker
base   acid   acid   base
                              products favored

(c)

⬡⬡-OH + CH₃O⁻ ⇌ ⬡⬡-O⁻ + CH₃OH

stronger   stronger   weaker   weaker
acid   base   base   acid
                              products favored

(d) ⬠-OH + KOH ⇌ H₂O + ⬠-O⁻ K⁺

                weaker   stronger
                base   acid
weaker   stronger
acid   base
reactants favored

(e) $(CH_3)_3CO^-$ + CH_3CH_2OH ⇌ $(CH_3)_3COH$ + $CH_3CH_2O^-$

stronger   stronger   weaker   weaker
base   acid   acid   base
                              products favored

(f) $(CH_3)_3CO^-$ + H_2O ⇌ $(CH_3)_3COH$ + HO^-

stronger   stronger   weaker   weaker
base   acid   acid   base
                              products favored

(g) KOH + CH_3CH_2OH ⇌ H_2O + $CH_3CH_2O^-$ K⁺

weaker   weaker   stronger   stronger
base   acid   acid   base
reactants favored

10-43

(a) ⎓⎓⎓⎓—CHO —NaBH₄/CH₃OH→ ⎓⎓⎓⎓—OH OR —1) LiAlH₄ / 2) H₃O⁺→

OR

⎓⎓⎓⎓—COOH —1) LiAlH₄ / 2) H₃O⁺→

OR

⎓⎓⎓⎓—C(O)OR —1) LiAlH₄ / 2) H₃O⁺→

224

(b) [structure: cyclohexyl propyl ketone] $\xrightarrow[\text{CH}_3\text{OH}]{\text{NaBH}_4}$ [structure: cyclohexyl alcohol] OR $\xrightarrow[\text{2) H}_3\text{O}^+]{\text{1) LiAlH}_4}$

(c) [structure: phenyl propyl ketone] $\xrightarrow[\text{CH}_3\text{OH}]{\text{NaBH}_4}$ [structure: phenyl alcohol] OR $\xrightarrow[\text{2) H}_3\text{O}^+]{\text{1) LiAlH}_4}$

(d) [structure: tetralone] $\xrightarrow[\text{CH}_3\text{OH}]{\text{NaBH}_4}$ [structure: tetralol] OR $\xrightarrow[\text{2) H}_3\text{O}^+]{\text{1) LiAlH}_4}$

(e) [structure: cyclooctanone ester] $\xrightarrow[\text{CH}_3\text{OH}]{\text{NaBH}_4}$ [structure: cyclooctanol ester OEt]

(f) [structure: cyclooctanone ester] $\xrightarrow[\text{2) H}_3\text{O}^+]{\text{1) LiAlH}_4}$ [structure: cyclooctane diol] $+ \; \text{HOCH}_2\text{CH}_3$

10-44 The goal is to synthesize the target compound (boxed) from starting materials of six carbons or fewer. The product has 12 carbons, so the logical "disconnection" in working backwards is two six carbon fragments which could be joined in a Grignard reaction. The best way to make epoxides is from the double bond, and double bonds are made from alcohols which are the products of Grignard reactions.

[reaction scheme: bromobenzene $\xrightarrow[\text{ether}]{\text{Mg}}$ phenylmagnesium bromide + cyclopentanecarbaldehyde $\xrightarrow{\text{H}_3\text{O}^+}$ alcohol $\xrightarrow[\text{H}_2\text{SO}_4]{\Delta}$ benzylidenecyclopentane $\xrightarrow{\text{MCPBA}}$ target (epoxide)]

target

10-45 All steps are reversible.

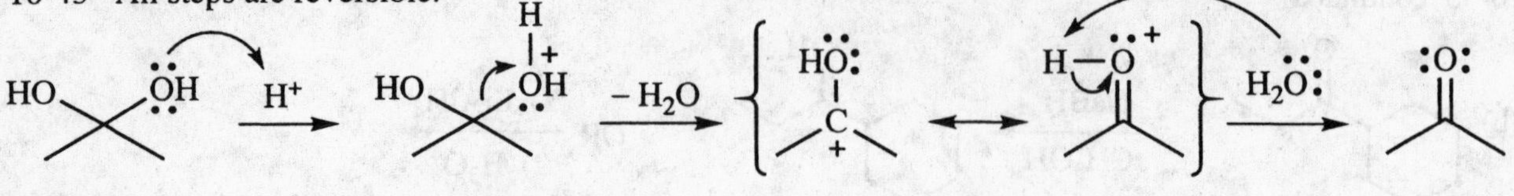

10-46 The symbol H—B represents a generic acid, where B⁻ is the conjugate base.

(a)

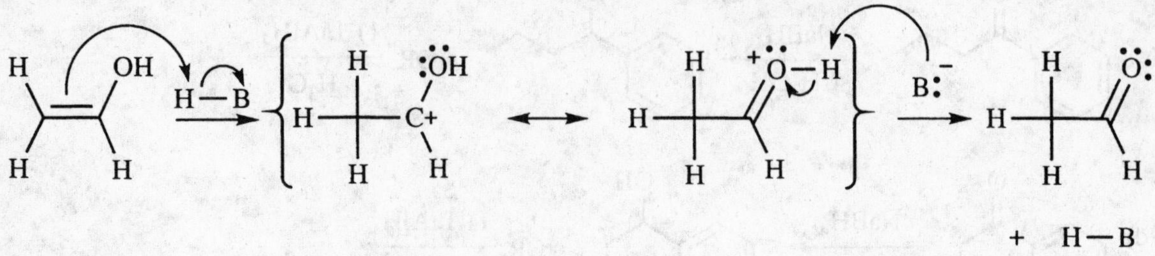

(b)

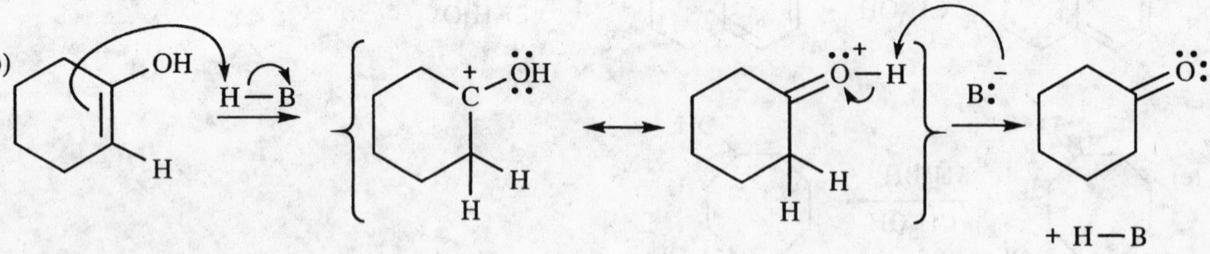

(c)

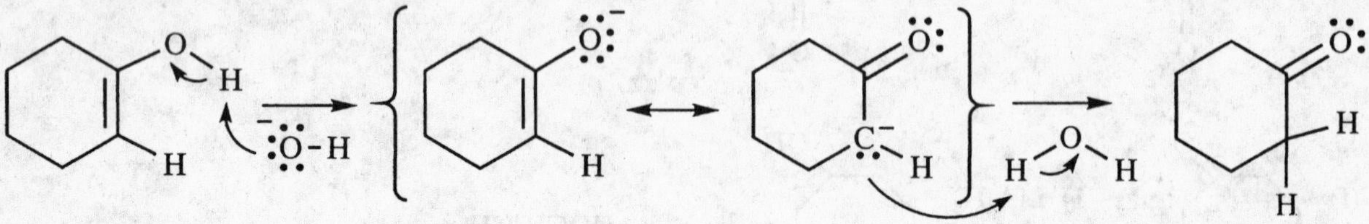

10-47

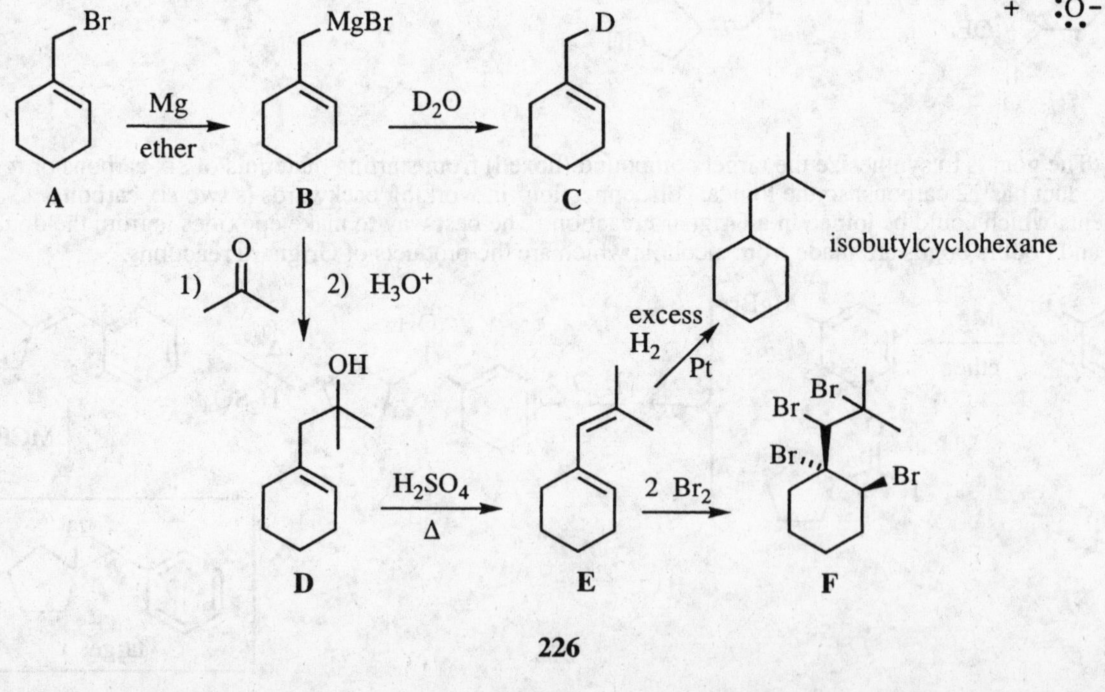

isobutylcyclohexane

10-48 This mechanism is similar to cleavage of the epoxide in ethylene oxide by Grignard reagents. The driving force for the reaction is relief of ring strain in the 4-membered cyclic ether, which is why it will undergo a Grignard reaction whereas most other ethers will not.

10-49 When mixtures of isomers can result, only the major product is shown.

10-50 The most important reactant in the deskunking mixture is hydrogen peroxide. Thiols are oxidized to structures having one, two, or three oxygens on the sulfur; all of these functional groups are acidic. The sodium bicarbonate is basic enough to ionize these acids, making them water soluble where the soap can wash them away.

227

10-51

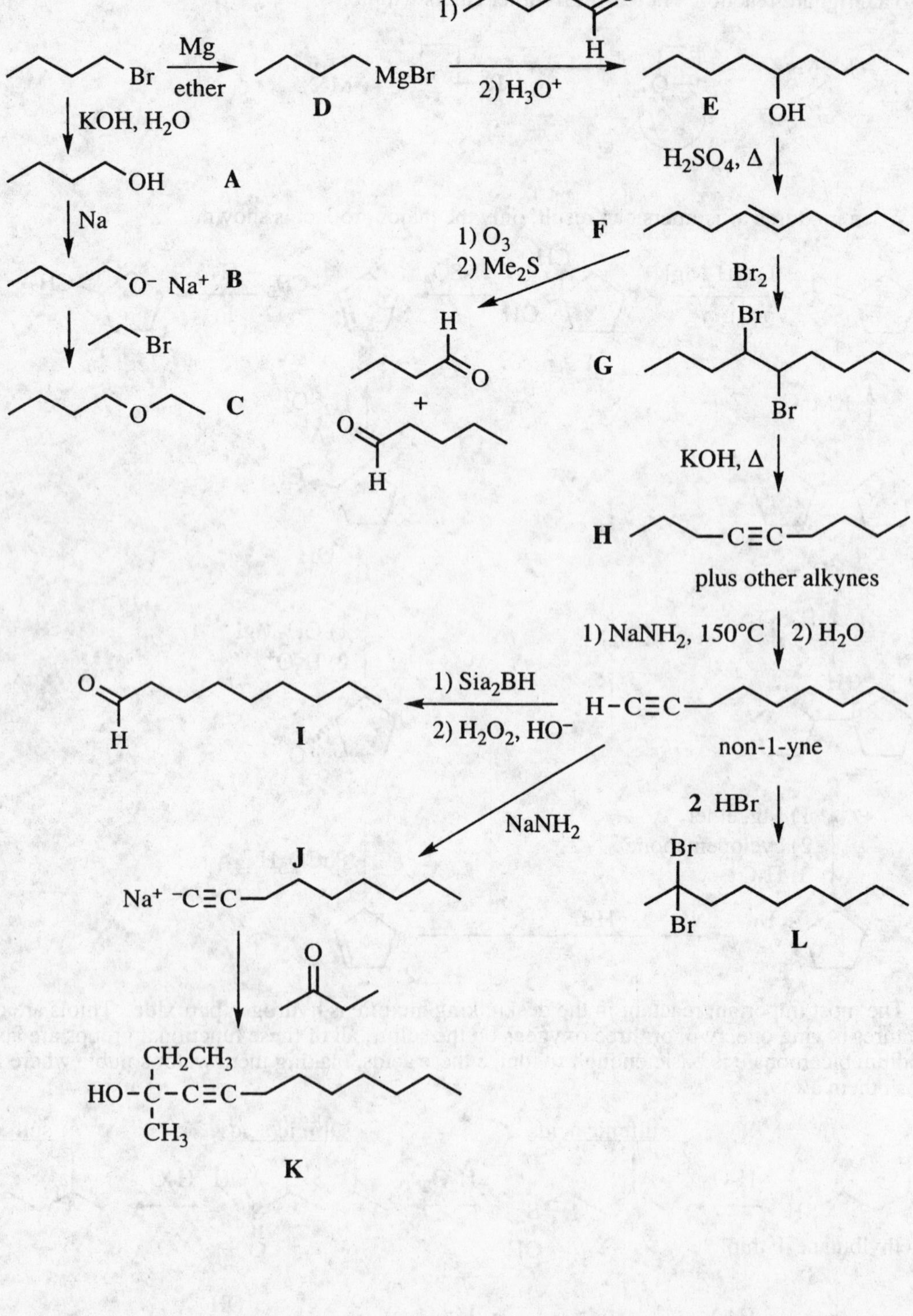

Mg
ether
1)
H
O
Br
MgBr
2) H3O+
D
E
OH
KOH, H2O
OH
A
H2SO4, Δ
Na
F
1) O3
2) Me2S
O− Na+
B
Br2
Br
H
Br
G
O
+
Br
C
O
KOH, Δ
H
H
C≡C
plus other alkynes
1) NaNH2, 150°C
2) H2O
O
1) Sia2BH
H−C≡C
H
I
2) H2O2, HO−
non-1-yne
2 HBr
NaNH2
Br
J
Na+ −C≡C
Br
L
O
CH2CH3
HO−C−C≡C
CH3
K

CHAPTER 11—REACTIONS OF ALCOHOLS

11-1
(a) both reactions are oxidations
(b) oxidation, oxidation, reduction, oxidation
(c) one carbon is oxidized and one carbon is reduced—no net change
(d) reduction: C—O is replaced by C—H
(e) neither oxidation nor reduction—the C still has two bonds to O
(f) oxidation (addition of X_2)
(g) neither oxidation nor reduction (addition of HX)
(h) neither oxidation nor reduction (elimination of H_2O)
(i) oxidation: adding an O to each carbon of the double bond
(j) the first reaction is oxidation as a new C—O bond is formed to each carbon of the alkene; the second reaction is neither oxidation nor reduction, as H_2O is added to the epoxide, and each carbon still has one bond to oxygen
(k) neither oxidation nor reduction: H—B is added in the first reaction, and B is replaced by O in the second reaction; overall, only H and OH are added, so there is no net oxidation nor reduction. (Note that in functional groups involving two carbons like alkenes or alkynes, both carbons have to be oxidized or reduced before the *net* change to the functional group is classified as oxidation or reduction.)

11-2

(a)

(b) no reaction $\xleftarrow{\ H_2CrO_4\ }$ $\xrightarrow{\ PCC\ }$ no reaction

(c)

(d) no reaction $\xleftarrow{\ H_2CrO_4\ }$ $\xrightarrow{\ PCC\ }$ no reaction

(e) no reaction $\xleftarrow{\ H_2CrO_4\ }$ $\xrightarrow{\ PCC\ }$ no reaction

(f) no reaction $\xleftarrow{\ H_2CrO_4\ }$ $H_3C-\overset{\overset{O}{\|}}{C}-OH$ $\xrightarrow{\ PCC\ }$ no reaction

(g) $H_3C-\overset{\overset{O}{\|}}{C}-OH$ $\xleftarrow{\ H_2CrO_4\ }$ CH_3CH_2OH $\xrightarrow{\ PCC\ }$ $H_3C-\overset{\overset{O}{\|}}{C}-H$

(h) $H_3C-\overset{\overset{O}{\|}}{C}-OH$ $\xleftarrow{\ H_2CrO_4\ }$ $H_3C-\overset{\overset{O}{\|}}{C}-H$ $\xrightarrow{\ PCC\ }$ no reaction

229

11-3

(a) A 1° alcohol loses two hydrogens when transformed to the aldehyde, and a 2° alcohol loses one hydrogen in forming a ketone; each alcohol is oxidized. DMSO loses an oxygen from the sulfur; it is clearly reduced. (If you got those two, you did the problem correctly.) To be rigorous, oxalyl chloride undergoes a disproportionation reaction: one C is oxidized to CO_2 and the other C is reduced to CO; however, the net effect on the elements in oxalyl chloride is "no change".

(b) Text section 8-15B shows that dimethyl sulfide reduces an ozonide. In the process, dimethyl sulfide is oxidized to DMSO.

11-4

(a) Dehydrogenation does not occur at 25° C—either: 1) there is a high kinetic barrier (a high activation energy) for this reaction, or 2) it is thermodynamically unfavorable, with $\Delta G > 0$. The latter possibility is supported by the fact that the reverse reaction (catalytic hydrogenation of a carbonyl) *is* spontaneous at 25° C (see text section 10-11C) and therefore has $\Delta G < 0$. This makes the question of kinetics academic—a reaction that cannot proceed must be uselessly slow.

(b) and (c) Kinetics will improve with increasing temperature for virtually all reactions, so both the hydrogenation and dehydrogenation reactions will go faster. In this case, however, the question is how to favor the dehydrogenation reaction. The answer is that thermodynamics will favor this reaction as the temperature is raised. The key is the fundamental thermodynamic equation $\Delta G = \Delta H - T\Delta S$. We can estimate that $\Delta H > 0$ since the product ketone plus hydrogen is less stable than the starting alcohol. Also, $\Delta S > 0$ since one molecule is converted to two: therefore, $-T\Delta S < 0$. At low temperature (25° C), ΔH dominates because T is so small, so $\Delta G > 0$. At a high enough temperature, the $-T\Delta S$ term will begin to overwhelm ΔH, and ΔG will become negative. For the reaction in question, this must be the case at 300° C.

11-5

(a)

(b) all three reagents give the same ketone product with a secondary alcohol

(c)

(d) all three reagents give no reaction with a tertiary alcohol

230

Note to the student: For simplicity, this book will use these standard laboratory methods of oxidation:

—PCC (pyridinium chlorochromate) to oxidize 1° alcohols to aldehydes;
—H_2CrO_4 (chromic acid) to oxidize 1° alcohols to carboxylic acids;
—CrO_3, H_2SO_4, acetone (Jones reagent) to oxidize 2° alcohols to ketones.

Understand that other choices are legitimate; for example, Swern oxidation works about as well as PCC in the preparation of aldehydes, and Collins reagent or PCC will oxidize a 2° alcohol to a ketone as well as chromic acid. If you have a question about the appropriateness of a reagent you choose, consult the table in the text before Problem 11-2.

11-6

(a)

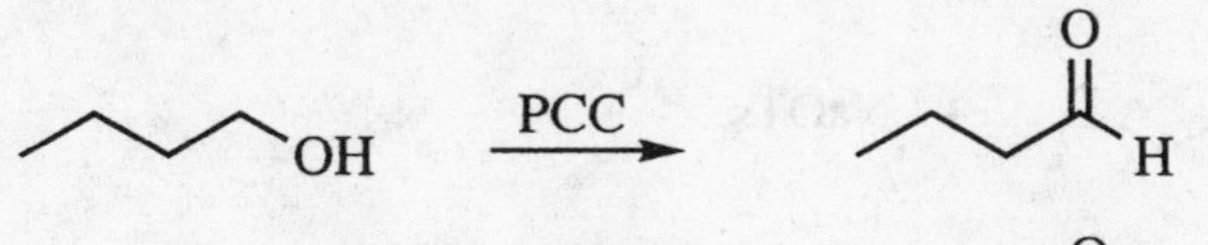

(b)

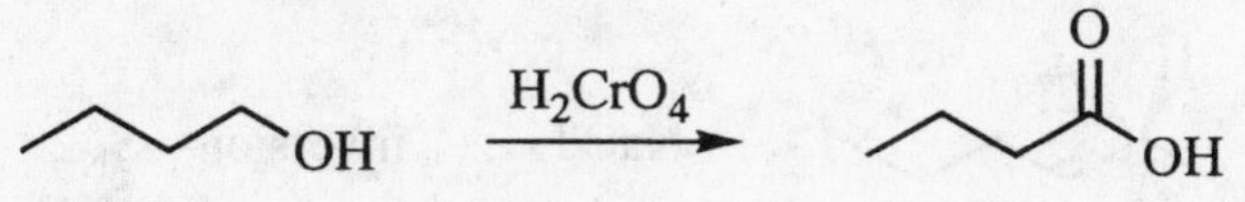

(c)

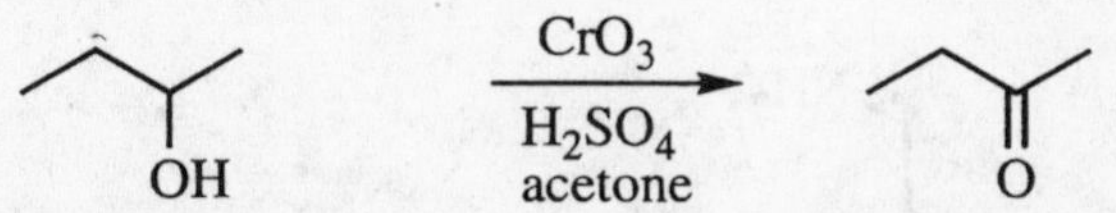

(d)

(e)

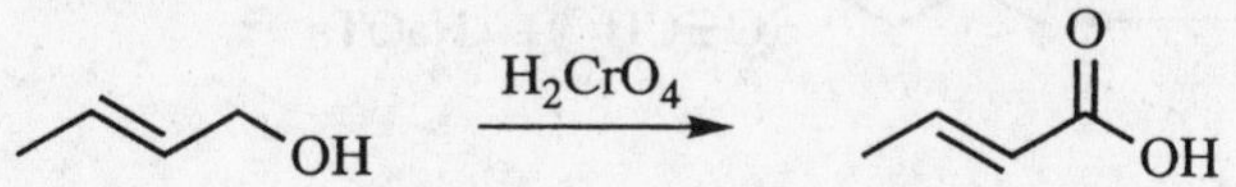

(f)

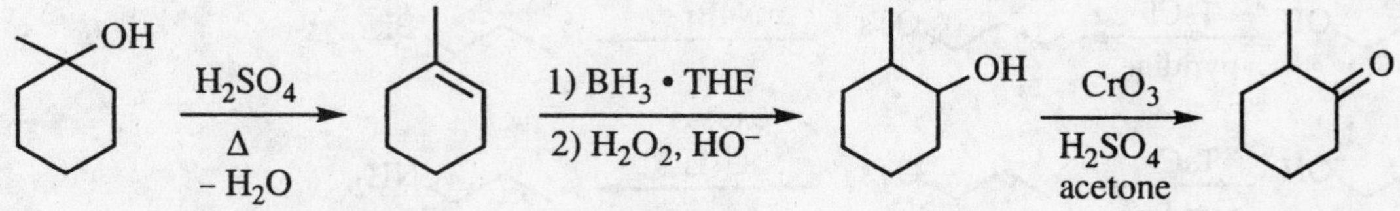

11-7 A chronic alcoholic has induced more ADH enzyme to be present to handle large amounts of imbibed ethanol, so requires more ethanol "antidote" molecules to act as a competitive inhibitor to "tie up" the extra enzyme molecules.

11-8

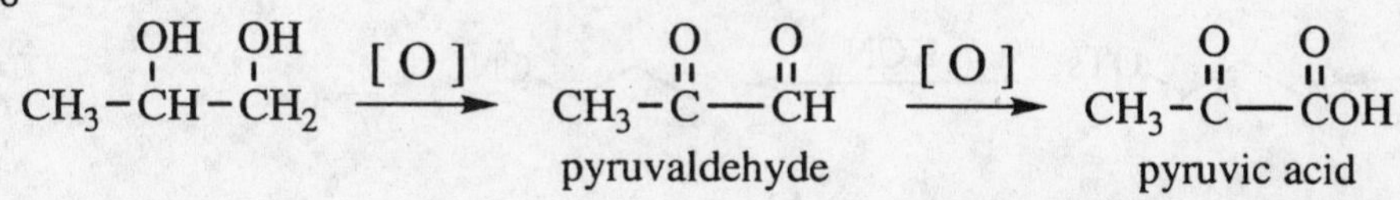

pyruvaldehyde

pyruvic acid

pyruvic acid is a normal metabolite
in the breakdown of glucose ("blood sugar")

11-9 From this problem on, "Ts" will refer to the "tosyl" or "p-toluenesulfonyl" group:

$$Ts \implies -\overset{O}{\underset{O}{\overset{\|}{\underset{\|}{S}}}}-\langle\text{benzene}\rangle-CH_3$$

(a) CH_3CH_2-OTs + $KO-\overset{CH_3}{\underset{CH_3}{\overset{|}{\underset{|}{C}}}}-CH_3$ $\longrightarrow$ $CH_3CH_2O-\overset{CH_3}{\underset{CH_3}{\overset{|}{\underset{|}{C}}}}-CH_3$ + KOTs

(E2 is also possible with this hindered base; the product would be ethylene, $CH_2=CH_2$.)

(b)   [structure] OTs + NaI $\longrightarrow$ [structure] I + NaOTs

(c)   [structure with TsO, H, labeled R] + NaCN $\longrightarrow$ [structure with H, CN, labeled S] + NaOTs   inversion—S_N2

(d)   [cyclohexyl-CH₂OTs] $\xrightarrow{NH_3}$ [cyclohexyl-CH₂$^+$NH₃ $^-$OTs] $\xrightarrow[NH_3]{excess}$ [cyclohexyl-CH₂NH₂] + $^+NH_4$ $^-$OTs

(e)   [structure] OTs + Na⁺ $^-$:C≡CH $\longrightarrow$ [structure] C≡CH + NaOTs

11-10

(a)   [structure] OH $\xrightarrow[pyridine]{TsCl}$ [structure] OTs $\xrightarrow{NaBr}$ [structure] Br

(b)   [structure] OH $\xrightarrow[pyridine]{TsCl}$ [structure] OTs $\xrightarrow[NH_3]{excess}$ [structure] NH₂

(c)   [structure] OH $\xrightarrow[pyridine]{TsCl}$ [structure] OTs $\xrightarrow{NaOCH_2CH_3}$ [structure] O

(d)   [structure] OH $\xrightarrow[pyridine]{TsCl}$ [structure] OTs $\xrightarrow{KCN}$ [structure] CN

11-11

(a)   [cyclohexyl]—CH₂OH $\xrightarrow[pyridine]{TsCl}$ [cyclohexyl]—CH₂OTs   OR   [cyclohexyl]—CH₂O$-\overset{O}{\underset{O}{\overset{\|}{\underset{\|}{S}}}}-\langle\text{benzene}\rangle-CH_3$

11-11 continued

(b)

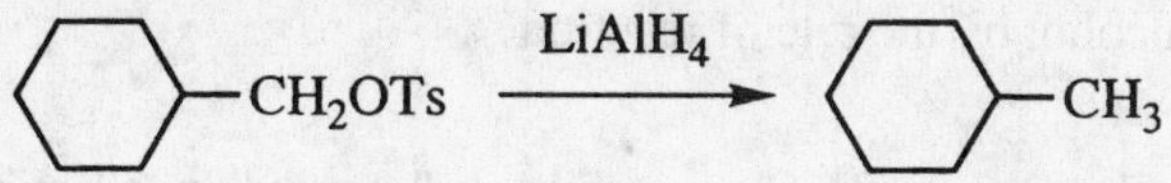

(c)

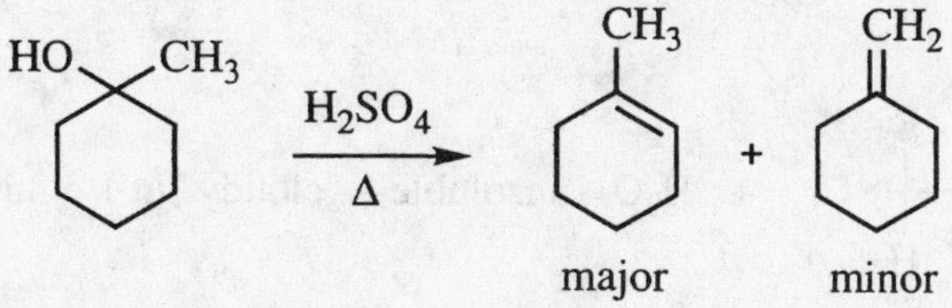

major   minor

(d)

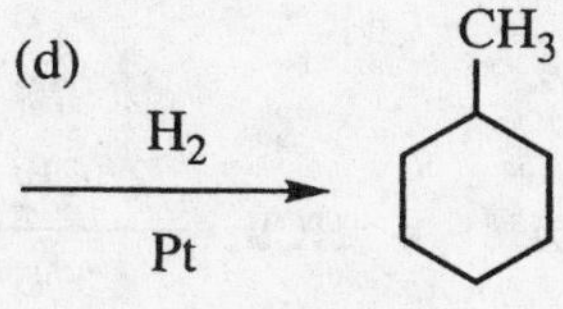

11-12

(a) either S_N1 or S_N2 on 2° alcohols

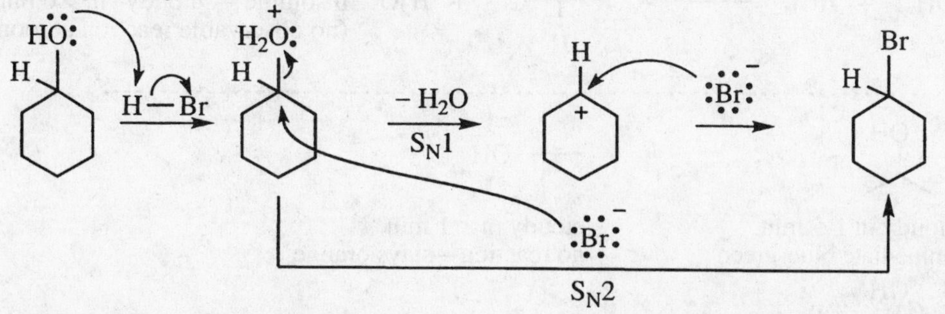

(b) S_N2 on 1° alcohols

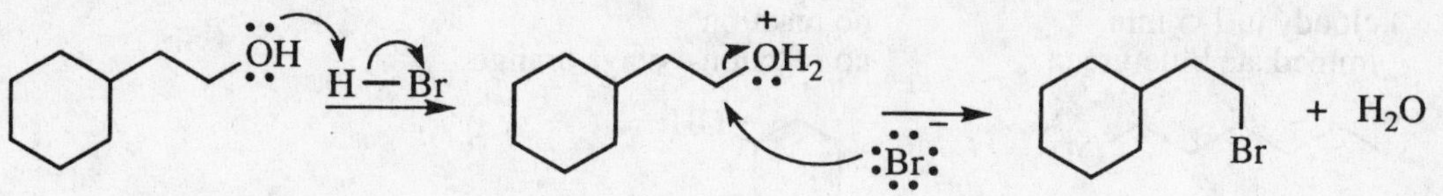

11-13

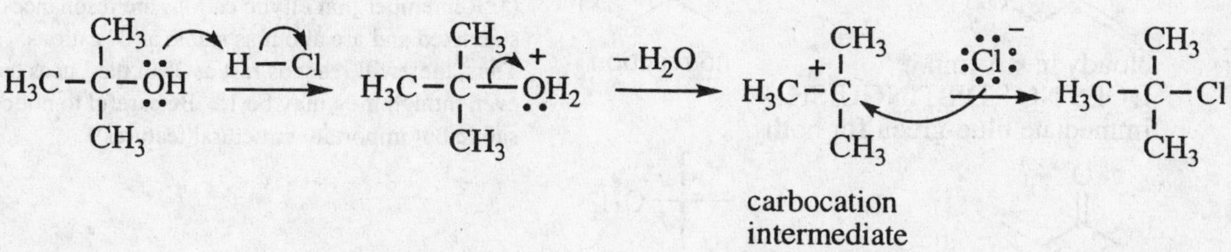

carbocation
intermediate

11-14 The two standard qualitative tests are:

1) <u>chromic acid</u>—distinguishes 3° alcohol from either 1° or 2°

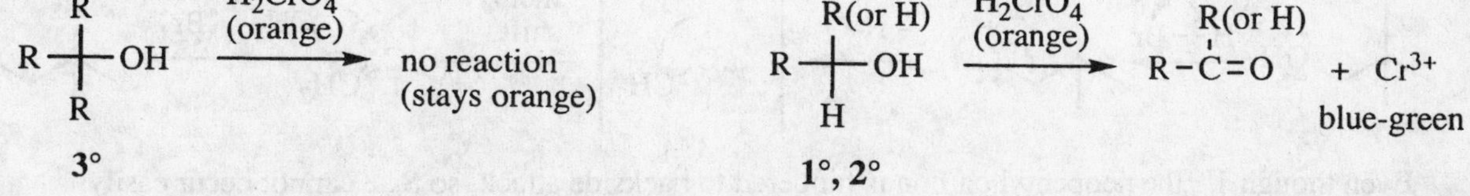

2) <u>Lucas test</u>—distinguishes 1° from 2° from 3° alcohol by the rate of reaction

3°   R—C(R)(R)—OH + HCl $\xrightarrow{ZnCl_2}$ R—C(R)(R)—Cl + H_2O insoluble—"cloudy" in < 1 minute
soluble

2°   R—C(R)(H)—OH + HCl $\xrightarrow{ZnCl_2}$ R—C(R)(H)—Cl + H_2O insoluble—"cloudy" in 1-5 minutes
soluble

1°   R—C(H)(H)—OH + HCl $\xrightarrow{ZnCl_2}$ R—C(H)(H)—Cl + H_2O insoluble—"cloudy" in > 6 minutes
soluble   (no observable reaction at room temp.)

(a)

Lucas: cloudy in 1-5 min.   cloudy in < 1 min.
H_2CrO_4:   immediate blue-green   no reaction—stays orange

(b)

Lucas: cloudy in 1-5 min.   no reaction
H_2CrO_4:   immediate blue-green   no reaction—stays orange

(c)

Lucas: no reaction   cloudy in 1-5 min.
H_2CrO_4:   DOES NOT DISTINGUISH—immediate blue-green for both

(d)

Lucas: cloudy in < 1 min. **   no reaction
H_2CrO_4:   DOES NOT DISTINGUISH—
            immediate blue-green for both

(**Remember that allylic cations are resonance-stabilized and are about as stable as 3° cations. Thus, they will react as fast as 3° in the Lucas test, even though they may be 1°. Be careful to notice subtle but important structural features!)

(e)

Lucas: no reaction   cloudy in < 1 min.
H_2CrO_4:   DOES NOT DISTINGUISH—stays orange for both

11-15

methyl shift

Even though 1°, the neopentyl carbon is hindered to backside attack, so S_N1 cannot occur easily. Instead, an S_N1 mechanism occurs, with rearrangement.

11-16

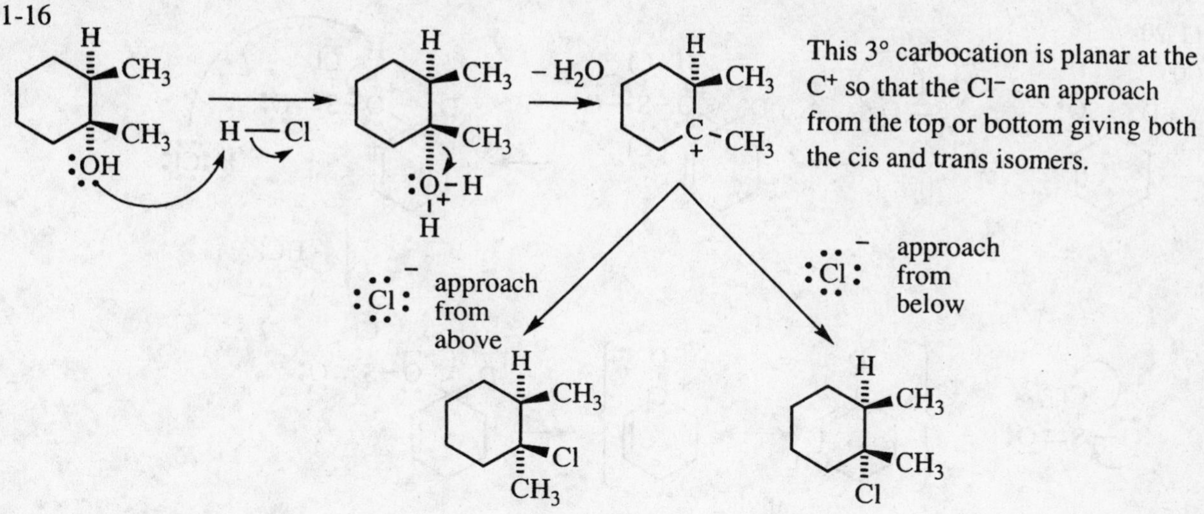

This 3° carbocation is planar at the C⁺ so that the Cl⁻ can approach from the top or bottom giving both the cis and trans isomers.

$-H_2O$

Cl⁻ approach from above

Cl⁻ approach from below

11-17

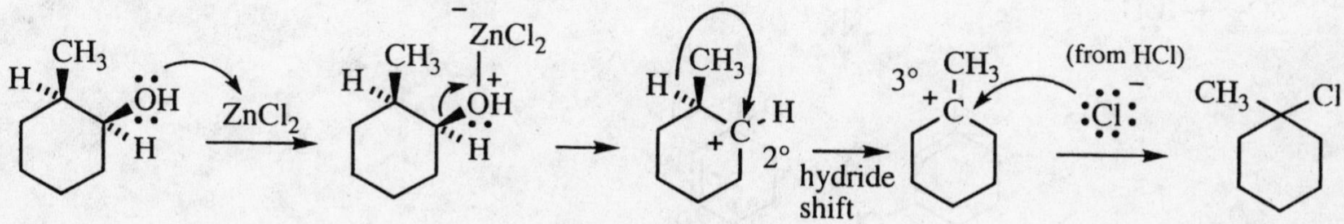

$ZnCl_2$

2° hydride shift

3°

(from HCl)

Cl⁻

11-18

3 (CH₃)₃CCH₂OH $+ PBr_3 \longrightarrow 3$ (CH₃)₃CCH₂Br $+ P(OH)_3$

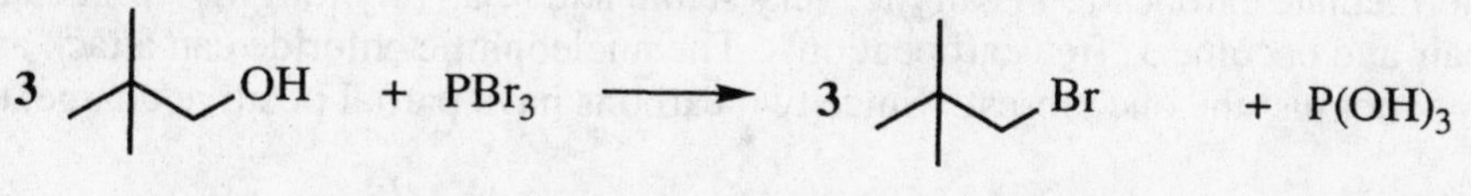

$6 \ CH_3(CH_2)_{14}CH_2OH + 2 \ P + 3 \ I_2 \longrightarrow 6 \ CH_3(CH_2)_{14}CH_2I + 2 \ P(OH)_3$

11-19

(a)

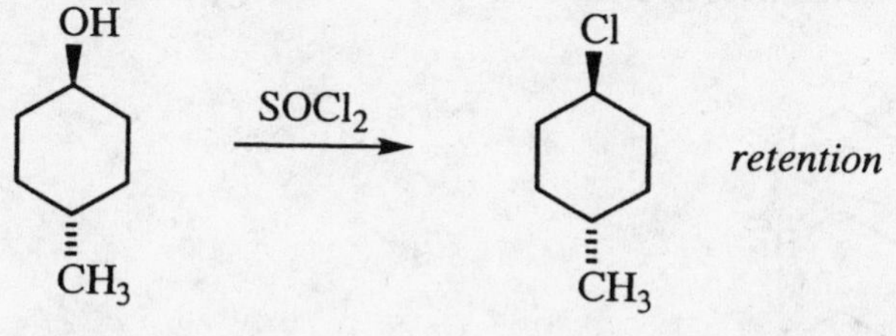

$SOCl_2$

retention

(b)

TsCl pyridine

NaCl

S_N2—*inversion*

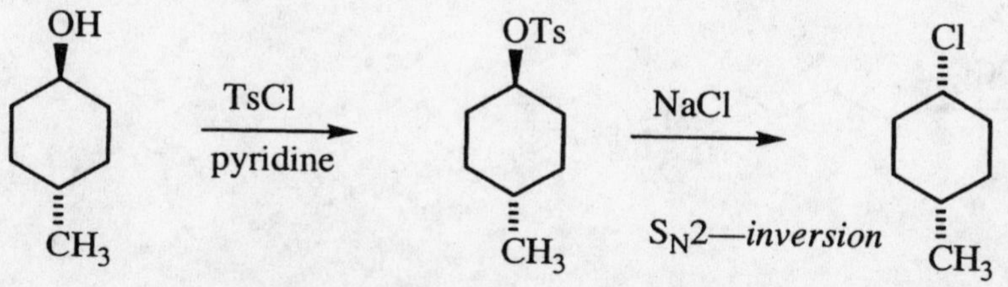

Another possible answer would be to use PCl_3.

11-20

(a)

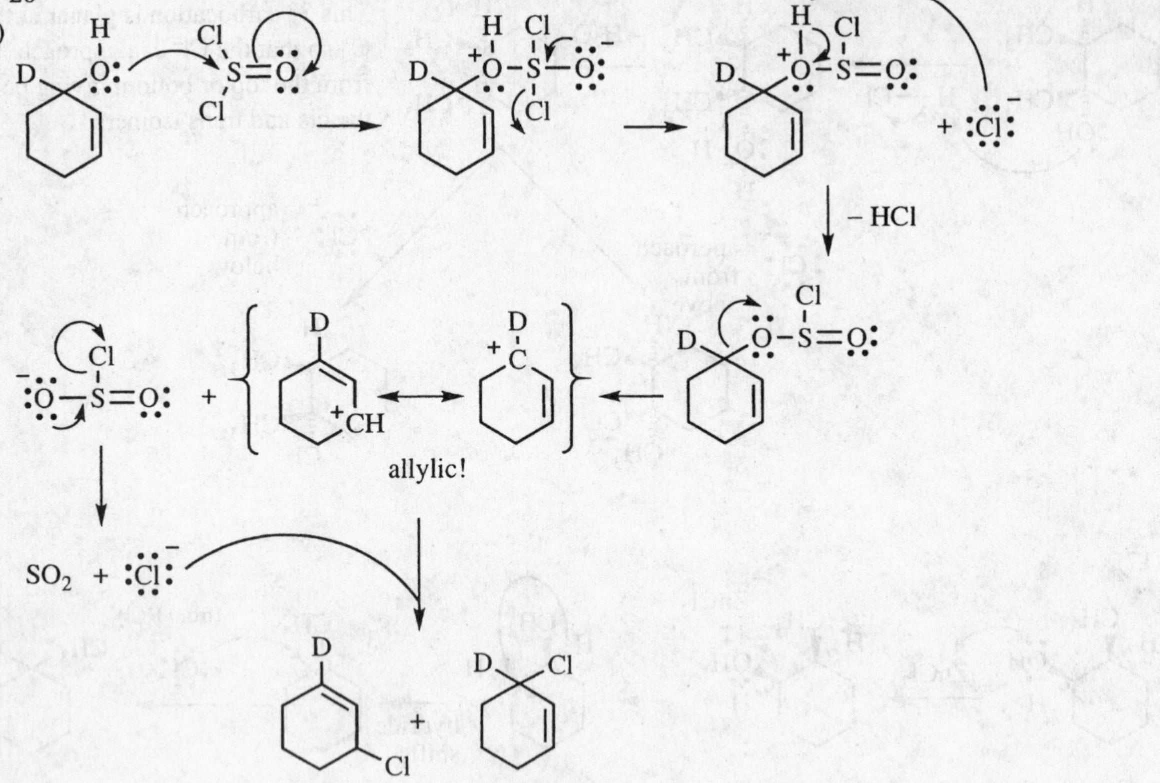

(b) The key is that the intermediate carbocation is allylic, very stable and relatively long-lived. It can therefore escape the ion pair and become a "free carbocation". The nucleophilic chloride can attack any carbon with positive charge, not just the one closest. Since two carbons have partial positive charge, two products result.

11-21

(a)

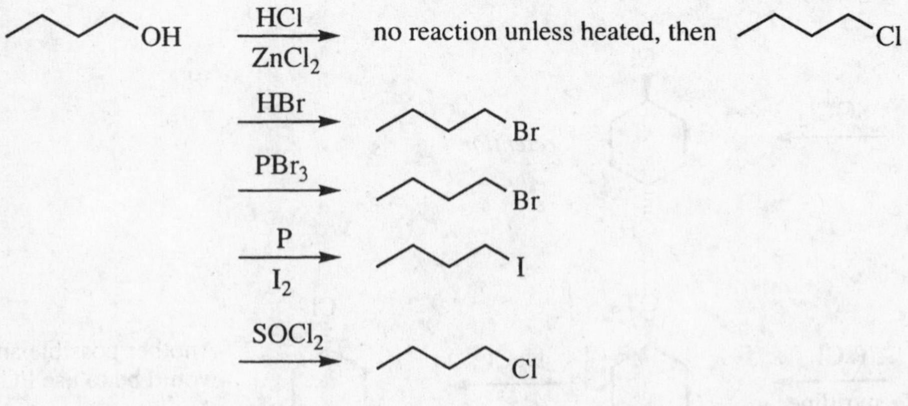

11-21 continued

(b) $\xrightarrow[\text{ZnCl}_2]{\text{HCl}}$

$\xrightarrow{\text{HBr}}$

$\xrightarrow{\text{PBr}_3}$

$\xrightarrow[\text{I}_2]{\text{P}}$

$\xrightarrow{\text{SOCl}_2}$

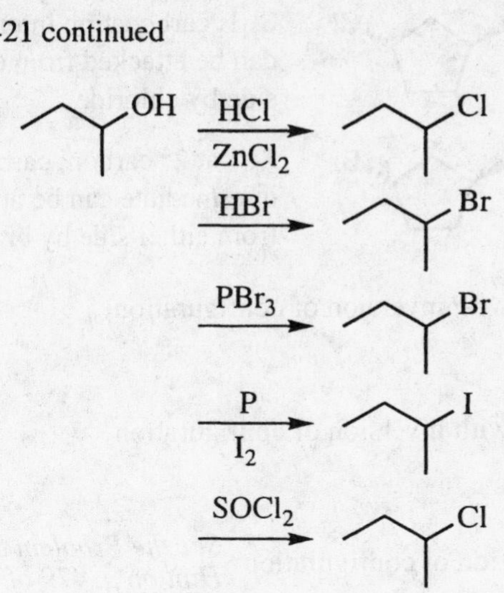

(c) $\xrightarrow[\text{ZnCl}_2]{\text{HCl}}$

$\xrightarrow{\text{HBr}}$

$\xrightarrow{\text{PBr}_3}$

(poor reaction on 3°)

$\xrightarrow[\text{I}_2]{\text{P}}$

(poor reaction on 3°)

$\xrightarrow{\text{SOCl}_2}$

(poor reaction on 3°)

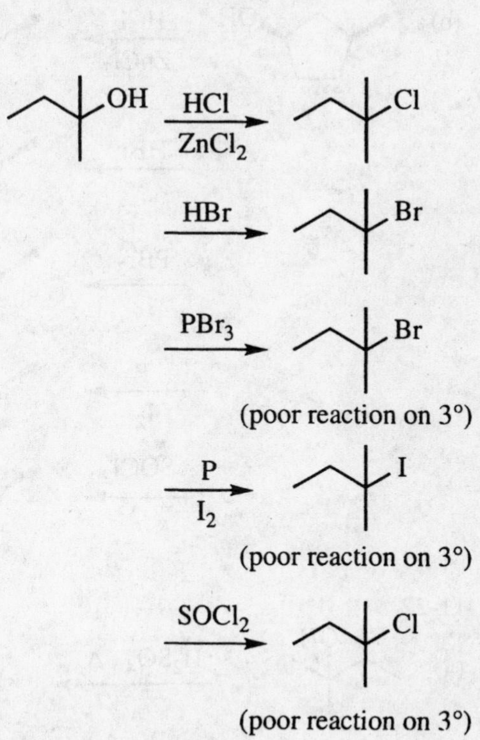

(d) $\xrightarrow[\text{ZnCl}_2]{\text{HCl}}$ no reaction unless heated, then S_N1—rearrangement

1°, neopentyl

$\xrightarrow{\text{HBr}}$ S_N2—minor (hindered) + S_N1—rearrangement

$\xrightarrow{\text{PBr}_3}$

$\xrightarrow[\text{I}_2]{\text{P}}$

$\xrightarrow{\text{SOCl}_2}$

237

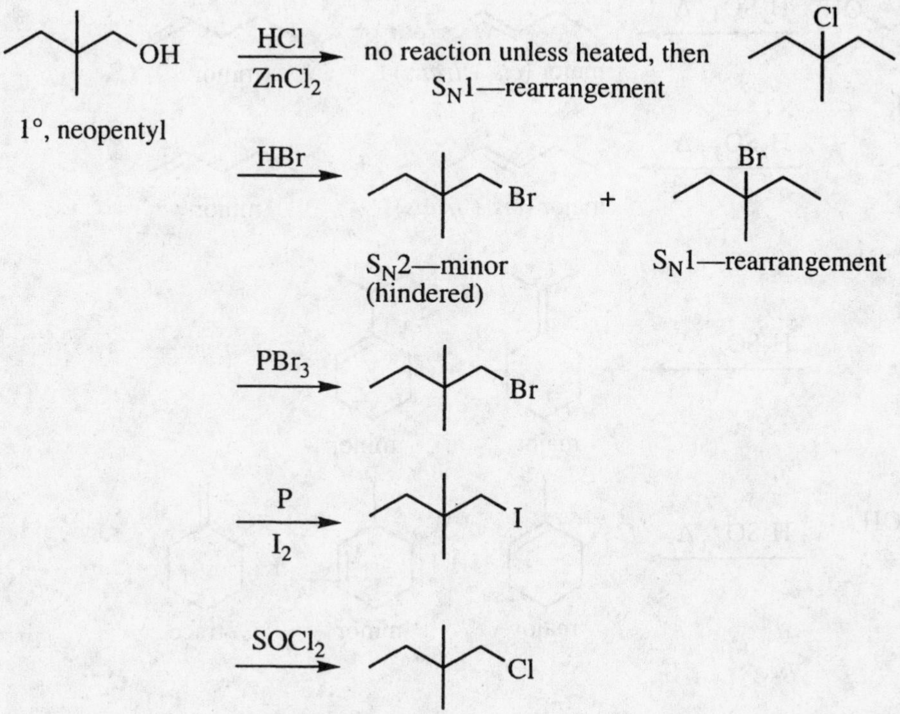

11-21 continued

(e)

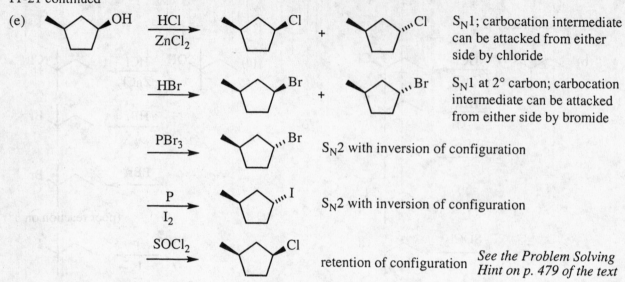

OH
HCl
ZnCl2
Cl
+
Cl

S_N1; carbocation intermediate can be attacked from either side by chloride

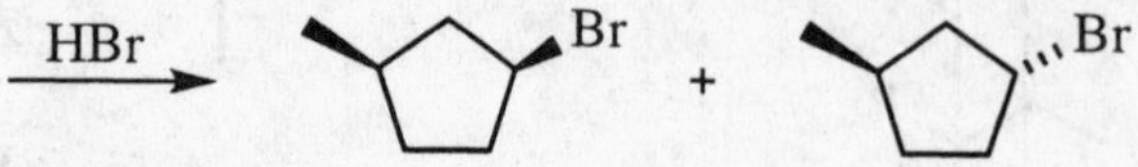
HBr
Br
+
Br

S_N1 at 2° carbon; carbocation intermediate can be attacked from either side by bromide

PBr3
Br

S_N2 with inversion of configuration

P
I2
I

S_N2 with inversion of configuration

SOCl2
Cl

retention of configuration *See the Problem Solving Hint on p. 479 of the text*

11-22

(a)

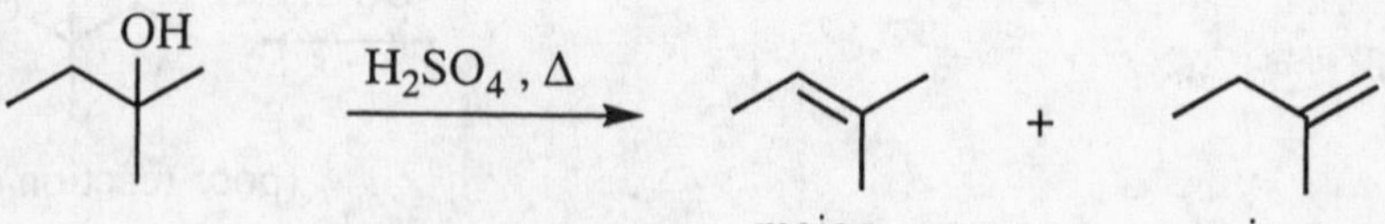
OH
H2SO4 , Δ
+

major minor

(b)

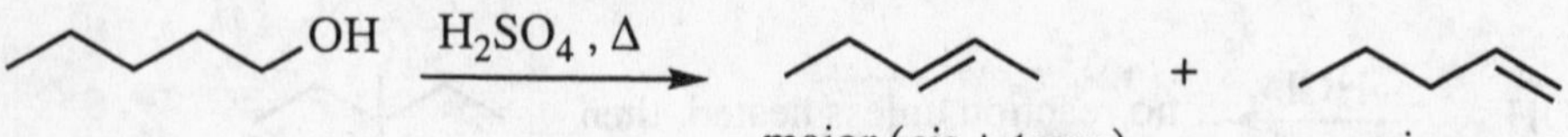
OH
H2SO4 , Δ
+

major (*cis* + *trans*) minor

(c)

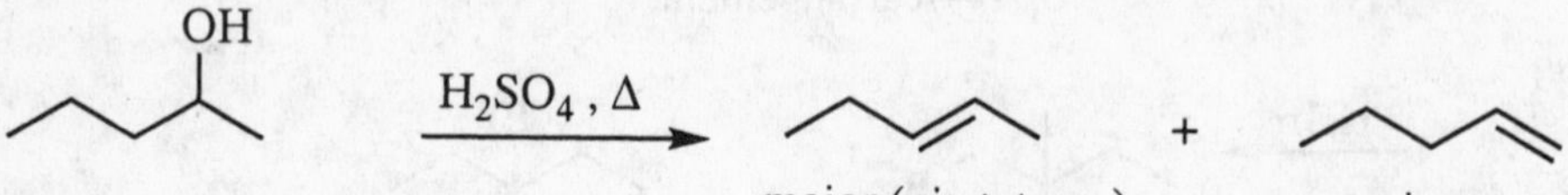
OH
H2SO4 , Δ
+

major (*cis* + *trans*) minor

(d)

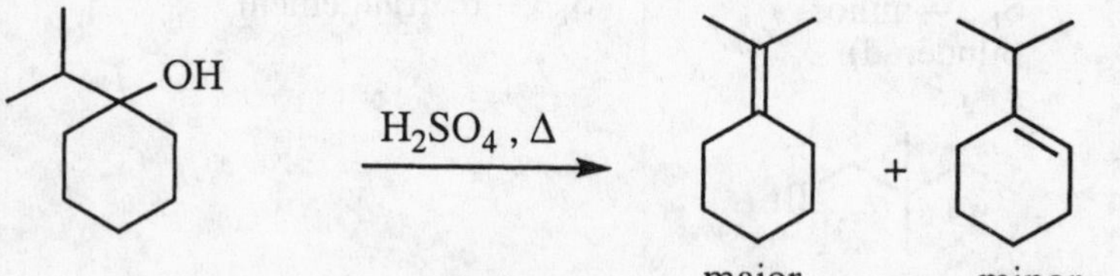
OH
H2SO4 , Δ
+

major minor

(e)

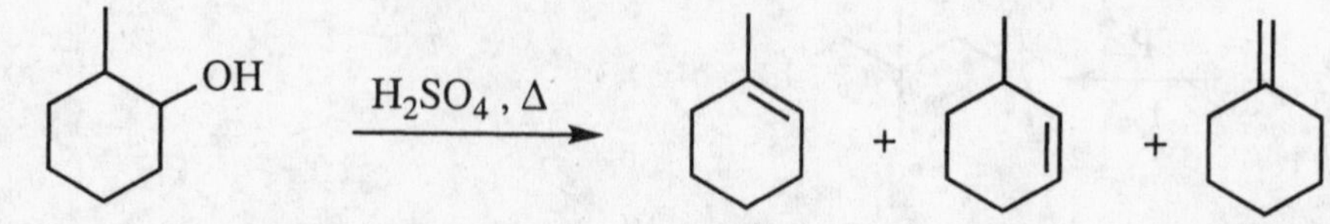
OH
H2SO4 , Δ
+
+

major minor trace

11-23

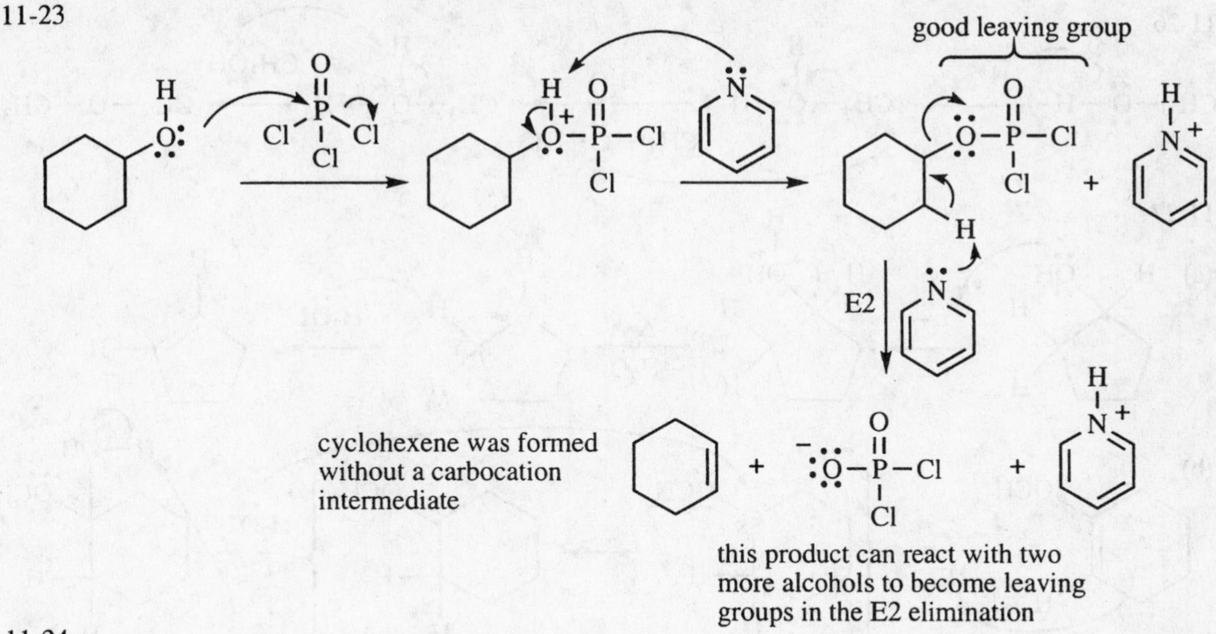

good leaving group

cyclohexene was formed
without a carbocation
intermediate

this product can react with two
more alcohols to become leaving
groups in the E2 elimination

11-24

Both mechanisms begin with protonation of the oxygen.

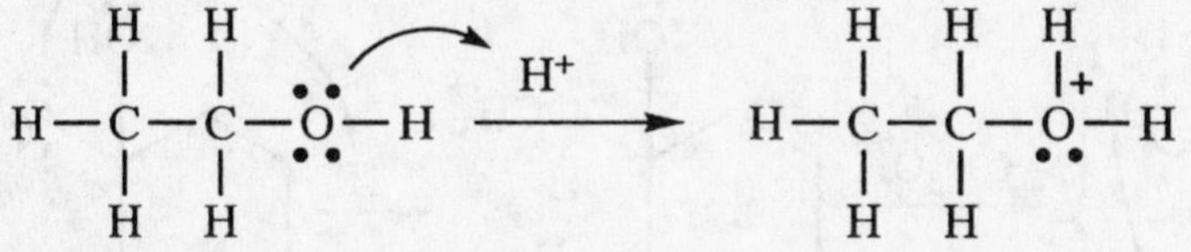

One mechanism involves another molecule of ethanol acting as a base, giving elimination.

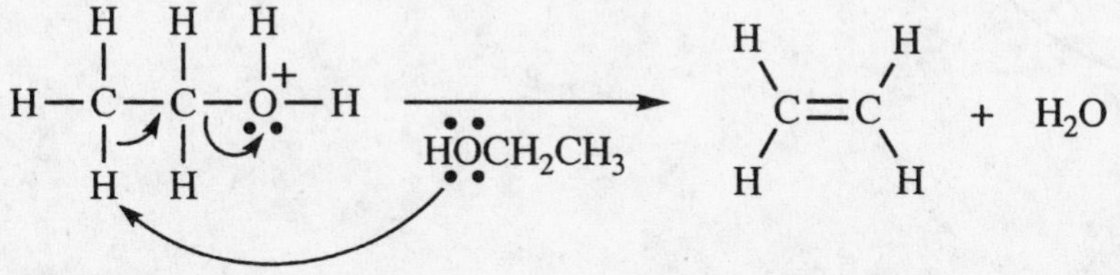

The other mechanism involves another molecule of ethanol acting as a nucleophile, giving substitution.

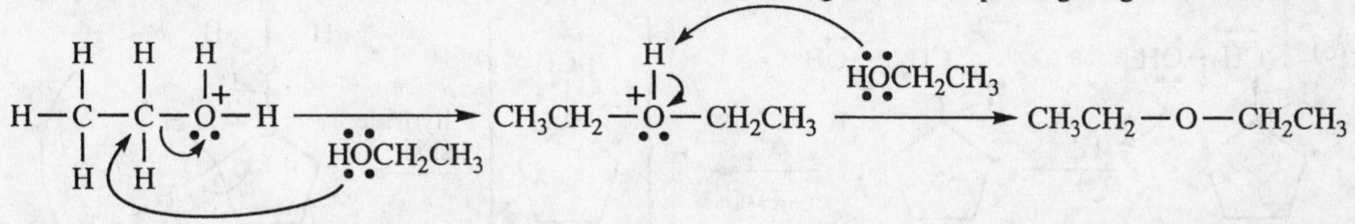

11-25 An equimolar mixture of methanol and ethanol would produce all three possible ethers. The difficulty in separating these compounds would preclude this method from being a practical route to any one of them. This method is practical only for symmetric ethers, that is, where both alkyl groups are identical.

$$CH_3CH_2OH + HOCH_3 \xrightarrow[\Delta]{H^+} H_2O + CH_3CH_2OCH_3 + CH_3OCH_3 + CH_3CH_2OCH_2CH_3$$

11-26

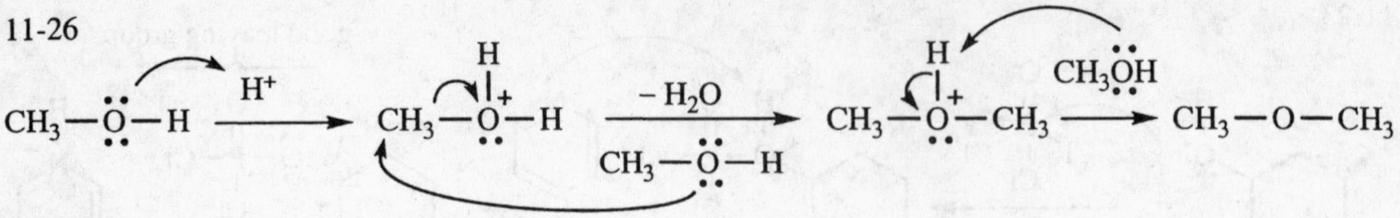

11-27

(a)

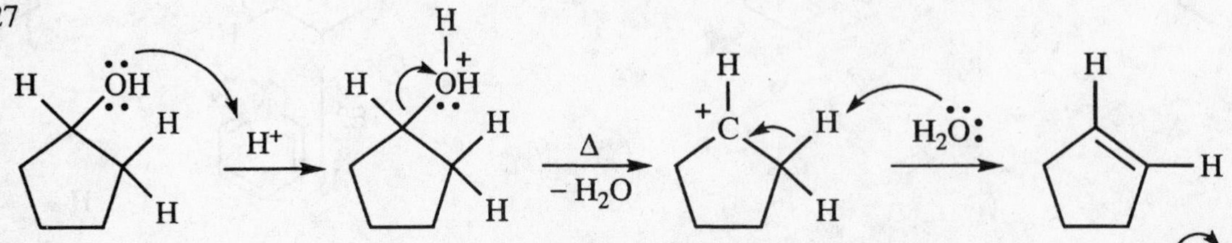

(b)

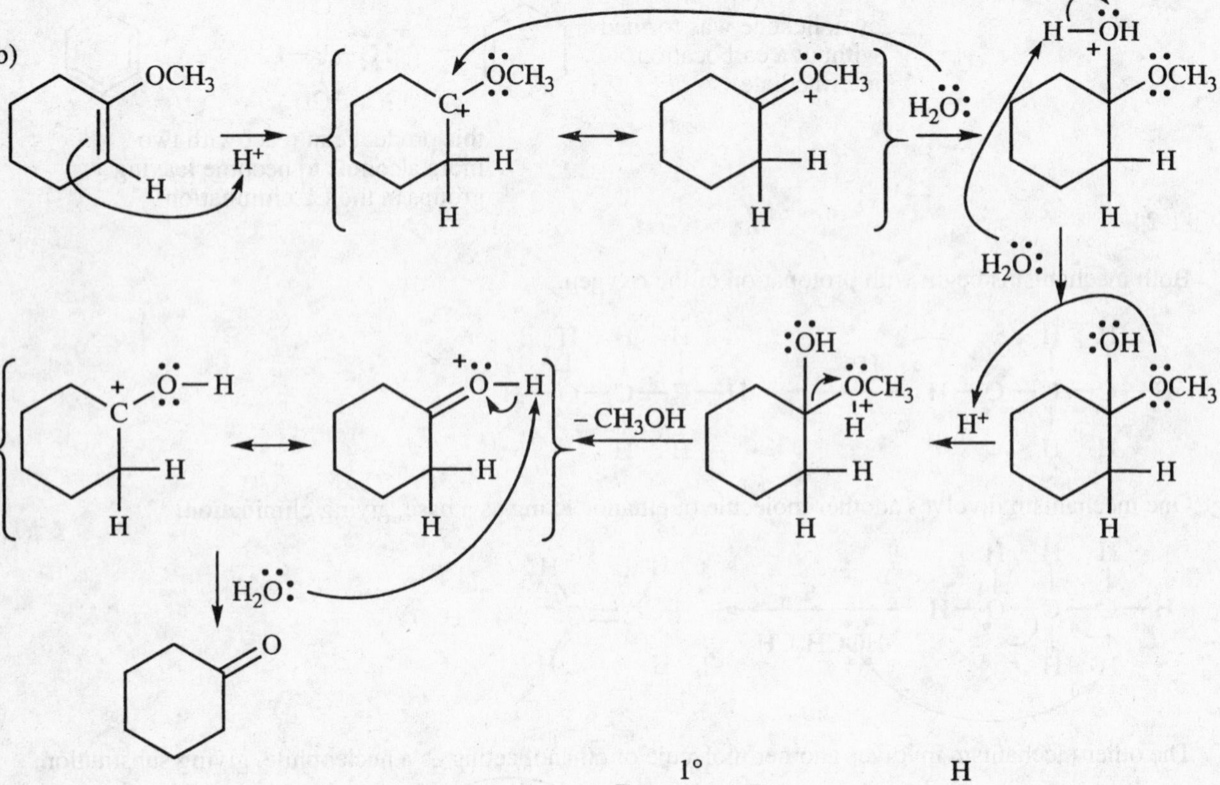

$-CH_3OH$

(c)

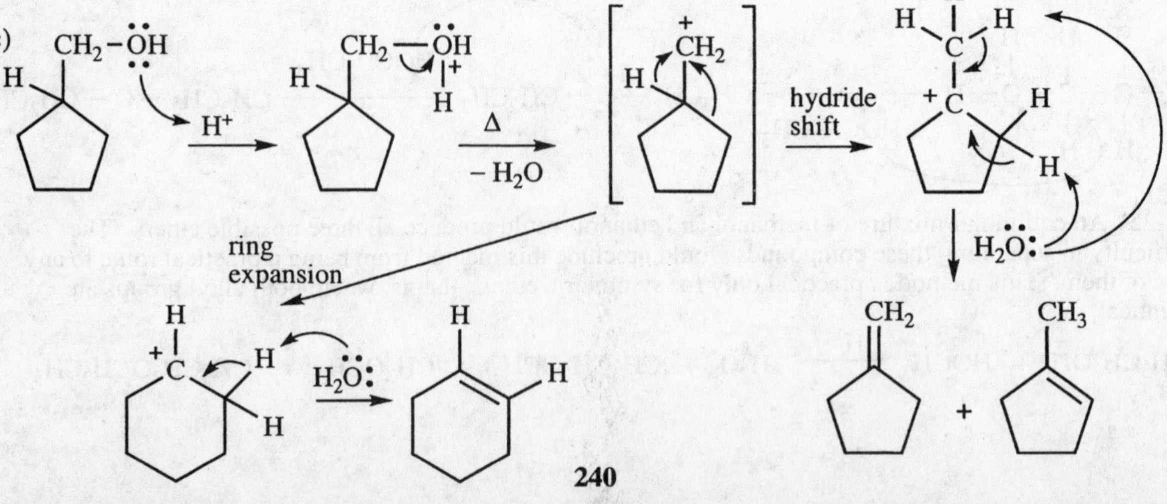

$1°$

hydride shift

ring expansion

11-27 continued

(d)

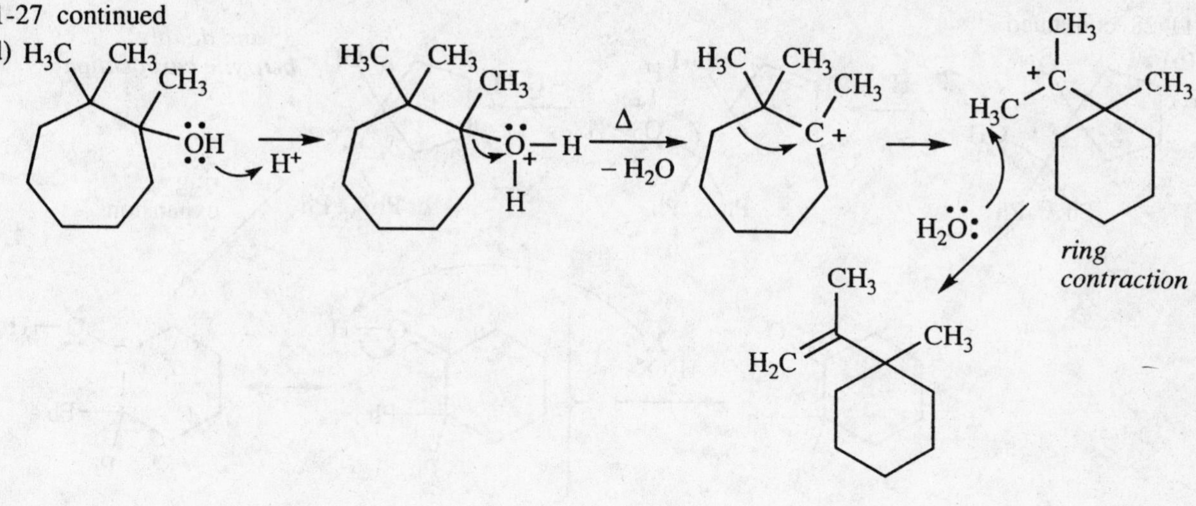

H₃C CH₃ CH₃ ... :ÖH → H⁺ → H₃C CH₃ CH₃ ... Ö⁺—H | H → Δ, −H₂O → H₃C CH₃ CH₃ ... C⁺ →

H₃C—C⁺(CH₃) CH₃ (ring contraction)

H₂Ö:

H₂C=C(CH₃) CH₃

ring contraction

11-28

(a)

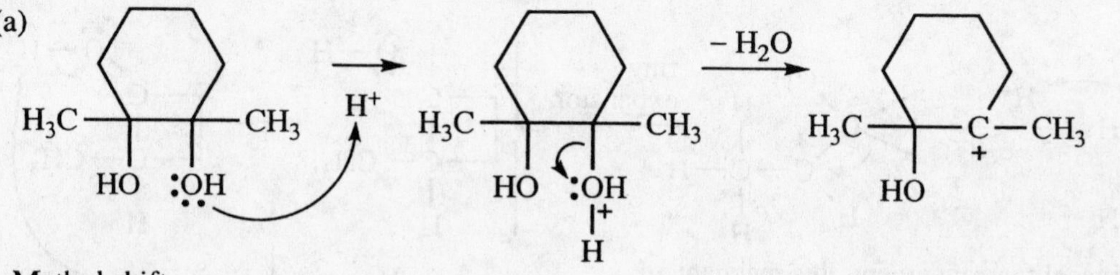

H₃C—C(OH) C(CH₃)(:ÖH) H⁺ → H₃C—C(OH) C(CH₃)(:ÖH⁺—H) → −H₂O → H₃C—C(OH) C⁺—CH₃

Methyl shift

H₃C—C(HO) C⁺—CH₃ → { ⁺C(HO:)(CH₃) — C—CH₃ ↔ H—O⁺(:)(CH₃) — C(CH₃) } → O=C — C(CH₃)(CH₃)

H₂Ö:

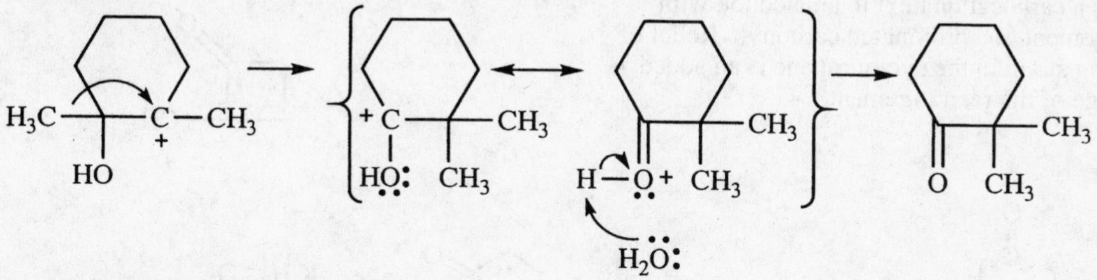

Alkyl shift—ring *contraction*

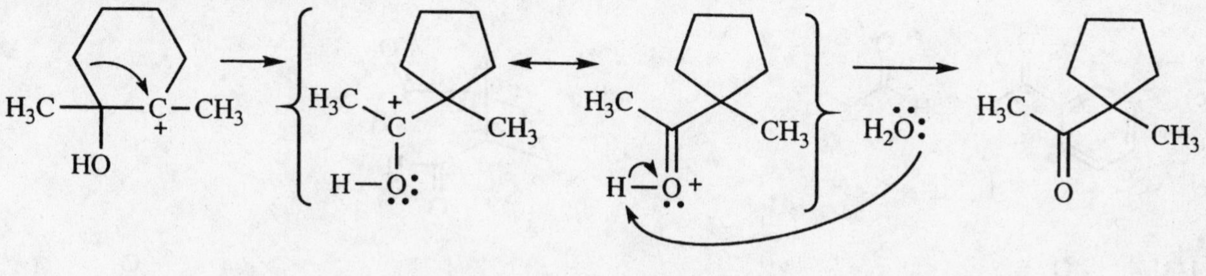

H₃C—C(HO) C⁺—CH₃ → { H₃C—C⁺(H—Ö:) CH₃ ↔ H₃C—C(H—O⁺:) CH₃ } H₂Ö: → H₃C—C(=O) — C(CH₃)(CH₃)

11-28 continued

(b)

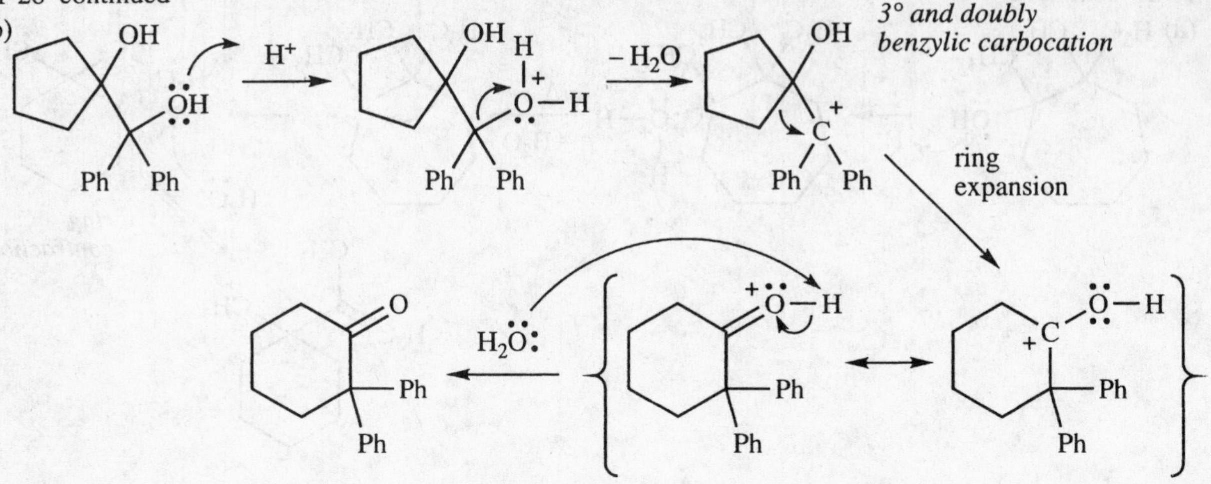

H$^+$ $-H_2O$ *3° and doubly benzylic carbocation*

ring expansion

11-29

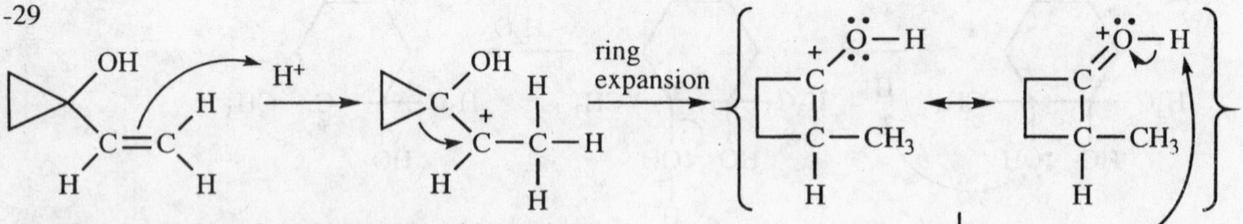

Similar to the pinacol rearrangement, this mechanism involves a carbocation next to an alcohol, with rearrangement to a protonated carbonyl. Relief of some ring strain in the cyclopropane is an added advantage of the rearrangement.

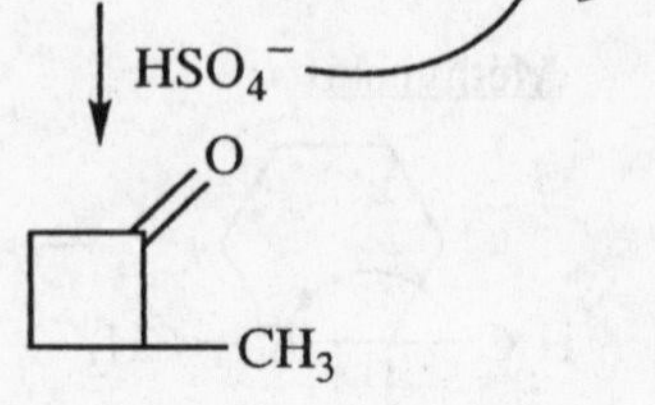

ring expansion HSO_4^-

11-30

(a)   **2** $H_3C-\overset{\overset{\displaystyle O}{\|}}{C}-H$

(b) $+$ $CH_2=O$

(c) $+$

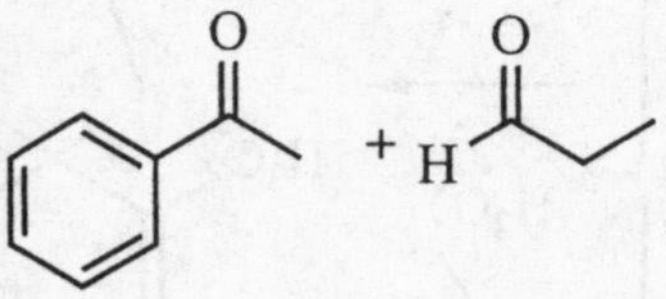

(d)

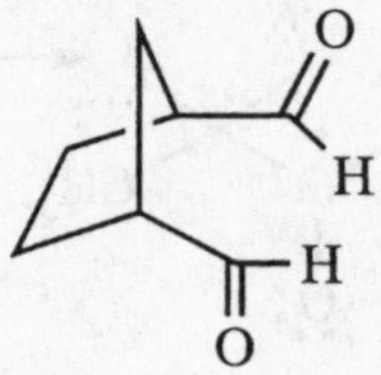

11-31

(a) $Cl-\overset{\overset{\displaystyle O}{\|}}{C}-CH_2CH_2CH_3$ $+$ $HOCH_2CH_2CH_3$

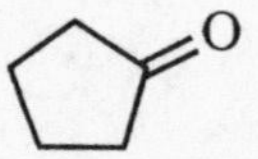

(b) $CH_3(CH_2)_3OH$ $+$ $Cl-\overset{\overset{\displaystyle O}{\|}}{C}CH_2CH_3$

(c) H_3C-⟨benzene⟩$-OH$ $+$ $Cl-\overset{\overset{\displaystyle O}{\|}}{C}CH(CH_3)_2$

(d)   ⟨cyclopropyl⟩$-OH$ $+$ $Cl-\overset{\overset{\displaystyle O}{\|}}{C}-$⟨Ph⟩

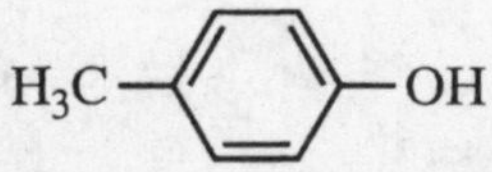

242

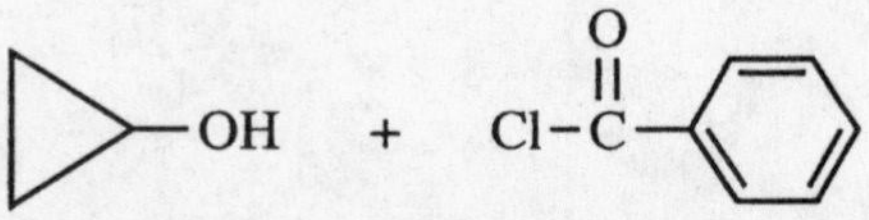

11-32

11-33 Proton transfer (acid-base) reactions are much faster than almost any other reaction. Methoxide will act as a base and remove a proton from the oxygen much faster than methoxide will act as a nucleophile and displace water.

11-34

(a)

(b) There are two problems with this attempted bimolecular dehydration. First, all three possible ether combinations of cyclohexanol and ethanol would be produced. Second, hot sulfuric acid are the conditions for dehydrating secondary alcohols like cyclohexanol, so elimination would compete with substitution.

11-35

(a) What the student did:

sodium *(S)*-2-butoxide   *(S)*-2-ethoxybutane

The product also has the *S* configuration, not the *R*. Why? The substitution is indeed an S_N2 reaction, but the *substitution did not take place at the chiral center*, so the configuration of the starting material is retained, not inverted.

(b) There are two ways to make *(R)*-2-ethoxybutane. Start with *(R)*-2-butanol, make the anion, and substitute on ethyl tosylate similar to part (a), or do an S_N2 inversion at the chiral center of *(S)*-2-butanol. S_N2 works better at 1° carbons so the former method would be preferred to the latter.

(c) This is not the optimum method because it requires S_N2 at a 2° carbon, as discussed in part (b).

11-36

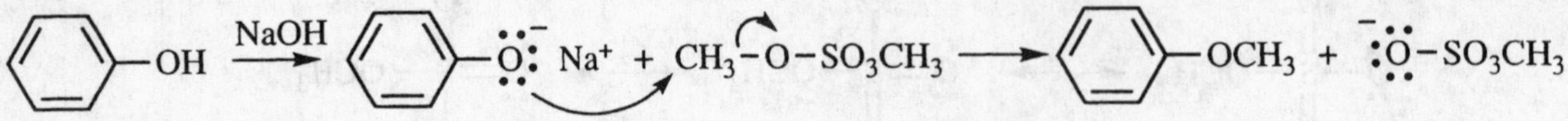

11-37

(a) $CH_3CH_2-\overset{O}{\overset{\|}{C}}-OH$ $\xrightarrow{SOCl_2}$ $CH_3CH_2-\overset{O}{\overset{\|}{C}}-Cl$ $\xrightarrow[2)\ H_3O^+]{1)\ \textbf{2}\ CH_3CH_2MgBr}$ $CH_3CH_2-\overset{CH_2CH_3}{\underset{OH}{\overset{|}{\underset{|}{C}}}}-CH_2CH_3$

$\downarrow H_2SO_4$

$CH_3\overset{CH_2CH_3}{\underset{}{\overset{|}{C}H-\underset{}{C}H}}CH_2CH_3 \xleftarrow[2)\ H_2O_2,\ HO^-]{1)\ BH_3\bullet THF}$ $CH_3CH=\overset{CH_2CH_3}{\overset{|}{C}}-CH_2CH_3$

(with OH on the product: $CH_3CHCHCH_2CH_3$ with $\overset{CH_2CH_3}{|}$ and $\underset{OH}{|}$)

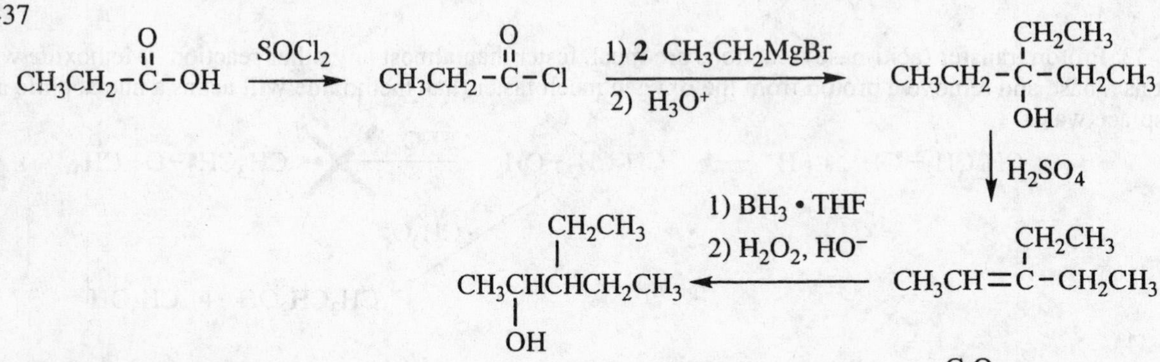

(b) $CH_3CH_2CH_2OH$ $\xrightarrow{PCC}$ $CH_3CH_2-\overset{O}{\overset{\|}{C}}-H$ $\xrightarrow[2)\ H_3O^+]{1)\ CH_3CH_2MgBr}$ [OH structure] $\xrightarrow[acetone]{\overset{CrO_3}{H_2SO_4}}$ [ketone structure]

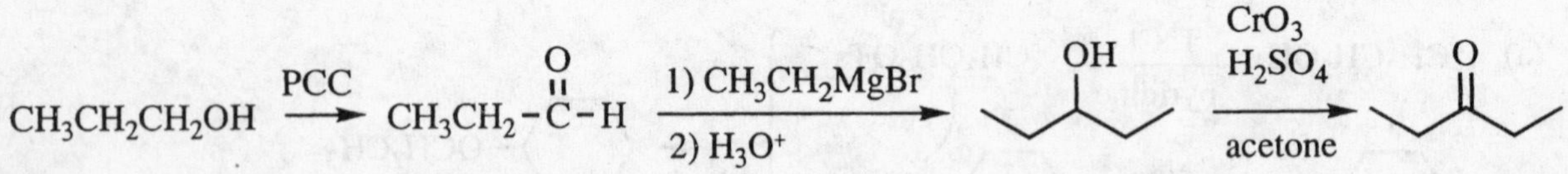

11-38

(a)

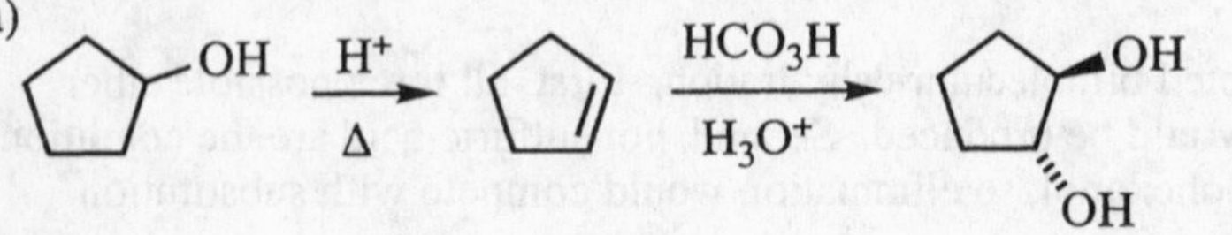

any peroxy acid can be used to form the epoxide which is cleaved to the *trans*-diol in aqueous acid

(b)

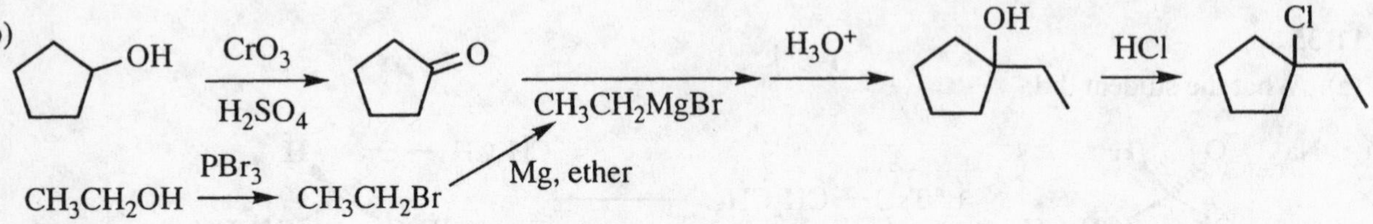

(c)

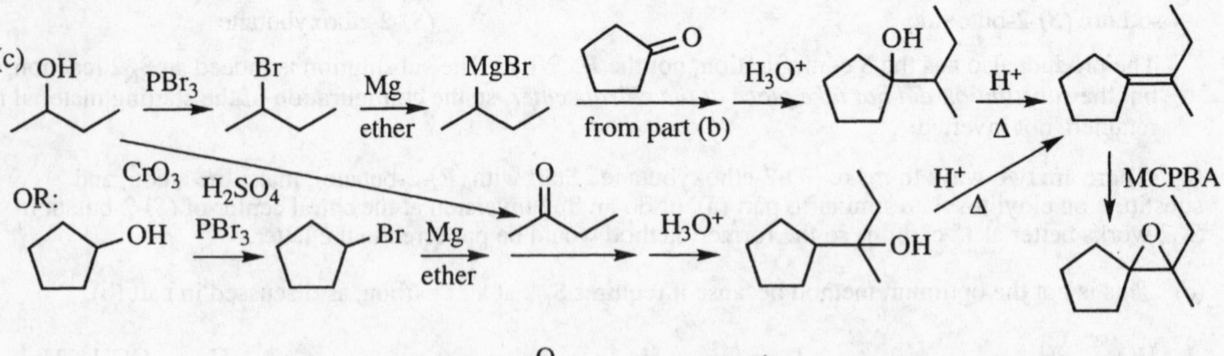

OR:

(d)

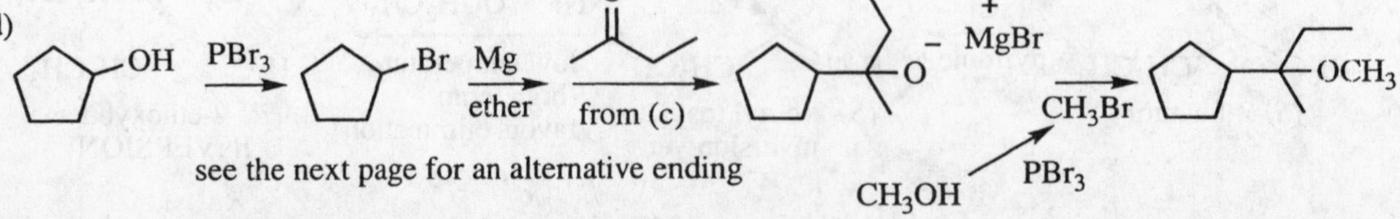

see the next page for an alternative ending

244

11-38 continued
(d) continued from the previous page

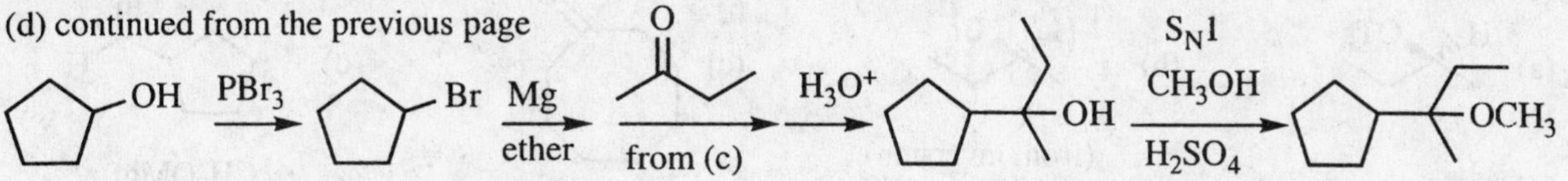

(e) There are several possible combinations of Grignard reactions on aldehydes or ketones. This is one example. Your example may be different and still be correct.

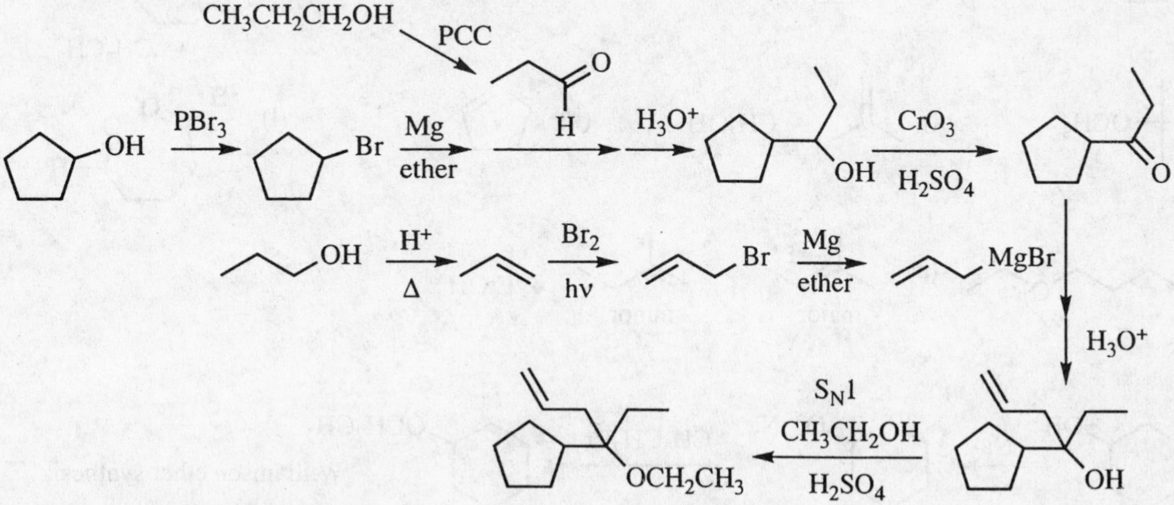

11-39 Please refer to solution 1-20, page 12 of this Solutions Manual.

11-40

(a)

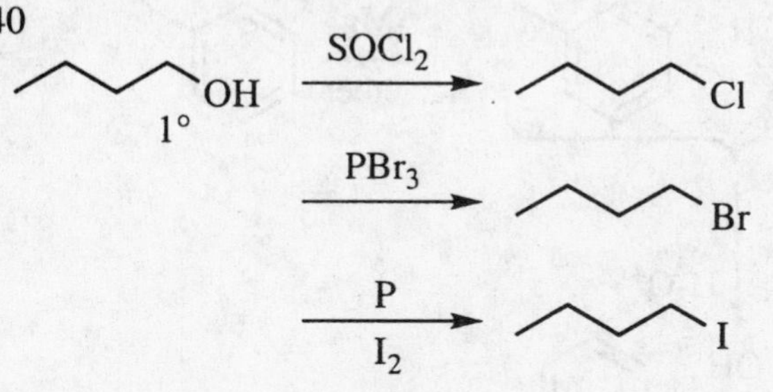

(b)

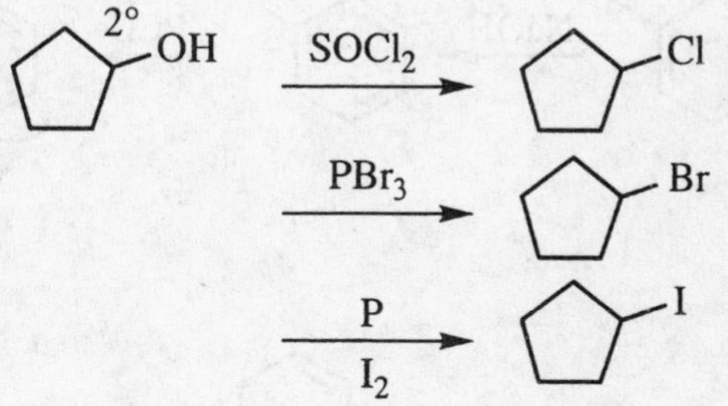

(c)

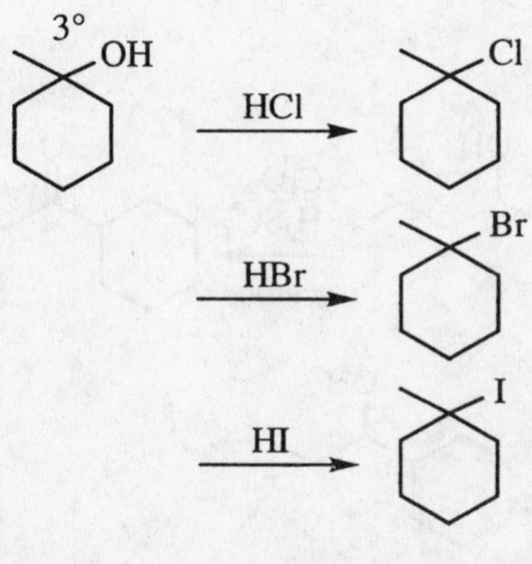

(d)

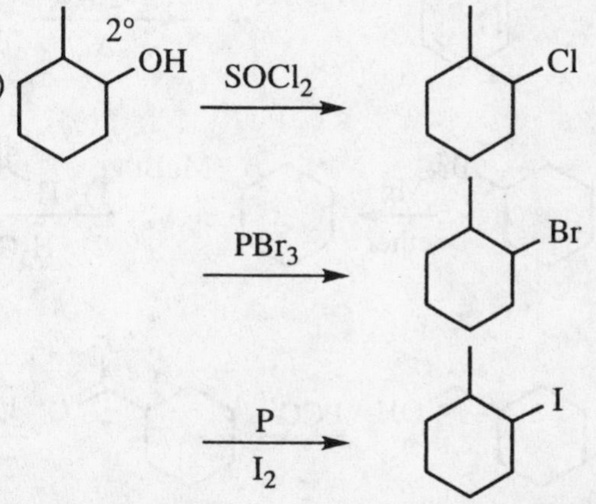

245

11-41

(a) 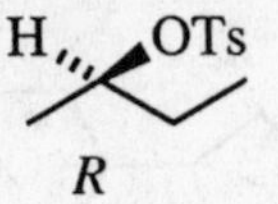
H,,,OTs
R

(b) 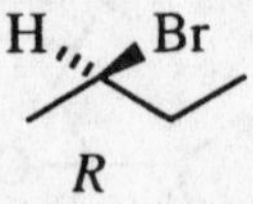
H,,,Br
R
(from inversion)

(c)

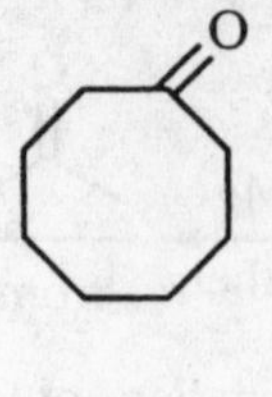

(d)

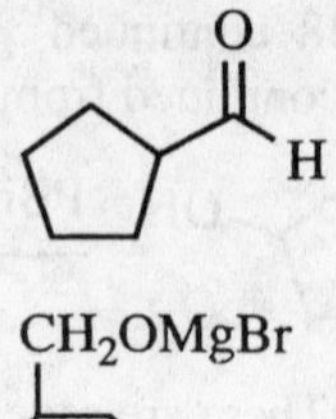

(e) 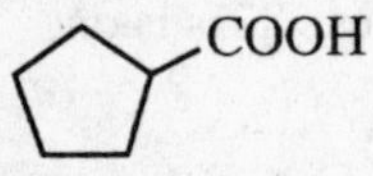 COOH

(f) 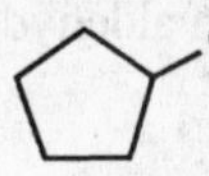 Cl

(g) 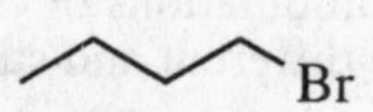 Br

(h)
CH$_2$OMgBr

+ CH$_3$CH$_3$

(i) 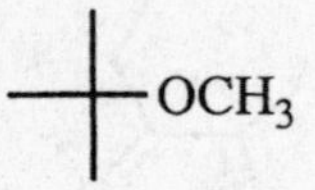 $\rightarrow$OCH$_3$

(j) + CH$_3$OH

(k)

(l)

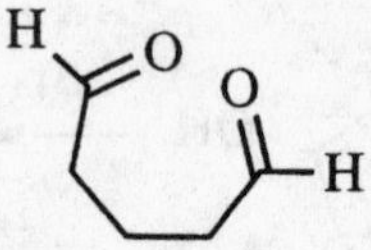

(m)

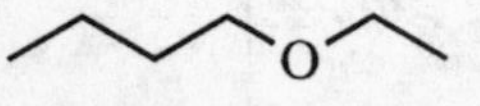

(n) 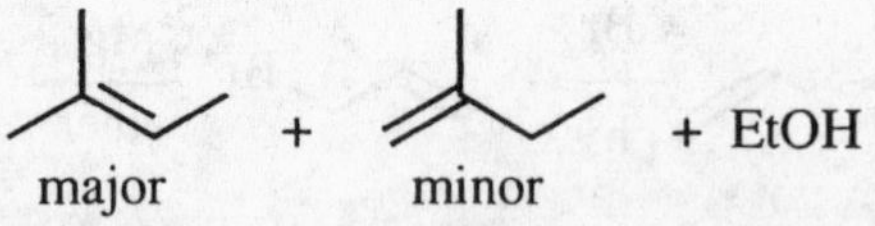 + + EtOH
    major   minor

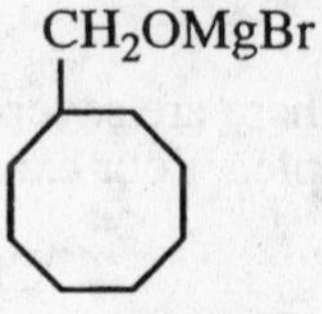

11-42

(a) 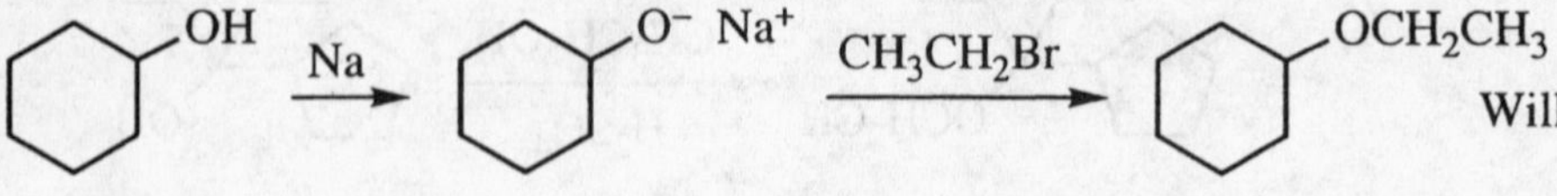 OH $\xrightarrow{\text{Na}}$ O$^-$ Na$^+$ $\xrightarrow{\text{CH}_3\text{CH}_2\text{Br}}$ OCH$_2$CH$_3$ Williamson ether synthesis

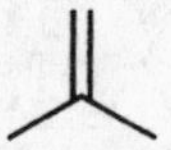

(b) 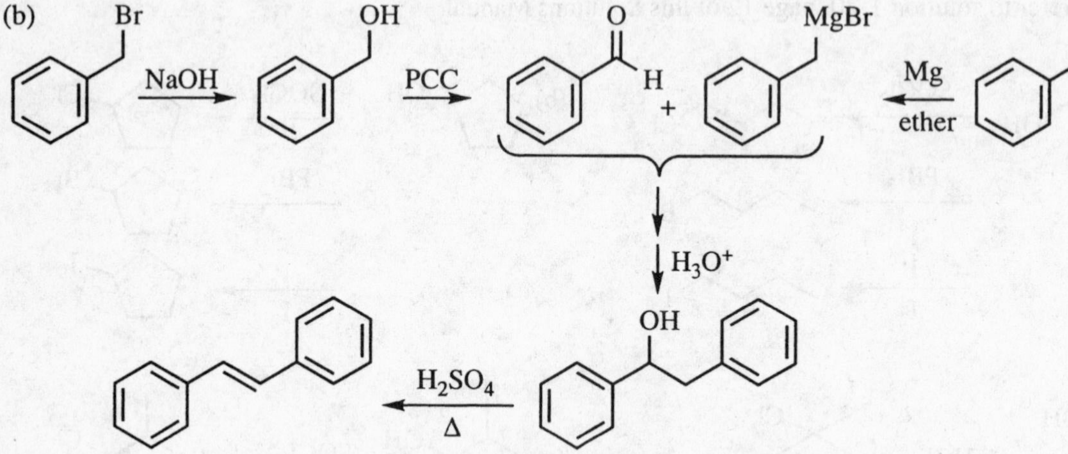 Br $\xrightarrow{\text{NaOH}}$ OH $\xrightarrow{\text{PCC}}$ + MgBr $\xleftarrow[\text{ether}]{\text{Mg}}$ Br

$\downarrow$ H$_3$O$^+$

$\xleftarrow[\Delta]{\text{H}_2\text{SO}_4}$ OH

(c) 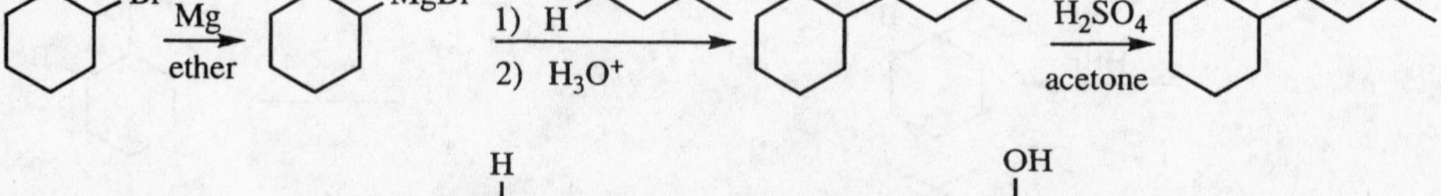 Br $\xrightarrow[\text{ether}]{\text{Mg}}$ MgBr $\xrightarrow[\text{2) H}_3\text{O}^+]{\text{1) H}}$ OH $\xrightarrow[\text{acetone}]{\text{CrO}_3 \ \text{H}_2\text{SO}_4}$

(d) 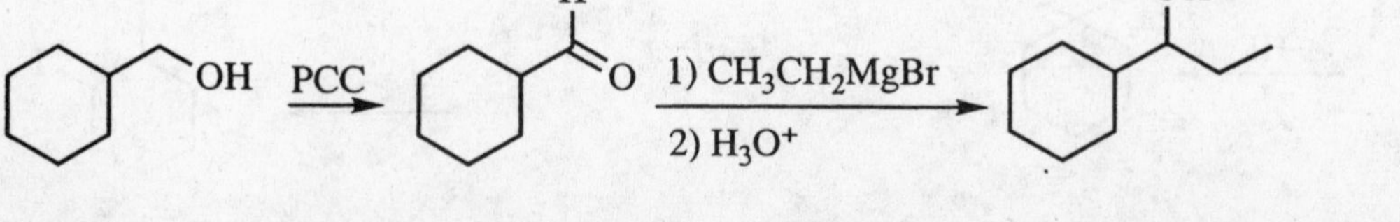 OH $\xrightarrow{\text{PCC}}$ $\xrightarrow[\text{2) H}_3\text{O}^+]{\text{1) CH}_3\text{CH}_2\text{MgBr}}$ OH

246

11-43 Major product for each reaction is shown.

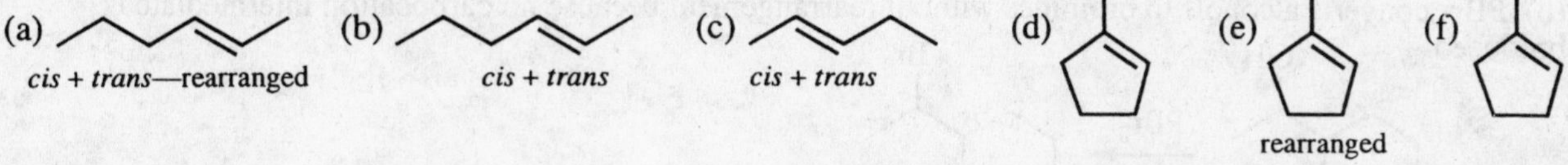
(a)
cis + trans—rearranged
(b)
cis + trans
(c)
cis + trans
(d)
(e)
(f)
rearranged

Note that (d), (e), and (f), produce the same alkene.

11-44

(a) $CH_3CH_2CH_2COOCH_3$ (b)

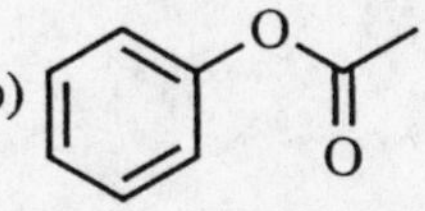
O
O

(c)

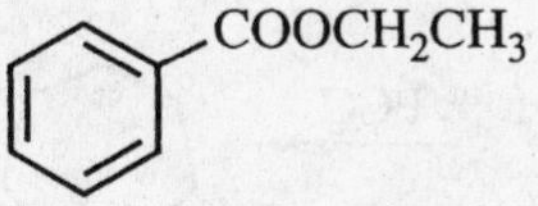
COOCH$_2$CH$_3$

(d) $CH_3CH_2O-P(=O)(OCH_2CH_3)-OH$

(e) CH_3ONO_2

11-45

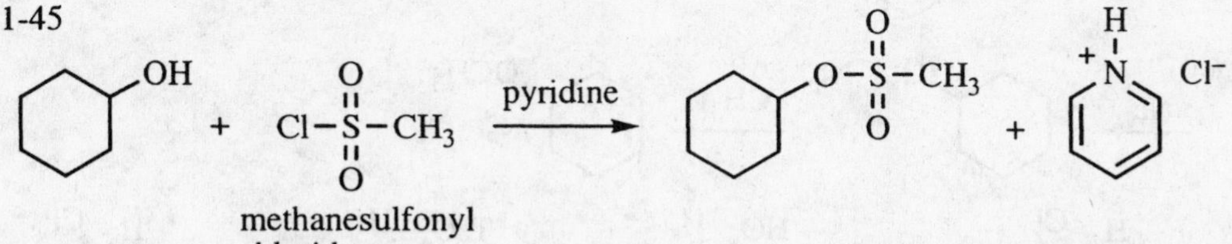
OH
Cl–S–CH$_3$
pyridine
O–S–CH$_3$
Cl$^-$

methanesulfonyl chloride

11-46

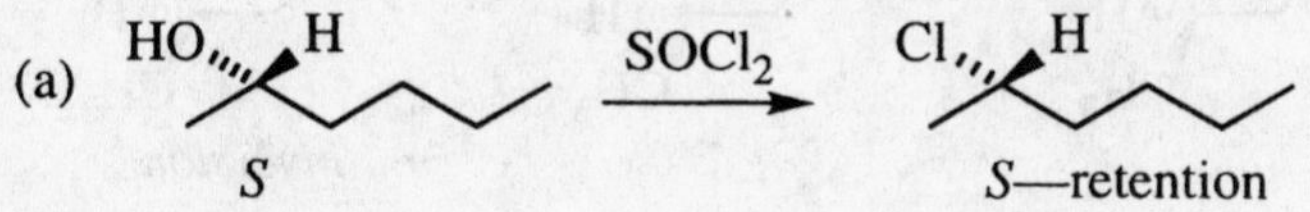
(a)
HO
H
S
SOCl$_2$
Cl
H
S—retention

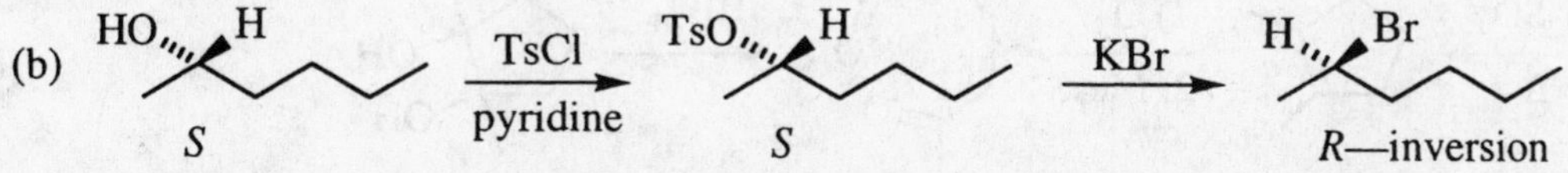
(b)
HO
H
S
TsCl
pyridine
TsO
H
S
KBr
H
Br
R—inversion

Alternatively, PBr_3 could be used.

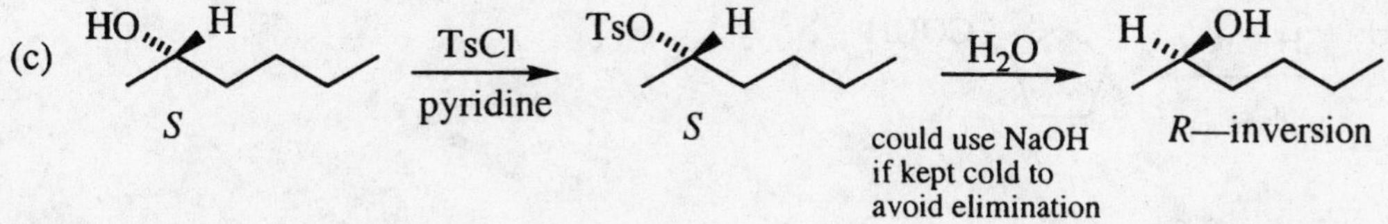
(c)
HO
H
S
TsCl
pyridine
TsO
H
S
H$_2$O
H
OH
R—inversion
could use NaOH
if kept cold to
avoid elimination

11-47

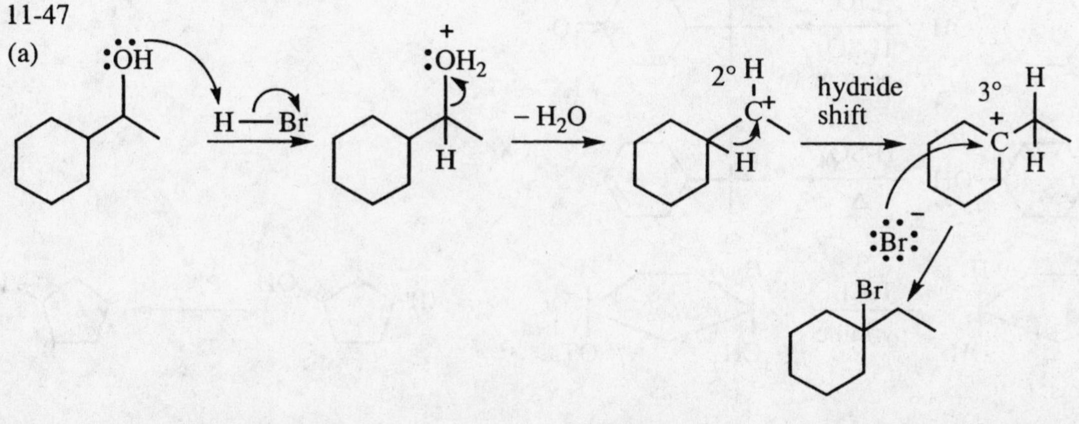
(a)
:ÖH
H—Br
:OH$_2$
H
– H$_2$O
2° H
C+
H
hydride
shift
3°
H
C
H
:Br:
Br

11-47 continued

(b) PBr$_3$ converts alcohols to bromides without rearrangement because no carbocation intermediate is produced.

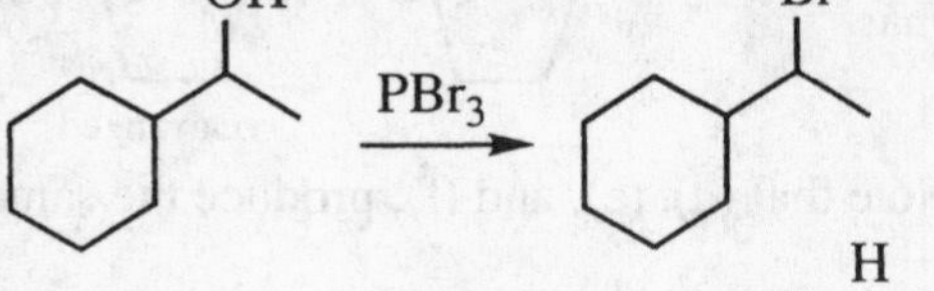

11-48

(a)

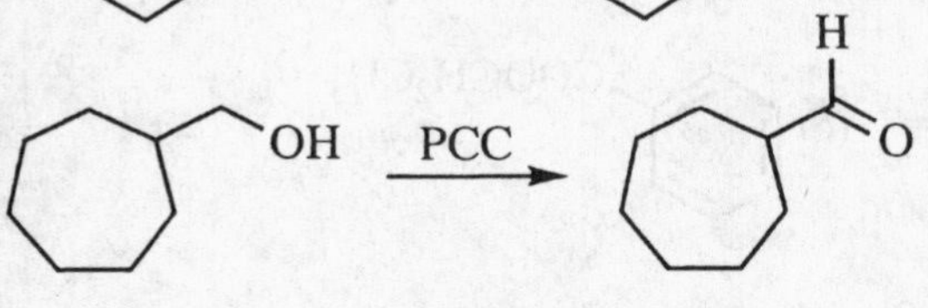

(b)

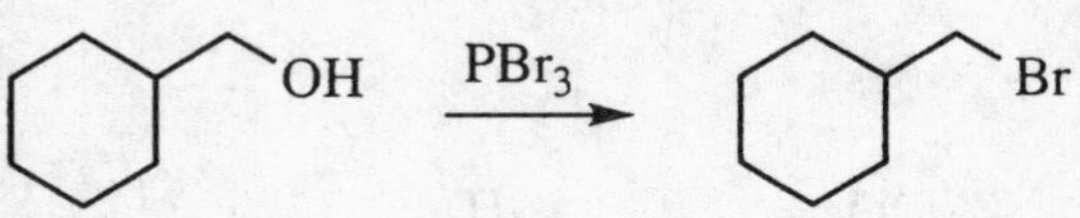

(c)

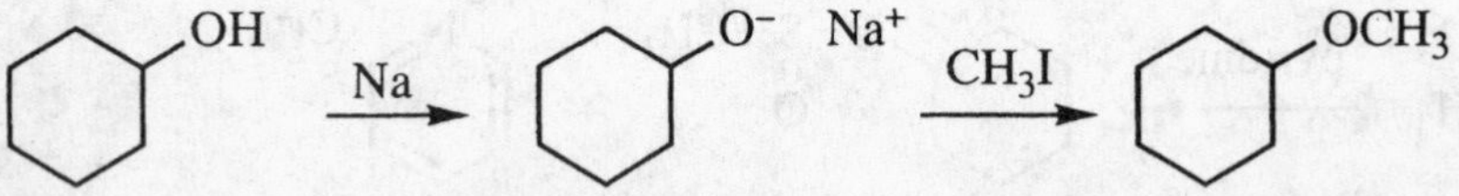

(d)

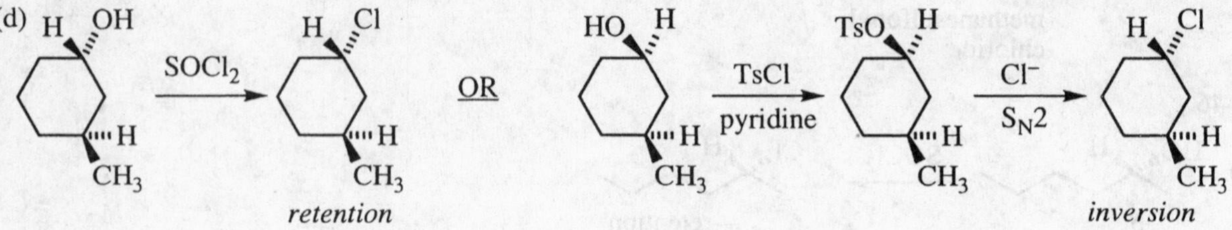

retention   OR   *inversion*

(e)

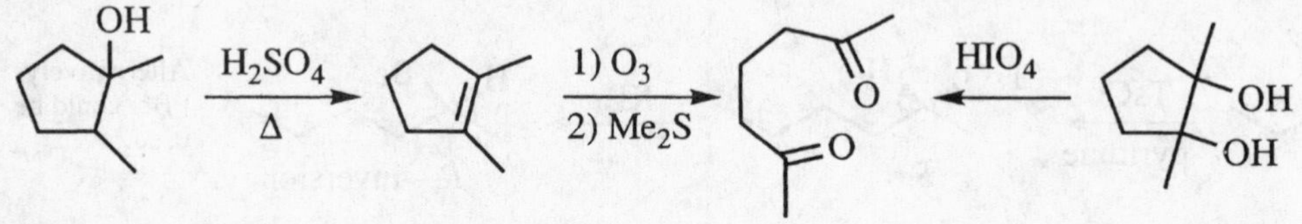

(f)

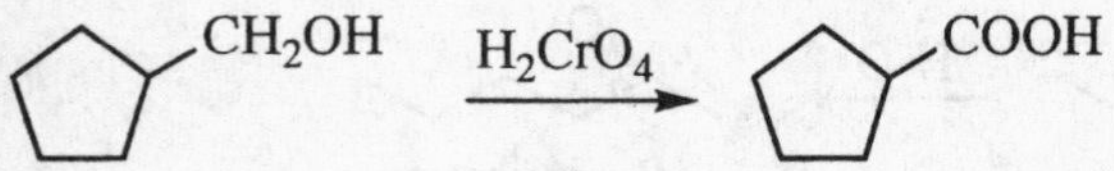

(g)

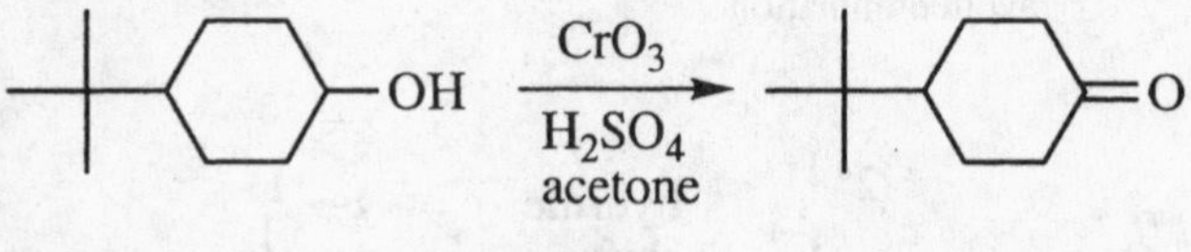

(h)

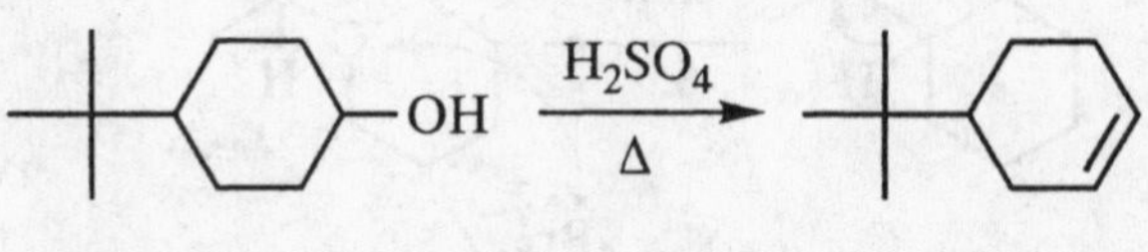

(i)

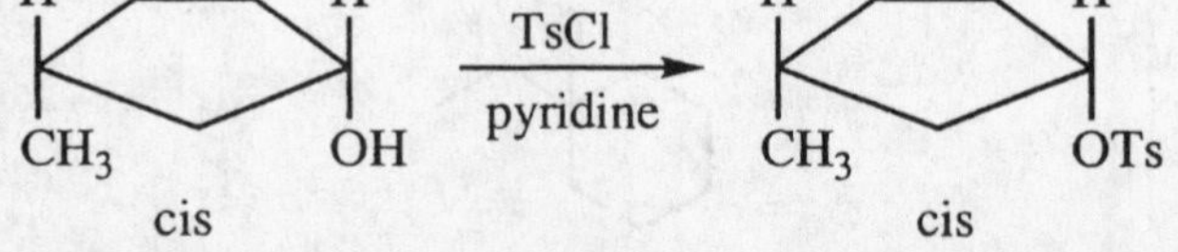

cis   cis

(j)

248

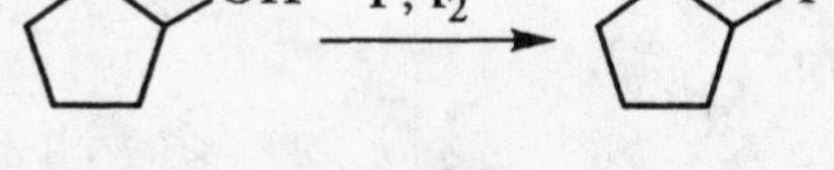

11-49

(a) 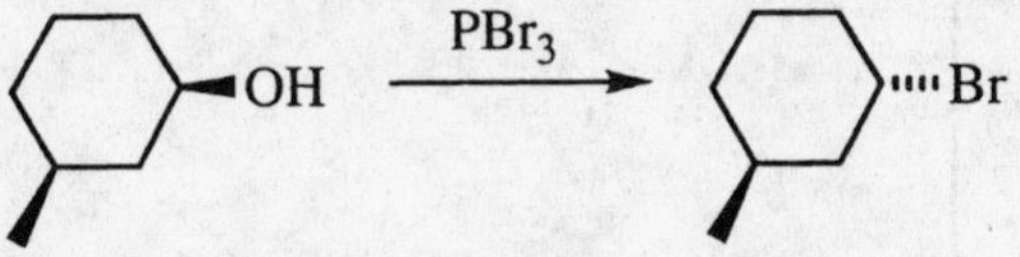 [cyclohexane structure with OH] → PBr₃ → [cyclohexane with Br] *inversion*

$$\text{OH} \xrightarrow{\text{PBr}_3} \text{Br} \quad \textit{inversion}$$

(b) 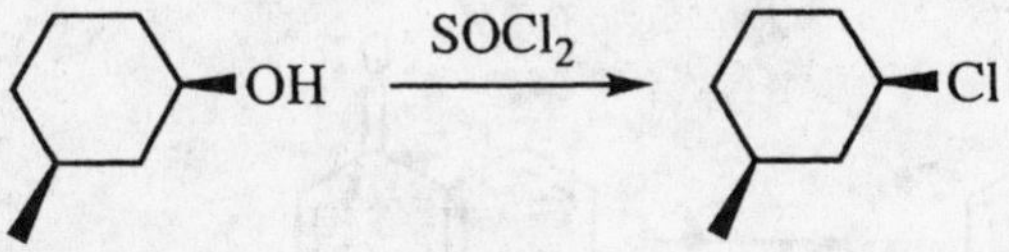
$$\text{OH} \xrightarrow{\text{SOCl}_2} \text{Cl} \quad \textit{retention}$$

(c) 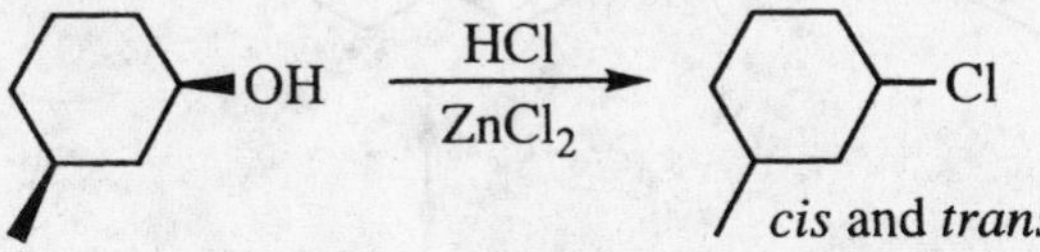
$$\text{OH} \xrightarrow[\text{ZnCl}_2]{\text{HCl}} \text{Cl} \quad S_N1$$
cis and *trans*

(d) 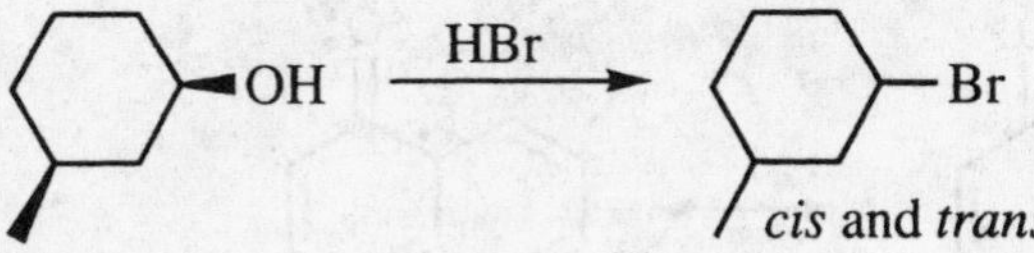
$$\text{OH} \xrightarrow{\text{HBr}} \text{Br} \quad S_N1$$
cis and *trans*

(e) 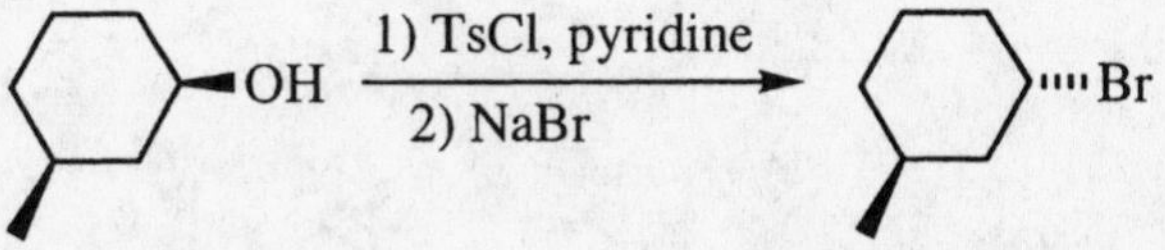
$$\text{OH} \xrightarrow[\text{2) NaBr}]{\text{1) TsCl, pyridine}} \text{Br} \quad \textit{inversion}, S_N2$$

11-50

(a)  [butanol structure] ; [2-butanol structure, OH]

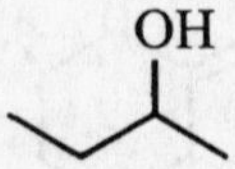

Lucas:　　no reaction　　　　　　cloudy in 1-5 min.

(b) 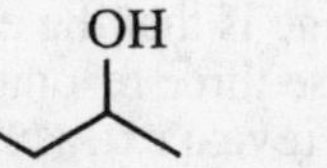 [2-methyl-1-propanol, OH] ; [2-methyl-2-butanol, OH]

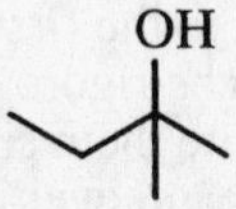

Lucas:　　cloudy in 1-5 min.　　　cloudy in < 1 min.
H₂CrO₄:　immediate blue-green　no reaction—stays orange

(c) 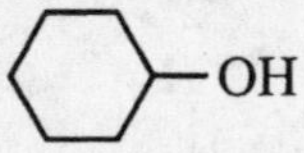 [cyclohexanol, OH] ; [cyclohexene]

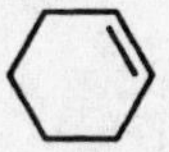

Lucas:　　cloudy in 1-5 min.　　　no reaction
H₂CrO₄:　immediate blue green　no reaction—stays orange

(d) 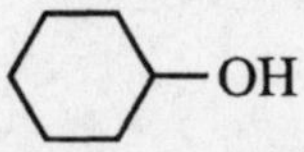 [cyclohexanol, OH] ; [cyclohexanone, =O]

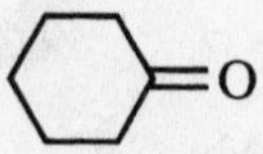

Lucas:　　cloudy in 1-5 min.　　　no reaction
H₂CrO₄:　immediate blue green　no reaction—stays orange

(e) 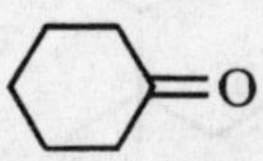 [cyclohexanone, =O] ; [1-methylcyclohexanol, OH]

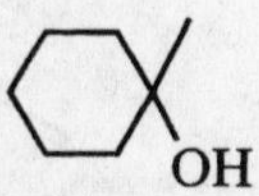

Lucas:　　no reaction　　　　　　cloudy in < 1 min.

249

11-51

(a)

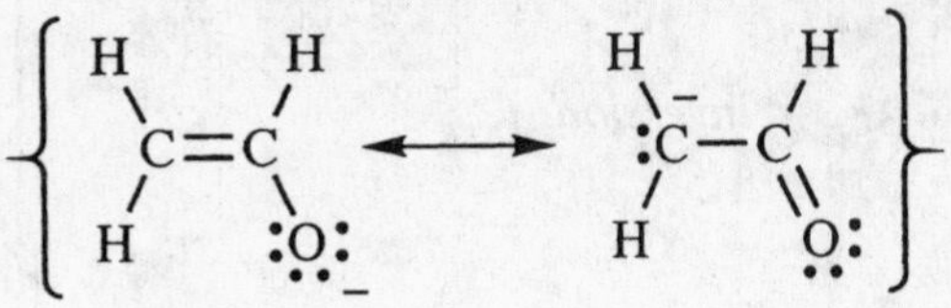

(b)

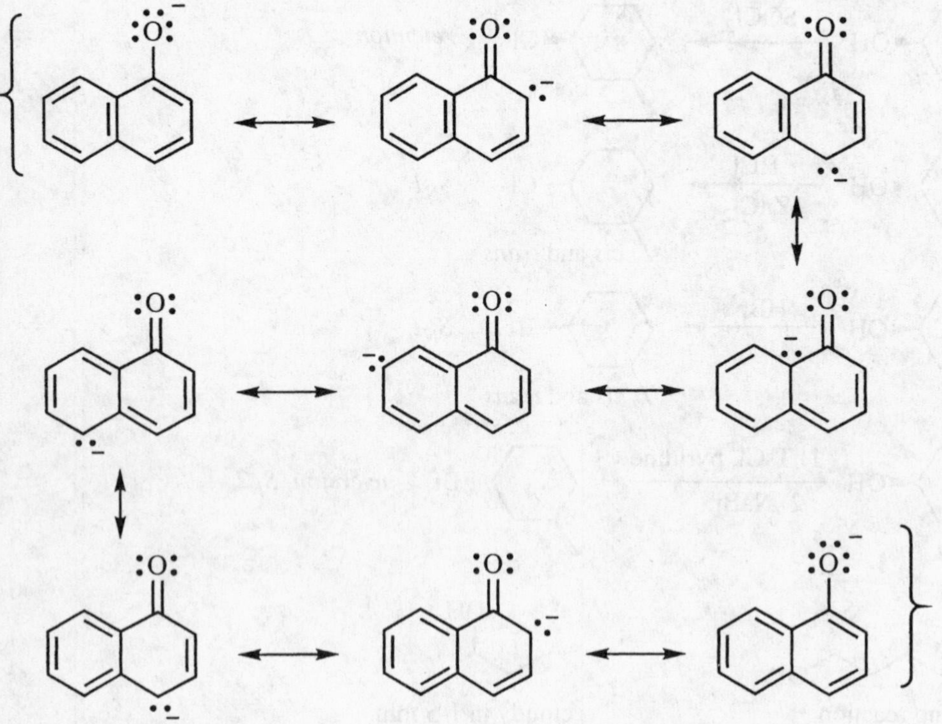

The last three resonance forms are similar to the first three; the change is that the electrons are shown in alternate positions in the benzene ring. To be rigorously correct, these three resonance forms should be included, but most chemists would not write them since they do not reveal extra charge delocalization; understand that they would still be significant, even if not written with the others.

(c)

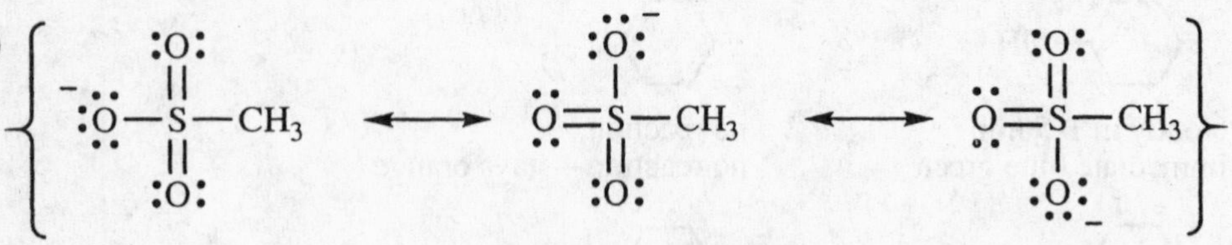

11-52

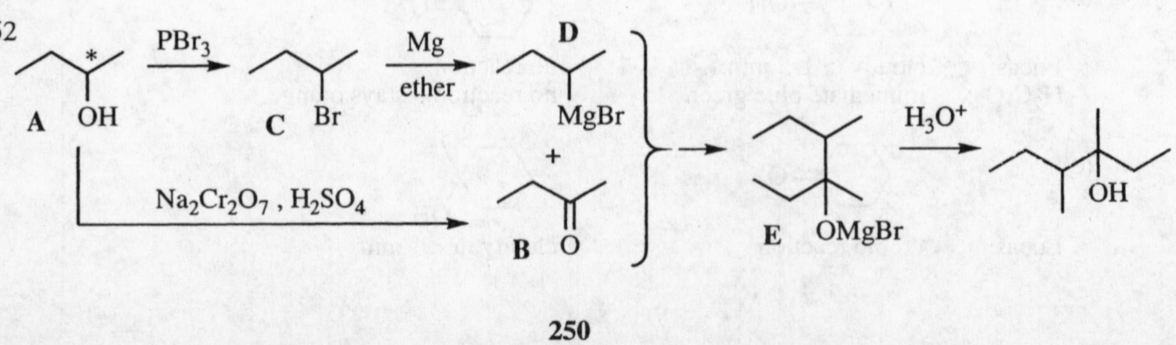

250

11-53

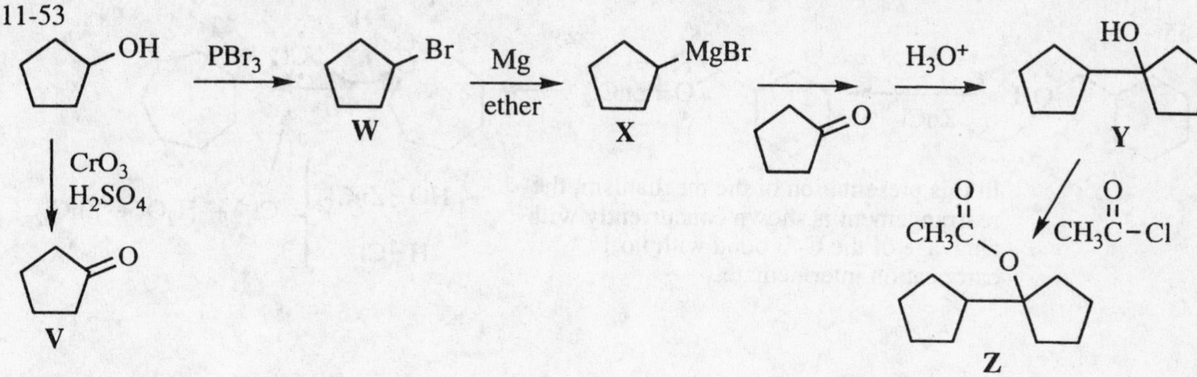

11-54

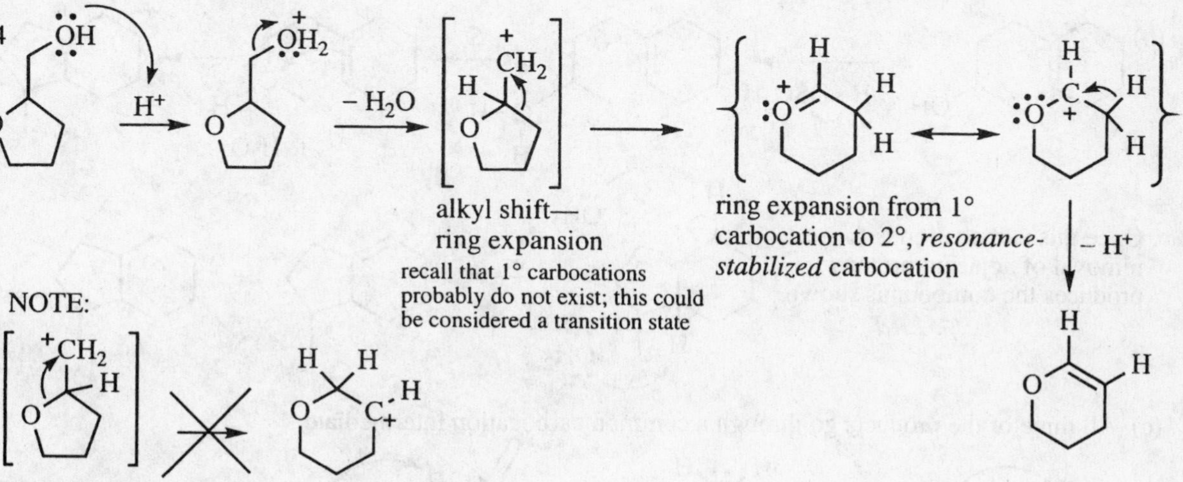

alkyl shift—
ring expansion

recall that 1° carbocations
probably do not exist; this could
be considered a transition state

ring expansion from 1°
carbocation to 2°, *resonance-stabilized* carbocation

NOTE:

The migration directly above does NOT occur
as the cation produced is not resonance-stabilized.

An alternative mechanism could be proposed: protonate the ring oxygen, open the ring to a 2°
carbocation followed by a hydride shift to a resonance-stabilized cation, ring closure, and dehydration.

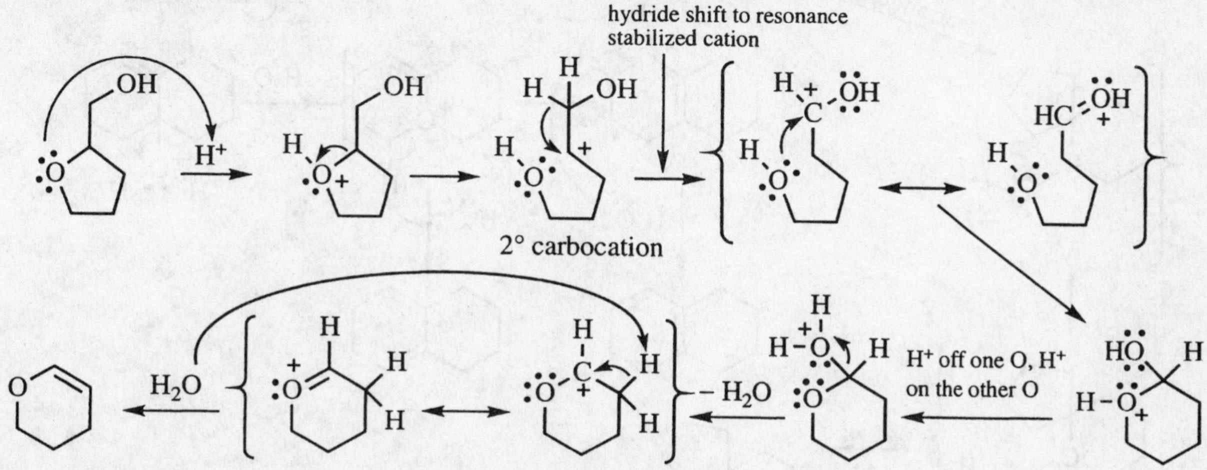

hydride shift to resonance
stabilized cation

2° carbocation

H⁺ off one O, H⁺
on the other O

11-55

(a)

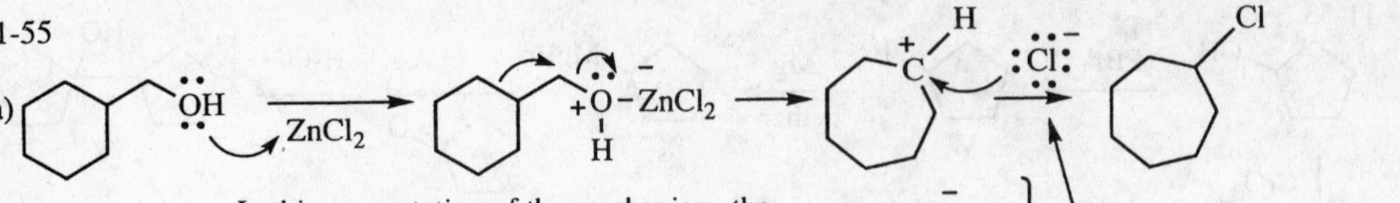

In this presentation of the mechanism, the rearrangement is shown concurrently with cleavage of the C-O bond with no 1° carbocation intermediate.

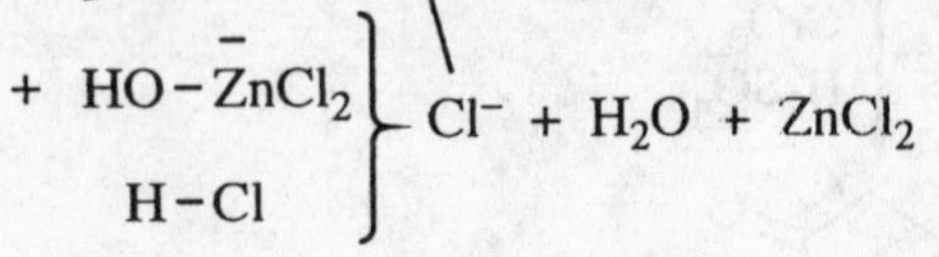

(b)

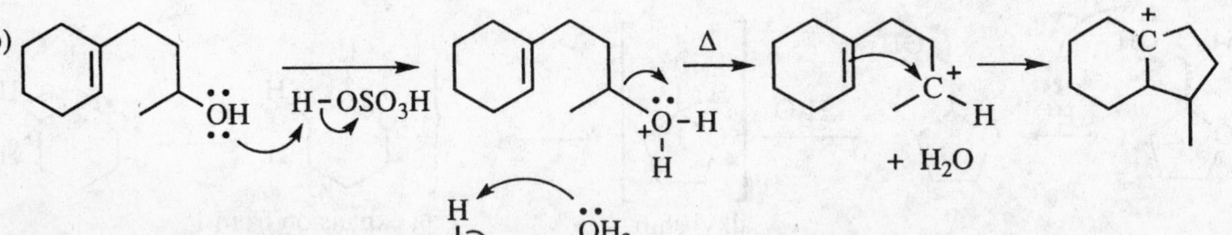

Once this carbocation is formed, removal of adjacent protons produces the compounds shown.

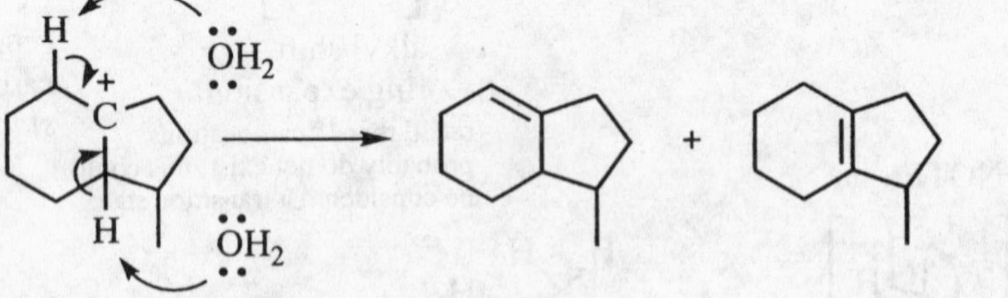

(c) All three of the products go through a common carbocation intermediate.

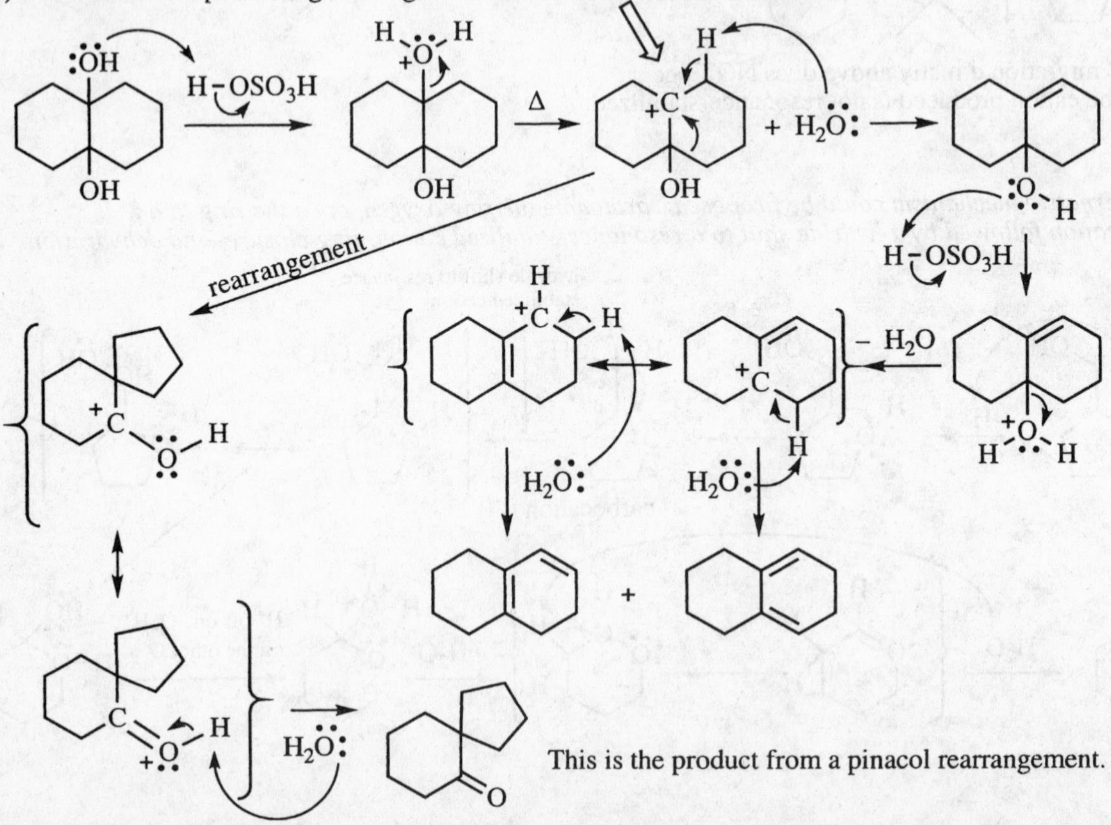

This is the product from a pinacol rearrangement.

252

11-56

(a)
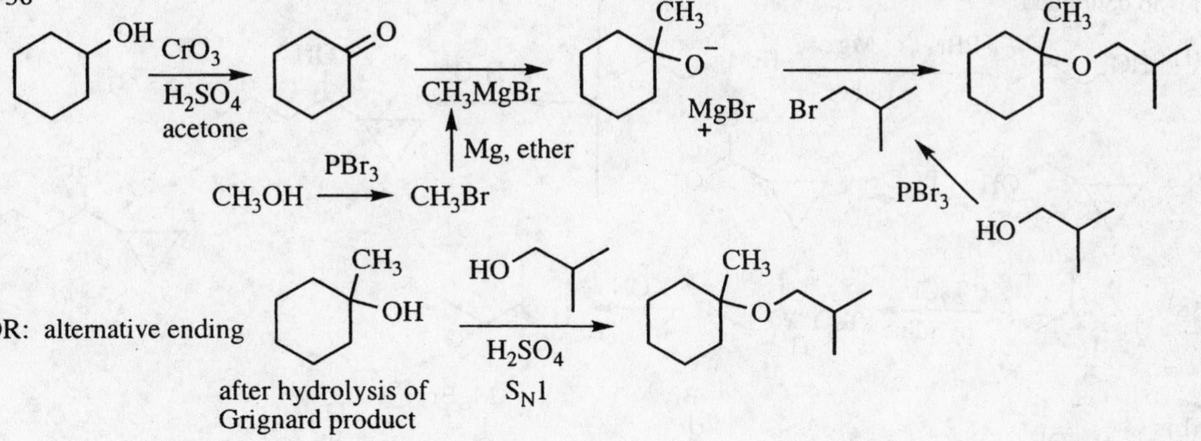

OH
CrO3
H2SO4
acetone
O
CH3MgBr
Mg, ether
PBr3
CH3OH
CH3Br
CH3
O−
MgBr
+
Br
PBr3
HO
CH3
O
OR: alternative ending
CH3
OH
HO
H2SO4
SN1
CH3
O
after hydrolysis of
Grignard product

(b)
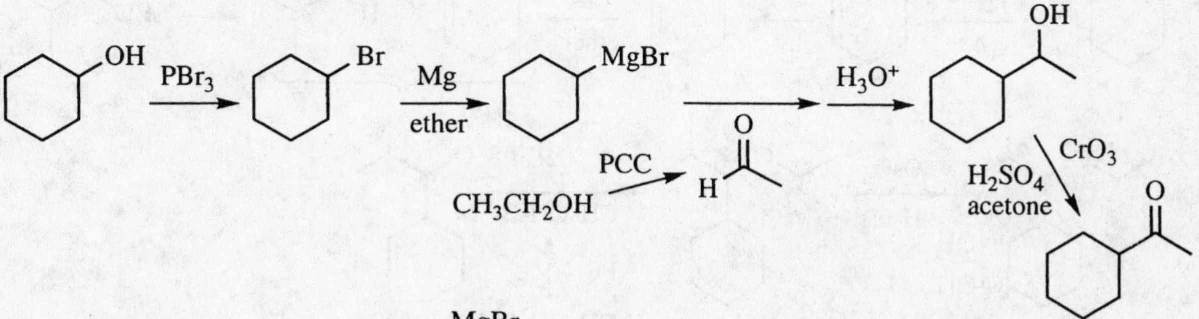

OH
PBr3
Br
Mg
ether
MgBr
H3O+
OH
PCC
CH3CH2OH
O
H
CrO3
H2SO4
acetone
O

(c)
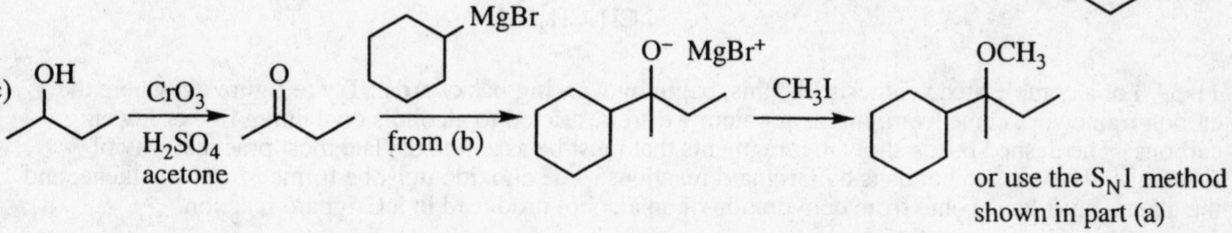

OH
CrO3
H2SO4
acetone
O
MgBr
from (b)
O− MgBr+
CH3I
OCH3
or use the SN1 method
shown in part (a)

(d)
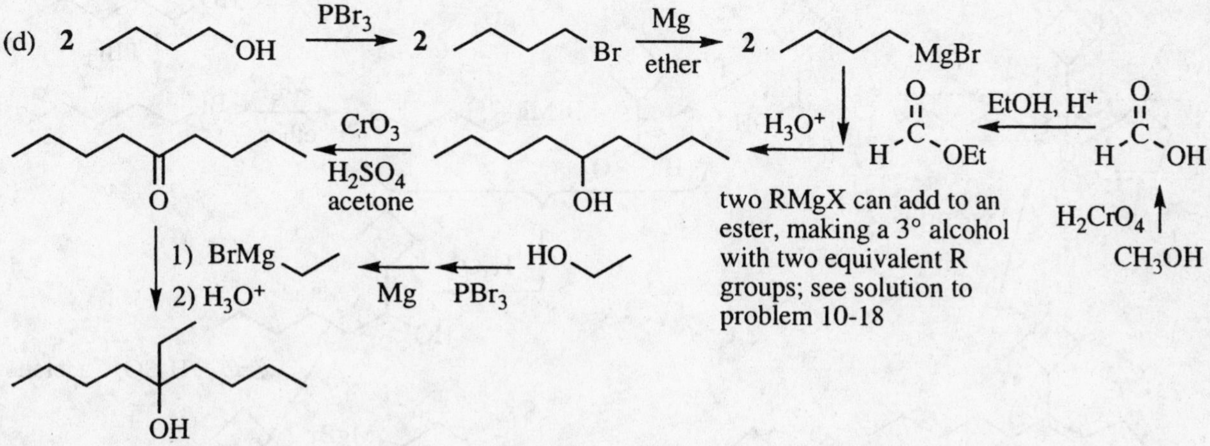

2
OH
PBr3
2
Br
Mg
ether
2
MgBr
O
C
H
OEt
EtOH, H+
O
C
H
OH
H2CrO4
CH3OH
H3O+
OH
CrO3
H2SO4
acetone
O
two RMgX can add to an
ester, making a 3° alcohol
with two equivalent R
groups; see solution to
problem 10-18
1) BrMg
2) H3O+
Mg
PBr3
HO
OH

(e)
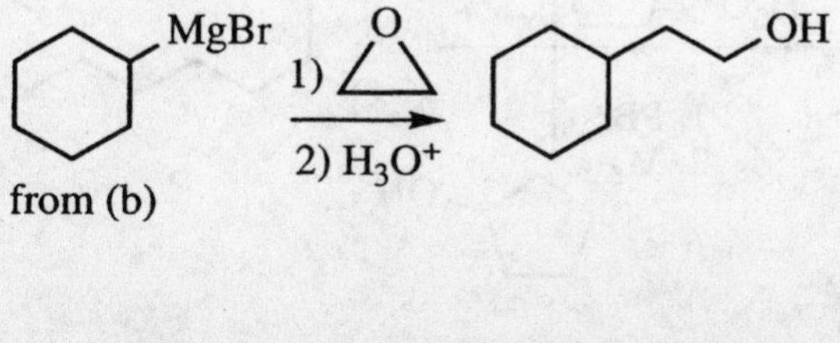

MgBr
from (b)
1)
O
2) H3O+
OH

11-56 continued

(f)

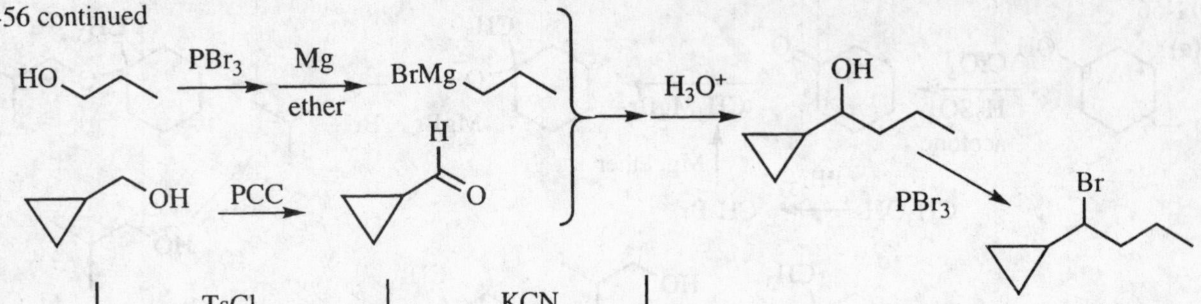

(g)

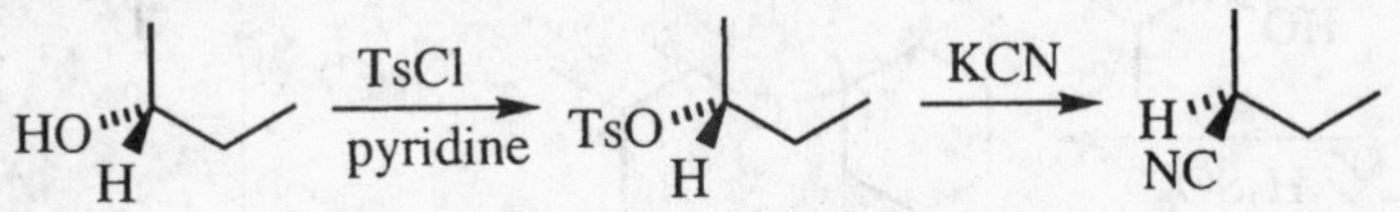

(h)

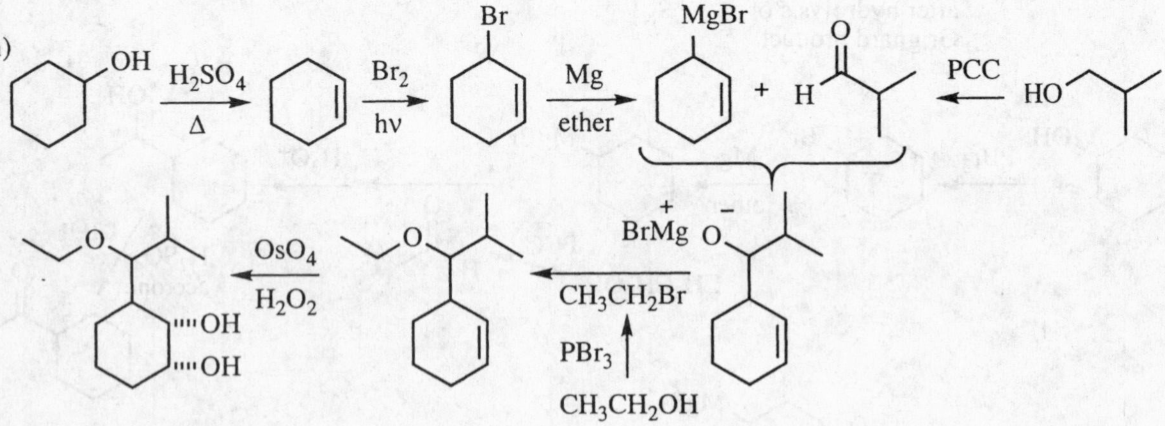

11-57 For a complicated synthesis like this, begin by working backwards. Try to figure out where the carbon framework came from; in this problem we are restricted to alcohols containing five or fewer carbons. The dashed boxes show the fragments that must be assembled. The most practical way of forming carbon-carbon bonds is by Grignard reactions. The epoxide must be formed from an alkene, and the alkene must have come from dehydration of an alcohol produced in a Grignard reaction.

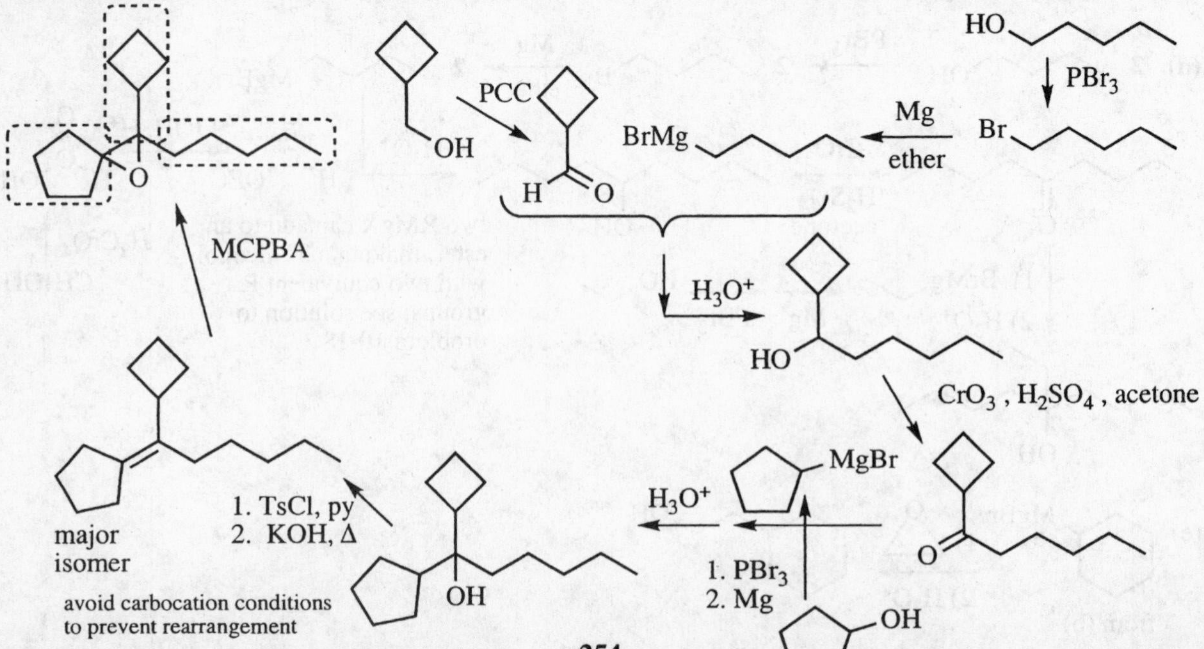

major isomer

avoid carbocation conditions
to prevent rearrangement

(a) Both of these pseudo-syntheses suffer from the misconception that incompatible reagents or conditions can co-exist. In the first example, the S_N1 conditions of ionization cannot exist with the S_N2 conditions of sodium methoxide. The tertiary carbocation in the first step would not wait around long enough for the sodium methoxide to be added in the second step. (The irony is that the first step by itself, the solvolysis of *t*-butyl bromide in methanol, would give the desired product without the sodium methoxide.)

In the second reaction, the acidic conditions of the first step in which the alcohol is protonated are incompatible with the basic conditions of the second step. If basic sodium methoxide were added to the sulfuric acid solution, the instantaneous acid-base neutralization would give methanol, sodium sulfate, and the starting alcohol. No reaction on the alcohol would occur.

(b)

S_N1 solvolysis conditions

Several synthetic sequences are possible for the second synthesis.

11-59

Compound X : —must be a 1° or 2° alcohol with an alkene; no reaction with Lucas leads to a 1° alcohol; can't be allylic as this would give a positive Lucas test

Compound Y : —must be a cyclic ether, not an alcohol and not an alkene; other isomers of cyclic ethers possible

11-60

this reaction works fine, but wait.......

NO REACTION!
cannot do an S_N2 reaction

The Williamson ether synthesis is an S_N2 displacement of a leaving group by an alkoxide ion. There are two reasons why this tosylate cannot unergo an S_N2 reaction. First, backside attack cannot occur because the back side of the bridgehead carbon is blocked by the other bridgehead. Second, the bridgehead carbon cannot undergo inversion because of the constraints of the bridged ring system.

backside attack is blocked

side view

this carbon cannot invert which is required in the S_N2 mechanism

side view

alternative synthesis: $R-OH \xrightarrow{Na} R-O^- Na^+ \xrightarrow{CH_3I} R-OCH_3$

11-61 Let's begin by considering the facts.

The axial alcohol is oxidized ten times as fast as the equatorial alcohol. (In the olden days, this observation was used as evidence suggesting the stereochemistry of a ring alcohol.)

Second, it is known that the oxidation occurs in two steps: 1) formation of the chromate ester; and 2) loss of H and chromate to form the C=O. Let's look at each mechanism.

AXIAL

EQUATORIAL

So what do we know about these systems? We know that substituents are more stable in the equatorial position than in the axial position because any group at the axial position has 1,3-diaxial interactions. So what if Step 1 were the rate-limiting step? We would expect that the equatorial chromate ester would form faster than the axial chromate ester; since this is contrary to what the data show, Step 1 is not likely to be rate-limiting. How about Step 2? If the elimination is rate limiting, we would expect the approach of the base (probably water) to the equatorial hydrogen (axial chromate ester) would be faster than the approach of the base to the axial hydrogen (equatorial chromate ester). Moreover, the axial ester is more motivated to leave due to steric congestion associated with such a large group. This is consistent with the relative rates of reaction from experiment. Thus, it is reasonable to conclude that the second step of the mechanism is rate-limiting.

11-62
(a)

continued on next page

256

11-62 (a) continued

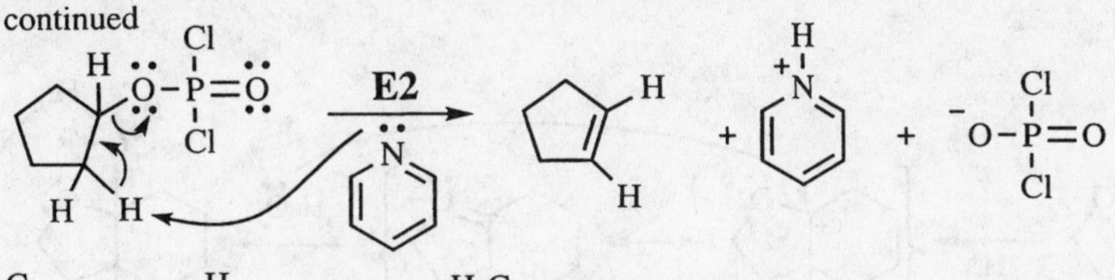

$$\text{E2} \qquad + \qquad + $$

(b)

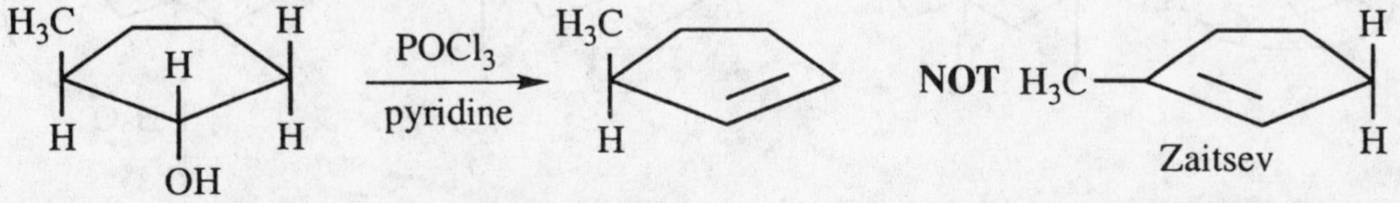

$$\xrightarrow[\text{pyridine}]{\text{POCl}_3}$$

NOT H$_3$C

Zaitsev

There must be a stereochemical requirement in this elimination. If the Saytzeff alkene is not produced because the methyl group is *trans* to the leaving group, then the H and the leaving group must be *trans* and the elimination must be anti—the characteristic stereochemistry of E2 elimination. This evidence differentiates between the two possibilities in part (a).

11-63

(a)

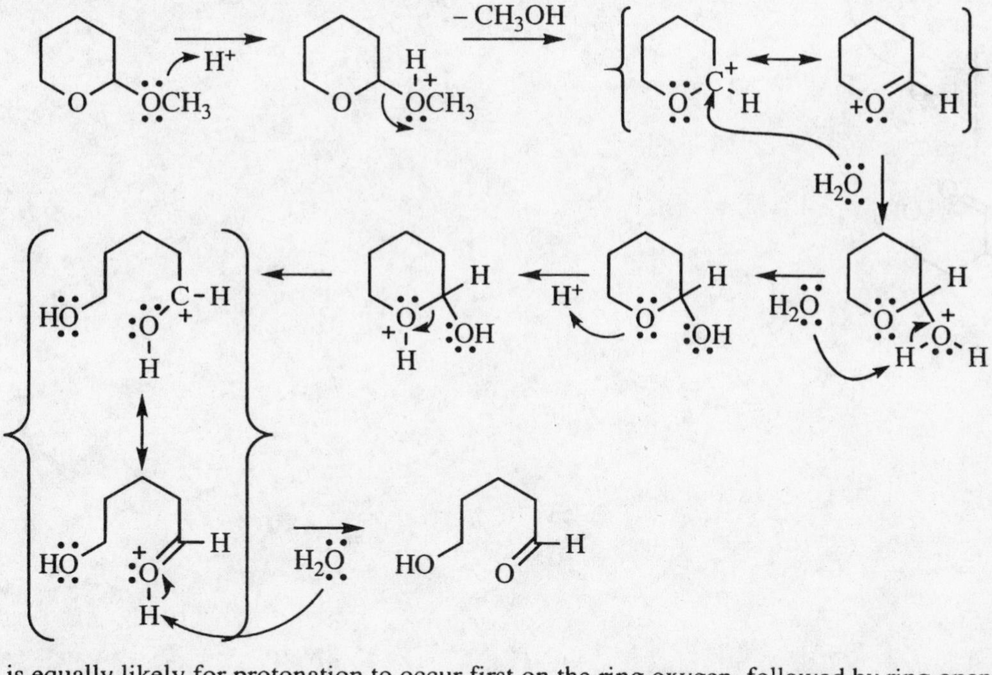

$$\xrightarrow{\text{H}^+} \qquad \xrightarrow{-\text{CH}_3\text{OH}}$$

It is equally likely for protonation to occur first on the ring oxygen, followed by ring opening, then replacement of OCH$_3$ by water.

Dr. Kantorowski suggests this alternative. He and I will arm wrestle to determine which mechanism is correct.

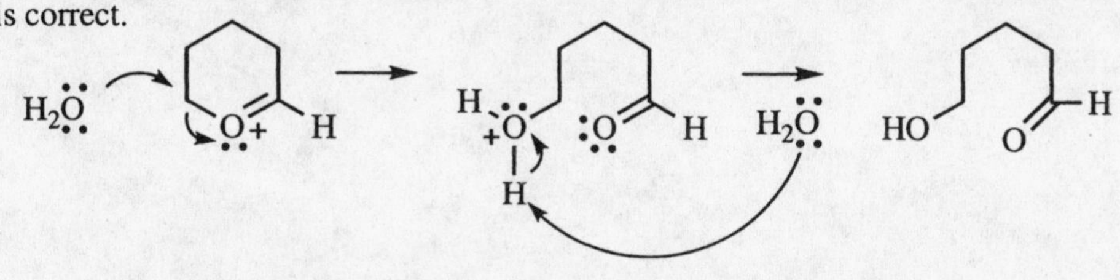

257

11-63 continued

(b)

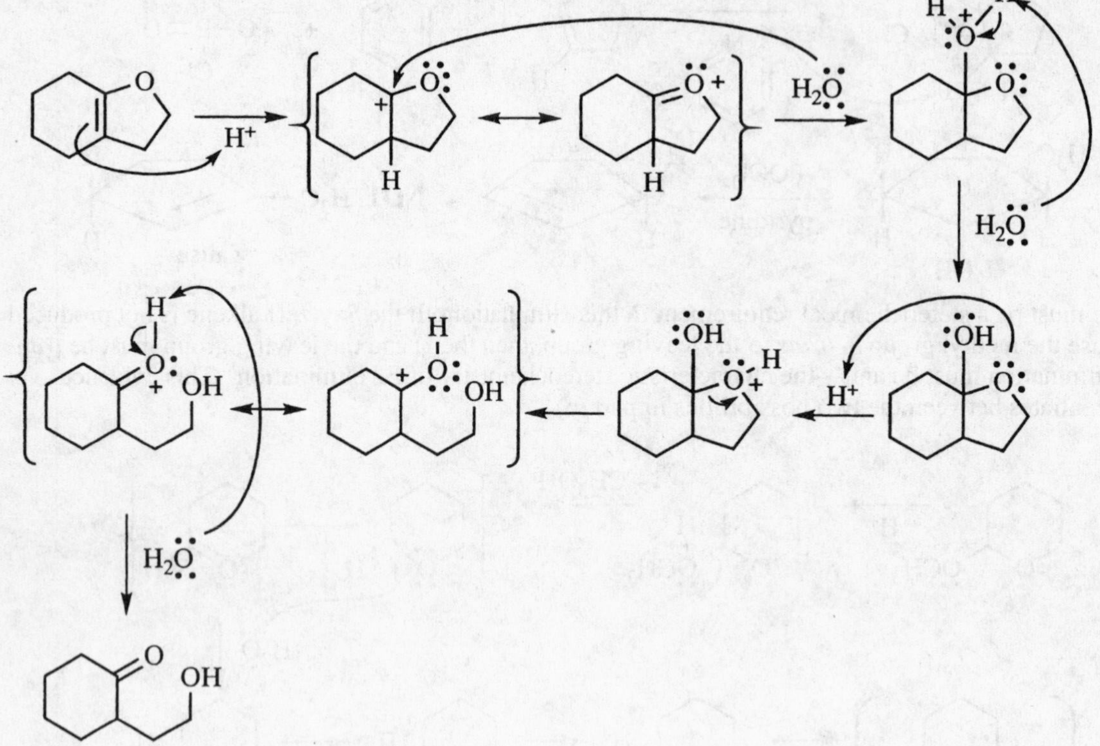

CHAPTER 12—INFRARED SPECTROSCOPY AND MASS SPECTROMETRY

See p. 270 for some useful web sites with infrared and mass spectra.

12-1 The table is completed by recognizing that: $(\overline{\nu})(\lambda) = 10,000$

$\overline{\nu}$ (cm^{-1})	4000	**3300**	**3003**	**2198**	1700	1640	1600	400
λ (μm)	2.50	3.03	3.33	4.55	**5.88**	**6.10**	**6.25**	25.0

12-2 In general, only bonds with dipole moments will have an IR absorption.

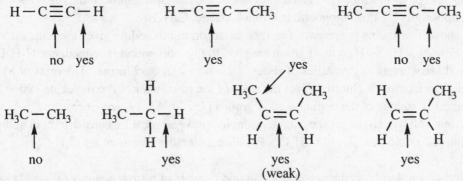

12-3
(a) alkene: C=C at 1640 cm^{-1} , =C—H at 3080 cm^{-1}
(b) alkane: no peaks indicating sp or sp^2 carbons present
(c) This IR shows more than one group. There is a terminal alkyne shown by: C≡C at 2100 cm^{-1} , ≡C—H at 3300 cm^{-1} . These signals indicate an aromatic hydrocarbon as well: =C—H at 3050 cm^{-1}, C=C at 1600 cm^{-1} .

12-4
(a) 2° amine, R—NH—R: one peak at 3300 cm^{-1} indicates an N—H bond; this spectrum also shows a C=C at 1640 cm^{-1}
(b) carboxylic acid: the extremely broad absorption in the 2500-3500 cm^{-1} range, with a "shoulder" around 2500-2700 cm^{-1} , and a C=O at 1710 cm^{-1} , are compelling evidence for a carboxylic acid
(c) alcohol: strong, broad O—H at 3330 cm^{-1}

12-5
(a) conjugated ketone: the small peak at 3030 cm^{-1} suggests =C—H, and the strong peak at 1685 cm^{-1} is consistent with a ketone conjugated with the alkene. The C=C is indicated by a very small peak around 1600 cm^{-1} .
(b) ester: the C=O absorption at 1738 cm^{-1} (higher than the ketone's 1710 cm^{-1}), in conjunction with the strong C—O at 1200 cm^{-1}, points to an ester
(c) amide: the two peaks at 3160-3360 cm^{-1} are likely to be an NH$_2$ group; the strong peak at 1640 cm^{-1} is too strong for an alkene, so it must be a different type of C=X, in this case a C=O, so low because it is part of an amide

12-6
(a) The small peak at 1642 cm^{-1} indicates a C=C, consistent with the =C—H at 3080 cm^{-1} . This appears to be a simple alkene.
(b) The strong absorption at 1691 cm^{-1} is unmistakably a C=O. The smaller peak at 1626 cm^{-1} indicates a C=C, probably conjugated with the C=O. The two peaks at 2712 cm^{-1} and at 2814 cm^{-1} represent H—C=O confirming that this is an aldehyde.

259

12-6 continued

(c) The strong peak at 1650 cm^{-1} is C=C, probably conjugated as it is unusually strong. The 1703 cm^{-1} peak appears to be a conjugated C=O, undeniably a carboxylic acid because of the strong, broad O—H absorption from 2400-3400 cm^{-1}.

(d) The C=O absorption at 1742 cm^{-1} coupled with C—O at 1220 cm^{-1} suggest an ester. The small peak at 1604 cm^{-1}, peaks above 3000 cm^{-1}, and peaks in the 600-800 cm^{-1} region indicate a benzene ring.

12-7

(a) The M and M+2 peaks of equal intensity identify the presence of bromine. The mass of M (156) minus the weight of the lighter isotope of bromine (79) gives the mass of the rest of the molecule: $156 - 79 = 77$. The C_6H_5 (phenyl) group weighs 77; this compound is bromobenzene, C_6H_5Br.

(b) The m/z 127 peak shows that iodine is present. The molecular ion minus iodine gives the remainder of the molecule: $156 - 127 = 29$. The C_2H_5 (ethyl) group weighs 29; this compound is iodoethane, C_2H_5I.

(c) The M and M+2 peaks have relative intensities of about 3:1, a sure sign of chlorine. The mass of M minus the mass of the lighter isotope of chlorine gives the mass of the remainder of the molecule: $90 - 35 = 55$. A fragment of mass 55 is not one of the common alkyl groups (15, 29, 43, 57, *etc.*, increasing in increments of 14 mass units (CH_2)), so the presence of an atom like oxygen must be considered. In addition to the chlorine atom, mass 55 could be C_4H_7 or C_3H_3O. Possible molecular formulas are C_4H_7Cl or C_3H_3ClO.

(d) The odd-mass molecular ion indicates the presence of an odd number of nitrogen atoms (always begin by assuming *one* nitrogen). The rest of the molecule must be: $115 - 14 = 101$; this is most likely C_7H_{17}. The formula $C_7H_{17}N$ is the correct formula of a molecule with no elements of unsaturation. The seven carbons probably include alkyl groups like ethyl or propyl or isopropyl.

12-8 Recall that radicals are not detected in mass spectrometry; only positively-charged ions are detected.

12-9

260

12-10 The molecular weight of each isomer is 116 g/mole, so the molecular ion appears at m/z 116. The left half of each structure is the same; loss of a three carbon radical gives a stabilized cation, each with m/z 73:

$\left\{H_2C \overset{+}{\underset{\cdot\cdot}{\overset{\cdot\cdot}{O}}} \diagdown\diagup \longleftrightarrow H_2C \underset{\cdot\cdot}{\overset{\cdot\cdot}{O}} \overset{+}{\diagdown\diagup}\right\}$ $\left\{H_2C \overset{+}{\underset{\cdot\cdot}{\overset{\cdot\cdot}{O}}}\diagup\diagdown \longleftrightarrow H_2C \underset{\cdot\cdot}{\overset{\cdot\cdot}{O}}\overset{+}{\diagup\diagdown}\right\}$

m/z 73   m/z 73

Where the two structures differ is in the alpha-cleavage on the right side of the oxygen. Alpha-cleavage on the left structure loses two carbons, whereas alpha-cleavage on the right structure loses only one carbon.

$\left[\diagdown\diagup\diagdown O \diagup\diagdown\right]^{+\cdot}$ $\left[\diagdown\diagup\diagdown O \diagup\diagdown\diagup\right]^{+\cdot}$

alpha-cleavage loss of $\overset{\cdot}{C}H_2CH_3$ alpha-cleavage loss of $\overset{\cdot}{C}H_3$

$\left\{\diagdown\diagup\diagdown\underset{\cdot\cdot}{\overset{\cdot\cdot}{O}}\overset{\cdot-CH_2}{+} \longleftrightarrow \diagdown\diagup\diagdown\underset{\cdot\cdot}{\overset{+}{O}}=CH_2\right\}$ $\left\{\diagdown\diagup\diagdown\underset{\cdot\cdot}{\overset{\cdot\cdot}{O}}\overset{H}{\underset{+}{\overset{|}{C}}}\diagdown \longleftrightarrow \diagdown\diagup\diagdown\underset{\cdot\cdot}{\overset{+}{O}}=\overset{H}{C}\diagdown\right\}$

m/z 87   m/z 101

12-11 2,6-Dimethylheptan-4-ol, $C_9H_{20}O$, has molecular weight 144. The highest mass peak at 126 is *not* the molecular ion, but rather is the loss of water (18) from the molecular ion.

$\left[\diagup\diagdown\underset{OH}{\overset{|}{\diagup}}\diagdown\diagup\right]^{+\cdot} \longrightarrow H_2O + \left[\diagup\diagdown\diagup=\diagdown\diagup\right]^{+\cdot}$

m/z 144   m/z 126

The peak at m/z 111 is loss of another 15 (CH_3) from the fragment of m/z 126. This is called allylic cleavage; it generates a 2°, allylic, resonance-stabilized carbocation.

$\left[\diagup\diagdown\diagup=\diagdown\diagup\diagdown\right]^{+\cdot} \longrightarrow \left\{\diagdown\diagup\diagdown\diagup=\diagdown\overset{+}{\diagup} \longleftrightarrow \diagdown\diagup\diagdown\overset{+}{\diagup}\diagdown=\diagup\right\} + \overset{\cdot}{C}H_3$

m/z 126   m/z 111

The peak at m/z 87 results from fragmentation on one side of the alcohol:

$\left[\diagup\diagdown\underset{OH}{\overset{|}{\diagup}}\diagdown\diagup\right]^{+\cdot}$ alpha-cleavage $\left\{\diagdown\diagup\diagdown\overset{:\overset{\cdot\cdot}{O}H}{\underset{+}{\diagup}} \longleftrightarrow \diagdown\diagup\diagdown\diagup=\overset{+\overset{\cdot\cdot}{O}H}{}\right\} + \diagup\diagdown$

m/z 144   m/z 87   mass 57
                                    resonance-stabilized

12-12 Please refer to solution 1-20, page 12 of this Solutions Manual.

12-13 Divide the numbers into 10,000 to arrive at the answer.

(a) 1603 cm^{-1} (b) 2959 cm^{-1} (c) 1709 cm^{-1} (d) 1739 cm^{-1} (e) 2212 cm^{-1} (f) 3300 cm^{-1}

12-14

(a) $\overset{H}{\underset{H}{\diagup}}C=C\overset{CH_2CH_3}{\underset{H}{\diagdown}}$ or $O=C\overset{CH_2CH_3}{\underset{H}{\diagdown}}$

            1660 cm^{-1} 1710 cm^{-1}
                              stronger absorption—larger dipole

261

12-14 continued

(b)

$$\underset{H}{\overset{H}{}}C=C\underset{H}{\overset{CH_2CH_3}{}}$$

1660 cm^{-1}

or

$$\underset{H}{\overset{H}{}}C=C\underset{H}{\overset{OCH_2CH_3}{}}$$

1640 cm^{-1}
stronger absorption—larger dipole

(c)

$$\overset{\cdot\cdot}{\underset{H}{N}}=C\underset{H}{\overset{CH_2CH_3}{}}$$

1660 cm^{-1}
stronger absorption—
larger dipole

or

$$\underset{H}{\overset{H}{}}C=C\underset{H}{\overset{CH_2CH_3}{}}$$

1660 cm^{-1}

(d)

$$\underset{H_3C}{\overset{H}{}}C=C\underset{H}{\overset{CH_3}{}}$$

1660 cm^{-1}
(no dipole moment)

or

$$\underset{H}{\overset{H}{}}C=C\underset{H}{\overset{CH_2CH_3}{}}$$

1660 cm^{-1}
stronger absorption—larger dipole

12-15

(a)

$$\underset{H_3C}{\overset{H_3C}{}}C=C\underset{CH_3}{\overset{CH_3}{}}$$

1660 cm^{-1}
weak or non-existent

and

3000-3100 cm^{-1}

$$\underset{H}{\overset{H}{}}C=C\underset{CH(CH_3)_2}{\overset{CH_3}{}}$$

1660 cm^{-1}
moderate intensity

(b)

1620 cm^{-1}

conjugated

and

1645 cm^{-1}

not conjugated

(c) both carbonyls show strong absorptions around 1710 cm^{-1}

$$CH_3(CH_2)_3 - \overset{O}{\overset{\|}{C}} - H$$

2700-2800 cm^{-1}
two small peaks

and

$$CH_3(CH_2)_2 - \overset{O}{\overset{\|}{C}} - CH_3$$

(d)

O—H 3300 cm^{-1}
broad, strong

1200 cm^{-1}

and

1710 cm^{-1}
strong

12-15 continued

(e) CH₃(CH₂)₅ - C≡C - H   ← 3300 cm⁻¹   and   CH₃(CH₂)₆ - C≡N

$CH_3(CH_2)_5 - C{\equiv}C - H$ ← 3300 cm⁻¹

↑ 2100-2200 cm⁻¹
weak to moderate intensity

↑ 2200-2300 cm⁻¹
moderate to strong intensity

(f) both carbonyls show strong absorptions around 1710 cm⁻¹

3300 cm⁻¹
broad, strong

$CH_3CH_2CH_2 - \overset{\displaystyle O}{\underset{\displaystyle ||}{C}} - OH$

and

$CH_3 - CH - CH_2 - \overset{\displaystyle O}{\underset{\displaystyle ||}{C}} - H$

2500-3500 cm⁻¹
very broad

↑ 2700-2800 cm⁻¹
two small peaks

(g) 1650 cm⁻¹

$CH_3CH_2CH_2 - \overset{\displaystyle O}{\underset{\displaystyle ||}{C}} - N - H$

and 1710 cm⁻¹

$CH_3CH_2 - \overset{\displaystyle O}{\underset{\displaystyle ||}{C}} - CH_2CH_3$

H
3300 cm⁻¹
two peaks

12-16

(a) 1700 cm⁻¹

$\overset{\displaystyle O}{\underset{\displaystyle ||}{C}} - OH$

H - C = C
H   CH₃
2400-3400 cm⁻¹
1640 cm⁻¹

(b) 1715 cm⁻¹

$CH_3 - CH \cdot \overset{\displaystyle O}{\underset{\displaystyle ||}{C}} - CH_3$
          |
         CH₃

(c) 3000-3100 cm⁻¹

H

CH₂ - C≡N

2250 cm⁻¹

1600 cm⁻¹

(d) 3000-3100 cm⁻¹

H   H
         3400 cm⁻¹

N

CH₂CH₃

2900-3000 cm⁻¹

1600 cm⁻¹

12-17

(a)
$\left[CH_3 \overset{\overset{\displaystyle 71}{CH_3}}{\underset{\underset{\displaystyle 43}{CH}}{|}} CH_2CH_2CH_3 \right]^{+\bullet}$
m/z 86

⟶

CH₃
|
+CH — CH₂CH₂CH₃
m/z 71

⟶

CH₃
|
CH₃ — CH   m/z 43
        +

263

12-17 continued

(b) $[CH_3 \cdot CH = C(CH_3) - CH_2 \cdot CH_2CH_3]^{+\cdot}$ $\longrightarrow$ $\{CH_3 \cdot CH = C(CH_3) - \overset{+}{C}H_2 \longleftrightarrow CH_3 - \overset{+}{C}H - C(CH_3) = CH_2\}$

69

m/z 98—allylic cleavage   m/z 69

(c) $\begin{bmatrix} \overset{87}{\overset{|}{\underset{|}{OH}}} & CH_3 \\ CH_3 + CH + CH_2 - CH \cdot CH_3 \\ 45 \end{bmatrix}^{+}$
   alpha-cleavage
$\{:\ddot{O}H \ldots +CH - CH_2 - CH(CH_3) \cdot CH_3 \longleftrightarrow \overset{+}{\ddot{O}}H \ldots CH - CH_2 - CH(CH_3) \cdot CH_3\}$

m/z 102   m/z 87

↓ − H_2O alpha-cleavage

$[CH_3 \cdot CH = CH - CH(CH_3) \cdot CH_3]^{+\cdot}$ $\{:\ddot{O}H \ldots CH_3 - \overset{+}{C}H \longleftrightarrow \overset{+}{\ddot{O}}H \ldots CH_3 - CH\}$

m/z 84   m/z 45

(d) [benzene ring with CH(CH₃) substituent; labels 77, 57, 91, 43]$^{+\cdot}$ $\longrightarrow$ benzyl cation ($\overset{+}{C}H_2$)   rearrange   tropylium ion

m/z 134   benzyl cation   m/z 91

The *tropylium ion* is a characteristic fragment from phenylalkanes.

[phenyl cation] C^+ m/z 77

$HC^+(CH_3)(CH_3)$ m/z 43

$H_2\overset{+}{C}$–CH(CH₃)₂   rearrange $H_3C-\overset{+}{C}(CH_3)-CH_3$ with CH_3 m/z 57

(e) [cyclohexyl–O–C(CH₃)₃ structure; labels 85, 43, 129]$^{+\cdot}$ $\longrightarrow$ [cyclohexyl cation] $+CH$ m/z 85

$HC^+(CH_3)(CH_3)$ m/z 43

m/z 144

↓ alpha-cleavage

$\{$cyclohexyl$-\overset{..}{\underset{..}{O}}-\overset{+}{C}H(CH_3) \longleftrightarrow$ cyclohexyl$-\overset{+}{\underset{..}{O}}=CH(CH_3)\}$

m/z 129

264

12-18

(a) $[\text{...}]^{+\bullet}$ m/z 114 (with labels 71, 57, 85) $\longrightarrow$

$\sim\sim\overset{+}{CH_2}$ m/z 85 $+$ $H_2\overset{\bullet}{C}-CH_3$ mass 29

$\sim\sim\overset{+}{CH_2}$ m/z 71 $+$ $H_2\overset{\bullet}{C}\sim$ mass 43

$\sim\overset{+}{CH_2}$ m/z 57 $+$ $H_2\overset{\bullet}{C}\sim$ mass 57

(b) $\left[\begin{array}{c}CH_3\\ \bigcirc\end{array}\right]^{+\bullet}$ m/z 98 $\longrightarrow$ $\overset{H}{\underset{+}{\bigcirc}}C$ m/z 83 $+ \bullet CH_3$ mass 15

(c) $\left[CH_3-\underset{\underset{CH_3}{|}}{C}=CH-CH_2{+}CH_3\right]^{+\bullet}$ m/z 84 (label 69) $\longrightarrow$

$\left\{CH_3-\underset{\underset{CH_3}{|}}{C}=CH-\overset{+}{CH_2} \longleftrightarrow CH_3-\underset{\underset{+}{|}}{\overset{\overset{CH_3}{|}}{C}}-CH=CH_2\right\}$ m/z 69 $+ \bullet CH_3$ mass 15

(d) $\left[\begin{array}{c}OH\\ |\\ CH_2{+}CH_2-CH_2CH_2CH_3\end{array}\right]^{+\bullet}$ m/z 88 (label 31) $\longrightarrow$

$\left\{\underset{\overset{+}{CH_2}}{\overset{:\overset{\bullet\bullet}{O}H}{|}} \longleftrightarrow \underset{CH_2}{\overset{+\overset{\bullet\bullet}{O}H}{||}}\right\}$ m/z 31 $+ \bullet CH_2CH_2CH_2CH_3$ mass 57

$\downarrow -H_2O$

$\left[CH_2=CH-CH_2{+}CH_2{+}CH_3\right]^{+\bullet}$ m/z 70 (label 41, 55) $\longrightarrow$ $CH_2=CH-CH_2-\overset{+}{CH_2}$ m/z 55 $+ \bullet CH_3$ mass 15

$\downarrow$

$\left\{CH_2=CH-\overset{+}{CH_2} \longleftrightarrow \overset{+}{CH_2}-CH=CH_2\right\}$ m/z 41 $+ \bullet CH_2CH_3$ mass 29

(e) $\left[\begin{array}{c}H\\ \bigcirc-N\end{array}\right]^{+\bullet}$ m/z 121 (label 77, 106) $\longrightarrow$ $\bigcirc C^+$ m/z 77 $+ \bullet NHCH_2CH_3$ mass 44

continued on next page

265

12-18 (e) continued

alpha-cleavage → {...} + •CH₃ mass 15

m/z 121 → m/z 106

12-19

(a) The characteristic frequencies of the OH absorption and the C=C absorption will indicate the presence or absence of the groups. A spectrum with an absorption around 3300 cm⁻¹ will have some cyclohexanol in it; if that same spectrum also has a peak at 1645 cm⁻¹, then the sample will also contain some cyclohexene. Pure samples will have peaks representative of only one of the compounds and not the other. Note that *quantitation* of the two compounds would be very difficult by IR because the strength of absorptions are very different. Usually, other methods are used in preference to IR for quantitative measurements.

3300 cm⁻¹ broad, strong

1200 cm⁻¹

1645 cm⁻¹ moderate

(b) Mass spectrometry can be misleading with alcohols. Usually, alcohols dehydrate in the inlet system of a mass spectrometer, and the characteristic peaks observed in the mass spectrum are those of the alkene, not of the parent alcohol. For this particular analysis, mass spectrometry would be unreliable and perhaps misleading.

12-20

(a) The "student prep" compound must be 1-bromobutane. The most obvious feature of the mass spectrum is the pair of peaks at M and M+2 of approximately equal heights, characteristic of a bromine atom. Loss of bromine (79) from the molecular ion at 136 gives a mass of 57, C_4H_9, a butyl group. Which of the four possible butyl groups? The peaks at 107 (loss of 29, C_2H_5) and 93 (loss of 43, C_3H_7) are consistent with a linear chain, not a branched chain.

(b) The base peak at 57 is so strong because the carbon-halogen bond is the weakest in the molecule. Typically, loss of halogen is the dominant fragmentation in alkyl halides.

266

12-21

(a) Deuterium has twice the mass of hydrogen, but similar spring constant, k. Compare the frequency of C—D vibration to C—H vibration by setting up a ratio, changing only the mass (substitute 2m for m).

$$\frac{\nu_D}{\nu_H} = \frac{\sqrt{k/2m}}{\sqrt{k/m}} = \frac{\sqrt{1/2}\,\sqrt{k/m}}{\sqrt{k/m}} = \sqrt{1/2} = 0.707$$

$$\nu_D = 0.707\,\nu_H = 0.707\,(3000\ \text{cm}^{-1}) \approx \mathbf{2100\ cm^{-1}}$$

(b) The functional group most likely to be confused with a C—D stretch is the alkyne (carbon-carbon triple bond), which appears in the same region and is often very weak.

12-22

a 1° carbocation—not favorable

a 2° carbocation—reasonable

a 3° carbocation—the best

The most likely fragmentation of 2,2,3,3-tetramethylbutane will give a 3° carbocation, the most stable of the common alkyl cations. The molecular ion should be small or non-existent while m/z 57 is likely to be the base peak, whereas the molecular ion peaks will be more prominent for *n*-octane and for 3,4-dimethylhexane.

12-23

(a) The information that this mystery compound is a hydrocarbon makes interpreting the mass spectrum much easier. (It is relatively simple to tell if a compound has chlorine, bromine, or nitrogen by a mass spectrum, but oxygen is difficult to determine by mass spectrometry alone.) A hydrocarbon with molecular ion of 110 can have only 8 carbons (8 x 12 = 96) and 14 hydrogens. The formula C_8H_{14} has two elements of unsaturation.

(b) The IR will be useful in determining what the elements of unsaturation are. Cycloalkanes are generally not distinguishable in the IR. An alkene should have an absorption around 1600-1650 cm^{-1}; none is present in this IR. An alkyne should have a small, sharp peak around 2200 cm^{-1}—PRESENT AT 2120 cm^{-1}! Also, a sharp peak around 3300 cm^{-1} indicates a hydrogen on an alkyne, so the alkyne is at one end of the molecule. Both elements of unsaturation are accounted for by the alkyne.

12-23

(c) The only question is how are the other carbons arranged. The mass spectrum shows a progression of peaks from the molecular ion at 110 to 95 (loss of CH_3), to 81 (loss of C_2H_5), to 67 (loss of C_3H_7). The mass spectrum suggests it is a linear chain. The extra evidence that hydrogenation of the mystery compound gives n-octane verifies that the chain is linear. The original compound must be oct-1-yne.

(d) The base peak is so strong because the ion produced is stabilized by resonance.

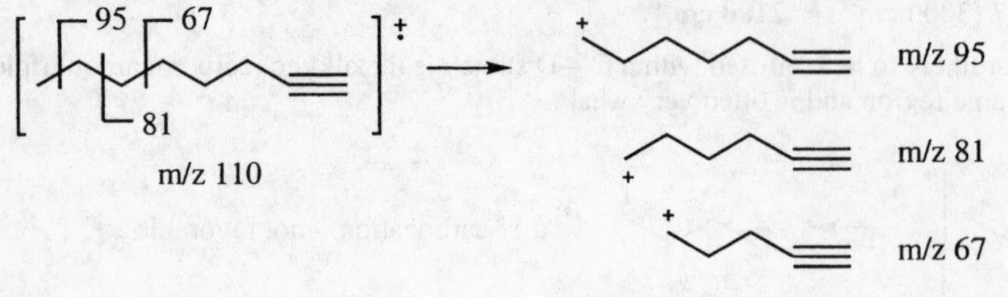

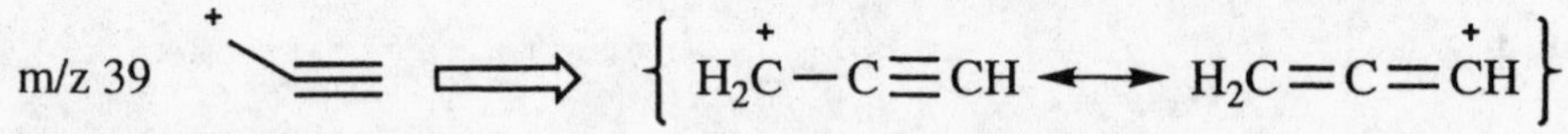

12-24

(a) and (b) The mass spec is consistent with the formula of the alkyne, C_8H_{14}, mass 110. The IR is not consistent with the alkyne, however. Often, symmetrically substituted alkynes have a miniscule $C\equiv C$ peak, so the fact that the IR does not show this peak does not prove that the alkyne is absent. The important evidence in the IR is the significant peak at 1620 cm^{-1}; this absorption is characteristic of a conjugated diene. Instead of the alkyne being formed, the reaction must have been a double elimination to the diene.

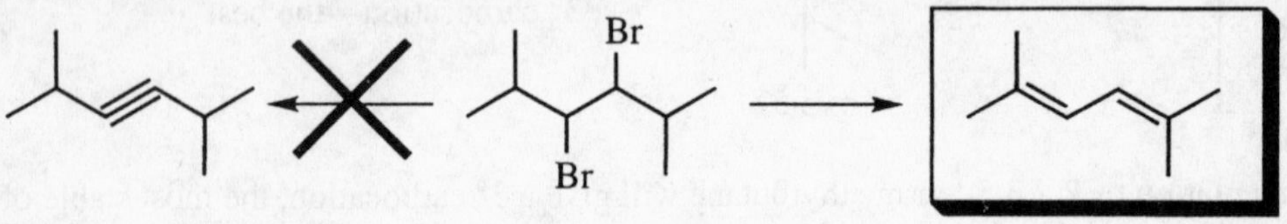

12-25

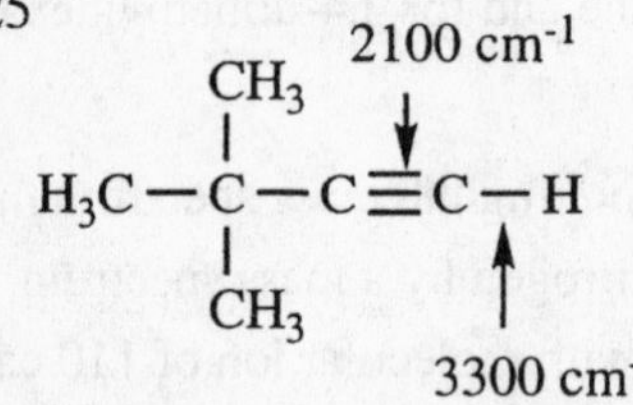

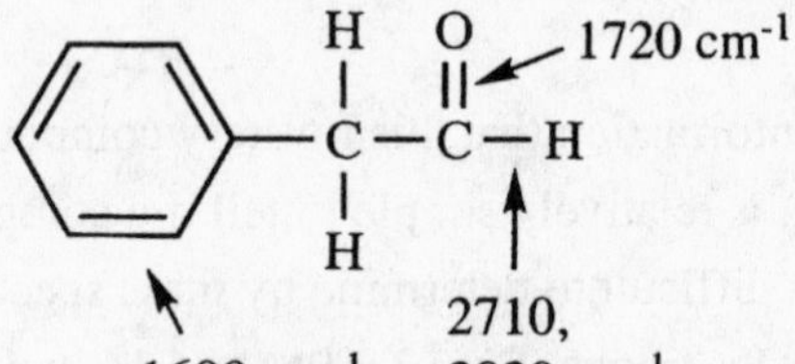

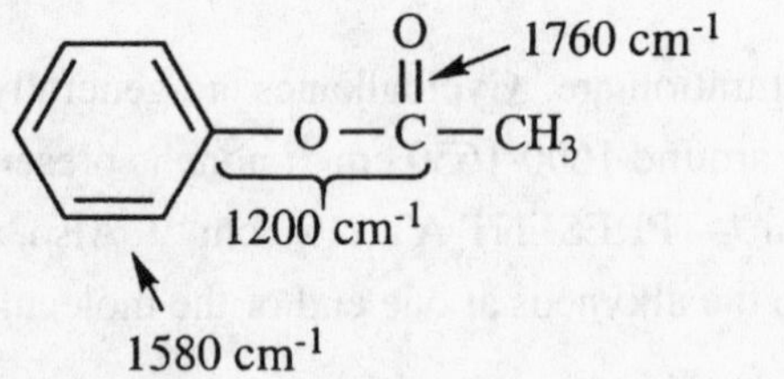

12-26
(a)

2-methylhexan-2-ol
$C_7H_{16}O$ mol. wt. 116

(b) The molecular ion is not visible in the spectrum. Alcohols typically dehydrate in the hot inlet system of the mass spectrometer, especially true for 3° alcohols that are the easiest type to dehydrate. The two fragmentations that produce a resonance-stabilized carbocation give the major peaks in the spectrum at m/z 59 and 101.

mass 116

m/z 59

m/z 101

12-27 The unknown compound has peaks in the mass spec at m/z 198, 155, 127, 71, and 43. It is helpful that the masses of two of the fragments, 155 and 43, sum to 198, as do the other two fragment masses, 127 and 71. We can say with certainty that the molecular ion is at m/z 198, and that the unknown is a relatively simple molecule with two main fragmentations.

This is a fairly high mass for a simple compound; some heavy group must be present. What is NOT present is N because of the even molecular ion mass, nor Cl nor Br because of the lack of isotope peaks, nor a phenyl group because of no peak at 77. The progression of alkyl group masses: 15, 29, **43**, 57, **71**, 85, 99—includes two of the peaks, so it appears that the unknown contains a propyl group and a pentyl group (the propyl could be part of the pentyl group). The 127 fragment is key; the fragment C_9H_{19} has this mass, but we would expect much more fragmentation from a nine carbon piece. There must be some other explanation for this 127 peak.

And there is! There is one piece—more specifically, one atom—that has mass 127: iodine! In all probability, the iodine atom is attached to a fragment of mass 71 which is C_5H_{11} , a pentyl group. We cannot tell with certainty what isomer it is, so unless there is some other evidence, let's propose a straight chain isomer, 1-iodopentane.

The 155 fragment probably has this bridged structure because iodine is so big:

12-28

(a)

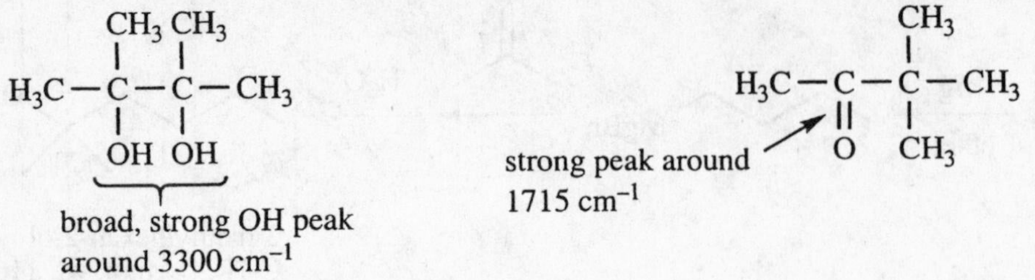

$H_3C - \underset{\underset{OH}{\overset{CH_3}{|}}}{C} - \underset{\underset{OH}{\overset{CH_3}{|}}}{C} - CH_3$

$\underbrace{}$
broad, strong OH peak
around 3300 cm^{-1}

$H_3C - \underset{\overset{\|}{O}}{C} - \underset{\overset{|}{CH_3}}{\overset{CH_3}{C}} - CH_3$

strong peak around
1715 cm^{-1}

(b)

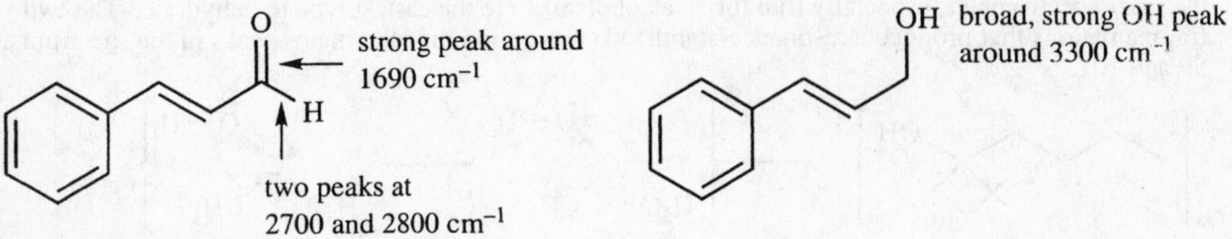

strong peak around
1690 cm^{-1}

two peaks at
2700 and 2800 cm^{-1}

OH broad, strong OH peak
around 3300 cm^{-1}

(c)

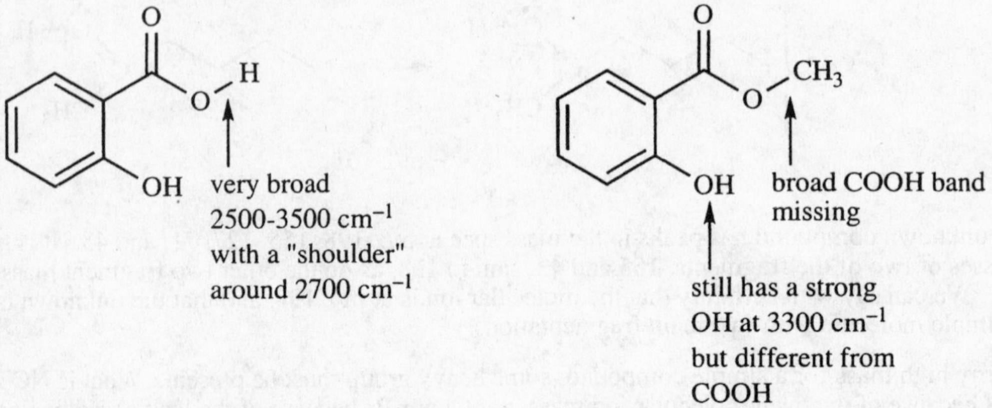

very broad
2500-3500 cm^{-1}
with a "shoulder"
around 2700 cm^{-1}

broad COOH band
missing

still has a strong
OH at 3300 cm^{-1}
but different from
COOH

If you wish to find IR spectra and mass spectra of common compounds, there are two web sites
that are very helpful. Entering a name or molecular formula will give isomers from which to
choose the desired structure and the IR or MS if available in their database.

http://webbook.nist.gov/
"NIST" is the U.S. National Institute of Standards and Technology.

http://www.aist.go.jp/RIODB/SDBS/menu-e.html
This is from the National Institute of Advanced Industrial Science and Technology of Japan.

A benzene ring can be written with three alternating double bonds or with a circle in the ring. All of the carbons and hydrogens in an unsubstituted benzene ring are equivalent, regardless of which symbolism is used.

          equivalent to

The Japanese web site listed at the bottom of p. 270 also gives proton and carbon NMR spectra.

Reminder: The word "spectrum" is singular; the word "spectra" is plural.

13-1

(a) $\dfrac{650 \text{ Hz}}{300 \times 10^6 \text{ Hz}} = 2.17 \times 10^{-6} = 2.17$ ppm downfield from TMS

(b) Difference in magnetic field $= 70{,}459$ gauss $\times\ (2.17 \times 10^{-6}) = 0.153$ gauss

(c) The chemical shift does not change with field strength: $\delta\ 2.17$ at both 60 MHz and 300 MHz.

(d) $(2.17 \text{ ppm}) \times (60 \text{ MHz}) = (2.17 \times 10^{-6}) \times (60 \times 10^6 \text{ Hz}) = 130$ Hz

13-2 Numbers are chemical shift values, in ppm, derived from Table 13-3 and the Appendix in the text. *Your predictions should be in the given range, or within 0.5-1.0 ppm of the given value.*

(a)
$$a = \delta\ 5\text{-}6$$
$$b = \delta\ 0.9$$

(b)
$$a = \delta\ 7.2$$
$$b = \delta\ 2.3$$

(c)
$$a = \delta\ 7.2$$
$$b = \delta\ 3.6$$

(d)
$$a = \delta\ 2\text{-}5$$
$$b = \delta\ 2.5$$
$$c = \delta\ 1\text{-}2$$

(e)
$$a = \delta\ 10\text{-}12$$
$$b \approx \delta\ 3.1$$
$$c = \delta\ 7.2$$

The hydrogens labeled "b" are between two deshielding functional groups, the deshielding effects of which are rougly additive. From Table 13-3, a CH next to C=O is at about 2.1; a CH next to a benzene ring is at about 2.3, that is, about 1.0 delta units further downfield than a CH_2 by itself. Add 2.1 to 1.0 to get an approximate value of 3.1; a safe prediction would be from about 2.8 to 3.8. The actual value is 3.6, within 0.5 of the predicted value.

(f)
$$a = \delta\ 3\text{-}4$$
$$b = \delta\ 1\text{-}2$$

(The hydrogens labeled "c" are not equivalent. They appear at roughly the same chemical shift because the substituent is neither strongly electron-donating nor withdrawing.)

13-3

(a) $\overset{c}{C}H_3\overset{b}{C}H_2\overset{a}{C}H_2Cl$

three types of H

(b) $\overset{b}{C}H_3\overset{a}{C}H\overset{b}{C}H_3$
$\;\;\;\;\;\;\;\;\;|$
$\;\;\;\;\;\;\;\;Cl$

two types of H

(c)
$$\overset{a}{C}H_3-\overset{\overset{\displaystyle CH_3\;a}{|}}{\underset{\underset{\displaystyle CH_3}{|}}{C}}-\overset{b}{C}H_2\overset{c}{C}H_3$$
$\;\;\;\;\;\;\;\;\;\;\;\;\;\;\;\;\;\;a$

three types of H

(d)
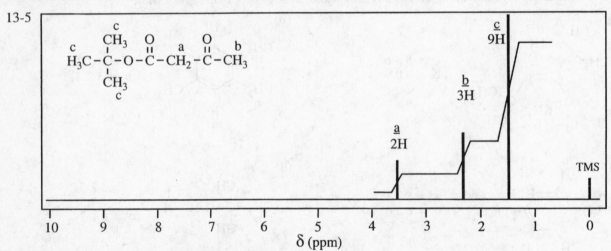

five types of H

13-4

(a)
[benzene ring with $\overset{a}{C}H_3$ at top, b and b on upper ring positions (H), c and c on lower positions (H), d at bottom (H)]

four types of H

(b) The three types of aromatic hydrogens appear in a relatively small space around δ 7.2. The signal is complex because all the peaks from the three types of hydrogens overlap.

Note: NMR spectra drawn in this Solutions Manual will represent peaks as single lines. These lines may not look like "real" peaks, but this avoids the problem of variation among spectrometers and printers. Individual spectra may look different from the ones presented here, but all of the important information will be contained in these representational spectra.

13-5

$$\overset{c}{H_3C}-\overset{\overset{\displaystyle \overset{c}{CH_3}}{|}}{\underset{\underset{\displaystyle \underset{c}{CH_3}}{|}}{C}}-O-\overset{\overset{\displaystyle O}{||}}{C}-\overset{a}{C}H_2-\overset{\overset{\displaystyle O}{||}}{C}-\overset{b}{C}H_3$$

[NMR spectrum with peaks labeled: a 2H (~3.5), b 3H (~2.3), c 9H (~1.4), TMS at 0]

δ (ppm)

13-6 The three spectra are identified with their structures. Data are given as chemical shift values, with the integration ratios of each peak given in parentheses.

Spectrum (a)

$$\overset{c}{C}H_3-\overset{\overset{\displaystyle \overset{b}{OH}}{|}}{\underset{\underset{\displaystyle \underset{c}{CH_3}}{|}}{C}}-C\equiv C-\overset{a}{H}$$

a = δ 2.4 (1) (1H)
b = δ 2.6 (1) (1H)
c = δ 1.5 (6) (6H)

Spectrum (b)

[benzene ring: a H at top positions, a H at bottom positions, b CH$_3$O— on left, b —OCH$_3$ on right]

a = δ 6.8 (2) (4H)
b = δ 3.7 (3) (6H)

Spectrum (c)

$$\overset{b}{C}H_3-\overset{\overset{\displaystyle \overset{b}{CH_3}\;a}{|}}{\underset{\underset{\displaystyle Br}{|}}{C}}-\overset{a}{C}H_2Br$$

a = δ 3.9 (1) (2H)
b = δ 1.9 (3) (6H)

(This is the compound in Problem 13-2(f). You may wish to check your answer to that question against the spectrum.)

13-7 Chemical shift values are approximate and may vary slightly from yours. The splitting and integration values should match exactly, however.

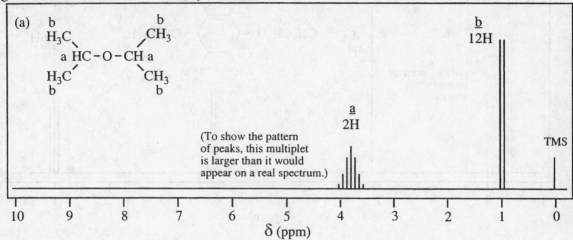

(a)

(To show the pattern of peaks, this multiplet is larger than it would appear on a real spectrum.)

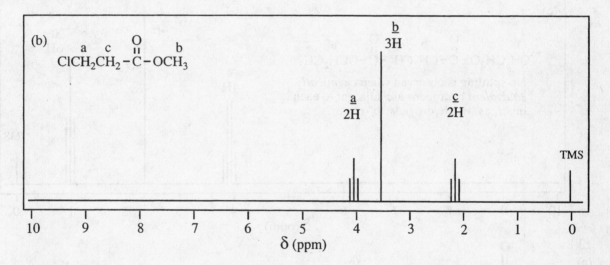

(b)

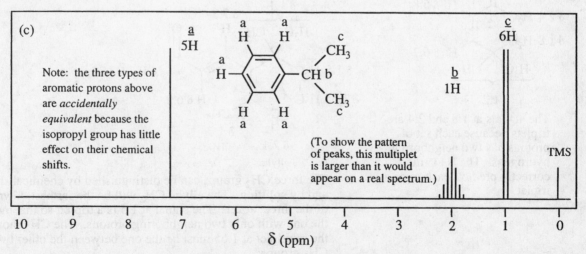

(c)

Note: the three types of aromatic protons above are *accidentally equivalent* because the isopropyl group has little effect on their chemical shifts.

(To show the pattern of peaks, this multiplet is larger than it would appear on a real spectrum.)

13-7 continued

(d)

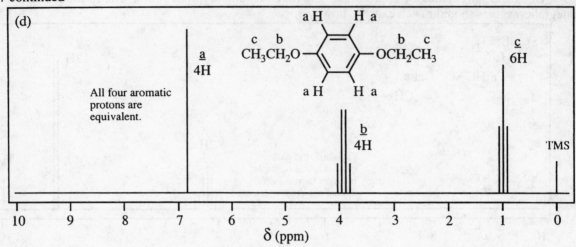

All four aromatic protons are equivalent.

a
4H

c b
CH₃CH₂O

a H   H a

a H   H a

b c
OCH₂CH₃

c
6H

b
4H

TMS

10 9 8 7 6 5 4 3 2 1 0
δ (ppm)

(e)

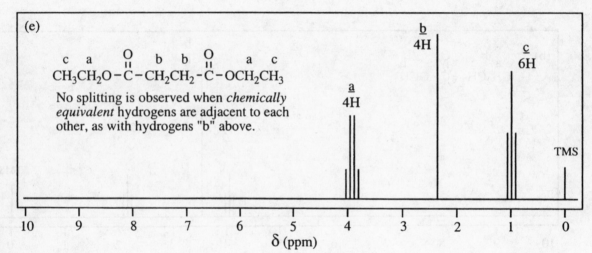

c a   O   b b   O   a c
CH₃CH₂O – C – CH₂CH₂ – C – OCH₂CH₃

No splitting is observed when *chemically equivalent* hydrogens are adjacent to each other, as with hydrogens "b" above.

b
4H

c
6H

a
4H

TMS

10 9 8 7 6 5 4 3 2 1 0
δ (ppm)

13-8

(a)

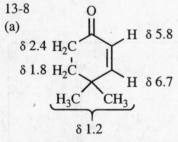

δ 2.4 H₂C
δ 1.8 H₂C

O

H δ 5.8

H δ 6.7

H₃C   CH₃

δ 1.2

The signals at 1.8 and 2.4 are triplets because each set of protons has two neighboring hydrogens. The N+1 rule correctly predicts each to be a triplet.

(b)

δ 1.1

H₃C   CH₃

δ 7.3
H

O

δ 1.5 H₂C

δ 1.65 H₂C
C
H₂

δ 2.1

allylic

CH₃

δ 1.7
allylic

CH₃ δ 2.3

H δ 6.2

The three CH₂ groups can be distinguished by chemical shift and by splitting. The allylic CH₂ will be the farthest downfield of the three, at 2.1. The signal at 1.5 is a triplet, so that must be the one with only two neighboring protons. The CH₂ showing the multiplet at 1.65 must be the one between the other two CH₂ groups.

274

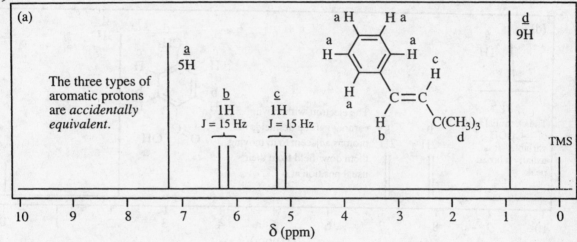

(a)

The three types of aromatic protons are *accidentally equivalent.*

a 5H

b 1H J = 15 Hz

c 1H J = 15 Hz

d 9H

TMS

δ (ppm)

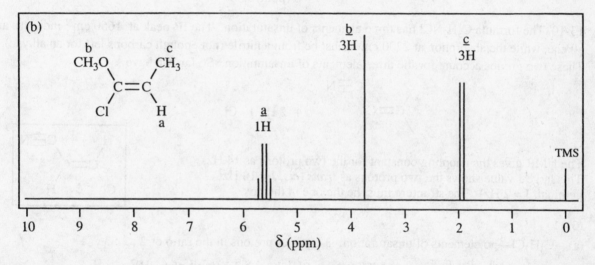

(b)

b CH₃O

c CH₃

Cl

H a

a 1H

b 3H

c 3H

TMS

δ (ppm)

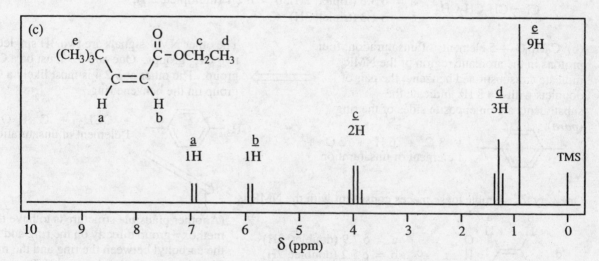

(c)

e (CH₃)₃C

c d C-OCH₂CH₃

H a

H b

a 1H

b 1H

c 2H

d 3H

e 9H

TMS

δ (ppm)

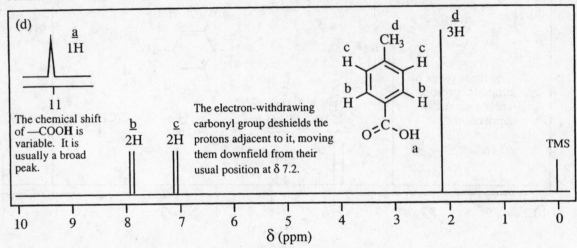

(d)

a
1H

11

The chemical shift of —COOH is variable. It is usually a broad peak.

b
2H

c
2H

The electron-withdrawing carbonyl group deshields the protons adjacent to it, moving them downfield from their usual position at δ 7.2.

d
CH₃

d
3H

TMS

δ (ppm)

13-10 The formula C_3H_2NCl has three elements of unsaturation. The IR peak at 1650 cm⁻¹ indicates an alkene, while the absorption at 2200 cm⁻¹ must be from a nitrile (not enough carbons left for an alkyne). These two groups account for the three elements of unsaturation. So far, we have:

$$C=C \qquad C\equiv N \qquad + 2H + Cl$$

The NMR gives the coupling constant for the two protons as 14 Hz. This large J value shows the two protons as *trans* (*cis*, J = 10 Hz; geminal, J = 2 Hz). The structure must be the one in the box.

13-11

(a) C_3H_7Cl—no elements of unsaturation; 3 types of protons in the ratio of 2 : 2 : 3 .

$$\overset{a}{\underset{}{Cl-CH_2}}\overset{b}{CH_2}\overset{c}{CH_3}$$

a = δ 3.8 (triplet, 2H); b = δ 2.1 (multiplet, 2H);
c = δ 1.3 (triplet, 3H)

(b) $C_9H_{10}O_2$—5 elements of unsaturation; four protons in the aromatic region of the NMR indicate a disubstituted benzene; the pair of doublets with J = 8 Hz indicate the substituents are on opposite sides of the ring (*para*).

The other NMR signals are two 3H singlets, two CH_3 groups. One at δ 3.9 must be a CH_3O group. The other at δ 2.4 is most likely a CH_3 group on the benzene ring.

+ 3 C + 6 H + 2 O + 1 element of unsaturation

CH_3—⟨ ⟩— + OCH_3 + C + O + 1 element of unsaturation

One way to assemble these pieces consistent with the NMR is:

b   a
H   H   O
d   ‖   c
CH₃—⟨ ⟩—C—OCH₃
H   H
b   a

a = δ 7.9 (doublet, 2H)
b = δ 7.2 (doublet, 2H)
c = δ 3.9 (singlet, 3H)
d = δ 2.4 (singlet, 3H)

(Another plausible structure is to have the methoxy group directly on the ring and to put the carbonyl between the ring and the methyl. This does not fit the chemical shift values quite as well as the above structure, as the methyl would appear around δ 2.1 or 2.2 instead of 2.4.)

13-12 H_c, δ 5.1

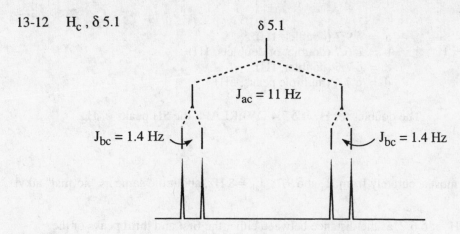

13-13

(a)

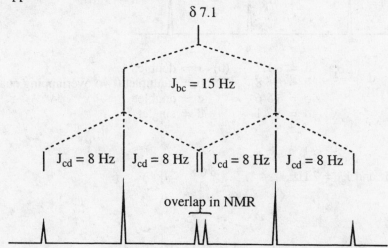

a = δ 12.1 (broad singlet, 1H)
b = δ 5.8 (doublet, 1H)
c = δ 7.1 (multiplet, 1H)
d = δ 2.2 (quartet, 2H)
e = δ 1.5 (sextet, 2H)
f = δ 0.9 (triplet, 3H)

(b) The vinyl proton at δ 7.1 is H_c; it is coupled with H_b and H_d, with two different coupling constants, J_{bc} and J_{cd}, respectively. The value of J_{bc} can be measured most precisely from the signal for H_b at δ 5.8; the two peaks are separated by about 15 Hz, corresponding to 0.05 ppm in a 300 MHz spectrum. The value of J_{cd} appears to be about the standard value 8 Hz, judging from the signal at δ 7.1. The splitting tree would thus appear:

13-14

(a)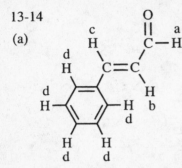

a = δ 9.7 (doublet, 1H)
b = δ 6.7 (doublet of doublets, 1H)
c = δ 7.5 (doublet, 1H)
d = δ 7.4 (multiple peaks, 5H)

The doublet for H_c at δ 7.4 OVERLAPS the 5H peaks of H_d.

(b) J_{ab} can be determined most accurately from H_a at δ 9.7: $J_{ab} \approx 8$ Hz, about the same as "normal" alkyl coupling.

J_{bc} can be measured from H_b at δ 6.7, as the distance between either the first and third peaks or the second and fourth peaks (see diagram below): $J_{bc} \approx 18$ Hz, about double the "normal" alkyl coupling.

(c)

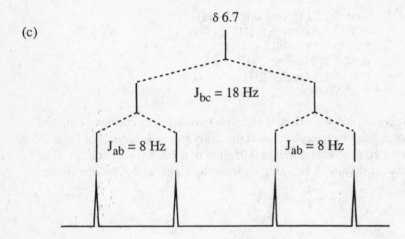

13-15

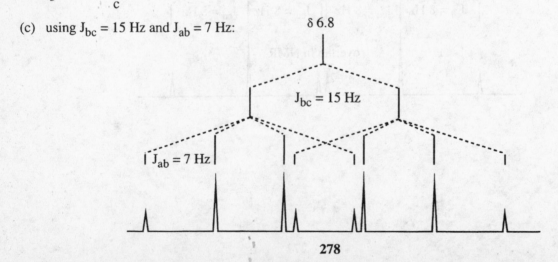

(a) a = δ 1.7   (b) a = doublet
     b ≈ δ 6.8   b = multiplet (two overlapping quartets—see part (c))
     c = δ 5-6   c = doublet
     d = δ 2.1   d = singlet

(c) using $J_{bc} = 15$ Hz and $J_{ab} = 7$ Hz:

13-16

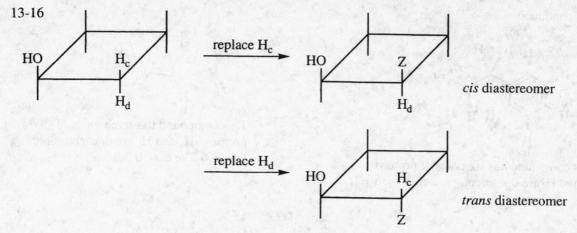

13-17

(a)

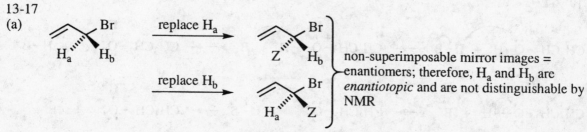

replace H_a

replace H_b

non-superimposable mirror images = enantiomers; therefore, H_a and H_b are *enantiotopic* and are not distinguishable by NMR

(b)

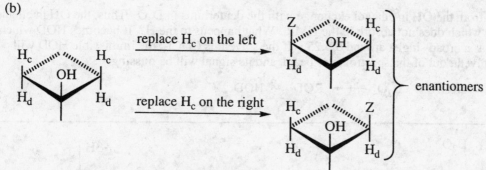

replace H_c on the left

replace H_c on the right

enantiomers

(c) The H_d protons are also enantiotopic.

13-18

(a)

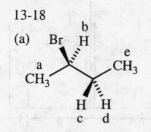

This compound has five types of protons. H_c and H_d are diastereotopic. $a = \delta\ 1.5$; $b = \delta\ 3.6$; $c,d = \delta\ 1.7$; $e = \delta\ 1.0$

(b)

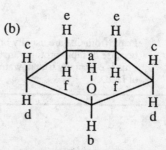

This compound has six types of protons. H_c and H_d are diastereotopic, as are H_e and H_f. $a = \delta\ 2-5$; $b = \delta\ 3.9$; $c,d = \delta\ 1.6$; $e,f = \delta\ 1.3$

13-18 continued

(c)

a
H   Br   d
H
b
H
          Br
c
H   H   H   H
a   H   e   f
    b

This compound has six types of protons.
H_e and H_f are diastereotopic. a,b,c = δ 7.2;
d ≈ δ 5.0; e,f = δ 3.6

(d)

a
H   Cl
C = C
H   H
b   c

This compound has three types of
protons. H_a and H_b are diastereotopic.
a,b = δ 5-6; c = δ 7-8

13-19

(a)

CH_3CH_2-Ö-H + H-B ⟶ CH_3CH_2-Ö-H + :B⁻ ⟶ CH_3CH_2-Ö: + H-B

(b)

CH_3CH_2-Ö-H + :B⁻ ⟶ CH_3CH_2-Ö:⁻ + H-B ⟶ CH_3CH_2-Ö: + :B⁻

13-20 The protons from the OH in ethanol exchange with the deuteriums in D_2O. Thus, the OH in ethanol
is replaced with OD which does not absorb in the NMR. What happens to the H? It becomes HOD which
can usually be seen as a broad singlet around δ 5.25. (If the solvent is $CDCl_3$, the immiscible HOD will
float on top of the solvent, out of the spectrometer beam, and its signal will be missing.)

$$ROH + D_2O \longrightarrow ROD + HOD$$

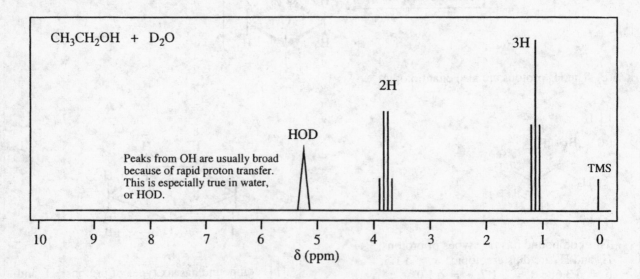

(a) The formula $C_4H_{10}O_2$ has no elements of unsaturation, so the oxygens must be alcohol or ether functional groups. The doublet at δ 1.2 represents 3H and must be a CH_3 next to a CH. The peaks centered at δ 1.65 integrating to 2H appear to be an uneven quartet and signify a CH_2 between two sets of non-equivalent protons; apparently the coupling constants between the non-equivalent protons are not equal, leading to a complicated pattern of overlapping peaks. The remaining five hydrogens appear in four groups of 1H, 1H, 1H, and 2H, between δ 3.7 and 4.2. The 2H multiplet at δ 3.75 is a CH_2 next to O, with complex splitting (doublet of doublets) due to diastereotopic neighbors. The 1H multiplet at δ 4.0 is a CH between many neighbors. The two 1H signals at δ 4.1-4.2 appear to be OH peaks in different environments, partially split by adjacent CH groups.

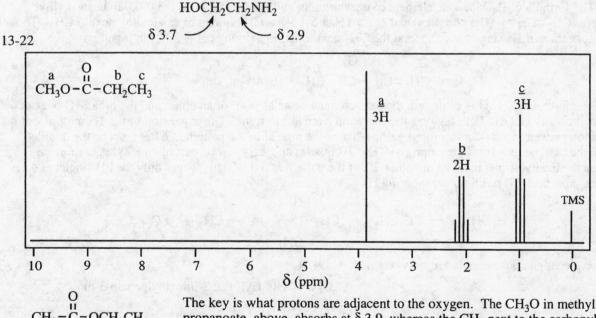

The 3-dimensional view shows clearly that the two Hs on a CH_2 are not equivalent

Of the two possible structures:

this one has no chiral centers and would not display such complex splitting

CORRECT!

If the possibility of intramolecular H-bonding is considered, the differences in the environments for each H become even more obvious.

H-bond

(b) The formula C_2H_7NO has no elements of unsaturation. The N must be an amine and the O must be an alcohol or ether. Two triplets, each with 2H, are certain to be $-CH_2CH_2-$. Since there are no carbons left, the N and O, with enough hydrogens to fill their valences, must go on the ends of this chain. The rapidly exchanging OH and NH_2 protons appear as a broad, 3H hump at δ 2.8.

$$HOCH_2CH_2NH_2$$
δ 3.7 $\qquad$ δ 2.9

a $\quad$ O $\quad$ b $\quad$ c
$$CH_3O-\overset{\overset{O}{\|}}{C}-CH_2CH_3$$

$\underline{a}$
3H

$\underline{c}$
3H

$\underline{b}$
2H

TMS

δ (ppm)

$$CH_3-\overset{\overset{O}{\|}}{C}-OCH_2CH_3$$
δ 2.05 $\qquad$ δ 4.1

The key is what protons are adjacent to the oxygen. The CH_3O in methyl propanoate, above, absorbs at δ 3.9, whereas the CH_2 next to the carbonyl absorbs at δ 2.2. This is in contrast to ethyl acetate, at the left.

13-23

$H_a = \delta\ 2.4$ (singlet, 1H)
$H_b = \delta\ 3.4$ (doublet, 2H)
$H_c = \delta\ 1.8$ (multiplet, 1H)
$H_d = \delta\ 0.9$ (doublet, 6H)

$$H-O-\overset{d}{\underset{b}{\overset{CH_3}{\underset{\underset{CH_3}{|}}{\overset{|}{C}}}}}-\overset{c}{\underset{}{C}}-H$$

13-24

(a) The formula $C_4H_8O_2$ has one element of unsaturation. The 1H singlet at δ 12.1 indicates carboxylic acid. The 1H multiplet and the 6H doublet scream isopropyl group.

$$\underset{CH_3}{\overset{CH_3}{}}CH-\overset{O}{\overset{||}{C}}-OH$$

(b) The formula $C_9H_{10}O$ has five elements of unsaturation. The 5H pattern between δ 7.2-7.4 indicates monosubstituted benzene. The peak at δ 9.85 is unmistakably an aldehyde, trying to be a triplet because it is weakly coupled to an adjacent CH_2. The two triplets at δ 2.7-3.0 are adjacent CH_2 groups.

$$\text{(benzene ring)}-\overset{H_2}{\underset{H_2}{C}}-\overset{O}{\overset{||}{C}}-H$$

(c) The formula $C_5H_8O_2$ has two elements of unsaturation. A 3H singlet at δ 2.3 is probably a CH_3 next to carbonyl. The 2H quartet and the 3H triplet are certain to be ethyl; with the CH_2 at δ 2.7, this also appears to be next to carbonyl.

$$CH_3CH_2-\overset{O}{\overset{||}{C}}-\overset{O}{\overset{||}{C}}-CH_3$$

(d) The formula C_4H_8O has one element of unsaturation, and the signals from δ 5.0-6.0 indicate a vinyl pattern ($CH_2{=}CH-$). The complex quartet for 1H at δ 4.3 is a CH bonded to an alcohol, next to CH_3. The OH appears as a 1H singlet at δ 2.5, and the CH_3 next to CH is a doublet at δ 1.3. Put together:

$$CH_2{=}CH-\overset{OH}{\overset{|}{CH}}-CH_3 \qquad \text{but-3-en-2-ol}$$

(e) The formula $C_7H_{16}O$ is saturated; the oxygen must be an alcohol or an ether, and the broad 1H peak at δ 1.2 is probably an OH. Let's analyze the spectrum from left to right. The expansion of the 1H multiplet at δ 1.7 shows seven peaks—an isopropyl group! Six of the nine H in the pattern at δ 0.9 must be the doublet from the two methyls from the isopropyl. The 2H quartet at δ 1.5 must be part of an ethyl pattern, from which the methyl triplet must be the other 3H of the pattern at δ 0.9. This leaves only the 3H singlet at δ 1.1 which must be a CH_3 with no neighboring Hs.

$$-OH \ + \ -CH_2CH_3 \ + \ CH_3-\overset{H}{\underset{CH_3}{\overset{|}{\underset{|}{C}}}}- \ + \ -CH_3 \ + \ 1\ C$$

These pieces can be assembled in only one way:

$$HO-\overset{CH_3}{\underset{CH_2CH_3}{\overset{|}{\underset{|}{C}}}}-CH(CH_3)_2 \qquad \text{2,3-dimethylpentan-3-ol}$$

13-25 Chemical shift values are estimates from Figure 13-41 and from Appendix 1C, except in (c) and (d), where the values are exact.

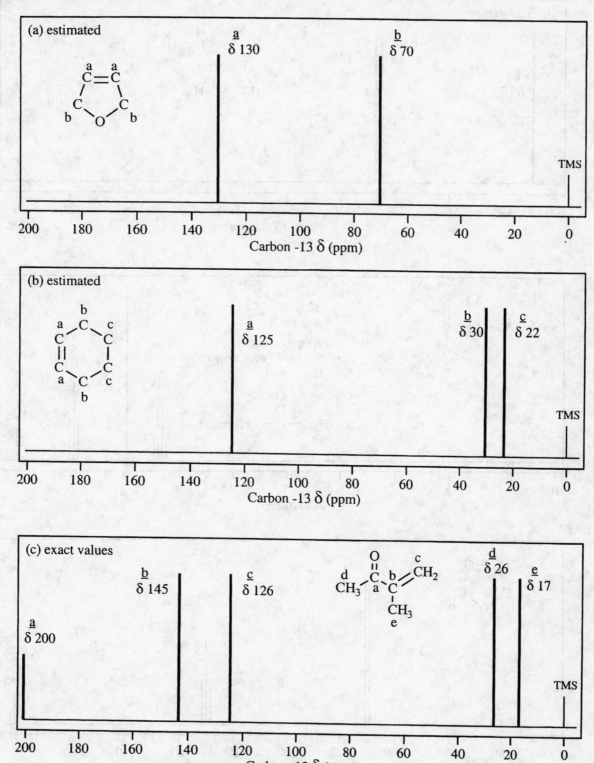

(a) estimated

a
δ 130

b
δ 70

TMS

200 180 160 140 120 100 80 60 40 20 0
Carbon -13 δ (ppm)

(b) estimated

a
δ 125

b
δ 30

c
δ 22

TMS

200 180 160 140 120 100 80 60 40 20 0
Carbon -13 δ (ppm)

(c) exact values

b
δ 145

c
δ 126

d
δ 26

e
δ 17

a
δ 200

TMS

200 180 160 140 120 100 80 60 40 20 0
Carbon -13 δ (ppm)

13-25 continued

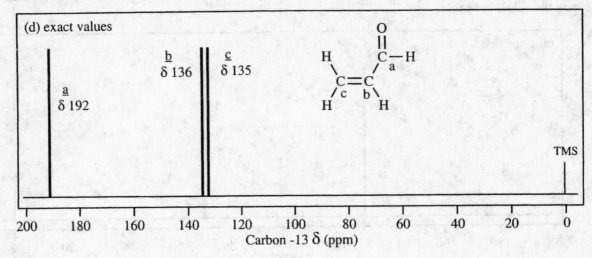

(d) exact values

a
δ 192

b
δ 136

c
δ 135

TMS

Carbon -13 δ (ppm)

13-26

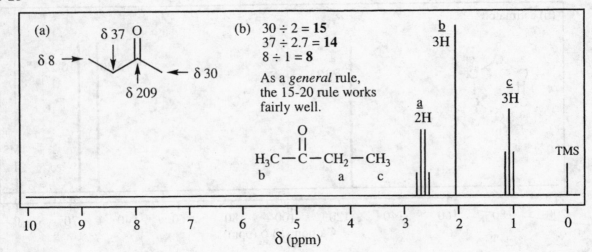

(a)

δ 8 →   δ 37 O
              ↘ ‖
                 ↗ ← δ 30
              δ 209

(b) 30 ÷ 2 = **15**
     37 ÷ 2.7 = **14**
     8 ÷ 1 = **8**

As a *general* rule,
the 15-20 rule works
fairly well.

            O
            ‖
H₃C — C — CH₂ — CH₃
b   a   c

b
3H

a
2H

c
3H

TMS

δ (ppm)

13-27

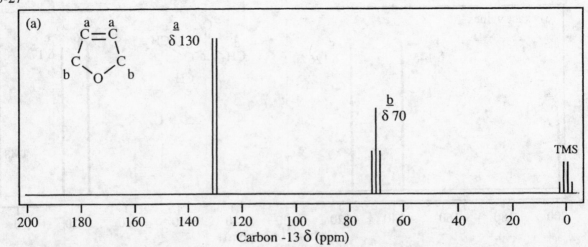

(a)

    a   a
    C = C
   /   \
  C   C
  b \ O / b

a
δ 130

b
δ 70

TMS

Carbon -13 δ (ppm)

13-27 continued

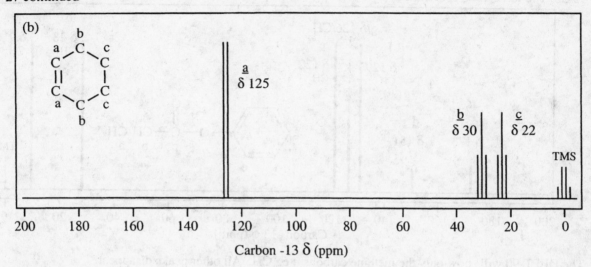

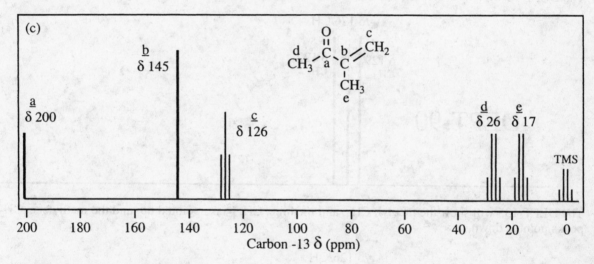

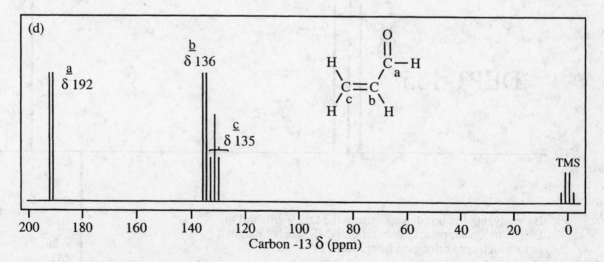

13-28 The full carbon spectrum of phenyl propanoate is presented below.

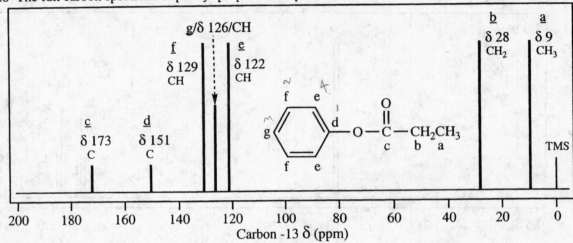

The DEPT-90 will show only the methine carbons, i.e., CH. All other peaks disappear.

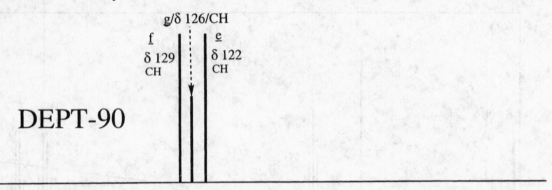

The DEPT-135 will show the methyl, CH_3, and methine, CH, peaks pointed up, and the methylene, CH_2, peaks pointed down.

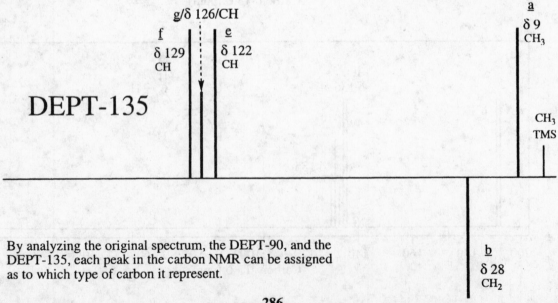

By analyzing the original spectrum, the DEPT-90, and the DEPT-135, each peak in the carbon NMR can be assigned as to which type of carbon it represent.

13-29

Since allyl bromide was the starting material, it is reasonable to expect the allyl group to be present in the impurity: the triplet at 115 is a $=CH_2$, the doublet at 138 is $=CH-$, and the triplet at 63 is a deshielded aliphatic CH_2; assembling the pieces forms an allyl group. The formula has changed from C_3H_5Br to C_3H_6O, so OH has replaced the Br.

$$H_2C=CH-CH_2OH$$

δ 115
triplet   δ 138
doublet   δ 63
triplet

Allyl bromide is easily hydrolyzed by water, probably an S_N1 process.

$$H_2C=CH-CH_2Br \xrightarrow{H_2O} H_2C=CH-CH_2OH + HBr$$

13-30

a = δ 180, singlet
b = δ 70, triplet
c = δ 28, triplet
d = δ 22, triplet

Two elements of unsaturation in $C_4H_6O_2$, one of which is a carbonyl, and no evidence of a C=C, prove that a ring must be present.

13-31

a = δ 128, doublet
b = δ 25, triplet
c = δ 23, triplet

Two elements of unsaturation in C_6H_{10} must be a C=C and a ring. Only three peaks indicates symmetry.

Using PBr_3 instead of $H_2SO_4/NaBr$ would give a higher yield of bromocyclohexane.

13-32   Compound 2

Mass spectrum: the molecular ion at m/z 136 shows a peak at 138 of about equal height, indicating a bromine atom is present: 136 – 79 = 57. The fragment at m/z 57 is the base peak; this fragment is most likely a butyl group, C_4H_9, so a likely molecular formula is C_4H_9Br.

Infrared spectrum: Notable for the absence of functional groups: no O—H, no N—H, no =C—H, no C=C, no C=O ⇒ most likely an alkyl bromide.

NMR spectrum: The 6H doublet at δ 1.0 suggests two CH_3's split by an adjacent H—an isopropyl group. The 2H doublet at δ 3.2 is a CH_2 between a CH and the Br.

Putting the pieces together gives isobutyl bromide.

$$Br-CH_2-CH \begin{matrix} CH_3 \\ \\ CH_3 \end{matrix}$$

13-33

The formula $C_9H_{11}Br$ indicates four elements of unsaturation, just enough for a benzene ring.

Here is the most accurate method for determining the number of protons per signal from integration values *when the total number of protons is known*. Add the integration heights: 4.4 cm + 13.0 cm + 6.7 cm = 24.1 cm. Divide by the total number of hydrogens: 24.1 cm ÷ 11H = 2.2 cm/H. Each 2.2 cm of integration height = 1H, so the ratio of hydrogens is 2 : 6 : 3.

The 2H singlet at δ 7.1 means that only two hydrogens remain on the benzene ring, that is, it has 4 substituents. The 6H singlet at δ 2.3 must be two CH_3's on the benzene ring in identical environments. The 3H singlet at δ 2.2 is another CH_3 in a slightly different environment from the first two. Substitution of the three CH_3's and the Br in the most symmetric way leads to the structures on the next page.

a   b
H   CH₃

c
CH₃ ——— Br

H   CH₃
a   b

a = δ 7.1 (singlet, 2H)
b = δ 2.3 (singlet, 6H)
c = δ 2.2 (singlet, 3H)

a second structure is also possible although it is less likely because the Br would probably deshield the Hs labeled "a" to about 7.3-7.4

b H₃C   H a

c
CH₃ ——— Br

b H₃C   H a

13-34 The numbers in italics indicate the number of peaks in each signal.

(a) $CH_3—CH_2—CCl_2—CH_3$
       3   *4*   *1*

(b) $CH_3—\overset{7}{CH}—OH$ (assume OH exchanging
              | *1*   rapidly—no splitting)
            CH_3
              2

(c) $CH_3—\overset{10}{CH}—CH_3$
              | *2*
            CH_3

(d)
H   H

H —   — CH₃
                1

H   H
      1

(all of these benzene H's
are accidentally equivalent
and do not split each other)

(e)
[ring with O] —CH₃
                2

13-35 Consult Appendix 1 in the text for chemical shift values. *Your predictions should be in the given range, or within 0.5 ppm of the given value.*

(a)
[benzene ring]—H   all at δ 7.2

(b)
[cyclohexane ring] H   all at δ 1.3
                   H

(c)   δ 1.6
       $CH_3—O—CH_2—CH_2—CH_2Cl$
       δ 3.4   δ 3.8   δ 3.2

(d) $CH_3—CH_2—C≡C—H$
       δ 1.2   δ 2.2   δ 2.5

(e)   O
                   ||
       $CH_3—CH_2—C—CH_3$
       δ 1.0   δ 2.5   δ 2.0

(f) CH_3 δ 4.3   δ 2-5
          \\
           CH—O—CH₂——CH₂—OH
          //   δ 3.8 δ 3.8
       CH_3
       δ 0.9

(g)   O
                 ||
       H   [benzene ring]—C—H
       δ 8.0   δ 9-10

       H   H
       δ 7.5 δ 8.0

The C=O has its strongest deshielding
effect at the adjacent H (ortho) and
across the ring (para). The remaining
H (meta) is less deshielded.

(h)   δ 6-7   O
                       ||
       $CH_3—CH=CH—C—H$
       δ 1.7   δ 5-6   δ 9-10

13-35 continued

(i)

HOOC—CH$_2$—CH$_2$—C(=O)—O—CH(CH$_3$)$_2$ ← δ 4.0
δ 10-12   δ 2.3   δ 2.3   CH$_3$ δ 1.1

(j)

δ 1.3 { ring structure } δ 4.5
δ 1.3   δ 1.7
             allylic

(k)

δ 7.2   δ 2.5   } δ 1.4
             benzylic

(l)

δ 6.0
δ 7.2   δ 3.0
             allylic and
             benzylic

13-36

c   a   c
CH$_3$—CH—CH$_3$ a = δ 4.0 (septet, 1H)
        | b = δ 2.5 (broad singlet, 1H) (rapidly exchanging)
        OH   c = δ 1.2 (doublet, 6H)
        b

13-37

(a) The chemical shift *in ppm* would not change: δ 4.00.

(b) Coupling constants do not change with field strength: J = 7 Hz, regardless of field strength.

(c) At 60 MHz, δ 4.00 = 4.00 ppm = $(4.00 \times 10^{-6}) \times (60 \times 10^6 \text{ Hz})$ = 240 Hz

The signal is 240 Hz downfield from TMS in a 60 MHz spectrum.

At 300 MHz, $(4.00 \times 10^{-6}) \times (300 \times 10^6 \text{ Hz})$ = 1200 Hz

The signal is 1200 Hz downfield from TMS in a 300 MHz spectrum.
Necessarily, 1200 Hz is exactly 5 times 240 Hz because 300 MHz is
exactly 5 times 60 MHz. They are directly proportional.

13-38

c   b   O   d
CH$_2$—CH$_2$—O—C(=O)—CH$_3$

a = δ 7.2-7.3 (multiplet, 5H)
b = δ 4.3 (triplet, 2H)
c = δ 2.9 (triplet, 2H)
d = δ 2.0 (singlet, 3H)

13-39

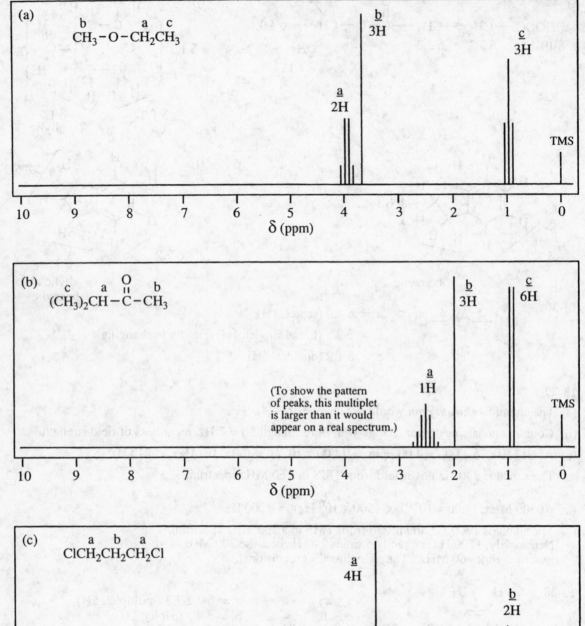

(a)

b   a c
CH₃−O−CH₂CH₃

$\underline{b}$ 3H

$\underline{a}$ 2H

$\underline{c}$ 3H

TMS

δ (ppm)

(b)

c   a   O   b
(CH₃)₂CH−C−CH₃

$\underline{b}$ 3H

$\underline{c}$ 6H

(To show the pattern
of peaks, this multiplet
is larger than it would
appear on a real spectrum.)

$\underline{a}$ 1H

TMS

δ (ppm)

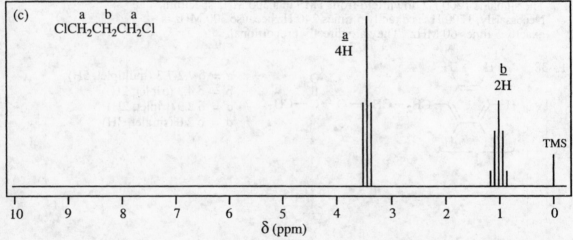

(c)

a   b   a
ClCH₂CH₂CH₂Cl

$\underline{a}$ 4H

$\underline{b}$ 2H

TMS

δ (ppm)

(d)

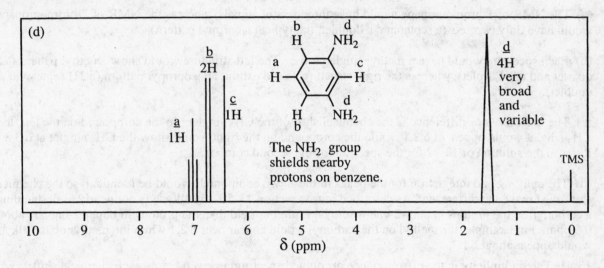

a
1H

b
2H

c
1H

d
4H
very
broad
and
variable

The NH$_2$ group
shields nearby
protons on benzene.

TMS

(e)

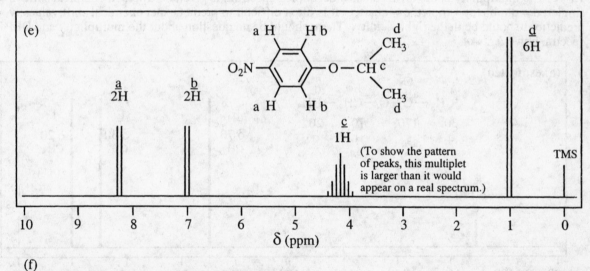

a
2H

b
2H

c
1H

d
6H

(To show the pattern
of peaks, this multiplet
is larger than it would
appear on a real spectrum.)

TMS

(f)

Signal (a) is split into a quartet because of the adjacent
CH$_3$ with J = 7 Hz. Each of those peaks is then split
into a doublet because of the coupling with the trans
H, J = 15 Hz. This is called a doublet of quartets, and
it is drawn here as two quartets. In a real spectrum,
these peaks would overlap and would not be a clean
doublet of quartets. See the splitting tree for problem
13-15.

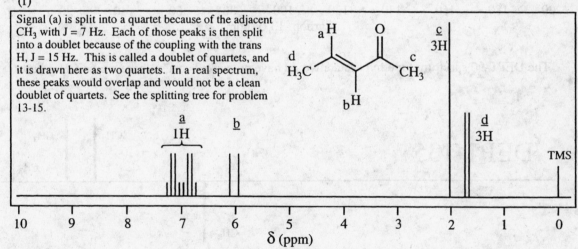

a
1H

b

c
3H

d
3H

TMS

13-40

(a) The NMR of 1-bromopropane would have three sets of signals, whereas the NMR of 2-bromopropane would have only two sets (a septet and a doublet, the typical isopropyl pattern).

(b) Each spectrum would have a methyl singlet at δ 2. The left structure would show an ethyl pattern (a 2H quartet and a 3H triplet), whereas the right structure would exhibit an isopropyl pattern (a 1H septet and a 6H doublet).

(c) The most obvious difference is the chemical shift of the CH_3 singlet. In the compound on the left, the CH_3 singlet would appear at δ 2.1, while the compound on the right would show the CH_3 singlet at δ 3.8. Refer to the solution of 13-22 for the spectrum of the second compound.

(d) The splitting and integration for the peaks in these two compounds would be identical, so the chemical shift must make the difference. As described in text section 13-5B, the alkyne is not nearly as deshielding as a carbonyl, so the protons in pent-2-yne would be farther upfield than the protons in butan-2-one, by about 0.5 ppm. For example, the methyl on the carbonyl would appear near δ 2.1 while the methyl on the alkyne would appear about δ 1.7.

13-41 The multiplicity in the off-resonance decoupled spectrum is given below each chemical shift: s = singlet; d = doublet; t = triplet; q = quartet. It is often difficult to predict exact chemical shift values; your predictions should be in the right vicinity. There should be no question about the multiplicity and DEPT spectra, however.

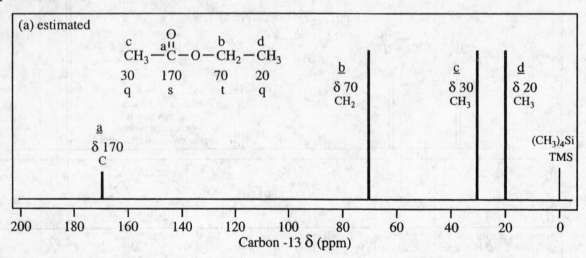

The DEPT-90 spectrum for ethyl acetate would have no peaks because there are no CH groups.

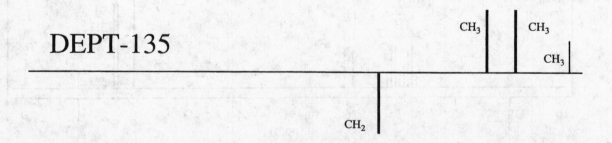

13-41 continued

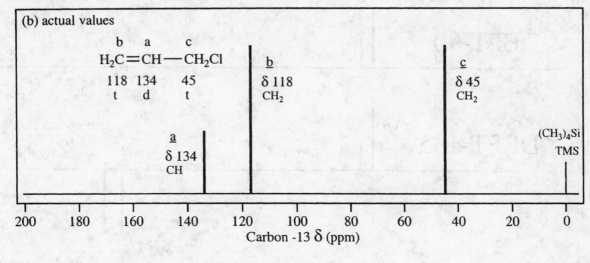

(b) actual values

b   a   c
$H_2C{=}CH{-}CH_2Cl$
118 134   45
 t   d   t

<u>b</u>
δ 118
CH_2

<u>c</u>
δ 45
CH_2

<u>a</u>
δ 134
CH

$(CH_3)_4Si$
TMS

Carbon -13 δ (ppm)

DEPT-90

CH

DEPT-135

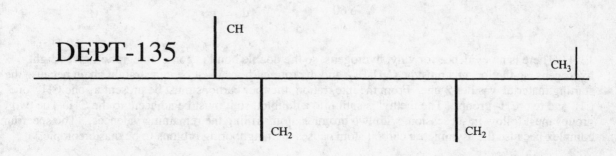

CH

CH_3

CH_2

CH_2

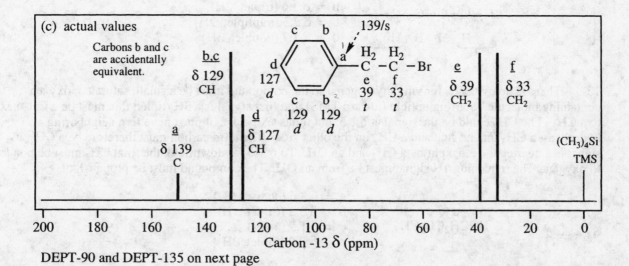

(c) actual values

Carbons b and c
are accidentally
equivalent.

<u>b,c</u>
δ 129
CH

<u>a</u>
δ 139
C

<u>d</u>
δ 127
CH

139/s

127
d

129
d

129
d

H_2 H_2
$C{-}C{-}Br$
e   f
39   33
t   *t*

<u>e</u>
δ 39
CH_2

<u>f</u>
δ 33
CH_2

$(CH_3)_4Si$
TMS

Carbon -13 δ (ppm)

DEPT-90 and DEPT-135 on next page

13-41 (c) continued

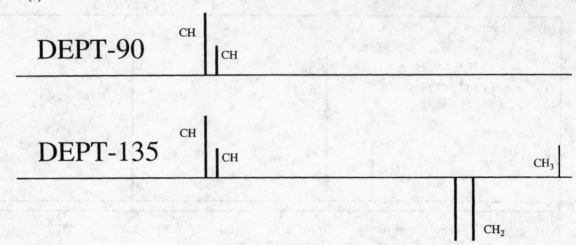

DEPT-90   CH | CH

DEPT-135   CH | CH   CH₃ | | CH₂

13-42 The multiplicity of the peaks in this off-resonance decoupled spectrum show two different CH's and a CH₃. There is only one way to assemble these pieces with three chlorines.

$$\delta 20 \rightarrow \quad CH_3 - \underset{\underset{\delta 60}{\overset{Cl}{|}}}{CH} - \underset{\underset{Cl}{\overset{Cl}{|}}}{CH}$$

q   δ 75
                              d
         δ 60
         d

13-43 There is no evidence for vinyl hydrogens, so the double bond is gone. Integration gives eight hydrogens, so the formula must be $C_4H_8Br_2$, and the four carbons must be in a straight chain because the starting material was but-2-ene. From the integration, the four carbons must be present as one CH_3, one CH, and two CH_2 groups. The methyl is split into a doublet, so it must be adjacent to the CH. The two CH_2 groups must follow in succession, with two bromine atoms filling the remaining valences. (The spectrum is complex because the asymmetric carbon atom causes the neighboring protons to be diastereotopic.)

```
    H  Br  H  Br          a = δ 4.3 (sextet, 1H)
    |  |   |  |            b = δ 3.6 (triplet, 2H)
H - C- C - C- C - H        c = δ 2.3 (multiplet, 2H)
    |  |   |  |            d = δ 1.7 (doublet, 3H)
    H  H   H  H
  └─┘ └──────┘
   d  a  c   b
```

13-44 There is no evidence for vinyl hydrogens, so the compound must be a small, saturated, oxygen-containing molecule. Starting upfield (toward TMS), the first signal is a 3H triplet; this must be a CH_3 next to a CH_2. The CH_2 could be the signal at δ 1.5, but it has six peaks: it must have five neighboring hydrogens, a CH_3 on one side and a CH_2 on the other side. The third carbon must therefore be a CH_2; its signal is a quartet at δ 3.6, split by a CH_2 and an OH. To be so far downfield, the final CH_2 must be bonded to oxygen. The remaining 1H signal must be from an OH. The compound must be propan-1-ol.

                    a = δ 3.6 (quartet, 2H)
     b   a   c   d   b = δ 3.2 (triplet, 1H)
    HOCH₂CH₂CH₃   c = δ 1.5 (6 peaks, 2H)
                    d = δ 0.9 (triplet, 3H)

13-45

(a)

a = δ 5.2 (quartet, 1H)   d = δ 4.7 (singlet, 2H)
b = δ 1.7 (singlet, 3H)   e = δ 2.0 (quartet, 2H)
c = δ 1.6 (singlet, 3H)   f = δ 1.7 (singlet, 3H)
d = δ 1.5 (doublet, 3H)   g = δ 1.0 (triplet, 3H)

(b) With NaOH as base, the more highly substituted alkene, Isomer A, would be expected to predominate—
the Zaitsev Rule. With KO-*t*-Bu as a hindered, bulky base, the less substituted alkene, Isomer B, would
predominate (the Hofmann product).

13-46 "Nuclear waste" is comprised of radioactive products from either nuclear reactions, for example,
from electrical generating stations powered by nuclear reactors, or residue from medical or scientific studies
using radioactive nuclides as therapeutic agents (like iodine for thyroid treatment) or as molecular tracers
(carbon-14, tritium H-3, phosphorus-32, nitrogen-15, and many others). The physical technique of *nuclear*
magnetic resonance neither uses nor generates any radioactive elements, and does not generate "nuclear
waste". (Some people assume that the medical application of NMR, medical resonance imaging or MRI,
purposely dropped the word "nuclear" from the technique to avoid the confusion between "nuclear" and
"radioactive".)

13-47

Mass spectrum: The molecular ion of m/z 117 suggests the presence of an odd number of nitrogens.

Infrared spectrum: No NH or OH appears. Hydrogens bonded to both sp^2 and sp^3 carbon are indicated
around 3000 cm^{-1}. The characteristic C≡N peak appears at 2250 cm^{-1} and aromatic C=C is suggested by
the peak at 1600 cm^{-1}.

NMR spectrum: Five aromatic protons are shown in the NMR at δ 7.3. A CH$_2$ singlet appears at δ 3.7.
Assemble the pieces:

13-48 This is a challenging problem, despite the molecule being relatively small.

Mass spectrum: The molecular ion at 96 suggests no Cl, Br, or N. The molecule must have seven carbons or fewer.

Infrared spectrum: The dominant functional group peak is at 1685 cm^{-1}, a carbonyl that is conjugated with C=C (lower wavenumber than normal, very intense peak). The presence of an oxygen and a molecular ion of 96 lead to a formula of C_6H_8O, with three elements of unsaturation, a C=O and one or two C=C.

Carbon NMR spectrum: The six peaks show, by chemical shift, one carbonyl carbon (196), two alkene carbons (129, 151), and three aliphatic carbons (23, 26, 36). By off-resonance decoupling multiplicity, the groups are: three CH_2 groups, two alkene CH groups, and carbonyl.

Since the structure has one carbonyl and only two alkene carbons, the third element of unsaturation must be a ring.

$$C=O$$
$$C=C$$
$$H \quad H$$

CH_2 + CH_2 + CH_2 + 1 ring

Since the structure has no methyl group, and no $H_2C=$, all of the carbons must be included in the ring. The only way these pieces can fit together is in cyclohex-2-enone. Notice that the proton NMR was unnecessary to determine the structure, fortunately, since the HNMR was not easily interpreted except for the two alkene hydrogens; the hydrogen on carbon-2 appears as the doublet at 6.0.

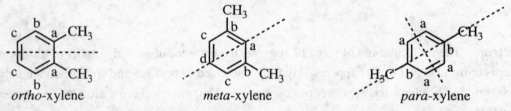

cyclohex-2-enone

The mystery mass spec peak at m/z 68 comes from a fragmentation that will be discussed later; it is called a retro-Diels-Alder fragmentation.

13-49 The key to the carbon NMR lies in the symmetry of these structures.

ortho-xylene   *meta*-xylene   *para*-xylene

(a) In each molecule, the methyl carbons are equivalent, giving one signal in the CNMR. Considering the ring carbons, the symmetry of the structures shows that *ortho*-xylene would have 3 carbon signals from the ring (total of 4 peaks), *meta*-xylene would have 4 carbon signals from the ring (total of 5 peaks), and *para*-xylene would have only 2 carbon signals from the ring (total of 3 peaks). These compounds would be instantly identifiable simply by the number of peaks in the carbon NMR.

(b) The proton NMR would be a completely different problem. Unless the substituent on the benzene ring is moderately electron-withdrawing or donating, the ring protons absorb at roughly the same position. A methyl group has essentially no electronic effect on the ring hydrogens, so while the *para* isomer would give a clean singlet because all its ring protons are equivalent, the *ortho* and *meta* isomers would have only slightly broadened singlets for their proton signals. (Only a very high field NMR, 500 MHz or higher, would be able to distinguish these isomers in the proton NMR.)

13-50 (a), (b) and (c) The six isomers are drawn here. Below each structure is the number of proton signals and the number of carbon signals. (Note that splitting patterns would give even more clues in the proton NMR.)

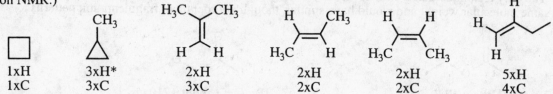

1xH	3xH*	2xH	2xH	2xH	5xH
1xC	3xC	3xC	2xC	2xC	4xC

*Rings always present challenges in stereochemistry. When viewed in three dimensions, it becomes apparent that the two hydrogens on a CH_2 are not equivalent: on each CH_2, one H is cis to the methyl and one H is trans to the methyl. These are diastereotopic protons. A more correct answer to part (b) would be four types of protons; whether all four could be distinguished in the NMR is a harder question to answer. For the purpose of this problem, whether it is 3 or 4 types of H does not matter because either one, in combination with three types of carbon, will distinguish it from the other 5 structures.

(d) Two types of H and three types of C can be only one isomer: 2-methylpropene (isobutylene). (The only isomers that would not be distinguished from each other would be *cis*- and *trans*-but-2-ene.)

13-51 For DEPT to be useful in distinguishing isomers, there should be different numbers of CH_3, CH_2, CH, and C signals in the carbon NMRs.

A

B

C

6 C signals

one CH_3 peak
four CH_2 peaks
one CH peak

5 C signals

one CH_3 peak
two CH_2 peaks
two CH peaks

5 C signals

one CH_3 peak
two CH_2 peaks
one CH peak
one C peak

A carbon NMR could not distinguish isomers **B** and **C**, and if there is any overlap in the signals, it might have a difficult time distinguishing **A**. Fortunately, the patterns that would appear in the DEPT-90 would distinguish **B** (two CH peaks up) from the other two (1 CH peak up), and the DEPT-135 would instantly distinguish **A** (4 CH_2 peaks down) from **C** (2 CH_2 peaks down).

13-52

(a) MS or IR could not easily distinguish these isomers: same molecular weight and same functional group. They would give dramatically different proton and carbon NMRs however.

OCH_3

2 singlets in HNMR
3 singlets in CNMR

OCH_2CH_3

4 signals, all with splitting, in HNMR
4 singlets in CNMR

13-52 continued

(b) The only technique that would not readily distinguish these isomers would be MS because they have the same molecular weight and would have similar, though not identical, fragmentation patterns.

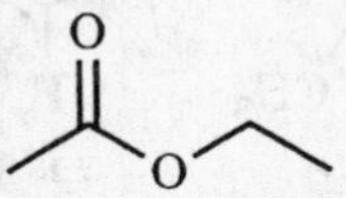

IR: C=O about 1730 cm^{-1}
HNMR: methyl singlet and ethyl pattern
CNMR: ester C=O about δ 170

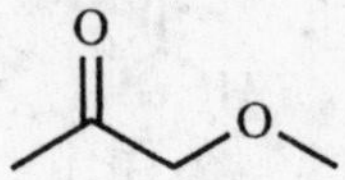

IR: C=O about 1710 cm^{-1}
HNMR: 3 singlets
CNMR: ketone C=O about δ 200

(c) The big winner here is MS: they have different molecular weights, plus the Cl has the two isotope peaks that make a Cl atom easily distinguished. The other techniques would have minor differences and would require having a detailed table of frequencies or chemical shifts to determine which is which.

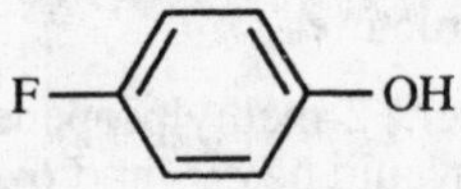

M^+ = m/z 112

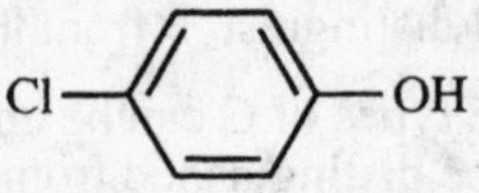

M^+ = m/z 128, 130

298

14-1

The four solvents decrease in polarity in this order: water, ethanol, ethyl ether, and dichloromethane. The three solutes decrease in polarity in this order: sodium acetate, 2-naphthol, and naphthalene. The guiding principle in determining solubility is, "Like dissolves like." Compounds of similar polarity will dissolve (in) each other. Thus, sodium acetate will dissolve in water, will dissolve only slightly in ethanol, and will be virtually insoluble in ethyl ether and dichloromethane. 2-Naphthol will be insoluble in water, somewhat soluble in ethanol, and soluble in ether and dichloromethane. Naphthalene will be insoluble in water, partially soluble in ethanol, and soluble in ethyl ether and dichloromethane. (Actual solubilities are difficult to predict, but you should be able to predict *trends*.)

14-2

Oxygen shares one of its electron pairs with aluminum; oxygen is the Lewis base, and aluminum is the Lewis acid. An oxygen atom with three bonds and one unshared pair has a positive formal charge. An aluminum atom with four bonds has a negative formal charge.

14-3

The crown ether has two effects on KMnO$_4$: first, it makes KMnO$_4$ much more soluble in benzene; second, it holds the potassium ion tightly, making the permanganate more available for reaction. Chemists call this a "naked anion" because it is not complexed with solvent molecules.

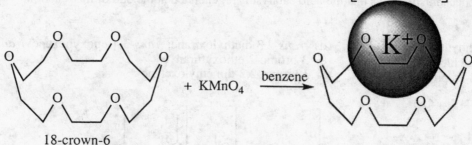

18-crown-6
a "crown ether"

Please see the note on p. 136 of this Solutions Manual regarding placement of position numbers.

14-4 IUPAC name first; then common name (see Appendix 1 in this Solutions Manual for a summary of IUPAC nomenclature)

(a) methoxycyclopropane; cyclopropyl methyl ether
(b) 2-ethoxypropane; ethyl isopropyl ether
(c) 1-chloro-2-methoxyethane; 2-chloroethyl methyl ether
(d) 2-ethoxy-2,3-dimethylpentane; no common name
(e) 2-*t*-butoxybutane; *sec*-butyl *t*-butyl ether
(f) *trans*-2-methoxycyclohexan-1-ol; no common name

14-5

(a)

The alcohol is ethane-1,2-diol; the common name is ethylene glycol.

299

14-5 continued

(b)

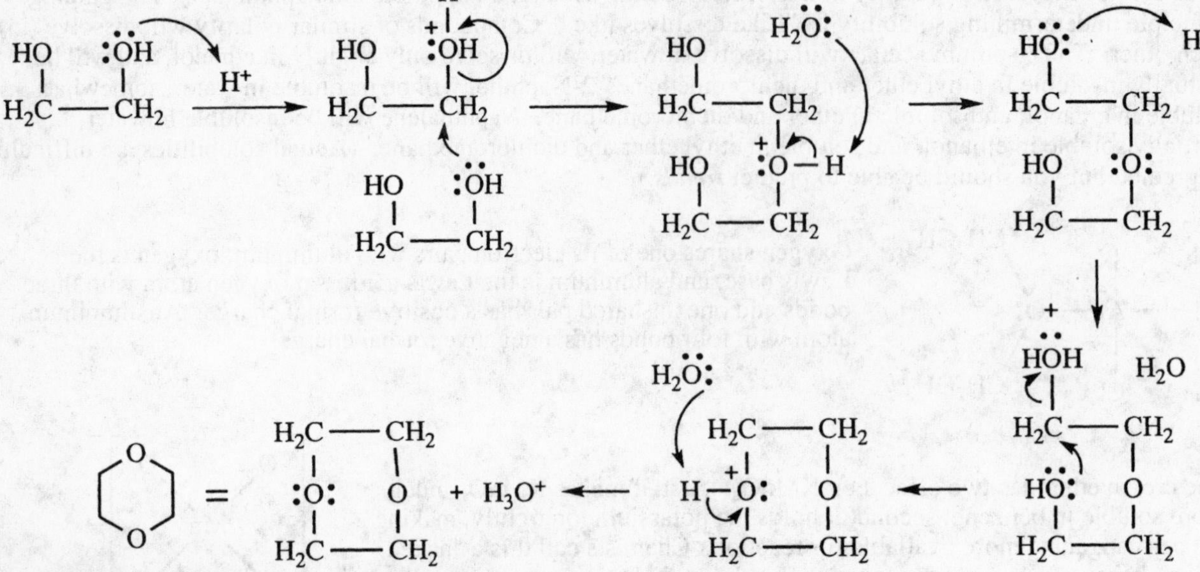

The mechanism shows that the acid catalyst is regenerated at the end of the reaction.

14-6
(a) dihydropyran
(b) 2-chloro-1,4-dioxane
(c) 3-isopropylpyran
(d) *trans*-2,3-diethyloxirane; *trans*-3,4-epoxyhexane; *trans*-3-hexene oxide
(e) 3-bromo-2-ethoxyfuran
(f) 3-bromo-2,2-dimethyloxetane

14-7

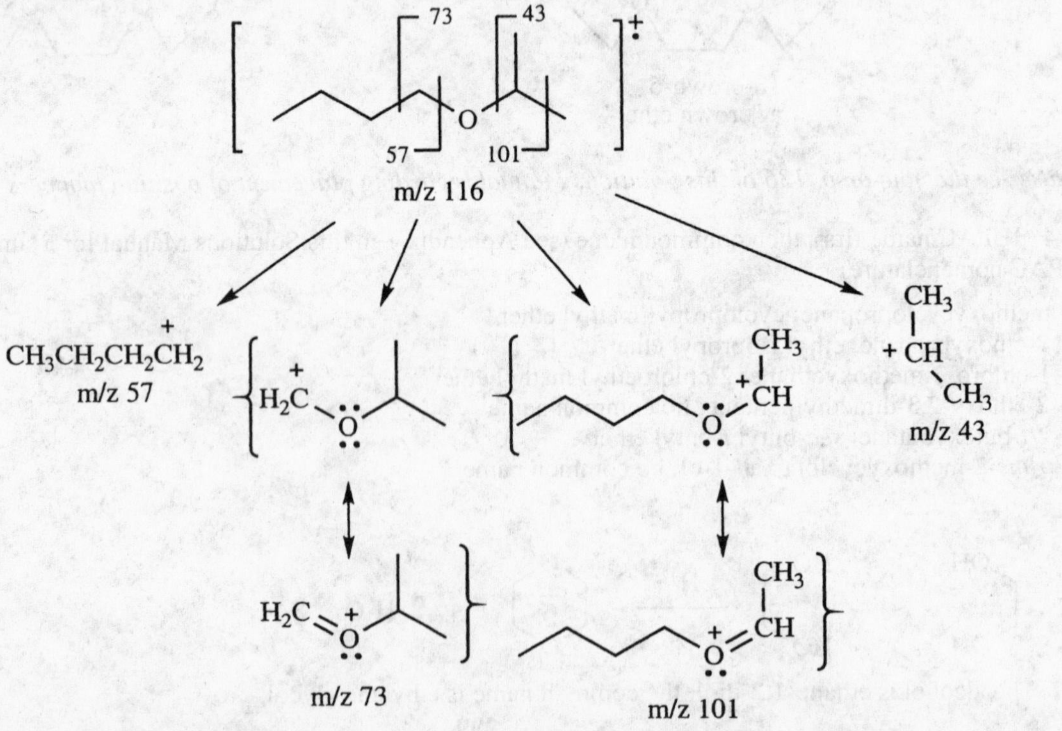

300

14-8 S_N2 reactions, including the Williamson ether synthesis, work best when the nucleophile attacks a 1° or methyl carbon. Instead of attempting to form the bond from oxygen to the 2° carbon on the ring, form the bond from oxygen to the 1° carbon of the butyl group.

The OH must first be transformed into a good leaving group: either a tosylate, or one of the halides (not fluoride).

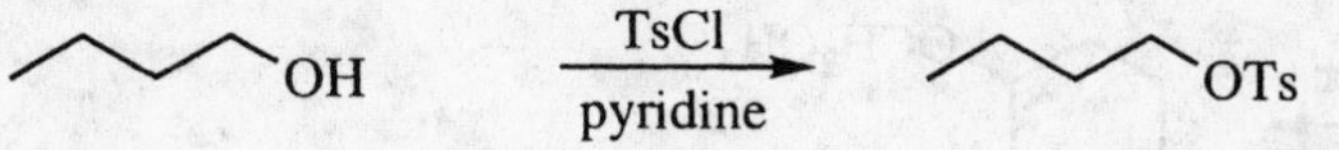

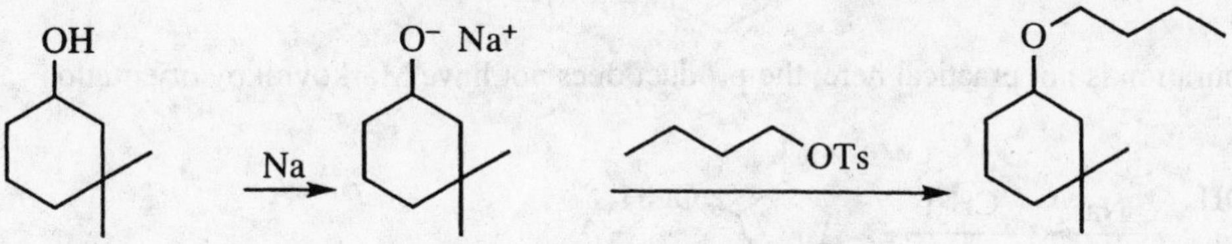

3-butoxy-1,1-dimethylcyclohexane

14-9

(a)

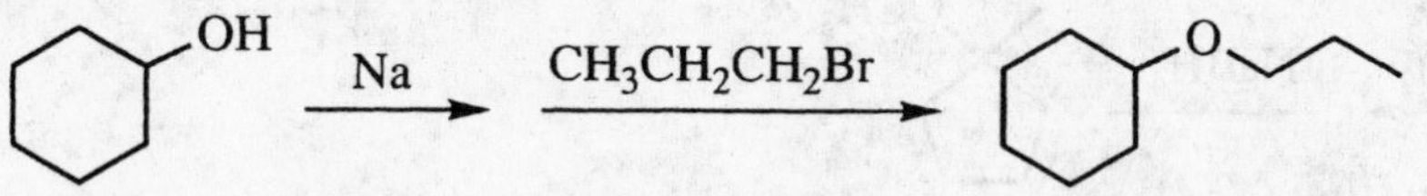

(b)

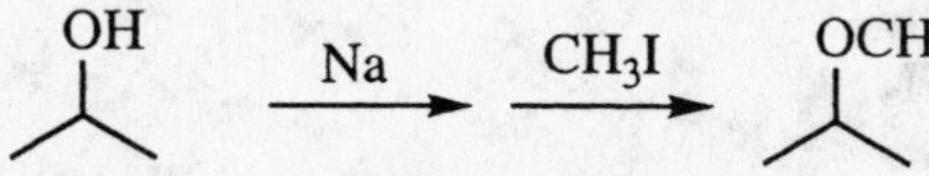

(c)

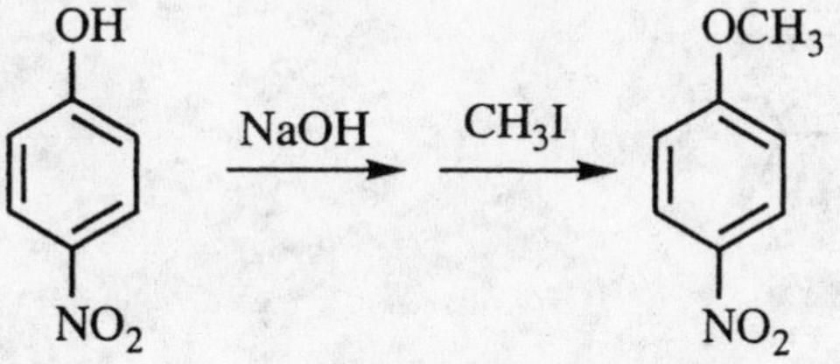

(d) $CH_3CH_2CH_2OH \xrightarrow{Na} \xrightarrow{CH_3CH_2Br} CH_3CH_2CH_2OCH_2CH_3$

$CH_3CH_2OH \xrightarrow{Na} \xrightarrow{CH_3CH_2CH_2Br} CH_3CH_2CH_2OCH_2CH_3$

(e)

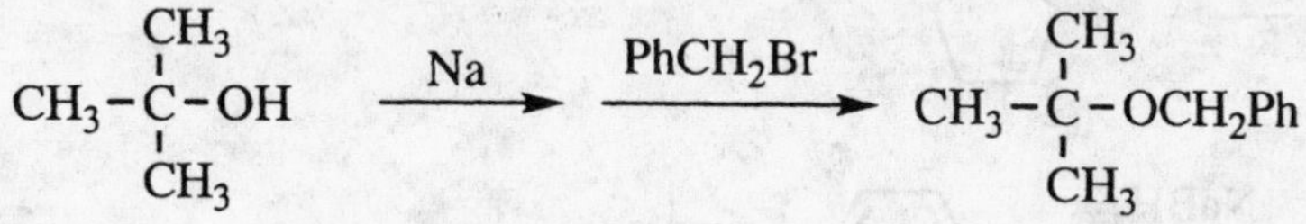

14-10

(a) (1)

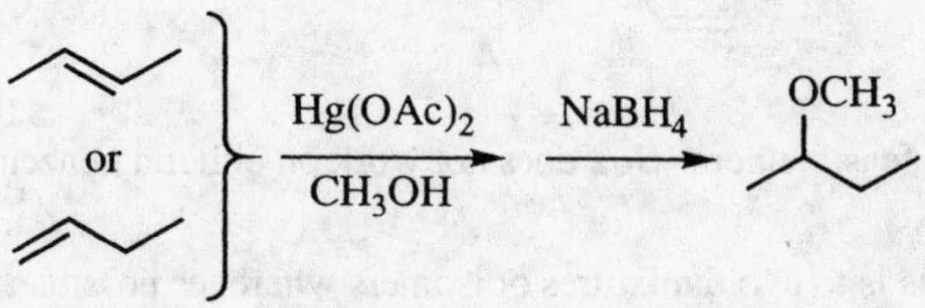

(2)

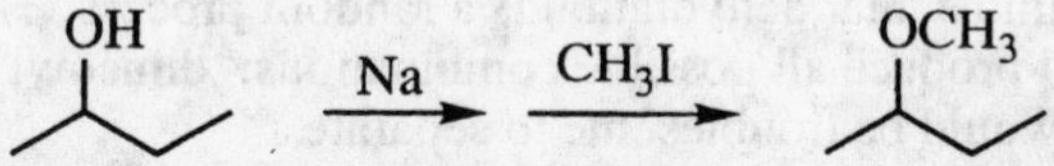

14-10 continued

(b) (1)

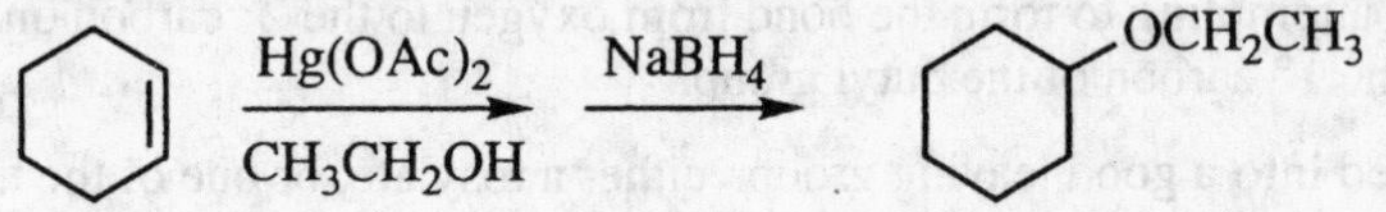

(2)

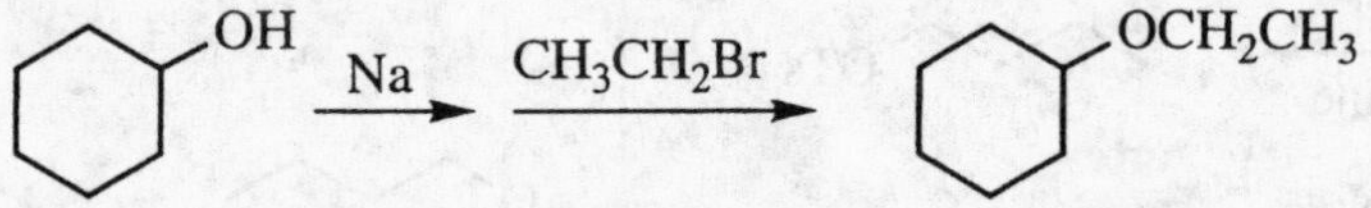

(c) (1) Alkoxymercuration is not practical here; the product does not have Markovnikov orientation.

(2)

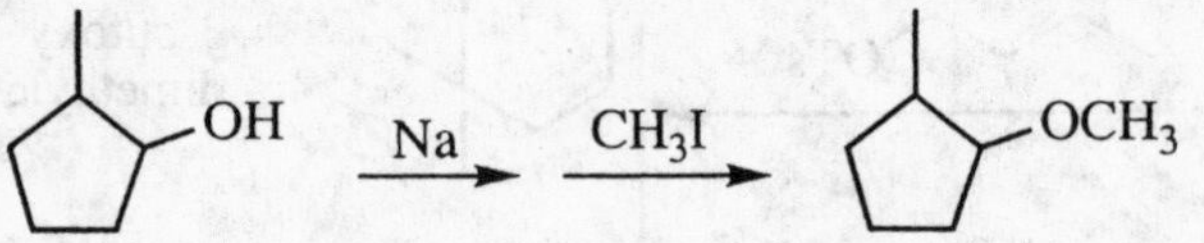

(d) (1)

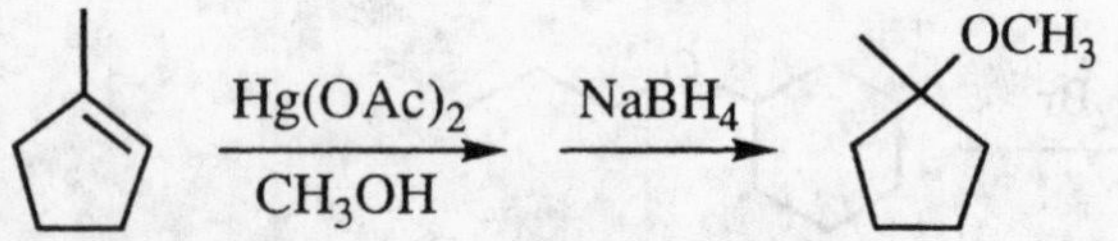

(2)

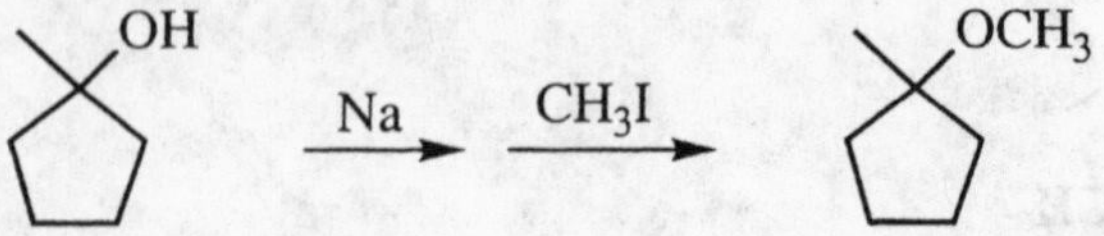

(e) (1)

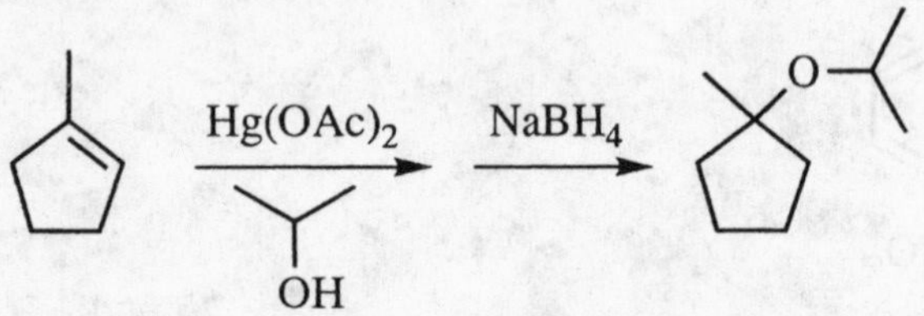

(2) Williamson ether synthesis would give a poor yield of product as the halide is on a 2° carbon.

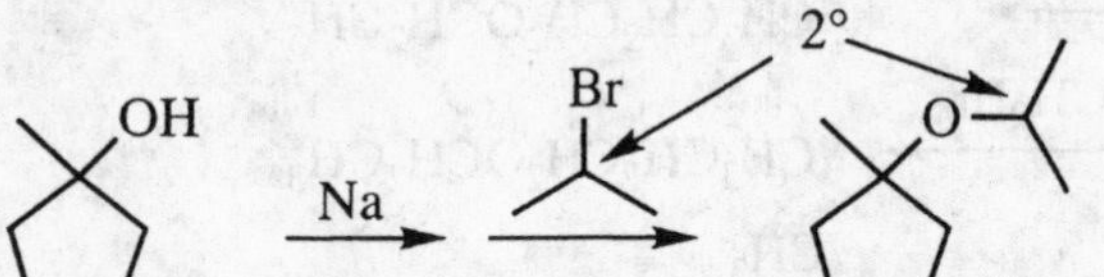

(f) (1)

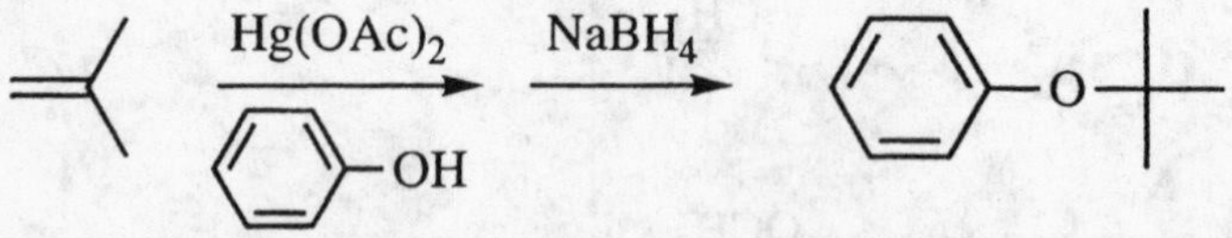

(2) Williamson ether synthesis is not feasible here. S_N2 does not work on either a benzene or a 3° halide.

14-11 An important principle of synthesis is to avoid mixtures of isomers wherever possible; minimizing separations increases recovery of products. Bimolecular dehydration is a random process. Heating a mixture of ethanol and methanol with acid will produce all possible combinations: dimethyl ether, ethyl methyl ether, and diethyl ether. This mixture would be troublesome to separate.

14-12

Ether formation

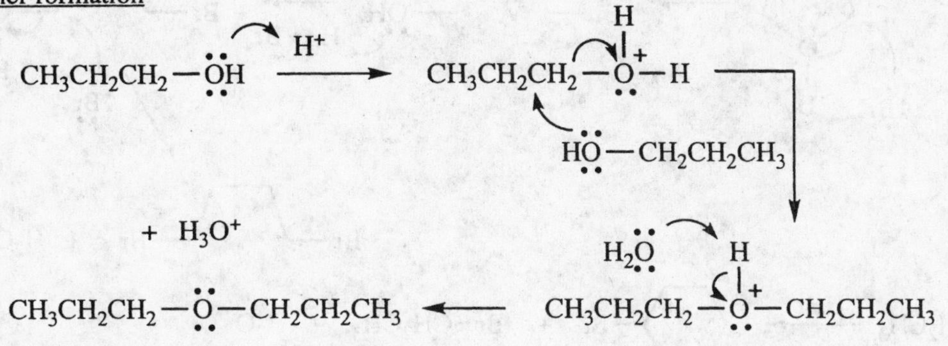

Dehydration

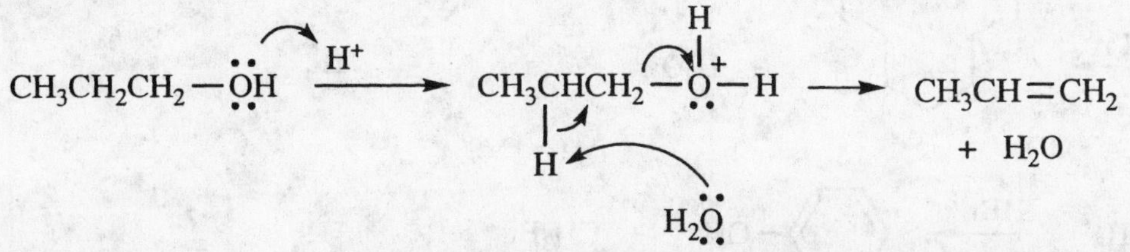

Remember $\Delta G = \Delta H - T\Delta S$? Thermodynamics of a reaction depend on the sign and magnitude of ΔG. As temperature increases, the entropy term grows in importance. In ether formation, the ΔS is small because two molecules of alcohol give one molecule of ether plus one molecule of water—no net change in the number of molecules. In dehydration, however, one molecule of alcohol generates one molecule of alkene plus one molecule of water—a large increase in entropy. So $T\Delta S$ is more important for dehydration than for ether formation. As temperature increases, the competition will shift toward more dehydration.

14-13

(a) This symmetrical ether at 1° carbons could be produced in good yield by bimolecular dehydration.

(b) This unsymmetrical ether could not be produced in high yield by bimolecular dehydration. Williamson synthesis would be preferred.

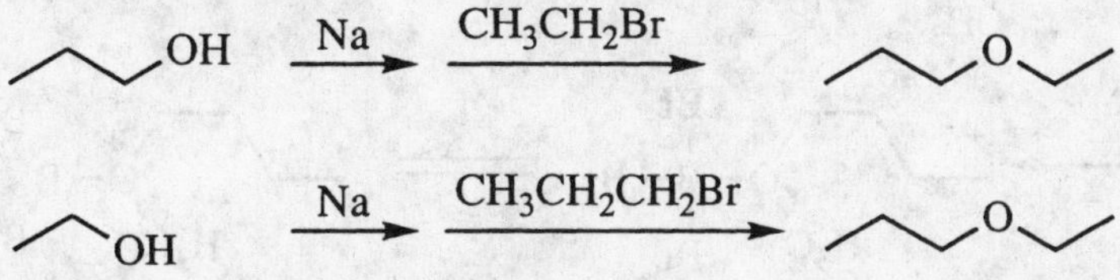

(c) Even though this ether is symmetrical, both carbons are 2°, so bimolecular dehydration would give low yields. Unimolecular dehydration to give alkenes would be the dominant pathway. Alkoxymercuration-demercuration is the preferred route.

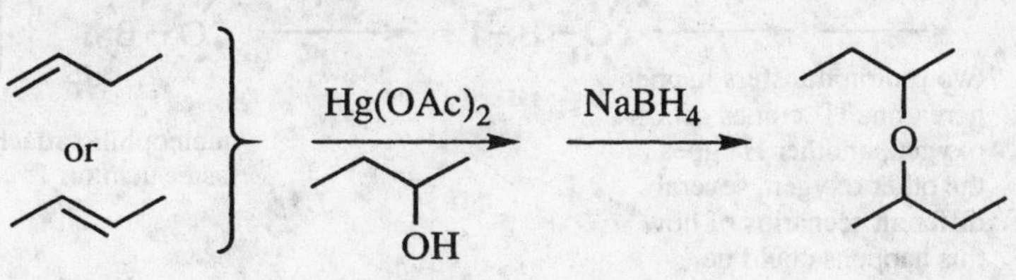

14-14

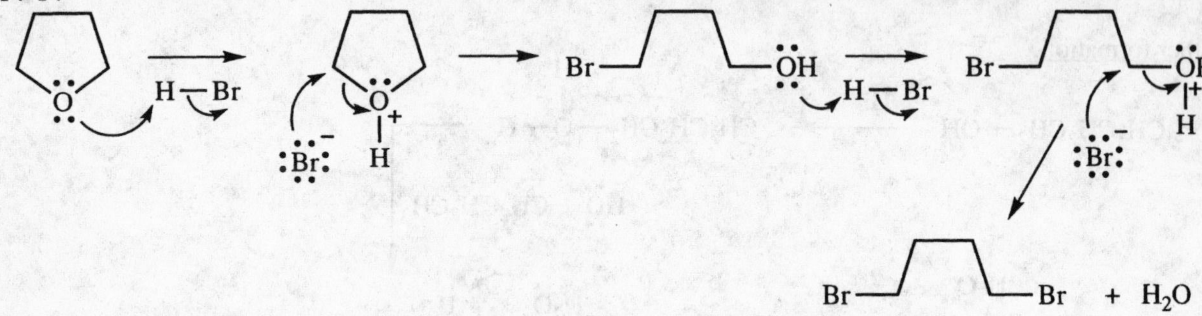

14-15

(a)

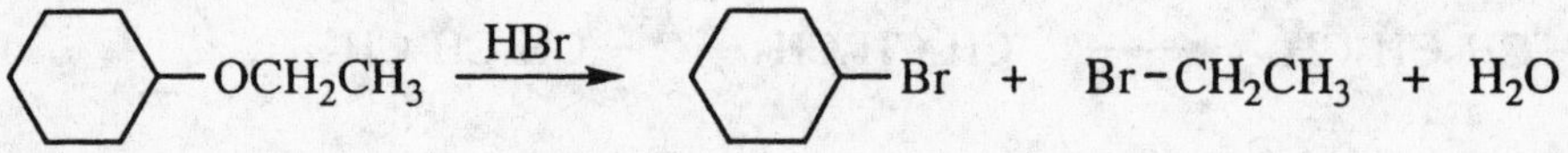

$$\text{cyclohexyl}-OCH_2CH_3 \xrightarrow{HBr} \text{cyclohexyl}-Br + Br-CH_2CH_3 + H_2O$$

(b)

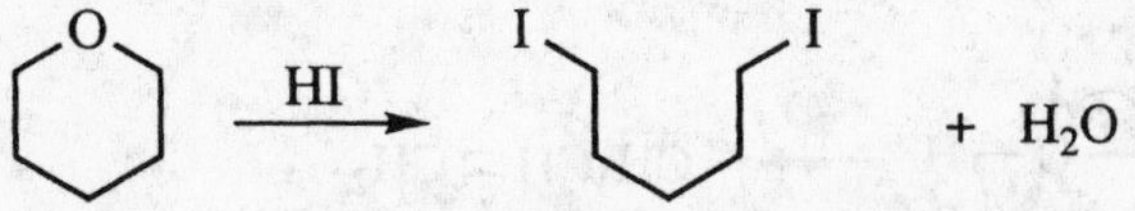

$$\text{(tetrahydropyran)} \xrightarrow{HI} I-\text{chain}-I + H_2O$$

(c)

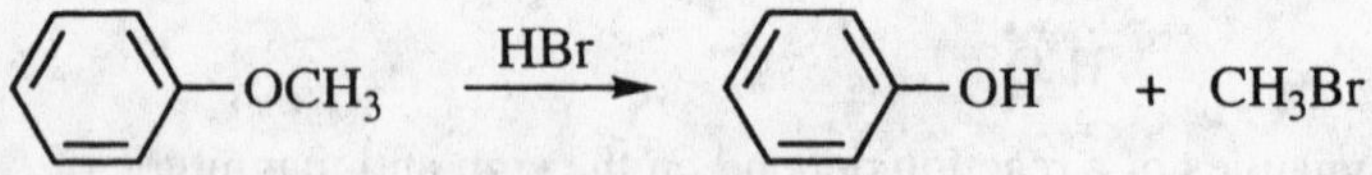

$$\text{Ph}-OCH_3 \xrightarrow{HBr} \text{Ph}-OH + CH_3Br$$

(d)

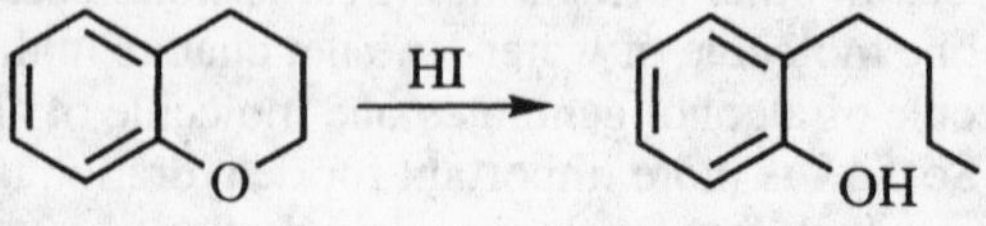

$$\text{(chroman)} \xrightarrow{HI} \text{o-substituted phenol with } CH_2CH_2I$$

(e)

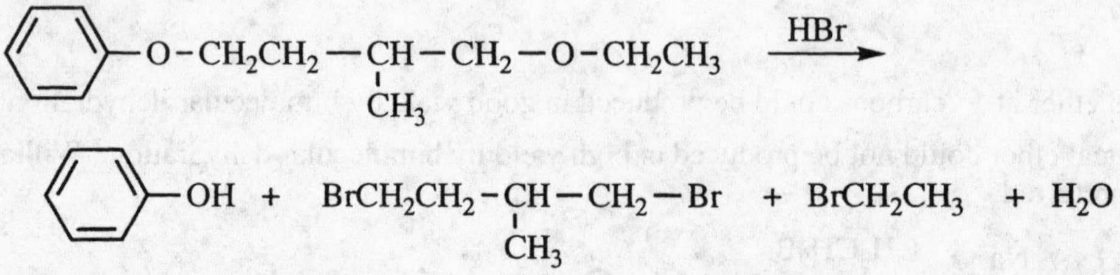

$$\text{Ph}-O-CH_2CH_2-\underset{\underset{CH_3}{|}}{CH}-CH_2-O-CH_2CH_3 \xrightarrow{HBr}$$

$$\text{Ph}-OH + BrCH_2CH_2-\underset{\underset{CH_3}{|}}{CH}-CH_2-Br + BrCH_2CH_3 + H_2O$$

14-16

nucleophilic attach on CH_3 faster than on 1° carbon

two proton transfers happen here: one H^+ comes off of oxygen, another H^+ goes on the other oxygen; several different scenarios of how this happens could be proposed

BuOH +

HOBBr$_2$

$\downarrow$ 2 H$_2$O

B(OH)$_3$ + 2 HBr

304

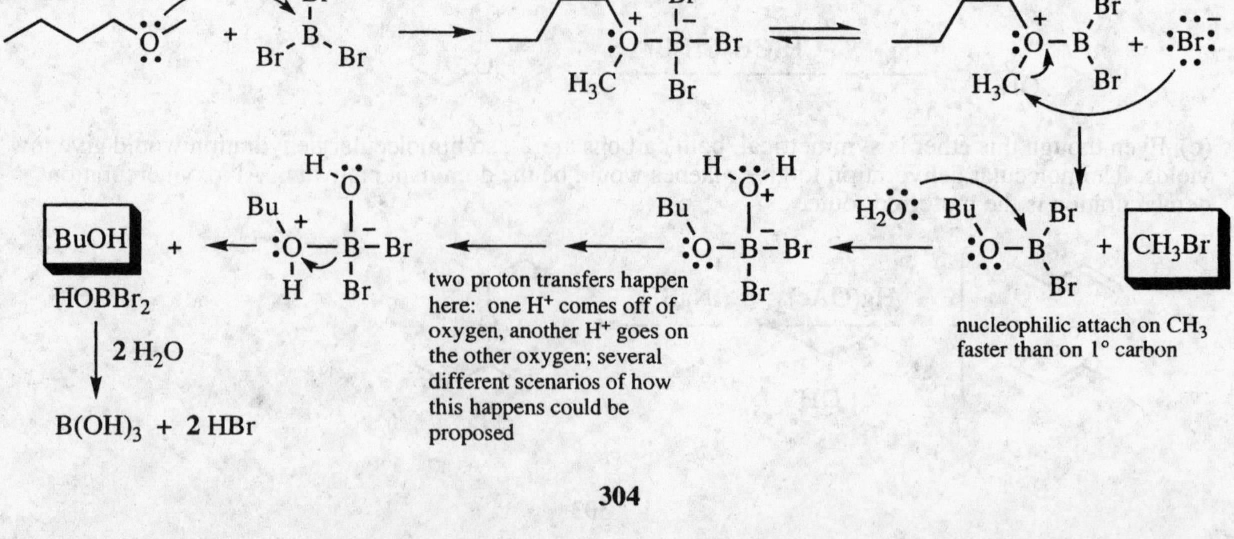

14-17 Begin by transforming the alcohols into good leaving groups like halides or tosylates:

$$\text{OH} \xrightarrow{\text{PBr}_3} \text{Br} \xrightarrow{\text{NaSH}} \text{SH} \xrightarrow{\text{NaOH}} \text{S}^- \text{ Na}^+$$

$$\underset{\text{OH}}{\bigvee} \xrightarrow[\text{pyridine}]{\text{TsCl}} \underset{\text{OTs}}{\bigvee} \longrightarrow \underset{\text{S}}{\bigvee}$$

14-18 The sulfur at the center of mustard gas is an excellent nucleophile, and chloride is a decent leaving group. Sulfur can do in *internal* nucleophilic subsitution to make a reactive sulfonium salt and the sulfur equivalent of an epoxide.

(a)

very reactive
alkylating agent

Cl ⟶ S ⟶ Cl $\xrightarrow{S_N i}$ S+ ⟶ Cl ⟶ HN ⟶ S ⟶ Cl

inactivated enzyme

:NH₂

(b) NaOCl is a powerful oxidizing agent. It oxidizes sulfur to a sulfoxide or more likely a sulfone, either of which is no longer nucleophilic, preventing formation of the cyclic sulfonium salt.

Cl ⟶ S ⟶ Cl $\xrightarrow{\text{NaOCl}}$ Cl ⟶ S+ ⟶ Cl   +   Cl ⟶ S++ ⟶ Cl

O⁻   O⁻

sulfoxide   sulfone

14-19

Generally, chemists prefer the peroxyacid method of epoxide formation to the halohydrin method. Reactions (a) and (b) show the peroxyacid method, but the halohydrin method could also be used.

(a) $\xrightarrow[\text{CH}_2\text{Cl}_2]{\text{MCPBA}}$

(b) $\underset{\text{Ph}}{\overset{\text{HO}}{\bigwedge}} \xrightarrow[\Delta]{\text{H}_2\text{SO}_4} \underset{\text{Ph}}{\bigvee} \xrightarrow[\text{CH}_2\text{Cl}_2]{\text{MCPBA}} \underset{\text{Ph}}{\bigvee}$

(c) $\underset{\text{Cl}}{\bigvee} \xrightarrow{\text{BH}_3 \cdot \text{THF}} \xrightarrow[\text{HO}^-]{\text{H}_2\text{O}_2} \underset{\text{Cl}}{\overset{\text{HO}}{\bigvee}} \xrightarrow{\text{NaOH}} \bigcirc$

(d) $\underset{\text{Cl}}{\bigvee} \xrightarrow[\text{H}_2\text{O}]{\text{Hg(OAc)}_2} \xrightarrow{\text{NaBH}_4} \underset{\text{Cl}}{\overset{\text{OH}}{\bigvee}} \xrightarrow{\text{NaOH}} \bigcirc$

(e) $\underset{\text{Cl}}{\bigvee}\text{OH} \xrightarrow{\text{NaOH}} \bigvee$

305

14-20

(a) 1) *tert*-Butyl hydroperoxide is the oxidizing agent. The $(CH_3)_3COOH$ contains the O—O bond just like a peroxyacid. 2) Diethyl tartrate has two asymmetric carbons and is the source of asymmetry; its function is to create a chiral transition state that is of lowest energy, leading to only one enantiomer of product. This process is called *chirality transfer*. 3) The function of the titanium (IV) isopropoxide is to act as the glue that holds all of the reagents together. The titanium holds an oxygen from each reactant— geraniol, *t*-BuOOH, and diethyl tartrate—and tethers them so that they react together, rather than just having them in solution and hoping that they will eventually collide.

(b) All three reactants are required to make Sharpless epoxidation work, but the key to *enantioselective* epoxidation is the chiral molecule, diethyl tartrate. When it complexes (or *chelates*) with titanium, it forms a large structure that is also chiral. As the *t*-BuOOH and geraniol approach the complex, the steric requirements of the complex allow the approach in one preferred orientation. When the reaction between the alkene and *t*-BuOOH occurs, it occurs preferentially from one face of the alkene, leading to one major stereoisiomer of the epoxide. Without the chiral diethyl tartrate in the complex, the alkene could approach from one side just as easily as the other, and a racemic mixture would be formed.

(c) Using the enantiomer of diethyl L-tartrate, called diethyl D-tartrate, would give exactly the opposite stereochemical results.

diethyl D-tartrate, the enantiomer of diethyl L-tartrate

14-21

IDENTICAL—MESO

stereochemistry shown in Newman projections:

306

14-21 continued

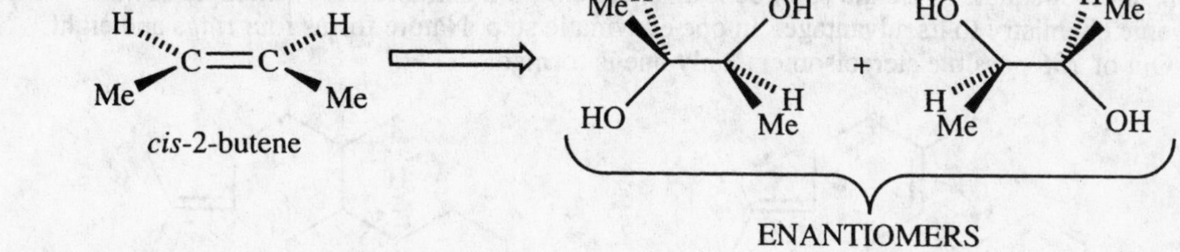

cis-2-butene

ENANTIOMERS

stereochemistry shown in Newman projections:

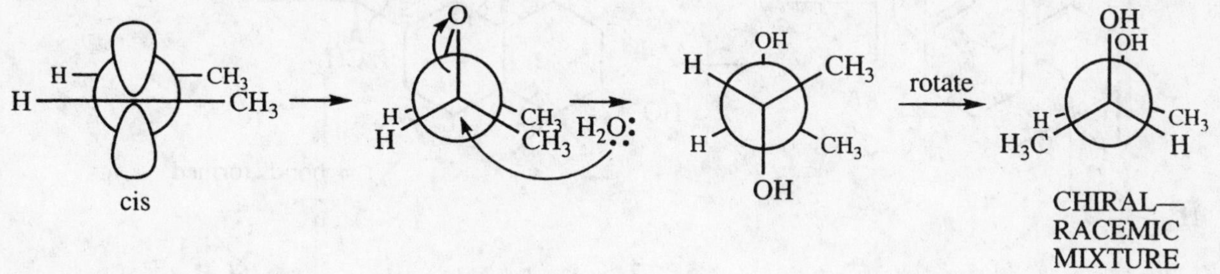

cis

rotate

CHIRAL—
RACEMIC
MIXTURE

14-22

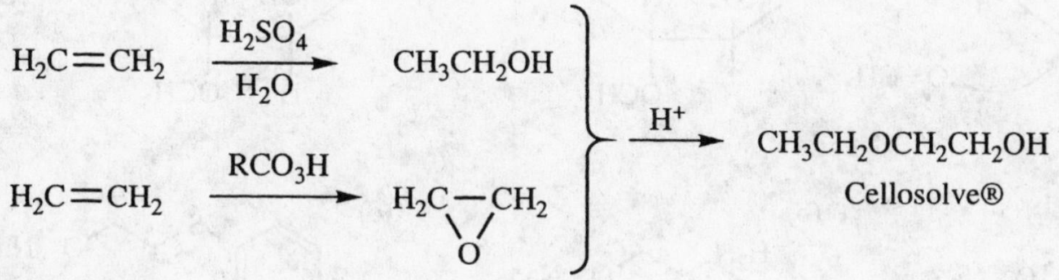

$$H_2C=CH_2 \xrightarrow[\text{H}_2\text{O}]{\text{H}_2\text{SO}_4} CH_3CH_2OH$$

$$H_2C=CH_2 \xrightarrow{RCO_3H} H_2C\overset{O}{-}CH_2$$

$$\xrightarrow{H^+} CH_3CH_2OCH_2CH_2OH$$
Cellosolve®

14-23

<u>anhydrous HBr</u>—only Br⁻ nucleophiles present

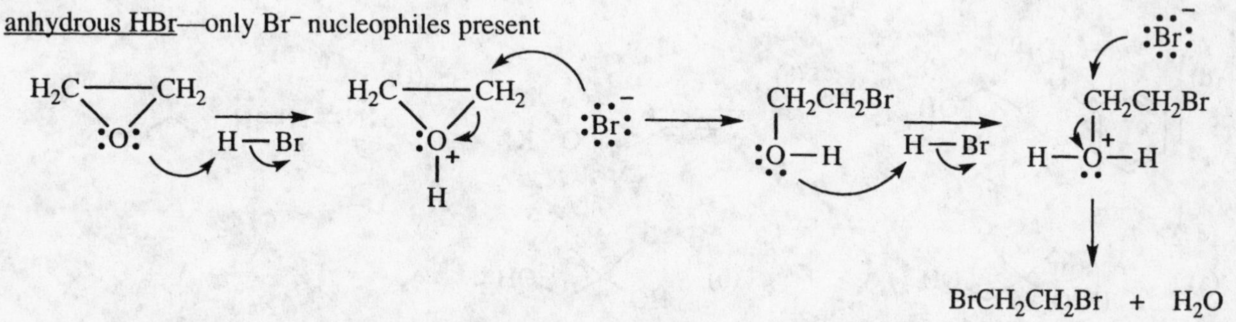

$$BrCH_2CH_2Br + H_2O$$

<u>aqueous HBr</u>—many more H_2O nucleophiles than Br⁻ nucleophiles

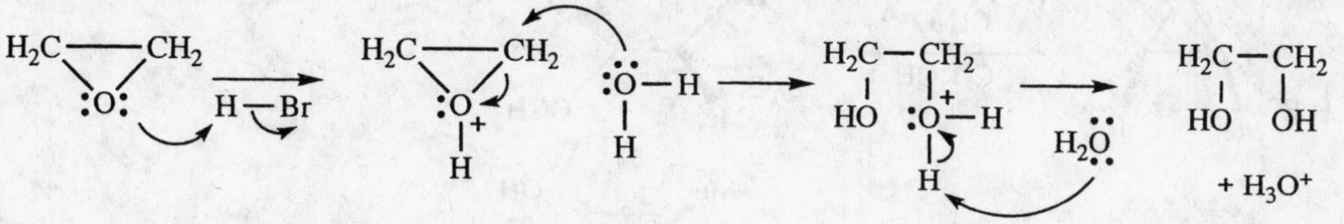

$$+ H_3O^+$$

307

14-24 The cyclization of squalene via the epoxide is an excellent (and extraordinary) example of how Nature uses organic chemistry to its advantage. In one enzymatic step, Nature forms four rings and eight chiral centers! Out of 256 possible stereoisomers, only one is formed!

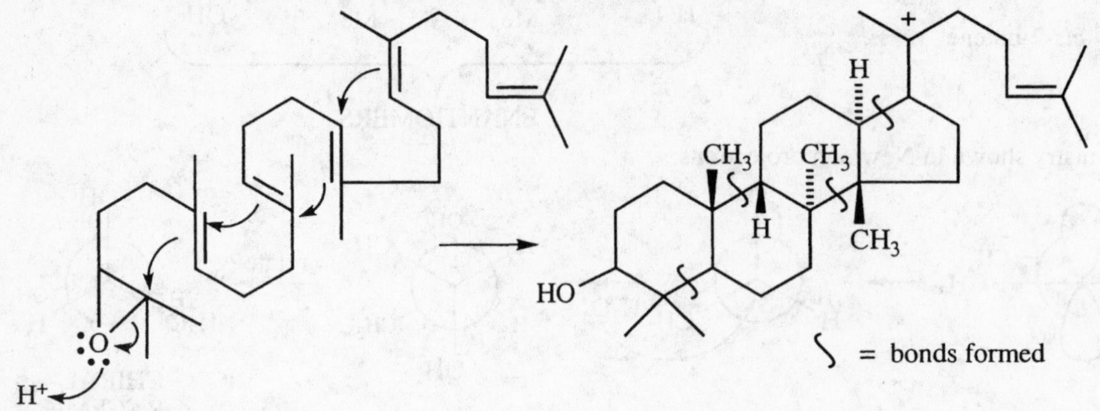

ζ = bonds formed

14-25

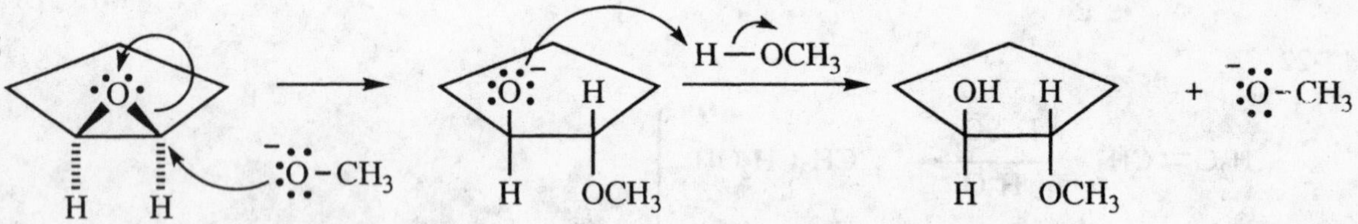

14-26

(a)

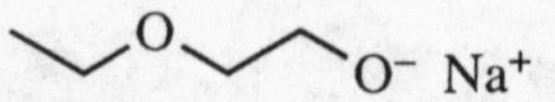

(b)

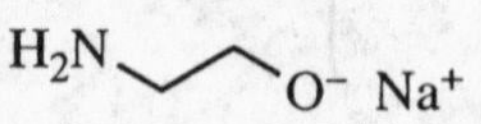

(c)

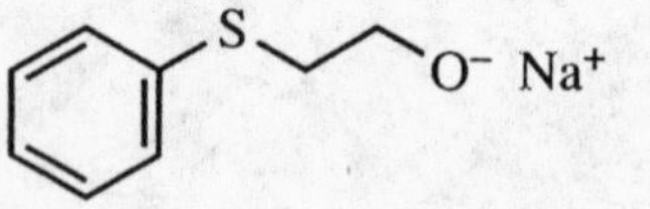

(d)

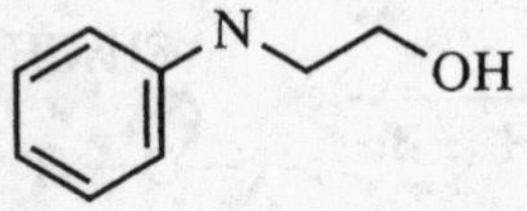

(e)

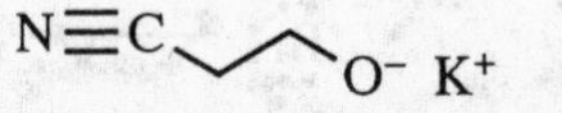

(f)

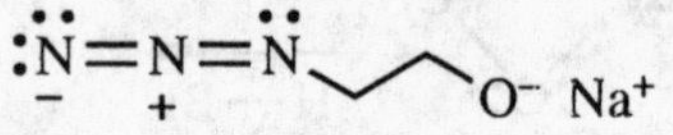

14-27

(a)

(b)

(c)

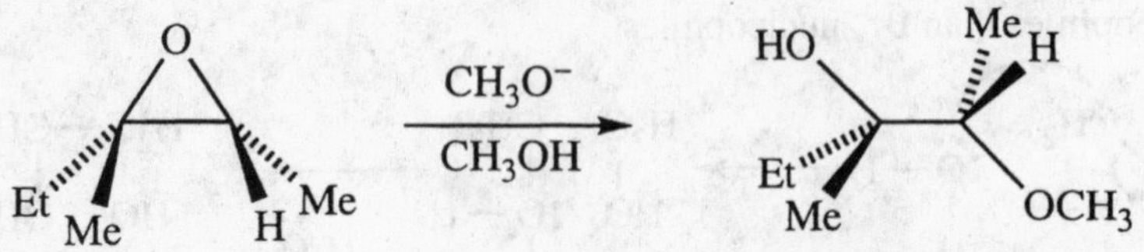

(d)

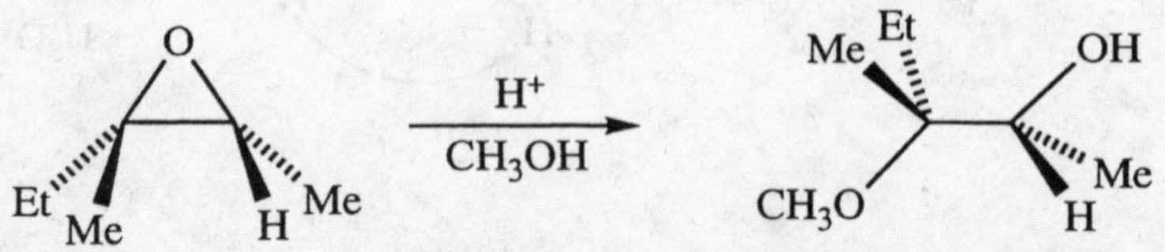

14-28 Newly formed bonds are shown in bold.

(a)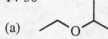

(b) (structure)

(c) (structure)

14-29 Please refer to solution 1-20, page 12 of this Solutions Manual.

14-30

(a) (structure)   (b) (structure)   (c) (structure)

(d) (structure)   (e) (structure)   (f) (structure)

(g) (structure)   (h) 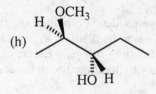   (i) (structure)

14-31

(a) *sec*-butyl isopropyl ether
(b) *t*-butyl isobutyl ether
(c) ethyl phenyl ether
(d) chloromethyl *n*-propyl ether

(e) *trans*-cyclohexene glycol
(f) cyclopentyl methyl ether
(g) propylene oxide
(h) cyclopentene oxide

14-32

(a) 2-methoxypropan-1-ol
(b) ethoxybenzene or phenoxyethane
(c) methoxycyclopentane
(d) 2,2-dimethoxycyclopentan-1-ol
(e) *trans*-1-methoxy-2-methylcyclohexane

(f) *trans*-3-chloro-1,2-epoxycycloheptane
(g) *trans*-1-methoxy-1,2-epoxybutane; or, *trans*-2-ethyl-3-methoxyoxirane
(h) 3-bromooxetane
(i) 1,3-dioxane

14-33

(a) (structure) + (structure) + H_2O

(b) (structure) + (structure) + H_2O

(c) and (d)   no reaction

(e) (structure) $-OH + CH_3CH_2I$

(f) (structure)

(g) (structure)

(h) (structure)

(i) (structure)

Keep at low temperature to minimize the competing E2 reaction.

14-33 continued

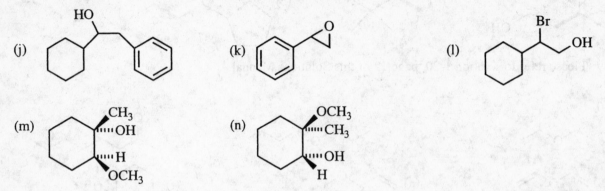

(j)   HO

(k)   O

(l)   Br   OH

(m)   CH₃   ''''OH   ''''H   OCH₃

(n)   OCH₃   ''''CH₃   ''''OH   H

14-34

(a) On long-term exposure to air, ethers form peroxides. Peroxides are explosive when concentrated or heated. (For exactly this reason, ethers should *never* be distilled to dryness.)

(b) Peroxide formation can be prevented by excluding oxygen. Ethers can be checked for the presence of peroxides, and peroxides can be destroyed safely by treatment with reducing agents.

14-35

(a) Beginning with *(R)*-butan-2-ol and producing the *(R)* sulfide requires two inversions of configuration.

$$HO\ H \quad \xrightarrow{PBr_3} \quad H\ Br \quad \xrightarrow{Na^+\ {}^-SCH_3} \quad H_3C-S\ H$$

$$R \qquad\qquad S \qquad\qquad\qquad R$$

An alternative approach would be to make the tosylate, displace with chloride or bromide (S$_N$2 with inversion), then do a second inversion with NaSCH$_3$.

(b) Synthesis of the *(S)* isomer directly requires only one inversion.

$$HO\ H \quad \xrightarrow[\text{pyridine}]{TsCl} \quad TsO\ H \quad \xrightarrow{Na^+\ {}^-SCH_3} \quad H\ S-CH_3$$

$$R \qquad\qquad\qquad R \qquad\qquad\qquad S$$

14-36

(a)

$$\left[\ \overset{43}{}\ O\ \overset{73}{}\ \right]^{\ddagger} \longrightarrow \left\{\ \ddot{\overset{..}{O}}{}^{+}\!-\!CH_2 \longleftrightarrow \ddot{\overset{..+}{O}}\!=\!CH_2\ \right\}$$

molecular ion m/z 102

m/z 73

$$CH_3CH_2CH_2{}^+ \quad m/z\ 43$$

14-36 continued
(b)

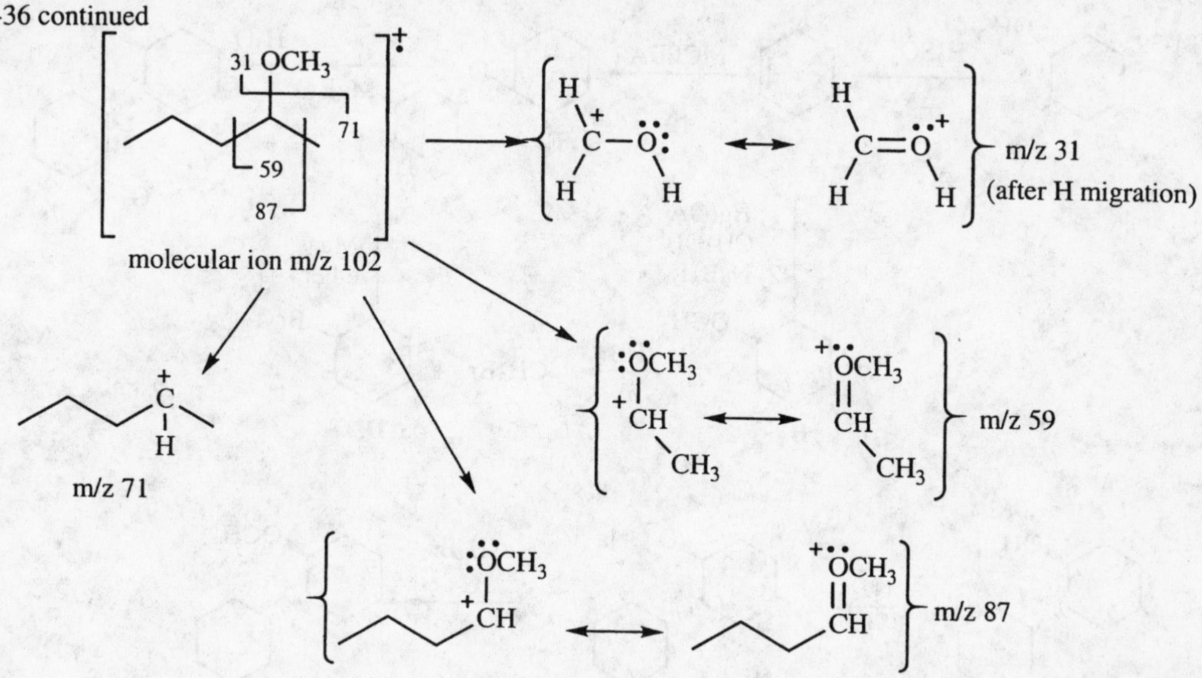

31 OCH₃

71

59

87

molecular ion m/z 102

m/z 31
(after H migration)

m/z 71

m/z 59

m/z 87

14-37

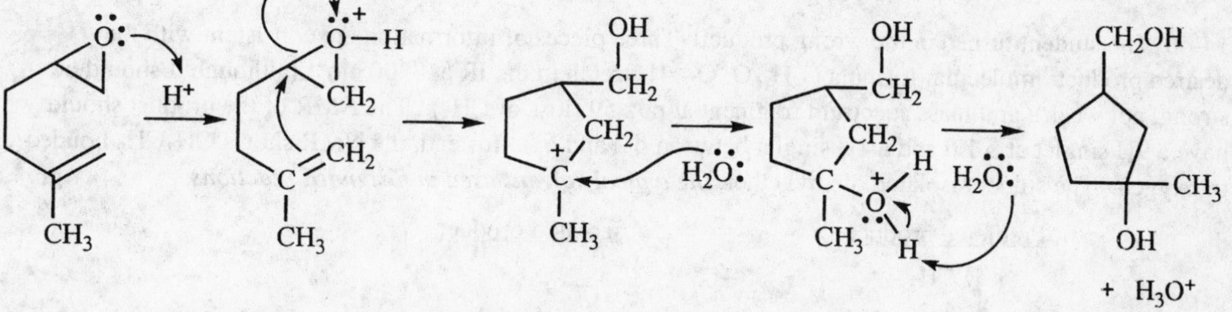

H⁺

+ H₃O⁺

14-38

(a)

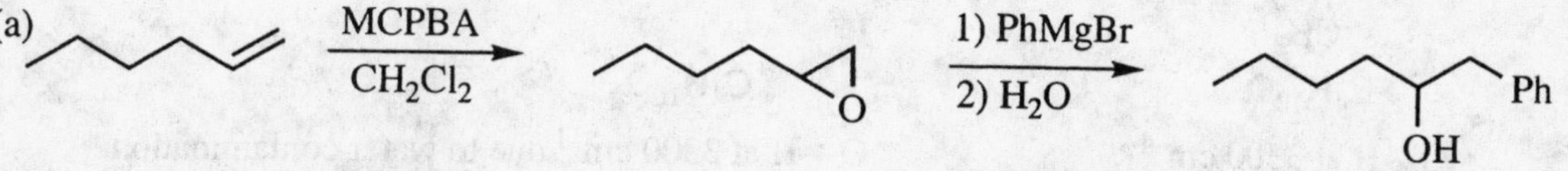

$\dfrac{\text{MCPBA}}{\text{CH}_2\text{Cl}_2}$

$\dfrac{\text{1) PhMgBr}}{\text{2) H}_2\text{O}}$

(b)

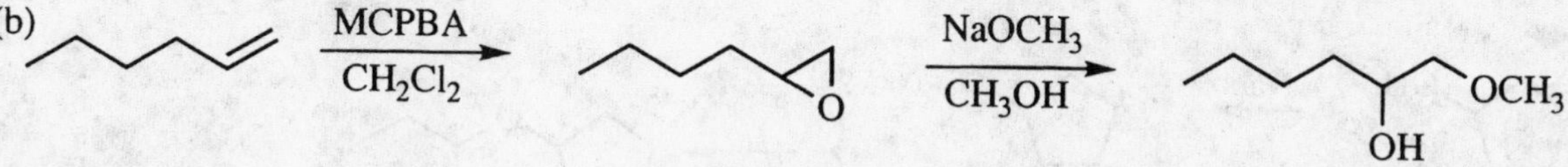

$\dfrac{\text{MCPBA}}{\text{CH}_2\text{Cl}_2}$

$\dfrac{\text{NaOCH}_3}{\text{CH}_3\text{OH}}$

(c)

$\dfrac{\text{MCPBA}}{\text{CH}_2\text{Cl}_2}$

$\dfrac{\text{H}^+}{\text{CH}_3\text{OH}}$

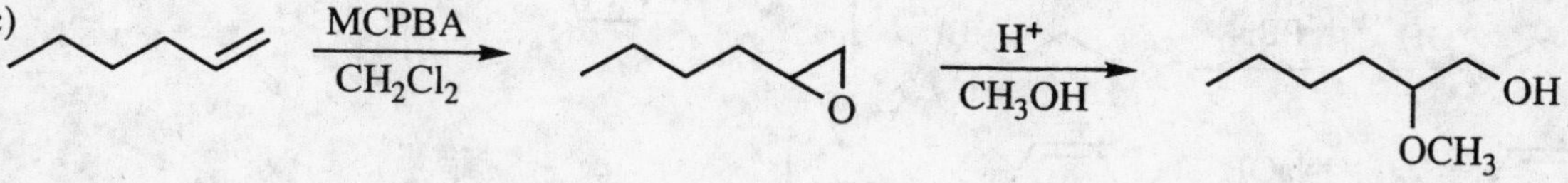

14-39

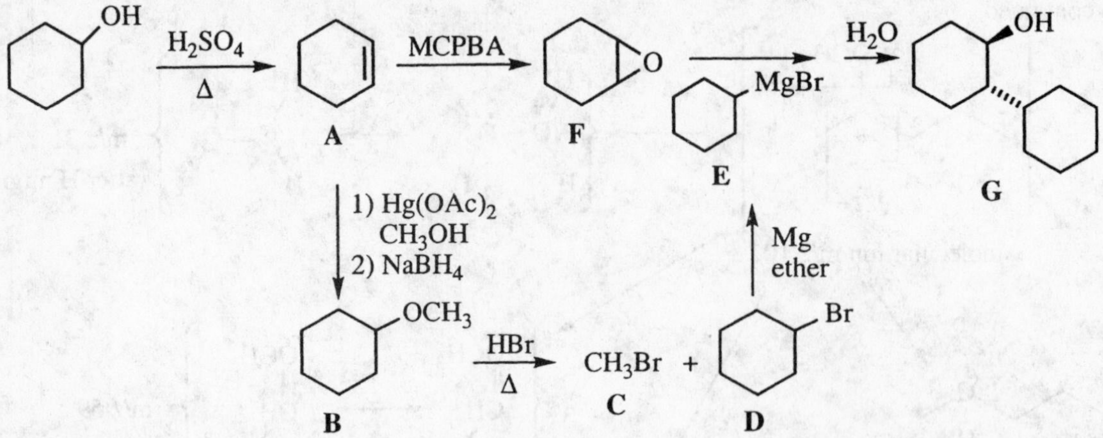

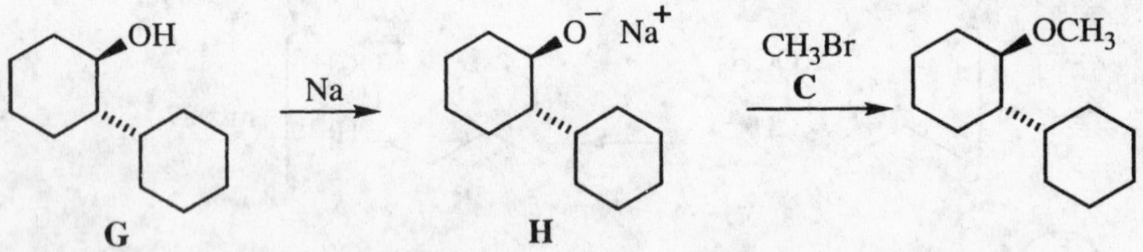

14-40 The student turned in the wrong product! Three pieces of information are consistent with the desired product: molecular formula $C_4H_{10}O$; O—H stretch in the IR at 3300 cm^{-1} (although it should be strong, not weak); and mass spectrum fragment at m/z 59 (loss of CH_3). The NMR of the product should have a 9H singlet at δ 1.0 and a 1H singlet between δ 2 and δ 5. Instead, the NMR shows CH_3CH_2 bonded to oxygen. The student isolated diethyl ether, *the typical **solvent** used in Grignard reactions.*

Predicted product	Isolated product
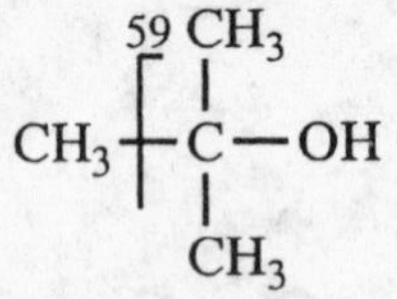	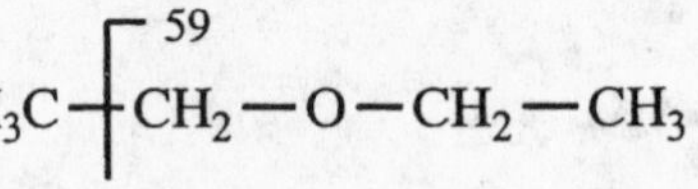
$C_4H_{10}O$	$C_4H_{10}O$
O—H at 3300 cm^{-1}	O—H at 3300 cm^{-1} due to water contamination

14-41

(a)

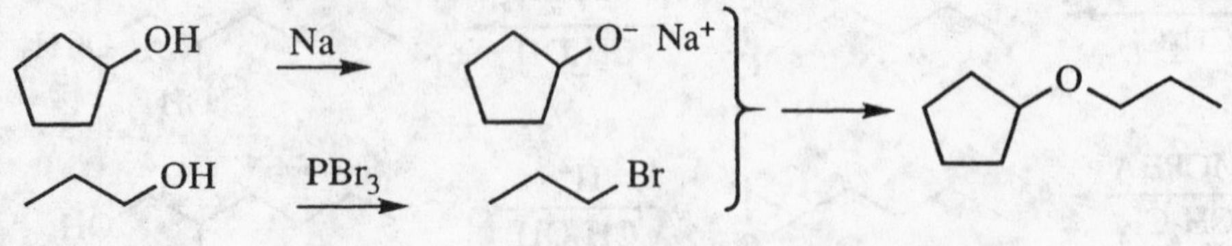

(b)

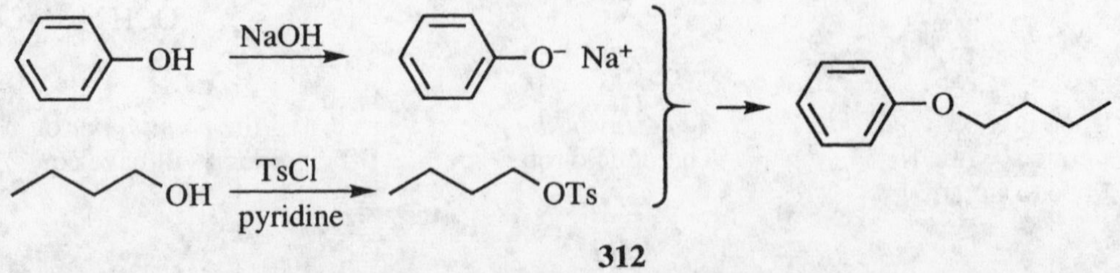

14-41 continued

(c)

$$\xrightarrow[\text{CH}_3\text{OH}]{\text{Hg(OAc)}_2} \xrightarrow{\text{NaBH}_4}$$

(product with OCH₃)

(d)

$$\xrightarrow[\text{2) H}_2\text{O}_2,\ \text{HO}^-]{\text{1) BH}_3 \cdot \text{THF}}$$

(product with OH)

↓ HBr, ROOR

(product with Br) $\xrightarrow{\text{NaOCH}_3}$ (product with OCH₃) $\xleftarrow[\text{2) CH}_3\text{I}]{\text{1) Na}}$

(e)

$$\xrightarrow[\text{CH}_3\text{CH}_2\text{OH}]{\text{Hg(OAc)}_2} \xrightarrow{\text{NaBH}_4}$$

(cyclohexane with OCH₂CH₃)

(f)

$$\xrightarrow[\text{CH}_2\text{Cl}_2]{\text{MCPBA}}$$

(epoxide product, O)

14-42 In the first sequence, no bond is broken to the chiral center, so the configuration of the product is the same as the configuration of the starting material.

$$[\alpha]_D = -8.24°$$

$$\xrightarrow{\text{Na}}$$ (H, O⁻ Na⁺) $$\xrightarrow{\text{CH}_3\text{CH}_2\text{I}}$$ (H, OCH₂CH₃) $\quad [\alpha]_D = -15.6°$

(Assume the enantiomer shown is levorotatory.)

In the second reaction sequence, however, bonds to the chiral carbon are broken twice, so the stereochemistry of each process must be considered.

(mechanism scheme showing tosylation with Cl–SO₂–C₆H₄–CH₃, pyridine, **RETENTION OF CONFIGURATION**)

Undoubtedly, some E2 products will also form in this reaction.

$$\text{CH}_3\text{CH}_2\overset{..}{\underset{..}{\text{O}}}{:}^- \quad \text{Na}^+$$

↓ S_N2—INVERSION

(product CH₃CH₂O, H)

The second sequence involves *retention* followed by *inversion*, thereby producing the *enantiomer* of the 2-ethoxyoctane generated by the first sequence. The optical rotation of the final product will have equal magnitude but opposite sign, $[\alpha]_D = +15.6°$.

14-43

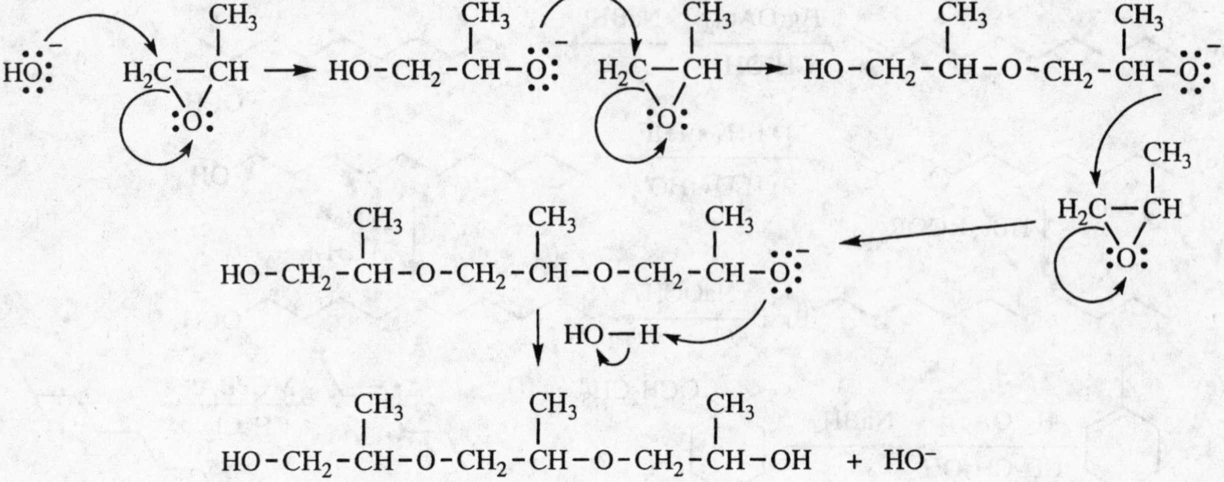

14-44 The text uses HA to indicate an acid; A⁻ is the conjugate base.

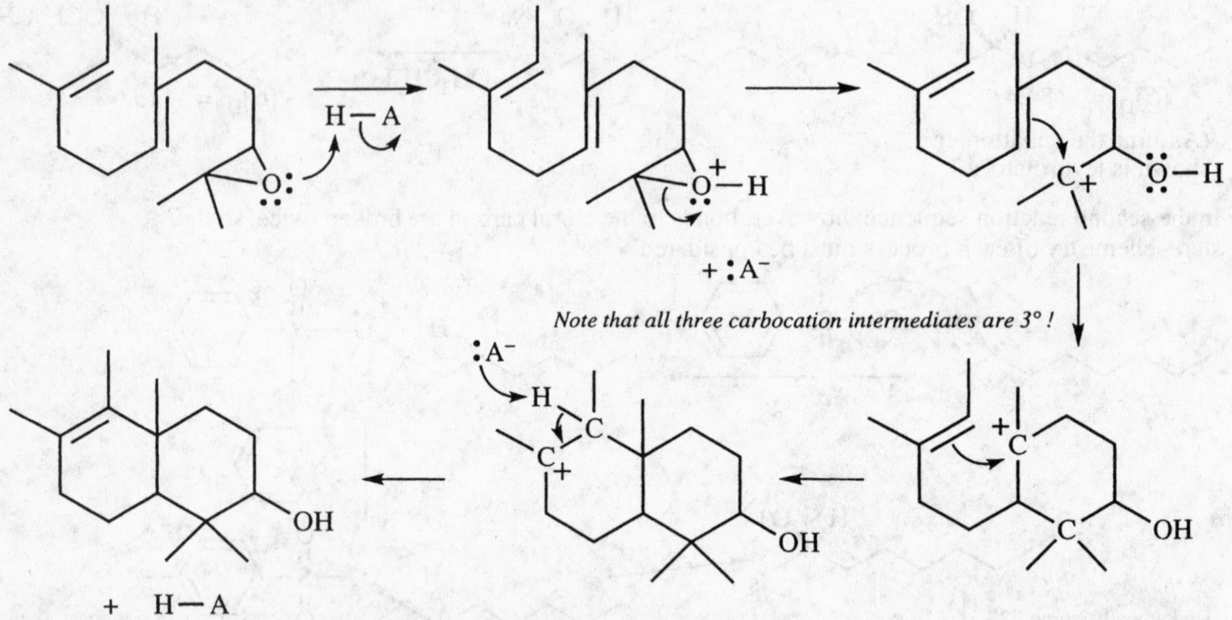

+ :A⁻

Note that all three carbocation intermediates are 3° !

+ H—A

This process resembles the cyclization of squalene oxide to lanosterol. (See the solution to problem 14-24.)
In fact, pharmaceutical synthesis of steroids uses the same type of reaction called a "biomimetic cyclization".

14-45

(a)

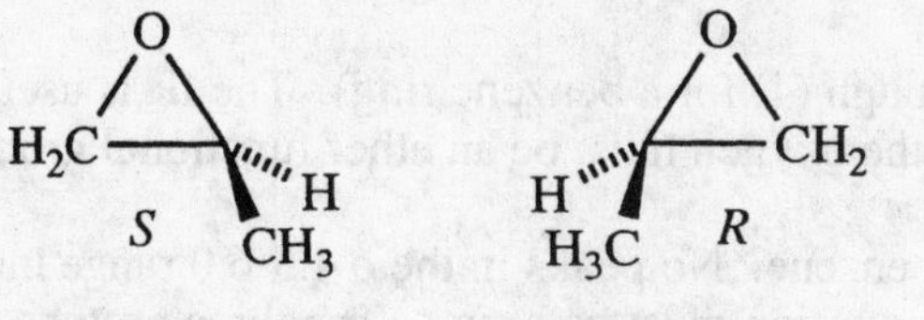

(b)

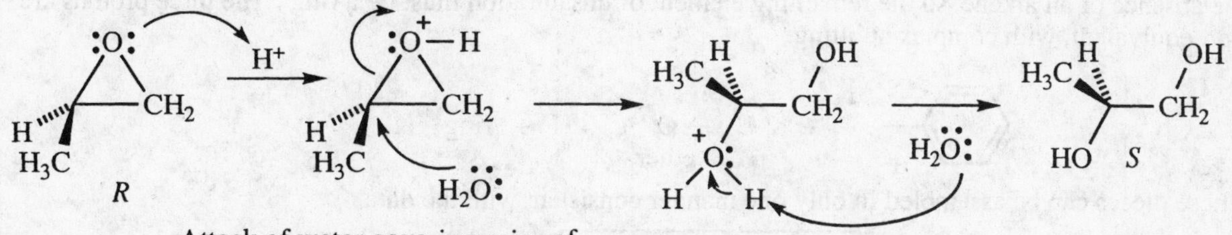

Attack of water gave *inversion* of
configuration at the chiral center; *R* became *S*.

(c)

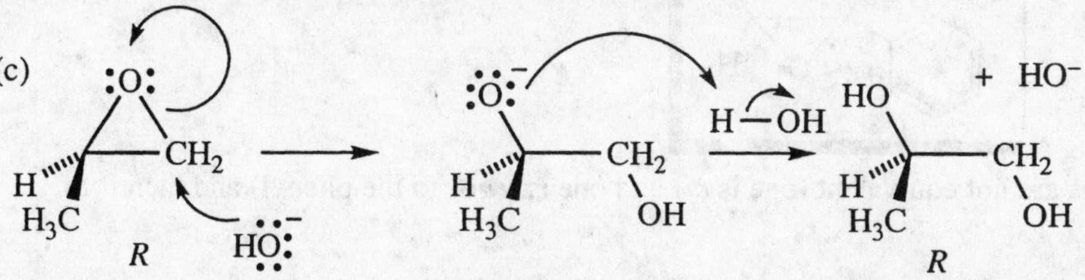

No bond to the chiral center
was broken. Configuration
is retained; *R* stays as *R*.

(d) The difference in these mechanisms lies in where the nucleophile attacks. Attack at the chiral carbon gives inversion; attack at the achiral carbon retains the configuration at the chiral carbon. These products are enantiomers and must necessarily have optical rotations of opposite sign.

14-46   methyl cellosolve $CH_3OCH_2CH_2OH$

To begin, what can be said about methyl cellosolve? Its molecular weight is 76; its IR would show C--O in the 1000-1200 cm^{-1} region and a strong O—H around 3300 cm^{-1}; and its NMR would show four sets of signals in the ratio of 3:2:2:1.

The unknown has molecular weight 134; this is double the weight of methyl cellosolve, minus 18 (water). The IR shows no OH, only ether C—O. The NMR shows no OH, only H—C—O in the ratio of 3:2:2. Apparently, two molecules of methyl cellosolve have combined in an acid-catalyzed, bimolecular dehydration.

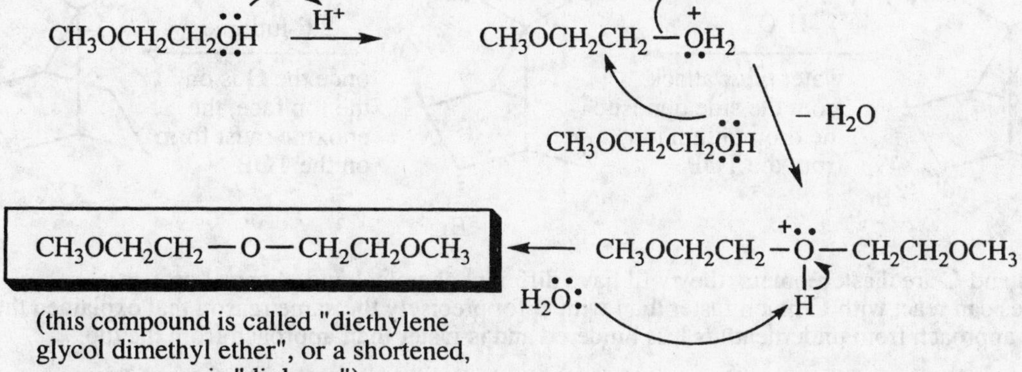

(this compound is called "diethylene
glycol dimethyl ether", or a shortened,
common name is "diglyme")

14-47

The formula C_8H_8O has five elements of unsaturation (enough (4) for a benzene ring). The IR is useful for what is does *not* show. There is neither OH nor C=O, so the oxygen must be an ether functional group.

The NMR shows a 5H signal at δ 7.2, a monosubstituted benzene. No peaks in the δ 4.5-6.0 range indicate the absence of an alkene, so the remaining element of unsaturation must be a ring. The three protons are non-equivalent, with complex splitting.

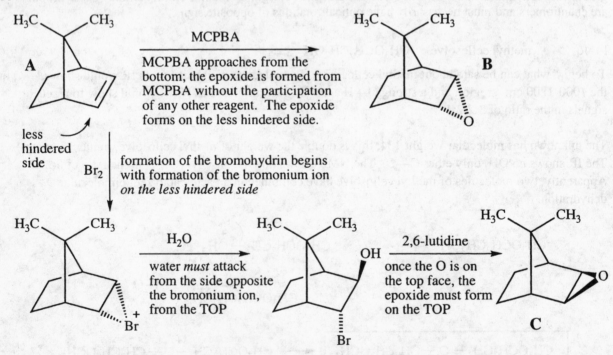

These pieces can be assembled in only one manner consistent with the data.

(Note that the CH_2 hydrogens are not equivalent (one is *cis* and one is *trans* to the phenyl) and therefore have distinct chemical shifts.)

14-48 The key concept is that reagents always go to the less hindered side of a molecule first. In this case, the "underneath" side is less hindered; the "top" side has a CH_3 hovering over the double bond and approach from the top will be much more difficult, and therefore slower, than approach from underneath.

A

less hindered side

MCPBA

MCPBA approaches from the bottom; the epoxide is formed from MCPBA without the participation of any other reagent. The epoxide forms on the less hindered side.

B

Br_2

formation of the bromohydrin begins with formation of the bromonium ion *on the less hindered side*

H_2O

water *must* attack from the side opposite the bromonium ion, from the TOP

2,6-lutidine

once the O is on the top face, the epoxide must form on the TOP

C

Epoxides **B** and **C** are diastereomers; they will have different chemical and physical properties. Nucleophiles can react with **C** much faster than with **B** for precisely the same reason that explained their formation: approach from underneath is less hindered and is faster than approach from the top.

316

CHAPTER 15—CONJUGATED SYSTEMS, ORBITAL SYMMETRY, AND

ULTRAVIOLET SPECTROSCOPY

15-1 Look for: 1) the number of double bonds to be hydrogenated—the fewer C=C, the smaller the ΔH; 2) conjugation—the more conjugated, the more stable, the lower the ΔH; 3) degree of substitution of the alkenes—the more substituted, the more stable, the lower the ΔH.

(a)

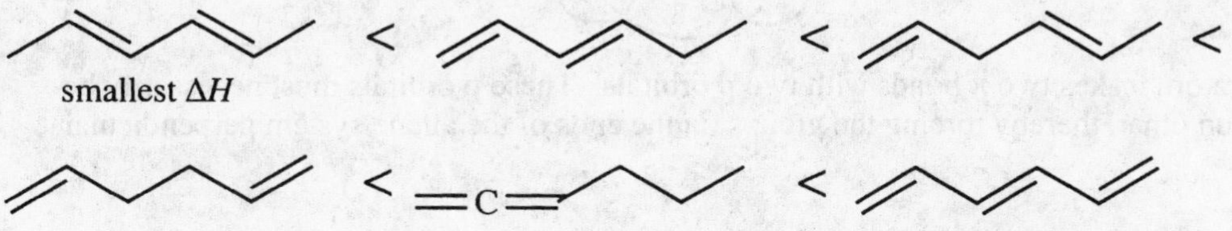

smallest ΔH

biggest ΔH

(b)

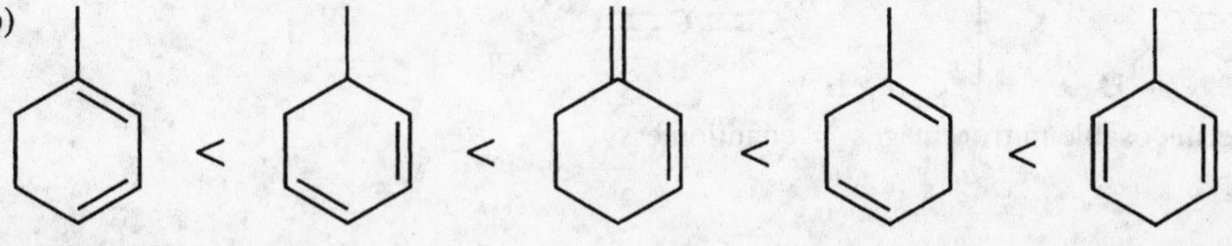

smallest ΔH biggest ΔH

15-2 Reminder: H—B is used to symbolize the general form for an acid, that is, a protonated base; B^- is the conjugate base.

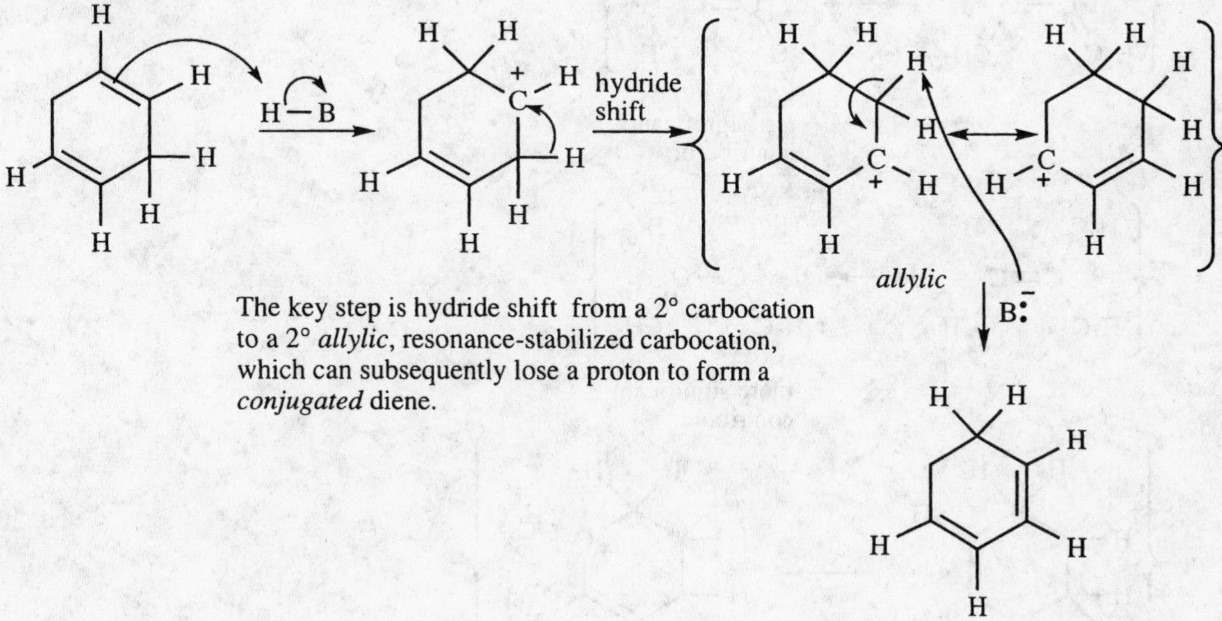

The key step is hydride shift from a 2° carbocation to a 2° *allylic*, resonance-stabilized carbocation, which can subsequently lose a proton to form a *conjugated* diene.

317

15-3 (You may wish to refer to problem 2-6.) <u>orbital picture</u>

(a)

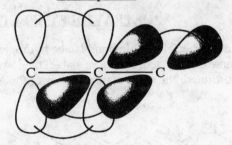

$$H \overset{\cdots}{\underset{H}{}} \overset{1}{C} = \overset{2}{C} = \overset{3}{C} \overset{H}{\underset{H}{}}$$

sp

The central carbon atom makes two π bonds with two p orbitals. These p orbitals must necessarily be perpendicular to each other, thereby forcing the groups on the ends of the allene system perpendicular.

(b)

mirror

$$H \overset{\cdots}{\underset{Cl}{}} C = C = C \overset{Cl}{\underset{H}{}} \qquad \qquad Cl \overset{}{\underset{H}{}} C = C = C \overset{H}{\underset{Cl}{}}$$

non-superimposable mirror images = enantiomers

15-4 Carbocation stability depends on conjugation (benzylic, allylic), then on degree of carbon carrying the positive charge.

$$\left\{ \quad \overset{H}{\underset{H_2C}{}} C - \overset{H}{\underset{CH_3}{}} C + \quad \longleftrightarrow \quad \overset{H}{\underset{H_2C}{}} C = \overset{H}{\underset{CH_3}{}} C \quad \right\}$$

2°   1°
less significant
contributor

$$\left\{ \quad \overset{H_3C}{\underset{H_3C}{}} C = \overset{H}{\underset{CH_2}{}} C \quad \longleftrightarrow \quad \overset{H_3C}{\underset{H_3C}{}} +C - \overset{H}{\underset{CH_2}{}} C \quad \right\}$$

1°   3°
more significant
contributor

2°   equivalent to the
first resonance form

318

15-5

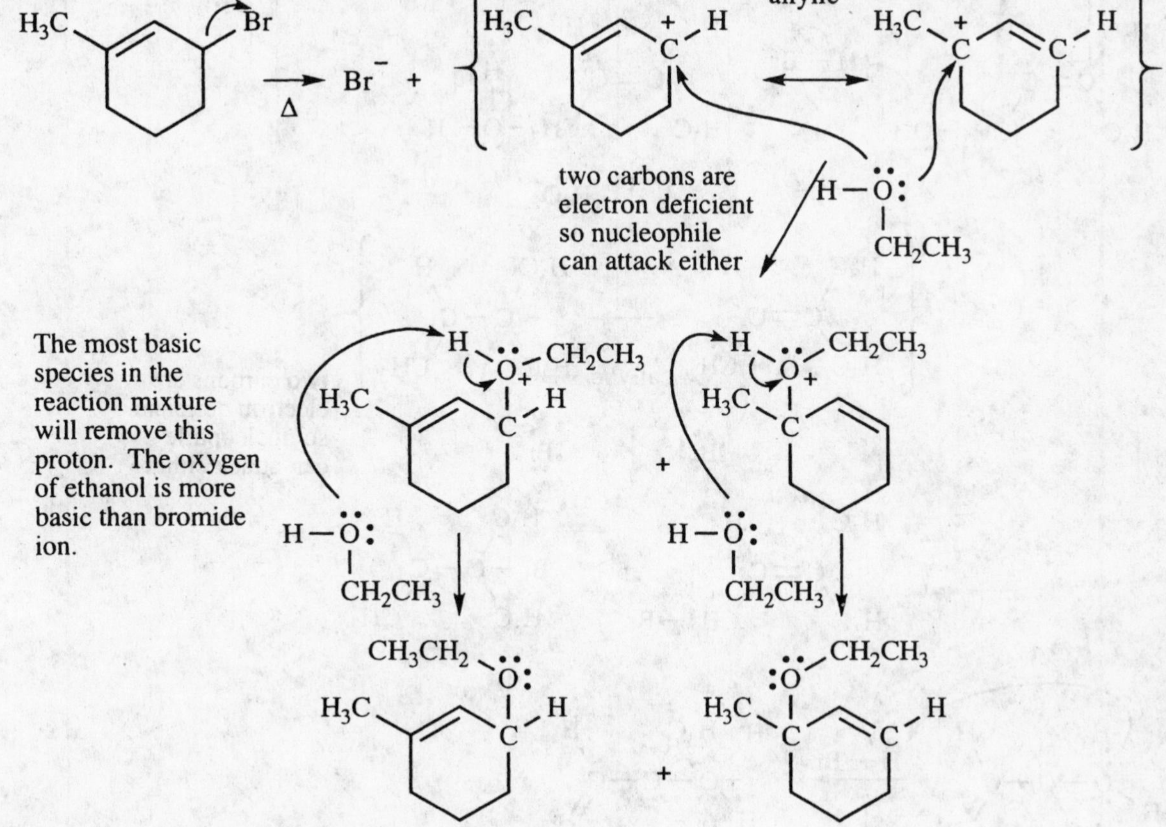
allylic
Δ
two carbons are electron deficient so nucleophile can attack either
The most basic species in the reaction mixture will remove this proton. The oxygen of ethanol is more basic than bromide ion.

15-6

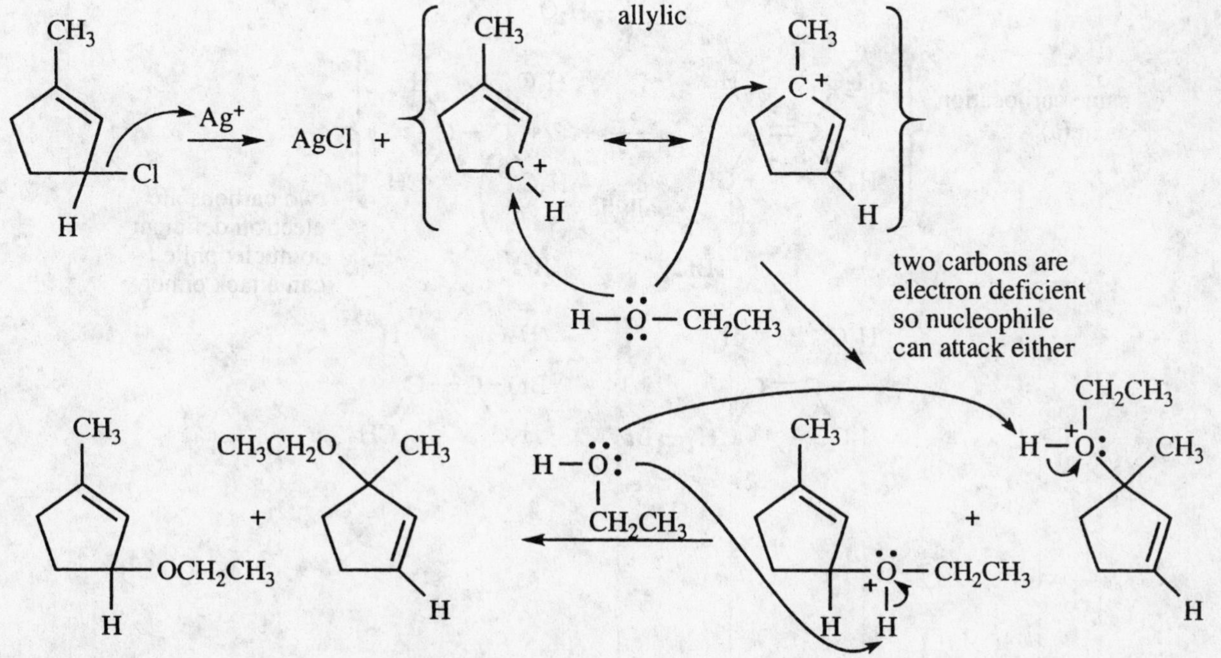
Ag+
AgCl +
allylic
two carbons are electron deficient so nucleophile can attack either

15-7

(a)

(b)

320

15-7 continued

(c)

$$H_2C=C(H)-C(H)=CH_2 \xrightarrow{Br-Br}$$

$$\left\{ \begin{array}{c} \underset{H}{\overset{Br}{|}}H_2C-\overset{+}{C}-C(H)=CH_2 \quad \longleftrightarrow \quad \underset{H}{\overset{Br}{|}}H_2C-C(H)=C-\overset{+}{C}H_2 \end{array} \right\}$$

allylic

$:\ddot{\underset{..}{Br}}:^{-}$

two carbons are electron deficient so nucleophile can attack either

$$\underset{H}{\overset{Br\;\;Br}{|\;\;\;|}}H_2C-C-C(H)=CH_2 \quad + \quad \underset{H\;\;H}{\overset{Br\;\;\;\;\;\;Br}{|\;\;\;\;\;\;\;\;|}}H_2C-C=C-CH_2$$

While the bromonium ion mechanism is typical for isolated alkenes, the greater stability of the resonance-stabilized carbocation will make it the lower energy intermediate for conjugated systems.

(d)

$$H_3C-C(H)=C(H)-\overset{Cl}{\underset{|}{C}}H_2 \xrightarrow[-\;AgCl]{Ag^+}$$

$$\left\{ \begin{array}{c} H_3C-C(H)=C(H)-\overset{+}{C}H_2 \quad \longleftrightarrow \quad H_3C-\overset{+}{C}(H)-C(H)=CH_2 \end{array} \right\}$$

allylic

$H_2\ddot{\underset{..}{O}}$

two carbons are electron deficient so nucleophile can attack either

$$H-\overset{..\;+}{\underset{..}{O}}-H \qquad\qquad H-\overset{..\;+}{\underset{..}{O}}-H$$

$$H_3C-C(H)=C(H)-CH_2 \quad + \quad H_3C-C(H)-C(H)=CH_2$$

$H_2\ddot{\underset{..}{O}} \qquad\qquad H_2\ddot{\underset{..}{O}}$

$$\underset{H\;\;H}{\overset{OH}{|}}H_3C-C=C-CH_2 \quad + \quad \underset{H\;\;H}{\overset{OH}{|}}H_3C-C-C=CH_2$$

15-7 continued

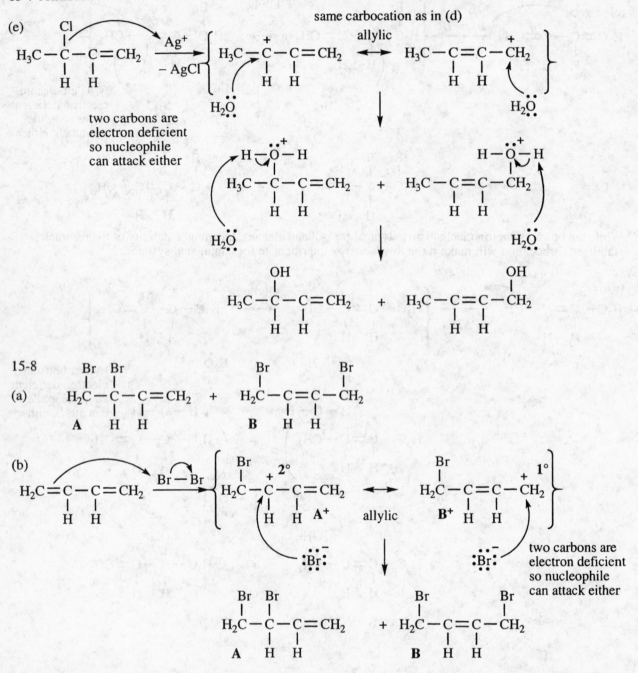

(c) The resonance form **A⁺**, which eventually leads to product **A**, has positive charge on a 2° carbon and is a more significant resonance contributor than structure **B⁺**. With greater positive charge on the 2° carbon than on the 1° carbon, we would expect bromide ion attack on the 2° carbon to have lower activation energy. Therefore, **A** must be the *kinetic* product. At higher temperature, however, the last step becomes reversible, and the stability of the products becomes the dominant factor in determining product ratios. As **B** has a disubstituted alkene whereas **A** is only monosubstituted, it is reasonable that **B** is the major, *thermodynamic* product at 60° C.

15-8 continued

(d) At 60° C, ionization of **A** would lead to the same allylic carbocation as shown in (b), which would give the same product ratio as formation of **A** and **B** from butadiene.

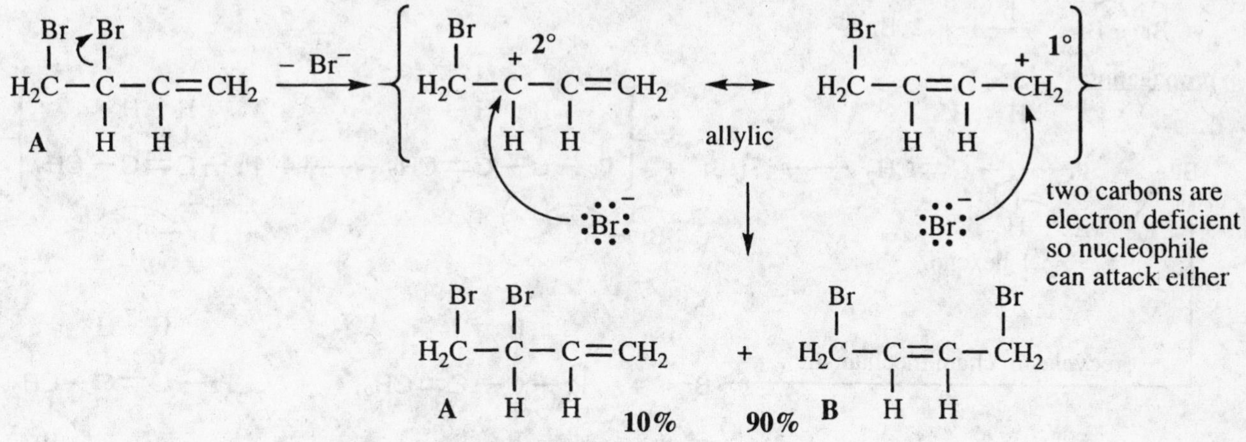

15-9

(a)

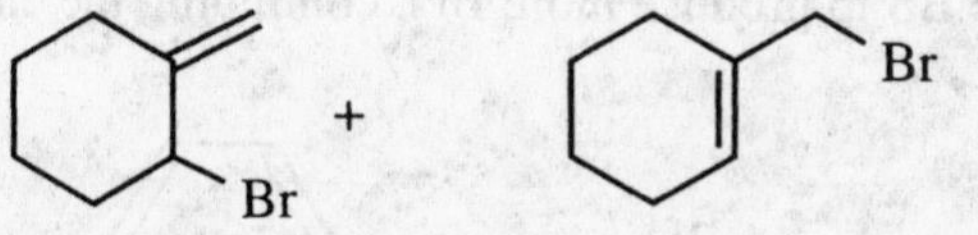

(b)

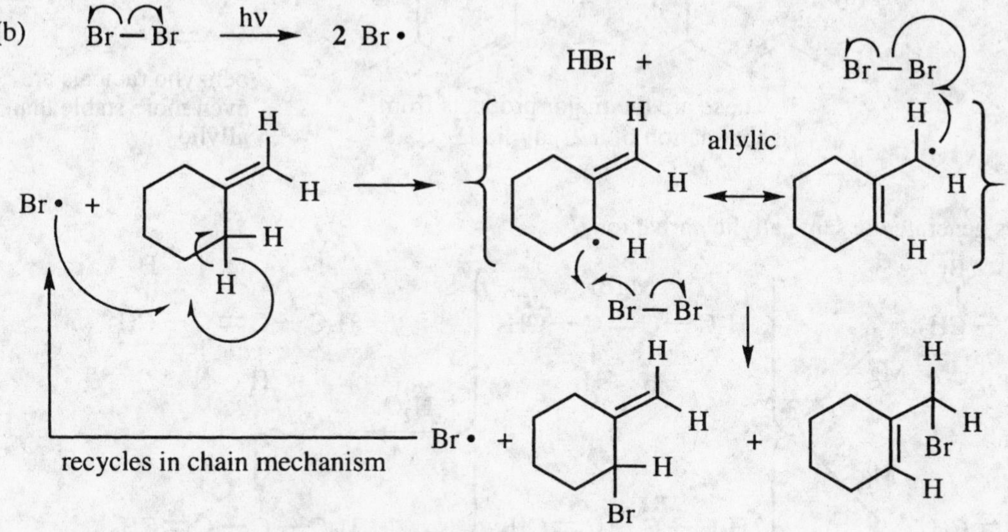

recycles in chain mechanism

15-10 ("Pr" is the abbreviation for *n*-propyl, used below.)

NBS generates a low concentration of Br_2

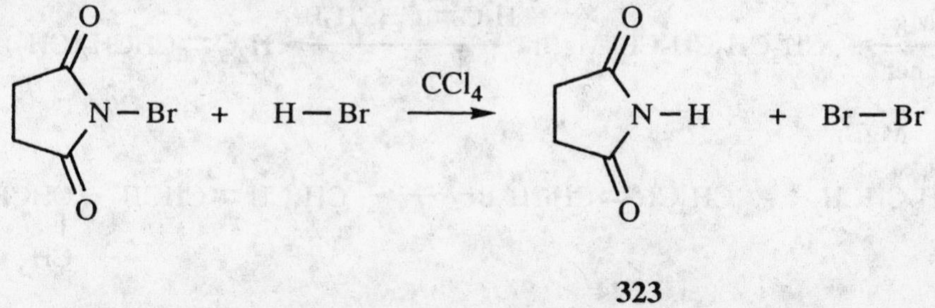

323

15-10 continued

<u>initiation</u>

$$Br - Br \xrightarrow{h\nu} 2 \; Br \cdot$$

<u>propagation</u>

Br • + Pr—C—C=CH₂ ⟶ HBr + { Pr—C—C=CH₂ ⟷ Pr—C=C—CH₂ }

1-hexene

recycles in chain mechanism

Br • + Pr—C—C=CH₂ + Pr—C=C—CH₂
              | |
              Br   Br

The HBr generated in the propagation step combines with NBS to produce more Br_2, continuing the chain mechanism.

15-11

(a) [cyclopentenyl bromide structure] Br

(b) [two allylic bromide structures] Br + Br

These are the major products from abstraction of a 2° allylic H.

(c) [benzyl bromide structure] CH₂Br

benzylic radicals are even more stable than allylic

15-12 Both halides generate the same allylic carbanion.

Br
|
H₃C—C=C—CH₂   Mg
      | |
      H   H

Br
|
H₃C—C—C=CH₂   Mg
    | |
    H   H

{ MgBr⁺
  H₃C—C=C—⋅⋅CH₂⁻
        | |
        H   H
  ↕
  H₃C—⋅⋅C⁻—C=CH₂
       | |
       H   H }

H₂O

H
|
H₃C—C=C—CH₂
      | |
      H   H

+

H
|
H₃C—C—C=CH₂
    | |
    H   H

15-13

(a) $CH_3CH_2CH_2CH_2Br \xrightarrow[\text{ether}]{Mg} CH_3CH_2CH_2CH_2MgBr \xrightarrow{H_2C=CHCH_2Br} H_2C=CHCH_2(CH_2)_3CH_3$

(b)
Br
|
$CH_3CHCH_3 \xrightarrow[\text{ether}]{Mg}$
MgBr
|
CH_3CHCH_3 + $CH_3CH=CHCH_2Br \longrightarrow CH_3CH=CHCH_2—CHCH_3$
                                                                    |
                                                                    CH₃

324

15-13 continued

(c) $CH_3CH_2CH_2Br \xrightarrow[\text{ether}]{Mg} CH_3CH_2CH_2MgBr$

This synthesis could also be performed sequentially.

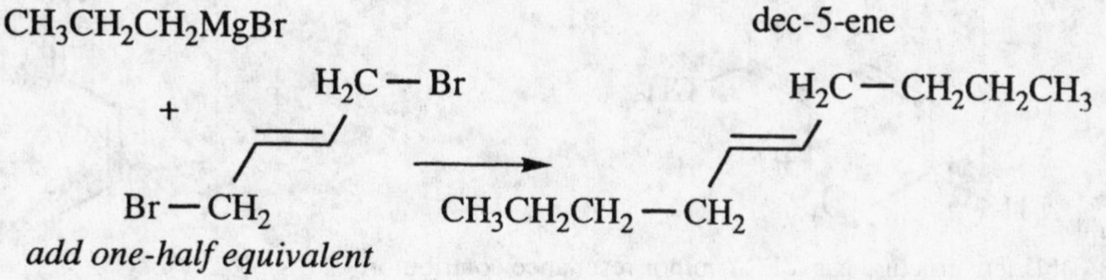

dec-5-ene
+
H₂C — Br
Br — CH₂
H₂C — CH₂CH₂CH₃
CH₃CH₂CH₂ — CH₂
add one-half equivalent

15-14

(a)

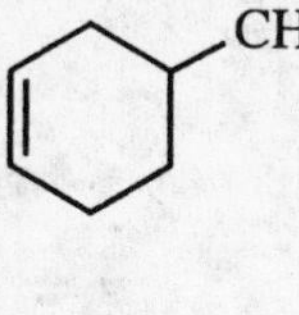

CHO

(b)

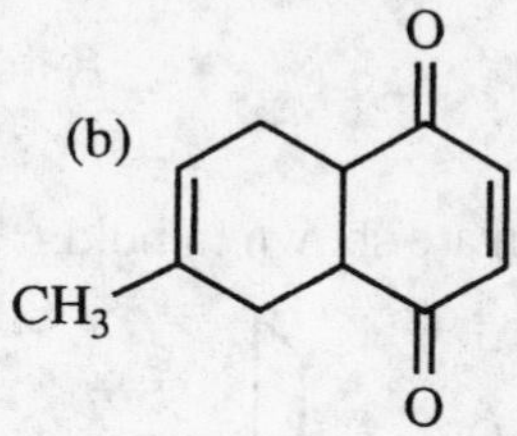

O
CH₃
O

(c)

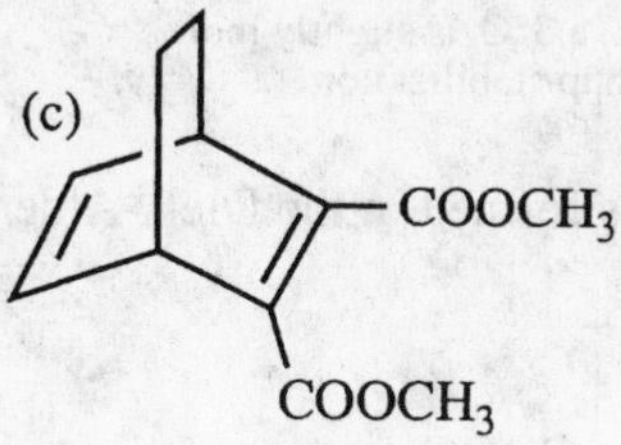

COOCH₃
COOCH₃

(d)

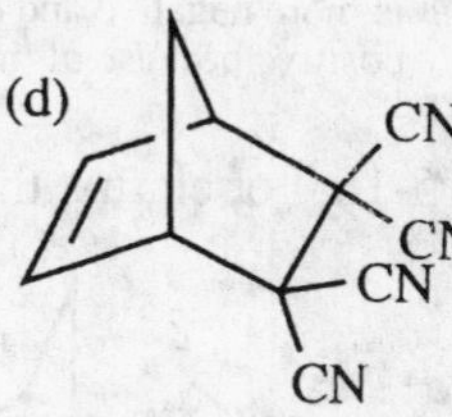

CN
CN
CN
CN

(e)

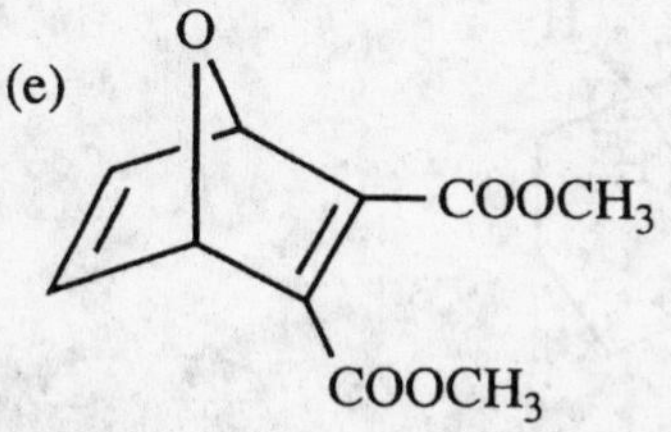

O
COOCH₃
COOCH₃

(f)

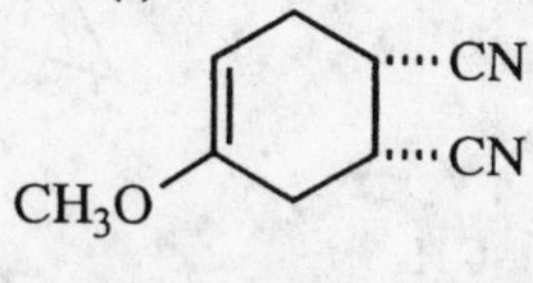

CN
CN
CH₃O

15-15

(a)

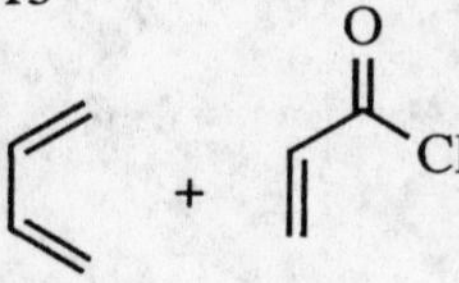

+
O
CH₃

(b)

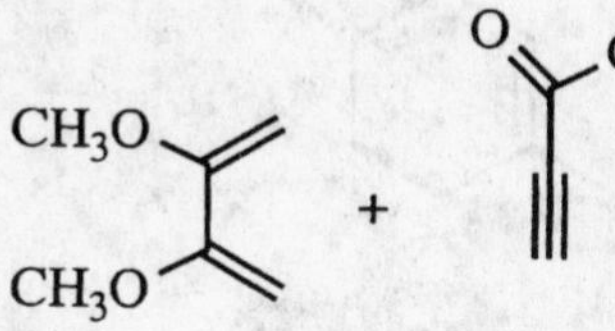

CH₃O
CH₃O
+
O
OEt

(c)

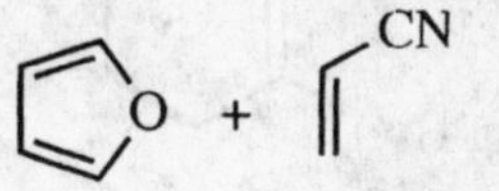

O
+
CN

(d)

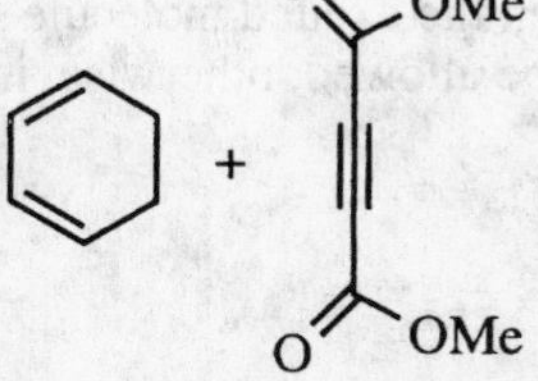

+
O
OMe
O
OMe

(e)

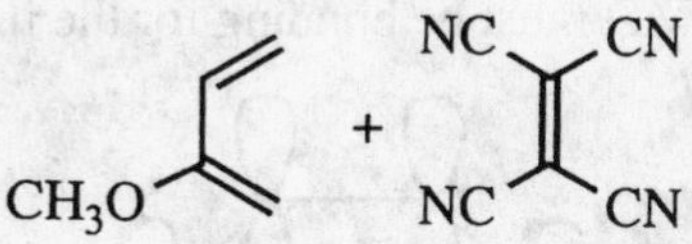

CH₃O
+
NC
CN
NC
CN

(f)

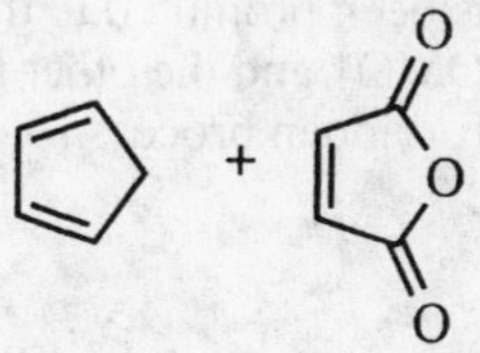

+
O
O
O

15-16

(a)

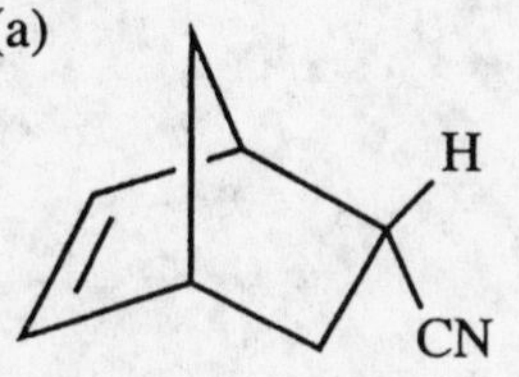

H
CN

(b)

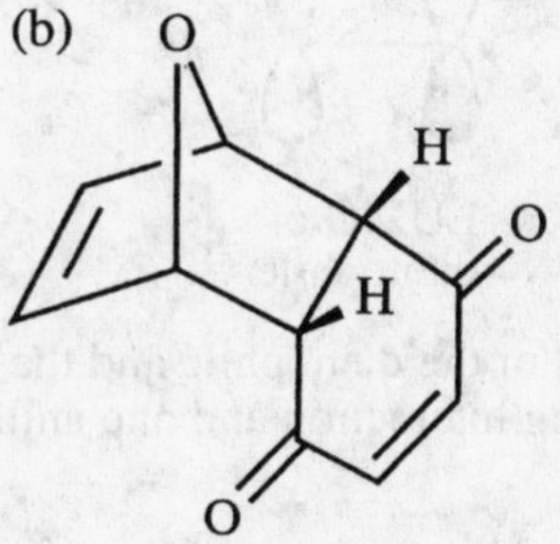

O
H
O
H
O

(c)

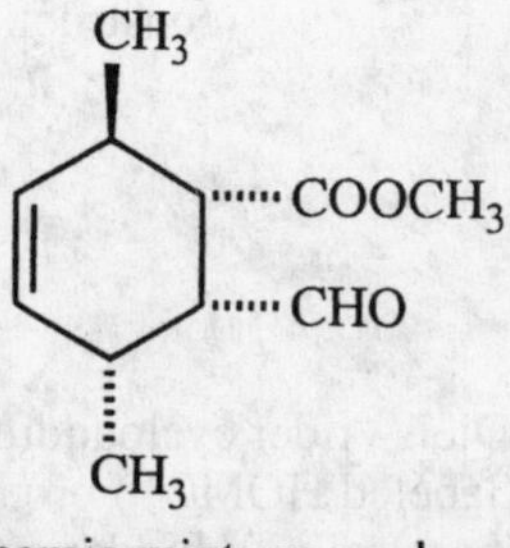

CH₃
COOCH₃
CHO
CH₃

racemic mixture; wedge and dashed bonds show relative stereochemistry

15-17 These structures show the alignment of diene and dienophile in the Diels-Alder transition state, leading to 1,4-orientation in (a) and 1,2-orientation in (b).

(a)

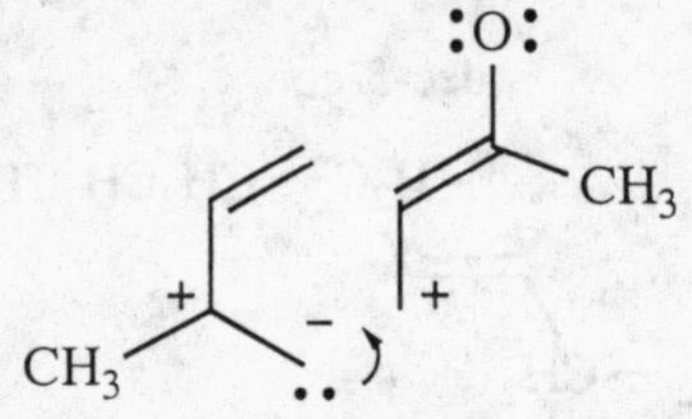

this left structure is a VERY minor resonance contributor; however, it explains the orientation for the diene as carbon-1 is more negative and carbon-2, a 3°C, is slightly more positive because of methyl group stabilization

(b)

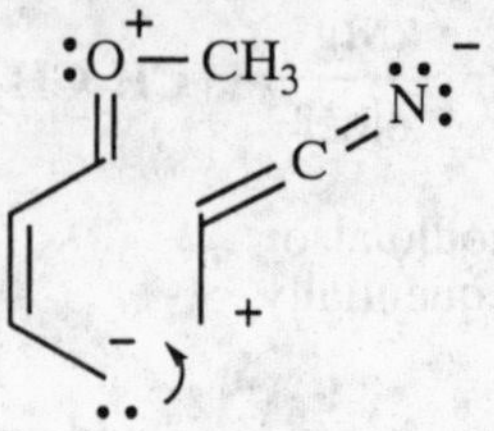

15-18 For clarity, the bonds formed in the Diels-Alder reaction are shown in bold.

(a)

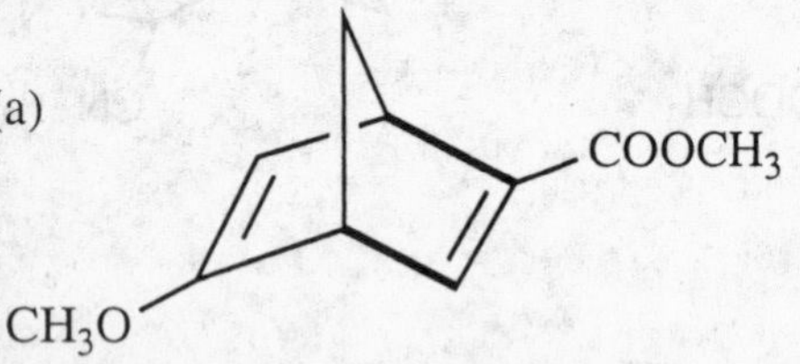

(b)

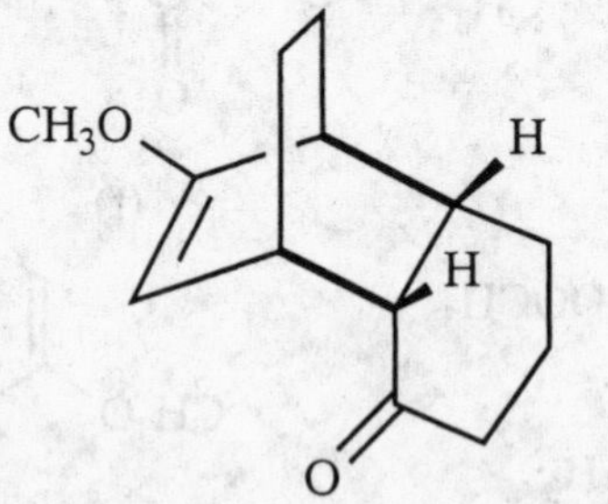

(c)

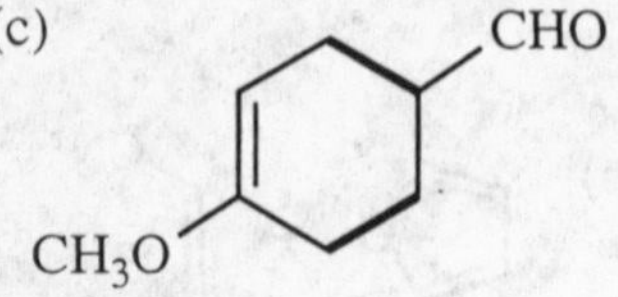

(d)

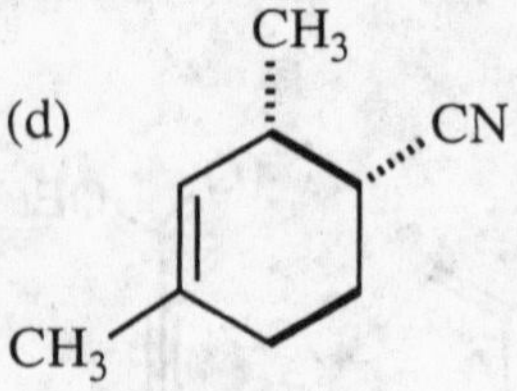

15-19 For a photochemically *allowed* process, one molecule must use an excited state in which an electron has been promoted to the first antibonding orbital. All orbital interactions between the excited molecule's HOMO* and the other molecule's LUMO must be bonding for the interaction to be allowed; otherwise, it is a forbidden process.

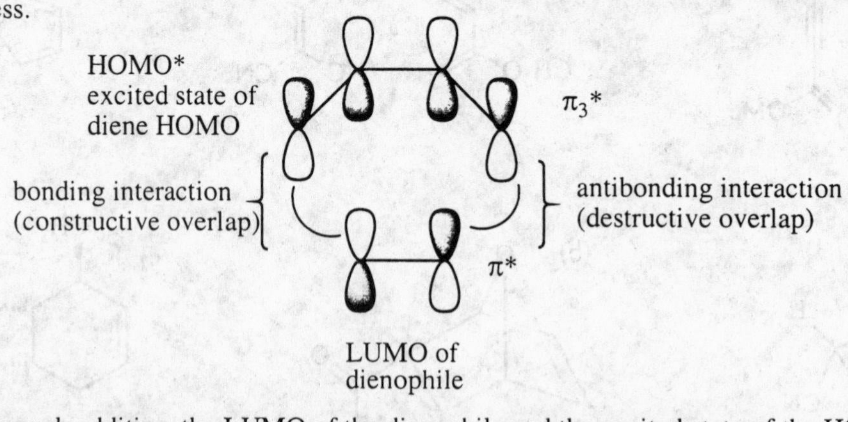

HOMO*
excited state of
diene HOMO

π_3*

bonding interaction
(constructive overlap)

antibonding interaction
(destructive overlap)

π*

LUMO of
dienophile

In the Diels-Alder cycloaddition, the LUMO of the dienophile and the excited state of the HOMO of the diene (labeled HOMO*) produce one bonding interaction and one antibonding interaction. Thus, this is a photochemically forbidden process.

15-20 For a [4 + 4] cycloaddition:

(a)

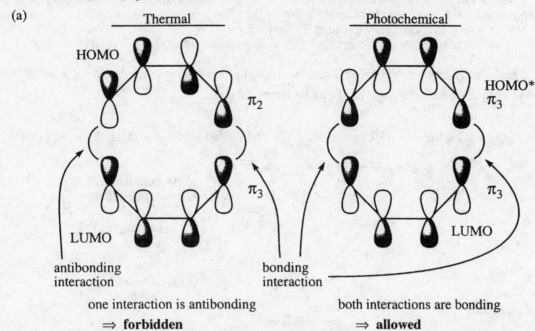

Thermal

HOMO
π_2

LUMO
π_3

antibonding
interaction

bonding
interaction

one interaction is antibonding
⇒ **forbidden**

Photochemical

HOMO*
π_3

π_3

LUMO

both interactions are bonding
⇒ **allowed**

(b) A [4 + 4] cycloaddition is not thermally allowed, but a [4 + 2] (Diels-Alder) is!

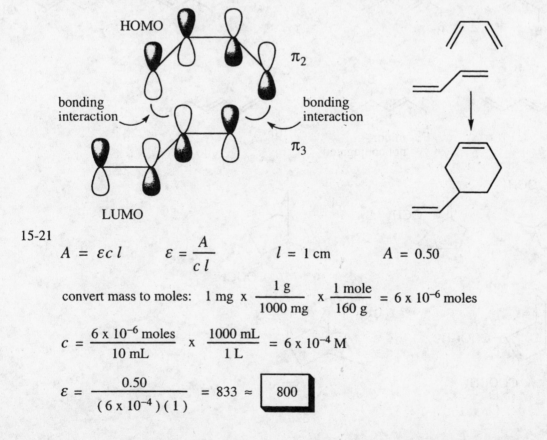

HOMO
π_2

bonding
interaction

bonding
interaction

π_3

LUMO

15-21

$$A = \varepsilon c\, l \qquad \varepsilon = \frac{A}{c\, l} \qquad l = 1 \text{ cm} \qquad A = 0.50$$

convert mass to moles: $1 \text{ mg} \times \dfrac{1 \text{ g}}{1000 \text{ mg}} \times \dfrac{1 \text{ mole}}{160 \text{ g}} = 6 \times 10^{-6} \text{ moles}$

$$c = \frac{6 \times 10^{-6} \text{ moles}}{10 \text{ mL}} \times \frac{1000 \text{ mL}}{1 \text{ L}} = 6 \times 10^{-4} \text{ M}$$

$$\varepsilon = \frac{0.50}{(6 \times 10^{-4})(1)} = 833 \approx \boxed{800}$$

15-22

(a) 353 nm: a conjugated tetraene—must have highest absorption maximum among these compounds;
(b) 313 nm: closest to the bicyclic conjugated triene in Table 15-2; the diene is in a more substituted ring, so it is not surprising for the maximum to be slightly higher than 304 nm;
(c) 232 nm: similar to 3-methylenecyclohexene in Table 15-2;
(d) 273 nm: like cyclohexa-1,3-diene (256 nm) + 2 alkyl substituents (2 x 5 nm) = predicted value of 266 nm;
(e) 237 nm: like 3-methylenecyclohexene (232 nm) + 1 alkyl group (5 nm) = predicted value of 237 nm

15-23 Please refer to solution 1-20, page 12 of this Solutions Manual.

15-24

(a) isolated   (b) conjugated   (c) cumulated   (d) conjugated   (e) conjugated

(f) cumulated (1,2) and conjugated (2,4)

$$H_2C=C=C\overset{\displaystyle H}{\underset{\displaystyle H}{\,}}\,C\overset{\displaystyle H}{\,}=CH_2$$

15-25

(a)

(b)
one equivalent of HCl

(c)

(d) (with Br substituent) + (with Br substituent)

(e) Br—(with OH) + Br—(with OH)

(f) (Br, Br isomer) + (Br, Br isomer) minor—not conjugated + (Br isomer)

(g) (ring with OCH₃ and methyl) + (ring with methylene and OCH₃)

(h) (bicyclic with COOCH₃)

(i) (bicyclic with two COOCH₃)

(j) (oxabicyclic with two C≡N)

328

15-26 Grignard reactions are performed in ether solvent. For clarity, the bonds formed are shown in bold.

(a)

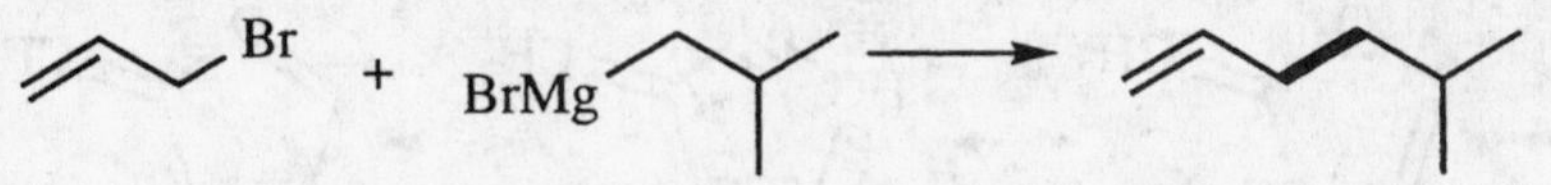

(b)

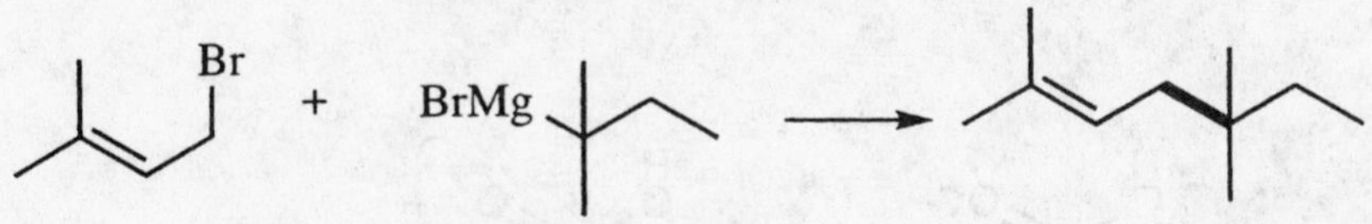

(c)

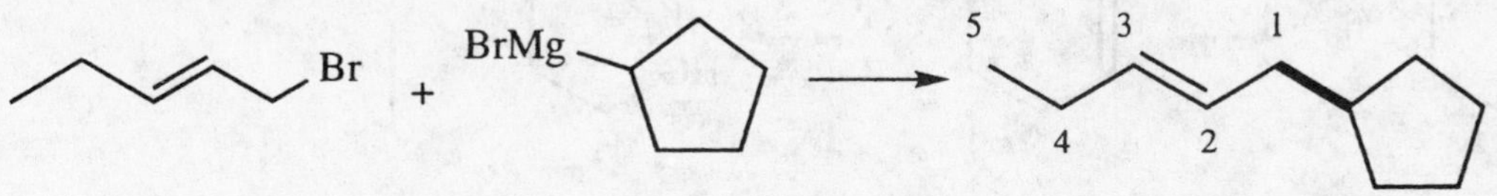

15-27

(a)

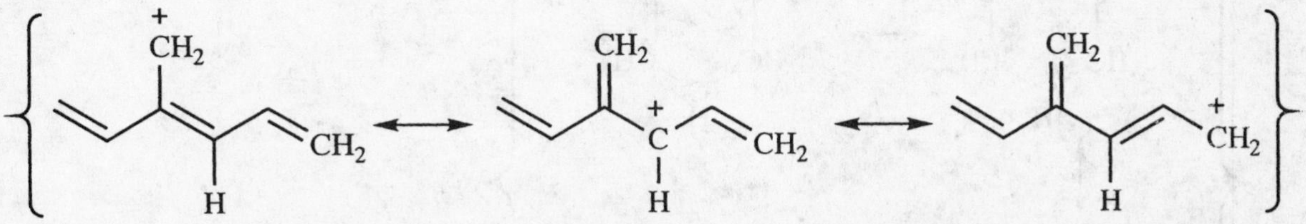

(b)

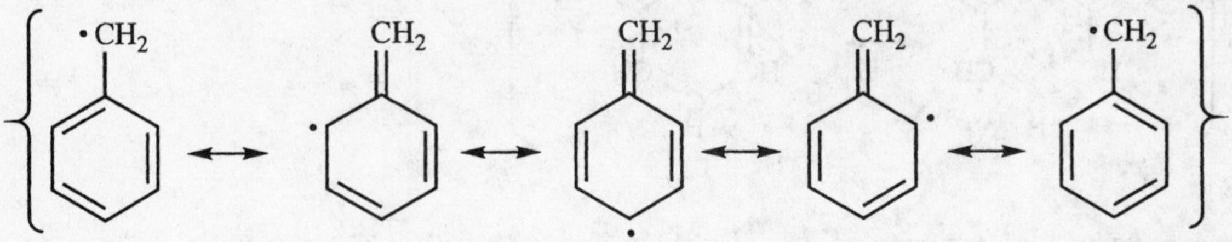

(c)

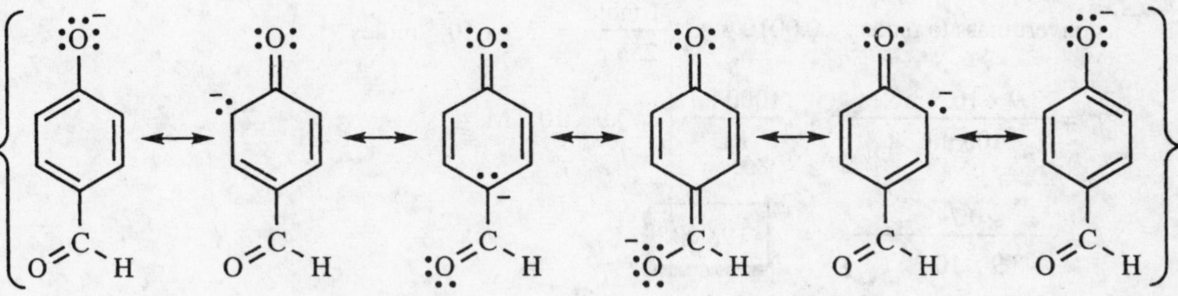

(d)

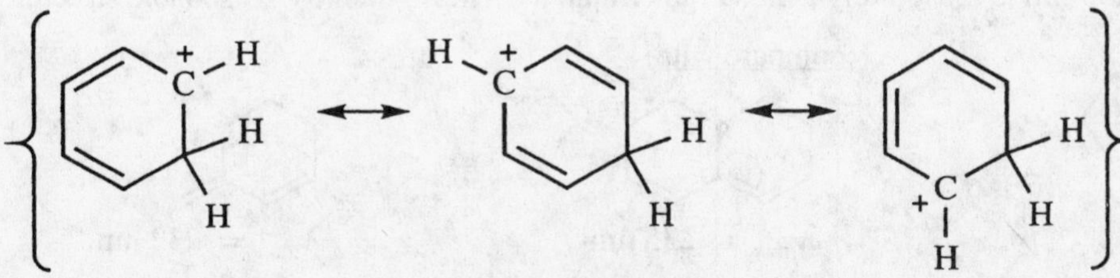

329

15-27 continued

(e)

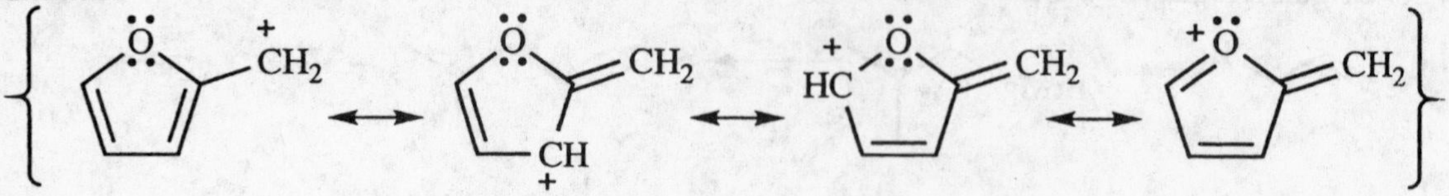

$\left\{ \text{resonance structures} \right\}$

(f)

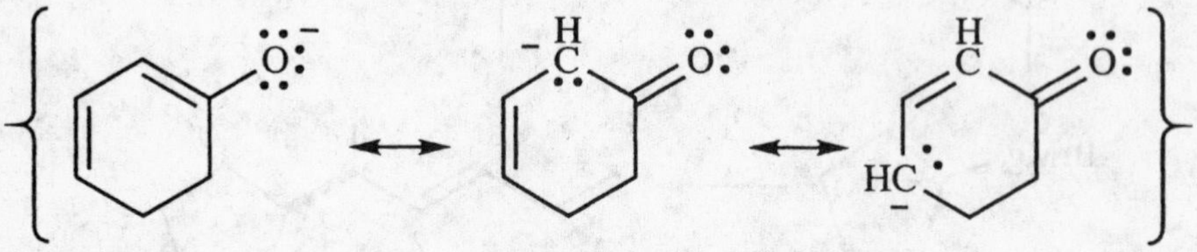

$\left\{ \text{resonance structures} \right\}$

(g)

$\left\{ \text{resonance structures} \right\}$

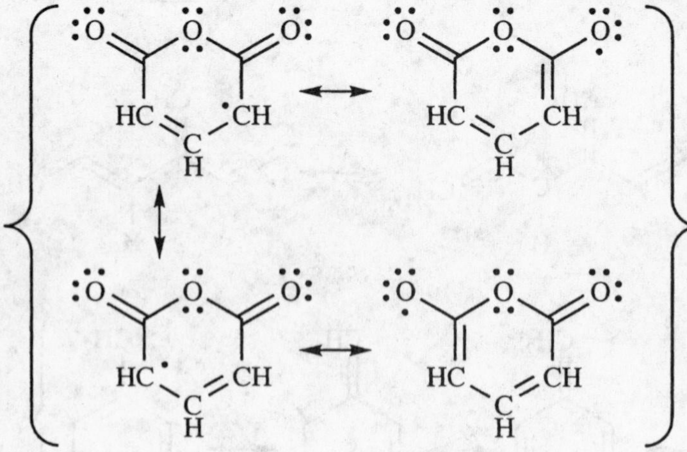

15-28

(a) $A = \varepsilon c \, l$ $\qquad$ $\varepsilon = \dfrac{A}{c \, l}$ $\qquad$ $l = 1 \text{ cm}$ $\qquad$ $A = 0.74$

convert mass to moles: $\quad 0.0010 \text{ g} \times \dfrac{1 \text{ mole}}{255 \text{ g}} = 3.9 \times 10^{-6} \text{ moles}$

$c = \dfrac{3.9 \times 10^{-6} \text{ moles}}{100 \text{ mL}} \times \dfrac{1000 \text{ mL}}{1 \text{ L}} = 3.9 \times 10^{-5} \text{ M}$

$\varepsilon = \dfrac{0.74}{(3.9 \times 10^{-5})(1)} \approx \boxed{19{,}000}$

(b) This large value of ε could only come from a conjugated system, eliminating the first structure. The absorption maximum at 235 nm is most likely a diene rather than a triene. The most reasonable structure is:

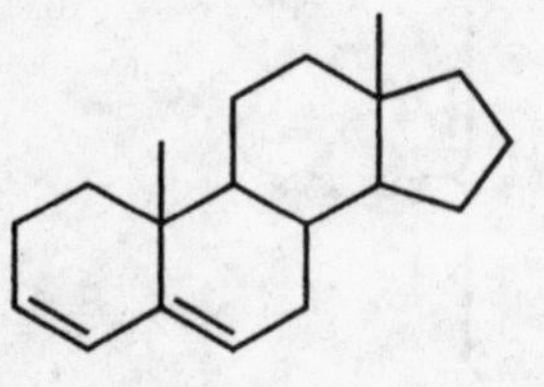

compare with:

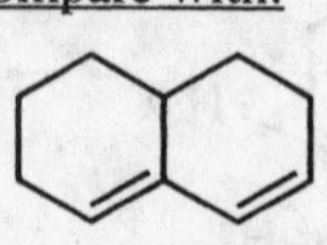

$\lambda_{\text{max}} = 235 \text{ nm}$

Solved Problem 15-3

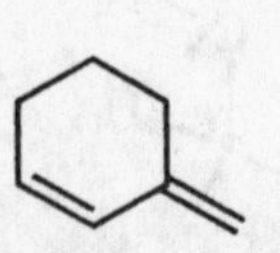

$\lambda_{\text{max}} = 232 \text{ nm}$

Table 15-2

15-29

(a)

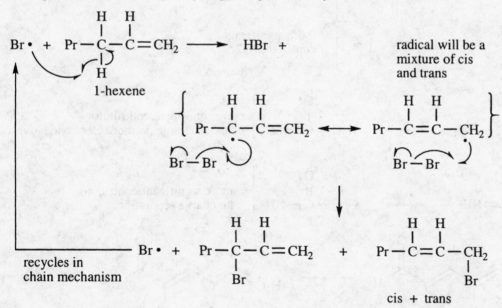

(b)   ("Pr" is the abbreviation for *n*-propyl, used below.)

NBS generates a low concentration of Br$_2$

initiation

propagation

first step is abstraction of allylic hydrogen to generate allylic radical

recycles in chain mechanism

15-30

(a)   (b)   (c)

15-30 continued

(d) [structure: cyclohexene ring with OCH$_3$ and CHO substituents]

(e) [structure: cyclohexene ring with CH$_3$, CN, CN, CH$_3$ substituents]

(f) [structure: cyclohexene ring with CH$_3$, CN, CN, CH$_3$ substituents]

(Note: Yield of the product in (f) would be small or zero due to severe steric interaction in the *s-cis* conformation of the diene. See Figure 15-16.)

15-31

(a)
$$\left\{ \; H_2C = \overset{+}{C} - \text{(with H below)} \quad \longleftrightarrow \quad \overset{+}{H_2C} - = \text{(1°)} \; \right\}$$

more significant
contributor: 2°

(b)
$$\left\{ \text{1° } \overset{+}{CH_2} \text{ ... cyclohexene structure } \longleftrightarrow \text{ } CH_2 \text{ ... } \overset{+}{C} \text{ 2°} \right\}$$

more significant
contributor: 2°

(c)
$$\left\{ \overset{:\ddot{O}:}{\underset{}{H_2\ddot{C}^- - C - CH_3}} \quad \longleftrightarrow \quad \overset{:\ddot{O}:^-}{\underset{}{H_2C = C - CH_3}} \right\}$$
more significant contributor—
negative charge on more electronegative atom

(d)
$$\left\{ \overset{:\ddot{O}:^-}{\underset{}{CH_3 - C = \overset{+}{NH_2}}} \quad \longleftrightarrow \quad \overset{:O:}{\underset{}{CH_3 - C - \ddot{N}H_2}} \right\}$$
more significant contributor—
no charge separation

(e)
$$\left\{ \; \overset{-}{H_2\ddot{C}} \text{...} = \text{...} = \text{...} \overset{O}{\overset{||}{C}} - H \quad \longleftrightarrow \quad H_2C = \text{...} \overset{-}{\ddot{}} \text{...} = \text{...} \overset{O}{\overset{||}{C}} - H \; \right\}$$

$$\left\{ \; H_2C = \text{...} = \text{...} \overset{:\ddot{O}:^-}{=} - H \quad \longleftrightarrow \quad H_2C = \text{...} \overset{-}{\ddot{}} \text{...} \overset{:O:}{=} - H \; \right\}$$

most significant contributor—
negative charge on more electronegative atom

332

15-31 continued

(f)

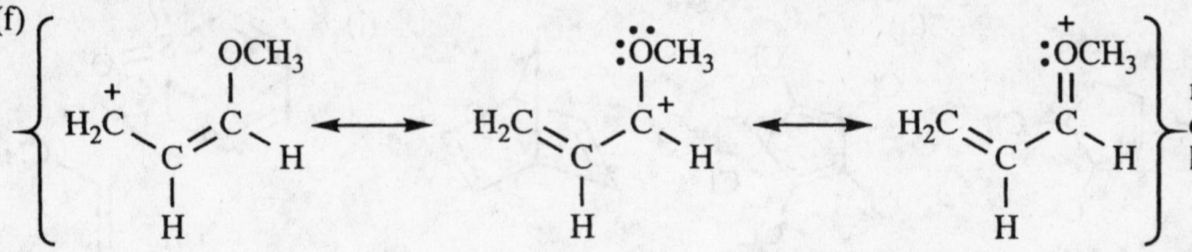

OCH3
H2C
C
C
H
H
:ÖCH3
H2C
C+
C
H
H
:ÖCH3
H2C
C
C
H
H

most significant contributor—all atoms have full octets

(g)

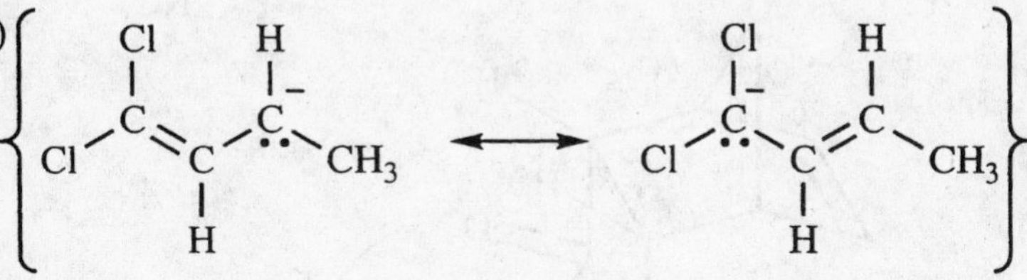

Cl
H
Cl
C
C
C
CH3
H
Cl
H
Cl
C
C
C
CH3
H

more significant contributor—negative charge can be stabilized by electronegative chlorines

15-32

(a) The absorption at 1630 cm^{-1} suggests a conjugated alkene. The higher temperature allowed for migration of the double bond.

(b)

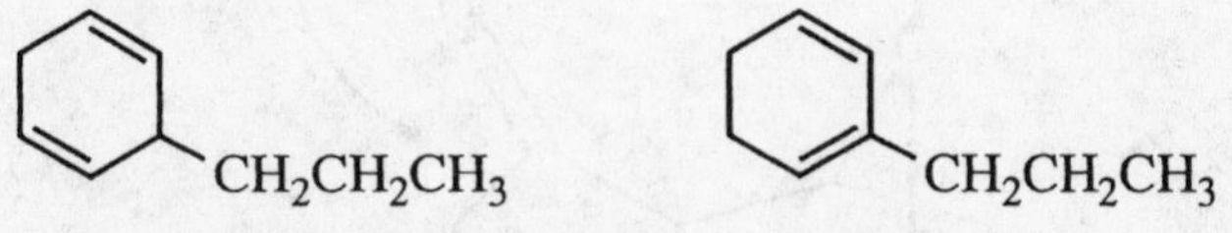

CH2CH2CH3
CH2CH2CH3

desired actual

(c)

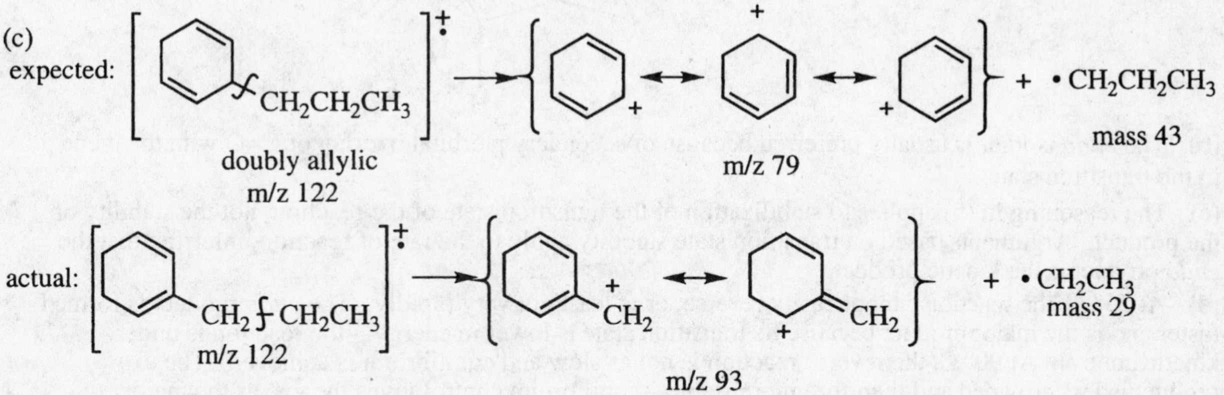

expected:
CH2CH2CH3
doubly allylic
m/z 122
+ • CH2CH2CH3
mass 43
m/z 79
actual:
CH2 CH2CH3
m/z 122
CH2
CH2
+ • CH2CH3
mass 29
m/z 93

15-33

(a)

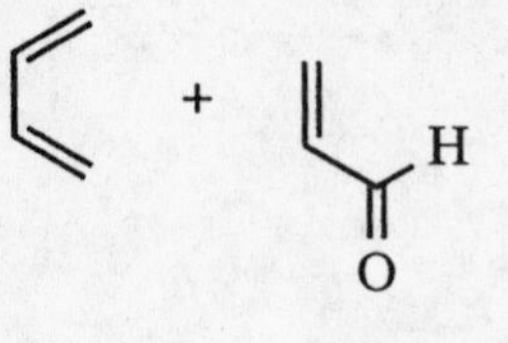

H
O

(b)

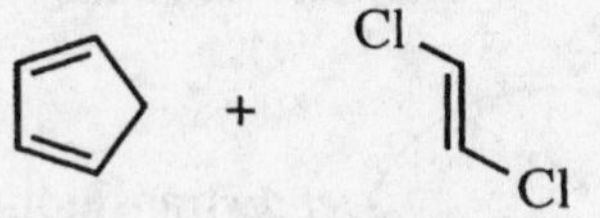

Cl
Cl

(c)

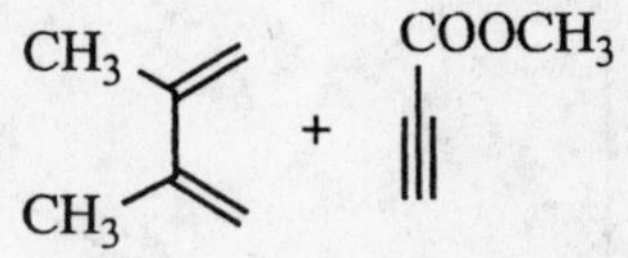

CH3
CH3
COOCH3

(d)

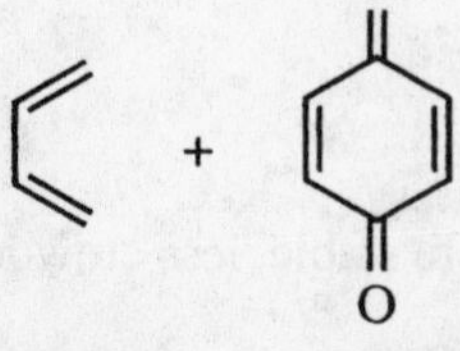

O
O

(e)

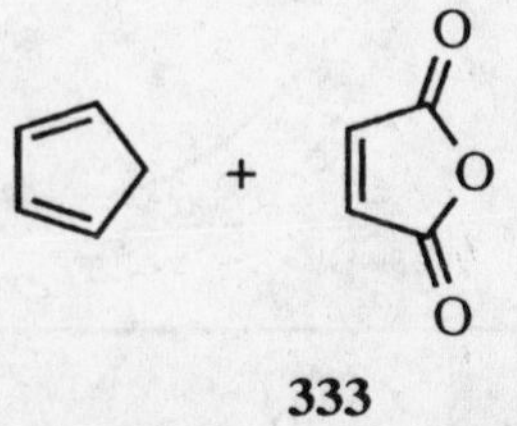

O
O
O

(f)

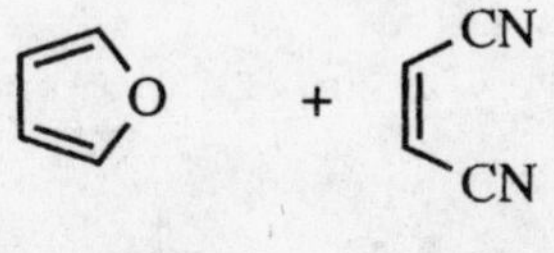

O
CN
CN

15-33 continued

(g)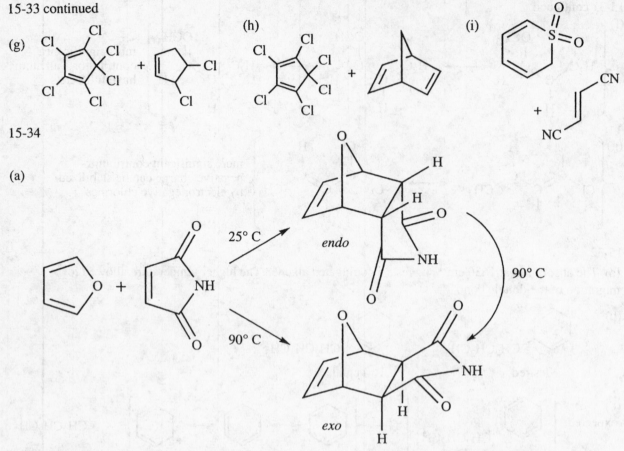

(h)

(i)

15-34

(a)

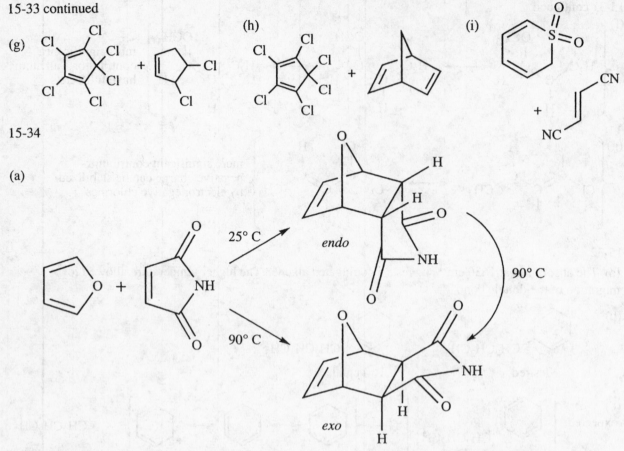

25° C

endo

90° C

90° C

exo

(b) The *endo* isomer is usually preferred because of secondary p orbital overlap of C=O with the diene in the transition state.

(c) The reasoning in (b) applies to stabilization of the transition state of the reaction, not the stability of the product. Arguments based on transition state stability apply to the rate of reaction, inferring that the *endo* product is the kinetic product.

(d) At 25° C, the reaction cannot easily reverse, or at least not very rapidly. The *endo* product is formed faster and is the major product because its transition state is lower in energy—the reaction is under kinetic control. At 90° C, the reverse reaction is not as slow and equilibrium is achieved. The *exo* product is less crowded and therefore more stable—equilibrium control gives the *exo* as the major product.

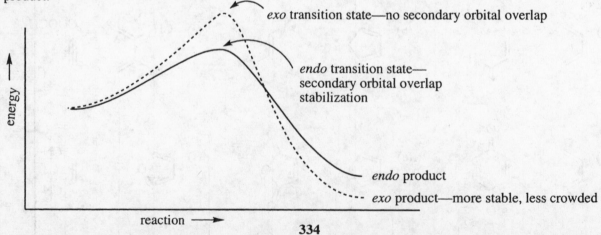

exo transition state—no secondary orbital overlap

endo transition state—
secondary orbital overlap
stabilization

energy

endo product

exo product—more stable, less crowded

reaction ⟶

334

15-35 Nodes are
represented by dashed lines.

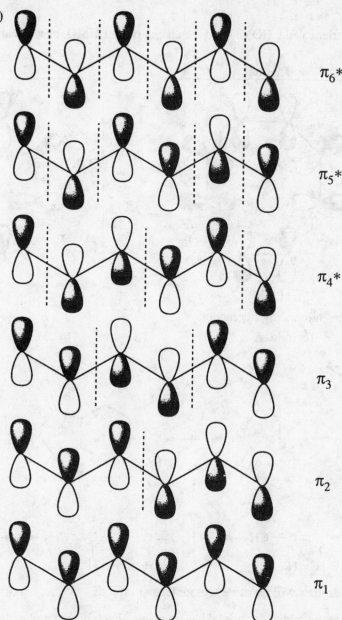

15-35 continued

(d) Whether the triene is the HOMO and the alkene is the LUMO, or *vice versa*, the answer will be the same.

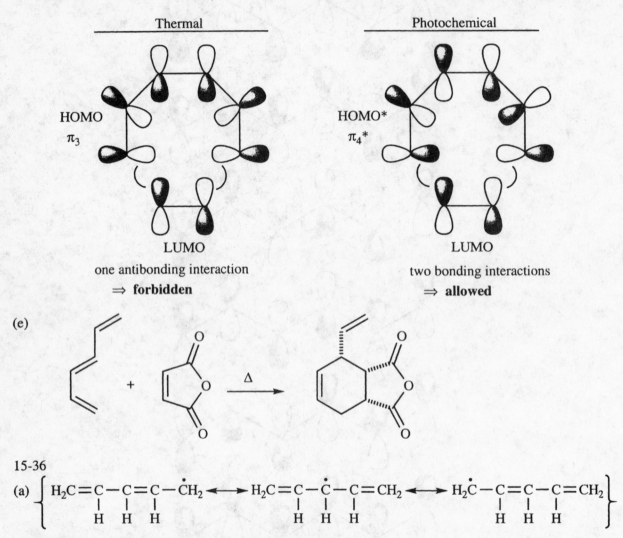

Thermal

HOMO
π_3

LUMO

one antibonding interaction
⟹ **forbidden**

Photochemical

HOMO*
π_4*

LUMO

two bonding interactions
⟹ **allowed**

(e)

15-36

(a) $\left\{ H_2C=C-C=C-\overset{\bullet}{C}H_2 \longleftrightarrow H_2C=C-\overset{\bullet}{C}-C=CH_2 \longleftrightarrow H_2\overset{\bullet}{C}-C=C-C=CH_2 \right\}$
with H, H, H below each.

(b) Five atomic orbitals will generate five molecular orbitals.

(c) The lowest energy molecular orbital has no nodes. Each higher molecular orbital will have one more node, so the fifth molecular orbital will have four nodes.

15-36 continued

(d) Nodes are represented by dashed lines.

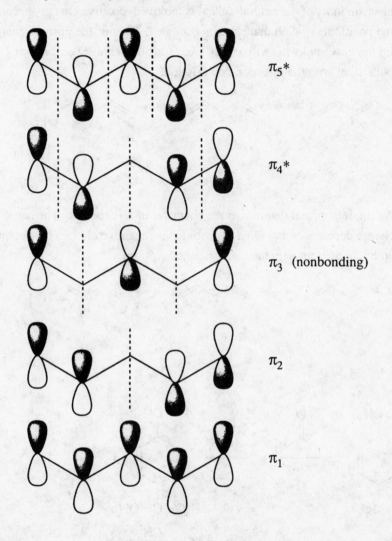

$\pi_5{}^*$

$\pi_4{}^*$

π_3 (nonbonding)

π_2

π_1

(e)

$\pi_5{}^*$

$\pi_4{}^*$

π_3

π_2

π_1

(f) The HOMO, π_3, contains an unpaired electron giving this species its radical character. The HOMO is a non-bonding orbital with lobes only on carbons 1, 3, and 5, consistent with the resonance picture.

15-36 continued

(g)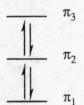

π_5^* ——

π_4^* ——

π_3 ——

π_2 ↑↓

π_1 ↑↓

Again, it is π_3 that determines the character of this species. When the single electron in π_3 of the neutral radical is removed, positive charge appears only in the position(s) which that electron occupied. That is, the positive charge depends on the now *empty* π_3, with *empty* lobes (positive charge) on carbons 1, 3, and 5, consistent with the resonance description.

(h)

π_5^* ——

π_4^* ↑↓

π_3 ↑↓

π_2 ↑↓

π_1 ↑↓

Again, it is π_3 that determines the character of this species. The negative charge depends on the *filled* π_3, with lobes (negative charge) on carbons 1, 3, and 5, consistent with the resonance description.

15-37
(a)

$$CH_3 \xrightarrow[\text{ether}]{Mg} CH_3 (MgBr) \xrightarrow{\text{(acetone)}} \xrightarrow{H_3O^+} CH_3 \quad A$$

$$H_3C-\underset{\underset{CH_3}{|}}{\overset{}{C}}-OH$$

(b) Use Appendix 3 to predict λ_{max} values. Alkyl substituents are circled.

transoid cyclic diene = 217 nm
4 alkyl groups = 20 nm
exocyclic alkene = 5 nm
TOTAL = 242 nm

desired product

cisoid cyclic diene = 253 nm
3 alkyl groups = 15 nm
TOTAL = 268 nm

cisoid cyclic diene = 253 nm
4 alkyl groups = 20 nm
TOTAL = 273 nm—AHA!

actual product **B**

338

15-37 continued

(c)

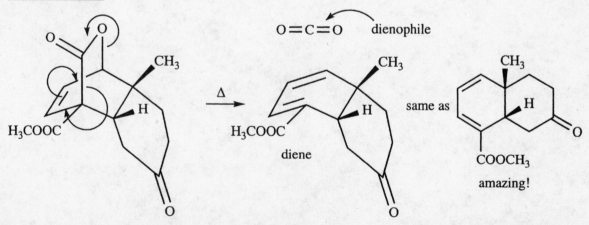

B is produced in preference to the other cisoid diene above because **B**'s diene system is more highly substituted and therefore more stable

15-38 It is stunningly clever reactions like this that earned E. J. Corey his Nobel Prize.

<u>Diels-Alder</u>

imagine the diene on top of the dienophile in the transition state, leading to the *endo* adduct

endo product

<u>retro-Diels-Alder</u>

O=C=O   dienophile

diene

same as

amazing!

339

Note: The representation of benzene with a circle to represent the π system is fine for questions of nomenclature, properties, isomers, and reactions. For questions of mechanism or reactivity, however, the representation with three alternating double bonds (the Kekulé picture) is more informative. For clarity and consistency, this Solutions Manual will use the Kekulé form exclusively.

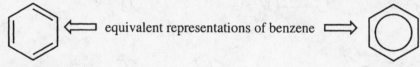

Kekulé form used in
the Solutions Manual

16-1

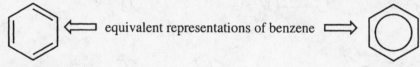

16-2 All values are per mole.

(a)

	benzene	−208 kJ	− 49.8 kcal
−	1,4-cyclohexadiene	−240 kJ	− 57.4 kcal
	ΔH = + 32 kJ	+ 7.6 kcal

(b)

	benzene	−208 kJ	− 49.8 kcal
−	cyclohexene	−120 kJ	− 28.6 kcal
	ΔH = − 88 kJ	− 21.2 kcal

(c)

	1,3-cyclohexadiene	−232 kJ	− 55.4 kcal
−	cyclohexene	−120 kJ	− 28.6 kcal
	ΔH = − 112 kJ	−26.8 kcal

16-3

(a)

16-3 continued

(b)

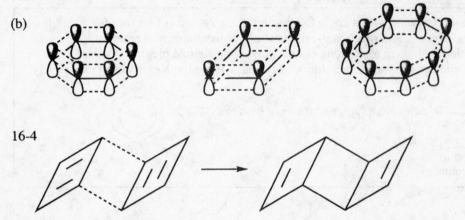

16-4

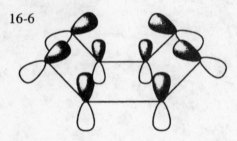

16-5 Figure 16-8 shows that the first 3 pairs of electrons are in three bonding molecular orbitals of cyclooctatetraene. Electrons 7 and 8, however, are located in two different nonbonding orbitals. As in cyclobutadiene, a planar cyclooctatetraene is predicted to be a diradical, a particularly unstable electron configuration.

16-6

Models show that the angles between p orbitals on adjacent π bonds approach 90°.

16-7
(a) nonaromatic: internal hydrogens prevent planarity
(b) nonaromatic: not all atoms in the ring have a p orbital
(c) aromatic: [14]annulene
(d) aromatic: also a [14]annulene in the outer ring: the internal alkene is not part of the aromatic system

16-8 Azulene satisfies all the criteria for aromaticity, and it has a Huckel number of π electrons: 10. Both heptalene (12 π electrons) and pentalene (8 π electrons) are antiaromatic.

16-9

(a)

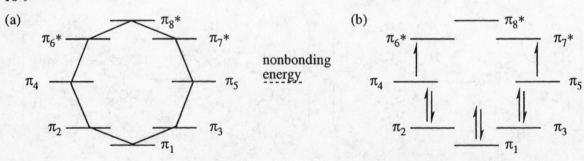

nonbonding energy

(b)

This electronic configuration is antiaromatic.

16-9 continued

(c)

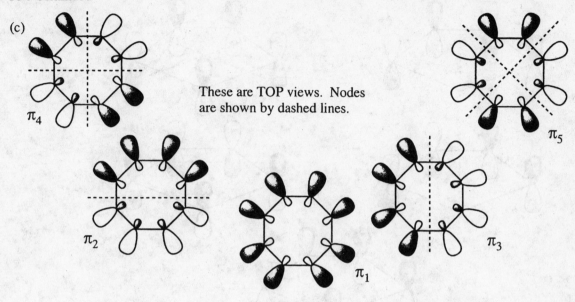

These are TOP views. Nodes are shown by dashed lines.

16-10

(a)

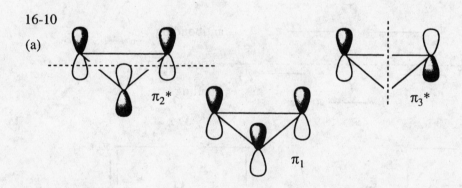

(b)

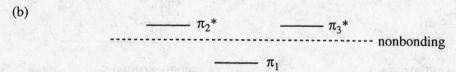

π_1 is bonding; π_2* and π_3* are antibonding.

(c)

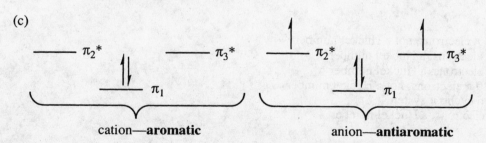

16-11

(a)

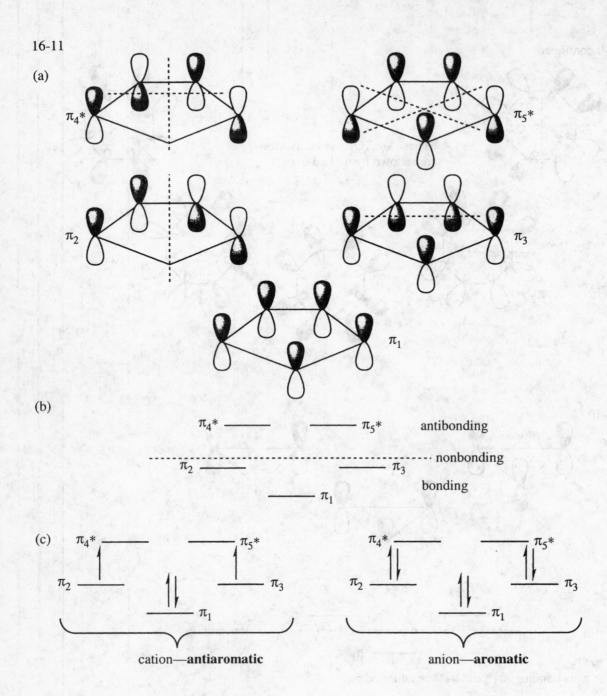

(b)

π_4* ——————— ——————— π_5* antibonding

—   nonbonding

π_2 ——————   ——————— π_3

bonding

——————— π_1

(c)

π_4* ———   ——— π_5*

π_2 ———   ——— π_3

——— π_1

cation—**antiaromatic**

π_4* ———   ——— π_5*

π_2 ———   ——— π_3

——— π_1

anion—**aromatic**

16-12

(a) antiaromatic: 8 π electrons, not a Huckel number
(b) aromatic: 10 π electrons, a Huckel number
(c) aromatic: 18 π electrons, a Huckel number
(d) antiaromatic: 20 π electrons, not a Huckel number
(e) nonaromatic: no cyclic π system
(f) aromatic: 18 π electrons, a Huckel number

16-13 The reason for the dipole can be seen in a resonance form distributing the electrons to give each ring 6 π electrons. This resonance picture gives one ring a negative charge and the other ring a positive charge.

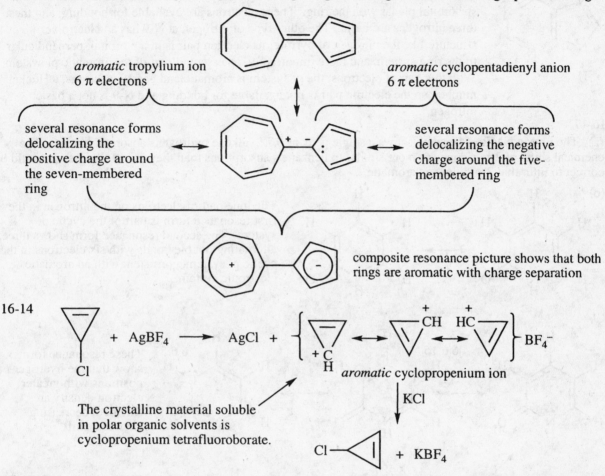

aromatic tropylium ion
6 π electrons

aromatic cyclopentadienyl anion
6 π electrons

several resonance forms delocalizing the positive charge around the seven-membered ring

several resonance forms delocalizing the negative charge around the five-membered ring

composite resonance picture shows that both rings are aromatic with charge separation

16-14

$+ \text{AgBF}_4 \longrightarrow \text{AgCl} +$ ... *aromatic* cyclopropenium ion BF_4^-

The crystalline material soluble in polar organic solvents is cyclopropenium tetrafluoroborate.

KCl

Cl— ... $+ \text{KBF}_4$

16-15 Draw resonance forms showing the carbonyl polarization, leaving a positive charge on the carbonyl carbon.

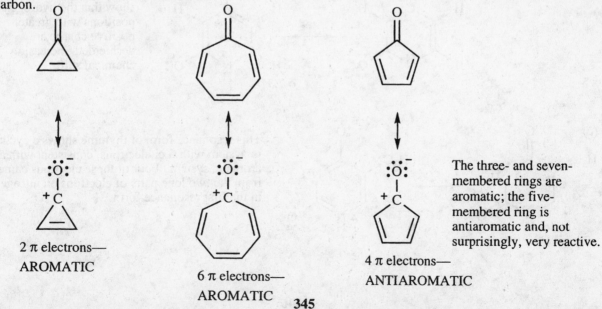

2 π electrons—
AROMATIC

6 π electrons—
AROMATIC

4 π electrons—
ANTIAROMATIC

The three- and seven-membered rings are aromatic; the five-membered ring is antiaromatic and, not surprisingly, very reactive.

16-16

The structure of purine shows two types of nitrogens. One type (N-1, N-3, and N-7) has an electronic structure like the nitrogen in pyridine; the pair of electrons is in an sp² orbital planar with the ring. These electrons are available for bonding, and these three nitrogens are basic. The other type of nitrogen at N-9 has an electronic structure like the nitrogen of pyrrole; its electron pair is in a p orbital, perpendicular to the ring system, and more importantly, an essential part of the aromatic pi system. With this pair of electrons, the pi system is aromatic and has 10 electrons, a Huckel number, so the electron pair is not available for bonding and N-9 is not a basic nitrogen.

16-17

(a)   The proton NMR of benzene shows a single peak at δ 7.2; alkene hydrogens absorb at δ 4.5-6. The chemical shifts of 2-pyridone are more similar to benzene's absorptions than they are to alkenes. It would be correct to infer that 2-pyridone is aromatic.

(b)

The lone pair of electrons on the nitrogen in the first resonance form is part of the cyclic pi system. The second resonance form shows three alternating double bonds with six electrons in the cyclic pi system, consistent with an aromatic electronic system.

δ 6.15   δ 6.57 These resonance forms show that the hydrogens at positions with greater electron density are shielded, decreasing chemical shift.

H δ 7.26

δ 7.31 These resonance forms show that the hydrogens at positions with greater positive charge are deshielded, increasing chemical shift.

(c)

This resonance form of thymine shows a cyclic pi system with 6 pi electrons, consistent with an aromatic system. Four of these electrons came from the two lone pairs of electrons on nitrogens in the first resonance form.

16-18

(a)

aromatic: 4 π
electrons in
double bonds
plus one pair
from oxygen

(b)

aromatic: 4 π
electrons in
double bonds
plus one pair
from sulfur

(c)

not aromatic:
no cyclic π
system
because of sp³
carbon

(d)

aromatic: cation on carbon-
4 indicates an empty p
orbital; two π bonds plus a
pair of electrons from
oxygen makes a 6 π
electron system

(e)

aromatic: resonance form
shows "push-pull" of
electrons from one O to
the other, making a cyclic
π system with 6 electrons

(f)

not aromatic:
no cyclic π
system
because of sp³
carbon

(g)

aromatic: resonance form shows
electron pair from N making a cyclic
π system with 6 electrons

16-19

Borazole is a non-carbon equivalent of
benzene. Each boron is hybridized in its
normal sp². Each nitrogen is also sp² with its
pair of electrons in its p orbital. The system
has six π electrons in 6 p orbitals—aromatic!

16-20
(a)

anthracene

phenanthrene

347

16-20 continued

(b) anthracene

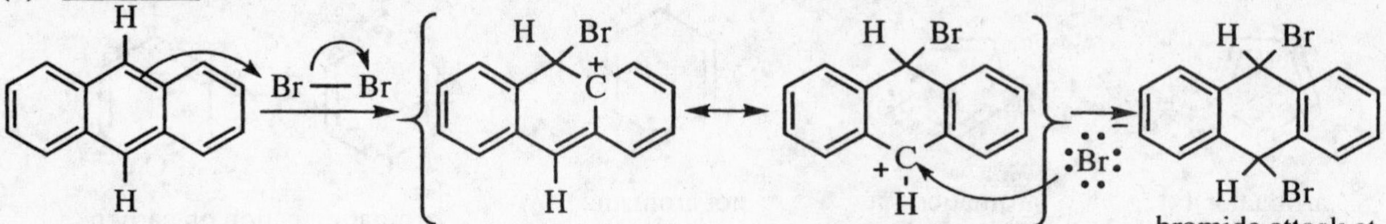

plus many other
resonance forms

bromide attack at
C-10 leaves
two aromatic rings

phenanthrene

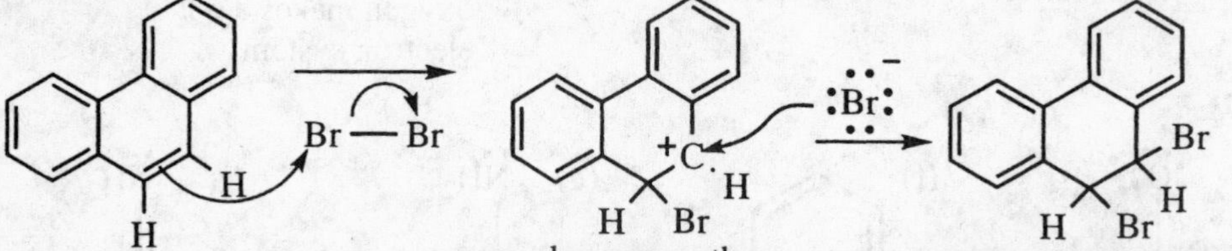

plus many other
resonance forms

bromide attack at
C-9 leaves two
aromatic rings

cis and *trans*

(c) A typical addition of bromine occurs with a bromonium ion intermediate which can give only anti addition. Addition of bromine to phenanthrene, however, generates a free carbocation because the carbocation is benzylic, stabilized by resonance over two rings. In the second step of the mechanism, bromide nucleophile can attack either side of the carbocation giving a mixture of *cis* and *trans* products.

(d)

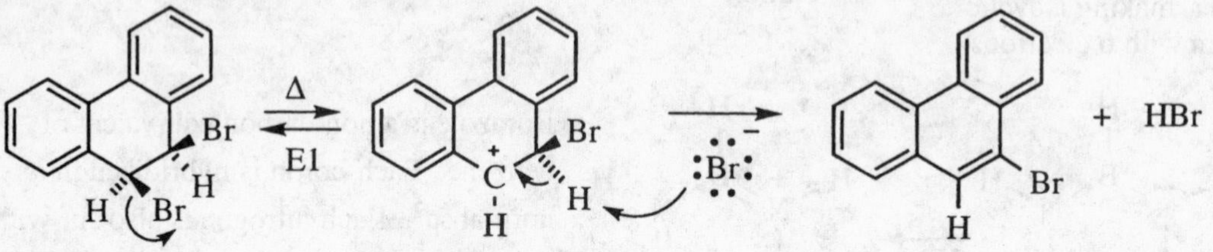

+ HBr

resonance-stabilized

16-21

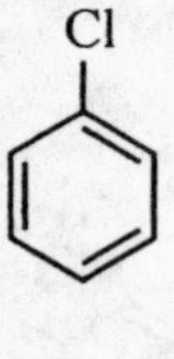

chlorobenzene

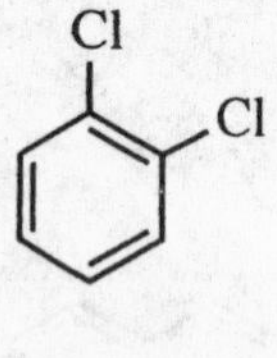

o-dichlorobenzene
(1,2-dichlorobenzene)

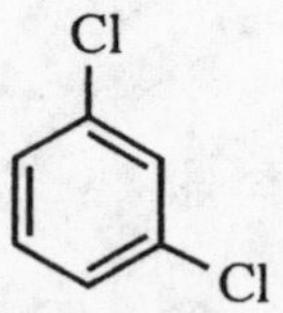

m-dichlorobenzene
(1,3-dichlorobenzene)

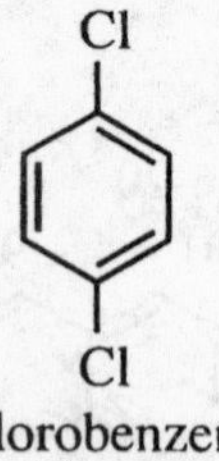

p-dichlorobenzene
(1,4-dichlorobenzene)

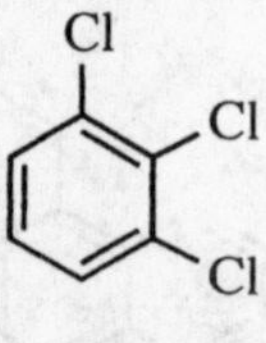

1,2,3-trichlorobenzene

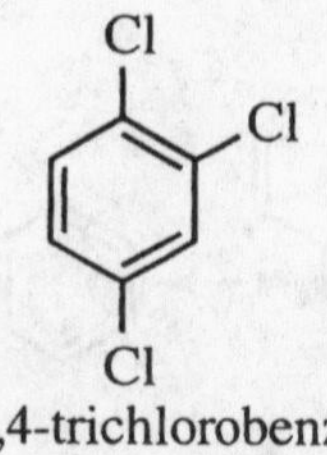

1,2,4-trichlorobenzene

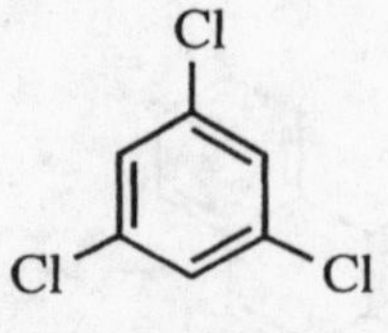

1,3,5-trichlorobenzene

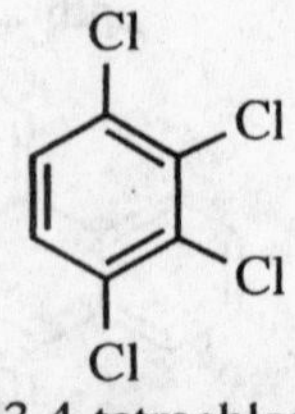

1,2,3,4-tetrachloro-
benzene

16-21 continued

Cl Cl
Cl Cl
Cl Cl

1,2,3,5-tetrachloro-benzene

Cl Cl
Cl
Cl

1,2,4,5-tetrachloro-benzene

Cl Cl
Cl
Cl Cl

1,2,3,4,5-pentachloro-benzene

Cl Cl
Cl Cl
Cl Cl

1,2,3,4,5,6-hexachloro-benzene

16-22

(a) fluorobenzene
(b) 4-phenylbut-1-yne
(c) *m*-methylphenol, or 3-methylphenol (common name: *m*-cresol)

(d) *o*-nitrostyrene
(e) *p*-bromobenzoic acid, or 4-bromobenzoic acid
(f) isopropoxybenzene, or isopropyl phenyl ether

(g) 3,4-dinitrophenol
(h) benzyl ethyl ether, or benzoxyethane, or (ethoxymethyl)benzene, or α-ethoxytoluene

16-23 These examples are representative. Your examples may be different from these and still be correct.

(a) ⌬—O—CH₃

methyl phenyl ether or methoxybenzene or anisole

(b) CH₃—⌬—SO₃H

p-toluenesulfonic acid

(c) ⌬—Li

phenyllithium

(d) O₂N—⌬—OH

These are *phenols*. This is 4-nitrophenol.

(e)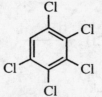

diphenylmethanol

(f)

1,3-dibromo-2-phenylbenzene

(g) CH₃CH₂O—⌬—CH₂OH

3-ethoxybenzyl alcohol or 3-ethoxyphenylmethanol

16-24

⁺CH₂ ↔ CH₂ ↔ CH₂ ↔ CH₂ ↔ ⁺CH₂

349

16-25

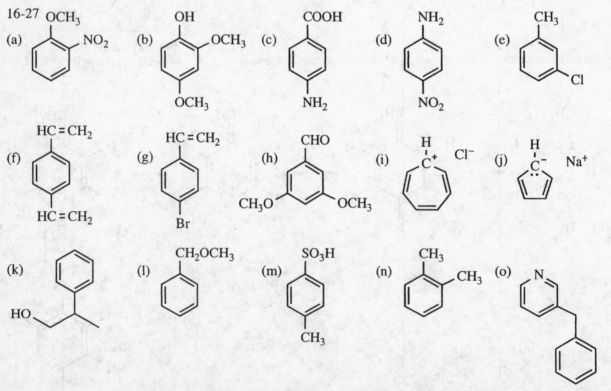

λ_{max} 220 nm (strong)
258 nm (weak)

− H₂O

plus resonance forms with positive
charge on the benzene ring

λ_{max} 250 nm (strong)
290 nm (weak)

electronic systems with extended
conjugation absorb at longer wavelength

16-26 Please refer to solution 1-20, page 12 of this Solutions Manual.

16-27

(a) OCH₃, NO₂

(b) OH, OCH₃, OCH₃

(c) COOH, NH₂

(d) NH₂, NO₂

(e) CH₃, Cl

(f) HC=CH₂, HC=CH₂

(g) HC=CH₂, Br

(h) CHO, CH₃O, OCH₃

(i) H, C⁺, Cl⁻

(j) H, C⁻, Na⁺

(k) HO

(l) CH₂OCH₃

(m) SO₃H, CH₃

(n) CH₃, CH₃

(o) N

16-28

(a) 1,2-dichlorobenzene (*ortho*)
(b) 4-nitroanisole (*para*)
(c) 2,3-dibromobenzoic acid
(d) 2,7-dimethoxynaphthalene

(e) 3-chlorobenzoic acid (*meta*)
(f) 2,4,6-trichlorophenol
(g) 2-*sec*-butylbenzaldehyde (*ortho*)
(h) cyclopropenium tetrafluoroborate

16-29

toluene

o-xylene

m-xylene

p-xylene

1,2,3-trimethylbenzene

1,2,4-trimethylbenzene

1,3,5-trimethylbenzene
(common name: mesitylene)

16-30 Aromaticity is one of the strongest stabilizing forces in organic molecules. The cyclopentadienyl system is stabilized in the anion form where it has 6 π electrons, a Huckel number. The question then becomes: which of the four structures can lose a proton to become aromatic?

While the first, third, and fourth structures can lose protons from sp^3 carbons to give resonance-stabilized anions, only the second structure can make a cyclopentadienide anion. It will lose a proton most easily of these four structures which, by definition, means it is the strongest acid.

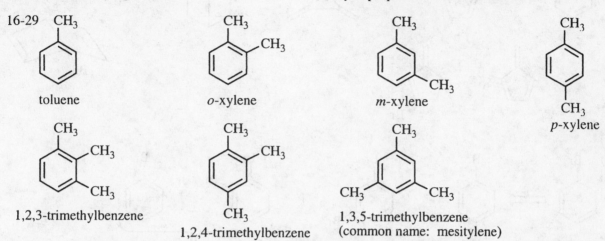

AROMATIC

16-31

(a)

16-31 continued

(b)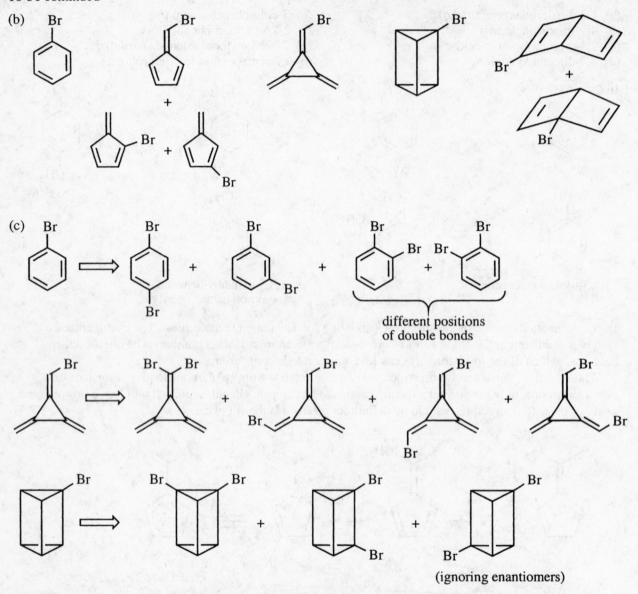

(c)

different positions
of double bonds

(ignoring enantiomers)

(d) The only structure consistent with three isomers of dibromobenzene is the prism structure, called Ladenburg benzene. It also gives no test for alkenes, consistent with the behavior of benzene. (Kekulé defended his structure by claiming that the "two" structures of *ortho*-dibromobenzene were rapidly interconverted, equilibrating so quickly that they could never be separated.)

(e) We now know that three- and four-membered rings are the least stable, but this fact was unknown to chemists during the mid-1800s when the benzene controversy was raging. Ladenburg benzene has two three-membered rings and three four-membered rings (of which only four of the rings are independent), which we would predict to be unstable. (In fact, the structure has been synthesized. Called *prismane*, it is NOT aromatic, but rather, is very reactive toward addition reactions.)

16-32

(a)

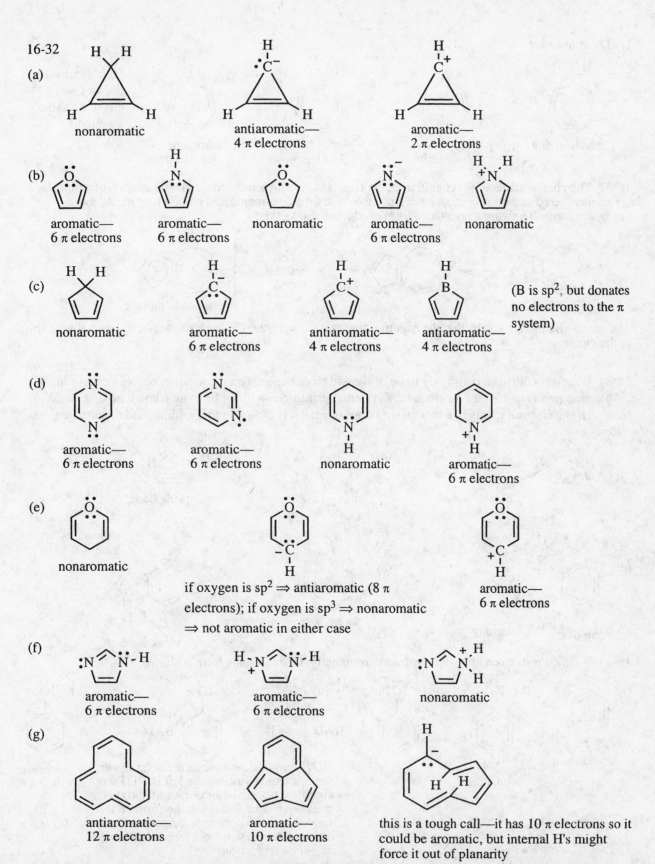

nonaromatic

antiaromatic—
4 π electrons

aromatic—
2 π electrons

(b)

aromatic—
6 π electrons

aromatic—
6 π electrons

nonaromatic

aromatic—
6 π electrons

nonaromatic

(c)

nonaromatic

aromatic—
6 π electrons

antiaromatic—
4 π electrons

antiaromatic—
4 π electrons

(B is sp², but donates
no electrons to the π
system)

(d)

aromatic—
6 π electrons

aromatic—
6 π electrons

nonaromatic

aromatic—
6 π electrons

(e)

nonaromatic

if oxygen is sp² ⇒ antiaromatic (8 π
electrons); if oxygen is sp³ ⇒ nonaromatic
⇒ not aromatic in either case

aromatic—
6 π electrons

(f)

aromatic—
6 π electrons

aromatic—
6 π electrons

nonaromatic

(g)

antiaromatic—
12 π electrons

aromatic—
10 π electrons

this is a tough call—it has 10 π electrons so it
could be aromatic, but internal H's might
force it out of planarity

353

16-32 continued

(h)

nonaromatic

aromatic—
6 π electrons

antiaromatic—
8 π electrons

aromatic—
6 π electrons

(B is sp², but
donates no
electrons to the
π system)

16-33 The clue to azulene is recognition of the five- and seven-membered rings. To attain aromaticity, a seven-membered carbon ring must have a positive charge; a five-membered carbon ring must have a negative charge. Drawing a resonance form of azulene shows this:

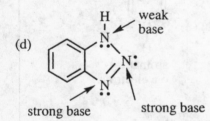

other
resonance
forms

composite picture

The composite picture shows that the negative charge is concentrated in the five-membered ring, giving rise to the dipole.

16-34 Whether a nitrogen is strongly basic or weakly basic depends on the location of its electron pair. If the electron pair is needed for an aromatic π system, the nitrogen will not be basic (shown here as "weak base"). If the electron pair is in either an sp² or sp³ orbital, it is available for bonding, and the nitrogen is a "strong base".

(a)

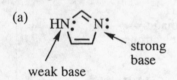

strong
base

weak base

(b)

strong base

(c)

strong base

(d)

weak
base

strong base   strong base

(e)

strong base

(f)

strong
base

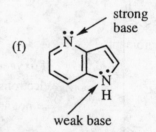

weak base

16-35 Where a resonance form demonstrates aromaticity, the resonance form is shown.

(a)

aromatic

(b)

aromatic

(c)

aromatic

(d)

BAD

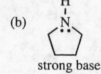

BAD

NOT aromatic—although it can be drawn, the resonance form on the left is NOT a significant contributor because the oxygen does not have a full octet; the form on the right shows the correct polarization of the carbonyl, but it's still not aromatic because of only 4 π electrons

(e)

aromatic

354

16-35 continued

(f)
aromatic;
not basic

(g)
aromatic;
N is not
basic but
the O⁻ is a
weak base

(h) + ← not basic
← basic
aromatic

(i) ← basic
sp³ ← basic
NOT aromatic
because of sp³
carbon; both N are
basic, although the
top one is conjugated,
lowering its basicity

(j) ← basic
← basic
aromatic; bottom N is
not basic because it
has donated its
electron pair to make
the ring aromatic

(k)
aromatic;
not basic

(l)
NOT aromatic: if
both N and O are
sp², then the pi
system has 8 e⁻;
N will be basic

(m)
NOT aromatic: if O
is sp², then the pi
system has 8 e⁻

(2) Nitrogens whose electrons are needed to complete the aromatic π system will not be basic. The only N that are more basic than water are the ones indicated in parts (h), (i), (j), and (l).

16-36

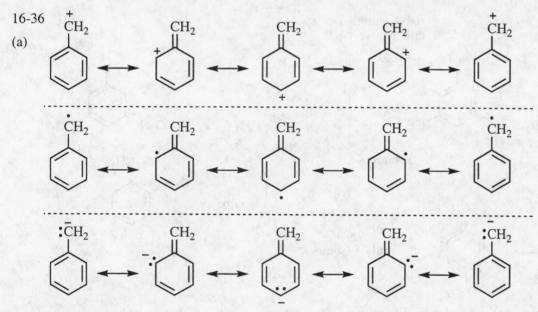

(b) initiation

$$Br-Br \xrightarrow{h\nu} 2\ Br\cdot$$

propagation steps on the next page

355

16-36 continued

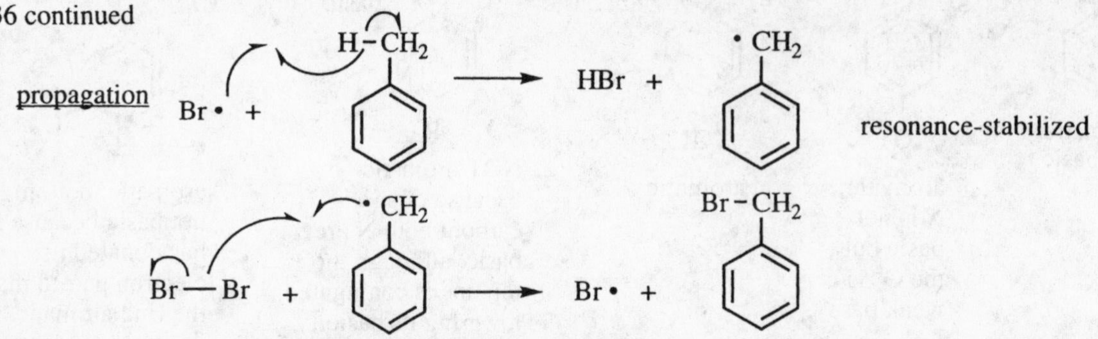

propagation

resonance-stabilized

(c) Both reactions are S_N2 on primary carbons, but the one at the benzylic carbon occurs faster. In the transition state of S_N2, as the nucleophile is approaching the carbon and the leaving group is departing, the electron density resembles that of a p orbital. As such, it can be stabilized through overlap with the π system of the benzene ring.

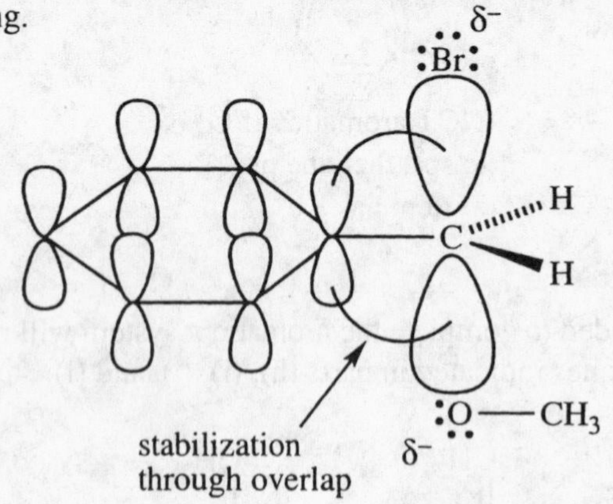

stabilization through overlap

16-37

(a)

3 isomers   +   +

same as

(b)

only 1 isomer

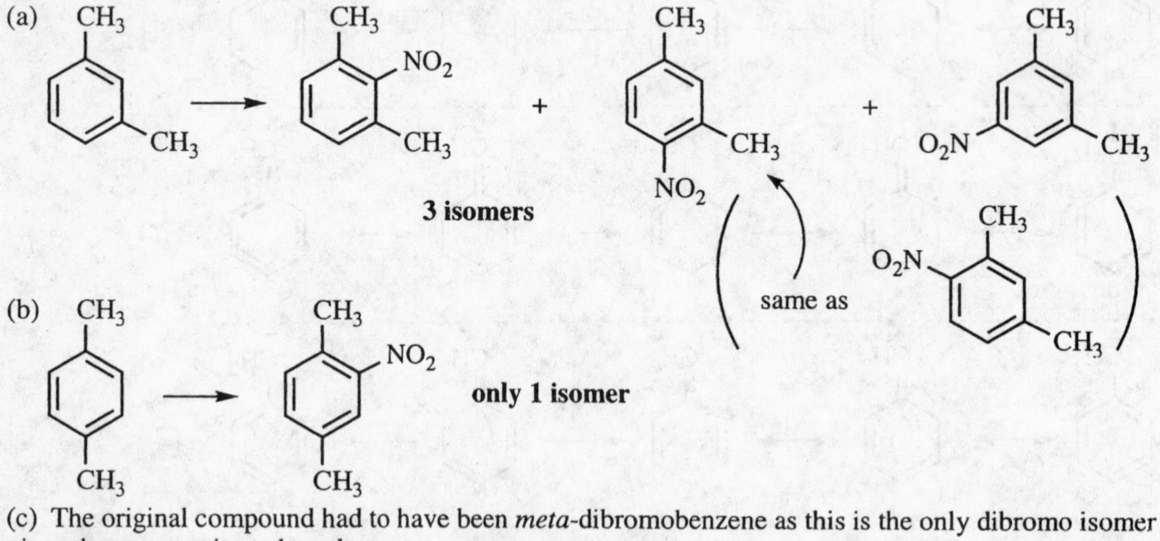

(c) The original compound had to have been *meta*-dibromobenzene as this is the only dibromo isomer that gives three mononitrated products.

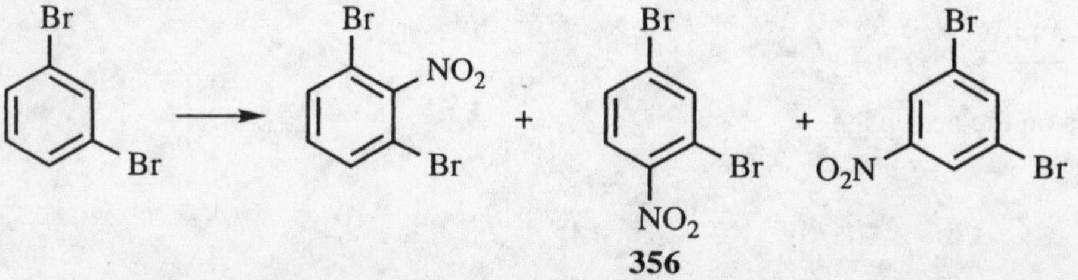

356

16-38

(a) The formula C_8H_7OCl has five elements of unsaturation, probably a benzene ring (4) plus either a double bond or a ring. The IR suggests a conjugated carbonyl at 1690 cm^{-1} and an aromatic ring at 1602 cm^{-1}. The NMR shows a total of five aromatic protons, indicating a monosubstituted benzene. A 2H singlet at δ 4.7 is a deshielded methylene.

(b) The mass spectral evidence of molecular ion peaks of 1 : 1 intensity at 184 and 186 shows the presence of a bromine atom. The m/z 184 minus 79 for bromine gives a mass of 105 for the rest of the molecule, which is about a benzene ring plus two carbons and a few hydrogens. The NMR shows four aromatic hydrogens in a typical *para* pattern (two doublets), indicating a *para*-disubstituted benzene. The 2H quartet and 3H triplet are characteristic of an ethyl group.

16-39

(a)

like the ends of a conjugated diene

(b)

Diels-Alder product

new sigma bonds shown in bold

16-40

(a) No, biphenyl is not fused. The rings must share two atoms to be labeled "fused".
(b) There are 12 π electrons in biphenyl compared with 10 for naphthalene.
(c) Biphenyl has 6 "double bonds". An isolated alkene releases 120 kJ/mole upon hydrogenation.

predicted: 6 x 120 kJ/mole (28.6 kcal/mole) ≈ 720 kJ/mole (172 kcal/mole)

observed: 418 kJ/mole (100 kcal/mole)

resonance energy: 302 kJ/mole (72 kcal/mole)

(d) On a "per ring" basis, biphenyl is 302 ÷ 2 = 151 kJ/mole, the same as the value for benzene. Naphthalene's resonance energy is 252 kJ/mole (60 kcal/mole); on a "per ring" basis, naphthalene has only 126 kJ/mole of stabilization per ring. This is consistent with the greater reactivity of naphthalene compared with benzene. In fact, the more fused rings, the lower the resonance energy per ring, and the more reactive the compound. (Refer to Problem 16-20.)

16-41 Two protons are removed from sp^3 carbons to make sp^2 carbons and to generate a π system with 10 π electrons.

16-42

(a) 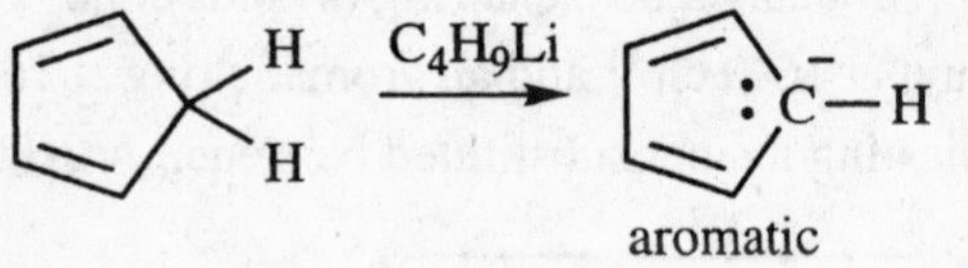 aromatic

(b) 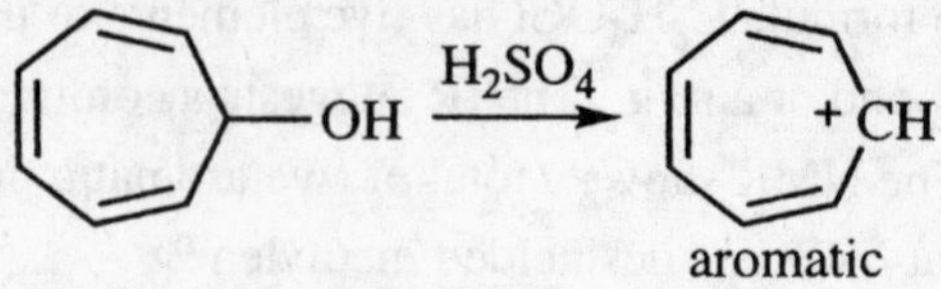 aromatic

(c) 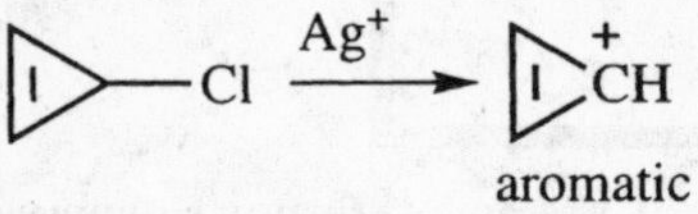 aromatic

(d) 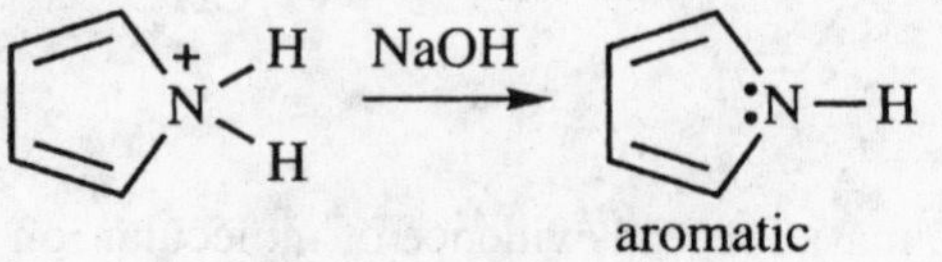 aromatic

(e) 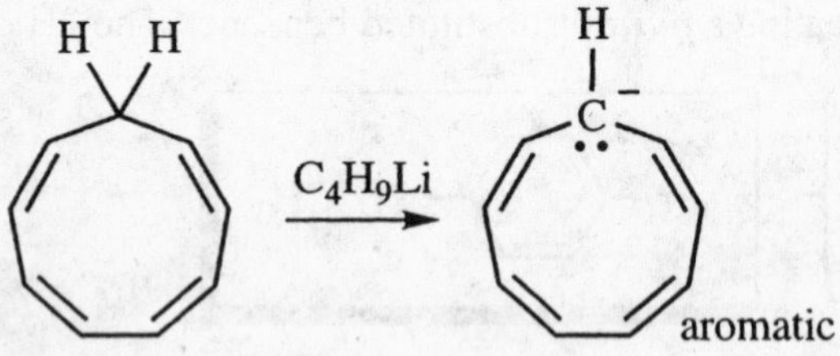 aromatic

(f) 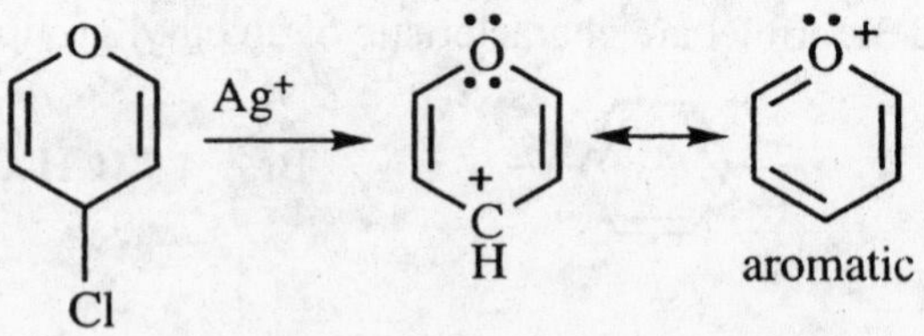 aromatic

16-43 These four bases can be aromatic, partially aromatic, or aromatic in a tautomeric form. In other words, aromaticity plays an important role in the chemistry of all four structures. (Only electron pairs involved in the important resonance are shown.) For part (b), nitrogens that are basic are denoted by **B**. Those that are not basic are shown as **NB**. Note that some nitrogens change depending on the tautomeric form.

(a) and (c)

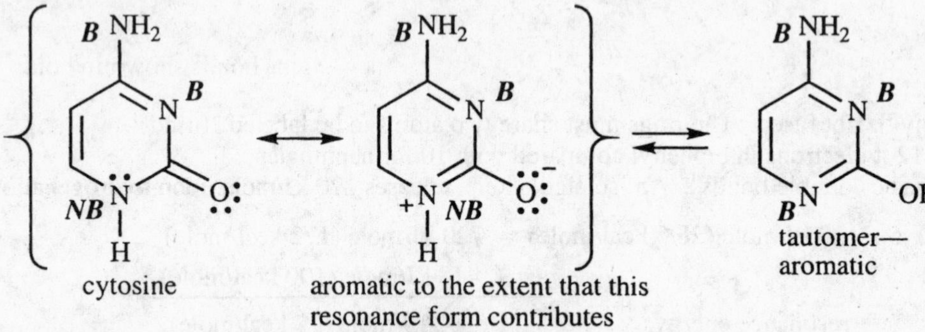

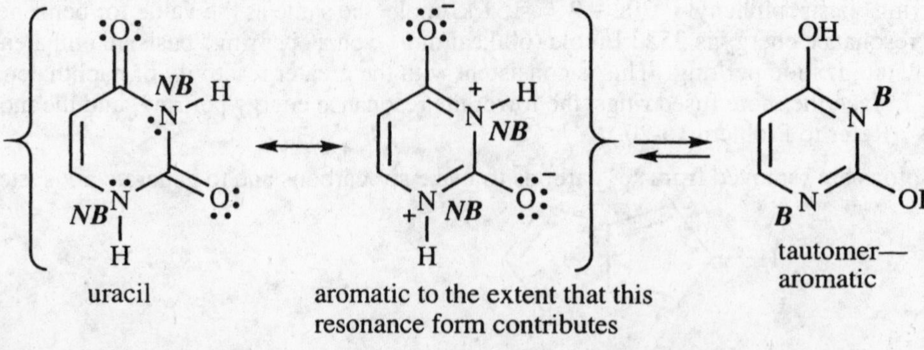

cytosine

aromatic to the extent that this resonance form contributes

tautomer—aromatic

uracil

aromatic to the extent that this resonance form contributes

tautomer—aromatic

continued on the next page

16-43 continued

guanine ↔ aromatic to the extent that this resonance form contributes ⇌ tautomer— fully aromatic

- -

fully aromatic

adenine

16-44

(a) Antiaromatic—only 4 π electrons.

(b) This molecule is electronically equivalent to cyclobutadiene. Cyclobutadiene is unstable and undergoes a Diels-Alder reaction with another molecule of itself. The *t*-butyl groups prevent dimerization by blocking approach of any other molecule.

(c) Yes, the nitrogen should be basic. The pair of electrons on the nitrogen is in an sp^2 orbital and is not part of the π system.

(d)

Analysis of structure **1** shows the three *t*-butyl groups in unique environments in relation to the nitrogen. We would expect three different signals in the NMR, as is observed at –110° C. Why do signals coalesce as the temperature is increased? Two of the *t*-butyl groups become equivalent—which two? Most likely, they are **a** and **c** that become equivalent as they are symmetric around the nitrogen. But they are *not* equivalent in structure **1**—what is happening here?

What must happen is an equilibration between structures **1** and **2**, very slow at –110° C, but very fast at room temperature, faster than the NMR can differentiate. So the signal which has coalesced is an average of **a** and **a'** and **c** and **c'**. (This type of low temperature NMR experiment is also used to differentiate axial and equatorial hydrogens on a cyclohexane.)

The NMR data prove that **1** is not aromatic, and that **1** and **2** are isomers, not resonance forms. If **1** were aromatic, then **a** and **c** would have identical NMR signals at all temperatures.

359

16-45

Mass spectrum: Molecular ion at 150; base peak at 135, M − 15, is loss of methyl.

Infrared spectrum: The broad peak at 3500 cm^{-1} is OH; thymol must be an alcohol. The peak at 1620 cm^{-1} suggests an aromatic compound.

NMR spectrum: The singlet at δ 4.8 is OH; it disappears upon shaking with D_2O. The 6H doublet at δ 1.2 and the 1H multiplet at δ 3.2 are an isopropyl group, apparently on the benzene ring. A 3H singlet at δ 2.3 is a methyl group, also on the benzene ring.

Analysis of the aromatic protons suggests the substitution pattern. The three aromatic hydrogens confirm that there are three substituents. The singlet at δ 6.5 is a proton between two substituents (no neighboring H's). The doublets at δ 6.75 and δ 7.1 are ortho hydrogens, splitting each other.

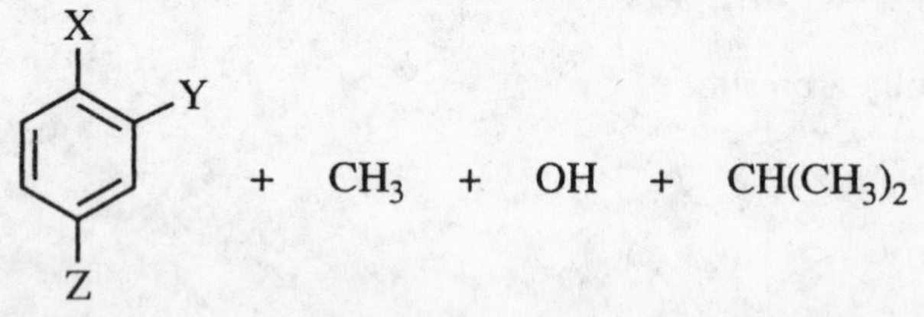

Several isomeric combinations are consistent with the spectra (although the single H giving δ 6.5 suggests that either Y or Z is the OH group—an OH on a benzene ring shields hydrogens ortho to it, moving them upfield). The structure of thymol is:

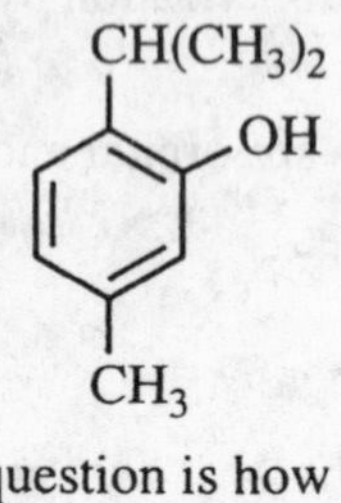

The final question is how the molecule fragments in the mass spectrometer:

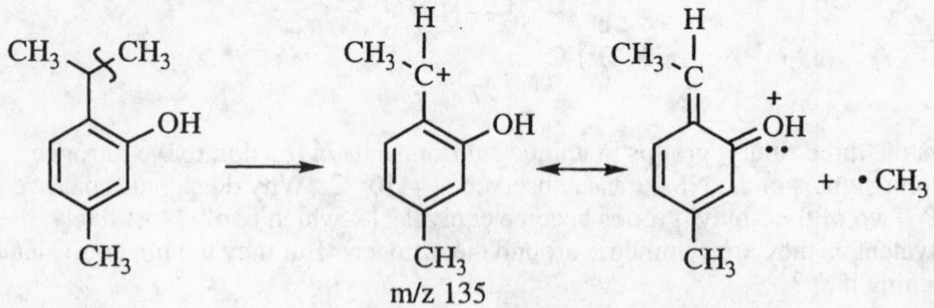

resonance stabilization of this benzylic cation includes forms with
positive charge on three ring carbons and on oxygen (shown)

16-46

Mass spectrum: Molecular ion at 170; two prominent peaks are M − 15 (loss of methyl) and M − 43 (as we shall see, most likely the loss of acetyl, CH_3CO).

Infrared spectrum: The two most significant peaks are at 1680 cm^{-1} (conjugated carbonyl) and 1600 cm^{-1} (aromatic C=C).

NMR spectrum: A 3H singlet at δ 2.7 is methyl next to a carbonyl, shifted slightly downfield by an aromatic ring. The other signals are seven aromatic protons. The 1H at d 8.7 is a deshielded proton next to a carbonyl. Since there is only one, the carbonyl can have only one neighboring hydrogen.

Conclusions:

$$CH_3-\overset{\overset{\displaystyle O}{\|}}{C}- \quad + \text{ m/z 127 including 7H} \Rightarrow \text{mass 120 for carbons} \Rightarrow 10\ C$$

The fragment $C_{10}H_7$ is almost certainly a naphthalene. The correct isomer (box) is indicated by the NMR.

This isomer would would have two deshielded protons in the NMR.

16-47

Although all carbons in hexahelicene are sp^2, the molecule is not flat. Because of the curvature of the ring system, one end of the molecule has to sit on top of the other end—the carbons and the hydrogens would bump into each other if they tried to occupy the same plane. In other words, the molecule is the beginning of a spiral. An "upward" spiral is the nonsuperimposable mirror image of a "downward" spiral, so the molecule is chiral and therefore optically active.

The magnitude of the optical rotation is extraordinary: it is one of the largest rotations ever recorded. In general, alkanes have small rotations and aromatic compounds have large rotations, so it is reasonable to expect that it is the interaction of plane-polarized light (electromagnetic radiation) with the electrons in the twisted pi system (which can also be considered as having wave properties) that causes this enormous rotation.

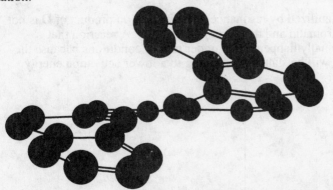

three-dimensional picture of hexahelicene showing the twist in the system of six rings

16-48 The key concept in parts (a)-(c) is that an aromatic product is created.

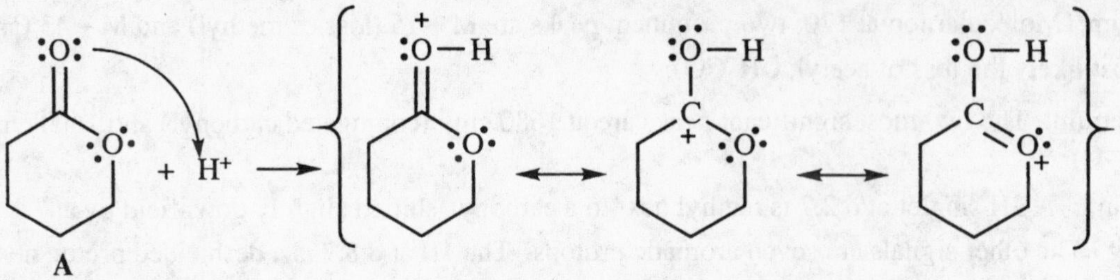

(a)

The protonated carbonyl gives a resonance-stabilized cation. Protonation of the singly bonded oxygen does not generate a resonance-stabilized product.

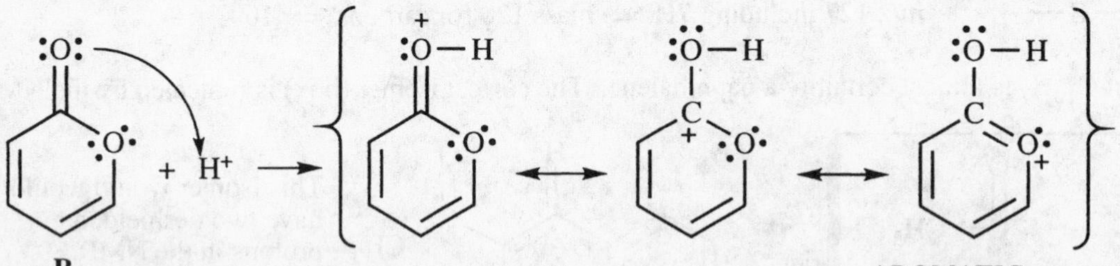

AROMATIC

The last resonance form shows that the cation produced is aromatic and therefore more stable than the corresponding nonaromatic ion. In this second reaction, the products are more favored than in the first reaction which is interpreted as the reactant **B** being more basic than **A**.

(b)

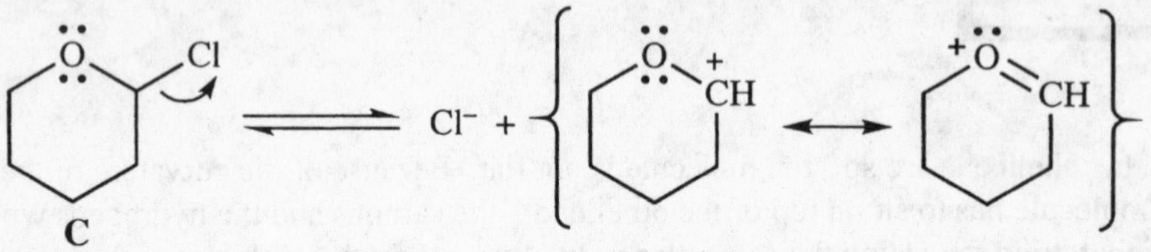

AROMATIC

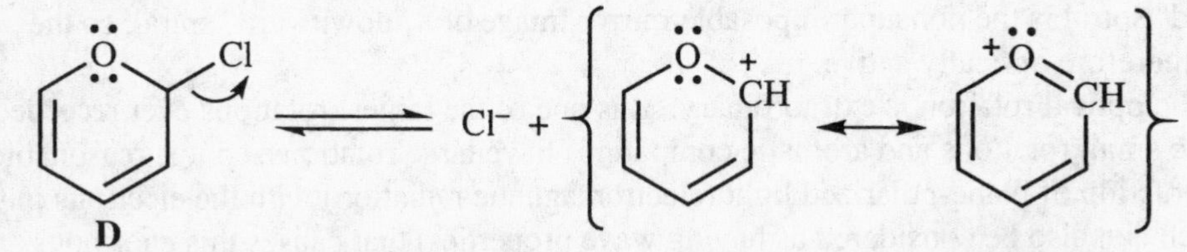

The product from ionization of **C** is stabilized by resonance. The ionization product of **D** is not only resonance-stabilized but is also aromatic and therefore more stable. A reaction that produces a more stable product will usually happen faster under milder conditions because the transition state leading to that product will be stabilized, leading to a lower activation energy.

16-48 continued

(c)

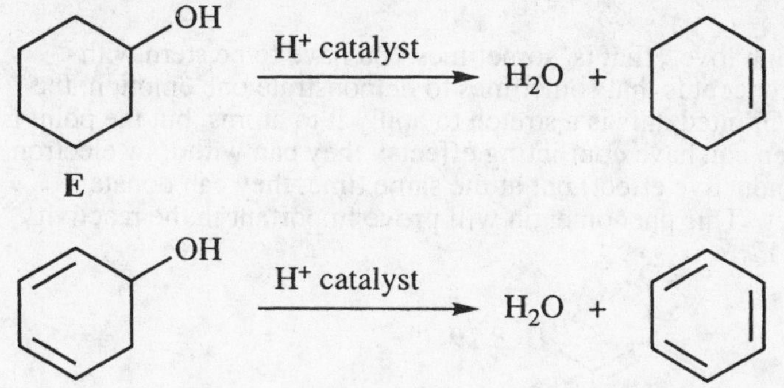

Dehydration of **F** produces an aromatic product that is more stable than the product from **E**. A reaction that produces a more stable product will usually happen faster under milder conditions because the transition state leading to that product will be stabilized, leading to a lower activation energy.

(d)

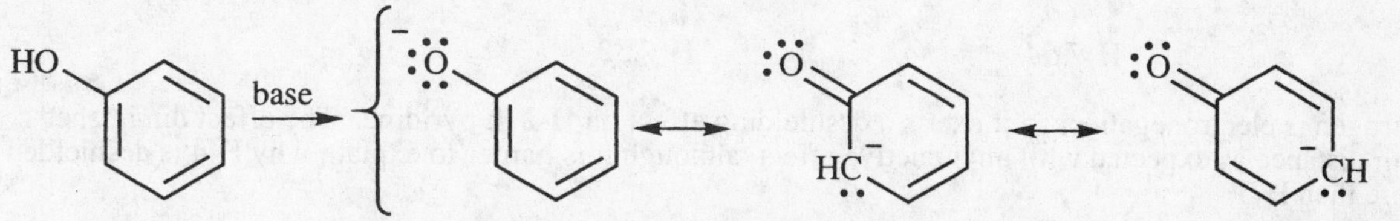

phenol

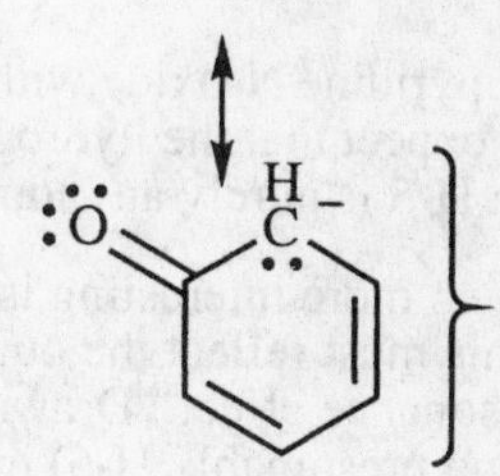

Resonance stabilization of the phenoxide anion shows the negative charge distributed over the one para and two ortho carbons.

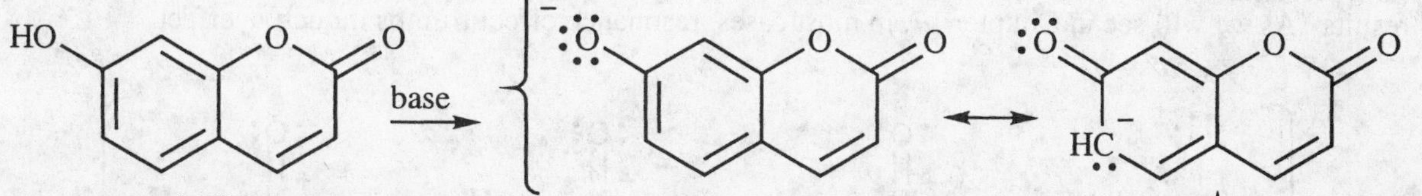

umbelliferone

(Umbelliferone is one of about 1000 coumarins isolated from plants, primarily the families of Angiosperms: Fabaceae, Asteraceae, Apiaceae, and Rutaceae.)

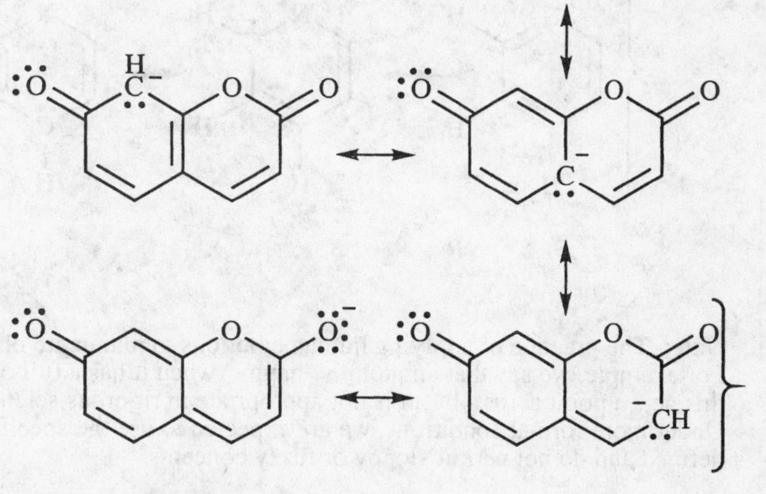

Resonance stabilization of the anion of umbelliferone gives not only the same three forms as the phenoxide anion, but in addition, gives an extra resonance form with (-) charge on a carbon, and the most significant resonance contributor, another form with the (-) charge on the other carbonyl oxygen. This anion is much more stable than phenoxide which we interpret as enhanced acidity of the starting material, umbelliferone. In fact, the pKa of umbelliferone is 7.7 while phenol is about 10.

16-49 Humulon (sometimes spelled humulone), even though highly resonance stabilized, cannot be aromatic because the carbon shown at the bottom of the ring is tetrahedral and must be sp^3 hybridized. A ring is aromatic only when all of the atoms in the ring have a p orbital which the sp^3 carbon does not.

16-50 Are you familiar with the concept "tough love", that is, sometimes you have to be stern with someone you love for their own good? The concept is that sometimes to demonstrate one emotion, the behavior has to appear exactly the opposite. Granted, this is a stretch to apply it to atoms, but the point is that sometimes atoms like oxygen and nitrogen can have conflicting effects: they can withdraw electron density by their strong electronegativity (an inductive effect) but at the same time, they can donate electron density through their resonance effect. This phenomenon will prove important in the reactivity of substituted benzenes described in Chapter 17.

pyridine

H 8.60
H 7.25
H 7.64

pyridine N-oxide

H 8.19
H 7.40
H 7.32

Nitrogen is electronegative, so it exerts a deshielding effect on H-2 in pyridine. The effect diminishes with distance as expected with an inductive effect, although it is harder to explain why H-4 is deshielded more than H-3.

In pyridine N-oxide, with an even more electronegative oxygen attached to the N, it would be reasonable to expect that the hydrogens would be deshielded. This is the case with H-3; we can infer that the effect on H-3 is purely an inductive effect.

Even more interesting is the *shielding* effect on H-2 and H-4; these chemical shifts are shifted *upfield*. This must reflect the other side of oxygen's personality, the donation of electron density through a resonance effect. Drawing the resonance forms clearly shows that the electron density at H-2 and H-4 (and presumably H-6) increases through this donation by resonance, in perfect agreement with the NMR results. As we will see in Chapter 17, in most cases, resonance effect trumps inductive effect.

Note: The practice of applying human emotions to inanimate objects is called "anthropomorphizing". For example, we say that an atom is "happy" when it has a full octet of electrons. In casual conversation, this gets a point across, but it is not appropriate in rigorous scientific terms, for example, on exams. Under more formal conditions, we are expected to use the specific terms of science because they are well defined and do not permit sloppy or fuzzy concepts.

As our equipment technician says: "Don't anthropomorphize computers. They hate that."

CHAPTER 17—REACTIONS OF AROMATIC COMPOUNDS

The representation of benzene with a circle to represent the π system is fine for questions of nomenclature, properties, isomers, and reactions. For questions of mechanism or reactivity, however, the representation with three alternating double bonds (the Kekulé picture) is more informative. For clarity and consistency, this Solutions Manual will use the Kekulé form exclusively.

17-1

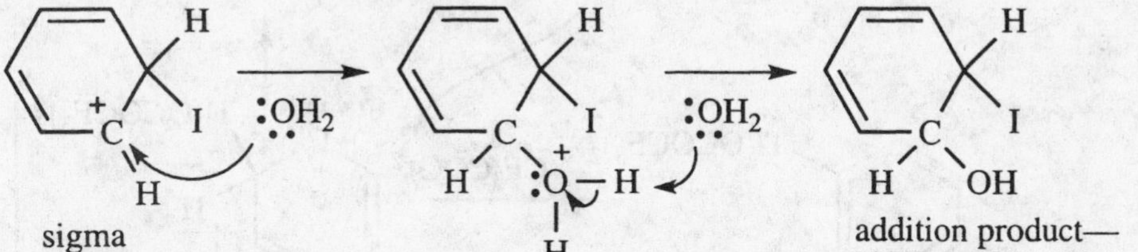

sigma complex

addition product—
NOT AROMATIC

While the addition of water to the sigma complex can be shown in a reasonable mechanism, the product is not aromatic. Thus, it has lost the 152 kJ/mol (36 kcal/mole) of resonance stabilization energy. The addition reaction is not favorable energetically, and substitution prevails.

17-2

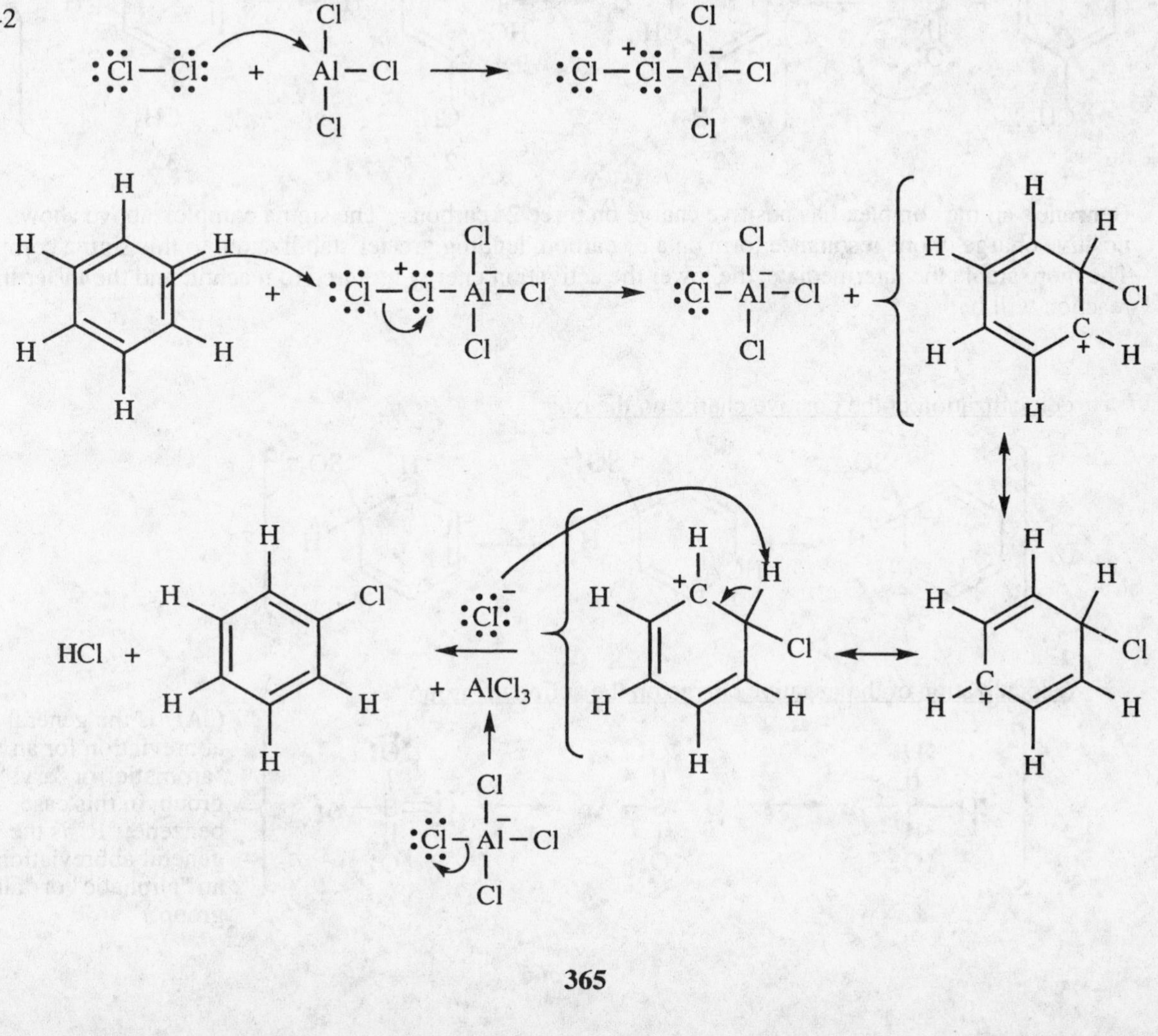

17-3 Like most heavy metals, thallium is highly toxic and should not be used on breakfast cereal.

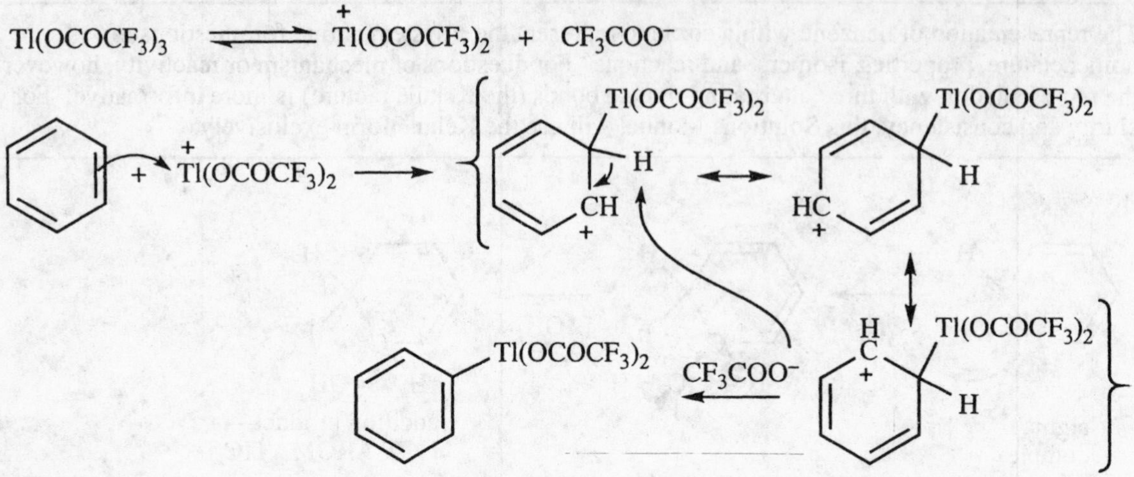

$$Tl(OCOCF_3)_3 \rightleftharpoons \overset{+}{Tl}(OCOCF_3)_2 + CF_3COO^-$$

17-4

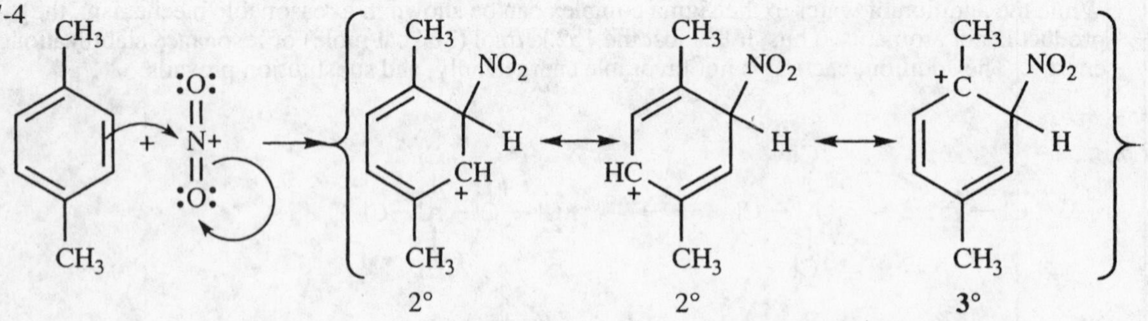

Benzene's sigma complex has positive charge on three 2° carbons. The sigma complex above shows positive charge in one resonance form on a 3° carbon, lending greater stabilization to this sigma complex. The more stable the intermediate, the lower the activation energy required to reach it, and the faster the reaction will be.

17-5   <u>delocalization of the positive charge on the ring</u>

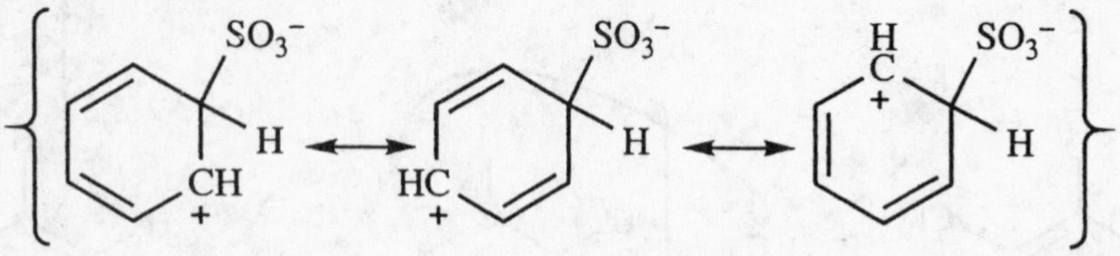

<u>delocalization of the negative charge on the sulfonate group</u>

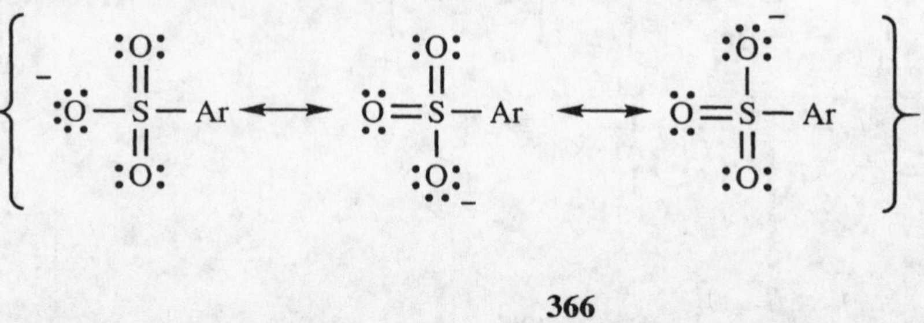

("Ar" is the general abbreviation for an "aromatic" or "aryl" group, in this case, benzene; "R" is the general abbreviation for an "aliphatic" or "alkyl" group.)

366

17-6

(a) The key to electrophilic aromatic substitution lies in the stability of the sigma complex. When the electrophile bonds at ortho or para positions of ethylbenzene, the positive charge is shared by the 3° carbon with the ethyl group. Bonding of the electrophile at the meta position lends no particular advantage because the positive charge in the sigma complex is never adjacent to, and therefore never stabilized by, the ethyl group.

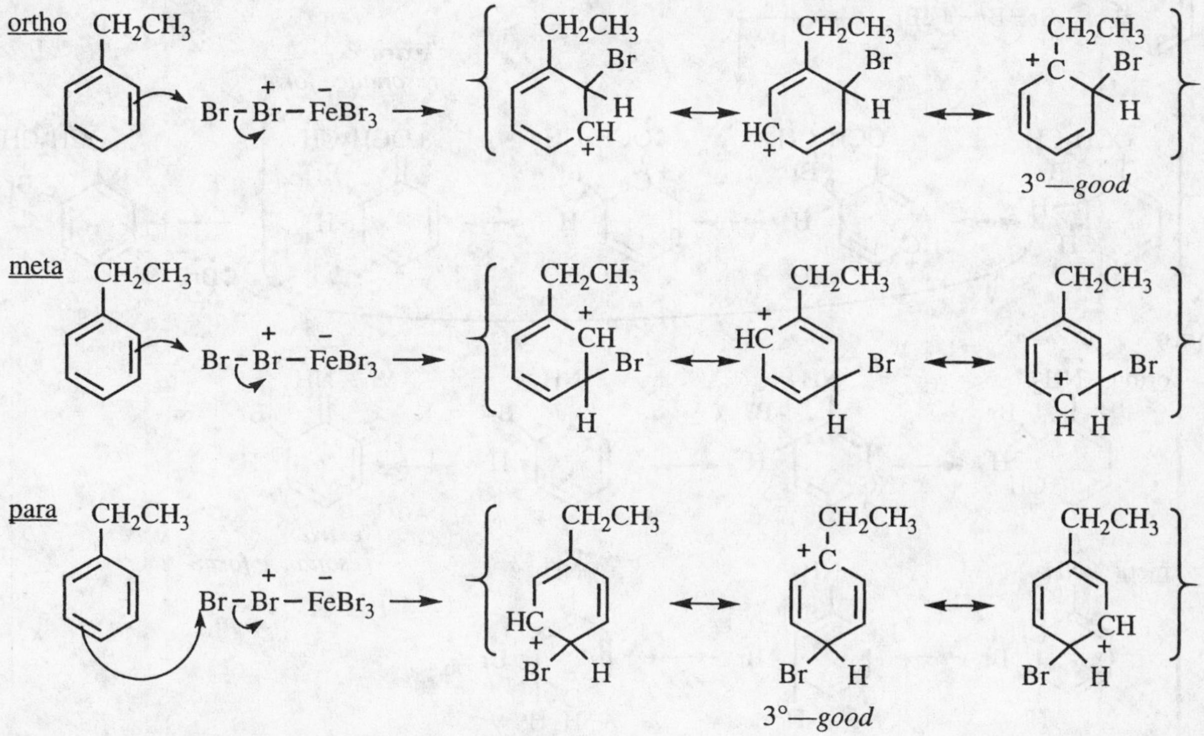

ortho

meta

para

(b) Electrophilic attack on *p*-xylene gives an intermediate in which only one of the three resonance forms is stabilized by a substituent (see the solution to Problem 17-4). *m*-Xylene, however, is stabilized in two of its three resonance forms. A more stable intermediate gives a faster reaction.

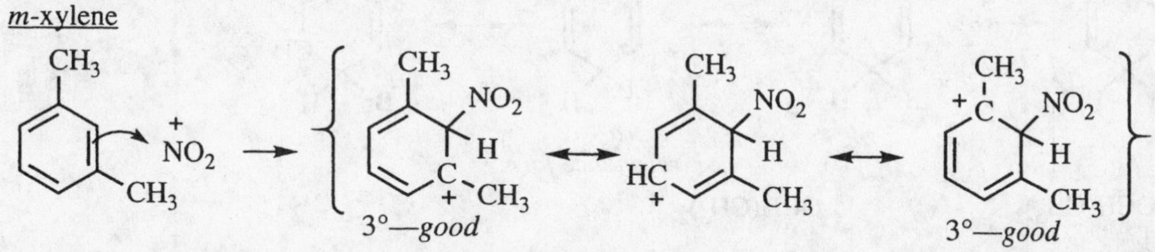

m-xylene

17-7 For ortho and para attack, the positive charge in the sigma complex can be shared by resonance with the vinyl group. This cannot happen with meta attack because the positive charge is never adjacent to the vinyl group. (Ortho attack is shown; para attack gives an intermediate with positive charge on the same carbons.)

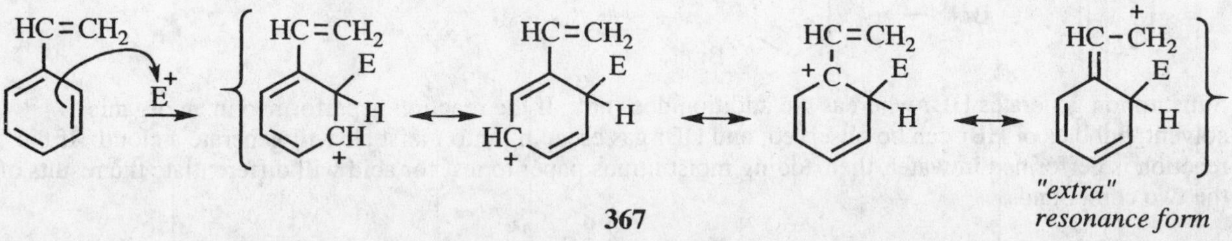

"extra" *resonance form*

17-8 Attack at only ortho and para positions (not meta) places the positive charge on the carbon with the ethoxy group, where the ethoxy group can stabilize the positive charge by resonance donation of a lone pair of electrons. (Ortho attack is shown; para attack gives a similar intermediate.)

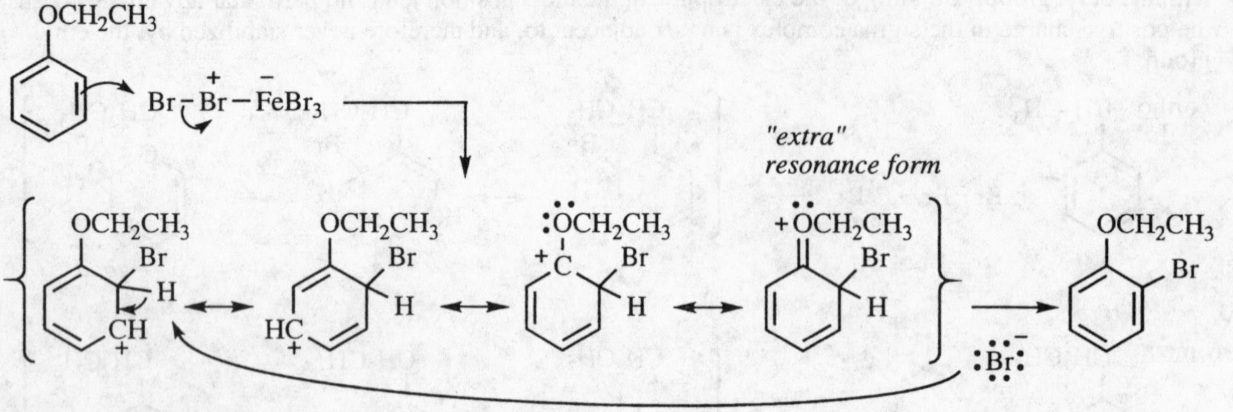

17-9

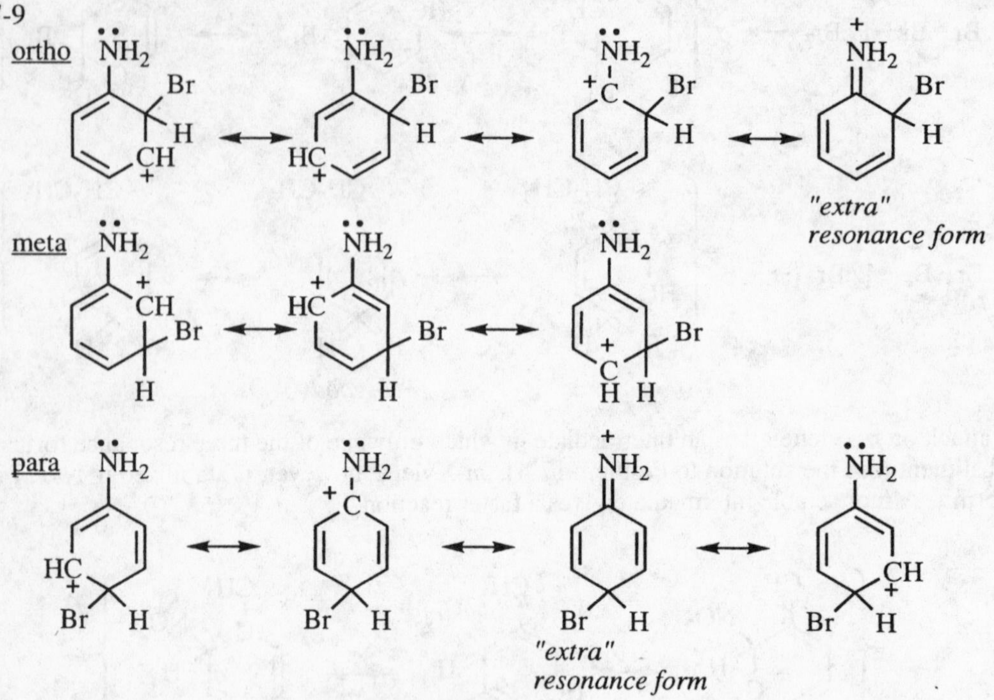

ortho

"extra"
resonance form

meta

para

"extra"
resonance form

17-10

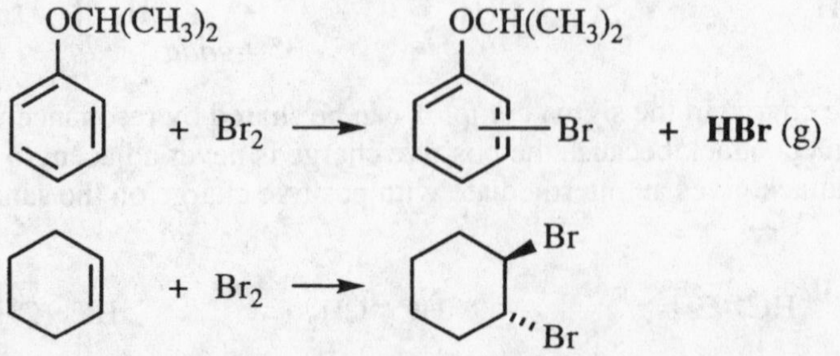

$+ \; Br_2 \longrightarrow$

$+ \; \textbf{HBr} \; (g)$

$+ \; Br_2 \longrightarrow$

Substitution generates HBr whereas the addition does not. If the reaction is performed in an organic solvent, bubbles of HBr can be observed, and HBr gas escaping into moist air will generate a cloud. If the reaction is performed in water, then adding moist litmus paper to test for acid will differentiate the results of the two compounds.

17-11

(a) Nitration is performed with nitric acid and a sulfuric acid catalyst. In strong acid, amines in general, including aniline, are protonated.

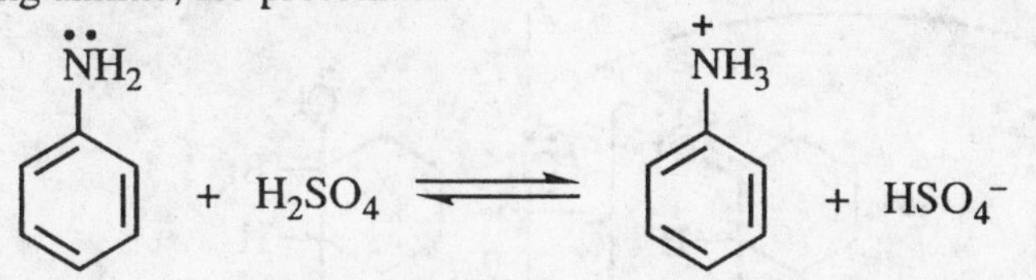

(b) The NH_2 group is a strongly activating ortho,para-director. In acid, however, it exists as the protonated ammonium ion—a strongly **deactivating meta-director**. The strongly acidic nitrating mixture itself forces the reaction to be slower.

(c) The acetyl group removes some of the electron density from the nitrogen, making it much less basic; the nitrogen of this amide is not protonated under the reaction conditions. The N retains enough electron density to share with the benzene ring, so the $NHCOCH_3$ group is still an activating ortho,para-director, though weaker than NH_2.

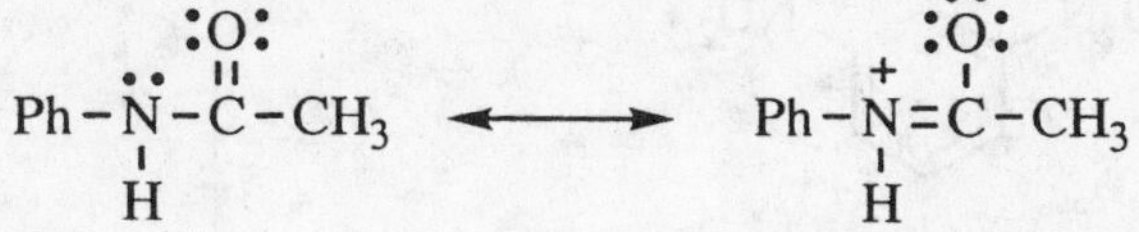

17-12 Nitronium ion attack at the ortho and para positions places positive charge on the carbon adjacent to the bromine, allowing resonance stabilization by an unshared electron pair from the bromine. Meta attack does not give a stabilized intermediate.

ortho

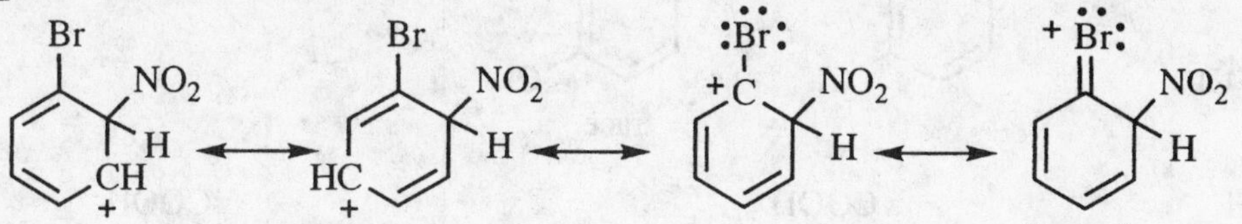

"extra"
resonance form

meta

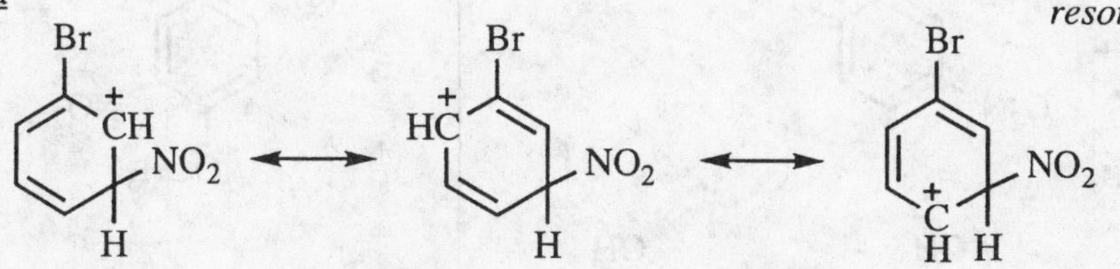

para

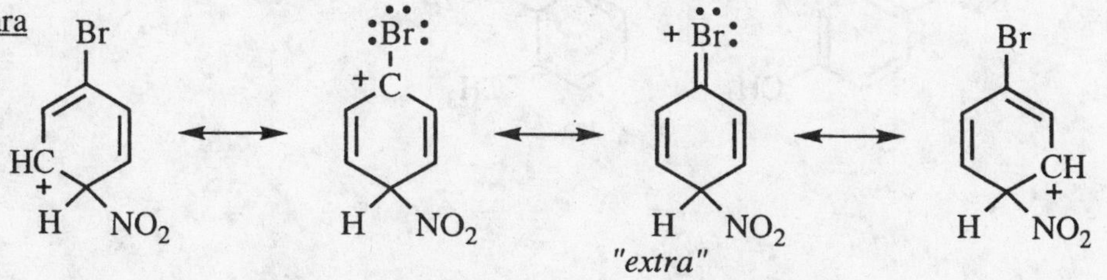

"extra"
resonance form

17-13

(a)

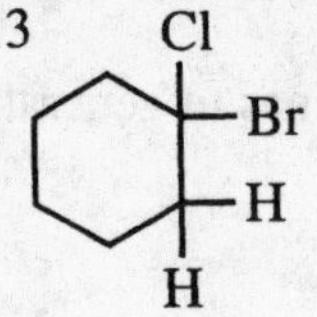

(b)

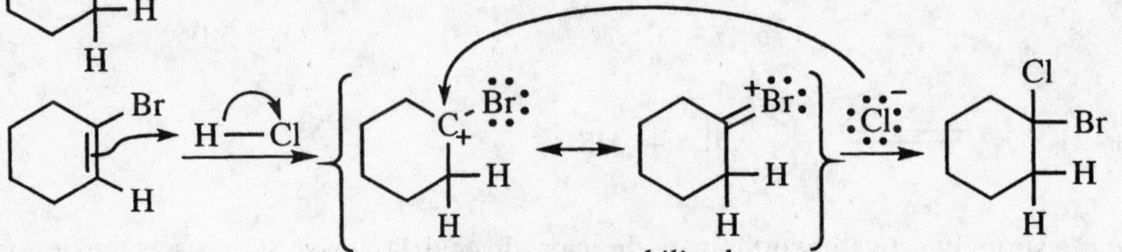

resonance-stabilized

(c) A bromine atom can stabilize positive charge by sharing a pair of electrons. Bromine can do this in any cationic species, whether from electrophilic addition or electrophilic aromatic substitution.

17-14

(a)

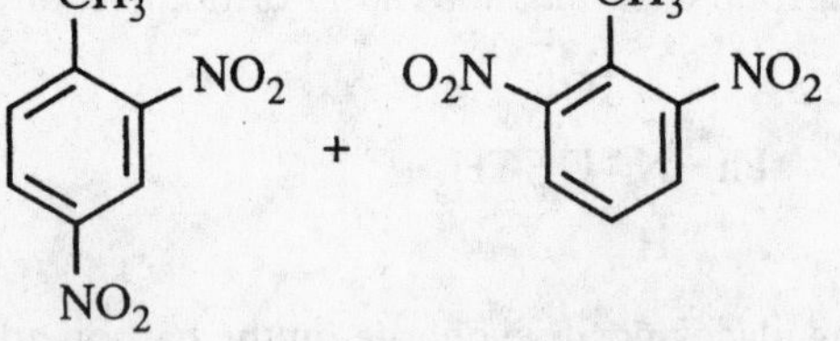

(b)

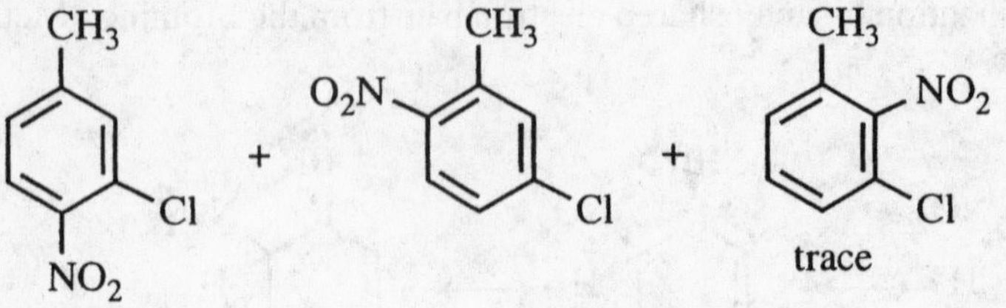

trace

(c)

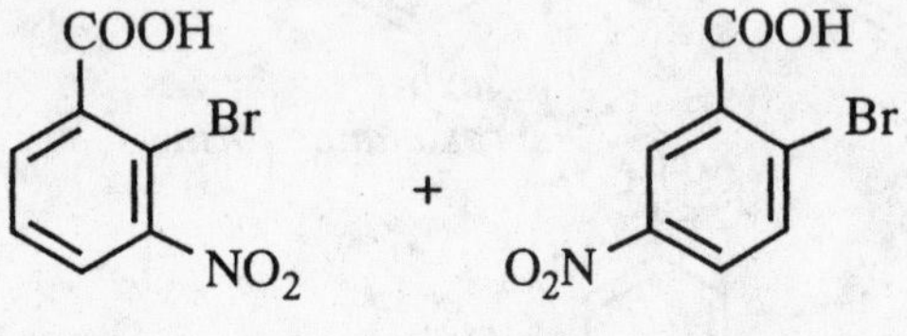

(d)

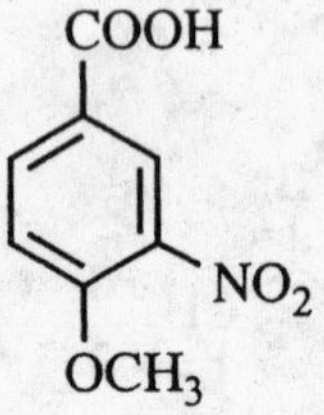

(e)

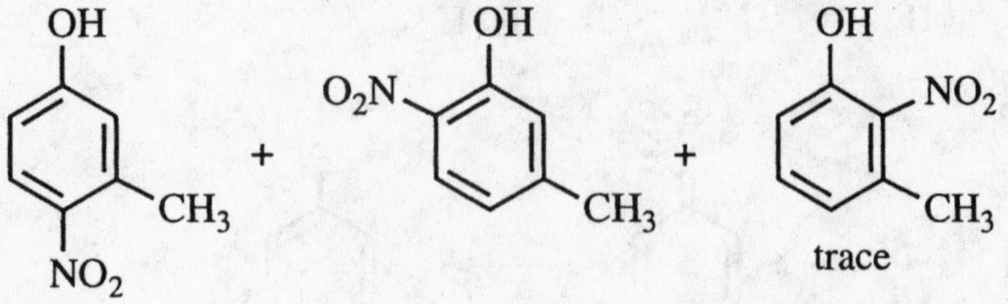

trace

17-15

(a)

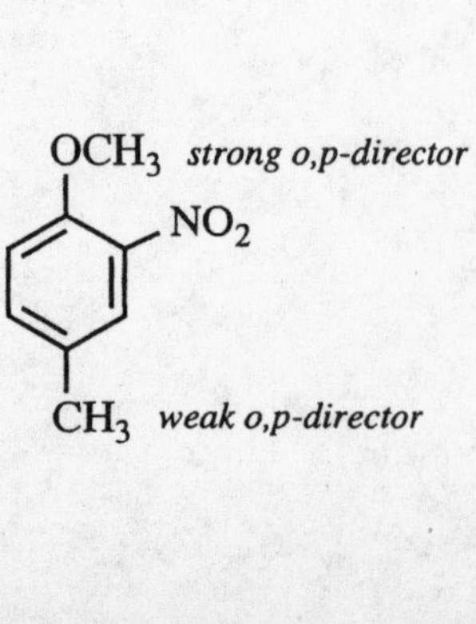

OCH_3 *strong o,p-director*

NO_2

CH_3 *weak o,p-director*

17-15 continued

(b)

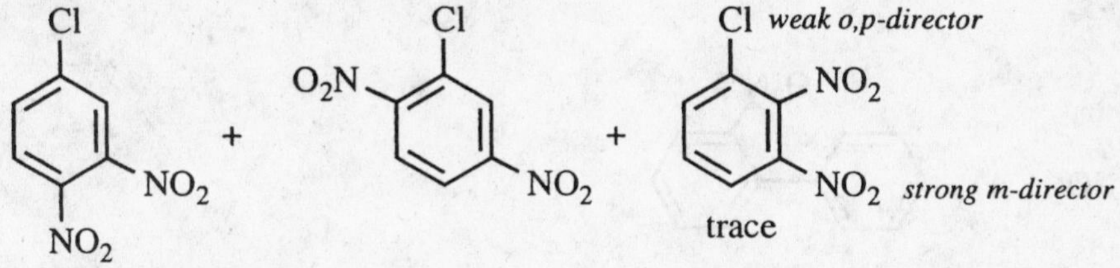

Cl / NO$_2$ / NO$_2$

+

O$_2$N / Cl / NO$_2$

+

Cl *weak o,p-director* / NO$_2$ / NO$_2$ *strong m-director*
trace

(c)

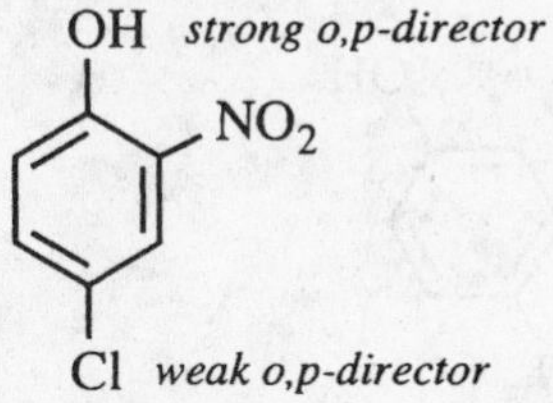

OH *strong o,p-director* / NO$_2$ / Cl *weak o,p-director*

(d)

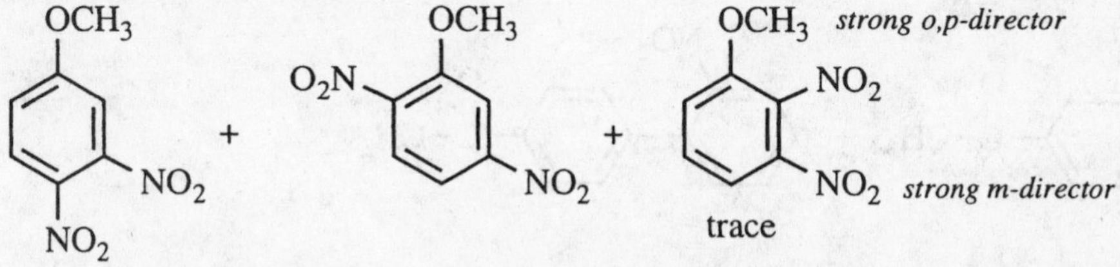

OCH$_3$ / NO$_2$ / NO$_2$

+

O$_2$N / OCH$_3$ / NO$_2$

+

OCH$_3$ *strong o,p-director* / NO$_2$ / NO$_2$ *strong m-director*
trace

(e)

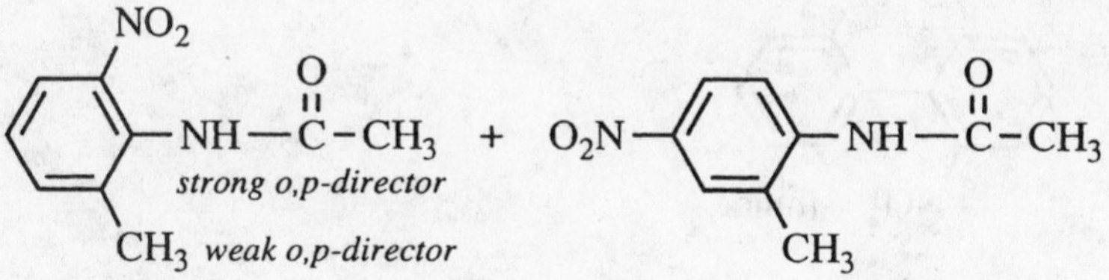

NO$_2$ / —NH—C(=O)—CH$_3$ *strong o,p-director* / CH$_3$ *weak o,p-director*

+

O$_2$N / —NH—C(=O)—CH$_3$ / CH$_3$

(f)

CH$_3$—C(=O)—NH— *strong o,p-director* ... O$_2$N ... —C(=O)—NH$_2$ *strong m-director*

activating   deactivating

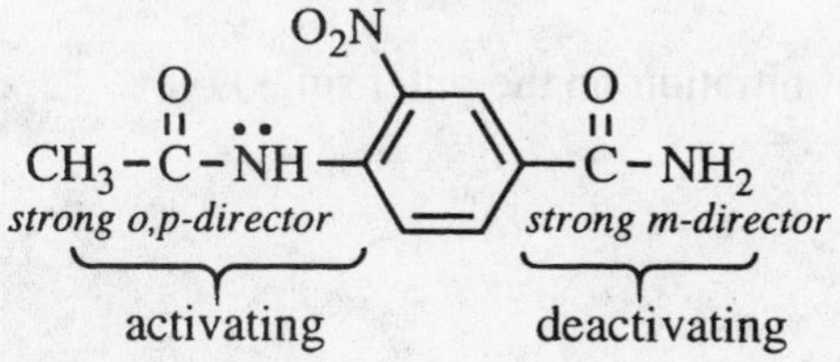

17-16

(a) Sigma complex of ortho attack—the phenyl substituent stabilizes positive charge by resonance:

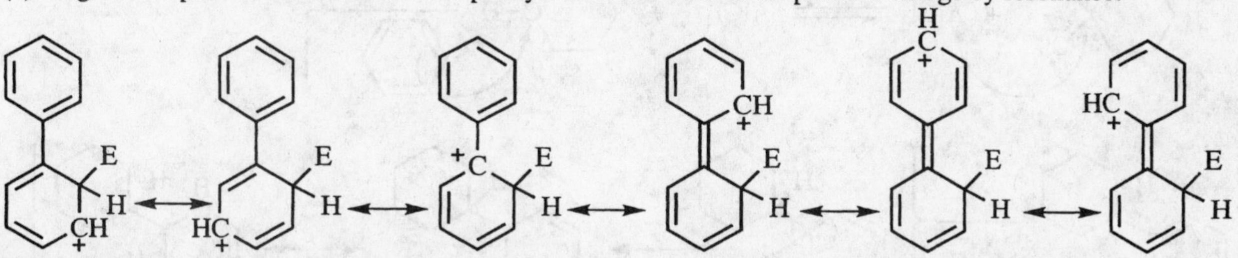

Para attack gives similar stabilization. Meta attack does not permit delocalization of the positive charge on the phenyl substituent.

371

17-16 continued

(b)

(i)

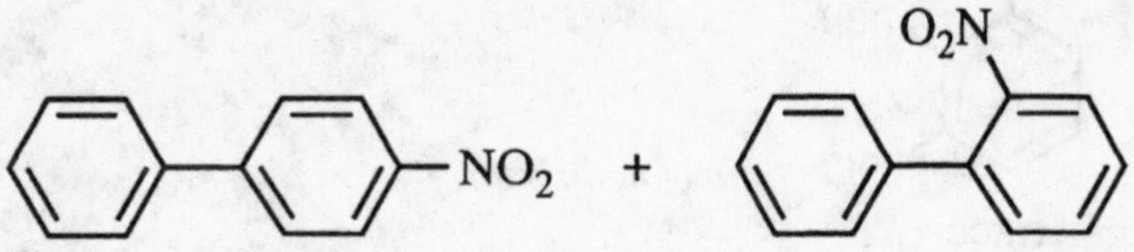

(ii)

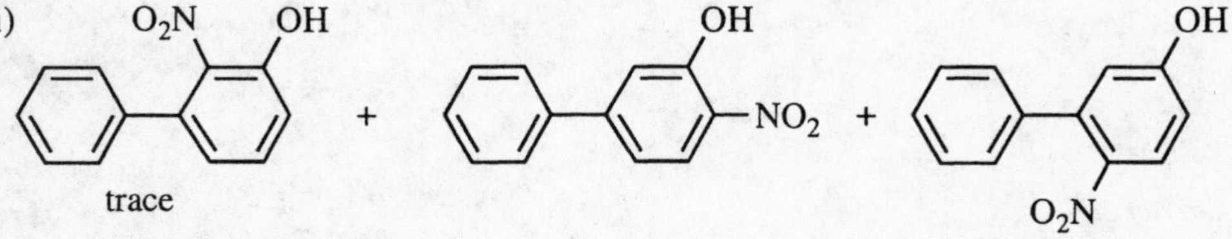

trace

(iii)

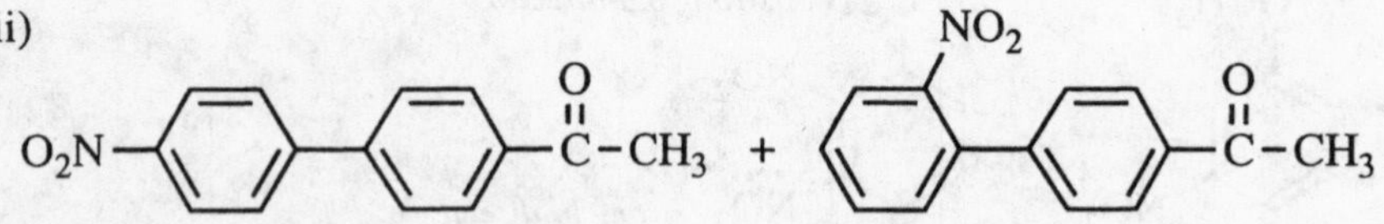

(iv)

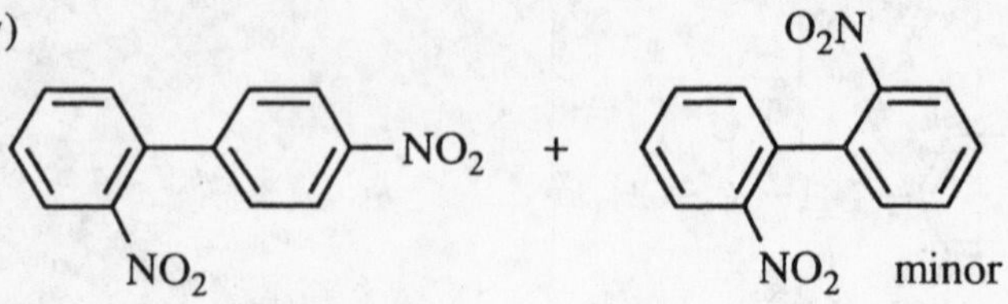

NO₂   minor

(v)

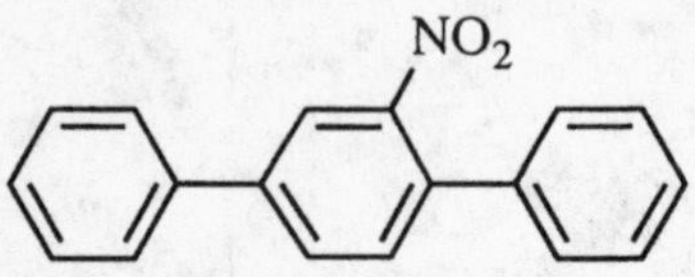

major (minor amounts of nitration on the outer rings)

17-17

(a)

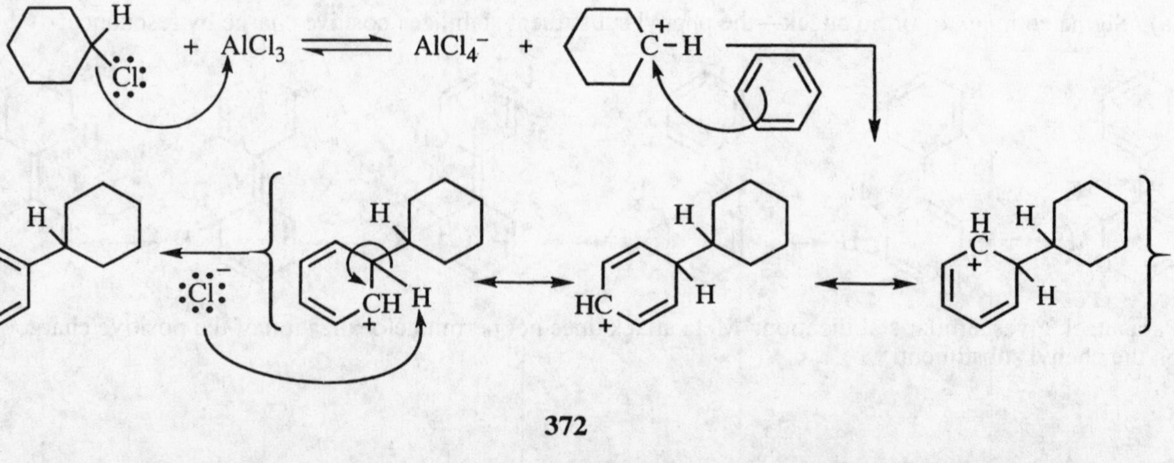

372

17-17 continued

(b)

CH_3—$\overset{..}{\underset{..}{Cl}}$: + $AlCl_3$ $\rightleftharpoons$ CH_3—$\overset{+\,..}{\underset{..}{Cl}}$—$\bar{A}lCl_3$

(benzene ring)—OCH_3

para isomer
also formed by
similar mechanism

(resonance structures showing electrophilic aromatic substitution with OCH₃ and CH₃ substituents leading to product with OCH₃ and CH₃)

(c)

CH_3—$\overset{CH_3}{\underset{CH_3}{C}}$—$CHCH_3$ $\xrightarrow{AlCl_3}$ CH_3—$\overset{CH_3}{\underset{CH_3}{C}}$—$\overset{+}{C}HCH_3$ $AlCl_4^-$ $\xrightarrow[\text{shift}]{\text{methyl}}$ CH_3—$\overset{\overset{3°}{CH_3}}{\underset{CH_3}{\overset{+}{C}}}$—$CHCH_3$

$2°$

(isopropylbenzene + carbocation → resonance structures of arenium ion with CH₃—C—CH(CH₃)₂ and CH₃ substituents)

$\overset{CH(CH_3)_2}{\underset{CH_3}{\underset{|}{C}}}$—$CH_3$ on para-isopropyl benzene ring

Only a small amount of the ortho isomer might be produced as steric
interactions will discourage this approach path of the electrophile.

17-18

(a)

(cyclohexene) + HF $\longrightarrow$ (cyclohexyl cation) (+ benzene) $\longrightarrow$ (cyclohexylbenzene)

(b)

CH_3—$\overset{CH_3}{\underset{CH_3}{C}}$—OH + BF_3 $\longrightarrow$ CH_3—$\overset{CH_3}{\underset{CH_3}{C}}$+ (+ benzene) $\longrightarrow$ CH_3—$\overset{CH_3}{\underset{CH_3}{C}}$—(phenyl)

373

17-18 continued

(c)

$CH_2=C$ (with CH_3 groups) $+ HF \longrightarrow CH_3-C^+$ (with CH_3 groups) $\longrightarrow CH_3-C(CH_3)_2$—(benzene ring)—$C(CH_3)_2-CH_3$

(d) $HO-CHCH_3$ (with CH_3) $+ BF_3 \longrightarrow {}^+CHCH_3$ (with CH_3) $\longrightarrow$ (para-isopropyltoluene) $+$ (ortho-isopropyltoluene)

17-19 In (a), (b), and (d), the electrophile has rearranged.

(a) $CH_3-C(CH_3)_2$—(benzene ring)

(b) (para-sec-butyltoluene) $+$ (ortho-sec-butyltoluene)

(c) No reaction: nitrobenzene is too deactivated for the Friedel-Crafts reaction to succeed.

(d) (benzene)—$C(CH_3)_2-CH(CH_3)_2$

17-20

(a) (benzene) $+ CH_3CH_2CH_2CH_2Br \xrightarrow{AlCl_3}$ (sec-butylbenzene)   plus over-alkylation products

(b) gives desired product
(c) gives desired product, plus ortho isomer; use excess bromobenzene to avoid overalkylation
(d) gives desired product, plus ortho isomer
(e) gives desired product

17-21

(a) (benzene) (excess) $\xrightarrow[AlCl_3]{(CH_3)_3CCl}$ (tert-butylbenzene, $C(CH_3)_3$) $\xrightarrow[H_2SO_4]{HNO_3}$ (para-nitro-tert-butylbenzene, $C(CH_3)_3$ and NO_2)   (separate from ortho, although steric hindrance would prevent much ortho substitution)

(b) (benzene) (excess) $\xrightarrow[AlCl_3]{CH_3I}$ (toluene, CH_3) $\xrightarrow[H_2SO_4]{SO_3}$ (para-toluenesulfonic acid, CH_3 and SO_3H)   (separate from ortho)

374

17-21 continued

(c)

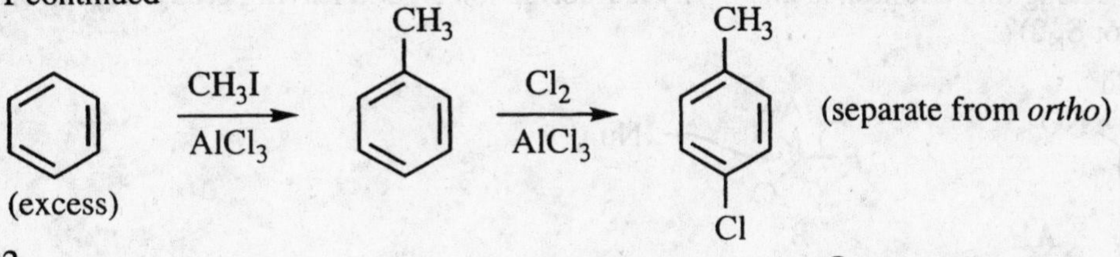

17-22

(a)

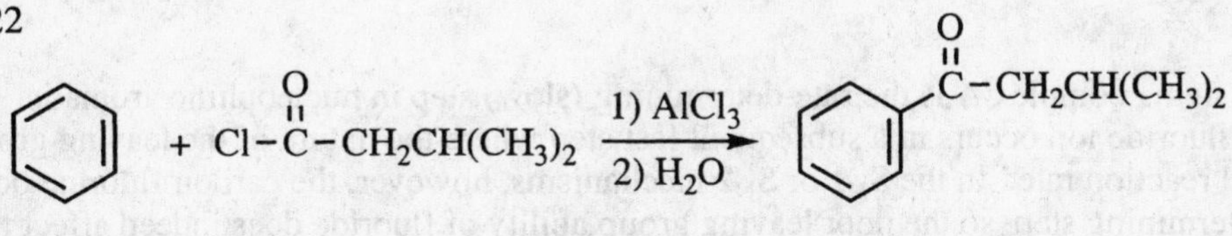

(b)

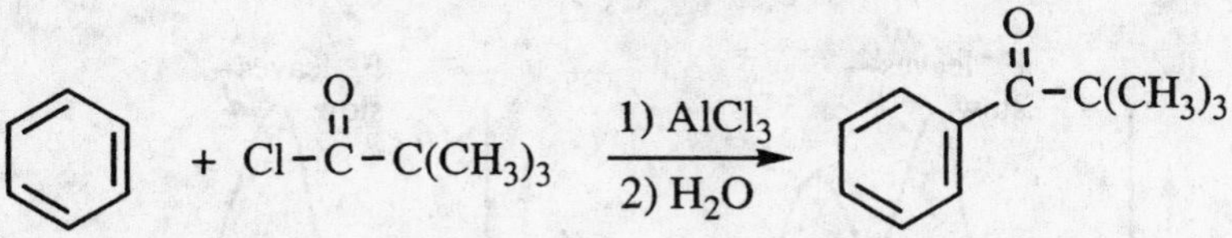

(c)

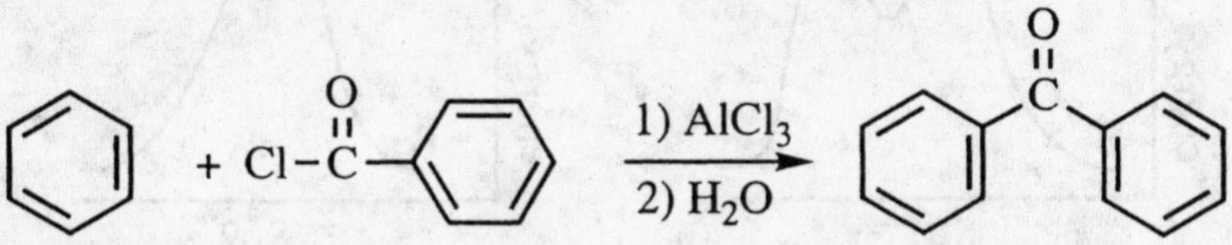

(d)

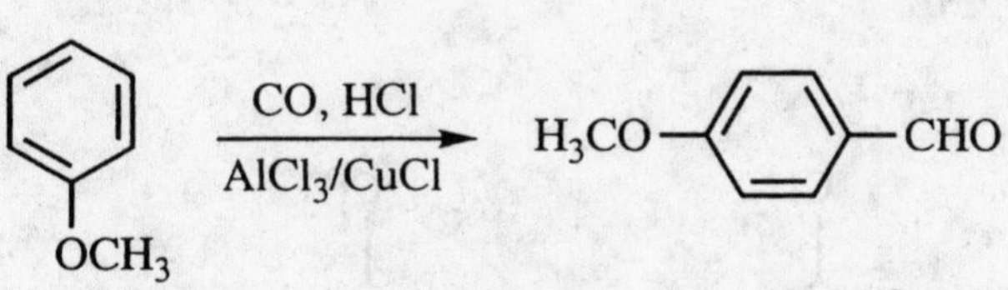

(e)

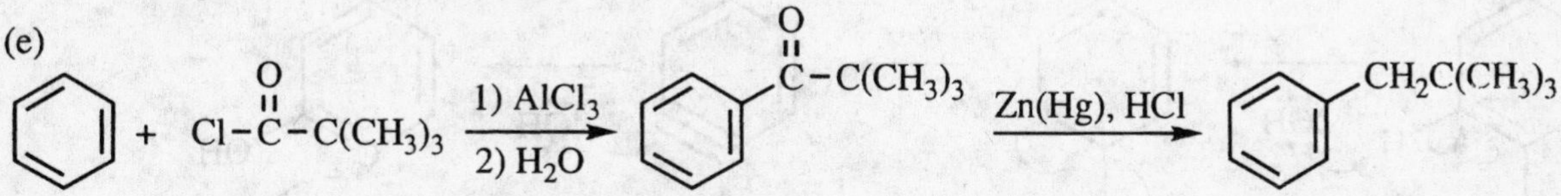

(f)

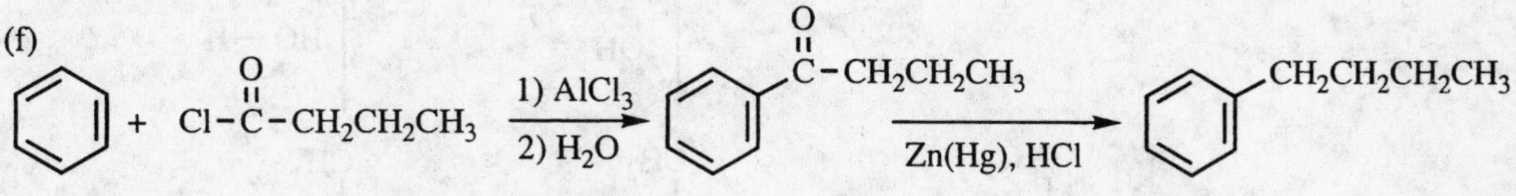

17-23 Another way of asking this question is this: Why is fluoride ion a good leaving group from **A** but not from **B** (either by S_N1 or S_N2)?

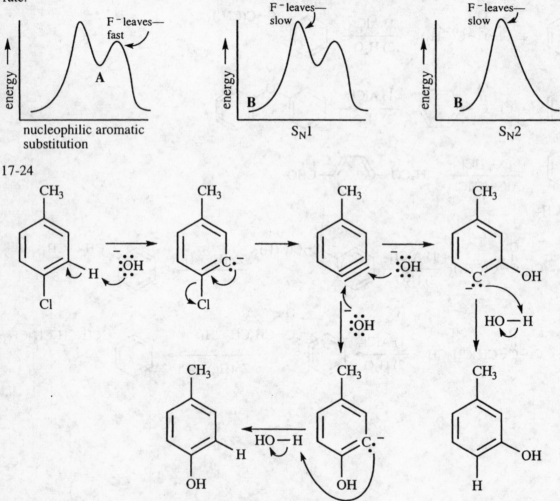

Formation of the anionic sigma complex **A** is the rate-determining (slow) step in nucleophilic aromatic substitution. The loss of fluoride ion occurs in a subsequent fast step where the nature of the leaving group does not affect the overall reaction rate. In the S_N1 or S_N2 mechanisms, however, the carbon-fluorine bond is breaking in the rate-determining step, so the poor leaving group ability of fluoride does indeed affect the rate.

17-24

17-25

(a)

(b)

17-25 continued

(c)

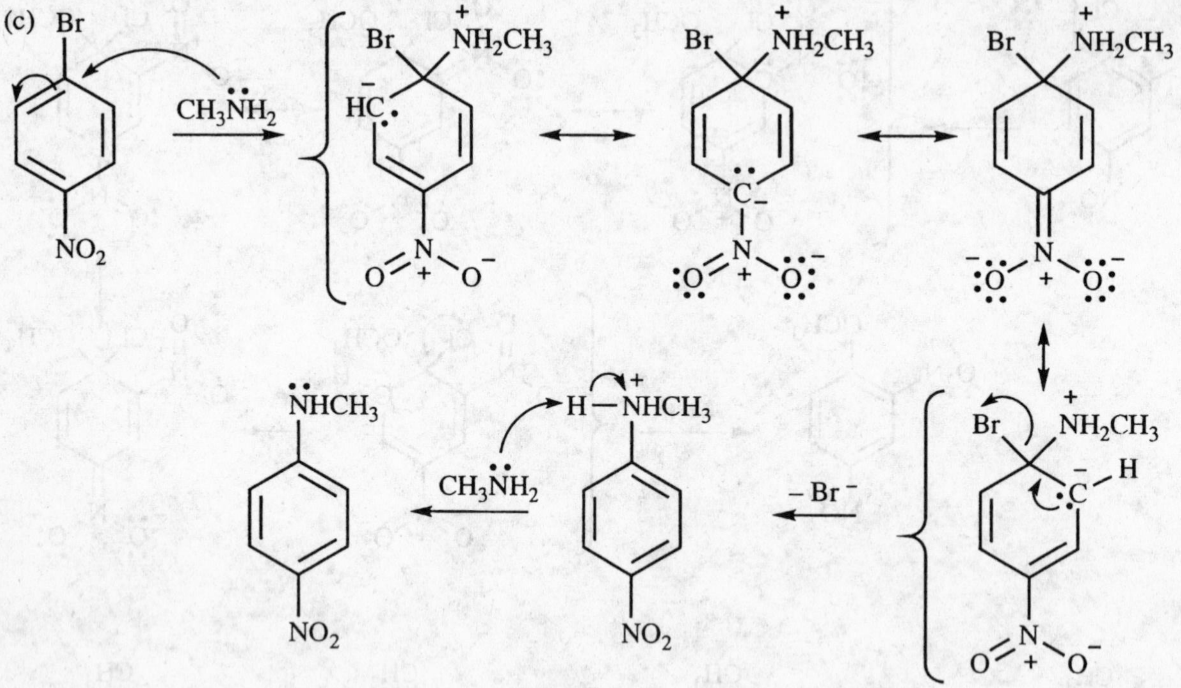

(d)

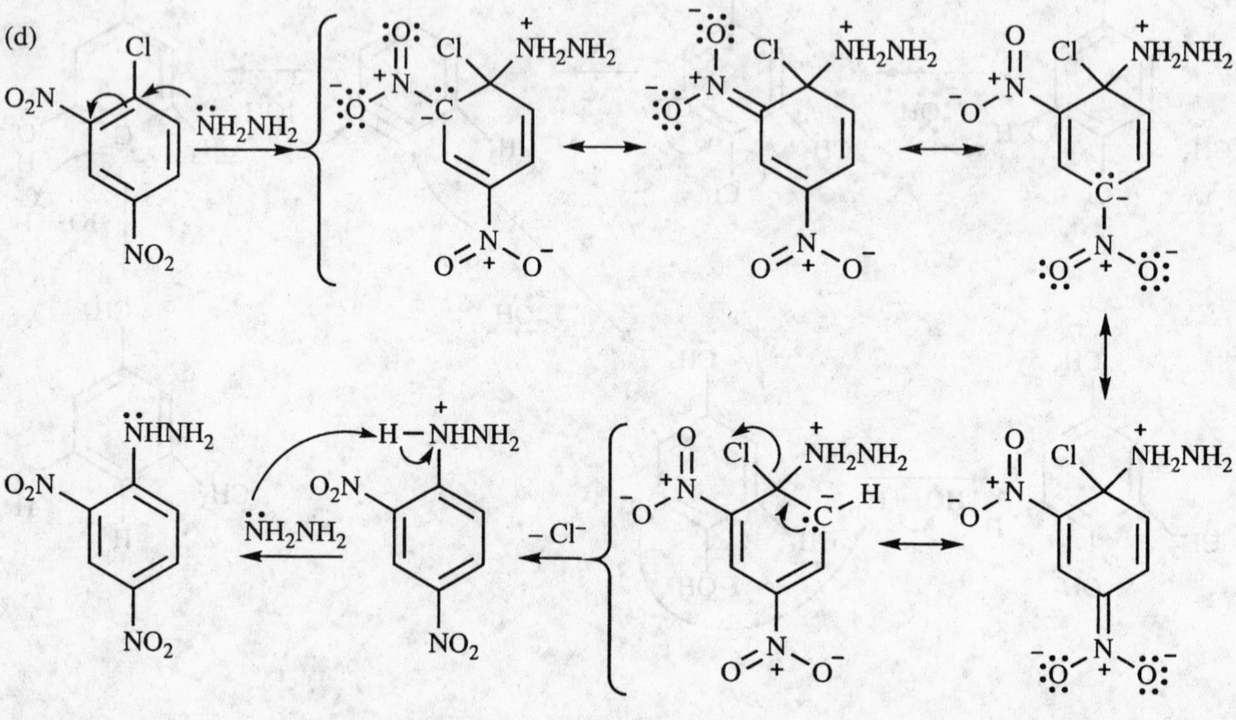

17-26 Assume an acidic workup to each of these reactions to produce the phenol, not the phenoxide ion.

(a)

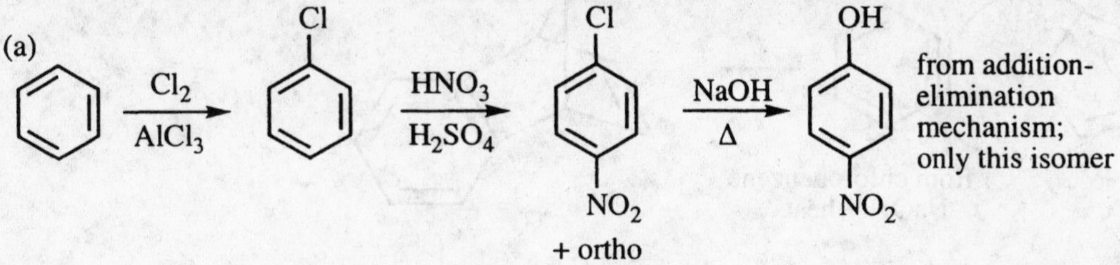

$\xrightarrow[\text{AlCl}_3]{\text{Cl}_2}$ $\xrightarrow[\text{H}_2\text{SO}_4]{\text{HNO}_3}$ $\xrightarrow[\Delta]{\text{NaOH}}$ from addition-elimination mechanism; only this isomer

+ ortho

(b)

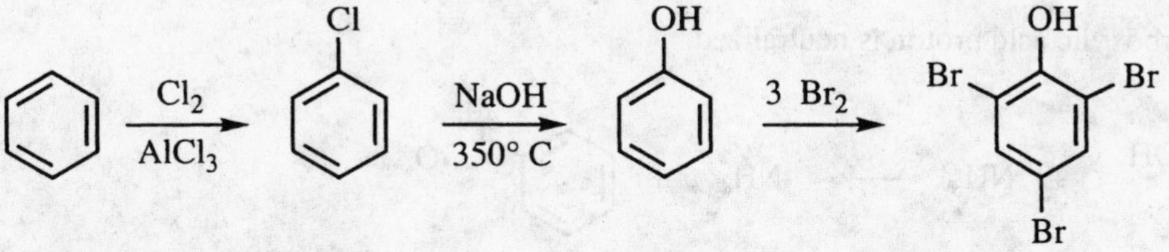

$\xrightarrow[\text{AlCl}_3]{\text{Cl}_2}$ $\xrightarrow[350°\text{C}]{\text{NaOH}}$ $\xrightarrow{3 \ \text{Br}_2}$

(c)

$\xrightarrow[\text{AlCl}_3]{2 \ \text{Cl}_2}$ $\xrightarrow[\Delta]{\text{NaOH}}$ + via benzyne mechanism

+ ortho

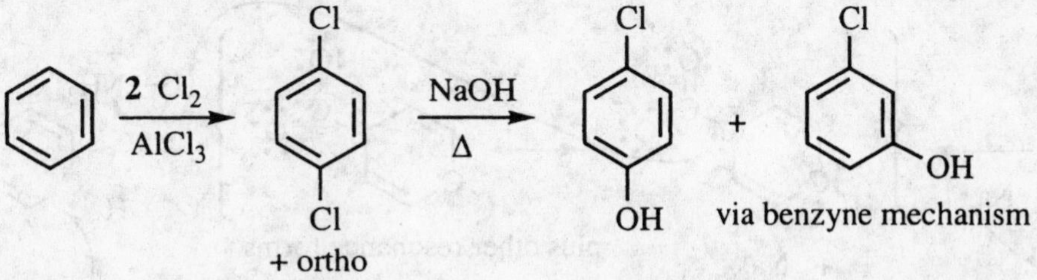

(d)

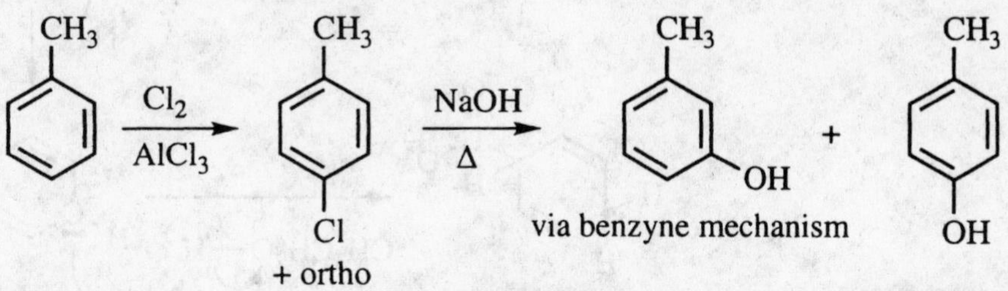

$\xrightarrow[\text{AlCl}_3]{\text{Cl}_2}$ $\xrightarrow[\Delta]{\text{NaOH}}$ via benzyne mechanism +

+ ortho

(e)

$\xrightarrow[\text{AlCl}_3]{\text{CH}_3\text{CH}_2\text{CH}_2-\overset{\text{O}}{\overset{\|}{\text{C}}}-\text{Cl}}$ $\xrightarrow[\text{HCl}]{\text{Zn(Hg)}}$ $\xrightarrow[\text{AlCl}_3]{\text{Cl}_2}$

$\xleftarrow[\Delta]{\text{NaOH}}$ + ortho

+ via benzyne mechanism

379

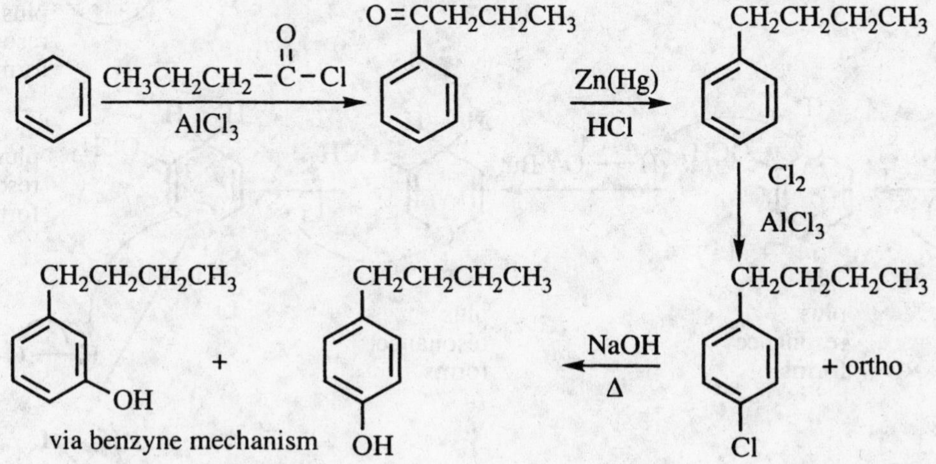

17-27

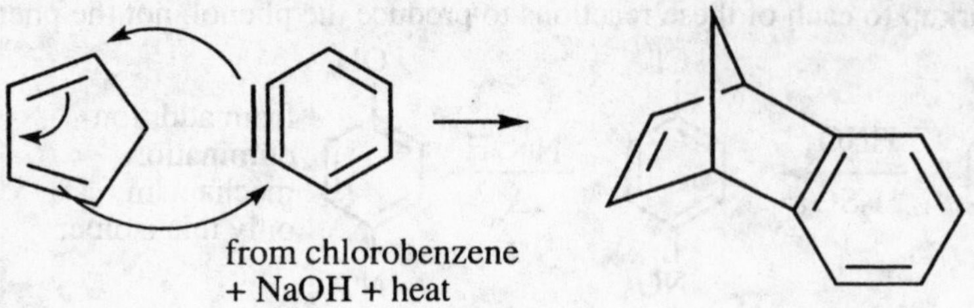

from chlorobenzene
+ NaOH + heat

17-28

(a) First, the carboxylic acid proton is neutralized.

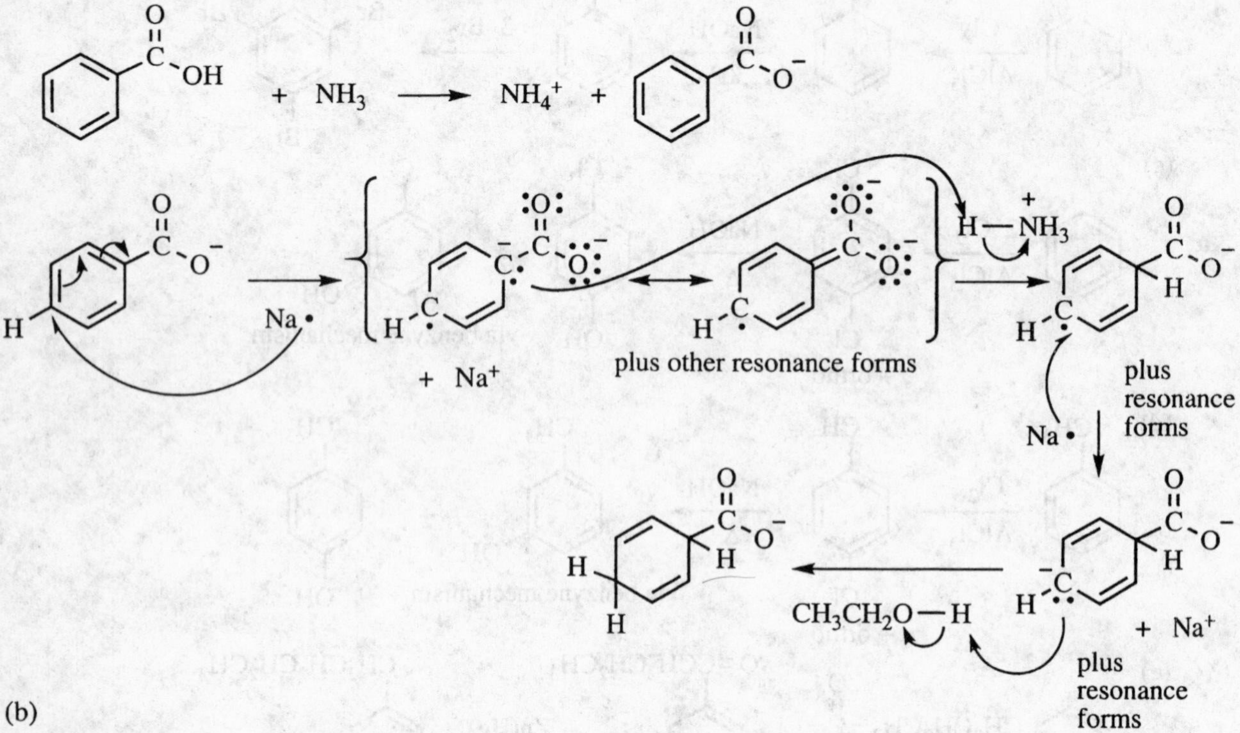

(b)

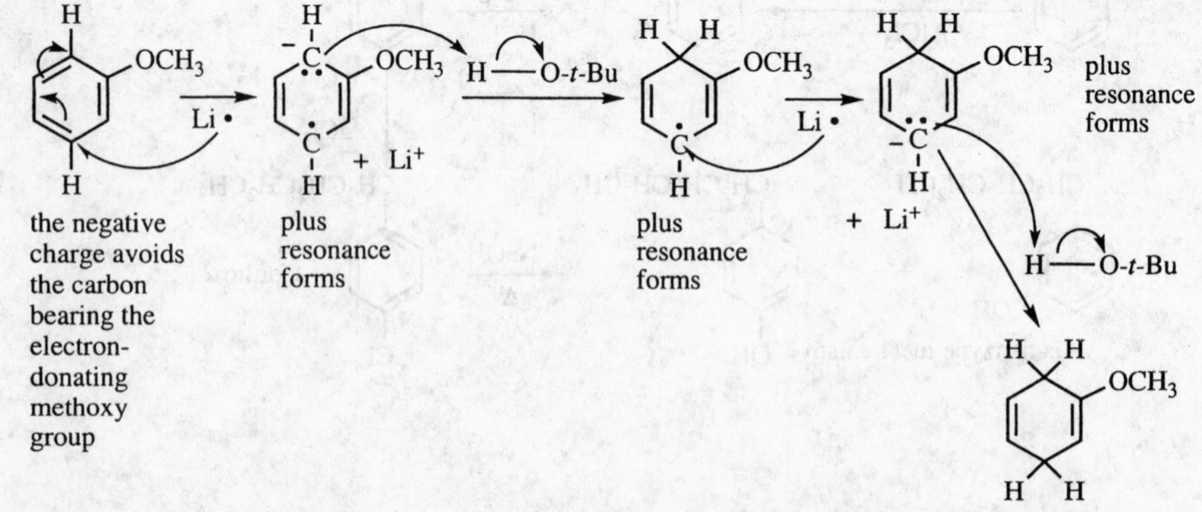

the negative
charge avoids
the carbon
bearing the
electron-
donating
methoxy
group

plus
resonance
forms

plus
resonance
forms

plus
resonance
forms

17-29

(a)

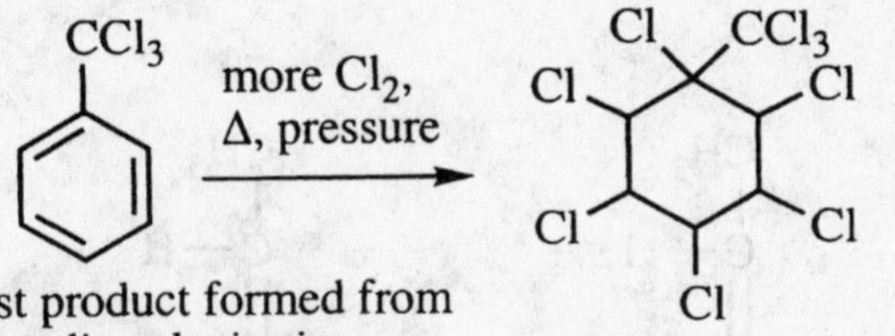

first product formed from benzylic substitution

$\xrightarrow{\text{more Cl}_2, \Delta, \text{ pressure}}$

addition of chlorine to the π system of the ring, giving a mixture of stereoisomers

(b)

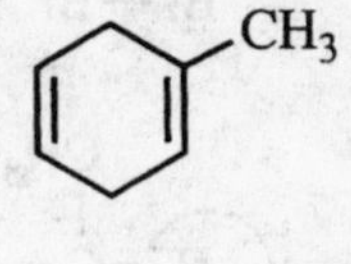

(c)

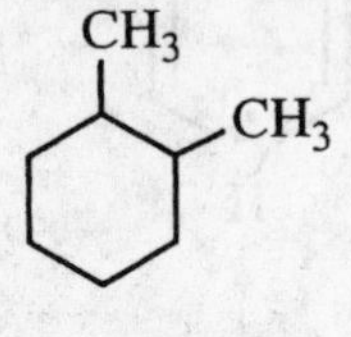

cis + trans

(d)

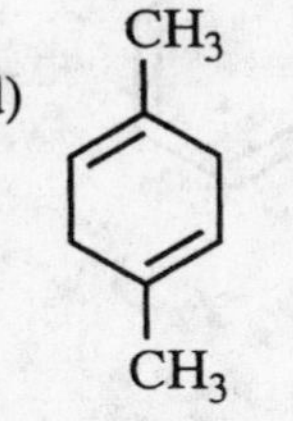

(e) 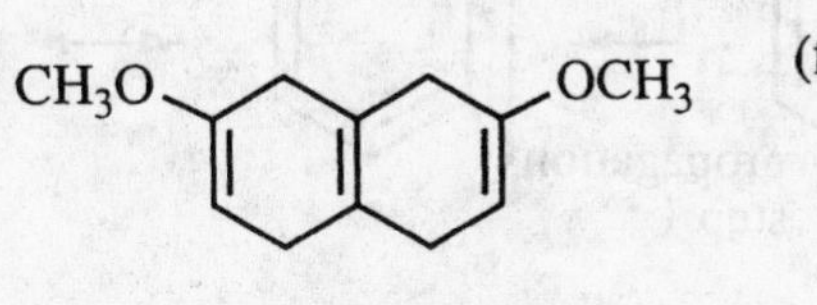 CH₃O ⸺ OCH₃

(f)

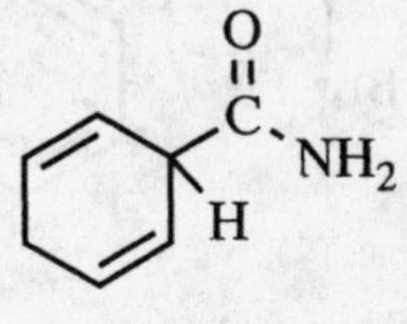

17-30

(a)

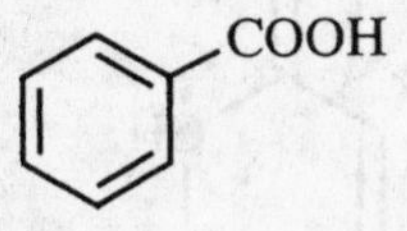

(b)

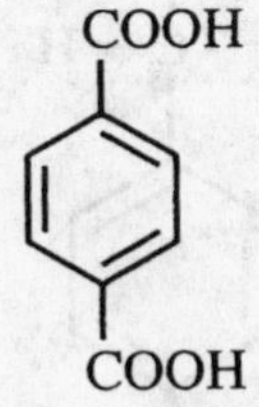

(c)

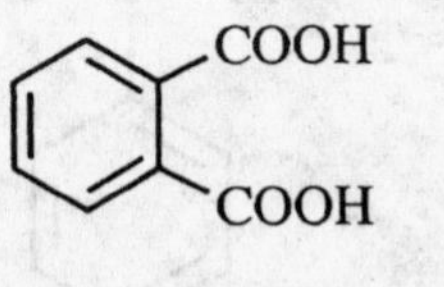

17-31

$\text{Cl—Cl} \xrightarrow{h\nu} 2\ \text{Cl} \cdot$ initiation

propagation step 1

propagation step 2

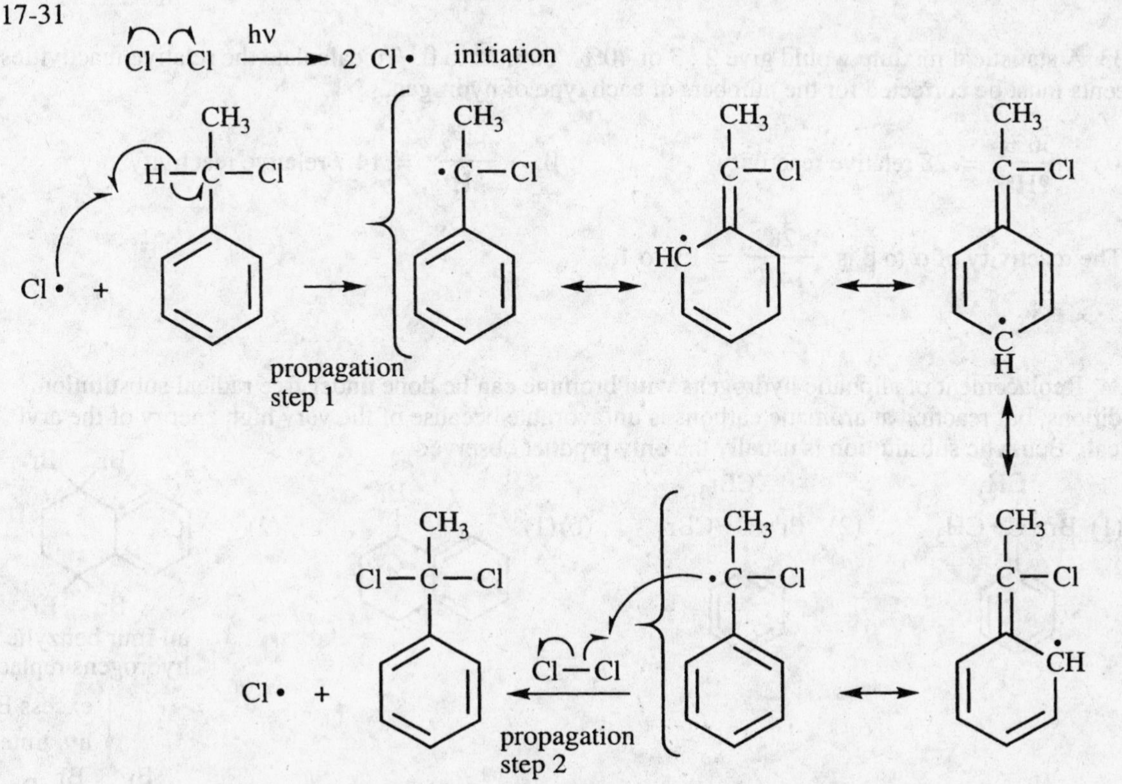

17-32

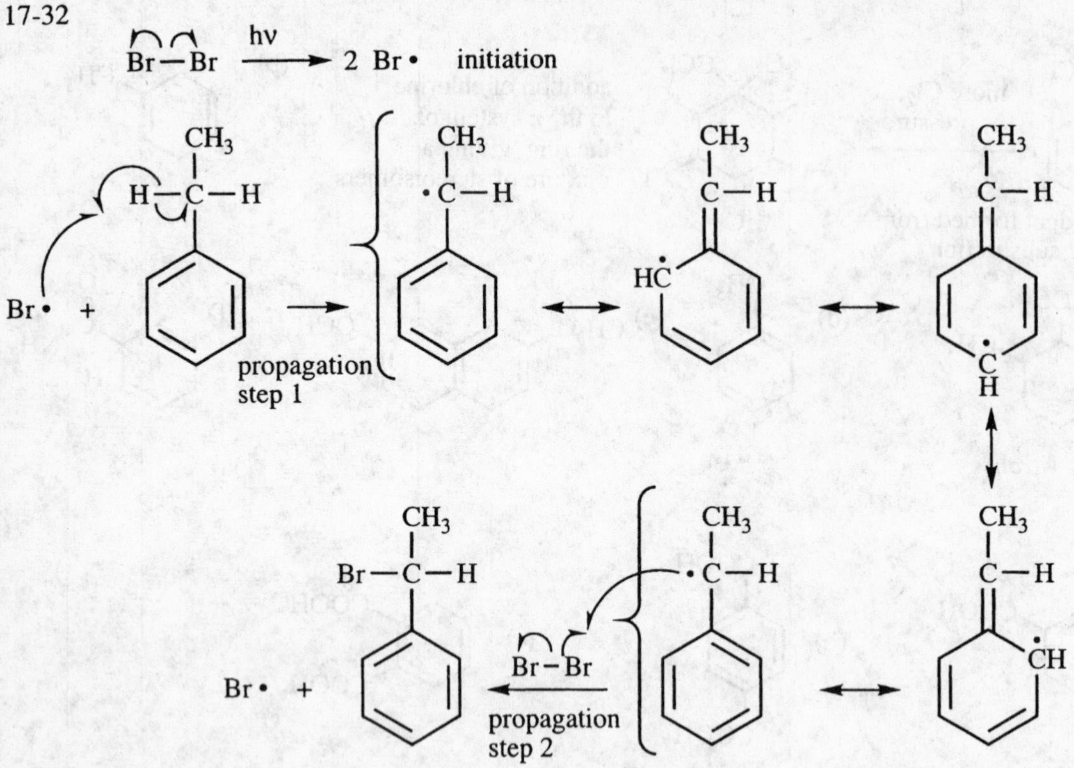

17-33 A statistical mixture would give 2 : 3 or 40% : 60% α to β. To calculate the relative reactivities, the percents must be corrected for the numbers of each type of hydrogen.

α: $\dfrac{56\%}{2H}$ = 28 relative reactivity   β: $\dfrac{44\%}{3H}$ = 14.7 relative reactivity

The reactivity of α to β is $\dfrac{28}{14.7}$ = 1.9 to 1

17-34 Replacement of aliphatic hydrogens with bromine can be done under free radical substitution conditions, but reaction at aromatic carbons is unfavorable because of the very high energy of the aryl radical. Benzylic substitution is usually the only product observed.

(a) (1)

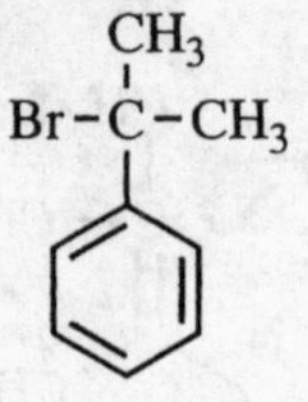

(2)

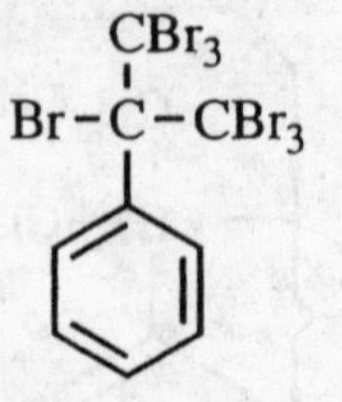

(b)(1)

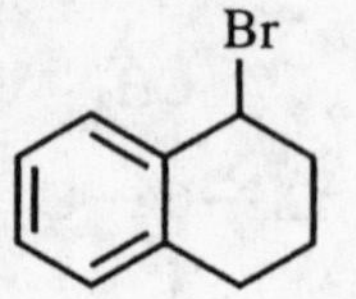

(2)

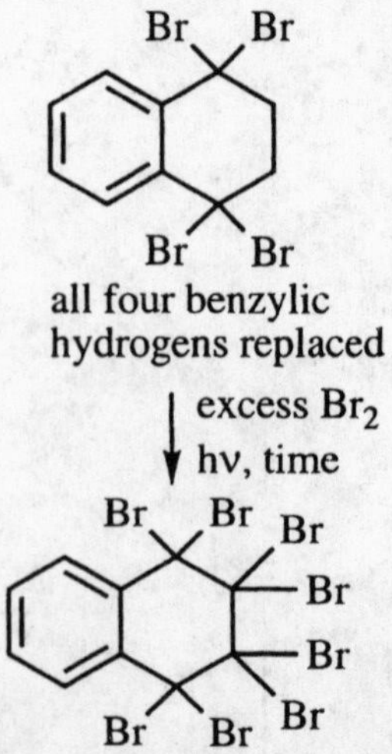

all four benzylic hydrogens replaced

excess Br_2 hv, time

17-35

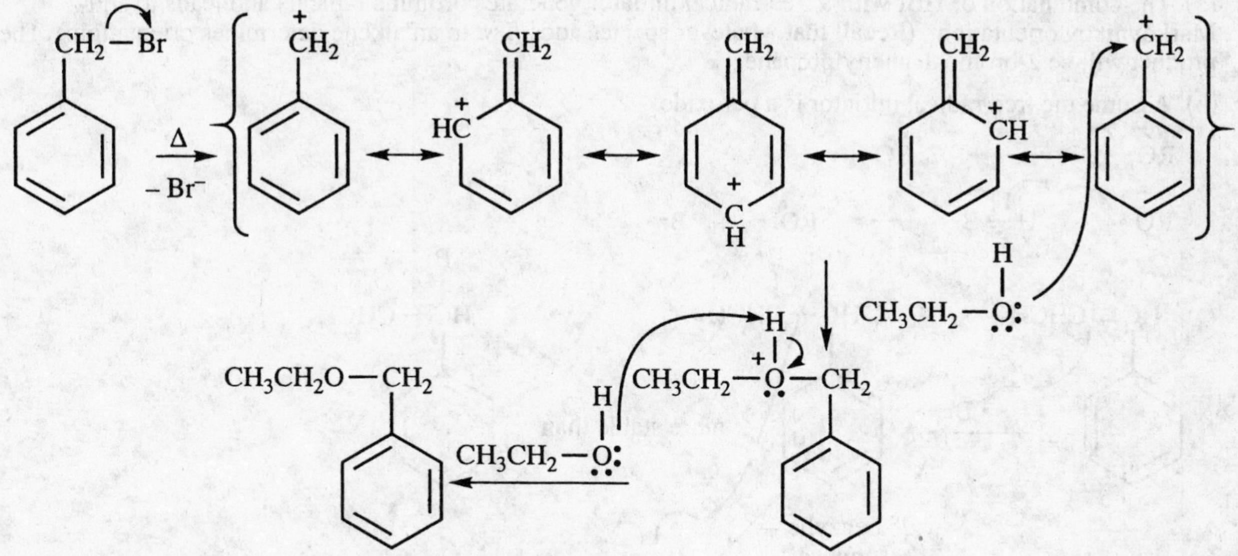

17-36

(a) Benzylic cations are stabilized by resonance and are much more stable than regular alkyl cations. The product is 1-bromo-1-phenylpropane.

(b)

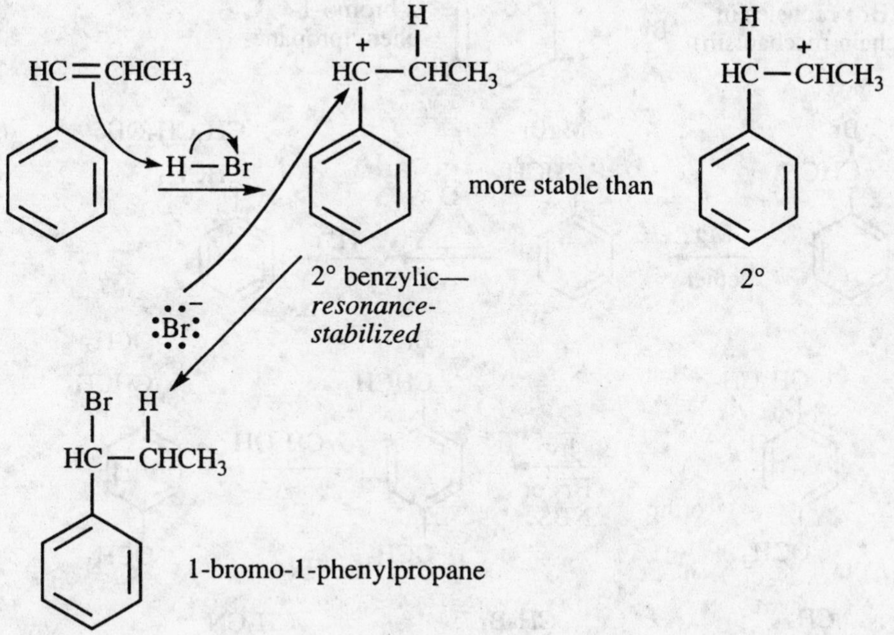

2° benzylic—
resonance-stabilized

more stable than

2°

1-bromo-1-phenylpropane

17-37

(a) The combination of HBr with a free-radical initiator generates bromine radicals and leads to anti-Markovnikov orientation. (Recall that whatever species adds *first* to an alkene determines orientation.) The product will be 2-bromo-1-phenylpropane.

(b) Assume the free-radical initiator is a peroxide.

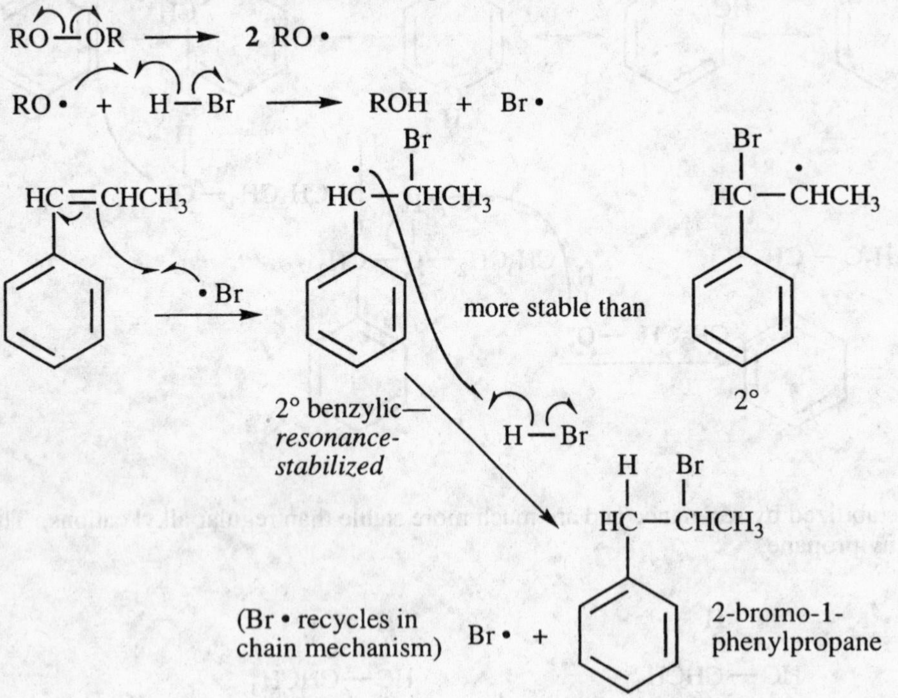

$$RO - OR \longrightarrow 2\ RO\cdot$$

$$RO\cdot\ +\ H - Br \longrightarrow ROH\ +\ Br\cdot$$

HC=CHCH₃ $\xrightarrow{\cdot Br}$ HC—CHCH₃ (Br)   more stable than   HC—CHCH₃ (Br)

2° benzylic—
resonance-stabilized

H—Br

2°

H Br
HC—CHCH₃

(Br· recycles in chain mechanism)   Br· +

2-bromo-1-phenylpropane

17-38

(a) HC=CH₂ $\xrightarrow{HBr}$ CHCH₃ (Br) $\xrightarrow[\text{ether}]{Mg}$ CHCH₃ (MgBr) $\xrightarrow{\triangle O}$ $\xrightarrow{H^+}$ CHCH₃ (CH₂CH₂OH)

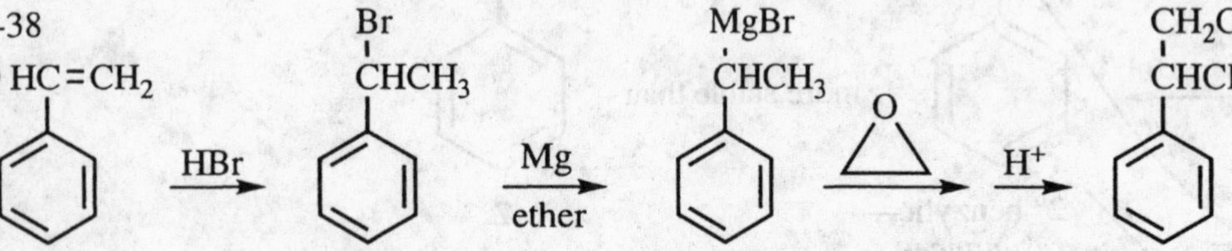

(b) $\xrightarrow[\text{AlCl}_3]{CH_3CH_2Br}$ CH₂CH₃ (OCH₃) + ortho $\xrightarrow[\text{Br}_2 \text{ or} \atop \text{NBS}]{h\nu}$ CHCH₃ (Br, OCH₃) $\xrightarrow[\Delta]{CH_3OH}$ CHCH₃ (OCH₃, OCH₃)

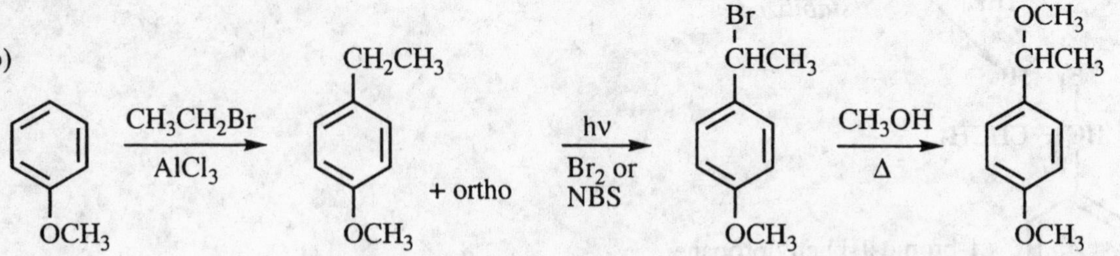

(c) CH₃ $\xrightarrow[\text{H}_2\text{SO}_4]{HNO_3}$ CH₃ (NO₂) + ortho $\xrightarrow[\text{Br}_2 \text{ or} \atop \text{NBS}]{h\nu}$ CH₂Br (NO₂) $\xrightarrow{NaCN}$ CH₂CN (NO₂)

384

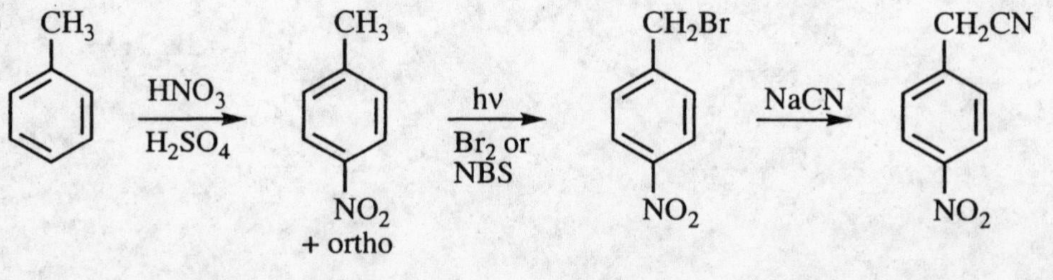

17-39

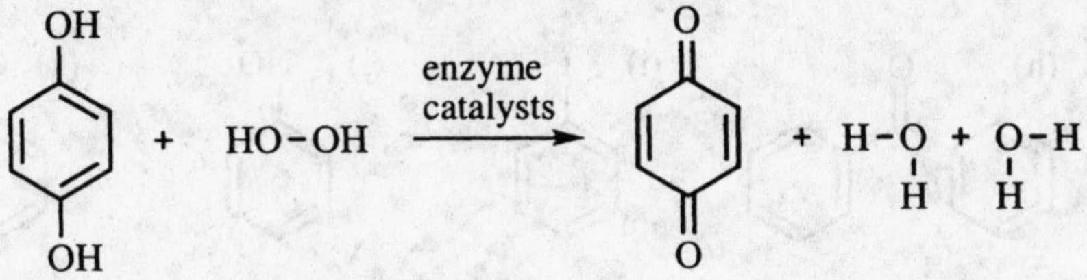

17-40

(a) OCH$_2$CH$_3$ ... CH$_3$

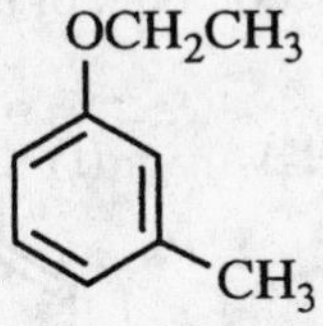

(b) O–C(=O)–CH$_3$ ... CH$_3$

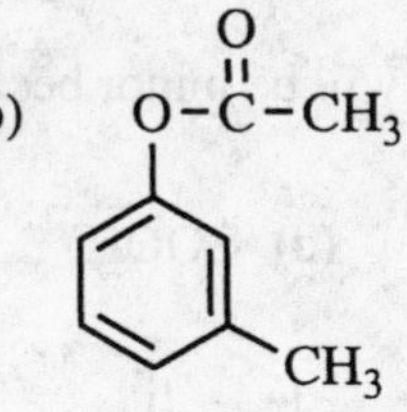

(c) OH ... CH$_3$, Br + Br, OH ... CH$_3$

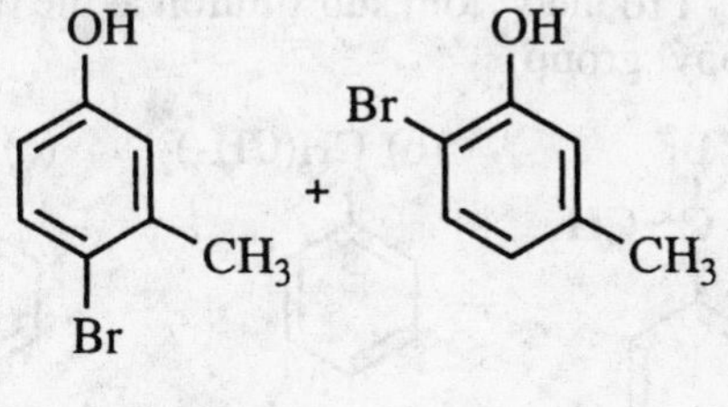

(d) OH, Br, Br, CBr$_3$, Br

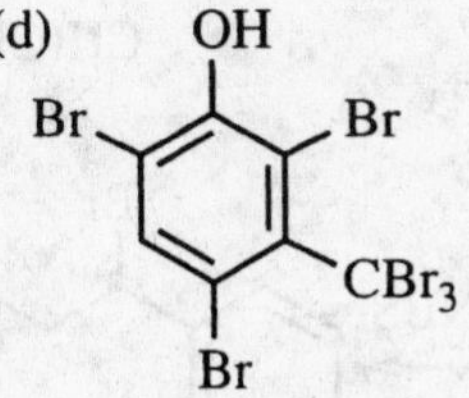

(e) O ... CH$_3$ ... O

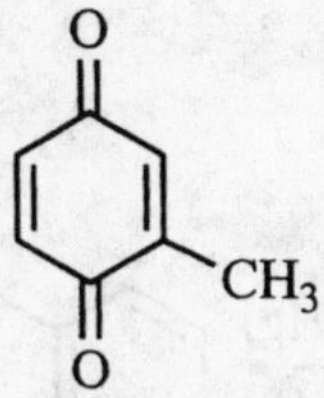

(f) (CH$_3$)$_3$C ... OH ... CH$_3$, C(CH$_3$)$_3$

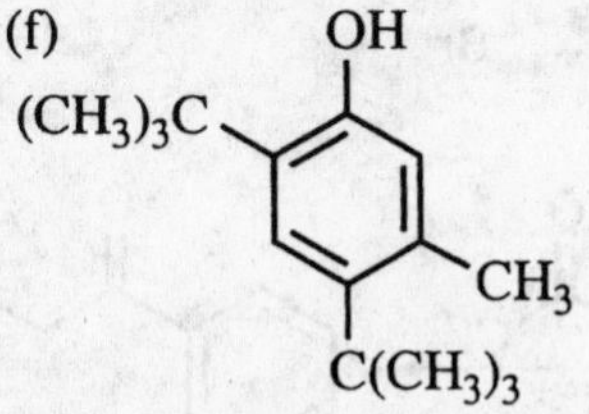

17-41

(a)

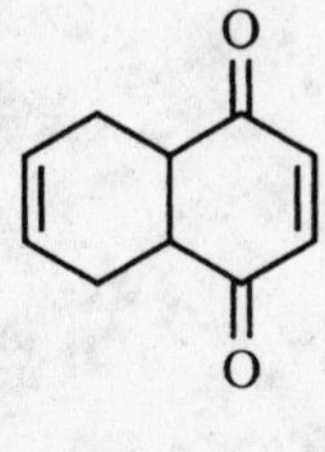

(b)

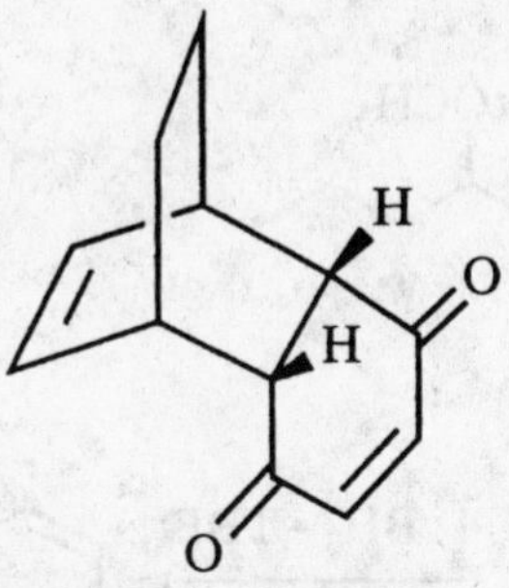

(c)

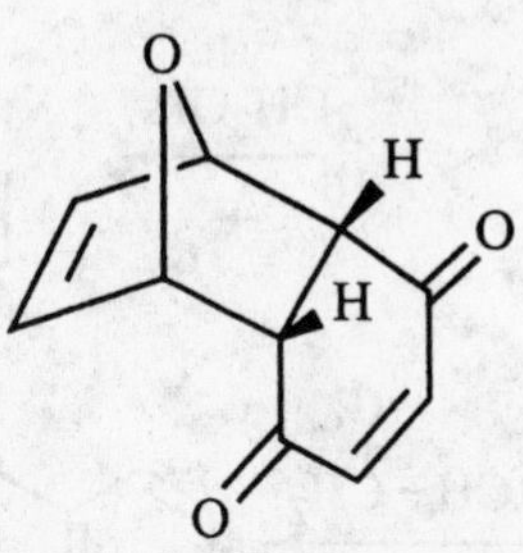

17-42

OH $\xrightarrow{\text{3 Br}_2}$ Br, OH, Br, Br $\xrightarrow{\text{Br}_2}$ O, Br, Br, Br, Br

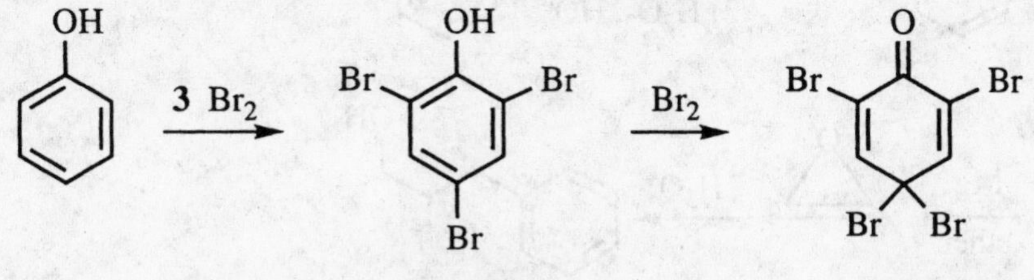

C$_6$H$_3$OBr$_3$ C$_6$H$_2$OBr$_4$

17-43 Please refer to solution 1-20, page 12 of this Solutions Manual.

17-44

(a) C(CH$_3$)$_3$ (b) CH$_3$–CHCH$_2$CH$_3$ (c) C(CH$_3$)$_3$ (d) Br (e) C(CH$_3$)$_3$ (f) SO$_3$H

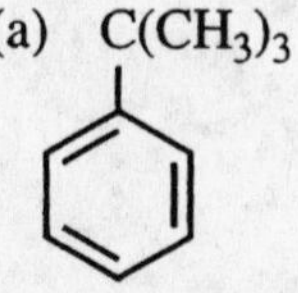

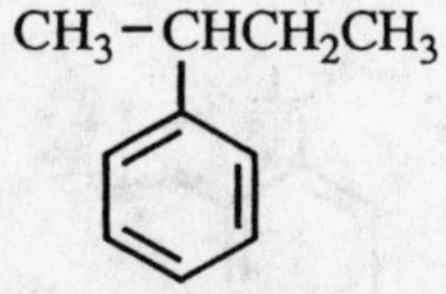

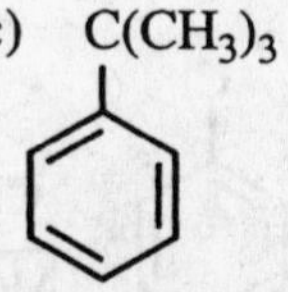

rearrangement   rearrangement

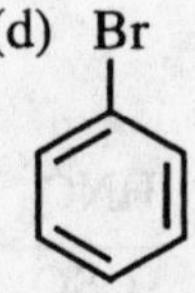

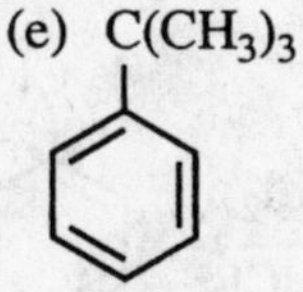

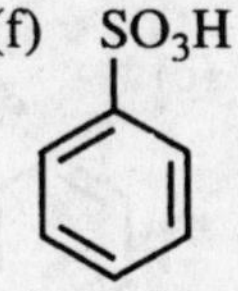

17-44 continued

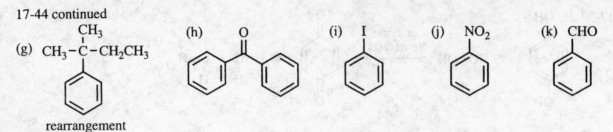

(g) $CH_3-\overset{\overset{\displaystyle CH_3}{|}}{\underset{|}{C}}-CH_2CH_3$ (with phenyl)

rearrangement

(h) diphenyl ketone (O)

(i) I (iodobenzene)

(j) NO₂ (nitrobenzene)

(k) CHO (benzaldehyde)

17-45 Products from substitution at the ortho position will be minor because of the steric bulk of the isopropyl group.

(a) $CH_3-\overset{\overset{\displaystyle Br}{|}}{\underset{|}{C}}-CH_3$ (with phenyl)

(b) CH(CH₃)₂ benzene with Br para

(c) CH(CH₃)₂ benzene with SO₃H para

(d) COOH benzene

(e) CH(CH₃)₂ benzene with $\overset{O}{\underset{||}{C}}-CH_3$ para

(f) CH(CH₃)₂ benzene with CH(CH₃)₂ para

17-46

(a) benzene $\xrightarrow[\text{AlCl}_3]{\overset{O}{\underset{||}{C}}\text{Cl (butanoyl)}}$ phenyl ketone $\xrightarrow[\text{HCl}]{\text{Zn(Hg)}}$ butylbenzene $\xrightarrow{\text{NBS}}$ benzene with CHBr chain

(b) benzene with CHBr chain, from (a) $\xrightarrow[\Delta]{\text{CH}_3\text{OH}}$ benzene with OCH₃ chain

(c) benzene $\xrightarrow[\text{AlCl}_3]{\text{CH}_2=\text{CH}-\overset{\overset{\displaystyle Cl}{|}}{\text{CH}}_2}$ allylbenzene $\xrightarrow[\text{2) H}_2\text{O}_2,\ \text{HO}^-]{\text{1) BH}_3\cdot\text{THF}}$ benzene with propyl–OH

OR CH₃ toluene $\xrightarrow[\text{or NBS}]{\text{Br}_2,\ h\nu}$ CH₂Br benzyl bromide $\xrightarrow[\text{ether}]{\text{Mg}}$ (epoxide) $\xrightarrow{\text{H}_3\text{O}^+}$ benzene with propyl–OH

(d) benzene $\xrightarrow[\text{AlCl}_3]{\text{Cl}_2}$ Cl chlorobenzene $\xrightarrow[350°C]{\text{NaOH}}$ OH phenol $\xrightarrow[\text{CH}_3\text{CH}_2\text{Br}]{\text{NaOH}}$ OCH₂CH₃

(e) benzene $\xrightarrow[\text{AlCl}_3]{\text{Cl}_2}$ Cl chlorobenzene $\xrightarrow[\text{H}_2\text{SO}_4]{\text{HNO}_3}$ Cl benzene with NO₂ para $\xrightarrow[\text{AlCl}_3]{\text{Cl}_2}$ Cl,Cl benzene with NO₂

17-46 continued

(f) three methods

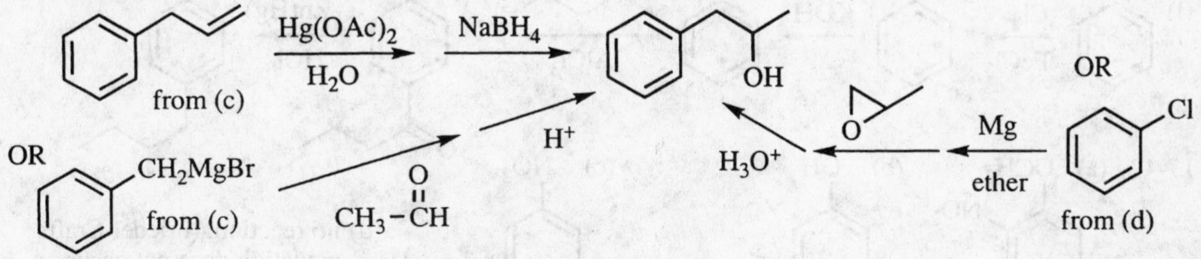

(g)

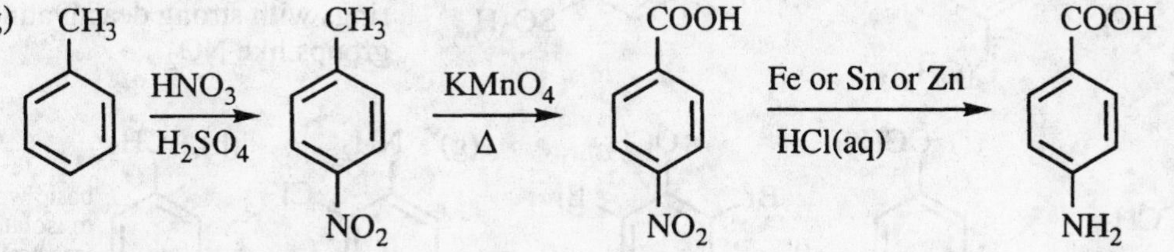

(h) two possible methods

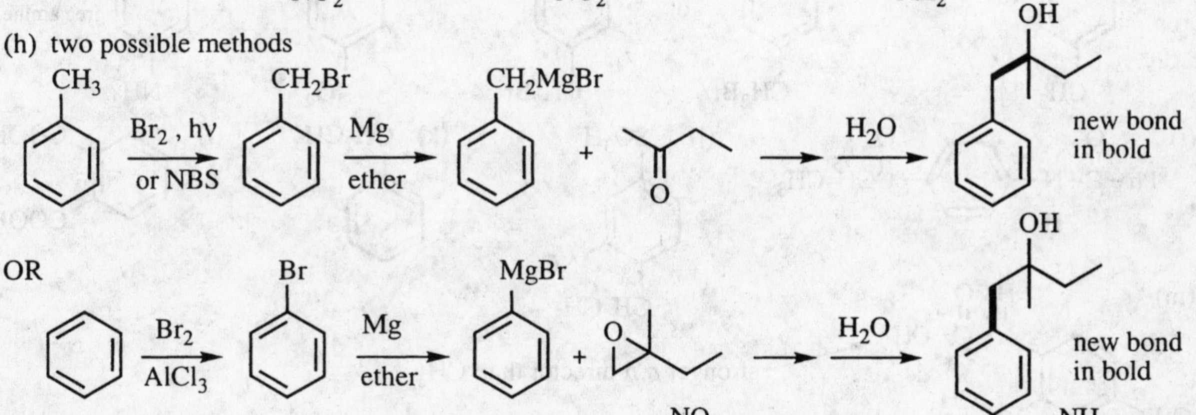

OR

(i)

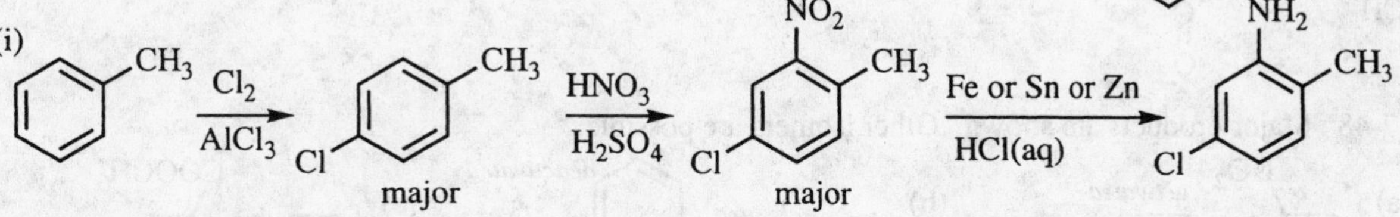

(j)

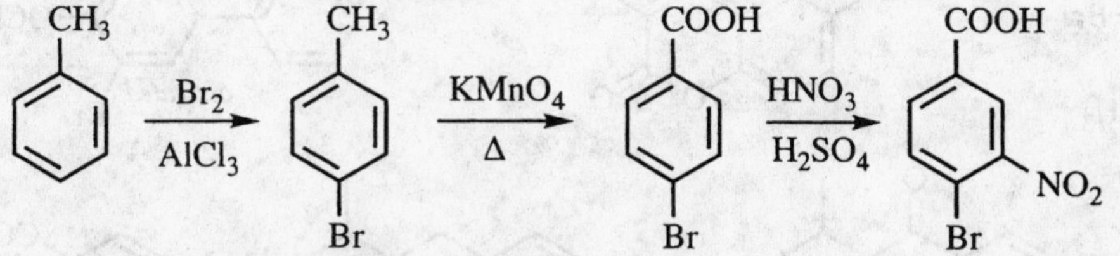

(k)

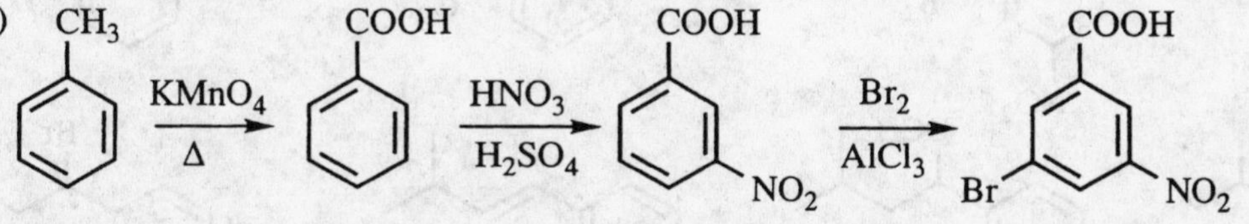

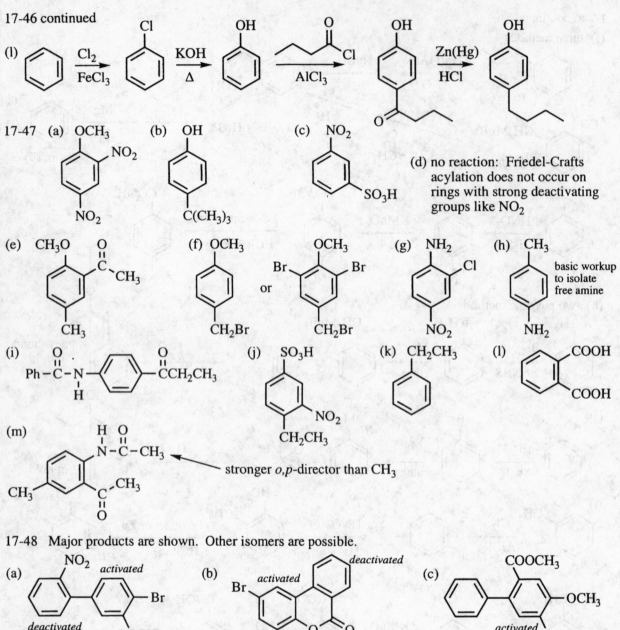

17-46 continued

(l)

17-47 (a) (b) (c) (d) no reaction: Friedel-Crafts acylation does not occur on rings with strong deactivating groups like NO_2

(e) (f) or (g) (h) basic workup to isolate free amine

(i) (j) (k) (l)

(m) stronger *o,p*-director than CH_3

17-48 Major products are shown. Other isomers are possible.

(a) *activated* *deactivated* (b) *deactivated* *activated* (c) *activated*

17-49 A B C D

E F G H (same as E)

17-50

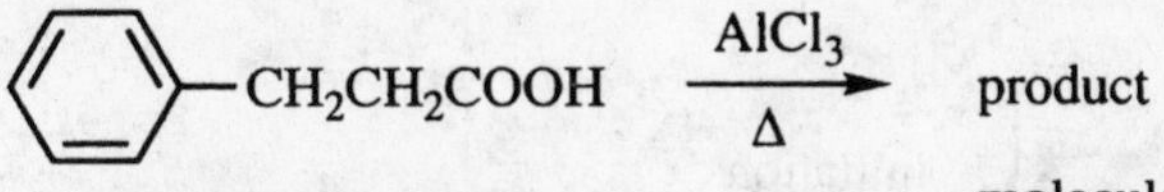

starting material
molecular weight 150

product

molecular weight 132 =
loss of 18 = loss of H₂O

IR spectrum: The dominant peak is the carbonyl at 1710 cm⁻¹. No COOH stretch.

NMR spectrum: The splitting is complicated but the integration is helpful. In the region of δ 2.6-3.2, there are two signals each with integration value of 2H; these must be the two adjacent methylenes, CH₂CH₂. The aromatic region from δ 7.3-7.8 has integration of 4H, so the ring must be disubstituted.

Carbon NMR: Four of the aromatic signals are tall, indicating C—H; two are short, with no attached hydrogens, also showing a disubstituted benzene. Also indicated are carbons in a carbonyl and two methylenes.

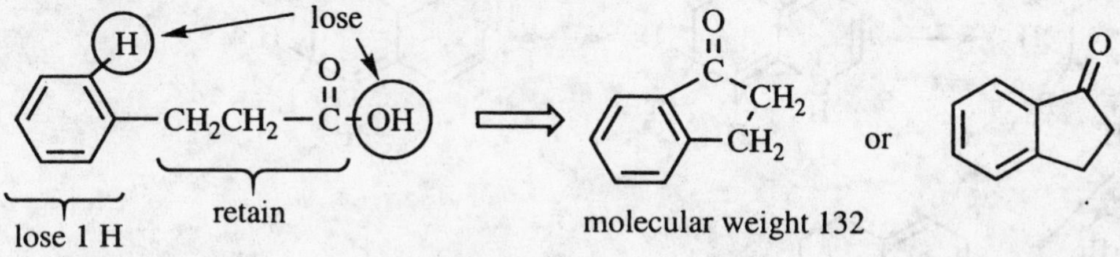

The product must be the cyclized ketone, formed in an intramolecular Friedel-Crafts acylation.

17-51

(a)

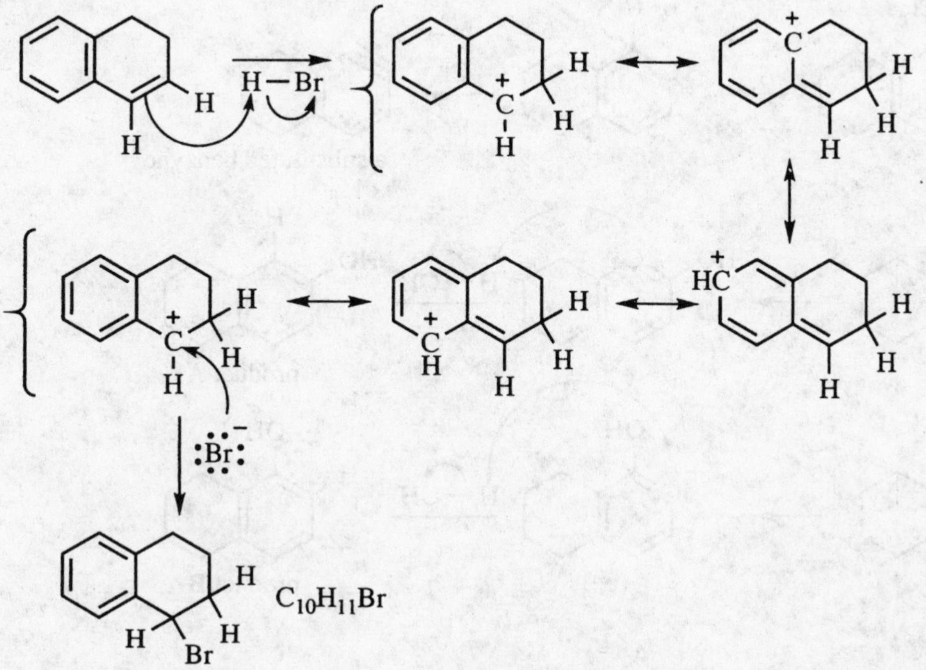

17-51 continued

(b) Assume the free-radical initiator is a peroxide.

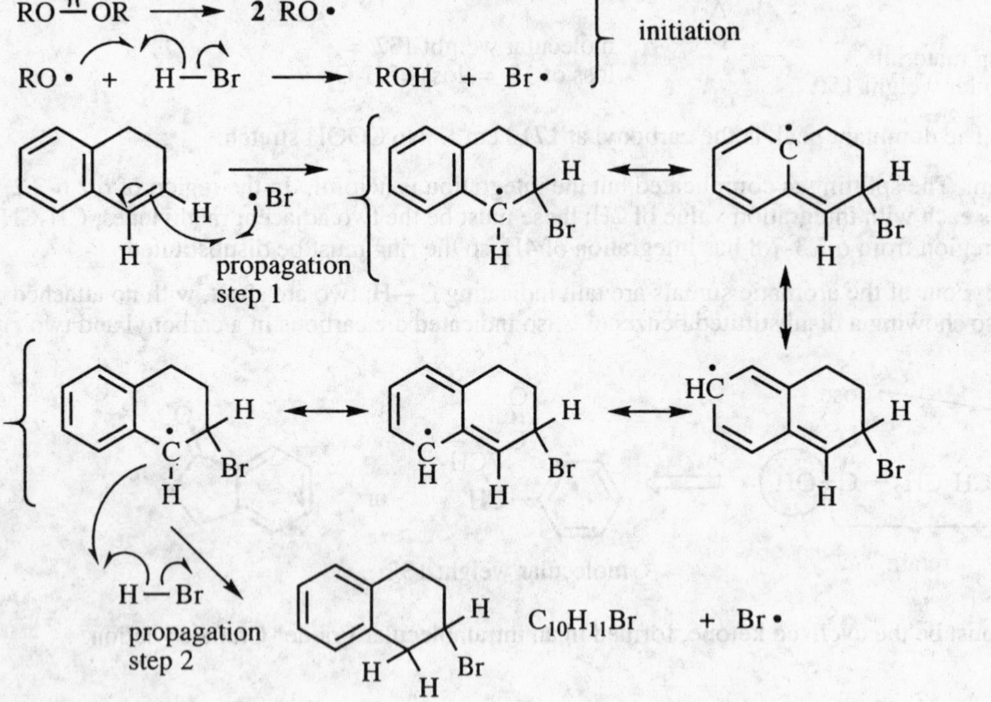

$$RO{-}OR \longrightarrow 2\ RO\cdot$$

$$RO\cdot\ +\ H{-}Br \longrightarrow ROH\ +\ Br\cdot$$

initiation

propagation step 1

propagation step 2

$C_{10}H_{11}Br$ + Br·

17-52

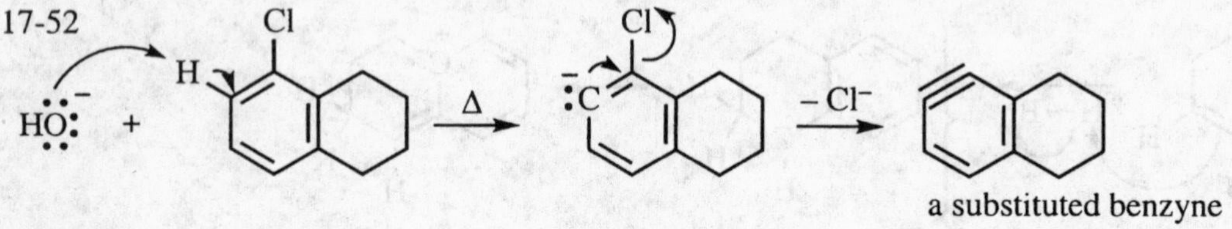

a substituted benzyne

Attack A

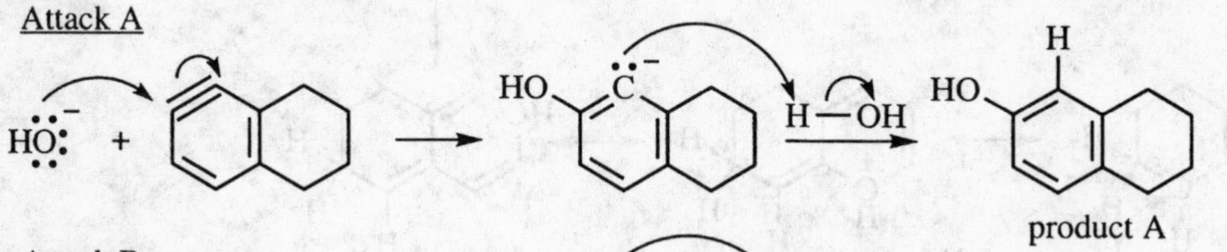

product A

Attack B

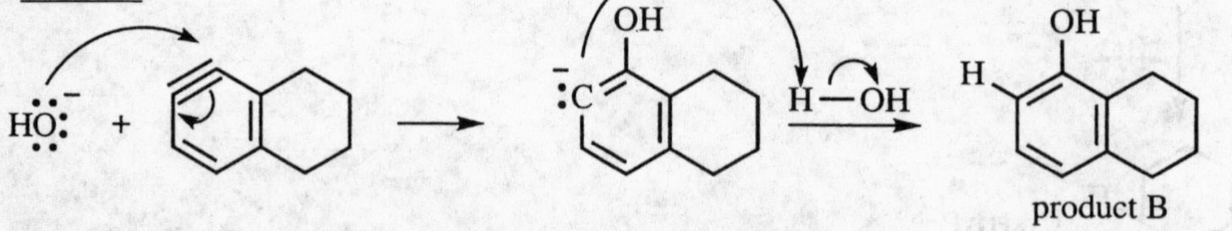

product B

17-53 The electron-withdrawing carbonyl group stabilizes the adjacent negative charge.

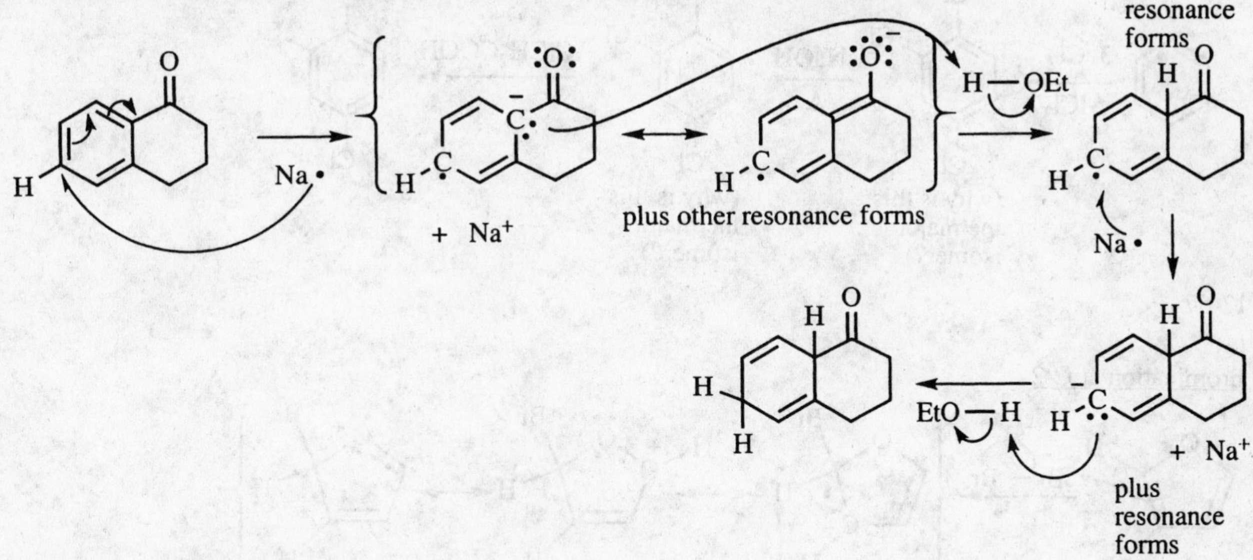

17-54

(a) NO₂

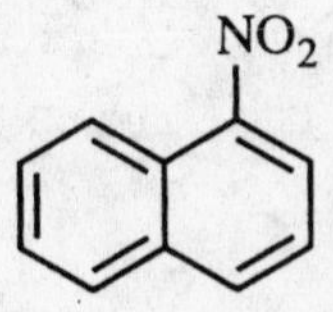

(b) Br

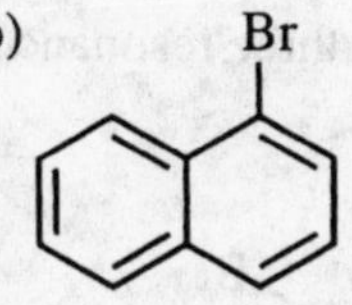

(c)

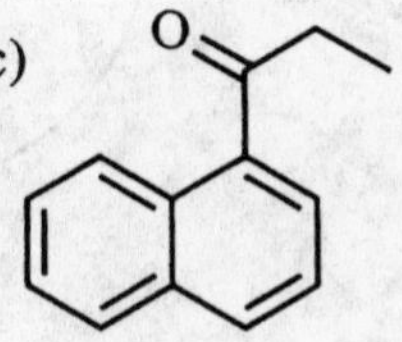

(d) C(CH₃)₃

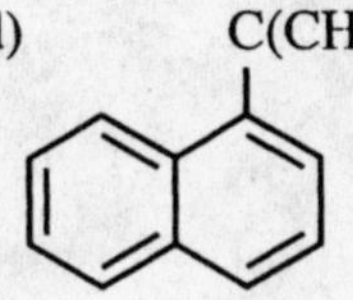

(e)

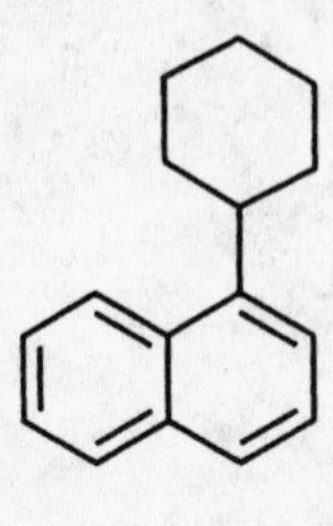

(f) SO₃H

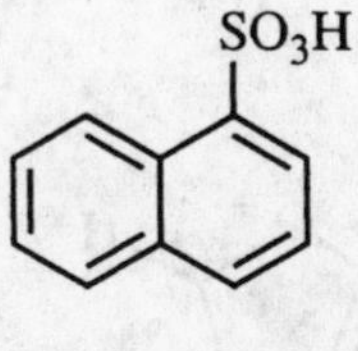

17-55

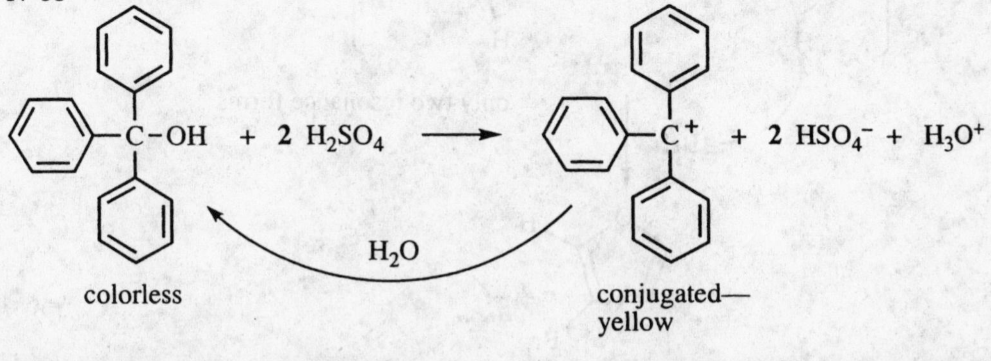

colorless   conjugated—
                             yellow

Concentrated sulfuric acid "dehydrates" the alcohol, producing a highly conjugated, colored carbocation, and protonates the water to prevent the reverse reaction. Upon adding more water, however, there are too many water molecules for the acid to protonate, and triphenylmethanol is regenerated.

17-56

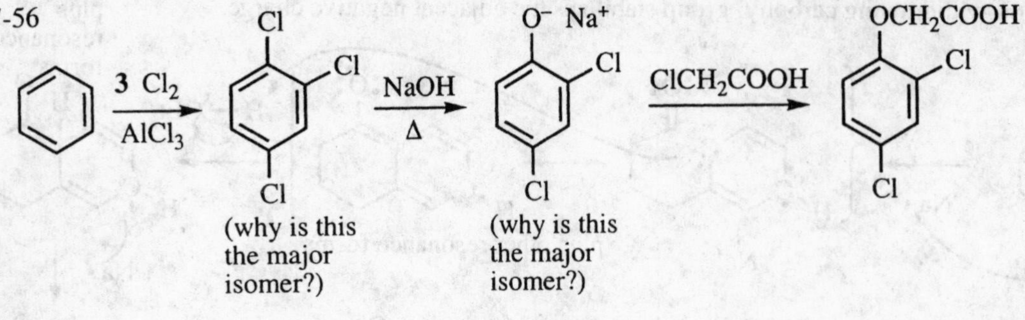

(why is this the major isomer?)

(why is this the major isomer?)

17-57

(a)

bromination at C-2

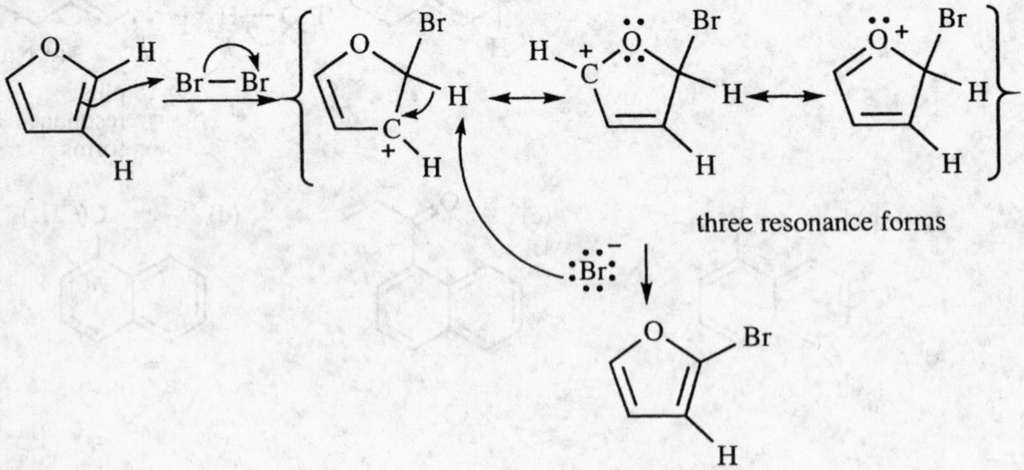

three resonance forms

bromination at C-3

only two resonance forms

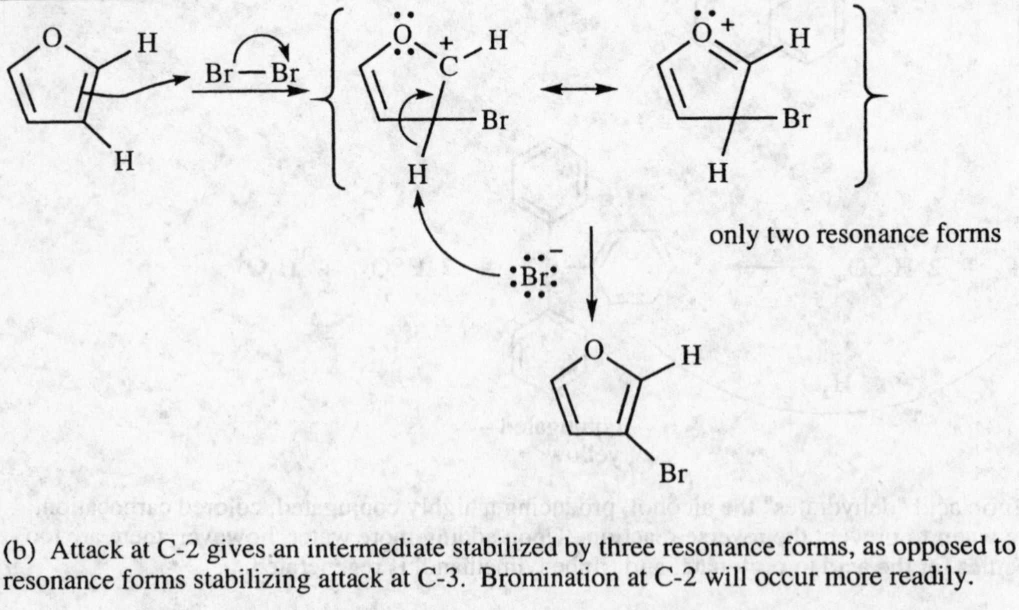

(b) Attack at C-2 gives an intermediate stabilized by three resonance forms, as opposed to only two resonance forms stabilizing attack at C-3. Bromination at C-2 will occur more readily.

17-58

abbreviate PhCH$_2$—CHCOOH as R
(NH$_2$ above CHCOOH) ... NH$_2$ above R

phenylalanine

−F$^-$

O$_2$N ... NO$_2$

393

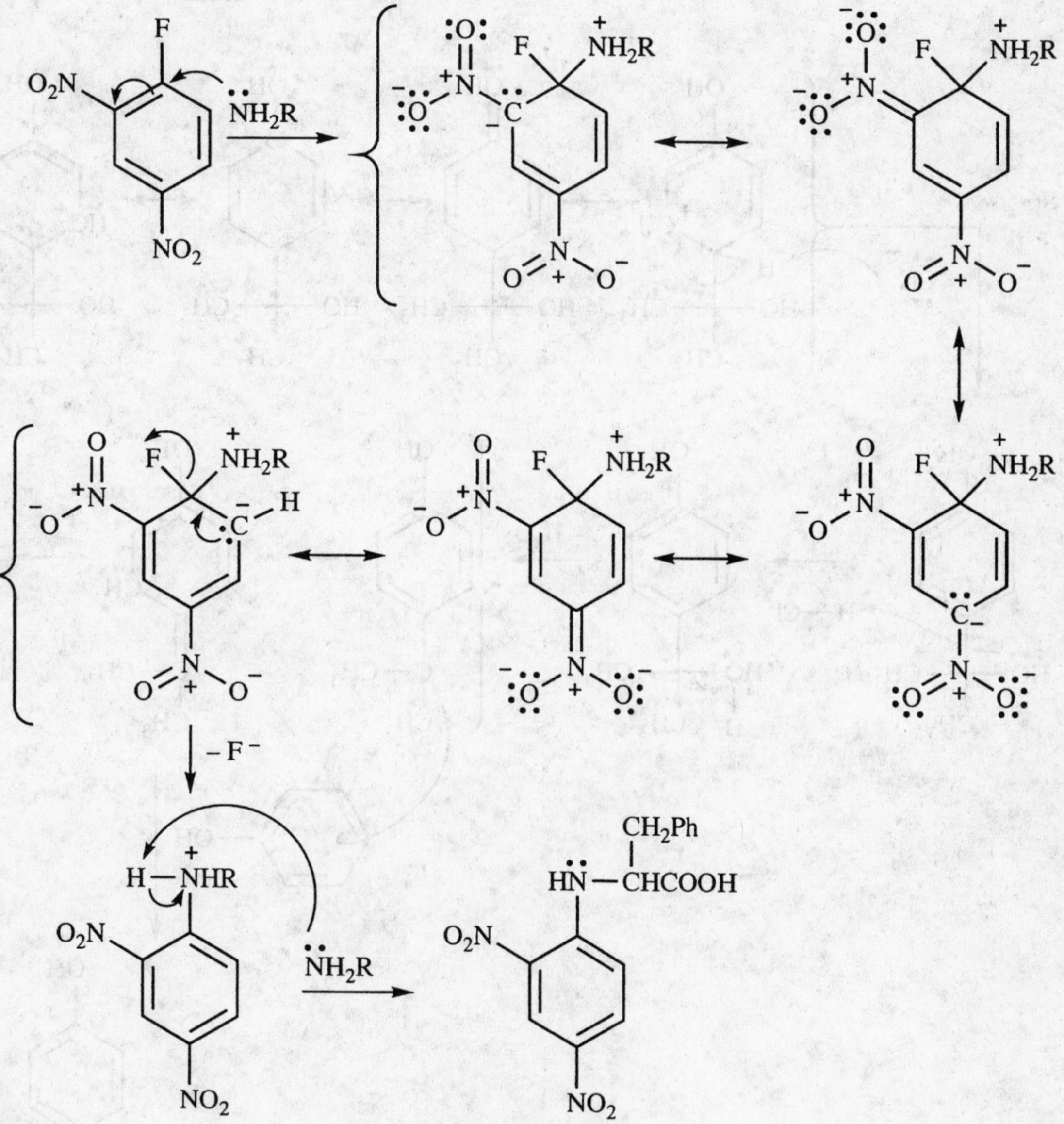

17-59

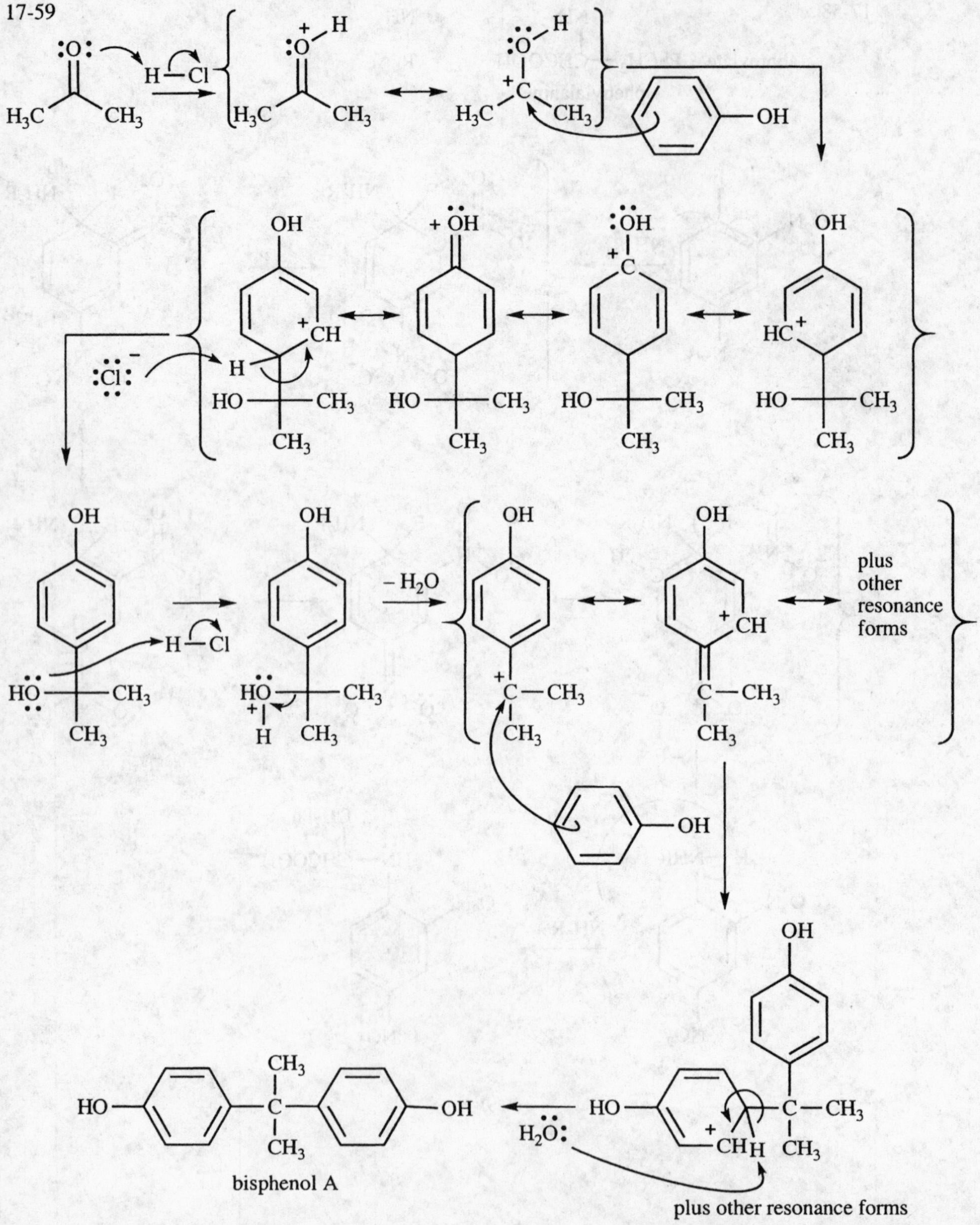

− H₂O
plus other resonance forms
bisphenol A
plus other resonance forms

17-60

(a) This is an example of kinetic versus thermodynamic control of a reaction. At low temperature, the kinetic product predominates: in this case, almost a 1 : 1 mixture of ortho and para. These two isomers must be formed at approximately equal rates at 0° C. At 100° C, however, enough energy is provided for the *desulfonation* to occur rapidly; the large excess of the para isomer indicates the para is more stable, even though it is formed initially at the same rate as the ortho.

(b) The product from the 0° C reaction will equilibrate as it is warmed, and at 100° C will produce the same ratio of products as the reaction which was run initially at 100° C.

17-61

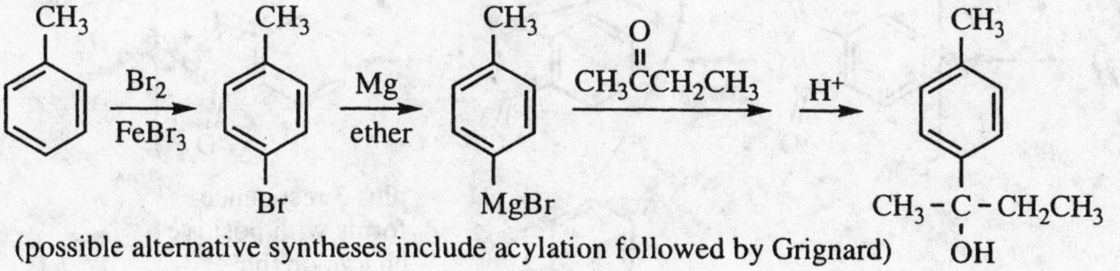

(possible alternative syntheses include acylation followed by Grignard)

17-62

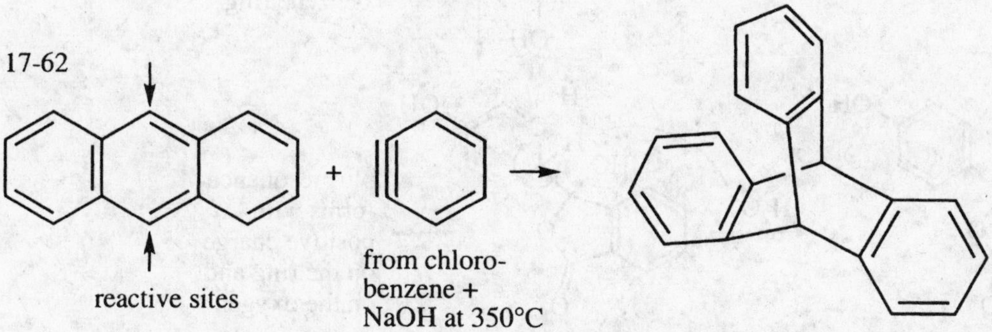

reactive sites

from chloro-
benzene +
NaOH at 350°C

As we saw in Chapter 16, the carbons of the center ring of anthracene are susceptible to electrophilic addition, leaving two isolated benzene rings on the ends. Benzyne is such a reactive dienophile that the reluctant anthracene is forced into a Diels-Alder reaction.

17-63

(a)

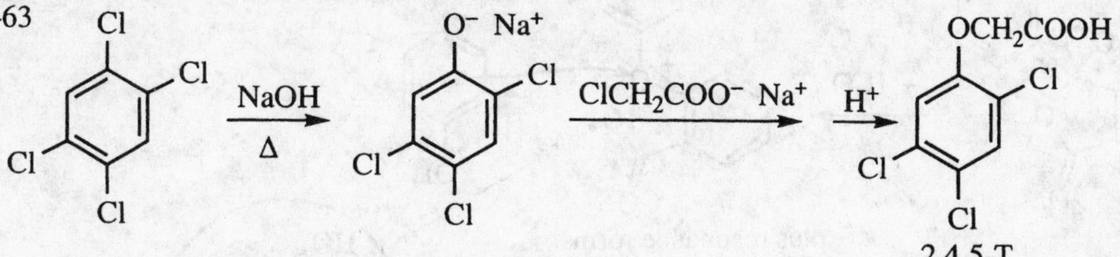

2,4,5-T

(b)

Δ 2 Cl$^-$ +

TCDD

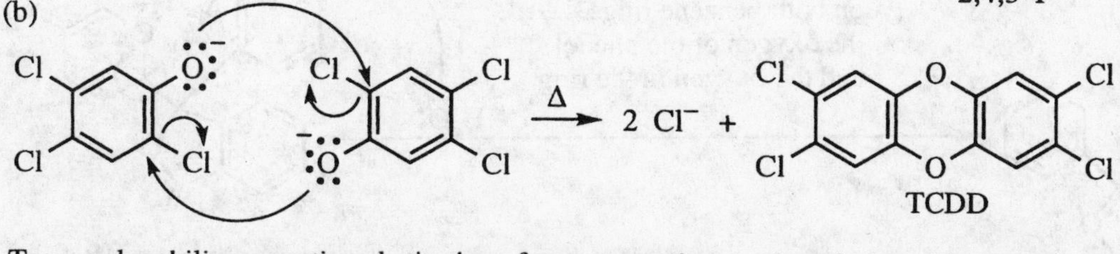

(This is the compound used to poison Ukrainian political leader, Boris Yushchenko.)

Two nucleophilic aromatic substitutions form a new six-membered ring. (Though not shown here, this reaction would follow the standard addition-elimination mechanism.)

17-63 continued

(c) To minimize formation of TCDD during synthesis: 1) keep the solutions dilute; 2) avoid high temperature; 3) replace chloroacetate with a more reactive molecule like bromoacetate or iodoacetate; 4) add an excess of the haloacetate.

To separate TCDD from 2,4,5-T at the end of the synthesis, take advantage of the acidic properties of 2,4,5-T. The 2,4,5-T will dissolve in an aqueous solution of a weak base like NaHCO$_3$. The TCDD will remain insoluble and can be filtered or extracted into an organic solvent like ether or dichloromethane. The 2,4,5-T can be precipitated from aqueous solution by adding acid.

17-64
(a)

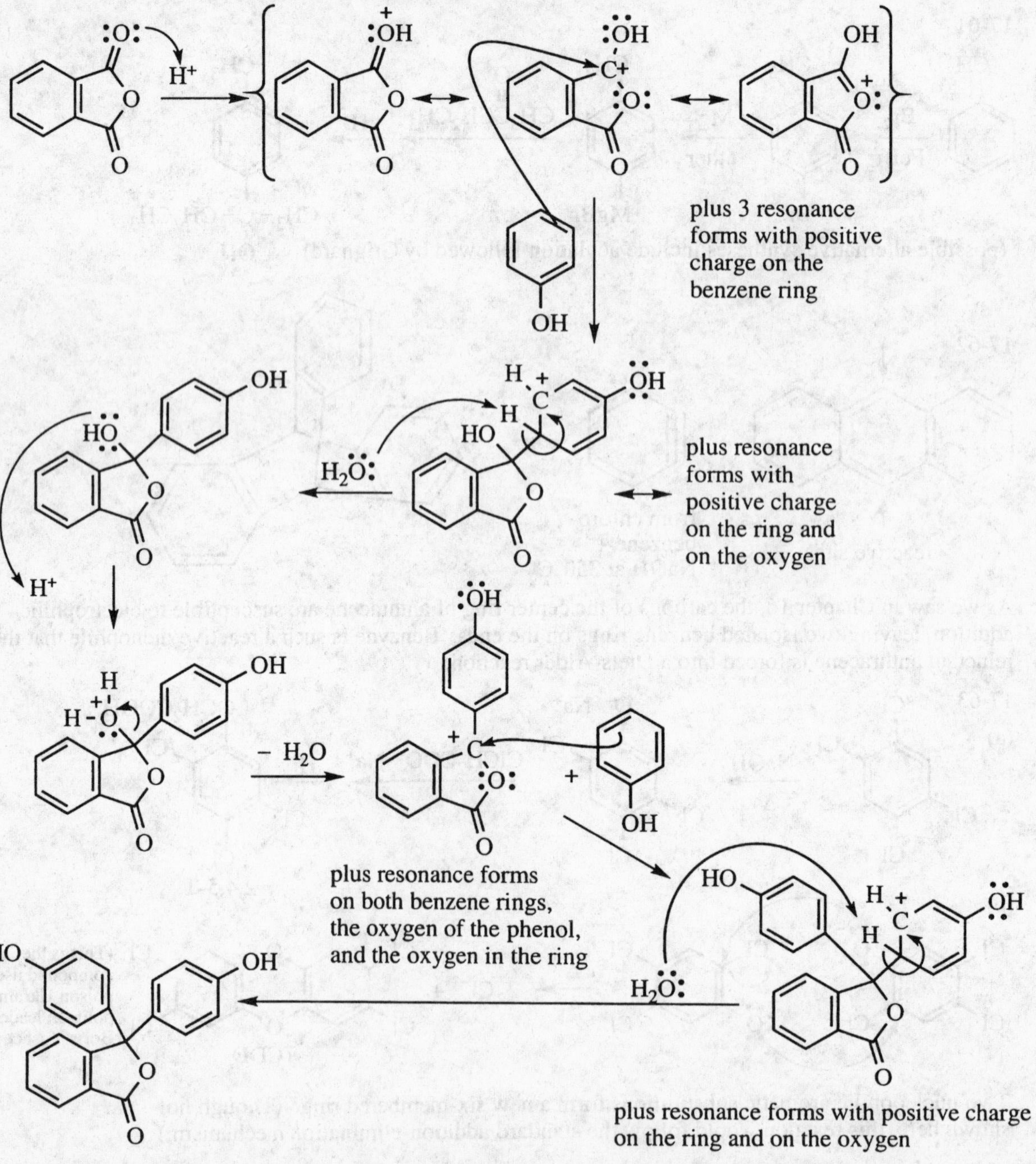

plus 3 resonance forms with positive charge on the benzene ring

plus resonance forms with positive charge on the ring and on the oxygen

plus resonance forms on both benzene rings, the oxygen of the phenol, and the oxygen in the ring

plus resonance forms with positive charge on the ring and on the oxygen

17-64 continued

(b)

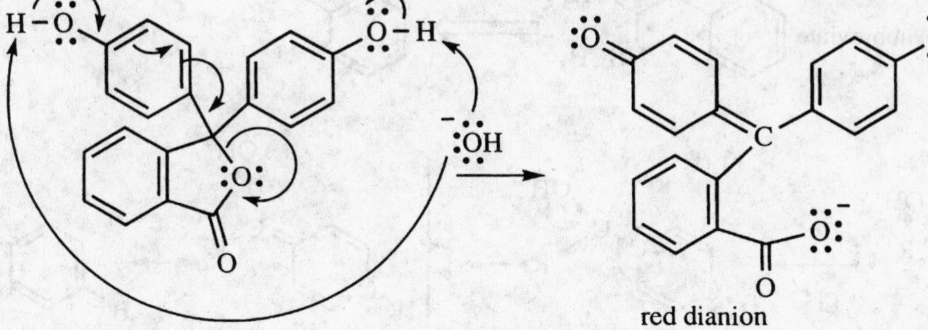

red dianion

(c)

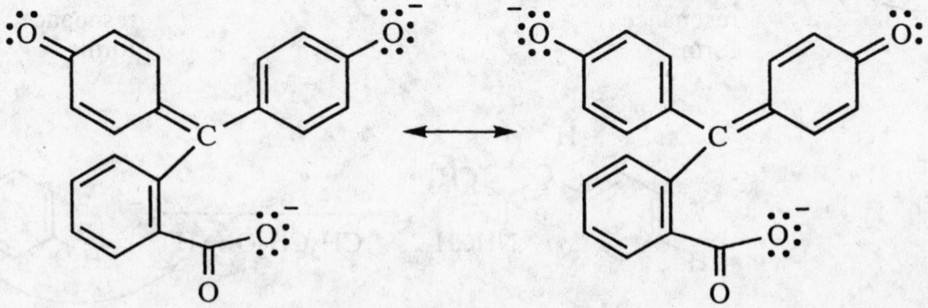

17-65

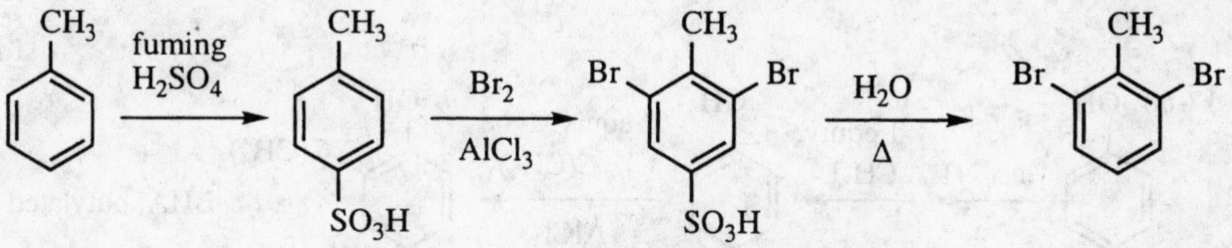

para position blocked

17-66 A benzyne must have been generated from the Grignard reagent.

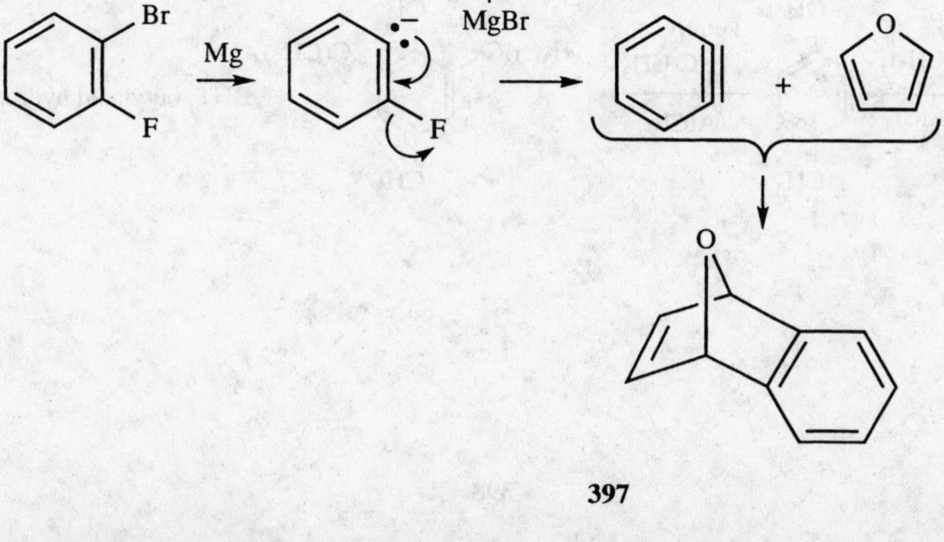

397

17-67 The intermediate anion forces the loss of hydroxide.

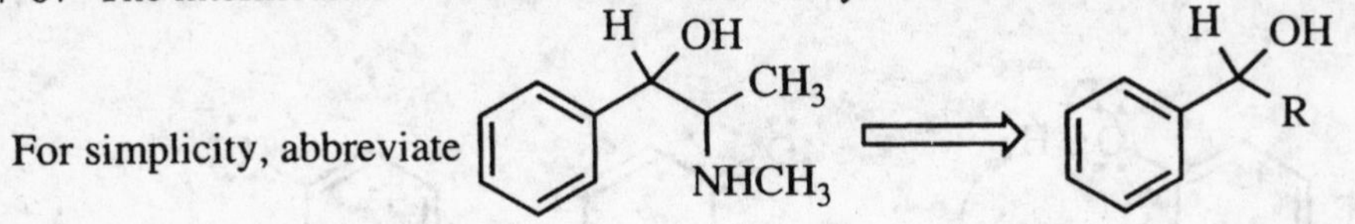

For simplicity, abbreviate
H
OH
CH3
NHCH3
H
OH
R

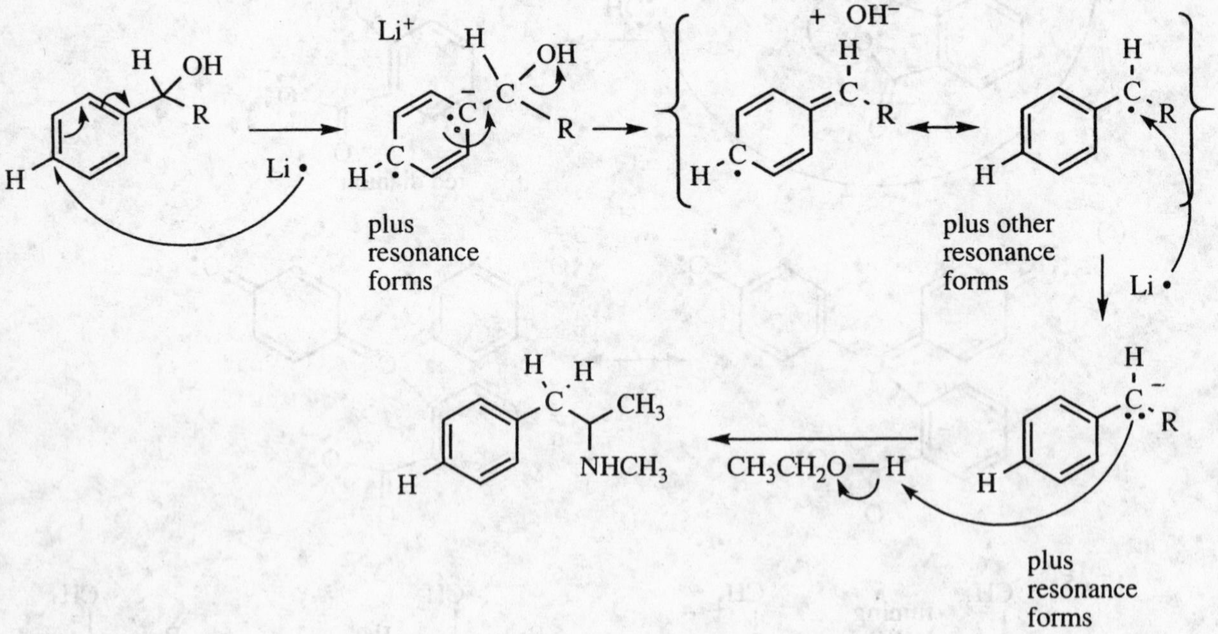

H
OH
R
H
Li
Li+
H
OH
C
C
R
H
C
plus
resonance
forms
+ OH−
H
C
R
H
C
H
C
R
H
plus other
resonance
forms
Li
H
C
R
H
CH3CH2O — H
H H
C
CH3
NHCH3
H
plus
resonance
forms

17-68

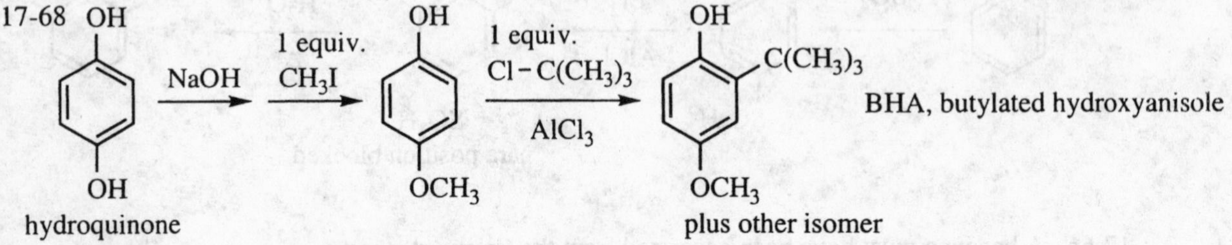

OH
OH
hydroquinone
NaOH
1 equiv.
CH3I
OH
OCH3
1 equiv.
Cl − C(CH3)3
AlCl3
OH
C(CH3)3
OCH3
plus other isomer
BHA, butylated hydroxyanisole

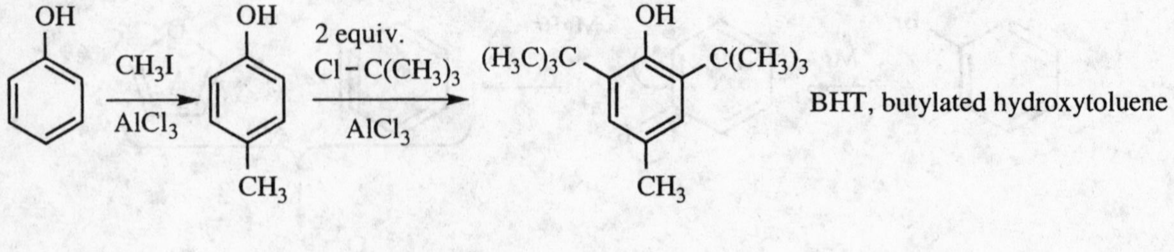

OH
CH3I
AlCl3
OH
CH3
2 equiv.
Cl − C(CH3)3
AlCl3
OH
(H3C)3C
C(CH3)3
CH3
BHT, butylated hydroxytoluene

17-69 Solve the problem by writing the mechanism. (See the solution to problem 17-25(a) for an identical mechanism.)

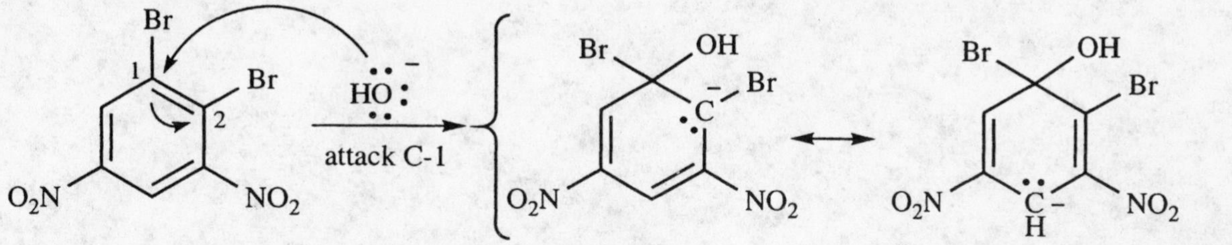

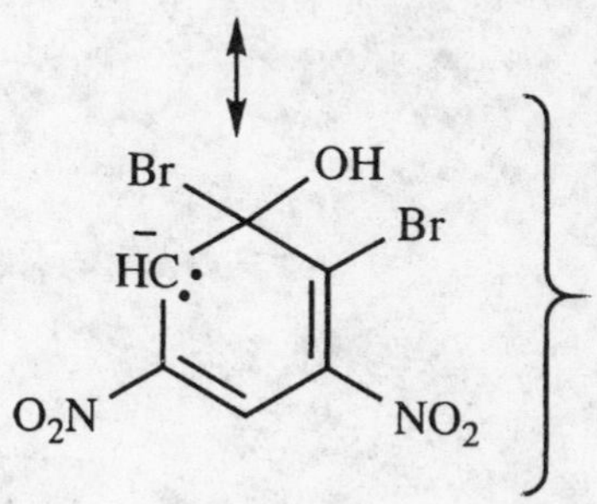

Hydroxide attack on C-1 puts the negative charge on carbons that do not have the NO_2 groups, so these anions are not stabilized.

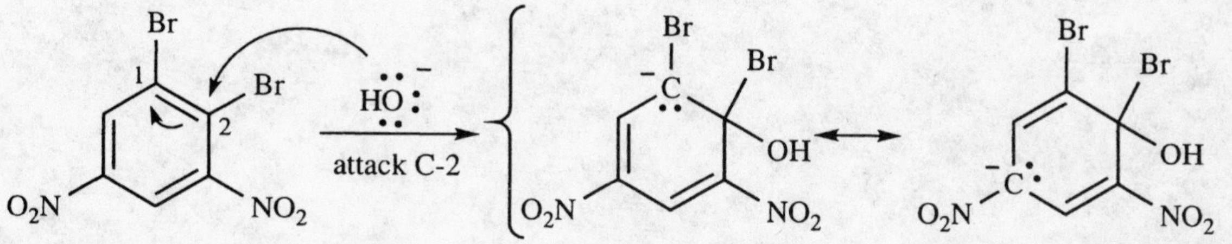

attack C-1

Hydroxide attack on C-2 puts the negative charge on carbons with nitro groups, thereby increasing the stabilization by delocalizing the negative charge. This intermediate is formed preferentially.

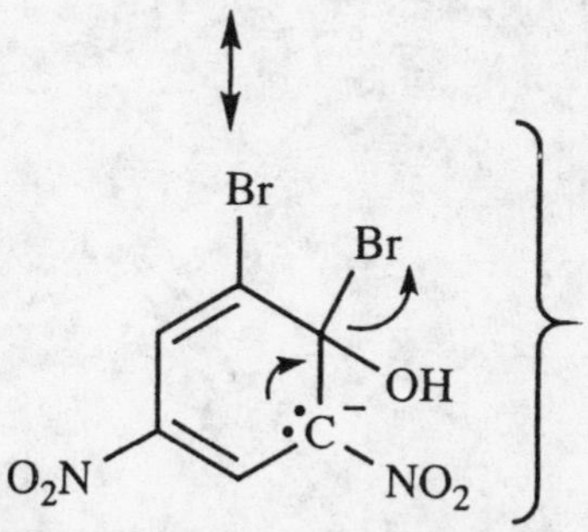

attack C-2

also stabilized by resonance onto the nitro group

also stabilized by resonance onto the nitro group

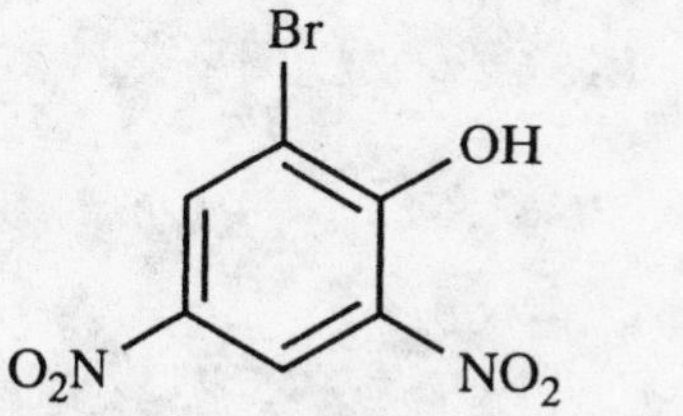

only product formed

CHAPTER 18—KETONES AND ALDEHYDES

18-1

(a) 5-hydroxyhexan-3-one; ethyl β-hydroxypropyl ketone

(b) 3-phenylbutanal; β-phenylbutyraldehyde

(c) *trans*-2-methoxycyclohexanecarbaldehyde (or *(R,R)* if you named this enantiomer); no common name

(d) 6,6-dimethylcyclohexa-2,4-dienone; no common name

18-2

(a) $C_9H_{10}O \Rightarrow$ 5 elements of unsaturation

1H doublet (very small coupling constant) at δ 9.7 $\Rightarrow$ aldehyde hydrogen, next to CH

5H multiple peaks at δ 7.2-7.4 $\Rightarrow$ monosubstituted benzene

1H multiplet at δ 3.6 and 3H doublet at δ 1.4 $\Rightarrow$ $CHCH_3$

The splitting of the hydrogen on carbon-2, next to the aldehyde, is worth examining. In its overall shape, it looks like a quartet due to the splitting from the adjacent CH_3. A closer examination of the peaks shows that each peak of the quartet is split into two peaks: this is due to the splitting from the aldehyde hydrogen. The aldehyde hydrogen and the methyl hydrogens are not equivalent, so it is to be expected that the coupling constants will not be equal. If a hydrogen is coupled to different neighboring hydrogens by different coupling constants, they must be considered separately, just as you would by drawing a splitting tree for each type of adjacent hydrogen.

(b) $C_8H_8O \Rightarrow$ 5 elements of unsaturation

cluster of 4 peaks at δ 128-145 $\Rightarrow$ mono- or para-substituted benzene ring

peak at δ 197 $\Rightarrow$ carbonyl carbon (the small peak height suggests a ketone rather than an aldehyde)

peak at δ 26 $\Rightarrow$ methyl next to carbonyl or benzene

18-3 A compound has to have a hydrogen on a γ carbon (or other atom) in order for the McLafferty rearrangement to occur. Butan-2-one has no γ-hydrogen.

18-4

m/z 128

m/z 43

$^+CH_2CH_2CH_2CH_2CH_2CH_3$

m/z 85

m/z 113

McLafferty rearrangement

m/z 58

18-5

The first value is the π to π^*; the second value is the n to π^*. The values are approximate.

(a) < 200 nm; 280 nm; this simple ketone should have values similar to acetone

(b) 230 nm; 310 nm; conjugated system (210) plus 2 alkyl groups (20) = 230; the value of 310 nm is similar to Figure 18-7: ketone (280 nm base value) plus 30 nm for the conjugated double bond = 310 nm

(c) 280 nm; 360 nm; conjugated system (210) plus 1 extra double bond (30) plus 4 alkyl groups (40) = 280; similar reasoning for the other transition, starting with an average base value of 290

(d) 270 nm; 350 nm; same as in (c) except only 3 alkyl groups instead of 4

REMINDERS ABOUT SYNTHESIS PROBLEMS:
1. There may be more than one legitimate approach to a synthesis, especially as the list of reactions gets longer.
2. Begin your analysis by comparing the target to the starting material. If the product has more carbons than the reactant, you will need to use one of the small number of reactions that form carbon-carbon bonds.
3. Where possible, work backwards from the target back to the starting material.
4. KNOW THE REACTIONS. There is no better test of whether you know the reactions than attempting synthesis problems.

18-6 All three target molecules in this problem have more than six carbons, so all answers will include carbon-carbon-bond-forming reactions. So far, there are three types of reactions that form carbon-carbon bonds: the Grignard reaction, S_N2 substitution by an acetylide ion, and the Friedel-Crafts reactions (alkylation and acylation) on benzene.

(a)

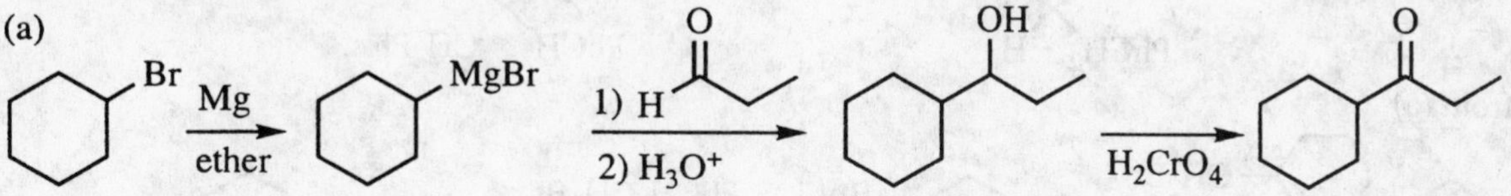

(b)

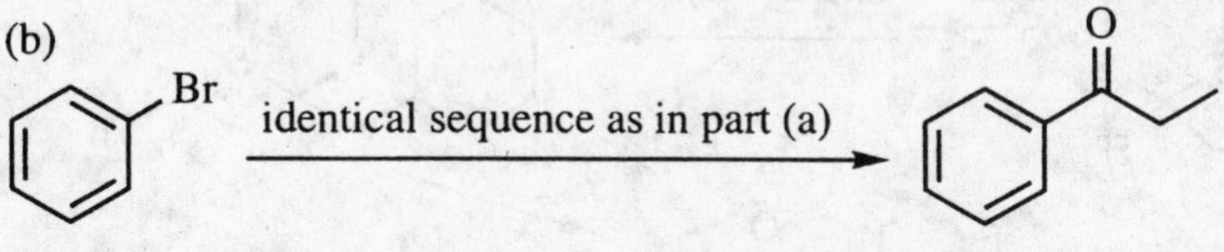

OR

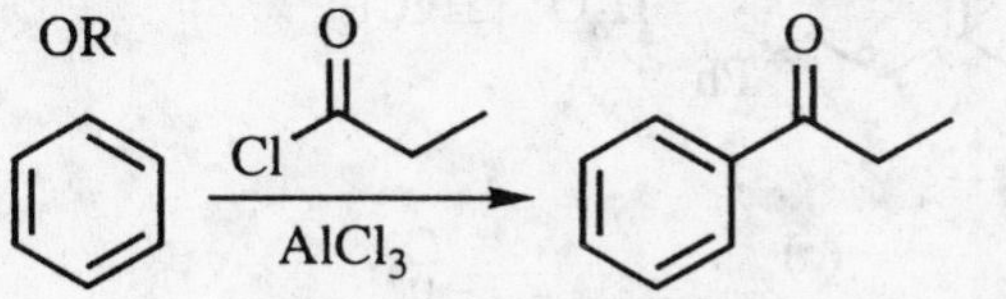

this method using Friedel-Crafts acylation is more efficient as it is only one step

(c) a synthesis as in part (a) could also be used here

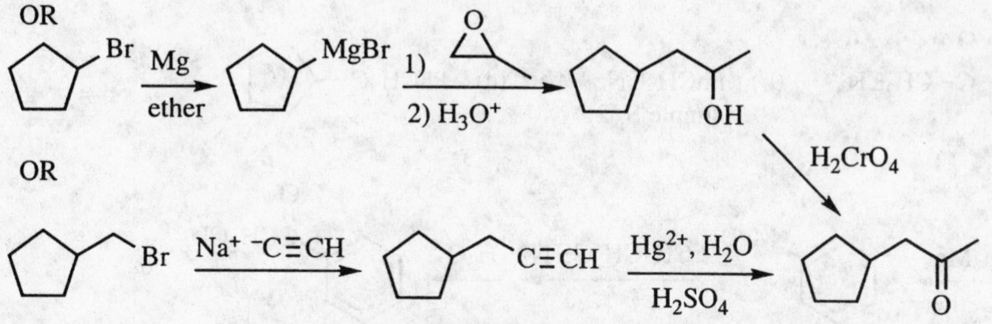

OR

OR

18-7

(a)

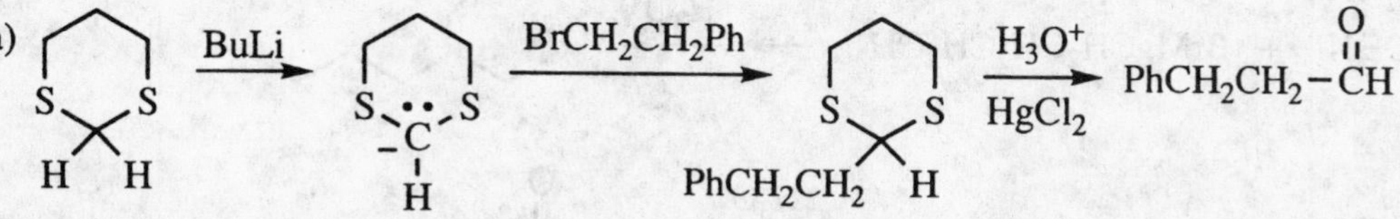

(b)

from (a)

403

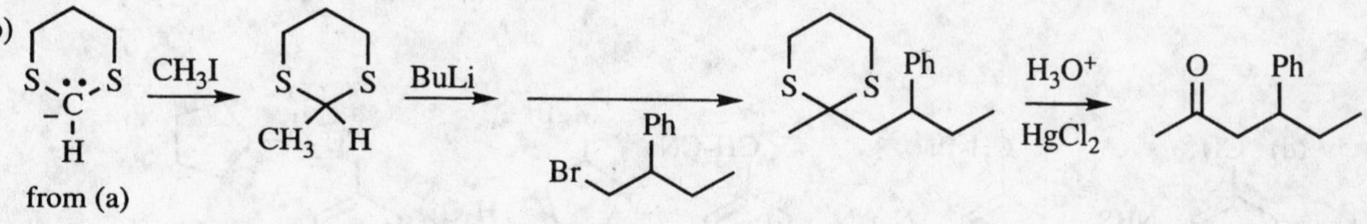

18-7 continued

(c) from (a)

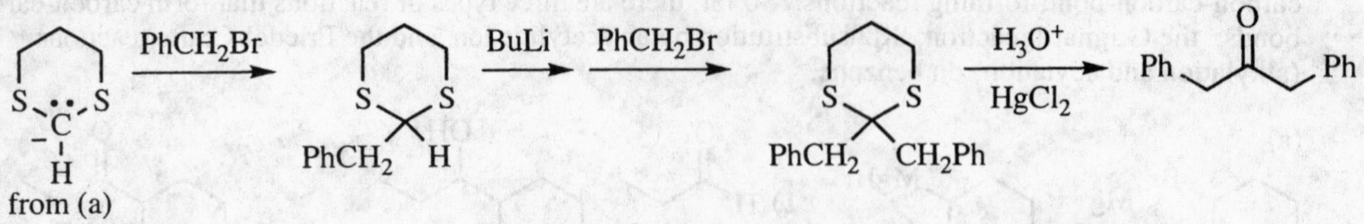

(d) from (a)

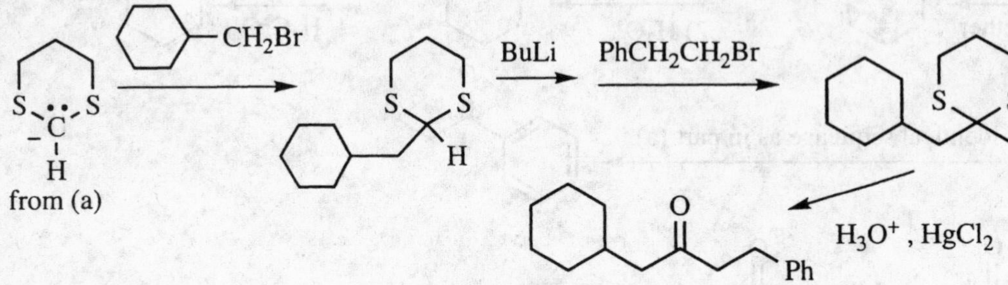

18-8

(a)

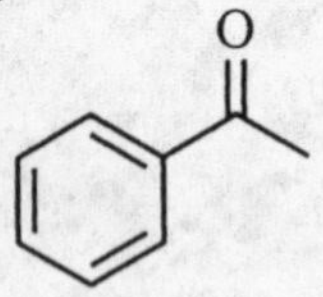

(b)

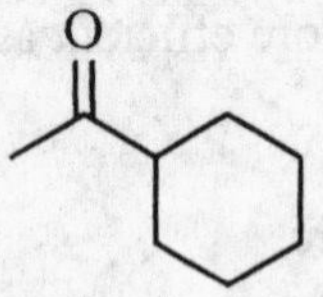

(c) $CH_3(CH_2)_3-\overset{\overset{\displaystyle O}{\|}}{C}-CH_2CH_3$

18-9

(a) $CH_3(CH_2)_3-\overset{\overset{\displaystyle O}{\|}}{C}-CH_2CH_3$

(b) $PhCH_2CN$
(simple S_N2)

(c)

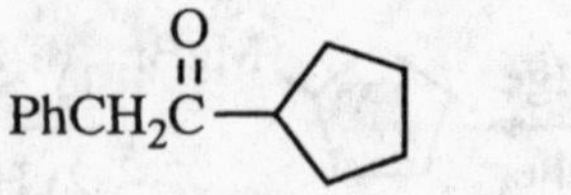

18-10

(a)

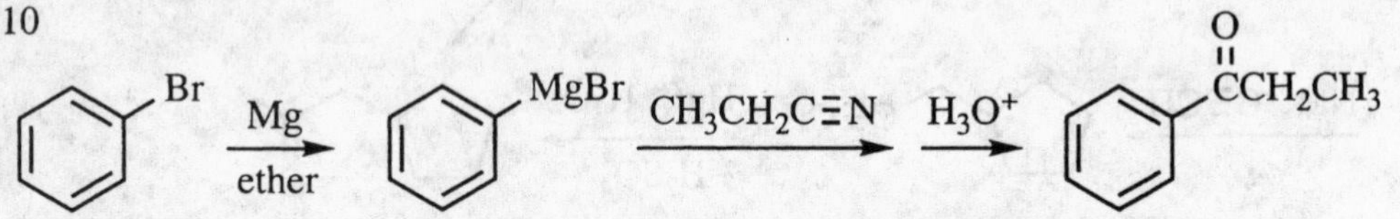

(b) $CH_3CH_2C\equiv N$ + $BrMgCH_2CH_2CH_2CH_3$ $\xrightarrow{}$ $\xrightarrow{H_3O^+}$

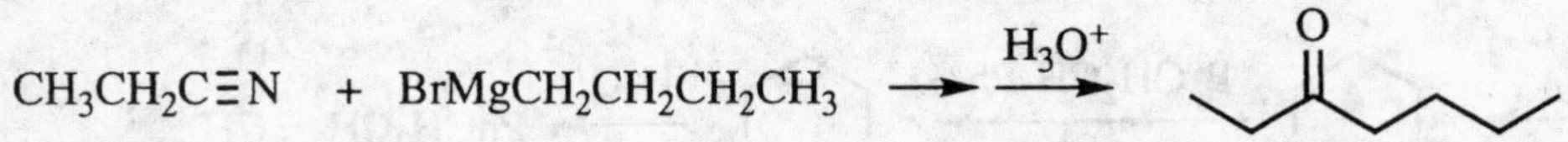

(c) $CH_3(CH_2)_3COOH$ + 2 CH_3CH_2Li $\xrightarrow{H_3O^+}$

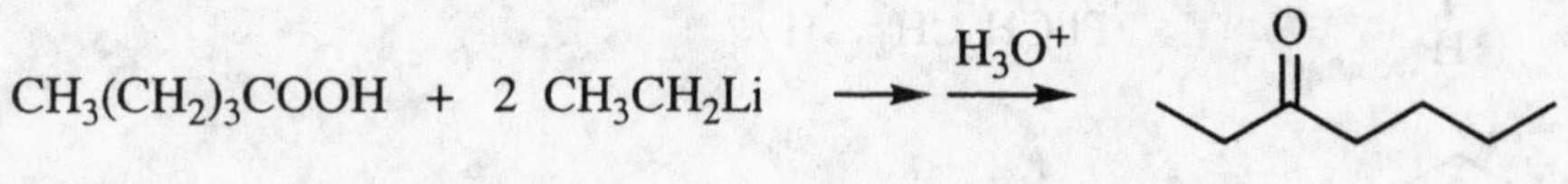

(d)

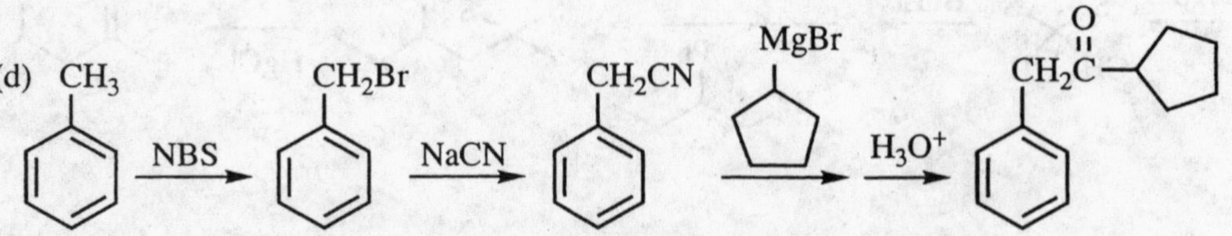

18-11

(a) 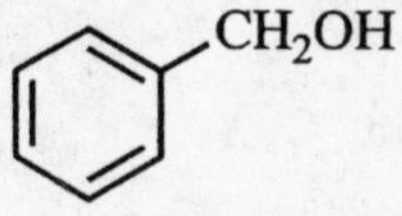 CH_2OH

(b) 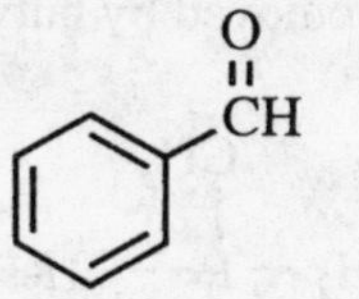 $\overset{O}{\underset{||}{C}}H$

(c) OH

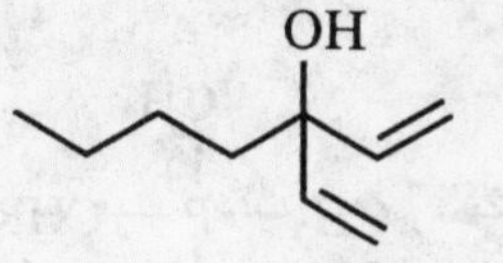

(d)

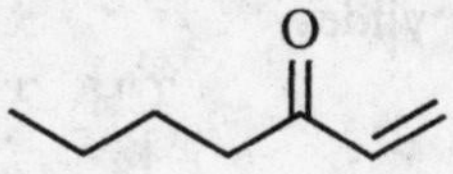

18-12 Review the reminders on p. 402. There are often more than one correct way to do syntheses, but a more direct route with fewer steps is usually better.

(a)

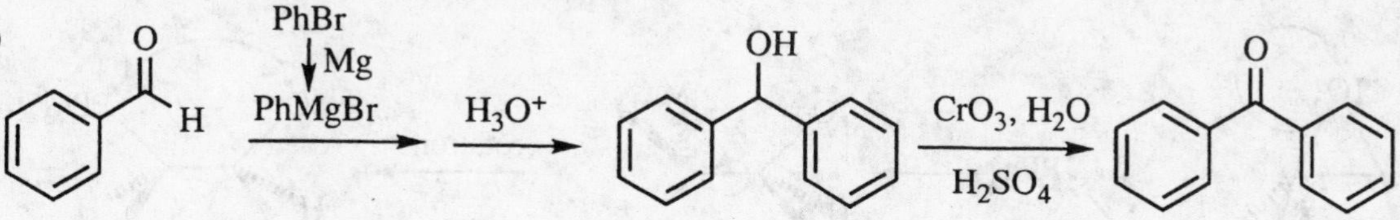

(b)

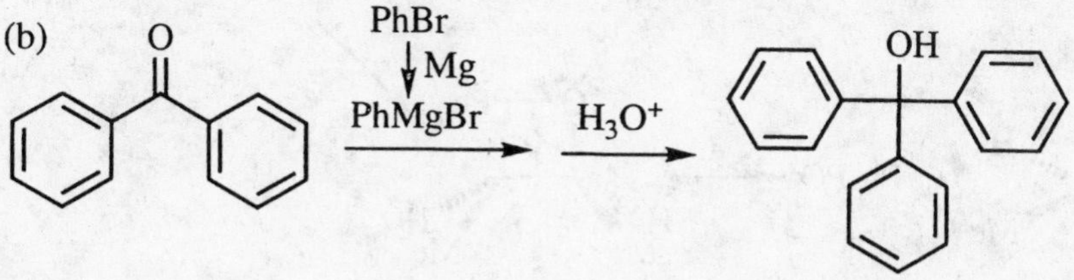

(c)

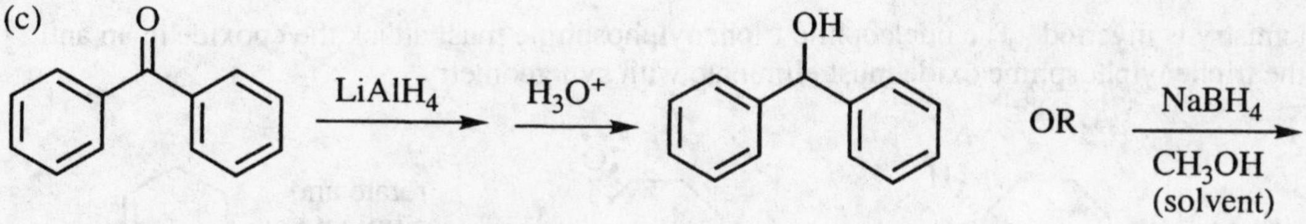

(d)

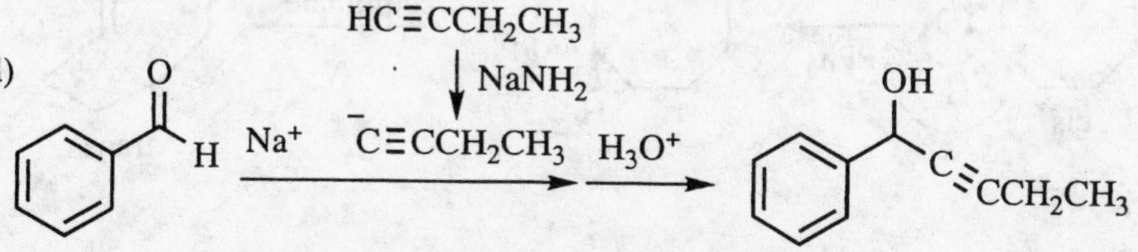

18-13 The triacetoxyborohydride ion is similar to borohydride, BH_4^-, where three acetoxy groups have replaced three hydrides.

(a)

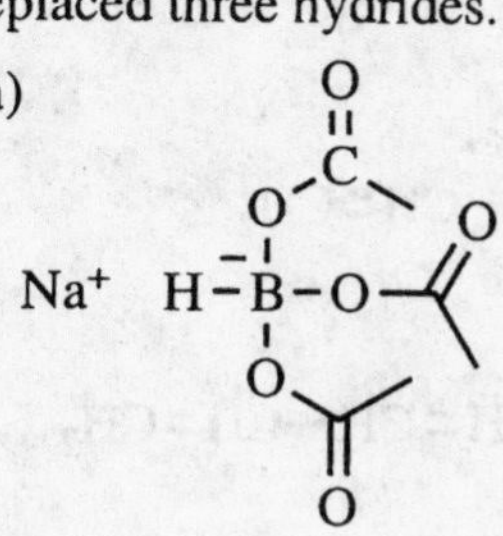

(b)

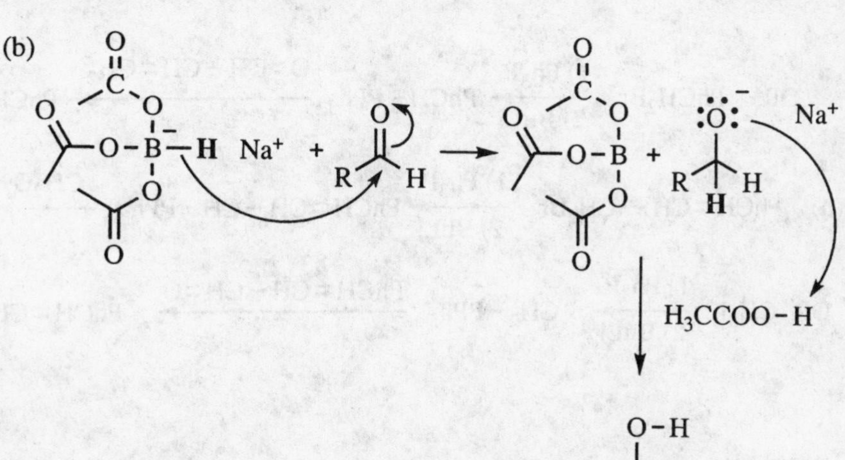

18-14 Trimethylphosphine has α-hydrogens that could be removed by butyllithium, generating undesired ylides.

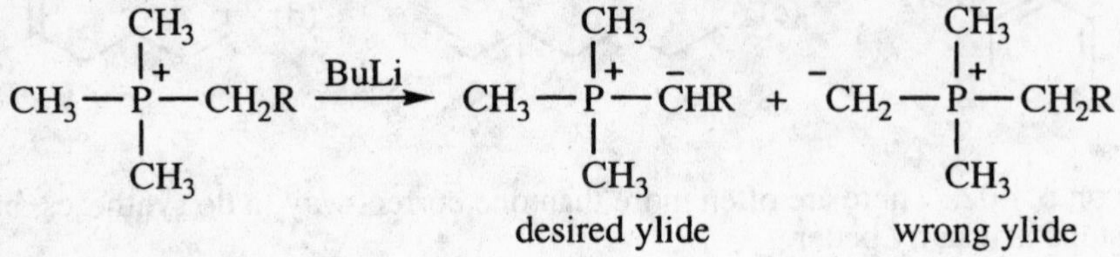

desired ylide   wrong ylide

18-15

(a)

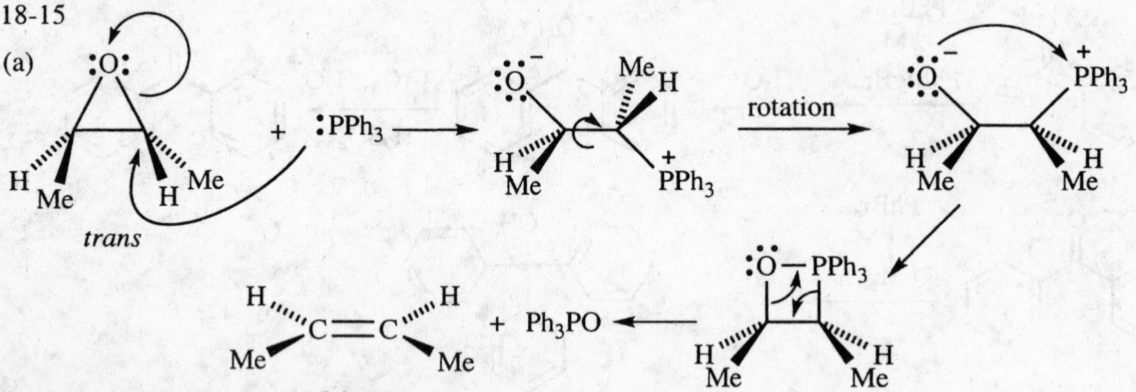

cis-2-butene

The stereochemistry is inverted. The nucleophile triphenylphosphine must attack the epoxide in an anti fashion, yet the triphenylphosphine oxide must eliminate with syn geometry.

(b)

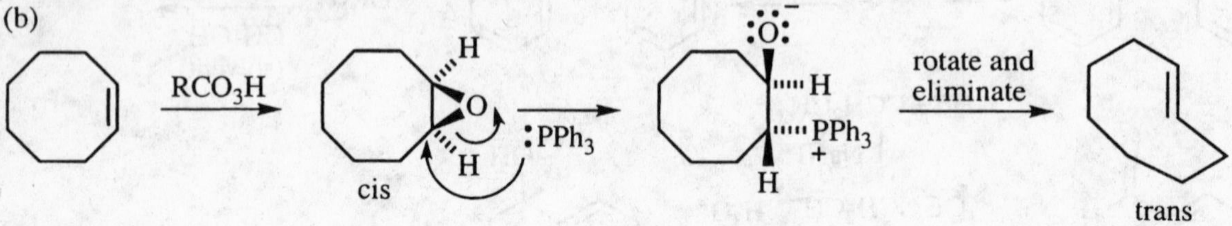

18-16

(a) $CH_2=CHCH_2Br$ $\xrightarrow[\text{2) BuLi}]{\text{1) Ph}_3\text{P}}$ $CH_2=CHCH\overset{-}{-}\overset{+}{PPh_3}$ $\xrightarrow{\text{PhCHO}}$ $PhCH=CH-CH=CH_2$

OR $PhCH_2Br$ $\xrightarrow[\text{2) BuLi}]{\text{1) Ph}_3\text{P}}$ $Ph\overset{-}{CH}-\overset{+}{PPh_3}$ $\xrightarrow{O=CH-CH=CH_2}$ $PhCH=CH-CH=CH_2$

(b) $PhCH=CH-CH_2Br$ $\xrightarrow[\text{2) BuLi}]{\text{1) Ph}_3\text{P}}$ $PhCH=CH-\overset{-}{CH}-\overset{+}{PPh_3}$ $\xrightarrow{CH_2O}$ $PhCH=CH-CH=CH_2$

OR CH_3I $\xrightarrow[\text{2) BuLi}]{\text{1) Ph}_3\text{P}}$ $\overset{-}{CH_2}-\overset{+}{PPh_3}$ $\xrightarrow{PhCH=CH-CH=O}$ $PhCH=CH-CH=CH_2$

18-17 Many alkenes can be synthesized by two different Wittig reactions (as in the previous problem). The ones shown here form the phosphonium salt from the less hindered alkyl halide.

(a) $PhCH_2Br \xrightarrow[\text{2) BuLi}]{\text{1) Ph}_3\text{P}} PhCH^- - \overset{+}{P}Ph_3 + \underset{H_3C \quad CH_3}{\overset{O}{\underset{\|}{C}}} \longrightarrow PhCH=C(CH_3)_2$

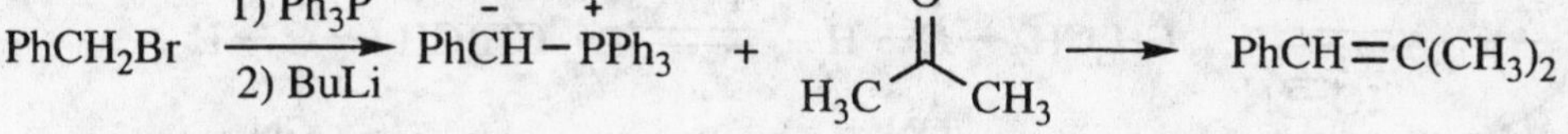

(b) $CH_3I \xrightarrow[\text{2) BuLi}]{\text{1) Ph}_3\text{P}} CH_2^- - \overset{+}{P}Ph_3 + \underset{Ph \quad CH_3}{\overset{O}{\underset{\|}{C}}} \longrightarrow \underset{Ph \quad CH_3}{\overset{CH_2}{\underset{\|}{C}}}$

(c)

$PhCH_2Br \xrightarrow[\text{2) BuLi}]{\text{1) Ph}_3\text{P}} PhCH^- - \overset{+}{P}Ph_3 \xrightarrow{PhCH=CH-CH=O} PhCH=CH-CH=CHPh$

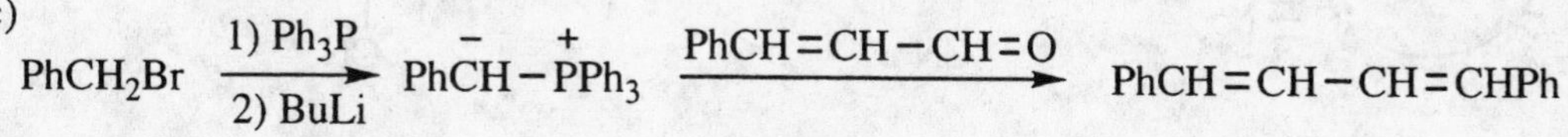

(d) $CH_3I \xrightarrow[\text{2) BuLi}]{\text{1) Ph}_3\text{P}} CH_2^- - \overset{+}{P}Ph_3 + \text{(cyclopentanone)} \longrightarrow \text{(methylenecyclopentane)}$

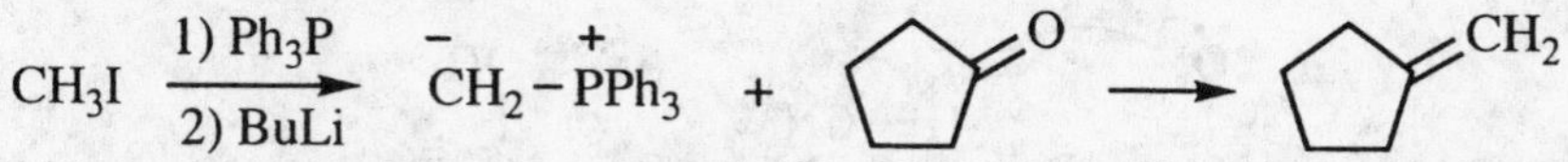

(e) $CH_3CH_2Br \xrightarrow[\text{2) BuLi}]{\text{1) Ph}_3\text{P}} CH_3CH^- - \overset{+}{P}Ph_3 + \text{(cyclohexanone)} \longrightarrow \text{(ethylidenecyclohexane =CHCH}_3\text{)}$

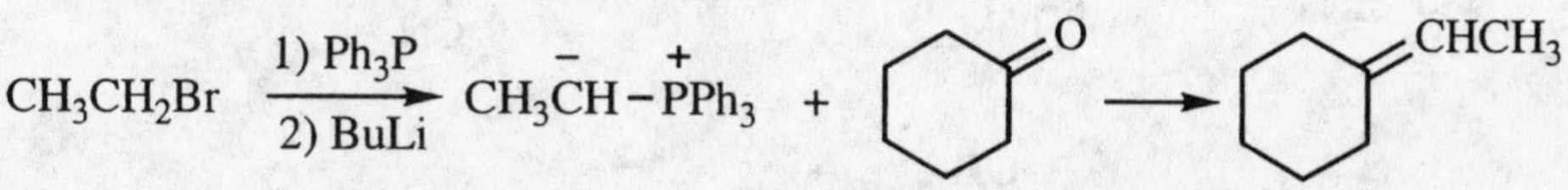

18-18

(a) $Cl_3C-\overset{\overset{\displaystyle :O:}{\|}}{C}-H \underset{}{\overset{H^+}{\rightleftharpoons}} \left\{ Cl_3C-\overset{\overset{\displaystyle \overset{+}{O}-H}{\|}}{C}-H \longleftrightarrow Cl_3C-\overset{\overset{\displaystyle \overset{..}{O}-H}{}}{\underset{+}{C}}-H \right\} \underset{}{\overset{}{\rightleftharpoons}} Cl_3C-\overset{\overset{\displaystyle OH}{|}}{C}-H$

$H_2\overset{..}{O}:$ $H-\overset{+}{\underset{|}{O}}-H$

$Cl_3C-\overset{\overset{\displaystyle OH}{|}}{\underset{\underset{\displaystyle OH}{|}}{C}}-H \rightleftharpoons$ $H_2\overset{..}{O}:$

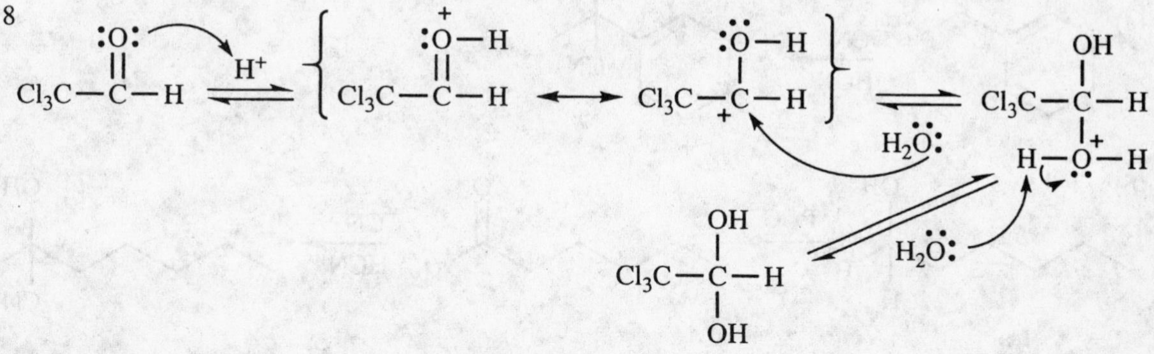

(b) $CH_3-\overset{\overset{\displaystyle :O:}{\|}}{C}-CH_3 \underset{:\overset{..}{O}H}{\overset{}{\rightleftharpoons}} CH_3-\overset{\overset{\displaystyle :\overset{-}{\overset{..}{O}}:}{|}}{\underset{\underset{\displaystyle OH}{|}}{C}}-CH_3 \underset{}{\overset{H-OH}{\rightleftharpoons}} CH_3-\overset{\overset{\displaystyle OH}{|}}{\underset{\underset{\displaystyle OH}{|}}{C}}-CH_3 + HO^-$

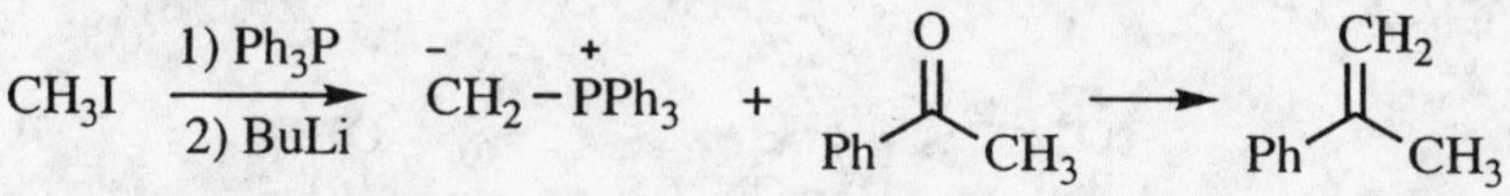

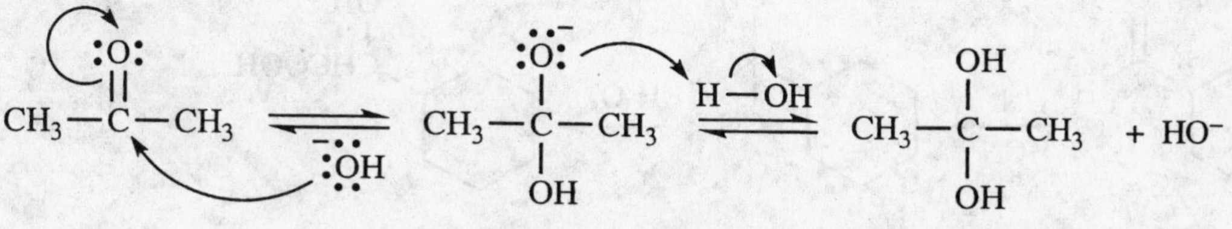

18-19

$\text{(cyclohexanone)} < \text{(2-bromocyclohexanone, Br)} < \text{(cyclohexanecarbaldehyde, CHO)} < \text{(1-bromocyclohexanecarbaldehyde, CHO, Br)}$

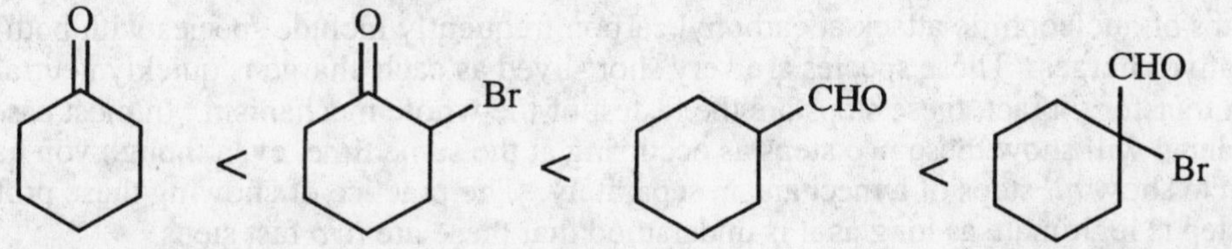

least amount
of hydrate

greatest amount
of hydrate

18-20

(a)

$$CH_3CH_2-\overset{\displaystyle :\!O\!:}{\underset{}{\overset{\|}{C}}}-H \;\rightleftharpoons\; \underset{\underset{:C\equiv N}{}}{} \; CH_3CH_2-\overset{\displaystyle :\!\overset{-}{O}\!:}{\underset{CN}{\overset{|}{C}}}-H \;\xrightarrow{H-CN}\rightleftharpoons\; CH_3CH_2-\overset{\displaystyle OH}{\underset{CN}{\overset{|}{C}}}-H$$

(b)

$$CH_3CH_2-\overset{\displaystyle :\!O\!:}{\underset{}{\overset{\|}{C}}}-CH_3 \;\rightleftharpoons\; \underset{\underset{:C\equiv N}{}}{} \; CH_3CH_2-\overset{\displaystyle :\!\overset{-}{O}\!:}{\underset{CN}{\overset{|}{C}}}-CH_3 \;\xrightarrow{H-CN}\rightleftharpoons\; CH_3CH_2-\overset{\displaystyle OH}{\underset{CN}{\overset{|}{C}}}-CH_3$$

(c)

$$t\text{-Bu}-\overset{\displaystyle :\!O\!:}{\underset{}{\overset{\|}{C}}}-t\text{-Bu} \;\rightleftharpoons\; \underset{\underset{:C\equiv N}{}}{} \; t\text{-Bu}-\overset{\displaystyle :\!\overset{-}{O}\!:}{\underset{CN}{\overset{|}{C}}}-t\text{-Bu} \;\xrightarrow{H-CN}\rightleftharpoons\; t\text{-Bu}-\overset{\displaystyle OH}{\underset{CN}{\overset{|}{C}}}-t\text{-Bu}$$

18-21

(a)

Acetophenone $\xrightarrow[\text{HCN}]{^-CN}$ 2-phenyl-2-hydroxypropanenitrile (C with OH, CH₃, CN substituents on benzene ring)

(b)

1-hexanol (secondary alcohol with OH, H) $\xrightarrow{\text{PCC}}$ aldehyde (with O, H) $\xrightarrow[\text{HCN}]{^-CN}$ cyanohydrin (with OH, H, CN)

(c)

cyclohexanecarbaldehyde $\xrightarrow[\text{HCN}]{H\;^-CN}$ cyclohexyl cyanohydrin (OH, H, CN) $\xrightarrow{H_3O^+}$ cyclohexyl–CH(OH)–CHCOOH

Note: Mechanisms of nucleophilic attack at carbonyl carbon frequently include species with both positive and negative charges. These species are very short-lived as each charge is quickly neutralized by a rapid proton transfer; in fact, these steps are the fastest of the whole mechanism. In most cases, this Solutions Manual will show these *two* steps as occurring at the same time, even though you have been admonished to show *all* steps of a mechanism separately. The practice of showing these proton transfers in one step is legitimate as long as it is understood that these are *two* fast steps.

18-22

(a)

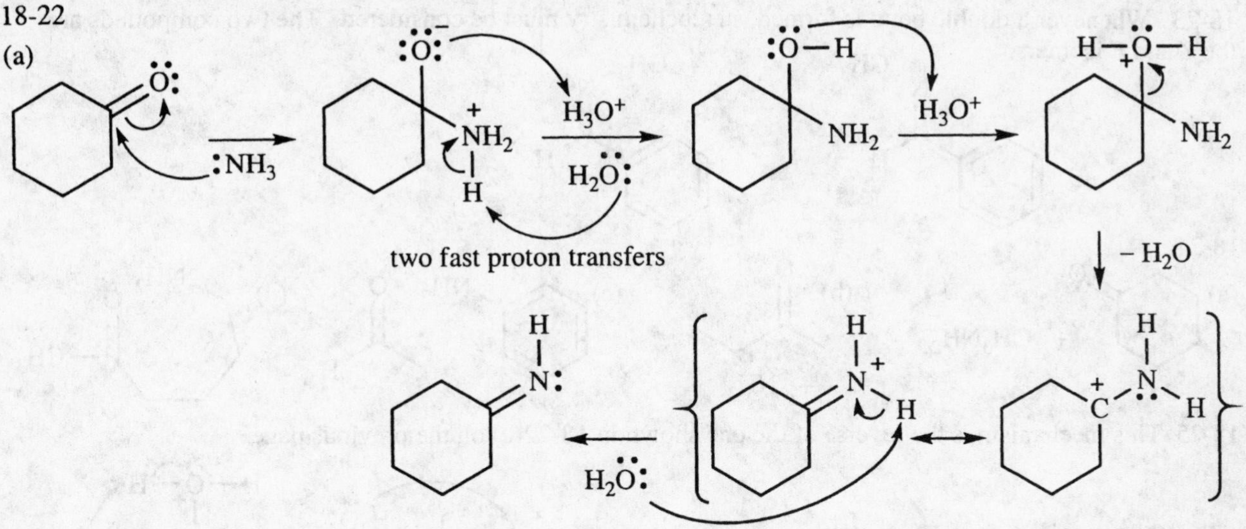

two fast proton transfers

$- H_2O$

$H_2O:$

(b)

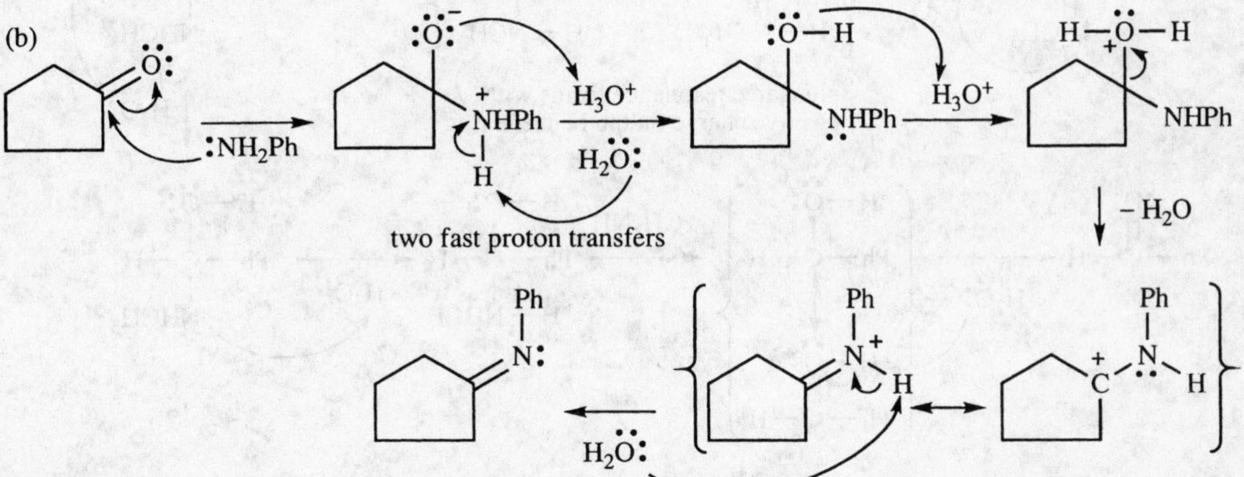

two fast proton transfers

$- H_2O$

$H_2O:$

(c)

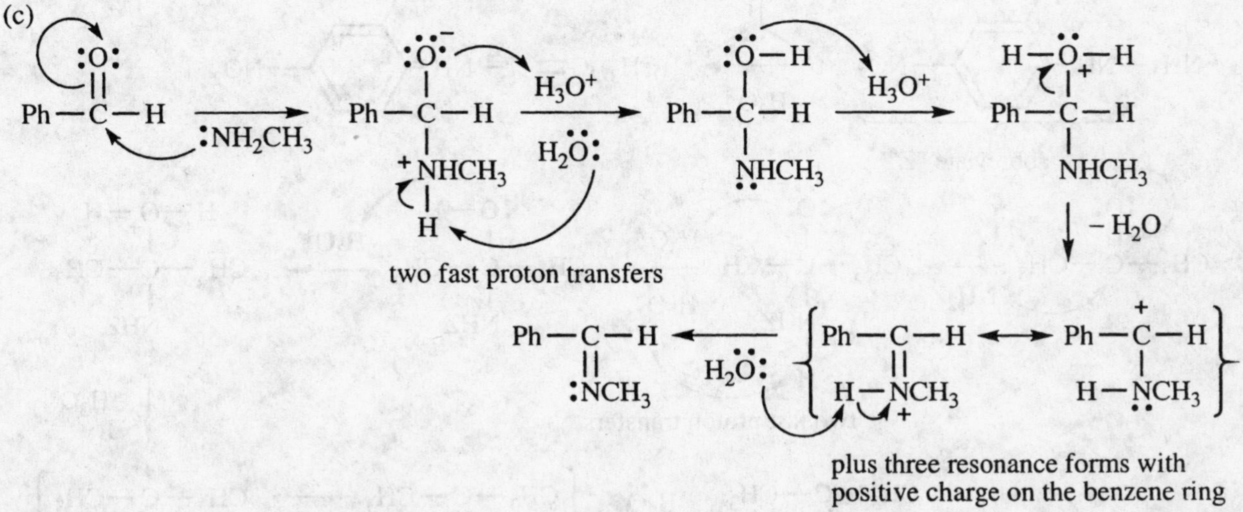

two fast proton transfers

$- H_2O$

$H_2O:$

plus three resonance forms with
positive charge on the benzene ring

18-23 Whenever a double bond is formed, stereochemistry must be considered. The two compounds are the Z and E isomers.

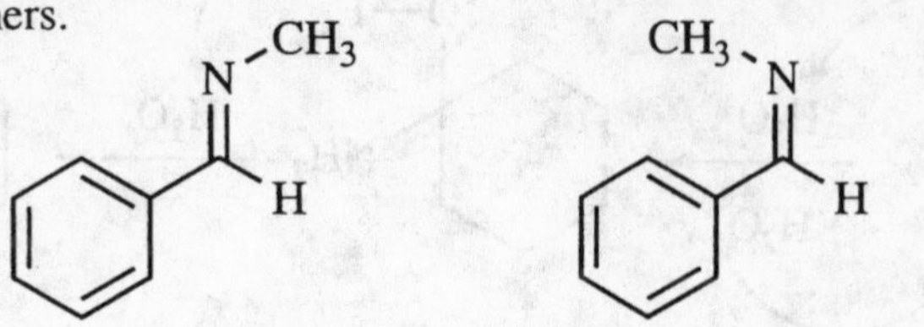

18-24

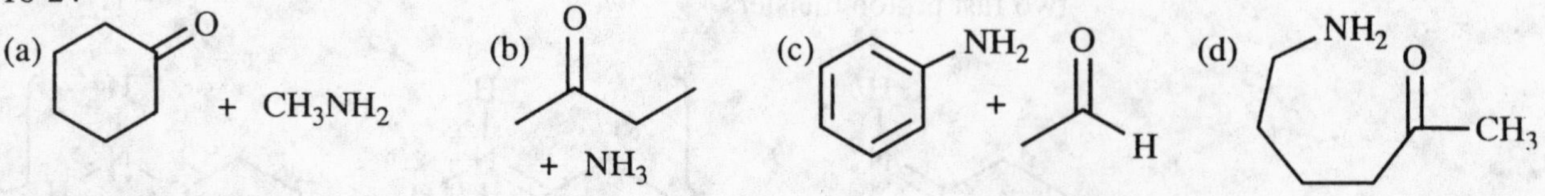

(a) cyclohexanone + CH_3NH_2

(b) + NH_3

(c) + (aldehyde)

(d)

18-25 This mechanism is the reverse of the one shown in 18-22(c) on the previous page.

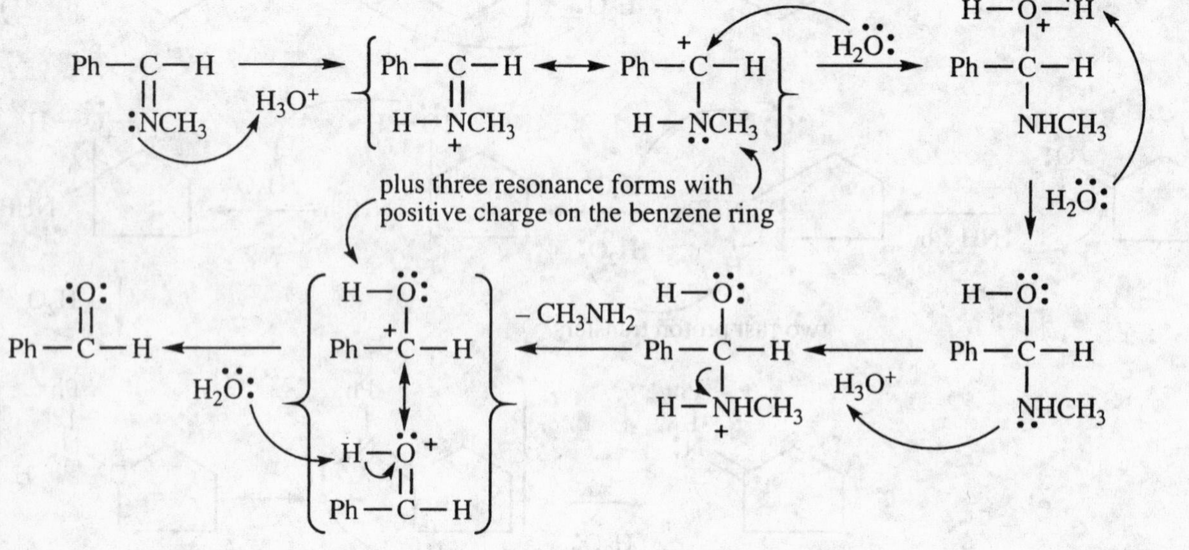

plus three resonance forms with positive charge on the benzene ring

18-26

abbreviate "Z"

two fast proton transfers

410

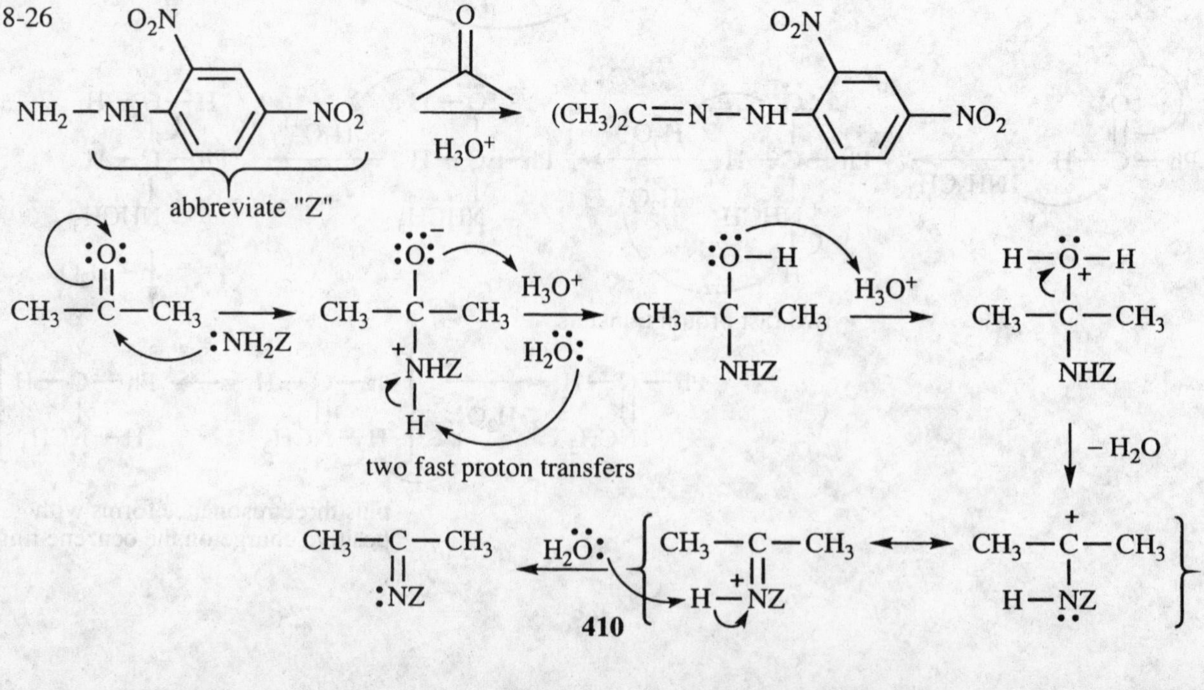

18-27

(a) 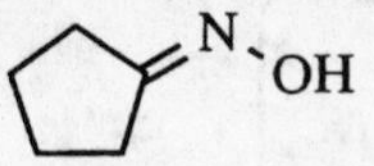

(b) 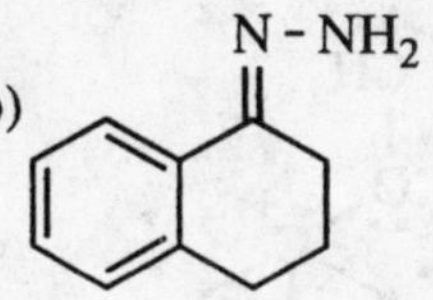

(c) 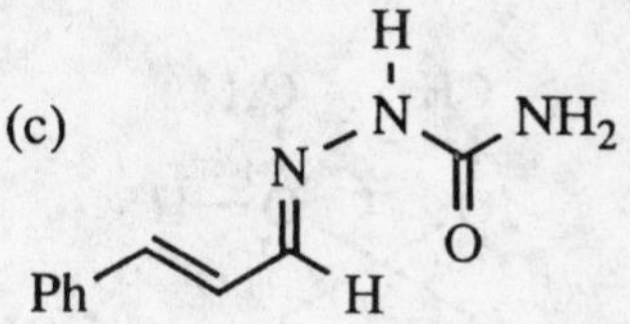

(d) 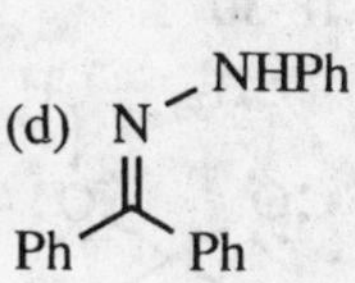

18-28

(a) PhCHO + $H_2NNH-\overset{O}{\overset{\|}{C}}-NH_2$

(b) 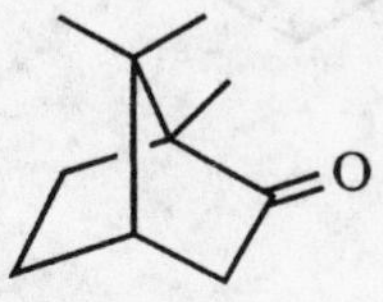 + H_2NOH

(c) 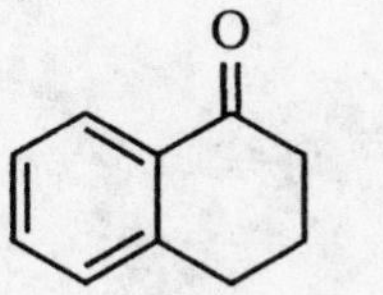 + $H_2N-NHPh$

(d) 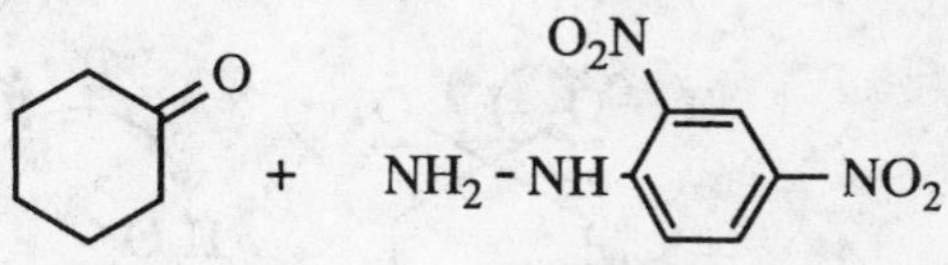

(e) 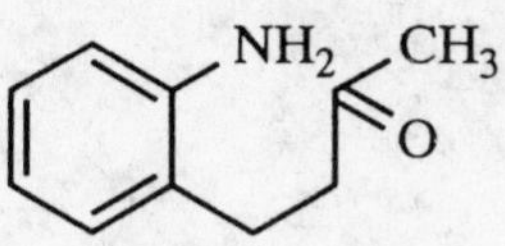

18-29

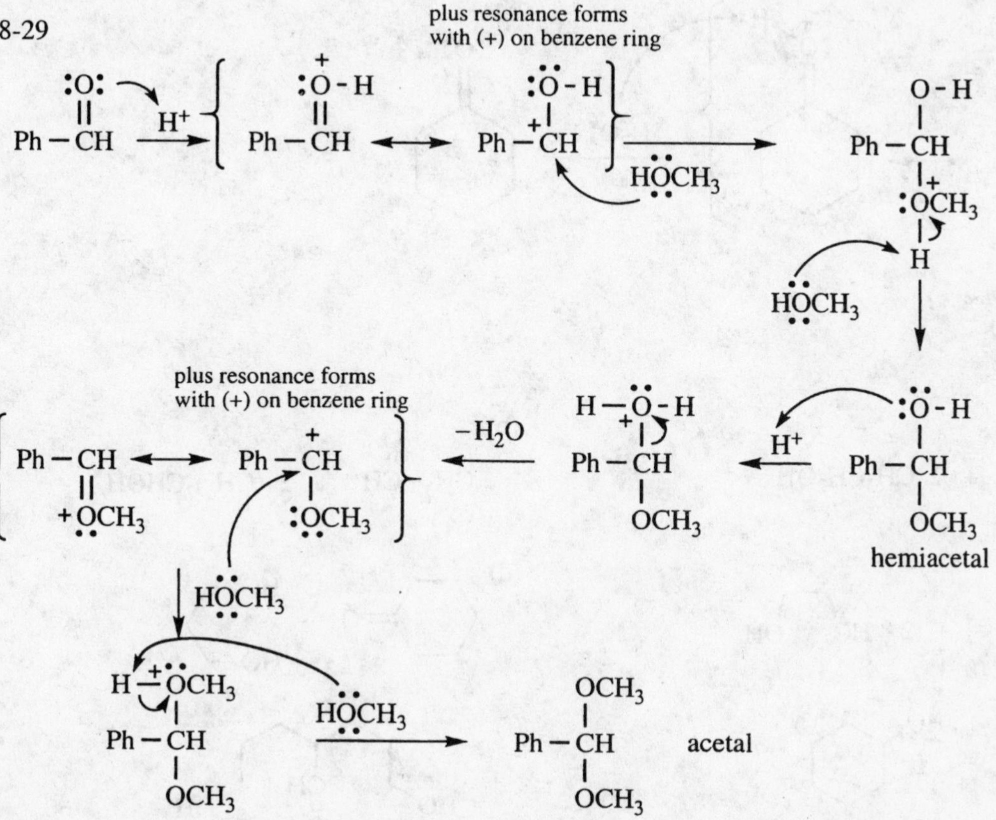
plus resonance forms
with (+) on benzene ring
hemiacetal
plus resonance forms
with (+) on benzene ring
acetal

18-30

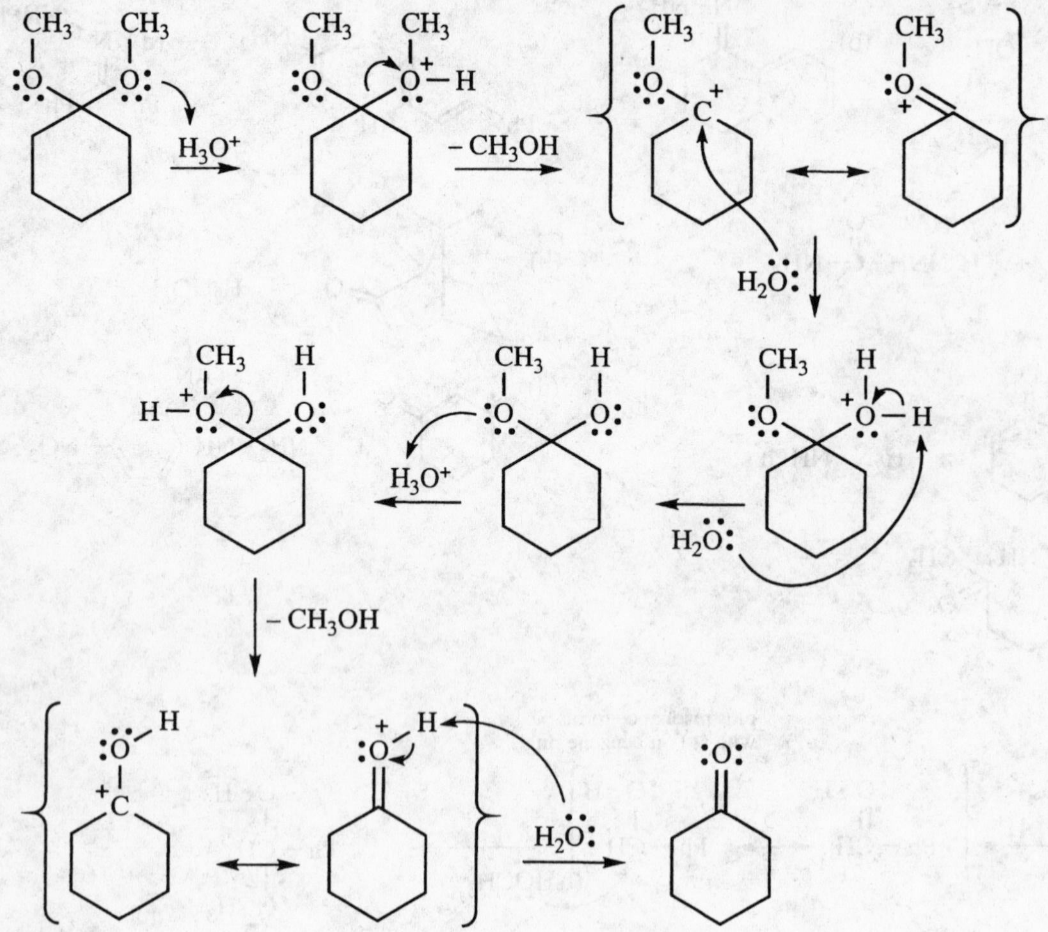

18-31

(a)

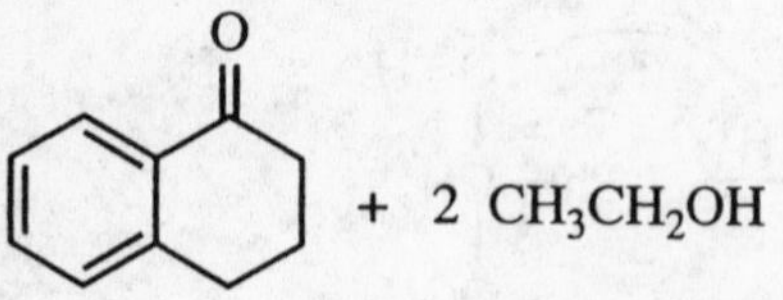

+ 2 CH₃CH₂OH

(b)

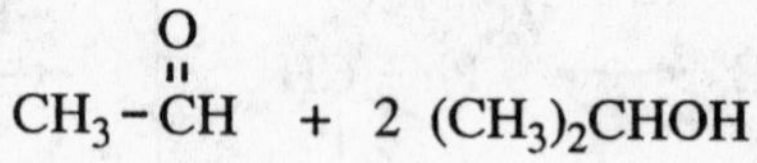

CH_3-CH (with =O) + 2 $(CH_3)_2CHOH$

(c)

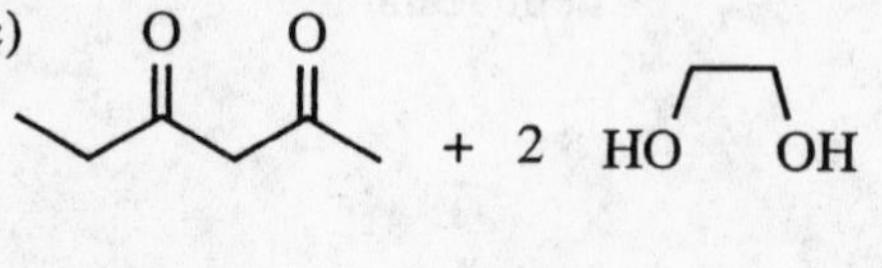

+ 2 HO⌒OH

(d)

=O + HO⌒OH (HO and HO)

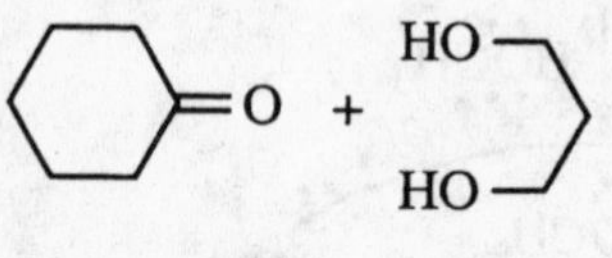

(e)

+

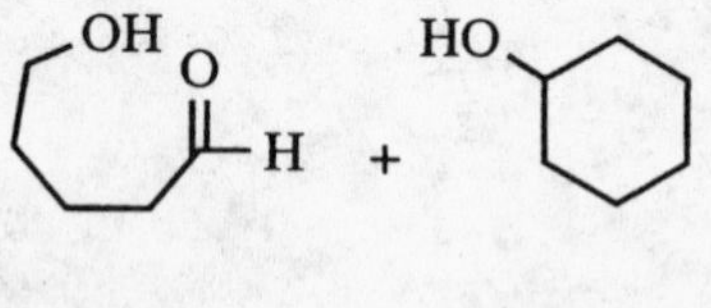

(f)

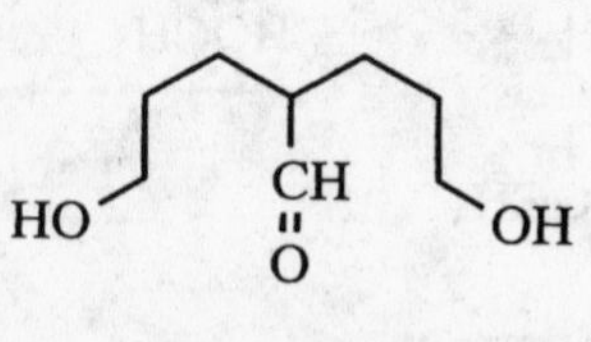

18-32

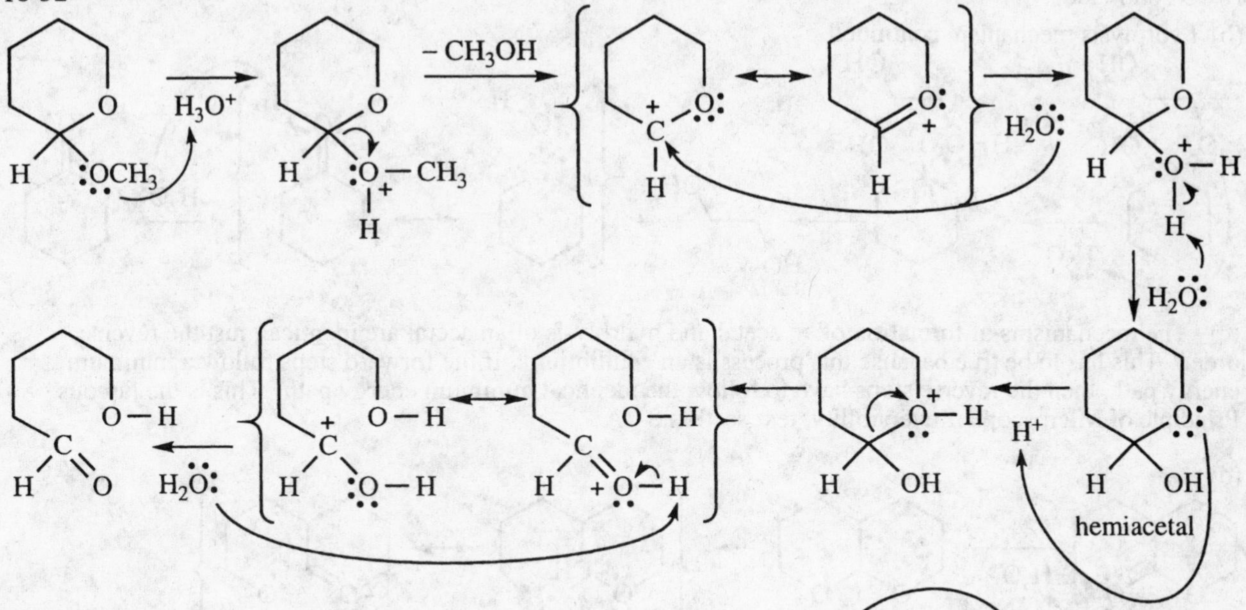

18-33

(a)

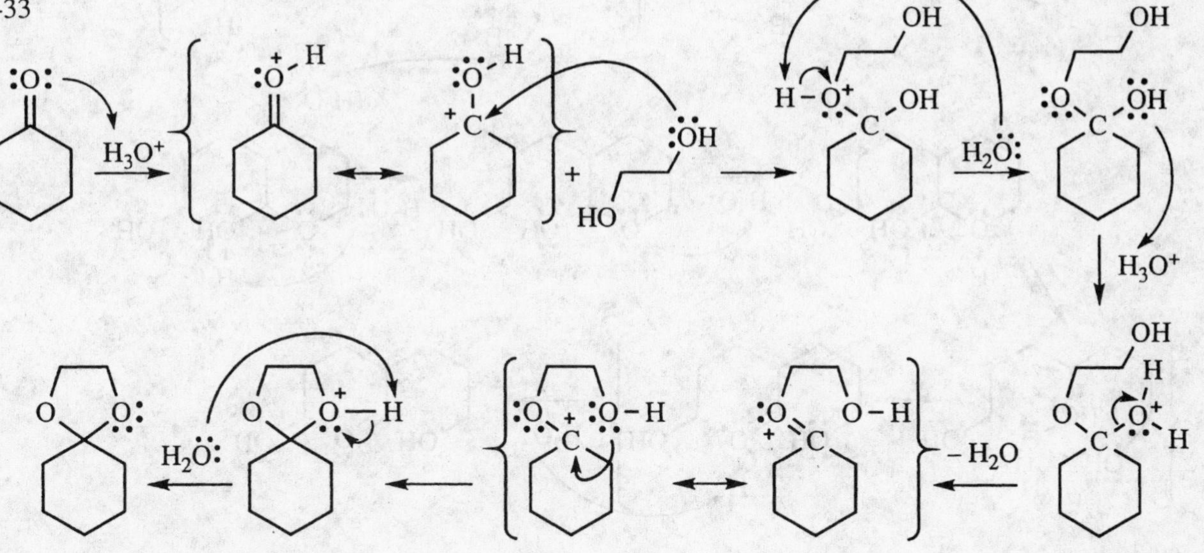

(b)

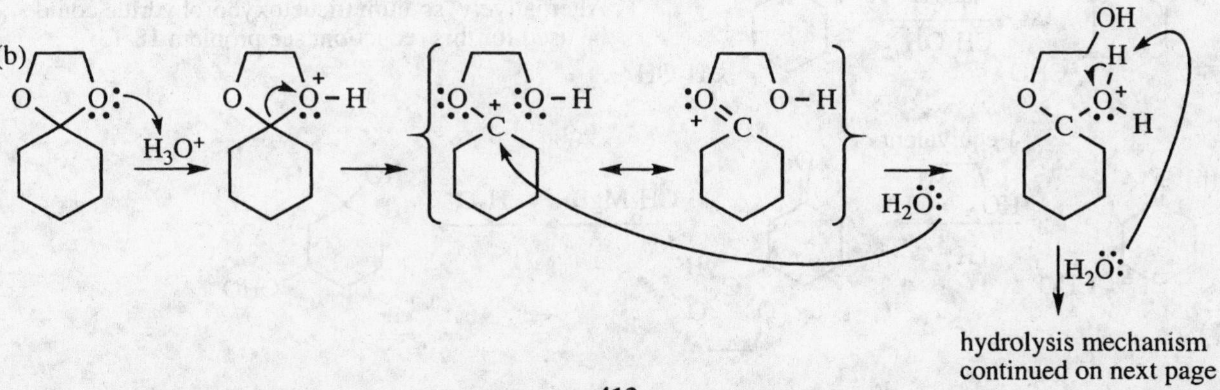

hydrolysis mechanism
continued on next page

413

18-33 continued

(b) hydrolysis mechanism continued

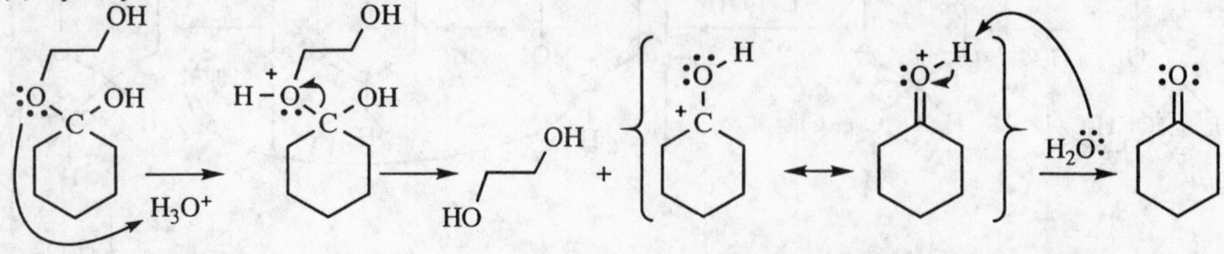

(c) The mechanisms of formation of an acetal and hydrolysis of an acetal are identical, just the reverse order. This has to be true because this process is an equilibrium: if the forward steps follow a minimum energy path, then the reverse steps have to follow the identical minimum energy path. This is the famous Principle of Microscopic Reversibility, text section 8-4A.

(d)

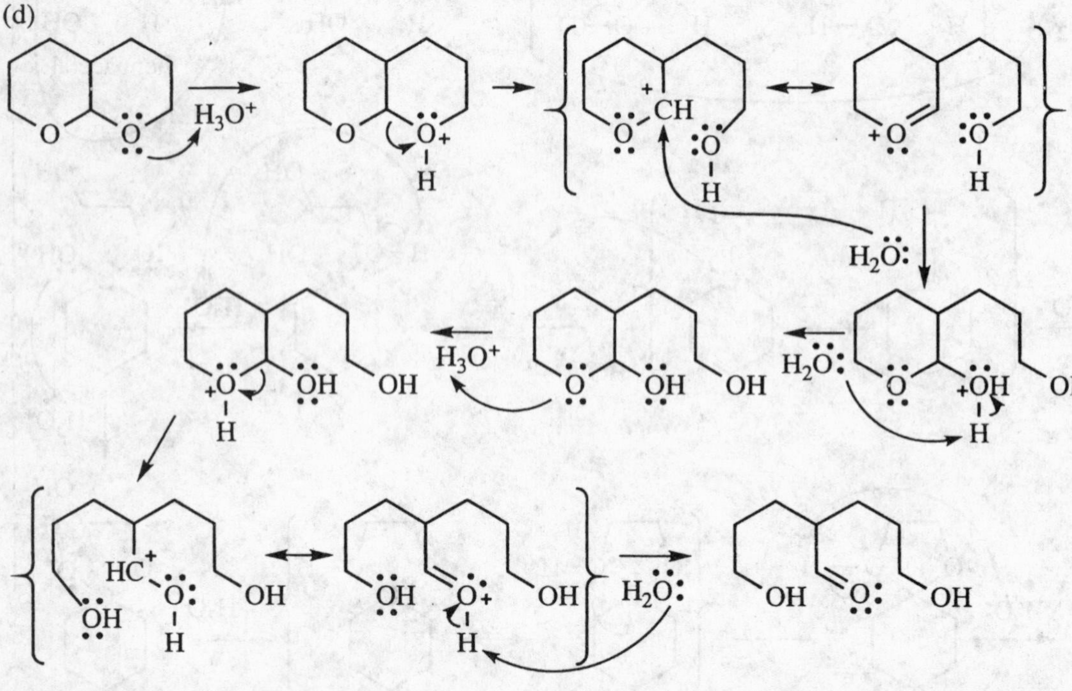

18-34

(a)

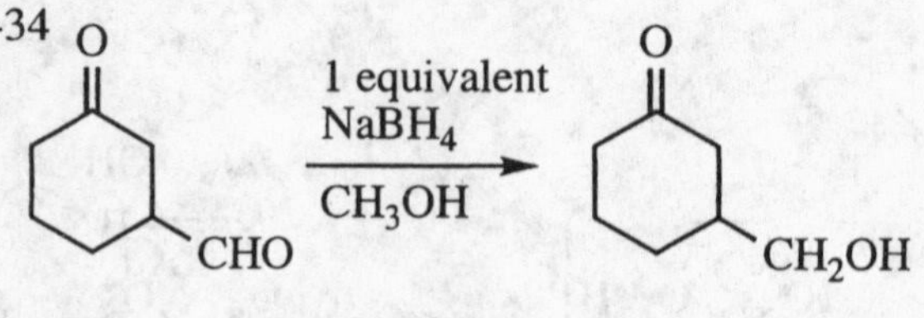

Aldehydes are reduced faster than ketones (keeping this reaction cold will increase selectivity). Alternatively, sodium triacetoxyborohydride could be used for this reduction; see problem 18-13.

(b)

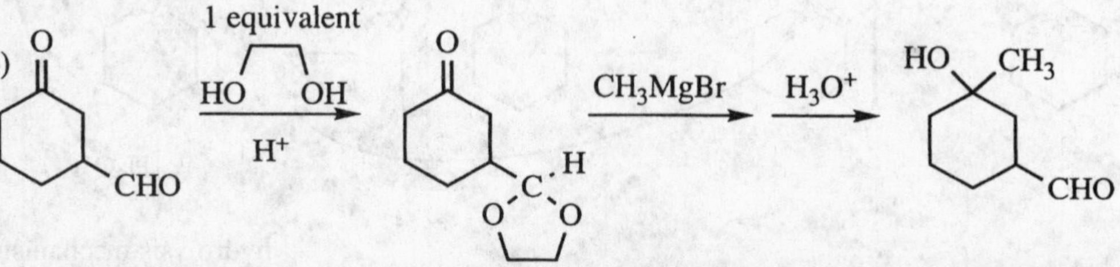

414

18-34 continued

(c)

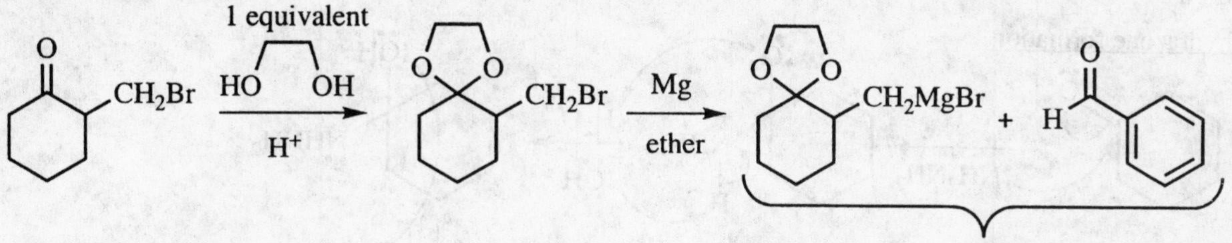

The last step protonates the oxygen, dehydrates the alcohol, and hydrolyzes the acetal.

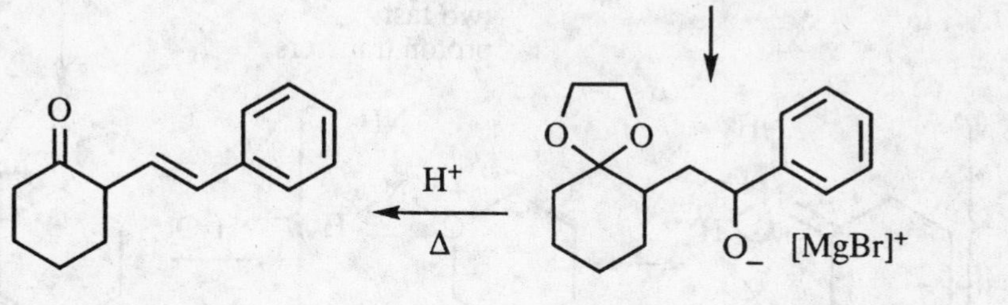

(d)

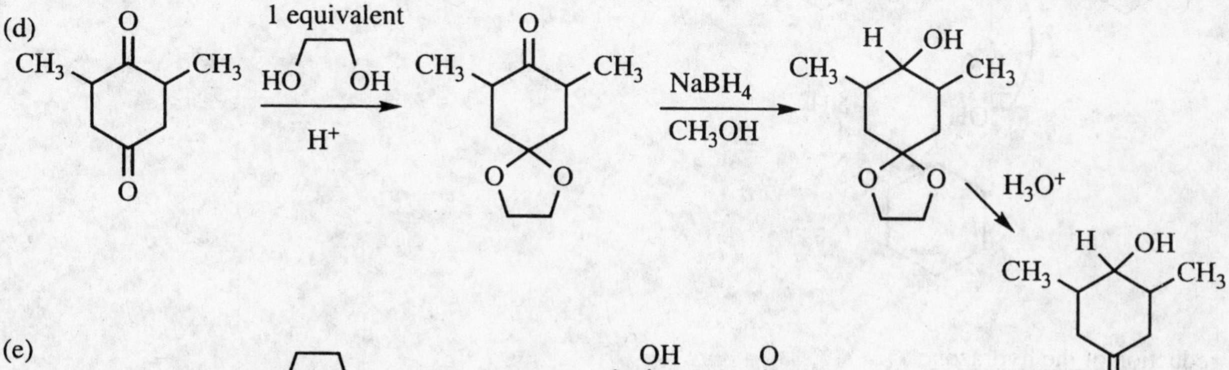

(e)

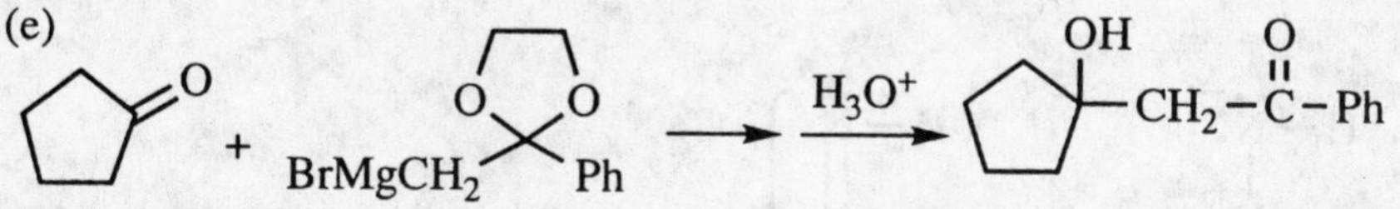

(f)

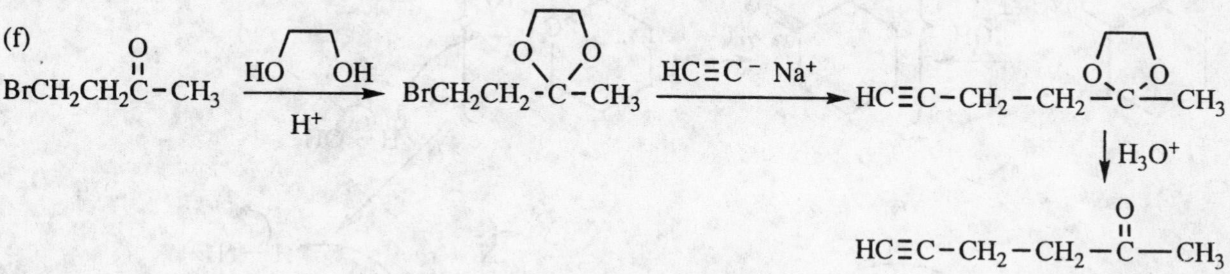

$BrCH_2CH_2\overset{O}{\overset{\|}{C}}-CH_3 \xrightarrow[H^+]{HO \quad OH} BrCH_2CH_2-\overset{O \quad O}{C}-CH_3 \xrightarrow{HC\equiv C^- \ Na^+} HC\equiv C-CH_2-CH_2-\overset{O \quad O}{C}-CH_3$

$\downarrow H_3O^+$

$HC\equiv C-CH_2-CH_2-\overset{O}{\overset{\|}{C}}-CH_3$

18-35

(a)

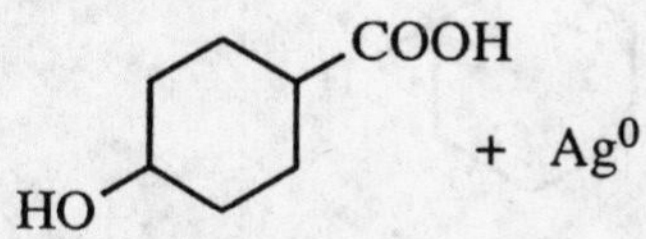

+ Ag⁰

after adding H⁺

(b)

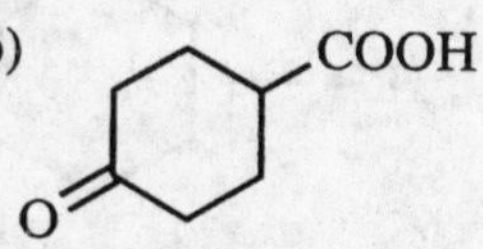

(c)

+ Ag⁰

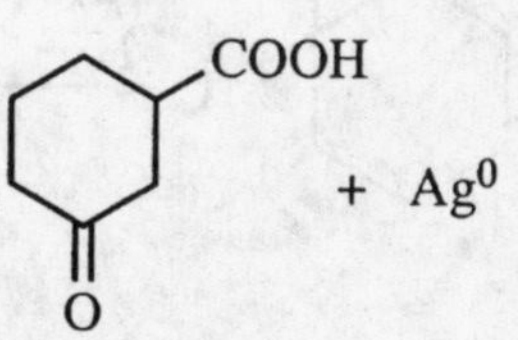

after adding H⁺

(d)

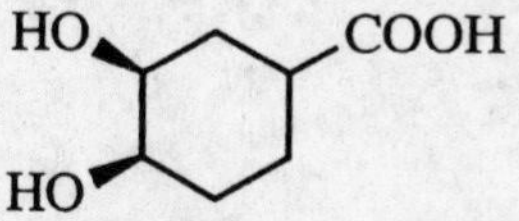

18-36

hydrazone formation

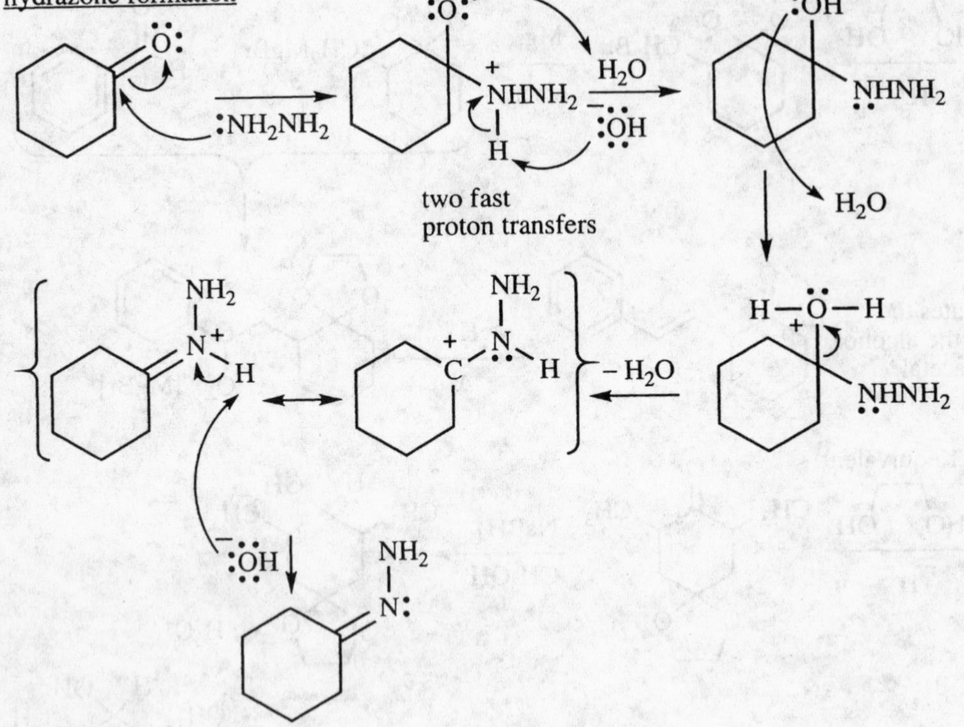

two fast
proton transfers

reduction of the hydrazone

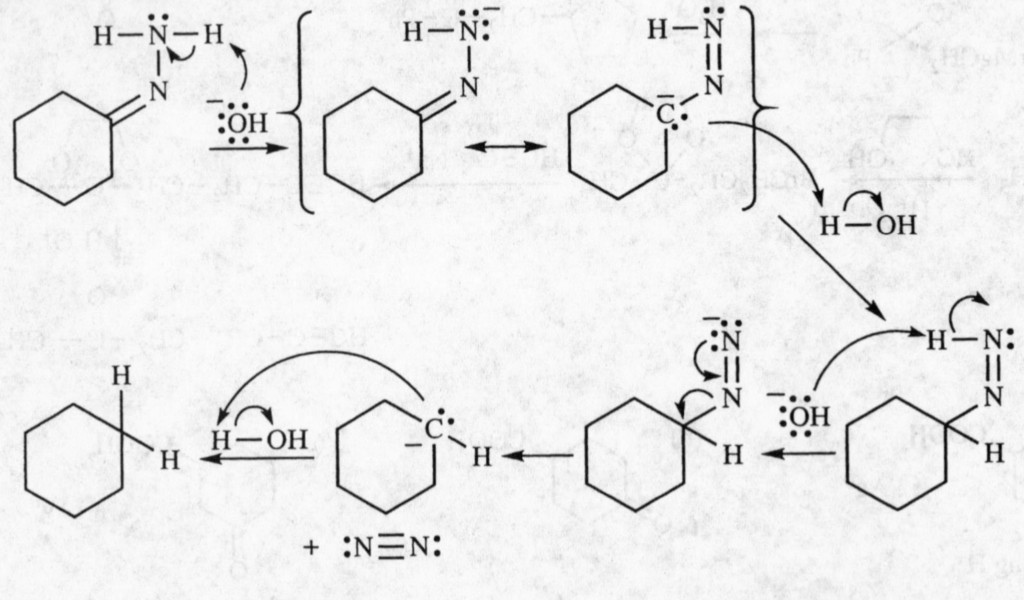

18-37

(a)

(b)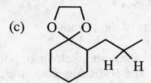

(c) (structure)

(d) (structure)

18-38   Please refer to solution 1-20, page 12 of this Solutions Manual.

18-39   IUPAC names first; then common names. Please see the note on p. 136 of this Solutions Manual regarding placement of position numbers.

(a) heptan-2-one; methyl *n*-pentyl ketone
(b) heptan-4-one; di-*n*-propyl ketone
(c) heptanal; no simple common name
(d) benzophenone; diphenyl ketone
(e) butanal; butyraldehyde
(f) propanone; acetone (IUPAC accepts "acetone")

(g) 4-bromo-2-methylhexanal; no common name
(h) 3-phenylprop-2-enal; cinnamaldehyde
(i) hexa-2,4-dienal; no common name
(j) 3-oxopentanal; no common name
(k) 3-oxocyclopentanecarbaldehyde; no common name
(l) *cis*-2,4-dimethylcyclopentanone; no common name

18-40 In order of increasing equilibrium constant for hydration:

$$CH_3-\overset{O}{\overset{\|}{C}}-CH_3 \quad < \quad CH_3-\overset{O}{\overset{\|}{C}}-CH_2Cl \quad < \quad CH_3-\overset{O}{\overset{\|}{C}}-H \quad < \quad ClCH_2-\overset{O}{\overset{\|}{C}}-H \quad < \quad H-\overset{O}{\overset{\|}{C}}-H$$

least amount
of hydration

greatest amount
of hydration

18-41

(NMR spectrum)

$$H-\overset{O}{\overset{\|}{\underset{a}{C}}}-\overset{b}{CH_2}-\overset{CH_3}{\underset{CH_3}{\overset{|}{\underset{|}{C}}}}-CH_3 \Big\}\underline{c}$$

a
1H

b
2H

c
9H

TMS

δ (ppm)

18-42

(structure: 3-methylcyclohex-2-enone with CH_3)

By comparison with similarly substituted molecules shown in the text:

$\pi \rightarrow \pi^*$ base value (210) plus 3 alkyl groups (30) = 240 nm
$n \rightarrow \pi^*$ 300-320 nm

417

18-43

$C_6H_{10}O_2$ indicates two elements of unsaturation.

The IR absorption at 1708 cm^{-1} suggests a ketone, or possibly two ketones since there are two oxygens and two elements of unsaturation. The NMR singlets in the ratio of 2 : 3 indicate a highly symmetric molecule. The singlet at δ 2.15 is probably methyl next to carbonyl, and the singlet at δ 2.67 integrating to two is likely to be CH_2 on the other side of the carbonyl.

Since the molecular formula is double this fragment, the molecule must be twice the fragment.

Two questions arise. Why is the integration 2 : 3 and not 4 : 6? Integration provides a *ratio*, not absolute numbers, of hydrogens. Why don't the two methylenes show splitting? Adjacent, *identical* hydrogens, with identical chemical shifts, do not split each other; the signals for ethane or cyclohexane appear as singlets.

18-44 The formula $C_{10}H_{12}O$ indicates 5 elements of unsaturation. A solid 2,4-DNP derivative suggests an aldehyde or a ketone, but a negative Tollens test precludes the possibility of an aldehyde; therefore, the unknown must be a ketone.

The NMR shows the typical ethyl pattern at δ 1.0 (3H, triplet) and δ 2.5 (2H, quartet), and a monosubstituted benzene at δ 7.3 (5H, multiplet). The singlet at δ 3.7 is a CH_2, but quite far downfield, apparently deshielded by two groups. Assemble the pieces:

18-45

(a)

(b)

(c)

$$\left[\begin{array}{c} \overset{H}{\underset{CH}{O}} \overset{\cdot CH_3}{\underset{CH}{CH}} \\ H_3C-\overset{C}{\underset{H_2}{C}}-\overset{C}{\underset{CH}{C}}\cdot CH_3 \end{array} \right]^{\ddagger}$$

m/z 114

$$\longrightarrow \left[\begin{array}{c} O\cdots H \\ \overset{\parallel}{\underset{}{}} \\ H_3C-C=CH_2 \end{array} \right]^{\ddagger}$$

m/z 58

$$+ \begin{array}{c} CHCH_3 \\ \parallel \\ CHCH_3 \end{array}$$

mass 56

18-46

18-47

A molecular ion of m/z 70 means a fairly small molecule. A solid semicarbazone derivative and a negative Tollens test indicate a ketone. The carbonyl (CO) has mass 28, so 70 – 28 = 42, enough mass for only 3 more carbons. The molecular formula is probably C_4H_6O (mass 70); with two elements of unsaturation, we can infer the presence of a double bond or a ring in addition to the carbonyl.

The IR shows a strong peak at 1790 cm^{-1}, indicative of a ketone in a small ring. No peak in the 1600-1650 cm^{-1} region shows the absence of an alkene. The only possibilities for a small ring ketone containing four carbons are these:

A

B

The HNMR can distinguish these. No methyl doublet appears in the NMR spectrum, ruling out **B**. The NMR does show a 4H triplet at δ 3.1; this signal comes from the two methylenes (C-2 and C-4) adjacent to the carbonyl, split by the two hydrogens on C-3. The signal for the methylene at C-3 appears at δ 2.0, roughly a quintet because of splitting by four neighboring protons.

The unknown is cyclobutanone, **A**. The symmetry indicated by the carbon NMR rules out structure **B**. The IR absorption of the carbonyl at 1790 cm^{-1} is characteristic of small ring ketones; ring strain strengthens the carbon-oxygen double bond, increasing its frequency of vibration. (See Section 12-9 in the text.)

18-48

(a) The conjugated diene has a maximum at 235 nm (see Solved Problem 15-3) and the ketone has a maximum at about 237 nm, so the π to π* transition cannot be used to differentiate the compounds.
(b) The ketone has an n to π* transition around 315 nm that the diene cannot have.

18-49

(a) acetal
(old: ketal)

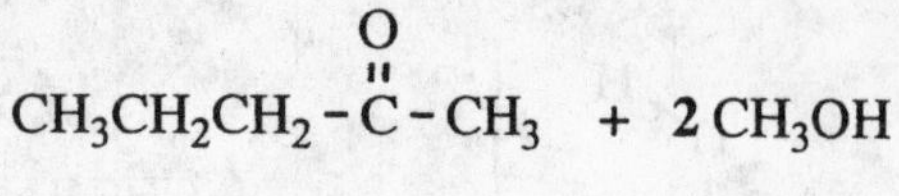

(e) acetal

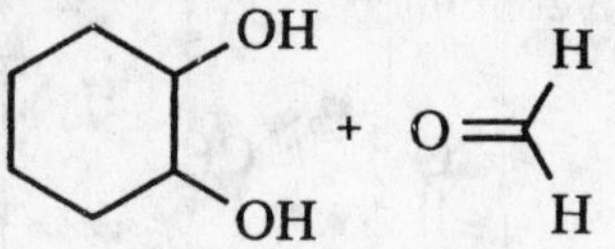

(b) hemiacetal
(old: hemiketal)

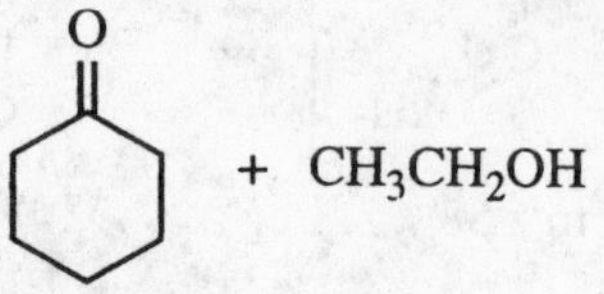

(f) diether, inert to hydrolysis

(c) acetal

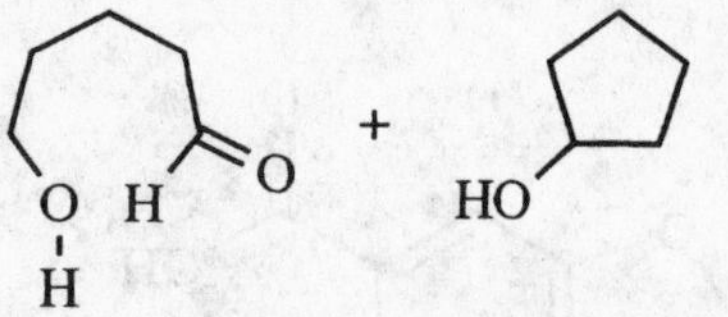

(g) imine(s)

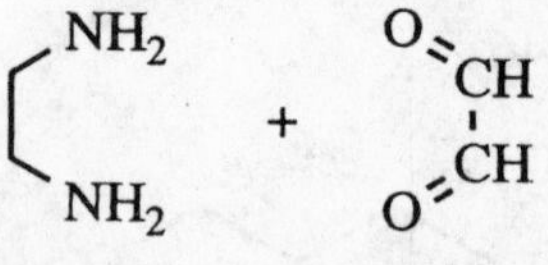

(d) acetal
(old: ketal)

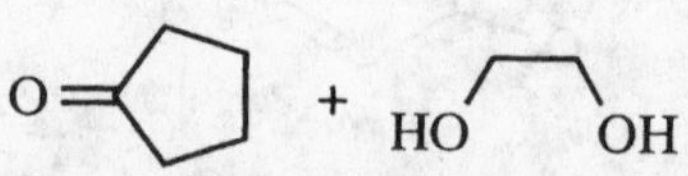

(h) imine

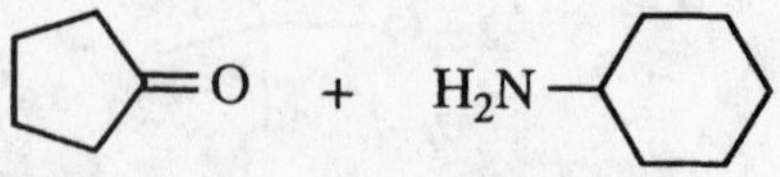

18-50

(a)

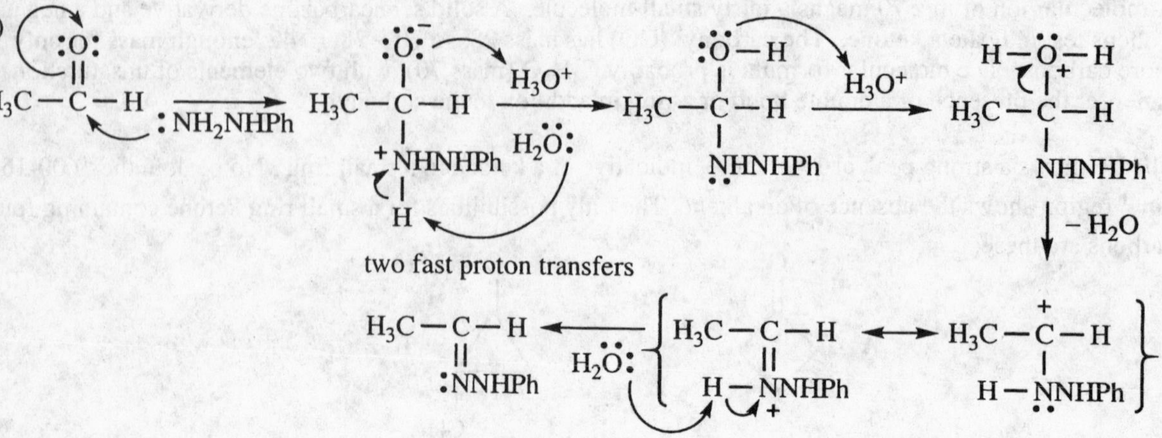

(b)

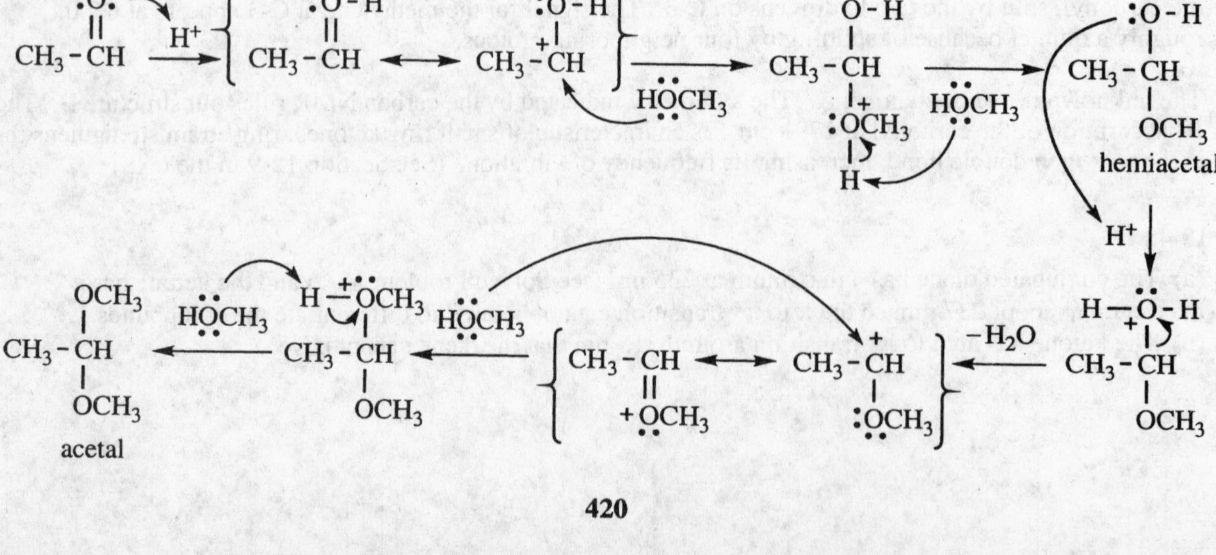

18-50 continued

(c)

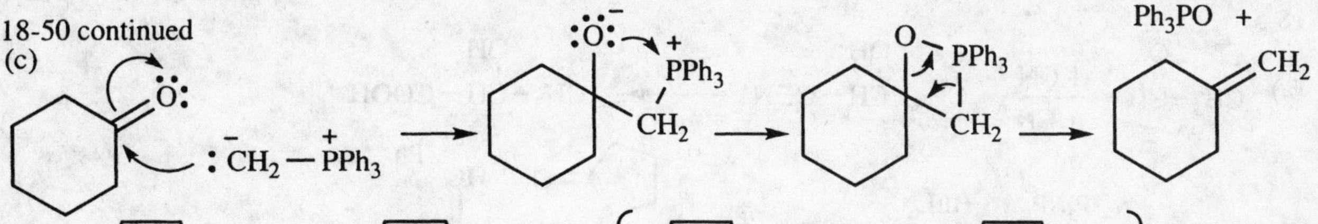

(d)

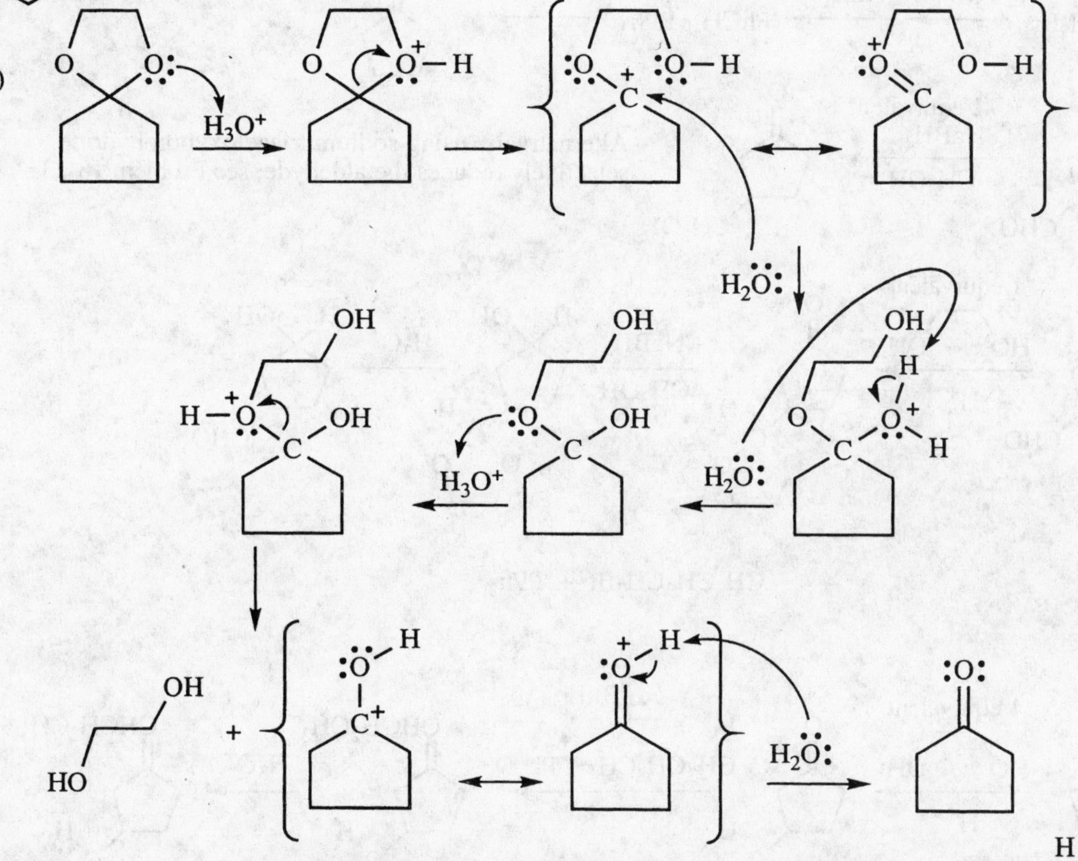

(e)

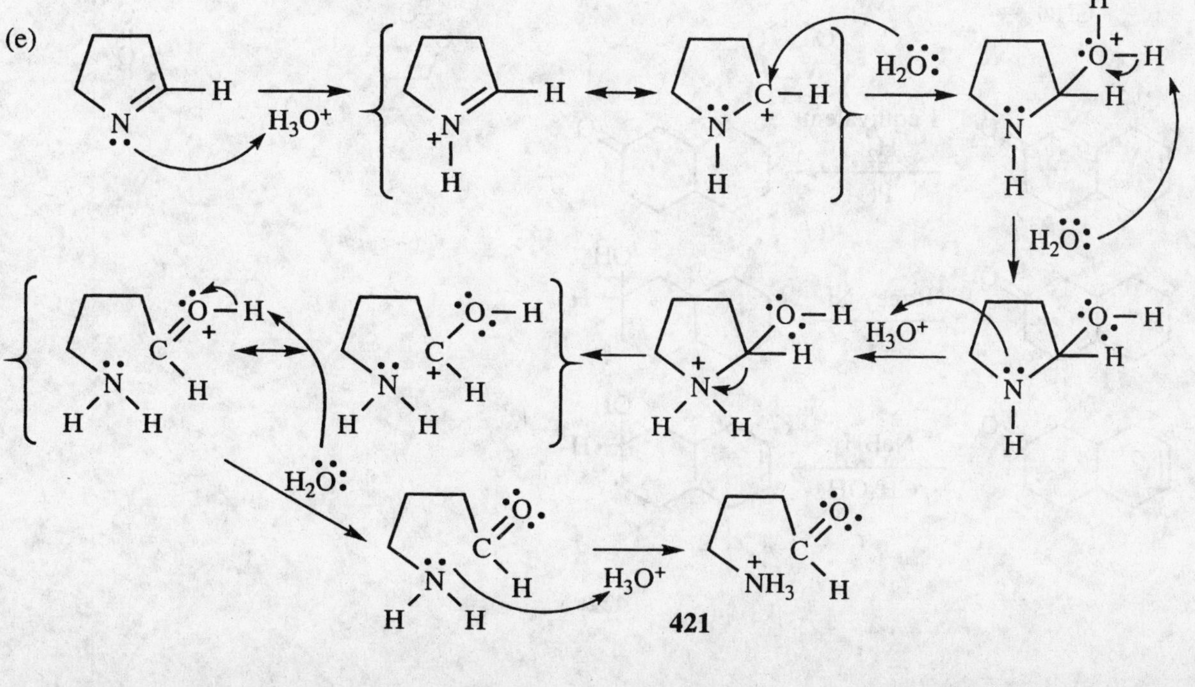

421

18-51

(a) $CH_3-\overset{O}{\overset{\|}{C}}H$ $\xrightarrow[\text{HCN}]{\text{KCN}}$ $CH_3-\overset{OH}{\underset{|}{C}}H-C\equiv N$ $\xrightarrow{H_3O^+}$ $CH_3-\overset{OH}{\underset{|}{C}}H-COOH$

(b) 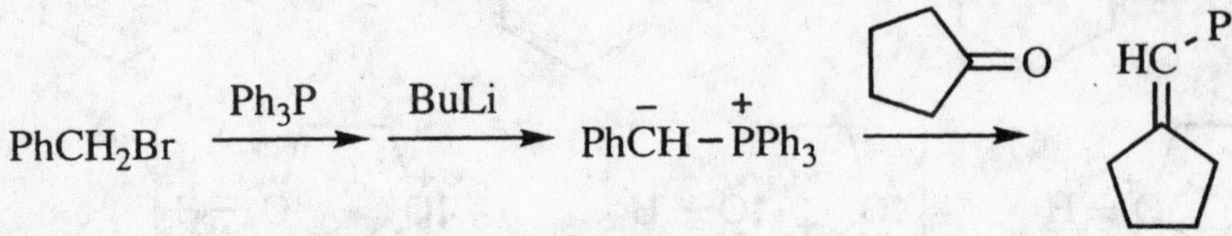 $PhCH_2Br$ $\xrightarrow{Ph_3P}$ $\xrightarrow{BuLi}$ $Ph\overset{-}{C}H-\overset{+}{P}Ph_3$ $\longrightarrow$

(c) $\xrightarrow[\text{CH}_3\text{OH}]{\substack{\text{1 equivalent} \\ \text{NaBH}_4}}$ Alternatively, using sodium triacetoxyborohydride selectively reduces the aldehyde; see Problem 18-13.

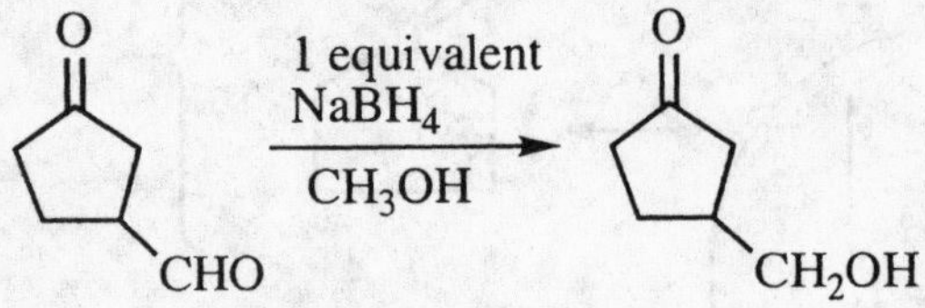

(d) 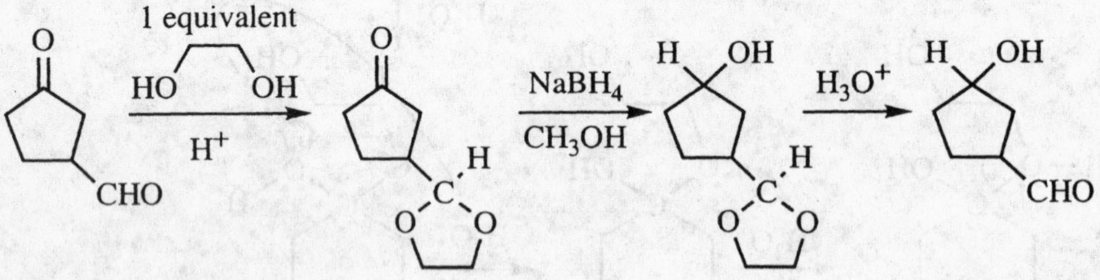 $\xrightarrow[\text{H}^+]{\substack{\text{1 equivalent} \\ \text{HO} \quad \text{OH}}}$ $\xrightarrow[\text{CH}_3\text{OH}]{\text{NaBH}_4}$ $\xrightarrow{H_3O^+}$

$CH_3CH_2CH_2Br + PPh_3$

$\downarrow BuLi$

(e) 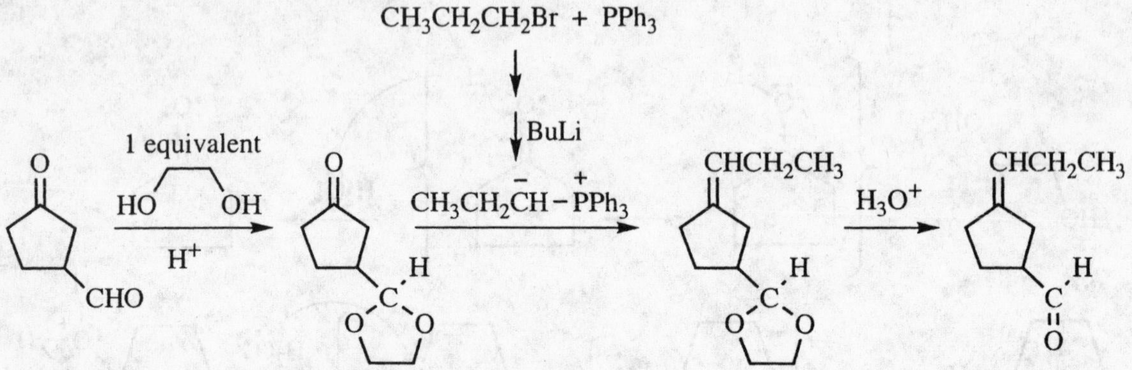 $\xrightarrow[\text{H}^+]{\substack{\text{1 equivalent} \\ \text{HO} \quad \text{OH}}}$ $\xrightarrow{CH_3CH_2\overset{-}{C}H-\overset{+}{P}Ph_3}$ $\xrightarrow{H_3O^+}$

(f) 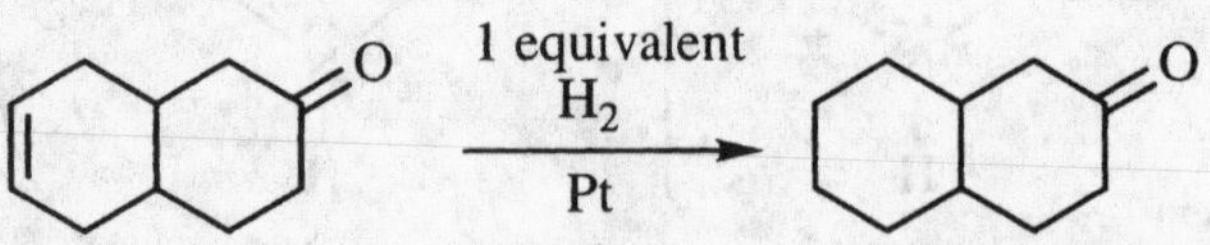 $\xrightarrow[\text{Pt}]{\substack{\text{1 equivalent} \\ \text{H}_2}}$

(g) 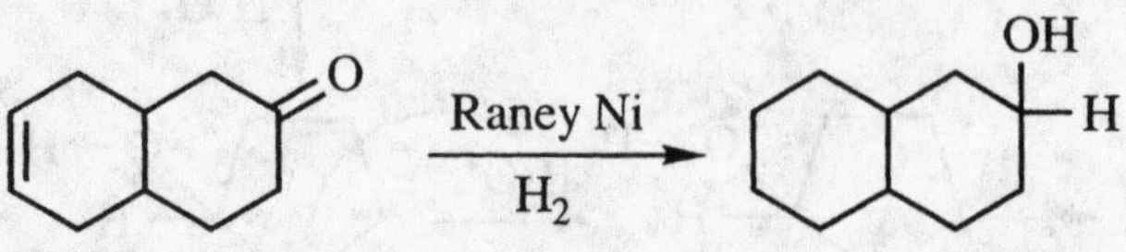 $\xrightarrow[\text{H}_2]{\text{Raney Ni}}$

(h) $\xrightarrow[\text{CH}_3\text{OH}]{\text{NaBH}_4}$

422

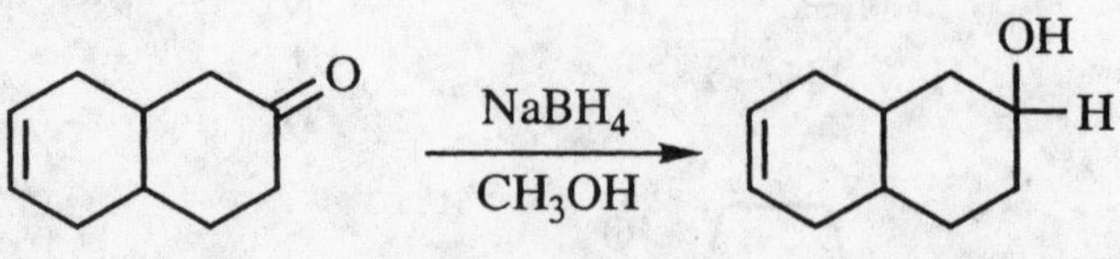

18-52 All of these reactions would be acid-catalyzed.

(a) 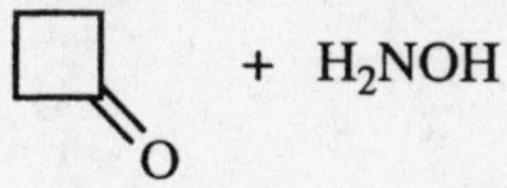 [cyclobutanone structure] + H₂NOH

(b) 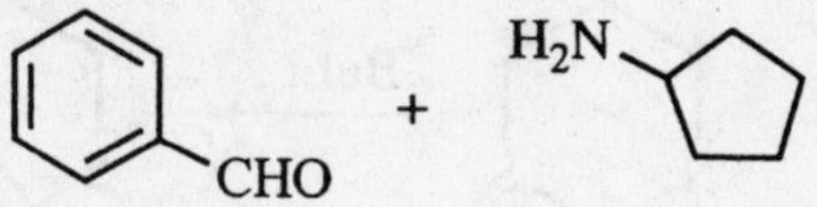 [benzaldehyde with CHO] + H₂N–[cyclopentane]

(c) 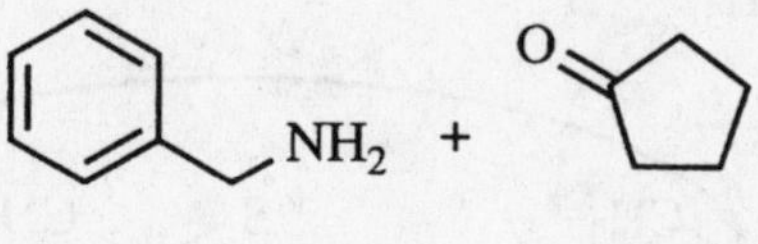 [benzyl with NH₂] + [cyclopentanone structure]

(d) 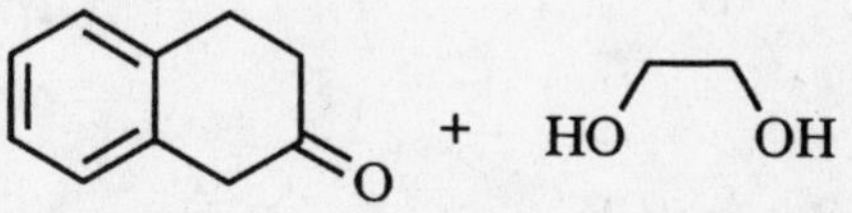 [tetralone bicyclic structure with O] + HO–CH₂CH₂–OH

(e) 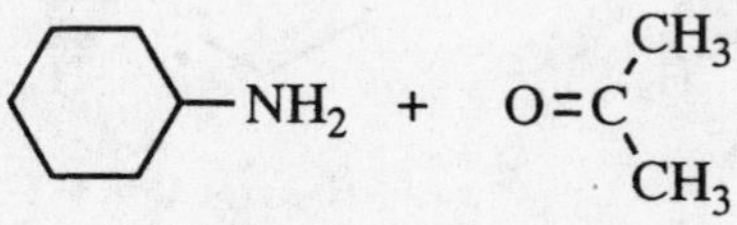 [cyclohexyl]–NH₂ + O=C(CH₃)(CH₃)

(f) 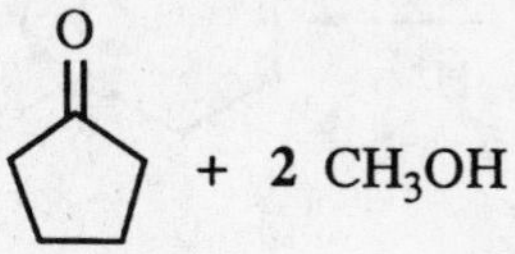 [cyclopentanone structure] + 2 CH₃OH

18-53

(a) 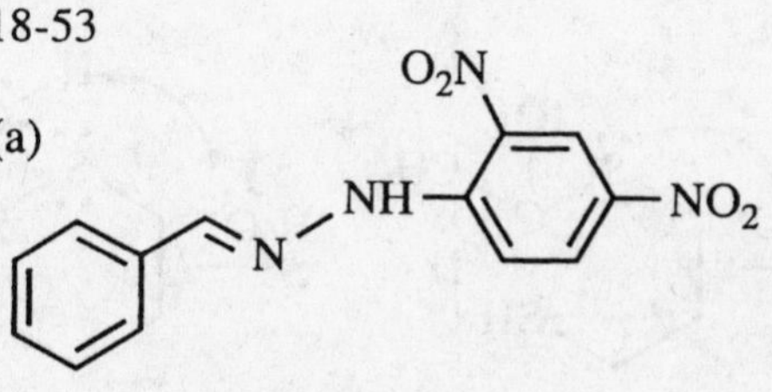 [benzaldehyde phenylhydrazone structure with 2,4-dinitrophenyl: O₂N, NO₂ groups]

(b) 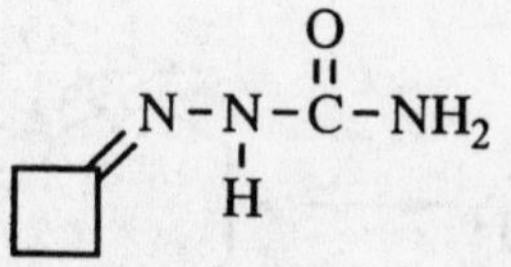 [cyclobutanone semicarbazone: N–N(H)–C(=O)–NH₂]

(c) [cyclopropane with =NOH]

(d) [dioxolane ring with propyl and ethyl substituents]

(e) CH₃–CH(OCH₃)(OCH₃)

(f) H–CH(OH)(OCH₃)

(g) [phenyl]–C(=N–CH₂CH₃)(CH₂CH₃)

(h) [1,3-dithiane ring with CH₃CH₂ and H substituents]

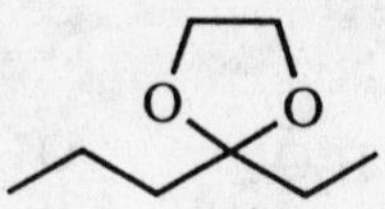

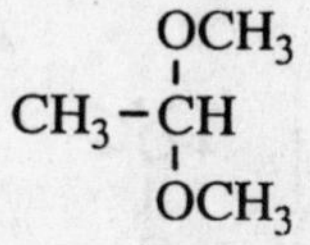

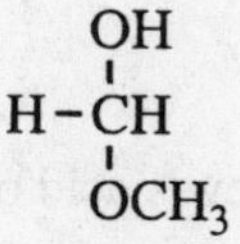

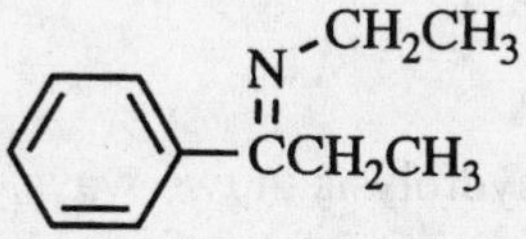

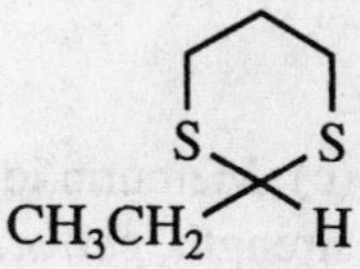

18-54

(a)

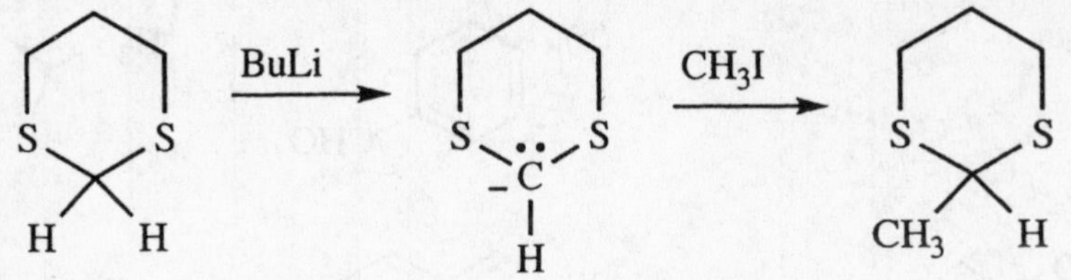

(b)

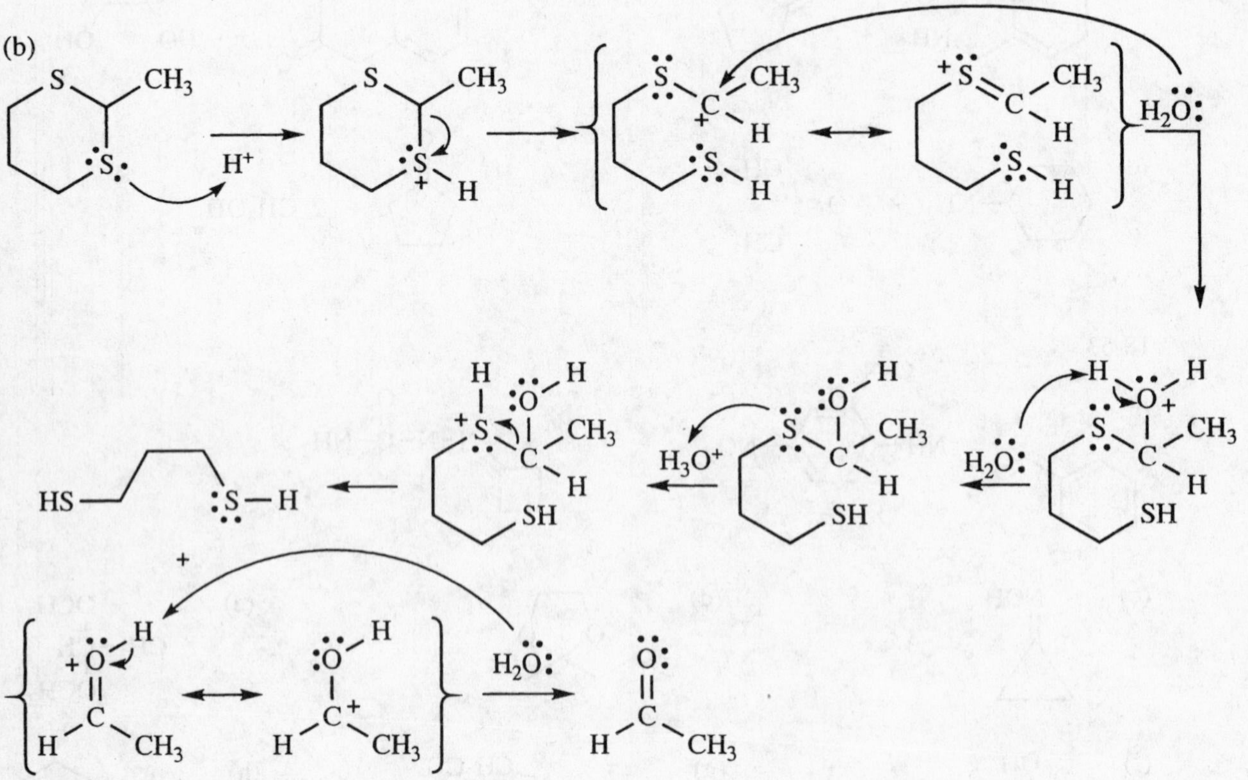

(c) Mercuric ion, Hg^{2+}, assists the hydrolysis in two ways. First, mercuric ion is a Lewis acid of moderate strength, performing the same function as a proton from a protic acid.

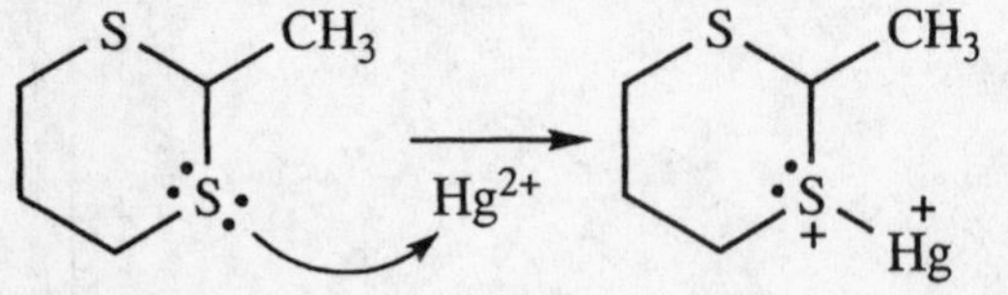

The effectiveness of Hg^{2+} as a Lewis acid is partly due to its charge: even complexed with a sulfur, the mercury atom still has a positive charge, attracting the sulfur's electrons.

A second explanation for mercuric ion's effectiveness in hydrolysis of thioacetals lies in the complex formed between the ion and the two sulfur atoms. This stable complex effectively removes HSCH$_2$CH$_2$SH from the equilibrium, shifting the equilibrium to product. (An example of whose principle? His initials are "Le Châtelier".)

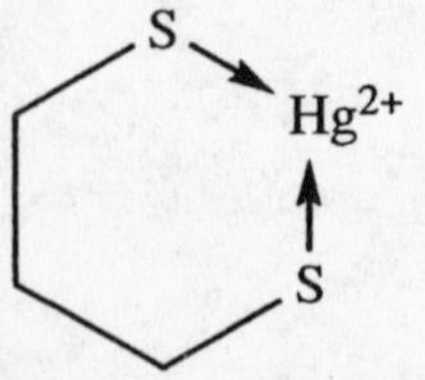

Thiols are often referred to as "mercaptans" because of their ability to CAPTure MERcury.

424

18-55 The key to this problem is understanding that the relative proximity of the two oxygens can dramatically affect their chemistry.

1,2-dioxane

The "second" isomer described: two oxygens connected by a sigma bond are a peroxide. The O—O bond is easily cleaved to give radicals. In the presence of organic compounds, radical reactions can be explosive.

1,3-dioxane

The "third" isomer described: two oxygens bonded to the same sp³ carbon constitute an acetal which is hydrolyzed in aqueous acid. See the mechanism below.

1,4-dioxane

The "first" isomer described: an excellent solvent (although toxic), these oxygens are far enough apart to act independently. It is a simple ether.

<u>Mechanism of acetal hydrolysis</u>

18-56

(a)
NCH₃

(b)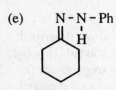
CH₃O   OCH₃

(c)
NOH

(d)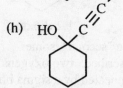

(e)
N–N–Ph
   |
   H

(f)
Ph   OH

(g) no reaction

(h)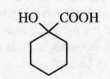
HO   C≡C–H

(i)

(j)
CH₂

(k) Na⁺ ⁻O   C≡N

(l) HO   COOH

18-57 The new bond to carbon comes from the NaBH₄ or NaBD₄, shown in bold below. The new bond to oxygen comes from the protic solvent.

(a)
$$CH_3-\overset{O}{\overset{\|}{C}}-CH_2CH_3 \xrightarrow{NaB\mathbf{D}_4} \xrightarrow{H_2O} CH_3-\underset{\mathbf{D}}{\overset{OH}{\underset{|}{\overset{|}{C}}}}-CH_2CH_3$$

(b)
$$CH_3-\overset{O}{\overset{\|}{C}}-CH_2CH_3 \xrightarrow{NaB\mathbf{D}_4} \xrightarrow{D_2O} CH_3-\underset{\mathbf{D}}{\overset{OD}{\underset{|}{\overset{|}{C}}}}-CH_2CH_3$$

(c)
$$CH_3-\overset{O}{\overset{\|}{C}}-CH_2CH_3 \xrightarrow{NaBH_4} \xrightarrow{D_2O} CH_3-\underset{\mathbf{H}}{\overset{OD}{\underset{|}{\overset{|}{C}}}}-CH_2CH_3$$

18-58 While hydride is a small group, the actual chemical species supplying it, AlH₄⁻, is fairly large, so it prefers to approach from the less hindered side of the molecule, that is, the side opposite the methyl. This forces the oxygen to go to the same side as the methyl, producing the *cis* isomer as the major product.

less hindered face

H·Al·H
   |
   H

H⁺

cis major product

18-59

(a)

$$CH_3Br \xrightarrow{Ph_3P} \xrightarrow{BuLi} \ ^-CH_2-\overset{+}{P}Ph_3 \ \xrightarrow{\text{(cyclohexanone)}} \ \text{(methylenecyclohexane)}$$

An exocyclic double bond is less stable than an endocyclic double bond.

methylenecyclohexane

(b) The difficulty in synthesizing methylenecyclohexane from cyclohexanone without using the Wittig reaction rests in the stability of the double bond inside the ring (endocyclic) versus outside the ring (exocyclic).

cyclohexanone + $CH_3MgBr \longrightarrow \xrightarrow{H_2O}$ 1-methylcyclohexanol $\xrightarrow[\Delta]{H^+}$ 1-methylcyclohexene (major) + methylenecyclohexane (minor)

A dehydration following the E1 mechanism passing through a carbocation intermediate will give the more substituted, endocyclic double bond as the major product. The only chance of making the exocyclic double bond as the major product is to do an E2 elimination using a bulky base to give Hofmann orientation (Chapter 7).

$\xrightarrow[\text{cold}]{HBr}$ 1-bromo-1-methylcyclohexane $\xrightarrow[\text{E2}]{K^+ \ ^-O\text{-}t\text{-}Bu}$ 1-methylcyclohexene (minor) + methylenecyclohexane (major)

18-60

(a) benzene $\xrightarrow[\text{AlCl}_3]{Cl-\overset{O}{\overset{\|}{C}}CH_2CH_2CH_3}$ (phenyl propyl ketone) $\xrightarrow[\text{HCl}]{Zn(Hg)}$ (butylbenzene)

(b) benzonitrile (C≡N) $\xrightarrow[\text{ether}]{CH_3CH_2MgBr} \xrightarrow{H_3O^+}$ $\overset{O}{\overset{\|}{C}}-CH_2CH_3$

(c) benzene $\xrightarrow[\text{AlCl}_3]{Cl_2}$ (chlorobenzene, Cl) $\xrightarrow[\text{350° C}]{NaOH}$ (phenol, OH) $\xrightarrow[\text{2) CH}_3I]{\text{1) NaOH}}$ (anisole, OCH_3) $\xrightarrow[\text{CuCl}]{\substack{CO \\ HCl \\ AlCl_3}}$ (p-methoxybenzaldehyde, OCH_3 / CHO)

(d) (HO— with ring and chain) $\xrightarrow{H_2CrO_4}$ (HO-C=O acid) $\xrightarrow{\text{conc. } H_2SO_4}$ (α-tetralone)

Alternatively, the acid chloride could be made with SOCl_2, then cyclized by Friedel-Crafts acylation with AlCl_3.

18-61

(a)

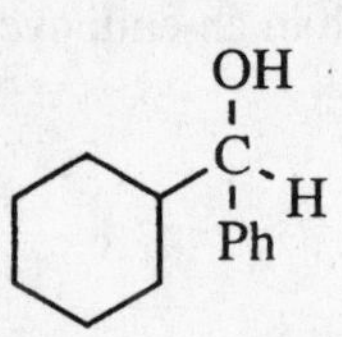

(b)

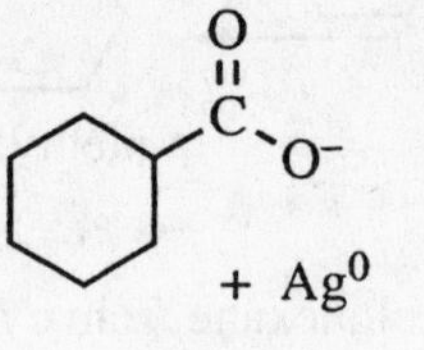

+ Ag⁰

(c)

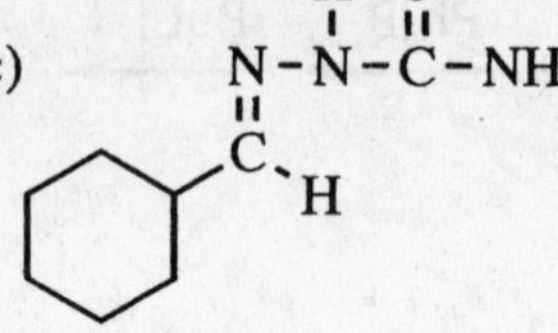

(d)

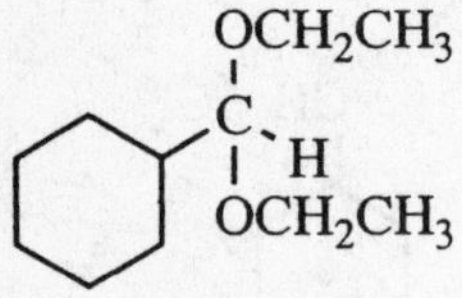

(e)

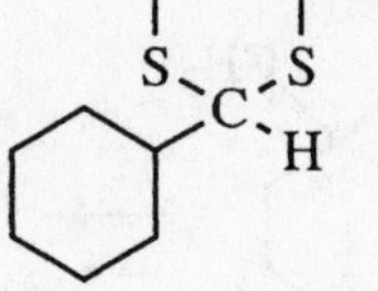

(f)

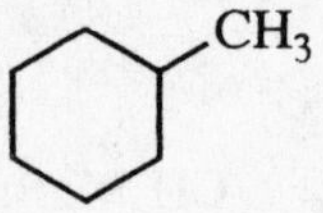

18-62

(a)

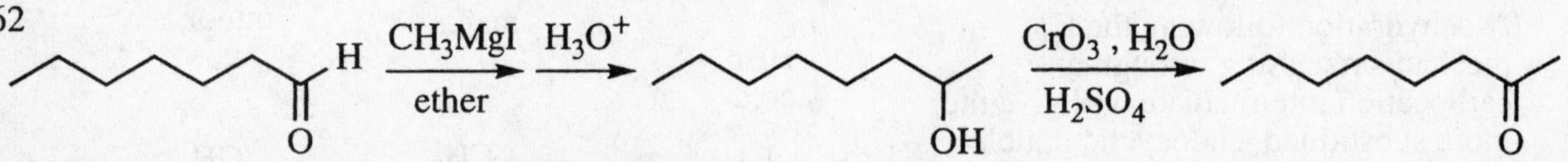

$$\xrightarrow[\text{ether}]{CH_3MgI} \xrightarrow{H_3O^+}$$

$$\xrightarrow[H_2SO_4]{CrO_3,\ H_2O}$$

(b)

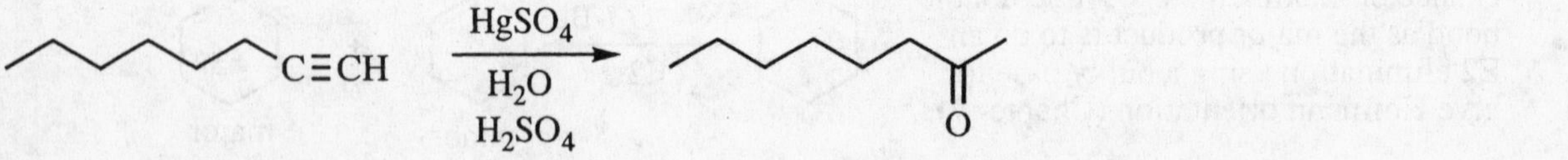

$$\xrightarrow[H_2SO_4]{HgSO_4 \quad H_2O}$$

(c)

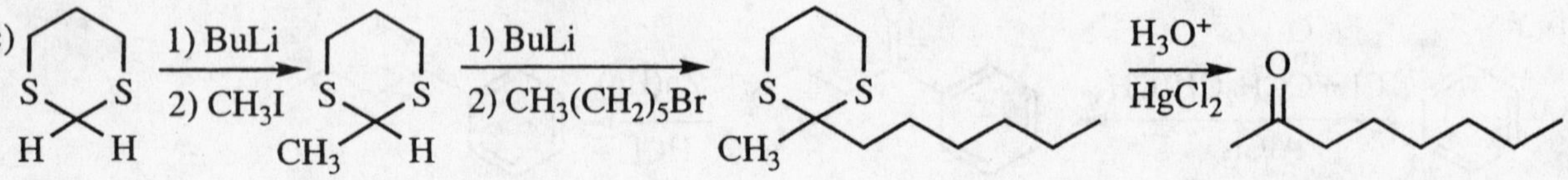

$$\xrightarrow[\text{2) CH}_3\text{I}]{\text{1) BuLi}}$$

$$\xrightarrow[\text{2) CH}_3(CH_2)_5Br]{\text{1) BuLi}}$$

$$\xrightarrow[HgCl_2]{H_3O^+}$$

(d)

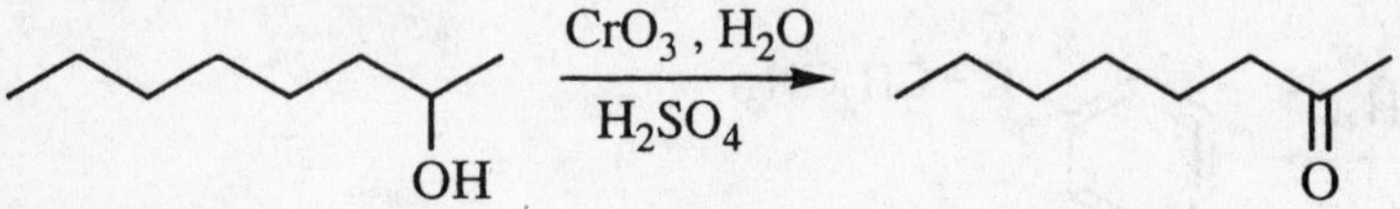

$$\xrightarrow[H_2SO_4]{CrO_3,\ H_2O}$$

(e)

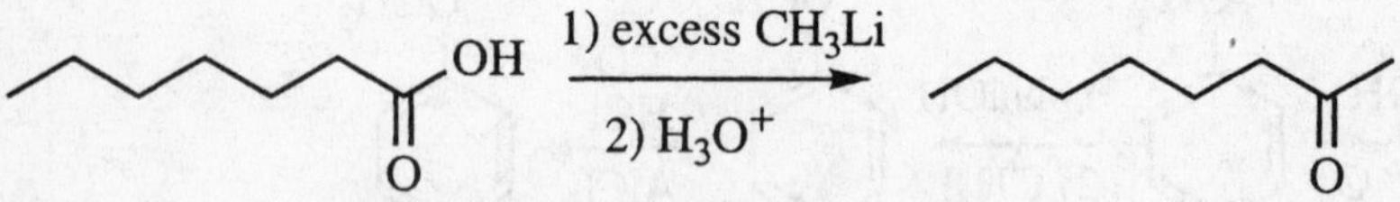

$$\xrightarrow[\text{2) } H_3O^+]{\text{1) excess } CH_3Li}$$

(f)

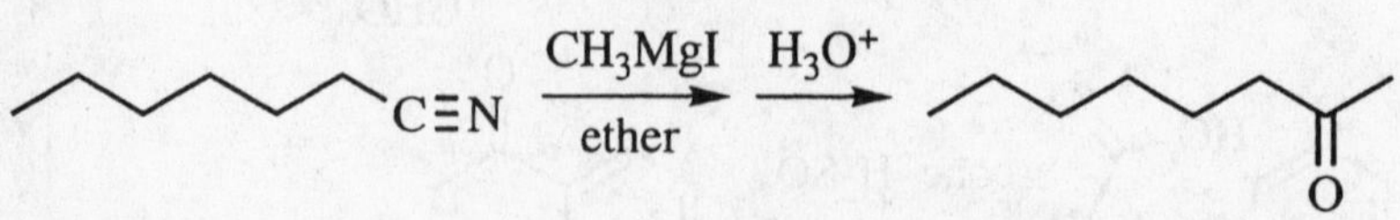

$$\xrightarrow[\text{ether}]{CH_3MgI \quad H_3O^+}$$

(g)

$$\xrightarrow[\text{2) } Me_2S]{\text{1) } O_3}$$

$$+$$

428

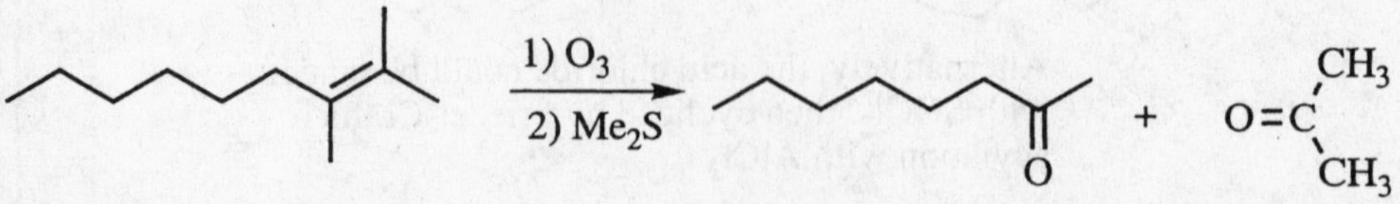

18-63

(a)

$\underset{\text{(octanol)}}{\text{CH}_3(\text{CH}_2)_6\text{CH}_2\text{OH}}$ $\xrightarrow{\text{PCC}}$ (aldehyde, O=CH–)

(b)

(alkene with =CH$_2$) $\xrightarrow[\text{2) Me}_2\text{S}]{\text{1) O}_3}$ (aldehyde)

(c)

(terminal alkyne, C≡CH) $\xrightarrow[\text{2) H}_2\text{O}_2, \text{HO}^-]{\text{1) Sia}_2\text{BH}}$ (aldehyde)

(d)

(1,3-dithiane, S–CH$_2$–S with H, H) $\xrightarrow[\text{2) CH}_3(\text{CH}_2)_6\text{Br}]{\text{1) BuLi}}$ (alkylated dithiane, H) $\xrightarrow[\text{HgCl}_2]{\text{H}_3\text{O}^+}$ (aldehyde)

(e)

(1,1-dichloro compound, Cl, Cl) $\xrightarrow[\text{H}_2\text{O}]{\text{KOH}}$ (aldehyde)

(This reaction needs a solvent like THF to keep all reactants in solution.)

(f)

(carboxylic acid, OH) $\xrightarrow[\text{2) H}_3\text{O}^+]{\text{1) LiAlH}_4}$ (primary alcohol, OH) $\xrightarrow{\text{PCC}}$ (aldehyde)

OR

(carboxylic acid, OH) $\xrightarrow{\text{SOCl}_2}$ (acid chloride, Cl) $\xrightarrow{\text{LiAl(O}t\text{-Bu})_3\text{H}}$ (aldehyde)

18-64

(a) (ketone)

(b) (ketone) + (ketone)

(c) (ketone)

(d) (cyclic ketone, O)

(e) (cyclic ketone with CH$_3$) + (cyclic ketone with CH$_3$)

18-65

(a) ketone: no reaction
(b) aldehyde: positive
(c) enol of an aldehyde—tautomerizes to aldehyde in base: positive
(d) hemiacetal of an aldehyde in equilibrium with the aldehyde in base: positive
(e) acetal—stable in base: no reaction
(f) hemiacetal of an aldehyde in equilibrium with the aldehyde in base: positive

18-66 The structure of **A** can be deduced from its reaction with **J** and **K**. What is common to both products of these reactions is the heptan-2-ol part; the reactions must be Grignard reactions with heptan-2-one, so **A** must be heptan-2-one.

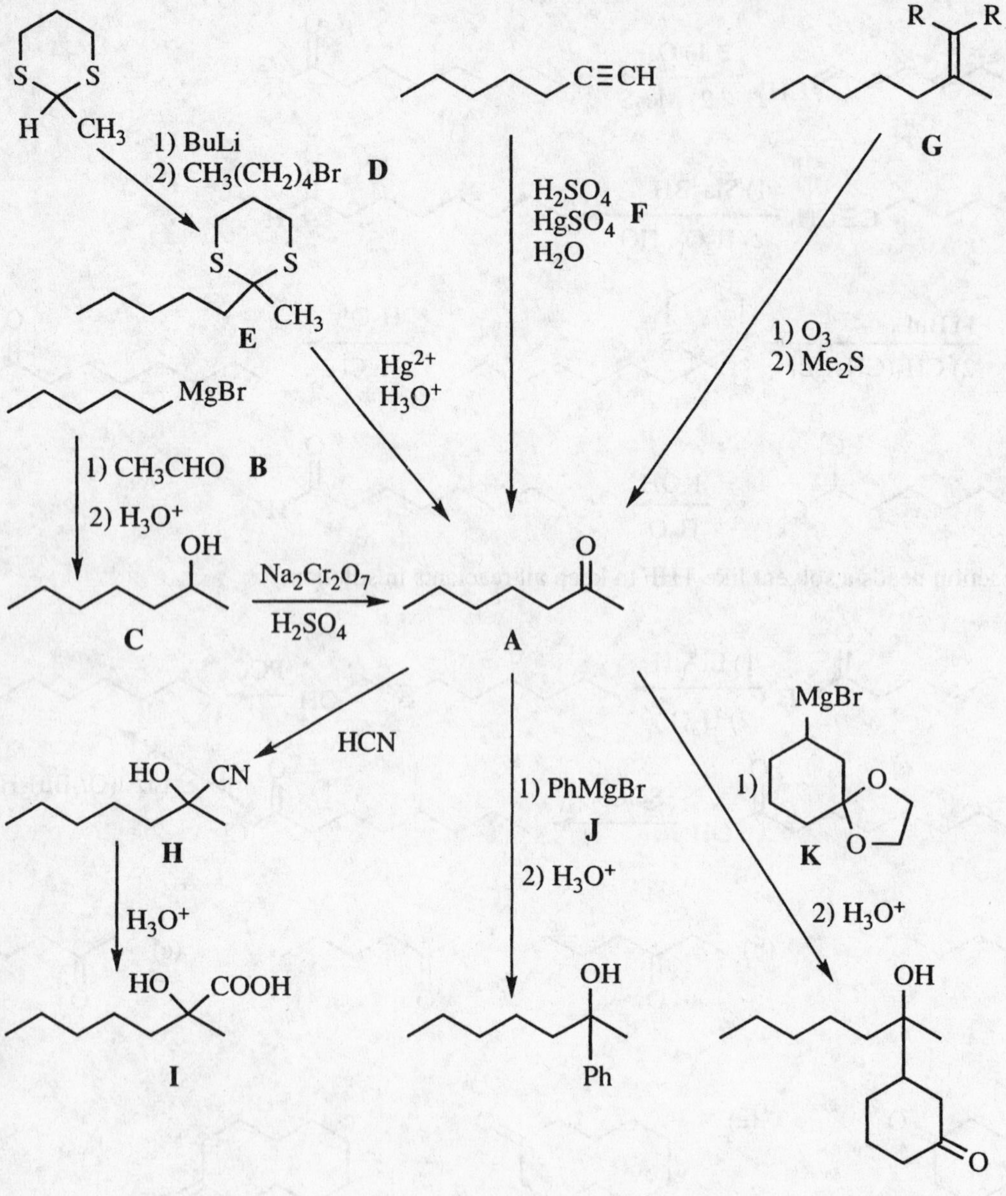

1) BuLi
2) CH₃(CH₂)₄Br **D**

S S
H CH₃

S S
CH₃
E

MgBr

1) CH₃CHO **B**
2) H₃O⁺

OH
C

Na₂Cr₂O₇
H₂SO₄

Hg²⁺
H₃O⁺

C≡CH

H₂SO₄
HgSO₄
H₂O **F**

R R

G

1) O₃
2) Me₂S

O
A

HCN

HO CN
H

H₃O⁺

HO COOH
I

1) PhMgBr
J
2) H₃O⁺

OH
Ph

MgBr

1)
O O
K

2) H₃O⁺

OH

O

430

18-67 The very strong π to π* absorption at 225 nm in the UV spectrum suggests a conjugated ketone or aldehyde. The IR confirms this: strong, conjugated carbonyl at 1690 cm^{-1} and small alkene at 1610 cm^{-1}. The absence of peaks at 2700-2800 cm^{-1} shows that the unknown is not an aldehyde.

The molecular ion at 96 leads to the molecular formula:

$$C=C-\overset{\overset{\displaystyle O}{\|}}{C}-C$$

mass 64

$$\begin{array}{r} 96 \\ -64 \\ \hline \end{array}$$

32 mass units ⇒ add 2 carbons and 8 hydrogens

molecular formula = C_6H_8O = 3 elements of unsaturation

Two elements of unsaturation are accounted for in the enone. The other one is likely a ring.

The NMR shows two vinyl hydrogens. The doublet at δ 6.0 says that the two hydrogens are on neighboring carbons (two peaks = one neighboring H).

doublet: 1 neighboring H

$$C-\overset{\overset{\displaystyle H}{|}}{C}=\overset{\overset{\displaystyle H}{|}}{C}-\overset{\overset{\displaystyle O}{\|}}{C}-C \quad + \ 1\,C \ + \ 6\,H \ + \ 1\ ring$$

No methyls are apparent in the NMR, so the 6H group of peaks at δ 2.0-2.4 is most likely 3 CH$_2$ groups. Combining the pieces:

The mass spectral fragmentation can be explained by a "retro" or reverse Diels-Alder fragmentation:

m/z 96 → m/z 68 + $\overset{CH_2}{\underset{CH_2}{\|}}$ loss of 28

In the HNMR, one of the vinyl hydrogens appears at δ 7.0. This is typical of an α,β-unsaturated carbonyl because of the resonance form that shows deshielding of the β-hydrogen.

$$C-\underset{\beta}{C}=\underset{\alpha}{C}-\overset{\overset{\displaystyle O}{\|}}{C}-C \quad \longleftrightarrow \quad C-\underset{\beta}{\overset{+}{C}}-\underset{\alpha}{C}=C-C \quad (\delta\,7.0)$$

18-68 Building a model will help visualize this problem.

(a)

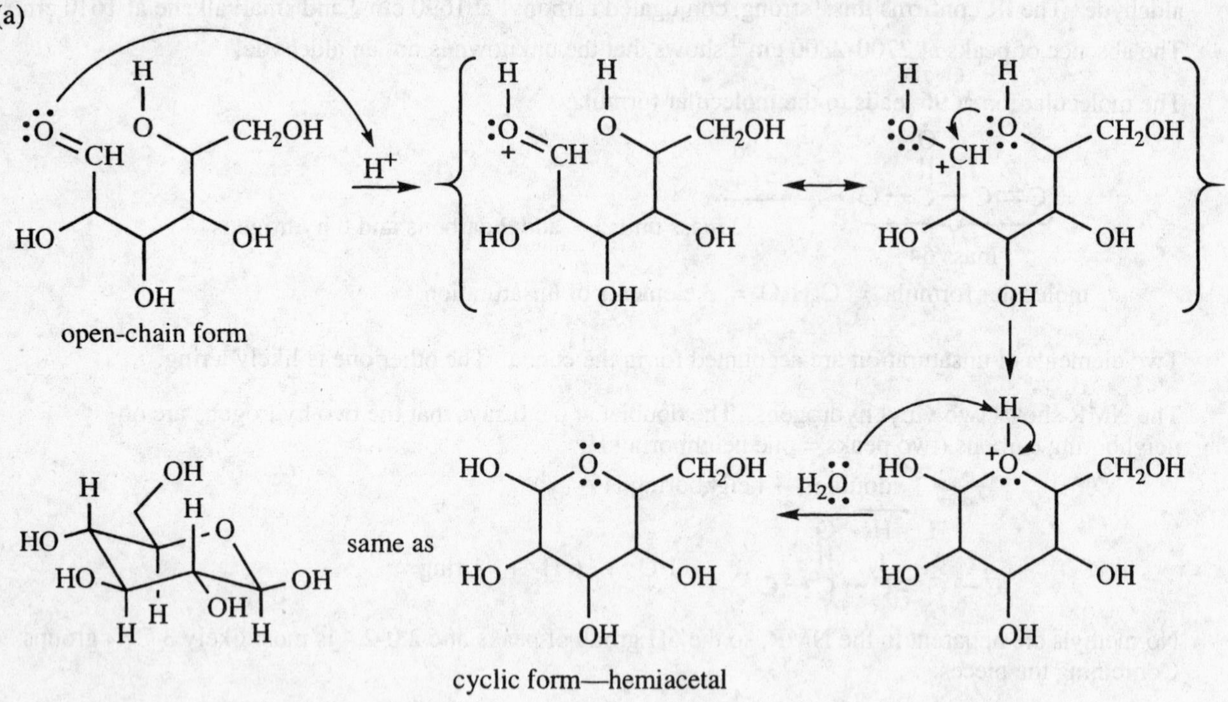

open-chain form

cyclic form—hemiacetal

same as

(b) Yes, the cyclic form of glucose will give a positive Tollens test. In the basic solution of the Tollens test, the hemiacetal is in equilibrium with the open-chain aldehyde with the cyclic form in much larger concentration. However, it is the open-chain aldehyde that reacts with silver ion, so even though there is only a small amount of open-chain form present at any given time, as more of the open-chain form is oxidized by silver ion, more cyclic form will open to replace the consumed open-chain form. Eventually all of the cyclic form will be dragged kicking and screaming through the open-chain form to be oxidized to the carboxylate. Le Châtelier's Principle strikes again!

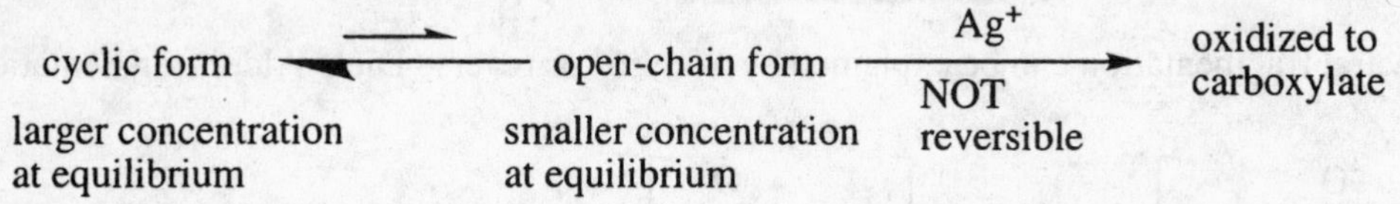

cyclic form ⇌ open-chain form $\xrightarrow[\text{NOT reversible}]{Ag^+}$ oxidized to carboxylate

larger concentration at equilibrium

smaller concentration at equilibrium

18-69

(a)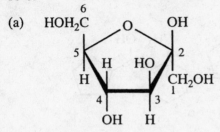

Any carbon with two oxygens bonded to it with single bonds belongs to the acetal family. If one of the oxygen groups is an OH, then the functional group is a hemiacetal. Thus, the functional group at C-2 is a hemiacetal. (The old name for this group is hemiketal as it came from a ketone.)

(b) Models will help. Ignore stereochemistry for the mechanism.

same as

18-70 Recall that "dilute acid" means an aqueous solution, and aqueous acid will remove acetals.

18-71

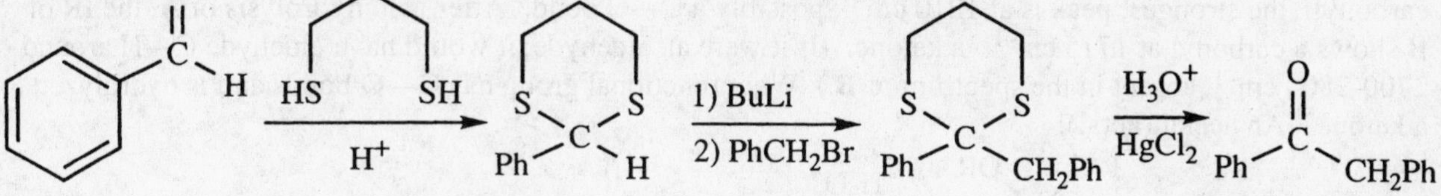

18-72

(a)

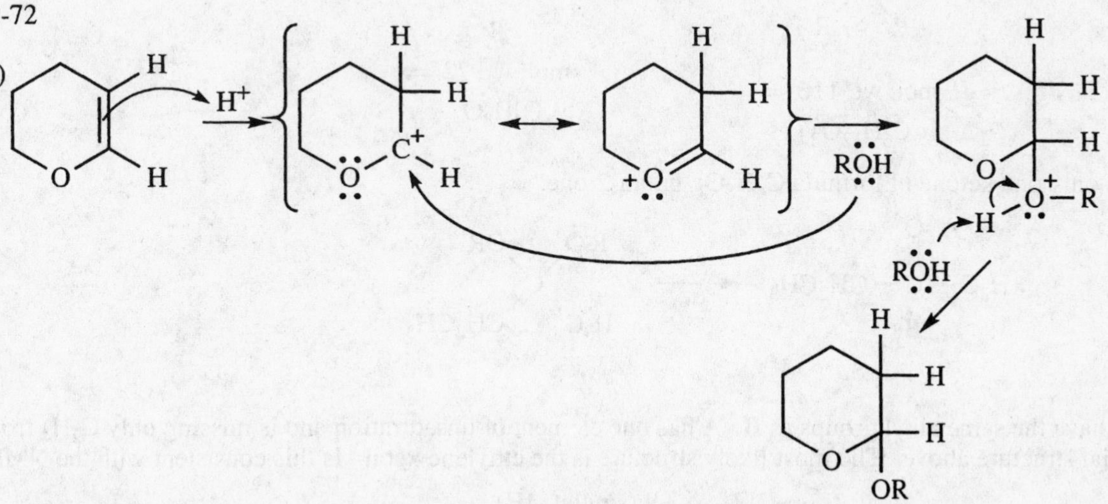

(b) This is not an ether, but rather an acetal, stable to base but reactive with aqueous acid.

(c)

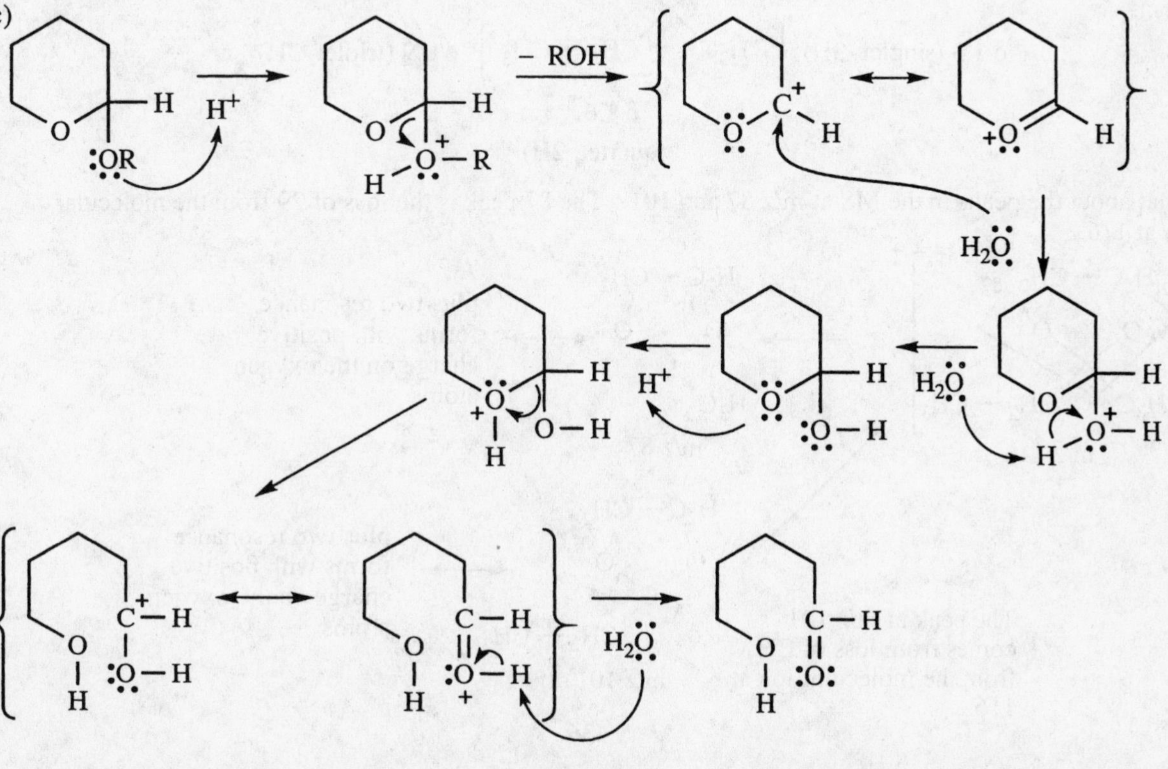

(a) First, deduce what functional groups are present in **A** and **B**. The IR of **A** shows no alkene and no carbonyl: the strongest peak is at 1210 cm^{-1}, possibly a C—O bond. After acid hydrolysis of **A**, the IR of **B** shows a carbonyl at 1715 cm^{-1}: a ketone. (If it were an aldehyde, it would have aldehyde C—H around $2700\text{-}2800 \text{ cm}^{-1}$, absent in the spectrum of **B**.) What functional group has C—O bonds and is hydrolyzed to a ketone? An acetal (ketal)!

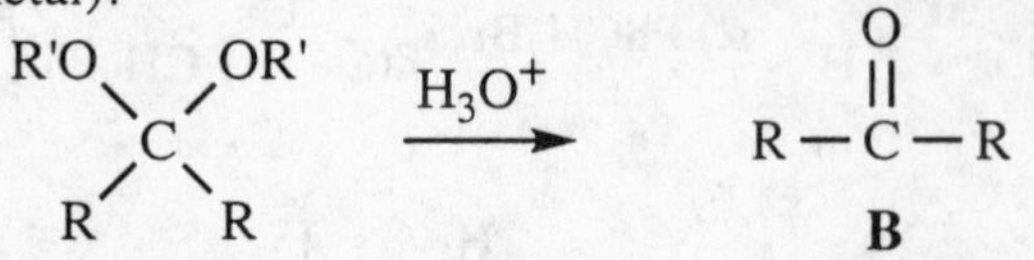

$$R'O \diagdown \underset{\underset{\displaystyle R \quad R}{|}}{C} \diagup OR'$$

A
mol. wt. 116 $\Rightarrow$
$C_6H_{12}O_2$

$\xrightarrow{H_3O^+}$

$$\overset{\displaystyle O}{\overset{\displaystyle \|}{R-C-R}}$$

B
mol. wt. 72 $\Rightarrow$
C_4H_8O

There is only one ketone of formula C_4H_8O: butan-2-one.

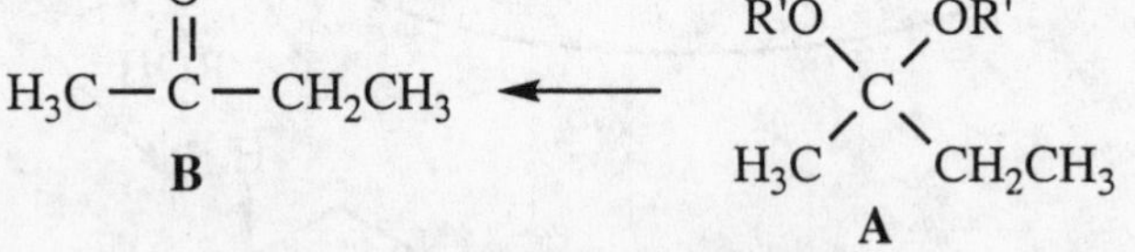

$$H_3C - \overset{\overset{\displaystyle O}{\|}}{C} - CH_2CH_3$$

B

$\longleftarrow$

$$R'O \diagdown \underset{\underset{\displaystyle H_3C \quad CH_2CH_3}{}}{C} \diagup OR'$$

A

A must have the same alkyl groups as **B**. **A** has one element of unsaturation and is missing only C_2H_4 from the partial structure above. The most likely structure is the ethylene ketal. Is this consistent with the NMR?

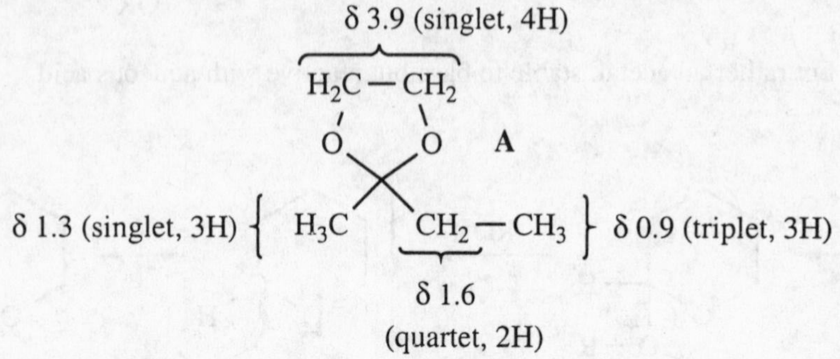

δ 3.9 (singlet, 4H)

$$\overbrace{H_2C - CH_2}$$

$$O \diagdown \qquad \diagup O \quad \textbf{A}$$

δ 1.3 (singlet, 3H) $\left\{ \quad H_3C \qquad \underbrace{CH_2} - CH_3 \quad \right\}$ δ 0.9 (triplet, 3H)

δ 1.6
(quartet, 2H)

What about the peaks in the MS at m/z 87 and 101? The 87 peak is the loss of 29 from the molecular ion at 116.

$$\left[\begin{array}{c} {}_{101}H_2C - CH_2 \; {}_{87} \\ O \diagdown \qquad \diagup O \\ H_3C \qquad CH_2 - CH_3 \end{array} \right]^{+\bullet}$$

m/z 116

$\longrightarrow$

$$\begin{array}{c} H_2C - CH_2 \\ O \diagdown \qquad \diagup O \\ \overset{\displaystyle |}{\underset{\displaystyle C+}{}} \\ H_3C \end{array}$$

m/z 87

$\longleftrightarrow$ plus two resonance forms with positive charge on the oxygen atoms

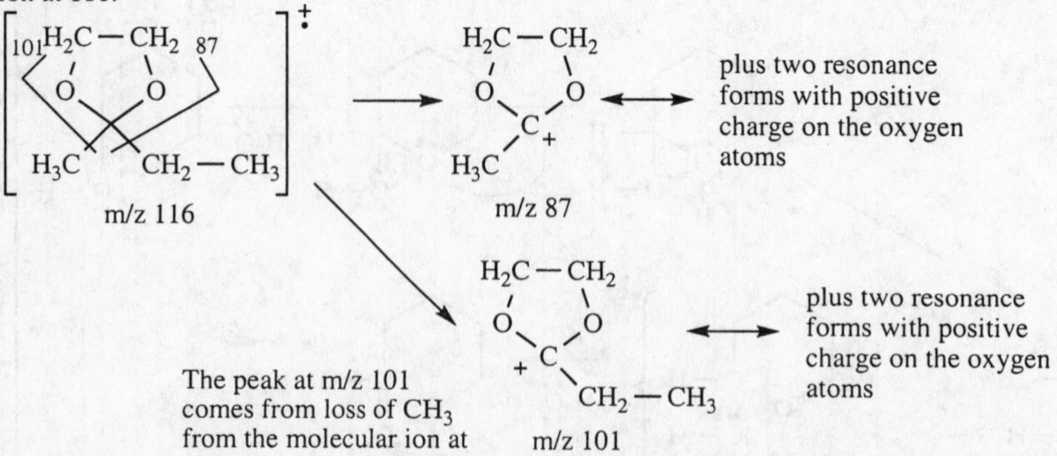

The peak at m/z 101 comes from loss of CH_3 from the molecular ion at 116.

$$\begin{array}{c} H_2C - CH_2 \\ O \diagdown \qquad \diagup O \\ +\overset{\displaystyle C}{\underset{\displaystyle CH_2 - CH_3}{}} \end{array}$$

m/z 101

$\longleftrightarrow$ plus two resonance forms with positive charge on the oxygen atoms

18-73 continued

(b)

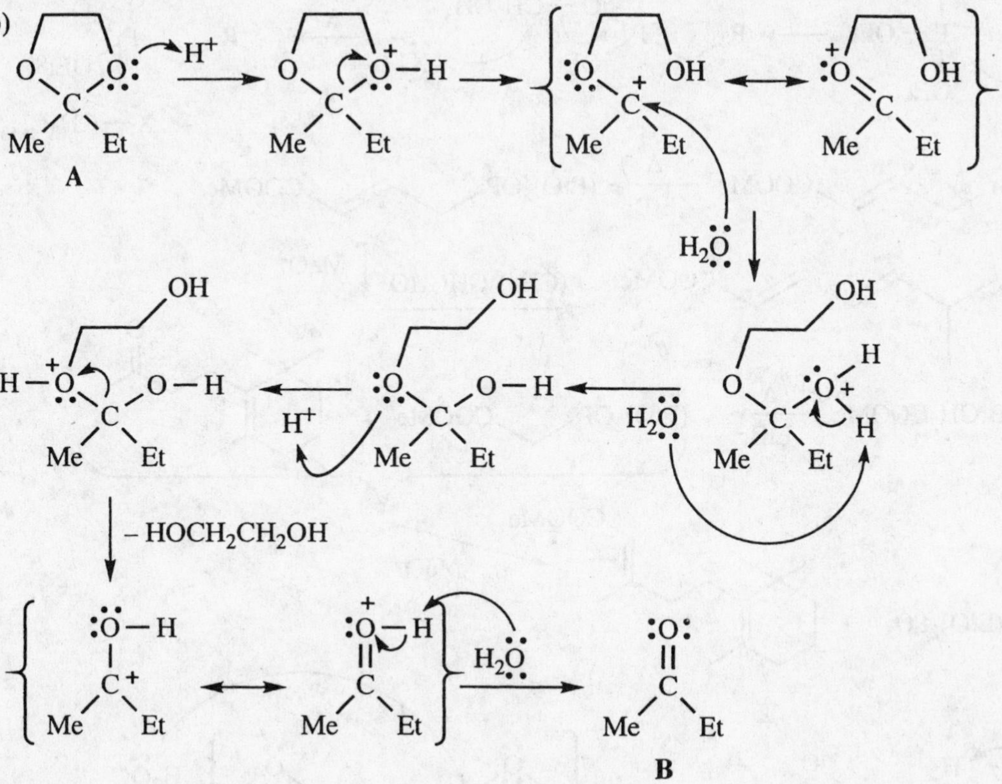

18-74 The strong UV absorption at 220 nm indicates a conjugated aldehyde or ketone. The IR shows a strong carbonyl at 1690 cm^{-1}, alkene at 1625 cm^{-1}, and two peaks at 2720 cm^{-1} and 2810 cm^{-1} —aldehyde!

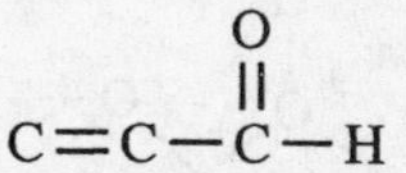

$$C=C-\overset{\overset{\displaystyle O}{\|}}{C}-H$$

The NMR shows the aldehyde proton at δ 9.5 split into a doublet, so it has one neighboring H. There are only two vinyl protons, so there must be an alkyl group coming off the β carbon:

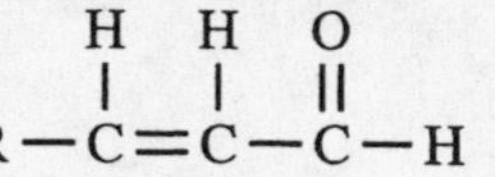

$$R-\overset{\overset{\displaystyle H}{|}}{C}=\overset{\overset{\displaystyle H}{|}}{C}-\overset{\overset{\displaystyle O}{\|}}{C}-H$$

The only other NMR signal is a 3H doublet: R must be methyl.

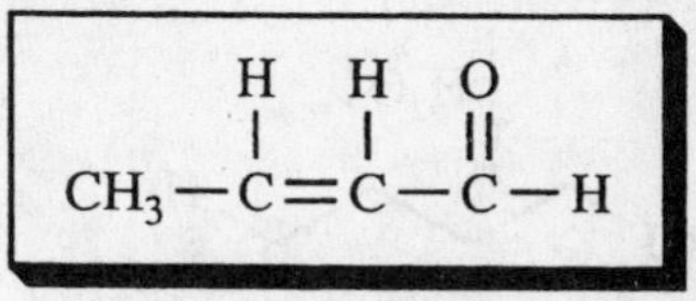

$$CH_3-\overset{\overset{\displaystyle H}{|}}{C}=\overset{\overset{\displaystyle H}{|}}{C}-\overset{\overset{\displaystyle O}{\|}}{C}-H$$

"crotonaldehyde"

18-75

(a)

$$(EtO)_2PO_2^- \ + $$

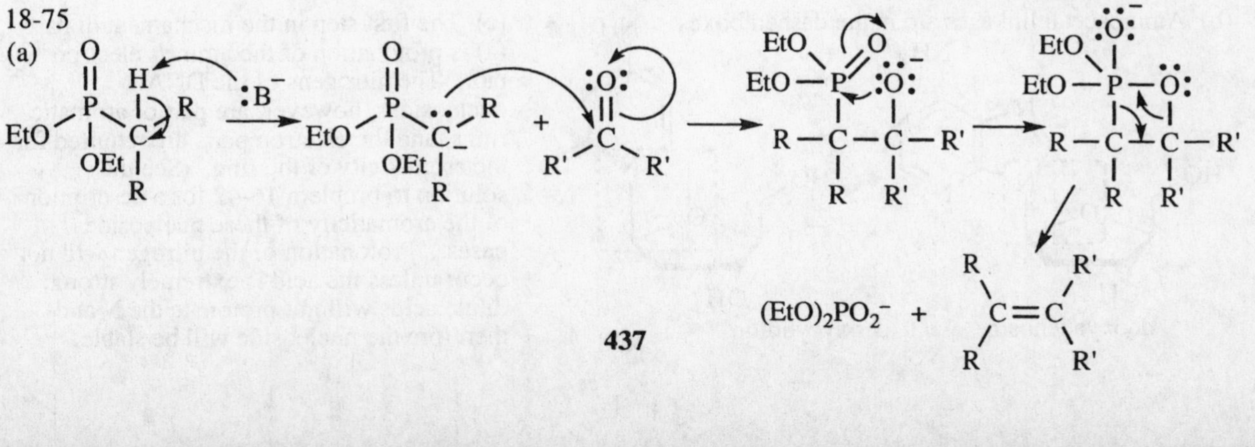

18-75 continued

(b)

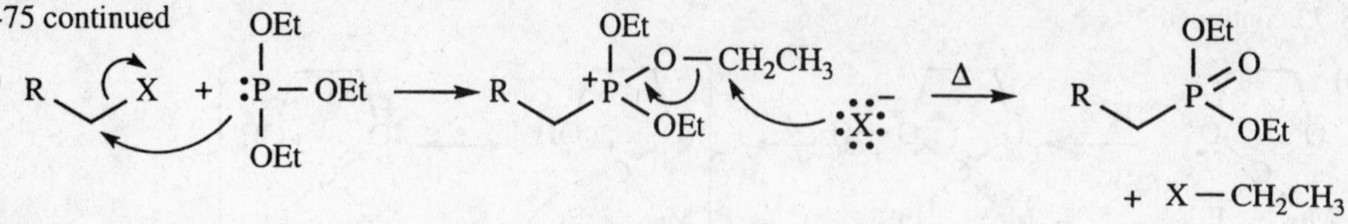

$$R-CH_2-X + :P(OEt)_3 \longrightarrow \text{...} \xrightarrow{\Delta} R-CH_2-\overset{O}{\underset{OEt}{\overset{|}{P}}}-OEt + X-CH_2CH_3$$

(c)

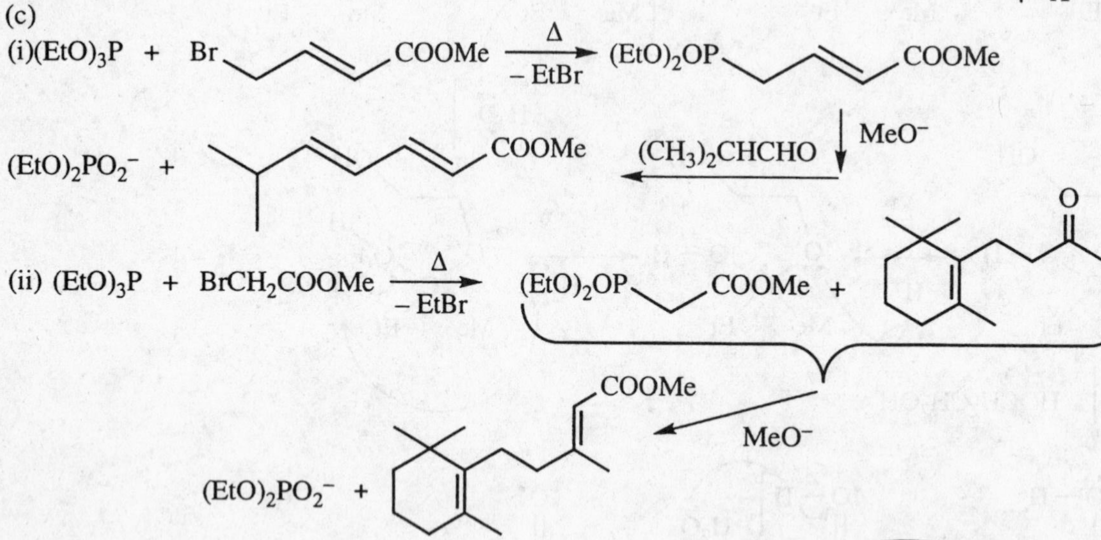

(i) $(EtO)_3P$ + Br~~~COOMe $\xrightarrow[-\,EtBr]{\Delta}$ $(EtO)_2OP$~~~COOMe

$\downarrow$ MeO⁻ (CH₃)₂CHCHO

$(EtO)_2PO_2^-$ + (isopropyl diene)~~~COOMe

(ii) $(EtO)_3P$ + $BrCH_2COOMe$ $\xrightarrow[-\,EtBr]{\Delta}$ $(EtO)_2OP$~~~COOMe + (cyclohexenyl ketone)

$\downarrow$ MeO⁻

$(EtO)_2PO_2^-$ + (cyclohexenyl chain COOMe product)

18-76

(a)

HO~~~CHO
(with O at bottom)

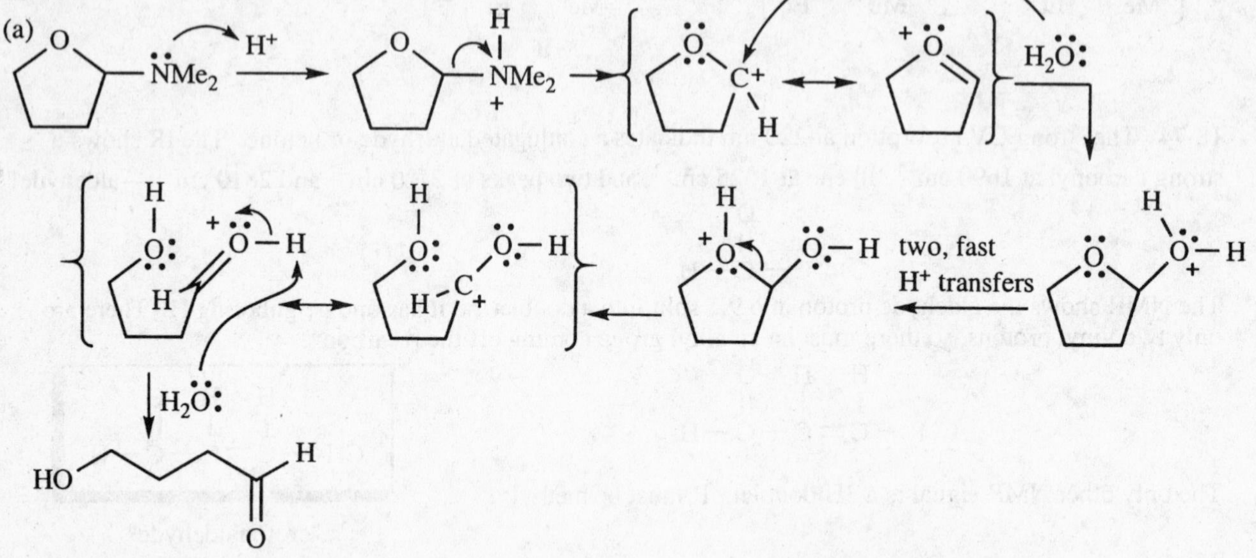

(b) Aminoacetal linkages are in the dashed boxes.

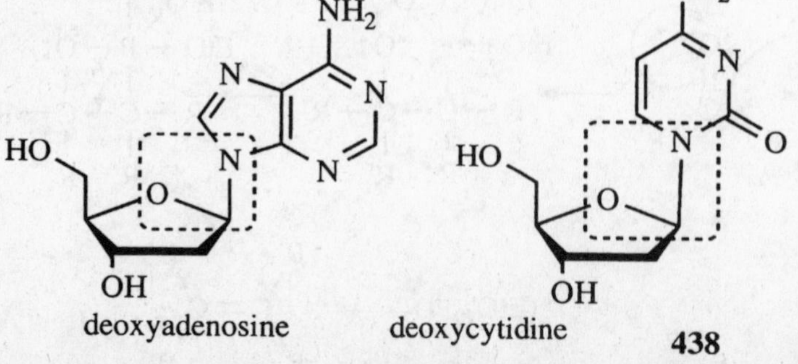

deoxyadenosine

deoxycytidine

(c) The first step in the mechanism in part (a) is protonation of the amine's electron pair. The nitrogens of the DNA nucleosides, however, are part of aromatic rings, and the electron pairs are required for the aromaticity of the ring. (See the solution to problem 16-42 for a description of the aromaticity of these nucleoside bases.) Protonation of the nitrogen will not occur unless the acid is extremely strong; dilute acids will not protonate the N and therefore the nucleoside will be stable.

438

19-1 These compounds satisfy the criteria for aromaticity (planar, cyclic π system, and the Huckel number of $4n + 2$ π electrons): pyrrole, imidazole, indole, pyridine, 2-methylpyridine, pyrimidine, and purine. The systems with 6 π electrons are: pyrrole, imidazole, pyridine, 2-methylpyridine, and pyrimidine. The systems with 10 π electrons are: indole and purine. The other nitrogen heterocycles shown are not aromatic because they do not have cyclic π systems.

19-2

(a)

$$H_3C-\underset{\underset{CH_3}{|}}{\overset{\overset{CH_3}{|}}{C}}-NH_2$$

(b)

$$CH_3-\underset{\underset{}{|}}{\overset{\overset{NH_2}{|}}{C}}HCHO$$

(c)

(d)

(e) $H_3C \underset{N}{\diagdown} CH_2CH_3$

(f)

19-3

(a) pentan-2-amine
(b) *N*-methylbutan-2-amine
(c) 3-aminophenol (or *meta-*)
(d) 3-methylpyrrole
(e) *trans*-cyclopentane-1,2-diamine
(f) *cis*-3-aminocyclohexanecarbaldehyde

19-4

(a) resolvable: there are two asymmetric carbons; carbon does not invert
(b) not resolvable: the nitrogen is free to invert
(c) not resolvable: it is symmetric
(d) not resolvable: even though the nitrogen is quaternary, one of the groups is a proton which can exchange rapidly, allowing for inversion
(e) resolvable: the nitrogen is quaternary and cannot invert when bonded to carbons

19-5 In order of increasing boiling point (increasing intermolecular hydrogen bonding):
(a) triethylamine and *n*-propyl ether have the same b.p. < di-*n*-propylamine
(b) dimethyl ether < dimethylamine < ethanol
(c) trimethylamine < diethylamine < diisopropylamine

19-6 Listed in order of increasing basicity. (See Appendix 2 for a discussion of acidity and basicity.)

(a) $PhNH_2$ < NH_3 < CH_3NH_2 < NaOH
(b) *p*-nitroaniline < aniline < *p*-methylaniline (*p*-toluidine)
(c) pyrrole < aniline < pyridine
(d) 3-nitropyrrole < pyrrole < imidazole

19-7

(a) secondary amine: one peak in the 3200-3400 cm^{-1} region, indicating NH

(b) primary amine: two peaks in the 3200-3400 cm^{-1} region, indicating NH_2

(c) alcohol: strong, broad peak around 3400 cm^{-1}

19-8 A compound with formula $C_4H_{11}N$ has no elements of unsaturation. The proton NMR shows five types of H, with the NH_2 appearing as a broad peak at δ 1.15, meaning that there are four different groups of hydrogens on the four carbons. The carbon NMR also shows four carbons, so there is no symmetry in this structure; that is, it does not contain a *t*-butyl group or an isopropyl group.

The multiplet farthest downfield is a CH deshielded by the nitrogen; integration shows it to be one H. There is a 2H multiplet at δ 1.35, the broad NH_2 peak at δ 1.15, a 3H doublet at δ 1.05, and a 3H triplet at δ 0.90. The latter two signals must represent methyl groups next to a CH and a CH_2 respectively. So far:

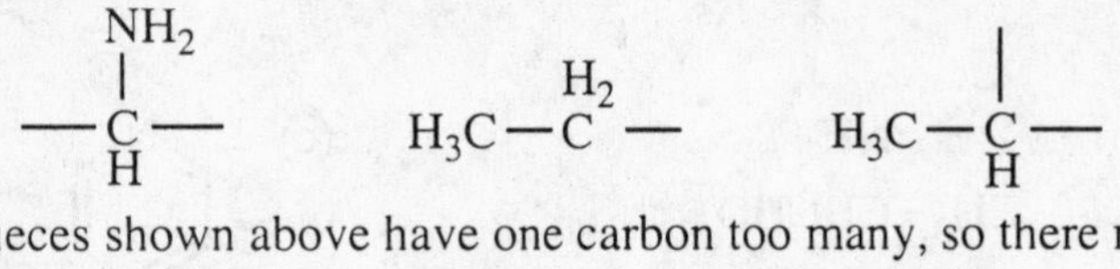

The pieces shown above have one carbon too many, so there must be one carbon that is duplicated: the only possible one is the CH, and the structure reveals itself.

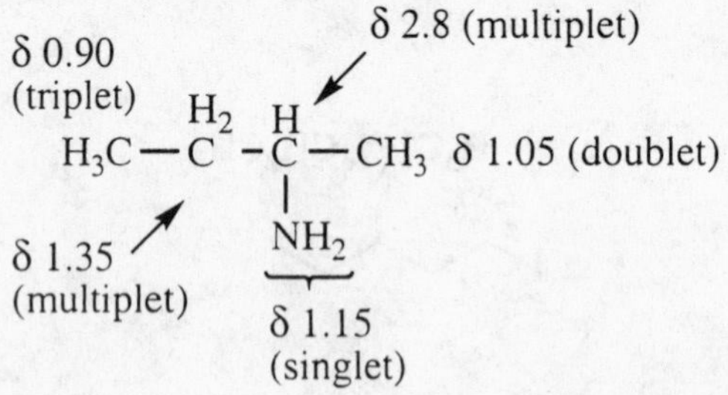

19-9

(a)

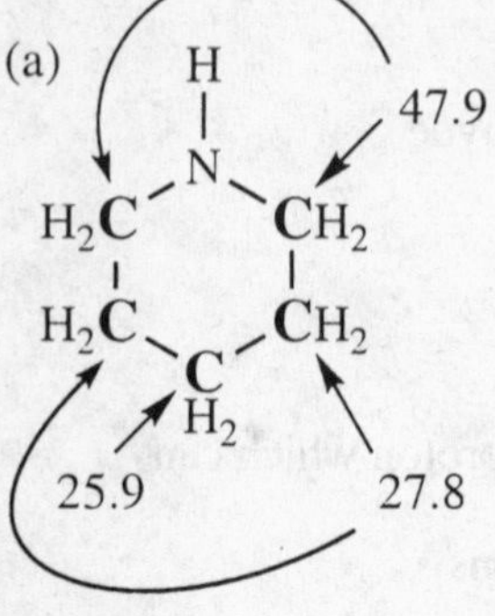

(b) $41.0 \longrightarrow CH_3$
$CH_3CH_2 - N - CH_2CH_3$
51.1
12.4
(An older printing of the text used values of 13.8, 47.5, and 58.2. The values shown here are taken from a spectrum.)

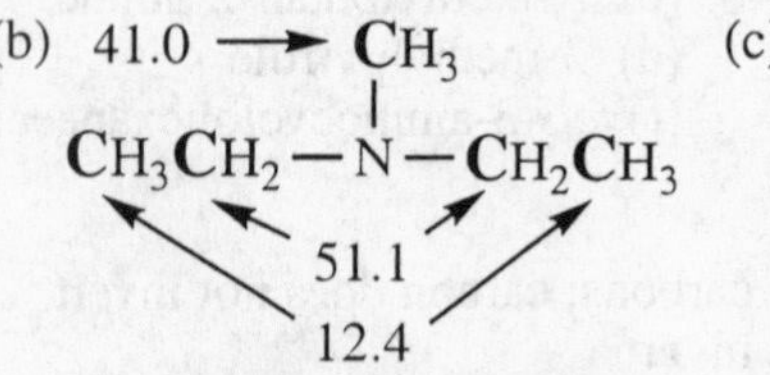

(c) 44.7
$CH_3CH_2 - \overset{\overset{\displaystyle O}{\|}}{CH}$
7.9   201.9

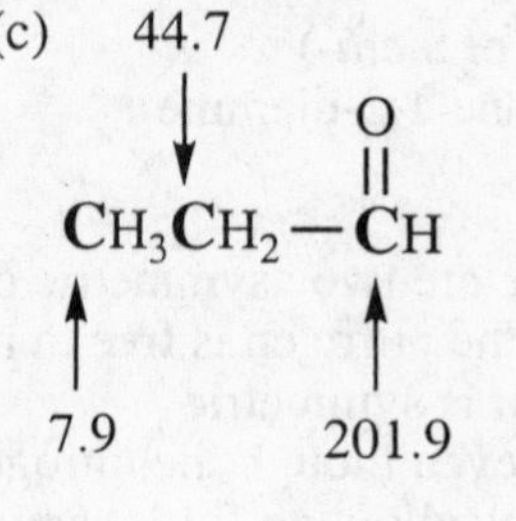

(d) 25.8
$CH_3CH_2CH_2OH$
10.0   63.6

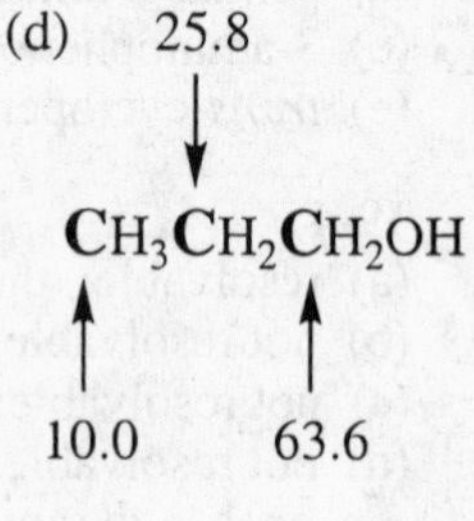

19-10

(a)

m/z 87

$+ \ CH_3\dot{C}H_2$
mass 29

m/z 58

(b)

m/z 72

$+ \ \cdot CH_3$
mass 15

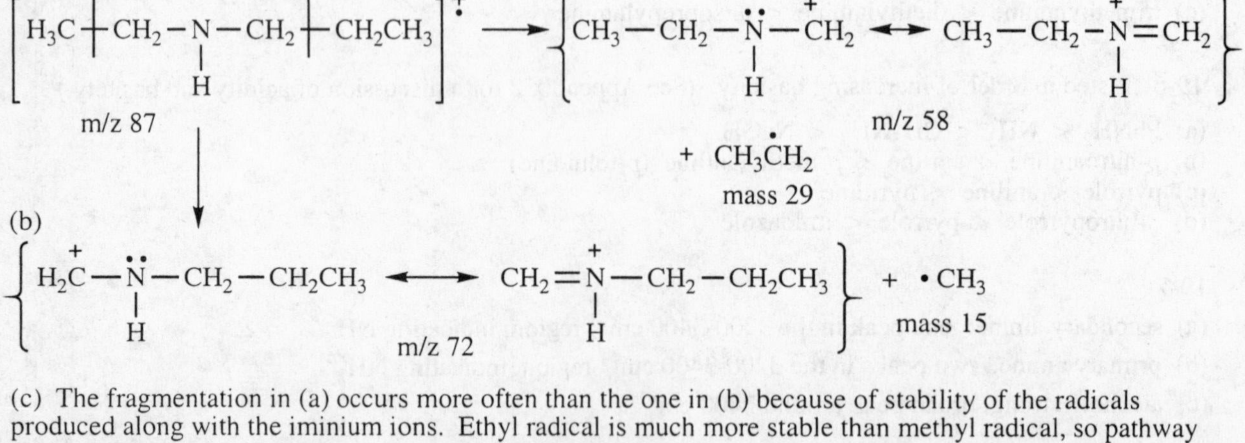

(c) The fragmentation in (a) occurs more often than the one in (b) because of stability of the radicals produced along with the iminium ions. Ethyl radical is much more stable than methyl radical, so pathway (a) is preferred.

19-11 Nitration at the 4-position of pyridine is not observed for the same reason that nitration at the 2-position is not observed: the intermediate puts some positive character on an electron-deficient nitrogen, and electronegative nitrogen hates that. (It is important to distinguish this type of positive nitrogen without a complete octet of electrons, from the quaternary nitrogen, also positively charged but with a full octet. It is the number of electrons around atoms that is most important; the charge itself is less important.)

GOOD: (quaternary nitrogen structure)   VERY BAD: (electron-deficient nitrogen structure)

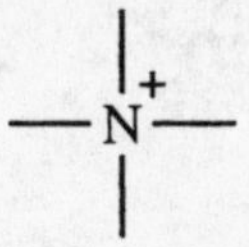

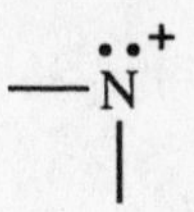

mechanism

(reaction mechanism scheme showing pyridine nitration at the 4-position with resonance structures)

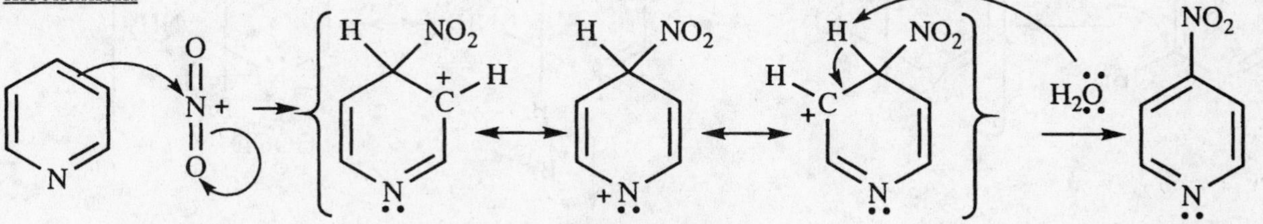

VERY BAD—
N does not have
an octet of electrons

not produced
because of
unfavorable
intermediate

19-12 Any electrophilic attack, including sulfonation, is preferred at the 3-position of pyridine because the intermediate is more stable than the intermediate from attack at either the 2-position or the 4-position. (Resonance forms of the sulfonate group are not shown, but remember that they are important!)

The N of pyridine is basic, and in the strong acid mixture, it will be protonated as shown here. That is part of the reason that pyridine is so sluggish to react: the ring already has a positive charge, so attack of an electrophile is slowed.

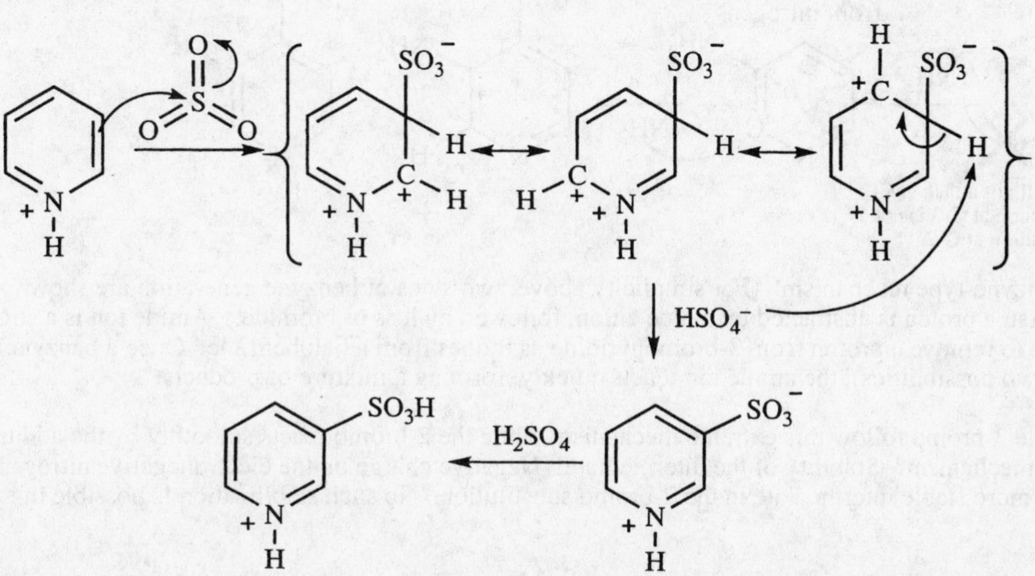

(reaction mechanism scheme showing pyridine sulfonation at the 3-position with resonance structures, HSO_4^-, H_2SO_4)

GOOD

(a)

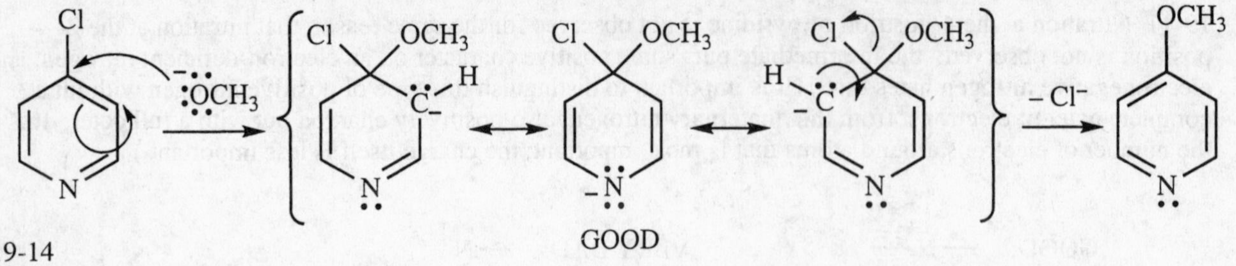

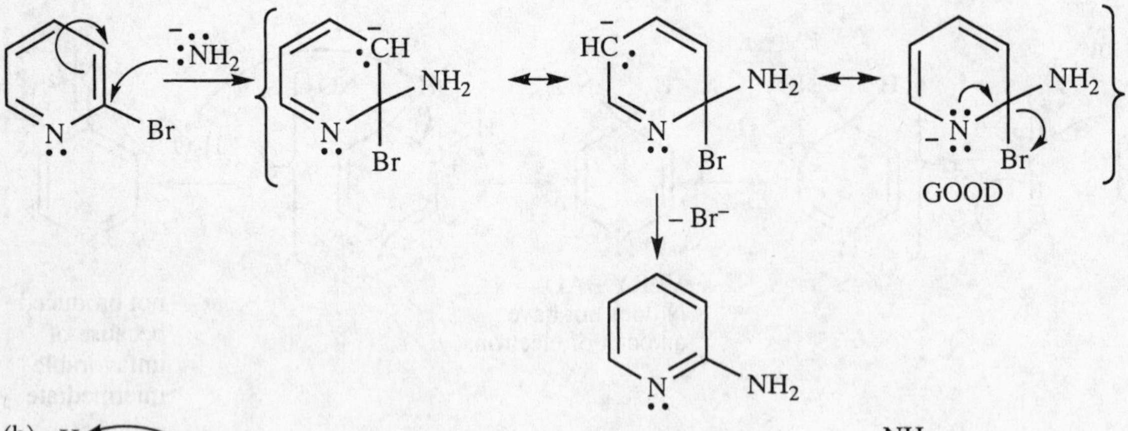

GOOD

(b)

stabilized by induction from nitrogen

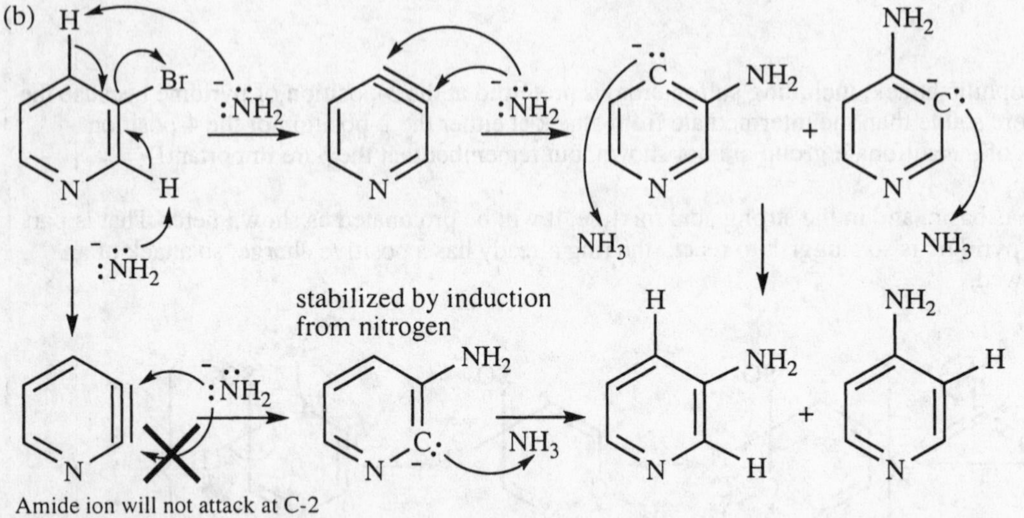

Amide ion will not attack at C-2 because the anion at C-3 is not as stable as the anion at C-2.

This is a benzyne-type mechanism. (For simplicity above, two steps of benzyne generation are shown as one step: first, a proton is abstracted by amide anion, followed by loss of bromide.) Amide ion is a strong enough base to remove a proton from 3-bromopyridine as it does from a halobenzene. Once a benzyne is generated (two possibilities), the amide ion reacts quickly, forming a mixture of products.

Why does the 3-bromo follow this extreme mechanism while the 2-bromo reacts smoothly by the addition-elimination mechanism? Stability of the intermediate! Negative charge on the electronegative nitrogen makes for a more stable intermediate in the 2-bromo substitution. No such stabilization is possible in the 3-bromo case.

19-15

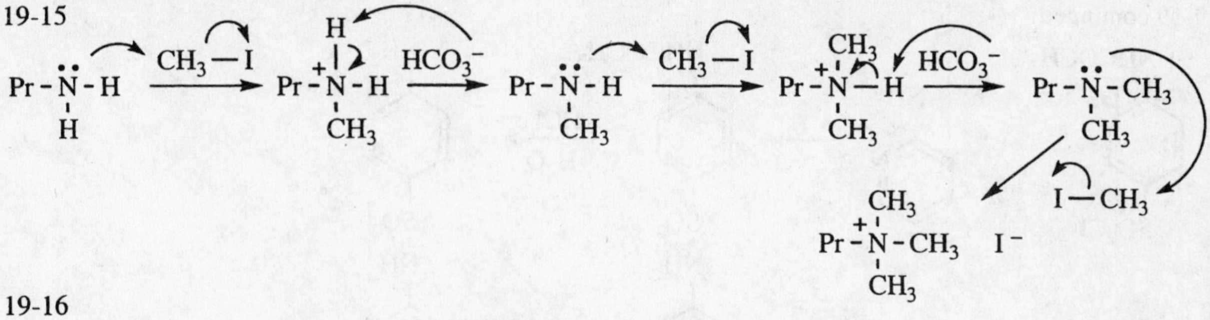

Pr–N–H $\xrightarrow{CH_3-I}$ Pr–N–H $\xrightarrow{HCO_3^-}$ Pr–N–H $\xrightarrow{CH_3-I}$ Pr–N–H $\xrightarrow{HCO_3^-}$ Pr–N–CH$_3$

$$Pr \overset{CH_3}{\underset{CH_3}{-\overset{+}{N}-}} CH_3 \quad I^-$$

19-16

(a) $PhCH_2NH_2$ + excess CH_3I $\xrightarrow{NaHCO_3}$ $PhCH_2\overset{+}{N}(CH_3)_3$ I^-

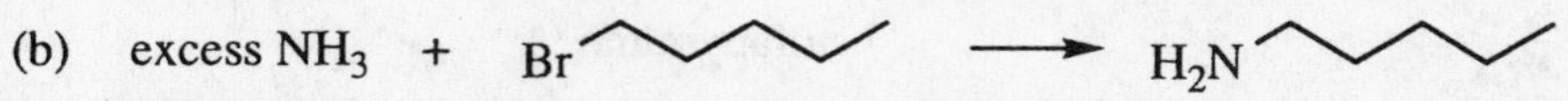

(b)   excess NH_3 + Br⟍⟋⟍⟋ ⟶ H_2N⟍⟋⟍⟋

(c)   excess NH_3 + $PhCH_2Br$ ⟶ $PhCH_2NH_2$

19-17

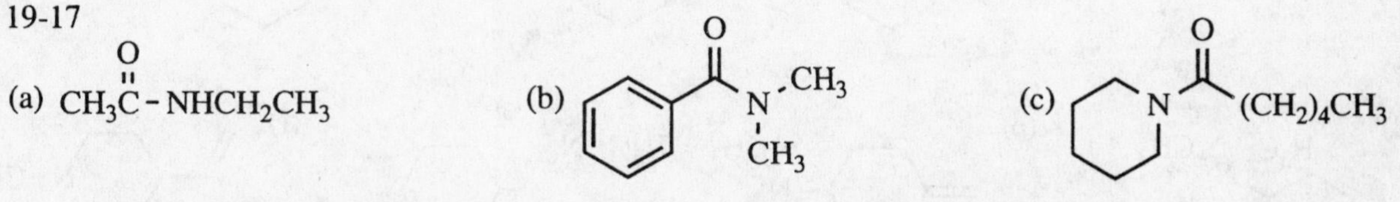

(a) $CH_3\overset{O}{\overset{\|}{C}}-NHCH_2CH_3$

(b) $\text{PhC}(=O)-N(CH_3)_2$ (N,N-dimethylbenzamide)

(c) piperidine-N–C(=O)(CH$_2$)$_4$CH$_3$

19-18 If the amino group were not protected, it would do a nucleophilic substitution on chlorosulfonic acid. Later in the sequence, this group could not be removed without cleaving the other sulfonamide group.

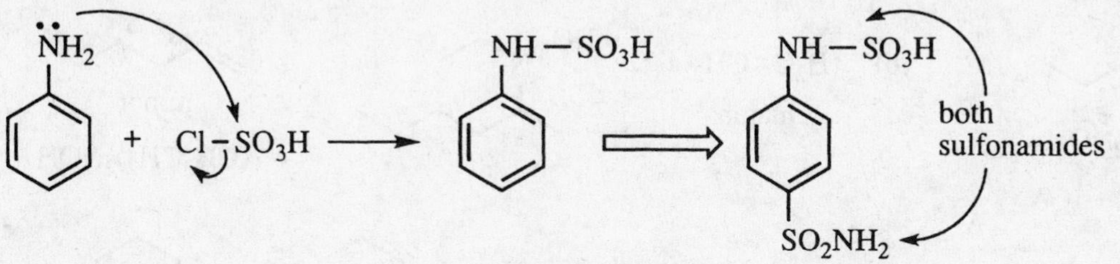

NH$_2$ (aniline) + Cl–SO$_3$H ⟶ NH–SO$_3$H (phenyl) ⟹ NH–SO$_3$H with SO$_2$NH$_2$ — both sulfonamides

19-19

NHCOCH$_3$ (with SO$_2$Cl) + thiazole-NH$_2$ ⟶ NHCOCH$_3$ (with SO$_2$–NH–thiazole) $\xrightarrow[\Delta]{\substack{HCl \\ H_2O}}$ NH$_2$ (with SO$_2$–NH–thiazole)

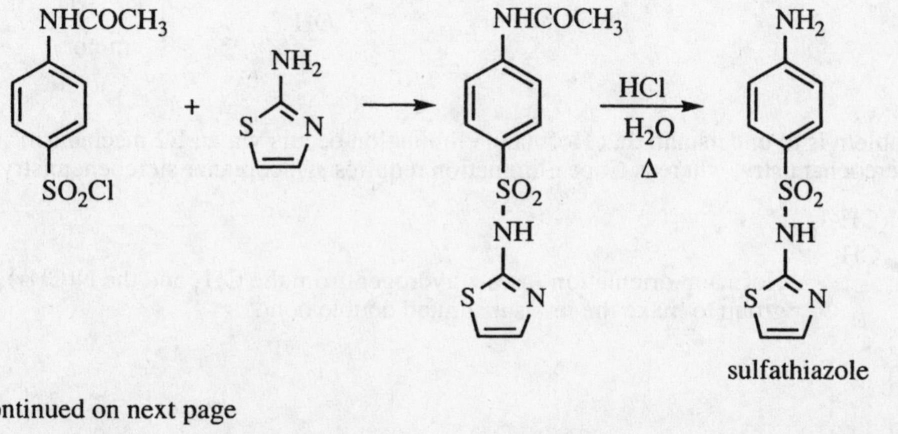

sulfathiazole

continued on next page

19-19 continued

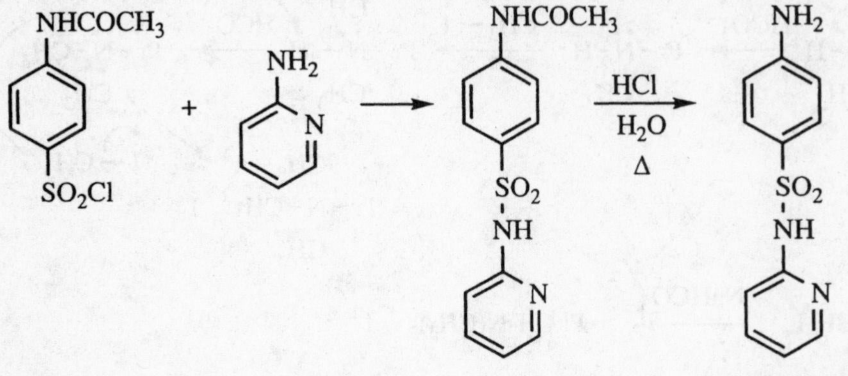

sulfapyridine

19-20

(a)

(b)

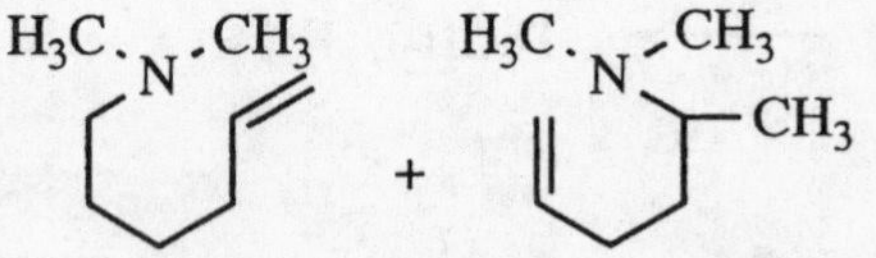

(c) 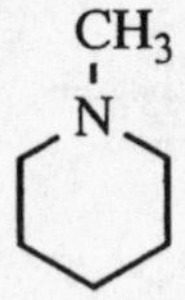 + $H_2C=CH_2$

(d)

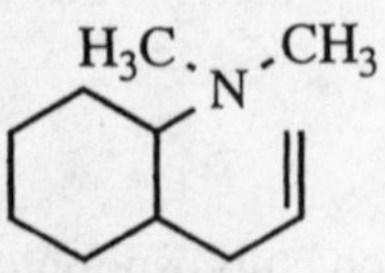

(e)

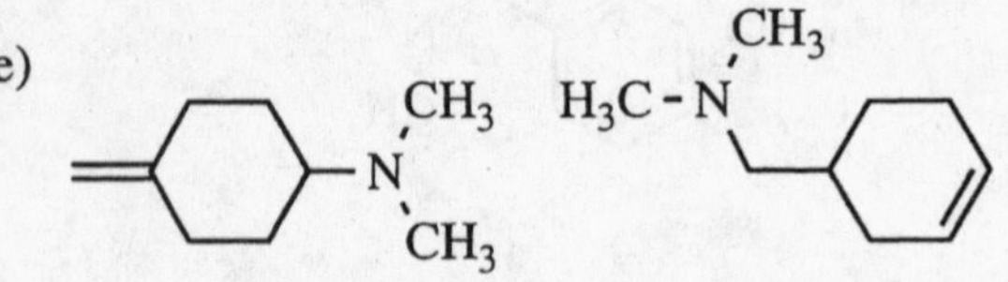

(f)

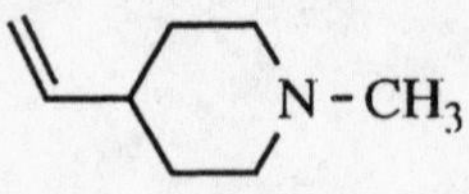

19-21 Orientation of the Cope elimination is similar to Hofmann elimination: the *less* substituted alkene is the major product.

(a)  + $(CH_3)_2NOH$

(b) $H_2C=CH_2$ + 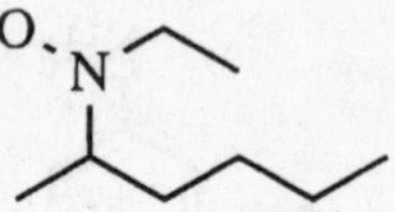 +

major   minor

+ $(CH_3CH_2)_2NOH$

(c) 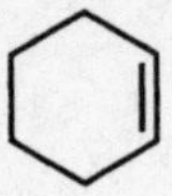 + $(CH_3)_2NOH$

(d) $H_2C=CH_2$ + 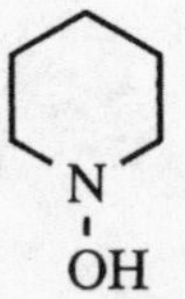 +

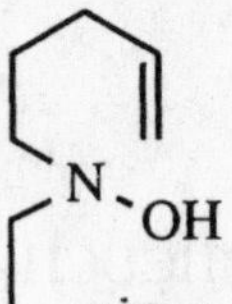

minor

19-22 The key to this problem is to understand that Hofmann elimination occurs via an E2 mechanism requiring *anti* coplanar stereochemistry, whereas Cope elimination requires *syn* coplanar stereochemistry.

(a)

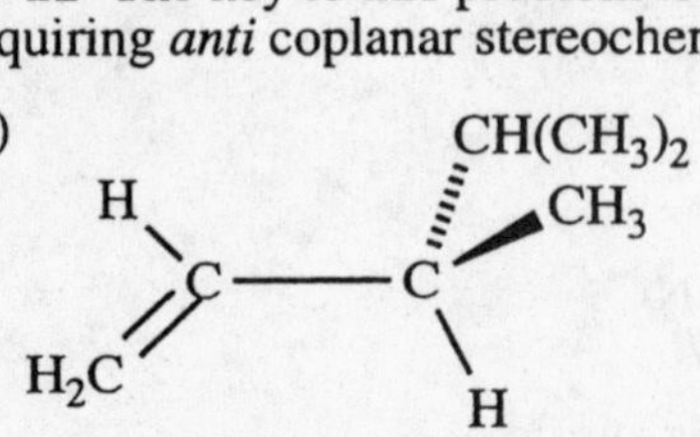

Hofmann orientation loses a hydrogen from the CH_3 and the $N(CH_3)_3$ group to make the less substituted double bond

19-22 continued

(b) Hofmann elimination

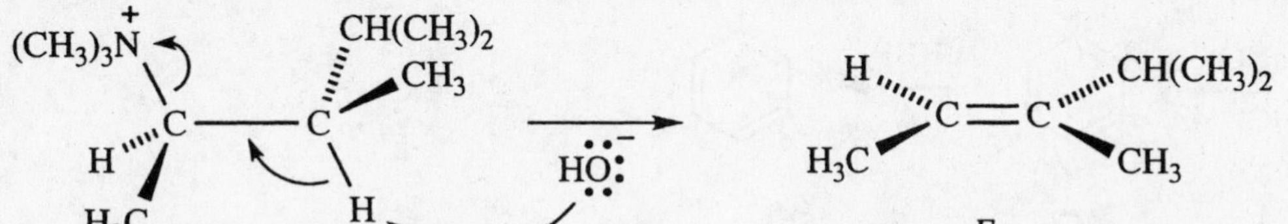

E
(Saytzeff product—more highly substituted)

Cope elimination

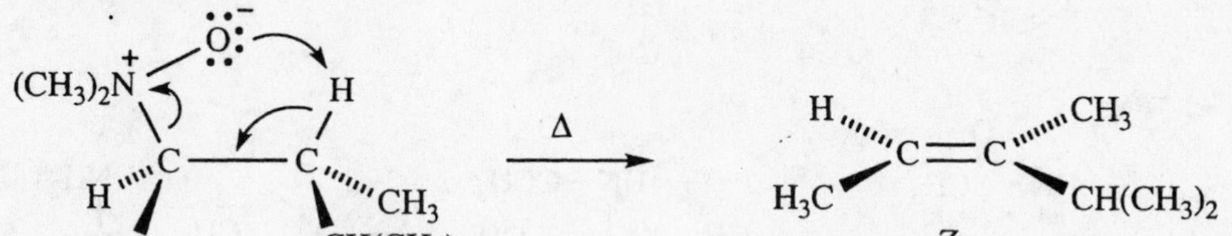

Z
(Saytzeff product—more highly substituted)

19-23

(a)

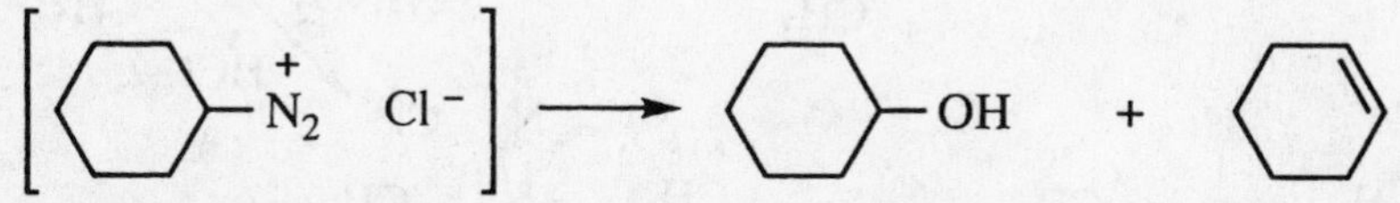

Aliphatic diazonium ions are very unstable, rapidly decomposing to carbocations.

(b)   (c) 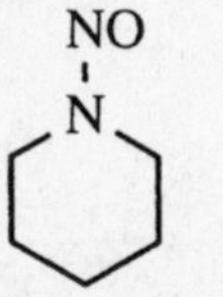   (d)

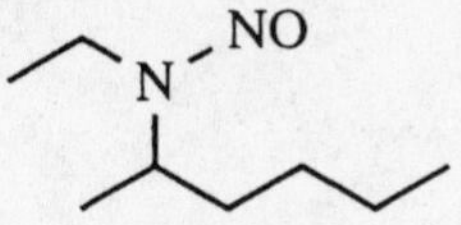

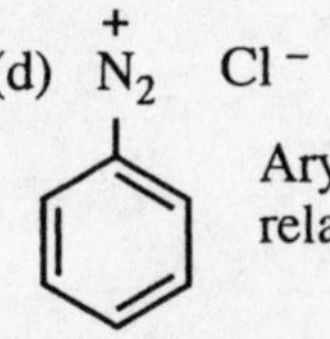

Aryl diazonium ions are relatively stable if kept cold.

19-24 The diazonium ion can do aromatic substitution like any other electrophile.

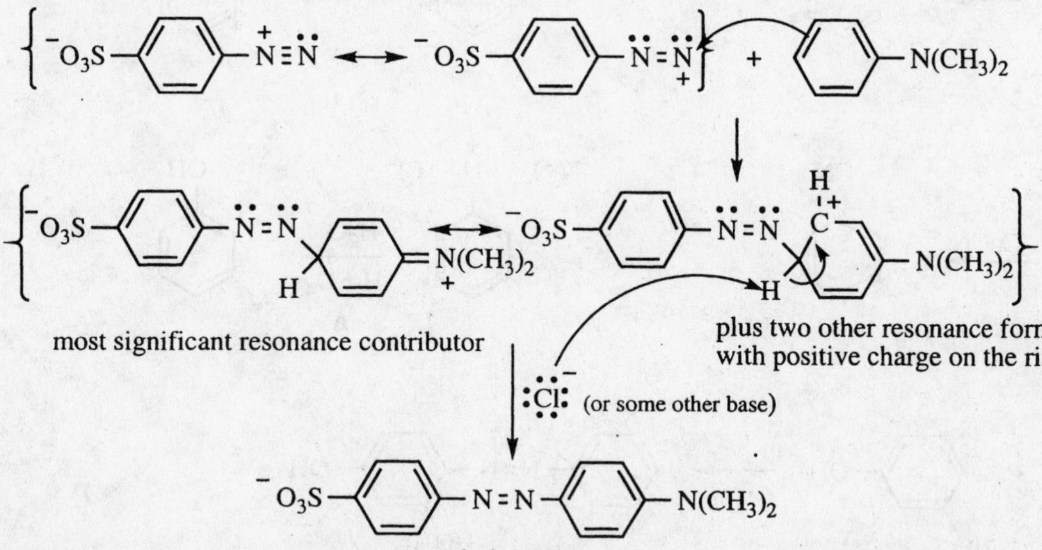

most significant resonance contributor

plus two other resonance forms
with positive charge on the ring

:Cl:⁻ (or some other base)

methyl orange   445

19-25

(a)

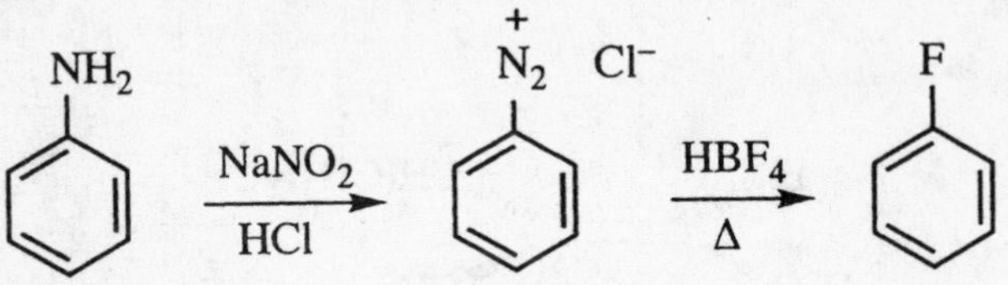

(b)

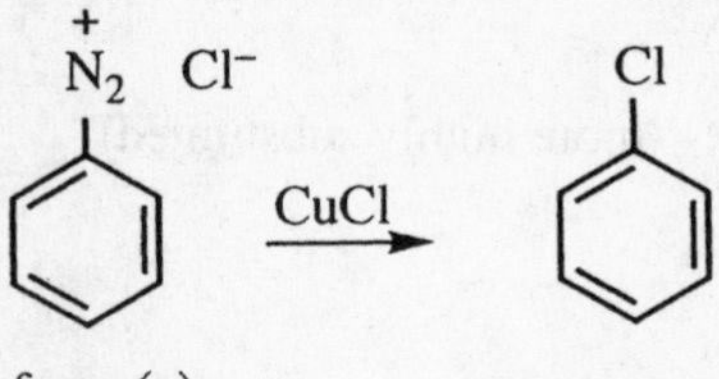

from (a)

(c)

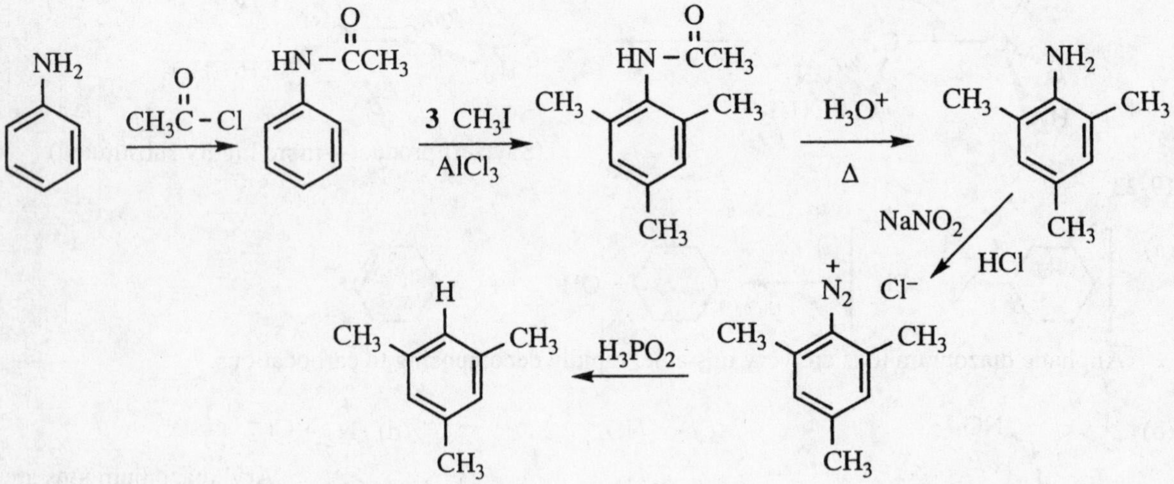

(d)

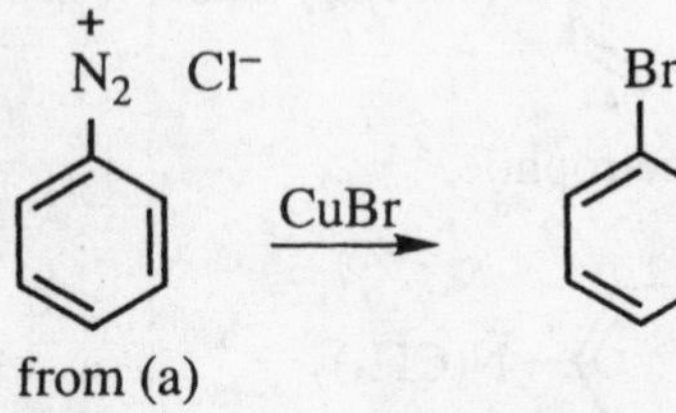

from (a)

(e)

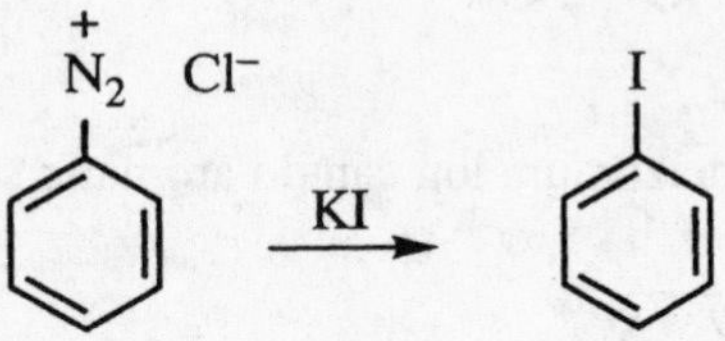

(f)

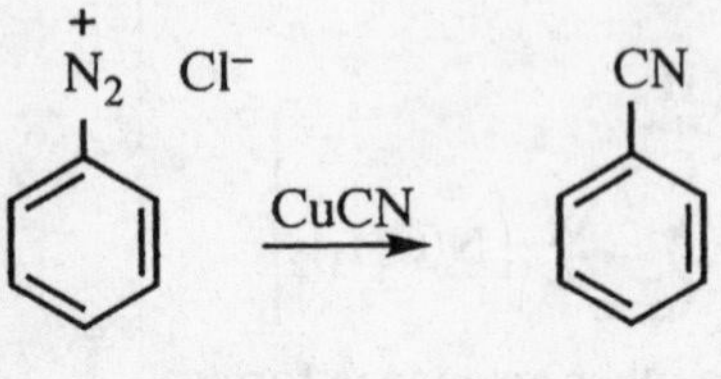

(g)

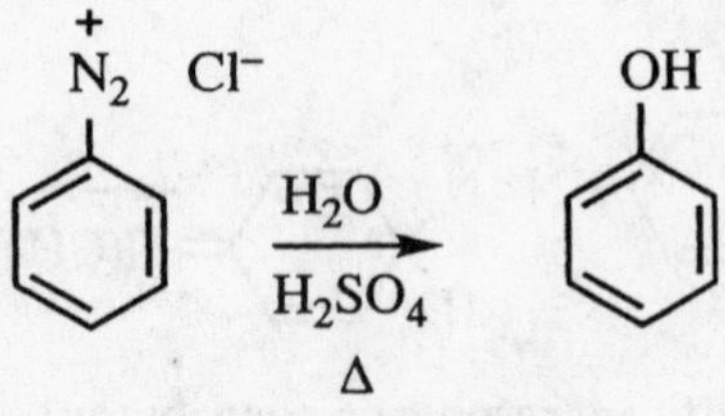

(h)

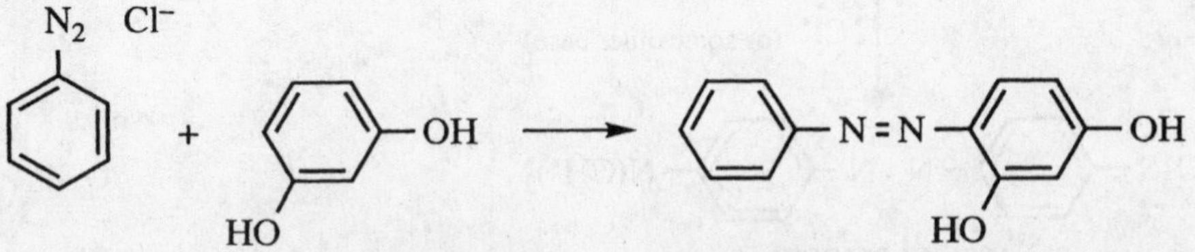

446

19-26 General guidelines for choice of reagent for reductive amination: use $LiAlH_4$ when the imine or oxime is isolated. Use $Na(CH_3COO)_3BH$ in solution when the imine or iminium ion is not isolated. Alternatively, catalytic hydrogenation works in most cases.

(a)

$$\text{(benzaldehyde)} + CH_3NH_2 \xrightarrow{Na(CH_3COO)_3BH} \text{(PhCH_2NHCH_3)}$$

(b)

$$PhCH_2-\overset{O}{\underset{}{C}}-CH_3 \xrightarrow[H^+]{H_2NOH} PhCH_2-\overset{NOH}{\underset{}{C}}-CH_3 \xrightarrow[2)\ H_2O]{1)\ LiAlH_4} PhCH_2-\overset{NH_2}{\underset{H}{C}}-CH_3$$

(c)

$$\text{(piperidine, N-H)} + \overset{O}{\underset{Ph\ \ \ H}{C}} \xrightarrow{Na(CH_3COO)_3BH} \text{(N-CH_2Ph piperidine)}$$

(d)

$$\text{(cyclohexanone)} + \text{(aniline NH}_2\text{)} \xrightarrow{H^+} \text{(N-Ph imine)} \xrightarrow[2)\ H_2O]{1)\ LiAlH_4} \text{(NHPh cyclohexyl)}$$

(e)

$$\text{(cyclohexanone)} \xrightarrow[H^+]{H_2NOH} \text{(NOH oxime)} \xrightarrow[2)\ H_2O]{1)\ LiAlH_4} \text{(NH}_2\text{ cyclohexylamine)}$$

(f)

$$\text{(piperidine NH)} + O=\text{(cyclopentanone)} \xrightarrow{Na(CH_3COO)_3BH} \text{(N-cyclopentyl piperidine)} \xleftarrow{Na(CH_3COO)_3BH} \left\{ \begin{array}{c} \text{(dialdehyde with two CHO)} \\ + \\ H_2N\text{-(cyclopentyl)} \end{array} \right. \text{OR}$$

19-27
(a)

$$\text{(piperidine N-H)} + \overset{O}{\underset{Cl}{C}}CH_2CH_3 \longrightarrow \text{(N-C(=O)CH_2CH_3 piperidine)} \xrightarrow[2)\ H_2O]{1)\ LiAlH_4} \text{(N-propyl piperidine)}$$

(b)

$$\text{(aniline NH}_2\text{)} + \overset{O}{\underset{Cl}{C}}\text{(Ph)} \longrightarrow \text{(PhNH-C(=O)-Ph)} \xrightarrow[2)\ H_2O]{1)\ LiAlH_4} \text{(PhNH-CH}_2\text{-Ph)}$$

19-28 Use a large excess of ammonia to avoid multiple alkylations of each nitrogen.

$$\text{(structure)} \quad + \quad NH_3 \quad \longrightarrow \quad \text{(structure)} NH_2$$
$$\text{excess}$$

19-29

(a)

$$\xrightarrow{BrCH_2Ph} \quad N-CH_2Ph \quad \xrightarrow[\Delta]{NH_2NH_2} \quad H_2NCH_2Ph$$

(b)

$$\xrightarrow{Br(CH_2)_5CH_3} \quad N-(CH_2)_5CH_3 \quad \xrightarrow[\Delta]{NH_2NH_2} \quad H_2N(CH_2)_5CH_3$$

(c)

$$\xrightarrow[\substack{\text{must use anion} \\ \text{to avoid protonating} \\ \text{the phthalimide anion}}]{Br(CH_2)_3COO^-} \quad N-(CH_2)_3COO^- \quad \xrightarrow[\text{2) } H^+]{\text{1) } NH_2NH_2 \text{, } \Delta} \quad H_2N(CH_2)_3COOH$$

19-30 Assume that $LiAlH_4$ or H_2/catalyst can be used interchangeably.

(a) $\quad PhCH_2Br \ + \ NaN_3 \ \longrightarrow \ PhCH_2N_3 \ \xrightarrow[Pt]{H_2} \ PhCH_2NH_2$

(b)

$$\xrightarrow{KCN} \quad \text{(C}\equiv\text{N)} \quad \xrightarrow[\text{2) } H_2O]{\text{1) } LiAlH_4} \quad \text{(NH}_2\text{)}$$

(c)

$$\text{(structure) } OH \quad \xrightarrow[\text{2) } H_2O]{\text{1) } LiAlH_4} \quad \text{(structure) } OH \quad \xrightarrow[\text{pyridine}]{TsCl} \quad \text{(structure) } OTs$$

$$\downarrow NaN_3$$

$$\text{(structure) } NH_2 \quad \xleftarrow[\text{2) } H_2O]{\text{1) } LiAlH_4} \quad \text{(structure) } N_3$$

OR

$$\text{(structure) } OH \quad \xrightarrow{SOCl_2} \quad \text{(structure) } Cl \quad \xrightarrow{NH_3} \quad \text{(structure) } NH_2$$

$$\text{1) } LiAlH_4 \ \downarrow \ \text{2) } H_2O$$

$$\text{(structure) } NH_2$$

(d)

$$\text{(structure) } OTs \quad \xrightarrow{KCN} \quad \text{(structure) C}\equiv\text{N} \quad \xrightarrow[\text{2) } H_2O]{\text{1) } LiAlH_4} \quad \text{(structure) } NH_2$$
$$\text{from (c)}$$

19-30 continued

(e)

$$\underset{R}{\text{Br}\ \text{H}} \xrightarrow[\substack{S_N2- \\ \text{inversion}}]{\text{NaN}_3} \underset{}{\text{H}\ \text{N}_3} \xrightarrow[\text{2) H}_2\text{O}]{\text{1) LiAlH}_4} \underset{S}{\text{H}\ \text{NH}_2}$$

(f)

$$\underset{R}{\text{Br}\ \text{H}} \xrightarrow[\substack{S_N2- \\ \text{inversion}}]{\text{NaCN}} \underset{}{\text{H}\ \text{C}\equiv\text{N}} \xrightarrow[\text{2) H}_2\text{O}]{\text{1) LiAlH}_4} \underset{S}{\text{H}\ \text{CH}_2\text{NH}_2}$$

(g)

$$\underset{}{\text{O}} \xrightarrow[\text{HCN}]{\text{KCN}} \underset{}{\text{HO}\ \text{CN}} \xrightarrow[\text{2) H}_2\text{O}]{\text{1) LiAlH}_4} \underset{}{\text{HO}\ \text{CH}_2\text{NH}_2}$$

19-31 To reduce nitroaromatics, the reducing reagents (H_2 plus a metal catalyst, or a metal plus HCl) can be used virtually interchangeably. Assume a workup in base to give the free amine final product.

(a)

$$\bigcirc \xrightarrow[\text{H}_2\text{SO}_4]{\text{HNO}_3} \underset{}{\text{NO}_2} \xrightarrow[\text{HCl}]{\text{Sn}} \underset{}{\text{NH}_2}$$

(b)

$$\bigcirc \xrightarrow[\text{FeBr}_3]{\text{Br}_2} \underset{}{\text{Br}} \xrightarrow[\text{H}_2\text{SO}_4]{\text{HNO}_3} \underset{\substack{\text{NO}_2 \\ + \text{ ortho}}}{\text{Br}} \xrightarrow[\text{HCl}]{\text{Fe}} \underset{\text{NH}_2}{\text{Br}}$$

(c)

$$\underset{\text{from (a)}}{\text{NO}_2} \xrightarrow[\text{FeBr}_3]{\text{Br}_2} \underset{\text{Br}}{\text{NO}_2} \xrightarrow[\text{HCl}]{\text{Sn}} \underset{\text{Br}}{\text{NH}_2}$$

(d)

$$\underset{}{\text{CH}_3} \xrightarrow[\substack{\text{H}_2\text{O} \\ \Delta}]{\text{KMnO}_4} \underset{}{\text{COOH}} \xrightarrow[\text{H}_2\text{SO}_4]{\text{HNO}_3} \underset{\text{NO}_2}{\text{COOH}} \xrightarrow[\text{HCl}]{\text{Fe}} \underset{\text{NH}_2}{\text{COOH}}$$

19-32

$$\text{PhCH}_2-\underset{\underset{\text{CH}_3}{|}}{\overset{\overset{\text{CH}_3}{|}}{\text{C}}}-\overset{\overset{\text{O}}{||}}{\text{C}}-\underset{\underset{\text{H}}{|}}{\text{N}}-\text{H} \quad :\ddot{\text{O}}\text{H}^- \rightarrow \left\{ \text{R}-\overset{\overset{\text{:O:}}{||}}{\text{C}}-\underset{\underset{\text{H}}{|}}{\ddot{\text{N}}}:^- \longleftrightarrow \text{R}-\overset{\overset{\text{:O:}^-}{||}}{\text{C}}=\underset{\underset{\text{H}}{|}}{\ddot{\text{N}}}: \right\} \quad \text{Br}-\text{Br} \rightarrow \text{R}-\overset{\overset{\text{O}}{||}}{\text{C}}-\underset{}{\ddot{\text{N}}}-\text{Br}$$

R

mechanism continued from previous page

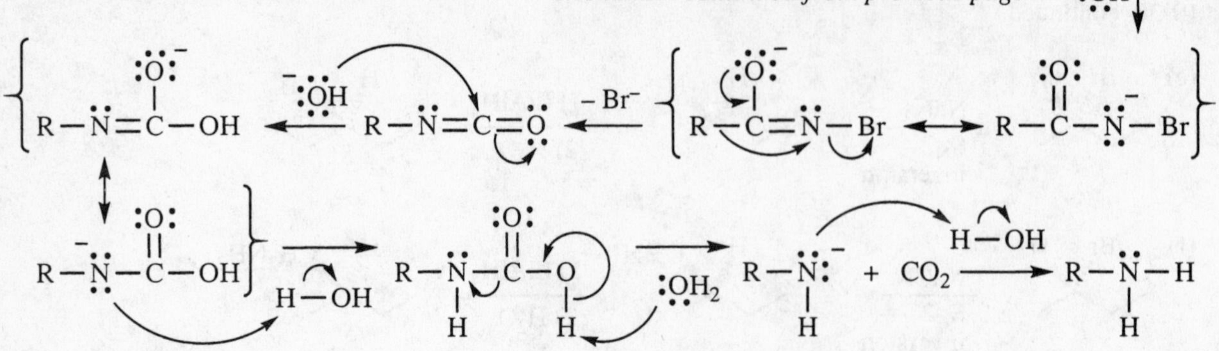

19-33

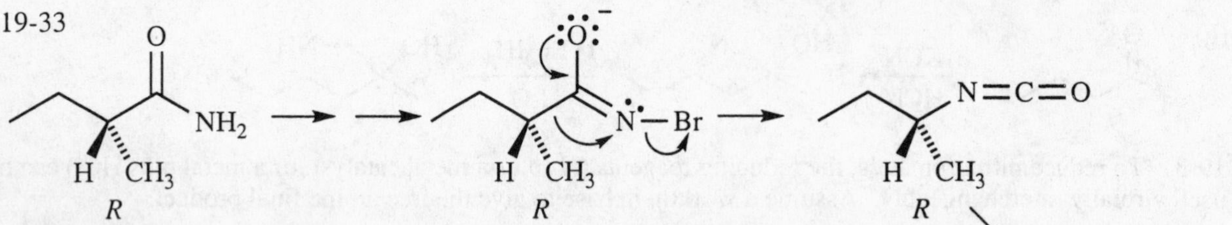

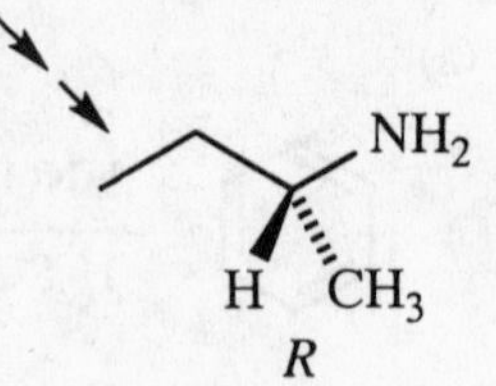

Stereochemistry at a chiral carbon is lost if the carbon goes through a planar intermediate, *e.g.*, either a carbocation or a free radical. Configuration at a chiral carbon can be inverted during substitution by a nucleophile from the side opposite the leaving group. However, when the carbon retains all four pairs of electrons, as in this Hofmann rearrangement, it retains its configuration.

19-34

(a) The acyl azide of the Curtius rearrangement is similar to the N-bromoamide of the Hofmann rearrangement in that both have an amide nitrogen with a good leaving group attached. Subsequent alkyl migration to the isocyanate and hydrolysis through the carbamic acid to the amine are identical in both mechanisms.

(b) The leaving group in the Curtius rearrangement is N_2 gas, one of the best leaving groups known "to man or beast", as we used to say.

(c)

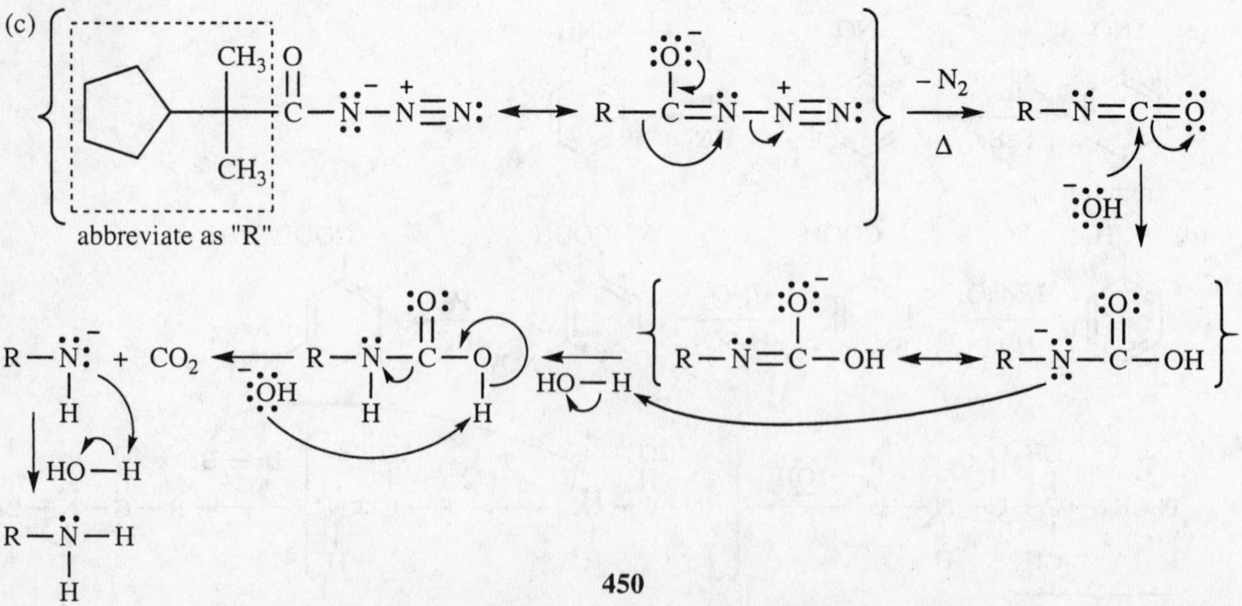

19-35 Please refer to solution 1-20, page 12 of this Solutions Manual.

19-36
(a) primary amine; 2,2-dimethylpropan-1-amine, or neopentylamine
(b) secondary amine; N-methylpropan-2-amine, or isopropylmethylamine
(c) tertiary heterocyclic amine and a nitro group; 3-nitropyridine
(d) quaternary heterocyclic ammonium ion; N,N-dimethylpiperidinium iodide
(e) tertiary aromatic amine oxide; N-ethyl-N-methylaniline oxide
(f) tertiary aromatic amine; N-ethyl-N-methylaniline
(g) tertiary heterocyclic ammonium ion; pyridinium chloride
(h) secondary amine; N,4-diethylhexan-3-amine

19-37 Shown in order of increasing basicity. In sets a-c, the aliphatic amine is the strongest base.

(a) $\text{Ph}-\overset{\overset{\text{H}}{|}}{\text{N}}-\text{Ph}$ < PhNH$_2$ < ⬡—NH$_2$ aliphatic amine is the strongest base

(b) pyrrole < pyridine < piperidine aliphatic amine is the strongest base; pyrrole's aromaticity would be lost if protonated

(c) pyrrole ⟨NH⟩ < HN⟨⟩N < ⟨⟩NH aliphatic amine is the strongest base; pyrrole's aromaticity would be lost if protonated

(d) O$_2$N—⟨⟩—NH$_2$ < ⟨⟩—NH$_2$ < H$_3$C—⟨⟩—NH$_2$ Basicity is a measure of the ability to donate the pair of electrons on the amine. Electron-withdrawing groups like NO$_2$ decrease basicity, while electron-donating groups like CH$_3$ increase basicity.

(e) ⟨⟩—C(=O)NH$_2$ < ⟨⟩—NH$_2$ < ⟨⟩—CH$_2$NH$_2$ aliphatic amine is the strongest base (amides are not basic)

(f) H$_3$C—⟨⟩—N(i-Pr)$_2$ < H$_2$N—⟨⟩—N(i-Pr)$_2$ < ⟨⟩—N(i-Pr)$_2$ (with CH$_3$)

Congratulations if you got this one right! CH$_3$ is only slightly electron-donating by induction so the first structure puts the least electron density on the N. The second structure is more basic (by about 1 pK unit) because NH$_2$ is electron-donating by resonance. The last structure is the strongest base for a completely different reason: steric inhibition of resonance. (See the last topic in Appendix Two of this Manual.) The ortho methyl interferes with the isopropyl substituents on N and it forces the NR$_2$ out of planarity with the ring, so that the electron pair on the N is not conjugated with the aromatic pi system. Essentially, this N becomes an aliphatic amine. The pK$_b$ of 2-methyl-N,N-diethylaniline is 6.9, about 2.5 pK units more basic than aniline. It is not as strong a base as a tertiary aliphatic amine because the aromatic ring is somewhat electron withdrawing by induction, yet the ortho methyl reduces the pi overlap of the nitrogen's electrons to make the N significantly more basic.

19-38

(a) not resolvable: planar

(b) resolvable: asymmetric carbon

(c) not resolvable: symmetric

(d) resolvable: nitrogen inversion is very slow

(e) not resolvable: symmetric

(f) resolvable: asymmetric nitrogen, unable to invert

(g) not resolvable in conditions where the proton on N can exchange

(h) resolvable: asymmetric nitrogen, unable to invert

19-39
The values of pK_b of amines or pK_a of the conjugate acids can be obtained from Table 19-3. The side of the reaction with the weaker acid and base will be favored at equilibrium. (See Appendix 2 for a discussion of acidity and basicity.)

(a)

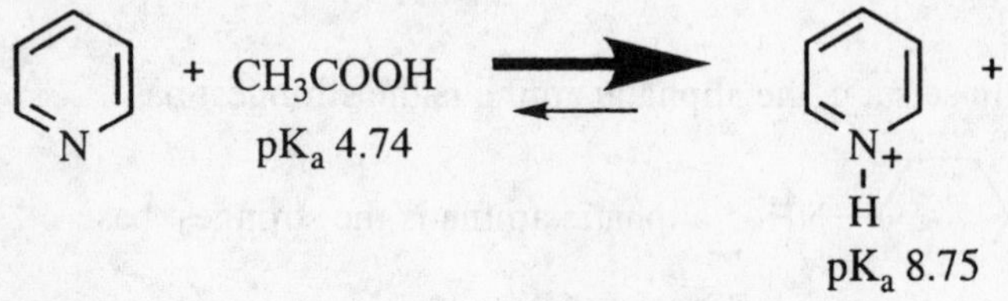

pyridine + CH_3COOH (pK_a 4.74) ⇌ pyridinium-H (pK_a 8.75) + CH_3COO^- *products are favored*

(b)

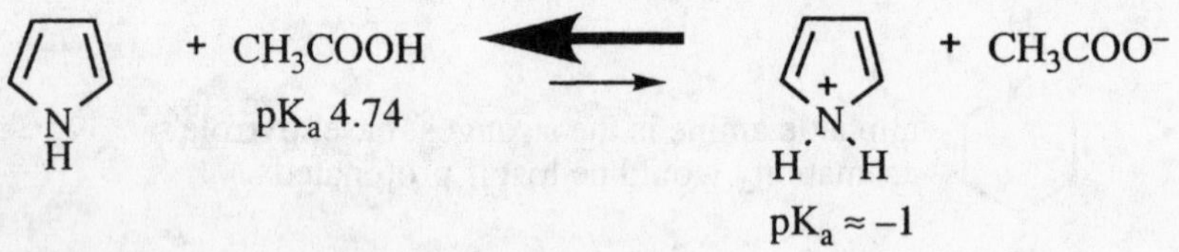

pyrrole + CH_3COOH (pK_a 4.74) ⇌ pyrrolium ($pK_a \approx -1$) + CH_3COO^- *reactants are favored*

(c)

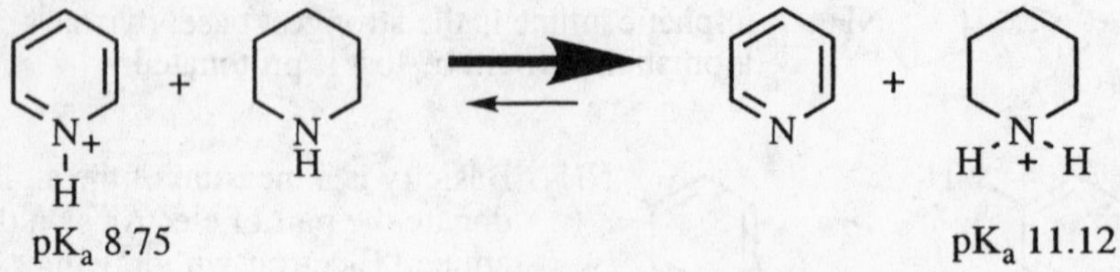

pyridinium (pK_a 8.75) + piperidine ⇌ pyridine + piperidinium (pK_a 11.12)   *products are favored*

(d)

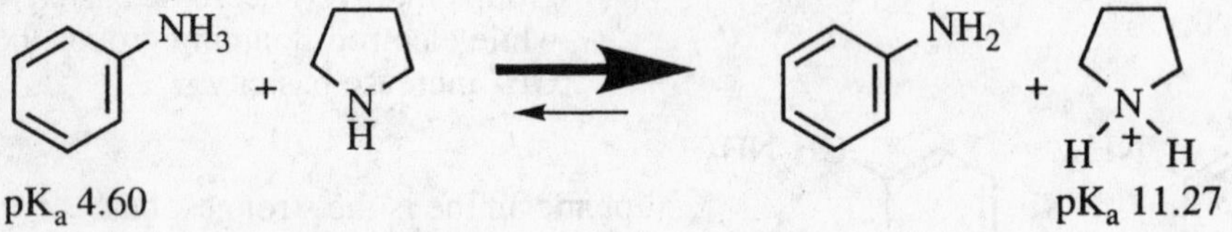

anilinium (pK_a 4.60) + pyrrolidine ⇌ aniline + pyrrolidinium (pK_a 11.27)   *products are favored*

19-40

(a) $PhCH_2CH_2NH_2$

(b) $H_2N(CH_2)_4NH_2$

(c)

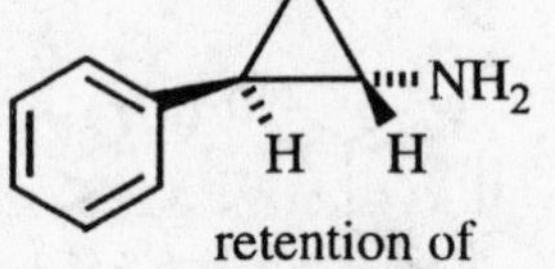

retention of configuration

19-41

(a) $PhCH_2CH_2CH_2NH_2$

(b)

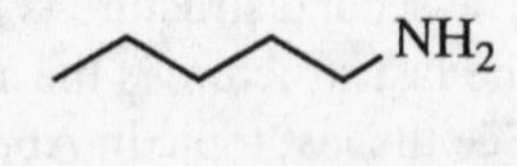

(c)

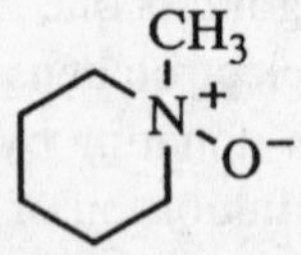

(d)

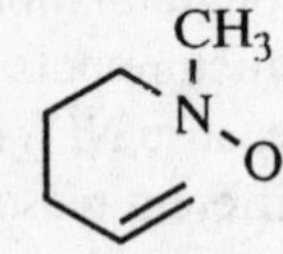

(e)

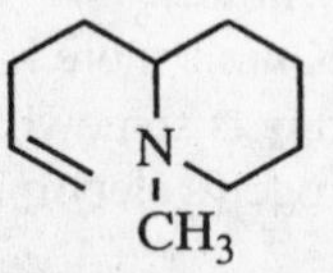

(f)

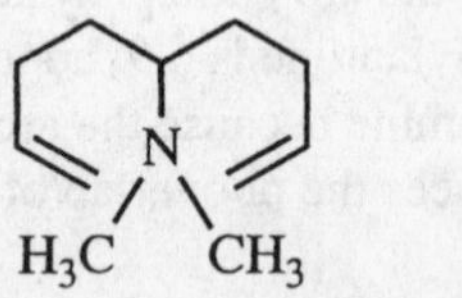

(g)

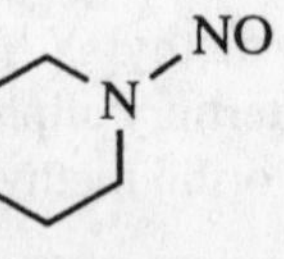

(h)

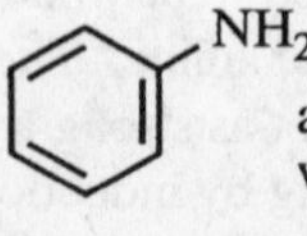

after workup with base

19-41 continued

(i)

cyclohexyl-CH$_2$-C(=O)-NHCH$_3$

(j)

cyclohexyl-CH$_2$CH$_2$NHCH$_3$

(k)

NHCH$_3$

CH$_3$(CH$_2$)$_3$CHCH$_2$CH$_3$

(l)

CH$_2$NH$_2$

PhCH$_2$CHCH$_3$

(m)

(C$_2$H$_5$)$_2$N-CH(CH$_3$)CH$_2$CH$_3$

(n)

pyridine with OCH$_2$CH$_3$

(o)

benzene ring with H and NO$_2$

(p)

HO—C(Et)—CN $\xrightarrow{\text{LiAlH}_4}$ $\xrightarrow{\text{H}_2\text{O}}$ HO—C(Et)—CH$_2$NH$_2$

19-42

(a) CH$_3$—C$_6$H$_4$—NH$_2$ $\xrightarrow[\text{HCl}]{\text{NaNO}_2}$ CH$_3$—C$_6$H$_4$—$\overset{+}{\text{N}}_2$ Cl$^-$ $\xrightarrow{\text{CuCN}}$ CH$_3$—C$_6$H$_4$—CN

(b) CH$_3$—C$_6$H$_4$—CN $\xrightarrow[\text{2) H}_2\text{O}]{\text{1) LiAlH}_4}$ CH$_3$—C$_6$H$_4$—CH$_2$NH$_2$

from (a)

(c) CH$_3$—C$_6$H$_4$—$\overset{+}{\text{N}}_2$ Cl$^-$ $\xrightarrow{\text{KI}}$ CH$_3$—C$_6$H$_4$—I

from (a)

(d) CH$_3$—C$_6$H$_4$—$\overset{+}{\text{N}}_2$ Cl$^-$ $\xrightarrow[\text{H}_2\text{O}]{\text{H}_2\text{SO}_4}$ CH$_3$—C$_6$H$_4$—OH

from (a)

(e) CH$_3$—C$_6$H$_4$—NH$_2$ $\xrightarrow{\text{CH}_3\text{C(=O)-Cl}}$ CH$_3$—C$_6$H$_4$—NHCOCH$_3$ $\xrightarrow[\text{H}_2\text{SO}_4]{\text{HNO}_3}$ CH$_3$, O$_2$N—C$_6$H$_3$—NHCOCH$_3$ (+ isomer)

after workup with base CH$_3$, O$_2$N—C$_6$H$_3$—NH$_2$ $\xleftarrow[\Delta]{\text{H}_3\text{O}^+}$

(f) CH$_3$—C$_6$H$_4$—NH$_2$ + O=cyclopentanone $\xrightarrow{\text{Na(CH}_3\text{COO)}_3\text{BH}}$ CH$_3$—C$_6$H$_4$—N(H)—cyclopentyl

19-43 This fragmentation is favorable because the iminium ion produced is stabilized by resonance. Also, there are three possible cleavages that give the same ion. Both factors combine to make the cleavage facile, at the expense of the molecular ion.

$$\left[CH_3 - \underset{\underset{CH_3}{|}}{\overset{\overset{CH_3}{|}}{C}} - NH_2 \right]^{+\cdot} \longrightarrow \cdot CH_3 + \left\{ \underset{\underset{CH_3}{|}}{\overset{\overset{CH_3}{|}}{\overset{+}{C}}} - \ddot{N}H_2 \longleftrightarrow \underset{\underset{CH_3}{|}}{\overset{\overset{CH_3}{|}}{C}} = \overset{+}{N}H_2 \right\}$$

m/z 73   mass 15   m/z 58

19-44

(a) benzene $\xrightarrow[H_2SO_4]{HNO_3}$ nitrobenzene $\xrightarrow[HCl]{Fe}$ aniline $\longrightarrow$ N-cyclopentanecarbonyl anilide (with Cl–C(=O)–cyclopentane reagent)

(b) aniline + Cl–S(=O)₂–C₆H₄–CH₃ $\longrightarrow$ N-(4-methylphenylsulfonyl)aniline

(c) piperidine (NH) $\xrightarrow[\text{CH}_3\text{I}]{\text{excess}}$ piperidinium $N^+(CH_3)_2$ $\xrightarrow[\Delta]{Ag_2O}$ pentenyl-$N(CH_3)_2$

(d) cycloheptanone $\xrightarrow[H^+]{H_2NOH}$ oxime (NOH) $\xrightarrow[\text{2) } H_2O]{\text{1) LiAlH}_4}$ cycloheptylamine (NH_2)

(e) N–CH₃ pyrrolidine $\xrightarrow{H_2O_2}$ N-oxide (N^+–O⁻, CH₃)

(f) N–CH₃ pyrrolidine $\xrightarrow{H_2O_2}$ N-oxide (N^+–O⁻, CH₃) $\xrightarrow{\Delta}$ butenyl-N(OH)(CH₃)

(g) CH_3-(m-C₆H₄)-COOH $\xrightarrow{SOCl_2}$ CH_3-(m-C₆H₄)-C(=O)Cl $\xrightarrow{HN(CH_2CH_3)_2}$ CH_3-(m-C₆H₄)-C(=O)-N(CH₂CH₃)₂

19-45 The problem restricts the starting materials to six carbons or fewer. Always choose starting materials with as many of the necessary functional groups as possible.

(a)

HO—⟨benzene⟩—NH₂ $\xrightarrow[\text{2) CH}_3\text{CH}_2\text{Br}]{\text{1) NaOH}}$ EtO—⟨benzene⟩—NH₂ $\xrightarrow{\text{H}_3\text{C}-\overset{\text{O}}{\overset{\|}{\text{C}}}-\text{Cl}}$ EtO—⟨benzene⟩—$\overset{\text{H}}{\underset{}{\text{N}}}-\overset{\text{O}}{\overset{\|}{\text{C}}}-\text{CH}_3$

phenacetin

(b)

⟨benzene⟩—Br $\xrightarrow[\text{ether}]{\text{Mg}}$ $\text{H}_2\text{C}\overset{\text{O}}{-}\text{CHCH}_3$ $\xrightarrow{\text{H}_3\text{O}^+}$ ⟨benzene⟩—$\text{CH}_2-\overset{\text{OH}}{\underset{}{\text{CHCH}_3}}$ $\xrightarrow[\substack{\text{H}_2\text{SO}_4 \\ \text{H}_2\text{O}}]{\text{CrO}_3}$ ⟨benzene⟩—$\text{CH}_2-\overset{\text{O}}{\overset{\|}{\text{C}}}\text{CH}_3$

$\xrightarrow[\text{Na(CH}_3\text{COO)}_3\text{BH}]{\text{CH}_3\text{NH}_2}$

⟨benzene⟩—$\text{CH}_2-\overset{\text{NHCH}_3}{\underset{}{\text{CHCH}_3}}$

methamphetamine

(c)

HO⟨benzene⟩HO $\xrightarrow[\text{AlCl}_3]{\text{CH}_3\text{I}}$ HO⟨benzene⟩HO—CH_3 $\xrightarrow{\text{NBS}}$ HO⟨benzene⟩HO—CH_2Br

$\downarrow$ KCN

HO⟨benzene⟩HO—$\text{CH}_2\text{CH}_2\text{NH}_2$ $\xleftarrow{\text{H}_2\text{, Pt}}$ HO⟨benzene⟩HO—CH_2CN

dopamine

19-46

(a)

19-46 continued

(b)

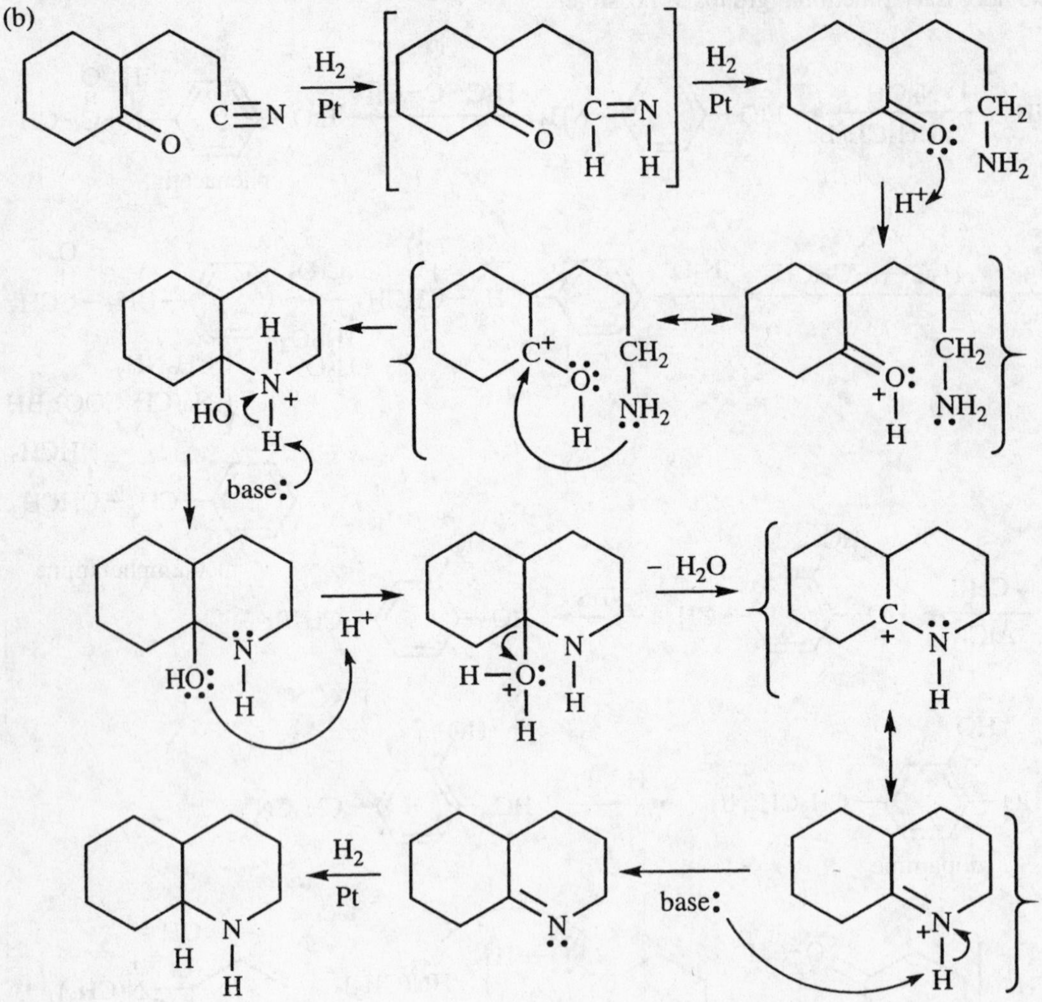

19-47

(a)

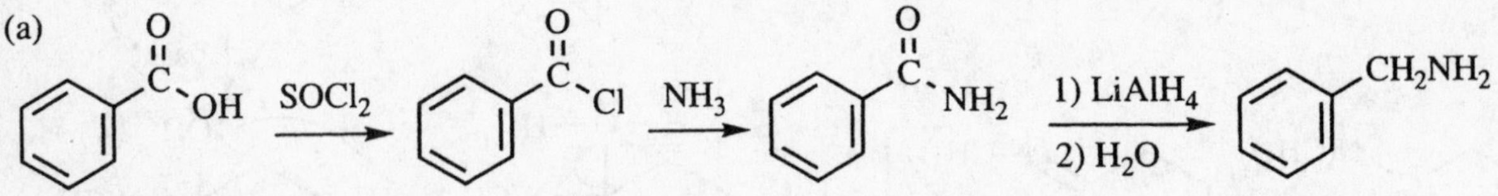

(b)

(c)

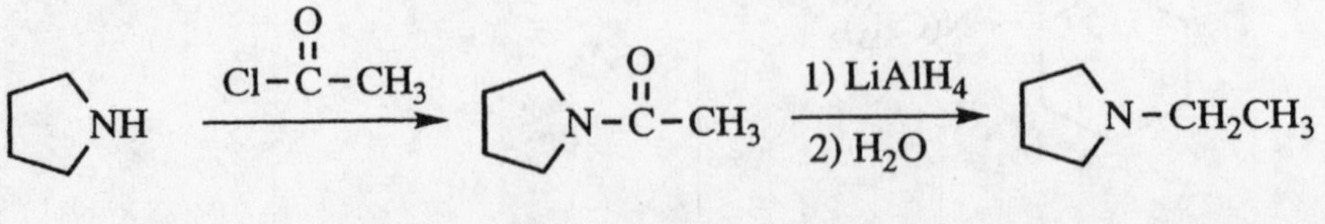

19-47 continued

(d)

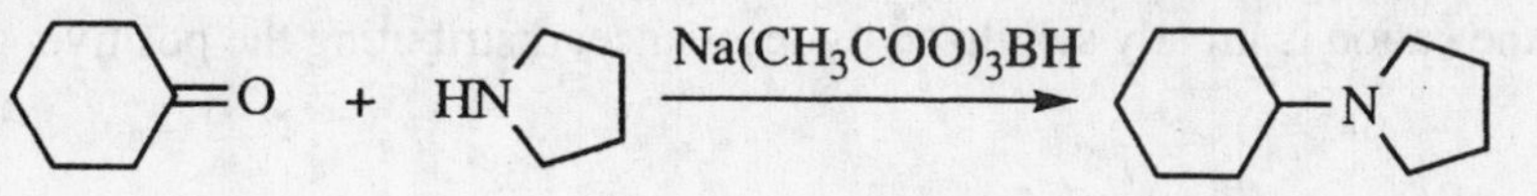

(e)

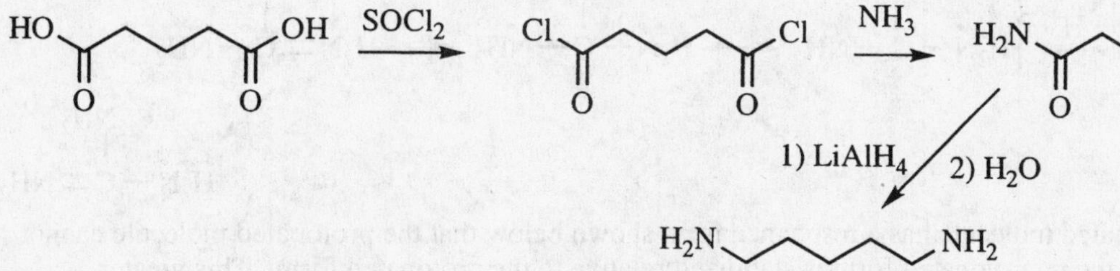

19-48

(a)

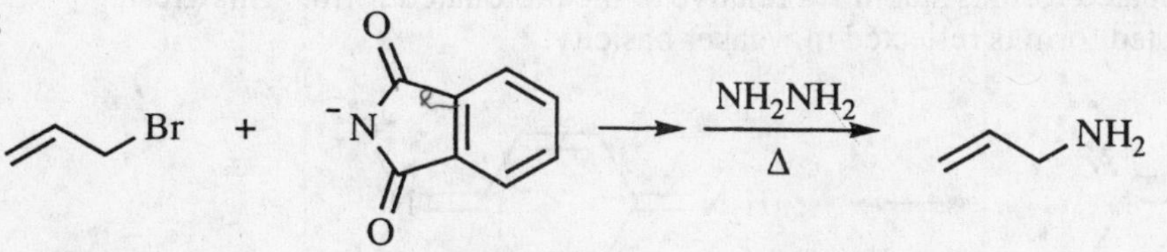

(b)

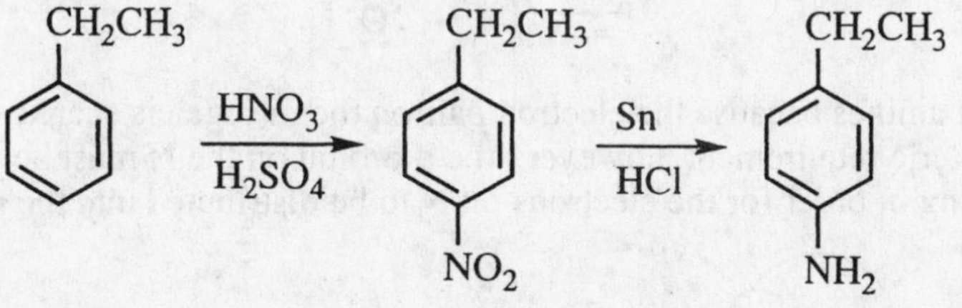

(c)

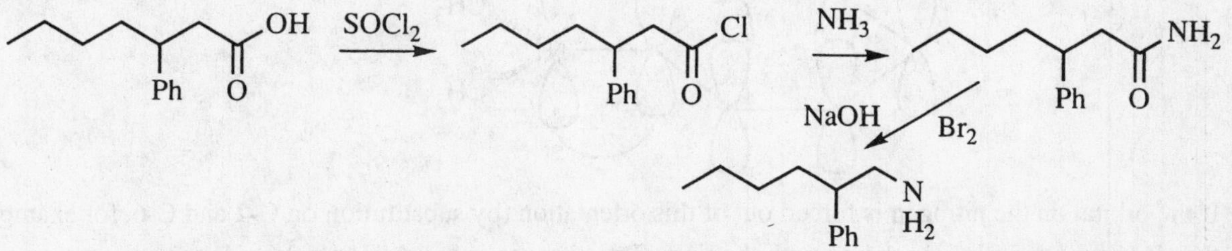

(d)

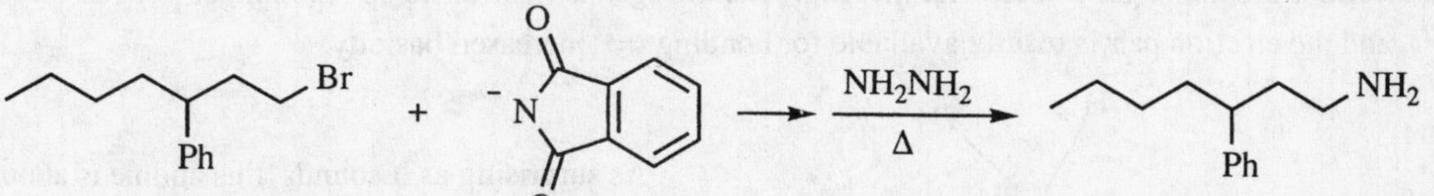

(e)

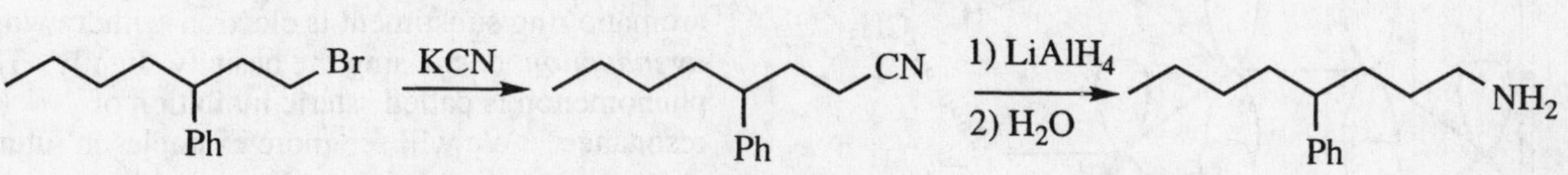

19-49

(a) When guanidine is protonated, the cation is greatly stabilized by resonance, distributing the positive charge over all atoms (except H):

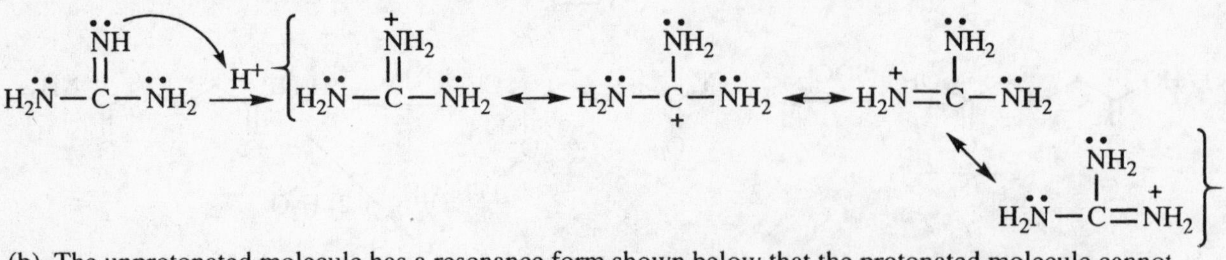

(b) The unprotonated molecule has a resonance form shown below that the protonated molecule cannot have. Therefore, the unprotonated form is stabilized relative to the protonated form. This greater stabilization of the unprotonated form is reflected in weaker basicity.

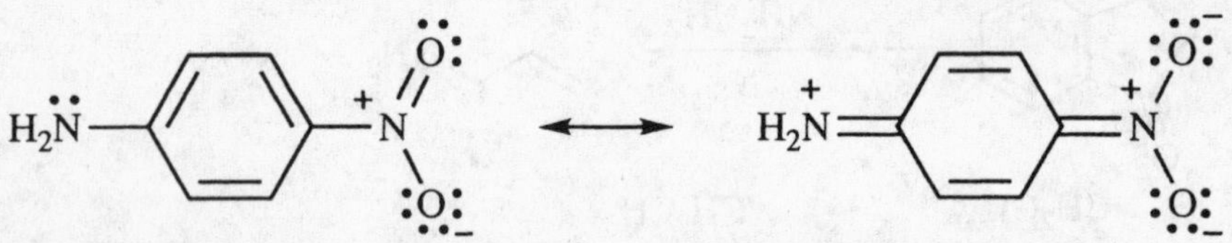

(c) Anilines are weaker bases than aliphatic amines because the electron pair on the nitrogen is shared with the ring, stabilizing the system. There is a steric requirement, however: the p orbital on the N must be parallel with the p orbitals on the benzene ring in order for the electrons on N to be distributed into the π system of the ring.

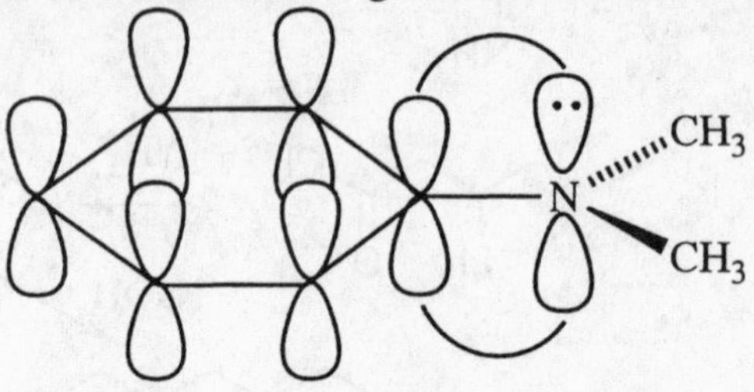

If the orbital on the nitrogen is forced out of this orientation (by substitution on C-2 and C-6, for example), the electrons are no longer shared with the ring. The nitrogen is hybridized sp^3 (no longer any reason to be sp^2), and the electron pair is readily available for bonding $\Rightarrow$ increased basicity.

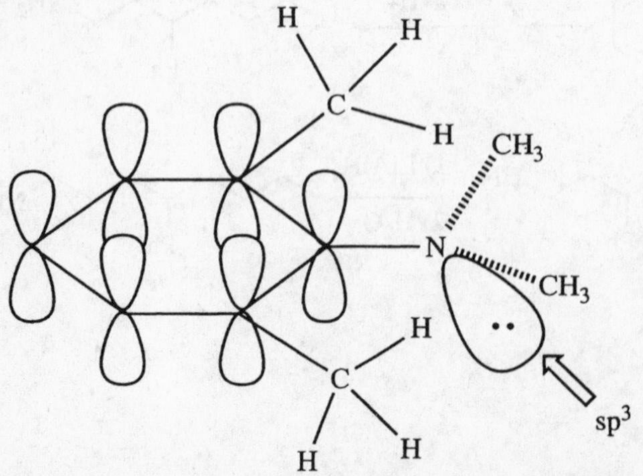

As surprising as it sounds, this aniline is about as basic as a tertiary aliphatic amine, except that the aromatic ring substituent is electron-withdrawing *by induction*, decreasing the basicity slightly. This phenomenon is called "steric inhibition of resonance". We will see more examples in future chapters. Also, it is the last topic in Appendix 2. (See another example in the solution to 19-37(f).)

19-50 In this problem, sodium triacetoxyborohydride will be represented as Na(AcO)$_3$BH.

(a)

$\diagdown\diagup\diagdown$OH $\xrightarrow[\text{pyridine}]{\text{TsCl}}$ $\diagdown\diagup\diagdown$OTs $\xrightarrow{\text{KCN}}$ $\diagdown\diagup\diagdown$CN $\xrightarrow[\text{2) H}_2\text{O}]{\text{1) LiAlH}_4}$ $\diagdown\diagup\diagdown\diagup$NH$_2$

(b)

$\diagdown\diagup\diagdown$OH $\xrightarrow{\text{PCC}}$ $\diagdown\diagup\diagdown$CHO $\xrightarrow[\text{Na(AcO)}_3\text{BH}]{\text{CH}_3\text{NH}_2}$ $\diagdown\diagup\diagdown$NHCH$_3$

(c)

$\diagdown\diagup$OH $\xrightarrow[\text{pyridine}]{\text{TsCl}}$ $\diagdown\diagup$OTs $\xrightarrow{\text{NH}_3}$ $\diagdown\diagup$NH$_2$

$\diagup\diagdown$OH $\xrightarrow[\text{H}_2\text{O}]{\text{CrO}_3 \atop \text{H}_2\text{SO}_4}$ $\diagup\diagdown$O $+$ $\diagdown\diagup$NH$_2$ $\xrightarrow{\text{Na(AcO)}_3\text{BH}}$ $\diagdown\diagup$HN $\diagup\diagdown$

$\xleftarrow{\text{Na(AcO)}_3\text{BH}}$ $\underset{\underset{\text{CH}_3\text{CH}_2\text{OH}}{\uparrow \text{PCC}}}{\text{CH}_3-\text{CH}=\text{O}}$

$\diagup\diagdown$N$\diagdown\diagup$

(d)

CH_3–Ph $\xrightarrow{\text{NBS}}$ CH$_2$Br–Ph $\xrightarrow{\text{NaOH}}$ CH$_2$OH–Ph $\xrightarrow{\text{PCC}}$ CHO–Ph

$\xleftarrow{\text{Na(AcO)}_3\text{BH}}$ $\underset{\diagdown\diagup\text{NH}_2 \text{ from (c)}}{}$ CH$_2$NHCH$_2$CH$_2$CH$_3$–Ph

(e)

CH$_3$–Ph $\xrightarrow{\text{KMnO}_4}$ COOH–Ph $\xrightarrow{\text{SOCl}_2}$ COCl–Ph $\xrightarrow{\text{NH}_3}$ C(=O)NH$_2$–Ph $\xrightarrow[\text{NaOH}]{\text{Br}_2}$ NH$_2$–Ph $\xleftarrow[\text{HCl}]{\text{NaNO}_2}$ $\overset{+}{\text{N}}_2$ Cl$^-$–Ph $\xrightarrow{\text{H}_3\text{O}^+}$ OH–Ph

Ph–N=N–Ph–OH

(f)

Ph $\xrightarrow[\text{AlCl}_3]{}$ (CH$_3$CH$_2$C(=O)Cl) Ph–C(=O)CH$_2$CH$_3$ $\xrightarrow[\text{H}_2\text{SO}_4]{\text{HNO}_3}$ Ph(NO$_2$)–C(=O)CH$_2$CH$_3$ $\xrightarrow[\substack{\text{HCl}\\\text{Clemmensen}\\\text{reduces both}}]{\text{Zn(Hg)}}$ Ph(NH$_2$)–CH$_2$CH$_2$CH$_3$

$\diagdown\diagup$OH $\xrightarrow{\text{H}_2\text{CrO}_4}$ $\xrightarrow{\text{SOCl}_2}$ $\diagdown\diagup$C(=O)Cl

459

19-50 continued

(g)

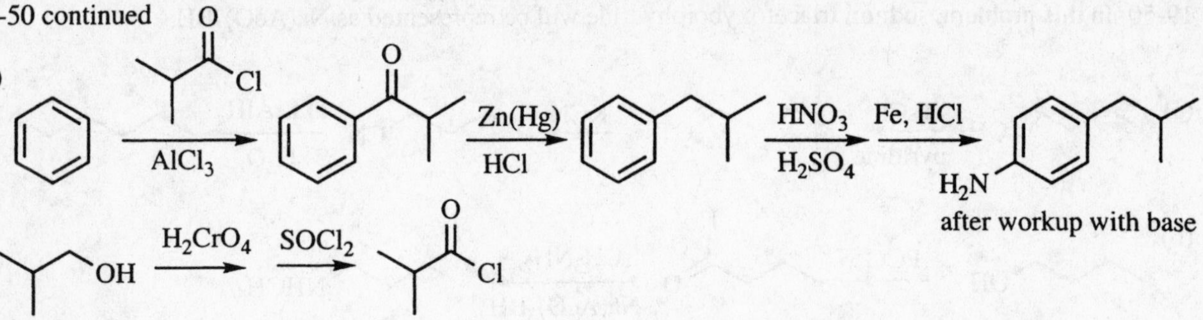

19-51

(a)

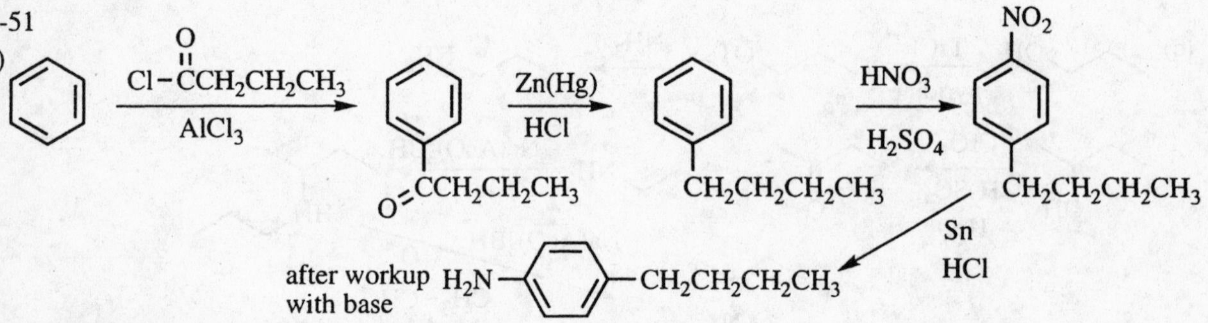

after workup with base

(b)

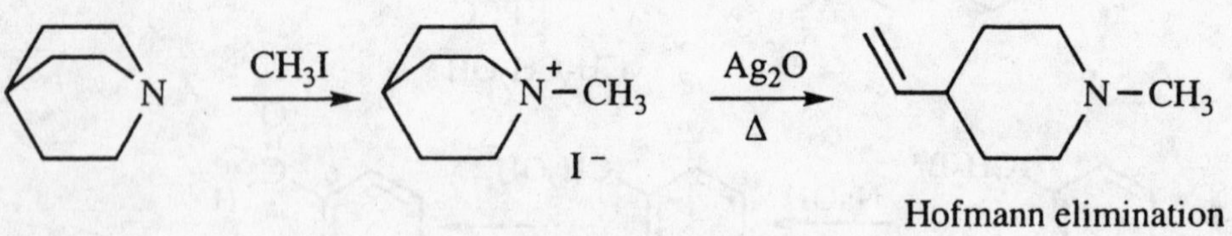

Hofmann elimination

(c)

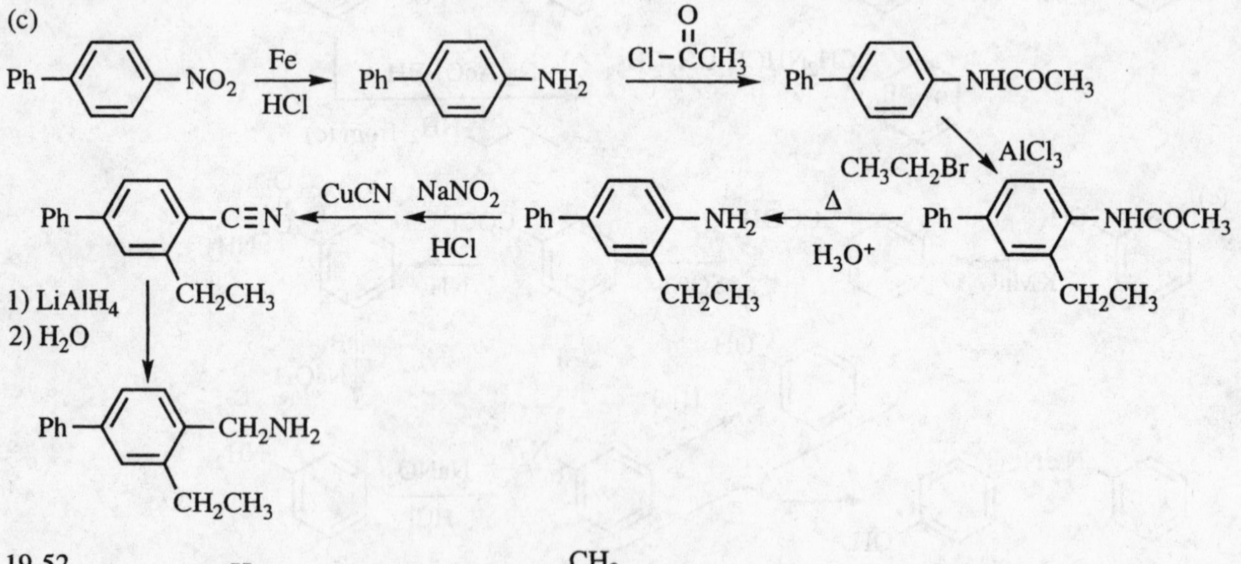

1) LiAlH₄
2) H₂O

19-52

coniine

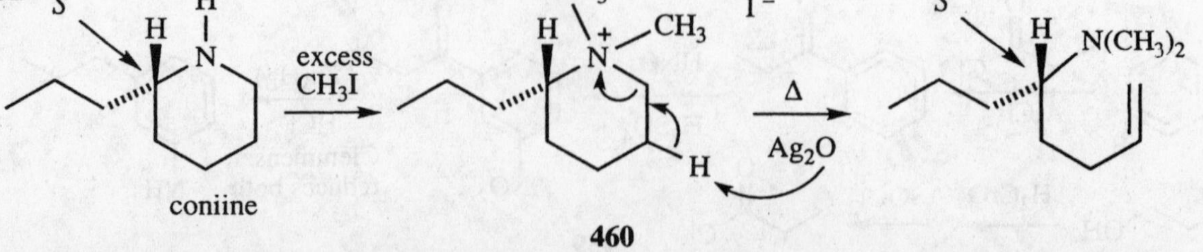

19-53
(a)

<u>unknown **X**</u>

—fishy odor $\Rightarrow$ amine

—molecular weight 101 $\Rightarrow$ odd number of nitrogens

$\Rightarrow$ if one nitrogen and no oxygen, the remainder is C_6H_{15}

<u>Mass spectrum:</u>

—fragment at 86 = M − 15 = loss of methyl $\Rightarrow$ the compound is likely to have this structural piece: $CH_3 \overline{} C - N$ α-cleavage

<u>IR spectrum:</u>

—no OH, no NH $\Rightarrow$ must be a 3° amine

—no C=O or C=C or C≡N or NO_2

<u>NMR spectrum:</u>

—only a triplet and quartet, integration about 3 : 2 $\Rightarrow$ ethyl group(s) only

assemble the evidence: $C_6H_{15}N$, 3° amine, only ethyl in the NMR:

$$CH_3CH_2 - N - CH_2CH_3$$
$$|$$
$$CH_2CH_3$$

(b) React the triethylamine with HCl. The pure salt is solid and odorless.

$$(CH_3CH_2)_3N \ + \ HCl \ \longrightarrow \ (CH_3CH_2)_3\overset{+}{N}H \quad Cl^-$$
salt

(c) Washing her clothing in dilute acid like vinegar (dilute acetic acid) or dilute HCl would form a water-soluble salt as shown in (b). Normal washing will remove the water-soluble salt.

19-54
(a)

(b) Both answers can be found in the resonance forms of the intermediate, in particular, the resonance form that shows the positive charge on the N. This is the major resonance contributor; what is special about it is that every atom has a full octet, the best of all possible conditions. That does not arise in the benzene intermediate, so it must be easier to form the intermediate from pyrrole than from benzene. Also, acylation at the 3-position puts positive charge at the 2-position and on the N, but never on the other side of the ring, so this substitution has only two resonance forms. The intermediate from acylation at the 3-position is therefore not as stable as the intermediate from acylation at the 2-position.

19-55

(a)

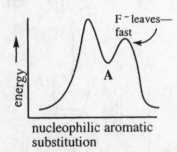

(b) Why is fluoride ion a good leaving group from **A** but not from **B** (either by S_N1 or S_N2)?

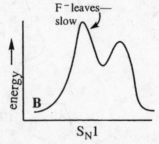

A   **B**

Formation of the anionic sigma complex **A** is the rate-determining (slow) step in nucleophilic aromatic substitution. The loss of fluoride ion occurs in a subsequent fast step where the nature of the leaving group does not affect the overall reaction rate. In the S_N1 or S_N2 mechanisms, however, the carbon-fluorine bond is breaking in the rate-determining step, so the poor leaving group ability of fluoride does indeed affect the rate.

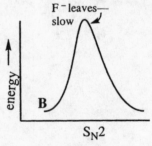

(c) Amines can act as nucleophiles as long as the electron pair on the N is available for bonding. The initial reactant, methylamine, CH_3NH_2, is a very reactive nucleophile. However, once the N is bonded to the benzene ring, the electron pair is delocalized onto the ring, especially with such strong electron-withdrawing groups like NO_2 in the ortho and para positions. The electrons on N are no longer available for bonding so there is no danger of it acting as a nucleophile in another reaction.

19-56   **Compound A**

Mass spectrum:

—molecular ion at 73 = odd mass = odd number of nitrogens;

if one nitrogen and no oxygen present ⇒ molecular formula $C_4H_{11}N$

—base peak at 44 is M - 29 ⇒ this fragment must be present:

$$
\left.\begin{array}{c}
-N- \\
| \\
-C\!\!+\!\!CH_2CH_3 \\
| \\
44 \\
\alpha\text{-cleavage}
\end{array}\right\} \text{EITHER} \quad
\begin{array}{c}
NH_2 \\
| \\
CH_3-C-CH_2CH_3 \\
| \\
H \quad \mathbf{1}
\end{array}
\ \text{OR}\
\begin{array}{c}
CH_3-NH \\
| \\
H-C-CH_2CH_3 \\
| \\
H \quad \mathbf{2}
\end{array}
$$

IR spectrum:

—two peaks around 3300 cm^{-1} indicate a 1° amine; no indication of oxygen

NMR spectrum:

—two exchangeable protons suggest NH_2 present

—1H multiplet at δ 2.8 means a CH—NH_2

The structure of **A** must be the same as **1** above:

$$
\mathbf{A} \quad
\begin{array}{c}
NH_2 \\
| \\
CH_3-C-CH_2CH_3 \\
| \\
H
\end{array}
$$

Compound B

an isomer of **A**, so its molecular formula must also be $C_4H_{11}N$

IR spectrum:

—only one peak at 3300 cm^{-1} ⇒ 2° amine

NMR spectrum:

—one exchangeable proton ⇒ NH

—two ethyls present

The structure of **B** must be:

$$
\mathbf{B} \quad
\begin{array}{c}
CH_3-CH_2-NCH_2CH_3 \\
| \\
H
\end{array}
$$

$$
\left[
\begin{array}{c}
58 \\
CH_3\!\!+\!\!CH_2-NCH_2CH_3 \\
| \\
H
\end{array}
\right]^{\stackrel{+}{\cdot}}
\longrightarrow
\begin{array}{c}
CH_2\!=\!\overset{+}{N}CH_2CH_3 \\
| \\
H \quad \text{resonance-stabilized} \\
m/z\ 58
\end{array}
$$

19-57

(a) The acid-catalyzed condensation of P2P (a controlled substance) with methylamine hydrochloride gives an imine which can be reduced to methamphetamine. The suspect was probably planning to use zinc in muriatic acid (dilute HCl) for the reduction.

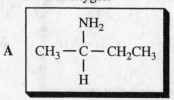

phenyl-2-propanone, P2P
(phenylacetone)   methamphetamine

(b) The jury acquitted the defendant on the charge of attempted manufacture of methamphetamine. There were legal problems with possible entrapment, plus the fact that he had never opened the bottle of the starting material. The defendant was convicted on several possession charges, however, and was awarded four years of institutional time to study organic chemistry.

19-58

Mass spectrum:

—molecular ion at 87 = odd mass = odd number of nitrogens present

—if one nitrogen and no oxygens ⇒ molecular formula $C_5H_{13}N$

—base peak at m/z 30 ⇒ structure must include this fragment

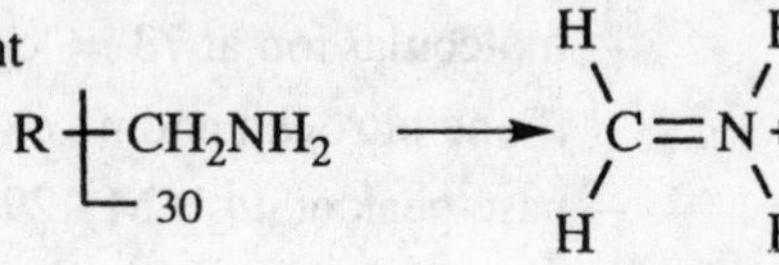

$$R\!-\!CH_2NH_2 \longrightarrow \begin{array}{c} H \quad\quad H \\ \backslash \quad / \\ C = N+ \\ / \quad\quad \backslash \\ H \quad\quad H \end{array}$$

m/z 30

IR spectrum:

—two peaks in the 3300-3400 cm^{-1} region ⇒ 1° amine

NMR spectrum:

—singlet at δ 0.9 for 9H must be a *t*-butyl group

—2H signal at δ 1.0 exchanges with D_2O ⇒ must be protons on N or O

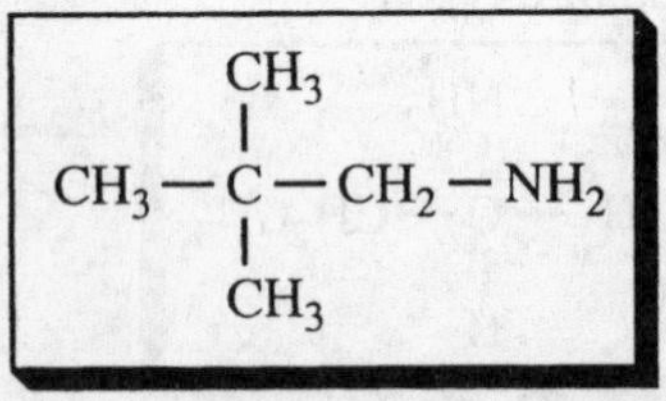

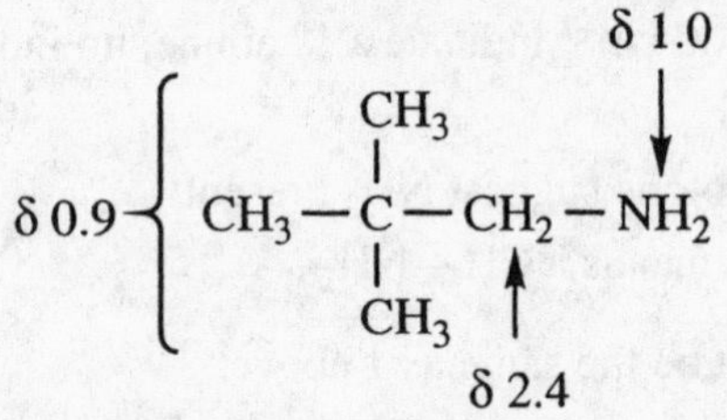

$$\boxed{\begin{array}{c} CH_3 \\ | \\ CH_3 - C - CH_2 - NH_2 \\ | \\ CH_3 \end{array}}$$

$$\delta\,0.9 \left\{ \begin{array}{c} \delta\,1.0 \\ \downarrow \\ CH_3 \\ | \\ CH_3 - C - CH_2 - NH_2 \\ | \\ CH_3 \;\uparrow \\ \delta\,2.4 \end{array} \right.$$

Note that the base peak in the MS arises from cleavage to give these two, relatively stable fragments:

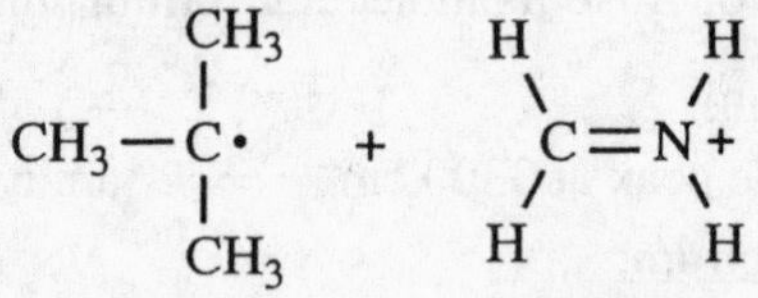

$$\begin{array}{c} CH_3 \\ | \\ CH_3 - C\cdot \\ | \\ CH_3 \end{array} \quad + \quad \begin{array}{c} H \quad\quad H \\ \backslash \quad / \\ C = N+ \\ / \quad\quad \backslash \\ H \quad\quad H \end{array}$$

m/z 30

19-59 (a tough problem)

molecular formula $C_{11}H_{16}N_2$ has 5 elements of unsaturation, enough for a benzene ring; no oxygens precludes NO_2 and amide; if C≡N is present, there are not enough elements of unsaturation left for a benzene ring, so benzene and C≡N are mutually exclusive

IR spectrum:

—one spike around 3300 cm^{-1} suggests a 2° amine

—no C≡N

—CH and C=C regions suggest an aromatic ring

Proton NMR spectrum:

—5H multiplet at δ 7.3 indicates a monosubstituted benzene ring (the fact that all the peaks are huddled around 7.3 precludes N being bonded to the ring)

—1H singlet at δ 2.0 is exchangeable ⇒ NH of secondary amine

—2H singlet at δ 3.5 is CH_2; the fact that it is so strongly deshielded and unsplit suggests that it is between a nitrogen and the benzene ring

fragments so far:

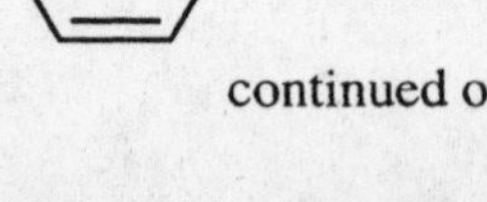

$$\langle\bigcirc\rangle\!-\!CH_2 - N \quad + \; NH \; + \; 4\,C \; + \; 8\,H \; + \; 1 \text{ element of unsaturation}$$

continued on next page   **464**

19-59 continued

Carbon NMR spectrum:

—four signals around δ 125-138 are the aromatic carbons

—the signal at δ 65 is the CH_2 bonded to the benzene

—the other 4 carbons come as two signals at δ 46 and δ 55; each is a triplet, so there are two sets of two equivalent CH_2 groups, each bonded to N to shift it downfield

fragments so far:

$$\text{Ph}-CH_2-N \quad + \quad NH \quad + \quad \frac{CH_2}{CH_2} \; + \; \frac{CH_2}{CH_2} \quad + \quad \text{1 element of unsaturation}$$

There is no evidence for an alkene in any of the spectra, so the remaining element of unsaturation must be a ring. The simplicity of the NMR spectra indicates a fairly symmetric compound.

Assemble the pieces:

$$\text{Ph}-CH_2-N \underset{\underset{\displaystyle C-C}{\underset{H_2 \; H_2}{|}}}{\overset{\overset{\displaystyle C-C}{\overset{H_2 \; H_2}{|}}}{}} NH$$

19-60

(a)

(b)

$$\text{Ph}-\overset{H}{N}- \quad \xrightarrow[Na(AcO)_3BH]{\overset{O}{\underset{}{H \; CH_3}}}$$

$$\text{Ph}-NH \quad \xrightarrow[Na(AcO)_3BH]{\overset{O}{\underset{}{H \; CH_2CH_3}}}$$

$$\text{Ph}-CH_2-\overset{O}{\underset{}{H}} \quad \xrightarrow[Na(AcO)_3BH]{} \quad HN$$

$$\overset{O}{\underset{Cl \quad CH_3}{}} \quad \text{Ph}-\overset{H}{N}- \quad \xrightarrow{LiAlH_4}$$

$$\xleftarrow{LiAlH_4} \quad \overset{O}{\underset{Cl \quad CH_2CH_3}{}} \quad \text{Ph}-NH$$

$$\xleftarrow{LiAlH_4} \quad \text{Ph}-\overset{O}{\underset{Cl}{}} \quad HN$$

Central product: $\text{Ph}-CH_2CH_2-N(CH_2CH_3)(CH_2CH_2CH_3)$

465

19-61 Not only is substitution at C-2 and C-4 the major products, but substitution occurs under surprisingly mild conditions.

Begin by drawing the resonance forms of pyridine N-oxide:

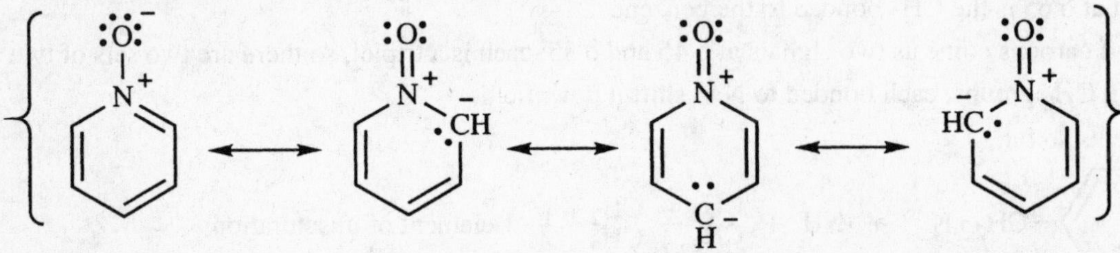

Resonance forms show that the electron density from the oxygen is distributed at C-2 and C-4; these positions would be the likely places for an electrophile to attack.

Resonance forms from electrophilic attack at C-2

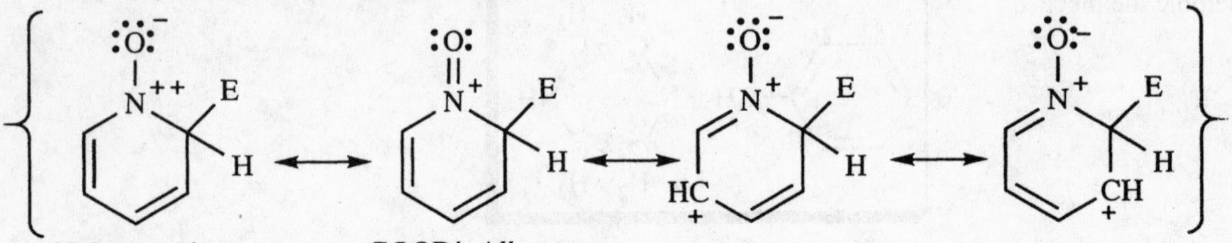

N does not have an octet; not a significant resonance contributor

GOOD! All atoms have octets.

Resonance forms from electrophilic attack at C-3

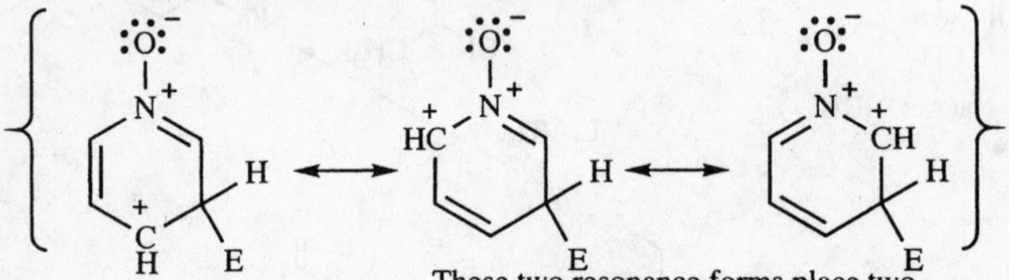

These two resonance forms place two positive charges on adjacent atoms—not good.

Resonance forms from electrophilic attack at C-4

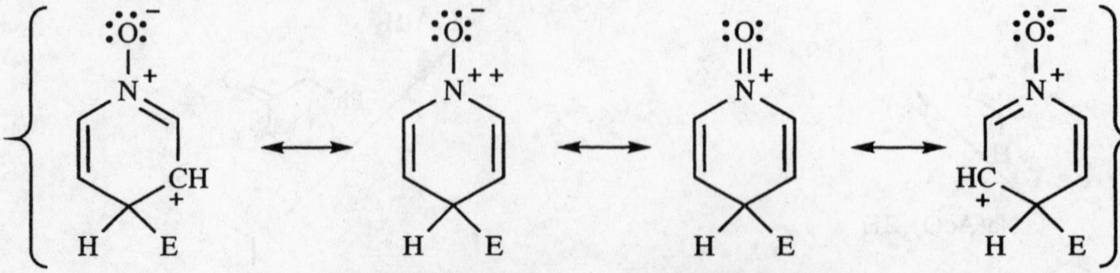

N does not have an octet; not a significant resonance contributor

GOOD! All atoms have octets.

continued on next page

19-61 continued

The resonance forms from electrophilic attack at C-3 are bad; only one of the three is a significant contributor, which means that there is not much resonance stabilization. When the electrophile attacks at C-2 or C-4, however, there are two forms that are good plus one great one that has all atoms with full octets. Clearly, attack at C-2 and C-4 give the most stable intermediates and will be the preferred sites of attack.

19-62

(a)

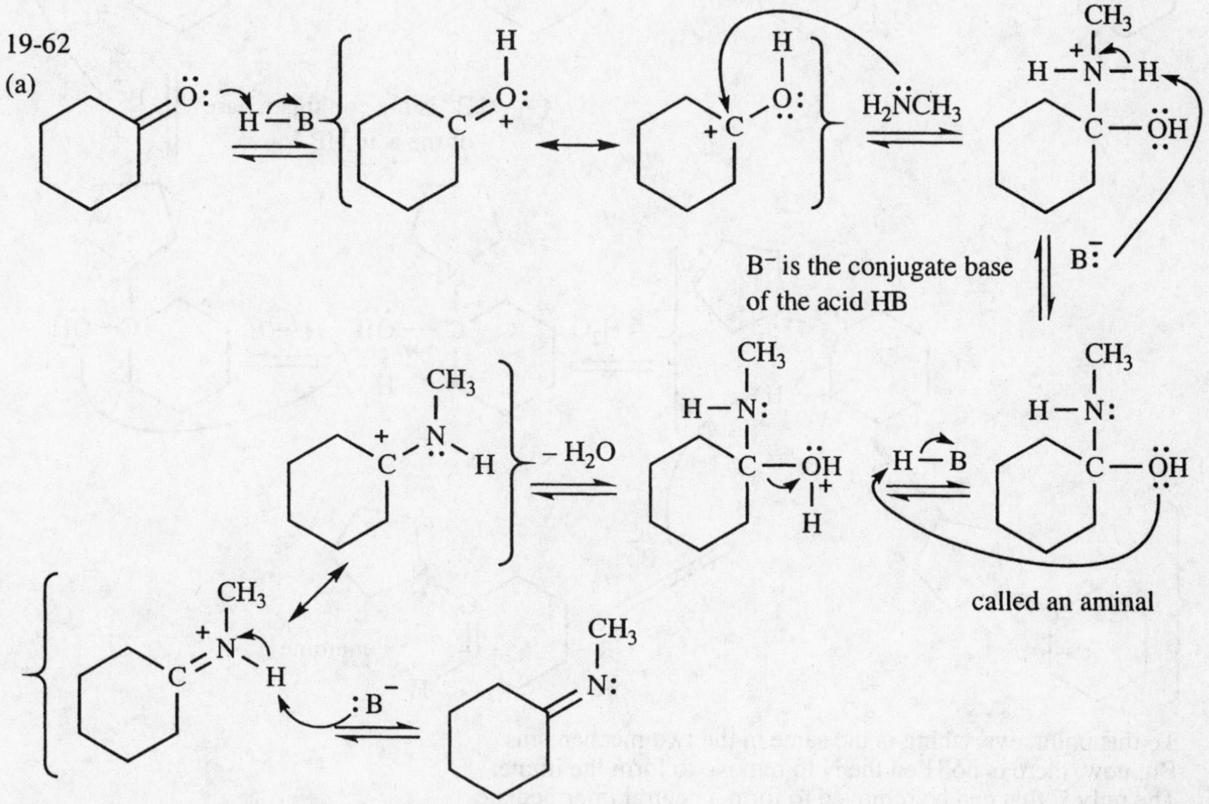

H—B
H₂N̈CH₃
B⁻ is the conjugate base of the acid HB
B:⁻
− H₂O
H—B
called an aminal
:B⁻

continued on next page

(b)

To this point, everything is the same in the two mechanisms.
But now, there is no H on the N to remove to form the imine.
The only H that can be removed to form a neutral intermediate
is the H on C next to the carbocation.

(c) A secondary amine has only one H to give which it loses in the first half of the mechanism to form the
neutral intermediate called an aminal, equivalent to a hemiacetal. In the second half of the mechanism, the H
on an adjacent carbon is removed to form the neutral product, the enamine. The type of product depends
entirely on whether the amine begins with one or two hydrogen atoms.

20-1

(a)
$$CH_3CH_2\overset{\overset{\displaystyle CH_3}{|}}{C}HCOOH$$

(b)
$$CH_3CH_2\overset{\overset{\displaystyle Br}{|}}{C}HCOOH$$

(c)
$$CH_3\overset{\overset{\displaystyle NH_2}{|}}{C}HCH_2CH_2COOH$$

(d)

(e)

(f)

(g)

(h)

(i)

(j)

(k)

(l)

20-2 IUPAC name first; then common name.

(a) 2-iodo-3-methylpentanoic acid; α-iodo-β-methylvaleric acid
(b) (Z)-3,4-dimethylhex-3-enoic acid
(c) 2,3-dinitrobenzoic acid; no common name
(d) *trans*-cyclohexane-1,2-dicarboxylic acid; (*trans*-hexahydrophthalic acid)
(e) 2-chlorobenzene-1,4-dicarboxylic acid; 2-chloroterephthalic acid
(f) 3-methylhexanedioic acid; β-methyladipic acid

20-3 Listed in order of increasing acid strength (weakest acid first). (See Appendix 2 for a review of acidity.)

(a)
$$CH_3CH_2COOH \quad < \quad CH_3-\overset{\overset{\displaystyle }{|}}{C}HCOOH \quad < \quad CH_3-\overset{\overset{\displaystyle Br}{|}}{\underset{\underset{\displaystyle Br}{|}}{C}}COOH$$
$$\qquad\qquad\qquad\qquad\qquad Br$$

The greater the number of electron-withdrawing substituents, the greater the stabilization of the carboxylate ion.

(b)
$$CH_3\overset{\overset{\displaystyle }{|}}{C}HCH_2CH_2COOH \quad < \quad CH_3CH_2\overset{\overset{\displaystyle }{|}}{C}CH_2COOH \quad < \quad CH_3CH_2CH_2\overset{\overset{\displaystyle }{|}}{C}HCOOH$$
$$\quad\;\; Br \qquad\qquad\qquad\qquad Br \qquad\qquad\qquad\qquad\quad Br$$

The closer the electron-withdrawing group, the greater the stabilization of the carboxylate ion.

(c)
$$CH_3CH_2COOH \;<\; CH_3-\overset{\overset{\displaystyle }{|}}{C}HCOOH \;<\; CH_3-\overset{\overset{\displaystyle }{|}}{C}HCOOH \;<\; CH_3-\overset{\overset{\displaystyle }{|}}{C}HCOOH$$
$$\qquad\qquad\qquad\qquad Cl \qquad\qquad\qquad C\equiv N \qquad\qquad\qquad NO_2$$

The stronger the electron-withdrawing effect of the substituent, the greater the stabilization of the carboxylate ion.

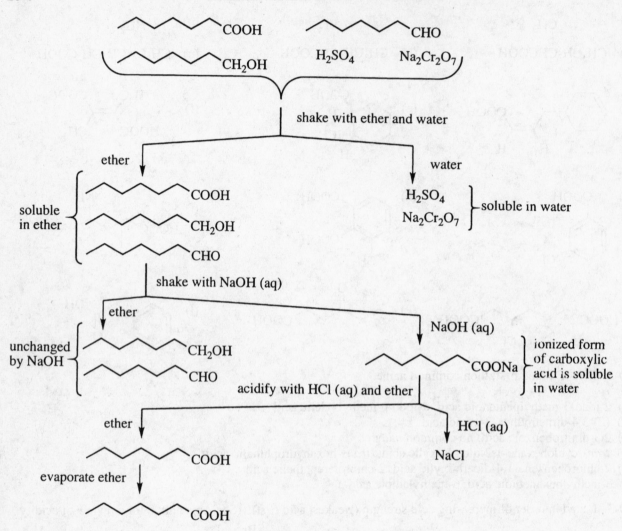

20-5 The principle used to separate a carboxylic acid (a stronger acid) from a phenol (a weaker acid) is to neutralize with a weak base ($NaHCO_3$), a base strong enough to ionize the stronger acid but not strong enough to ionize the weaker acid.

20-6 The reaction mixture includes the initial reactant, reagent, desired product, and the overoxidation product—not unusual for an organic reaction mixture.

(a)

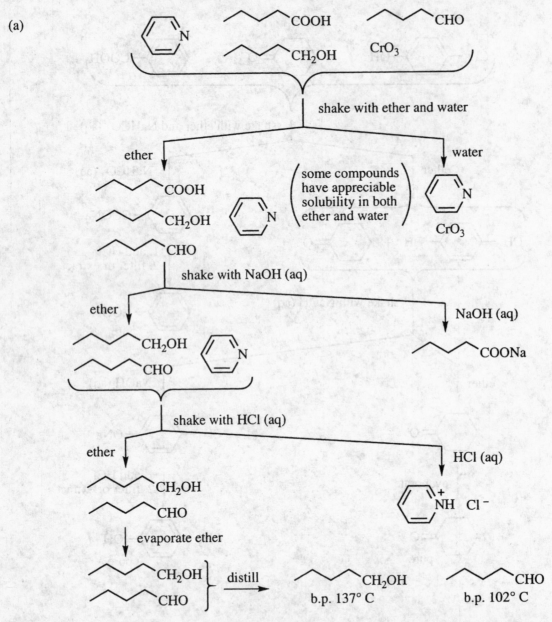

(b) Pentan-1-ol cannot be removed from pentanal by acid-base extraction. These two remaining products can be separated by distillation, the alcohol having the higher boiling point because of hydrogen bonding.

20-7 The COOH has a characteristic IR absorption: a broad peak from 3400-2400 cm^{-1}, with a "shoulder" around 2700 cm^{-1}. The carbonyl stretch at 1695 cm^{-1} is a little lower than the standard 1710 cm^{-1}, suggesting conjugation. The strong alkene absorption at 1650 cm^{-1} also suggests it is conjugated.

20-8

(a) The ethyl pattern is obvious: a 3H triplet at δ 1.15 and a 2H quartet at δ 2.4. The only other peak is the COOH at δ 11.9 (a 2.1 δ unit offset added to 9.8).

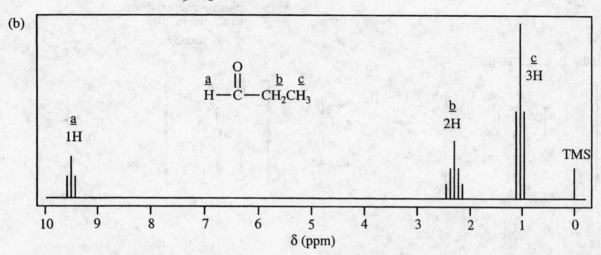

The multiplet between δ 2 and δ 3 is drawn as a pentet as though it were split equally by the aldehyde proton and the CH₂ group. These coupling constants are probably unequal, in which case the actual splitting pattern will be a complex multiplet.

(c) The chemical shift of the aldehyde proton is between δ 9-10, not as far downfield as the carboxylic acid proton. Also, the aldehyde proton is split into a triplet by the CH₂, unlike the COOH proton which always appears as a singlet. Finally, the CH₂ is split by an extra proton, so it will give a multiplet with complex splitting, instead of the quartet shown in the acid.

20-9

20-10

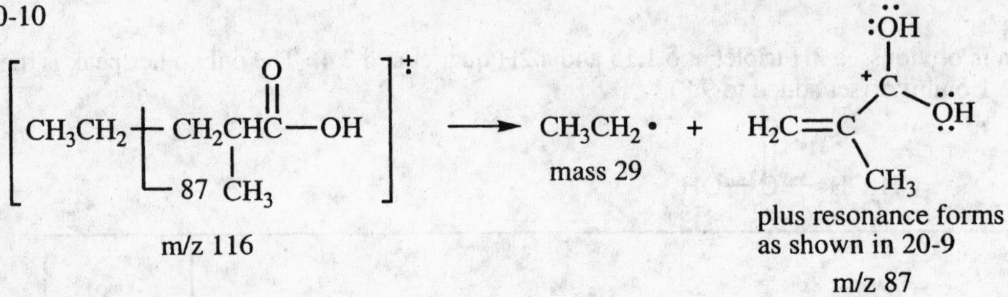

CH₃CH₂•
mass 29

plus resonance forms
as shown in 20-9
m/z 87

m/z 116

McLafferty rearrangement

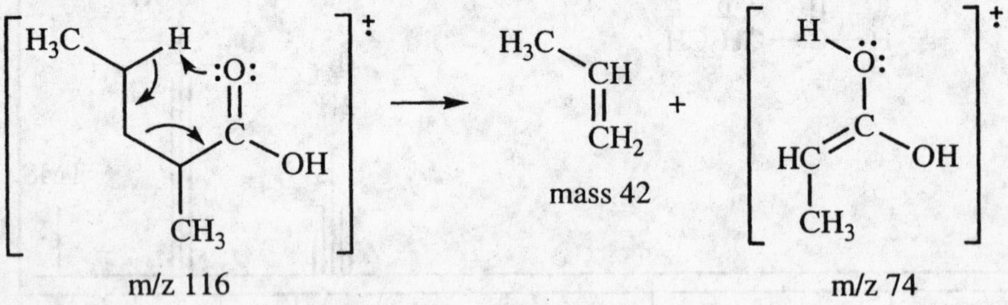

mass 42

m/z 116

m/z 74

20-11

(a)

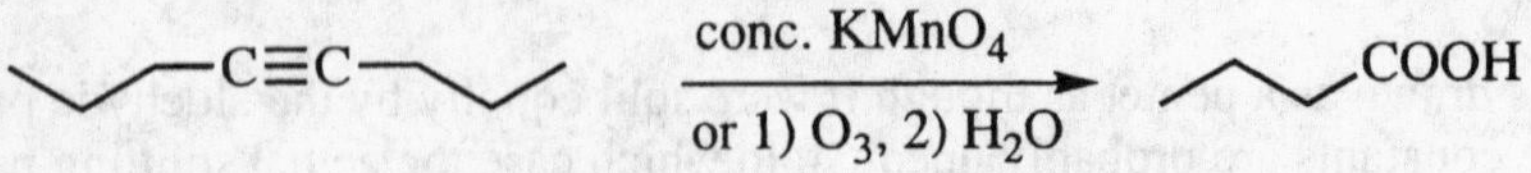

conc. KMnO₄
or 1) O₃, 2) H₂O

(b)

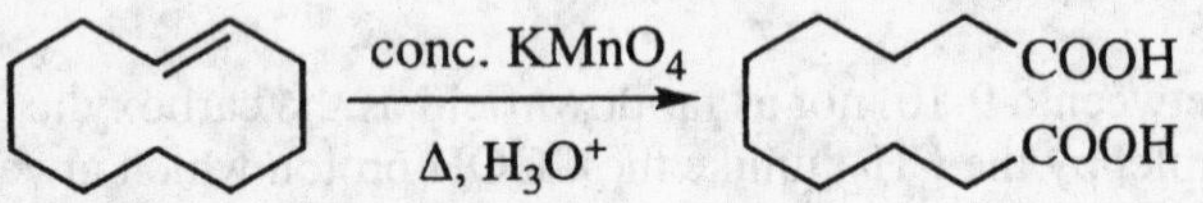

conc. KMnO₄
Δ, H₃O⁺

(c)

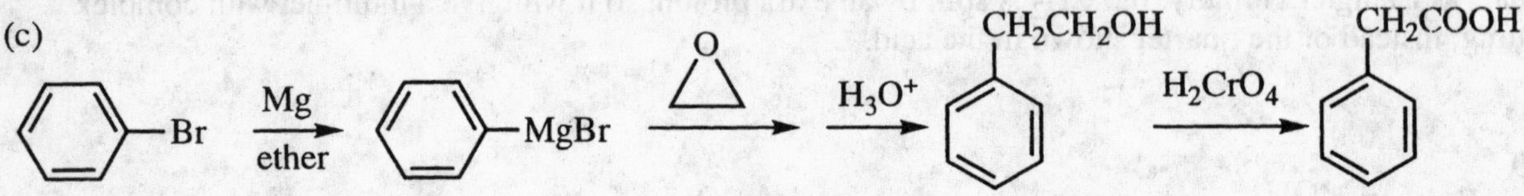

Mg
ether

H₃O⁺

H₂CrO₄

(d)

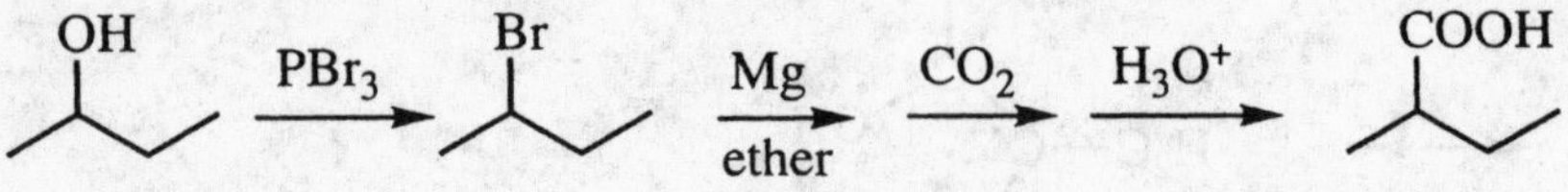

PBr₃ → Mg/ether → CO₂ → H₃O⁺

(e)

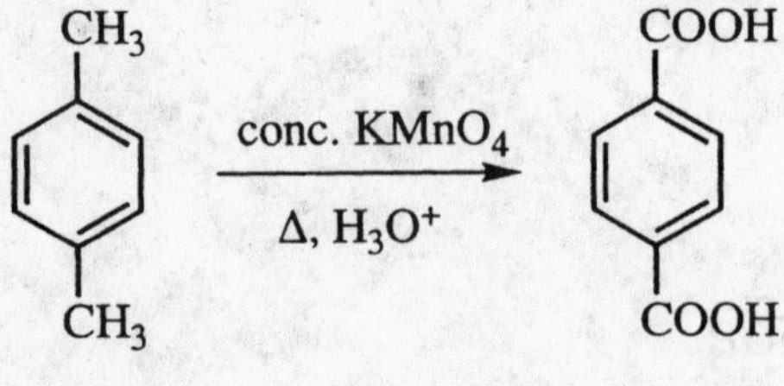

conc. KMnO₄
Δ, H₃O⁺

(f)

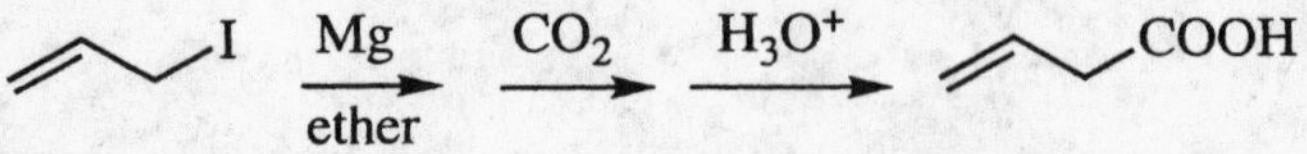

Mg
ether

CO₂

H₃O⁺

OR

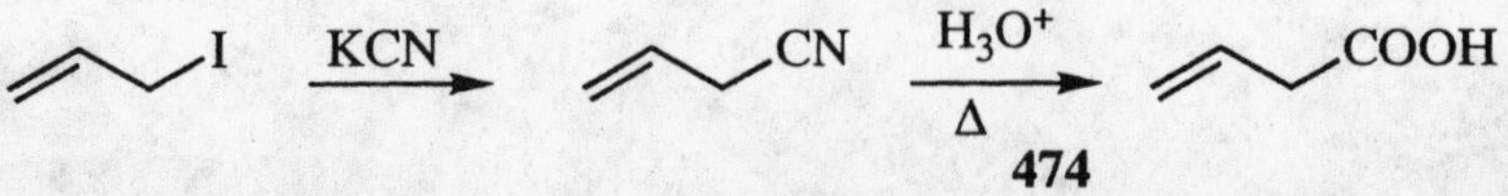

KCN

H₃O⁺
Δ

20-12 (a)

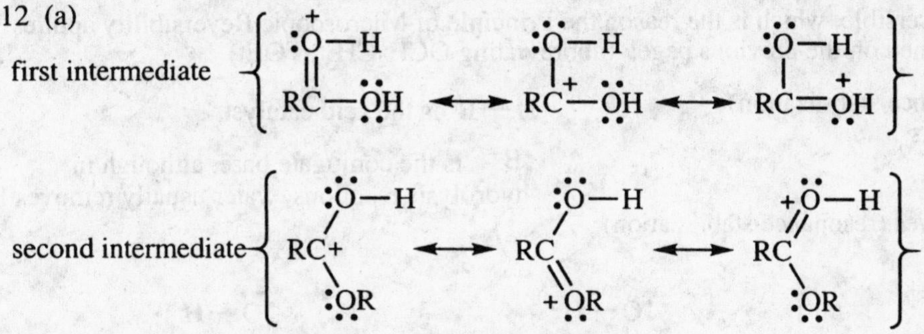

first intermediate $\left\{\begin{array}{c}\overset{+}{\underset{\|}{:\!O\!-\!H}}\\RC\!-\!\overset{..}{\underset{..}{O}}H\end{array}\longleftrightarrow\begin{array}{c}:\!\overset{..}{O}\!-\!H\\RC\overset{+}{-}\overset{..}{O}H\end{array}\longleftrightarrow\begin{array}{c}:\!\overset{..}{O}\!-\!H\\RC\!=\!\overset{+}{\underset{..}{O}}H\end{array}\right\}$

second intermediate $\left\{\begin{array}{c}:\!\overset{..}{O}\!-\!H\\RC^{+}\\:\!\overset{..}{O}R\end{array}\longleftrightarrow\begin{array}{c}:\!\overset{..}{O}\!-\!H\\RC\\\overset{+}{O}R\end{array}\longleftrightarrow\begin{array}{c}\overset{+}{:\!O}\!-\!H\\RC\\:\!\overset{..}{O}R\end{array}\right\}$

(b) The mechanism of acid-catalyzed nucleophilic acyl substitution may seem daunting, but it is simply a succession of steps that are already very familiar to you.

Typically, these mechanisms have six steps: four proton transfers (two on, two off), a nucleophilic attack, and a leaving group leaving, with a little resonance stabilization thrown in that makes the whole thing work. The six steps are labeled in the mechanism below:

Step A proton on (resonance stabilization)
Step B nucleophile attacks
Step C proton off
Step D proton on
Step E leaving group leaves (resonance stabilization)
Step F proton off

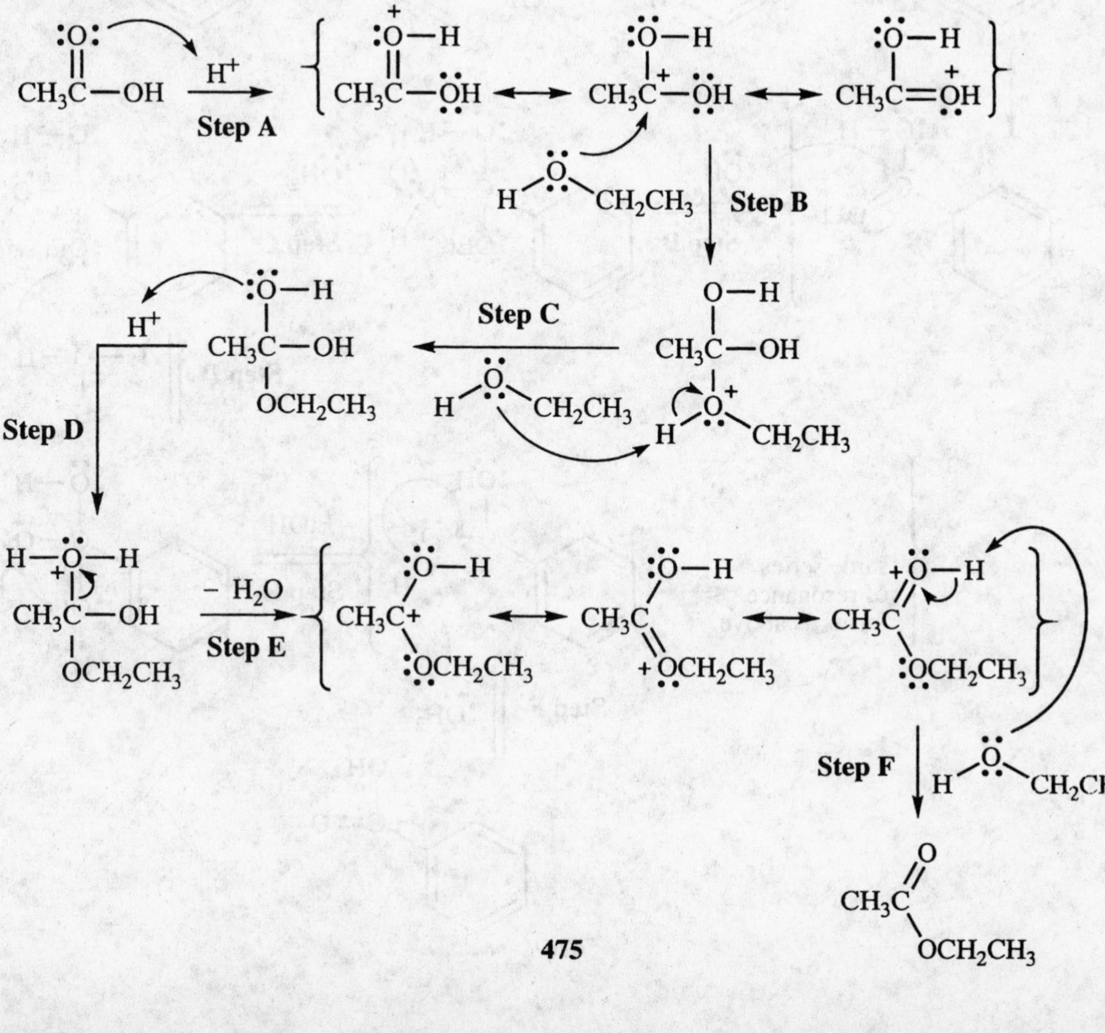

475

20-12 (c) All steps are reversible, which is the reason the Principle of Microscopic Reversibility applies. Applying the steps as outlined on the previous page: (abbreviating OCH_2CH_3 as OEt)

Step A proton on (resonance stabilization)
Step B nucleophile attacks
Step C proton off
Step D proton on
Step E leaving group leaves (resonance stabilization)
Step F proton off

H—B is the acid catalyst

:B⁻ is the conjugate base, although in hydrolysis reactions, water usually removes H⁺

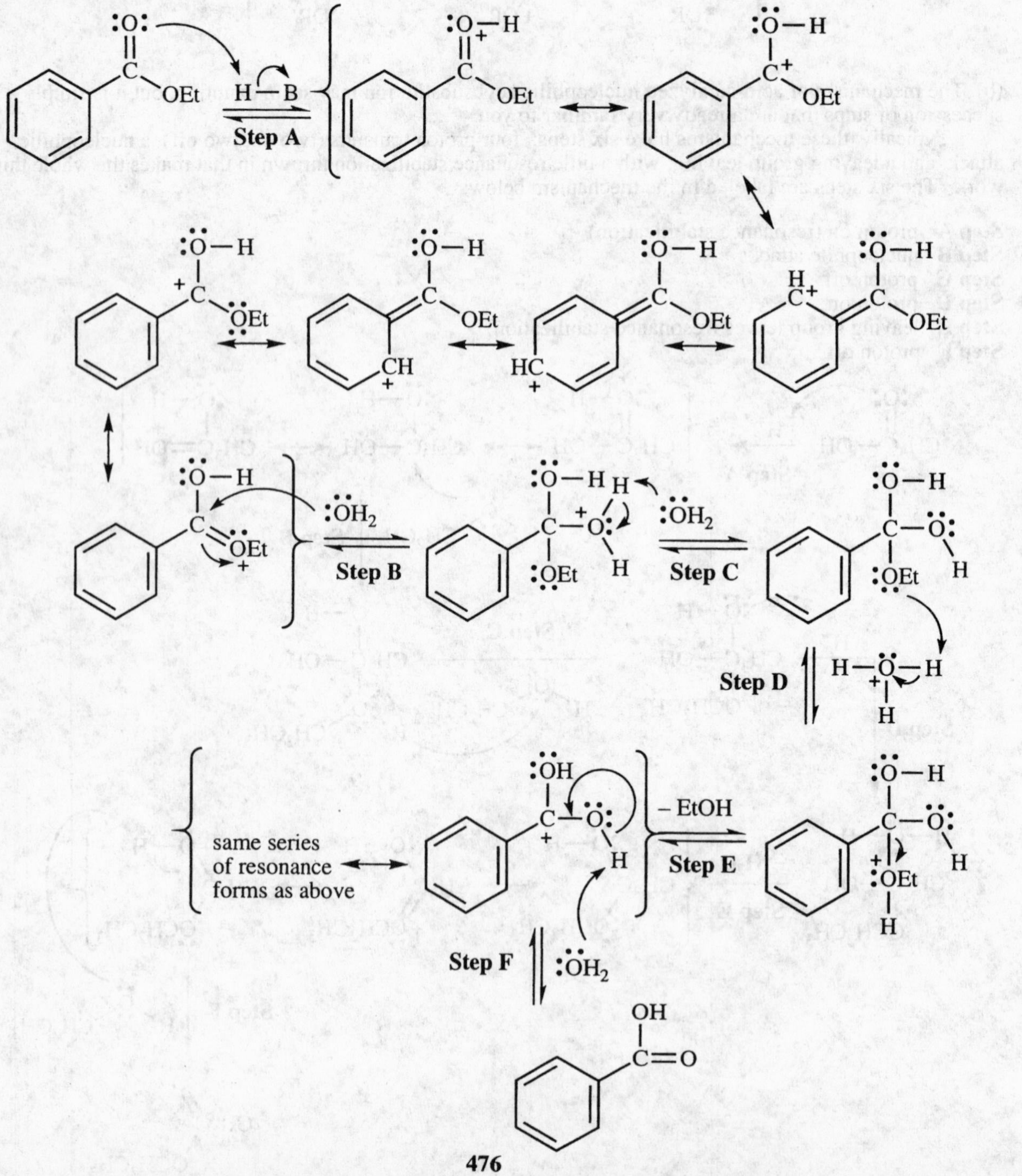

same series of resonance forms as above

− EtOH

476

20-13 For the sake of space in this problem, resonance forms will not be drawn, but remember that they are critical!

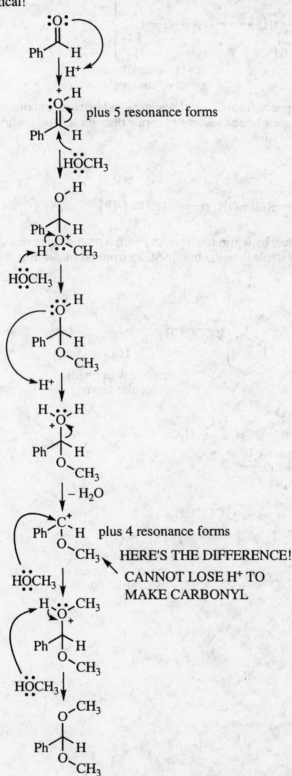

plus 5 resonance forms

plus 6 resonance forms

plus 4 resonance forms

HERE'S THE DIFFERENCE!
CANNOT LOSE H⁺ TO MAKE CARBONYL

plus 5 resonance forms

CAN LOSE H⁺ TO MAKE CARBONYL

20-14

(a)

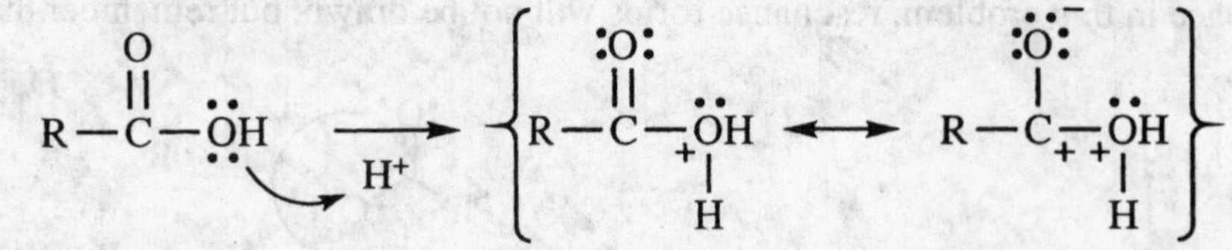
H+
BAD—two adjacent positive charges

(b) Protonation on the OH gives only two resonance forms, one of which is bad because of adjacent positive charges. Protonation on the C=O is good because of three resonance forms distributing the positive charge over three atoms, with no additional charge separation.

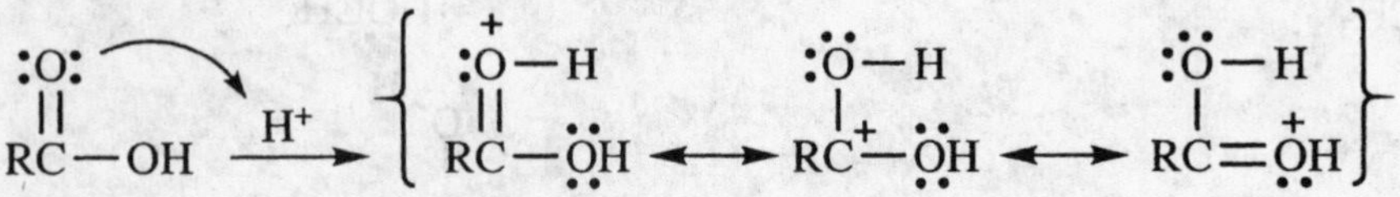
H+

(c) The carbonyl oxygen is more "basic" because, by definition, it reacts with a proton more readily. It does so because the intermediate it produces is more stable than the intermediate from protonation of the OH.

20-15

(a)

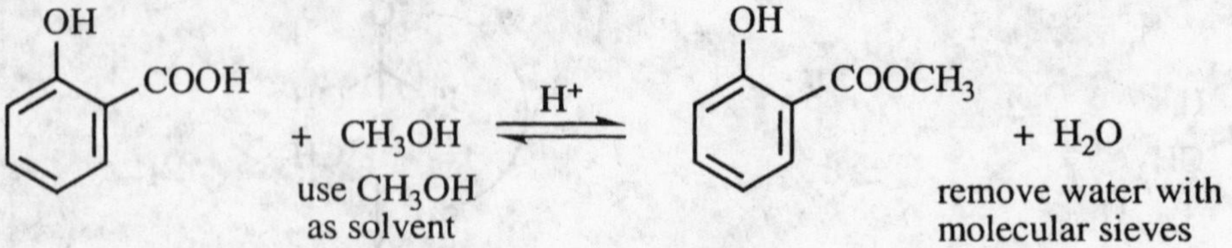
OH
COOH
+ CH3OH
H+
OH
COOCH3
+ H2O
use CH3OH
as solvent
remove water with
molecular sieves

(b)

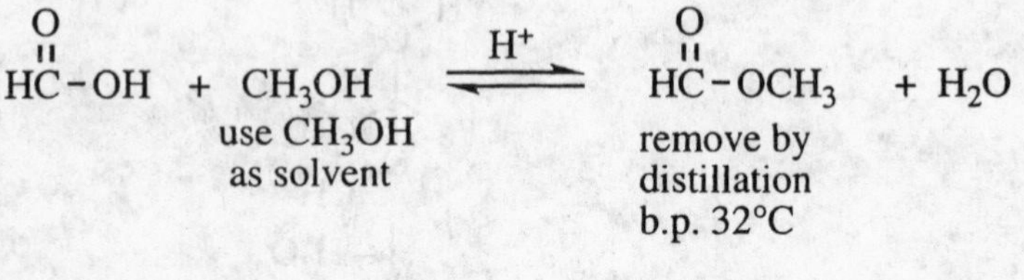
HC-OH + CH3OH
H+
HC-OCH3 + H2O
use CH3OH
as solvent
remove by
distillation
b.p. 32°C

(c)

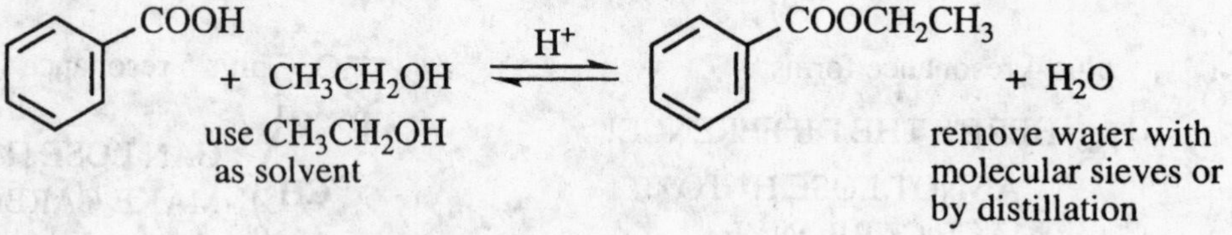
COOH
+ CH3CH2OH
H+
COOCH2CH3
+ H2O
use CH3CH2OH
as solvent
remove water with
molecular sieves or
by distillation

20-16 The asterisk ("*") denotes the ^{18}O isotope.

(a) and (b)

[reaction scheme showing protonation of acetic acid $CH_3C(=O)-OH$ by H^+ giving resonance forms of the protonated species:]

$$CH_3\overset{+}{C}(=\overset{..}{O}H)-\overset{..}{O}H \longleftrightarrow CH_3\overset{+}{C}(\overset{..}{O}H)-\overset{..}{O}H \longleftrightarrow CH_3C(\overset{..}{O}H)=\overset{+}{\overset{..}{O}}H$$

[attack by $H-\overset{..}{\overset{*}{O}}-CH_3$ methanol]

[tetrahedral intermediate $CH_3C(OH)_2(O^*CH_3)$ losing H^+]

[loss of H_2O from $CH_3C(OH)(O^*CH_3)$ giving resonance-stabilized cation]

$$CH_3\overset{+}{C}-\overset{..}{\overset{*}{O}}CH_3 \longleftrightarrow CH_3C=\overset{+}{\overset{*}{O}}CH_3 \longleftrightarrow CH_3C(\overset{+}{\overset{..}{O}}-H)-\overset{..}{\overset{*}{O}}CH_3$$

[final product:]

$$CH_3C(=O)-O^*CH_3$$

(c) The ^{18}O has two more neutrons, and therefore two more mass units, than ^{16}O. The instrument ideally suited to analyze compounds of different mass is the mass spectrometer.

20-17

(a)

first intermediate:

$$H-\overset{+}{C}(\overset{..}{O}-Et)(\overset{..}{O}-Et) \longleftrightarrow H-C(\overset{..}{O}-Et)(=\overset{+}{\overset{..}{O}}-Et) \longleftrightarrow H-C(=\overset{+}{\overset{..}{O}}-Et)(\overset{..}{O}-Et)$$

second intermediate:

$$H-\overset{+}{C}(\overset{..}{O}-H)(\overset{..}{O}-Et) \longleftrightarrow H-C(\overset{..}{O}-H)(=\overset{+}{\overset{..}{O}}-Et) \longleftrightarrow H-C(=\overset{+}{\overset{..}{O}}-H)(\overset{..}{O}-Et)$$

The more resonance forms that can be drawn to represent an intermediate, the more stable the intermediate. The more stable the intermediate, the more easily it can be formed, that is, under milder conditions. These intermediates are highly stabilized due to delocalization of the positive charge over the carbon and both oxygens. A trace of acid is all that is required to initiate this process.

20-17 continued

(b)

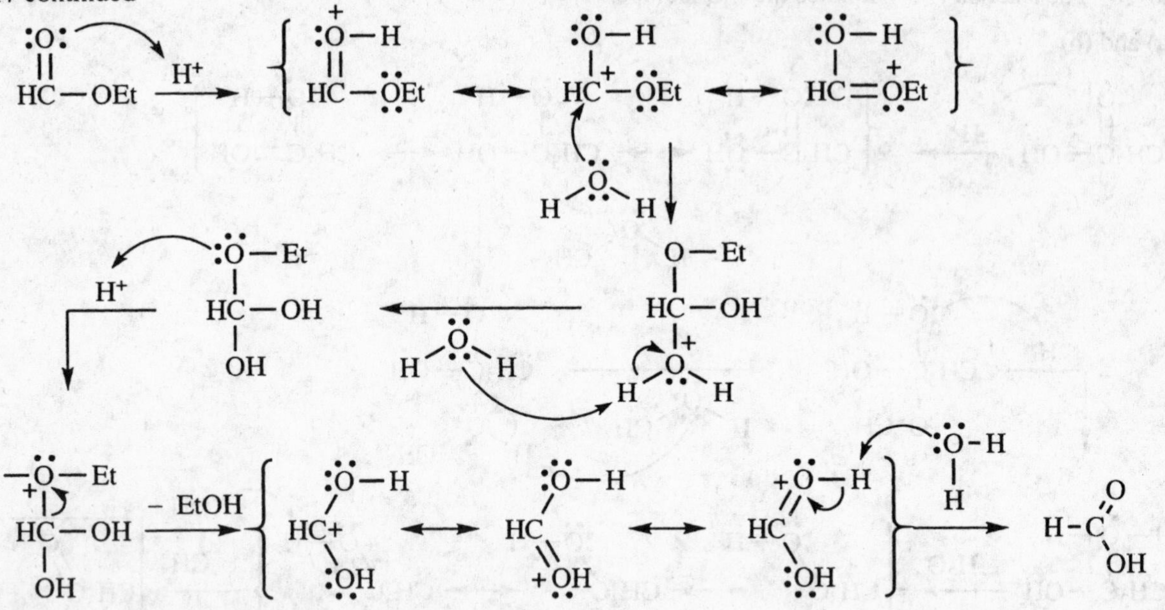

20-18

(a)

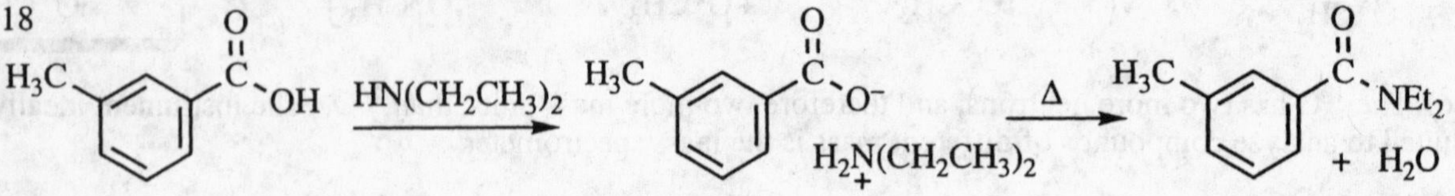

(b)

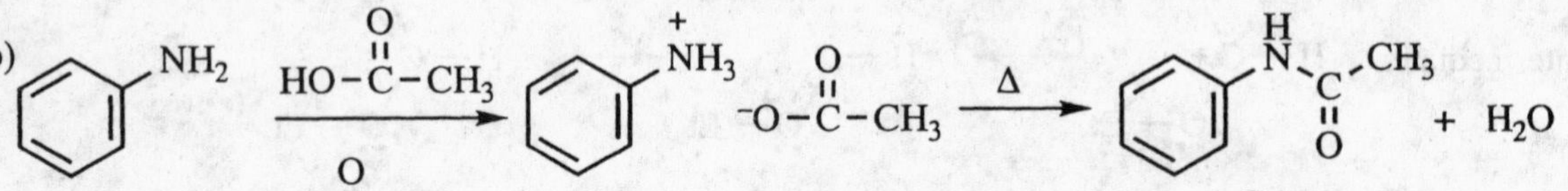

(c)

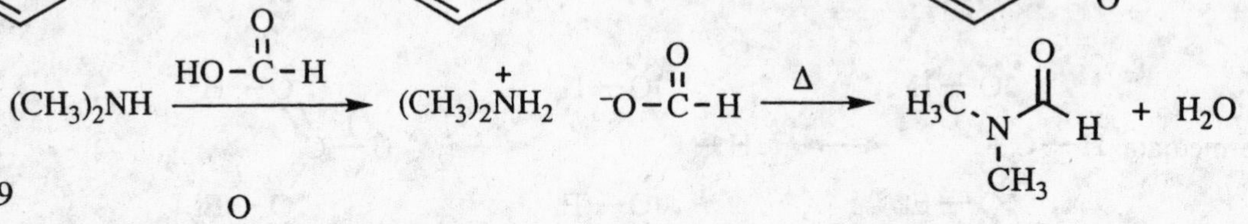

20-19

(a)

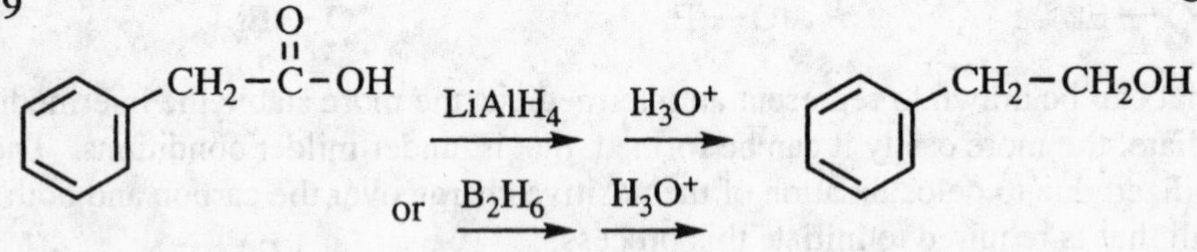

(b)

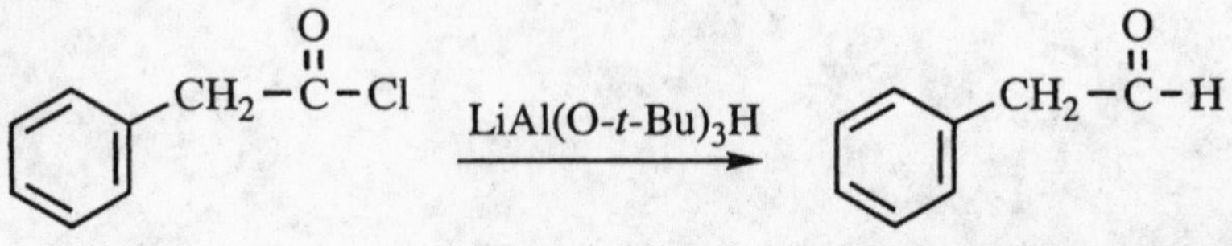

(c)

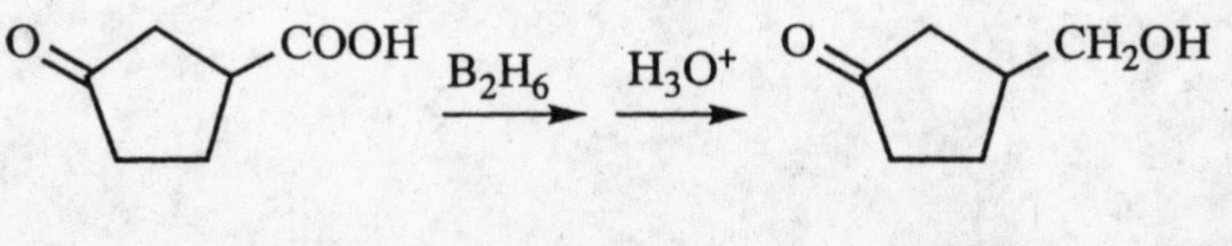

B_2H_6 selectively reduces a carboxylic acid in the presence of a ketone. Alternatively, protecting the ketone as an acetal, reducing the COOH, and removing the protecting group would also be possible but longer.

20-20

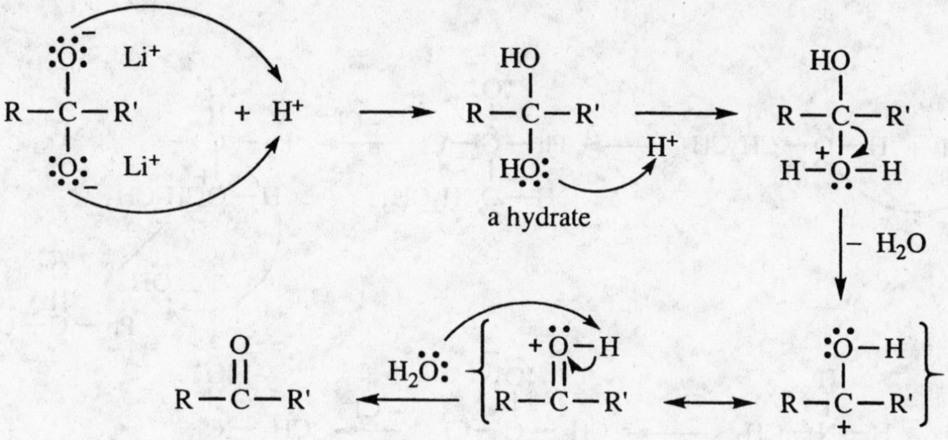

20-21

(a)

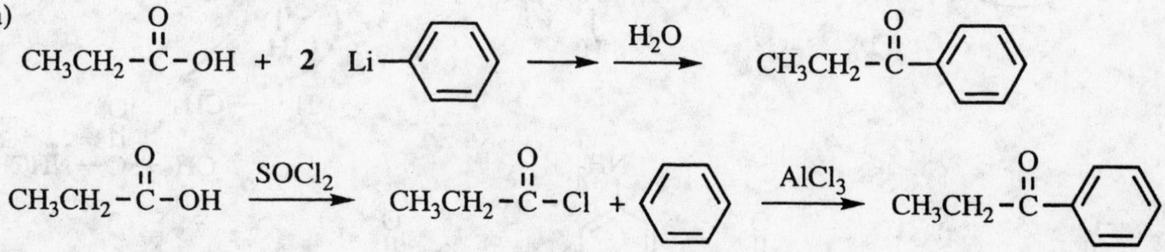

$$CH_3CH_2-\overset{\overset{\displaystyle O}{\|}}{C}-OH + 2\ Li-C_6H_5 \xrightarrow{\quad} \xrightarrow{H_2O} CH_3CH_2-\overset{\overset{\displaystyle O}{\|}}{C}-C_6H_5$$

$$CH_3CH_2-\overset{\overset{\displaystyle O}{\|}}{C}-OH \xrightarrow{SOCl_2} CH_3CH_2-\overset{\overset{\displaystyle O}{\|}}{C}-Cl + C_6H_6 \xrightarrow{AlCl_3} CH_3CH_2-\overset{\overset{\displaystyle O}{\|}}{C}-C_6H_5$$

(b)

$$C_6H_{11}-\overset{\overset{\displaystyle O}{\|}}{C}-OH + 2\ CH_3Li \xrightarrow{\quad} \xrightarrow{H_2O} C_6H_{11}-\overset{\overset{\displaystyle O}{\|}}{C}-CH_3$$

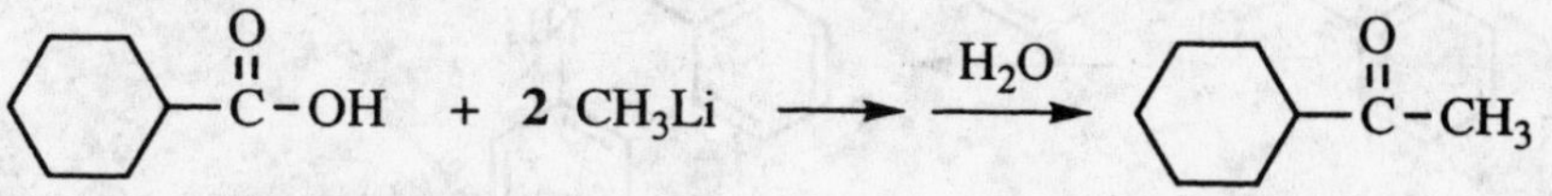

20-22

plus three other resonance forms with positive charge on the benzene ring

plus resonance forms

$$PhC(O)Cl + O=C=O + {}^-\overset{+}{C}\equiv O + Cl^-$$

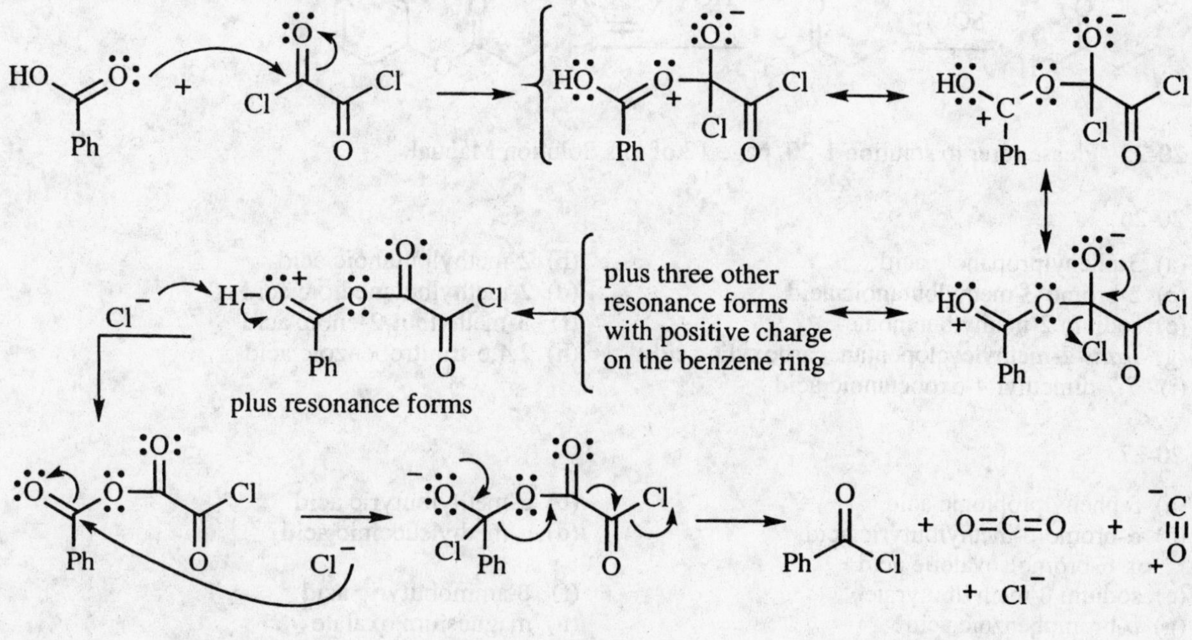

20-23

(a)

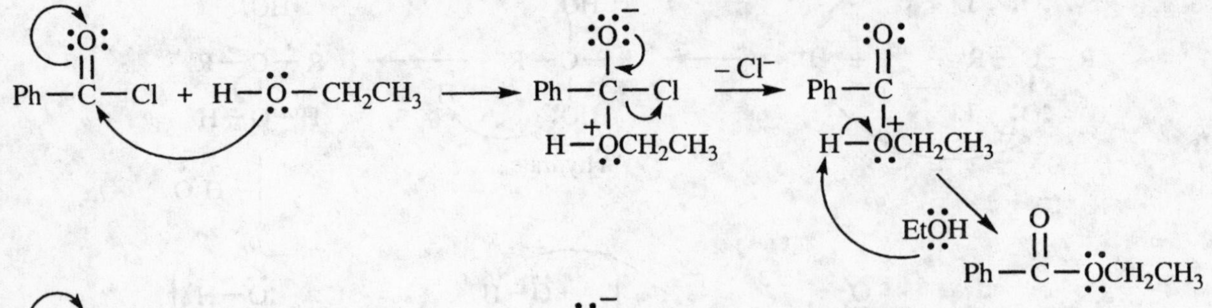

(b)

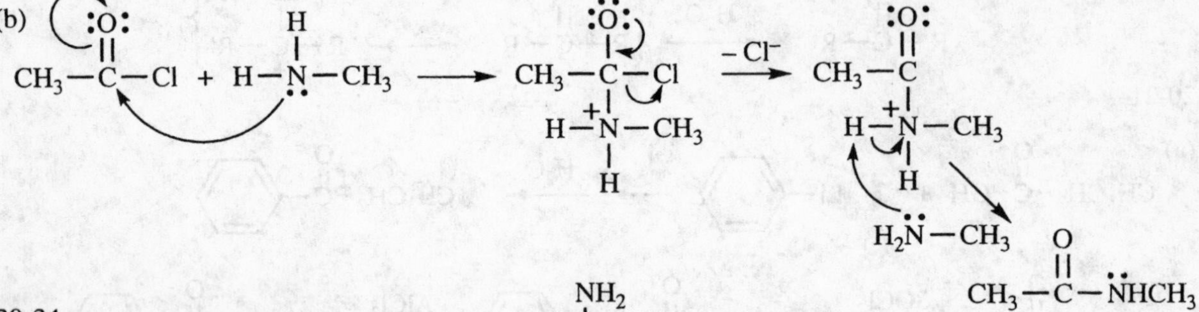

20-24

(a)

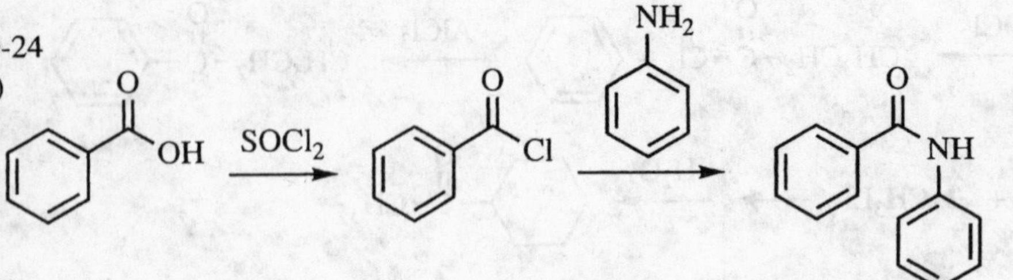

(b)

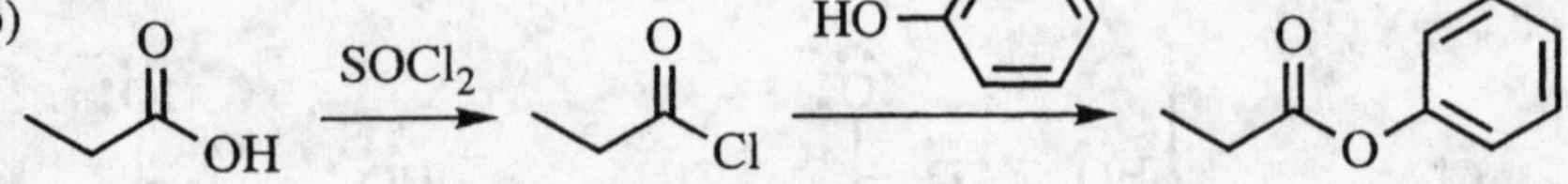

20-25   Please refer to solution 1-20, page 12 of this Solution Manual.

20-26

(a) 3-phenylpropanoic acid
(c) 2-bromo-3-methylbutanoic acid
(e) sodium 2-methylbutanoate
(g) *trans*-2-methylcyclopentanecarboxylic acid
(i) 7,7-dimethyl-4-oxooctanoic acid

(b) 2-methylbutanoic acid
(d) 2-methylbutanedioic acid
(f) 3-methylbut-2-enoic acid
(h) 2,4,6-trinitrobenzoic acid

20-27

(a) β-phenylpropionic acid
(c) α-bromo-β-methylbutyric acid,
     or α-bromoisovaleric acid
(e) sodium β-methylbutyrate
(g) *o*-bromobenzoic acid
(i) 4-methoxyphthalic acid

(b) α-methylbutyric acid
(d) α-methylsuccinic acid

(f) β-aminobutyric acid
(h) magnesium oxalate

20-28

(a)
$$CH_3-\overset{\overset{\displaystyle O}{\|}}{C}-OH$$

(b) benzene with COOH and COOH (phthalic acid)

(c)
$$\left(H-\overset{\overset{\displaystyle O}{\|}}{C}-O^- \right)_2 Mg^{2+}$$

(d)
$$HO-\overset{\overset{\displaystyle O}{\|}}{C}-CH_2-\overset{\overset{\displaystyle O}{\|}}{C}-OH$$ (malonic acid)

(e)
$$ClCH_2-\overset{\overset{\displaystyle O}{\|}}{C}-OH$$

(f)
$$CH_3-\overset{\overset{\displaystyle O}{\|}}{C}-Cl$$

(g)
$$\left(\text{long chain}-\overset{\overset{\displaystyle }{}}{C}(=O)-O^- \right)_2 Zn^{2+}$$

(h)
$$Ph-\overset{\overset{\displaystyle O}{\|}}{C}-O^- \ Na^+$$

(i)
$$FCH_2-\overset{\overset{\displaystyle O}{\|}}{C}-O^- \ Na^+$$

20-29 Weaker base listed first. (Weaker bases come from *stronger* conjugate acids.)

(a) $ClCH_2COO^- < CH_3COO^- < PhO^-$

(b) $CH_3COO^- \ Na^+ < HC\equiv C^- \ Na^+ < NaNH_2$

(c) $PhCOO^- \ Na^+ < PhO^- \ Na^+ < CH_3CH_2O^- \ Na^+$

(d) $CH_3COOH < CH_3CH_2OH < $ pyridine

20-30

(a) $CH_3COOH + NH_3 \longrightarrow CH_3COO^- \ {}^+NH_4$

(b)
phthalic acid (benzene-1,2-dicarboxylic acid) $+ \ 2 \ NaOH \longrightarrow$ disodium salt $+ \ 2 \ H_2O$

(c)
$$CH_3-\!\!\!\bigcirc\!\!\!-COOH + CF_3COO^- K^+ \longrightarrow \text{no reaction}$$

(d) $CH_3-\underset{\underset{\displaystyle Br}{|}}{C}HCOOH + CH_3CH_2COO^- \ Na^+ \longrightarrow CH_3-\underset{\underset{\displaystyle Br}{|}}{C}HCOO^- \ Na^+ + CH_3CH_2COOH$

(e)
$$Ph{-}COOH + Na^+ \ {}^-O{-}Ph \longrightarrow Ph{-}COO^- \ Na^+ + HO{-}Ph$$

20-31

$$CH_3\overset{\overset{\displaystyle O}{\|}}{C}-OCH_2CH_3 \quad < \quad \text{CH}_2=CH-O-CH_2CH_2OH \quad < \quad CH_3CH_2CH_2-\overset{\overset{\displaystyle O}{\|}}{C}-OH$$

lowest b.p. (77°C)   (b.p. 143°C)   highest b.p. (162°C)

The ester cannot form hydrogen bonds and will be the lowest boiling. The alcohol can form hydrogen bonds. The carboxylic acid forms two hydrogen bonds and boils as the dimer, the highest boiling among these three compounds.

20-32 Listed in order of increasing acidity (weakest acid first):

(a) ethanol < phenol < acetic acid
(b) acetic acid < chloroacetic acid < *p*-toluenesulfonic acid
(c) benzoic acid < *m*-nitrobenzoic acid < *o*-nitrobenzoic acid
(d) butyric acid < β-bromobutyric acid < α-bromobutyric acid

EWG = electron-withdrawing group

Acidity increases with:
1. closer proximity of EWG
2. great number of EWG
3. increasing strength (electronegativity) of EWG

(e) Br-cyclopentane-COOH < Cl-cyclopentane-COOH < F-cyclopentane-COOH

483

20-33 Acetic acid derivatives are often used as a test of electronic effects of a series of substituents: they are fairly easily synthesized (or are commercially available), and pK_a values are easily measured by titration.

Substituents on carbon-2 of acetic acid can express only an inductive effect; no resonance effect is possible because the CH_2 is sp^3 hybridized and no pi overlap is possible.

Two conclusions can be drawn from the given pK_a values. First, all four substituents are electron-withdrawing because all four substituted acids are stronger than acetic acid. Second, the magnitude of the electron-withdrawing effect increases in the order: $OH < Cl < CN < NO_2$. (It is always a safe assumption that nitro is the strongest electron-withdrawing group of all the common substituents.)

20-34

(a) Ascorbic acid is not a carboxylic acid. It is an example of a structure called an ene-diol where one of the OH groups is unusually acidic because of the adjacent carbonyl group. See part (c).

(b) Ascorbic acid, pK_a 4.71, is almost identical to the acidity of acetic acid, pK_a 4.74.

(c) The more acidic H will be the one that, when removed, gives the more stable conjugate base. *start here*

THREE resonance forms

more acidic

only two resonance forms

(d) In the slightly basic pH of physiological fluid, ascorbic acid is present as the conjugate base, ascorbate, whose structure can be represented by any of the three resonance forms on the left of part (c).

20-35

(a)

(b) CH_2COOH

(c)

(d) **2** $COOH$

(e) $COOH$

(f) $CH_3CH_2-\overset{\overset{\displaystyle Ph}{|}}{C}HCH_2OH$

(g)

(h) $COOH$ / $COOH$

(i)

(j)

(k)

484

20-36

(a)

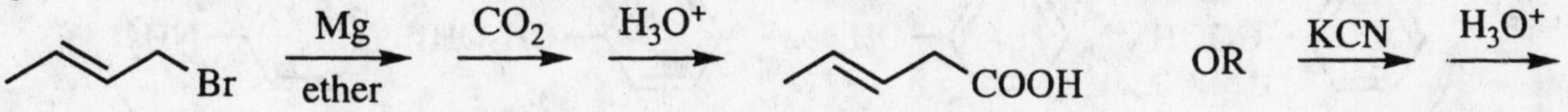

$CH_2=CH-CH_2-Br$ $\xrightarrow[\text{ether}]{\text{Mg}}$ $\xrightarrow{CO_2}$ $\xrightarrow{H_3O^+}$ ...COOH   OR $\xrightarrow{KCN}$ $\xrightarrow{H_3O^+}$

(b)

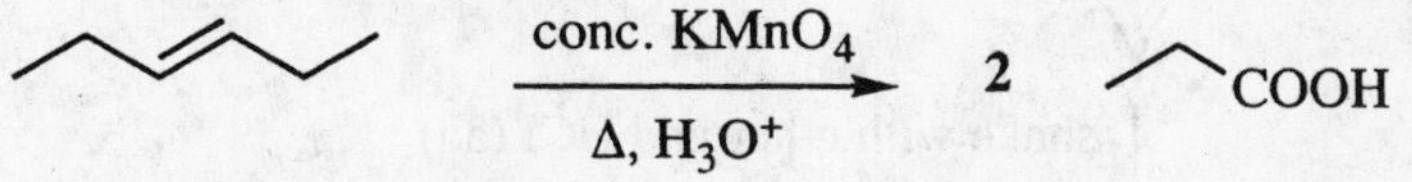

$\xrightarrow[\Delta,\ H_3O^+]{\text{conc. KMnO}_4}$ **2** ...COOH

(c)

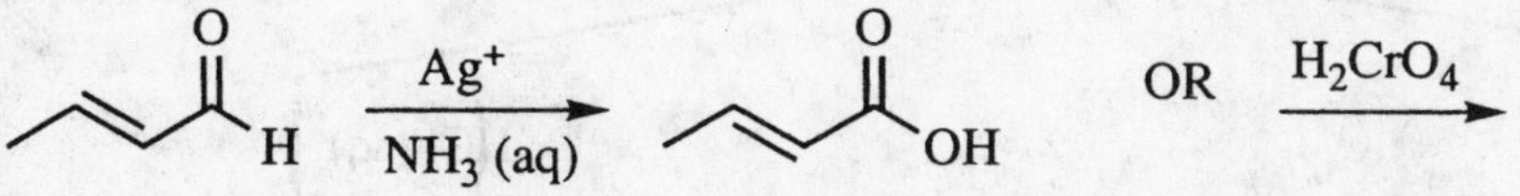

$\xrightarrow[\text{NH}_3\ (aq)]{\text{Ag}^+}$ ...OH   OR $\xrightarrow{H_2CrO_4}$

(d)

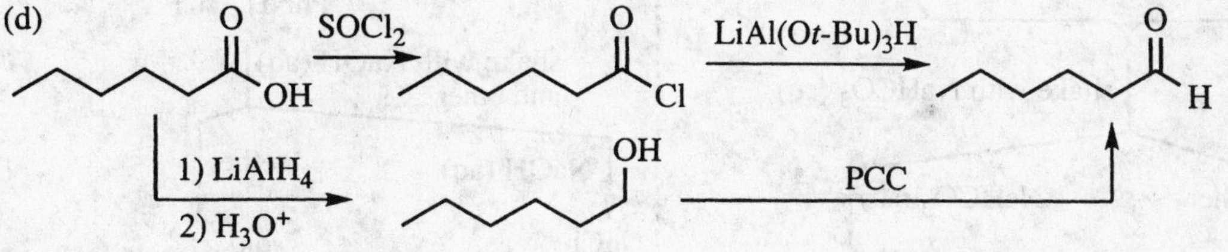

...OH $\xrightarrow{SOCl_2}$ ...Cl $\xrightarrow{\text{LiAl(O}t\text{-Bu)}_3\text{H}}$ ...H

$\xrightarrow[\text{2) }H_3O^+]{\text{1) LiAlH}_4}$ ...OH $\xrightarrow{\text{PCC}}$

(e)

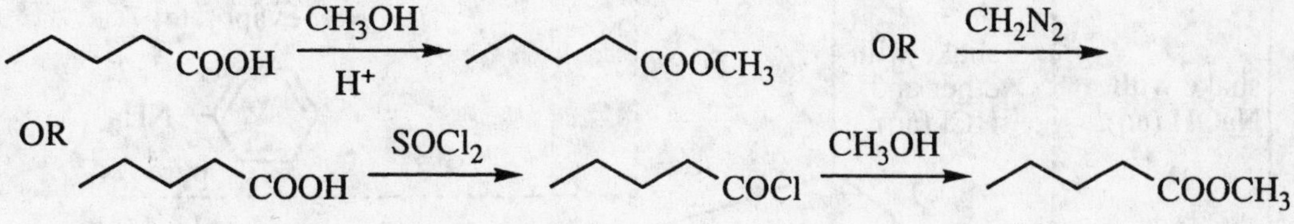

...COOH $\xrightarrow[\text{H}^+]{\text{CH}_3\text{OH}}$ ...COOCH$_3$ OR $\xrightarrow{\text{CH}_2\text{N}_2}$

OR   ...COOH $\xrightarrow{SOCl_2}$ ...COCl $\xrightarrow{\text{CH}_3\text{OH}}$ ...COOCH$_3$

(f)

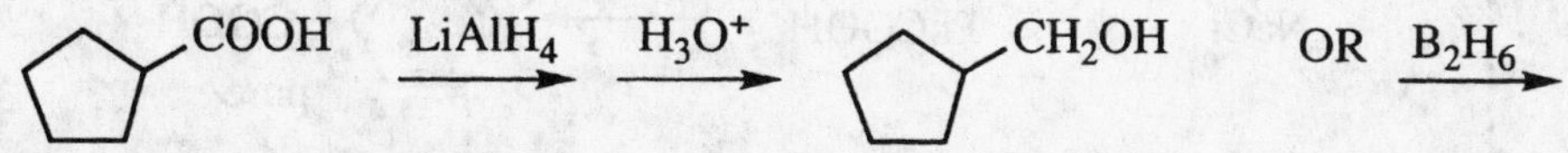

...COOH $\xrightarrow{\text{LiAlH}_4}$ $\xrightarrow{H_3O^+}$ ...CH$_2$OH   OR $\xrightarrow{B_2H_6}$

(g)

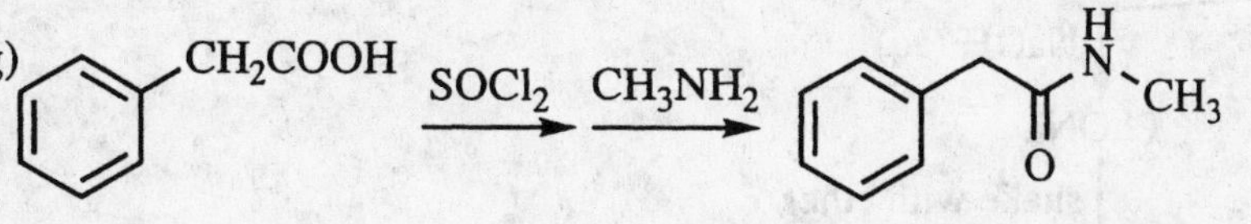

CH$_2$COOH $\xrightarrow{SOCl_2}$ $\xrightarrow{\text{CH}_3\text{NH}_2}$ ...

(h)

(Diels-Alder) $\xrightarrow[\text{Pt}]{H_2}$

485

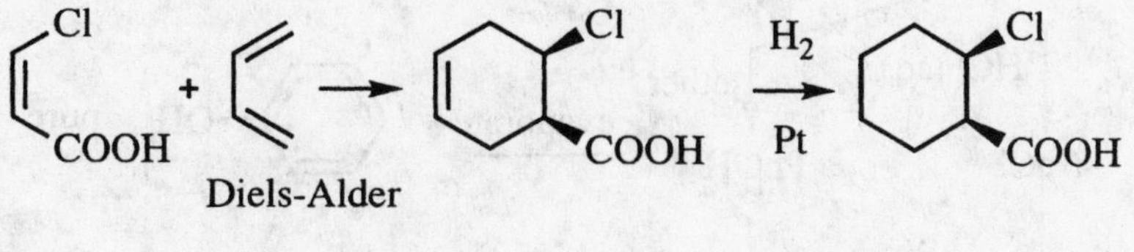

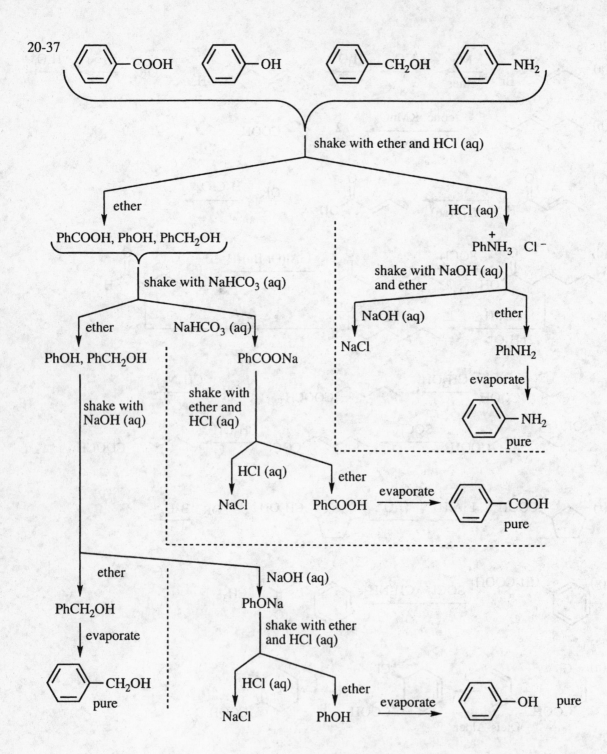

20-38

(a)

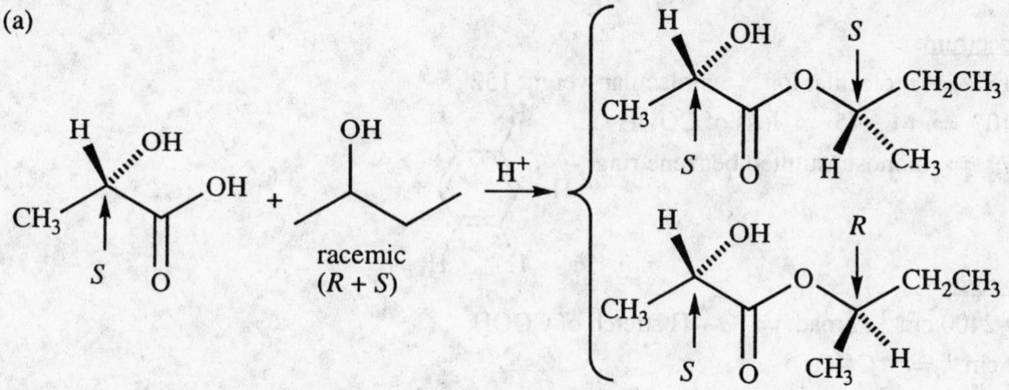

H
OH
OH
CH3
S
O
OH
+
racemic
(R + S)
H+
H
OH
S
O
CH2CH3
CH3
S
O
H
CH3
H
OH
R
O
CH2CH3
CH3
S
O
CH3
H

(b) Isomers which are *R,S* and *S,S* are diastereomers.

20-39

(a)

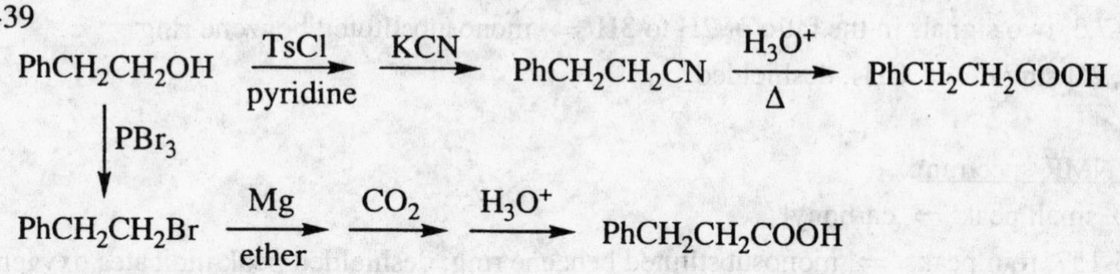

PhCH2CH2OH
TsCl
pyridine
KCN
PhCH2CH2CN
H3O+
Δ
PhCH2CH2COOH
PBr3
PhCH2CH2Br
Mg
ether
CO2
H3O+
PhCH2CH2COOH

(b)

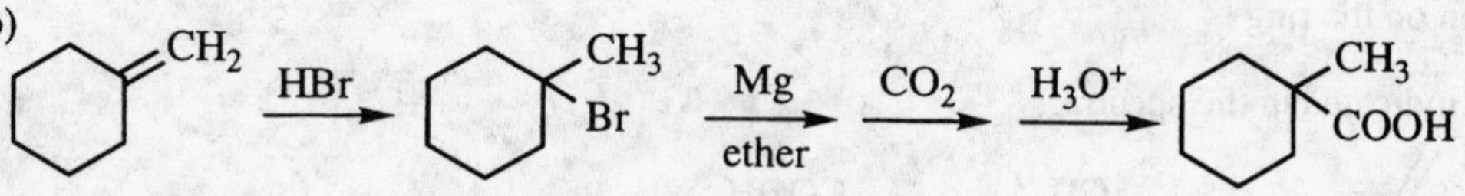

CH2
HBr
CH3
Br
Mg
ether
CO2
H3O+
CH3
COOH

(c)

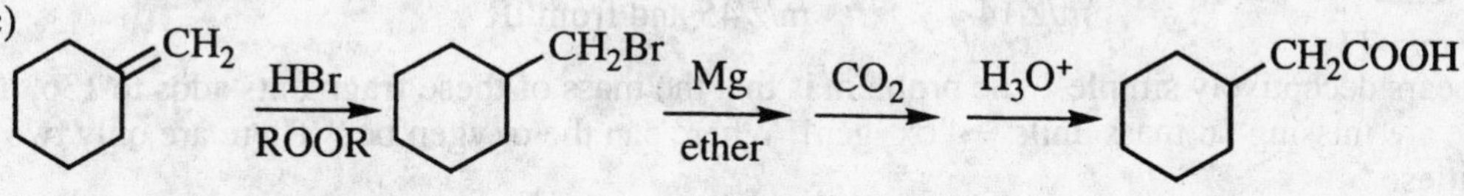

CH2
HBr
ROOR
CH2Br
Mg
ether
CO2
H3O+
CH2COOH

(d)

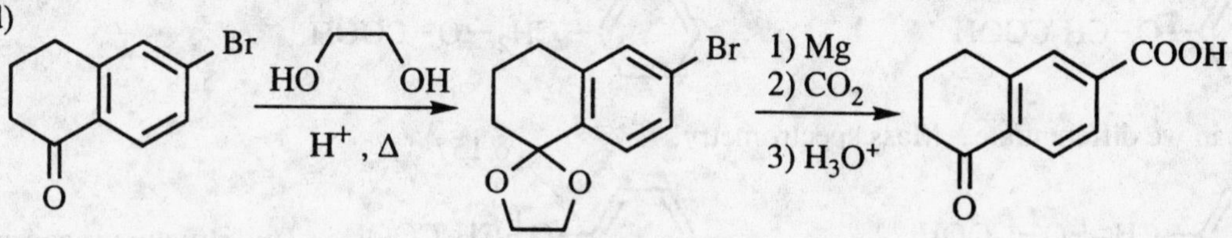

Br
O
HO
OH
H+, Δ
Br
O
O
1) Mg
2) CO2
3) H3O+
COOH
O

(e)

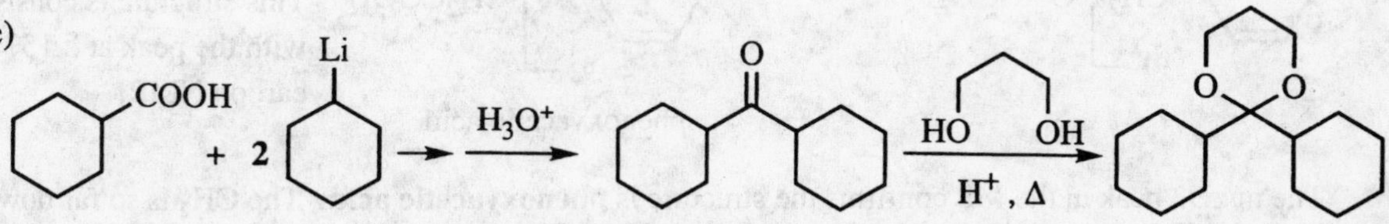

COOH
+ 2
Li
H3O+
O
HO
OH
H+, Δ
O
O

20-40

(a) Mass spectrum:

—m/z 152 $\Rightarrow$ molecular ion $\Rightarrow$ molecular weight 152

—m/z 107 $\Rightarrow$ M – 45 $\Rightarrow$ loss of COOH

—m/z 77 $\Rightarrow$ monosubstituted benzene ring,

IR spectrum:

—3400-2400 cm^{-1}, broad $\Rightarrow$ O—H stretch of COOH

—1700 cm^{-1} $\Rightarrow$ C=O

—1240 cm^{-1} $\Rightarrow$ C–O

—1600 cm^{-1} $\Rightarrow$ aromatic C=C

NMR spectrum:

—δ 6.8-7.3, two signals in the ratio of 2H to 3H $\Rightarrow$ monosubstituted benzene ring

—δ 4.6, 2H singlet $\Rightarrow$ CH$_2$, deshielded

Carbon NMR spectrum:

—δ 170, small peak $\Rightarrow$ carbonyl

—δ 115-157, four peaks $\Rightarrow$ monosubstituted benzene ring; deshielded peak indicates oxygen substitution on the ring

(b) Fragments indicated in the spectra:

CH$_2$ COOH

m/z 77   m/z 14   m/z 45 and from IR

This appears deceptively simple. The problem is that the mass of these fragments adds to 136, not 152—we are missing 16 mass units $\Rightarrow$ oxygen! Where can the oxygen be? There are only two possibilities:

How can we differentiate? Mass spectrometry!

phenoxyacetic acid

This structure is consistent with the peak at δ 157 in the carbon NMR.

The m/z 93 peak in the MS confirms the structure is **phenoxyacetic acid**. The CH$_2$ is so far downfield in the NMR because it is between two electron-withdrawing groups, the O and the COOH.

(c) The COOH proton is missing from the proton NMR. Either it is beyond 10 and the NMR was not scanned (unlikely), or the peak was broadened beyond detection because of hydrogen bonding with DMSO.

20-41

(a)

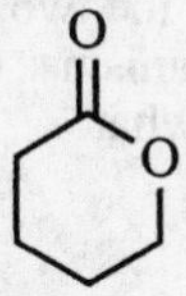

δ-valerolactone

(b)

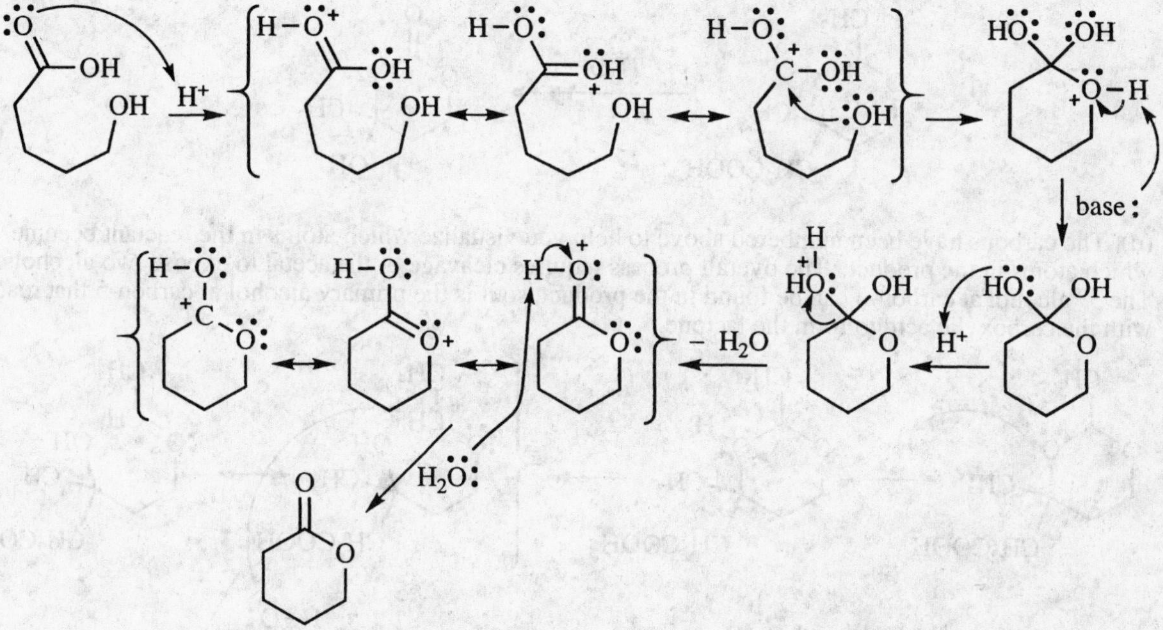

base:

20-42

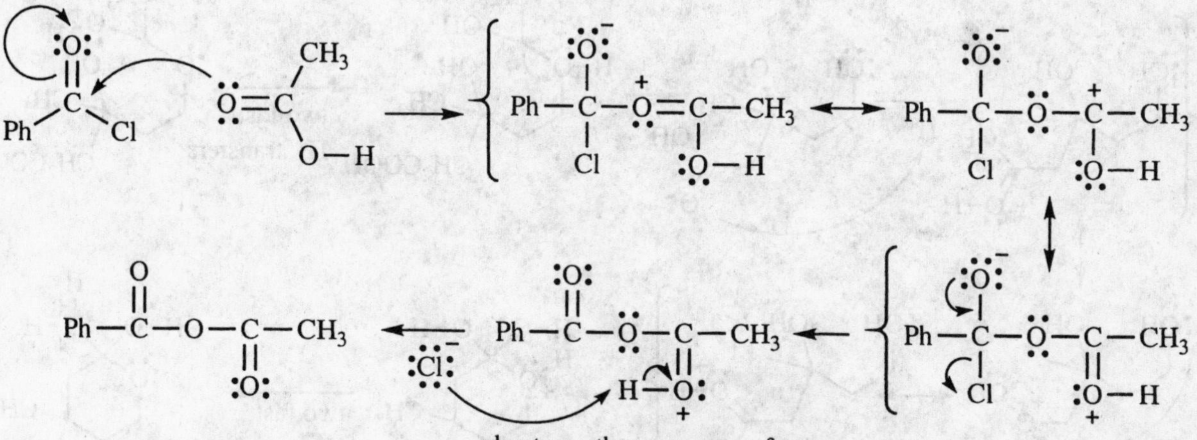

plus two other resonance forms

20-43

(a)

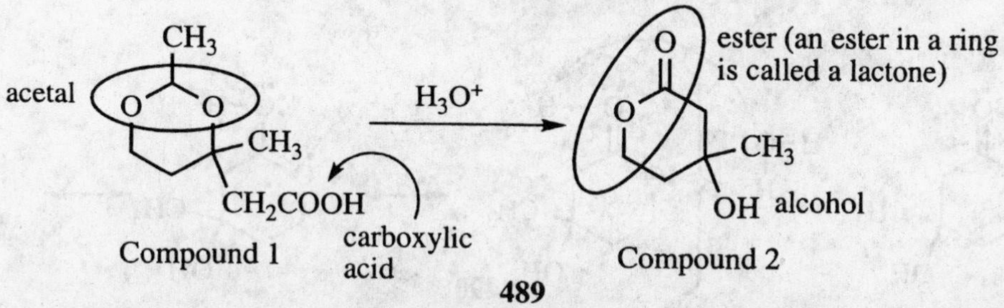

acetal

H_3O^+

ester (an ester in a ring is called a lactone)

Compound 1

carboxylic acid

CH$_2$COOH

CH$_3$

alcohol
OH

Compound 2

20-43 continued

(b) Compound 1 has 8 carbons, and Compound 2 has 6 carbons. Two carbons have been lost: the two carbons of the acetal have been cleaved. (This is the best way to figure out reactions and mechanisms: find out which atoms of the reactant have become which atoms of the product, then determine what bonds have been broken and formed.)

(c) Acetals are stable to base, so the acetal must have been cleaved when acid was added.

(d) The carbons have been numbered above to help you visualize which atoms in the reactant become which atoms in the product. The overall process requires cleavage of the acetal to expose two alcohols. The 3° alcohol at carbon-3 can be found in the product, so it is the primary alcohol at carbon-5 that reacts with the carboxylic acid to form the lactone.

20-44 (A more complete discussion of acidity and electronic effects can be found is Appendix 2.) A few words about the two types of electronic effects: induction and resonance. Inductive effects are a result of polarized σ bonds, usually because of electronegative atom substituents. Resonance effects work through π systems, requiring overlap of p orbitals to delocalize electrons.

All substituents have an inductive effect compared to hydrogen (the reference). Many groups also have a resonance effect; all that is required to have a resonance effect is that the atom or group have at least one p orbital for overlap.

The most interesting groups have both inductive and resonance effects. In such groups, how can we tell the direction of electron movement, that is, whether a group is electron-donating or electron-withdrawing? And do the resonance and inductive effects reinforce or conflict with each other? We can never "turn off" an inductive effect from a resonance effect; that is, any time a substituent is expressing its resonance effect, it is also expressing its inductive effect. We can minimize a group's inductive effect by moving it farther away; inductive effects decrease with distance. The other side of the coin is more accessible to the experimenter: we can "turn off" a resonance effect in order to isolate an inductive effect. We can do this by interrupting a conjugated π system by inserting an sp^3-hybridized atom, or by making resonance overlap impossible for steric reasons (steric inhibition of resonance).

These three problems are examples of separating inductive effects from resonance effects.

(a) and (b) In electrophilic aromatic substitution, the phenyl substituent is an ortho,para-director because it can stabilize the intermediate from electrophilic attack at the ortho and para positions. The phenyl substituent is electron-donating *by resonance*.

plus other resonance forms

BUT:

is a stronger acid than

The greater acidity of phenylacetic acid shows that the phenyl substituent is electron-withdrawing, thereby stabilizing the product carboxylate's negative charge. Does this contradict what was said above? Yes and no. What is different is that, since there is no p-orbital overlap between the phenyl group and the carboxyl group because of the CH_2 group in between, the increased acidity must be from a pure *inductive effect*. This structure isolates the inductive effect (which can't be "turned off") from the resonance effect of the phenyl group.

We can conclude three things: (1) phenyl is electron-withdrawing by induction; (2) phenyl is (in this case) electron-donating by resonance; (3) for phenyl, the resonance effect is stronger than the inductive effect (since it is an ortho,para-director).

20-44 continued

(c) The simpler case first—induction only:

$$CH_3O-CH_2-\overset{\overset{\displaystyle O}{\parallel}}{C}-O-H \quad \text{is a stronger acid than} \quad H-CH_2-\overset{\overset{\displaystyle O}{\parallel}}{C}-O-H$$

There is no resonance overlap between the methoxy group and the carboxyl group, so this is a pure inductive effect. The methoxy substituent increases the acidity, so methoxy must be electron-withdrawing by induction. This should come as no surprise as oxygen is the second most electronegative element.

The anomaly comes in the decreased acidity of 4-methoxybenzoic acid:

CH₃O—⟨benzene ring⟩—C(=O)—O—H   is a weaker acid than   H—⟨benzene ring⟩—C(=O)—O—H

Through resonance, a pair of electrons from the methoxy oxygen can be donated through the benzene ring to the carboxyl group—a stabilizing effect. However, this electron donation *destabilizes* the carboxylate anion as there is already a negative charge on the carboxyl group; the resonance donation intensifies the negative charge. Since the product of the equilibrium would be destabilized relative to the starting material, the proton donation would be less favorable, which we define as a weaker acid.

CH₃Ö—⟨benzene ring⟩—C(=O)—Ö—H ⟷ CH₃Ö⁺=⟨ring⟩=C(—Ö⁻)—Ö—H

Methoxy is another example of a group which is electron-withdrawing by induction but electron-donating by resonance.

(d) This problem gives three pieces of data to interpret:

(1) $$CH_3-CH_2-\overset{\overset{\displaystyle O}{\parallel}}{C}-O-H \quad \text{is a weaker acid than} \quad H-CH_2-\overset{\overset{\displaystyle O}{\parallel}}{C}-O-H$$

Interpretation: the methyl group is electron-donating by induction.

(2)

⟨benzene ring with CH₃ at top and CH₃ at bottom⟩—O—H   is a weaker acid than   ⟨benzene ring⟩—O—H

Interpretation: the methyl group is electron-donating by induction. This interpretation is consistent with (1), as expected, since methyl cannot have any resonance effect.

(3)

⟨benzene ring with CH₃ at top and CH₃ at bottom⟩—C(=O)—O—H   is a **stronger** acid than   ⟨benzene ring⟩—C(=O)—O—H

Interpretation: this is the anomaly. Contradictory to the data in (1) and (2), by putting on two methyl groups, the substituent seems to have become electron-withdrawing instead of electron-donating. How?

Quick! Turn the page!

Steric inhibition of resonance! In benzoic acid, the phenyl ring and the carboxyl group are all in the same plane, and benzene is able to donate electrons by resonance overlap through parallel p orbitals. This stabilizes the starting acid (and destabilizes the carboxylate anion) and makes the acid weaker than it would be without resonance.

plus other resonance forms

Putting substituents at the 2- and 6-positions prevents the carboxyl or carboxylate from coplanarity with the ring. Resonance is interrupted, and now the carboxyl group sees a phenyl substituent which cannot stabilize the acid through resonance; the stabilization of the acid is lost. At the same time, the *electron-withdrawing inductive effect* of the benzene ring stabilizes the carboxylate anion. These two effects work together to make this acid unusually strong. (Apparently, the slight electron-donating inductive effect of the methyls is overpowered by the stronger electron-withdrawing inductive effect of the benzene ring.)

COOH group is perpendicular to the plane of the benzene ring— no resonance interaction.

this three-dimensional view down the C-C bond between the COOH and the benzene ring shows that COOH is twisted out of the benzene plane

20-45

(a)

stock bottle   students' samples

(b) The spectrum of the students' samples shows the carboxylic acid present. Contact with oxygen from the air oxidized the sensitive aldehyde group to the acid.

(c) Storing the aldehyde in an inert atmosphere like nitrogen or argon prevents oxidation. Freshly prepared unknowns will avoid the problem.

20-46

20-47 Products are boxed.

(a) All steps are reversible in an acid-catalyzed ester hydrolysis. (abbreviating OCH_2CH_3 as OEt)

Step A proton on (resonance stabilization)
Step B nucleophile attacks
Step C proton off
Step D proton on
Step E leaving group leaves (resonance stabilization)
Step F proton off

H — B is the acid catalyst

:B⁻ is the conjugate base, although in hydrolysis reactions, water usually removes H⁺

20-47 continued

(b)

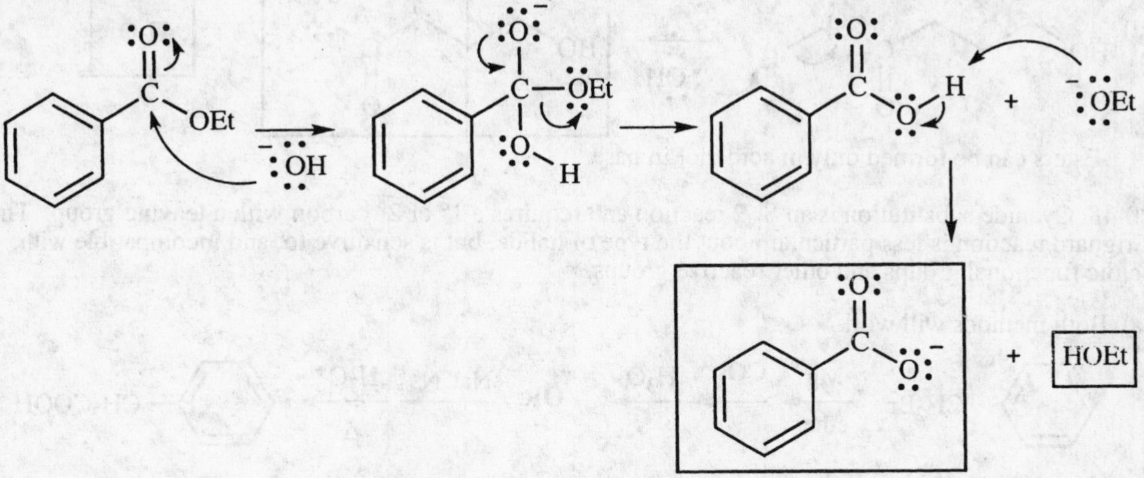

(c)

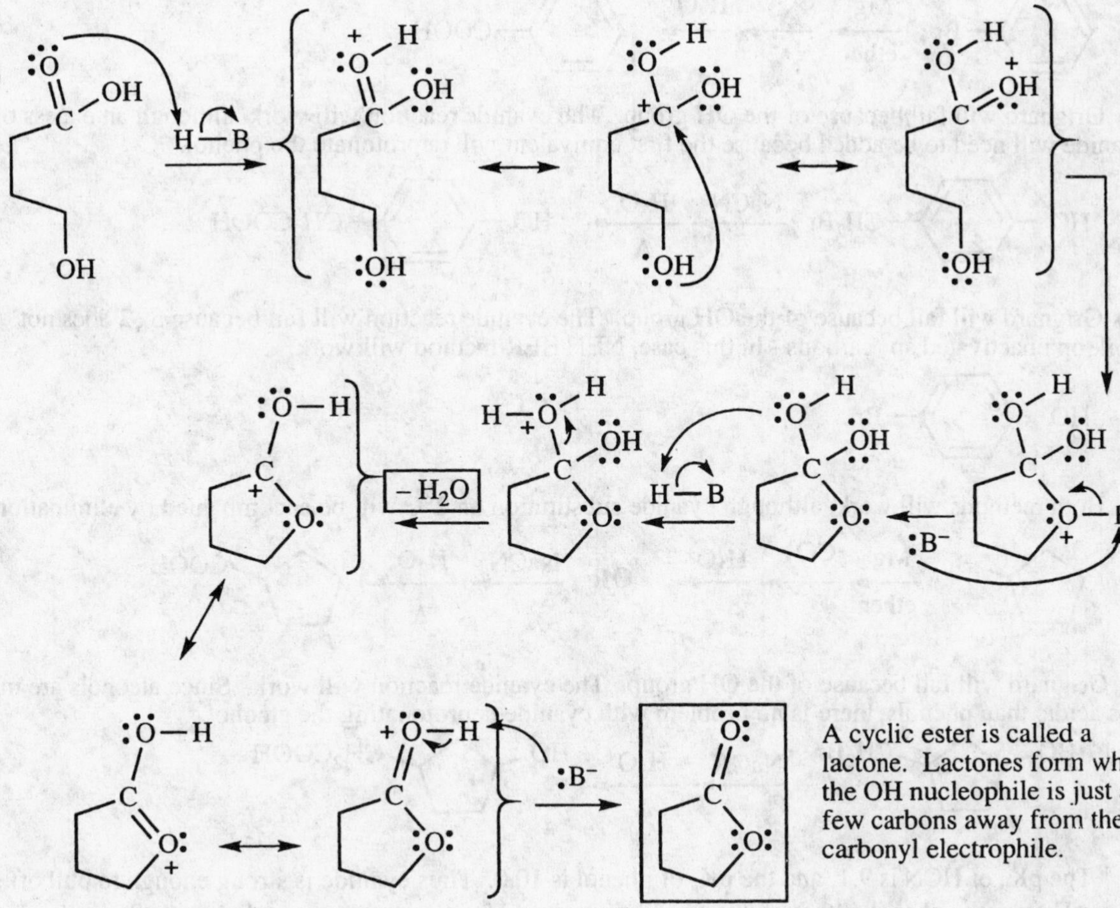

A cyclic ester is called a lactone. Lactones form when the OH nucleophile is just a few carbons away from the carbonyl electrophile.

20-47 continued

(d)

HO—CH₂CH₂CH₂—C(=O)—O—C—O—H with base ⁻:OH attacking, yielding HO—CH₂CH₂CH₂—C(=O)—O:⁻ + H₂O

Esters can be formed only in acid, not in base.

20-48 Cyanide substitution is an S$_N$2 reaction and requires a 1° or 2° carbon with a leaving group. The Grignard reaction is less particular about the type of halide, but is sensitive to, and incompatible with, acidic functional groups and other reactive groups.

(a) Both methods will work.

$$\text{Ph–CH}_2\text{Br} \xrightarrow[\text{ether}]{\text{Mg}} \xrightarrow{\text{CO}_2} \xrightarrow{\text{H}_3\text{O}^+} \quad \textbf{OR} \quad \xrightarrow{\text{NaCN}} \xrightarrow[\Delta]{\text{H}_3\text{O}^+} \text{Ph–CH}_2\text{COOH}$$

(b) Only Grignard will work. The S$_N$2 reaction does not work on unactivated benzene rings.

$$\text{Ph–Br} \xrightarrow[\text{ether}]{\text{Mg}} \xrightarrow{\text{CO}_2} \xrightarrow{\text{H}_3\text{O}^+} \text{Ph–COOH}$$

(c) Grignard will fail because of the OH group. The cyanide reaction will work, although an excess of cyanide will need to be added because the first equivalent will deprotonate the phenol.*

$$\text{HO–C}_6\text{H}_4\text{–CH}_2\text{Br} \xrightarrow{\text{NaCN}} \xrightarrow[\Delta]{\text{H}_3\text{O}^+} \text{HO–C}_6\text{H}_4\text{–CH}_2\text{COOH}$$

(d) Grignard will fail because of the OH group. The cyanide reaction will fail because S$_N$2 does not work on unactivated sp² carbons. In this case, NEITHER method will work.

HO–C₆H₄–Br

(e) Both methods will work, although cyanide substitution on 2°C will be accompanied by elimination.

$$\text{cyclopentyl–Br} \xrightarrow[\text{ether}]{\text{Mg}} \xrightarrow{\text{CO}_2} \xrightarrow{\text{H}_3\text{O}^+} \quad \textbf{OR} \quad \xrightarrow{\text{NaCN}} \xrightarrow[\Delta]{\text{H}_3\text{O}^+} \text{cyclopentyl–COOH}$$

(f) Grignard will fail because of the OH group. The cyanide reaction will work. Since alcohols are much less acidic than phenols, there is no problem with cyanide deprotonating the alcohol.*

$$\text{HO–cyclopentyl–CH}_2\text{Br} \xrightarrow{\text{NaCN}} \xrightarrow[\Delta]{\text{H}_3\text{O}^+} \text{HO–cyclopentyl–CH}_2\text{COOH}$$

* The pK_a of HCN is 9.1, and the pK_a of phenol is 10.0. Thus cyanide is strong enough to pull off some of the H from the phenol, although the equilibrium would favor cyanide ion and phenol. The pK_a of secondary alcohols is 16-18, so there is no chance that cyanide would deprotonate an alcohol.

$$^-\text{CN} + \text{HO–C}_6\text{H}_5 \rightleftharpoons \text{H–CN} + {}^-\text{O–C}_6\text{H}_5$$

The side with the weaker acid is favored.   pK_a 10.0   pK_a 9.1

496

20-47 continued

(d)

Esters can be formed only in acid, not in base.

20-48 Cyanide substitution is an S$_N$2 reaction and requires a 1° or 2° carbon with a leaving group. The Grignard reaction is less particular about the type of halide, but is sensitive to, and incompatible with, acidic functional groups and other reactive groups.

(a) Both methods will work.

(b) Only Grignard will work. The S$_N$2 reaction does not work on unactivated benzene rings.

(c) Grignard will fail because of the OH group. The cyanide reaction will work, although an excess of cyanide will need to be added because the first equivalent will deprotonate the phenol.*

(d) Grignard will fail because of the OH group. The cyanide reaction will fail because S$_N$2 does not work on unactivated sp² carbons. In this case, NEITHER method will work.

(e) Both methods will work, although cyanide substitution on 2°C will be accompanied by elimination.

(f) Grignard will fail because of the OH group. The cyanide reaction will work. Since alcohols are much less acidic than phenols, there is no problem with cyanide deprotonating the alcohol.*

* The pK_a of HCN is 9.1, and the pK_a of phenol is 10.0. Thus cyanide is strong enough to pull off some of the H from the phenol, although the equilibrium would favor cyanide ion and phenol. The pK_a of secondary alcohols is 16-18, so there is no chance that cyanide would deprotonate an alcohol.

The side with the weaker acid is favored.   pK_a 10.0   pK_a 9.1

20-49

(a)

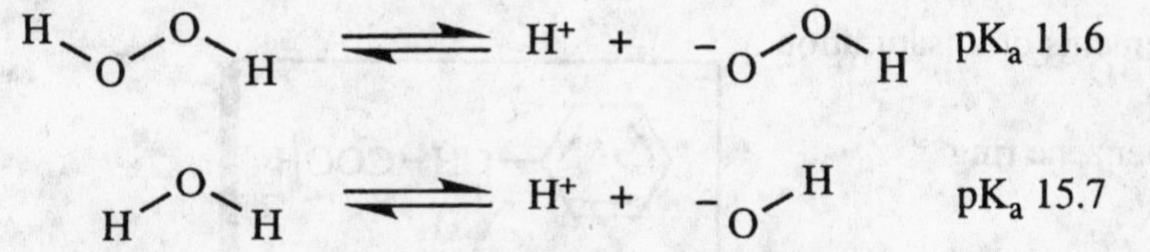

$$H\text{-}O\text{-}O\text{-}H \rightleftharpoons H^+ + {}^-O\text{-}O\text{-}H \quad pK_a\ 11.6$$

$$H\text{-}O\text{-}H \rightleftharpoons H^+ + {}^-O\text{-}H \quad pK_a\ 15.7$$

Hydrogen peroxide is four pK units (10^4 times) stronger acid than water, so the hydroperoxide anion, HOO$^-$, must be stabilized relative to hydroxide. This is from the *inductive effect* of the electronegative oxygen bonded to the O$^-$; by induction, the negative charge is distributed over both oxygens. The oxygen in hydroxide has to support the full negative charge with no delocalization.

(b)

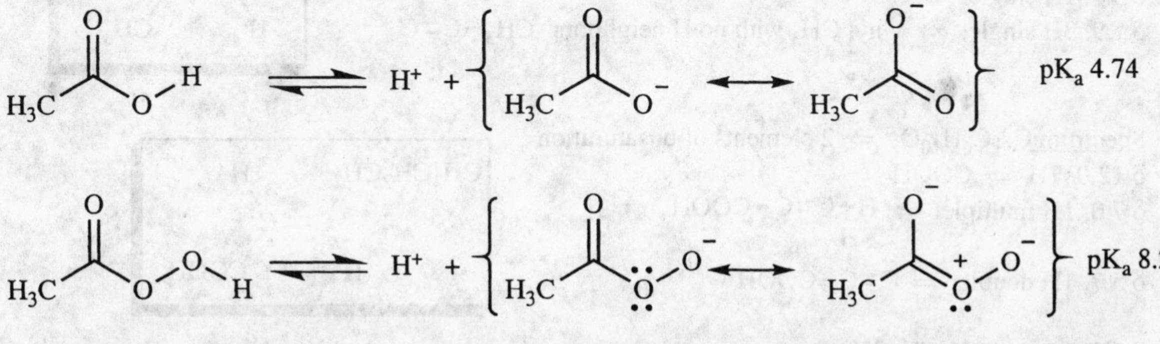

pK_a 4.74

pK_a 8.2

The reason that carboxylic acids are so acidic (over 10 pK units more acidic than alcohols) is because of the resonance stabilization of the carboxylate anion with two equivalent resonance forms in which all atoms have octets and the negative charge is on the more electronegative atom—the best of all possible resonance worlds. The peroxyacetate anion, however, cannot delocalize the negative charge onto the carbonyl oxygen; that negative charge is stuck out on the end oxygen like a wet nose on a frigid morning. There is some delocalization of the electron density onto the carbonyl, but with all the charge separation, this second form is a minor resonance contributor. This resonance does explain, however, why peroxyacetic acid is more acidic than hydrogen peroxide. It does not come close to acetic acid, though.

(c)

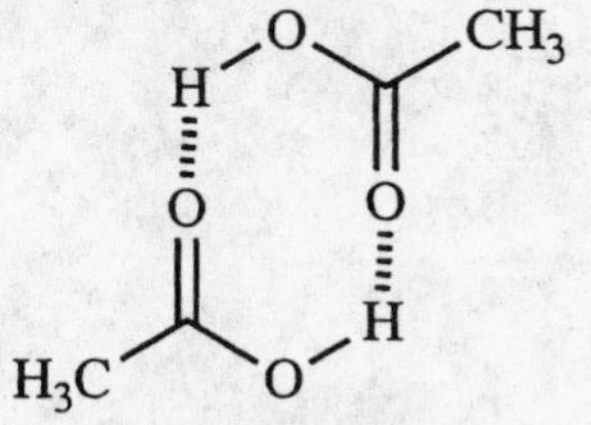

Carboxylic acids boil as the dimer, that is, two molecules are held tightly by hydrogen bonding. The dimer is an 8-membered ring with two hydrogen bonds as shown with dashed lines in the diagram. This works because the carbonyl oxygen has significant negative charge, and the H—O bond is weak because it is a relatively strong acid. The b.p. is 118°C.

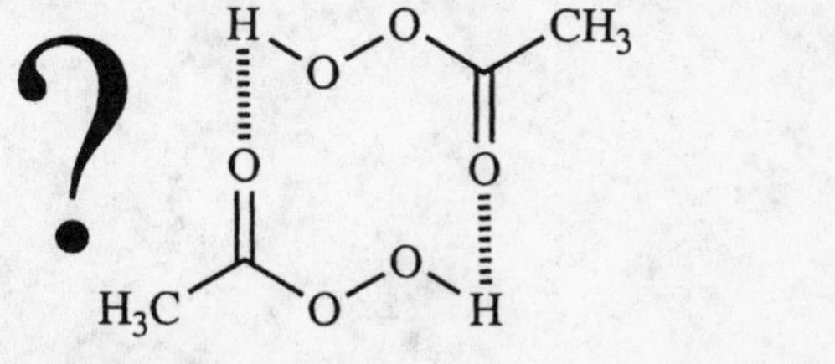

Do peroxyacids boil as the dimer? The author does not know, but there are three reasons to suspect that they do not. First, the b.p. is lower (105°C) instead of higher suggesting that they do not boil as a team but rather individually. Second, the dimer shown is a 10-membered ring—still possible but less likely than 8-membered. Third and most important, the electronic nature of the carbonyl group, as implied in the resonance forms in part (b), places less negative charge on the carbonyl oxygen, and the H is less acidic, suggesting that the hydrogren bonding is much less strong.

20-50

Spectrum A: $C_9H_{10}O_2 \Rightarrow$ 5 elements of unsaturation
δ 11.8, 1H $\Rightarrow$ COOH
δ 7.3, 5H $\Rightarrow$ monosubstituted benzene ring
δ 3.8, 1H quartet $\Rightarrow$ CHCH$_3$
δ 1.6, 3H doublet $\Rightarrow$ CHCH$_3$

Spectrum B: $C_4H_6O_2 \Rightarrow$ 2 elements of unsaturation
δ 12.1, 1H $\Rightarrow$ COOH
δ 6.2, 1H singlet $\Rightarrow$ H—C=C
δ 5.7, 1H singlet $\Rightarrow$ H—C=C
δ 1.9, 3H singlet $\Rightarrow$ vinyl CH$_3$ with no H neighbors   CH$_3$—C=C

Spectrum C: $C_6H_{10}O_2 \Rightarrow$ 2 elements of unsaturation
δ 12.0, 1H $\Rightarrow$ COOH
δ 7.0, 1H multiplet $\Rightarrow$ H-C=C-COOH
δ 5.7, 1H doublet $\Rightarrow$ C=C-COOH
                                  (with H on carbon)
δ 2.2-0.8 $\Rightarrow$ CH$_2$CH$_2$CH$_3$

must be *trans* due to large coupling
constant in doublet at δ 5.7

CHAPTER 21—CARBOXYLIC ACID DERIVATIVES

21-1 IUPAC name first; then common name

(a) isobutyl benzoate (both IUPAC and common)
(b) phenyl methanoate; phenyl formate
(c) methyl 2-phenylpropanoate; methyl α-phenylpropionate
(d) N-phenyl-3-methylbutanamide; β-methylbutyranilide
(e) N-benzylethanamide; N-benzylacetamide
(f) 3-hydroxybutanenitrile; β-hydroxybutyronitrile
(g) 3-methylbutanoyl bromide; isovaleryl bromide
(h) dichloroethanoyl chloride; dichloroacetyl chloride
(i) 2-methylpropanoic methanoic anhydride; isobutyric formic anhydride
(j) cyclopentyl cyclobutanecarboxylate (both IUPAC and common)
(k) 5-hydroxyhexanoic acid lactone; δ-caprolactone
(l) N-cyclopentylbenzamide (both IUPAC and common)
(m) propanedioic anhydride; malonic anhydride
(n) 1-hydroxycyclopentanecarbonitrile; cyclopentanone cyanohydrin
(o) cis-4-cyanocyclohexanecarboxylic acid; no common name
(p) 3-bromobenzoyl chloride; m-bromobenzoyl chloride
(q) N-methyl-5-aminoheptanoic acid lactam; no common name
(r) N-ethanoylpiperidine; N-acetylpiperidine

21-2 An aldehyde has a C—H absorption (usually 2 peaks) at 2700-2800 cm^{-1}. A carboxylic acid has a strong, broad absorption between 2400-3400 cm^{-1}. The spectrum of methyl benzoate has no peaks in this region.

21-3 The C—O single bond stretch in ethyl octanoate appears at 1170 cm^{-1}, while methyl benzoate shows this absorption at 1120 and 1280 cm^{-1}.

21-4

(a) acid chloride: single C=O peak at 1800 cm^{-1}; no other carbonyl comes so high

(b) primary amide: C=O at 1650 cm^{-1} and two N—H peaks between 3200-3400 cm^{-1}

(c) anhydride: two C=O absorptions at 1750 and 1820 cm^{-1}

21-5

(a) The formula C_3H_5NO has two elements of unsaturation. The IR spectrum shows two peaks between 3200-3400 cm^{-1}, an NH_2 group. The strong peak at 1670 cm^{-1} is a C=O, and the peak at 1610 cm^{-1} is a C=C. This accounts for all of the atoms.

The HNMR corroborates the assignment. The 1H multiplet at δ 5.8 is the vinyl H next to the carbonyl. The 2H multiplet at δ 6.3 is the vinyl hydrogen pair on carbon-3. The 2H singlet at δ 4.8 is the amide hydrogens. The CNMR confirms the structure: two vinyl carbons and a carbonyl.

(b) The formula $C_5H_8O_2$ has two elements of unsaturation. The IR spectrum shows no significant OH, so this compound is neither an alcohol nor a carboxylic acid. The strong peak at 1730 cm^{-1} is likely an ester carbonyl. The C—O appears between 1050-1250 cm^{-1}. The IR shows no C=C absorption, so the other element of unsaturation is likely a ring. The carbon NMR spectrum shows the carbonyl carbon at δ 171, the C—O carbon at δ 69, and three more carbons in the aliphatic region, but no carbons in the vinyl region between δ 100-150, so there can be no C=C. The proton NMR shows multiplets of 2H at δ 4.3 and 2.5, most likely CH_2 groups next to oxygen and carbonyl respectively.

The only structure with an ester, four CH_2 groups, and a ring, is δ-valerolactone:

21-6

(a)

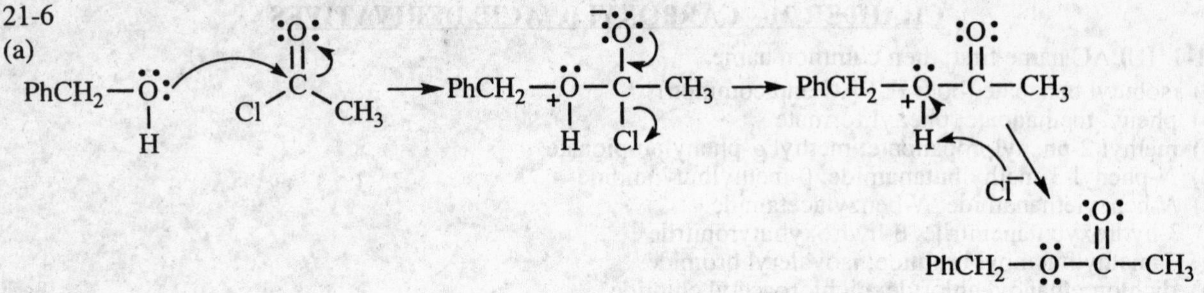

(b)

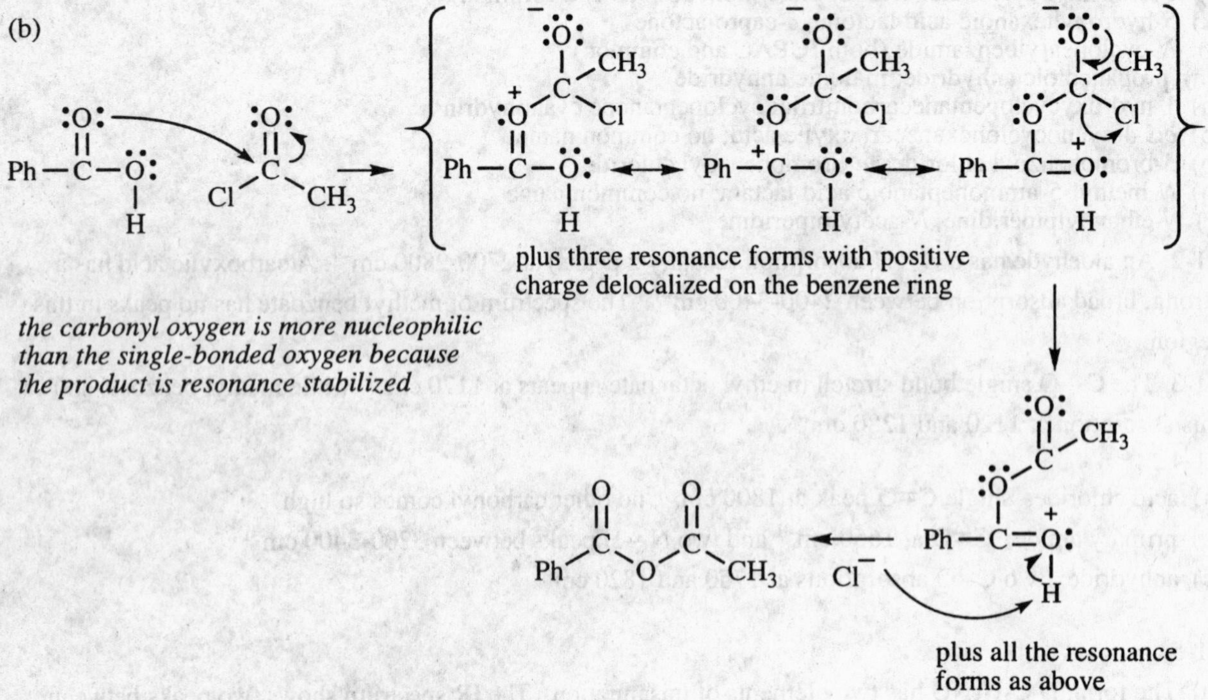

plus three resonance forms with positive
charge delocalized on the benzene ring

*the carbonyl oxygen is more nucleophilic
than the single-bonded oxygen because
the product is resonance stabilized*

plus all the resonance
forms as above

(c)

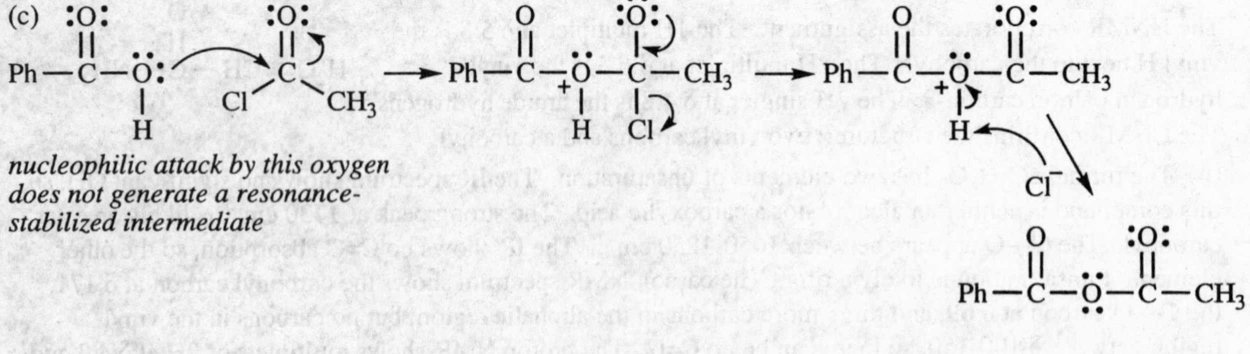

*nucleophilic attack by this oxygen
does not generate a resonance-
stabilized intermediate*

500

21-6 continued
(d)

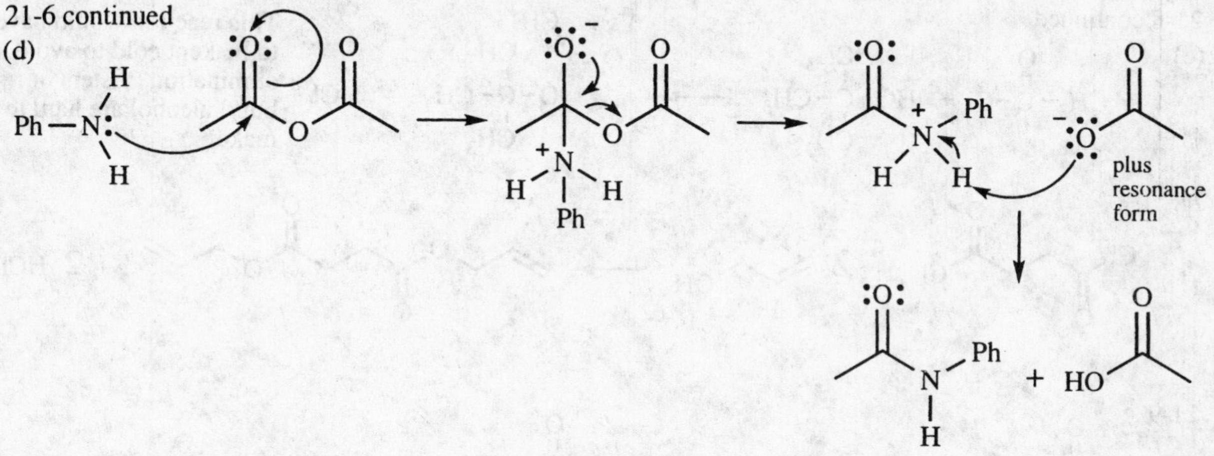

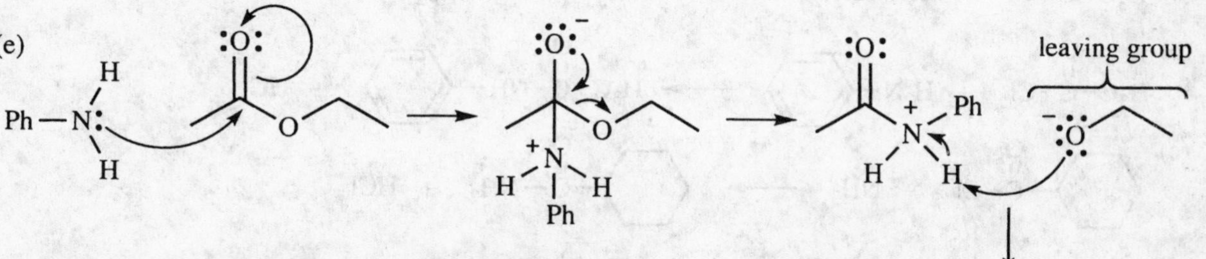

The leaving group is ethoxide ion, $CH_3CH_2O^-$, a very strong base. Ethoxide would never be a leaving group in an S_N2 reaction as it is too strong a base.

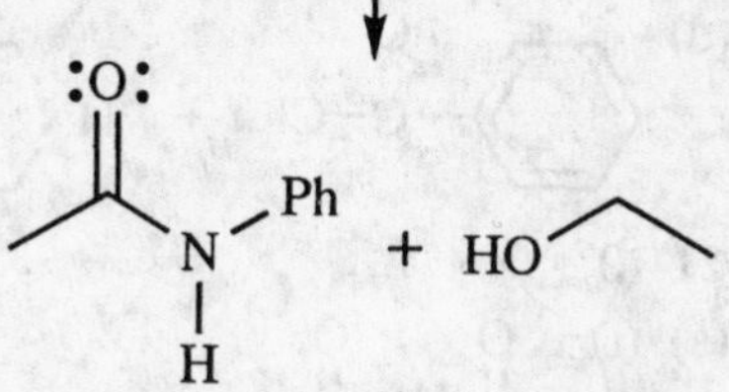

21-7 **Figure 21-9 is critical!** Reactions which go from a more reactive functional group to a less reactive functional group ("downhill reactions") will occur readily.
(a) amide to acid chloride will NOT occur—it is an "uphill" transformation
(b) acid chloride to amide will occur rapidly
(c) amide to ester will NOT occur—another "uphill" transformation
(d) acid chloride to anhydride will occur rapidly
(e) anhydride to amide will occur rapidly

21-8
(a) $CH_3CH_2-\overset{O}{\overset{\|}{C}}-Cl$ + $HOCH_2CH_3$ $\longrightarrow$ $CH_3CH_2-\overset{O}{\overset{\|}{C}}-OCH_2CH_3$ + HCl

(b)

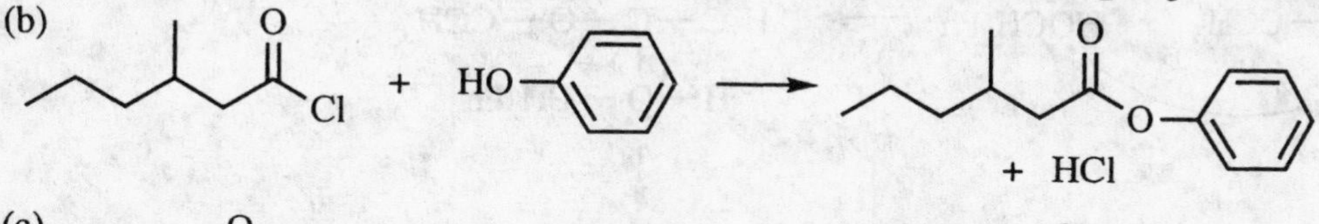

+ HCl

(c)

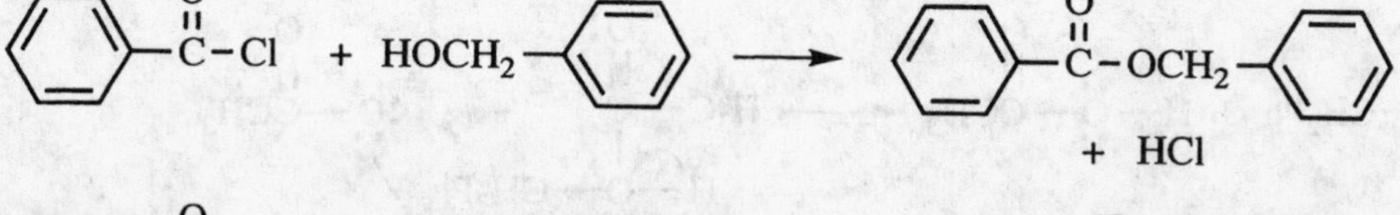

+ HCl

(d)

+ HCl

501

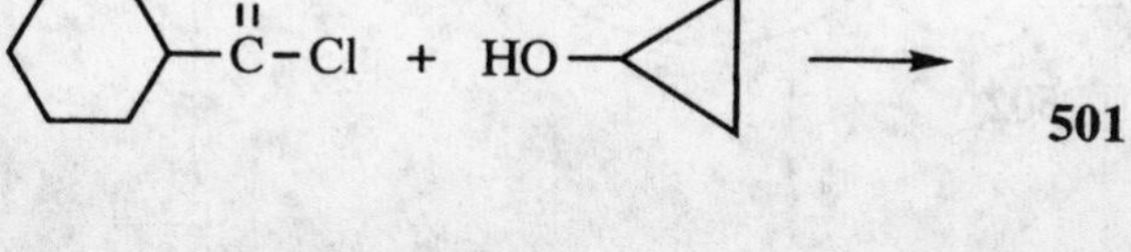

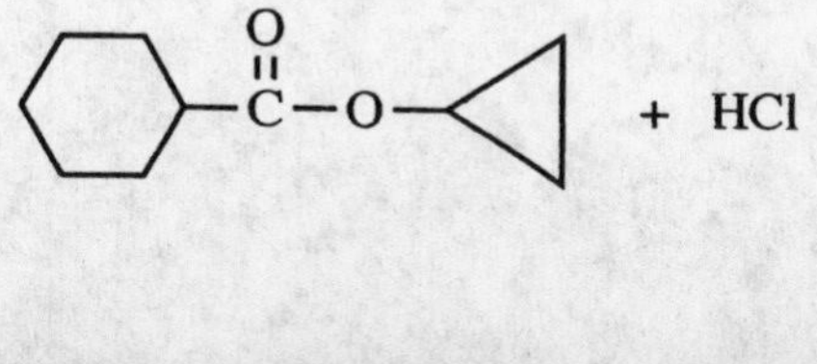

21-8 continued

(e)

$$CH_3-\overset{O}{\overset{\|}{C}}-Cl \ + \ HO-\overset{CH_3}{\underset{CH_3}{\overset{|}{\underset{|}{C}}}}-CH_3 \ \longrightarrow \ \overset{O=C-CH_3}{\underset{CH_3}{\overset{|}{\underset{|}{O-\overset{CH_3}{\underset{CH_3}{\overset{|}{\underset{|}{C}}}}-CH_3}}}} \ + \ HCl$$

This reaction would have to be kept cold to avoid elimination. Esters of *t*-butyl alcohol are hard to make.

(f)

$$\text{Cl}\overset{O}{\overset{\|}{C}}\text{—CH}_2\text{CH}_2\overset{O}{\overset{\|}{C}}\text{Cl} \ + \ 2 \ \diagdown\!\!\diagup\text{OH} \ \longrightarrow \ \text{allyl succinate diester} \ + \ 2 \ \text{HCl}$$

21-9

(a) $H_3C-\overset{O}{\overset{\|}{C}}-Cl \ + \ HN(CH_3)_2 \ \longrightarrow \ H_3C-\overset{O}{\overset{\|}{C}}-N(CH_3)_2 \ + \ HCl$

(b) $H_3C-\overset{O}{\overset{\|}{C}}-Cl \ + \ H_2N-\text{Ph} \ \longrightarrow \ H_3C-\overset{O}{\overset{\|}{C}}-NH-\text{Ph} \ + \ HCl$

(c) $\text{cyclohexyl}-\overset{O}{\overset{\|}{C}}-Cl \ + \ NH_3 \ \longrightarrow \ \text{cyclohexyl}-\overset{O}{\overset{\|}{C}}-NH_2 \ + \ HCl$

(d) $\text{Ph}-\overset{O}{\overset{\|}{C}}-Cl \ + \ HN(\text{piperidine}) \ \longrightarrow \ \text{Ph}-\overset{O}{\overset{\|}{C}}-N(\text{piperidine}) \ + \ HCl$

21-10

(a) (i) $H_3C-\overset{O}{\overset{\|}{C}}-O-\overset{O}{\overset{\|}{C}}-CH_3 \ + \ HOCH_2-\text{Ph} \ \longrightarrow \ H_3C-\overset{O}{\overset{\|}{C}}-OCH_2-\text{Ph} \ + \ HO-\overset{O}{\overset{\|}{C}}-CH_3$

(ii) $H_3C-\overset{O}{\overset{\|}{C}}-O-\overset{O}{\overset{\|}{C}}-CH_3 \ + \ HN(\text{diethyl}) \ \longrightarrow \ H_3C-\overset{O}{\overset{\|}{C}}-N(\text{diethyl}) \ + \ HO-\overset{O}{\overset{\|}{C}}-CH_3$

(b) (i)

$$H_3C-\overset{:O:}{\overset{\|}{C}}-O-\overset{O}{\overset{\|}{C}}CH_3 \ + \ H\ddot{O}CH_2Ph \ \longrightarrow \ \overset{:\overset{-}{\ddot{O}}:}{\underset{\underset{H-\overset{+}{\ddot{O}}-CH_2Ph}{|}}{H_3C-\overset{|}{C}-\overset{\cdot\cdot}{\ddot{O}}-\overset{O}{\overset{\|}{C}}CH_3}}$$

$$H_3C-\overset{:O:}{\overset{\|}{C}}-\overset{\cdot\cdot}{\ddot{O}}-CH_2Ph \ + \ H-\overset{\cdot\cdot}{\ddot{O}}-\overset{O}{\overset{\|}{C}}CH_3 \ \longleftarrow \ \underset{\underset{H-\overset{+}{\overset{|}{\ddot{O}}}-CH_2Ph}{|}}{H_3C-\overset{:O:}{\overset{\|}{C}}} \ + \ \overset{\cdot\cdot}{\overset{-}{:\ddot{O}}}-\overset{O}{\overset{\|}{C}}CH_3$$

502

21-10 (b) continued

(ii)

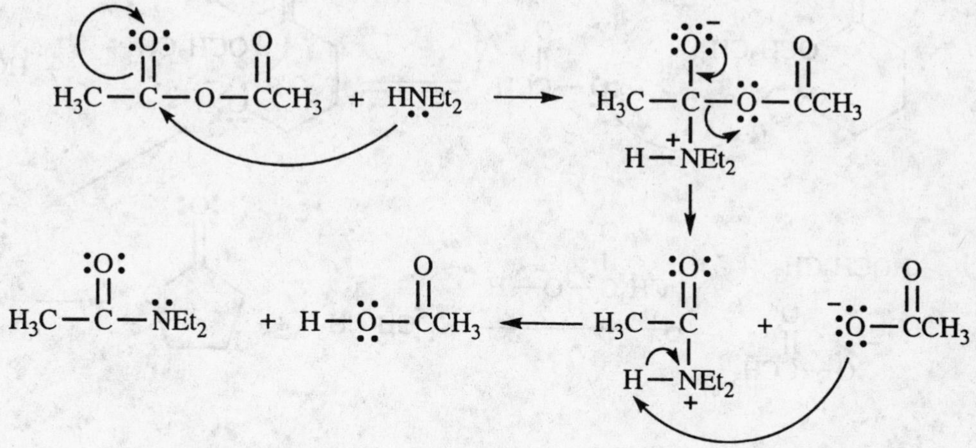

$$H_3C-C-O-CCH_3 \ + \ HNEt_2 \longrightarrow H_3C-\overset{+}{\underset{\underset{H-NEt_2}{|}}{C}}-O-CCH_3$$

$$H_3C-C-NEt_2 \ + \ H-O-CCH_3 \longleftarrow H_3C-\overset{\underset{H-\overset{+}{N}Et_2}{|}}{C} \ + \ {}^-O-CCH_3$$

21-11

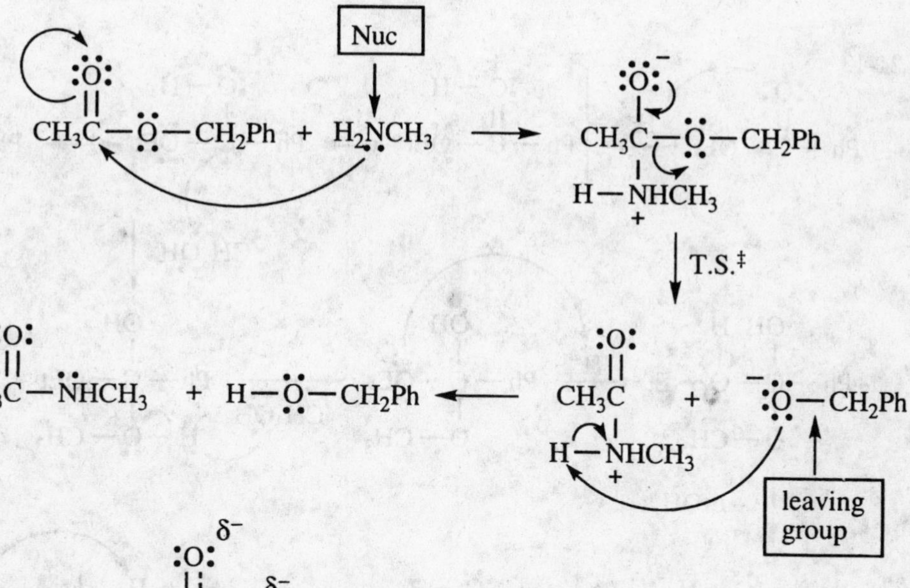

Nuc

$$CH_3C-O-CH_2Ph \ + \ H_2NCH_3 \longrightarrow CH_3\overset{\underset{H-\overset{+}{N}HCH_3}{|}}{C}-O-CH_2Ph$$

T.S.‡

$$CH_3C-NHCH_3 \ + \ H-O-CH_2Ph \longleftarrow CH_3C \ + \ {}^-O-CH_2Ph$$

leaving group

$$T.S.^\ddagger \Longrightarrow CH_3\overset{\underset{H-\overset{+}{N}HCH_3}{|}}{C}\cdots\overset{\delta^-}{O}-CH_2Ph$$

503

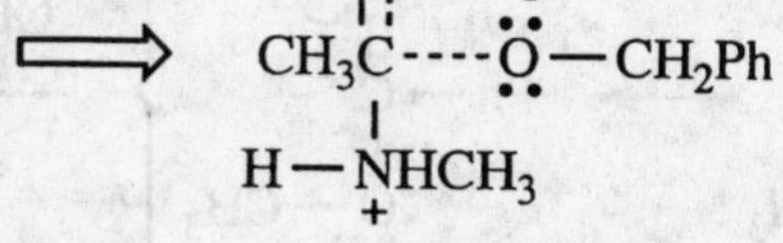

21-12

21-13

21-14

21-15

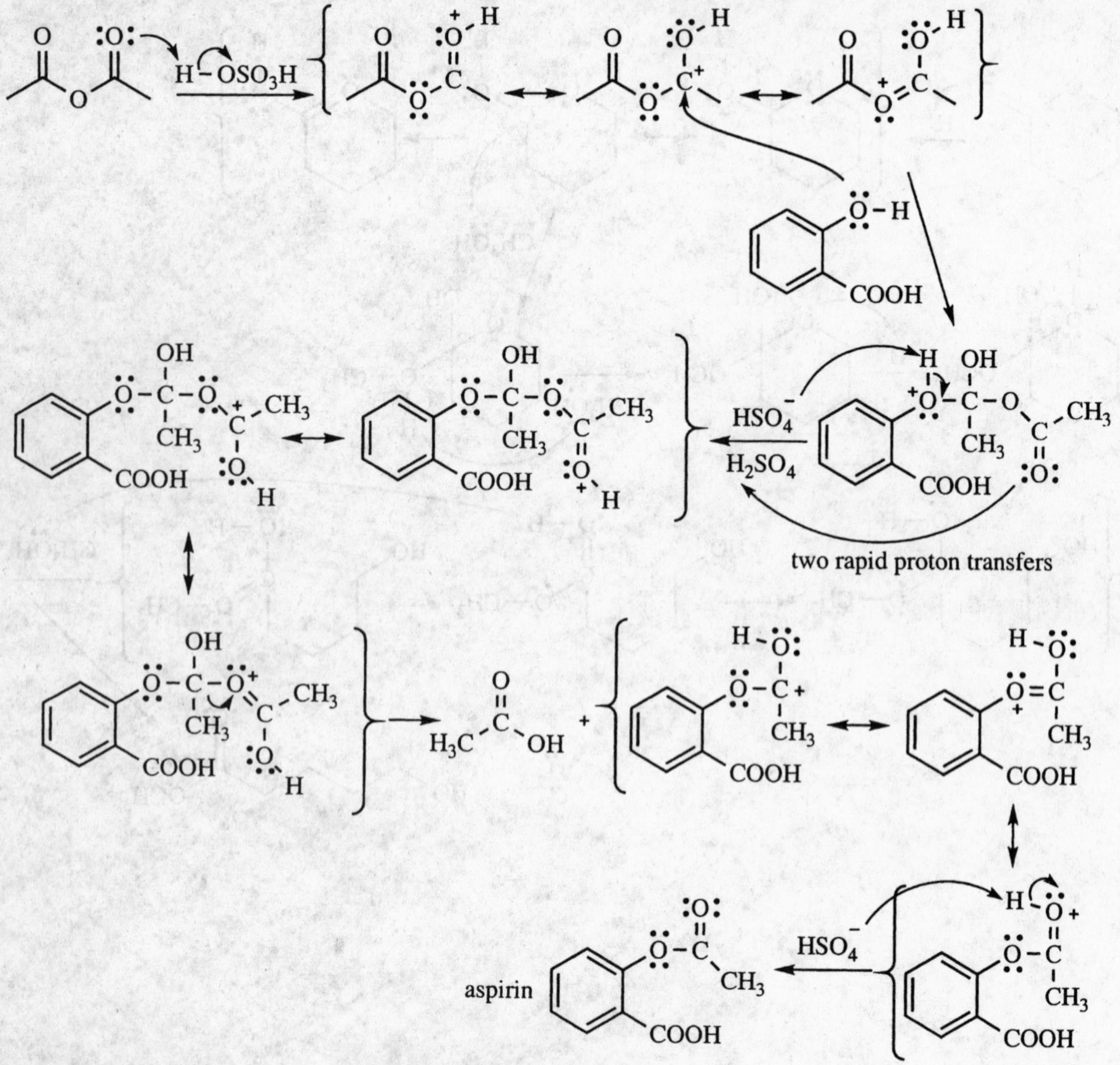
H−OSO$_3$H
COOH
HSO$_4^-$
H_2SO_4
two rapid proton transfers
H_3C OH
aspirin

21-16 The asterisk (*) will denote ^{18}O.

(a)

$$H_3C-\overset{\overset{\displaystyle :O:}{\|}}{C}-O^*R \longrightarrow H_3C-\overset{\overset{\displaystyle :\overset{-}{O}:}{|}}{\underset{\underset{\displaystyle :O-H}{|}}{C}}-\overset{..}{O}^*R \longrightarrow H_3C-\overset{\overset{\displaystyle :O:}{\|}}{C}-\overset{..}{\underset{\underset{\displaystyle H}{|}}{O}}: \; + \; :\overset{-}{O}^*R \longrightarrow H_3C-\overset{\overset{\displaystyle O}{\|}}{C}-O^-$$

with $:\ddot{O}H$ attacking.

$$O^*R = \quad {}^{18}\overset{..}{O}\underset{R}{\overset{\text{\scriptsize CH}_2\text{CH}_3}{\underset{\text{\scriptsize CH}_3}{\diagup\!\!\!\backslash\!\!\!\backslash H}}}$$

[boxed:] $H^{18}\overset{..}{O}\underset{R}{\overset{\text{\scriptsize CH}_2\text{CH}_3}{\underset{\text{\scriptsize CH}_3}{\diagup\!\!\!\backslash\!\!\!\backslash H}}}$ $+$

The alcohol product contains the ^{18}O label, with none in the carboxylate. The bond between ^{18}O and the tetrahedral carbon with (R) configuration did not break, so the configuration is retained.

(b) The products are identical regardless of mechanism.

$$H_3C-\overset{\overset{\displaystyle :O:}{\|}}{C}-O^*R \xrightarrow{H^+} \left\{ CH_3C \overset{\overset{\displaystyle :\overset{+}{O}-H}{\|}}{}-\overset{..}{O}^*R \longleftrightarrow CH_3C\overset{\overset{\displaystyle :\ddot{O}-H}{|}}{}-\overset{..}{\underset{+}{O}}^*R \longleftrightarrow CH_3C\overset{\overset{\displaystyle :\ddot{O}-H}{|}}{}=\overset{..}{\underset{+}{O}}^*R \right\}$$

$H_2\overset{..}{O}:$

$$CH_3\overset{\overset{\displaystyle OH\;H}{|}}{\underset{\underset{\displaystyle O-H}{|}}{C}}\overset{+}{O}^*R \xleftarrow{H^+} CH_3\overset{\overset{\displaystyle OH}{|}}{\underset{\underset{\displaystyle O-H}{|}}{C}}-\overset{..}{O}^*R \longleftarrow CH_3\overset{\overset{\displaystyle OH}{|}}{\underset{\underset{\displaystyle H-\overset{+}{\underset{..}{O}}-H}{|}}{C}}-O^*R$$

$H_2\overset{..}{O}:$

[boxed:] $- HO^*R$

$$\left\{ H_3C-\overset{\overset{\displaystyle :\ddot{O}-H}{|}}{\underset{\underset{\displaystyle :\ddot{O}-H}{|}}{C}}+ \longleftrightarrow H_3C-\overset{\overset{\displaystyle :\ddot{O}-H}{|}}{\underset{\underset{\displaystyle +\overset{}{O}-H}{\|}}{C}} \longleftrightarrow H_3C-\overset{\overset{\displaystyle +\overset{}{O}-H}{\|}}{\underset{\underset{\displaystyle :\ddot{O}-H}{|}}{C}} \right\} \xrightarrow{H_2\overset{..}{O}:} H_3C-\overset{\overset{\displaystyle :O:}{\|}}{C}-OH$$

acetic acid

(c) The ^{18}O has 2 more neutrons in its nucleus than ^{16}O. Mass spectra of these products would show the molecular ion of acetic acid at its standard value of m/z 60, whereas the molecular ion of 2-butanol would appear at m/z 76 instead of m/z 74, proving that the heavy isotope of oxygen went with the alcohol. This demonstrates that the bond between oxygen and the carbonyl carbon is broken, not the bond between the oxygen and the alkyl carbon.

To show if the alcohol was chiral or racemic, measuring its optical activity in a polarimeter and comparing to known values would prove its configuration. (The heavy oxygen isotope has a negligible effect on optical rotation.)

21-17

(a) A catalyst is defined as a chemical species that speeds a reaction but is not consumed in the reaction. In the acidic hydrolysis, acid is used in the first and fourth steps of the mechanism but is regenerated in the third and last steps. Acid is not consumed; the final concentration of acid is the same as the initial concentration. In the basic hydrolysis, however, the hydroxide that initially attacks the carbonyl is never regenerated. An alkoxide leaves from the carbonyl, but it quickly neutralizes the carboxylic acid. For every molecule of ester, one molecule of hydroxide is consumed; the base *promotes* the reaction but does not *catalyze* the reaction.

(b) Basic hydrolysis is not reversible. Once an ester molecule is hydrolyzed in base, the carboxylate cannot form an ester. Acid catalysis, however, is an equilibrium: the mixture will always contain some ester, and the yield will never be as high as in basic hydrolysis. Second, long chain fatty acids are not soluble in water until they are ionized; they are soluble only as their sodium salts (soap). Basic hydrolysis is preferred for higher yield and greater solubility of the product.

21-18

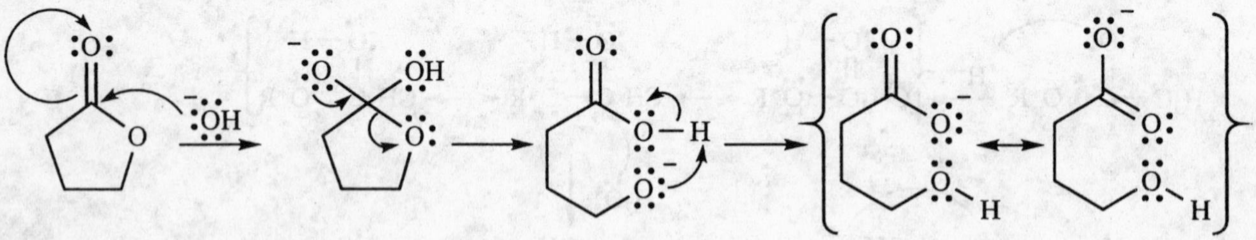

21-19

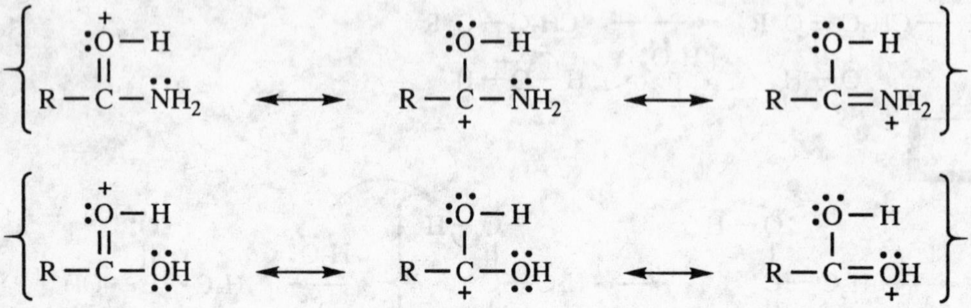

21-20

(a)

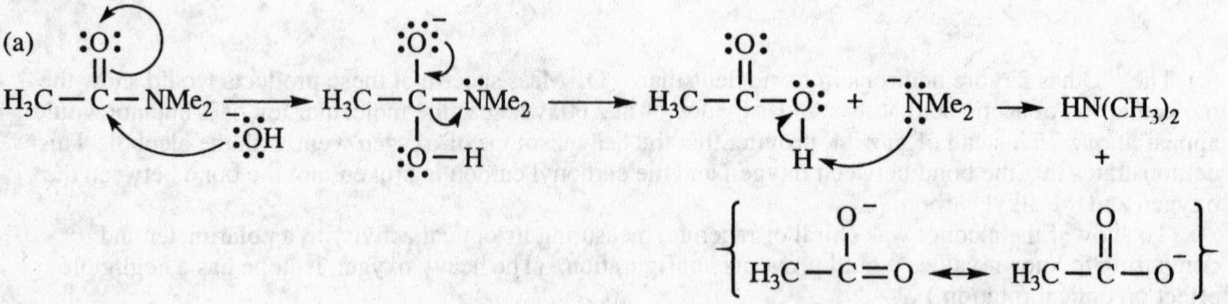

508

21-20 continued

(b)

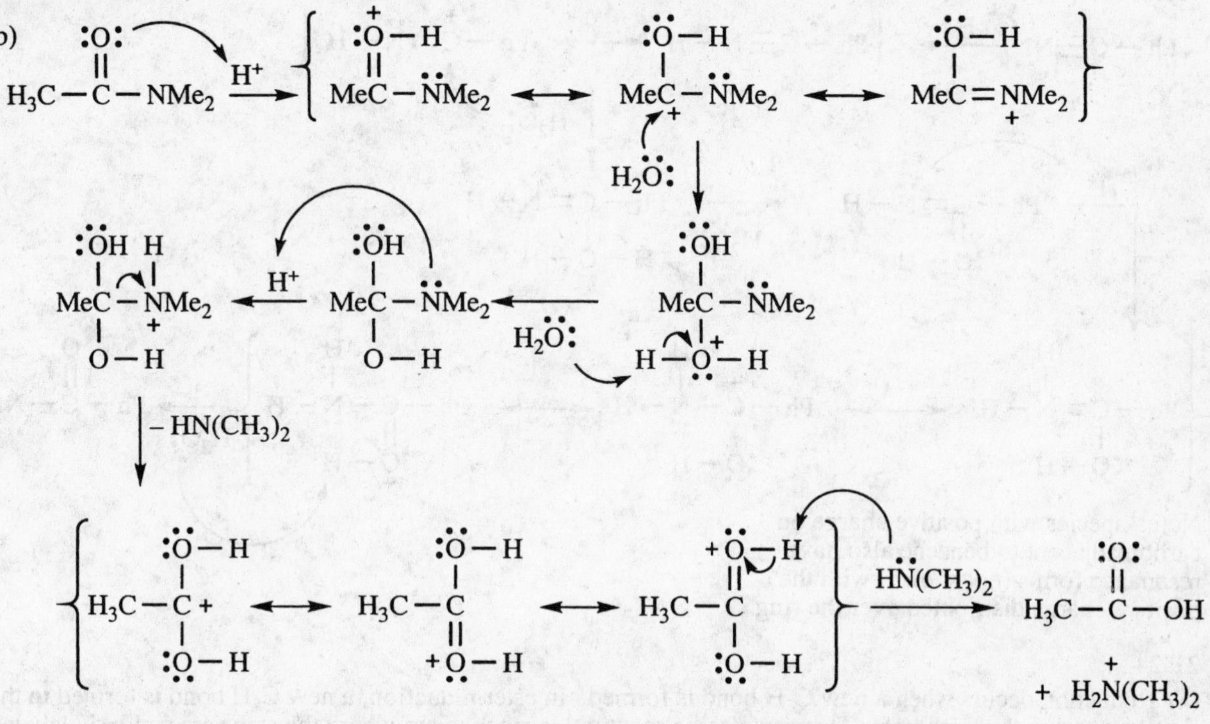

21-21 In the basic hydrolysis (21-20(a)), the step that drives the reaction to completion is the final step, the deprotonation of the carboxylic acid by the amide anion. In the acidic hydrolysis (21-20(b)), protonation of the amine by acid is exothermic and it prevents the reverse reaction by tying up the pair of electrons on the nitrogen so that the amine is no longer nucleophilic.

21-22

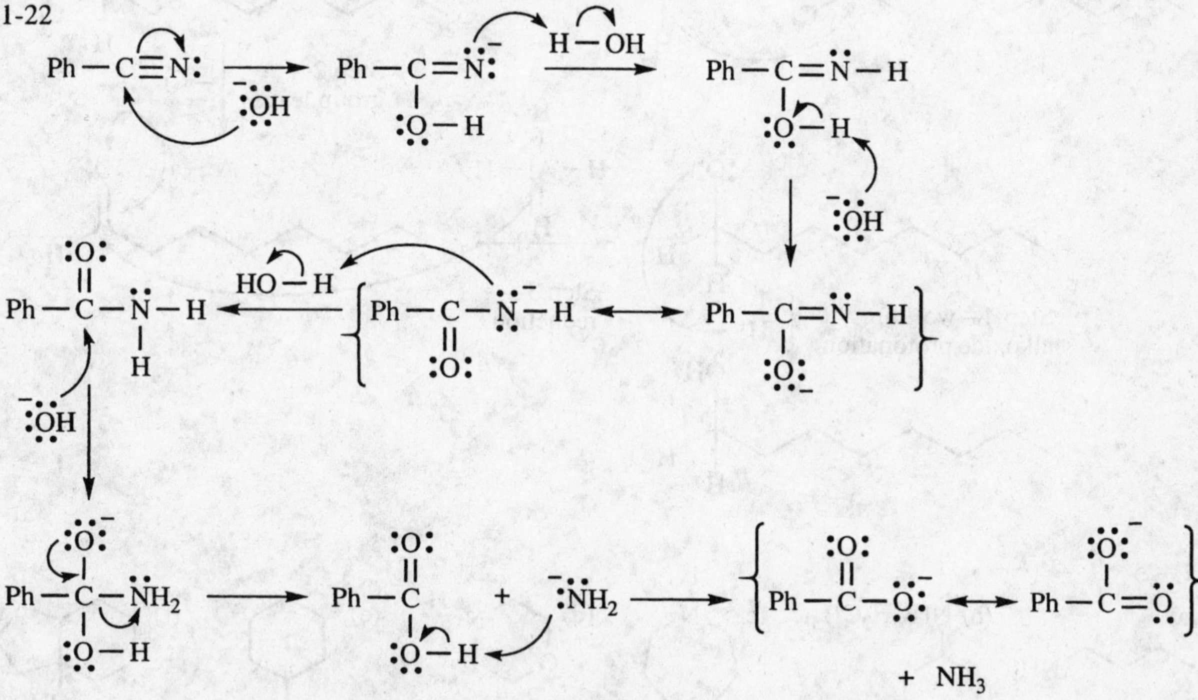

509

21-23

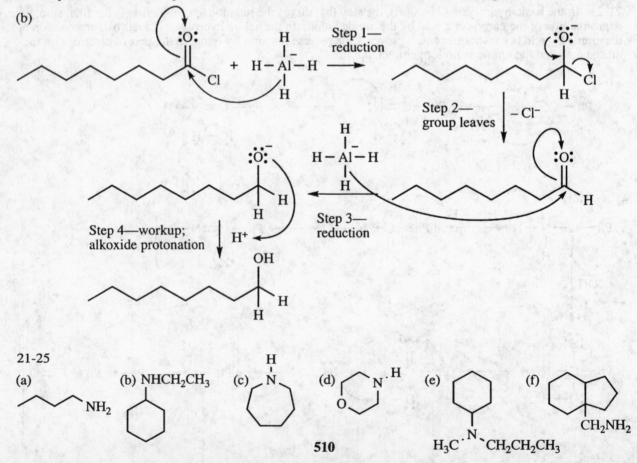

Note: species with positive charge on carbon adjacent to benzene also have resonance forms (not shown) with the positive charge distributed over the ring.

21-24

(a) Reduction occurs when a new C-H bond is formed. In ester reduction, a new C-H bond is formed in the first step and in the third step. This can also be seen in this mechanism where the steps are similarly labeled.

(b)

Step 1—reduction

Step 2—group leaves

Step 3—reduction

Step 4—workup; alkoxide protonation

21-25

(a) NH_2

(b) $NHCH_2CH_3$

(c) N—H

(d) O N—H

(e) H_3C—N—$CH_2CH_2CH_3$

(f) CH_2NH_2

21-26

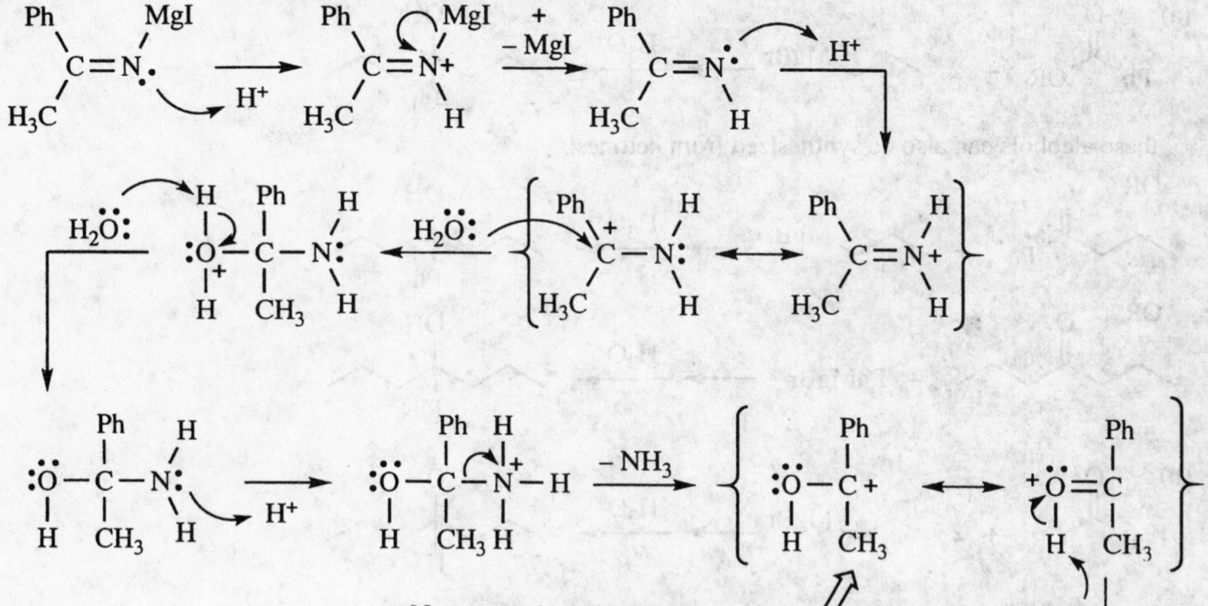

Note: species with positive charge on carbon adjacent to benzene also have resonance forms (not shown) with the positive charge distributed over the ring.

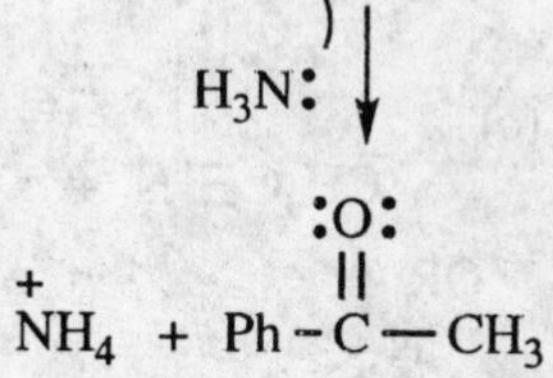

21-27

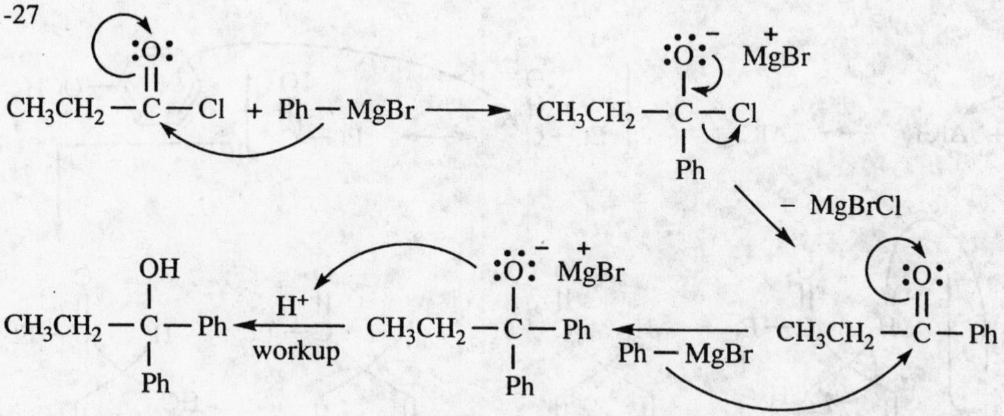

21-28

(a)

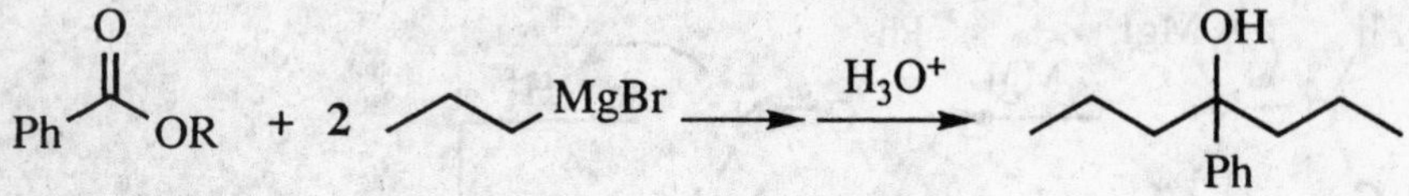

$Ph-C(=O)-OR$ + 2 (propyl)MgBr $\xrightarrow{\quad}$ $\xrightarrow{H_3O^+}$ (alcohol with OH, Ph)

these alcohols can also be synthesized from ketones:

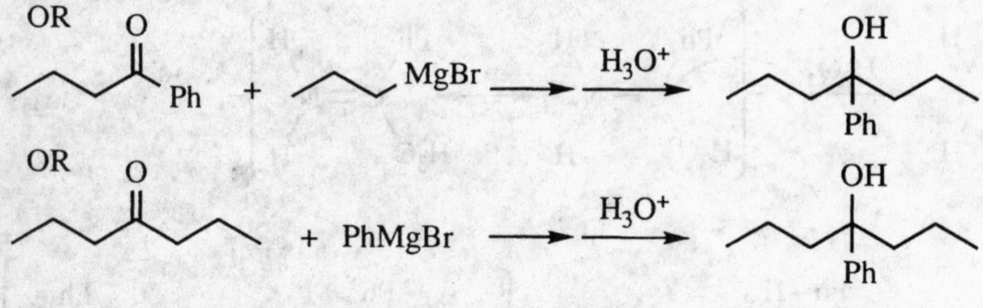

$\xrightarrow{\quad}$ ketone + (propyl)MgBr $\xrightarrow{\quad}$ $\xrightarrow{H_3O^+}$ (OH, Ph)

(ketone) + PhMgBr $\xrightarrow{\quad}$ $\xrightarrow{H_3O^+}$ (OH, Ph)

(b)

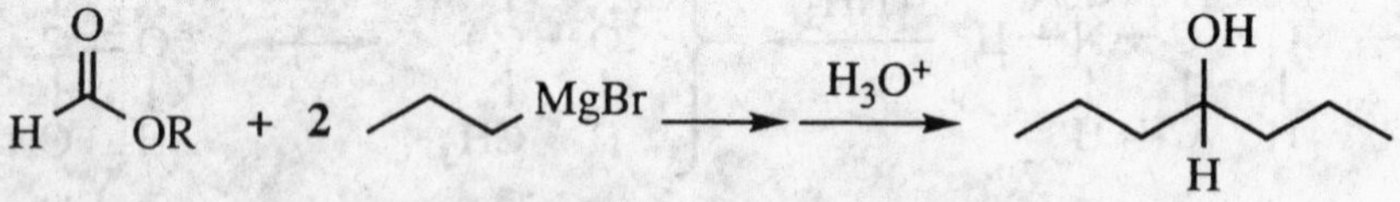

$H-C(=O)-OR$ + 2 (propyl)MgBr $\xrightarrow{\quad}$ $\xrightarrow{H_3O^+}$ (OH, H)

(c)

$CH_3C{\equiv}N$ + (propyl)MgBr $\xrightarrow{\quad}$ $\xrightarrow{H_3O^+}$ (ketone)

OR

$N{\equiv}C-$(propyl) + CH_3MgBr $\xrightarrow{\quad}$ $\xrightarrow{H_3O^+}$ (ketone)

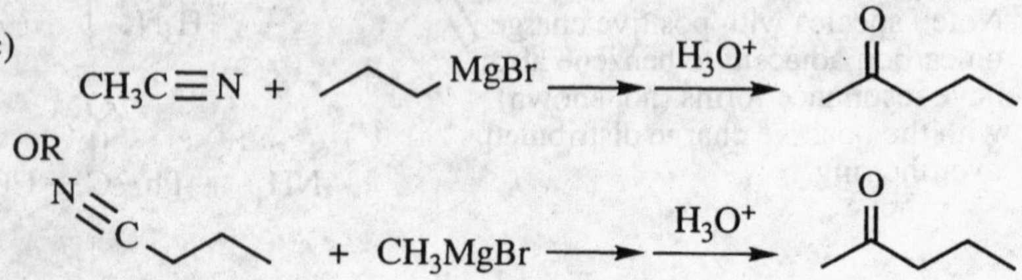

21-29

$Et-\overset{\ddot{O}}{\underset{}{C}}-Cl$ + $AlCl_3$ $\longrightarrow$ $AlCl_4^-$ + $\left\{ Et-\overset{\ddot{O}\colon}{\underset{+}{C}} \longleftrightarrow Et-\overset{\colon\overset{+}{O}}{C} \right\}$ (phenyl)$-OCH_3$

$AlCl_4^-$

$\left\{ \text{resonance structures with } H, C-Et, OCH_3 \right\}$

CH_3O-(phenyl)$-\overset{O}{C}CH_2CH_3$ + HCl + $AlCl_3$

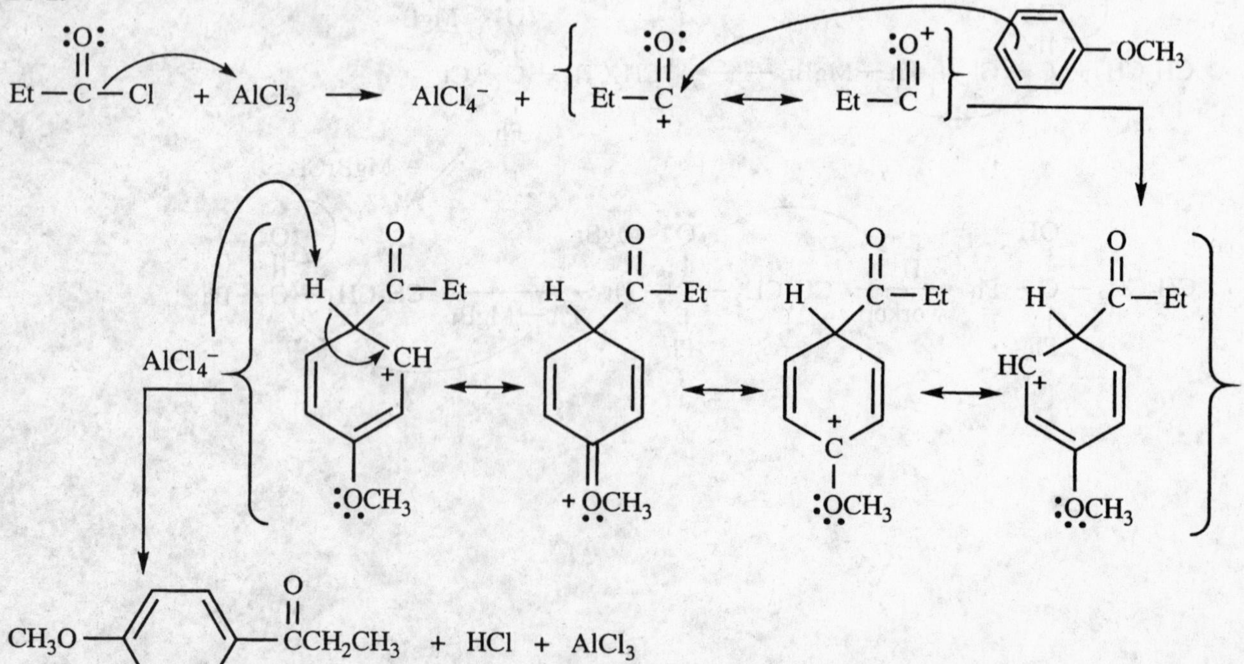

21-30

(a) [benzene] + CH$_3$C(=O)-Cl $\xrightarrow{\text{AlCl}_3}$ [C$_6$H$_5$-C(=O)-CH$_3$]

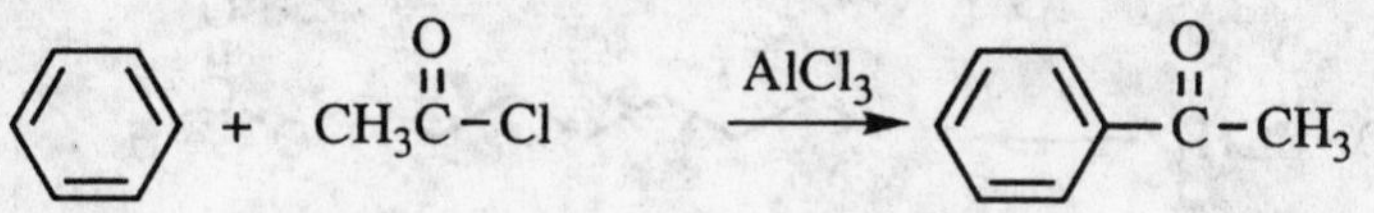

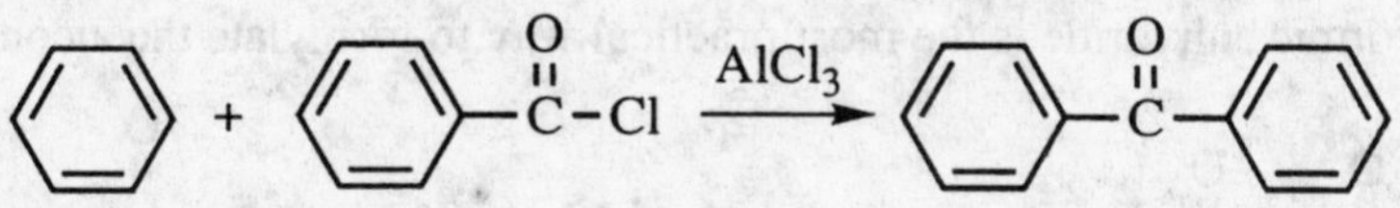

(b) [benzene] + [C$_6$H$_5$-C(=O)-Cl] $\xrightarrow{\text{AlCl}_3}$ [C$_6$H$_5$-C(=O)-C$_6$H$_5$]

(c) [benzene] + [CH$_3$CH$_2$CH$_2$-C(=O)-Cl] $\xrightarrow{\text{AlCl}_3}$ [C$_6$H$_5$-C(=O)-CH$_2$CH$_2$CH$_3$] $\xrightarrow[\text{HCl}]{\text{Zn(Hg)}}$ [C$_6$H$_5$-CH$_2$CH$_2$CH$_2$CH$_3$]

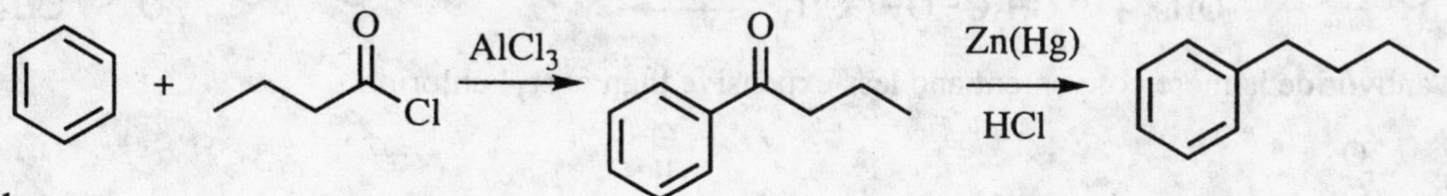

21-31

(a) (i) HC(=O)-O-C(=O)CH$_3$ + H$_2$N-[C$_6$H$_5$] $\longrightarrow$ HC(=O)-NH-[C$_6$H$_5$] + HO-C(=O)CH$_3$

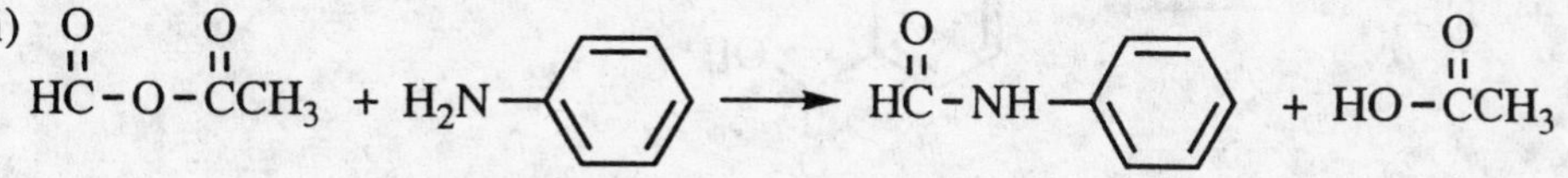

(ii) HC(=O)-O-C(=O)CH$_3$ + HOCH$_2$-[C$_6$H$_5$] $\longrightarrow$ HC(=O)-OCH$_2$-[C$_6$H$_5$] + HO-C(=O)CH$_3$

(b) (i)

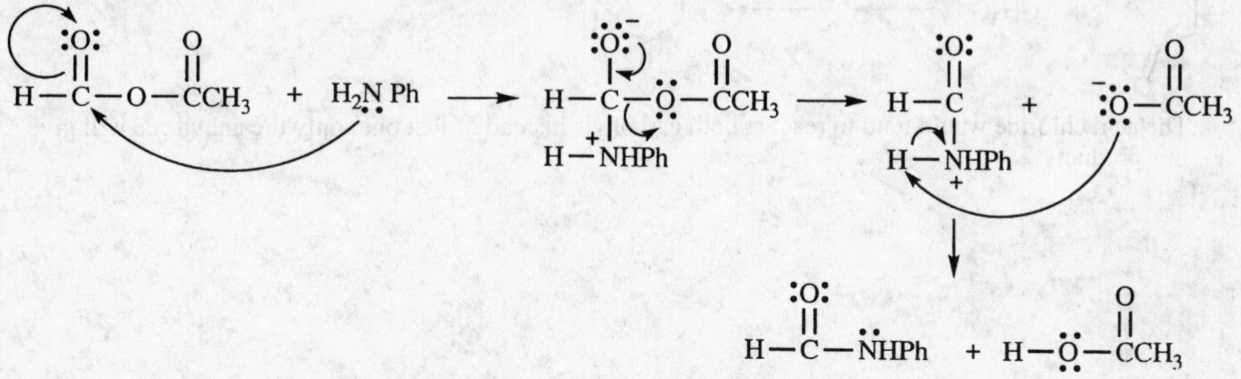

(ii)

513

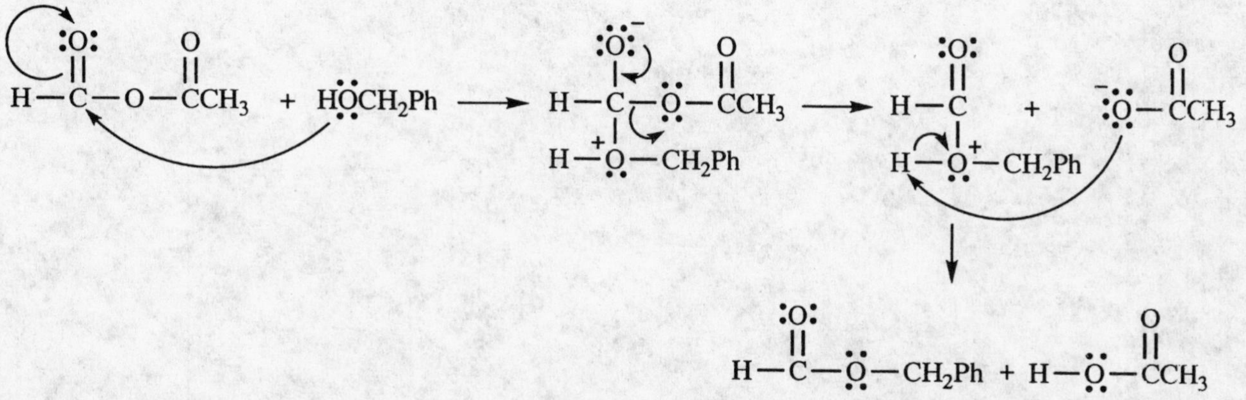

21-32

(a)

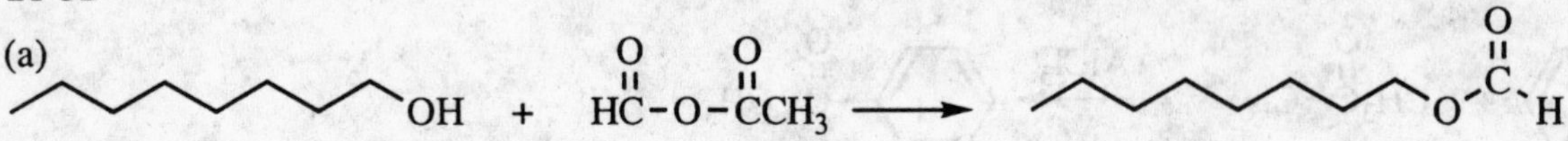

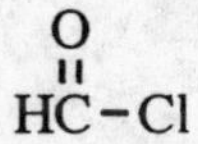

$$HC-Cl$$ does not exist, so acetic formic anhydride is the most practical way to formylate the alcohol.

(b)

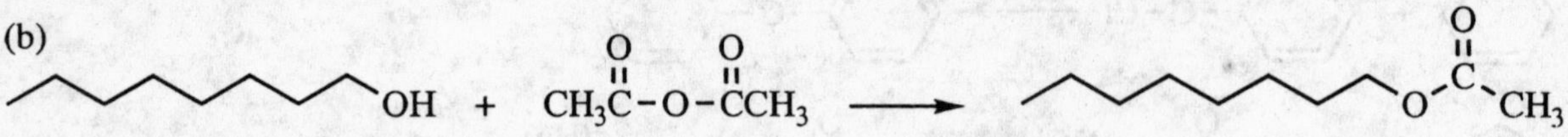

Acetic anhydride is more convenient and less expensive than acetyl chloride.

(c)

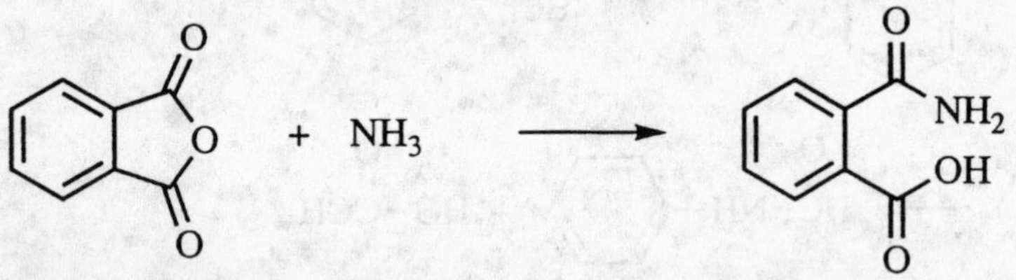

The acid chloride would tend to react at both carbonyls instead of just one; only the anhydride will give this product.

(d)

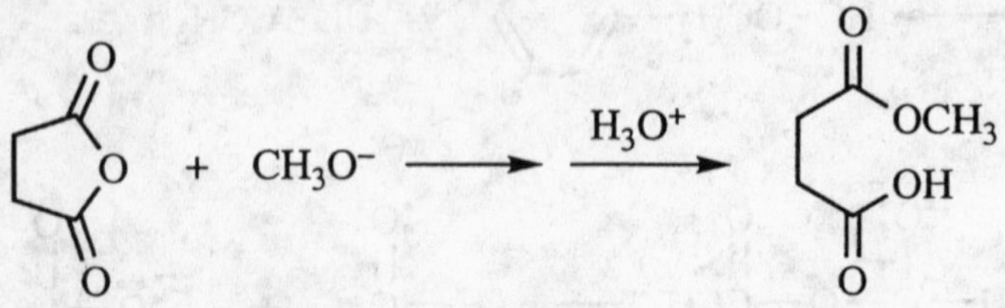

The acid chloride would tend to react at both carbonyls instead of just one; only the anhydride will give this product.

21-33

21-34

(a)

Generally, acetic anhydride is the optimum reagent for the preparation of acetate esters. Acetyl chloride would also react with the carboxylic acid to form a mixed anhydride.

(b)

Fischer esterification works well to prepare simple carboxylic esters. The diazomethane method would also react with the phenol, making the phenyl ether.

(c)

Fischer esterification would make the ester, but in the process, the acidic conditions would risk migrating the double bond into conjugation with the carbonyl group. The diazomethane reaction is run under neutral conditions where double bond migration will not occur.

21-35 Syntheses may have more than one correct approach.

(a)

$$Ph-\overset{O}{\underset{\|}{C}}-OCH_3 + 2\ PhMgBr \xrightarrow{ether} \xrightarrow{H_3O^+} Ph-\overset{OH}{\underset{\underset{Ph}{|}}{\overset{|}{C}}}-Ph$$

(b)

$$H-\overset{O}{\underset{\|}{C}}-OCH_2CH_3 + 2\ PhCH_2MgBr \xrightarrow{ether} \xrightarrow{H_3O^+} H-\overset{OH}{\underset{\underset{CH_2Ph}{|}}{\overset{|}{C}}}-CH_2Ph$$

(c)

$$Ph-\overset{O}{\underset{\|}{C}}-OCH_3 + H_2NCH_2CH_3 \longrightarrow Ph-\overset{O}{\underset{\|}{C}}-NHCH_2CH_3$$

(d)

$$H-\overset{O}{\underset{\|}{C}}-OCH_2CH_3 + 2\ PhMgBr \xrightarrow{ether} \xrightarrow{H_3O^+} H-\overset{OH}{\underset{\underset{Ph}{|}}{\overset{|}{C}}}-Ph$$

(e)

$$Ph-\overset{O}{\underset{\|}{C}}-OCH_3 + LiAlH_4 \xrightarrow{ether} \xrightarrow{H_3O^+} PhCH_2OH$$

(f)

$$Ph-\overset{O}{\underset{\|}{C}}-OCH_3 \xrightarrow{H_3O^+} Ph-\overset{O}{\underset{\|}{C}}-OH + CH_3OH$$

21-35 continued

(g) PhCH$_2$OH $\xrightarrow[\text{2) KCN}]{\text{1) TsCl, pyridine}}$ PhCH$_2$C≡N $\xrightarrow[\Delta]{\text{H}_3\text{O}^+}$ PhCH$_2$–C(=O)–OH $\xrightarrow{\text{H}_2\text{SO}_4}$ PhCH$_2$–C(=O)–O–CH(CH$_3$)(CH$_3$)

from (e)   ... HO–CH(CH$_3$)–CH$_3$ ... H$_3$C–CH–CH$_3$

(h)

PhCH$_2$–C(=O)–OCH(CH$_3$)$_2$ + **2** CH$_3$CH$_2$MgBr $\xrightarrow{\text{ether}}$ $\xrightarrow{\text{H}_3\text{O}^+}$ PhCH$_2$–C(OH)(CH$_2$CH$_3$)–CH$_2$CH$_3$

from (g)

(i) How to make an 8-carbon diol from an ester that has no more than 8 carbons? Make the ester a lactone!

[lactone structure, positions 1–8, O] + LiAlH$_4$ $\xrightarrow{\text{ether}}$ $\xrightarrow{\text{H}_3\text{O}^+}$ [8-carbon diol, positions 1–8, OH at 1 and 8]

21-36 There may be additional correct approaches to these problems.

(a)

[PhC(=O)OH] $\xrightarrow{\text{SOCl}_2}$ [PhC(=O)Cl] $\xrightarrow{\text{HN(CH}_3)_2}$ [PhC(=O)N(CH$_3$)$_2$] $\xrightarrow[\text{2) H}_2\text{O}]{\text{1) LiAlH}_4}$ [PhCH$_2$N(CH$_3$)$_2$]

(b)

[pyrrolidine, N–H] $\xrightarrow{\text{HC(=O)–O–C(=O)CH}_3}$ [N–C(=O)H pyrrolidine] $\xrightarrow{\text{LiAlH}_4 \quad \text{H}_2\text{O}}$ [N–CH$_3$ pyrrolidine]

21-37 Prepare the amide, then dehydrate.

(a)

[butanoic acid, C(=O)OH] $\xrightarrow{\text{SOCl}_2}$ [C(=O)Cl] $\xrightarrow{\text{NH}_3}$ [C(=O)NH$_2$] $\xrightarrow{\text{POCl}_3}$ [C≡N]

(b)

PhC(=O)–OH $\xrightarrow{\text{SOCl}_2}$ PhC(=O)–Cl $\xrightarrow{\text{NH}_3}$ PhC(=O)–NH$_2$ $\xrightarrow{\text{POCl}_3}$ PhC≡N

(c)

[cyclopentane C(=O)OH] $\xrightarrow{\text{SOCl}_2}$ [cyclopentane C(=O)Cl] $\xrightarrow{\text{NH}_3}$ [cyclopentane C(=O)NH$_2$] $\xrightarrow{\text{POCl}_3}$ [cyclopentane C≡N]

21-38

(a)

PhCH$_2$–C(=O)–OH $\xrightarrow{\text{SOCl}_2}$ PhCH$_2$–C(=O)–Cl $\xrightarrow{\text{NH}_3}$ PhCH$_2$–C(=O)–NH$_2$ $\xrightarrow{\text{POCl}_3}$ PhCH$_2$–C≡N

21-38 continued

(b)

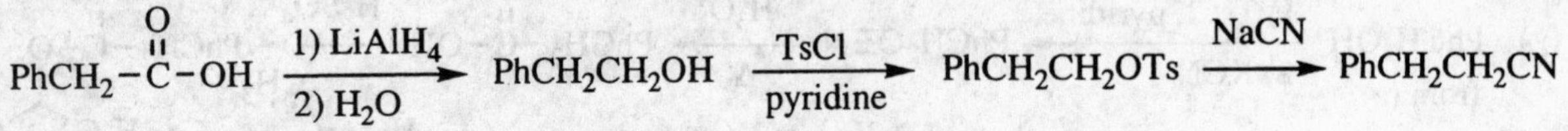

PhCH$_2$-$\overset{\displaystyle O}{\overset{\|}{C}}$-OH $\xrightarrow[\text{2) H}_2\text{O}]{\text{1) LiAlH}_4}$ PhCH$_2$CH$_2$OH $\xrightarrow[\text{pyridine}]{\text{TsCl}}$ PhCH$_2$CH$_2$OTs $\xrightarrow{\text{NaCN}}$ PhCH$_2$CH$_2$CN

(c)

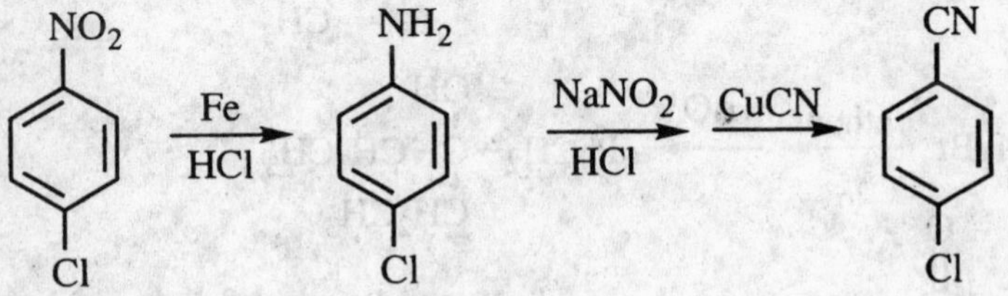

21-39

(a)

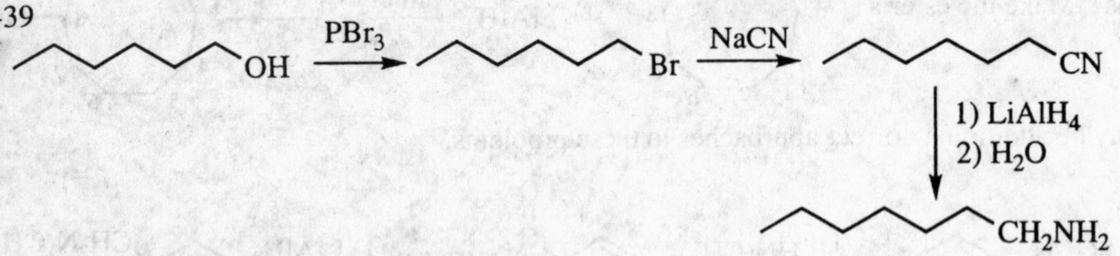

(b)

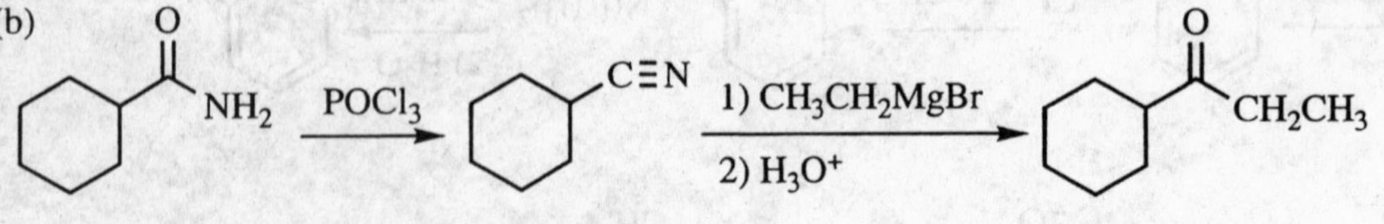

(c)

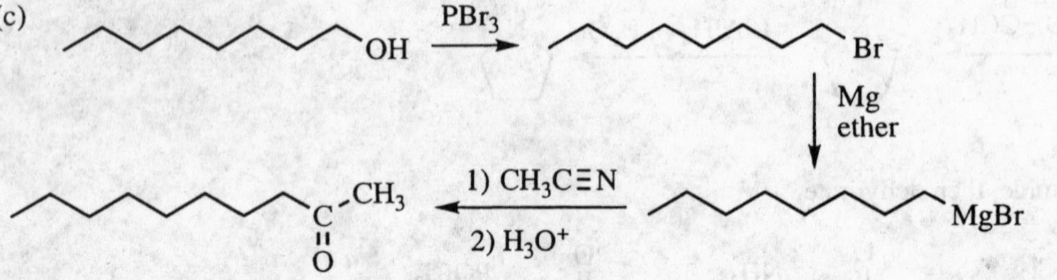

21-40

Ar =

H$_3$C$\diagdown$$\underset{\text{H}}{\text{N}}$$\overset{\displaystyle O}{\overset{\|}{C}}$$\diagup$O

Sevin
(carbaryl)

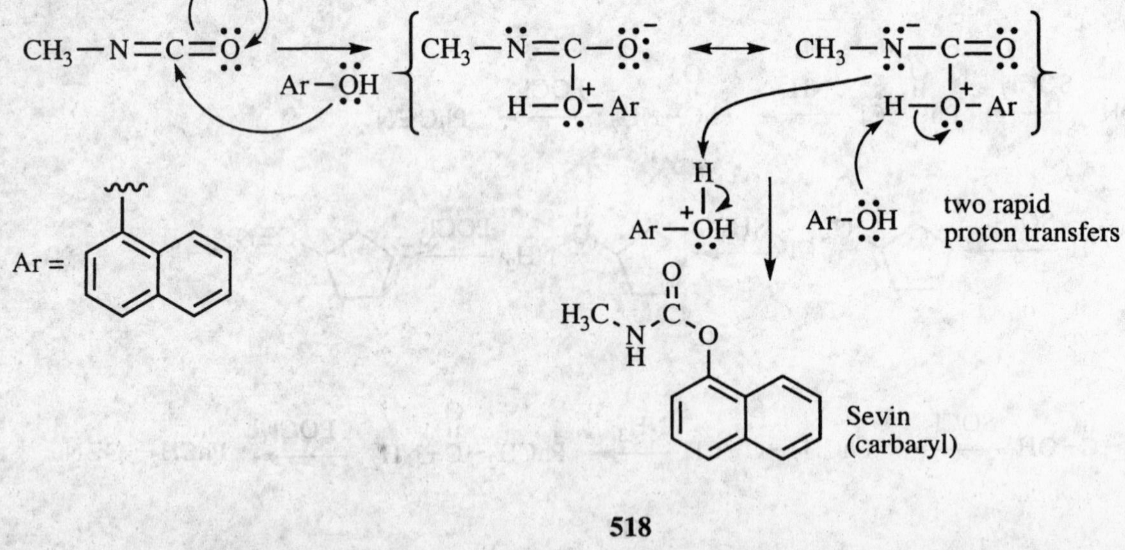

21-41

(a)

$$H_3O^+$$

(i) carbonate ester
(iii) not aromatic

(b)

$$H_3O^+$$

(i) thiolactone
(iii) not aromatic

(c)

$$H_3O^+$$

SH   SH $+ CO_2$

(i) thiocarbonate ester
(iii) not aromatic

(d)

$$\left\{ \quad HN \quad NH \quad \longleftrightarrow \quad HN \quad \overset{+}{N}H \quad \right\} \xrightarrow{H_3O^+} CO_2 + H_2N \quad NH_2 \longrightarrow H_2N \quad NH$$

enediamine   imine

(i) a substituted urea
(iii) AROMATIC—more easily seen in the resonance form shown

(The enediamine product would not be stable in aqueous acid. It would probably tautomerize to an imine, hydrolyze to ammonia and 2-aminoethanal, then polymerize.)

(e) At first glance, this AROMATIC compound does not appear to be an acid derivative. Like any enol, however, its tautomer must be considered.

$$\rightleftharpoons \quad \xrightarrow{H_3O^+} \quad \text{COOH} \quad \longrightarrow \quad \text{COOH} \quad \xrightarrow{H_3O^+} \quad \text{H} \quad \text{COOH}$$

lactam   enamine   imine $+ NH_3$

(f)

$$\left\{ \quad O \quad NH \quad \longleftrightarrow \quad \overset{..}{:}O \quad \overset{+}{N}H \quad \right\} \xrightarrow{H_3O^+} CO_2 + HO \quad NH_2 \longrightarrow \text{polymer}$$

see part (d)

(i) a carbamate or urethane
(iii) AROMATIC—more easily seen in the resonance form

21-42

(a) Carbamoyl phosphate is a mixed anhydride between carbamic acid and phosphoric acid. It would react easily with an amine to form an amide bond (technically, a urea), with phosphate as the leaving group.

(b) N-Carbamoylaspartate has a carbonyl with two nitrogens on either side; this group is a urea derivative.

(c) The NH_2 group on one end replaces the OH of a carboxylic acid on the other end; this reaction is a nucleophilic acyl substitution.

(d) Orotate is aromatic as can be seen readily in the tautomer. It is called a "pyrimidine base" because of its structural similarity to the pyrimidine ring.

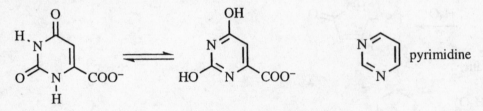

21-43 Please refer to solution 1-20, page 12 of this Solutions Manual.

21-44

(a) 3-methylpentanoyl chloride

(b) benzoic formic anhydride

(c) acetanilide; N-phenylethanamide

(d) N-methylbenzamide

(e) phenyl acetate; phenyl ethanoate

(f) methyl benzoate

(g) benzonitrile

(h) 4-phenylbutane nitrile; γ-phenylbutyronitrile

(i) dimethyl isophthalate, or dimethyl benzene-1,3-dicarboxylate

(j) N,N-diethyl-3-methylbenzamide

(k) 4-hydroxypentanoic acid lactone; γ-valerolactone

(l) 3-aminopentanoic acid lactam; β-valerolactam

21-45

(a)   O
       ‖
    PhC—OCH$_2$CH$_3$

(b)   O   O
       ‖   ‖
    PhC—O—CCH$_3$

(c)   O
       ‖
    PhC—N—⟨phenyl⟩
          |
          H

(d)   O
       ‖
    ⟨H$_3$CO—benzene⟩—C—Ph

(e)   OH
       |
    Ph—C—Ph
       |
       Ph

(f)   O
       ‖
    PhC—H

21-46 When a carboxylic acid is treated with a basic reagent, the base removes the acidic proton rather than attacking at the carbonyl (proton transfers are much faster than formation or cleavage of other types of bonds). Once the carboxylate anion is formed, the carbonyl is no longer susceptible to nucleophilic attack: nucleophiles do not attack sites of negative charge. By contrast, in acidic conditions, the protonated carbonyl has a positive charge and is activated to nucleophilic attack.

<u>basic conditions</u>

          O   O
          ‖   ‖
    R—C—OH   +   ⁻OR'   ⟶   R—C—O⁻   +   HOR'
                                      anion—not susceptible
                                      to nucleophilic attack

<u>acidic conditions</u>

          O   OH
          ‖ |
    R—C—OH   +   H⁺   ⟶   R—C—OH
                                         +
                                    rapidly attacked
                                    by R'OH nucleophile

21-47

(a)

(b)

(c) 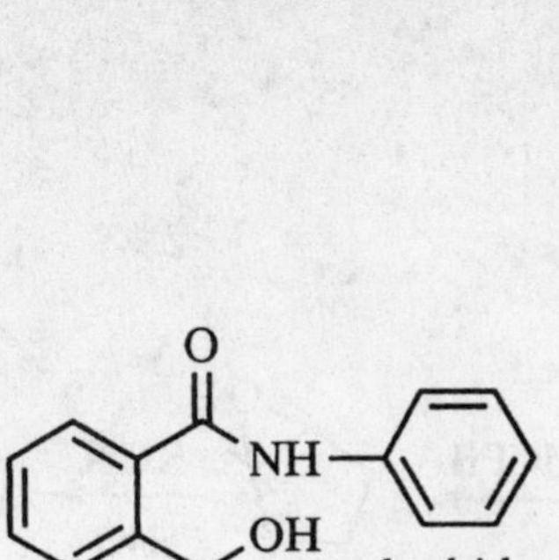
anhydrides react only once

(d)

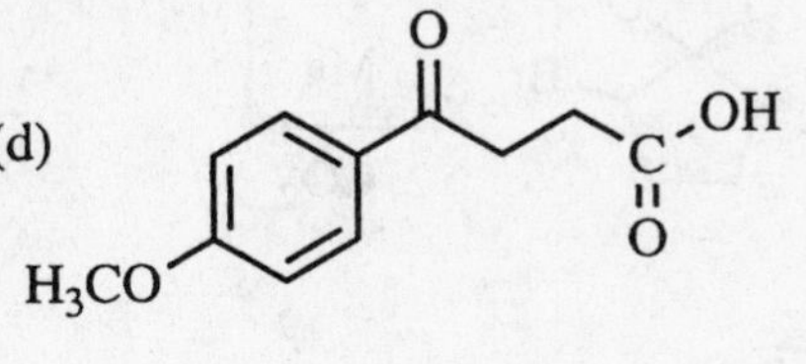

(e)
amines are more nucleophilic than alcohols 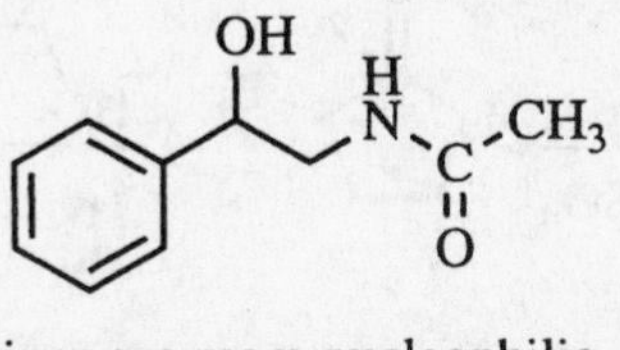

(f)

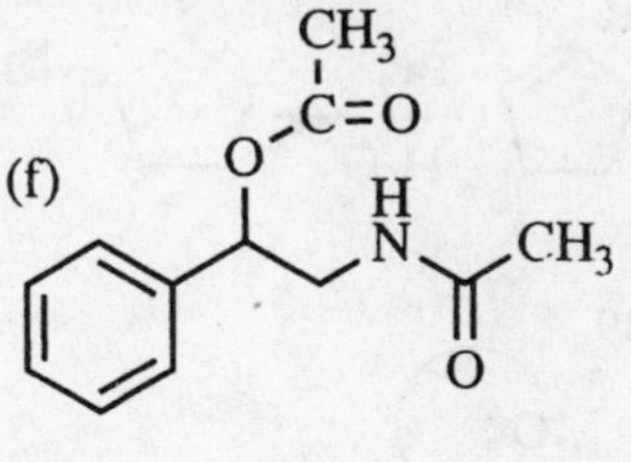

21-48

(a)

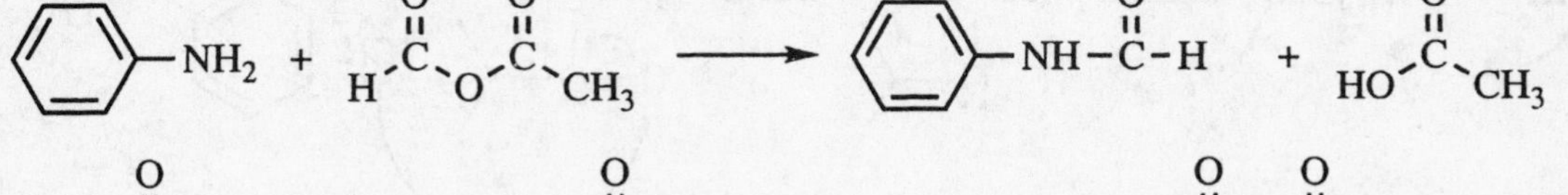

(b)

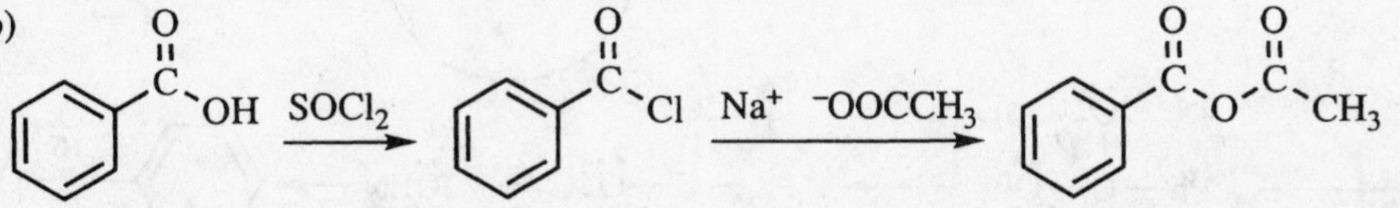

(c)

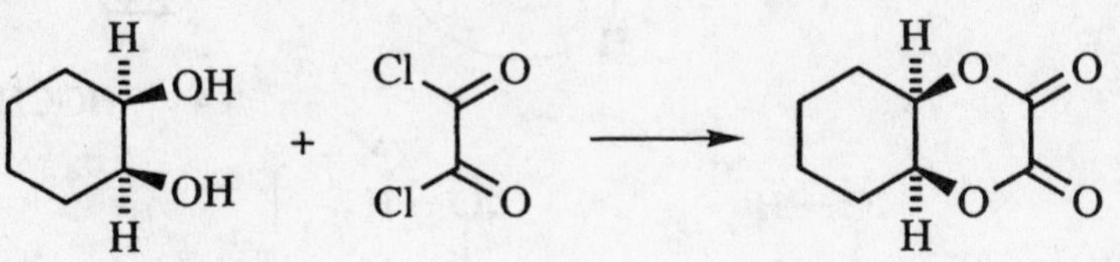

(d)

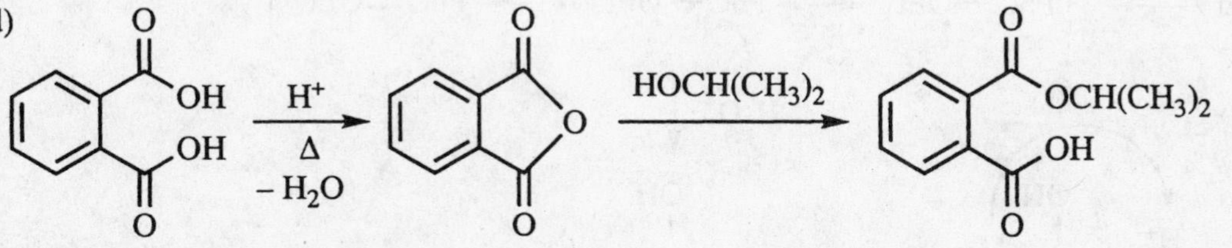

(e)

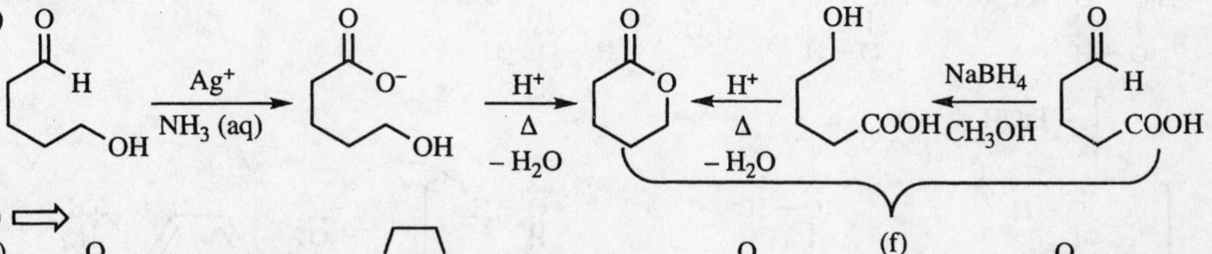

(f) ⟹

(g)

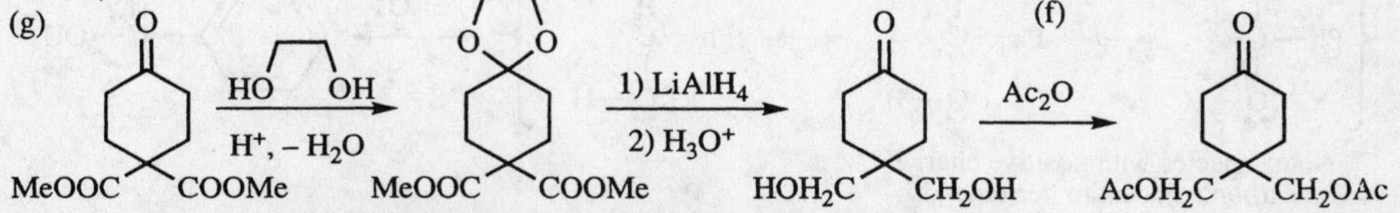

any ester where ethylene glycol displaces methanol will be reduced with LiAlH$_4$

aqueous acid workup removes ketal protecting group

521

21-48 continued

(h)

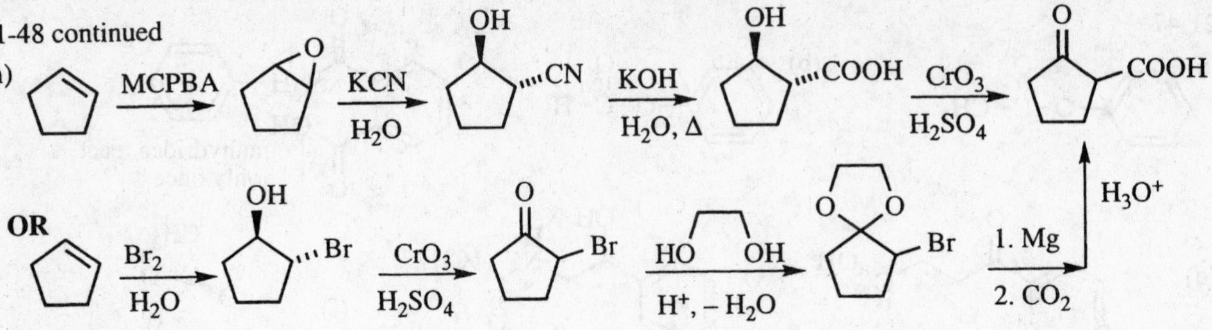

OR

21-49

(a)

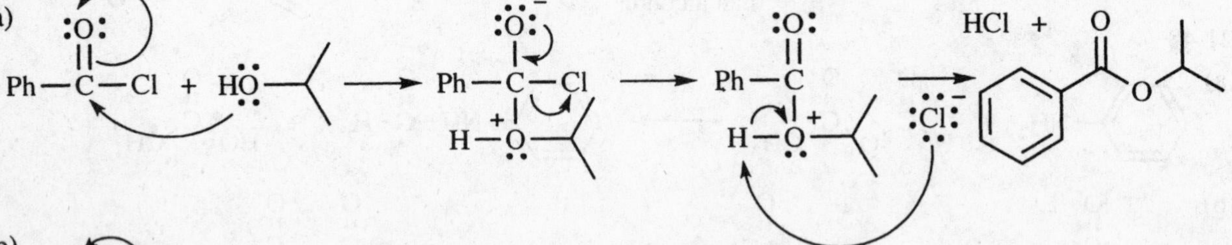

(b)

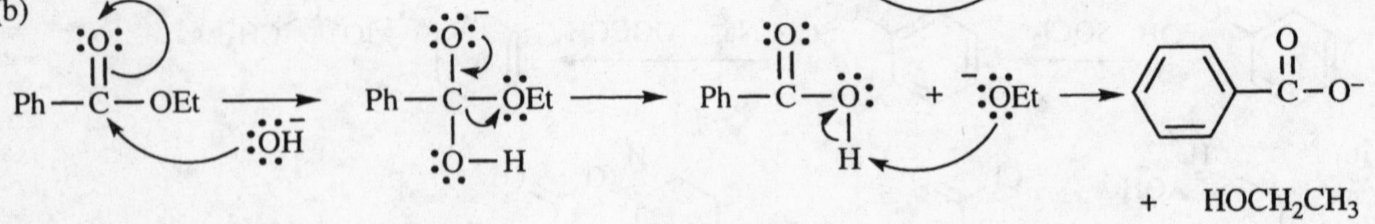

(c)

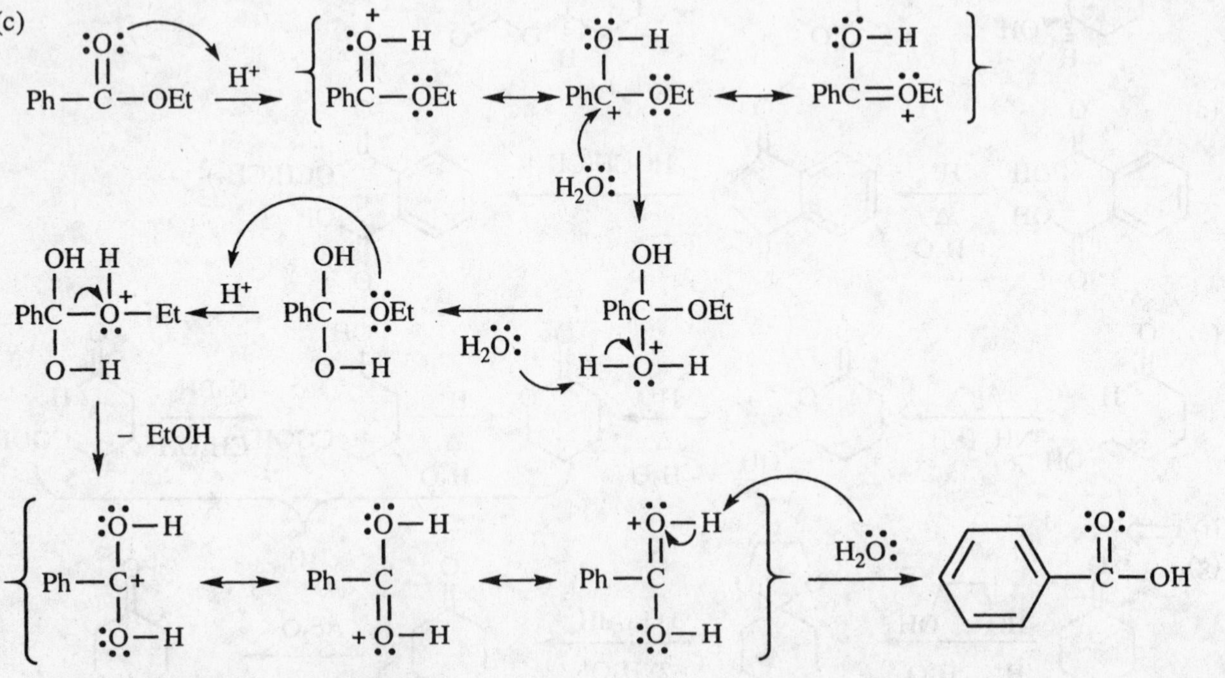

Note: species with positive charge on carbon adjacent to benzene also have resonance forms (not shown) with the positive charge distributed over the ring.

21-49 continued

(d)

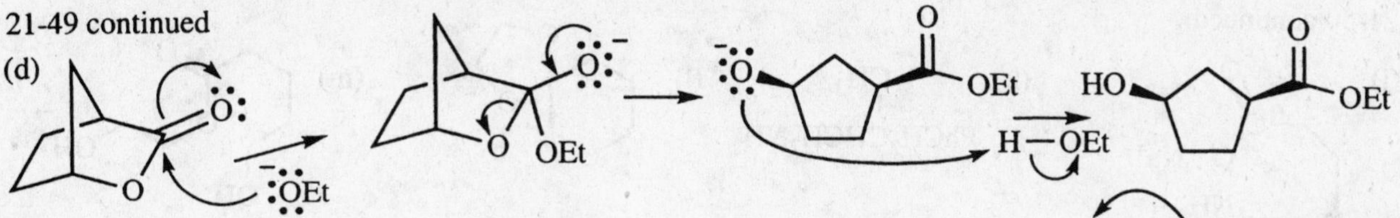

(e)

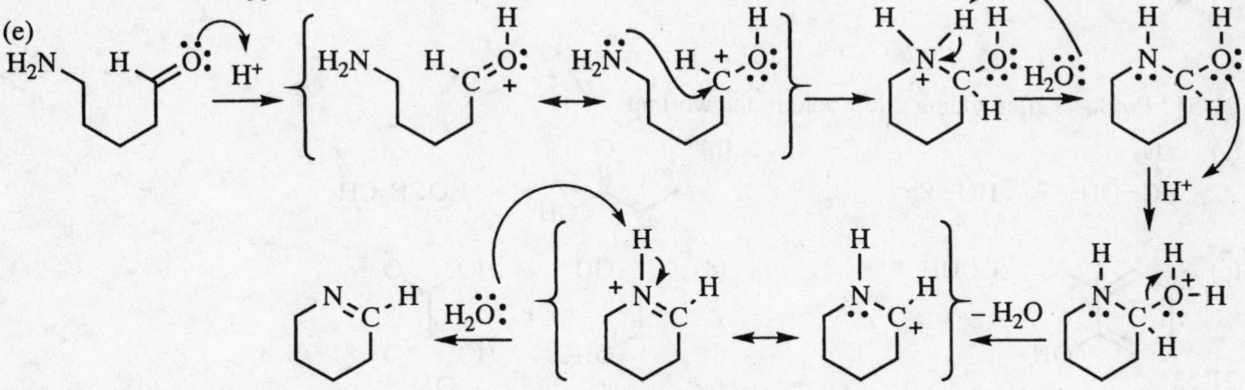

(f)

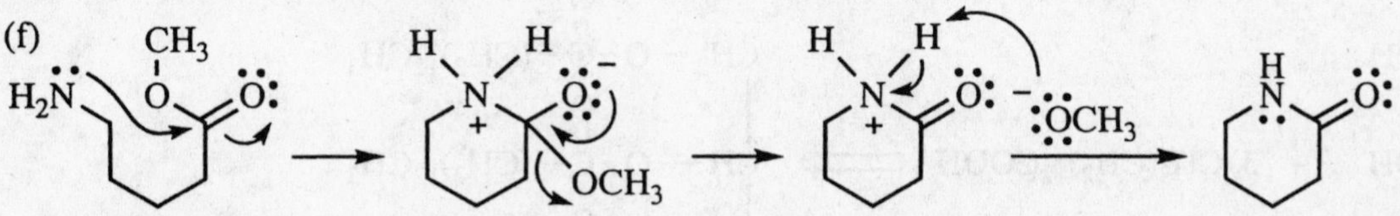

(g)

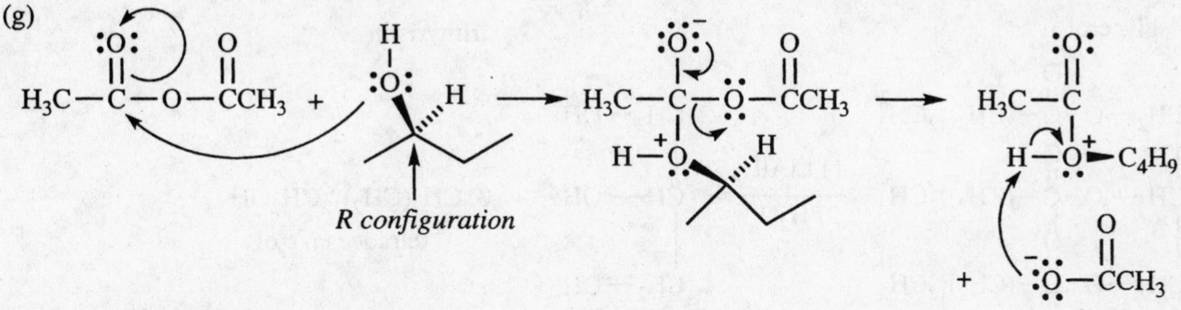

R configuration

No bond to the chiral center
is broken, so the
configuration is retained.

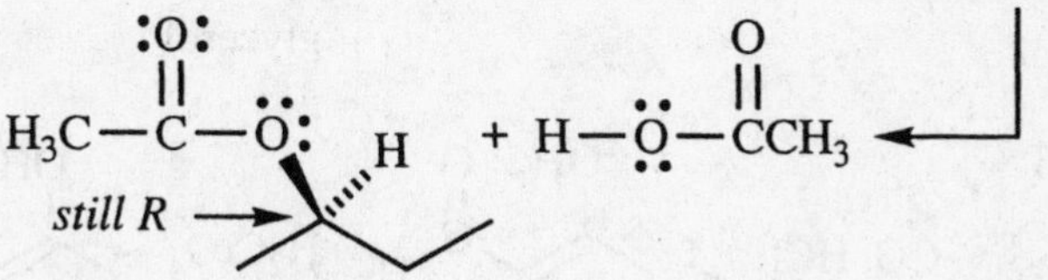

still R →

21-50

(a)

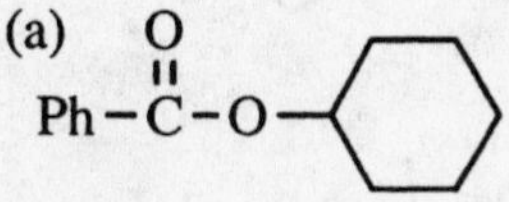

$Ph-\overset{O}{\underset{}{C}}-O-$

(b)

$\overset{O}{\underset{}{C}}-NHCH_3$

(c)

$Ph-\overset{O}{\underset{}{C}}-N$

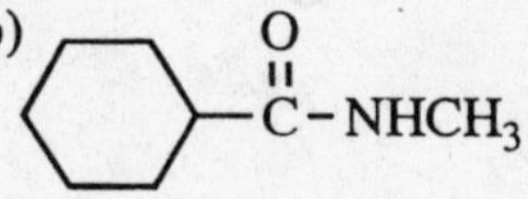

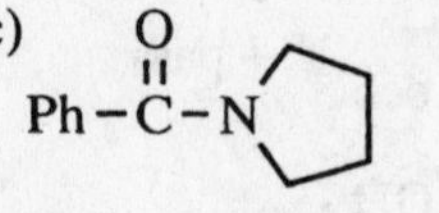

(d)

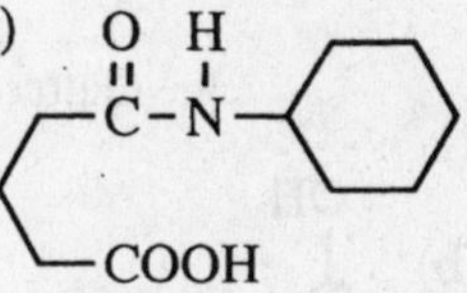

(e)
PhCH₂OH

(f)

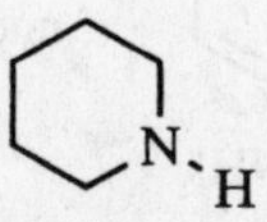

(g)

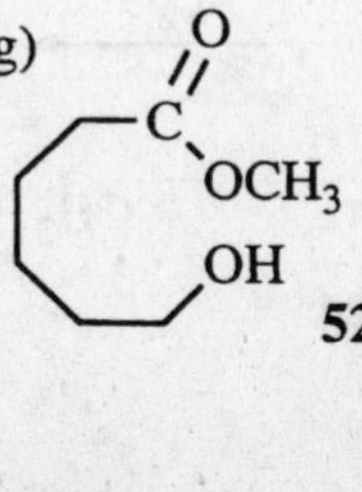

(h)

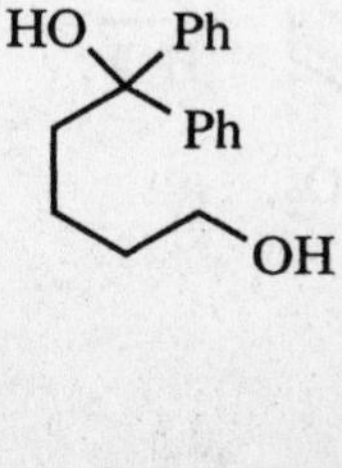

(i)

523

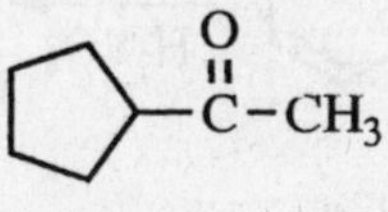

21-50 continued

(j)
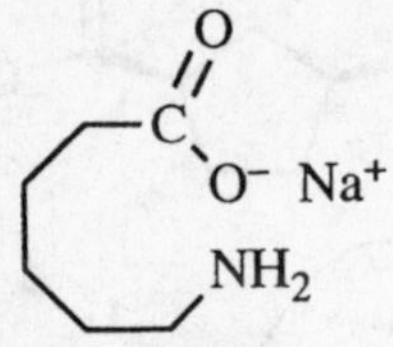

(k) $PhCH_2CH(CH_3)CH_2NH_2$

(l)
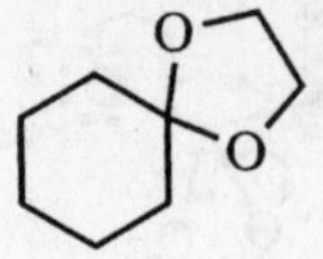

(m)
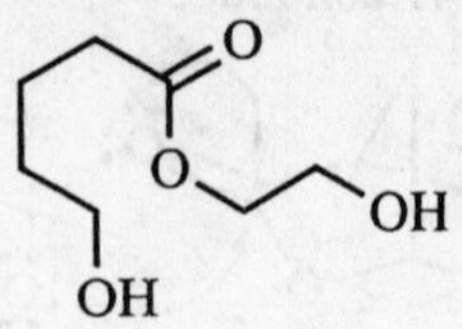

21-51 Products after adding dilute acid in the workup:

(a) $HC(=O)-OH$ + $HO-Ph$

(b)
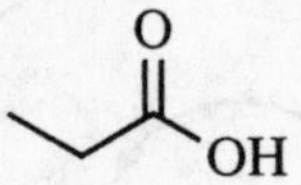

+ $HOCH_2CH_3$

(c)
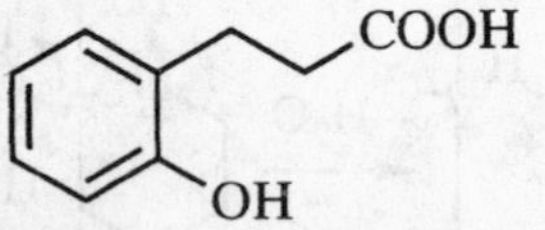

(d)
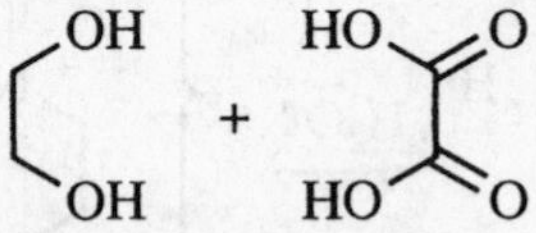

21-52

(a)
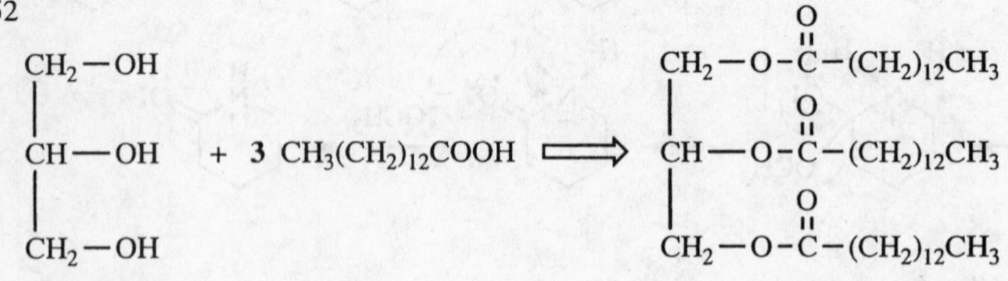

glycerol   trimyristin

(b)
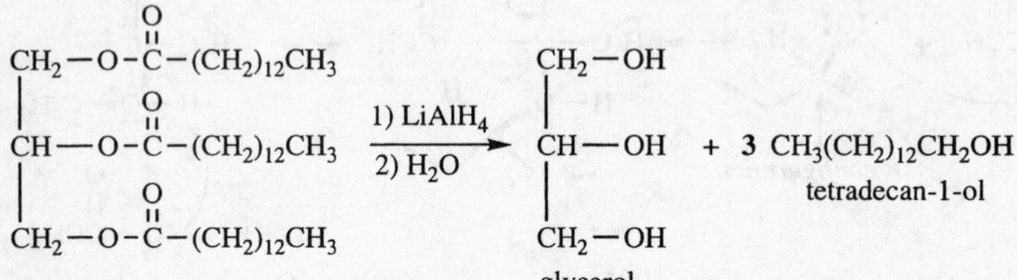

glycerol

21-53

(a)
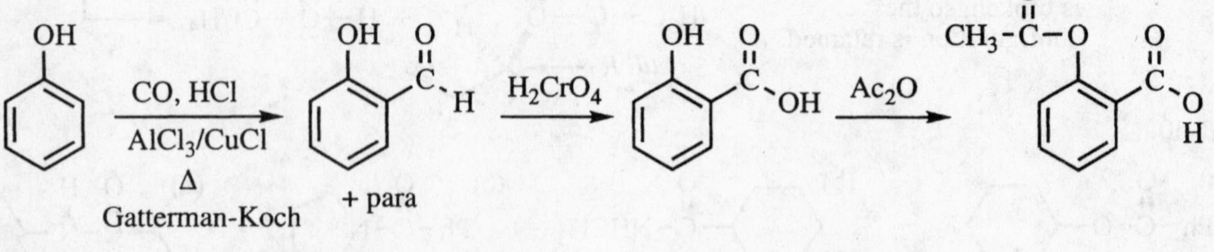

(b)
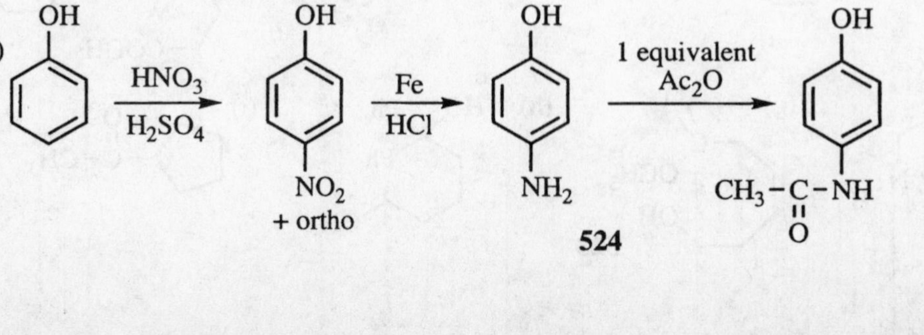

NH_2 is more nucleophilic than OH

21-54

(a)

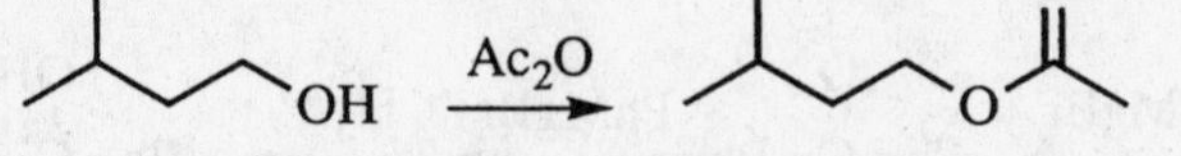

(b)

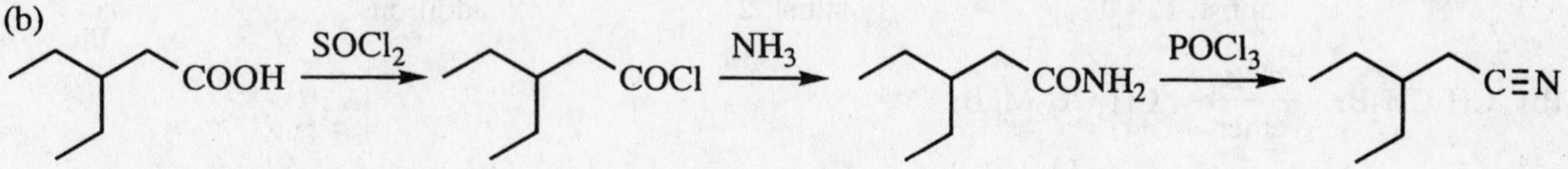

(c)

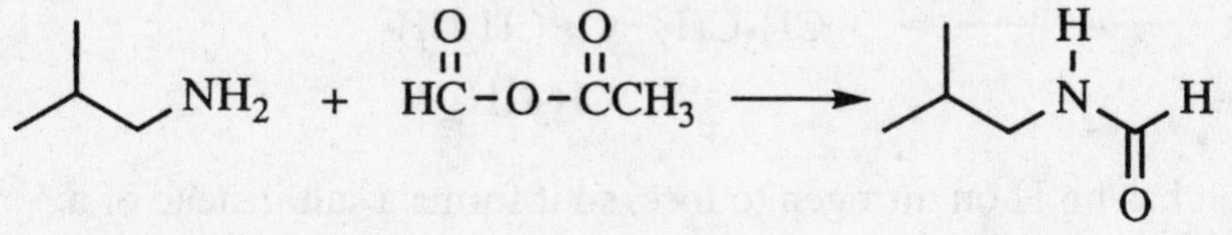

(d)

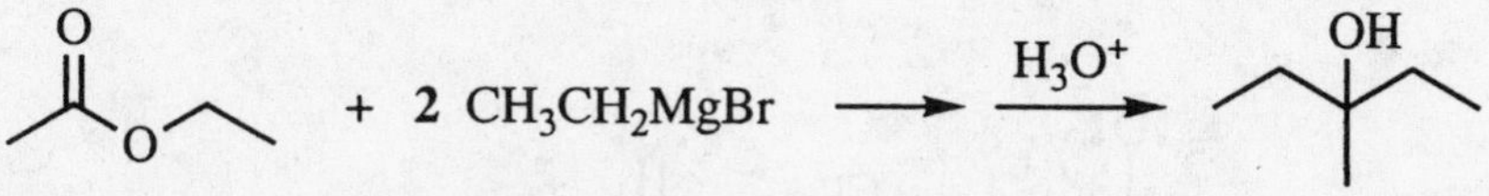

(e) See the solution to 21.36(b) for one method. Here is another: reductive alkylation.

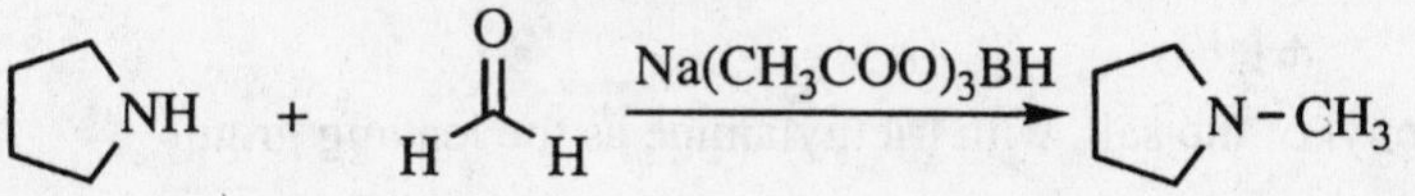

(f)

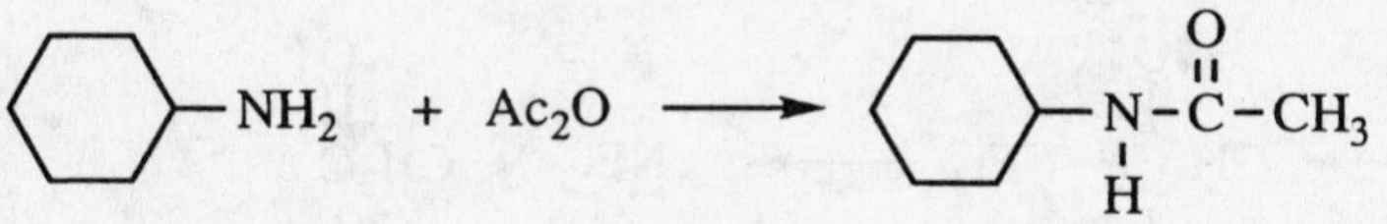

(g)

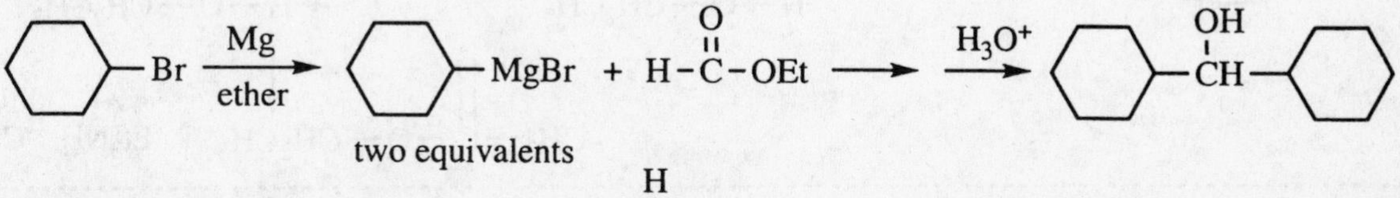

two equivalents

(h)

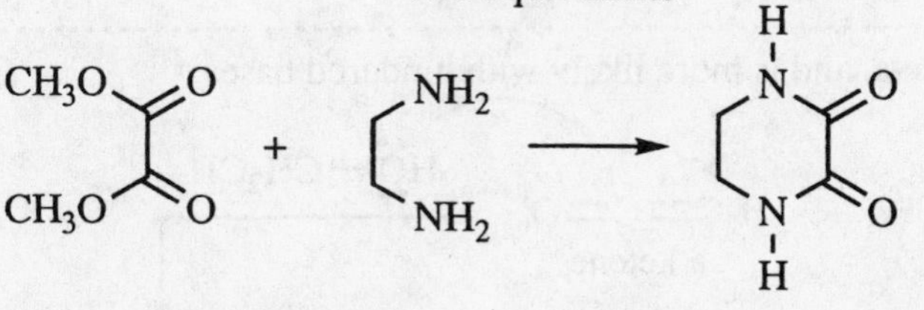

(i)

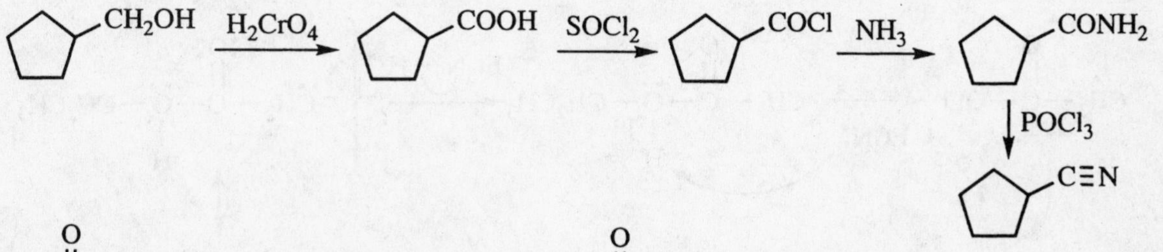

(j)

525

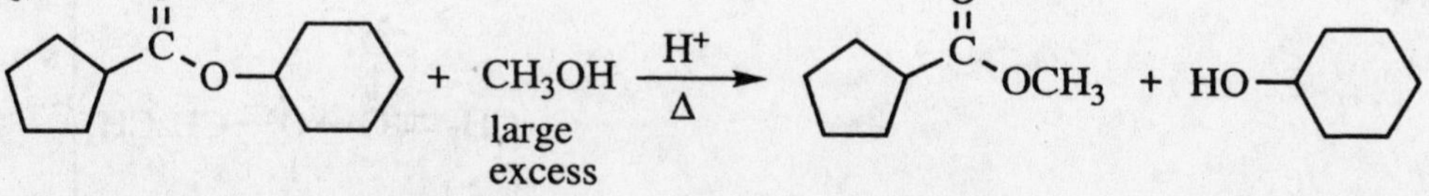

21-55 Diethyl carbonate has *two* leaving groups on the carbonyl. It can undergo *two* nucleophilic acyl substitutions, followed by one nucleophilic addition.

(a)

$$EtO-\overset{\overset{\displaystyle O}{\|}}{C}-OEt \xrightarrow[\text{subst. 1}]{PhMgBr} \left[Ph-\overset{\overset{\displaystyle O}{\|}}{C}-OEt \right] \xrightarrow[\text{subst. 2}]{PhMgBr} Ph-\overset{\overset{\displaystyle O}{\|}}{C}-Ph \xrightarrow[\text{addition}]{PhMgBr} \xrightarrow{H_3O^+} Ph-\overset{\overset{\displaystyle OH}{|}}{\underset{\underset{\displaystyle Ph}{|}}{C}}-Ph$$

(b) $CH_3CH_2Br \xrightarrow[\text{ether}]{Mg} CH_3CH_2MgBr$

$$3\ CH_3CH_2MgBr\ +\ EtO-\overset{\overset{\displaystyle O}{\|}}{C}-OEt \longrightarrow \xrightarrow{H_3O^+} CH_3CH_2-\overset{\overset{\displaystyle OH}{|}}{\underset{\underset{\displaystyle CH_2CH_3}{|}}{C}}-CH_2CH_3$$

21-56 Triethylamine is nucleophilic, but it has no H on nitrogen to lose, so it forms a salt instead of a stable amide.

$$CH_3\overset{\overset{\displaystyle :O:}{\|}}{C}-Cl\ +\ :NEt_3 \longrightarrow CH_3\overset{\overset{\displaystyle :\overset{-}{O}:}{|}}{\underset{\underset{\displaystyle +NEt_3}{|}}{C}}-Cl \xrightarrow{-Cl^-} CH_3\overset{\overset{\displaystyle :O:}{\|}}{C}-\overset{+}{N}Et_3$$

When ethanol is added, it attacks the carbonyl of the salt, with triethylamine as the leaving group.

$$CH_3\overset{\overset{\displaystyle :O:}{\|}}{C}-\overset{+}{N}Et_3\ +\ H\overset{..}{\underset{..}{O}}-CH_2CH_3 \longrightarrow CH_3\overset{\overset{\displaystyle :\overset{-}{O}:}{|}}{\underset{\underset{\displaystyle H-\overset{+}{\underset{..}{O}}-CH_2CH_3}{|}}{C}}-\overset{+}{N}Et_3 \longrightarrow \overset{..}{N}Et_3\ +\ CH_3\overset{\overset{\displaystyle :O:}{\|}}{C}$$

$$H-\overset{+}{\underset{..}{O}}-CH_2CH_3$$

$$CH_3-\overset{\overset{\displaystyle O}{\|}}{C}-O-CH_2CH_3\ +\ Et_3\overset{+}{N}H\ \ Cl^-$$

An alternate mechanism explains the same products, and is more likely with hindered bases:

$$H_2\overset{\underset{\underset{\displaystyle H}{|}}{|}}{C}-\overset{\overset{\displaystyle O}{\|}}{C}-Cl \xrightarrow{\text{like an E2}} Et_3\overset{+}{N}H\ \ Cl^-\ +\ H_2C=C=O\ +\ \underset{\text{a ketene}}{} \quad H\overset{..}{\underset{..}{O}}-CH_2CH_3$$

$$:NEt_3$$

$$CH_3-\overset{\overset{\displaystyle O}{\|}}{C}-OEt \xleftarrow[Et_3N:]{} CH_3-\overset{\overset{\displaystyle O}{\|}}{C}-\overset{+}{\underset{\underset{\displaystyle H}{|}}{\overset{..}{O}}}-CH_2CH_3 \xleftarrow{} Et_3\overset{+}{N}-H \quad \left\{ {}^-\overset{..}{C}H_2-\overset{\overset{\displaystyle :O:}{\|}}{C}-\overset{+}{\underset{\underset{\displaystyle H}{|}}{\overset{..}{O}}}-CH_2CH_3 \right.$$

$$CH_2=\overset{\overset{\displaystyle :\overset{-}{O}:}{|}}{C}-\overset{+}{\underset{\underset{\displaystyle H}{|}}{\overset{..}{O}}}-CH_2CH_3 \left. \right\}$$

21-57

(a)

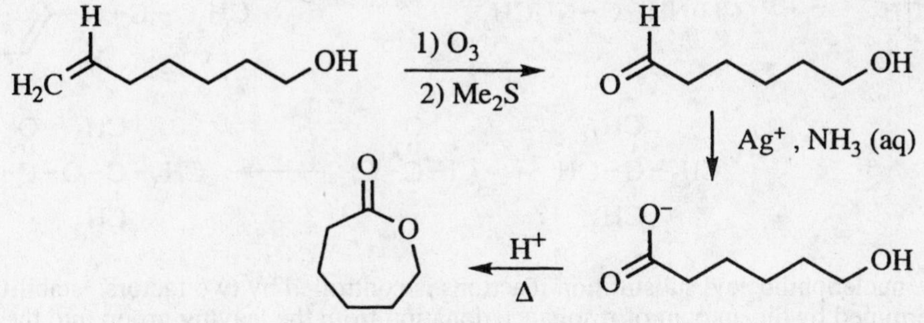

H_2C=...—OH $\xrightarrow[\text{2) Me}_2\text{S}]{\text{1) O}_3}$ H—(O=)...—OH

$\downarrow$ Ag^+, NH_3 (aq)

O (lactone) $\xleftarrow[\Delta]{H^+}$ O^-...—OH

(b)

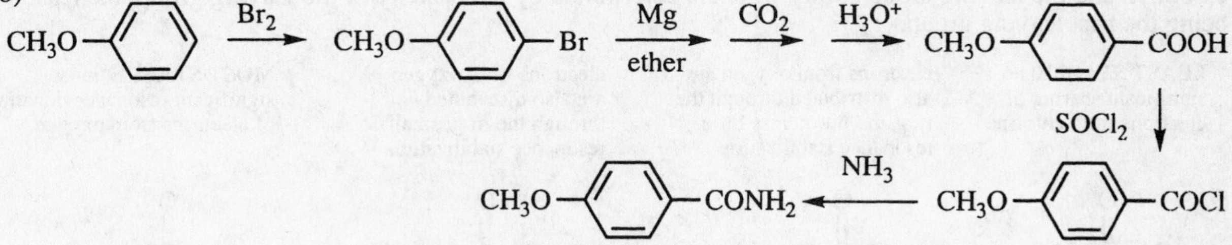

CH_3O—⟨benzene⟩ $\xrightarrow{\text{Br}_2}$ CH_3O—⟨benzene⟩—Br $\xrightarrow[\text{ether}]{\text{Mg}}$ $\xrightarrow{\text{CO}_2}$ $\xrightarrow{\text{H}_3\text{O}^+}$ CH_3O—⟨benzene⟩—COOH

$\downarrow$ $SOCl_2$

CH_3O—⟨benzene⟩—$CONH_2$ $\xleftarrow{\text{NH}_3}$ CH_3O—⟨benzene⟩—COCl

(c)

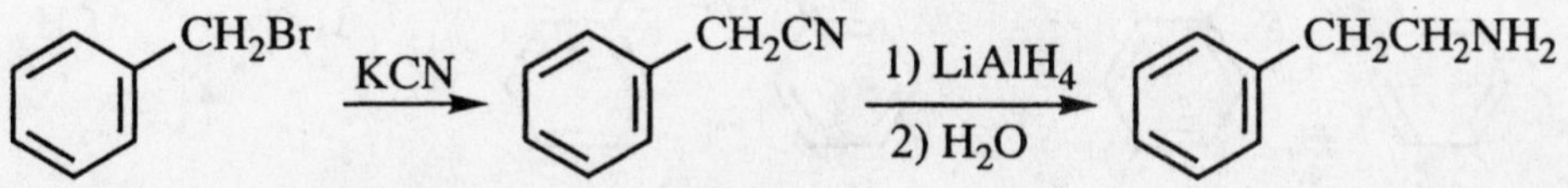

⟨benzene⟩—CH_2Br $\xrightarrow{\text{KCN}}$ ⟨benzene⟩—CH_2CN $\xrightarrow[\text{2) H}_2\text{O}]{\text{1) LiAlH}_4}$ ⟨benzene⟩—$CH_2CH_2NH_2$

(d)

COOH / HO, OH, OH benzene $\xrightarrow[\text{2) CH}_3\text{I excess}]{\text{1) NaOH excess}}$ COOH / CH_3O, OCH_3, OCH_3 benzene $\xrightarrow[\text{2) H}_2\text{O}]{\text{1) LiAlH}_4}$ CH_2OH / CH_3O, OCH_3, OCH_3 benzene

$\downarrow$ TsCl pyridine

$CH_2CH_2NH_2$ / CH_3O, OCH_3, OCH_3 $\xleftarrow[\text{2) H}_2\text{O}]{\text{1) LiAlH}_4}$ CH_2CN / CH_3O, OCH_3, OCH_3 $\xleftarrow{\text{KCN}}$ CH_2OTs / CH_3O, OCH_3, OCH_3

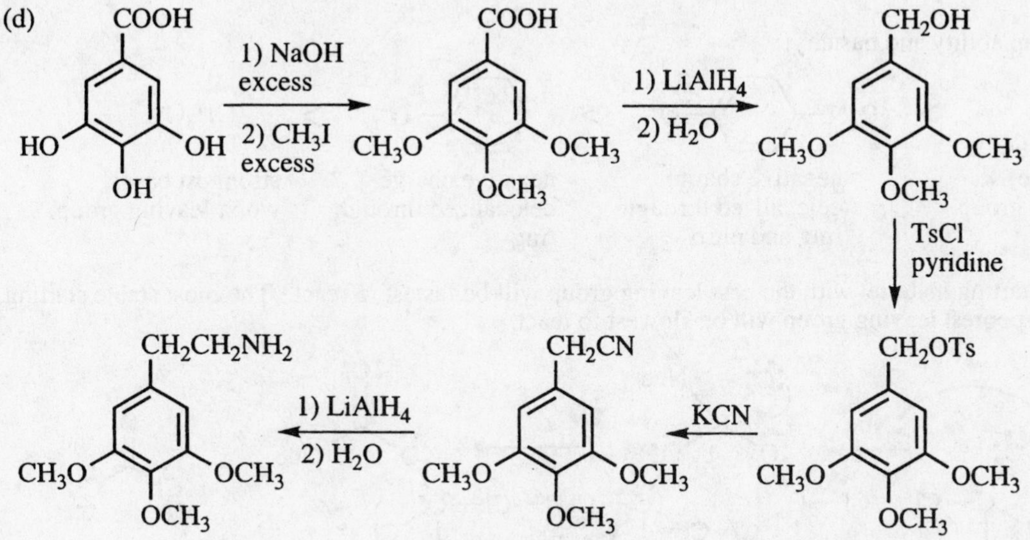

21-58

(a) $CH_3CH_2O-\overset{\overset{\displaystyle O}{\|}}{C}-OCH_2CH_3$

(b) $CH_3NH-\overset{\overset{\displaystyle O}{\|}}{C}-NHCH_3$

(c) $CH_3O-\overset{\overset{\displaystyle O}{\|}}{C}-\underset{\underset{\displaystyle H}{|}}{N}-C_6H_5$

(d) [cyclic carbonate: five-membered ring O–C(=O)–O with ethylene bridge]

(e) $CH_3-\overset{\overset{\displaystyle CH_3}{|}}{\underset{\underset{\displaystyle CH_3}{|}}{C}}-OH \;+\; Cl-\overset{\overset{\displaystyle O}{\|}}{C}-Cl \longrightarrow CH_3-\overset{\overset{\displaystyle CH_3}{|}}{\underset{\underset{\displaystyle CH_3}{|}}{C}}-O-\overset{\overset{\displaystyle O}{\|}}{C}-Cl$

21-59 (a) The rate of these nucleophilic acyl substitution reactions is controlled by two factors: stability of the starting material as determined by the amount of resonance donation from the leaving group into the carbonyl, and the leaving group ability which is determined by the basicity of the leaving group, the least basic being the best leaving group.

LEAST STABLE: no significant sharing of electrons from chlorine	electrons from oxygen are also distributed through the ring and nitro; very little resonance stabilization	electrons from oxygen are also distributed through the ring; small resonance stabilization	MOST STABLE: most significant resonance donation of electrons from oxygen

[structures, left to right: phosgene $Cl-C(=O)-Cl$; bis(4-nitrophenyl) carbonate; diphenyl carbonate; dimethyl carbonate]

leaving group ability and basicity:

$$Cl^- \;>\; O_2N\text{–}C_6H_4\text{–}O^- \;>\; C_6H_5\text{–}O^- \;>\; CH_3O^-$$

weakest base; best leaving group	negative charge delocalized through ring and nitro	negative charge delocalized through ring	strongest base; worst leaving group

The least stable starting material with the best leaving group will be fastest to react. The most stable starting material with the poorest leaving group will be slowest to react.

(b)

[mechanism: triphosgene reacting with :Nuc to form tetrahedral intermediate, collapsing to give product $Cl-C(=O)-O-Nuc$ plus phosgene $Cl-C(=O)-Cl$ + Cl^-]

triphosgene

The first step is the standard attack of a nucleophile like CH_3OH at the carbonyl carbon to make the tetrahedral intermediate. The second step is key: the collapse of the tetrahedral intermediate produces one equivalent of phosgene. Attack of a second nucleophile of the other side of triphosgene would release a latent (hidden or trapped) equivalent of phosgene from that side too. Thus, the equivalent of three molecules of phosgene are locked into the triphosgene molecule. Eventually, all six positions would be substituted with methanol producing three molecules of dimethyl carbonate for each molecule of triphosgene.

21-60

(a)
$$CH_3-N=C=O \ + \ H_2O \longrightarrow H_3C-\overset{\overset{H}{|}}{N}-\overset{\overset{\displaystyle O}{||}}{C}-OH \longrightarrow CH_3NH_2 \ (g) \ + \ CO_2 \ (g)$$

methyl isocyanate

a carbamic acid—unstable

Both of these reactions are exothermic. In a closed vessel like an industrial reactor, the production of gaseous products causes a large pressure increase, risking an explosion.

(b)

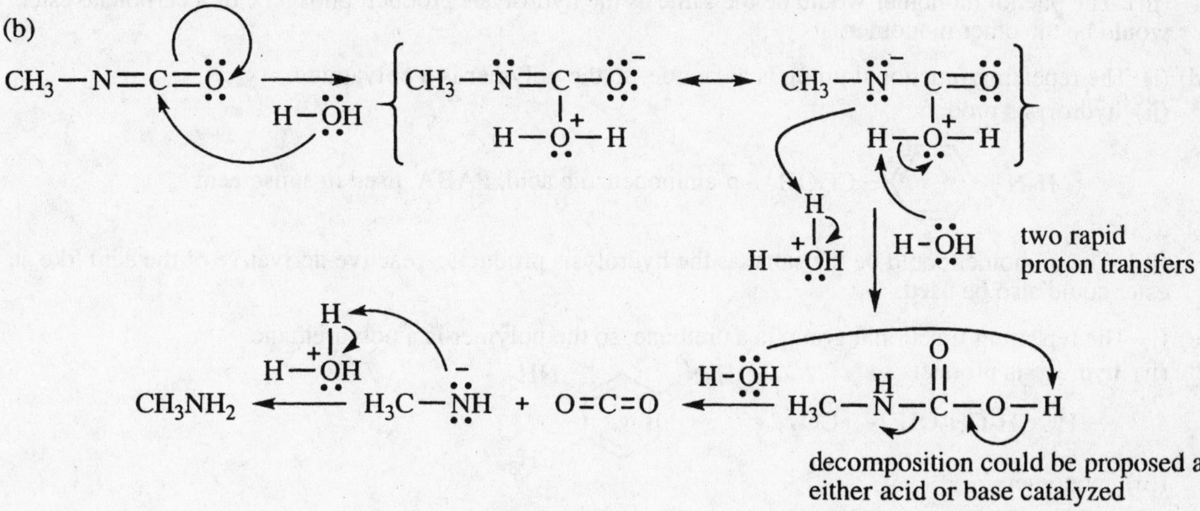

two rapid proton transfers

decomposition could be proposed as either acid or base catalyzed

(c)

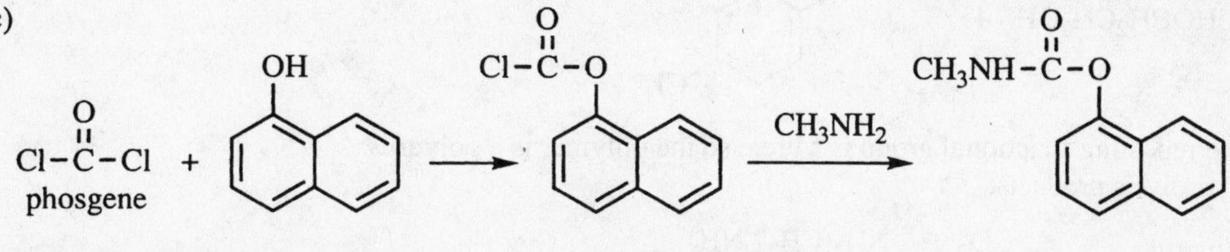

phosgene

21-61

(a) (i) The repeating functional group is an ester, so the polymer is a polyester (named Kodel®).

(ii) hydrolysis products:

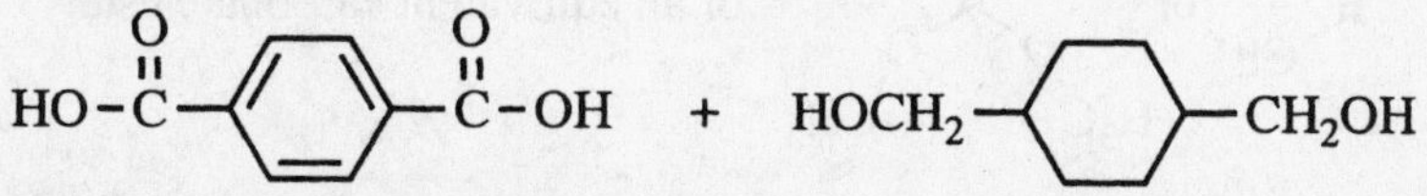

(iii) The monomers could be the same as the hydrolysis products, or else some reactive derivative of the dicarboxylic acid, like an acid chloride or an ester derivative.

(b)(i) The repeating functional group is an amide, so the polymer is a polyamide (named Nylon 6).

(ii) hydrolysis product:

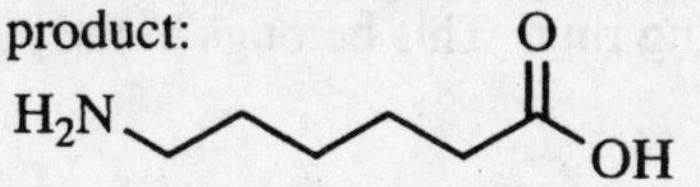

(iii) The monomer could be the same as the hydrolysis product, but in the polymer industry, the actual monomer used is the lactam shown at the right.

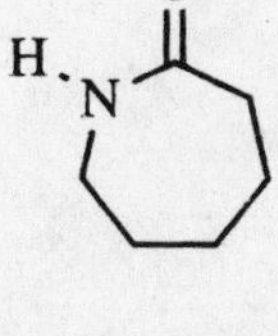

21-61 continued

(c) (i) The repeating functional group is a carbonate, so the polymer is a polycarbonate (named Lexan®).
(ii) hydrolysis products:

HO—〔benzene ring〕—C(CH₃)₂—〔benzene ring〕—OH + CO_2

(iii) The phenol monomer would be the same as the hydrolysis product; phosgene or a carbonate ester would be the other monomer.

(d) (i) The repeating functional group is an amide, so the polymer is a polyamide.
(ii) hydrolysis product:

H_2N—〔benzene ring〕—COOH   p-aminobenzoic acid, PABA, used in sunscreens

(iii) The monomer could be the same as the hydrolysis product; a reactive derivative of the acid like an ester could also be used.

(e) (i) The repeating functional group is a urethane, so the polymer is a polyurethane.
(ii) hydrolysis products:

$HOCH_2CH_2OH$ + CO_2 + 〔benzene ring with H_2N, NH_2, CH_3〕

(iii) monomers:

$HOCH_2CH_2OH$ + 〔O=C=N—benzene ring—N=C=O with CH_3〕

(f) (i) The repeating functional group is a urea, so the polymer is a polyurea.
(ii) hydrolysis products:

CO_2 + $NH_2(CH_2)_9NH_2$

(iii) monomers:

$NH_2(CH_2)_9NH_2$ + 〔Cl–CO–Cl〕 or 〔H_3C–O–CO–O–CH_3〕 or an equivalent carbonate ester

21-62

(a) Both structures are β-lactam antibiotics, a penicillin and a cephalosporin.
(b) "Cephalosporin N" has a 5-membered, sulfur-containing ring. This belongs in the penicillin class of antibiotics.

21-63 The rate of a reaction depends on its activation energy, that is, the difference in energy between starting material and the transition state. The transition state in saponification is similar in structure, and therefore in energy, to the tetrahedral intermediate:

$$
\underset{\text{tetrahedral intermediate}}{}
$$

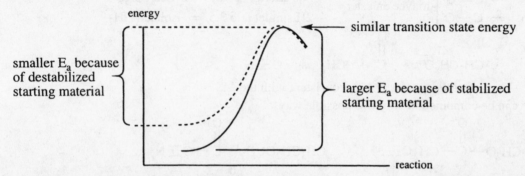

(structure labeled: tetrahedral intermediate)

The tetrahedral carbon has no resonance overlap with the benzene ring, so any resonance effect of a substituent on the ring will have very little influence on the energy of the transition state.

What will have a big influence on the activation energy is whether a substituent stabilizes or destabilizes the starting material. Anything that stabilizes the starting material will therefore increase the activation energy, slowing the reaction; anything that destabilizes the starting material will decrease the activation energy, speeding the reaction.

energy

similar transition state energy

smaller E_a because of destabilized starting material

larger E_a because of stabilized starting material

reaction

(a) One of the resonance forms of methyl *p*-nitrobenzoate has a positive charge on the benzene carbon adjacent to the positive carbonyl carbon. This resonance form destabilizes the starting material, lowering the activation energy, speeding the reaction.

poor resonance contributor, destabilizing the starting material; no effect in the transition state

(b) One of the resonance forms of methyl *p*-methoxybenzoate has all atoms with full octets, and negative charge on the most electronegative atom. This resonance form stabilizes the starting material, increasing the activation energy, slowing the reaction.

good resonance contributor, stabilizing the starting material; no effect in the transition state

21-64

21-65 A singlet at δ 2.15 is H on carbon next to carbonyl, the only type of proton in the compound. The IR spectrum shows no OH, and shows two carbonyl absorptions at high frequency, characteristic of an anhydride. Mass of the molecular ion at 102 proves that the anhydride must be acetic anhydride, a reagent commonly used in aspirin synthesis.

$$
\begin{array}{ccc}
 & O & \quad O \\
 & \parallel & \quad \parallel \\
 & C & \quad C \\
H_3C & \diagdown O \diagup & \diagdown CH_3
\end{array}
$$

Acetic anhydride can be disposed of by hydrolyzing (carefully! exothermic!) and neutralizing in aqueous base.

21-66

IR spectrum:	NMR spectrum:
—sharp spike at 2250 cm⁻¹ ⇒ C≡N	—triplet and quartet ⇒ CH_3CH_2
—1750 cm⁻¹ ⇒ C=O ⎫	—this quartet at δ 4.3 ⇒ CH_3CH_2O
—1200 cm⁻¹ ⇒ C—O ⎬ maybe an ester	—2H singlet at δ 3.5 ⇒ isolated CH_2

$$
\begin{array}{ccccccc}
 & & O & & & & \\
 & & \parallel & & & & \\
CH_3CH_2O & + & C & + & CH_2 & + & C\equiv N
\end{array}
$$

sum of the masses is 113, consistent with the MS

The fragments can be combined in only two possible ways:

$$
\begin{array}{cc}
\quad O & \quad\quad O \\
\quad \parallel & \quad\quad \parallel \\
CH_3CH_2O-C-CH_2C\equiv N & CH_3CH_2OCH_2-C-C\equiv N \\
\mathbf{A} & \mathbf{B}
\end{array}
$$

The NMR proves the structure to be **A**. If the structure were **B**, the CH_2 between oxygen and the carbonyl would come farther downfield than the CH_2 of the ethyl (deshielded by oxygen and carbonyl instead of by oxygen alone). As this is not the case, the structure cannot be **B**.

The peak in the mass spectrum at m/z 68 is due to α-cleavage of the ester:

$$
\left[\begin{array}{c} O \\ \parallel \\ CH_3CH_2O\!-\!\!\!-\!\!\!-C-CH_2C\equiv N \\ \quad\quad \llcorner 68 \end{array} \right]^{+\cdot} \longrightarrow \left\{ \begin{array}{c} :\ddot{O}: \quad\quad :O^+ \\ \parallel \quad\quad \parallel\parallel\parallel \\ +CCH_2CN \longleftrightarrow CCH_2CN \\ \quad\quad m/z\ 68 \end{array} \right\}
$$

m/z 113   + $CH_3CH_2O\cdot$ mass 45

21-67

The formula C_5H_9NO has 2 elements of unsaturation.

IR spectrum: The strongest peak at 1670 cm^{-1} comes low in the carbonyl region; in the absence of conjugation (no alkene peak observed), a carbonyl this low is almost certainly an amide. There is one broad peak in the NH/OH region, hinting at the likelihood of a secondary amide.

$$\underset{\underset{\displaystyle C}{\parallel}}{O} \quad \underset{\underset{\displaystyle N}{\mid}}{H}$$

—C—N—

HNMR spectrum: The broad peak at δ 7.55 is exchangeable with D_2O; this is an amide proton. A broad, 2H peak at δ 3.3 is a CH_2 next to nitrogen. A broad, 2H peak at δ 2.4 is a CH_2 next to carbonyl. The 4H peak at δ 1.8 is probably two more CH_2 groups. There appears to be coupling among these protons but it is not resolved enough to be useful for interpretation. This is often the case when the compound is cyclic, with restricted rotation around carbon-carbon bonds, giving *non-equivalent* (axial and equatorial) hydrogens on the same carbon.

$$CH_2 - \underset{\underset{\displaystyle C}{\parallel}}{O} - \underset{\underset{\displaystyle N}{\mid}}{H} - CH_2 \quad + \quad CH_2 \quad + \quad CH_2$$

CNMR spectrum: The peak at δ 175 is the C=O of the amide. All of the peaks between δ 25 and δ 50 are aliphatic sp^3 carbons, no sp^2 carbons, so the remaining element of unsaturation cannot be a C=C; it must be a ring. The carbon peak farthest downfield is the carbon adjacent to N.

The most consistent structure:

δ-valerolactam

21-68

<u>IR spectrum:</u> A strong carbonyl peak at 1720 cm^{-1}, in conjunction with the C—O peak at 1200 cm^{-1}, suggests the presence of an ester. An alkene peak appears at 1660 cm^{-1}.

$$C=C \qquad \overset{\overset{O}{\parallel}}{C}-O-C$$

<u>HNMR spectrum:</u> The typical ethyl pattern stands out: 3H triplet at δ 1.25 and 2H quartet at δ 4.2. The chemical shift of the CH$_2$ suggests it is bonded to an oxygen. The other groups are: a 3H doublet at δ 1.8, likely to be a CH$_3$ next to one H; a 1H doublet at δ 5.8, a vinyl hydrogen with one neighboring H; and a 1H multiplet at δ 6.9, another vinyl H with many neighbors, the far downfield chemical shift suggesting that it is beta to the carbonyl. The large coupling constant in the doublet at δ 5.8 shows that the two vinyl hydrogens are *trans*.

$$\underset{H}{\overset{H}{\diagdown}}C=C\diagdown \quad + \quad CH_3-\underset{\underset{H}{\mid}}{C} \quad + \quad OCH_2CH_3 \quad + \quad \overset{\overset{O}{\parallel}}{C}-O-C$$

There is only one possible way to assemble these pieces:

$$\begin{array}{c} \delta\,1.8,\,d\;\;H_3C \qquad\qquad H\;\;\delta\,5.8,\,d \\ \diagdown C=C\diagup \\ \delta\,6.9,\,m\;\;H \qquad\qquad C-OCH_2CH_3\;\;\delta\,1.25,\,t \\ \overset{\parallel}{O}\quad\;\delta\,4.2,\,q \end{array}$$

<u>CNMR spectrum:</u> The six unique carbons are unmistakeable: the C=O of the ester at δ 166; the two vinyl carbons at δ 144 (beta to C=O) and at δ 123 (alpha to C=O); the CH$_2$—O of the ester at δ 60; and the two methyls at δ 18 and at δ 14.

<u>Mass spectrum:</u> This structure has mass 114, consistent with the molecular ion.
Major fragmentations:

$$\left[\begin{array}{c} H_3C \qquad H \\ \diagdown C=C\diagup \\ H \qquad\;\; C \overset{99}{\underset{\mid}{+}}OCH_2 \Big| CH_3 \\ \overset{\parallel}{O} \\ 69 \end{array}\right]^{+\cdot} \longrightarrow CH_3CH_2O\cdot \;+\; \underset{\text{mass }45}{}$$

$$\begin{array}{c} H_3C \qquad H \\ \diagdown C=C\diagup \\ H \qquad\;\; C^+ \\ \overset{\parallel}{:O:} \end{array}$$

plus two other resonance forms

m/z 69

$$CH_3\cdot \;+\; \begin{array}{c} H_3C \qquad H \\ \diagdown C=C\diagup \\ H \qquad\;\; C-\overset{+}{\underset{\cdot\cdot}{O}}=CH_2 \\ \underset{O}{\overset{\parallel}{}} \end{array}$$ plus one other resonance form

m/z 99

21-69 If you solved this problem, put a gold star on your forehead.

The formula $C_6H_8O_3$ indicates 3 elements of unsaturation.

<u>IR spectrum</u>: The absence of strong OH peaks shows that the compound is neither an alcohol nor a carboxylic acid. There are two carbonyl absorptions: the one about 1770 cm^{-1} is likely a strained cyclic ester (reinforced with the C—O peak around 1150 cm^{-1}), while the one at 1720 cm^{-1} is probably a ketone. (An anhydride also has two peaks, but they are of higher frequency than the ones in this spectrum.)

$$\underset{\text{C}-\text{O}-\text{C}}{\overset{\displaystyle \overset{O}{\|}}{}} \qquad \underset{\text{C}-\text{C}-\text{C}}{\overset{\displaystyle \overset{O}{\|}}{}}$$

<u>Proton NMR spectrum</u>: The NMR shows four types of protons. The 2H triplet at δ 4.3 is a CH$_2$ group next to an oxygen on one side, with a CH$_2$ on the other. The 1H multiplet at δ 3.7 is also strongly deshielded (probably by two carbonyls), a CH next to a CH$_2$. The 3H singlet at δ 2.45 is a CH$_3$ on one of the carbonyls. The remaining two hydrogens are highly coupled, a CH$_2$ where the two hydrogens are not equivalent. There are no vinyl hydrogens (and no alkene carbon in the carbon NMR), so the remaining element of unsaturation must be a ring.

Assemble the pieces:

$$\underset{\text{C}-\text{O}-\text{CH}_2\text{CH}_2\text{CH}}{\overset{\displaystyle \overset{O}{\|}}{}} \quad + \quad \underset{\text{C}-\text{C}-\text{CH}_3}{\overset{\displaystyle \overset{O}{\|}}{}} \quad + \quad \text{1 ring}$$

On each carbon, the "up" hydrogen is cis to the acetyl group, while the "down" hydrogen is trans. Thus, the two hydrogens on each of these carbons are not equivalent, leading to complex splitting.

<u>Carbon NMR</u>:

δ 173 → C   δ 53   δ 200
δ 68 → C—C   δ 30
δ 24

CHAPTER 22—
CONDENSATIONS AND ALPHA SUBSTITUTIONS OF CARBONYL COMPOUNDS

22-1

(a)

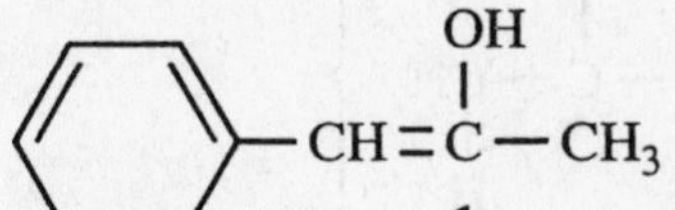

1

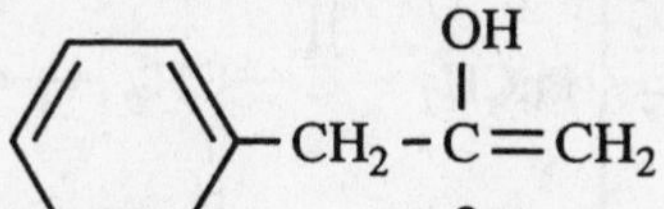

2

(b) Enol **1** will predominate at equilibrium as its double bond is conjugated with the benzene ring, making it more stable than **2**.

(c)

basic conditions forming enol 1

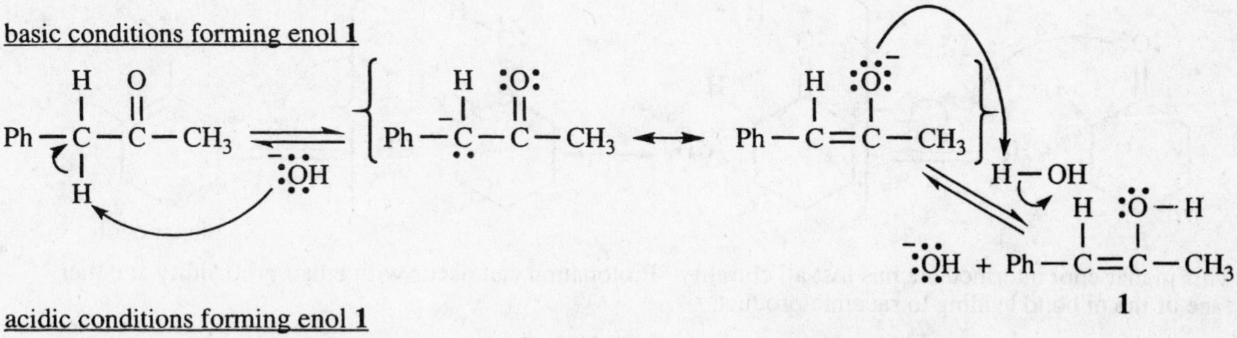

acidic conditions forming enol 1

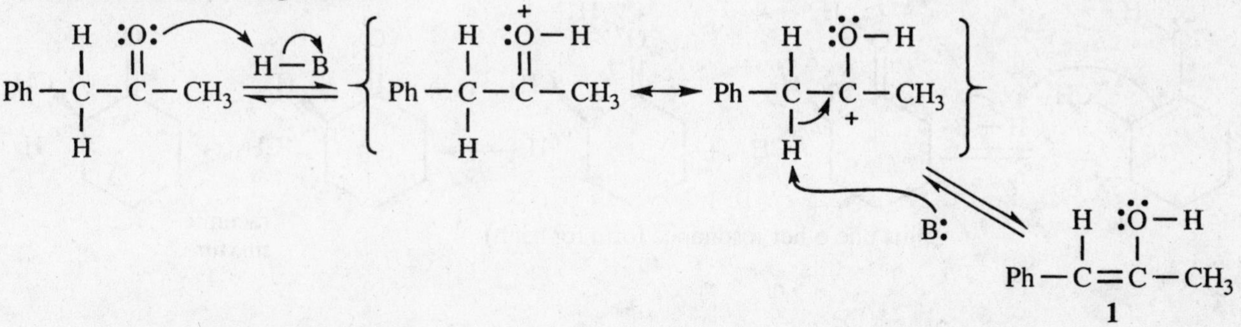

basic conditions forming enol 2

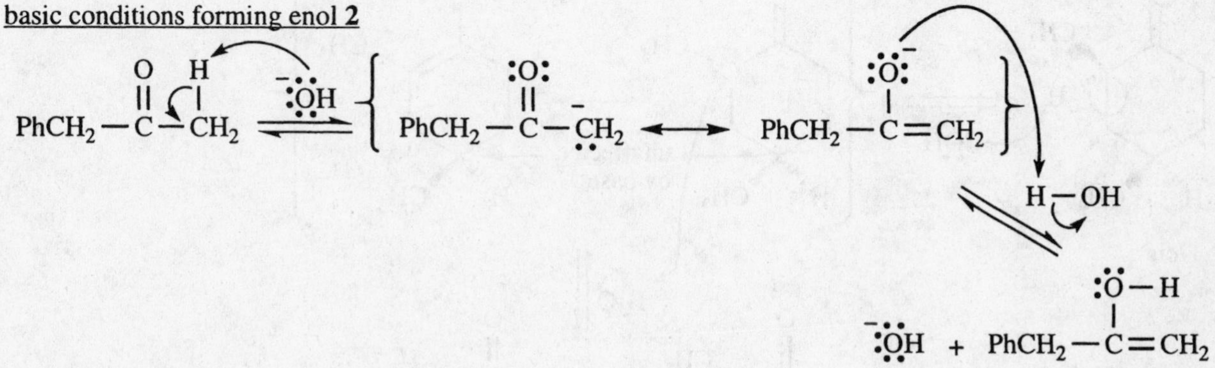

22-1 (c) continued

acidic conditions forming enol 2

22-2

This planar enol intermediate has lost all chirality. Protonation can occur with equal probability at either face of the pi bond leading to racemic product.

22-3

22-4

(a)

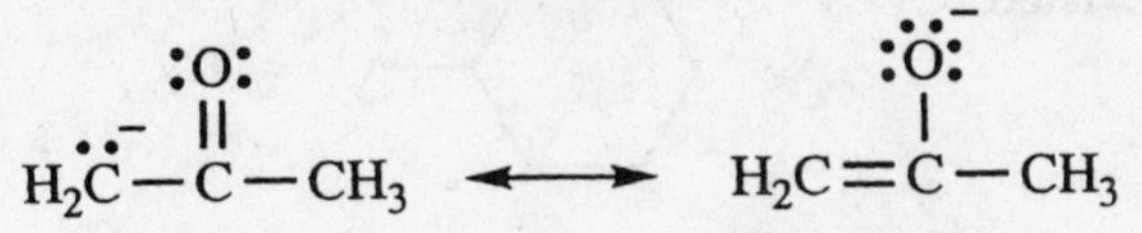

$$H_2\ddot{C}^- \overset{:\overset{..}{O}:}{-}C-CH_3 \longleftrightarrow H_2C=\overset{:\overset{..}{O}:^-}{C}-CH_3$$

(b)

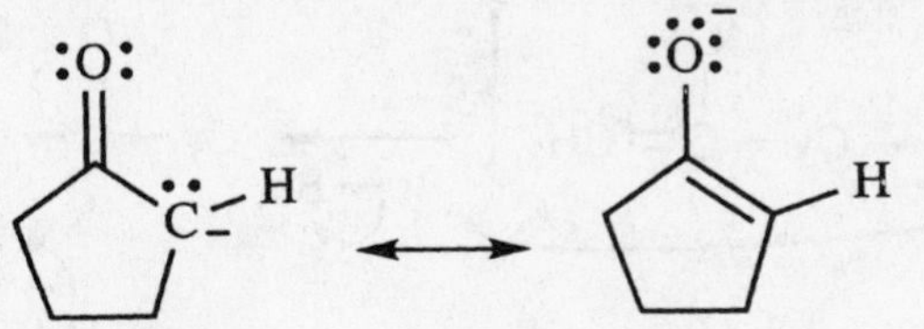

(c)

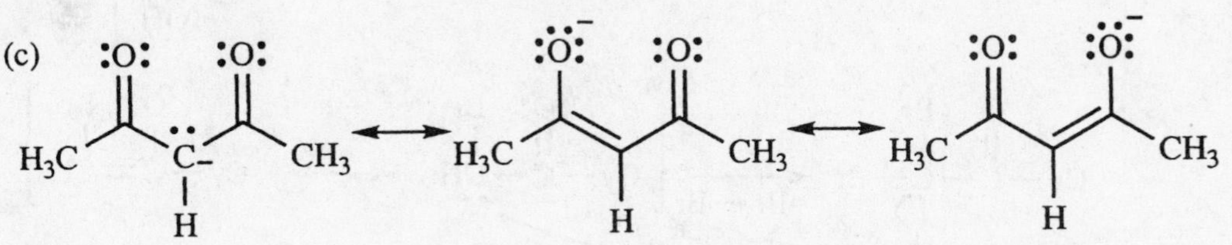

22-5

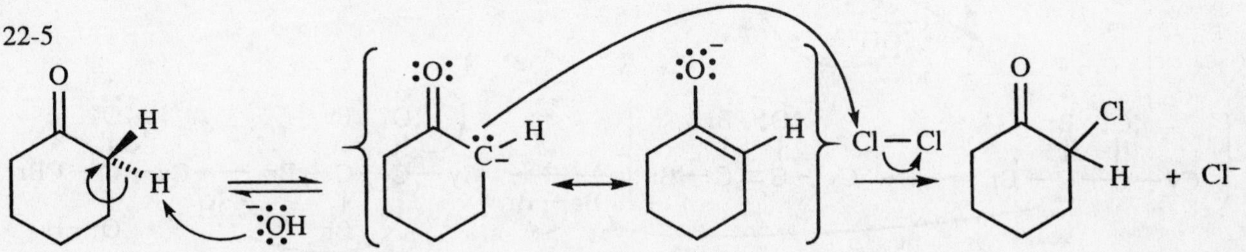

22-6

539

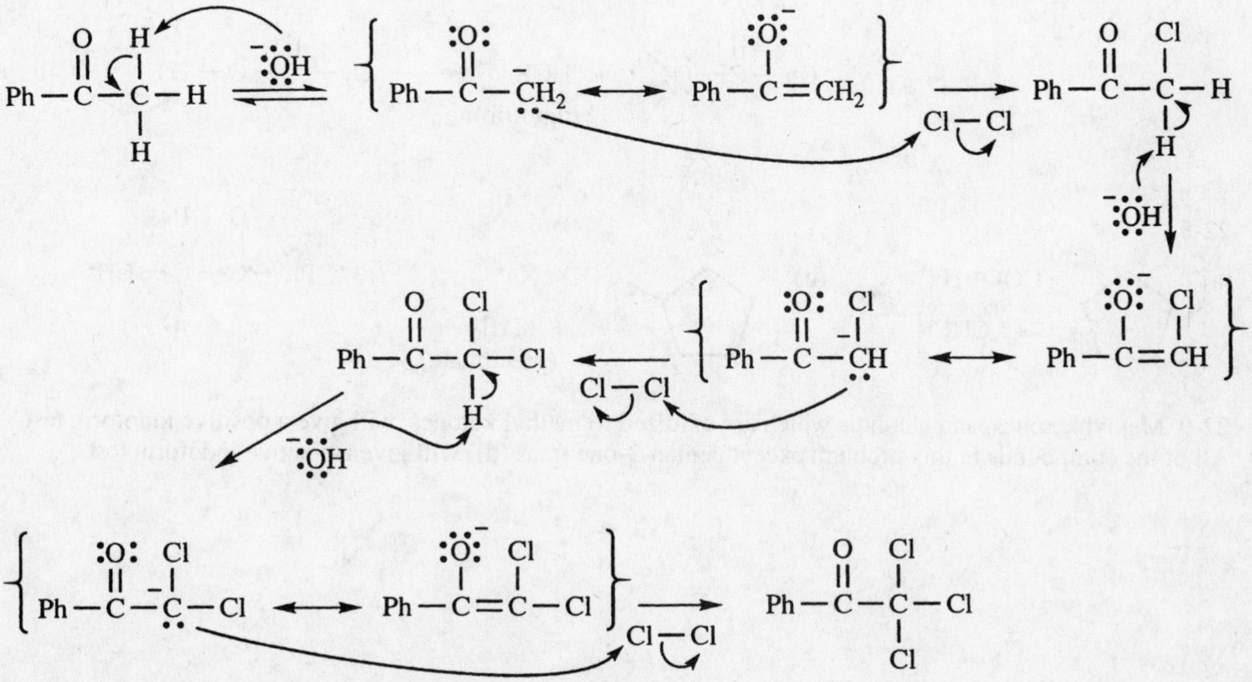

22-7 For this problem, the cyclohexyl group is abbreviated "Cy".

= Cy

$$
\text{Cy} - \underset{\underset{\text{H}}{|}}{\overset{\overset{\text{O}}{\parallel}}{\text{C}}} - \underset{}{\text{C}} - \text{H} \quad \overset{:\text{OH}}{\rightleftharpoons} \quad \left\{ \text{Cy} - \overset{\overset{:\text{O}:}{\parallel}}{\text{C}} - \overset{-}{\text{CH}_2} \longleftrightarrow \text{Cy} - \overset{\overset{:\ddot{\text{O}}:^{-}}{|}}{\text{C}} = \text{CH}_2 \right\} \overset{\text{Br}-\text{Br}}{\longrightarrow} \text{Cy} - \overset{\overset{\text{O}}{\parallel}}{\text{C}} - \overset{\overset{\text{Br}}{|}}{\underset{\underset{\text{H}}{|}}{\text{C}}} - \text{H}
$$

$$
\text{Cy} - \overset{\overset{\text{O}}{\parallel}}{\text{C}} - \overset{\overset{\text{Br}}{|}}{\underset{\underset{\text{H}}{|}}{\text{C}}} - \text{Br} \quad \overset{\text{Br}-\text{Br}}{\longleftarrow} \quad \left\{ \text{Cy} - \overset{\overset{:\text{O}:}{\parallel}}{\text{C}} - \overset{\overset{\text{Br}}{-|}}{\text{CH}} \longleftrightarrow \text{Cy} - \overset{\overset{:\ddot{\text{O}}:^{-}}{|}}{\text{C}} = \overset{\overset{\text{Br}}{|}}{\text{CH}} \right\}
$$

$$
\left\{ \text{Cy} - \overset{\overset{:\text{O}:}{\parallel}}{\text{C}} - \overset{\overset{\text{Br}}{-|}}{\text{C}} - \text{Br} \longleftrightarrow \text{Cy} - \overset{\overset{:\ddot{\text{O}}:^{-}}{|}}{\text{C}} = \overset{\overset{\text{Br}}{|}}{\text{C}} - \text{Br} \right\} \longrightarrow \text{Cy} - \overset{\overset{:\text{O}:}{\parallel}}{\text{C}} - \overset{\overset{\text{Br}}{|}}{\underset{\underset{\text{Br}}{|}}{\text{C}}} - \text{Br} \longrightarrow \text{Cy} - \overset{\overset{:\ddot{\text{O}}:^{-}}{|}}{\underset{\underset{\text{O}-\text{H}}{|}}{\text{C}}} - \text{CBr}_3
$$

$$
\text{Cy} - \overset{\overset{\text{O}}{\parallel}}{\text{C}} - \ddot{\text{O}}:^{-} \quad + \text{ HCBr}_3 \quad \longleftarrow \quad \text{Cy} - \overset{\overset{:\text{O}:}{\parallel}}{\text{C}} - \text{O} - \text{H} \quad + \quad ^{-}:\text{CBr}_3
$$
bromoform

22-8

(a) [cyclopentyl]—COO⁻ Na⁺ + CHCl₃

(b) [cyclopentyl]—COO⁻ Na⁺ + CHI₃ (precipitate)

(c) $\text{Ph} - \overset{\overset{\text{O}}{\parallel}}{\text{C}} - \overset{\overset{\text{Br}}{|}}{\underset{\underset{\text{Br}}{|}}{\text{C}}} - \text{CH}_3$

22-9 Methyl ketones, and alcohols which are oxidized to methyl ketones, will give a positive iodoform test. All of the compounds in this problem except pentan-3-one (part (d)) will give a positive iodoform test.

22-10

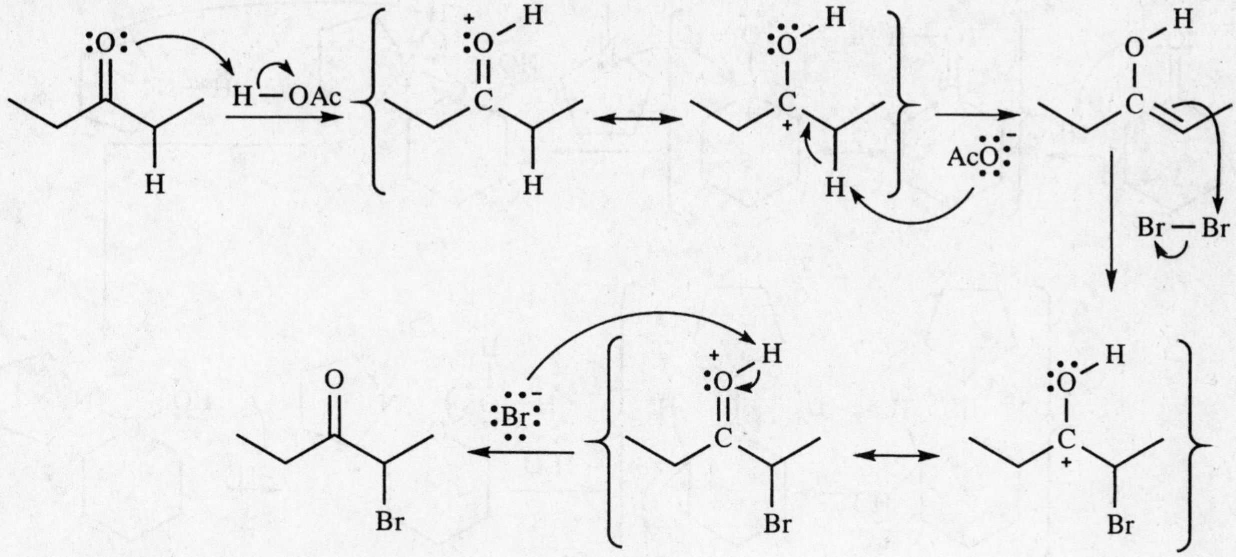

22-11

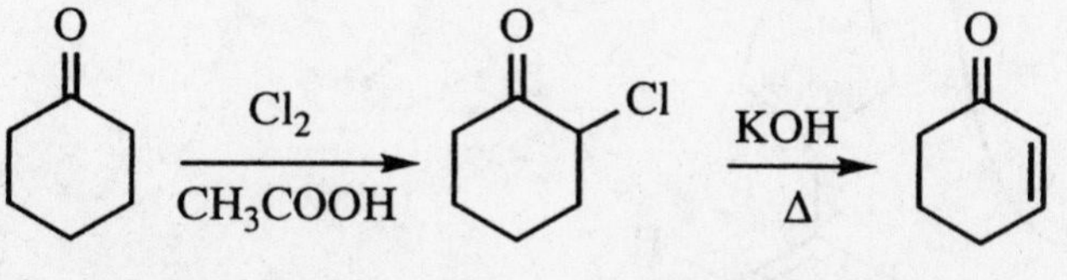

from Solved Problem 22-2   E2 elimination

22-12

(a)

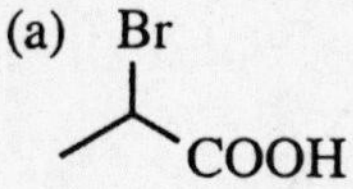

Br

COOH

(b) no reaction:
no α-hydrogen

(c)

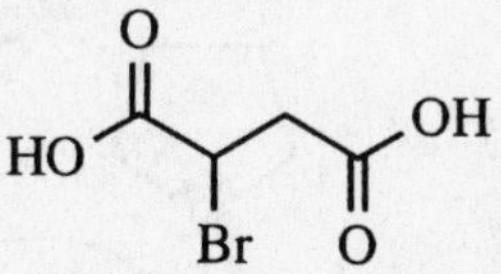

HO   OH

Br   O

(d) no reaction:
no α-hydrogen

22-13

(a)

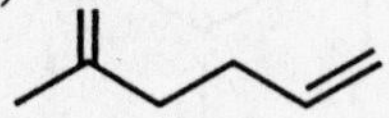

(b)

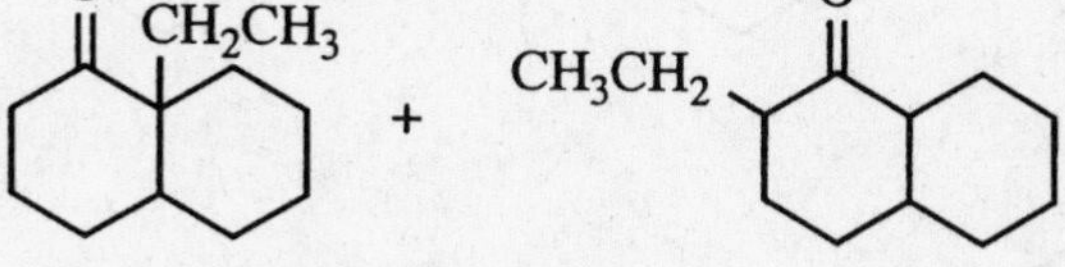

CH₂CH₃   +   CH₃CH₂

(c)

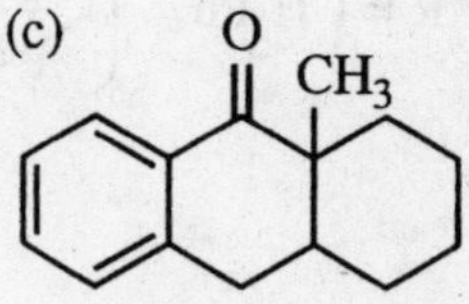

CH₃

22-14

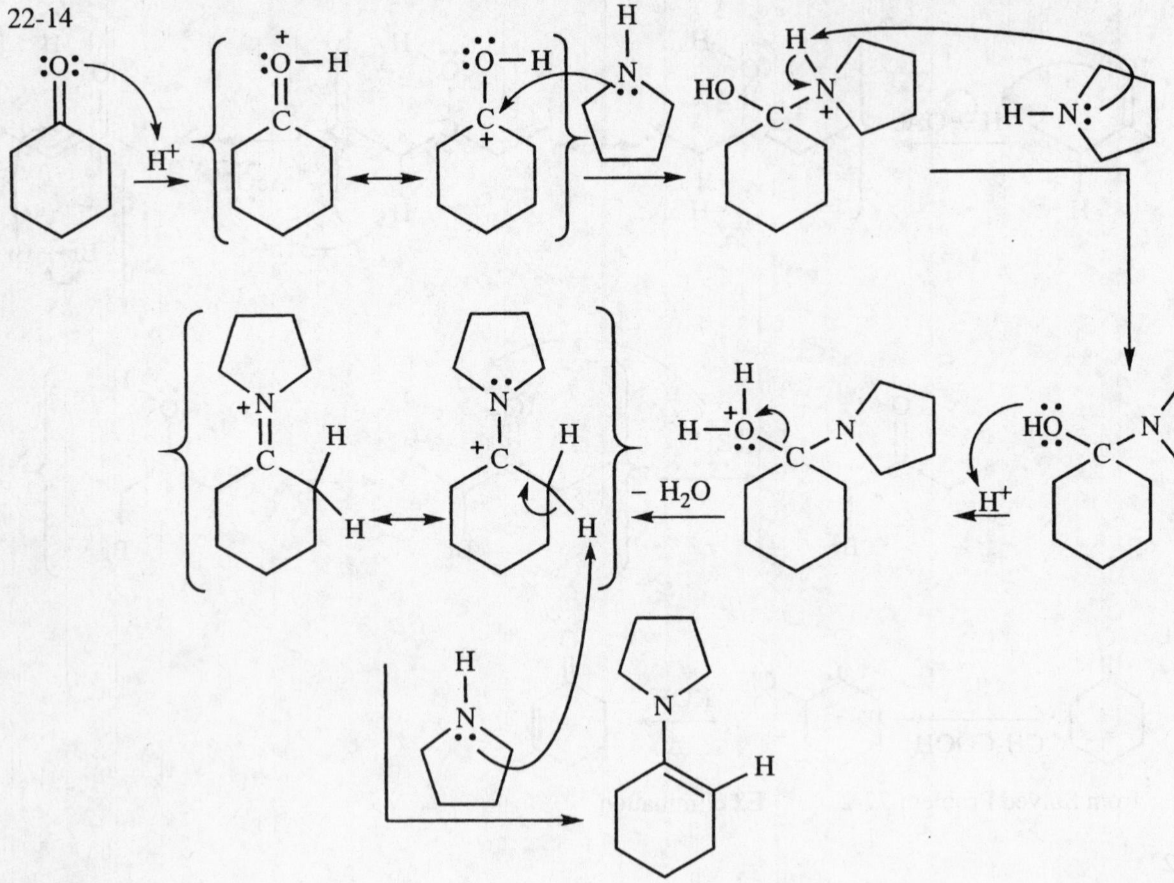

22-15

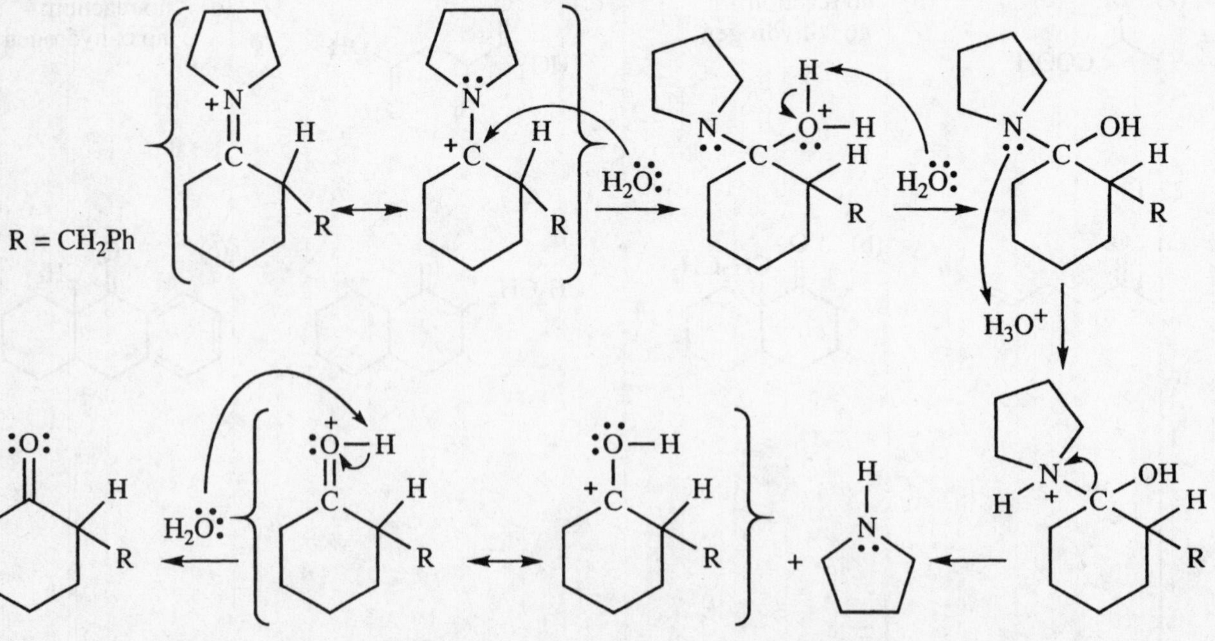

R = CH₂Ph

22-16

(a) 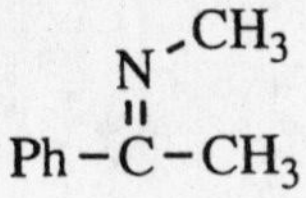 [structure: N-CH₃ imine, Ph-C-CH₃]

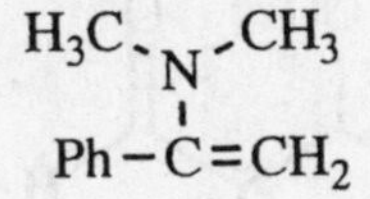

(b) [structure: H₃C-N-CH₃ enamine, Ph-C=CH₂]

(c) [structure: N-Ph imine of cyclohexanone]

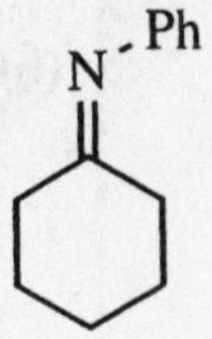

(d) 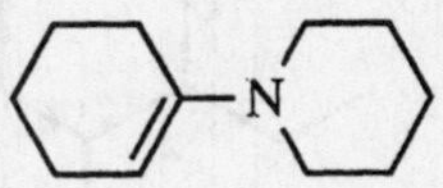 [structure: enamine, cyclohexenyl-N-piperidine]

22-17 Any 2° aliphatic amines can be used for this problem.

(a)

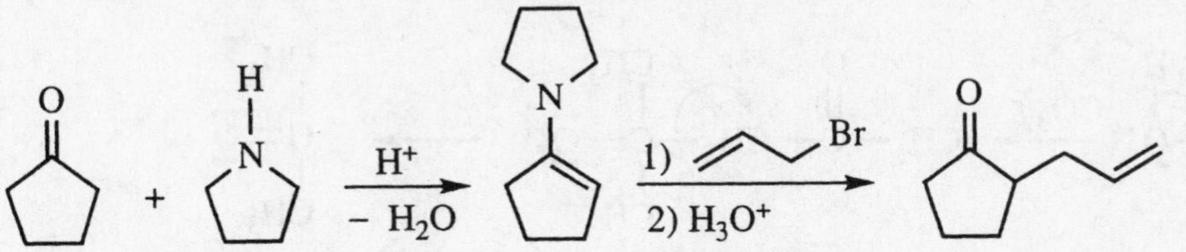

(b)

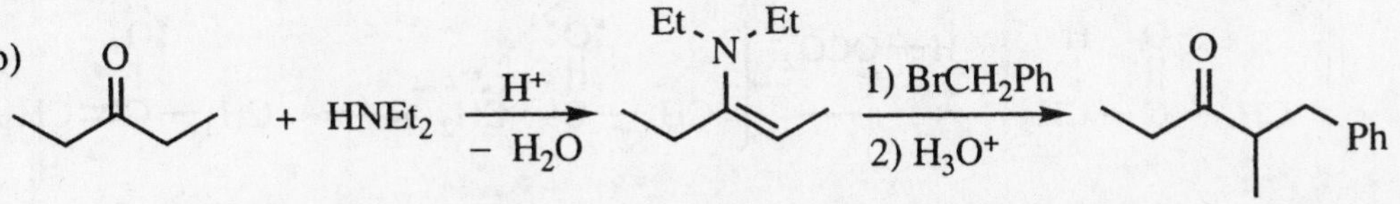

(c)

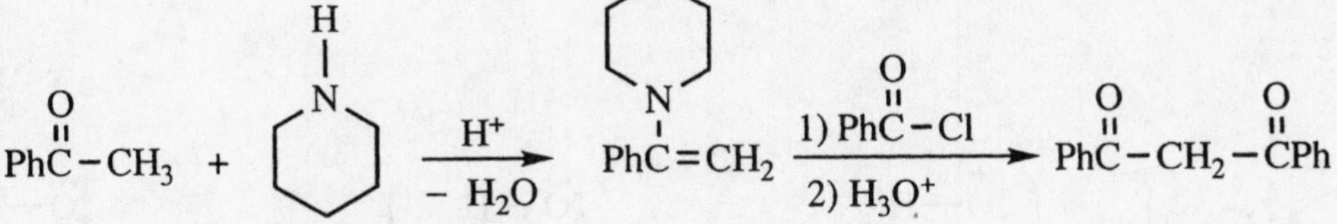

22-18

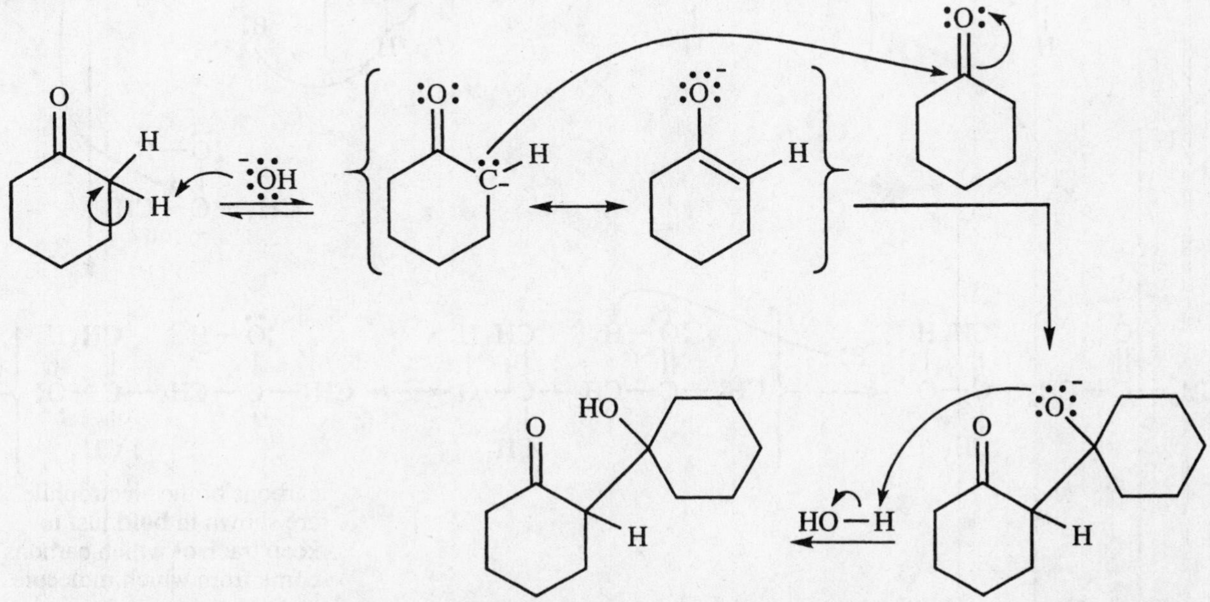

In general, the equilibrium in aldol condensations of ketones favors reactants rather than products. There is significant steric hindrance at both carbons with new bonds, so it is reasonable to conclude that this reaction of cyclohexanone would also favor reactants at equilibrium.

22-19

(a)

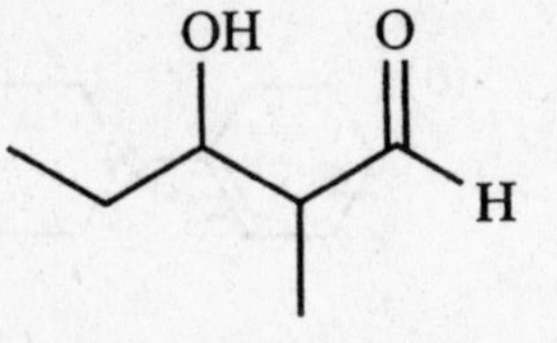

(b)

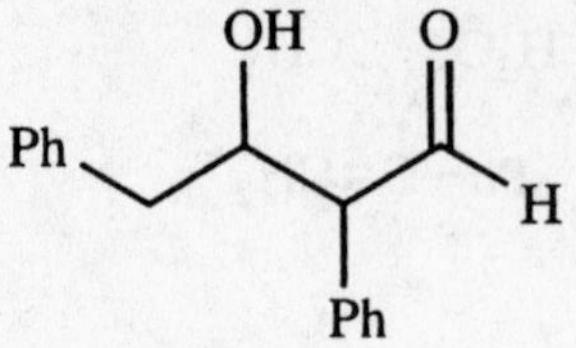

22-20 All the steps in the aldol condensation are reversible. Adding base to diacetone alcohol promoted the reverse aldol reaction. The equilibrium greatly favors acetone.

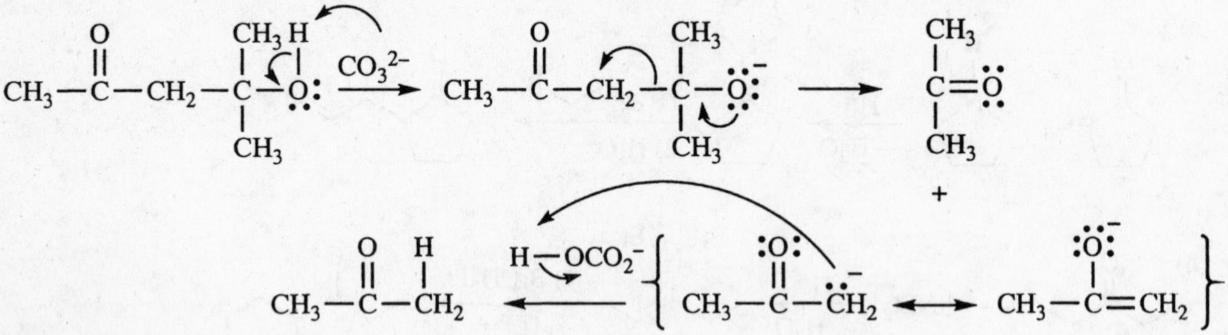

22-21

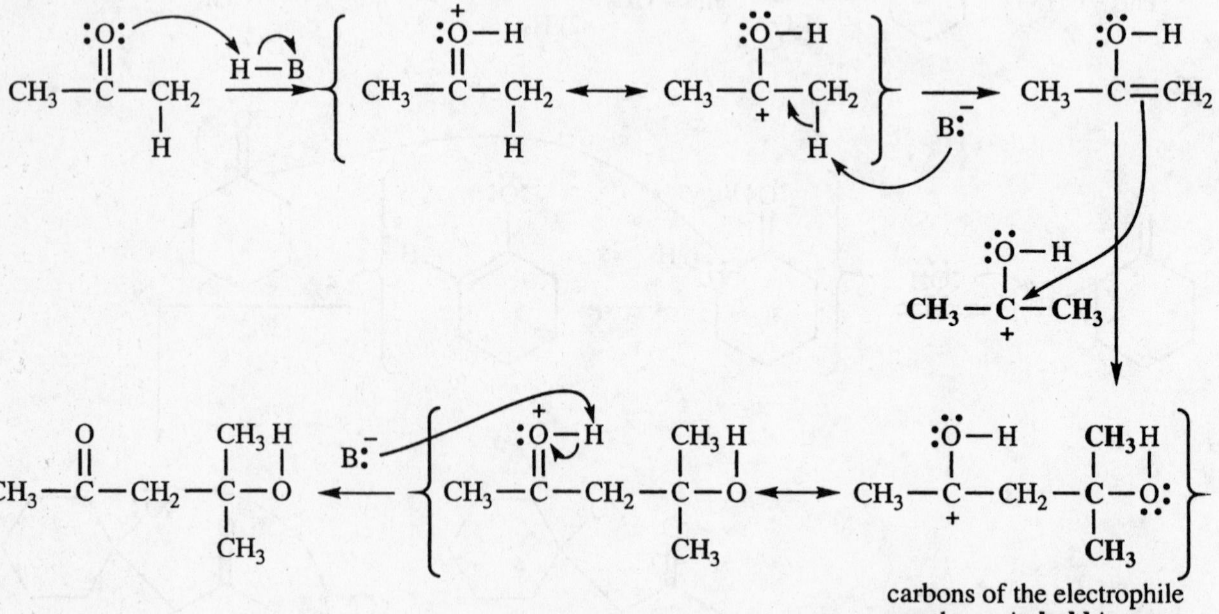

carbons of the electrophile
are shown in **bold** just to
keep track of which carbons
come from which molecule

22-22

(a) <u>acidic conditions</u>

$$H_3C-\overset{O}{\overset{\|}{C}}-\underset{H}{\overset{}{CH}}-\overset{CH_3}{\underset{CH_3}{\overset{|}{C}}}-\overset{H}{\overset{|}{\ddot{O}}}: \xrightarrow{\;H^+\;} H_3C-\overset{O}{\overset{\|}{C}}-\underset{H}{CH}-\overset{CH_3}{\underset{CH_3}{C}}-\overset{H}{\overset{+}{\ddot{O}}}-H$$

$$\downarrow\; -H_2O$$

$$H_3C-\overset{O}{\overset{\|}{C}}-CH=\overset{CH_3}{\underset{CH_3}{C}} \xleftarrow[H_2\ddot{O}:]{} H_3C-\overset{O}{\overset{\|}{C}}-\underset{H}{CH}-\overset{CH_3}{\underset{CH_3}{\overset{+}{C}}}$$

(b) <u>basic conditions</u>

$$H_3C-\overset{O}{\overset{\|}{C}}-\underset{H}{CH}-\overset{CH_3}{\underset{CH_3}{\overset{|}{C}}}-\overset{H}{O} \xrightarrow{\;^-:\ddot{O}:\;} \left\{ H_3C-\overset{:\ddot{O}:^-}{C}=CH-\overset{CH_3}{\underset{CH_3}{C}}-O \longleftrightarrow H_3C-\overset{:O:}{\overset{\|}{C}}-\overset{..}{CH}-\overset{CH_3}{\underset{CH_3}{C}}-\overset{H}{\ddot{O}:} \right\}$$

$$\downarrow$$

$$H_3C-\overset{O}{\overset{\|}{C}}-CH=\overset{CH_3}{\underset{CH_3}{C}} \;+\; {}^-:\ddot{O}H$$

22-23

$$H_3C-\overset{H}{\underset{H}{\overset{|}{C}}}-\overset{O}{\overset{\|}{C}}-H \xrightarrow{\;^-:\ddot{O}H\;} \left\{ H_3C-\overset{H}{\underset{..}{C}}-\overset{:O:}{\overset{\|}{C}}-H \longleftrightarrow H_3C-\overset{H}{C}=\overset{:\ddot{O}:^-}{C}-H \right\}$$

$$CH_3CH_2-\overset{:O:}{\overset{\|}{C}}-H$$

$$CH_3CH_2-\overset{HO}{\underset{H}{\overset{|}{C}}}-\overset{CH_3}{\underset{H}{\overset{|}{C}}}-\overset{O}{\overset{\|}{C}}-H \xleftarrow[HO-H]{} CH_3CH_2-\overset{:\ddot{O}:^-}{\underset{H}{\overset{|}{C}}}-\overset{CH_3}{\underset{H}{\overset{|}{C}}}-\overset{O}{\overset{\|}{C}}-H$$

$$\downarrow\; ^-:\ddot{O}H$$

$$\left\{ CH_3CH_2-\overset{H-O}{\underset{H}{\overset{|}{C}}}-\overset{CH_3}{\overset{|}{C}}-\overset{:O:}{\overset{\|}{C}}-H \longleftrightarrow CH_3CH_2-\overset{HO}{\underset{H}{\overset{|}{C}}}-\overset{CH_3}{\overset{|}{C}}=\overset{:\ddot{O}:^-}{C}-H \right\} \longrightarrow CH_3CH_2-\overset{CH_3}{\underset{H}{\overset{|}{C}}}=C-\overset{O}{\overset{\|}{C}}-H$$

545

22-24

(a)

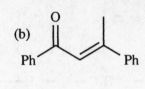

(b)

(c)

22-25

(a)

Step 1: carbon skeletons

Step 2: nucleophile generation

(2,2-Dimethylpropanal has no α-hydrogen.)

Step 3: nucleophilic attack

Step 4: conversion to final product

Step 5: combine Steps 2, 3, and 4 to complete the mechanism

546

22-25 continued

(b)

Step 1: carbon skeletons

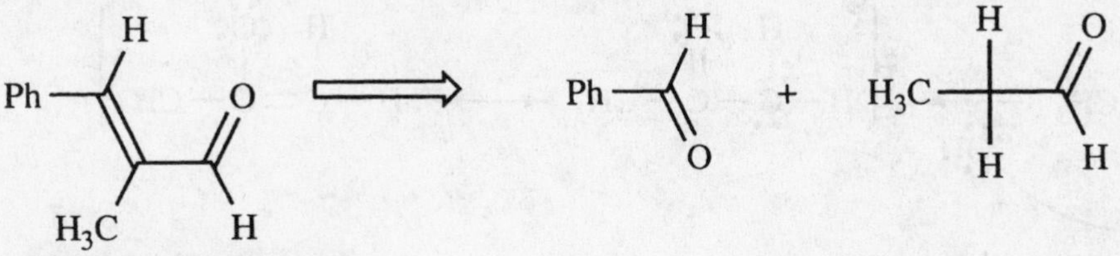

Step 2: nucleophile generation

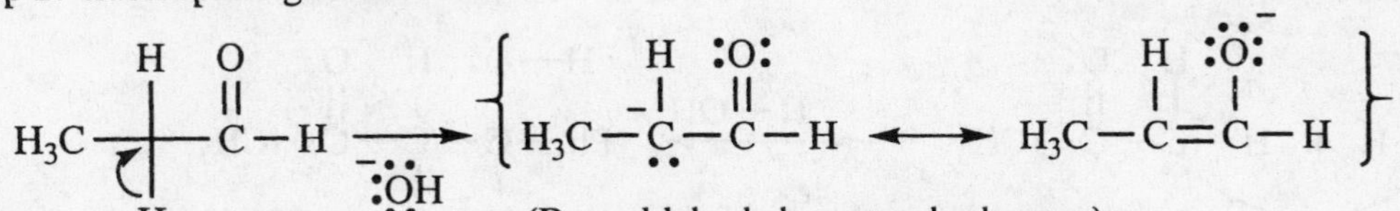

(Benzaldehyde has no α-hydrogen.)

Step 3: nucleophilic attack

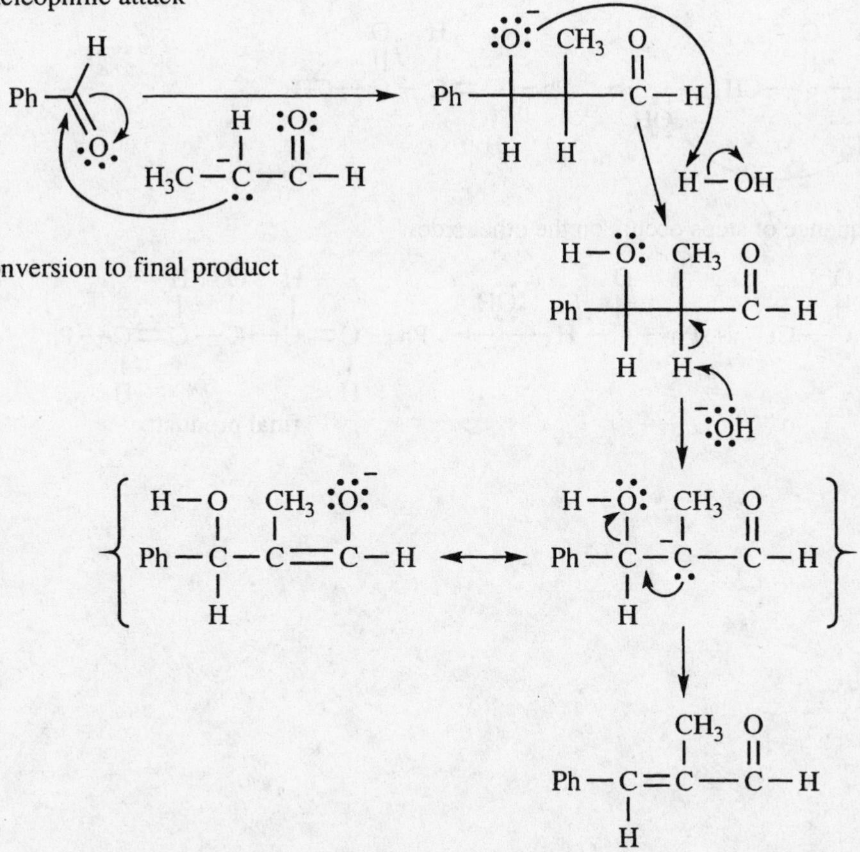

Step 4: conversion to final product

Step 5: combine Steps 2, 3, and 4 to complete the mechanism

22-26 This solution presents the sequence of reactions leading to the product, following the format of the Problem-Solving feature. This is not a complete mechanism.

Step 2: generation of the nucleophile

$$
\begin{array}{c}
\underset{\underset{H}{\displaystyle|}}{H-\underset{\underset{}{\overset{\displaystyle|}{C}}}{}}-\overset{\overset{\displaystyle O}{||}}{C}-CH_3 \quad \xrightarrow{\quad :\ddot{O}H\quad} \quad \left\{\ H-\overset{\overset{\displaystyle H}{\ }}{\underset{\underset{}{\ddot{C}}}{C}}-\overset{\overset{\displaystyle :\ddot{O}:}{||}}{C}-CH_3 \ \longleftrightarrow\ H-\overset{\overset{\displaystyle H}{\ }}{C}=\overset{\overset{\displaystyle :\ddot{O}:^{-}}{|}}{C}-CH_3\ \right\}
\end{array}
$$

Step 3: nucleophilic attack

$$
Ph-\overset{\overset{\displaystyle :O:}{||}}{C}-H \ +\ H-\overset{\overset{\displaystyle H}{|}}{\underset{\underset{}{\ddot{C}}}{C}}-\overset{\overset{\displaystyle :\ddot{O}:}{||}}{C}-CH_3 \ \xrightarrow{\quad H-OH\quad} \ Ph-\overset{\overset{\displaystyle H-\ddot{O}:}{|}}{\underset{\underset{H}{|}}{C}}-\overset{\overset{\displaystyle H}{|}}{\underset{\underset{H}{|}}{C}}-\overset{\overset{\displaystyle O}{||}}{C}-CH_3
$$

Step 4: dehydration

$$
Ph-\overset{\overset{\displaystyle H-O}{|}}{\underset{\underset{H}{|}}{C}}-\overset{\overset{\displaystyle H}{|}}{\underset{\underset{H}{|}}{C}}-\overset{\overset{\displaystyle O}{||}}{C}-CH_3 \ \xrightarrow{\quad :\ddot{O}H\quad} \ Ph-\overset{\overset{\displaystyle H}{|}}{C}=\overset{\overset{\displaystyle H}{|}}{\underset{\underset{H}{|}}{C}}-\overset{\overset{\displaystyle O}{||}}{C}-CH_3
$$

The same sequence of steps occurs on the other side.

$$
Ph-\overset{\overset{\displaystyle H}{|}}{\underset{\underset{H}{|}}{C}}=\overset{\overset{\displaystyle O}{\ }}{C}-\overset{\overset{\displaystyle O}{||}}{C}-CH_3 \ +\ Ph-\overset{\overset{\displaystyle O}{||}}{C}-H \ \xrightarrow{\quad :\ddot{O}H\quad} \ Ph-\overset{\overset{\displaystyle H}{|}}{\underset{\underset{H}{|}}{C}}=\overset{\overset{\displaystyle O}{\ }}{C}-\overset{\overset{\displaystyle O}{||}}{C}-\overset{\overset{\displaystyle H}{|}}{\underset{\underset{H}{|}}{C}}=C-Ph
$$

final product

22-27

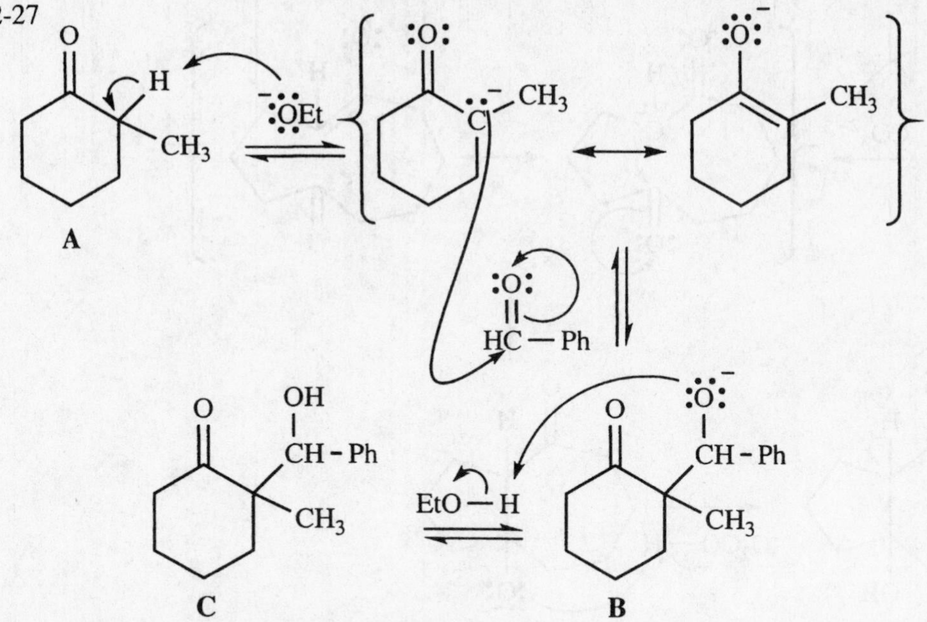

There are three problems with the reaction as shown:

1. Hydrogen on a 3° carbon (structure **A**) is less acidic than hydrogen on a 2° carbon. The 3° hydrogen will be removed at a slower rate than the 2° hydrogen.

2. Nucleophilic attack by the 3° carbon will be more hindered, and therefore slower, than attack by the 2° carbon. Structure **B** is quite hindered.

3. Once a normal aldol product is formed, dehydration gives a conjugated system which has great stability. The aldol product **C** cannot dehydrate because no α-hydrogen remains. Some **C** will form, but eventually the reverse-aldol process will return **C** to starting materials which, in turn, will react at the other α-carbon to produce the conjugated system. (This reason is the Kiss of Death for **C**.)

22-28

(a)

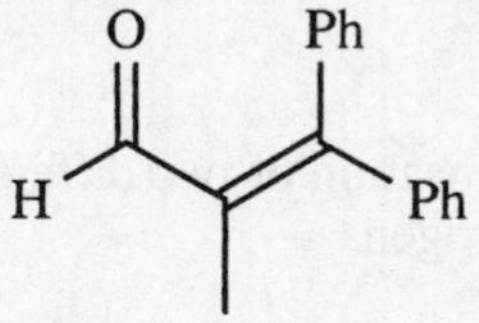

(b)

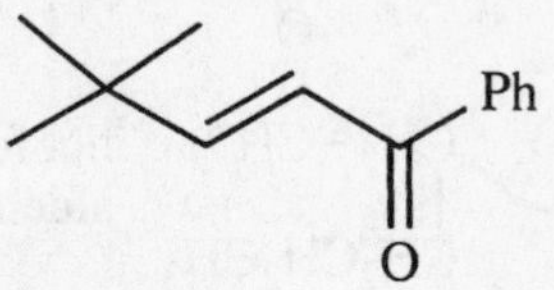

22-29

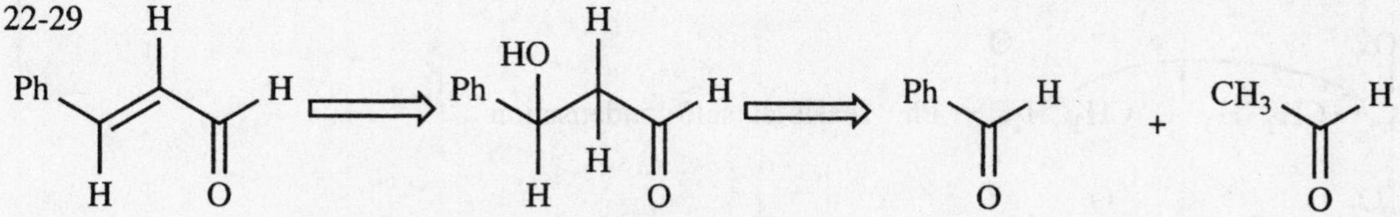

22-30

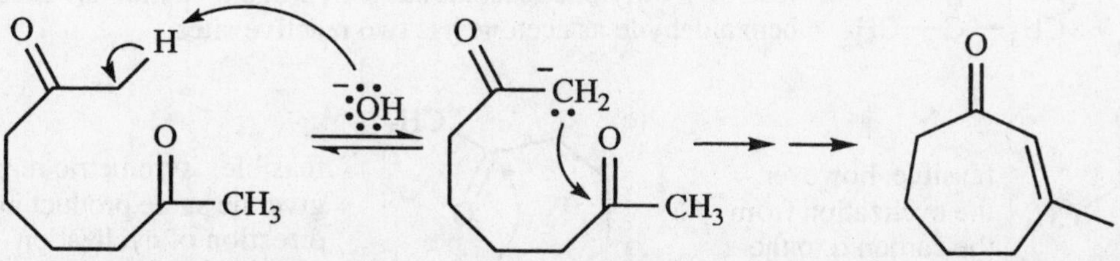

The formation of a seven-membered ring is unfavorable for entropy reasons: the farther apart the nucleophile and the electrophile, the harder time they will have finding each other. If the molecule has a possibility of forming a 5- or a 7-membered ring, it will almost always prefer to form the 5-membered ring.

22-31

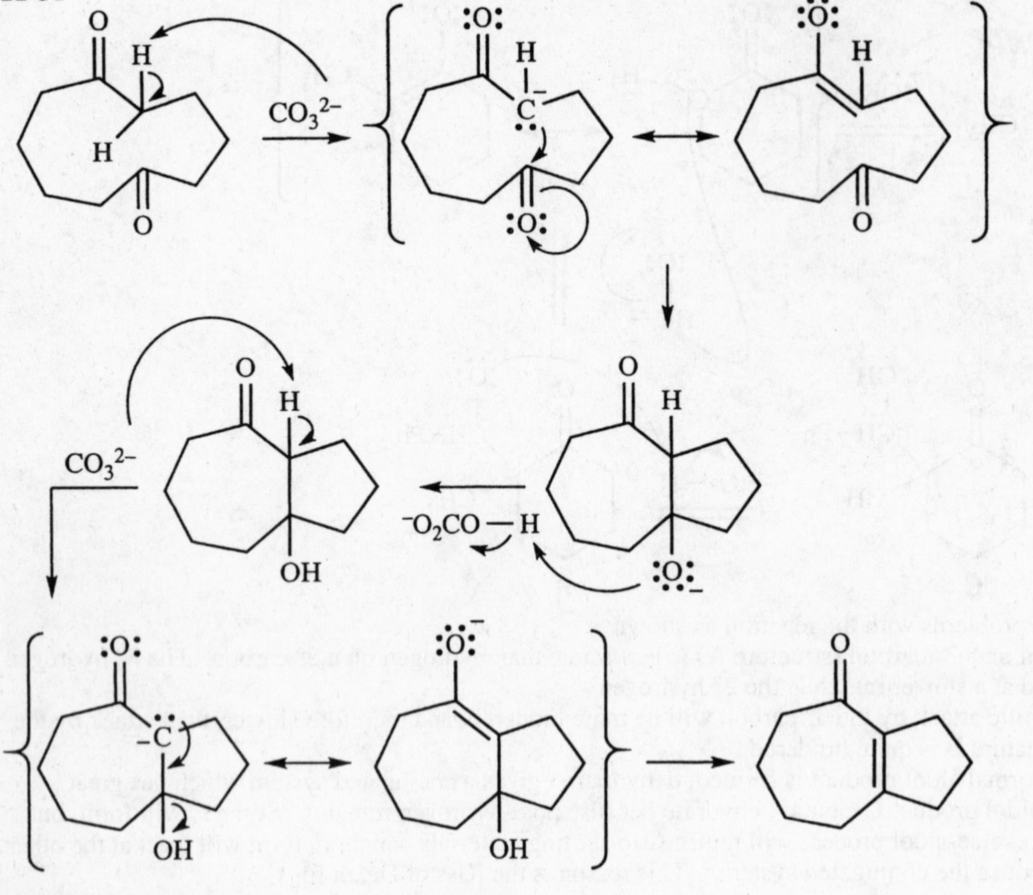

22-32

(a)

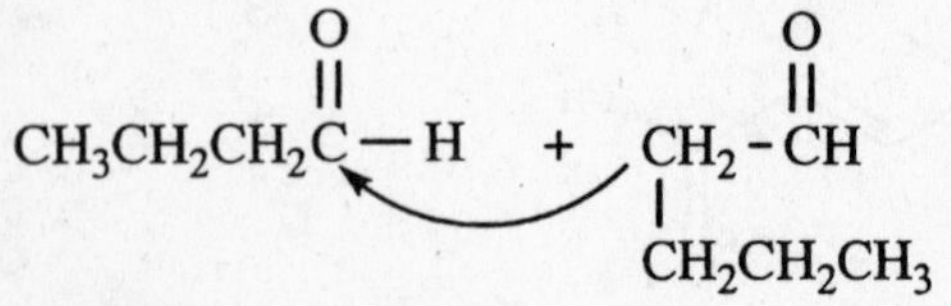

$CH_3CH_2CH_2\overset{O}{\underset{||}{C}}-H$ + $CH_2-\overset{O}{\underset{||}{CH}}$ not feasible: requires condensation of two different
aldehydes, each with α-hydrogens

$CH_2CH_2CH_3$

(b)

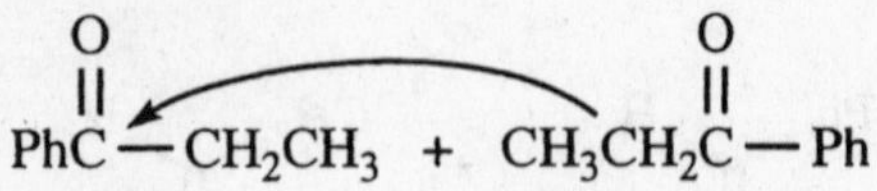

$Ph\overset{O}{\underset{||}{C}}-CH_2CH_3$ + $CH_3CH_2\overset{O}{\underset{||}{C}}-Ph$ feasible: self-condensation

(c)

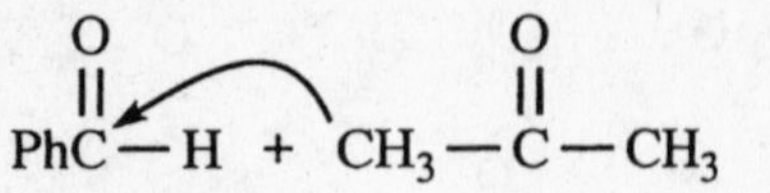

$Ph\overset{O}{\underset{||}{C}}-H$ + $CH_3-\overset{O}{\underset{||}{C}}-CH_3$ feasible: only one reactant has α-hydrogen; cannot use excess
benzaldehyde as acetone has two reactive sites

(d)

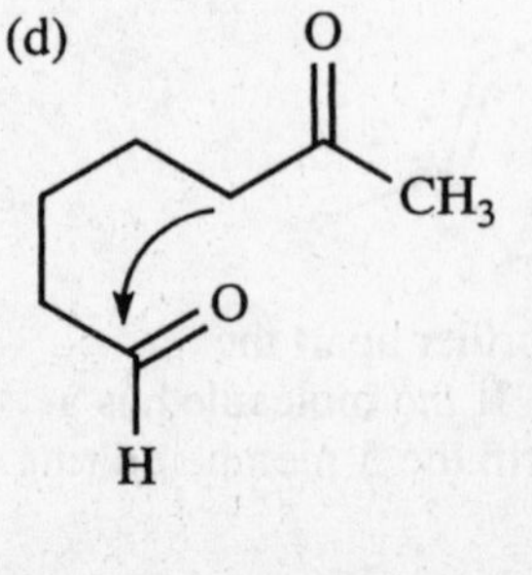

feasible; however, the cyclization from the carbon α to the aldehyde to the ketone carbonyl is also possible

(e)

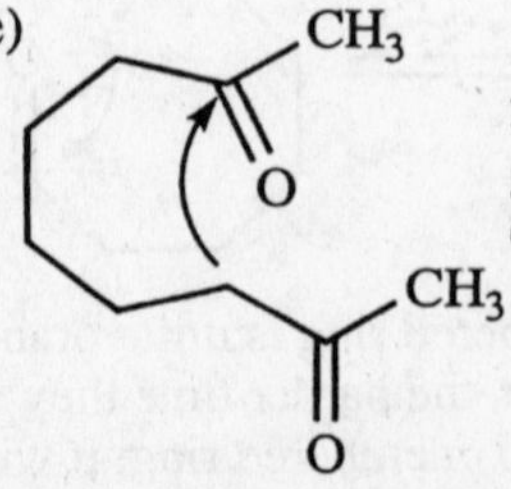

feasible: symmetric reagent will give the same product in either direction of cyclization

550

22-33 (a) and (b)

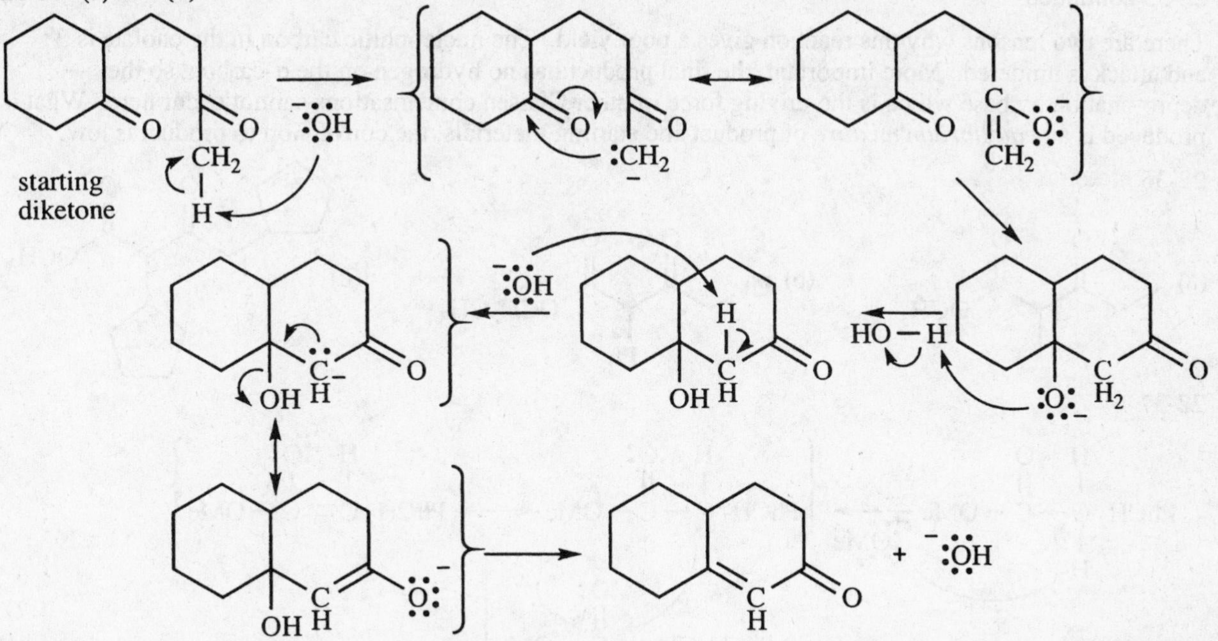

starting
diketone

22-34

(a) The side reaction with sodium methoxide is transesterification. The starting material, and therefore the product, would be a mixture of methyl and ethyl esters.

$$H_3C-\overset{\overset{\displaystyle O}{\|}}{C}-OCH_2CH_3 \;+\; NaOCH_3 \;\rightleftharpoons\; H_3C-\overset{\overset{\displaystyle O}{\|}}{C}-OCH_3 \;+\; NaOCH_2CH_3$$

(b) Sodium hydroxide would irreversibly saponify the ester, completely stopping the Claisen condensation as the carbonyl no longer has a leaving group attached to it.

$$H_3C-\overset{\overset{\displaystyle O}{\|}}{C}-OCH_2CH_3 \;+\; NaOH \;\longrightarrow\; H_3C-\overset{\overset{\displaystyle O}{\|}}{C}-O^-\,Na^+ \;+\; HOCH_2CH_3$$

22-35

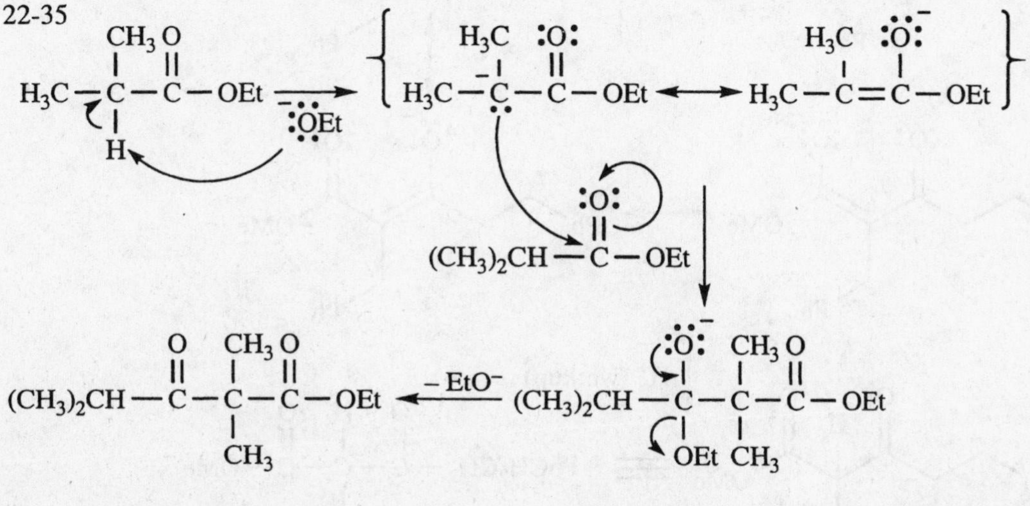

$-EtO^-$

explanation on next page

551

22-35 continued

There are two reasons why this reaction gives a poor yield. The nucleophilic carbon in the enolate is 3° and attack is hindered. More important, the final product has no hydrogen on the α-carbon, so the deprotonation by base which is the driving force in other Claisen condensations cannot occur here. What is produced is an *equilibrium mixture* of product and starting materials; the conversion to product is low.

22-36

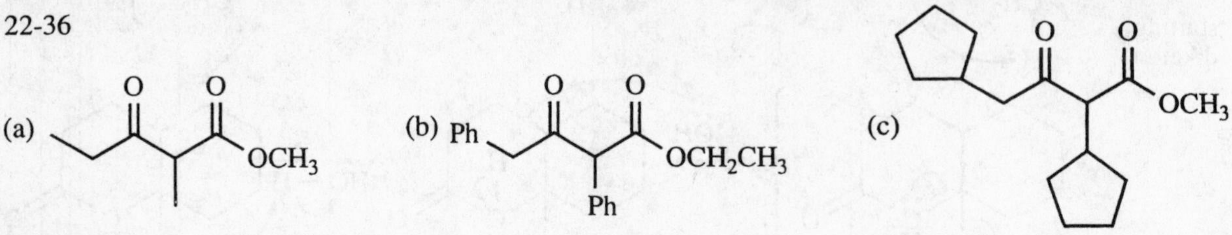

(a)

$$\underset{\text{O}}{\overset{\text{O}}{\parallel}} \quad \underset{\text{O}}{\overset{\text{O}}{\parallel}} \quad \text{OCH}_3$$

(b) Ph ... OCH$_2$CH$_3$ ... Ph

(c) ... OCH$_3$

22-37

PhCH$_2$CH$_2$—C—C—C—OMe
          |
        CH$_2$Ph

H$^+$ (workup)

≡ PhCH$_2$CH$_2$—C—C—C—OMe
              |
            CH$_2$Ph

552

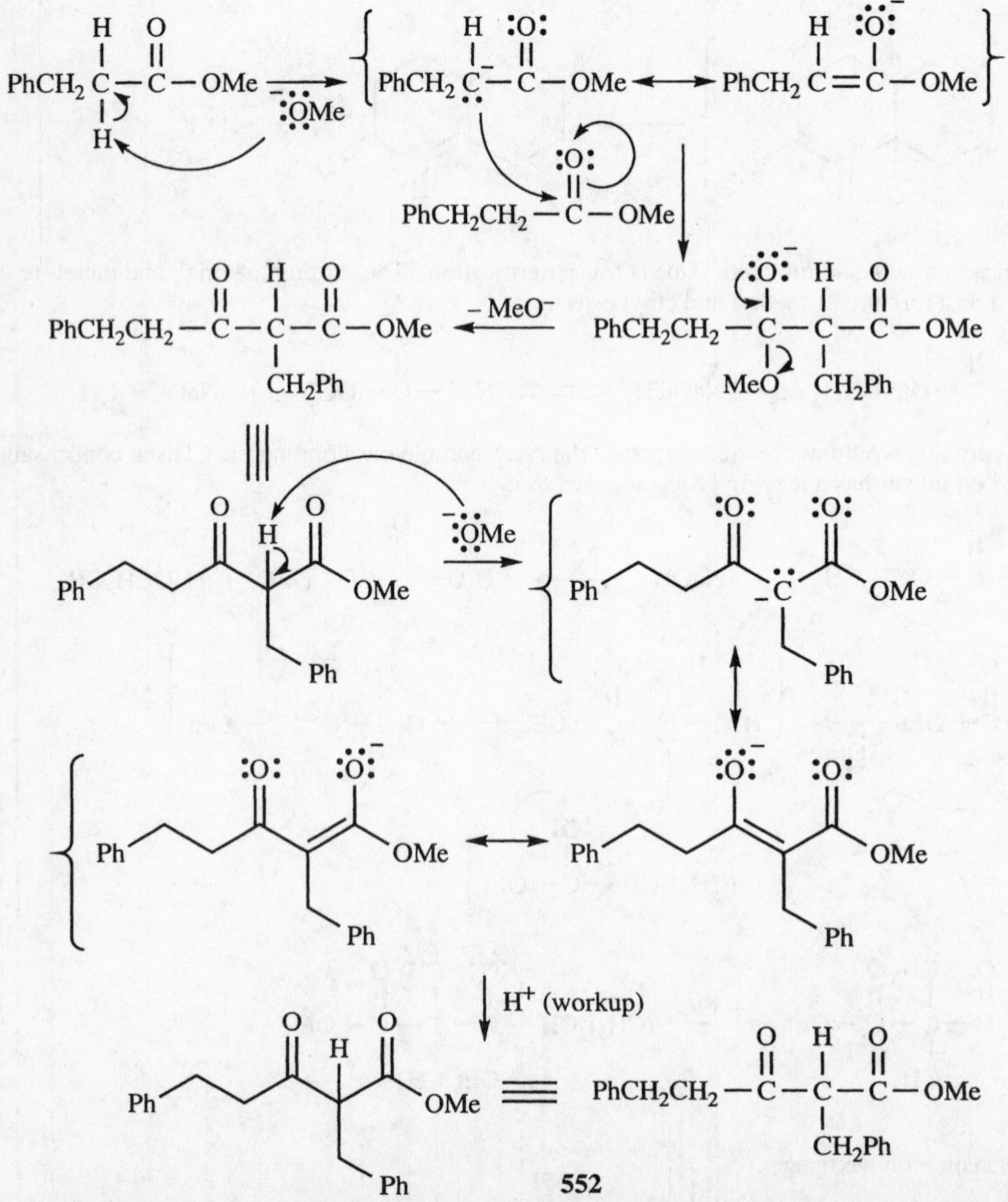

22-38

(a) $CH_3CH_2CH_2-\overset{\overset{\displaystyle O}{\|}}{C}-OCH_2CH_3$

(b) $PhCH_2\overset{\overset{\displaystyle O}{\|}}{C}-OCH_3$

(c) $CH_3\underset{\underset{\displaystyle CH_3}{|}}{CH}CH_2-\overset{\overset{\displaystyle O}{\|}}{C}-OCH_2CH_3$

This one would be difficult because the alpha-carbon is hindered.

22-39

This is the final product, after removal of the α-hydrogen by ethoxide, followed by reprotonation during the workup.

This is the final product, after removal of the α-hydrogen by ethoxide, followed by reprotonation during the workup.

22-40

(a) not possible by Dieckmann—not a β-keto ester

(b)

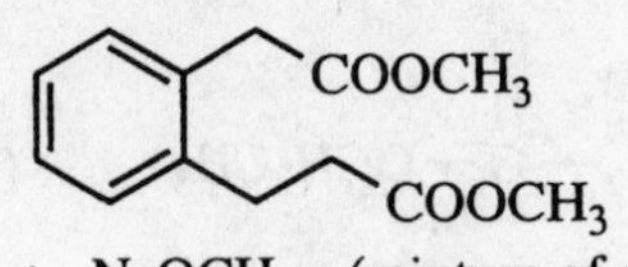

+ NaOCH$_3$ (mixture of products results)

(c)

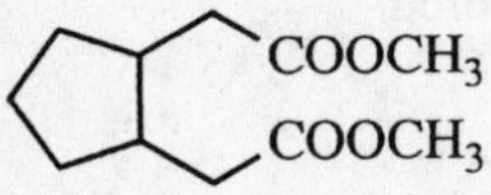

+ NaOCH$_3$

(d)

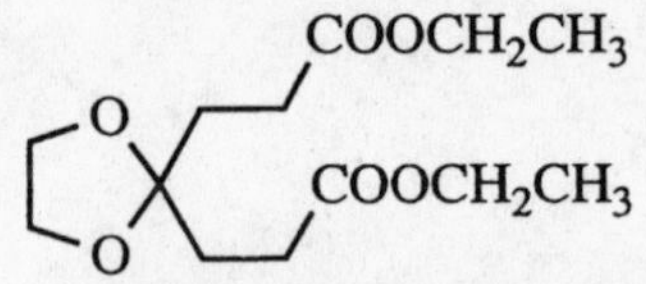

+ NaOCH$_2$CH$_3$

The protecting group is necessary to prevent aldol condensation. Aqueous acid workup removes the protecting group.

22-41

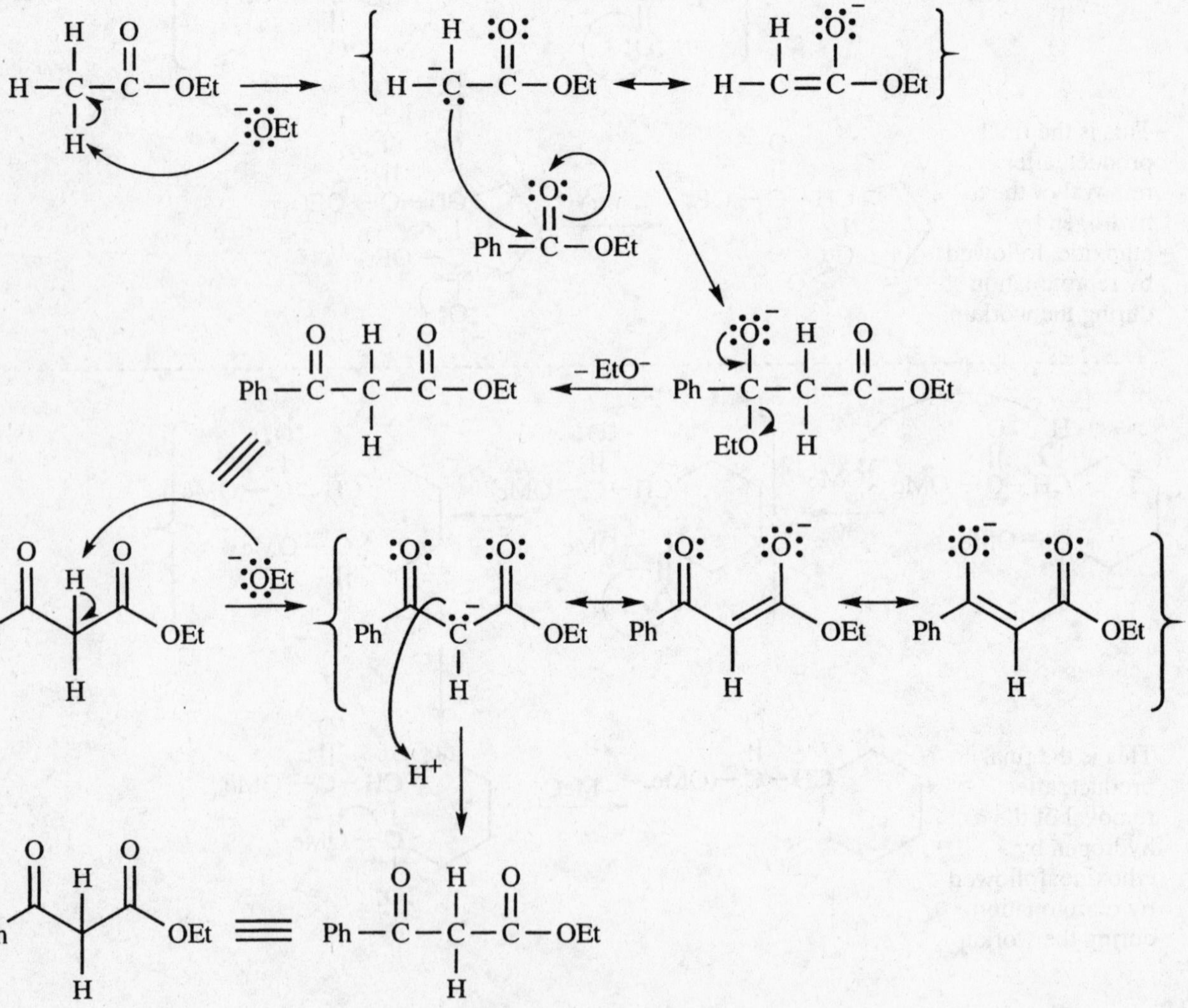

22-42

(a) Ph–C(=O)–CH(Ph)–C(=O)–OCH₃

(structure: Ph and Ph groups with two C=O, OCH₃)

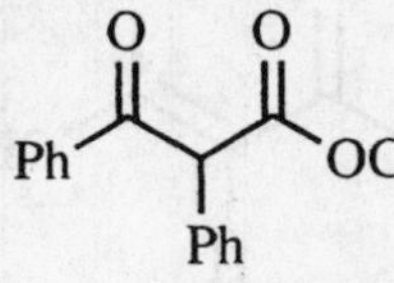

(b) CH₃–C(=O)–CH(Ph)–C(=O)–OCH₃ + Ph–CH₂–C(=O)–CH₂–C(=O)–OCH₃

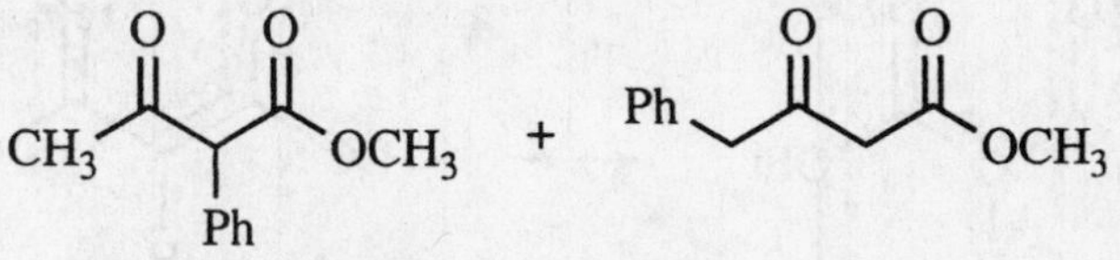

plus 2 self-condensation products—a poor choice because both esters have α-hydrogens

(c) EtO–C(=O)–CH₂–C(=O)–CH₂–C(=O)–OEt

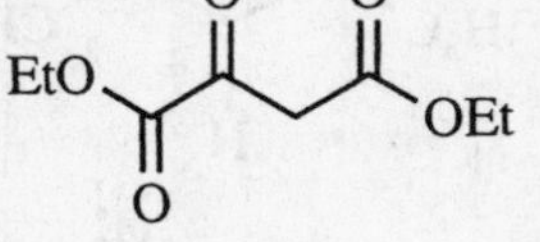

(d) EtO–C(=O)–CH(CH₃)–C(=O)–OEt

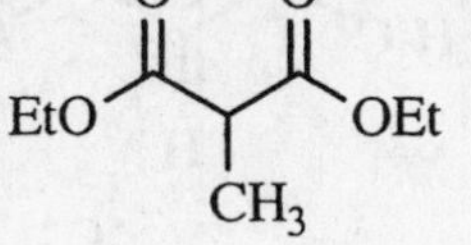

22-43

(a) $Ph-\overset{O}{\overset{\|}{C}}-OEt + CH_3CH_2-\overset{O}{\overset{\|}{C}}-OEt$

(b) $PhCH_2-\overset{O}{\overset{\|}{C}}-OMe + MeO-\overset{O}{\overset{\|}{C}}-\overset{O}{\overset{\|}{C}}-OMe$

(c) $EtO-\overset{O}{\overset{\|}{C}}-CH_2Ph + EtO-\overset{O}{\overset{\|}{C}}-OEt$

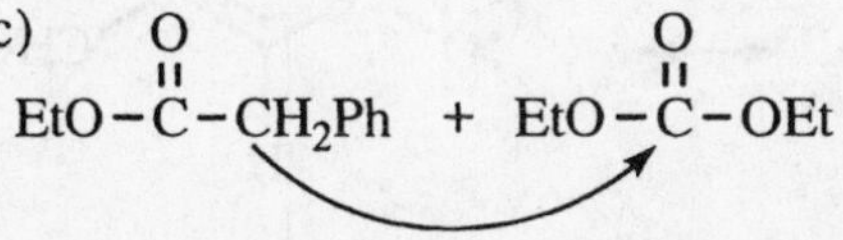

(d) $(CH_3)_3C-\overset{O}{\overset{\|}{C}}-OMe + CH_3CH_2CH_2CH_2\cdot\overset{O}{\overset{\|}{C}}-OMe$

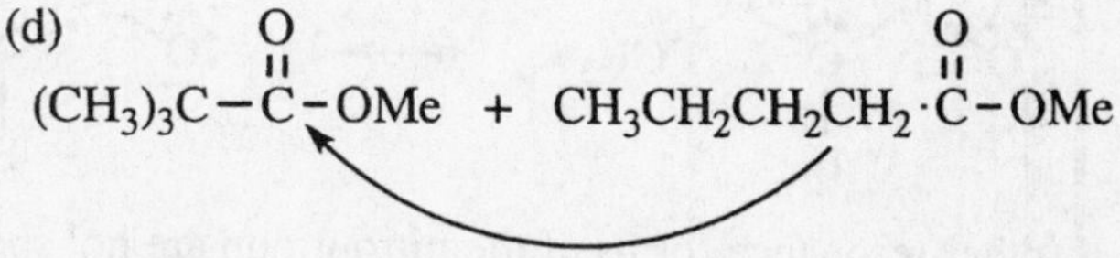

22-44

(a) (cyclohexanone with CHO) ⇌ (enol form)

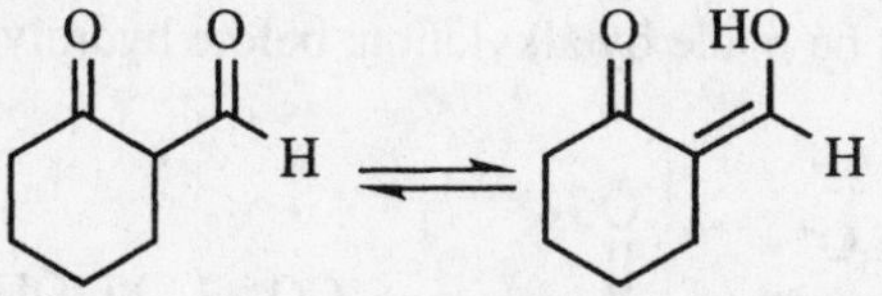

actually present in the enol form:

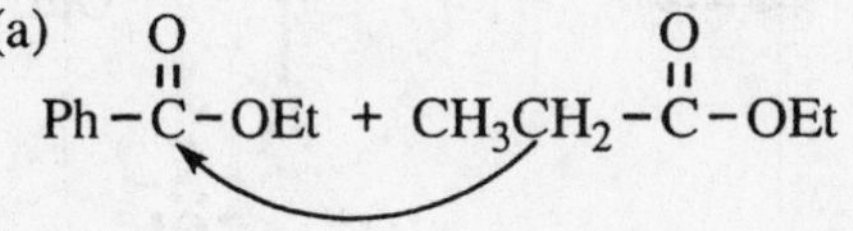

(b) CH₃–C(=O)–CH₂–C(=O)–CH₃

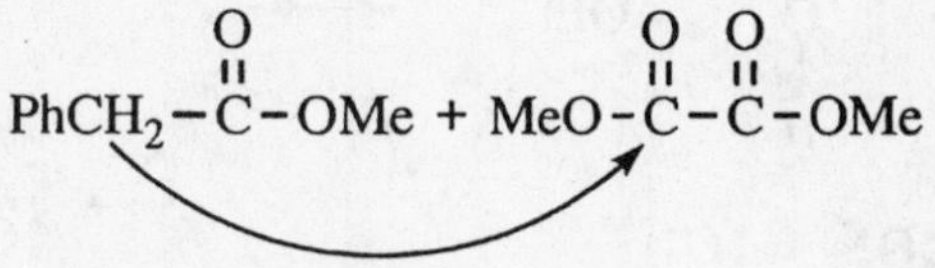

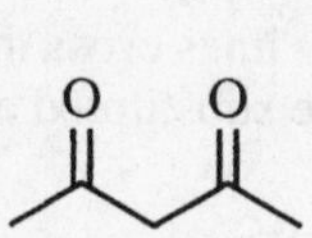

(c) (cyclopentane-1,3-dione)

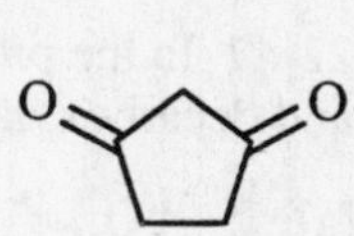

22-45

(a) two ways:

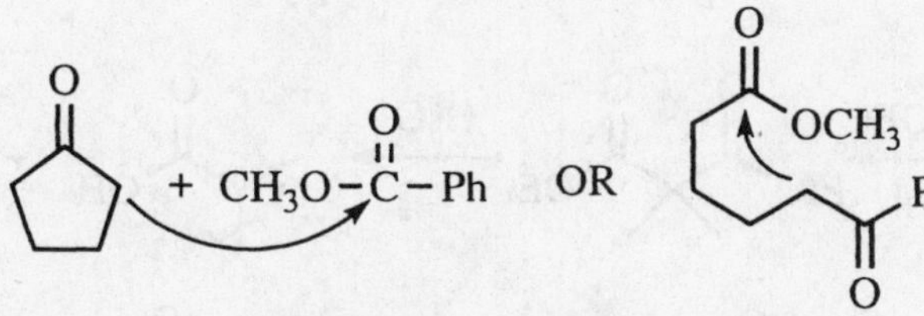

(cyclopentanone) + $CH_3O-\overset{O}{\overset{\|}{C}}-Ph$ OR (structure with OCH₃, Ph, two C=O)

(b) $CH_3CH_2-\overset{O}{\overset{\|}{C}}-CH_2CH_3 + CH_3CH_2O-\overset{O}{\overset{\|}{C}}-\overset{O}{\overset{\|}{C}}-OCH_2CH_3$

(c) (seven-membered ring ketone with ester OCH₂CH₃)

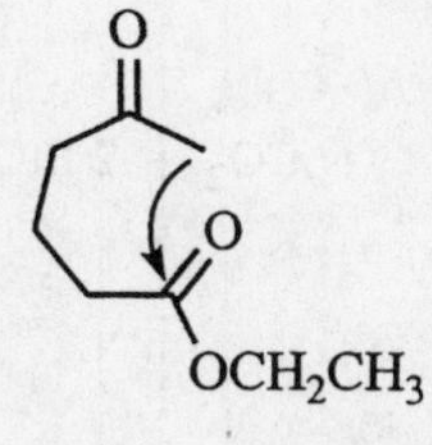

(d) two ways:

(cyclohexane-1,3-dione) + $CH_3CH_2O-\overset{O}{\overset{\|}{C}}-OCH_2CH_3$ OR (ring structure with OCH₂CH₃ groups)

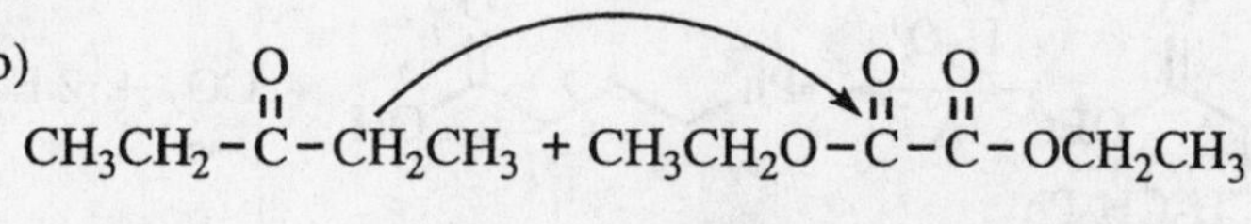

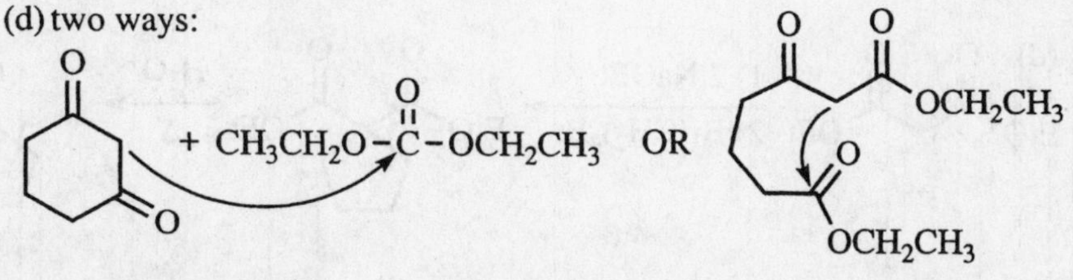

22-46

(a)

$$\text{H}_3\text{C}-\overset{:\text{O}:}{\overset{\|}{\text{C}}}-\overset{-}{\underset{\text{H}}{\text{C}}}-\overset{:\text{O}:}{\overset{\|}{\text{C}}}-\text{OEt} \longleftrightarrow \text{H}_3\text{C}-\overset{:\text{O}:^-}{\underset{\text{H}}{\text{C}}}=\overset{:\text{O}:}{\underset{}{\text{C}}}-\text{OEt} \longleftrightarrow \text{H}_3\text{C}-\overset{:\text{O}:}{\underset{\text{H}}{\text{C}}}=\overset{:\text{O}:^-}{\underset{}{\text{C}}}-\text{OEt}$$

(b)

$$\text{H}_3\text{C}-\overset{:\text{O}:}{\overset{\|}{\text{C}}}-\overset{-}{\underset{\text{H}}{\text{C}}}-\overset{:\text{O}:}{\overset{\|}{\text{C}}}-\text{CH}_3 \longleftrightarrow \text{H}_3\text{C}-\overset{:\text{O}:^-}{\underset{\text{H}}{\text{C}}}=\overset{:\text{O}:}{\underset{}{\text{C}}}-\text{CH}_3 \longleftrightarrow \text{H}_3\text{C}-\overset{:\text{O}:}{\underset{\text{H}}{\text{C}}}=\overset{:\text{O}:^-}{\underset{}{\text{C}}}-\text{CH}_3$$

(c)

$$\text{N}\equiv\text{C}-\overset{-}{\underset{\text{H}}{\text{C}}}-\overset{:\text{O}:}{\overset{\|}{\text{C}}}-\text{OEt} \longleftrightarrow :\overset{-}{\text{N}}=\text{C}=\overset{}{\underset{\text{H}}{\text{C}}}-\overset{:\text{O}:}{\overset{\|}{\text{C}}}-\text{OEt} \longleftrightarrow \text{N}\equiv\text{C}-\overset{}{\underset{\text{H}}{\text{C}}}=\overset{:\text{O}:^-}{\underset{}{\text{C}}}-\text{OEt}$$

(d)

$$:\overset{-}{\underset{}{\text{O}}}:-\overset{:\text{O}:}{\overset{+}{\text{N}}}-\overset{-}{\underset{\text{H}}{\text{C}}}-\overset{:\text{O}:}{\overset{\|}{\text{C}}}-\text{CH}_3 \longleftrightarrow :\overset{:\text{O}:^-}{\underset{}{\text{O}}}:\overset{+}{\text{N}}=\overset{}{\underset{\text{H}}{\text{C}}}-\overset{:\text{O}:}{\overset{\|}{\text{C}}}-\text{CH}_3 \longleftrightarrow :\overset{:\text{O}:}{\underset{}{\text{O}}}:\overset{+}{\text{N}}-\overset{}{\underset{\text{H}}{\text{C}}}=\overset{:\text{O}:^-}{\underset{}{\text{C}}}-\text{CH}_3$$

(other resonance forms of the nitro group are not shown)

22-47 In the products, the wavy lines cross the bonds that must be made by alkylation, before hydrolysis and decarboxylation produce the substituted acetic acid.

(a)

EtO–CO–CH$_2$–CO–OEt $\xrightarrow[\text{2) PhCH}_2\text{Br}]{\text{1) NaOEt}}$ EtO–CO–CH(CH$_2$Ph)–CO–OEt $\xrightarrow[\Delta]{\text{H}_3\text{O}^+}$ CH$_2$Ph–CH$_2$–COOH $+ \ CO_2 + 2\ EtOH$

(b)

EtO–CO–CH$_2$–CO–OEt $\xrightarrow[\text{2) CH}_3\text{I}]{\text{1) NaOEt}}$ $\xrightarrow[\text{2) CH}_3\text{I}]{\text{1) NaOEt}}$ EtO–CO–C(CH$_3$)$_2$–CO–OEt $\xrightarrow[\Delta]{\text{H}_3\text{O}^+}$ (CH$_3$)$_2$CH–COOH $+ \ CO_2 + 2\ EtOH$

(c)

EtO–CO–CH$_2$–CO–OEt $\xrightarrow[\text{2) Ph(CH}_2)_2\text{Br}]{\text{1) NaOEt}}$ EtO–CO–CH(CH$_2$CH$_2$Ph)–CO–OEt $\xrightarrow[\Delta]{\text{H}_3\text{O}^+}$ Ph–CH$_2$CH$_2$–CH$_2$–COOH $+ \ CO_2 + 2\ EtOH$

(d)

EtO–CO–CH$_2$–CO–OEt $\xrightarrow[\text{2) Br(CH}_2)_4\text{Br}]{\text{1) 2 NaOEt}}$ EtO–CO–C(cyclopentane)–CO–OEt $\xrightarrow[\Delta]{\text{H}_3\text{O}^+}$ cyclopentane–COOH $+ \ CO_2 + 2\ EtOH$

556

22-48

(a) Only two substituent groups plus a hydrogen atom can appear on the alpha carbon after decarboxylation at the end of the malonic ester synthesis. The product shown has three alkyl groups, so it cannot be made by malonic ester synthesis.

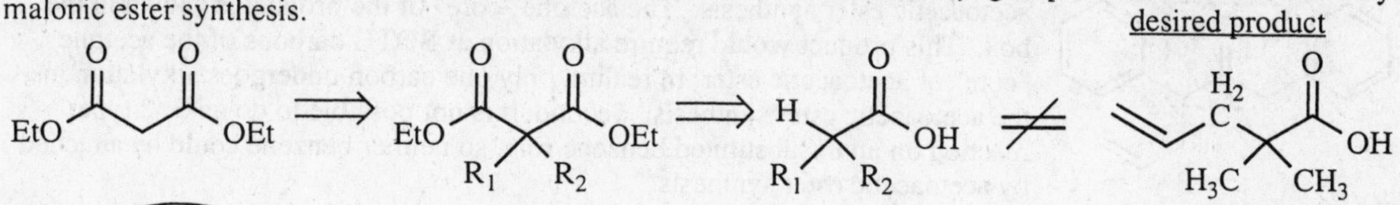

<u>desired product</u>

(b)

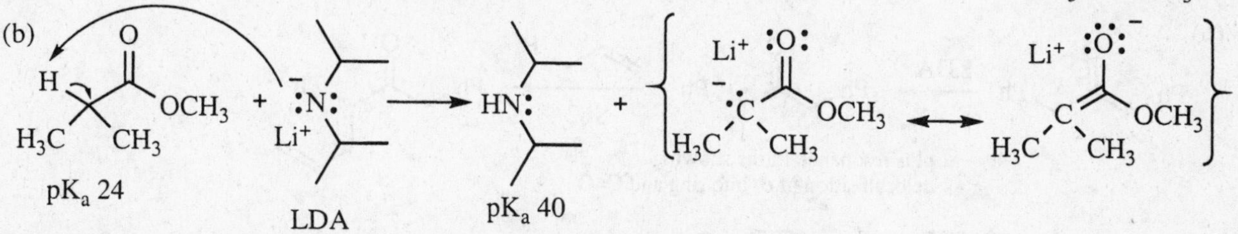

pK$_a$ 24

LDA

pK$_a$ 40

With such a large difference in pK$_a$ values, products are favored >> 99%.

(c)

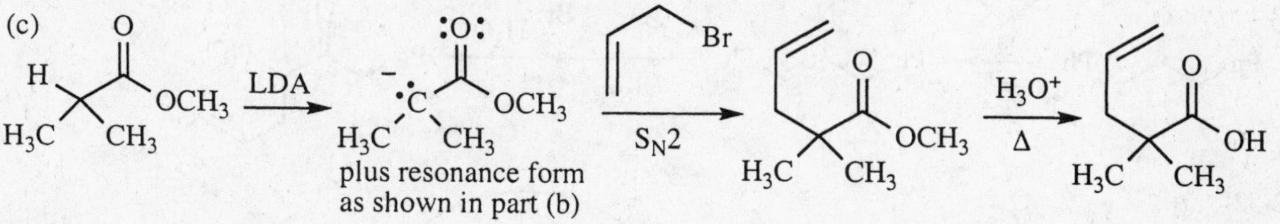

plus resonance form
as shown in part (b)

22-49

(a) PhCH$_2$CH$_2$ – C – CH$_3$ + CO$_2$ + EtOH

(b) 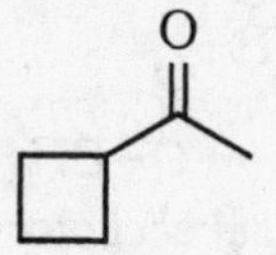 + CO$_2$ + MeOH

(c) 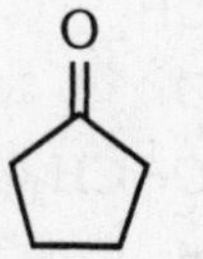 + CO$_2$ + EtOH

22-50 In the products, the wavy lines indicate the bonds that must be made by alkylation, before hydrolysis and decarboxylation produce the substituted acetone.

(a)

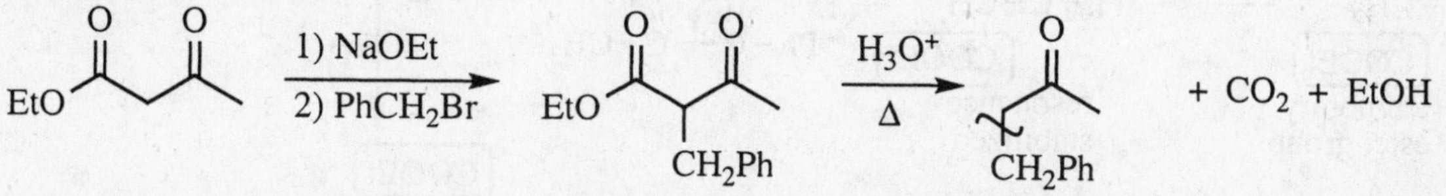

1) NaOEt
2) PhCH$_2$Br

H$_3$O$^+$
Δ

+ CO$_2$ + EtOH

(b)

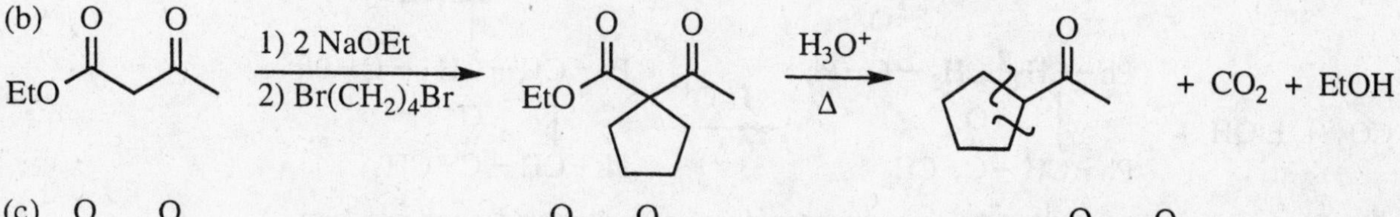

1) 2 NaOEt
2) Br(CH$_2$)$_4$Br

H$_3$O$^+$
Δ

+ CO$_2$ + EtOH

(c)

1) NaOEt
2) PhCH$_2$Br

1) NaOEt
2) Br⌁

H$_3$O$^+$
Δ

CO$_2$ + EtOH +

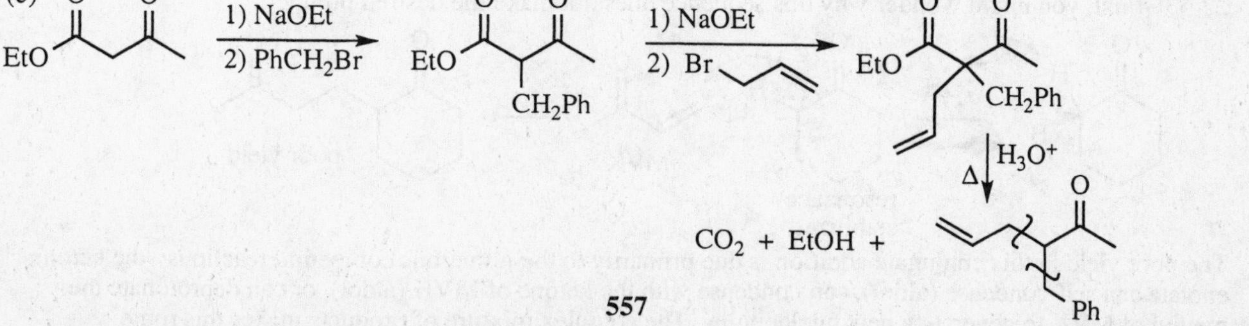

22-51

(a) There are two problems with attempting to make this compound by acetoacetic ester synthesis. The acetone "core" of the product is shown in the box. This product would require alkylation at BOTH carbons of the acetone "core" of acetoacetic ester; in reality, only one carbon undergoes alkylation in the acetoacetic ester synthesis. Second, it is not possible to do an S_N2 type reaction on an unsubstituted benzene ring, so neither benzene could be attached by acetoacetic ester synthesis.

(b)

Ph–CH$_2$–CO–CH$_2$–Ph $\xrightarrow{\text{LDA}}$ Ph–CH$_2$–CO–$\overset{-}{\underset{H}{C}}$–Ph $\xrightarrow[S_N2]{\text{(allyl Br)}}$ Ph–CH$_2$–CO–CH(Ph)–CH$_2$CH=CH$_2$

plus resonance forms showing
delocalization of e$^-$ into ring and C=O

(c)

Ph–CH$_2$–CO–CH$_2$–Ph $\xrightarrow[H^+]{\text{(pyrrolidine)}}$ (enamine) $\xrightarrow[S_N2]{\text{(allyl Br)}}$ $\xrightarrow{H_3O^+}$ Ph–CH$_2$–CO–CH(Ph)–CH$_2$CH=CH$_2$

22-52

$$\overset{\beta}{\text{Ph–CH}}\!-\!\overset{\gamma}{\text{CH}_2}\!-\!\overset{O}{\overset{\|}{C}}\!-\!\text{Ph}$$
$$|$$
$$\text{Ph–}\underset{\alpha}{\text{CH}}\!-\!\overset{O}{\overset{\|}{C}}\!-\!\text{CH}_3$$

came from ⟹

Ph–CH=C(–CO–CH$_3$)(Ph) **Michael acceptor**

+ $\overset{-}{C}$H$_2$–CO–Ph **Michael donor**

forward direction

Ph–CO–CH$_2$–[COOEt] $\xrightarrow{\text{NaOEt}}$ Ph–CO–$\overset{-}{C}$H–[COOEt] + Ph–CH=C(–CO–CH$_3$)(Ph) $\longrightarrow$

temporary ester group   resonance-stabilized

$\xrightarrow{H^+}$ Ph–CH(–CO–CH$_3$...)–CH(COOEt)–CO–Ph

[COOEt]
Ph–CH–CH–CO–Ph
|
Ph–CH–CO–CH$_3$

$\xrightarrow[\Delta]{H_3O^+}$

CO$_2$ + EtOH + Ph–CH$_2$–CH$_2$–CO–Ph
|
Ph–CH–CO–CH$_3$

22-53 First, you might wonder why this sequence does not make the desired product:

resonance-stabilized   MVK   poor yield

The poor yield in this conjugate addition is due primarily to the numerous competing reactions: the ketone enolate can self-condense (aldol), can condense with the ketone of MVK (aldol), or can deprotonate the methyl of MVK to generate a new nucleophile. The complex mixture of products makes this route practically useless. (continued on next page)

22-53 continued

What permits enamines (or other stabilized enolates) to work are: a) the certainty of which atom is the nucleophile, and b) the lack of self-condensation. Enamines can also do conjugate addition:

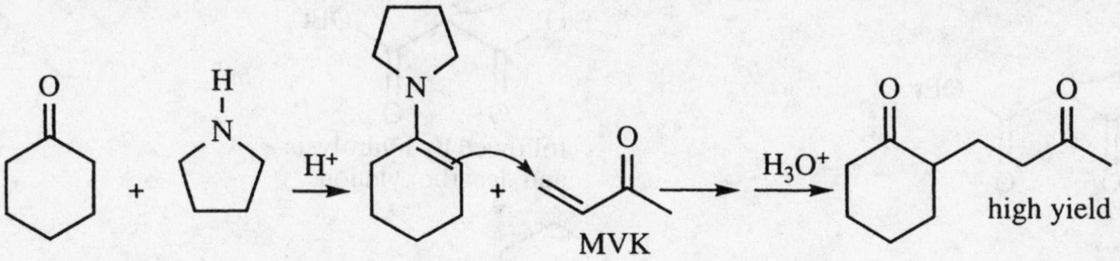
H⁺
MVK
H₃O⁺
high yield

22-54 The enolate of acetoacetic ester can be used in a Michael addition to an α,β-unsaturated ketone like MVK.

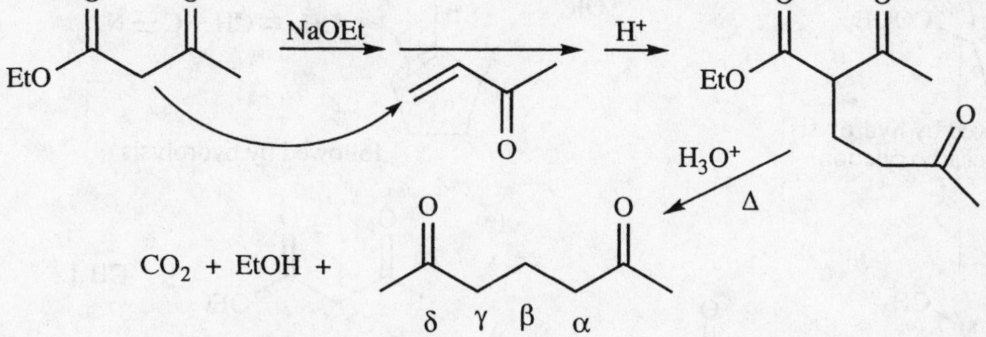
NaOEt
H⁺
EtO
H₃O⁺
Δ
CO₂ + EtOH +
δ γ β α

22-55

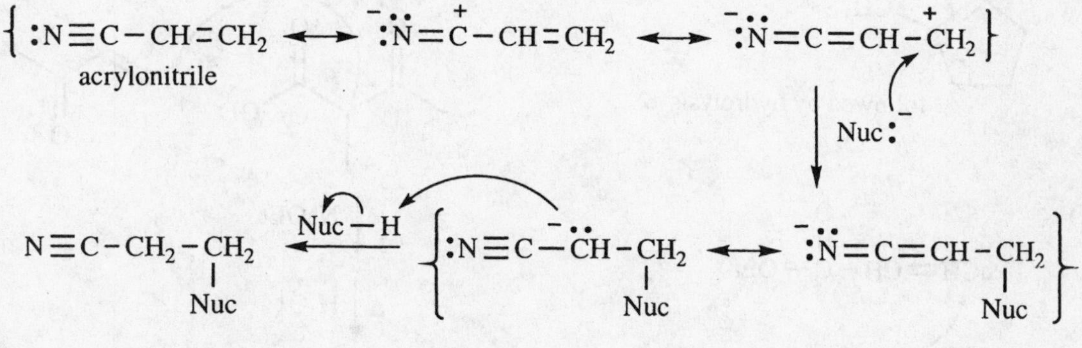
acrylonitrile
Nuc
Nuc—H

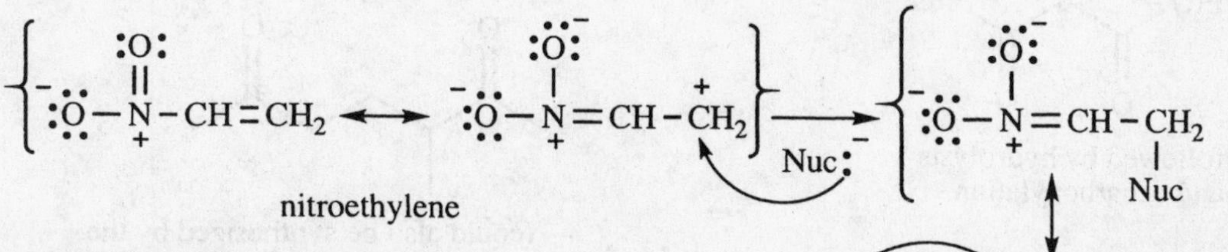
nitroethylene
Nuc
Nuc

(some resonance forms of the nitro group are not shown)

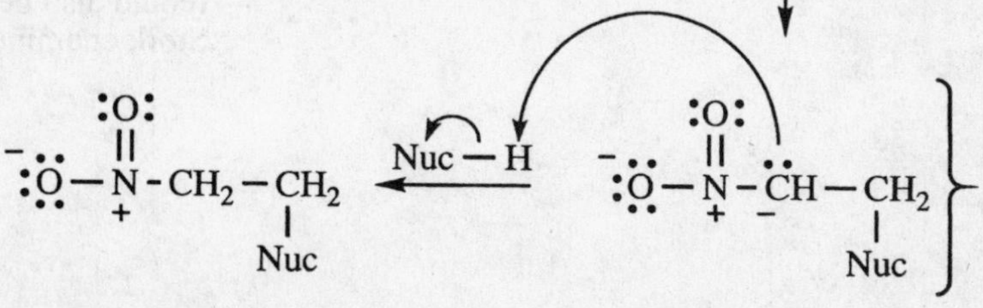
Nuc—H
Nuc

22-56

(a)

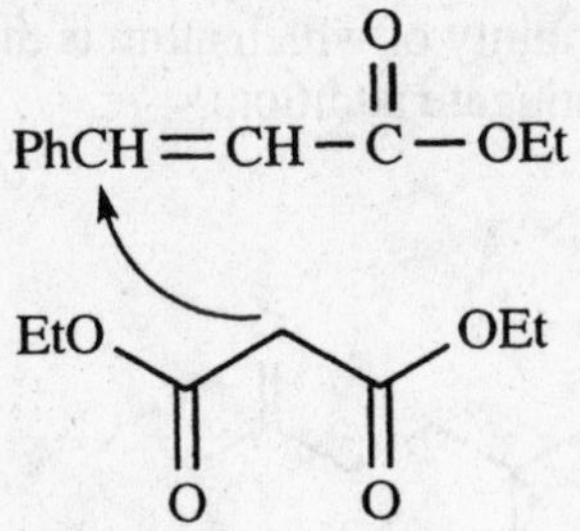

PhCH=CH−C−OEt (with O double bond)

EtO ... OEt (malonate, with two O double bonds)

(b)

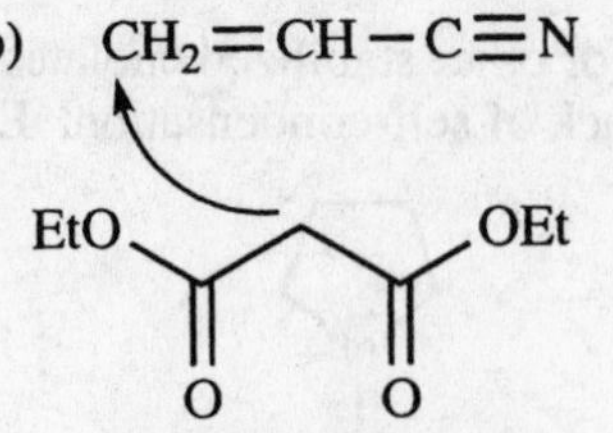

CH₂=CH−C≡N

EtO ... OEt (malonate, with two O double bonds)

followed by hydrolysis
and decarboxylation

(c) two ways

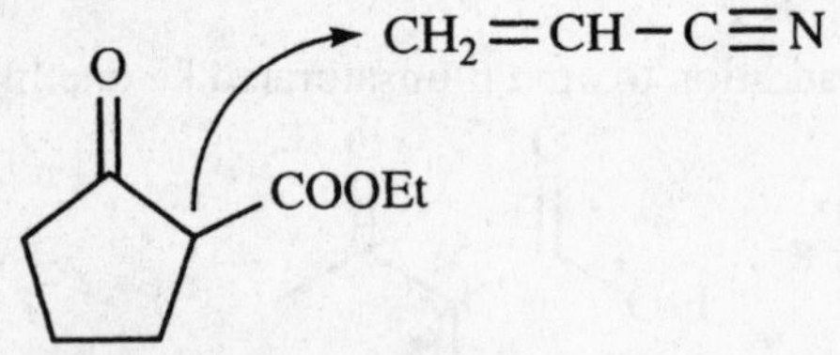

(cyclopentanone with COOEt) → CH₂=CH−C≡N

followed by hydrolysis
and decarboxylation

OR

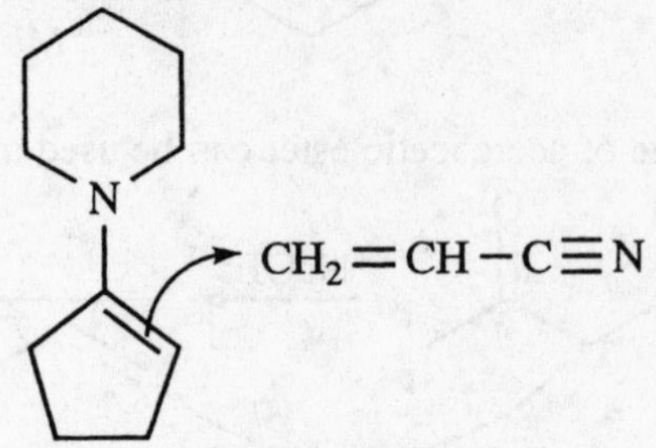

(piperidine enamine of cyclopentene) → CH₂=CH−C≡N

followed by hydrolysis

(d)

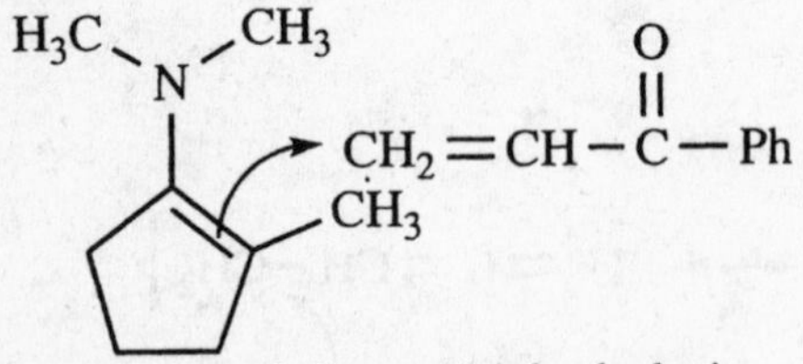

H₃C−N−CH₃ (dimethylenamine of cyclopentene with CH₃) → CH₂=CH−C−Ph (with O double bond)

followed by hydrolysis

(e)

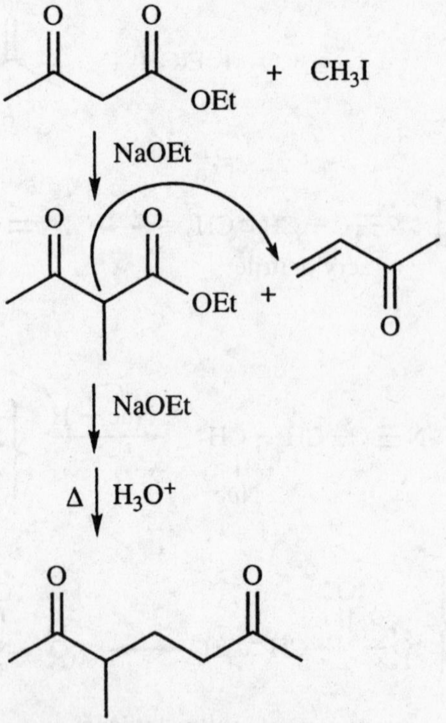

O   O
(ethyl acetoacetate)   OEt   +   CH₃I

│ NaOEt

(enolate attacks CH₃I) OEt   +   (methyl vinyl ketone)

│ NaOEt

Δ | H₃O⁺

O   O
(2-methyl-heptanedione)

(could also be synthesized by the
Stork enamine reactions)

(f)

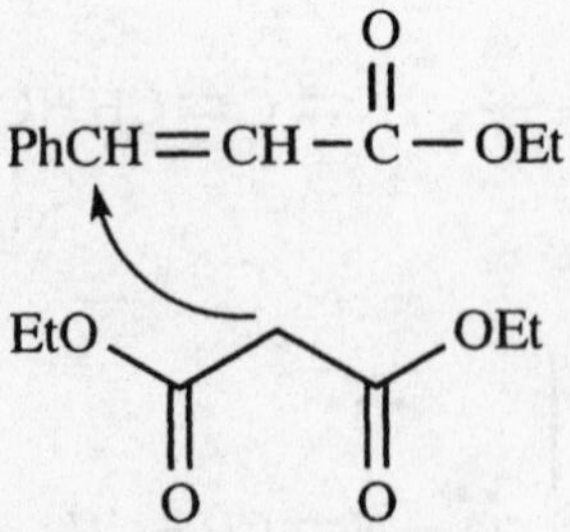

PhCH=CH−C−OEt (with O double bond)

EtO ... OEt (malonate, with two O double bonds)

followed by hydrolysis
and decarboxylation

560

22-57

Step 1: carbon skeleton

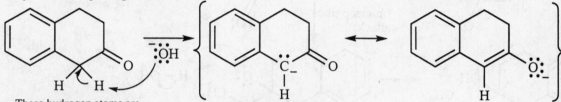

Step 2: nucleophile generation

These hydrogen atoms are more acidic than the other alpha hydrogens because the anion can be stabilized by the benzene ring.

plus resonance forms with negative charge on the benzene ring

Step 3: nucleophilic attack (Michael addition)

HO — H

continued on next page

22-57 continued

Step 4: conversion to final product

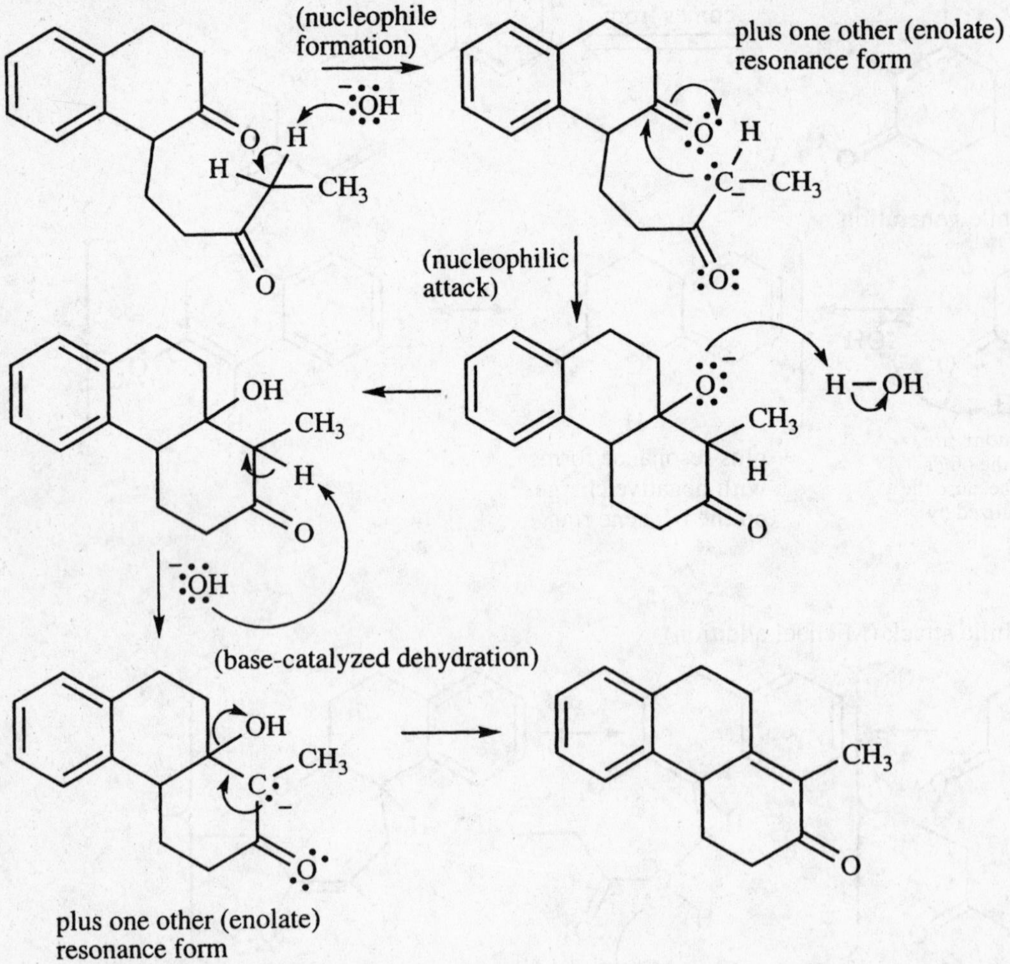

(nucleophile formation)

plus one other (enolate) resonance form

(nucleophilic attack)

(base-catalyzed dehydration)

plus one other (enolate)
resonance form

Step 5: The complete mechanism is the combination of Steps 2, 3, and 4. Notice that this mechanism is simply described by:
1) Enolate formation, followed by Michael addition;
2) Aldol condensation, followed by dehydration.

22-58

Step 1: carbon skeleton

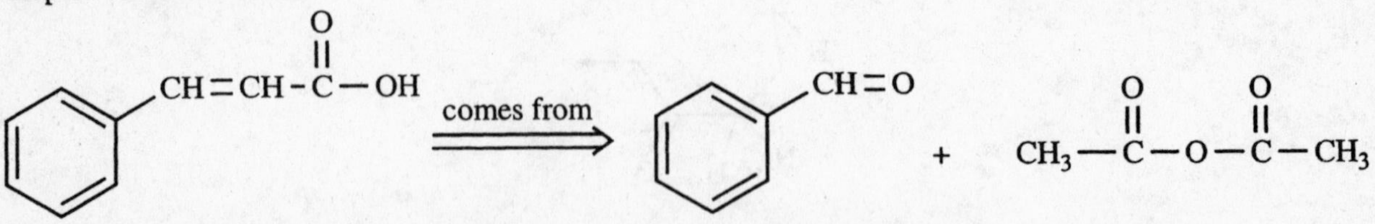

comes from

Step 2: nucleophile generation

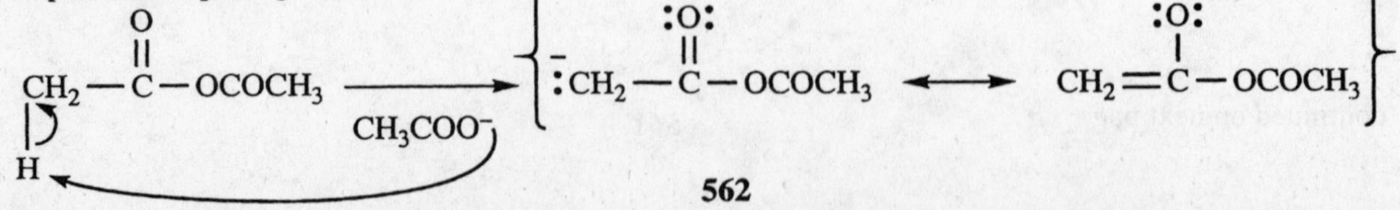

22-58 continued

Step 3: nucleophilic attack

$$Ph-\overset{\overset{\displaystyle :O:}{\|}}{C}-H \;+\; {}^{-}:CH_2-\overset{\overset{\displaystyle O}{\|}}{C}-OCOCH_3 \longrightarrow Ph-\overset{\overset{\displaystyle :\overset{-}{O}:}{|}}{CH}-CH_2-\overset{\overset{\displaystyle O}{\|}}{C}-OCOCH_3$$

$$Ph-\overset{\overset{\displaystyle OH}{|}}{CH}-CH_2-\overset{\overset{\displaystyle O}{\|}}{C}-OCOCH_3 \;\longrightarrow\; H-OAc$$

Step 4: conversion to final product

$$Ph-\overset{\overset{\displaystyle OH}{|}}{CH}-\overset{\overset{\displaystyle }{\underset{\displaystyle H}{CH}}}{}-\overset{\overset{\displaystyle O}{\|}}{C}-OCOCH_3 \;\xrightarrow{\;CH_3COO^-\;}\; Ph-\overset{\overset{\displaystyle OH}{|}}{CH}-\overset{-}{:}\underset{..}{CH}-\overset{\overset{\displaystyle :O:}{\|}}{C}-OCOCH_3$$

plus one other (enolate) resonance form

$$Ph-CH=CH-\overset{\overset{\displaystyle O}{\|}}{C}-OCOCH_3$$

<u>hydrolysis</u>

$$PhCH=CH-\overset{\overset{\displaystyle :O:}{\|}}{C}-O-\overset{\overset{\displaystyle O}{\|}}{C}-CH_3 \xrightarrow{\;H^+\;} \left\{ PhCH=CH-\overset{\overset{\displaystyle :\overset{+}{O}-H}{\|}}{C}-O-\overset{\overset{\displaystyle O}{\|}}{C}-CH_3 \right.$$

$$PhCH=CH-\overset{\overset{\displaystyle OH}{|}}{C}-O-\overset{\overset{\displaystyle :O:}{\|}}{C}-CH_3 \;\xrightarrow{}\; H^+$$

$$\longleftarrow PhCH=CH-\underset{\underset{\displaystyle +}{|}}{\overset{\overset{\displaystyle :\overset{..}{O}-H}{|}}{C}}-O-\overset{\overset{\displaystyle O}{\|}}{C}-CH_3 \left. \right\}$$

$$H_2\overset{..}{O}:$$

plus one other resonance form
with positive on the other oxygen

$$H-\overset{+}{\underset{\displaystyle ..}{O}}-H$$

$$H_2\overset{..}{O}:$$

two fast
proton transfers

$$PhCH=CH-\underset{\underset{\displaystyle OH}{|}}{\overset{\overset{\displaystyle OH}{|}}{C}}-O-\overset{\overset{\displaystyle :\overset{+}{O}-H}{\|}}{C}-CH_3 \;\longrightarrow\; O=\underset{\underset{\displaystyle CH_3}{|}}{C}-H \;+\; \left\{ PhCH=CH-\underset{\underset{\displaystyle :OH}{|}}{\overset{\overset{\displaystyle H-\overset{..}{O}:}{|}}{C}{}^+} \right.$$

plus two other
resonance forms

$$PhCH=CH-\overset{\overset{\displaystyle O}{\|}}{C}-OH \;\longleftarrow\; PhCH=CH-\underset{\underset{\displaystyle :OH}{|}}{\overset{\overset{\displaystyle H-\overset{+}{O}:}{\|}}{C}} \;\; H_2\overset{..}{O}:$$

plus one other resonance form
with positive on the other oxygen

Step 5: The complete mechanism is the combination of Steps 2, 3, and 4.

563

22-59 The Robinson annulation consists of a Michael addition followed by aldol cyclization with dehydration. In the retrosynthetic direction, disconnect the alkene formed in the aldol/dehydration, then disconnect the Michael addition to discover the reactants.

(a) aldol and dehydration forms the α,β double bond:

Michael addition forms a bond to the β' carbon:

(b) aldol and dehydration forms the α,β double bond:

Michael addition forms a bond to the β' carbon:

22-60 Please refer to solution 1-20, page 12 of this Solutions Manual.

22-61 The most acidic hydrogens are shown in boldface. (See Appendix 2 for a review of acidity.) (Braces around resonance forms omitted here.)

(a)

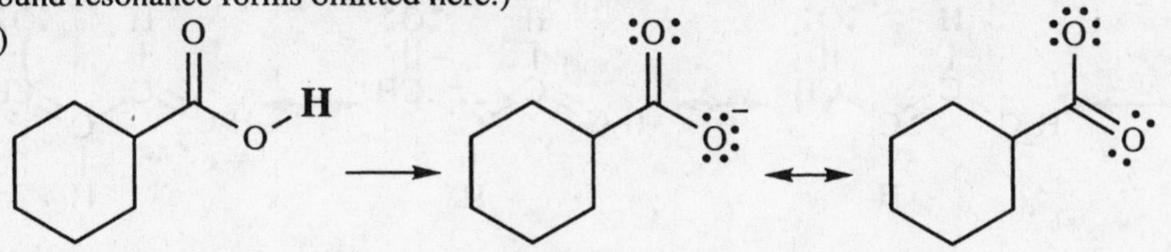

(b)

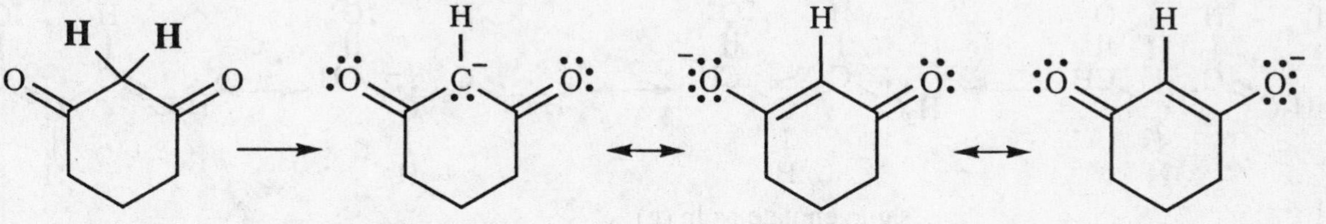

(c)

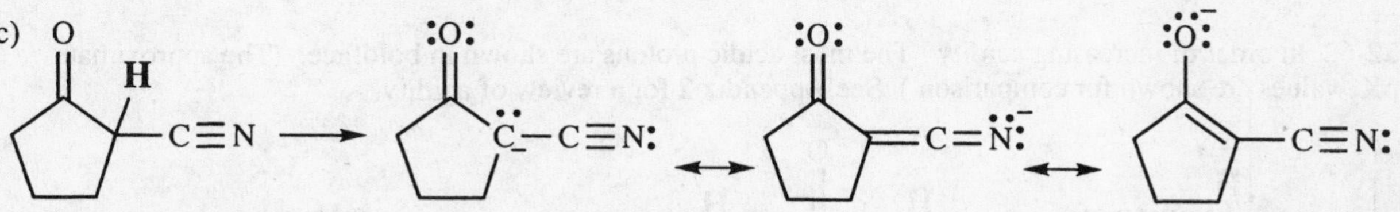

(d)

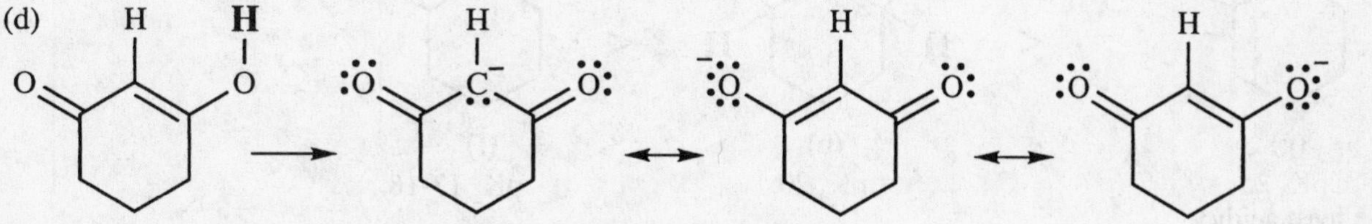

same enolate as in (b)

(e)

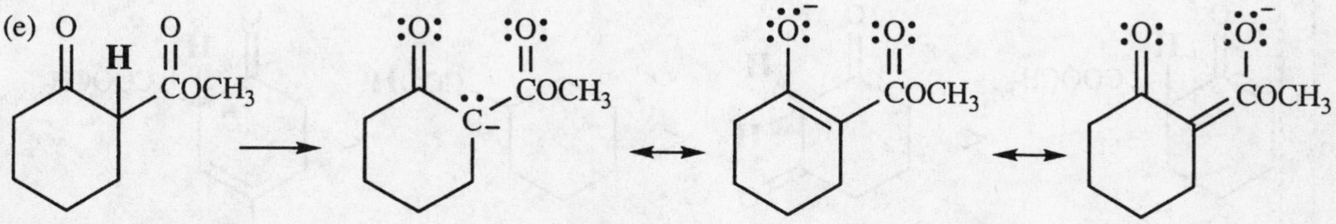

(f)

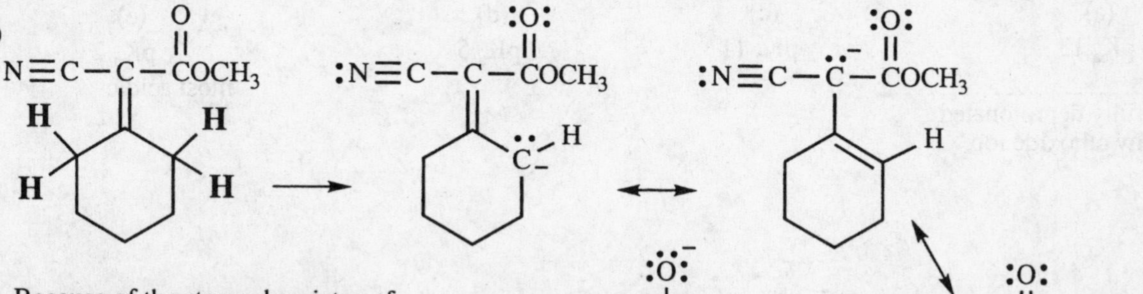

$N \equiv C - C - COCH_3$

Because of the stereochemistry of the double bond, the H atoms on the left side of the ring are not equivalent to those on the right side. However, the two enolates will be equivalent except for the double bond geometry.

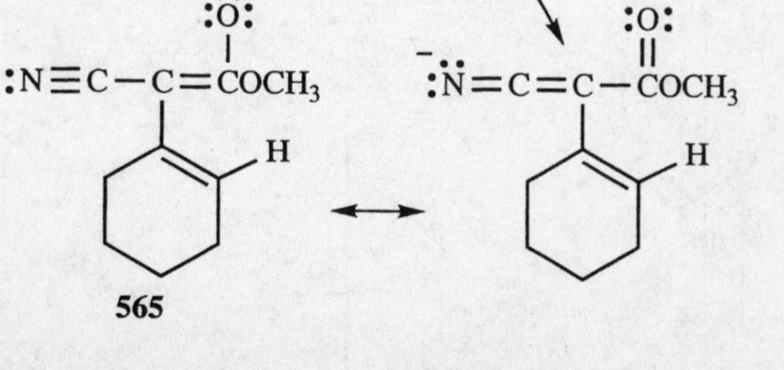

22-61 continued

(g)

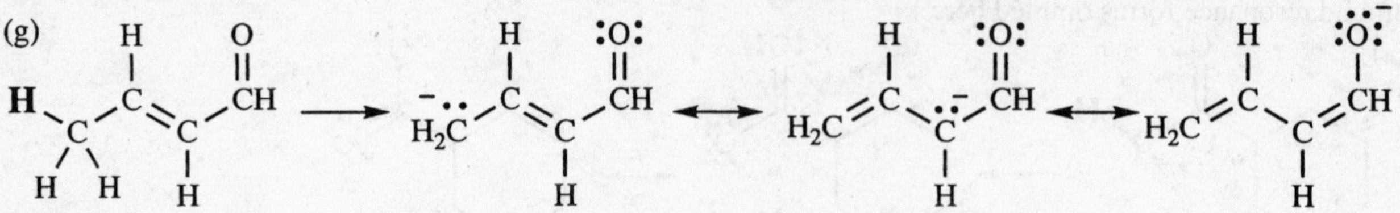

(h)

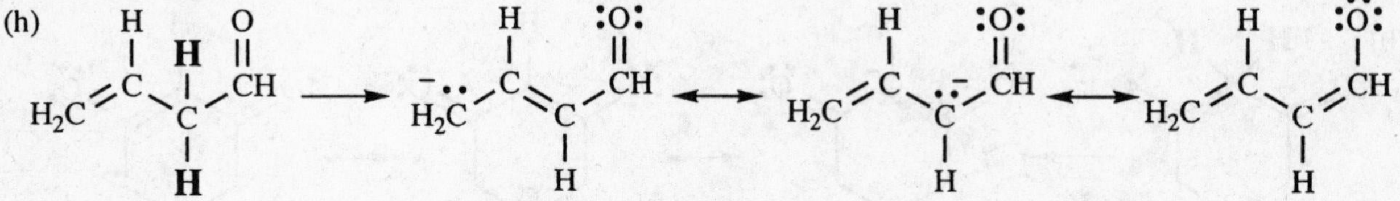

same enolate as in (g)

22-62 In order of increasing acidity. The most acidic protons are shown in boldface. (The approximate pK_a values are shown for comparison.) See Appendix 2 for a review of acidity.

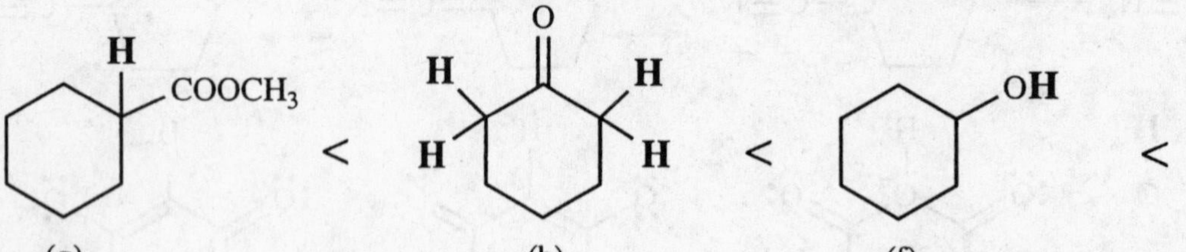

(g) pK_a 25 least acidic < (b) pK_a 20 < (f) pK_a 17-18 <

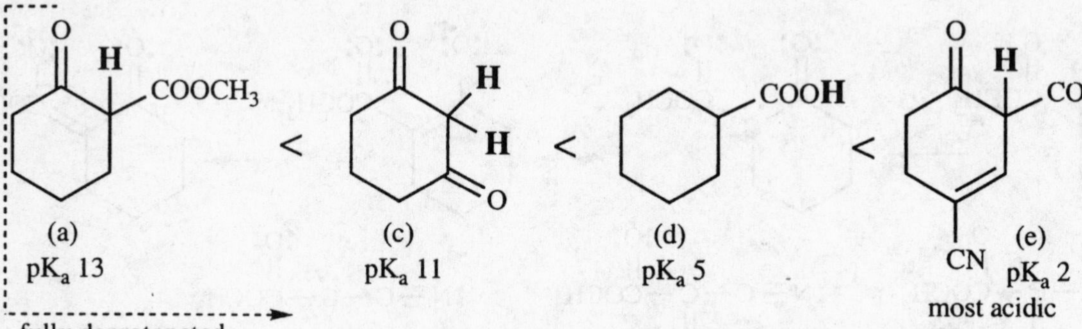

(a) pK_a 13 < (c) pK_a 11 < (d) pK_a 5 < (e) pK_a 2 most acidic

fully deprotonated by ethoxide ion

22-63

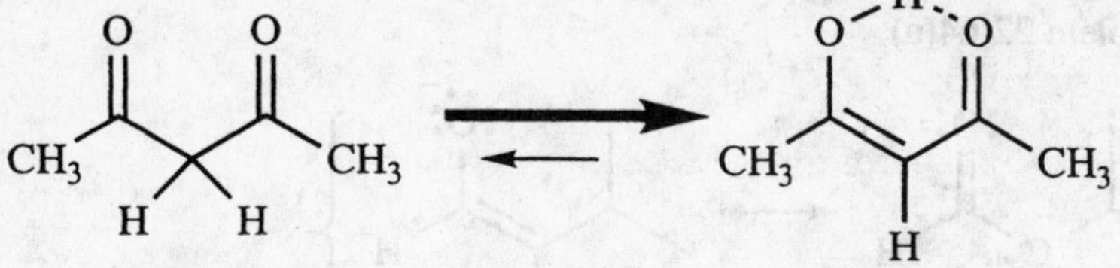

keto ⟶ ⟵ enol

The enol form is stable because of the conjugation and because of intramolecular hydrogen-bonding in a six-membered ring.

In dicarbonyl compounds in general, the weaker the electron-donating ability of the group G, the more it will exist in the enol form: aldehydes (G = H) are almost completely enolized, then ketones, esters, and finally amides which have virtually no enol content.

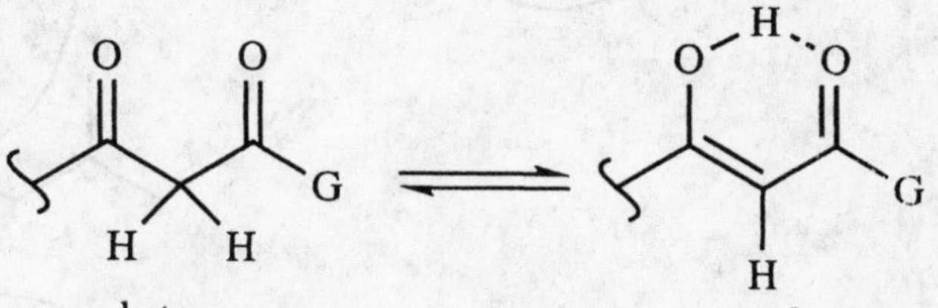

keto ⇌ enol

22-64 The wavy line lies across the bond formed in the aldol condensation.

(a) $-H_2O$

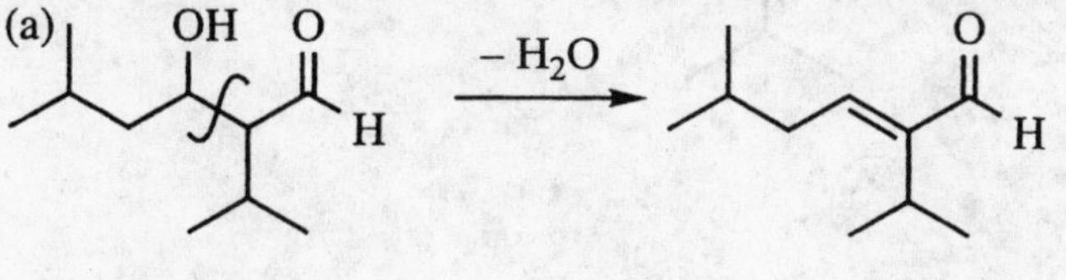

(b) $-H_2O$

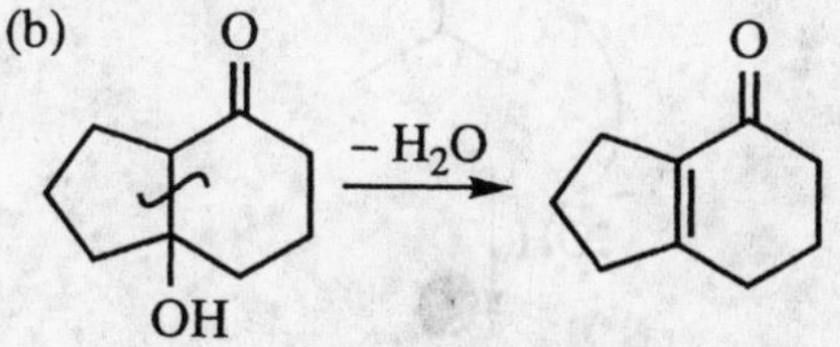

(c) $-2\,H_2O$

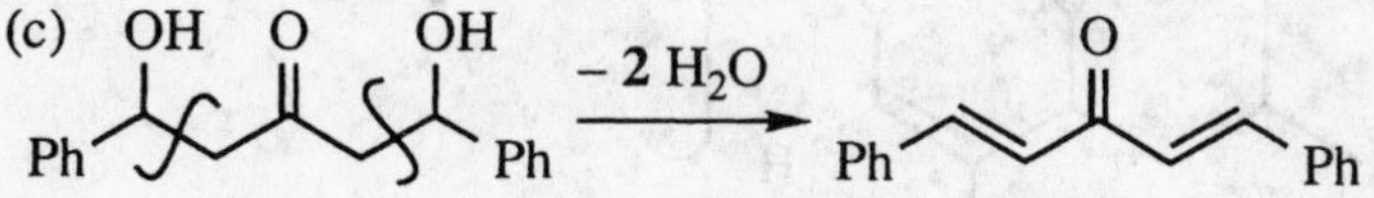

(d) $-H_2O$

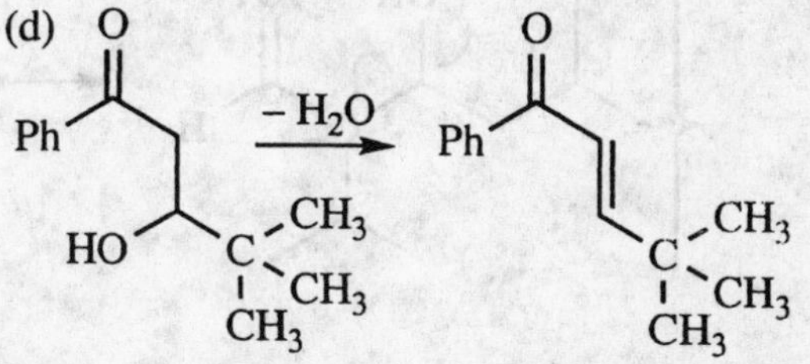

(e) $-H_2O$

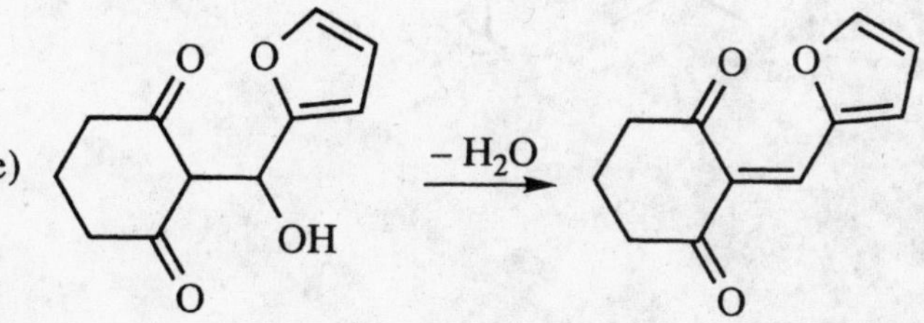

22-65 The wavy line lies across the bond formed in the Claisen condensation.

(a) (b) 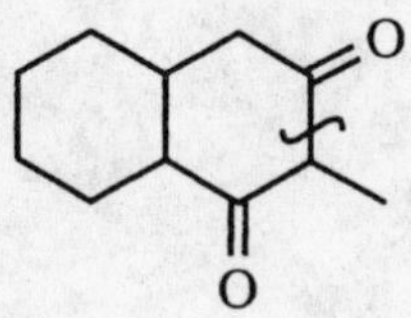 (c) 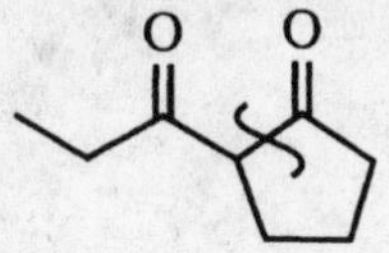 (d)

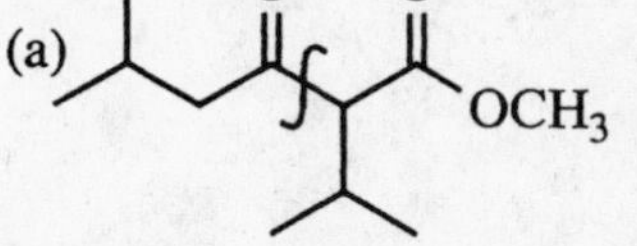

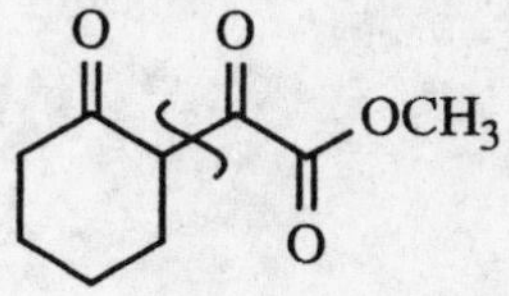

(e)

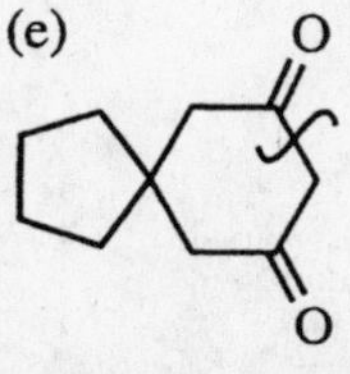

567

22-66

(a) mechanism of aldol condensation in problem 22-64(a)

22-66 continued

(b) mechanism of aldol condensation in problem 22-64(b)

22-66 continued

(c) mechanism of Claisen condensation in problem 22-65(a)

(this product will be deprotonated by methoxide,
but regenerated upon acidic workup)

(d) mechanism of Claisen condensation in problem 22-65(b)

(this product will be deprotonated by methoxide,
but regenerated upon acidic workup)

22-67 All products shown are after acidic workup.

(a) aldol self-condensation

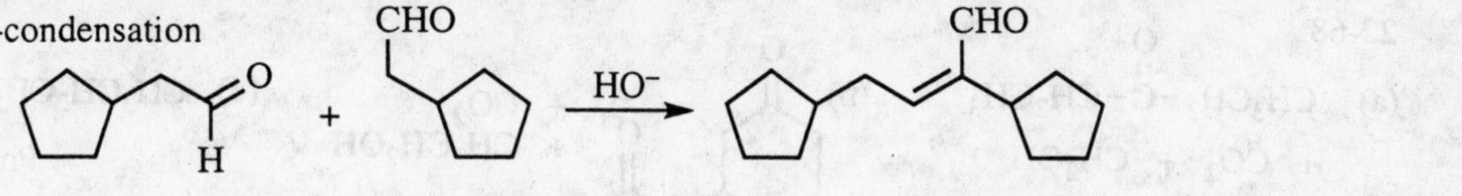

(b) Claisen self-condensation

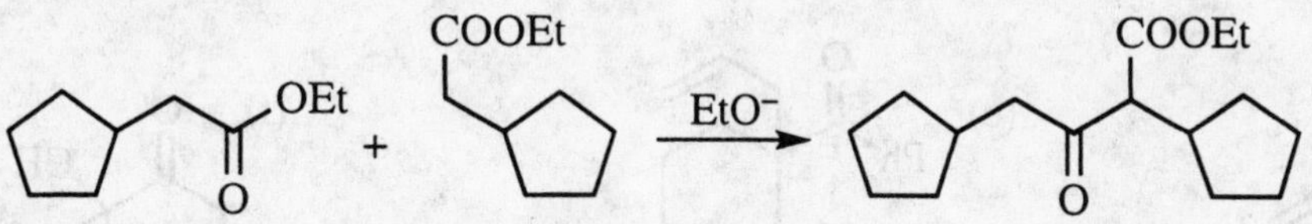

(c) aldol cyclization

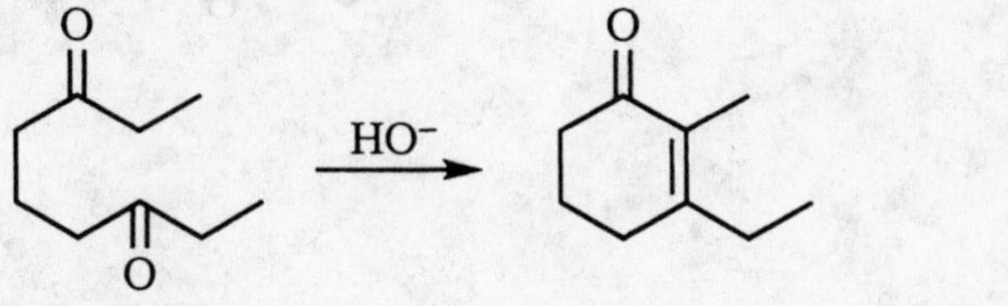

(d) mixed Claisen

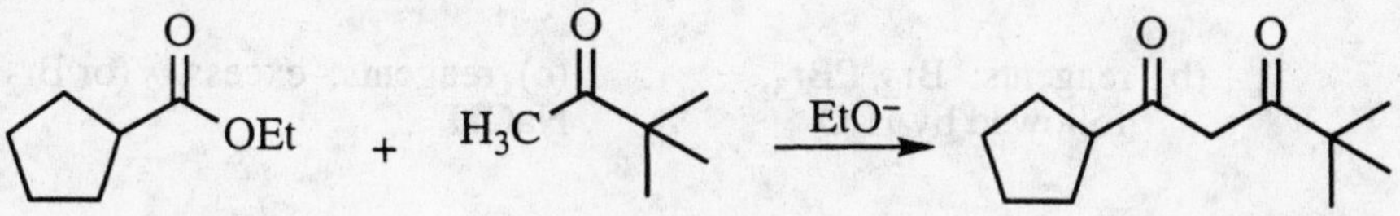

OR

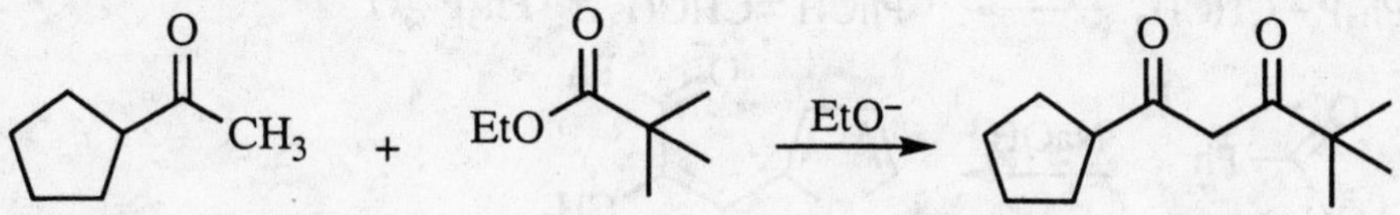

(e) mixed aldol

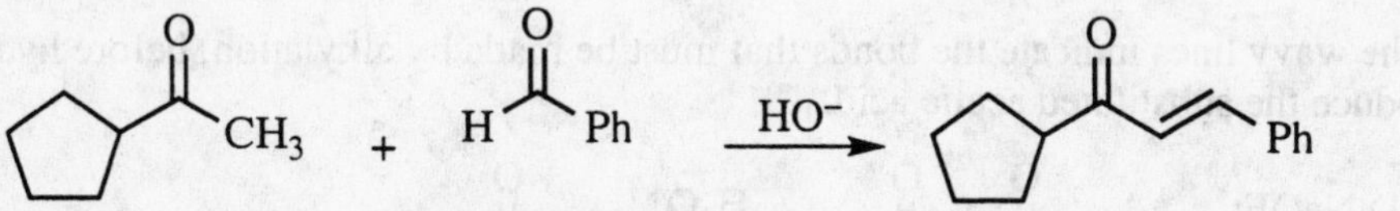

(f) enamine acylation or mixed Claisen

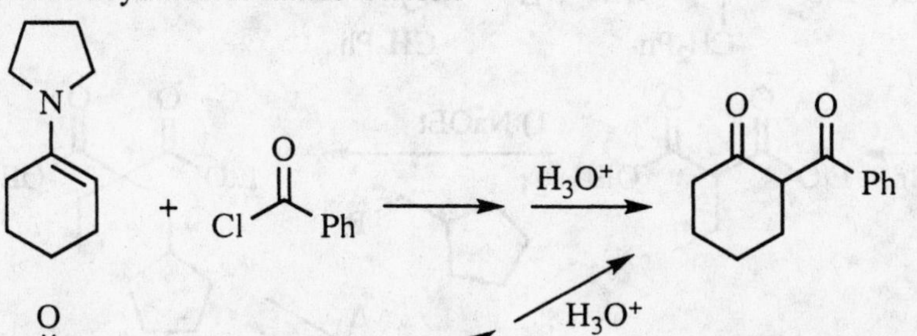

OR

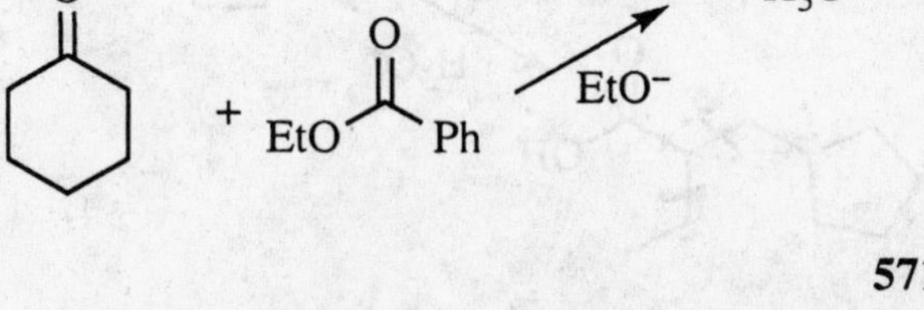

In practice, the mixed Claisen reactions starting from a ketone enolate plus an ester will give a considerable amount of aldol self-condensation.

571

22-68

(a) $CH_3CH_2-\overset{\overset{\displaystyle O}{\|}}{C}-CH_2CH_3$
   $+\ CO_2\ +\ CH_3OH$

(b) 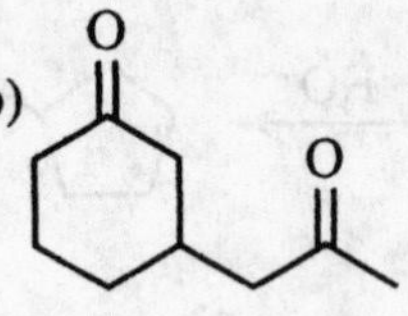 $+\ CO_2$
   $+\ CH_3CH_2OH$

(c)

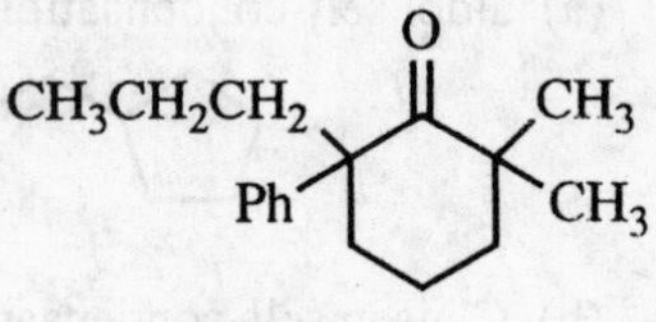

(d)

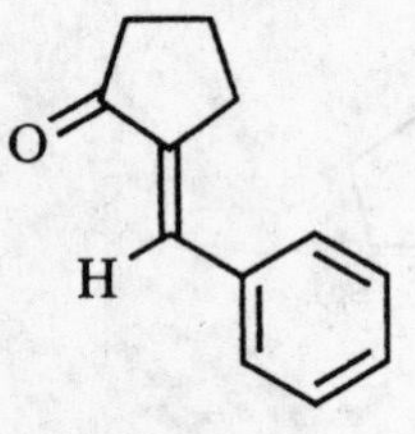

(e)

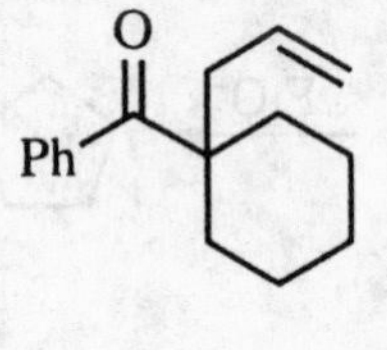

(f)

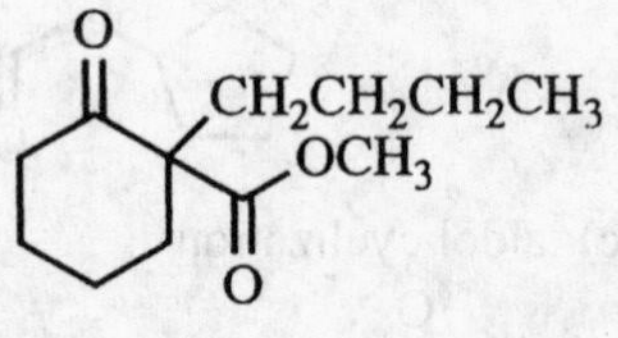

(g)

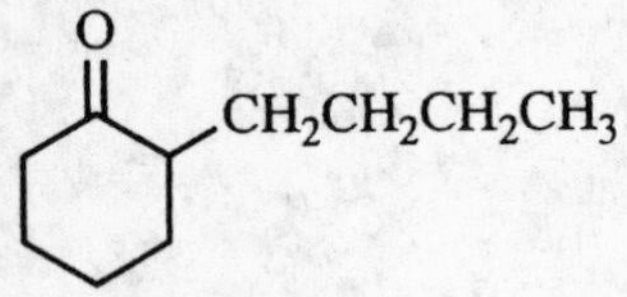

22-69

(a) reagents: Br_2, H^+

(b) reagents: Br_2, PBr_3, followed by H_2O

(c) reagents: excess I_2 (or Br_2 or Cl_2), NaOH

(d)
$Ph-\overset{\overset{\displaystyle O}{\|}}{C}-H\ +\ Ph_3\overset{+}{P}-\overset{-}{C}HCH_3\ \longrightarrow\ PhCH=CHCH_3\ +\ Ph_3P=O$

(e) $+$ $\xrightarrow{\text{NaOH}}$

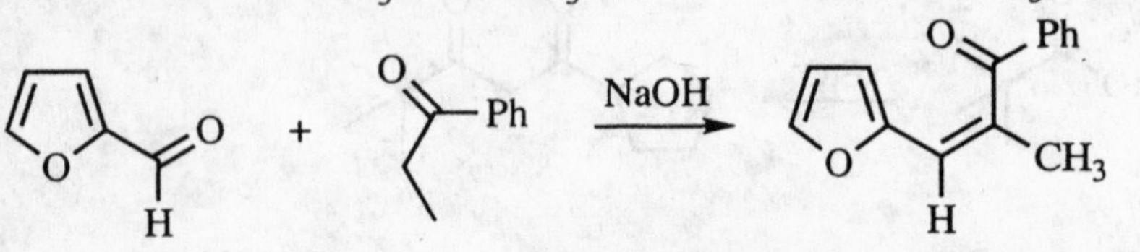

22-70 In the products, the wavy lines indicate the bonds that must be made by alkylation, before hydrolysis and decarboxylation produce the substituted acetic acid.

(a)

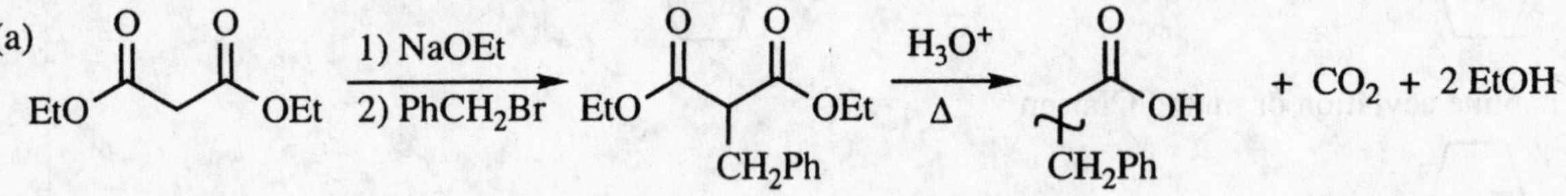

(b)

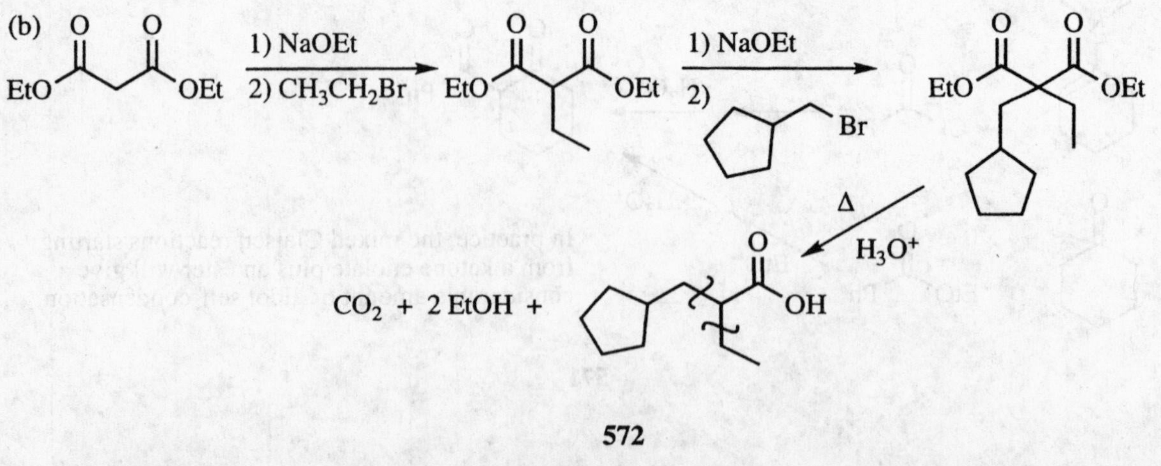

572

22-70 continued

(c)

EtO—C(=O)—CH₂—C(=O)—OEt $\xrightarrow[\text{2) Br(CH}_2)_5\text{Br}]{\text{1) 2 NaOEt}}$ [spirocyclic diester] $\xrightarrow[\Delta]{\text{H}_3\text{O}^+}$ [cyclohexane-COOH] + CO_2 + 2 EtOH

22-71 In the products, the wavy lines indicate the bonds that must be made by alkylation, before hydrolysis and decarboxylation produce the substituted acetone.

(a)

[ethyl acetoacetate] $\xrightarrow[\text{2) EtBr}]{\text{1) NaOEt}}$ [alkylated] $\xrightarrow[\text{2) CH}_2\text{Br(benzyl)}]{\text{1) NaOEt}}$ [dialkylated] $\xrightarrow[\Delta]{\text{H}_3\text{O}^+}$ [product] + CO_2 + EtOH

(b)

[ethyl acetoacetate] $\xrightarrow[\text{2) Br(CH}_2)_4\text{Br}]{\text{1) 2 NaOEt}}$ [cyclopentane diester/ketone] $\xrightarrow[\Delta]{\text{H}_3\text{O}^+}$ [product] + CO_2 + EtOH

(c) The acetoacetic ester synthesis makes substituted acetone, so where is the acetone in this product?

substituted acetone

The single bond to this substituted acetone can be made by the acetoacetic ester synthesis. How can we make the α,β double bond? Aldol condensation!

make by conjugate addition ⟶ [cyclohexenone with EtOOC] ⟵ make by aldol cyclization/dehydration

[ethyl acetoacetate] + [methyl vinyl ketone] $\xrightarrow{\text{NaOEt}}$ [Michael adduct] $\xrightarrow[\Delta]{\text{H}_3\text{O}^+}$ [diketone] $\xrightarrow{\text{NaOH}}$ [methylcyclohexenone]

573

22-72 These compounds are made by aldol condensations followed by other reactions. The key is to find the skeleton make by the aldol.

(a) Where is the possible α,β-unsaturated carbonyl in this skeleton?

$$PhCH_2CH_2-\overset{OH}{\underset{|}{C}}HPh \Longrightarrow PhCH \overset{}{\underset{}{\overset{|}{\not{}}}} CH-\overset{O}{\overset{||}{C}}-Ph \xrightarrow{\text{reverse aldol}} Ph-\overset{O}{\overset{||}{C}}-H \; + \; CH_3-\overset{O}{\overset{||}{C}}-Ph$$

forward synthesis

$$Ph-\overset{O}{\overset{||}{C}}-H \; + \; CH_3-\overset{O}{\overset{||}{C}}-Ph \xrightarrow{NaOH} PhCH=CH-\overset{O}{\overset{||}{C}}-Ph \xrightarrow{H_2, Pt} PhCH_2CH_2-\overset{OH}{\underset{|}{C}}HPh$$

(b) The aldol skeleton is not immediately apparent in this formidable product. What can we see from it? Most obvious is the β-dicarbonyl (β-ketoester) which we know to be a good nucleophile, capable of substitution or Michael addition. In this case, Michael addition is most likely as the site of attack is β to another carbonyl.

β-ketoester

AHA! The aldol product reveals itself.
(See the solution to 22-71 (c).)

forward synthesis

1) NaOCH₃
2) H⁺

(c) The key in this product is the α-nitroketone, the equivalent of a β-dicarbonyl system, capable of doing Michael addition to the β-carbon of the other carbonyl.

forward synthesis

NaOH · · · $\xrightarrow{H^+}$ · · · $\xrightarrow[Na_2CO_3]{}$

22-73

(a)

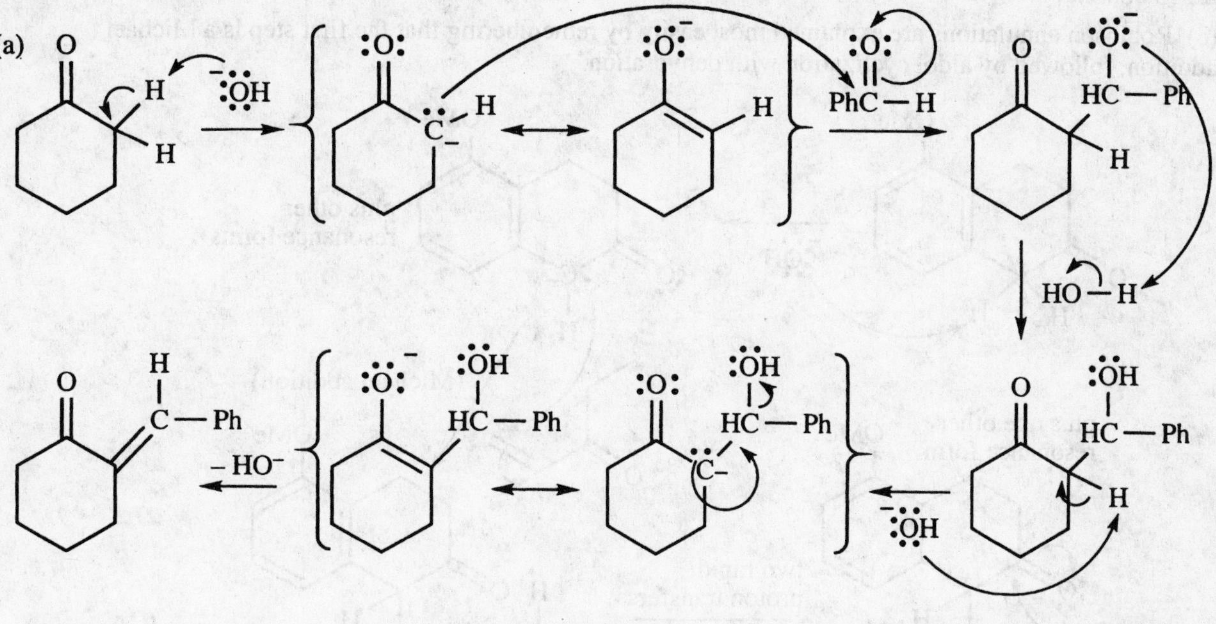

(b)

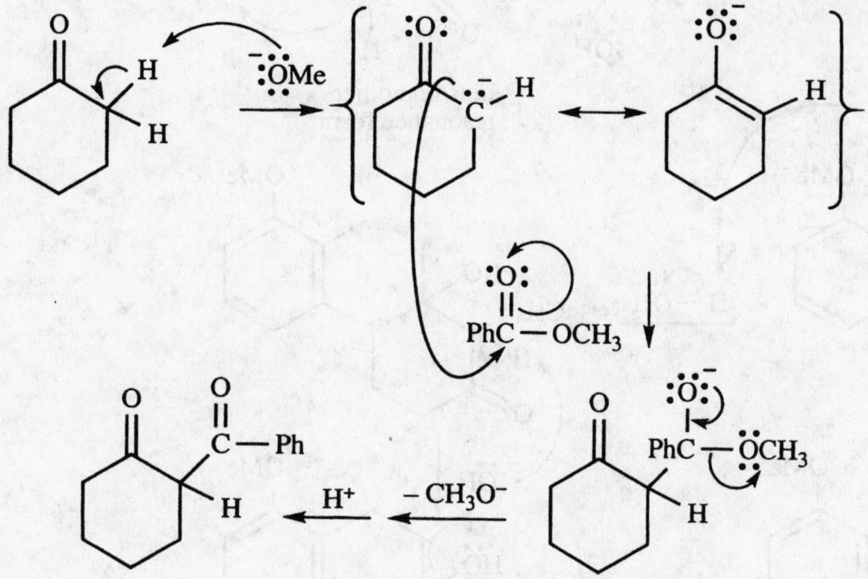

575

22-73 continued

(c) Robinson annulations are explained most easily by remembering that the first step is a Michael addition, followed by aldol cyclization with dehydration.

(d)

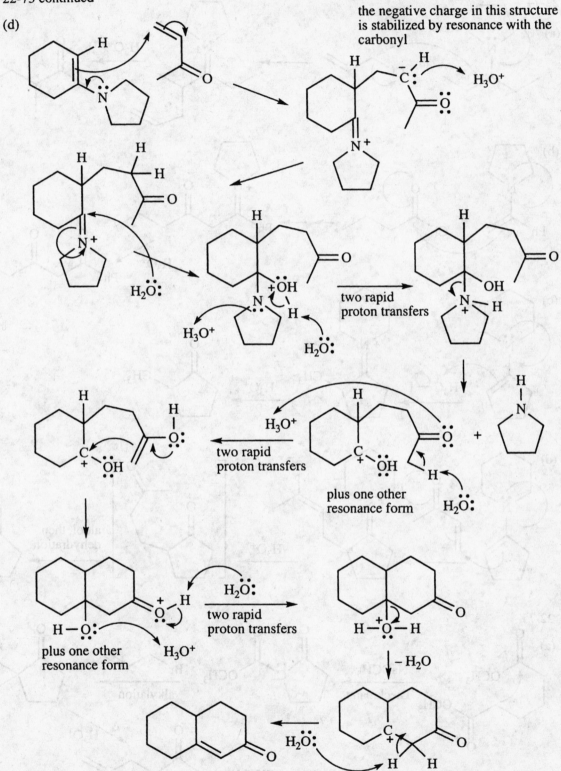

the negative charge in this structure is stabilized by resonance with the carbonyl

577

22-74

(a)

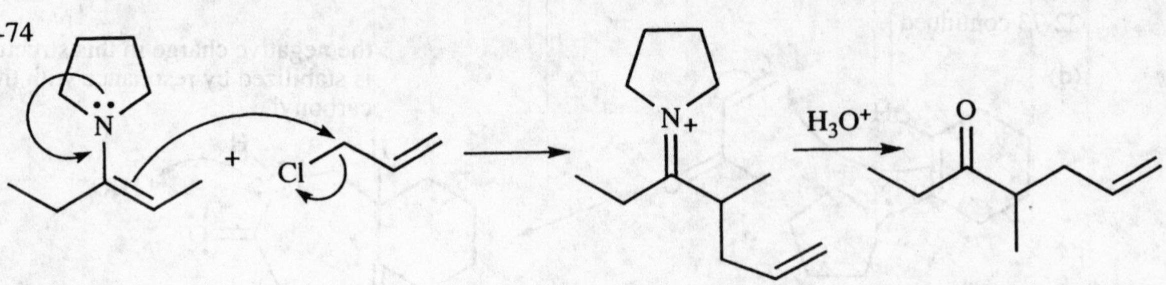

(b)

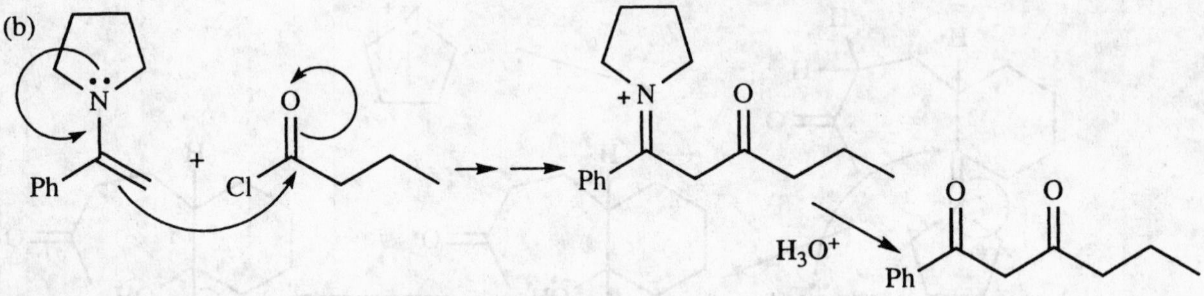

(c)

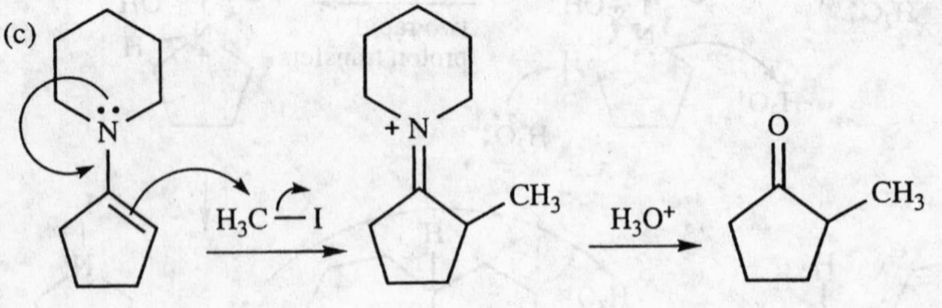

(d)

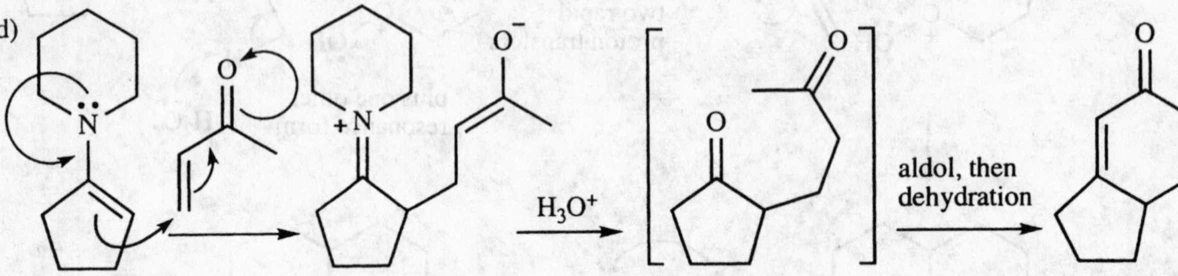

22-75

(a)

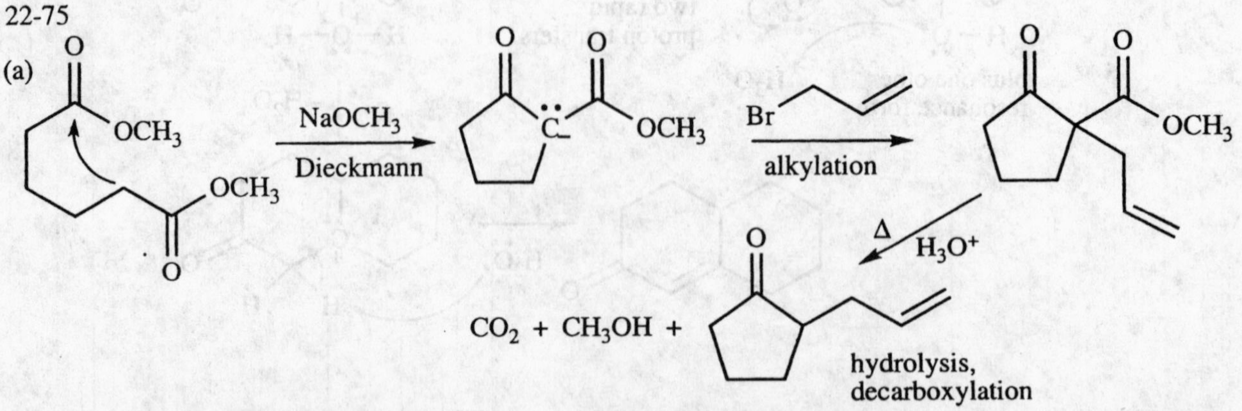

$$CO_2 + CH_3OH +$$

hydrolysis, decarboxylation

22-75 continuued

(b)

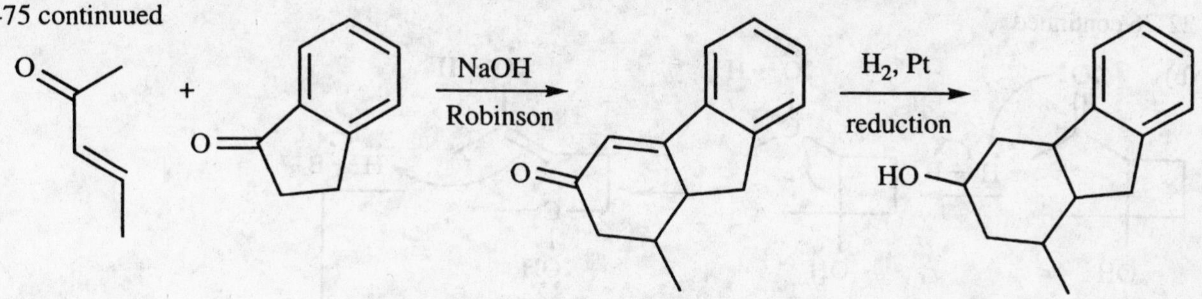

O
+
O
NaOH
Robinson
O
H2, Pt
reduction
HO

(c)

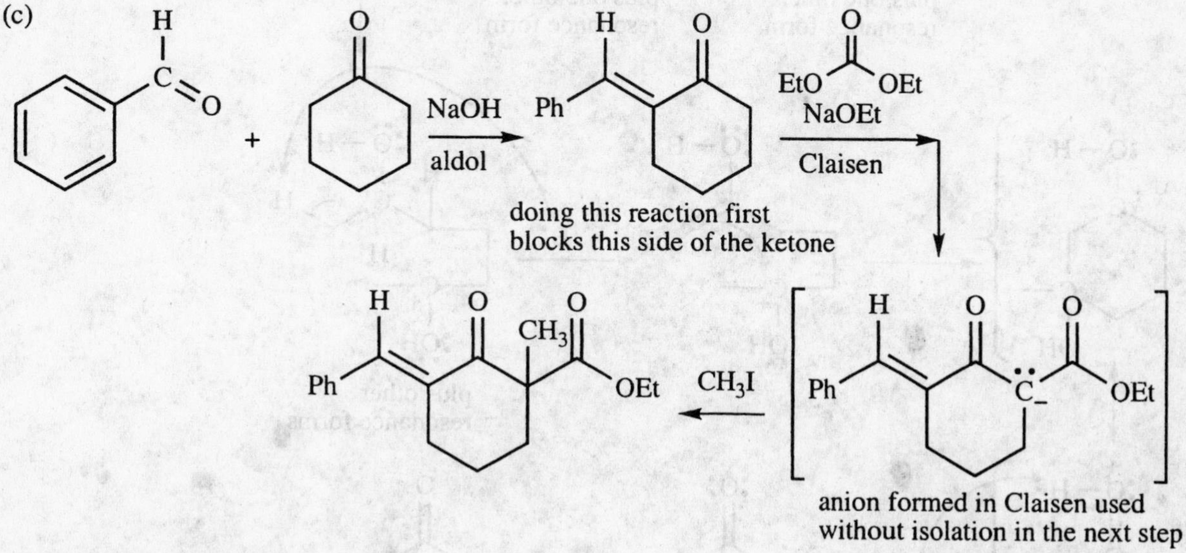

H
C
O
+
O
NaOH
aldol
Ph
H
O
doing this reaction first
blocks this side of the ketone
O
EtO
OEt
NaOEt
Claisen
H
O
O
Ph
C
OEt
anion formed in Claisen used
without isolation in the next step
CH3I
H
O
O
CH3
Ph
OEt

22-76

(a)

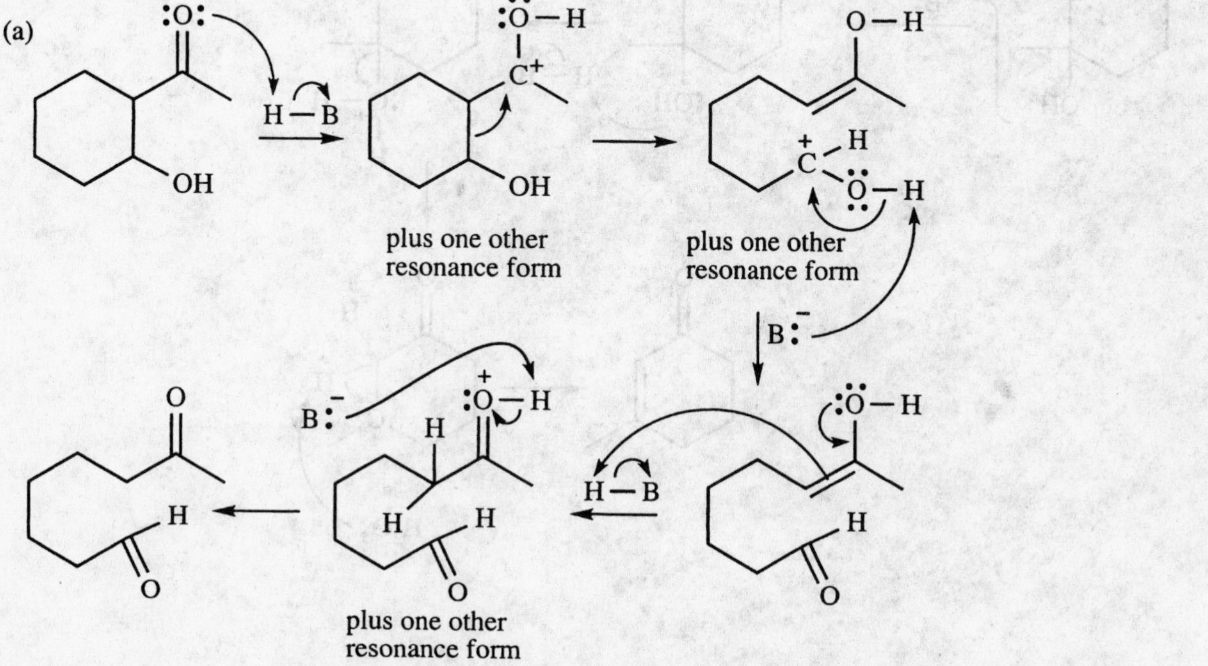

:O:
OH
H—B
:O—H
C+
OH
plus one other
resonance form
O—H
C
H
O—H
plus one other
resonance form
B:
:O—H
H
O
H—B
B:
+
:O—H
H
H
H
O
plus one other
resonance form
O
H
O

22-76 continued

(b)

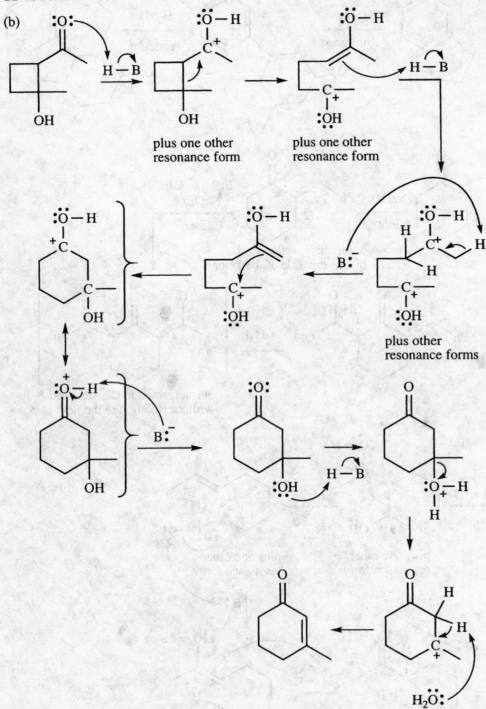

plus one other
resonance form

plus one other
resonance form

plus other
resonance forms

22-76 continued

(c)

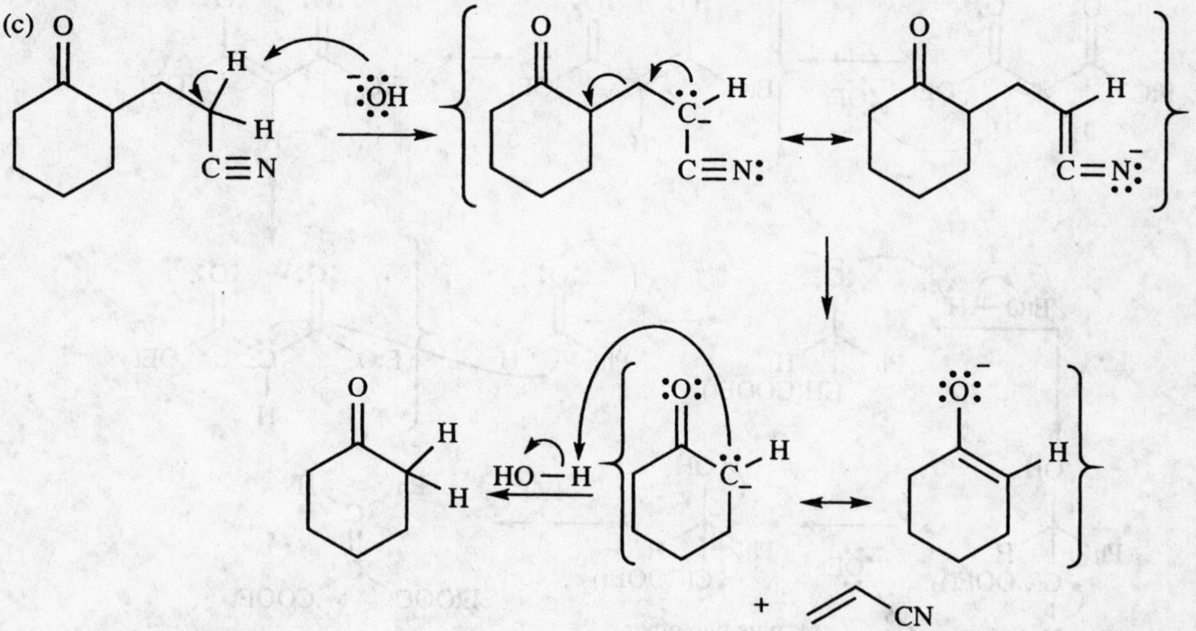

22-77 All of these Robinson annulations are catalyzed by NaOH.

(a)

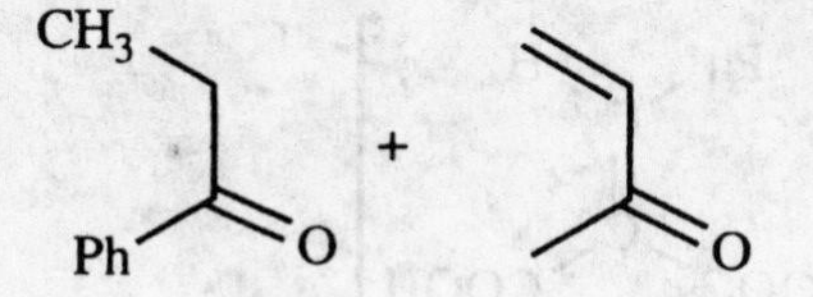

(b)

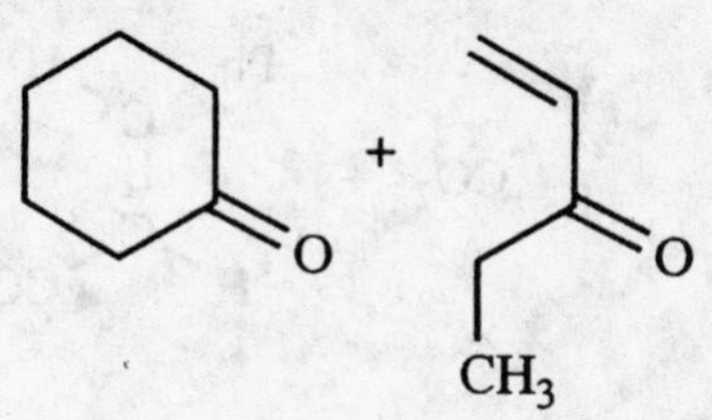

(c)

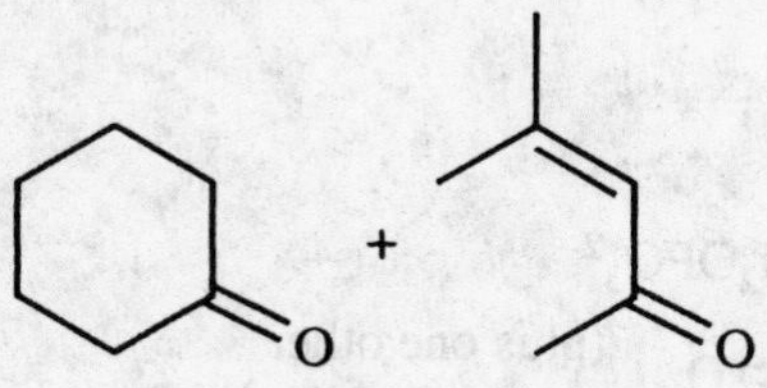

22-78

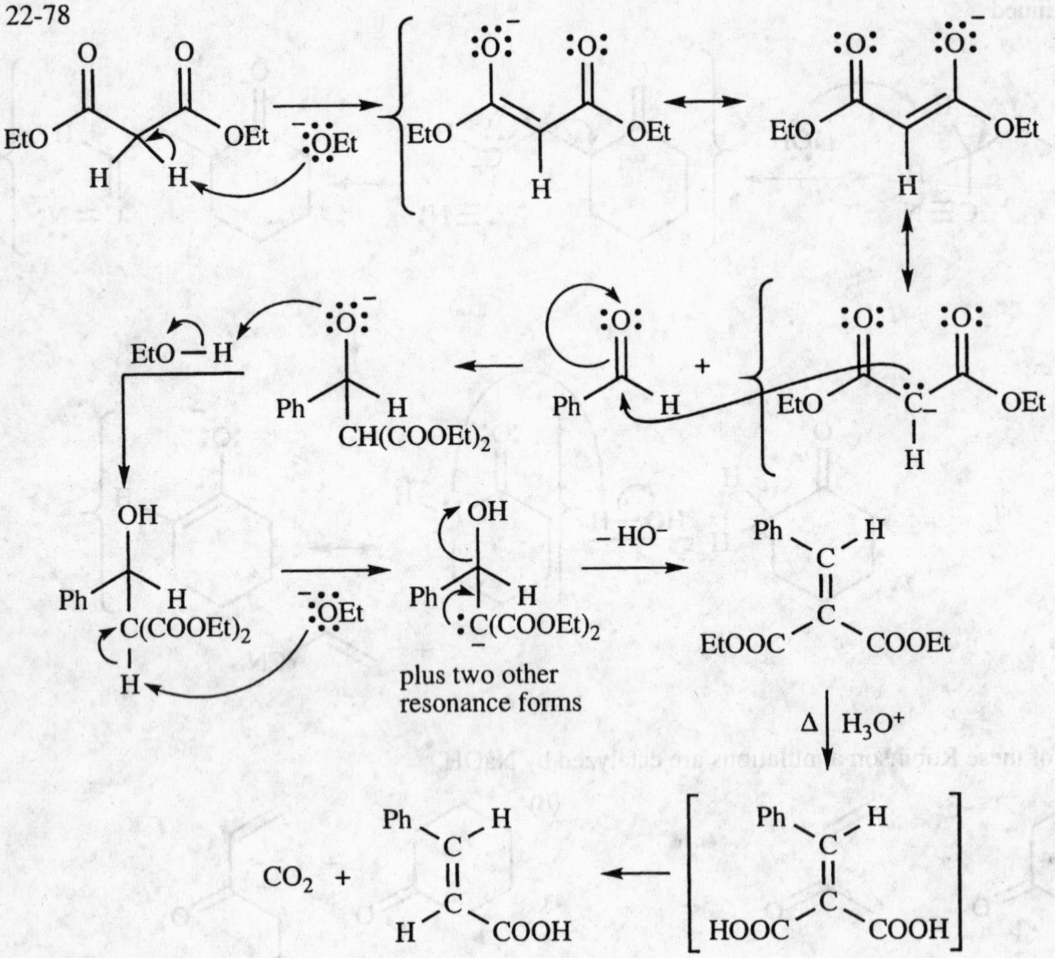

EtO
OEt
OEt
EtO—H
Ph
CH(COOEt)$_2$
C(COOEt)$_2$
– HO$^-$
plus two other resonance forms
EtOOC
COOEt
Δ
H$_3$O$^+$
CO_2 +
HOOC
COOH

22-79

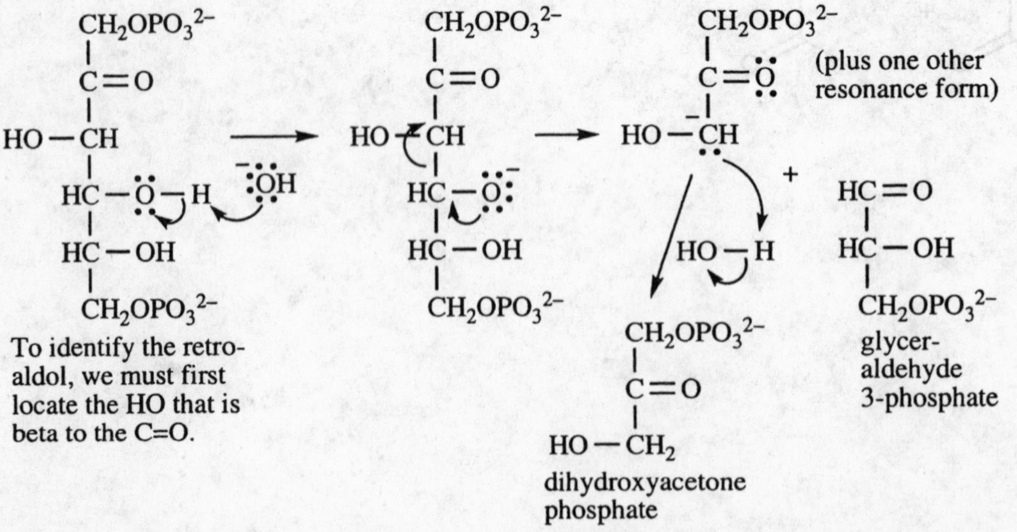

CH$_2$OPO$_3$$^{2-}$
C=O
HO—CH
HC—OH
(plus one other resonance form)
HC=O
glyceraldehyde 3-phosphate
dihydroxyacetone phosphate

To identify the retro-aldol, we must first locate the HO that is beta to the C=O.

22-80 This is an aldol condensation. **P** stands for a protein chain in this problem.

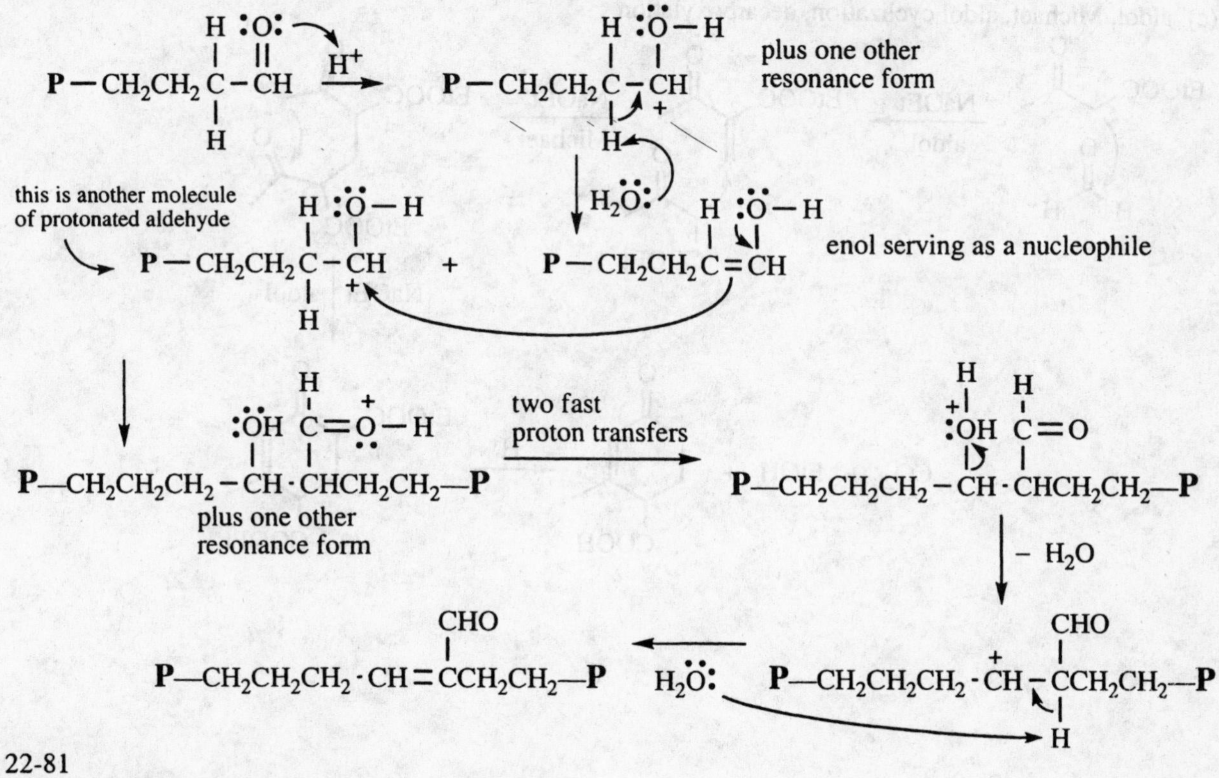

22-81

(a) aldol followed by Michael

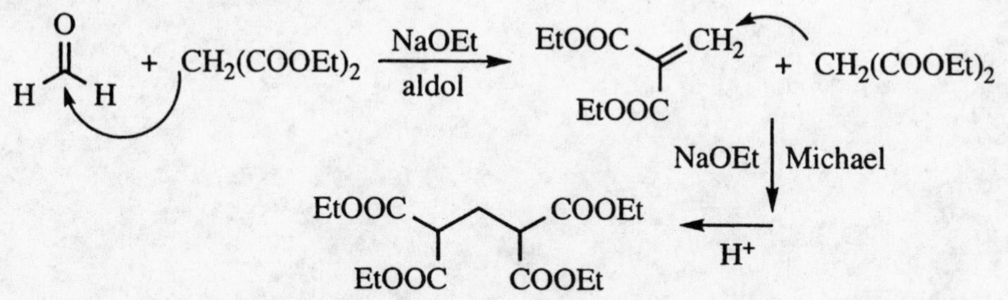

(b) Michael followed by Claisen condensation; hydrolysis and decarboxylation

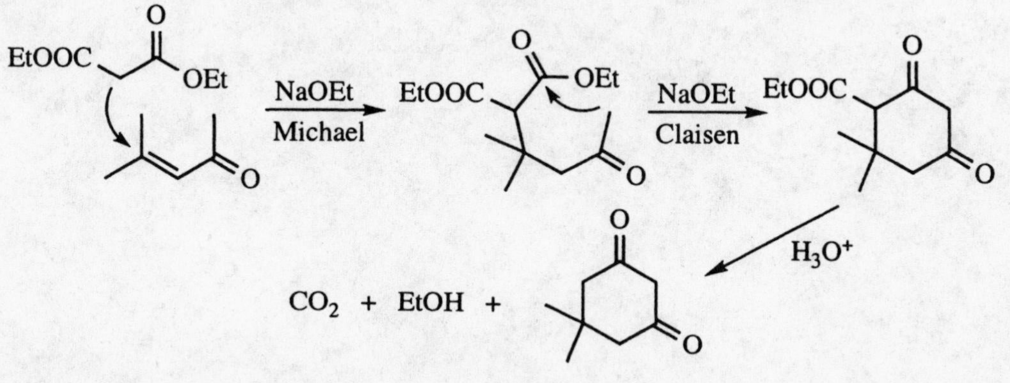

583

22-81 continued

(c) aldol, Michael, aldol cyclization, decarboxylation

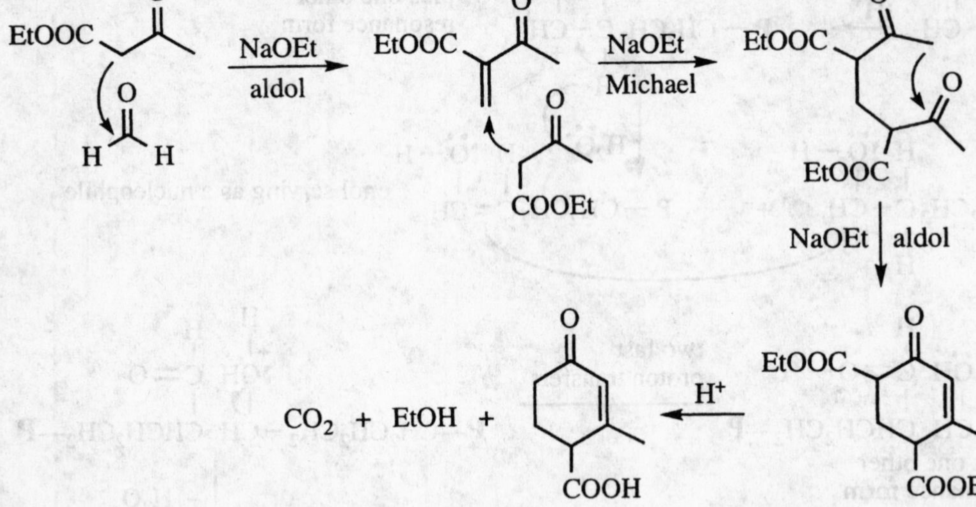

CO_2 + EtOH +

584

Reminder about Fischer projections, first introduced in Chapter 5, section 5-10: vertical bonds are equivalent to dashed bonds, going behind the plane of the paper, and horizontal bonds are equivalent to wedge bonds, coming toward the viewer.

Fischer projection:

$$CHO$$
$$HO \longrightarrow H$$
$$HO \longrightarrow H$$
$$CH_2OH$$

$\equiv$

$$O = C - H$$
$$HO \blacktriangleright C \blacktriangleleft H$$
$$HO \blacktriangleright C \blacktriangleleft H$$
$$CH_2OH$$

view from left side

view from right side

23-1

glucose

$$CHO$$
$$H \longrightarrow OH$$
$$HO \longrightarrow H$$
$$H \longrightarrow OH$$
$$H \longrightarrow OH$$
$$CH_2OH$$

mirror image

$$CHO$$
$$HO \longrightarrow H$$
$$H \longrightarrow OH$$
$$HO \longrightarrow H$$
$$HO \longrightarrow H$$
$$CH_2OH$$

fructose

$$CH_2OH$$
$$C = O$$
$$HO \longrightarrow H$$
$$H \longrightarrow OH$$
$$H \longrightarrow OH$$
$$CH_2OH$$

mirror image

$$CH_2OH$$
$$O = C$$
$$H \longrightarrow OH$$
$$HO \longrightarrow H$$
$$HO \longrightarrow H$$
$$CH_2OH$$

All four of these compounds are chiral and optically active.

23-2

(a)

$$CHO$$
$$H \longrightarrow OH$$
$$H \longrightarrow OH$$
$$CH_2OH$$

$$CHO$$
$$HO \longrightarrow H$$
$$HO \longrightarrow H$$
$$CH_2OH$$

$$CHO$$
$$H \longrightarrow OH$$
$$HO \longrightarrow H$$
$$CH_2OH$$

$$CHO$$
$$HO \longrightarrow H$$
$$H \longrightarrow OH$$
$$CH_2OH$$

two asymmetric carbons $\Rightarrow$ four stereoisomers (two pairs of enantiomers) if none are meso

23-2 continued

(b)

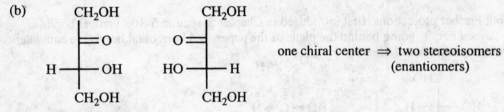

one chiral center ⇒ two stereoisomers
(enantiomers)

(c) An aldohexose has four chiral carbons and sixteen stereoisomers. A ketohexose has three chiral carbons and eight stereoisomers.

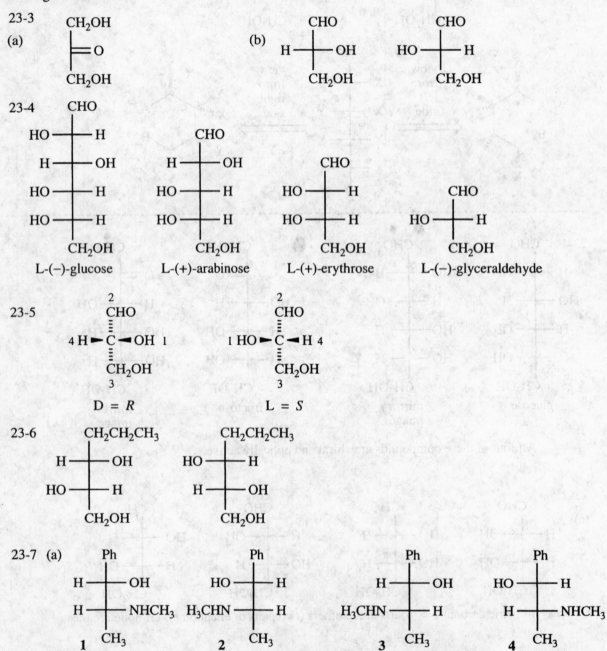

23-3

(a)

(b)

23-4

L-(−)-glucose L-(+)-arabinose L-(+)-erythrose L-(−)-glyceraldehyde

23-5

D = R L = S

23-6

23-7 (a)

1 2 3 4

23-7 (a) continued

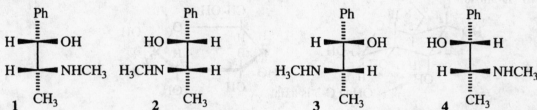

1   2   3   4

(b) Ephedrine is the *erythro* diastereomer, represented by structures **1** and **2**. Structures **3** and **4** represent pseudoephedrine, the *threo* diastereomer.

(c) The Fischer-Rosanoff convention assigns D and L to the configuration of the asymmetric carbon at the bottom of the Fischer projection. Structure **1** is D-ephedrine; structure **2** is L-ephedrine; structure **3** is L-pseudoephedrine; and structure **4** is D-pseudoephedrine.

(d) It is not possible to determine which is the (+) or (−) isomer of any compound just by looking at the structure. The only compound that has a direct correlation between the direction of optical rotation and its D or L designation is glyceraldehyde, about which the D and L system was designed. For all other compounds, optical rotation can only be determined by measurement in a polarimeter.

23-8

(a)
```
        CHO
  H ────── OH
  H ──3──── OH
  H ────── OH
  H ────── OH
       CH₂OH
```
D-allose

(b)
```
        CHO
 HO ──2── H
 HO ────── H
 HO ────── H
  H ────── OH
       CH₂OH
```
D-talose

(c)
```
        CHO
 HO ────── H
  H ──3──── OH
 HO ────── H
  H ────── OH
       CH₂OH
```
D-idose

(d)
```
        CHO
  H ────── OH
 HO ────── H
  H ──4──── OH
       CH₂OH
```
D-xylose

⟹

```
        CHO
  H ────── OH
 HO ────── H
 HO ────── H
       CH₂OH
```
L-arabinose

inversion at bottom chiral center ⇒ L-series sugar

23-9 D-mannose

```
      ¹CHO
 HO ─²── H
 HO ─³── H
  H ─⁴── OH
  H ─⁵── OH
     ⁶CH₂OH
```

rotate ⟹

(pyranose ring structure with labels: 6 CH₂OH, 5, 4 OH, HO, 3 H, 2 H, OH, H, O, HO)

587

23-10 D-allose

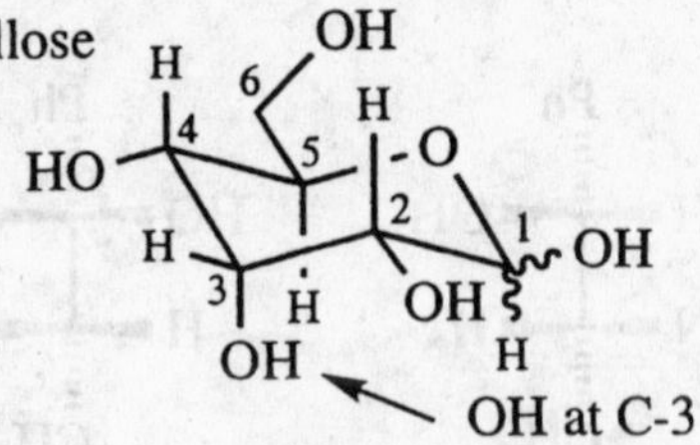

OH at C-3 is axial

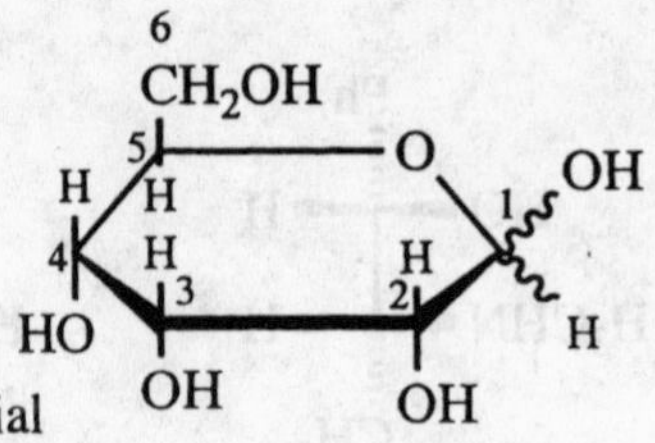

23-11 D-talopyranose

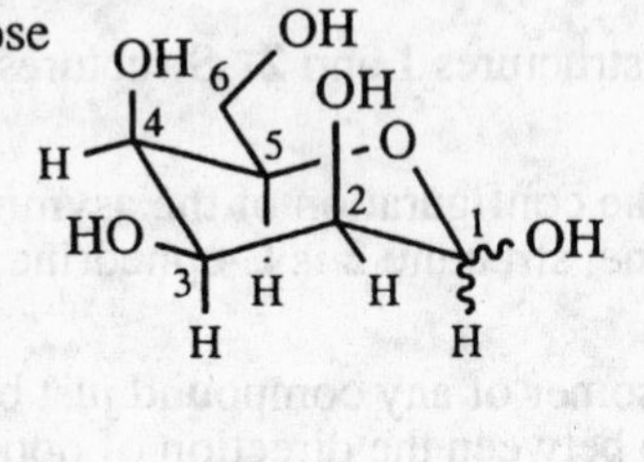

OH groups at C-2 and C-4 are axial

23-12

(a)

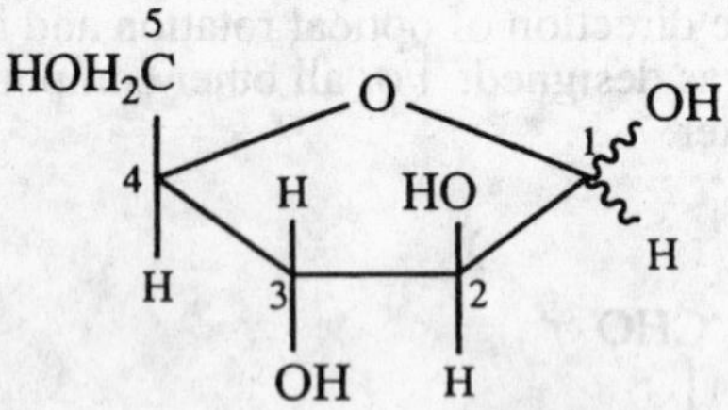

D-arabinofuranose

(b)

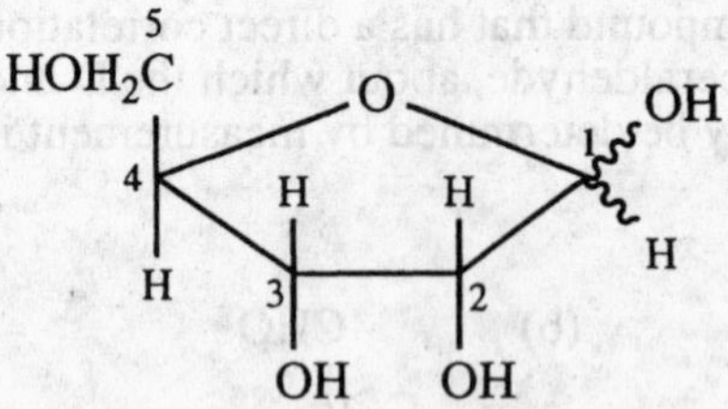

D-ribofuranose

23-13

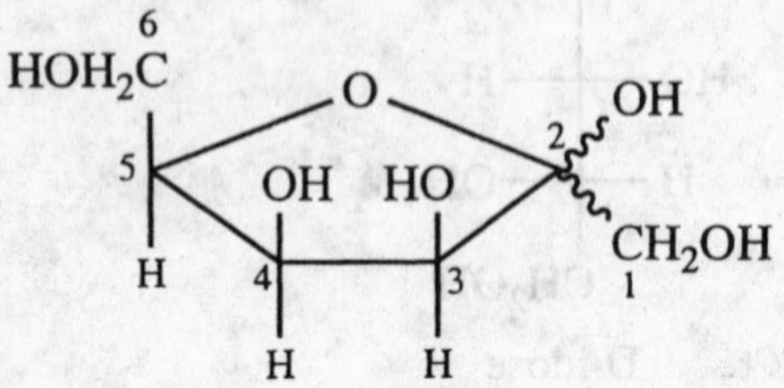

23-14

(a) α-D-mannopyranose

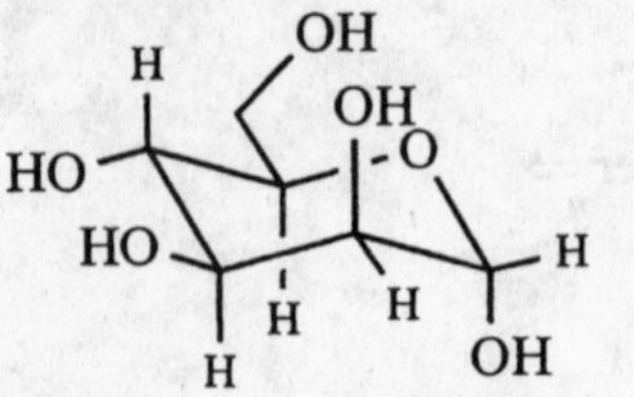

axial = α

(b) β-D-galactopyranose

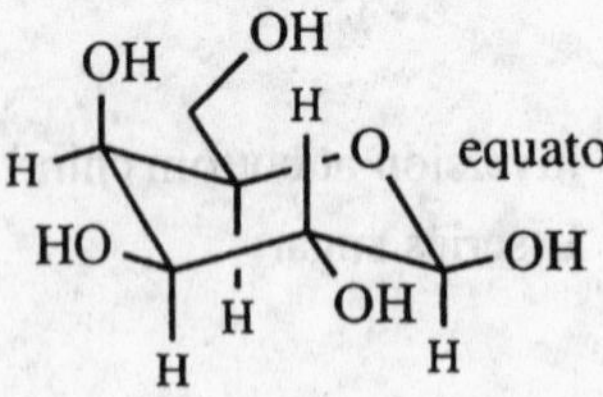

equatorial = β

(c) β-D-allopyranose

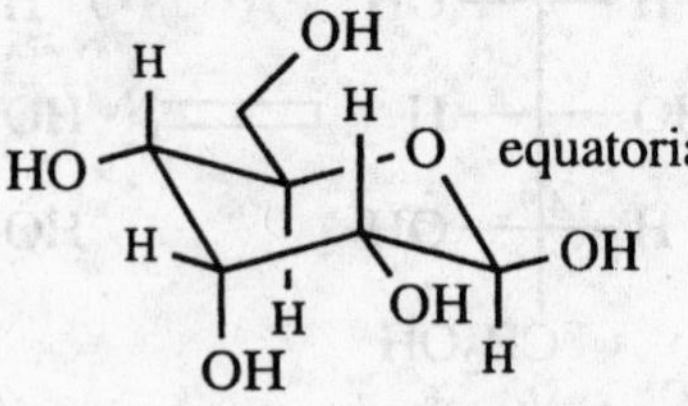

equatorial = β

(d) α-D-arabinofuranose

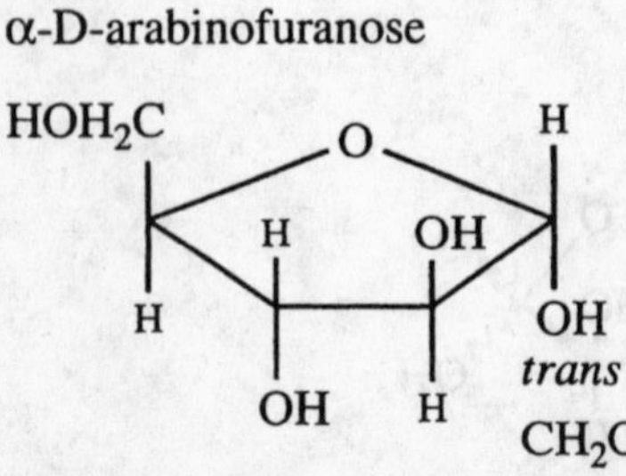

trans to
CH₂OH = α

(e) β-D-ribofuranose

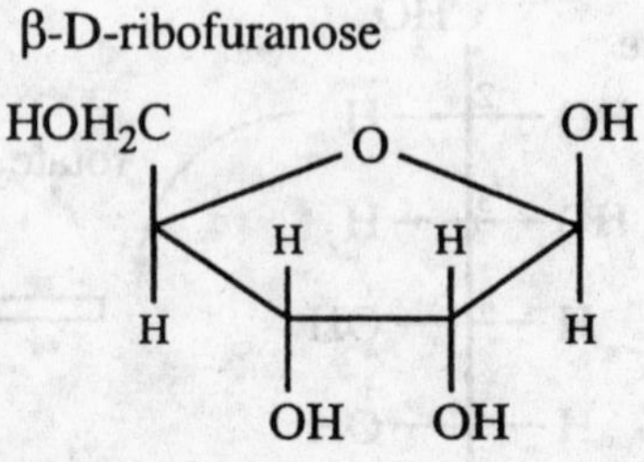

OH *cis* to CH₂OH
= β

23-15   a = fraction of galactose as the α anomer; b = fraction of galactose as the β anomer

$a\,(+150.7°) + b\,(+52.8°) = +80.2°$

$a + b = 1;\quad b = 1 - a$

$a\,(+150.7°) + (1-a)\,(+52.8°) = +80.2° \xRightarrow{\text{solve for "a"}} a = 0.28;\ b = 0.72$

The equilibrium mixture contains 28% of the α anomer and 72% of the β anomer.

23-16

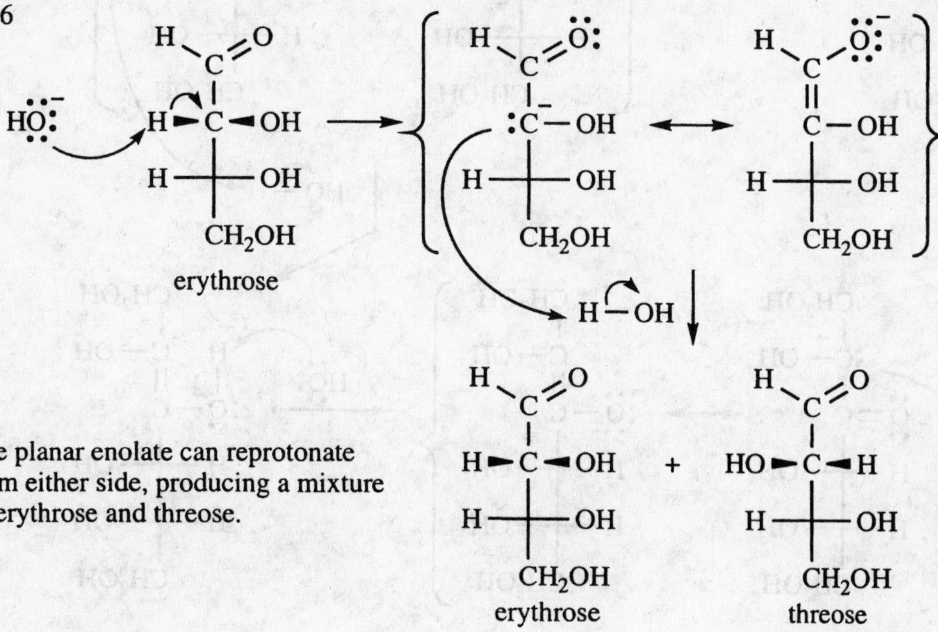

erythrose

The planar enolate can reprotonate from either side, producing a mixture of erythrose and threose.

erythrose   +   threose

23-17

fructose

589

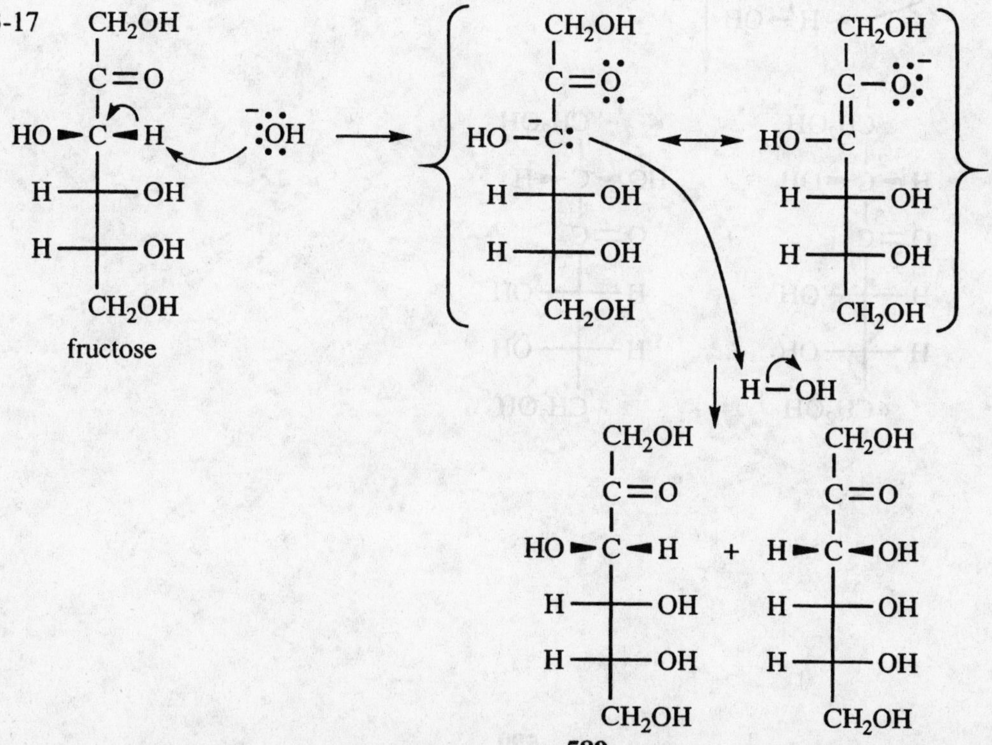

23-18

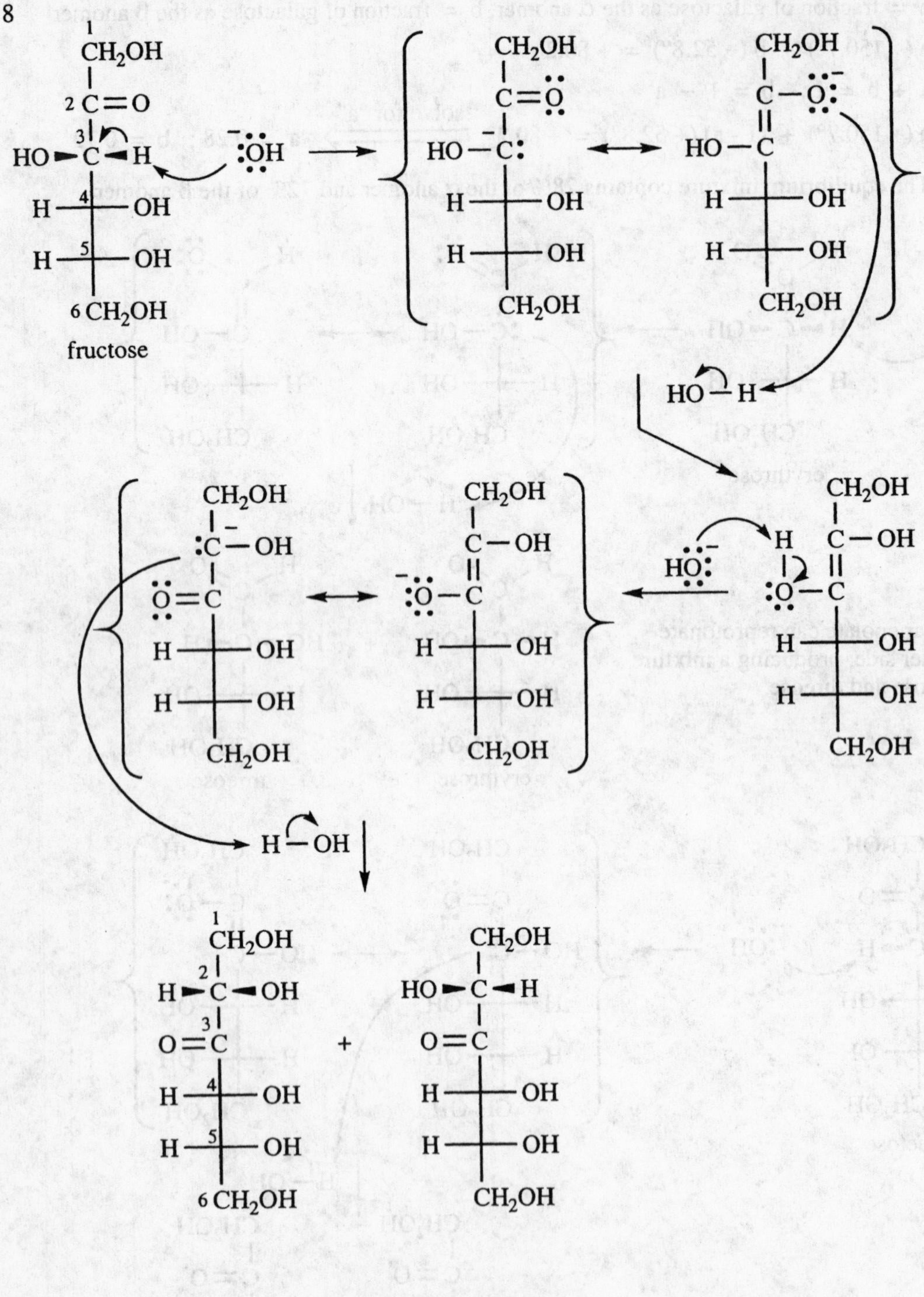

fructose

590

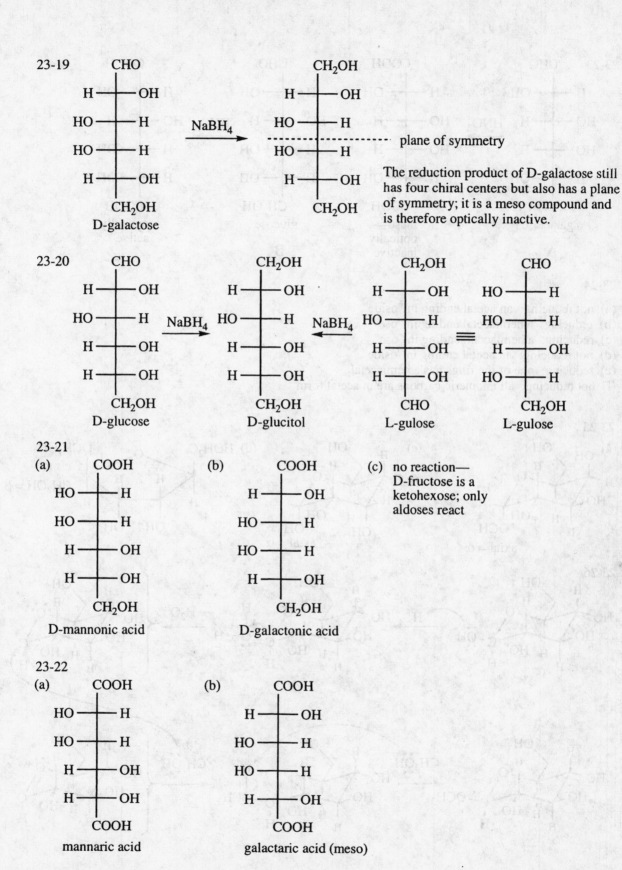

23-19

CHO
H——OH
HO——H
HO——H
H——OH
CH₂OH

D-galactose

$\xrightarrow{\text{NaBH}_4}$

CH₂OH
H——OH
HO——H
- - - - - - - - - - - plane of symmetry
HO——H
H——OH
CH₂OH

The reduction product of D-galactose still has four chiral centers but also has a plane of symmetry; it is a meso compound and is therefore optically inactive.

23-20

CHO
H——OH
HO——H
H——OH
H——OH
CH₂OH

D-glucose

$\xrightarrow{\text{NaBH}_4}$

CH₂OH
H——OH
HO——H
H——OH
H——OH
CH₂OH

D-glucitol

$\xleftarrow{\text{NaBH}_4}$

CH₂OH
H——OH
HO——H
H——OH
H——OH
CHO

L-gulose

$\equiv$

CHO
HO——H
HO——H
H——OH
HO——H
CH₂OH

L-gulose

23-21

(a)

COOH
HO——H
HO——H
H——OH
H——OH
CH₂OH

D-mannonic acid

(b)

COOH
H——OH
HO——H
HO——H
H——OH
CH₂OH

D-galactonic acid

(c) no reaction— D-fructose is a ketohexose; only aldoses react

23-22

(a)

COOH
HO——H
HO——H
H——OH
H——OH
COOH

mannaric acid

(b)

COOH
H——OH
HO——H
HO——H
H——OH
COOH

galactaric acid (meso)

23-23

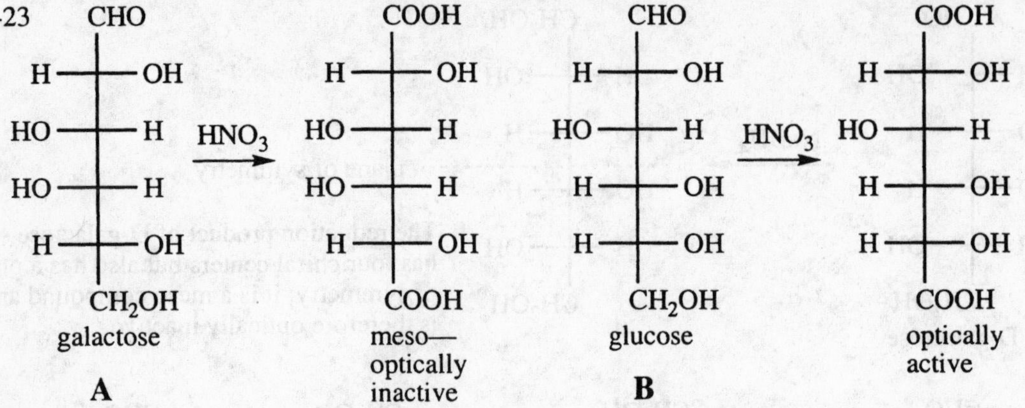

galactose **A** → (HNO₃) → meso— optically inactive

glucose **B** → (HNO₃) → optically active

23-24

(a) not reducing: an acetal ending in "oside"
(b) reducing: a hemiacetal ending in "ose"
(c) reducing: a hemiacetal ending in "ose"
(d) not reducing: an acetal ending in "oside"
(e) reducing: one of the rings has a hemiacetal
(f) not reducing: all anomeric carbons are in acetal form

23-25

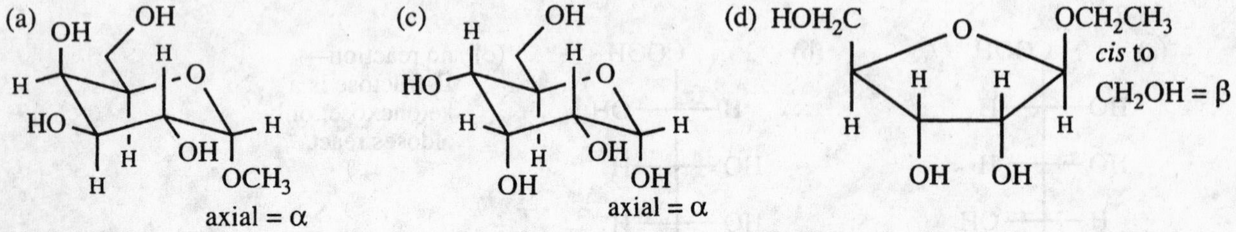

(a) axial = α

(c) axial = α

(d) *cis* to CH₂OH = β

23-26

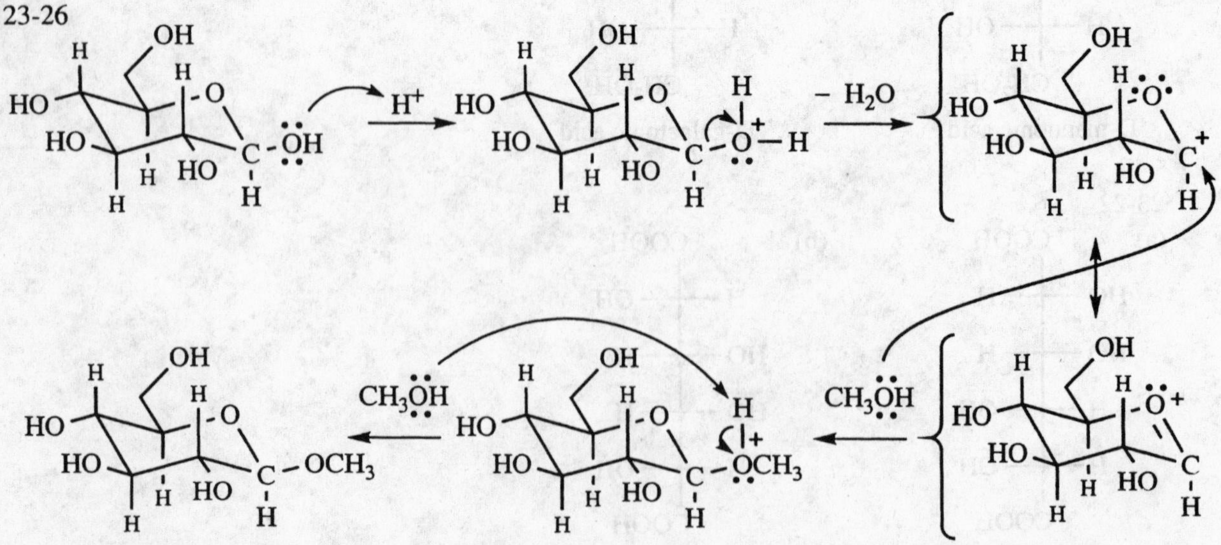

23-27

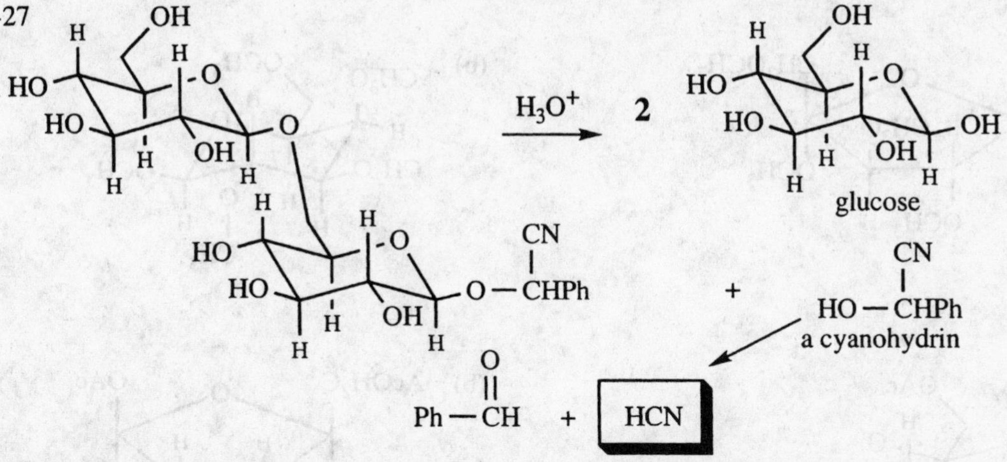

HCN is released from amygdalin. HCN is a potent cytotoxic (cell-killing) agent, particularly toxic to nerve cells.

23-28

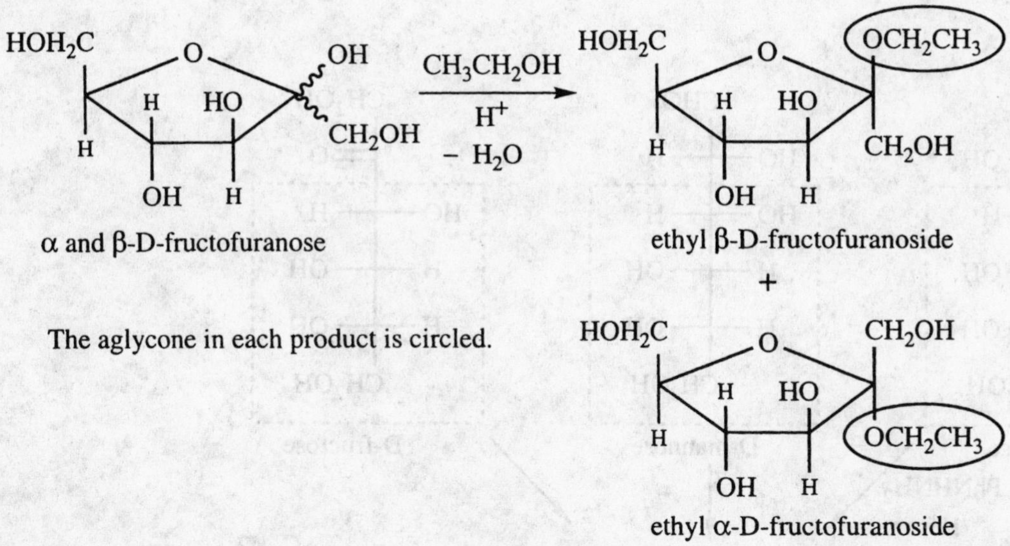

α and β-D-fructofuranose

ethyl β-D-fructofuranoside

The aglycone in each product is circled.

ethyl α-D-fructofuranoside

23-29

$+$ $^-OSO_3CH_3$

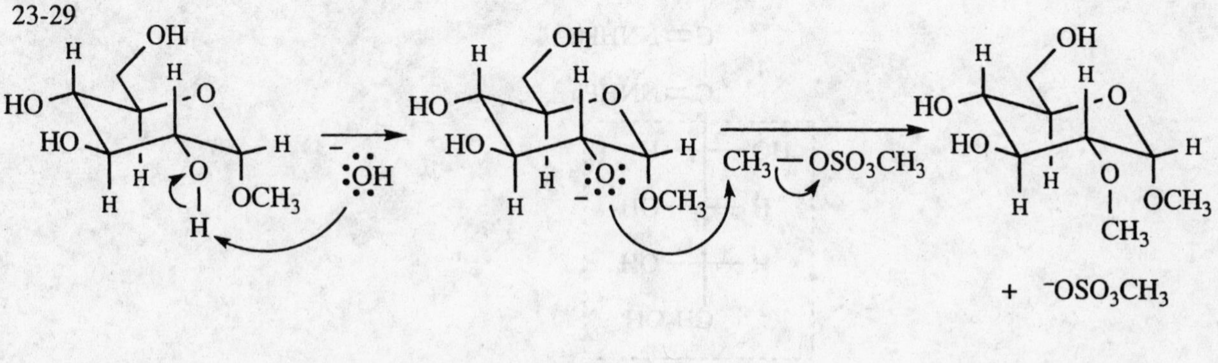

23-30

(a) CH$_3$OH$_2$C ... CH$_2$OCH$_3$
H CH$_3$O
OCH$_3$
H OCH$_3$
OCH$_3$ H

(b) CH$_3$O ... OCH$_3$
H
CH$_3$O
H OCH$_3$
H
OCH$_3$
H
CH$_3$O

23-31

(a) H OAc
H
AcO O
AcO H
H AcO
H OAc

(b) AcOH$_2$C O OAc
H H
H H
AcO OAc

23-32

(a)

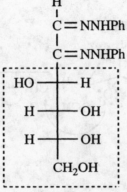

CHO	CHO	CH$_2$OH
H—OH	HO—H	=O
HO—H	HO—H	HO—H
H—OH	H—OH	H—OH
H—OH	H—OH	H—OH
CH$_2$OH	CH$_2$OH	CH$_2$OH
D-glucose	D-mannose	D-fructose

PhNHNH$_2$
H$^+$

H
C=NNHPh
C=NNHPh
HO—H
H—OH
H—OH
CH$_2$OH

23-32 continued

(b)

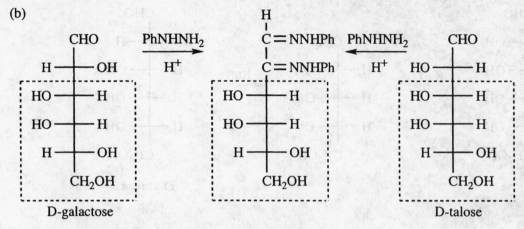

D-galactose

D-talose

D-Talose must be the C-2 epimer of D-galactose.

23-33 Reagents for the Ruff degradation are: 1. Br_2, H_2O; 2. H_2O_2, $Fe_2(SO_4)_3$.

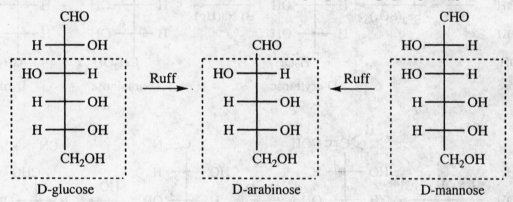

D-glucose

D-arabinose

D-mannose

23-34 Reagents for the Ruff degradation are: 1. Br_2, H_2O; 2. H_2O_2, $Fe_2(SO_4)_3$.

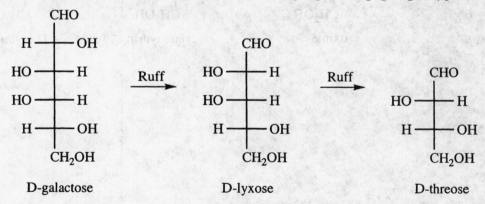

D-galactose

D-lyxose

D-threose

23-35   Reagents for the Ruff degradation are: 1. Br_2, H_2O; 2. H_2O_2, $Fe_2(SO_4)_3$.

$$
\begin{array}{c}
\text{CHO} \\
\text{H} \!-\!\!|\!-\! \text{OH} \\
\text{H} \!-\!\!|\!-\! \text{OH} \\
\text{H} \!-\!\!|\!-\! \text{OH} \\
\text{H} \!-\!\!|\!-\! \text{OH} \\
\text{CH}_2\text{OH}
\end{array}
\quad \xrightarrow{\ \text{Ruff}\ } \quad
\begin{array}{c}
\text{CHO} \\
\text{H} \!-\!\!|\!-\! \text{OH} \\
\text{H} \!-\!\!|\!-\! \text{OH} \\
\text{H} \!-\!\!|\!-\! \text{OH} \\
\text{CH}_2\text{OH}
\end{array}
\quad \xleftarrow{\ \text{Ruff}\ } \quad
\begin{array}{c}
\text{CHO} \\
\text{HO} \!-\!\!|\!-\! \text{H} \\
\text{H} \!-\!\!|\!-\! \text{OH} \\
\text{H} \!-\!\!|\!-\! \text{OH} \\
\text{H} \!-\!\!|\!-\! \text{OH} \\
\text{CH}_2\text{OH}
\end{array}
$$

D-allose   D-altrose

23-36

$$
\begin{array}{c}
\text{CHO} \\
\text{HO} \!-\!\!|\!-\! \text{H} \\
\text{H} \!-\!\!|\!-\! \text{OH} \\
\text{H} \!-\!\!|\!-\! \text{OH} \\
\text{CH}_2\text{OH}
\end{array}
\quad \xrightarrow[\ H_2O\ ]{\ Br_2\ } \ \xrightarrow[\ Fe_2(SO_4)_3\ ]{\ H_2O_2\ } \quad
\begin{array}{c}
\text{CHO} \\
\text{H} \!-\!\!|\!-\! \text{OH} \\
\text{H} \!-\!\!|\!-\! \text{OH} \\
\text{CH}_2\text{OH}
\end{array}
\quad
\begin{array}{l}
\text{1) HCN} \\
\text{2) } H_3O^+ \\
\text{3) Na(Hg)}
\end{array}
\quad
\begin{array}{c}
\text{CHO} \\
\text{HO} \!-\!\!|\!-\! \text{H} \\
\text{H} \!-\!\!|\!-\! \text{OH} \\
\text{H} \!-\!\!|\!-\! \text{OH} \\
\text{CH}_2\text{OH}
\end{array}
\ + \
\begin{array}{c}
\text{CHO} \\
\text{H} \!-\!\!|\!-\! \text{OH} \\
\text{H} \!-\!\!|\!-\! \text{OH} \\
\text{H} \!-\!\!|\!-\! \text{OH} \\
\text{CH}_2\text{OH}
\end{array}
$$

D-arabinose   D-erythrose   D-arabinose   D-ribose

23-37

$$
\begin{array}{c}
\text{CHO} \\
\text{HO} \!-\!\!|\!-\! \text{H} \\
\text{H} \!-\!\!|\!-\! \text{OH} \\
\text{H} \!-\!\!|\!-\! \text{OH} \\
\text{CH}_2\text{OH}
\end{array}
\xrightarrow{\ H_2NOH \cdot HCl\ }
\begin{array}{c}
\text{H}\!-\!\text{C}\!=\!\text{NOH} \\
\text{HO} \!-\!\!|\!-\! \text{H} \\
\text{H} \!-\!\!|\!-\! \text{OH} \\
\text{H} \!-\!\!|\!-\! \text{OH} \\
\text{CH}_2\text{OH}
\end{array}
\xrightarrow{\ Ac_2O\ }
\begin{array}{c}
\text{C}\!\equiv\!\text{N} \\
\text{HO} \!-\!\!|\!-\! \text{H} \\
\text{H} \!-\!\!|\!-\! \text{OH} \\
\text{H} \!-\!\!|\!-\! \text{OH} \\
\text{CH}_2\text{OH}
\end{array}
\xrightarrow[\ H_2O\ ]{\ HO^-\ }
\begin{array}{c}
\text{CN}^- \ + \\
\text{CHO} \\
\text{H} \!-\!\!|\!-\! \text{OH} \\
\text{H} \!-\!\!|\!-\! \text{OH} \\
\text{CH}_2\text{OH}
\end{array}
$$

D-arabinose   oxime   cyanohydrin   D-erythrose

23-38 Solve this problem by working backward from (+)-glyceraldehyde.

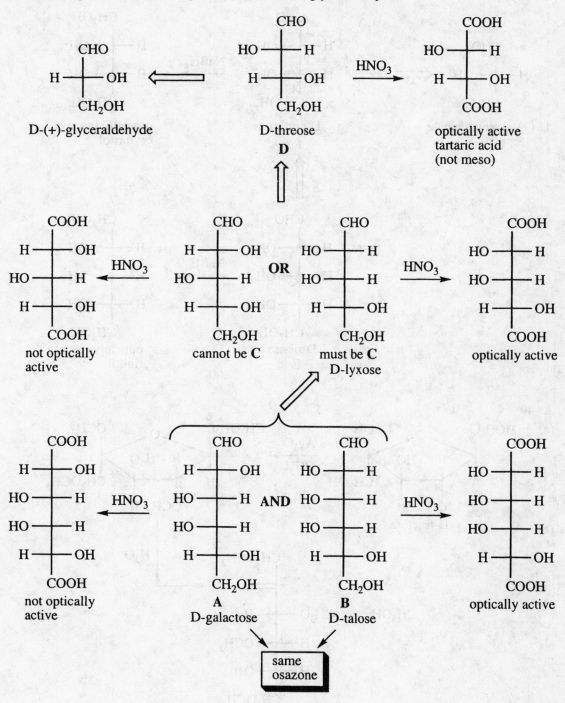

D-(+)-glyceraldehyde

D-threose
D

optically active
tartaric acid
(not meso)

not optically
active

cannot be **C**

OR

must be **C**
D-lyxose

optically active

not optically
active

A
D-galactose

AND

B
D-talose

optically active

same
osazone

23-39 Solve this problem by working backward from (+)-glyceraldehyde.

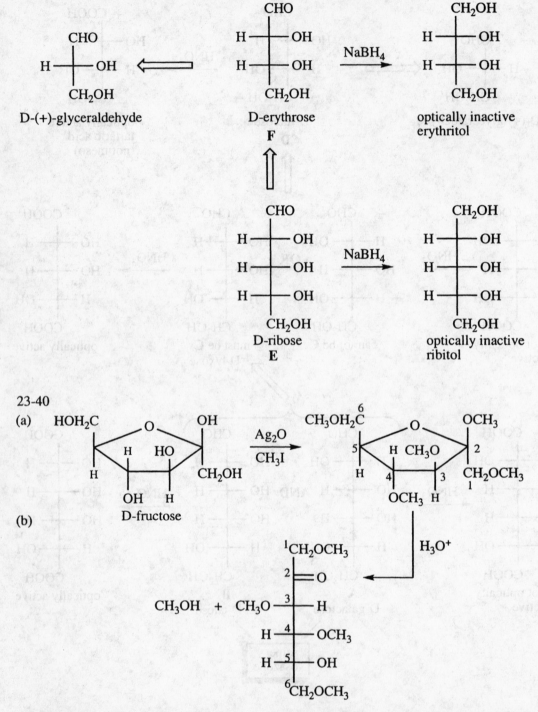

23-40
(a)

D-fructose

(b)

(c) Determining that the open chain form of fructose is a ketone at C-2 with a free OH at C-5 shows that fructose exists as a furanose hemiacetal.

23-41

(a)

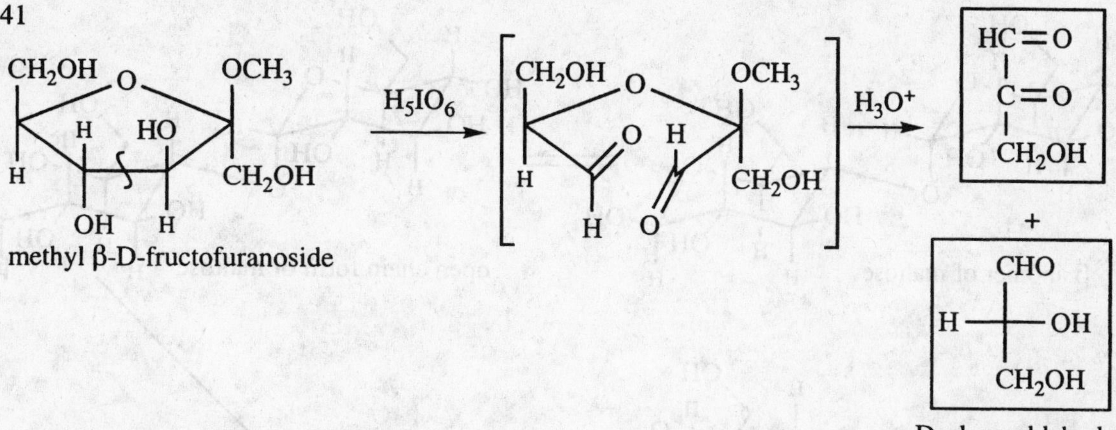

methyl β-D-fructofuranoside → (with H₅IO₆) → [intermediate] → (with H₃O⁺) →

$$HC=O, \quad C=O, \quad CH_2OH$$

+

D-glyceraldehyde

(b)

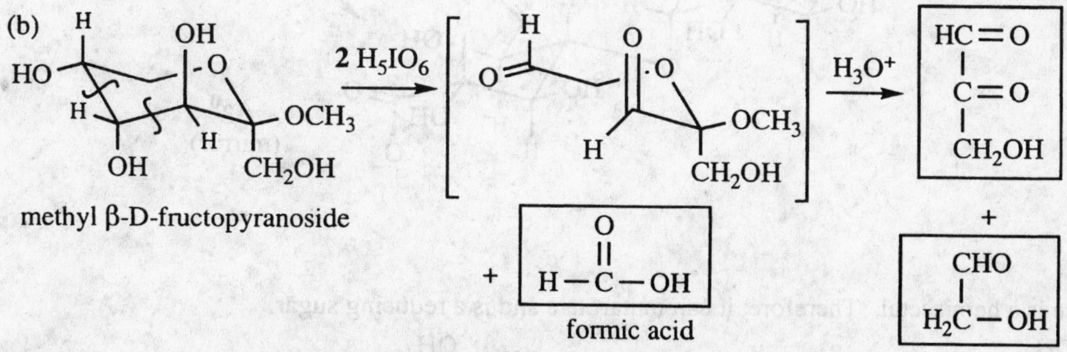

methyl β-D-fructopyranoside → (with 2 H₅IO₆) → [intermediate] → (with H₃O⁺) →

+ formic acid

Production of one equivalent of formic acid, and two fragments containing two and three carbons respectively, proves that the glycoside was in a six-membered ring.

(c) In periodic acid oxidation of an aldohexose glycoside, glyceraldehyde is generated from carbons 4, 5, and 6. If configuration at the middle carbon is D, that means that carbon-5 of the aldohexose must have had the D configuration. On the other hand, if the isolated glyceraldehyde had the L configuration, then the original aldohexose must have been an L sugar.

23-42

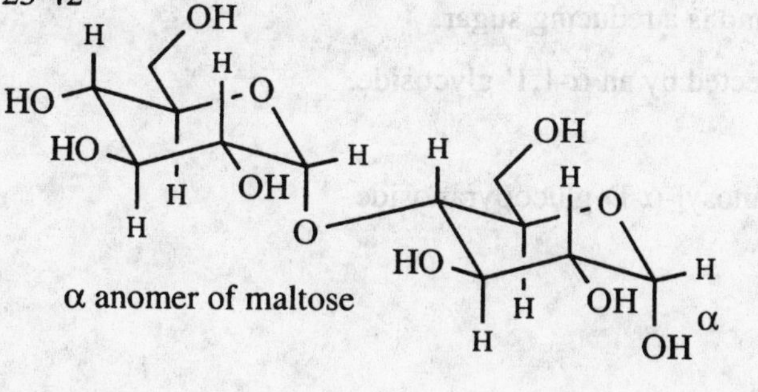

α anomer of maltose

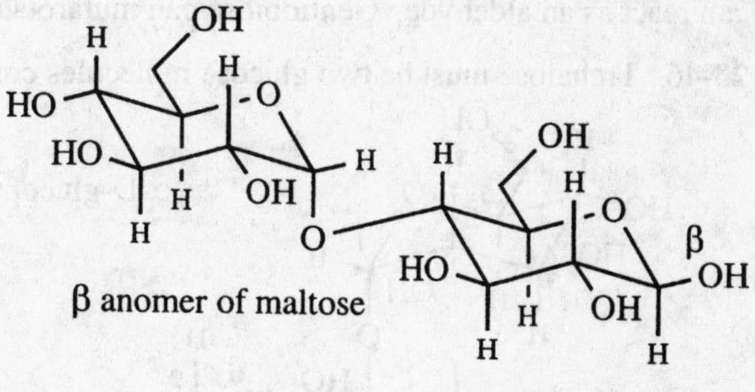

β anomer of maltose

599

23-43

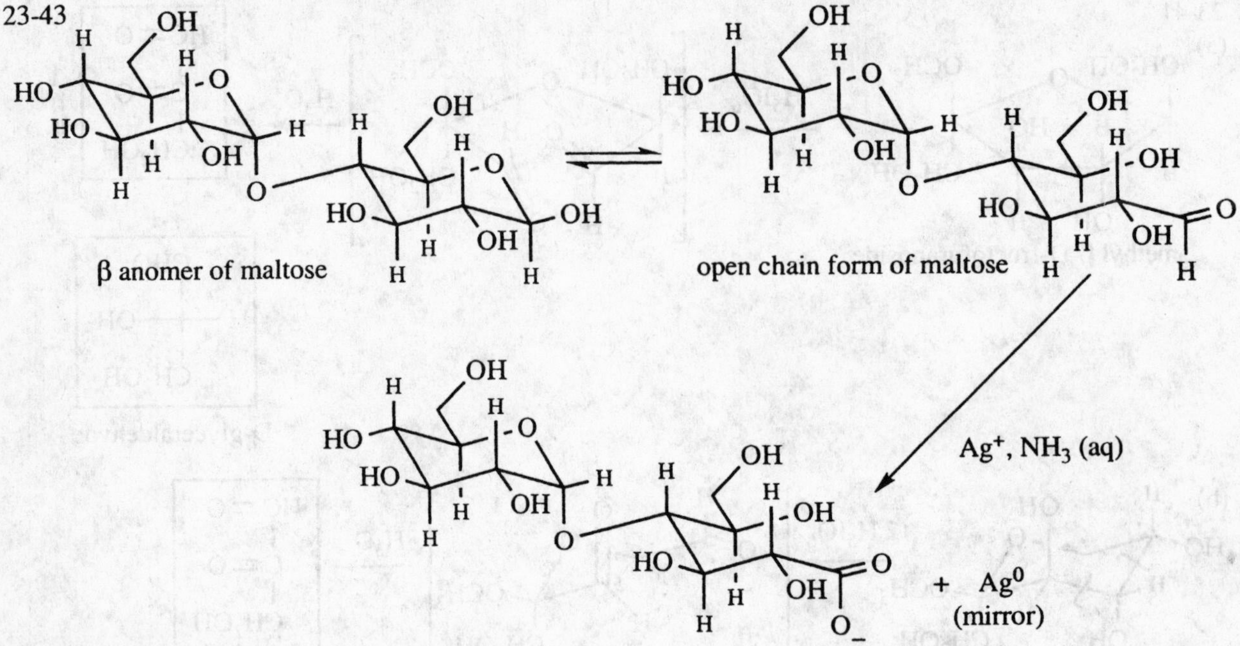

β anomer of maltose

open chain form of maltose

Ag⁺, NH₃ (aq)

+ Ag⁰
(mirror)

23-44 Lactose is a hemiacetal. Therefore, it can mutarotate and is a reducing sugar.

α anomer of lactose

β anomer of lactose

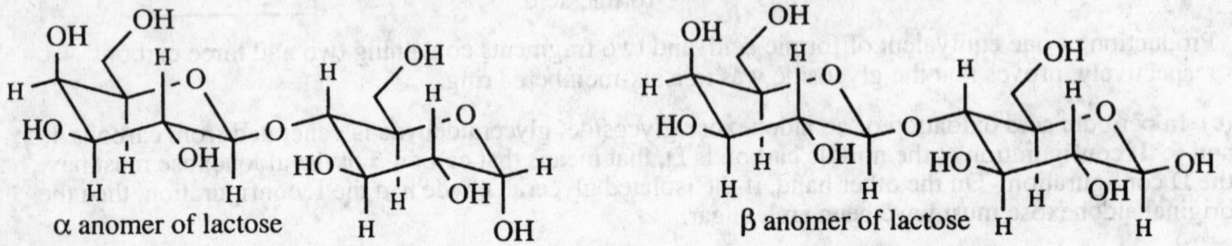

23-45 Gentiobiose is a hemiacetal; in water, the hemiacetal is in equilibrium with the open-chain form and can react as an aldehyde. Gentiobiose can mutarotate and is a reducing sugar.

23-46 Trehalose must be two glucose molecules connected by an α-1,1'-glycoside.

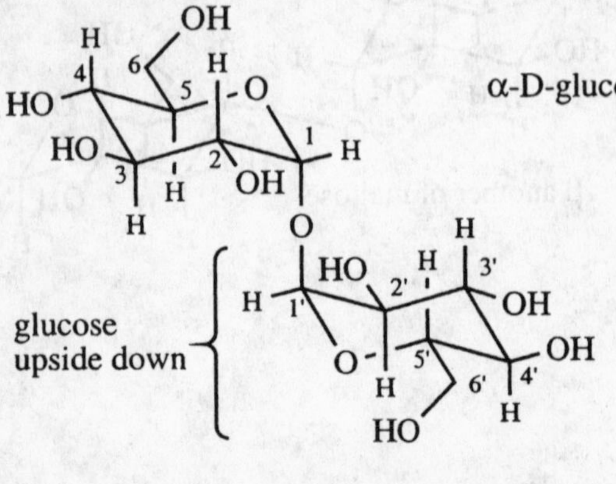

α-D-glucopyranosyl-α-D-glucopyranoside

glucose upside down

600

23-47 raffinose

melibiose = 6-O-(α-galactopyranosyl)-
D-glucopyranose

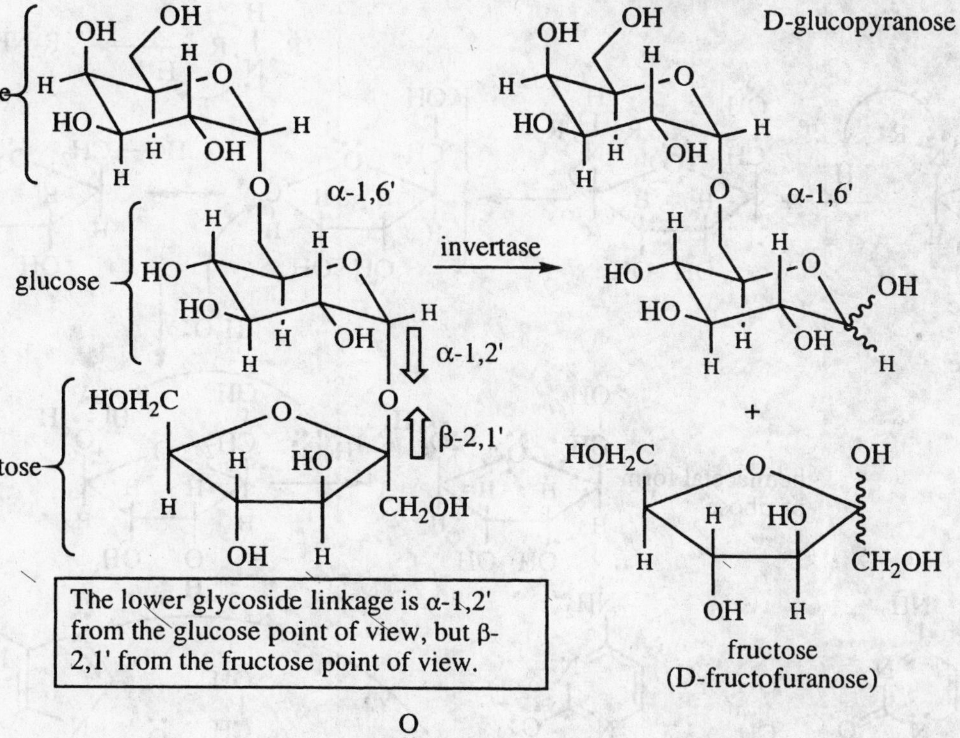

galactose

OH OH
H
H O
H
HO
H OH
H

glucose

α-1,6'

HO
HO
H OH
H
H O
H
H

invertase

fructose

HOH₂C O
H HO
H
H O
CH₂OH
OH H

α-1,2'

β-2,1'

The lower glycoside linkage is α-1,2'
from the glucose point of view, but β-
2,1' from the fructose point of view.

OH OH
H
H O
H
HO
H OH
H

α-1,6'

HO
HO
H OH
H
H O
H OH
H

+

HOH₂C O OH
H HO
H
H CH₂OH
OH H

fructose
(D-fructofuranose)

23-48 cellulose acetate

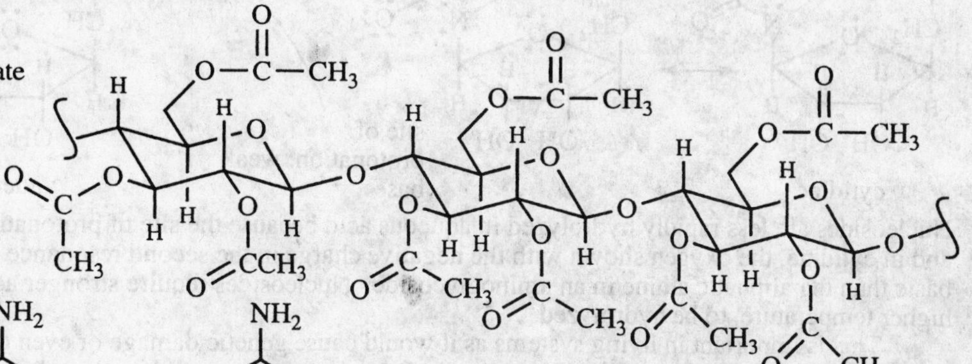

23-49 cytosine:

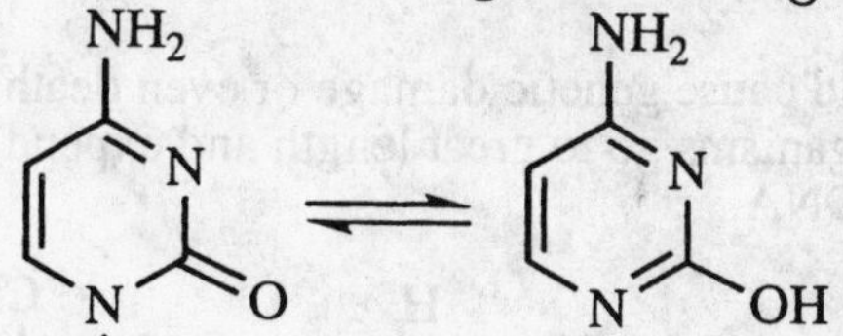

NH₂ ⇌ NH₂ (with N and OH)

uracil:

(series of four tautomeric structures connected by equilibrium arrows)

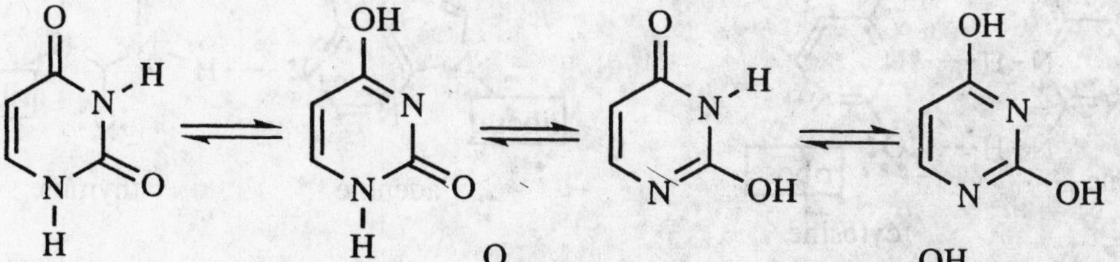

guanine:

(two tautomeric structures connected by equilibrium arrow)

601

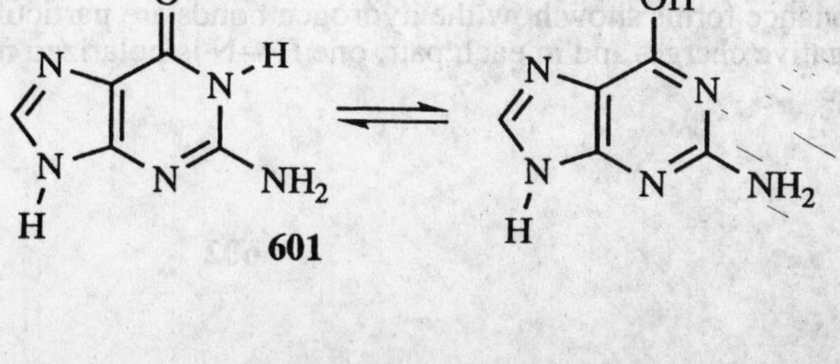

23-50 Aminoglycosides, including nucleosides, are similar to acetals: stable to base, cleaved by acid.

(a)

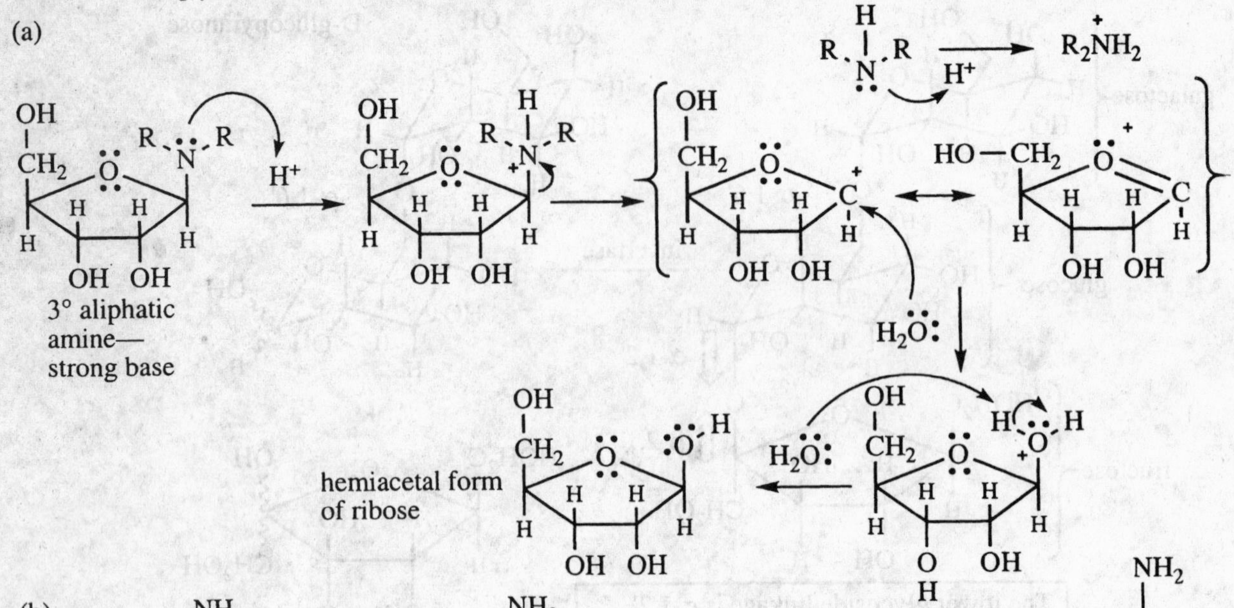

3° aliphatic
amine—
strong base

hemiacetal form
of ribose

(b)

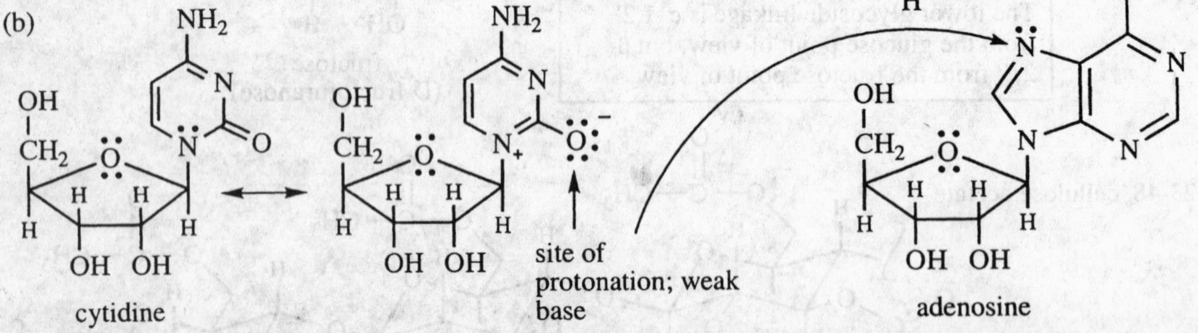

cytidine

site of
protonation; weak
base

adenosine

Nucleosides are less rapidly hydrolyzed in aqueous acid because the site of protonation (the N in adenosine, and in cytidine, the oxygen shown with the negative charge in the second resonance form) is much less basic than the aliphatic amine in an aminoglycoside. Nucleosides require stronger acid, or longer time and higher temperature, to be hydrolyzed.

This is important in living systems as it would cause genetic damage or even death of an organism if its DNA or RNA were too easily decomposed. Organisms go to great length and expend considerable energy to maintain the structural integrity of their DNA.

23-51

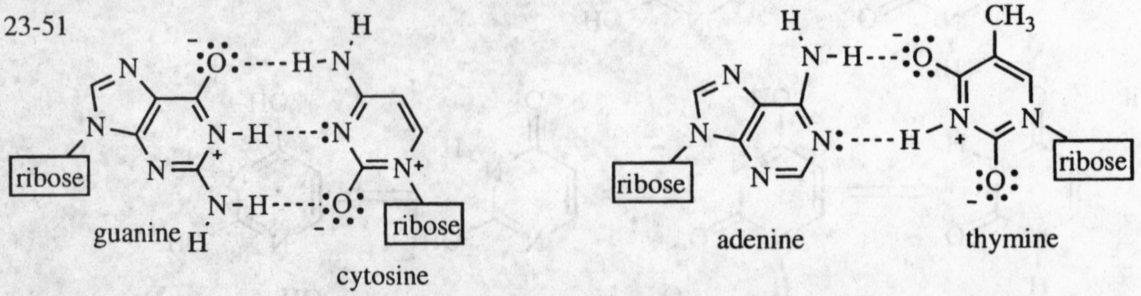

guanine

cytosine

adenine

thymine

The polar resonance forms show how the hydrogen bonds are particularly strong. Each oxygen has significant negative charge, and in each pair, one H—N is polarized more strongly because the N has positive charge.

23-52 Please refer to solution 1-20, page 12 of this Solutions Manual.

23-53

(a)

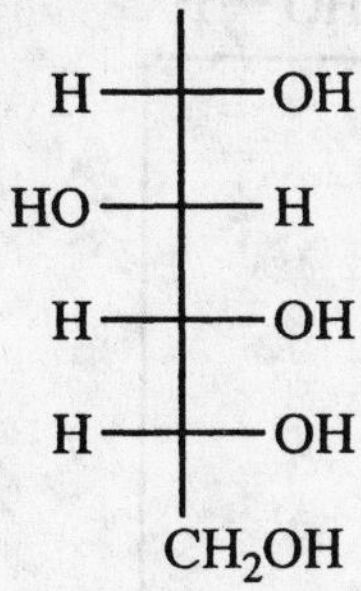

```
        CHO
   H ──┼── OH
  HO ──┼── H
   H ──┼── OH
   H ──┼── OH
       CH₂OH
```

(b) 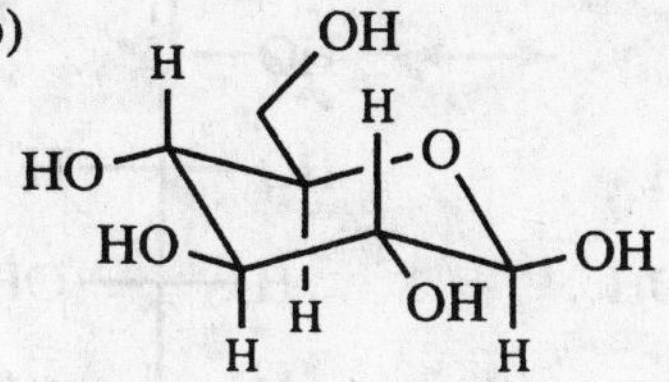 [cyclic sugar structure with OH, HO, O groups]

(c) 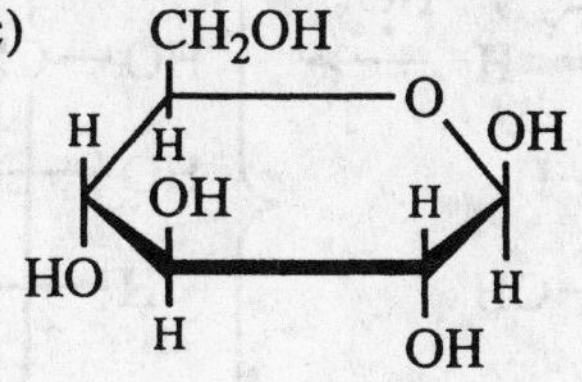 [cyclic sugar structure with CH₂OH, OH, HO groups]

23-54

(a) 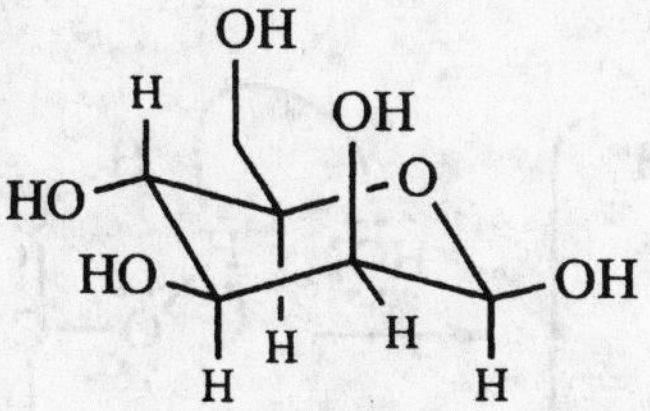 [cyclic sugar structure with OH, HO, HO, O, OH groups]

(b) 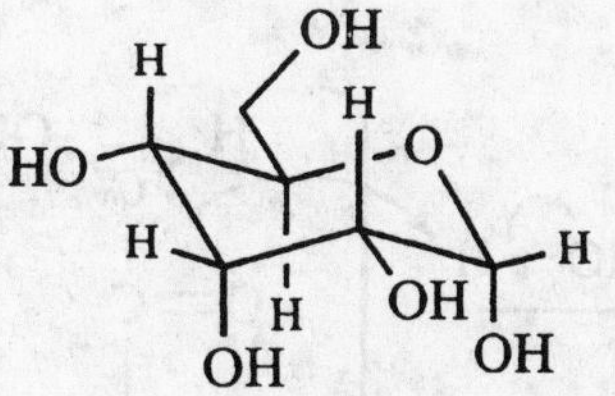 [cyclic sugar structure with OH, HO, O, OH groups]

(c) 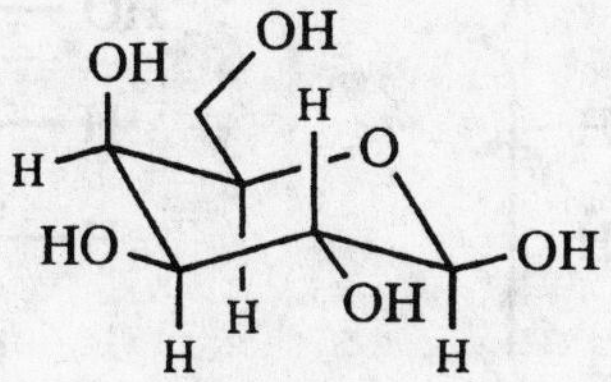 [cyclic sugar structure with OH, OH, HO, O, OH groups]

(d) 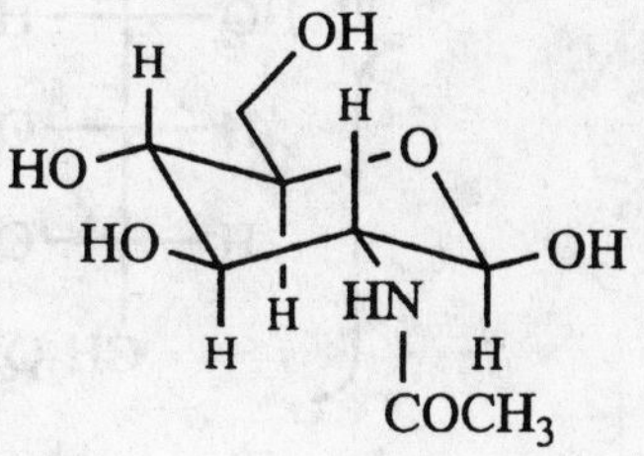 [cyclic sugar structure with OH, HO, HO, O, OH, HN, COCH₃ groups]

23-55

(a) D-aldohexose (D configuration, aldehyde, 6 carbons)
(b) D-aldopentose (D configuration, aldehyde, 5 carbons)
(c) L-ketohexose (L configuration, ketone, 6 carbons)
(d) L-aldohexose (L configuration, aldehyde, 6 carbons)
(e) D-ketopentose (D configuration, ketone, 5 carbons)
(f) L-aldotetrose (L configuration, aldehyde, 4 carbons)
(g) 2-acetamido D-aldohexose (D configuration, aldehyde, 6 carbons in chain, with acetamido group at C-2)

23-56

(a)

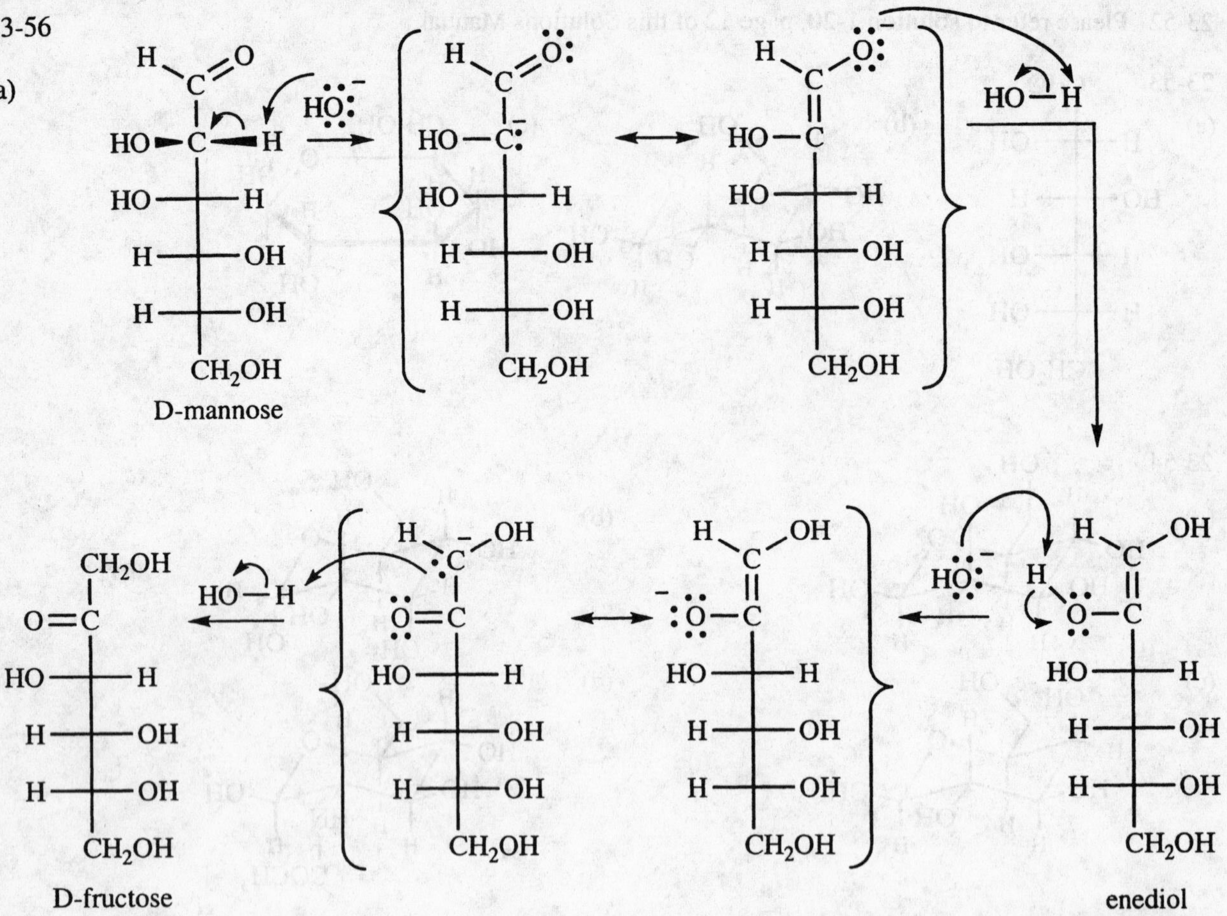

D-mannose
enediol
D-fructose

(b) The α isomer has the anomeric OH *trans* to the CH_2OH off of C-5. The β-anomer has these groups *cis*.

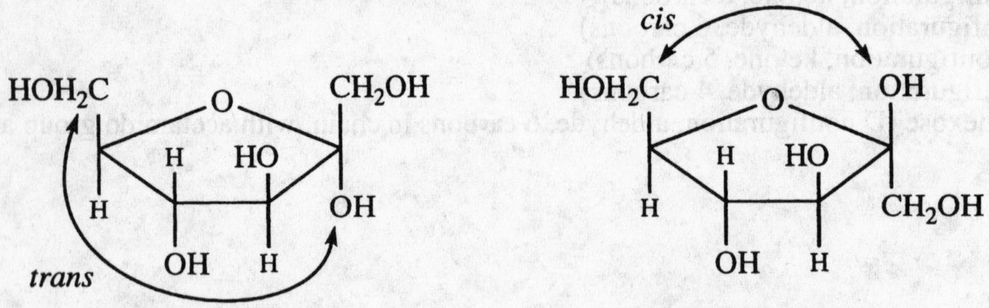

cis
trans

α-D-fructofuranose β-D-fructofuranose

23-57

(a)

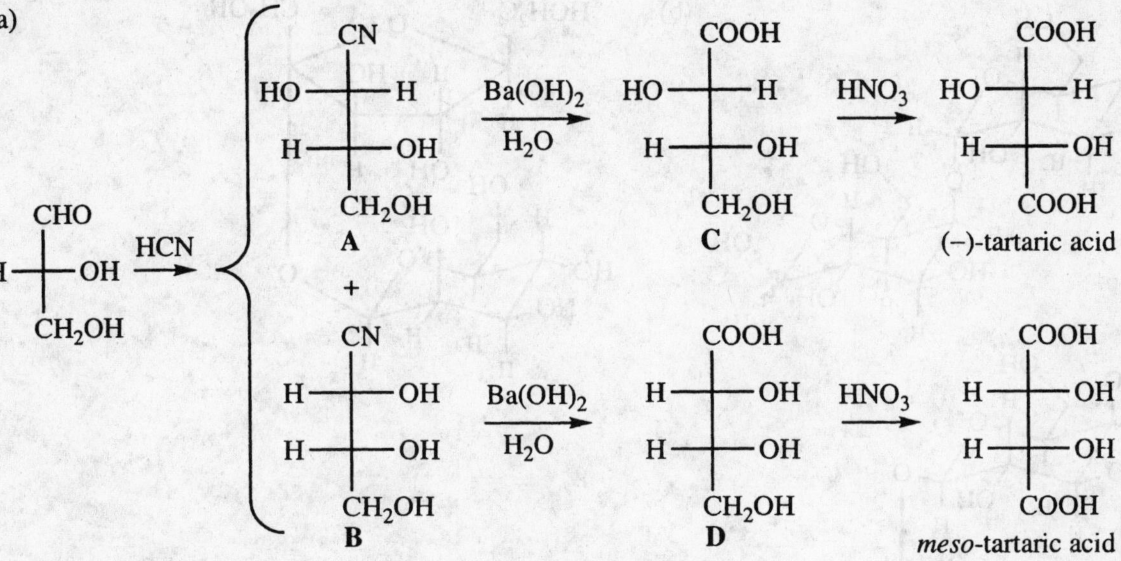

(b)

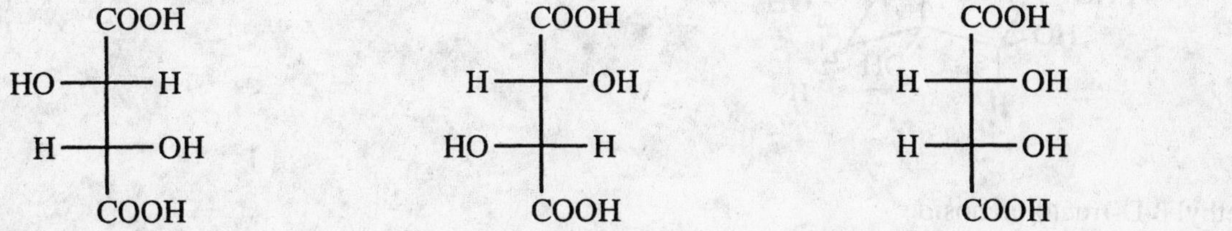

(*S*,*S*)-(−)-tartaric acid   (*R*,*R*)-(+)-tartaric acid   (*R*,*S*)-*meso*-tartaric acid

23-58

(a) D-(−)-ribose   (b) D-(+)-altrose   (c) L-(+)-erythrose   (d) L-(−)-galactose   (e) L-(+)-idose

23-59

(a)

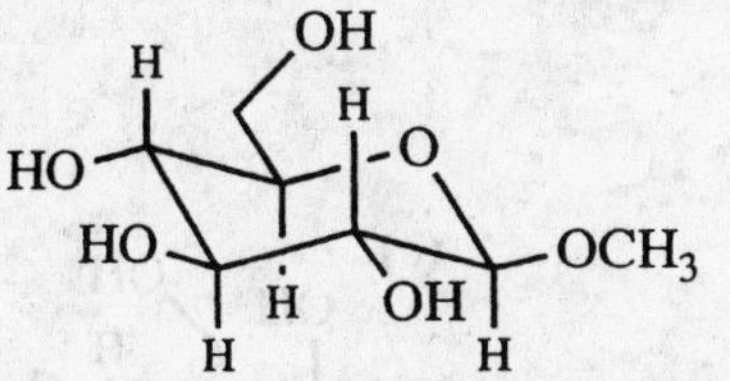

(b)

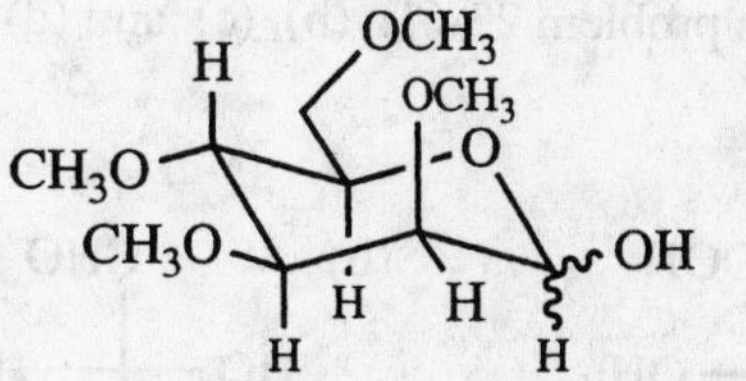

(c)

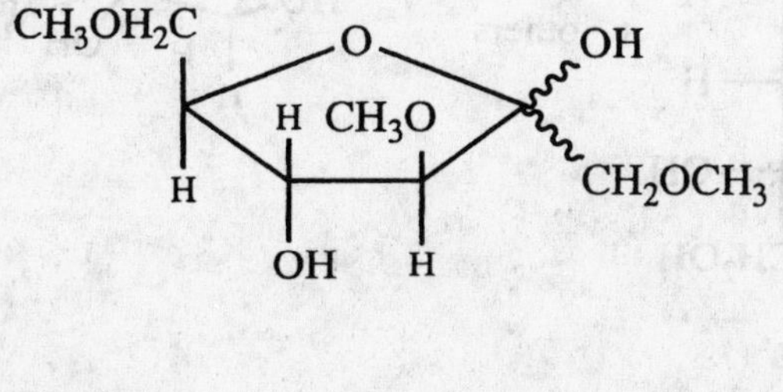

(d)

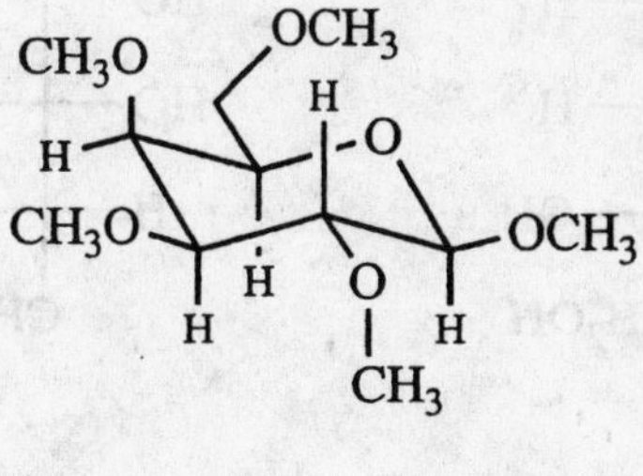

23-60

(a)

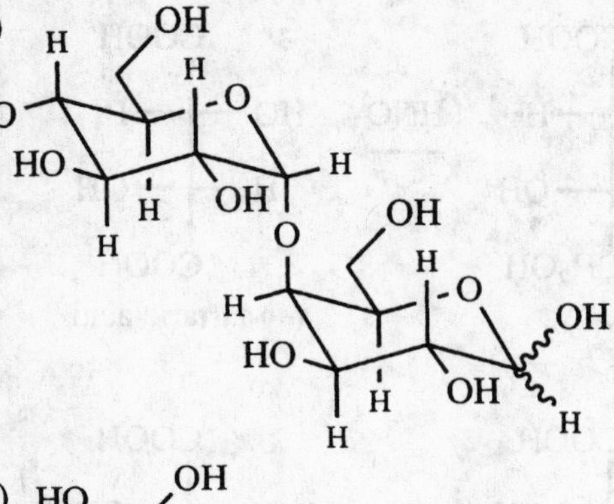

(b)

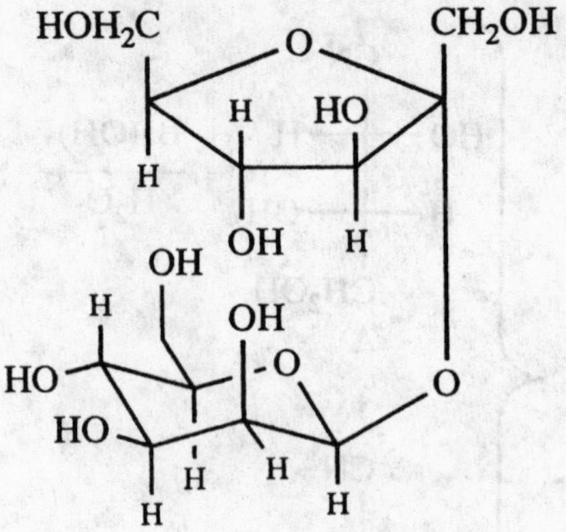

(c)

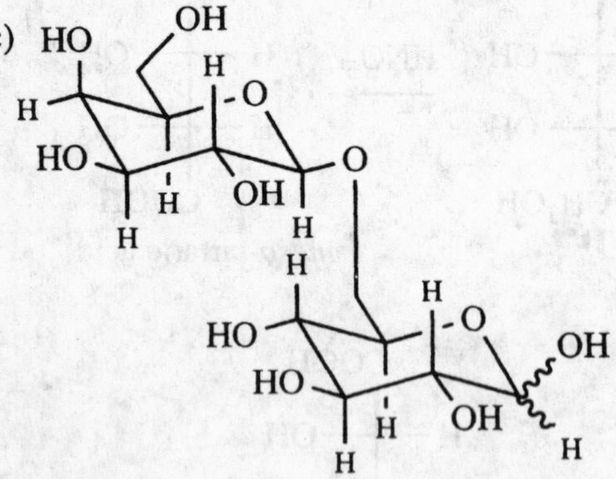

23-61

(a) methyl β-D-fructofuranoside
(b) 3,6-di-O-methyl-β-D-mannopyranose
(c) 4-O-(α-D-fructofuranosyl)-β-D-galactopyranose
(d) β-D-*N*-acetylgalactopyranosamine, or 2-acetamido-2-deoxy-β-D-galactopyranose

23-62 These are reducing sugars and would undergo mutarotation:

—in problem 23-59: (b) and (c);
—in problem 23-60: (a) and (c);
—in problem 23-61: (b), (c), and (d)

23-63

(a)

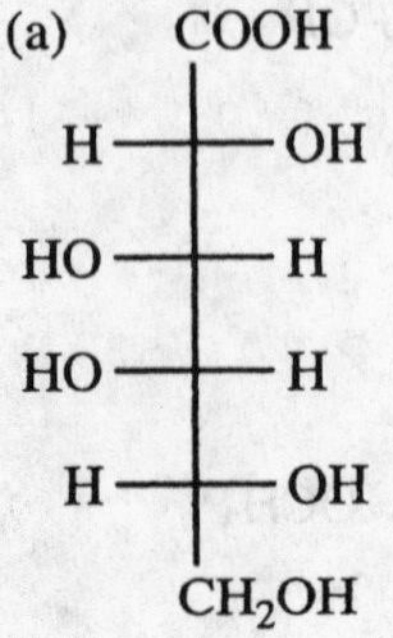

$$\begin{array}{c} COOH \\ H \rule{0pt}{0pt} OH \\ HO \rule{0pt}{0pt} H \\ HO \rule{0pt}{0pt} H \\ H \rule{0pt}{0pt} OH \\ CH_2OH \end{array}$$

(b)

$$\begin{array}{c} CHO \\ HO \rule{0pt}{0pt} H \\ HO \rule{0pt}{0pt} H \\ HO \rule{0pt}{0pt} H \\ H \rule{0pt}{0pt} OH \\ CH_2OH \end{array} \quad + \quad \begin{array}{c} CH_2OH \\ O \\ HO \rule{0pt}{0pt} H \\ HO \rule{0pt}{0pt} H \\ H \rule{0pt}{0pt} OH \\ CH_2OH \end{array} \quad + \text{ others}$$

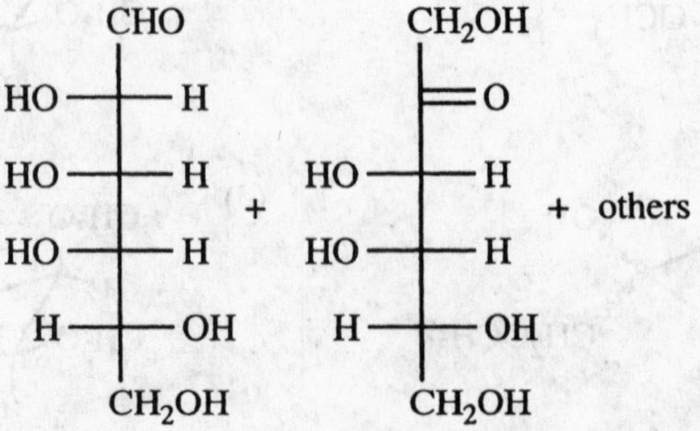

(c)

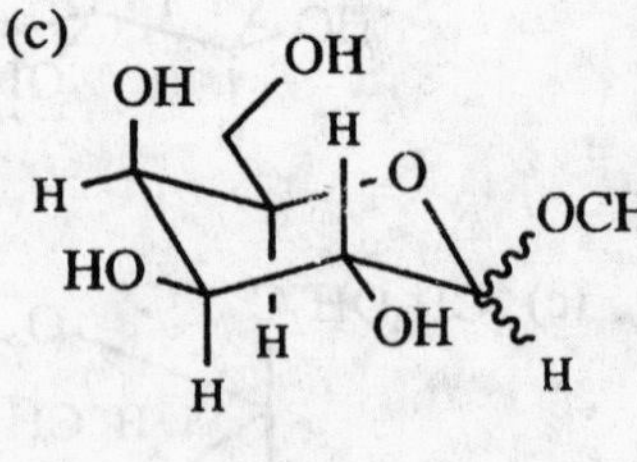

23-63 continued

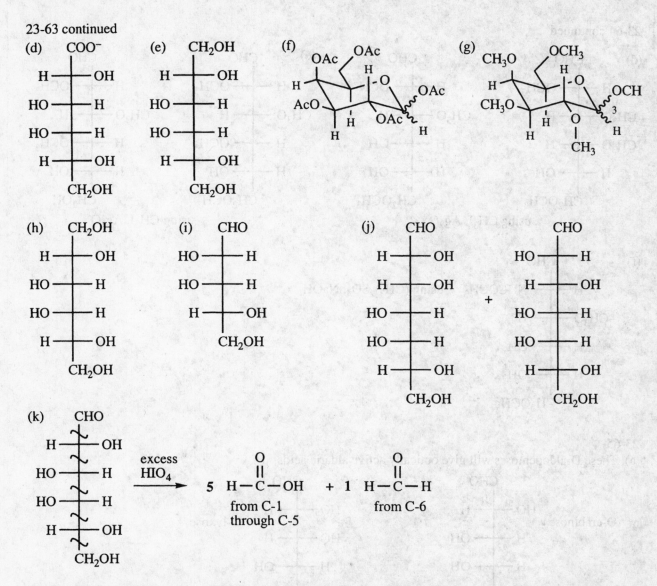

(d) COO⁻ / H—OH / HO—H / HO—H / H—OH / CH₂OH

(e) CH₂OH / H—OH / HO—H / HO—H / H—OH / CH₂OH

(f) structure with OAc groups

(g) structure with OCH₃ groups

(h) CH₂OH / H—OH / HO—H / HO—H / H—OH / CH₂OH

(i) CHO / HO—H / HO—H / H—OH / CH₂OH

(j) CHO / H—OH / H—OH / HO—H / HO—H / H—OH / CH₂OH + CHO / HO—H / H—OH / HO—H / HO—H / H—OH / CH₂OH

(k) CHO / H—OH / HO—H / HO—H / H—OH / CH₂OH

$\xrightarrow{\text{excess HIO}_4}$

5 H—C(=O)—OH + **1** H—C(=O)—H

from C-1 through C-5   from C-6

23-64 Use the milder reagent, CH_3I/Ag_2O, when the sugar is in the hemiacetal form; the mild conditions prevent isomerization. When the carbohydrate is present as an acetal (a glycoside), use the more basic reagent, $NaOH/(CH_3)_2SO_4$; an acetal is stable to basic conditions.

(a) CH₂OCH₃ / =O / CH₃O—H / H—OCH₃ / H—OH / CH₂OCH₃
using CH₃I, Ag₂O

(b) CHO / H—OCH₃ / CH₃O—H / H—OCH₃ / H—OH / CH₂OCH₃
using (CH₃)₂SO₄, NaOH

(c) CHO / H—OCH₃ / CH₃O—H / H—OCH₃ / H—OH / CH₂OCH₃ + CH₂OCH₃ / =O / CH₃O—H / H—OCH₃ / H—OH / CH₂OCH₃
using (CH₃)₂SO₄, NaOH

23-64 continued

(d)

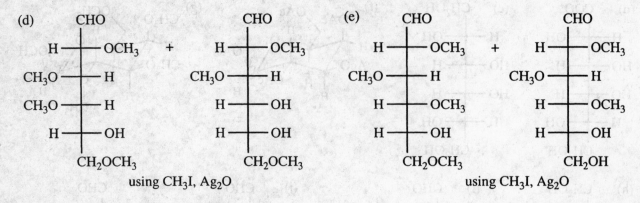

using CH_3I, Ag_2O

(e)

using CH_3I, Ag_2O

(f)

CHO

H —— $NHCOCH_3$ using $(CH_3)_2SO_4$, NaOH

CH_3O —— H

H —— OH

H —— OH

CH_2OCH_3

23-65

(a) These D-aldopentoses will give optically active aldaric acids.

D-arabinose

CHO

HO —— H

H —— OH

H —— OH

CH_2OH

D-lyxose

CHO

HO —— H

HO —— H

H —— OH

CH_2OH

(b) Only D-threose (of the aldotetroses) will give an optically active aldaric acid.

D-threose

CHO

HO —— H

H —— OH

CH_2OH

23-65 continued

(c) **X** is D-galactose.

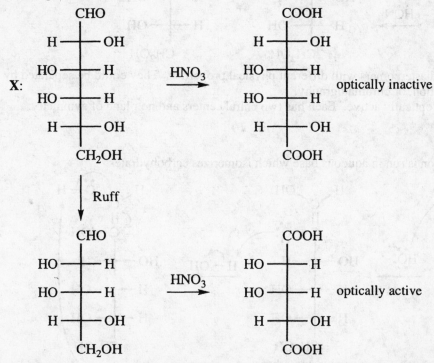

The other aldohexose that gives an optically inactive aldaric acid is D-allose, with all OH groups on the right side of the Fischer projection. Ruff degradation followed by nitric acid gives an optically *inactive* aldaric acid, however, so **X** cannot be D-allose.

(d) The optically active, five-carbon aldaric acid comes from the optically active pentose, not from the optically inactive, six-carbon aldaric acid. The principle is not violated.

(e)

D-threose

(S,S)-tartaric acid
optically active

23-66

(a)

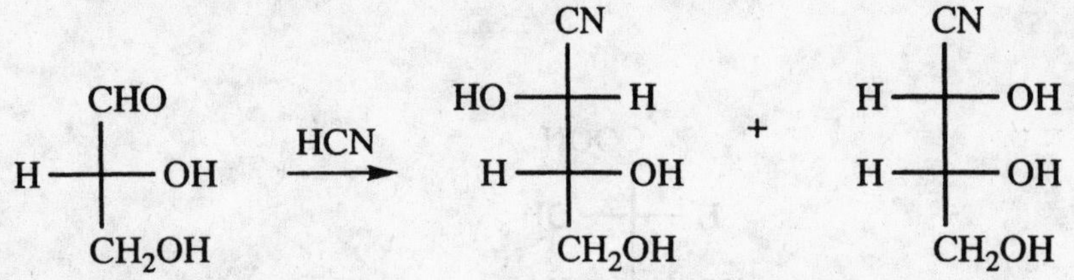

HCN

CHO / H—OH / CH₂OH → CN / HO—H / H—OH / CH₂OH + CN / H—OH / H—OH / CH₂OH

(b) The products are diastereomers with different physical properties. They could be separated by crystallization, distillation, or chromatography.

(c) Both products are optically active. Each has two chiral centers and no plane of symmetry.

23-67

(a) The Tollens reaction is run in aqueous base which isomerizes carbohydrates.

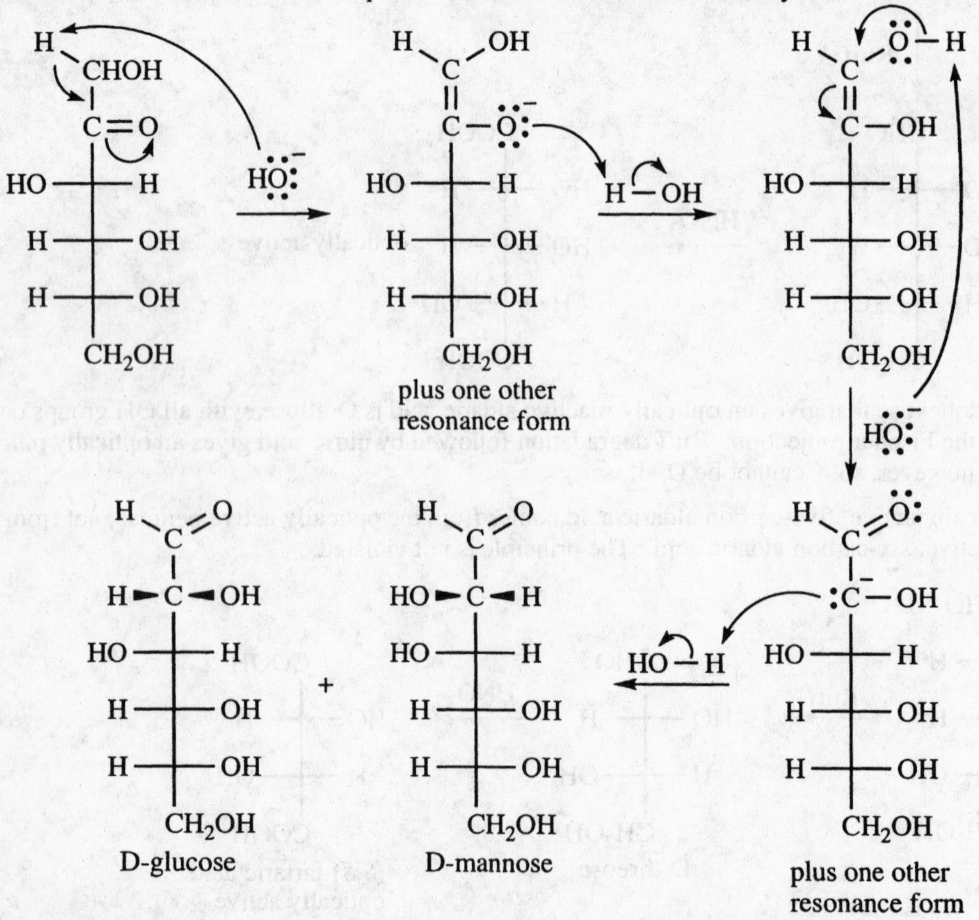

CHOH / C=O / HO—H / H—OH / H—OH / CH₂OH + HO⁻ → C(OH) / C—O⁻ / HO—H / H—OH / H—OH / CH₂OH

plus one other resonance form

H—OH / C / C—OH / HO—H / H—OH / H—OH / CH₂OH

D-glucose: H—C=O / H—C—OH / HO—H / H—OH / H—OH / CH₂OH

+

D-mannose: H—C=O / HO—C—H / HO—H / H—OH / H—OH / CH₂OH

H—C=O / :C—OH / HO—H / H—OH / H—OH / CH₂OH

plus one other resonance form

(b) Bromine water is acidic, not basic like the Tollens reagent. Carbohydrates isomerize quickly in base, but only very slowly in acid, so bromine water can oxidize without isomerization.

23-68 Tagatose is a monosaccharide, a ketohexose, that is found in the pyranose form.

(a)

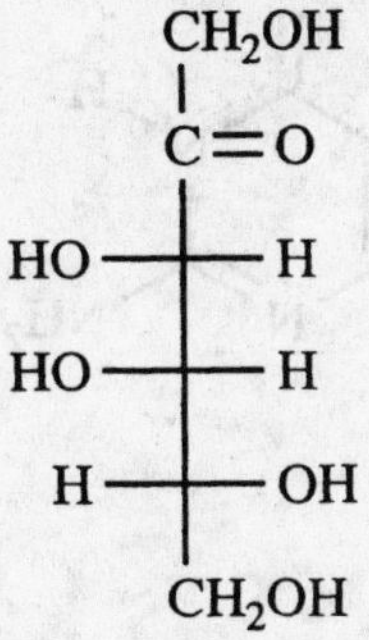

$$CH_2OH$$
$$C=O$$
$$HO—H$$
$$HO—H$$
$$H—OH$$
$$CH_2OH$$

(b)

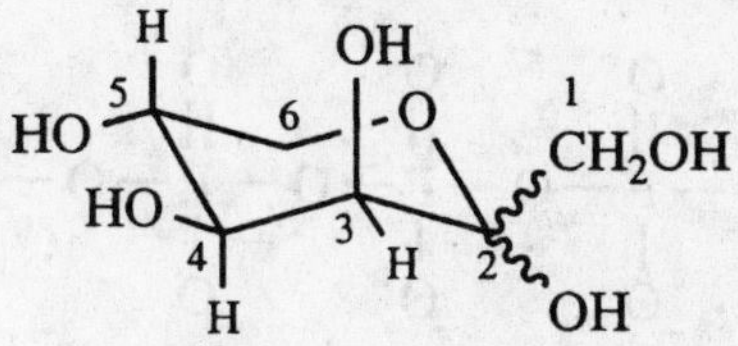

23-69

(a)

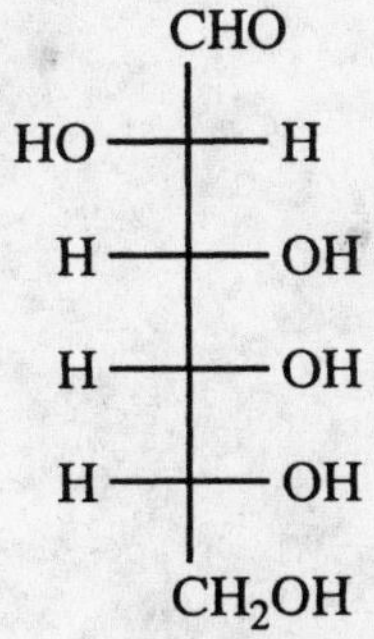

$$CHO$$
$$HO—H$$
$$H—OH$$
$$H—OH$$
$$H—OH$$
$$CH_2OH$$
D-altrose

(b)

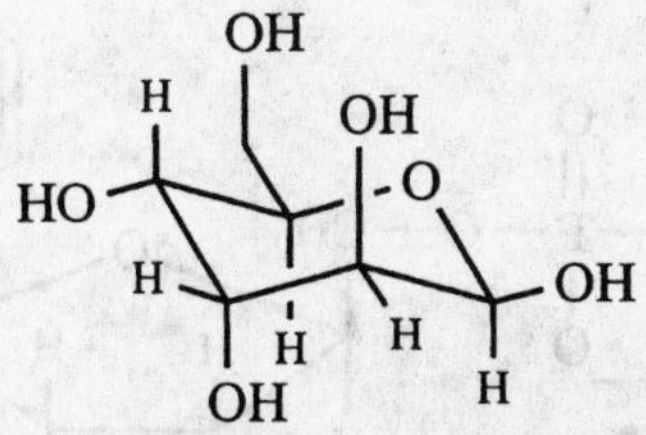

β-D-altropyranose

23-70

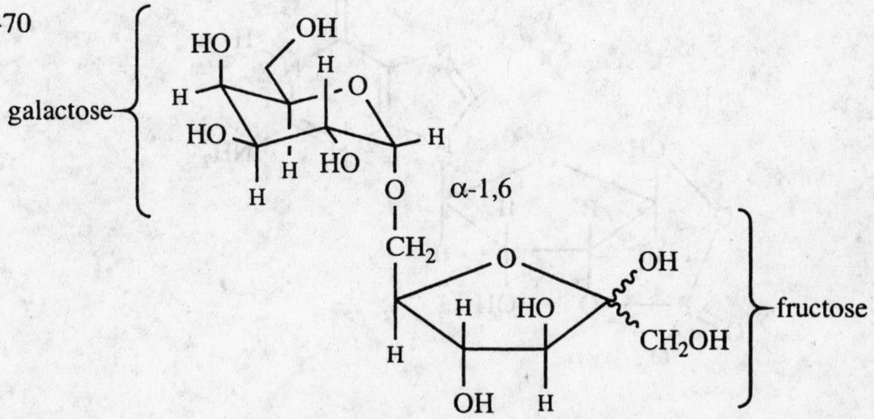

galactose

α-1,6

fructose

6-O-(α-D-galactopyranosyl)-D-fructofuranose

23-71

(a)

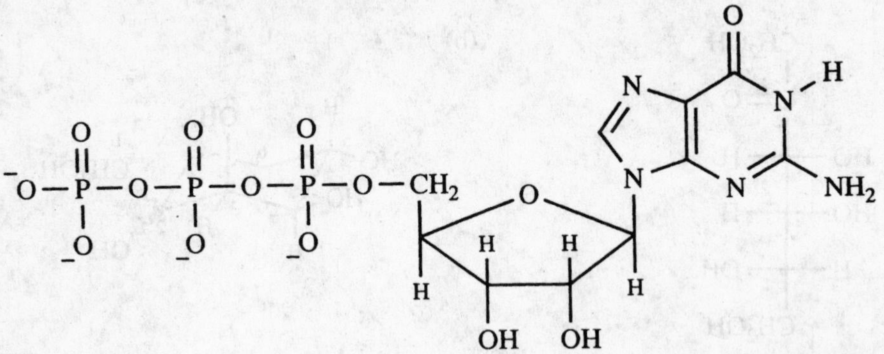

(b)

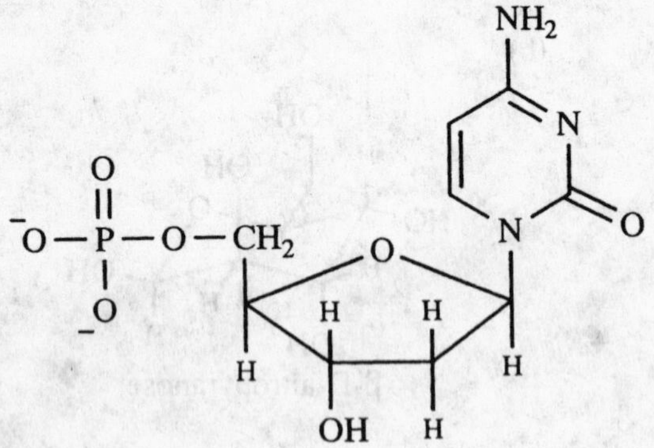

(c)

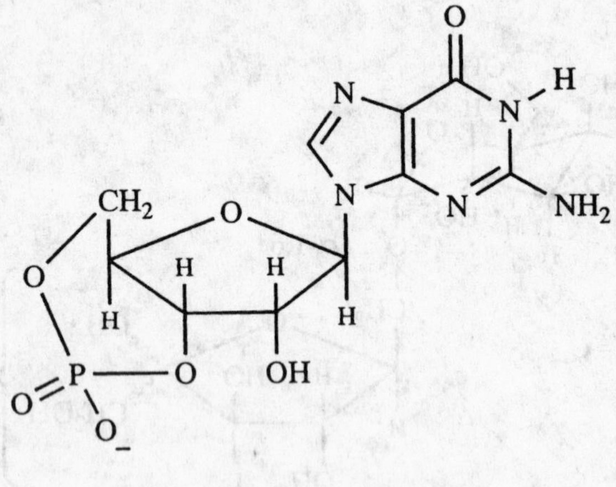

23-72 Bonds from C to H are omitted for simplicity.

23-73

(a) No, there is no relation between the amount of G and A.

(b) Yes, this must be true mathematically.

(c) Chargaff's rule must apply only to double-stranded DNA. For each G in one strand, there is a complementary C in the opposing strand, but there is no correlation between G and C *in the same strand*.

23-74 Bonds from C to H are omitted for simplicity.

2'-deoxythymidine
(abbreviated dT)

3'-azido-2',3'-dideoxythymidine
(AZT)

No phosphate can attach to the azide group, so
synthesis of the DNA chain is terminated.

AZT 5'-triphosphate

23-75 Recall from section 19-17 that nitrous acid is unstable, generating nitrosonium ion.

(a) $H-O-N=O + H^+ \longrightarrow H_2O + \overset{+}{N}=O$

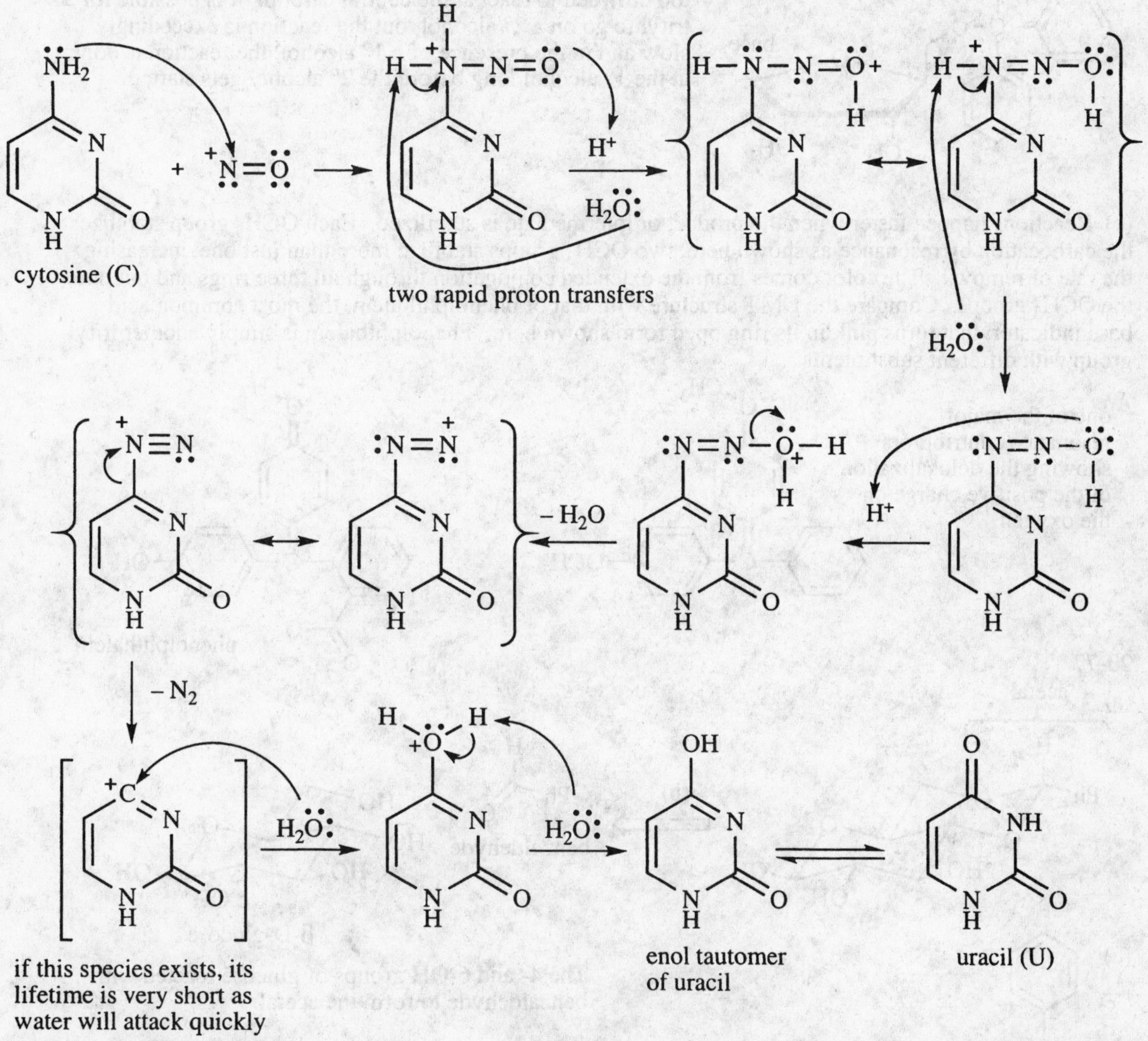

cytosine (C)

two rapid proton transfers

if this species exists, its
lifetime is very short as
water will attack quickly

enol tautomer
of uracil

uracil (U)

(b) In base pairing, cytosine pairs with guanine. If cytosine is converted to uracil, however, each replication will not carry the complement of cytosine (guanine) but instead will carry the complement of uracil (adenine). This is the definition of a mutation, where the wrong base is inserted in a nucleic acid chain.

(c) In RNA, the transformation of cytosine (C) to uracil (U) is not detected as a problem because U is a base normally found in RNA so it goes unrepaired. In DNA, however, thymine (with an extra methyl group) is used instead of uracil. If cytosine is diazotized to uracil, the DNA repair enzymes detect it as a mutation and correct it.

23-76

(a)

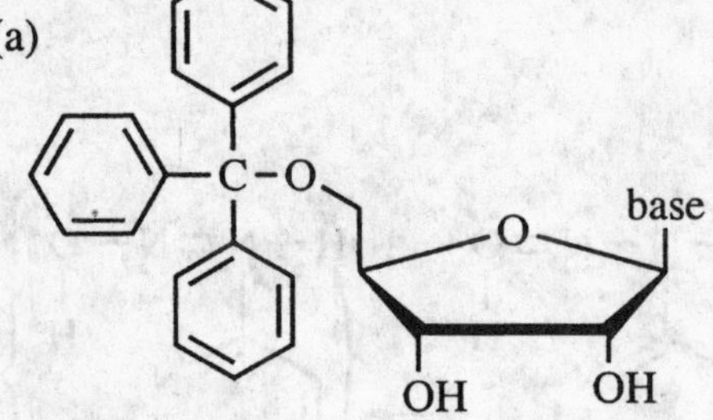

(b) Trityl groups are specific for 1° alcohols for steric reasons: the trityl group is so big that even a 2° alcohol is too crowded to react at the central carbon. It is possible for a trityl to go on a 2° alcohol, but the reaction is exceedingly slow and in the presence of a 1° alcohol, the reaction is done at the 1° alcohol long before the 2° alcohol gets started.

(c) Reactions happen faster when the product or intermediate is stabilized. Each OCH_3 group stabilizes the carbocation by resonance as shown here; two OCH_3 groups stabilize more than just one, increasing the rate of removal. The color comes from the extended conjugation through all three rings and out onto the OCH_3 groups. Compare the DMT structure with that of phenolphthalein, the most common acid base indicator, that turns pink in its ring open form shown here. Phenolphthalein is simply another trityl group with different substituents.

one of the major resonance contributors showing the delocalization of the positive charge on the oxygen

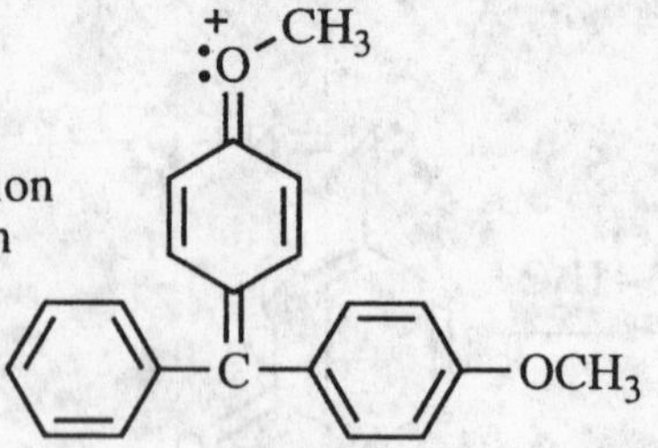

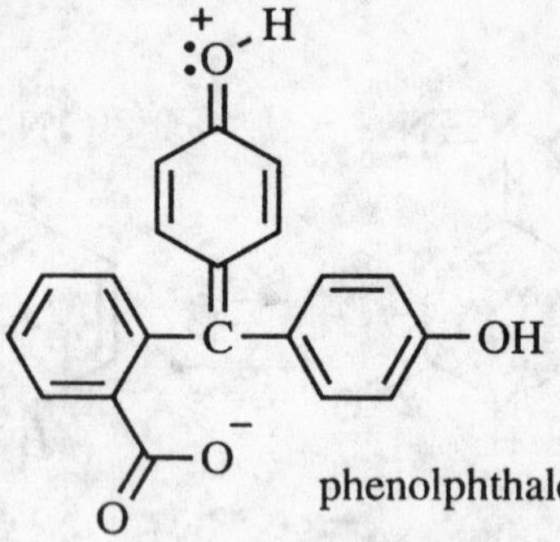

phenolphthalein

23-77

(a) acetal

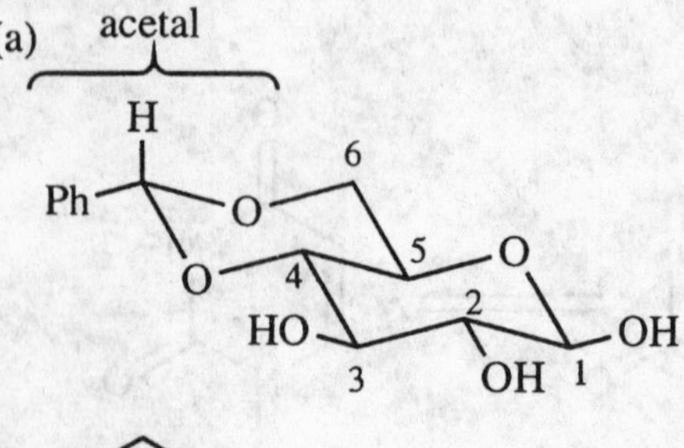

(b)

benzaldehyde

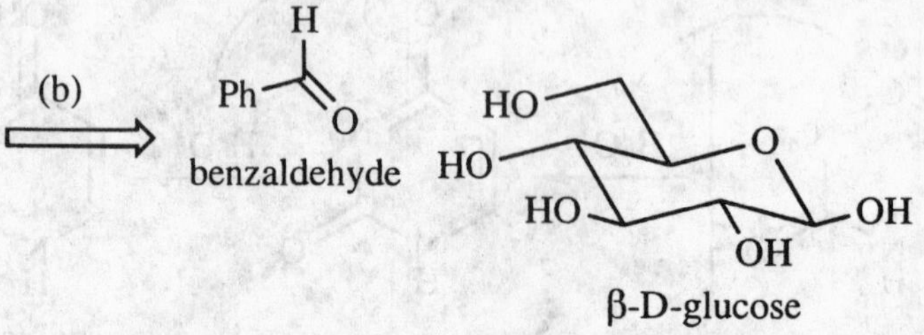

β-D-glucose

The 4- and 6-OH groups of glucose reacted with benzaldehyde to form the acetal.

(c)

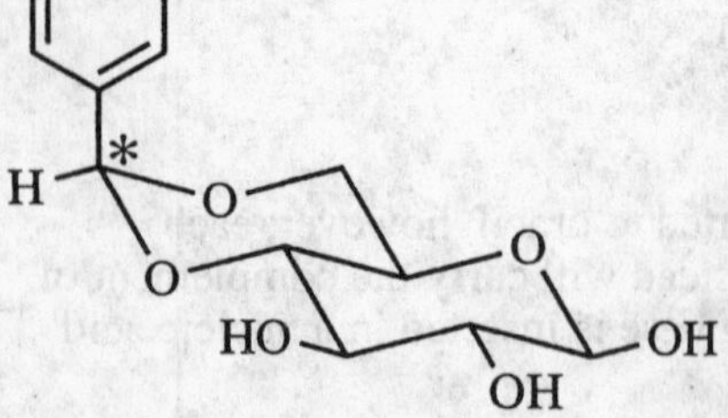

(c) and (d) The chiral center is marked with (*). This stereoisomer is a diastereomer since only one chiral center is inverted. This diastereomer puts the phenyl group in an axial position, definitely less stable than the structure shown in part (a). Only the product shown in (a) will isolated from this reaction.

(e)

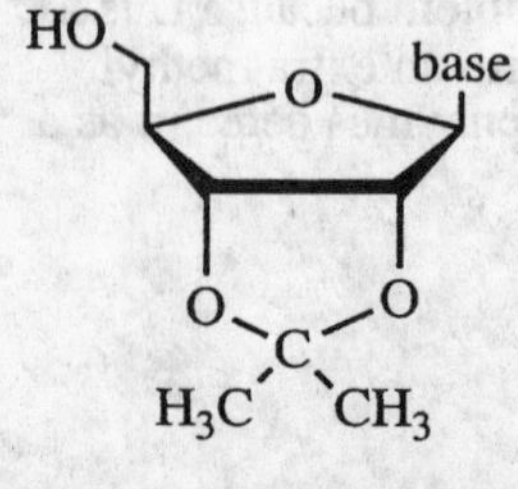

Only the 2'- and 3'-OH groups are close enough to form this cyclic acetal (ketal) from acetone, called an acetonide.

24-1

(a)

$$H_2N - \overset{\text{COOH}}{\underset{H}{\overset{|}{C}}} - CH_2Ph$$

(b)

$$H_2N - \overset{\text{COOH}}{\underset{H}{\overset{|}{C}}} - CH_2CH_2CH_2NH - \overset{NH}{\underset{||}{C}} - NH_2$$

(c)

$$H_2N - \overset{\text{COOH}}{\underset{CH_2OH}{\overset{|}{C}}} - H$$

(d)

$$H_2N - \overset{\text{COOH}}{\underset{H}{\overset{|}{C}}} - CH_2 - \text{(indole ring, NH)}$$

24-2

(a) The configurations around the asymmetric carbons of *(R)*-cysteine and *(S)*-alanine are the same. The *designation* of configuration changes because sulfur changes the priorities of the side chain and the COOH.

(S)-alanine with group priorities shown

$$H_2N \overset{2}{\underset{1}{-}} \overset{COOH}{\underset{4\,H}{\overset{|}{C}}} \overset{3}{-} CH_3$$

$$H_2N \overset{3}{\underset{1}{-}} \overset{COOH}{\underset{4\,H}{\overset{|}{C}}} \overset{2}{-} CH_2SH$$

(R)-cysteine with group priorities shown; note that the COOH and the CH$_2$SH priorities are reversed compared with *(S)*-alanine

(b) Fischer projections show that both *(S)*-alanine and *(R)*-cysteine are L-amino acids.

$$H_2N \overset{\text{COOH}}{\underset{CH_3}{-\!\!\!+\!\!\!-}} H$$

$$H_2N \overset{\text{COOH}}{\underset{CH_2SH}{-\!\!\!+\!\!\!-}} H$$

(S)-alanine
L-alanine

(R)-cysteine
L-cysteine

24-3 In their evolution, plants have needed to be more resourceful than animals in developing biochemical mechanisms for survival. Thus, plants make more of their own required compounds than animals do. The amino acid phenylalanine is produced by plants but required in the diet of mammals. To interfere with a plant's production of phenylalanine is fatal to the plant, but since humans do not produce phenylalanine, glyphosate is virtually non-toxic to us.

24-4 Here is a simple way of determining if a group will be protonated: *at solution pH below the group's pK$_a$ value, the group will be protonated; at pH higher than the group's pK$_a$ value, it will not be protonated.*

(a)

$$H_2N - \overset{H}{\underset{CH(CH_3)_2}{\overset{|}{C}}} - COO^-$$

(b)

$$H - \overset{+}{\underset{}{N}} \overset{H}{\underset{}{|}} - \overset{}{\underset{}{C}} - COOH \quad \text{(pyrrolidine ring)}$$

(c)

$$H_3\overset{+}{N} - \overset{H}{\underset{CH_2CH_2CH_2NH - \underset{+NH_2}{\overset{||}{C}} - NH_2}{\overset{|}{C}}} - COO^-$$

(d)

$$H_3\overset{+}{N} - \overset{H}{\underset{CH_2CH_2COO^-}{\overset{|}{C}}} - COO^-$$

24-4 continued

	alanine	lysine	aspartic acid

(e)

(i) pH 6

alanine:
$$H_3\overset{+}{N}-\underset{\underset{CH_3}{|}}{\overset{\overset{H}{|}}{C}}-COO^-$$

lysine:
$$H_3\overset{+}{N}-\underset{\underset{(CH_2)_4\overset{+}{N}H_3}{|}}{\overset{\overset{H}{|}}{C}}-COO^-$$

aspartic acid:
$$H_3\overset{+}{N}-\underset{\underset{CH_2COO^-}{|}}{\overset{\overset{H}{|}}{C}}-COO^-$$

(ii) pH 11

alanine:
$$H_2N-\underset{\underset{CH_3}{|}}{\overset{\overset{H}{|}}{C}}-COO^-$$

lysine:
$$H_2N-\underset{\underset{(CH_2)_4NH_2}{|}}{\overset{\overset{H}{|}}{C}}-COO^-$$

aspartic acid:
$$H_2N-\underset{\underset{CH_2COO^-}{|}}{\overset{\overset{H}{|}}{C}}-COO^-$$

(iii) pH 2

alanine:
$$H_3\overset{+}{N}-\underset{\underset{CH_3}{|}}{\overset{\overset{H}{|}}{C}}-COOH$$

lysine:
$$H_3\overset{+}{N}-\underset{\underset{(CH_2)_4\overset{+}{N}H_3}{|}}{\overset{\overset{H}{|}}{C}}-COOH$$

aspartic acid:
$$H_3\overset{+}{N}-\underset{\underset{CH_2COOH}{|}}{\overset{\overset{H}{|}}{C}}-COOH$$

24-5

[resonance structures of protonated guanidino group]

Protonation of the guanidino group gives a resonance-stabilized cation with all octets filled and the positive charge delocalized over three nitrogen atoms. Arginine's strongly basic isoelectric point reflects the unusual basicity of the guanidino group due to this resonance stabilization in the protonated form. (See Problems 1-39 and 19-49(a).)

24-6

tryptophan

$$H_2N-\underset{\underset{CH_2}{|}}{\overset{\overset{H}{|}}{C}}-COOH$$

not basic → N (like pyrrole)
|
H

histidine

$$H_2N-\underset{\underset{CH_2}{|}}{\overset{\overset{H}{|}}{C}}-COOH$$

basic → N (like pyridine)
NH → not basic (like pyrrole)

The basicity of any nitrogen depends on its electron pair's availability for bonding with a proton. In tryptophan, the nitrogen's electron pair is part of the aromatic π system; without this electron pair in the π system, the molecule would not be aromatic. Using this electron pair for bonding to a proton would therefore destroy the aromaticity—not a favorable process.

In the imidazole ring of histidine, the electron pair of one nitrogen is also part of the aromatic π system and is unavailable for bonding; this nitrogen is not basic. The electron pair on the other nitrogen, however, is in an sp^2 orbital available for bonding, and is about as basic as pyridine.

24-7 At pH 9.7, alanine (isoelectric point (IEP) 6.0) has a charge of –1 and will migrate to the anode. Lysine (IEP 9.7) is at its isoelectric point and will not move. Aspartic acid (IEP 2.8) has a charge of –2 and will also migrate to the anode, faster than alanine.

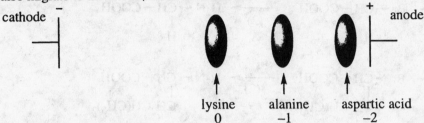

cathode ⁻

anode ⁺

lysine   alanine   aspartic acid
  0   –1   –2

24-8 At pH 6.0, tryptophan (IEP 5.9) has a charge of zero and will not migrate. Cysteine (IEP 5.0) has a partial negative charge and will move toward the anode. Histidine (IEP 7.6) has a partial positive charge and will move toward the cathode.

cathode ⁻

anode ⁺

histidine   tryptophan   cysteine

$$H_3N^+ - \overset{\overset{\displaystyle H}{|}}{C} - COO^-$$
$$CH_2$$

histidine (partially protonated imidazole ring)

$$H_3N^+ - \overset{\overset{\displaystyle H}{|}}{C} - COO^-$$
$$CH_2$$

tryptophan

$$H_3N^+ - \overset{\overset{\displaystyle H}{|}}{C} - COO^-$$
$$CH_2S^-$$

cysteine (partially deprotonated sulfur)

(The SH is more acidic than the NH_3^+ group.)

24-9

(a) $O=\overset{\overset{\displaystyle }{|}}{C}-COOH$ $\xrightarrow[\text{H}_2,\ \text{Pd}]{\text{NH}_3}$ $H_2N-\overset{|}{C}H-COOH$

       CH_3 CH_3

(b) $O=\overset{|}{C}-COOH$ $\xrightarrow[\text{H}_2,\ \text{Pd}]{\text{NH}_3}$ $H_2N-\overset{|}{C}H-COOH$

       $CH_2CH(CH_3)_2$ $CH_2CH(CH_3)_2$

(c) $O=\overset{|}{C}-COOH$ $\xrightarrow[\text{H}_2,\ \text{Pd}]{\text{NH}_3}$ $H_2N-\overset{|}{C}H-COOH$

       CH_2OH CH_2OH

(d) $O=\overset{|}{C}-COOH$ $\xrightarrow[\text{H}_2,\ \text{Pd}]{\text{NH}_3}$ $H_2N-\overset{|}{C}H-COOH$

       $CH_2CH_2CONH_2$ $CH_2CH_2CONH_2$

24-10 All of these reactions use: first arrow: (1) Br_2/PBr_3, followed by H_2O workup; second arrow: (2) excess NH_3, followed by neutralizing workup.

(a) $\underset{\overset{|}{H}}{H_2C}-COOH \xrightarrow{(1)} \underset{\overset{|}{H}}{Br-CH}-COOH \xrightarrow{(2)} \underset{\overset{|}{H}}{H_2N-CH}-COOH$

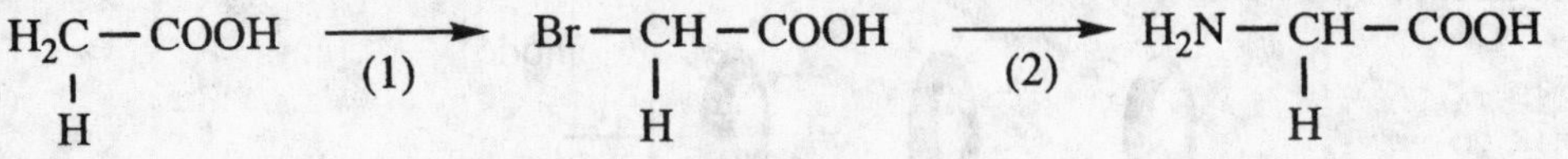

(b) $\underset{\overset{|}{CH_2CH(CH_3)_2}}{H_2C}-COOH \xrightarrow{(1)} \underset{\overset{|}{CH_2CH(CH_3)_2}}{Br-CH}-COOH \xrightarrow{(2)} \underset{\overset{|}{CH_2CH(CH_3)_2}}{H_2N-CH}-COOH$

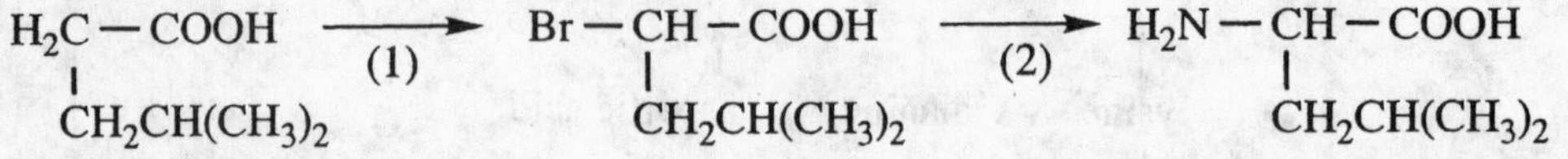

(c) $\underset{\overset{|}{CH(CH_3)_2}}{H_2C}-COOH \xrightarrow{(1)} \underset{\overset{|}{CH(CH_3)_2}}{Br-CH}-COOH \xrightarrow{(2)} \underset{\overset{|}{CH(CH_3)_2}}{H_2N-CH}-COOH$

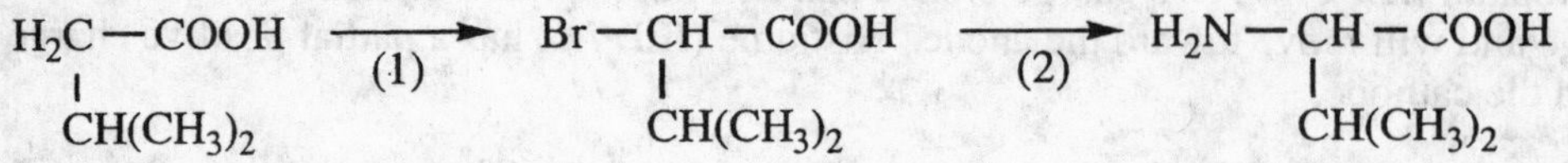

(d) $\underset{\overset{|}{CH_2CH_2COOH}}{H_2C}-COOH \xrightarrow{(1)} \underset{\overset{|}{CH_2CH_2COOH}}{Br-CH}-COOH \xrightarrow{(2)} \underset{\overset{|}{CH_2CH_2COOH}}{H_2N-CH}-COOH$

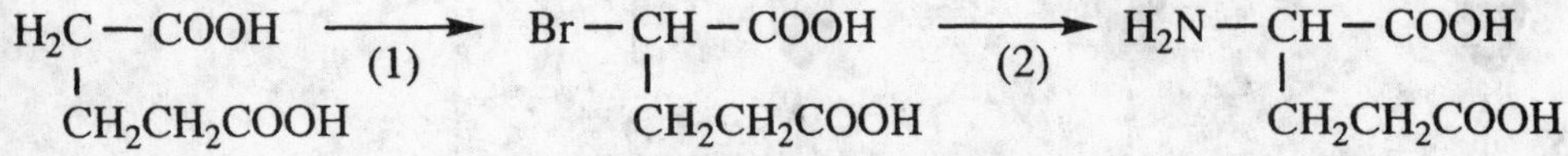

In part (d), care must be taken to avoid reaction α to the other COOH. In practice, this would be accomplished by using less than one-half mole of bromine per mole of the diacid.

24-11

(a)

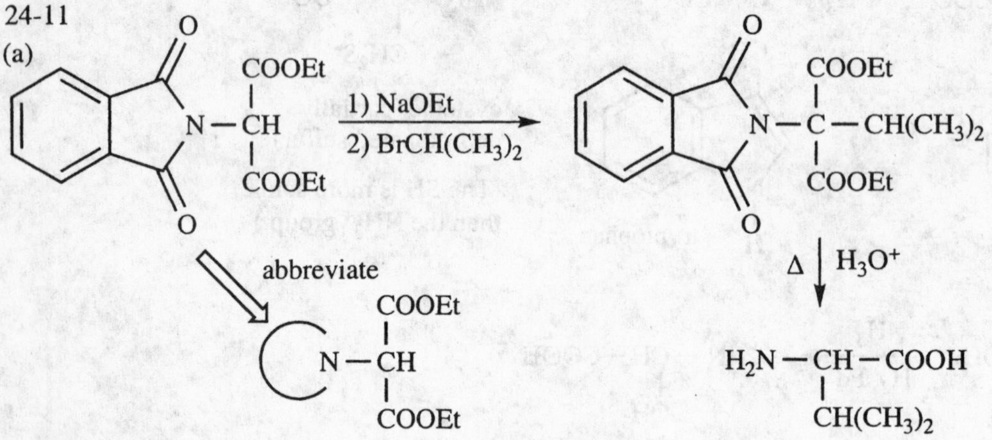

(b)

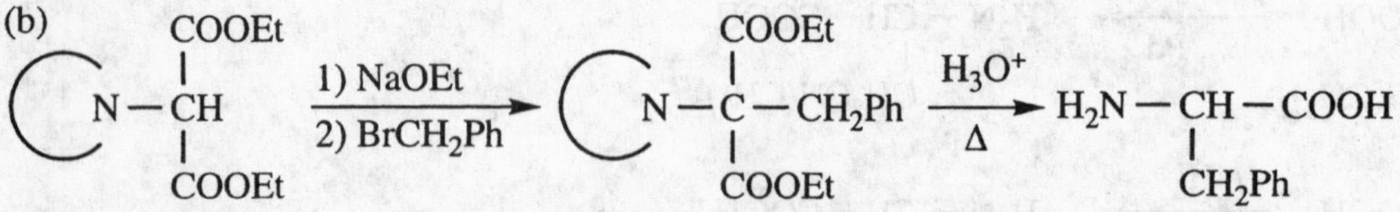

(c)

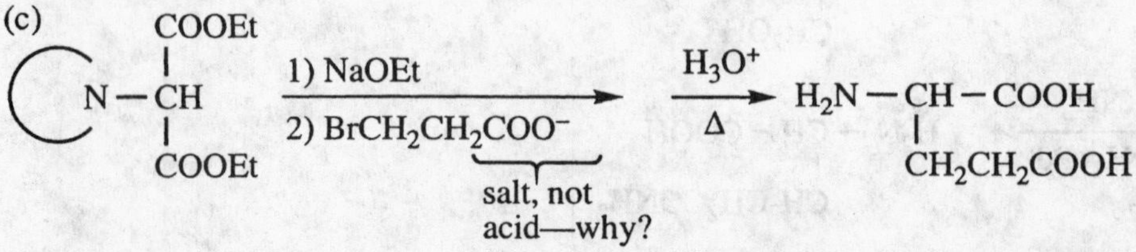

salt, not acid—why?

24-11 continued

(d)

$$\text{(ring)}N-\underset{\underset{\text{COOEt}}{|}}{\overset{\overset{\text{COOEt}}{|}}{CH}} \quad \xrightarrow[\text{2) BrCH}_2\text{CH(CH}_3)_2]{\text{1) NaOEt}} \quad \xrightarrow[\Delta]{\text{H}_3\text{O}^+} \quad H_2N-\underset{\underset{\text{CH}_2\text{CH(CH}_3)_2}{|}}{CH}-COOH$$

24-12

$$\text{AcNH}-\underset{\underset{\text{COOEt}}{|}}{\overset{\overset{\text{COOEt}}{|}}{CH}} \quad \xrightarrow[\text{2) BrCH}_2\text{Ph}]{\text{1) NaOEt}} \quad \text{AcNH}-\underset{\underset{\text{COOEt}}{|}}{\overset{\overset{\text{COOEt}}{|}}{C}}-CH_2Ph \quad \xrightarrow[\Delta]{\text{H}_3\text{O}^+} \quad H_2N-\underset{\underset{\text{CH}_2\text{Ph}}{|}}{CH}-COOH$$

acetamidomalonic
ester

24-13

(a)

$$\text{PhCH}_2-\overset{\overset{\text{O}}{\|}}{CH} \quad \xrightarrow[\text{H}_2\text{O}]{\text{NH}_3,\ \text{HCN}} \quad \text{PhCH}_2-\underset{\underset{\text{C}\equiv\text{N}}{|}}{\overset{\overset{\text{NH}_2}{|}}{CH}} \quad \xrightarrow[\Delta]{\text{H}_3\text{O}^+} \quad H_2N-\underset{\underset{\text{CH}_2\text{Ph}}{|}}{CH}-COOH$$

(b) While the solvent for the Strecker synthesis is water, the proton acceptor is ammonia and the proton donor is ammonium ion.

$$\text{PhCH}_2-\overset{\overset{:\ddot{O}:}{\|}}{CH} + :NH_3 \longrightarrow \text{PhCH}_2-\underset{\underset{\underset{H}{|}}{\overset{+}{NH_2}}}{\overset{\overset{:\ddot{O}:^-}{|}}{CH}} \quad \underset{:NH_3}{} \quad H-\overset{+}{NH_3}$$

$$\xrightarrow{\text{two fast proton transfers}}$$

$$\text{PhCH}_2-\underset{\underset{\text{NH}_2}{|}}{\overset{\overset{:\ddot{O}-H}{|}}{CH}} \quad \overset{+}{H_3N}-H \longleftarrow \text{PhCH}_2-\underset{\underset{\text{NH}_2}{|}}{\overset{\overset{H-\overset{+}{\ddot{O}}-H}{|}}{CH}} \quad \xrightarrow{-H_2O}$$

$$\left\{ \begin{array}{c} \text{PhCH}_2-\underset{\underset{:\text{NH}_2}{|}}{\overset{+}{CH}} \\ \updownarrow \\ \text{PhCH}_2-\underset{\underset{+\text{NH}_2}{\|}}{CH} \end{array} \right\}$$

this protonated
imine is rapidly
attacked by
cyanide nucleophile

$$\xrightarrow{\ :C\equiv N^-\ } \quad \text{PhCH}_2-\underset{\underset{\text{NH}_2}{|}}{\overset{\overset{\text{C}\equiv\text{N}}{|}}{CH}}$$

(abbreviate as

$$R-C\equiv N$$

on the next page)

24-13 (b) continued

mechanism of acid hydrolysis of the nitrile

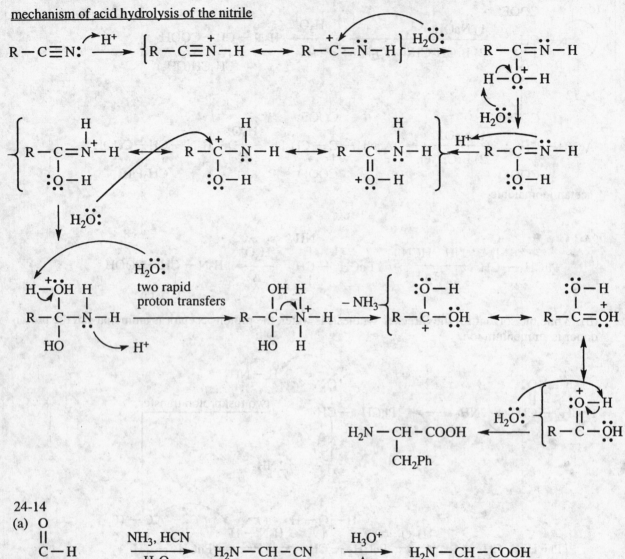

24-14

(a)

$$\underset{\substack{\text{C}-\text{H}\\|\\\text{CH}_2\text{CH}(\text{CH}_3)_2}}{\overset{\text{O}}{\overset{||}{}}} \xrightarrow[\text{H}_2\text{O}]{\text{NH}_3, \text{HCN}} \underset{\substack{|\\\text{CH}_2\text{CH}(\text{CH}_3)_2}}{\text{H}_2\text{N}-\text{CH}-\text{CN}} \xrightarrow[\Delta]{\text{H}_3\text{O}^+} \underset{\substack{|\\\text{CH}_2\text{CH}(\text{CH}_3)_2}}{\text{H}_2\text{N}-\text{CH}-\text{COOH}}$$

(b)

$$\underset{\substack{\text{C}-\text{H}\\|\\\text{H}}}{\overset{\text{O}}{\overset{||}{}}} \xrightarrow[\text{H}_2\text{O}]{\text{NH}_3, \text{HCN}} \underset{\substack{|\\\text{H}}}{\text{H}_2\text{N}-\text{CH}-\text{CN}} \xrightarrow[\Delta]{\text{H}_3\text{O}^+} \underset{\substack{|\\\text{H}}}{\text{H}_2\text{N}-\text{CH}-\text{COOH}}$$

(c)

$$\underset{\substack{\text{C}-\text{H}\\|\\\text{CH}(\text{CH}_3)_2}}{\overset{\text{O}}{\overset{||}{}}} \xrightarrow[\text{H}_2\text{O}]{\text{NH}_3, \text{HCN}} \underset{\substack{|\\\text{CH}(\text{CH}_3)_2}}{\text{H}_2\text{N}-\text{CH}-\text{CN}} \xrightarrow[\Delta]{\text{H}_3\text{O}^+} \underset{\substack{|\\\text{CH}(\text{CH}_3)_2}}{\text{H}_2\text{N}-\text{CH}-\text{COOH}}$$

24-15 In acid solution, the free amino acid will be protonated, with a positive charge, and probably soluble in water as are other organic ions. The acylated amino acid, however, is not basic since the nitrogen is present as an amide. In acid solution, the acylated amino acid is neutral and not soluble in water. Water extraction or ion-exchange chromatography (Figure 24-11) would be practical techniques to separate these compounds.

24-16

$$\text{abbreviate} \quad H_3\overset{+}{N}-CH-\overset{O}{\overset{\|}{C}}OEt \quad \text{as} \quad R-\overset{O}{\overset{\|}{C}}OEt$$
$$\hspace{5.5cm} | \\ \hspace{5.5cm} CH_2Ph$$

$$R-\overset{:\overset{\cdot\cdot}{O}:}{\overset{\|}{C}}-OEt \quad \xrightarrow{H^+} \quad R\overset{:\overset{\cdot\cdot}{O}H}{\underset{+}{\overset{|}{C}}}-\overset{\cdot\cdot}{\overset{\cdot\cdot}{O}}Et \quad \xrightarrow{H_2\overset{\cdot\cdot}{O}:} \quad RC\overset{OH}{\underset{H-\overset{+}{\underset{\cdot\cdot}{O}}-H}{\overset{|}{-}}}OEt \quad \longrightarrow \quad RC\overset{OH}{\underset{OH}{\overset{|}{-}}}\overset{\cdot\cdot}{\overset{\cdot\cdot}{O}}Et$$
$$\text{plus two other} \hspace{3cm} \overset{\curvearrowleft}{H_2\overset{\cdot\cdot}{O}:}$$
$$\text{resonance forms}$$

$$\hspace{5cm} \downarrow H^+$$

$$H_3\overset{+}{N}-CH-\overset{O}{\overset{\|}{C}}OH \quad \xleftarrow{H_2\overset{\cdot\cdot}{O}:} \quad RC\overset{:\overset{+}{O}-H}{\underset{:\overset{\cdot\cdot}{O}-H}{\overset{\|}{}}} \quad \xleftarrow{-EtOH} \quad RC\overset{OH}{\underset{O-H}{\overset{|}{\underset{H}{-}}}}\overset{H}{\overset{\curvearrowleft}{\underset{\cdot\cdot}{O}}}\overset{+}{-}Et$$
$$\hspace{1cm} | $$
$$\hspace{1cm} CH_2Ph$$
$$\text{(protonated form} \hspace{2cm} \text{plus two other}$$
$$\text{in acid solution)} \hspace{2cm} \text{resonance forms}$$

24-17

$$H_3\overset{+}{N}-CH-COO^- \quad \xrightarrow[H^+]{PhCH_2OH} \quad H_3\overset{+}{N}-CH-COOCH_2Ph \quad \xrightarrow{H_2, Pd} \quad H_3\overset{+}{N}-CH-COO^-$$
$$\hspace{1cm} | \hspace{5cm} | \hspace{5cm} |$$
$$\hspace{0.5cm} CH_2CH_2CONH_2 \hspace{3cm} CH_2CH_2CONH_2 \hspace{3cm} CH_2CH_2CONH_2$$
$$\hspace{10.5cm} + \ CH_3Ph$$

24-18

$$H_3\overset{+}{N}-CH-COO^- \quad \xrightarrow{PhCH_2O-\overset{O}{\overset{\|}{C}}-Cl} \quad PhCH_2O\overset{O}{\overset{\|}{C}}-\overset{H}{\overset{|}{N}}-CH-COOH$$
$$\hspace{1cm} | \hspace{7cm} |$$
$$\hspace{0.5cm} CH_2CH_2SCH_3 \hspace{5cm} CH_2CH_2SCH_3$$

$$\hspace{7cm} \downarrow H_2, Pd$$

$$PhCH_3 \ + \ CO_2 \ + \ H_3\overset{+}{N}-CH-COO^-$$
$$\hspace{6cm} |$$
$$\hspace{5.5cm} CH_2CH_2SCH_3$$

24-19

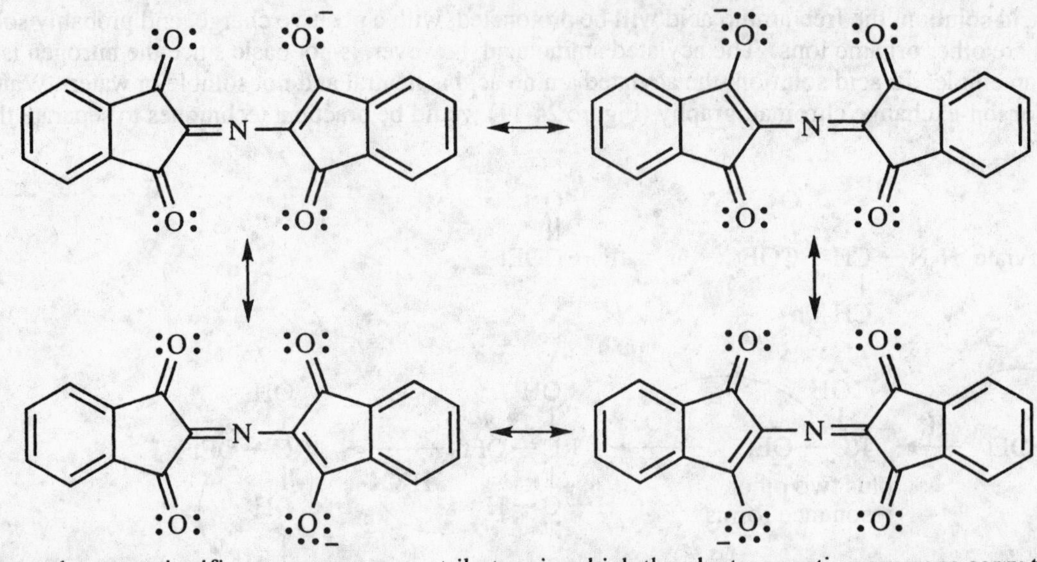

These are the most significant resonance contributors in which the electronegative oxygens carry the negative charge. There are also two other forms in which the negative charge is on the carbons bonded to the nitrogen, plus the usual resonance forms involving the alternate Kekulé structures of the benzene rings.

24-20

(a)

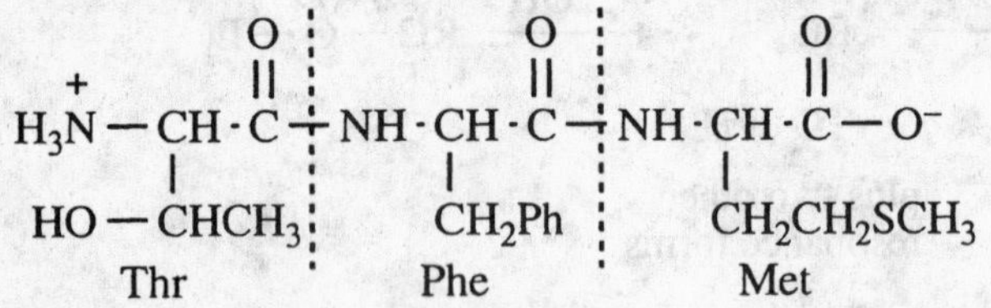

Thr — Phe — Met

(b)

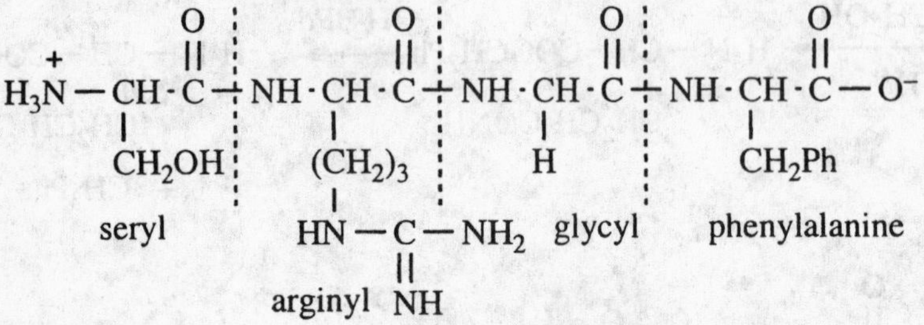

seryl — arginyl — glycyl — phenylalanine

(c)

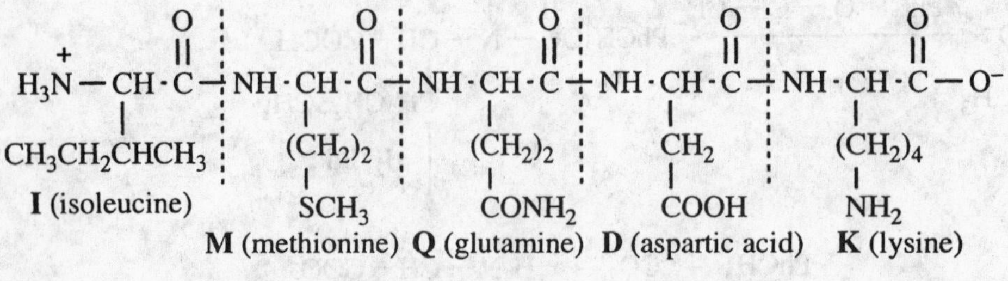

I (isoleucine) — **M** (methionine) — **Q** (glutamine) — **D** (aspartic acid) — **K** (lysine)

(d)

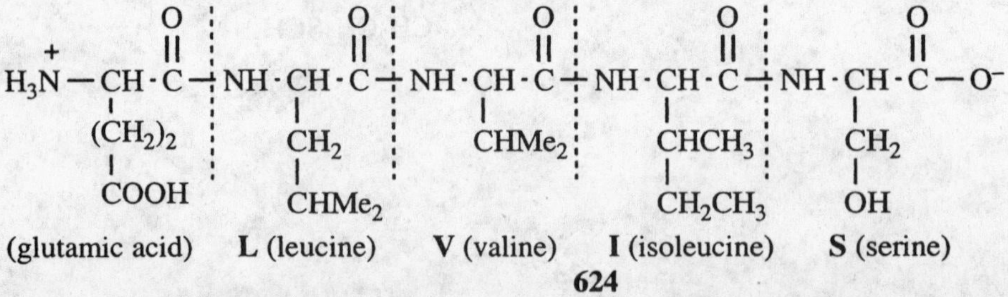

E (glutamic acid) — **L** (leucine) — **V** (valine) — **I** (isoleucine) — **S** (serine)

Try spelling your name in peptides!

24-21

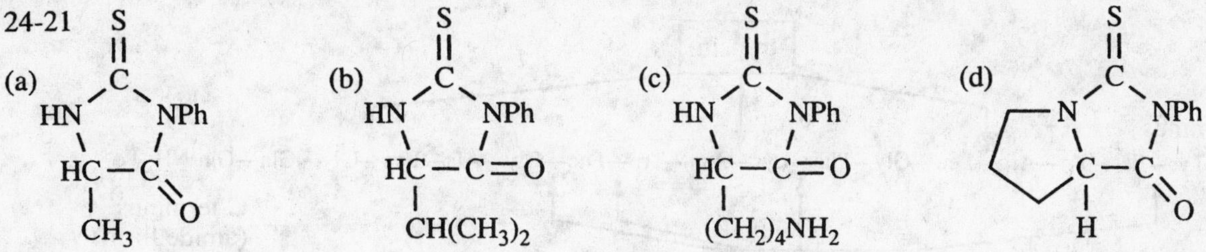

(a)

$$S$$
$$\parallel$$
$$C$$
HN⎯ ⎯NPh
| |
HC⎯C=O
|
CH_3

(b)

$$S$$
$$\parallel$$
$$C$$
HN⎯ ⎯NPh
| |
HC⎯C=O
|
$CH(CH_3)_2$

(c)

$$S$$
$$\parallel$$
$$C$$
HN⎯ ⎯NPh
| |
HC⎯C=O
|
$(CH_2)_4NH_2$

(d)

$$S$$
$$\parallel$$
$$C$$
N⎯ ⎯NPh
          |
C⎯C=O
|
H

24-22

__Step 3__

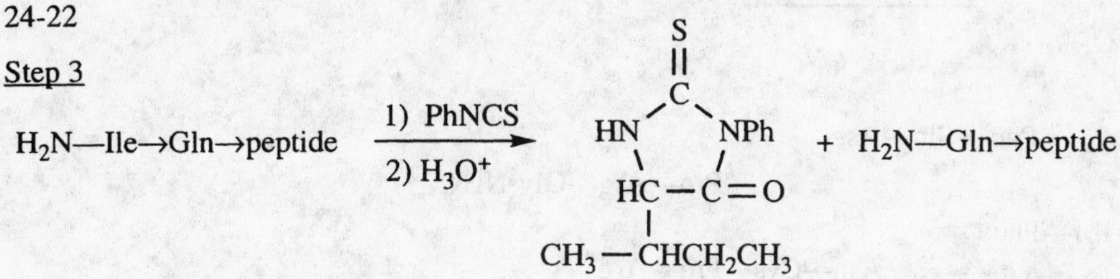

H_2N⎯Ile→Gln→peptide $\xrightarrow[\text{2) } H_3O^+]{\text{1) PhNCS}}$

$$S$$
$$\parallel$$
$$C$$
HN⎯ ⎯NPh
| |
HC⎯C=O
|
CH_3⎯$CHCH_2CH_3$

$+$ H_2N⎯Gln→peptide

__Step 4__

H_2N⎯Gln→peptide $\xrightarrow[\text{2) } H_3O^+]{\text{1) PhNCS}}$

$$S$$
$$\parallel$$
$$C$$
HN⎯ ⎯NPh
| |
HC⎯C=O
|
$CH_2CH_2CONH_2$

$+$ H_2N⎯peptide

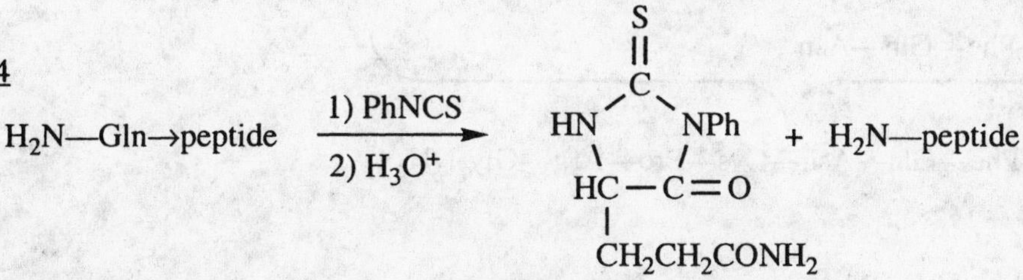

24-23 Abbreviate the N-terminus of the peptide chain as NH_2R .

(a) This is a nucleophilic aromatic substitution by the addition-elimination mechanism. The presence of two nitro groups makes this reaction feasible under mild conditions.

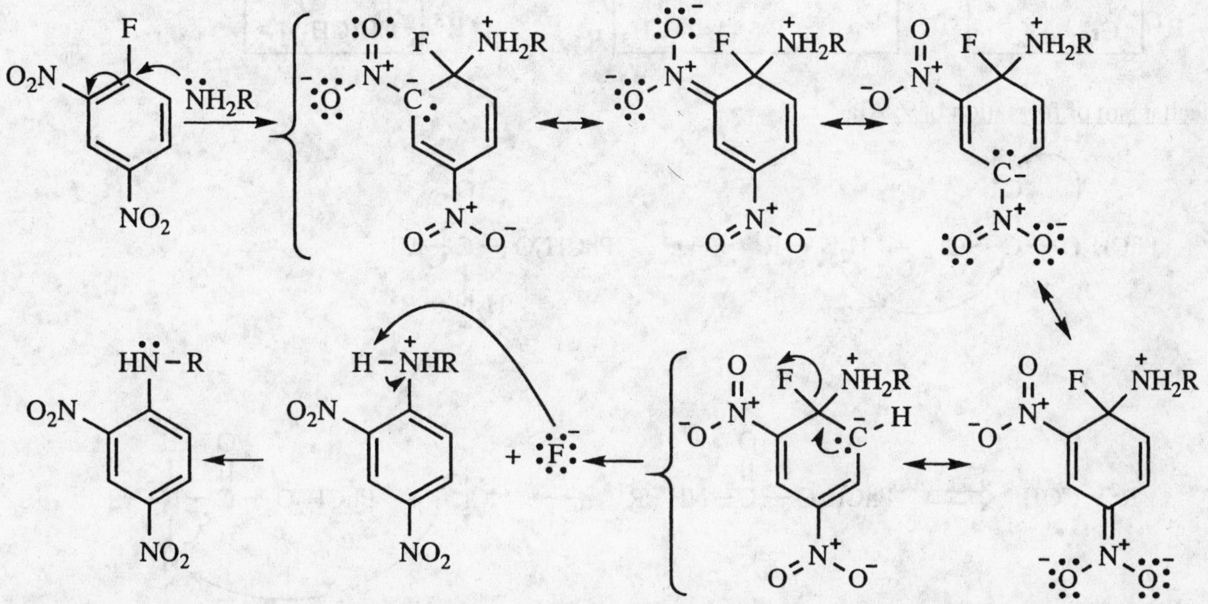

(b) The main drawback of the Sanger method is that only one amino acid is analyzed per sample of protein. The Edman degradation can usually analyze more than 20 amino acids per sample of protein.

24-24

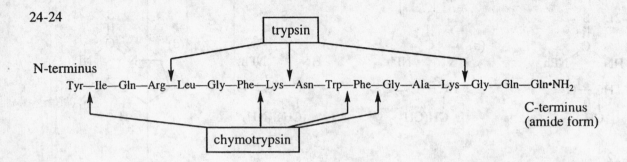

24-25   Phe—Gln—Asn

                                    Pro—Arg—Gly•NH₂

        Cys—Tyr—Phe

                        Asn—Cys—Pro—Arg

            Tyr—Phe—Gln—Asn

Cys—Tyr—Phe—Gln—Asn—Cys—Pro—Arg—Gly•NH₂

24-26 abbreviations used in this problem:

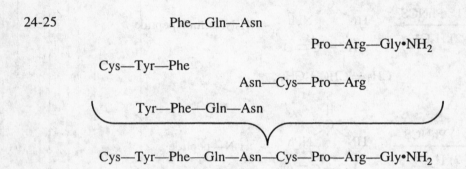

mechanism of formation of Z-Ala

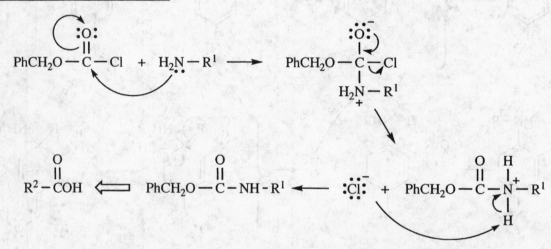

24-26 continued

two possible mechanisms of ethyl chloroformate activation

mechanism 1

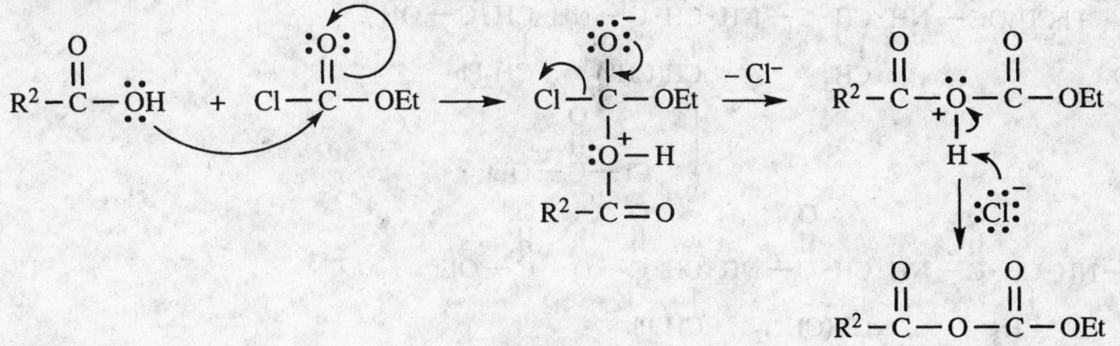
- Cl⁻

mechanism 2

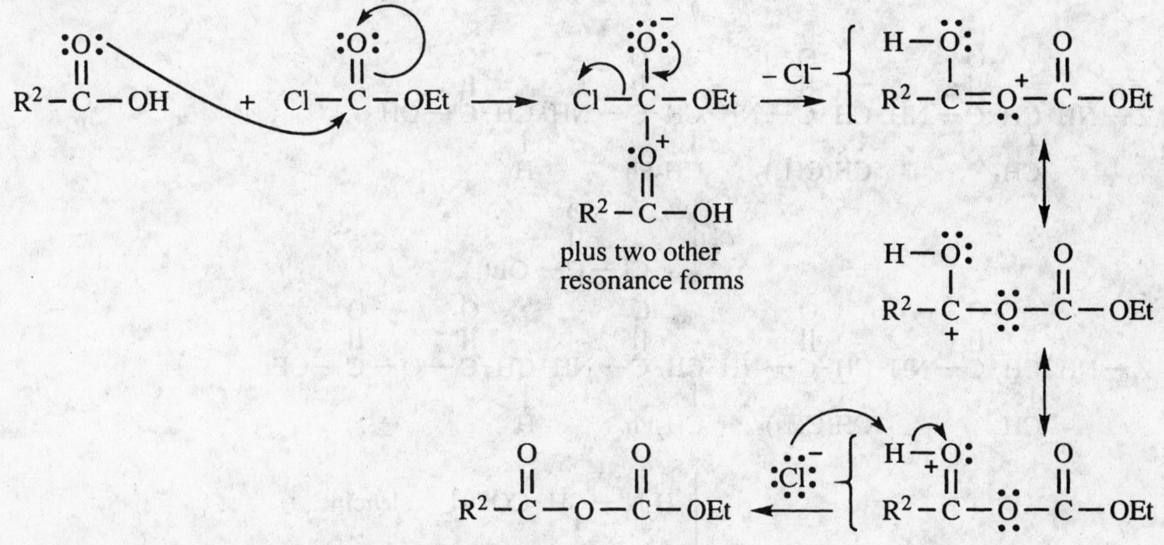
- Cl⁻
plus two other resonance forms

mechanism of the coupling with valine

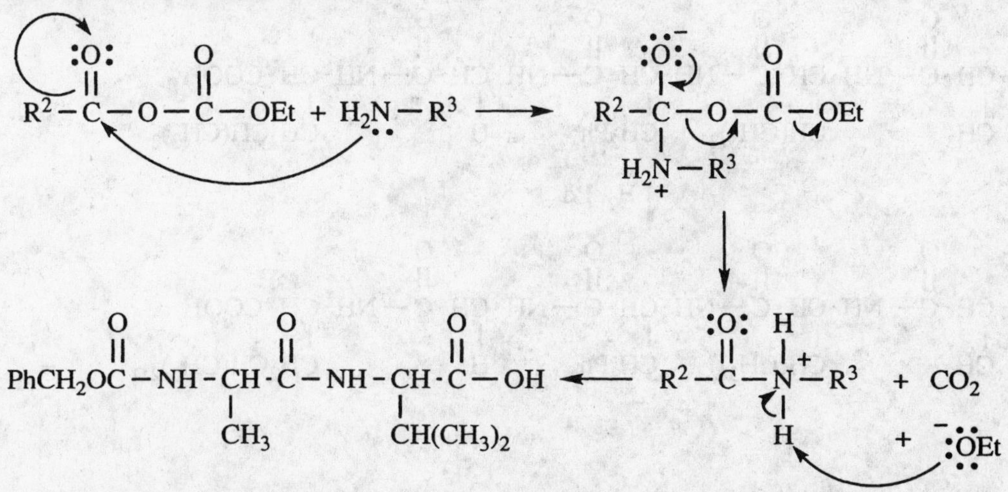

24-27

$$\overbrace{\hspace{2cm}}^{Z} \quad \overbrace{\hspace{2cm}}^{Ala} \quad \overbrace{\hspace{2cm}}^{Val} \quad \overbrace{\hspace{2cm}}^{Phe}$$

$$PhCH_2OC \overset{\overset{\displaystyle O}{\|}}{} - NH \cdot CH \cdot \overset{\overset{\displaystyle O}{\|}}{C} - NH \cdot CH \cdot \overset{\overset{\displaystyle O}{\|}}{C} - NH \cdot CH \cdot \overset{\overset{\displaystyle O}{\|}}{C} - OH$$
$$\underset{CH_3}{|}\underset{CH(CH_3)_2}{|}\underset{CH_2Ph}{|}$$

$$\Big\downarrow \quad Cl-\overset{\overset{\displaystyle O}{\|}}{C}-OEt$$

$$Z-NH \cdot CH \cdot \overset{\overset{\displaystyle O}{\|}}{C} - NH \cdot CH \cdot \overset{\overset{\displaystyle O}{\|}}{C} - NH \cdot CH \cdot \overset{\overset{\displaystyle O}{\|}}{C} - O - \overset{\overset{\displaystyle O}{\|}}{C} - OEt$$
$$\underset{CH_3}{|}\underset{CH(CH_3)_2}{|}\underset{CH_2Ph}{|}$$

$$\Big\downarrow \quad H_2NCH_2COOH \qquad glycine$$

$$Z-NH \cdot CH \cdot \overset{\overset{\displaystyle O}{\|}}{C} - NH \cdot CH \cdot \overset{\overset{\displaystyle O}{\|}}{C} - NH \cdot CH \cdot \overset{\overset{\displaystyle O}{\|}}{C} - NH \cdot CH \cdot \overset{\overset{\displaystyle O}{\|}}{C} - OH$$
$$\underset{CH_3}{|}\underset{CH(CH_3)_2}{|}\underset{CH_2Ph}{|}\underset{H}{|}$$

$$\Big\downarrow \quad Cl-\overset{\overset{\displaystyle O}{\|}}{C}-OEt$$

$$Z-NH \cdot CH \cdot \overset{\overset{\displaystyle O}{\|}}{C} - NH \cdot CH \cdot \overset{\overset{\displaystyle O}{\|}}{C} - NH \cdot CH \cdot \overset{\overset{\displaystyle O}{\|}}{C} - NH \cdot CH \cdot \overset{\overset{\displaystyle O}{\|}}{C} - O - \overset{\overset{\displaystyle O}{\|}}{C} - OEt$$
$$\underset{CH_3}{|}\underset{CH(CH_3)_2}{|}\underset{CH_2Ph}{|}\underset{H}{|}$$

$$\Big\downarrow \quad H_2N - \underset{\underset{CH_2CH(CH_3)_2}{|}}{CH} \cdot COOH \qquad leucine$$

$$Z-NH \cdot CH \cdot \overset{\overset{\displaystyle O}{\|}}{C} - NH \cdot CH \cdot \overset{\overset{\displaystyle O}{\|}}{C} - NH \cdot CH \cdot \overset{\overset{\displaystyle O}{\|}}{C} - NH \cdot CH \cdot \overset{\overset{\displaystyle O}{\|}}{C} - NH \cdot CH \cdot COOH$$
$$\underset{CH_3}{|}\underset{CH(CH_3)_2}{|}\underset{CH_2Ph}{|}\underset{H}{|}\underset{CH_2CH(CH_3)_2}{|}$$

$$\Big\downarrow \quad H_2, Pd$$

$$H_2N - CH \cdot \overset{\overset{\displaystyle O}{\|}}{C} - NH \cdot CH \cdot \overset{\overset{\displaystyle O}{\|}}{C} - NH \cdot CH \cdot \overset{\overset{\displaystyle O}{\|}}{C} - NH \cdot CH \cdot \overset{\overset{\displaystyle O}{\|}}{C} - NH \cdot CH \cdot COOH$$
$$\underset{CH_3}{|}\underset{CH(CH_3)_2}{|}\underset{CH_2Ph}{|}\underset{H}{|}\underset{CH_2CH(CH_3)_2}{|}$$

24-28

$$\underset{\text{(PhCH}_2\text{OC—Cl)}}{\overset{\overset{\displaystyle O}{\|}}{\text{PhCH}_2\text{OC—Cl}}} \;+\; \underset{\underset{\displaystyle \text{CH}_3-\text{CHCH}_2\text{CH}_3}{|}}{\text{H}_2\text{N}-\text{CH}\cdot\text{COOH}} \;\longrightarrow\; \underset{\underset{\displaystyle \text{CH}_3-\text{CHCH}_2\text{CH}_3}{|}}{\text{Z}-\text{NH}\cdot\text{CH}-\overset{\overset{\displaystyle O}{\|}}{\text{C}}-\text{OH}}$$

isoleucine

$$\Big\downarrow \;\; \text{Cl}-\overset{\overset{\displaystyle O}{\|}}{\text{C}}-\text{OEt}$$

$$\underset{\underset{\displaystyle \text{CH}_3-\text{CHCH}_2\text{CH}_3}{|}}{\text{Z}-\text{NH}\cdot\text{CH}-\overset{\overset{\displaystyle O}{\|}}{\text{C}}-\text{O}-\overset{\overset{\displaystyle O}{\|}}{\text{C}}-\text{OEt}}$$

$$\Big\downarrow \;\; \text{H}_2\text{NCH}_2\text{COOH} \qquad \text{glycine}$$

$$\underset{\underset{\displaystyle \text{CH}_3-\text{CHCH}_2\text{CH}_3 \quad\;\; \text{H}}{| \qquad\qquad |}}{\text{Z}-\text{NH}\cdot\text{CH}-\overset{\overset{\displaystyle O}{\|}}{\text{C}}-\text{NH}\cdot\text{CH}-\overset{\overset{\displaystyle O}{\|}}{\text{C}}-\text{OH}}$$

$$\Big\downarrow \;\; \text{Cl}-\overset{\overset{\displaystyle O}{\|}}{\text{C}}-\text{OEt}$$

$$\underset{\underset{\displaystyle \text{CH}_3-\text{CHCH}_2\text{CH}_3 \quad\;\; \text{H}}{| \qquad\qquad |}}{\text{Z}-\text{NH}\cdot\text{CH}-\overset{\overset{\displaystyle O}{\|}}{\text{C}}-\text{NH}\cdot\text{CH}-\overset{\overset{\displaystyle O}{\|}}{\text{C}}-\text{O}-\overset{\overset{\displaystyle O}{\|}}{\text{C}}-\text{OEt}}$$

$$\Big\downarrow \;\; \underset{\underset{\displaystyle \text{CH}_2\text{CONH}_2}{|}}{\text{H}_2\text{N}-\text{CH}\cdot\text{COOH}} \qquad \text{asparagine}$$

$$\underset{\underset{\displaystyle \text{CH}_3-\text{CHCH}_2\text{CH}_3 \quad\;\; \text{H} \qquad\quad \text{CH}_2\text{CONH}_2}{| \qquad\qquad | \qquad\qquad |}}{\text{Z}-\text{NH}\cdot\text{CH}-\overset{\overset{\displaystyle O}{\|}}{\text{C}}-\text{NH}\cdot\text{CH}-\overset{\overset{\displaystyle O}{\|}}{\text{C}}-\text{NH}\cdot\text{CH}\cdot\text{COOH}}$$

$$\Big\downarrow \;\; \text{H}_2,\,\text{Pd}$$

$$\underset{\underset{\displaystyle \text{CH}_3-\text{CHCH}_2\text{CH}_3 \quad\;\; \text{H} \qquad\quad \text{CH}_2\text{CONH}_2}{| \qquad\qquad | \qquad\qquad |}}{\text{H}_2\text{N}-\text{CH}-\overset{\overset{\displaystyle O}{\|}}{\text{C}}-\text{NH}\cdot\text{CH}-\overset{\overset{\displaystyle O}{\|}}{\text{C}}-\text{NH}\cdot\text{CH}\cdot\text{COOH}}$$

24-29 In this problem, "Cy" stands for "cyclohexyl".

$$\text{Cy}-\text{N}=\text{C}=\text{N}-\text{Cy} \Longrightarrow \text{Cy}-\text{N}=\text{C}=\text{N}-\text{Cy}$$
DCC

<u>mechanism</u>

$$\underset{\text{O}}{\overset{\parallel}{\text{CH}_3\text{C}}}-\text{O}-\text{H} + \text{H}_2\ddot{\text{N}}-\text{Ph} \longrightarrow \underset{\text{O}}{\overset{\parallel}{\text{CH}_3\text{C}}}-\ddot{\underset{..}{\text{O}}}:^- + \text{H}_3\overset{+}{\text{N}}-\text{Ph}$$

Cy—N=C=N—Cy

$$\left\{ \begin{array}{c} \underset{\text{O}}{\overset{\parallel}{\text{CH}_3\text{C}}}-\text{O}-\underset{\overset{\|}{\ddot{\text{N}}-\text{Cy}}}{\overset{:\ddot{\text{N}}^--\text{Cy}}{\text{C}}} \longleftrightarrow \underset{\text{O}}{\overset{\parallel}{\text{CH}_3\text{C}}}-\text{O}-\underset{\overset{|}{\ddot{\underset{..}{\text{N}}}^--\text{Cy}}}{\overset{\ddot{\text{N}}-\text{Cy}}{\text{C}}} \end{array} \right\}$$

plus other resonance forms

$$\underset{\overset{|}{\overset{+}{\text{H}_2\text{N}}-\text{Ph}}}{\overset{\text{H}}{}}$$

$$\underset{\text{CH}_3\text{C}}{\overset{:\ddot{\text{O}}:}{\parallel}}-\text{O}-\underset{\overset{\|}{\text{N}-\text{Cy}}}{\overset{\text{HN}-\text{Cy}}{\text{C}}} \xrightarrow{\text{H}_2\ddot{\text{N}}-\text{Ph}} \underset{\overset{|}{\overset{+}{\text{H}_2\text{N}}-\text{Ph}}}{\overset{:\ddot{\text{O}}:^-}{\underset{|}{\text{CH}_3\text{C}}}}-\text{O}-\underset{\overset{\|}{\text{N}-\text{Cy}}}{\overset{\text{HN}-\text{Cy}}{\text{C}}}$$

$$\underset{\text{H}}{\overset{\text{O}}{\underset{|}{\text{CH}_3\text{C}}}\overset{\parallel}{-}\overset{+}{\text{N}}\text{HPh}} + \underset{\text{resonance-stabilized}}{\text{Cy}-\overset{-}{\ddot{\underset{..}{\text{N}}}}-\underset{\overset{\parallel}{\text{O}}}{\text{C}}-\text{NH-Cy}}$$

$$\underset{\text{O}}{\overset{\parallel}{\text{CH}_3\text{C}}}-\text{NHPh} + \underset{\text{O}}{\text{Cy}-\text{NH}-\overset{\parallel}{\text{C}}-\text{NH}-\text{Cy}}$$
DCU

24-30

$\overbrace{\text{Boc}}$ $\overbrace{\text{Ala}}$ $\overbrace{\text{Val}}$ $\overbrace{\text{Phe}}$

$$\underset{\text{CH}_3}{\text{Me}_3\text{COC}-\text{NH}\cdot\text{CH}\cdot\overset{\text{O}}{\text{C}}}-\underset{\text{CH(CH}_3)_2}{\text{NH}\cdot\text{CH}\cdot\overset{\text{O}}{\text{C}}}-\underset{\text{CH}_2\text{Ph}}{\text{NH}\cdot\text{CH}\cdot\overset{\text{O}}{\text{C}}}-\text{O}-\text{P}$$

CF$_3$COOH

$$\overset{+}{\text{H}_3\text{N}}-\underset{\text{CH}_3}{\text{CH}\cdot\overset{\text{O}}{\text{C}}}-\underset{\text{CH(CH}_3)_2}{\text{NH}\cdot\text{CH}\cdot\overset{\text{O}}{\text{C}}}-\underset{\text{CH}_2\text{Ph}}{\text{NH}\cdot\text{CH}\cdot\overset{\text{O}}{\text{C}}}-\text{O}-\text{P}$$

DCC | Me$_3$COC$-$NHCH$_2$COOH   Boc—glycine

$$\text{Me}_3\text{COC}-\underset{\text{H}}{\text{NH}\cdot\text{CH}\cdot\overset{\text{O}}{\text{C}}}-\underset{\text{CH}_3}{\text{NH}\cdot\text{CH}\cdot\overset{\text{O}}{\text{C}}}-\underset{\text{CH(CH}_3)_2}{\text{NH}\cdot\text{CH}\cdot\overset{\text{O}}{\text{C}}}-\underset{\text{CH}_2\text{Ph}}{\text{NH}\cdot\text{CH}\cdot\overset{\text{O}}{\text{C}}}-\text{O}-\text{P}$$

CF$_3$COOH

$$\overset{+}{\text{H}_3\text{N}}-\underset{\text{H}}{\text{CH}\cdot\overset{\text{O}}{\text{C}}}-\underset{\text{CH}_3}{\text{NH}\cdot\text{CH}\cdot\overset{\text{O}}{\text{C}}}-\underset{\text{CH(CH}_3)_2}{\text{NH}\cdot\text{CH}\cdot\overset{\text{O}}{\text{C}}}-\underset{\text{CH}_2\text{Ph}}{\text{NH}\cdot\text{CH}\cdot\overset{\text{O}}{\text{C}}}-\text{O}-\text{P}$$

DCC | Me$_3$COC$-$NH·CH·COOH   Boc—leucine
                         CH$_2$CHMe$_2$

$$\text{Boc NH}\cdot\underset{\text{CH}_2\text{CHMe}_2}{\text{CH}\cdot\overset{\text{O}}{\text{C}}}-\underset{\text{H}}{\text{NH}\cdot\text{CH}\cdot\overset{\text{O}}{\text{C}}}-\underset{\text{CH}_3}{\text{NH}\cdot\text{CH}\cdot\overset{\text{O}}{\text{C}}}-\underset{\text{CH(CH}_3)_2}{\text{NH}\cdot\text{CH}\cdot\overset{\text{O}}{\text{C}}}-\underset{\text{CH}_2\text{Ph}}{\text{NH}\cdot\text{CH}\cdot\overset{\text{O}}{\text{C}}}-\text{O}-\text{P}$$

HF

$$\overset{+}{\text{H}_3\text{N}}-\underset{\text{CH}_2\text{CHMe}_2}{\text{CH}\cdot\overset{\text{O}}{\text{C}}}-\underset{\text{H}}{\text{NH}\cdot\text{CH}\cdot\overset{\text{O}}{\text{C}}}-\underset{\text{CH}_3}{\text{NH}\cdot\text{CH}\cdot\overset{\text{O}}{\text{C}}}-\underset{\text{CH(CH}_3)_2}{\text{NH}\cdot\text{CH}\cdot\overset{\text{O}}{\text{C}}}-\underset{\text{CH}_2\text{Ph}}{\text{NH}\cdot\text{CH}\cdot\overset{\text{O}}{\text{C}}}-\text{OH}$$

24-31

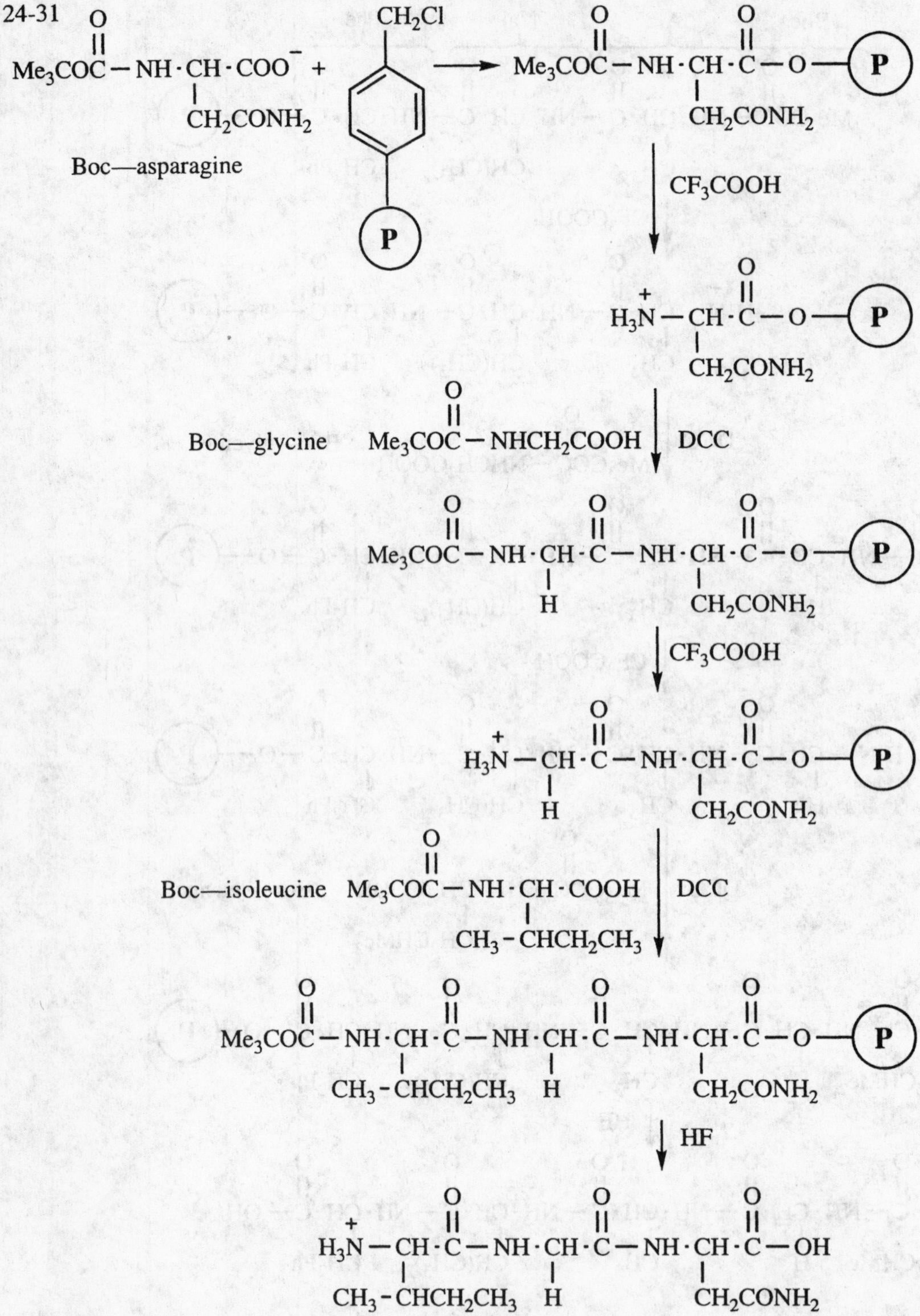

Me₃COC—NH·CH·COO⁻ + [CH₂Cl—benzene—P] ⟶ Me₃COC—NH·CH·C—O—P
         | |
      CH₂CONH₂   CH₂CONH₂

Boc—asparagine

│ CF₃COOH
▼

$\overset{+}{H_3N}$—CH·C—O—P
        |
     CH₂CONH₂

Boc—glycine   Me₃COC—NHCH₂COOH │ DCC
                                ▼

Me₃COC—NH·CH·C—NH·CH·C—O—P
           | |
           H   CH₂CONH₂

│ CF₃COOH
▼

$\overset{+}{H_3N}$—CH·C—NH·CH·C—O—P
        | |
        H   CH₂CONH₂

Boc—isoleucine   Me₃COC—NH·CH·COOH │ DCC
                          |
                      CH₃—CHCH₂CH₃
                                  ▼

Me₃COC—NH·CH·C—NH·CH·C—NH·CH·C—O—P
           | | |
      CH₃—CHCH₂CH₃ H   CH₂CONH₂

│ HF
▼

$\overset{+}{H_3N}$—CH·C—NH·CH·C—NH·CH·C—OH
        | | |
   CH₃—CHCH₂CH₃ H   CH₂CONH₂

24-32 Please refer to solution 1-20, page 12 of this Solutions Manual.

24-33

$$H_2N-CH\cdot\overset{\overset{\displaystyle O}{\|}}{C}-NH\cdot CH\cdot\overset{\overset{\displaystyle O}{\|}}{C}-NH\cdot CH\cdot\overset{\overset{\displaystyle O}{\|}}{C}-NH_2$$

with side chains: CH_2OH ; $(CH_2)_2$ / $CONH_2$; $CH_2CH_2SCH_3$

24-34

(a)

+ CO_2 + (isobutyraldehyde) CHO

(b)

$$H_2N-\underset{\underset{\displaystyle CH_3}{|}}{CH}COOH$$

(c)

$$CH_3\overset{\overset{\displaystyle O}{\|}}{C}-NH\cdot CH\cdot COOH$$
$$(CH_2)_4NHCOCH_3$$

(d)

L-proline + N-acetyl-D-proline

the enzyme does not recognize D amino acids

(e)

$$H_2N-CH\cdot CN$$
$$H_3C-CHCH_2CH_3$$

(f)

$$H_2N-CH\cdot COOH$$
$$H_3C-CHCH_2CH_3$$
isoleucine

(g)

$$Br-CH\cdot\overset{\overset{\displaystyle O}{\|}}{C}-Br$$
$$CH_2CH(CH_3)_2$$

with a water workup, the acid bromide would become COOH

(h)

$$H_2N-CH\cdot COOH \quad OR \quad H_2N-CH\cdot\overset{\overset{\displaystyle O}{\|}}{C}-NH_2$$
$$CH_2CH(CH_3)_2 \qquad\qquad\quad CH_2CH(CH_3)_2$$
leucine

another possible answer, depending on whether the acid bromide is hydrolyzed before adding ammonia

24-35

(a)

$$\overset{\overset{\displaystyle O}{\|}}{C}-COOH \xrightarrow[H_2,\ Pd]{NH_3} H_2N-CH\cdot COOH \quad valine$$
$$CH(CH_3)_2 \qquad\qquad CH(CH_3)_2$$

(b)

$$H_2C-COOH \xrightarrow[PBr_3]{Br_2} \xrightarrow{H_2O} Br-CH\cdot COOH \xrightarrow[\quad]{\overset{excess}{NH_3}} H_2N-CH\cdot COOH$$
$$H_3C-CHCH_2CH_3 \qquad\qquad H_3C-CHCH_2CH_3 \qquad\qquad H_3C-CHCH_2CH_3$$
isoleucine

633

24-35 continued

(c)

$$\underset{\underset{CH_2CH(CH_3)_2}{|}}{\overset{\overset{O}{\|}}{C-H}} \xrightarrow[\text{H}_2\text{O}]{\text{NH}_3,\ \text{HCN}} \underset{\underset{CH_2CH(CH_3)_2}{|}}{H_2N-CH\cdot CN} \xrightarrow[\Delta]{H_3O^+} \underset{\underset{CH_2CH(CH_3)_2}{|}}{H_2N-CH\cdot COOH}$$

leucine

(d)

phenylalanine

24-36

(a)

$$\underset{\underset{CH_3}{|}}{H_2N-CH\cdot COOH} + HOCH(CH_3)_2 \xrightarrow{H^+} \underset{\underset{CH_3}{|}}{H_2N-CH\cdot COOCH(CH_3)_2}$$

(b)

$$\underset{\underset{CH_3}{|}}{H_2N-CH\cdot COOH} + \overset{\overset{O}{\|}}{PhC}-Cl \xrightarrow{\text{(pyridine)}} \overset{\overset{O}{\|}}{PhC}-NH\cdot\underset{\underset{CH_3}{|}}{CH\cdot COOH}$$

(c)

$$\underset{\underset{CH_3}{|}}{H_2N-CH\cdot COOH} + \overset{\overset{O}{\|}}{PhCH_2OC}-Cl \longrightarrow \overset{\overset{O}{\|}}{PhCH_2OC}-NH-\underset{\underset{CH_3}{|}}{CH\cdot COOH}$$

(d)

$$\underset{\underset{CH_3}{|}}{H_2N-CH\cdot COOH} + \overset{\overset{O}{\|}}{Me_3COC}-O-\overset{\overset{O}{\|}}{COCMe_3} \longrightarrow \overset{\overset{O}{\|}}{Me_3COC}-NH-\underset{\underset{CH_3}{|}}{CH\cdot COOH}$$

24-37

$$\underset{\underset{CH_3}{|}}{\overset{\overset{COOH}{|}}{HO\blacktriangleright C\blacktriangleleft H}} \xrightarrow[\text{pyridine}]{\text{TsCl}} \underset{\underset{CH_3}{|}}{\overset{\overset{COOH}{|}}{TsO\blacktriangleright C\blacktriangleleft H}} \xrightarrow[\text{NH}_3]{\text{excess}} \underset{\underset{CH_3}{|}}{\overset{\overset{COOH}{|}}{H\blacktriangleright C\blacktriangleleft NH_2}}$$

D-alanine

24-38

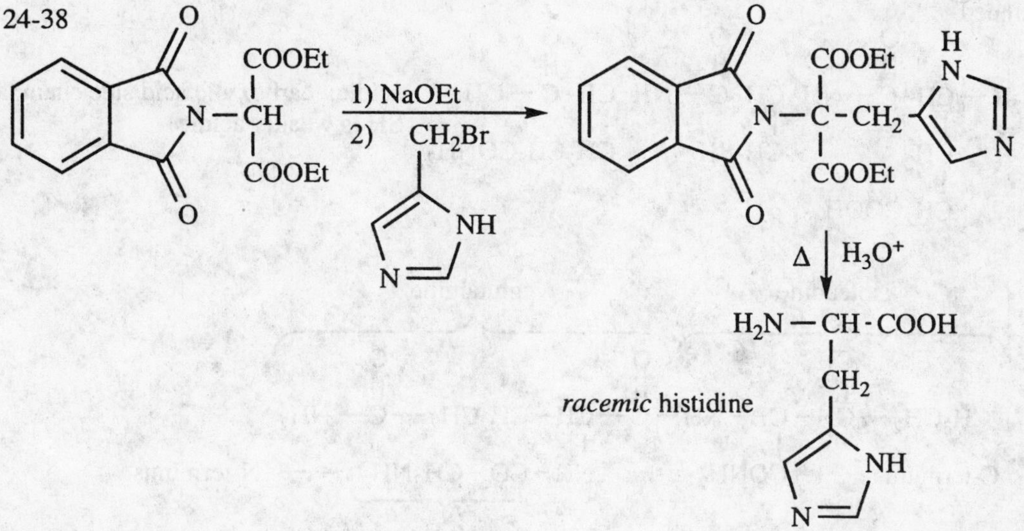

1) NaOEt
2) CH₂Br (imidazole)

Δ | H₃O⁺

$H_2N-CH \cdot COOH$

racemic histidine

24-39

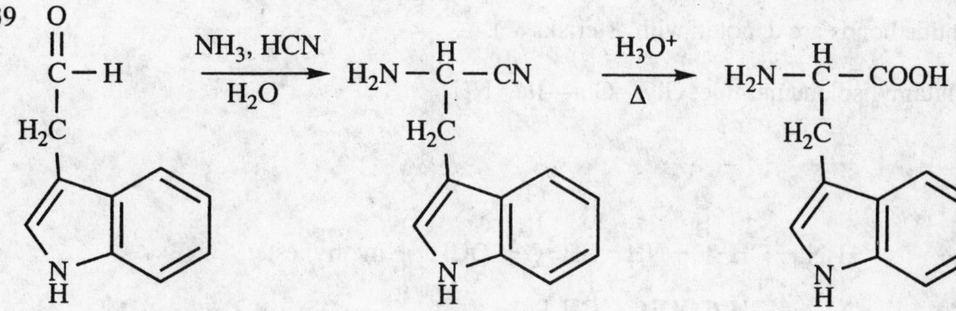

$\xrightarrow[H_2O]{NH_3,\ HCN}$

$H_2N-\overset{H}{\underset{H_2C}{C}}-CN$

$\xrightarrow[\Delta]{H_3O^+}$

$H_2N-\overset{H}{\underset{H_2C}{C}}-COOH$

racemic tryptophan

24-40

(a) $H_2N-CH\cdot C-NH\cdot CH\cdot C-OH$ neutral

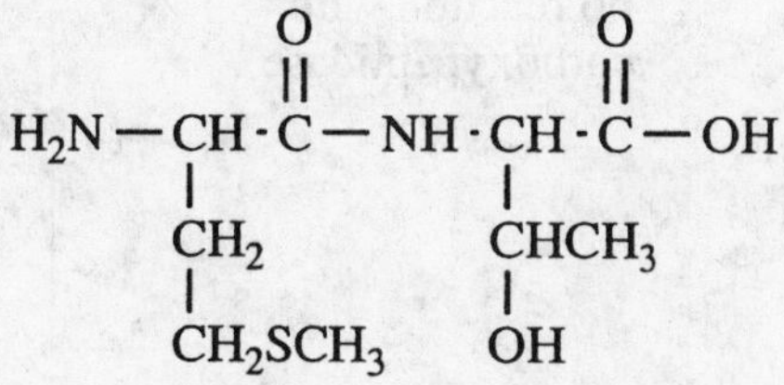

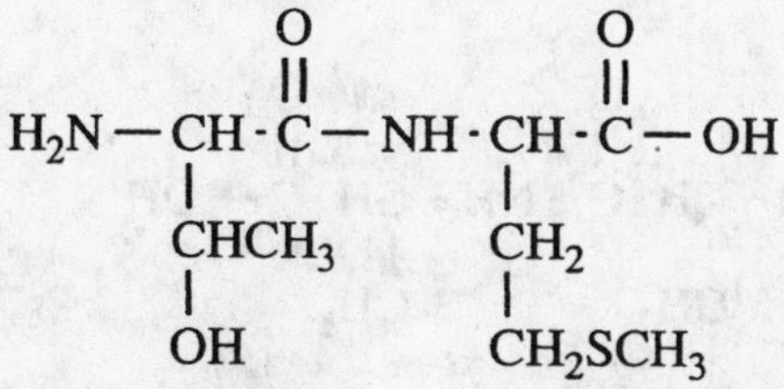

CH_2 $CHCH_3$

CH_2SCH_3 OH

(b) $H_2N-CH\cdot C-NH\cdot CH\cdot C-OH$ neutral

$CHCH_3$ CH_2

OH CH_2SCH_3

(c) $H_2N-CH\cdot C-NH\cdot CH\cdot C-NH\cdot CH\cdot C-OH$ basic: two basic side chains and one acidic side chain

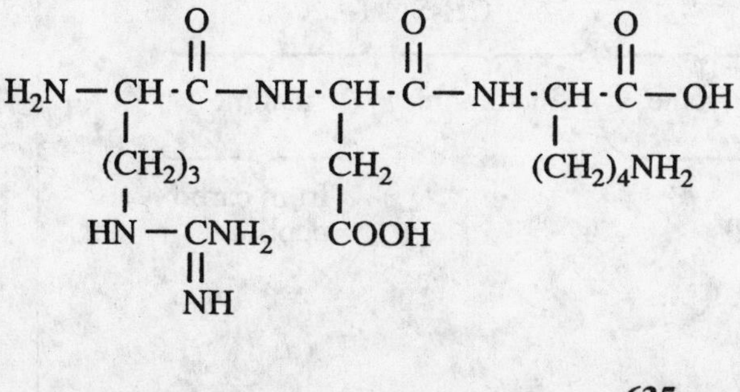

$(CH_2)_3$ CH_2 $(CH_2)_4NH_2$

$HN-CNH_2$ $COOH$

NH

635

24-40 continued

(d)

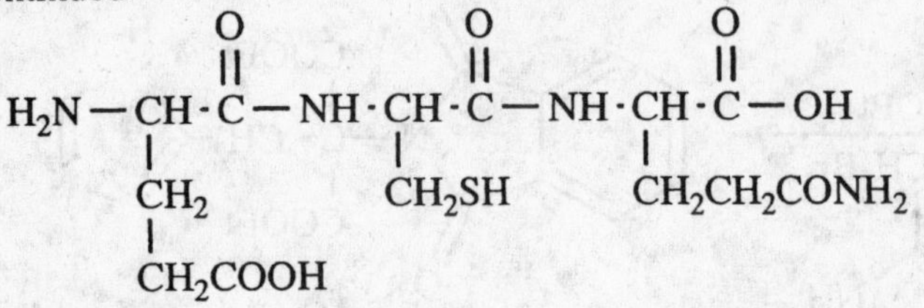

H₂N—CH·C(=O)—NH·CH·C(=O)—NH·CH·C(=O)—OH   acidic: carboxylic acid side chain, and the
     | | | SH is weakly acidic
     CH₂   CH₂SH   CH₂CH₂CONH₂
     |
     CH₂COOH

The side chains are: CH₂CH₂COOH → CH₂CH₂COOH

24-41

(a), (b), (c)

isoleucine ⎡ glutamine ⎤

CH₃CH₂—CH—CH—NH—*C(=O)—CH—CH₂CH₂—C(=O)—NH₂

with CH₃ branch on the isoleucine carbon; C-terminus → CONH₂; NH—*CO—CH₂NH₂ ← N-terminus (glycine)

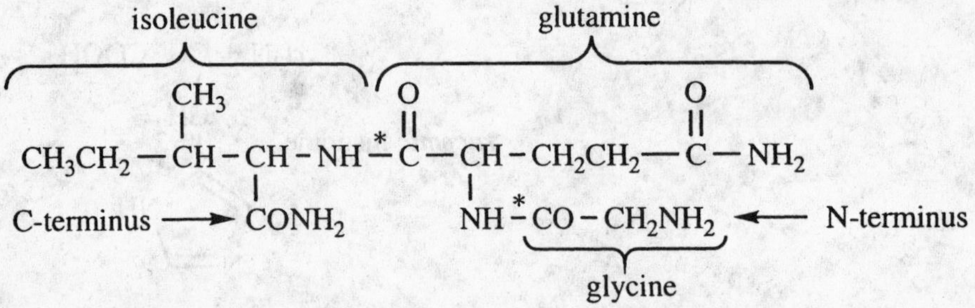

Peptide bonds are denoted with asterisks (*).

(d) glycylglutamylisoleucinamide; Gly—Gln—Ile • NH₂

24-42

Aspartame: H₂N—CH·C(=O)—NH—CH·C(=O)—OCH₃ ⎫ methyl ester
           | |
         CH₂COOH   CH₂Ph

aspartic acid (from Edman degradation) | phenylalanine

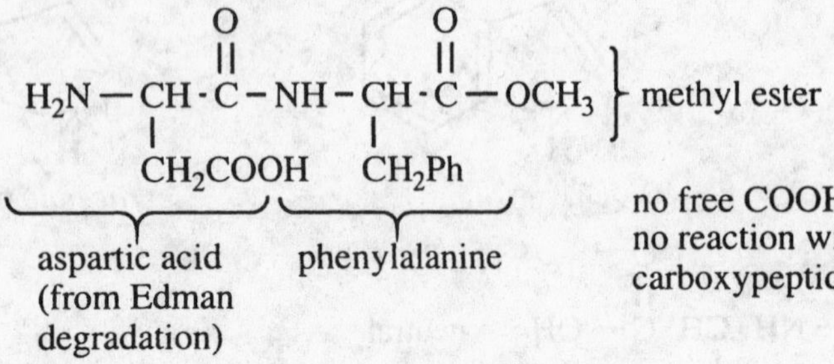

no free COOH ⇒ no reaction with carboxypeptidase

Aspartame is aspartylphenylalanine methyl ester.

24-43

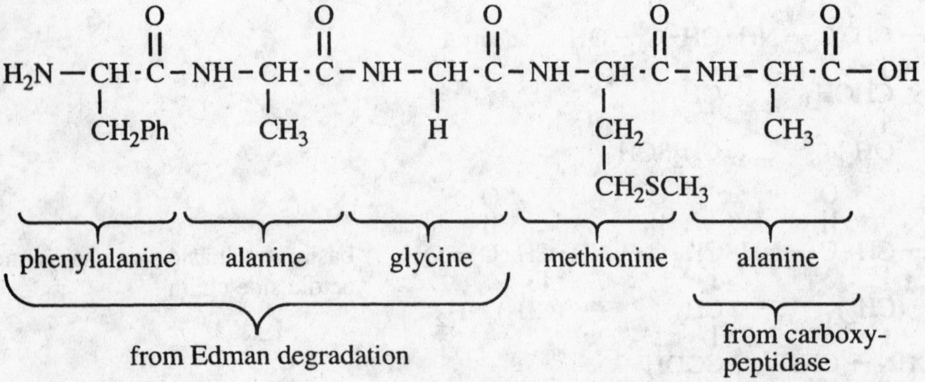

H₂N—CH·C(=O)—NH—CH·C(=O)—NH—CH·C(=O)—NH—CH·C(=O)—NH—CH·C(=O)—OH
    | | | | |
   CH₂Ph   CH₃   H   CH₂   CH₃
                            |
                          CH₂SCH₃

phenylalanine | alanine | glycine | methionine | alanine

from Edman degradation   from carboxy-peptidase

636

24-44

(a)   protect the N-terminus of the first amino acid

$$PhCH_2O\overset{\overset{\displaystyle O}{\|}}{C}-Cl \;+\; H_2N-\underset{\underset{\displaystyle CH_2CHMe_2}{|}}{CH}\cdot COOH \;\longrightarrow\; Z-NH\cdot\underset{\underset{\displaystyle CH_2CHMe_2}{|}}{CH}\cdot\overset{\overset{\displaystyle O}{\|}}{C}-OH$$

leucine

activate the C-terminus with $Cl-\overset{\overset{\displaystyle O}{\|}}{C}-OEt$

$$Z-NH\cdot\underset{\underset{\displaystyle CH_2CHMe_2}{|}}{CH}\cdot\overset{\overset{\displaystyle O}{\|}}{C}-O-\overset{\overset{\displaystyle O}{\|}}{C}-OEt$$

add the next amino acid $H_2N-\underset{\underset{\displaystyle CH_3}{|}}{CH}\cdot COOH$ alanine

$$Z-NH\cdot\underset{\underset{\displaystyle CH_2CHMe_2}{|}}{CH}\cdot\overset{\overset{\displaystyle O}{\|}}{C}-NH\cdot\underset{\underset{\displaystyle CH_3}{|}}{CH}\cdot\overset{\overset{\displaystyle O}{\|}}{C}-OH$$

activate the C-terminus $Cl-\overset{\overset{\displaystyle O}{\|}}{C}-OEt$

$$Z-NH\cdot\underset{\underset{\displaystyle CH_2CHMe_2}{|}}{CH}\cdot\overset{\overset{\displaystyle O}{\|}}{C}-NH\cdot\underset{\underset{\displaystyle CH_3}{|}}{CH}\cdot\overset{\overset{\displaystyle O}{\|}}{C}-O-\overset{\overset{\displaystyle O}{\|}}{C}-OEt$$

add the next amino acid $H_2N-\underset{\underset{\displaystyle CH_2Ph}{|}}{CH}\cdot COOH$ phenylalanine

$$Z-NH\cdot\underset{\underset{\displaystyle CH_2CHMe_2}{|}}{CH}\cdot\overset{\overset{\displaystyle O}{\|}}{C}-NH\cdot\underset{\underset{\displaystyle CH_3}{|}}{CH}\cdot\overset{\overset{\displaystyle O}{\|}}{C}-NH\cdot\underset{\underset{\displaystyle CH_2Ph}{|}}{CH}\cdot COOH$$

deprotect the N-terminus H_2, Pd

$$H_2N-\underset{\underset{\displaystyle CH_2CHMe_2}{|}}{CH}\cdot\overset{\overset{\displaystyle O}{\|}}{C}-NH\cdot\underset{\underset{\displaystyle CH_3}{|}}{CH}\cdot\overset{\overset{\displaystyle O}{\|}}{C}-NH\cdot\underset{\underset{\displaystyle CH_2Ph}{|}}{CH}\cdot COOH$$

24-44 continued
(b)

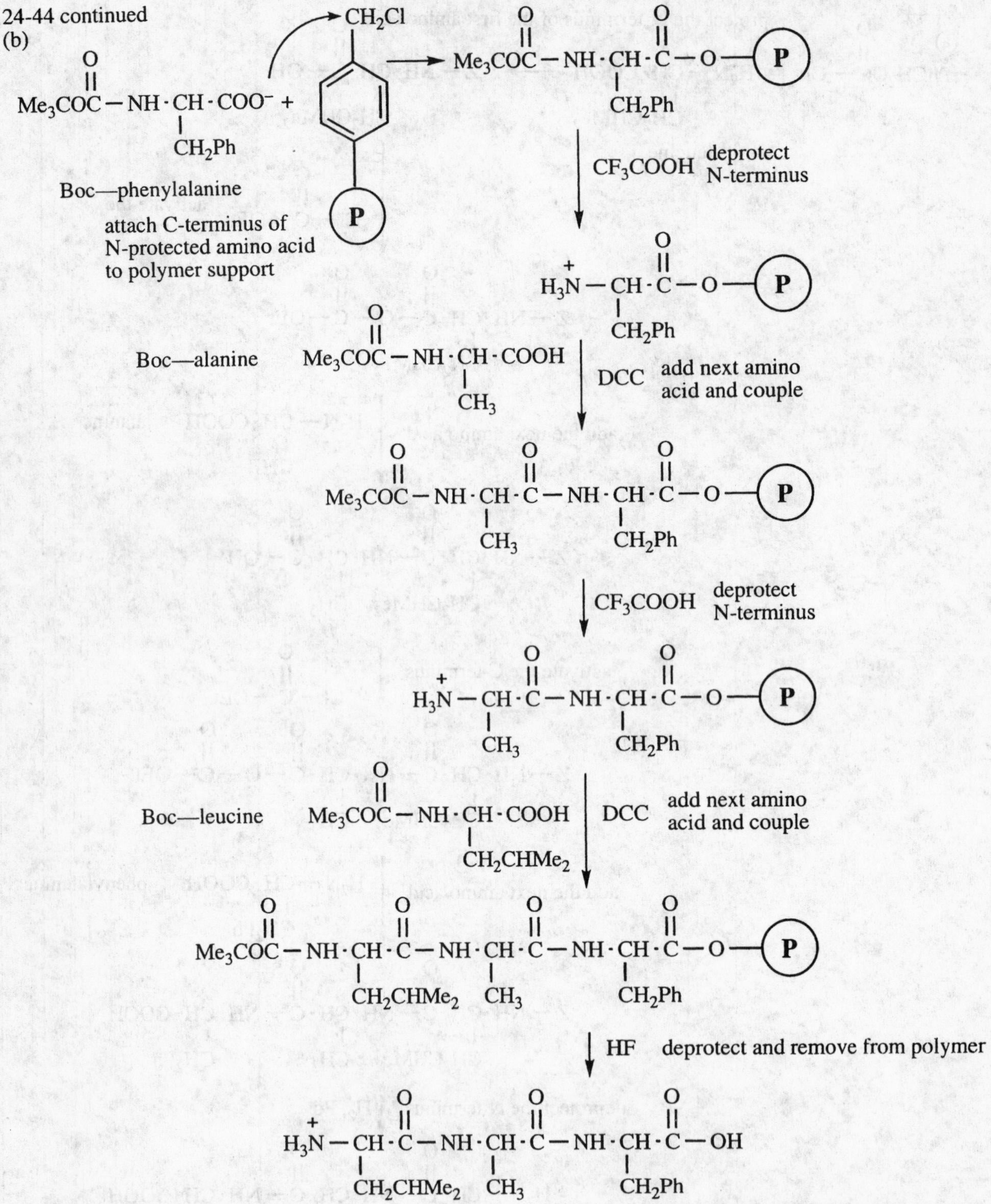

Me_3COC—$NH \cdot CH \cdot COO^-$ + [CH$_2$Cl on benzene ring with P] → Me_3COC—$NH \cdot CH \cdot C$—O—P

(with CH$_2$Ph substituents)

Boc—phenylalanine

attach C-terminus of
N-protected amino acid
to polymer support

CF$_3$COOH   deprotect N-terminus

$H_3\overset{+}{N}$—$CH \cdot C$—O—P
(CH$_2$Ph)

Boc—alanine Me_3COC—$NH \cdot CH \cdot COOH$ (CH$_3$)

DCC   add next amino acid and couple

Me_3COC—$NH \cdot CH \cdot C$—$NH \cdot CH \cdot C$—O—P
(CH$_3$) (CH$_2$Ph)

CF$_3$COOH   deprotect N-terminus

$H_3\overset{+}{N}$—$CH \cdot C$—$NH \cdot CH \cdot C$—O—P
(CH$_3$) (CH$_2$Ph)

Boc—leucine Me_3COC—$NH \cdot CH \cdot COOH$ (CH$_2$CHMe$_2$)

DCC   add next amino acid and couple

Me_3COC—$NH \cdot CH \cdot C$—$NH \cdot CH \cdot C$—$NH \cdot CH \cdot C$—O—P
(CH$_2$CHMe$_2$) (CH$_3$) (CH$_2$Ph)

HF   deprotect and remove from polymer

$H_3\overset{+}{N}$—$CH \cdot C$—$NH \cdot CH \cdot C$—$NH \cdot CH \cdot C$—OH
(CH$_2$CHMe$_2$) (CH$_3$) (CH$_2$Ph)

24-45

protect N-terminus

$$\text{PhCH}_2\text{O}\overset{\displaystyle O}{\overset{\|}{\text{C}}}-\text{Cl} \;+\; \text{H}_2\text{N}-\underset{\underset{\text{CH}_3}{|}}{\text{CH}}\cdot\text{COOH} \;\longrightarrow\; \text{Z}-\text{NH}\cdot\underset{\underset{\text{CH}_3}{|}}{\text{CH}}\cdot\overset{\displaystyle O}{\overset{\|}{\text{C}}}-\text{OH}$$

alanine

activate the C-terminus

$$\text{Cl}-\overset{\displaystyle O}{\overset{\|}{\text{C}}}-\text{OEt}$$

$$\text{Z}-\text{NH}\cdot\underset{\underset{\text{CH}_3}{|}}{\text{CH}}\cdot\overset{\displaystyle O}{\overset{\|}{\text{C}}}-\text{O}-\overset{\displaystyle O}{\overset{\|}{\text{C}}}-\text{OEt}$$

$$\text{H}_2\text{N}-\underset{\underset{\text{CH(CH}_3)_2}{|}}{\text{CH}}\cdot\text{COOH} \qquad \text{valine}$$

add the next amino acid

$$\text{Z}-\text{NH}\cdot\underset{\underset{\text{CH}_3}{|}}{\text{CH}}\cdot\overset{\displaystyle O}{\overset{\|}{\text{C}}}-\text{NH}\cdot\underset{\underset{\text{CH(CH}_3)_2}{|}}{\text{CH}}\cdot\overset{\displaystyle O}{\overset{\|}{\text{C}}}-\text{OH}$$

H_2, Pd deprotect the N-terminus

$$\text{H}_2\text{N}-\underset{\underset{\text{CH}_3}{|}}{\text{CH}}\cdot\overset{\displaystyle O}{\overset{\|}{\text{C}}}-\text{NH}\cdot\underset{\underset{\text{CH(CH}_3)_2}{|}}{\text{CH}}\cdot\overset{\displaystyle O}{\overset{\|}{\text{C}}}-\text{OH}$$

react the N-terminus of the dipeptide at the left with the N-protected, C-activated tripeptide below

$$\text{Z}-\text{NH}\cdot\underset{\underset{\text{H}_3\text{C}-\text{CHCH}_2\text{CH}_3}{|}}{\text{CH}}\cdot\overset{\displaystyle O}{\overset{\|}{\text{C}}}-\text{NH}\cdot\underset{\underset{\text{CH}_2\text{CHMe}_2}{|}}{\text{CH}}\cdot\overset{\displaystyle O}{\overset{\|}{\text{C}}}-\text{NH}\cdot\underset{\underset{\text{CH}_2\text{Ph}}{|}}{\text{CH}}\cdot\overset{\displaystyle O}{\overset{\|}{\text{C}}}-\text{O}-\overset{\displaystyle O}{\overset{\|}{\text{C}}}\text{OEt}$$

$$\text{Z}-\text{NH}\cdot\underset{\underset{\text{H}_3\text{C}-\text{CHCH}_2\text{CH}_3}{|}}{\text{CH}}\cdot\overset{\displaystyle O}{\overset{\|}{\text{C}}}-\text{NH}\cdot\underset{\underset{\text{CH}_2\text{CHMe}_2}{|}}{\text{CH}}\cdot\overset{\displaystyle O}{\overset{\|}{\text{C}}}-\text{NH}\cdot\underset{\underset{\text{CH}_2\text{Ph}}{|}}{\text{CH}}\cdot\overset{\displaystyle O}{\overset{\|}{\text{C}}}-\text{NH}\cdot\underset{\underset{\text{CH}_3}{|}}{\text{CH}}\cdot\overset{\displaystyle O}{\overset{\|}{\text{C}}}-\text{NH}\cdot\underset{\underset{\text{CH(CH}_3)_2}{|}}{\text{CH}}\cdot\text{COOH}$$

H_2, Pd deprotect the N-terminus

$$\text{H}_2\text{N}-\underset{\substack{| \\ \text{H}_3\text{C}-\text{CHCH}_2\text{CH}_3 \\ \text{Ile}}}{\text{CH}}\cdot\overset{\displaystyle O}{\overset{\|}{\text{C}}}-\text{NH}\cdot\underset{\substack{| \\ \text{CH}_2\text{CHMe}_2 \\ \text{Leu}}}{\text{CH}}\cdot\overset{\displaystyle O}{\overset{\|}{\text{C}}}-\text{NH}\cdot\underset{\substack{| \\ \text{CH}_2\text{Ph} \\ \text{Phe}}}{\text{CH}}\cdot\overset{\displaystyle O}{\overset{\|}{\text{C}}}-\text{NH}\cdot\underset{\substack{| \\ \text{CH}_3 \\ \text{Ala}}}{\text{CH}}\cdot\overset{\displaystyle O}{\overset{\|}{\text{C}}}-\text{NH}\cdot\underset{\substack{| \\ \text{CH(CH}_3)_2 \\ \text{Val}}}{\text{CH}}\cdot\text{COOH}$$

24-46

(a) There are two possible sources of ammonia in the hydrolysate. The C-terminus could have been present as the amide instead of the carboxyl, or the glutamic acid could have been present as its amide, glutamine.

(b) The C-terminus is present as the amide. The N-terminus is present as the lactam (cyclic amide) combining the amino group with the carboxyl group of the glutamic acid side chain.

(c) The fact that hydrolysis does not release ammonia implies that the C-terminus is not an amide. Yet, carboxypeptidase treatment gives no reaction, showing that the C-terminus is not a free carboxyl group. Also, treatment with phenyl isothiocyanate gives no reaction, suggesting no free amine at the N-terminus. The most plausible explanation is that the N-terminus has reacted with the C-terminus to produce a cyclic amide, a lactam. (These large rings, called macrocycles, are often found in nature as hormones or antibiotics.)

24-47

(a) Lipoic acid is a mild oxidizing agent. In the process of oxidizing another reactant, lipoic acid is reduced.

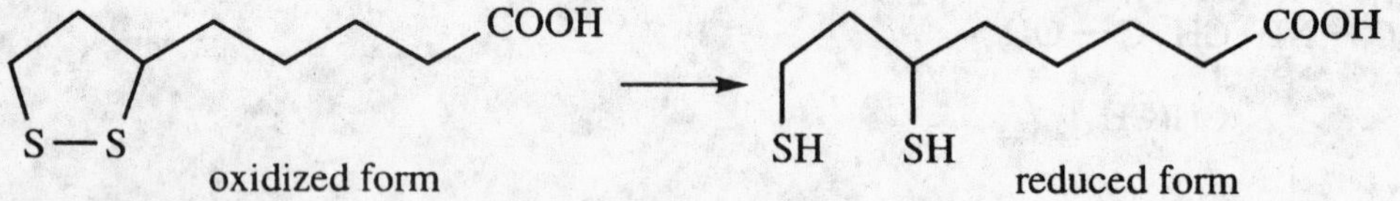

oxidized form → reduced form

(b)

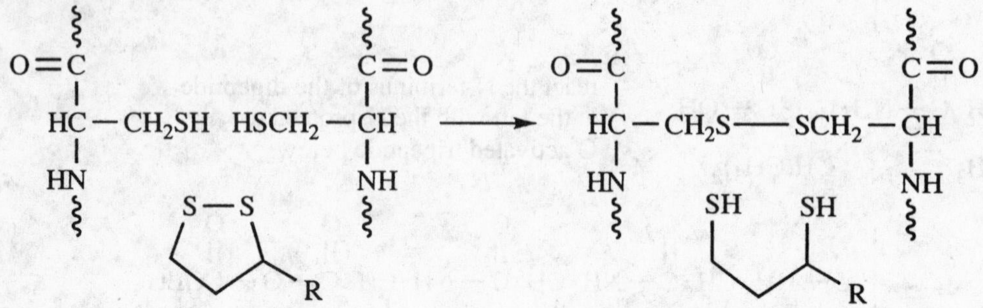

(c)

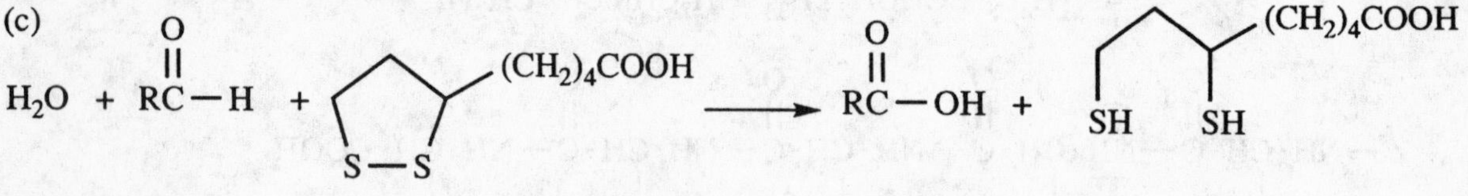

24-48

(a) histidine:

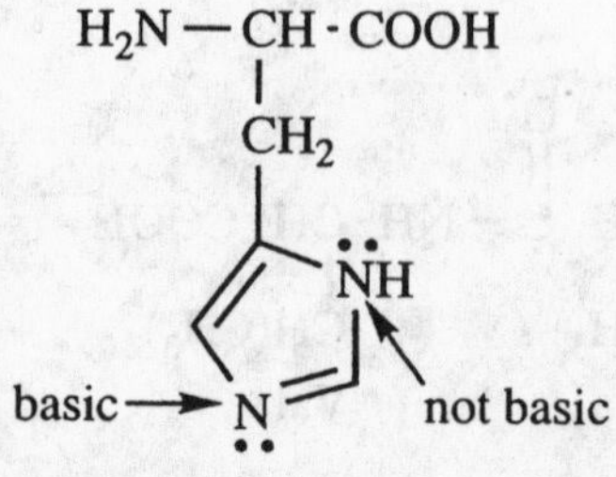

basic → N   not basic

24-48 continued

(b)

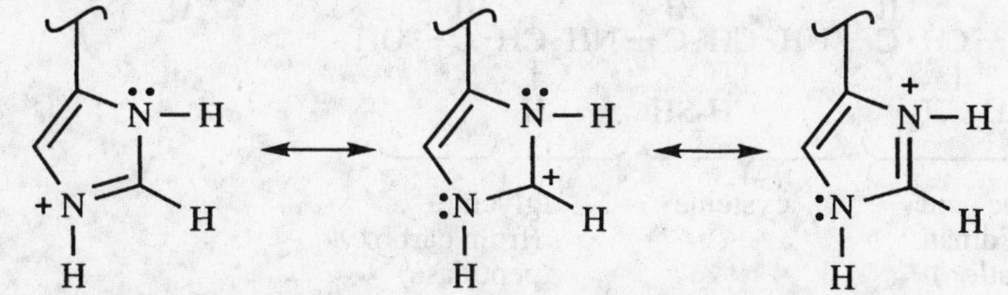

In the protonated imidazole, the two N's are similar in structure, and both NH groups are acidic.

(c)

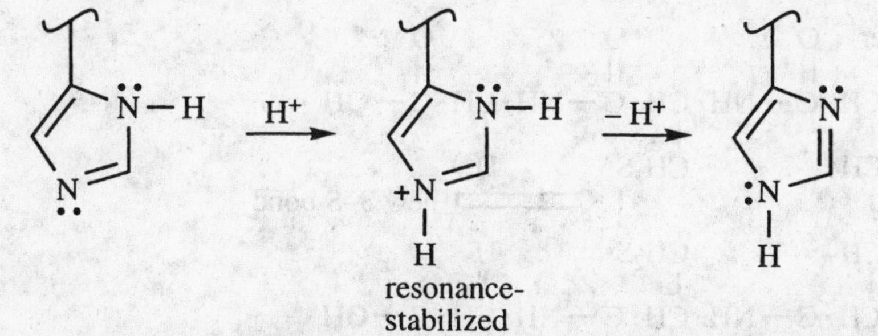
H+
– H+
resonance-
stabilized

We usually think of protonation-deprotonation reactions occurring in solution where protons can move with solvent molecules. In an enzyme active-site, there is no "solvent", so there must be another mechanism for movement of protons. Often, conformational changes in the protein will move atoms closer or farther. Histidine serves the function of moving a proton toward or away from a particular site by using its different nitrogens in concert as a proton acceptor and a proton donor.

24-49 The high isoelectric point suggests a strongly basic side chain as in lysine. The N—CH_2 bond in the side chain of arginine is likely to have remained intact during the metabolism. (Can you propose a likely mechanism for this reaction?)

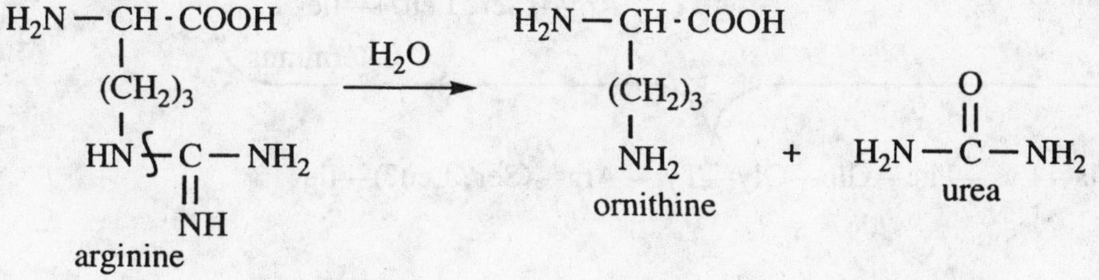
H2O
arginine
ornithine
urea

24-50

(a) glutathione:

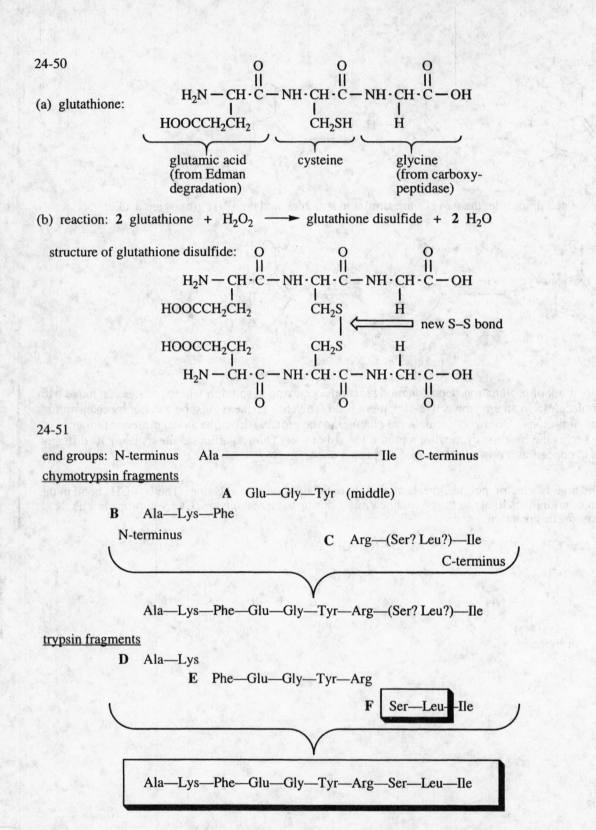

$H_2N-CH \cdot C(=O)-NH \cdot CH \cdot C(=O)-NH \cdot CH \cdot C(=O)-OH$

HOOCCH₂CH₂   CH₂SH   H

glutamic acid (from Edman degradation) | cysteine | glycine (from carboxy-peptidase)

(b) reaction: **2** glutathione + H_2O_2 ⟶ glutathione disulfide + **2** H_2O

structure of glutathione disulfide:

$H_2N-CH \cdot C(=O)-NH \cdot CH \cdot C(=O)-NH \cdot CH \cdot C(=O)-OH$

HOOCCH₂CH₂   CH₂S   H

⟸ new S–S bond

HOOCCH₂CH₂   CH₂S   H

$H_2N-CH \cdot C(=O)-NH \cdot CH \cdot C(=O)-NH \cdot CH \cdot C(=O)-OH$

24-51

end groups: N-terminus   Ala ——————————— Ile   C-terminus

<u>chymotrypsin fragments</u>

A   Glu—Gly—Tyr   (middle)

B   Ala—Lys—Phe
N-terminus

C   Arg—(Ser? Leu?)—Ile
C-terminus

Ala—Lys—Phe—Glu—Gly—Tyr—Arg—(Ser? Leu?)—Ile

<u>trypsin fragments</u>

D   Ala—Lys

E   Phe—Glu—Gly—Tyr—Arg

F   Ser—Leu—Ile

Ala—Lys—Phe—Glu—Gly—Tyr—Arg—Ser—Leu—Ile

24-52

(a) One important factor in ester reactivity is the ability of the alkoxy group to leave, and the main factor in determining leaving group ability is the stabilization of the anion. An NHS ester is more reactive than an alkyl ester because the anion R_2NO^- has an electron-withdrawing group on the O^-, thereby distributing the negative charge over two atoms instead of just one. A simple alkyl ester has the full negative charge on the oxygen with nowhere to go, RO^-.

(b)

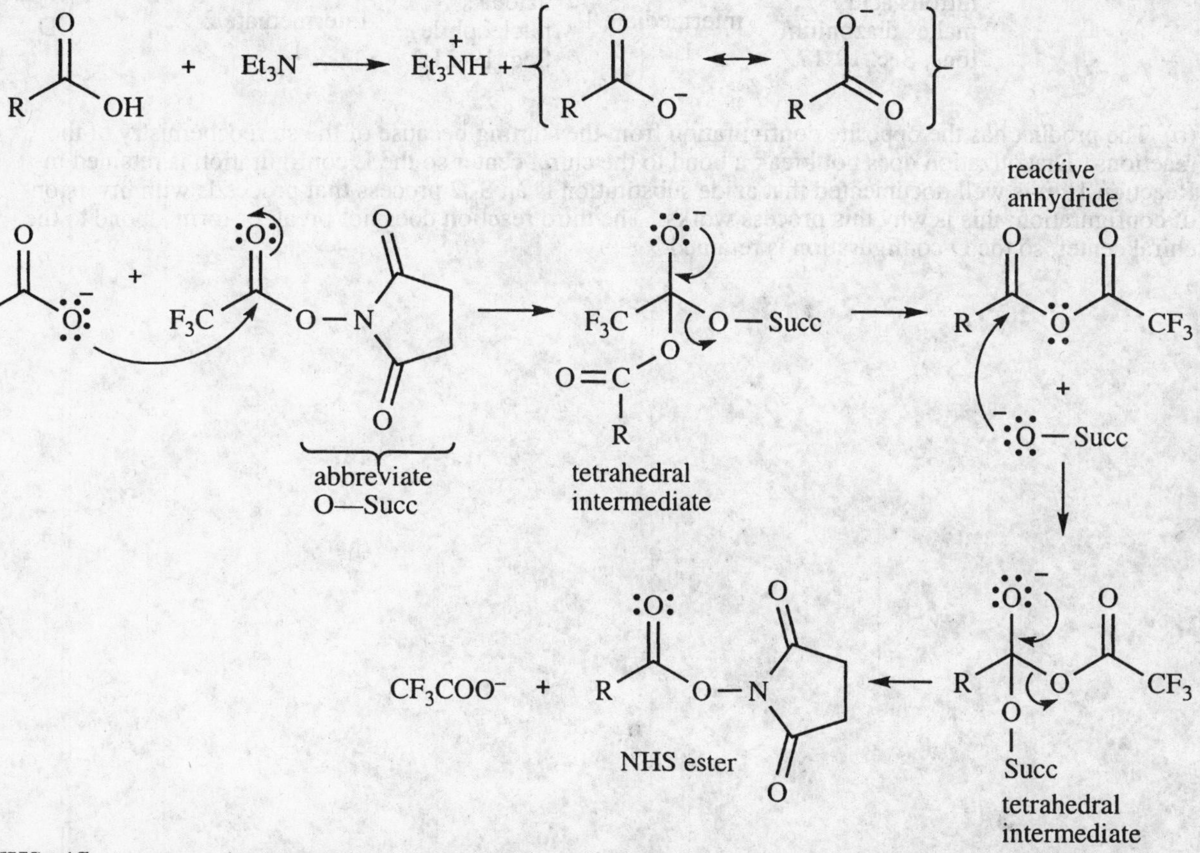

NHS trifluoroacetate is an amazing reagent. Not only does is activate the carboxylic acid through a mixed anhydride to form an ester under mild conditions, but the NHS leaving group is also the nucleophile that forms an ester that is both stable enough to work with, yet easily reactive when it needs to be. Perfect!

(c)

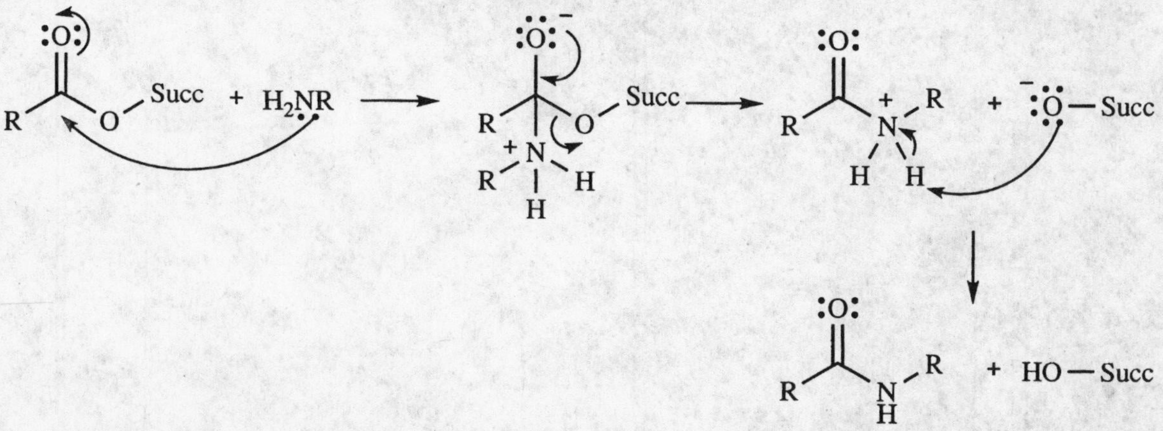

643

24-53

(a)

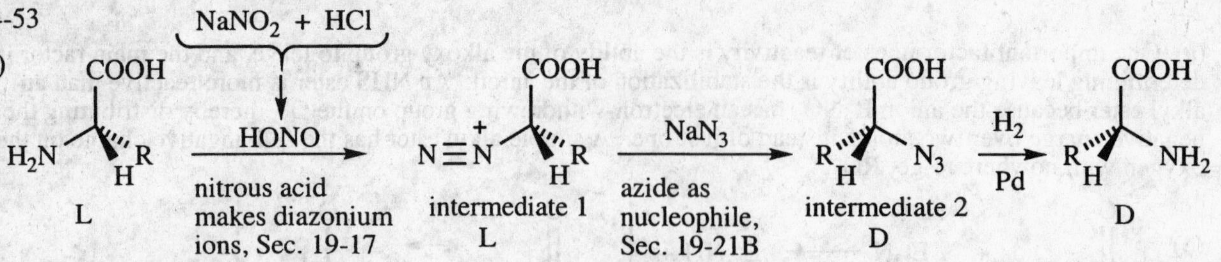

(b) The product has the opposite configuration from the starting because of the stereochemistry of the reactions. Diazotization does not break a bond to the chiral center so the L configuration is retained in Reaction 1. It is well documented that azide substitution is an S_N2 process that proceeds with inversion of configuration; this is why this process works. The third reaction does not break or form a bond to the chiral center, so the D configuration is retained.

25-1

$$CH_2-O-\overset{\overset{\displaystyle O}{\|}}{C}-(CH_2)_{12}CH_3$$

trimyristin $CH-O-\overset{\overset{\displaystyle O}{\|}}{C}-(CH_2)_{12}CH_3$

$$CH_2-O-\overset{\overset{\displaystyle O}{\|}}{C}-(CH_2)_{12}CH_3$$

25-2

all cis

$$CH_2-O-\overset{\overset{\displaystyle O}{\|}}{C}-(CH_2)_7CH=CH(CH_2)_7CH_3$$

$$CH-O-\overset{\overset{\displaystyle O}{\|}}{C}-(CH_2)_7CH=CH(CH_2)_7CH_3 \xrightarrow[\text{Ni}]{\substack{\text{excess} \\ H_2}}$$

$$CH_2-O-\overset{\overset{\displaystyle O}{\|}}{C}-(CH_2)_7CH=CH(CH_2)_7CH_3$$

triolein, m.p. –4° C
(liquid at room temperature)

$$CH_2-O-\overset{\overset{\displaystyle O}{\|}}{C}-(CH_2)_{16}CH_3$$

$$CH-O-\overset{\overset{\displaystyle O}{\|}}{C}-(CH_2)_{16}CH_3$$

$$CH_2-O-\overset{\overset{\displaystyle O}{\|}}{C}-(CH_2)_{16}CH_3$$

tristearin, m.p. 72° C
(solid at room temperature)

25-3

(a)

$$2 \ CH_3(CH_2)_{16}-\overset{\overset{\displaystyle O}{\|}}{C}-O^- \ Na^+ + \ Ca^{2+} \longrightarrow \left[CH_3(CH_2)_{16}-\overset{\overset{\displaystyle O}{\|}}{C}-O \right]_2 Ca \ + \ 2 \ Na^+$$

(b)

$$2 \ CH_3(CH_2)_{16}-\overset{\overset{\displaystyle O}{\|}}{C}-O^- \ Na^+ + \ Mg^{2+} \longrightarrow \left[CH_3(CH_2)_{16}-\overset{\overset{\displaystyle O}{\|}}{C}-O \right]_2 Mg \ + \ 2 \ Na^+$$

(c)

$$3 \ CH_3(CH_2)_{16}-\overset{\overset{\displaystyle O}{\|}}{C}-O^- \ Na^+ + \ Fe^{3+} \longrightarrow \left[CH_3(CH_2)_{16}-\overset{\overset{\displaystyle O}{\|}}{C}-O \right]_3 Fe \ + \ 3 \ Na^+$$

25-4

(a) Both sodium carbonate (its old name is "washing soda") and sodium phosphate will increase the pH above 6, so that the carboxyl group of the soap molecule will remain ionized, thus preventing precipitation.

(b) In the presence of calcium, magnesium, and ferric ions, the carboxylate group of soap will form precipitates called "hard-water scum", or as scientists label it, "bathtub ring". Both carbonate and phosphate ions will form complexes or precipitates with these cations, thereby preventing precipitation of the soap from solution.

25-5

abbreviate

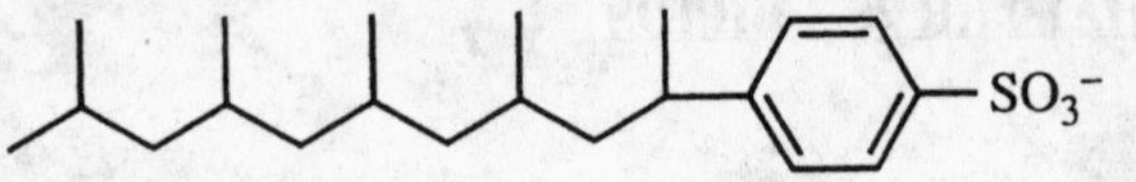

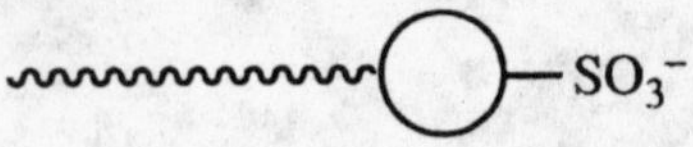

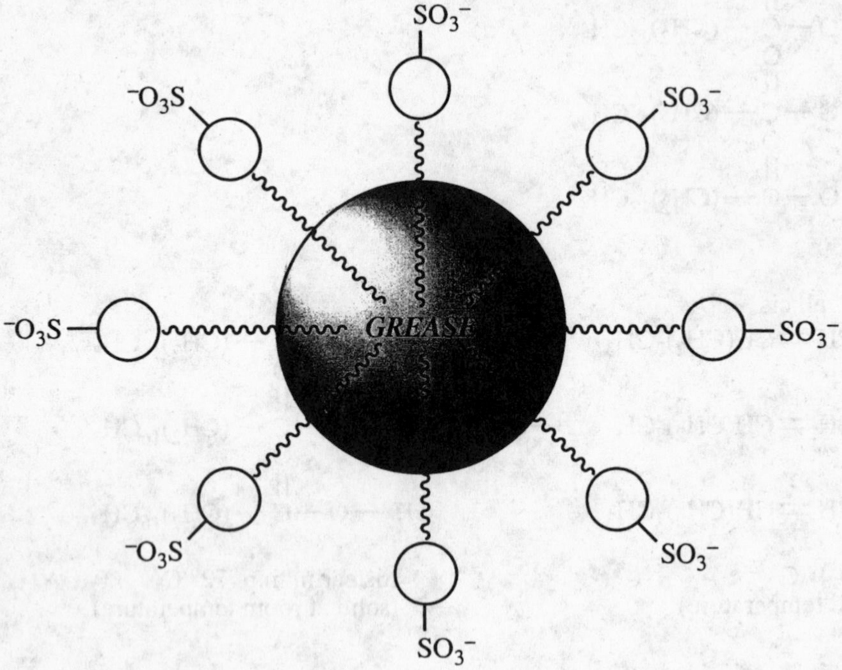

25-6 In each structure, the hydrophilic portion is circled. The uncircled part is hydrophobic.

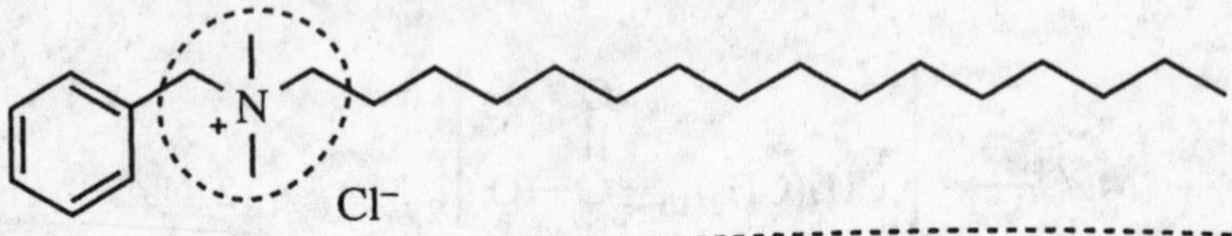

benzalkonium chloride

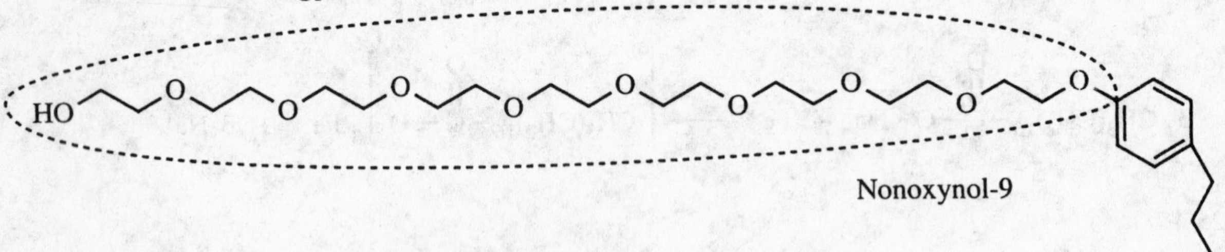

Nonoxynol-9

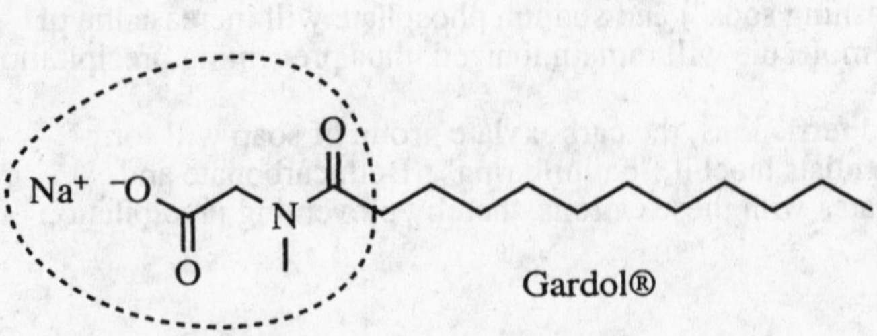

Gardol®

25-7

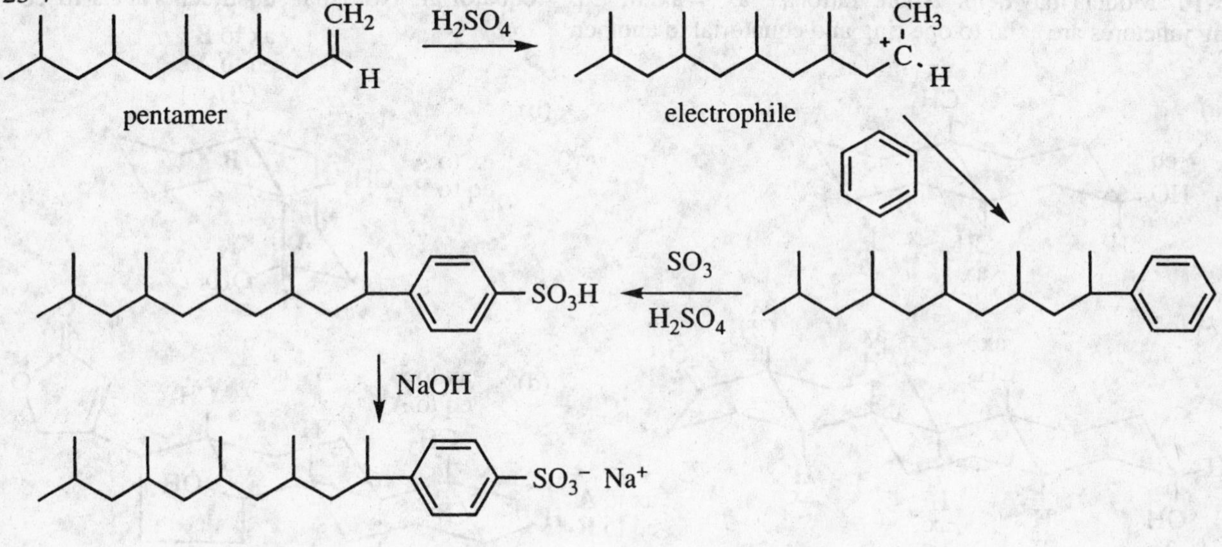

25-8

(a)

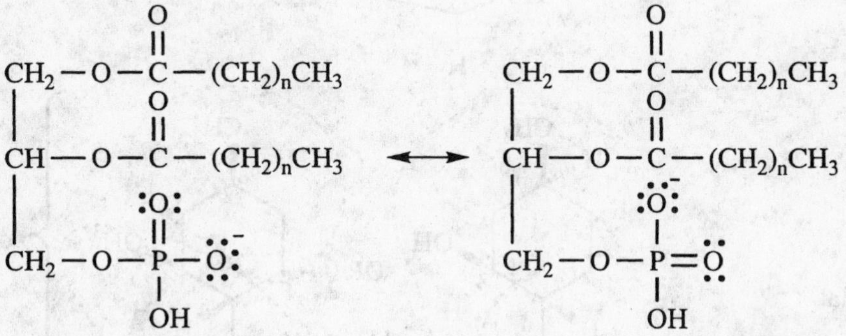

(b)

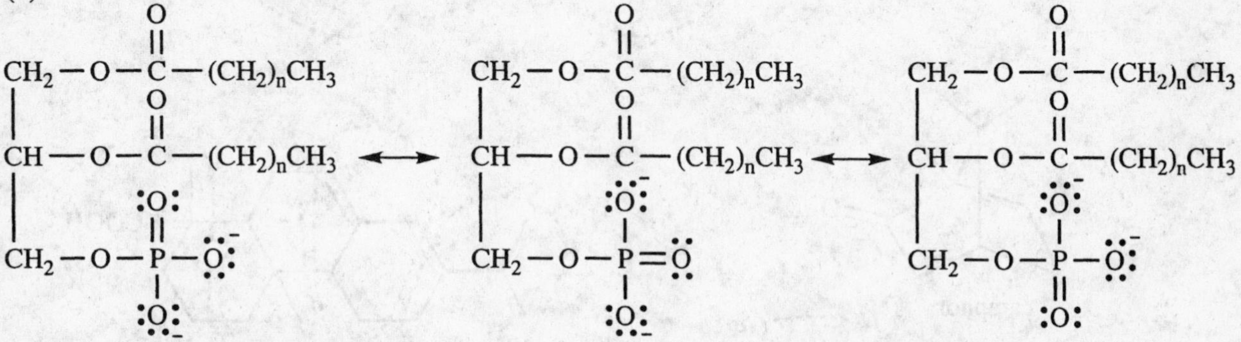

25-9 Estradiol is a phenol and can be ionized with aqueous NaOH. Testosterone does not have any hydrogens acidic enough to react with NaOH. Treatment of a solution of estradiol and testosterone in organic solvent with aqueous base will extract the phenoxide form of estradiol into the aqueous layer, leaving testosterone in the organic layer. Acidification of the aqueous base will precipitate estradiol which can be filtered. Evaporation of the organic solvent will leave testosterone.

647

25-10 Models may help. Abbreviations: "ax" = axial; "eq" = equatorial. Note that substituents at *cis*-fused ring junctures are axial to one ring and equatorial to another.

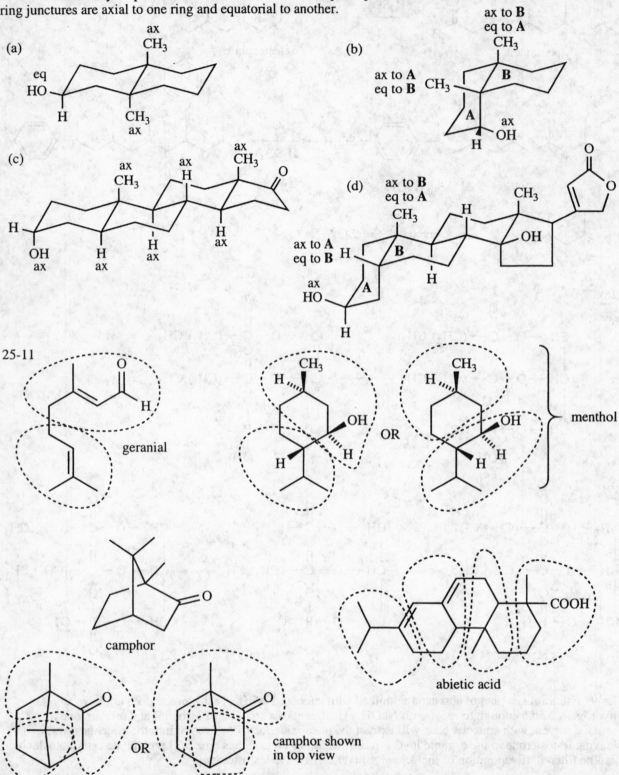

(a)

ax
CH₃
eq
HO
H
CH₃
ax

(b)

ax to **B**
eq to **A**
CH₃
ax to **A**
eq to **B** CH₃
B
A
ax
H OH

(c)

ax
CH₃
ax
H
ax
CH₃
O
CH₃
H
OH
ax
H
ax
H
ax
H
ax

(d)

ax to **B**
eq to **A**
CH₃
CH₃
H
OH
ax to **A**
eq to **B** H
B
A
H
ax
HO
H

25-11

O
H
geranial

CH₃
H
OH
H
H
OR
CH₃
H
OH
H
H
menthol

camphor

O

camphor shown in top view
OR

COOH
abietic acid

648

25-12 β-carotene

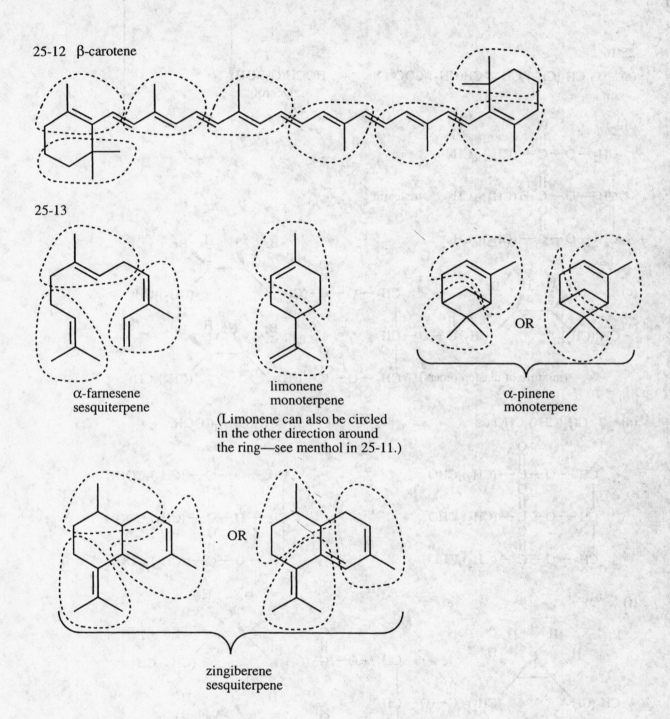

25-13

α-farnesene
sesquiterpene

limonene
monoterpene

(Limonene can also be circled
in the other direction around
the ring—see menthol in 25-11.)

OR

α-pinene
monoterpene

OR

zingiberene
sesquiterpene

25-14 Please refer to problem 1-20, page 12 of this Solutions Manual.

25-15

(a) triglyceride   (b) synthetic detergent   (c) wax   (d) sesquiterpene   (e) steroid

25-16

(a) **3** $CH_3(CH_2)_7CH=CH(CH_2)_7COO^-\ Na^+$ + $HOCH(CH_2OH)_2$

   soap   glycerol

(b)

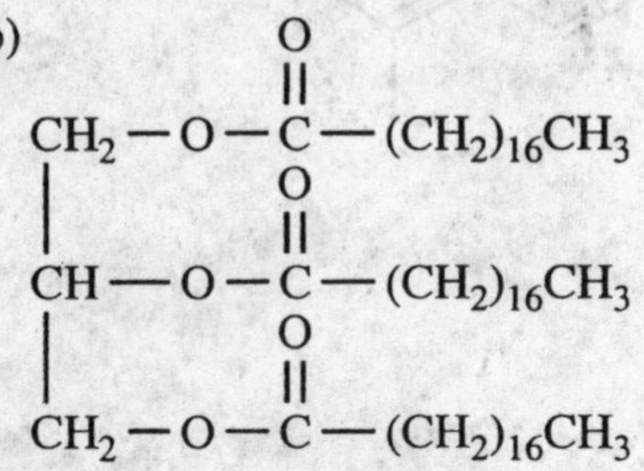

$$CH_2-O-\overset{\overset{\displaystyle O}{\|}}{C}-(CH_2)_{16}CH_3$$
$$CH-O-\overset{\overset{\displaystyle O}{\|}}{C}-(CH_2)_{16}CH_3 \quad \text{tristearin}$$
$$CH_2-O-\overset{\overset{\displaystyle O}{\|}}{C}-(CH_2)_{16}CH_3$$

(c)

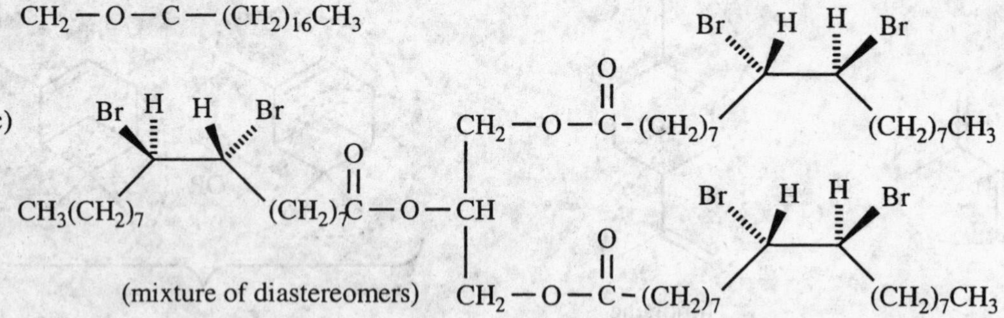

(mixture of diastereomers)

(d) **3** $CH_3(CH_2)_7CHO$ +

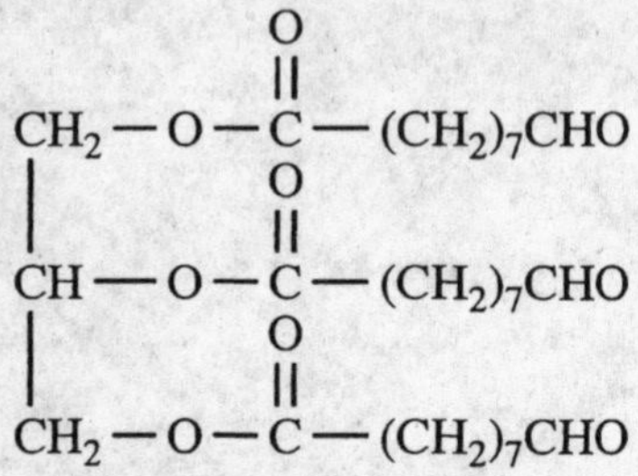

$$CH_2-O-\overset{\overset{\displaystyle O}{\|}}{C}-(CH_2)_7CHO$$
$$CH-O-\overset{\overset{\displaystyle O}{\|}}{C}-(CH_2)_7CHO$$
$$CH_2-O-\overset{\overset{\displaystyle O}{\|}}{C}-(CH_2)_7CHO$$

(e) **3** $CH_3(CH_2)_7COOH$ +

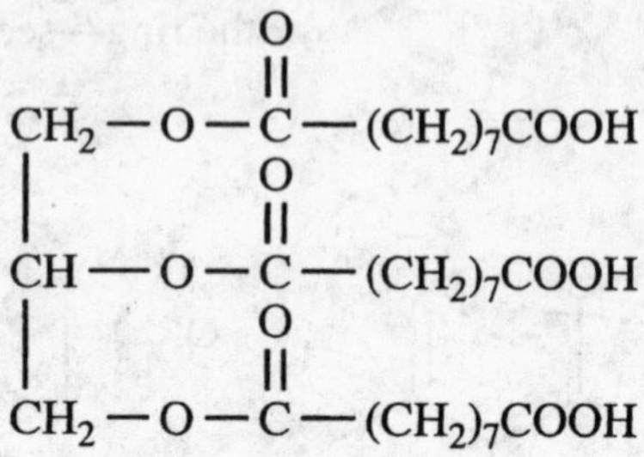

$$CH_2-O-\overset{\overset{\displaystyle O}{\|}}{C}-(CH_2)_7COOH$$
$$CH-O-\overset{\overset{\displaystyle O}{\|}}{C}-(CH_2)_7COOH$$
$$CH_2-O-\overset{\overset{\displaystyle O}{\|}}{C}-(CH_2)_7COOH$$

(f)

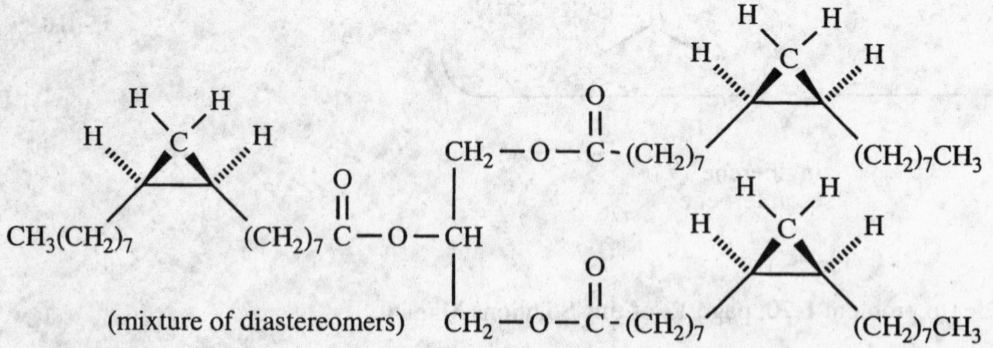

(mixture of diastereomers)

(g) **3** $CH_3(CH_2)_7CH=CH(CH_2)_7CH_2OH$ + $HOCH(CH_2OH)_2$

                                        glycerol

25-17

(a) $CH_3(CH_2)_7CH=CH(CH_2)_7COOH \xrightarrow{\text{H}_2 \atop \text{Ni}} \xrightarrow{\text{1) LiAlH}_4 \atop \text{2) H}_3\text{O}^+} CH_3(CH_2)_{16}CH_2OH$

(b) $CH_3(CH_2)_7CH=CH(CH_2)_7COOH \xrightarrow{\text{H}_2 \atop \text{Ni}} CH_3(CH_2)_{16}COOH$

(c) $\underset{\text{from (b)}}{CH_3(CH_2)_{16}COOH} + \underset{\text{from (a)}}{HOCH_2(CH_2)_{16}CH_3} \xrightarrow{\text{H}^+ \atop \Delta} CH_3(CH_2)_{16}COOCH_2(CH_2)_{16}CH_3$

(d) $CH_3(CH_2)_7CH=CH(CH_2)_7COOH \xrightarrow{\text{1) O}_3 \atop \text{2) Me}_2\text{S}} \underset{\text{nonanal}}{CH_3(CH_2)_7CH=O} + O=CH(CH_2)_7COOH$

(e) $CH_3(CH_2)_7CH=CH(CH_2)_7COOH \xrightarrow{\text{KMnO}_4 \atop \text{H}_2\text{O}, \Delta} CH_3(CH_2)_7COOH + \underset{\text{nonanedioic acid}}{HOOC(CH_2)_7COOH}$

(f) 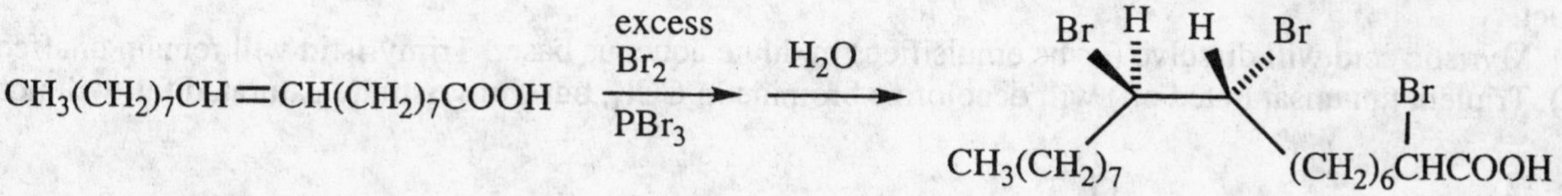 $CH_3(CH_2)_7CH=CH(CH_2)_7COOH \xrightarrow[\text{PBr}_3]{\text{excess} \atop \text{Br}_2} \xrightarrow{\text{H}_2\text{O}}$

25-18

(a)

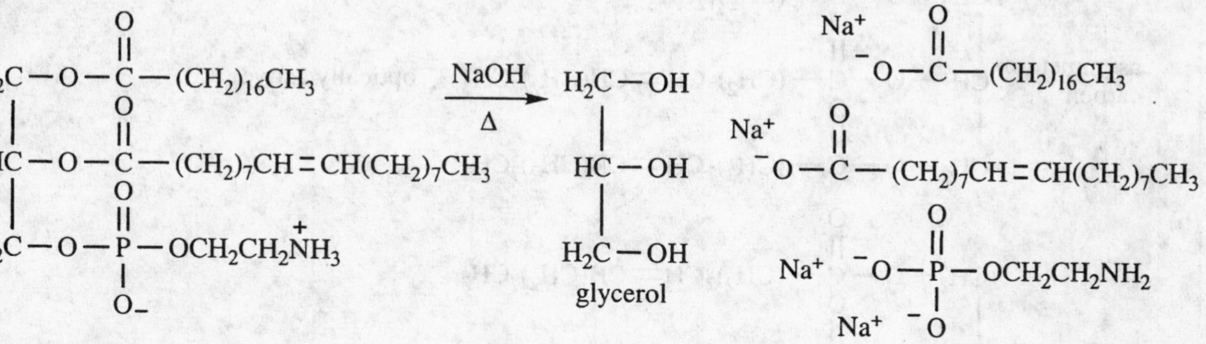

(b)

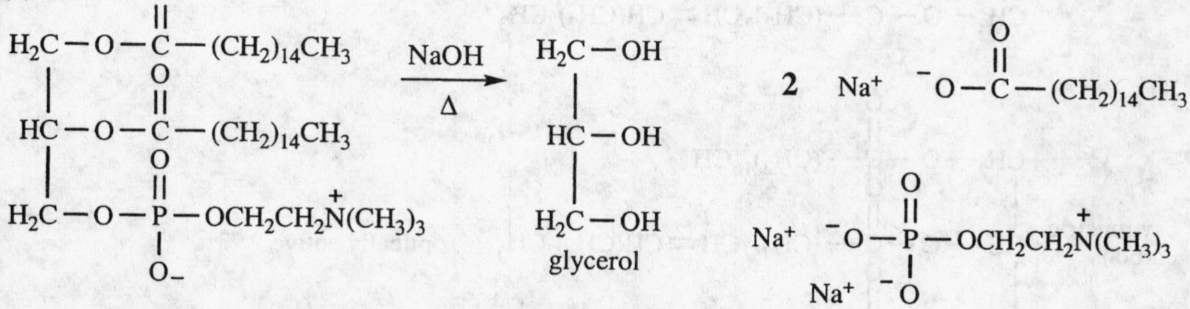

25-19

$$CH_2-O-\overset{\overset{\displaystyle O}{\|}}{C}-(CH_2)_{16}CH_3$$
$$CH-O-\overset{\overset{\displaystyle O}{\|}}{C}-(CH_2)_{16}CH_3 \quad\xrightarrow[\text{2) }H_3O^+]{\text{1) LiAlH}_4}\quad$$
$$CH_2-O-\overset{\overset{\displaystyle O}{\|}}{C}-(CH_2)_{16}CH_3$$

$$\begin{array}{c}CH_2OH\\ |\\ CHOH\\ |\\ CH_2OH\end{array} \quad+\quad 3 \quad CH_3(CH_2)_{16}CH_2OH$$

$$\downarrow \begin{array}{l}SO_3\\ H_2SO_4\\ \text{(cold)}\end{array}$$

$$CH_3(CH_2)_{16}CH_2OSO_3^-\ Na^+ \quad\xleftarrow{\text{NaOH}}\quad CH_3(CH_2)_{16}CH_2OSO_3H$$

25-20 Reagents in parts (a), (b), and (d) would react with alkenes. If both samples contained alkenes, these reagents could not distinguish the samples. Saponification (part (c)), however, is a reaction of an ester, so only the vegetable oil would react, not the hydrocarbon oil mixture.

25-21 (a) Add an aqueous solution of calcium ion or magnesium ion. Sodium stearate will produce a precipitate, while the sulfonate will not precipitate.
(b) Beeswax, an ester, can be saponified with NaOH. Paraffin wax is a solid mixture of alkanes and will not react.
(c) Myristic acid will dissolve (or be emulsified) in dilute aqueous base. Trimyristin will remain unaffected.
(d) Triolein (an unsaturated oil) will decolorize bromine in CCl_4, but trimyristin (a saturated fat) will not.

25-22

(a)

$$\begin{array}{l}CH_2-O-\overset{\overset{\displaystyle O}{\|}}{C}-(CH_2)_{12}CH_3\\ |\\ \overset{*}{CH}-O-\overset{\overset{\displaystyle O}{\|}}{C}-(CH_2)_7CH=CH(CH_2)_7CH_3\\ |\\ CH_2-O-\overset{\overset{\displaystyle O}{\|}}{C}-(CH_2)_7CH=CH(CH_2)_7CH_3\end{array}$$

asymmetric carbon   *   optically active

(b)

$$\begin{array}{l}CH_2-O-\overset{\overset{\displaystyle O}{\|}}{C}-(CH_2)_7CH=CH(CH_2)_7CH_3\\ |\\ CH-O-\overset{\overset{\displaystyle O}{\|}}{C}-(CH_2)_{12}CH_3\\ |\\ CH_2-O-\overset{\overset{\displaystyle O}{\|}}{C}-(CH_2)_7CH=CH(CH_2)_7CH_3\end{array}$$

not optically active; symmetric

25-23

$$\begin{array}{l}CH_2-O-\overset{\overset{\displaystyle O}{\|}}{C}-(CH_2)_{16}CH_3\\ |\\ \overset{*}{CH}-O-\overset{\overset{\displaystyle O}{\|}}{C}-(CH_2)_7CH=CH(CH_2)_7CH_3\\ |\\ CH_2-O-\overset{\overset{\displaystyle O}{\|}}{C}-(CH_2)_7CH=CH(CH_2)_7CH_3\end{array}$$

asymmetric carbon   *   optically active

25-23 continued

(a)

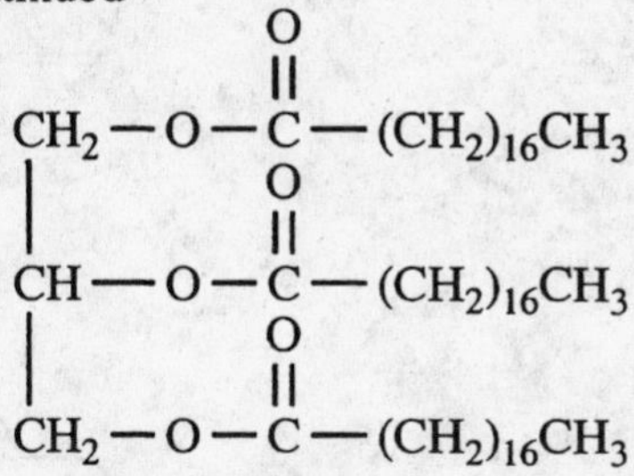

$$CH_2-O-\overset{\overset{\displaystyle O}{\|}}{C}-(CH_2)_{16}CH_3$$
$$CH-O-\overset{\overset{\displaystyle O}{\|}}{C}-(CH_2)_{16}CH_3$$
$$CH_2-O-\overset{\overset{\displaystyle O}{\|}}{C}-(CH_2)_{16}CH_3$$

not optically active

(b)

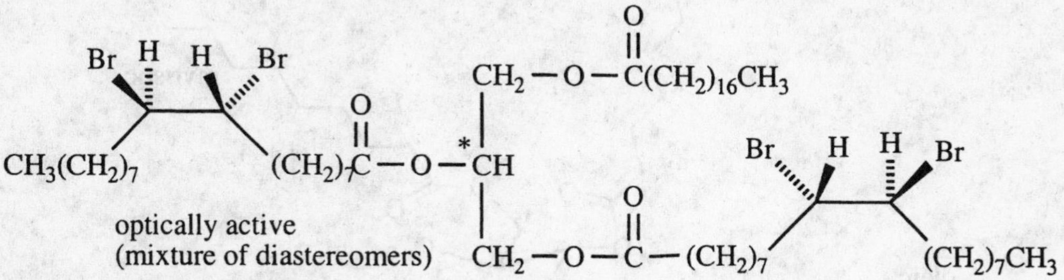

optically active
(mixture of diastereomers)

(c) products are not optically active

$$HOCH(CH_2OH)_2 \ + \ Na^+ \ ^-OOC(CH_2)_{16}CH_3 \ + \ 2 \ \ Na^+ \ ^-OOC(CH_2)_7CH=CH(CH_2)_7CH_3$$
glycerol

(d)

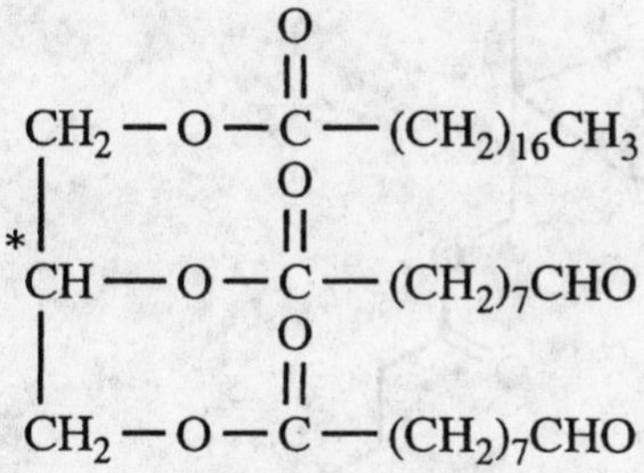

$$CH_2-O-\overset{\overset{\displaystyle O}{\|}}{C}-(CH_2)_{16}CH_3$$
$$\overset{*}{C}H-O-\overset{\overset{\displaystyle O}{\|}}{C}-(CH_2)_7CHO \ \ + \ 2 \ \ O=CH(CH_2)_7CH_3$$
$$CH_2-O-\overset{\overset{\displaystyle O}{\|}}{C}-(CH_2)_7CHO$$

not optically active

optically active

25-24 The products in (b), (c) and (f) are mixtures of stereoisomers.

(a) (b) 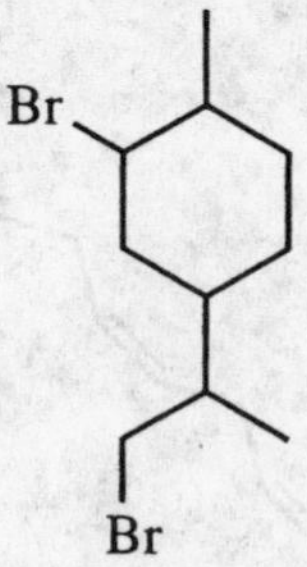 (c) 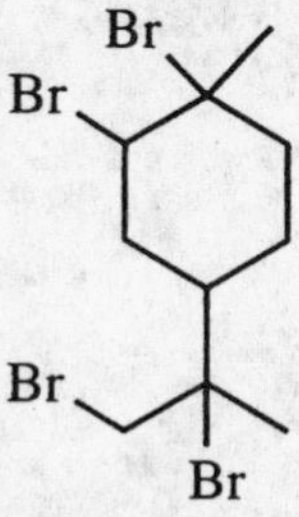 (d) (e)

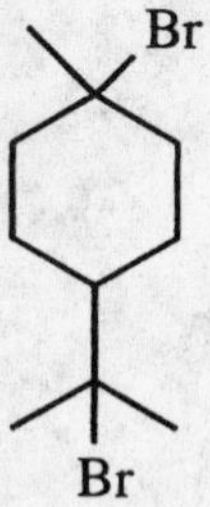

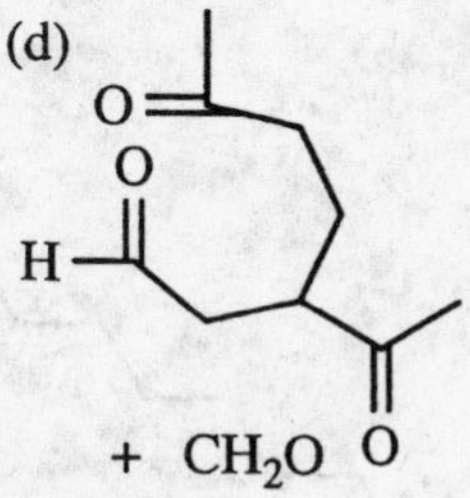

+ CH$_2$O

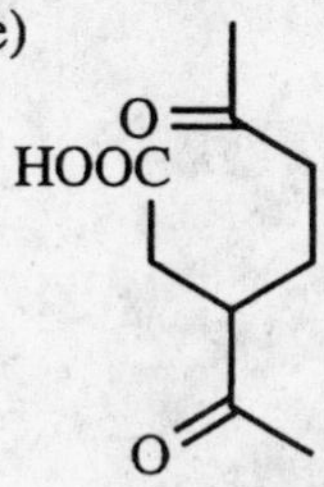

+ CO$_2$

(f)

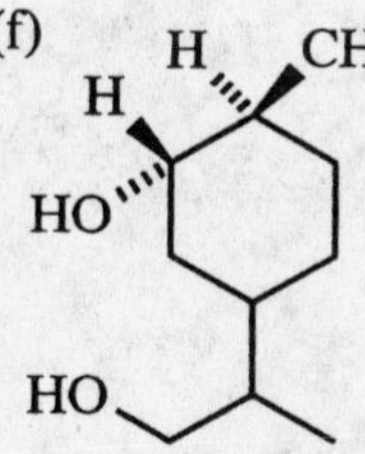

25-25 Is it any wonder that Olestra cannot be digested!

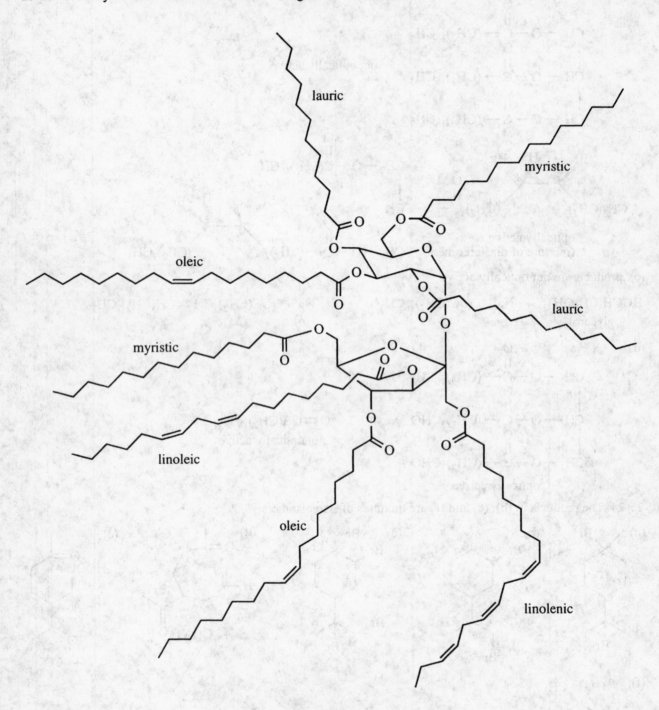

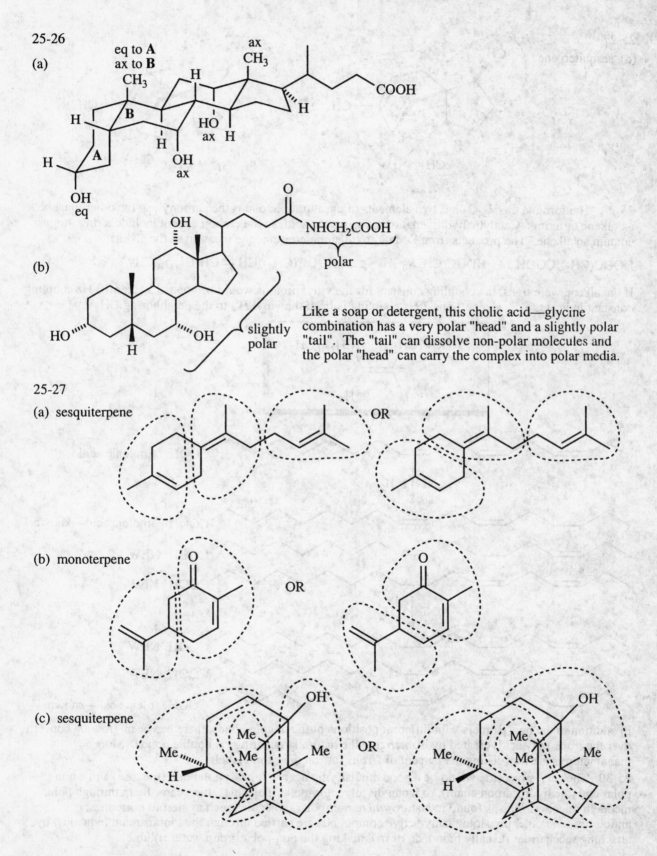

25-26

(a)

eq to **A**
ax to **B**

CH₃

ax
CH₃

H

H

B

H

HO
ax

H

H

H

A

OH
ax

H

OH
eq

COOH

(b)

OH

NHCH₂COOH

}

polar

HO

H

OH

}

slightly
polar

Like a soap or detergent, this cholic acid—glycine combination has a very polar "head" and a slightly polar "tail". The "tail" can dissolve non-polar molecules and the polar "head" can carry the complex into polar media.

25-27

(a) sesquiterpene

OR

(b) monoterpene

O

OR

O

(c) sesquiterpene

OH

Me

Me

Me

Me

OR

OH

Me

Me

Me

Me

H

H

25-27

(d) sesquiterpene

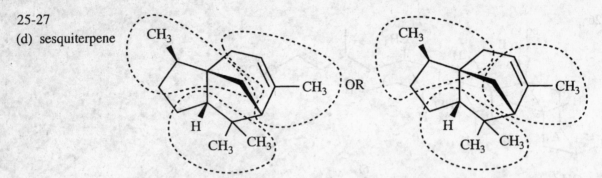

25-28

The formula $C_{18}H_{34}O_2$ has two elements of unsaturation; one is the carbonyl, so the other must be an alkene or a ring. Catalytic hydrogenation gives stearic acid, so the carbon cannot include a ring; it must contain an alkene. The products from $KMnO_4$ oxidation determine the location of the alkene:

$$HOOC(CH_2)_4COOH \ + \ HOOC(CH_2)_{10}CH_3 \implies HOOC(CH_2)_4CH=CH(CH_2)_{10}CH_3$$

If the alkene were trans, the coupling constant for the vinyl protons would be about 15 Hz; a 10 Hz coupling constant indicates a cis alkene. The 7 Hz coupling is from the vinyl H's to the neighboring CH_2's.

$$
\begin{array}{c}
HOOC(CH_2)_4 \qquad (CH_2)_{10}CH_3 \\
\diagdown \qquad \diagup \\
C = C \\
\diagup \qquad \diagdown \\
H \qquad\qquad H
\end{array}
$$

25-29

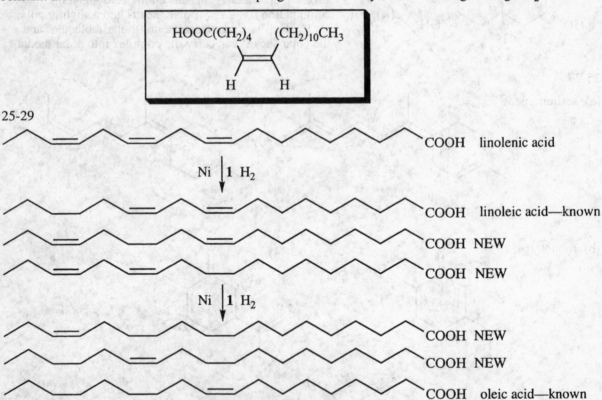

In addition to the new isomers with different positions of the double bonds, there has been growing concern over the *trans* fatty acids created by isomerizing the naturally occuring *cis* double bonds. More manufacturers are now listing the percent of "*trans* fat" on their food labels.

25-30 The cetyl glycoside would be a good emulsifying agent. It has a polar end (glucose) and a non-polar end (the hydrocarbon chain), so it can dissolve non-polar molecules, then carry them through polar media in micelles. This is found in Nature where non-polar molecules such as steroid hormones, antibiotics, and other physiologically-active compounds are carried through the bloodstream (aqueous) by attaching saccharides (usually mono, di, or tri), making the non-polar group water soluble.

25-31 (a) Four of these components in Vicks Vapo-Rub® are terpenes. The only one that is not is decane; it is not comprised of isoprene units despite having the correct number of carbons for a monterpene.

Three of the four terpenes have had their isoprene units indicated in previous problems; in each of those three, there are two possible ways to assign the isoprene units, so those pictures will not be duplicated here.

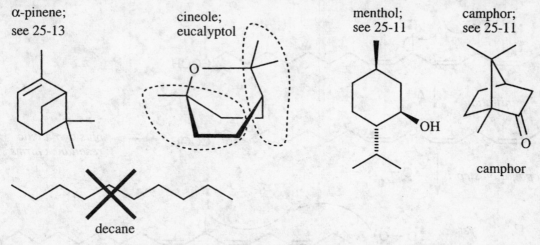

α-pinene;
see 25-13

cineole;
eucalyptol

menthol;
see 25-11

camphor;
see 25-11

camphor

decane

(b) Vicks Vapo-Rub® must be optically active as it contains four optically active terpenes.

25-32

(a) Of the two, only nepetalactone is a terpene. The other has only 8 carbons and terpenes must be in multiples of 5 carbons.

(b) Of the two, only the second is aromatic as can be readily seen in one of the resonance forms.

(c) Each compound is cleaved with NaOH (aq) to give an enolate that tautomerizes to the more stable keto form.

25-33

(a) High temperature and the diradical O_2 molecule strongly suggest a radical mechanism.

(b) Radical stability follows the order: benzylic > allylic > 3° > 2° > 1°. A radical at C-11 would be *doubly* allylic making it a prime site for radical reaction.

(c)

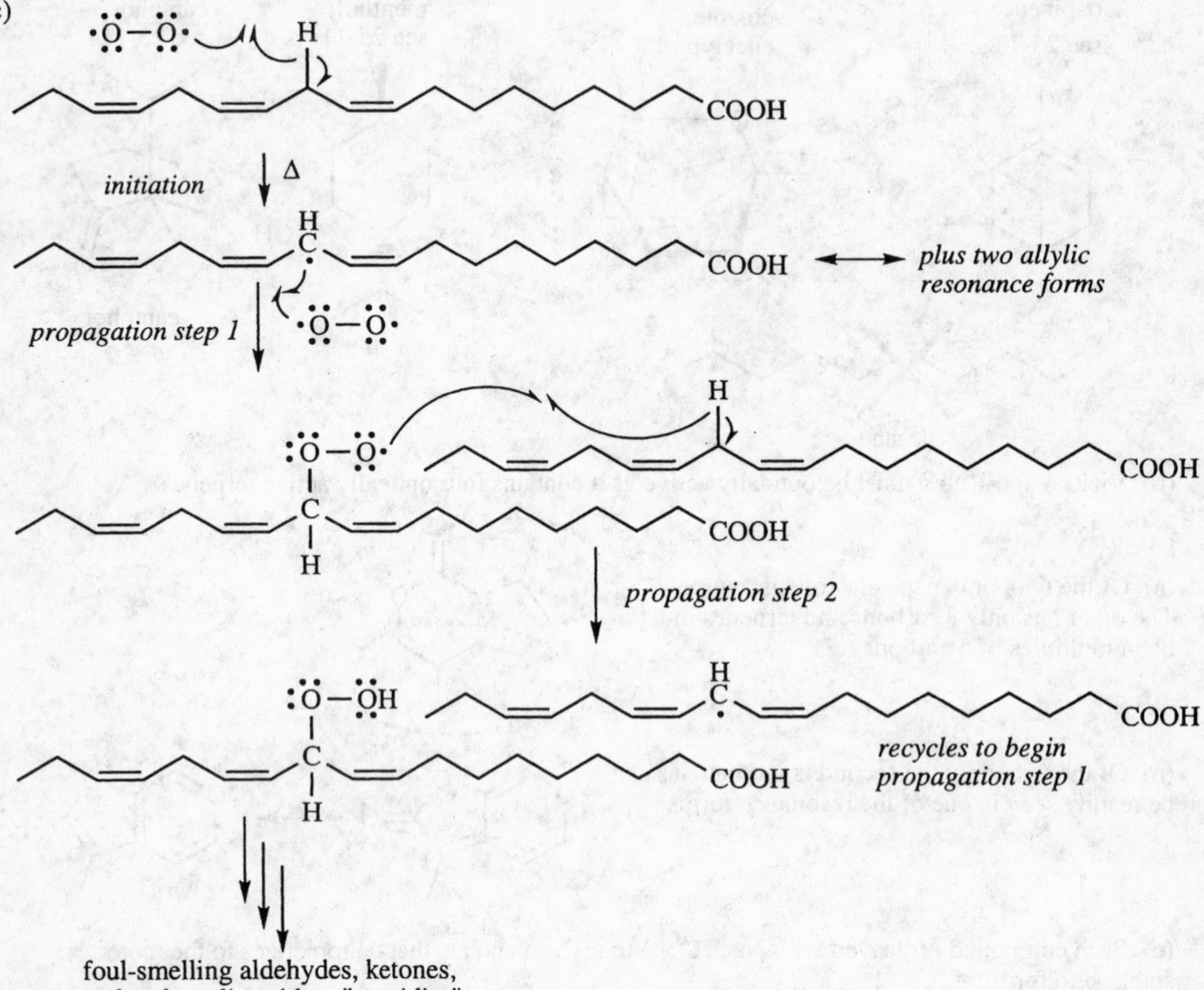

H
COOH
Δ
initiation
H
C
COOH
plus two allylic resonance forms
propagation step 1
H
COOH
C
H
COOH
propagation step 2
OH
H
C
COOH
C
H
COOH
recycles to begin propagation step 1
foul-smelling aldehydes, ketones, and carboxylic acids = "rancidity"

(d) Antioxidants are molecules that stop the free radical chain mechanism. In each of the two cases here, abstraction of the phenolic H makes an oxygen radical that is highly stabilized by resonance, so stable that it does not continue the free radical chain process. It only takes a small amount of antioxidant to prevent the chain mechanism. Interestingly, in breakfast cereals, the antioxidant is usually put in the plastic bag that the cereal is packaged in, rather than in the food itself. BHA and BHT can also be used directly in food as there is no evidence that they are harmful to humans.

$(H_3C)_3C$
OCH_3
$(CH_3)_3C$
$C(CH_3)_3$
CH_3

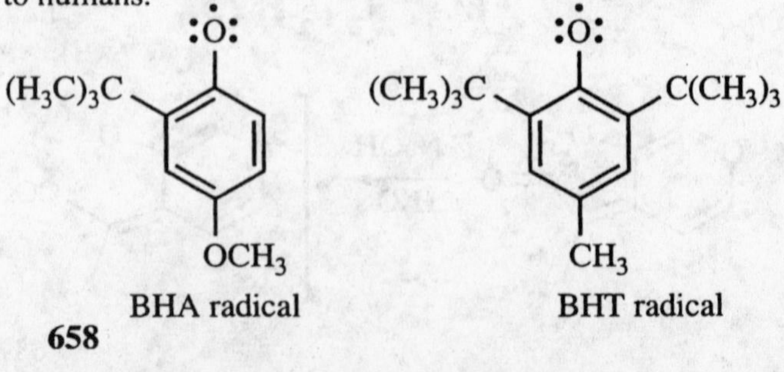

BHA radical BHT radical

CHAPTER 26—SYNTHETIC POLYMERS

Note: In this chapter, the "wavy bond" symbol means the continuation of a polymer chain.

26-1

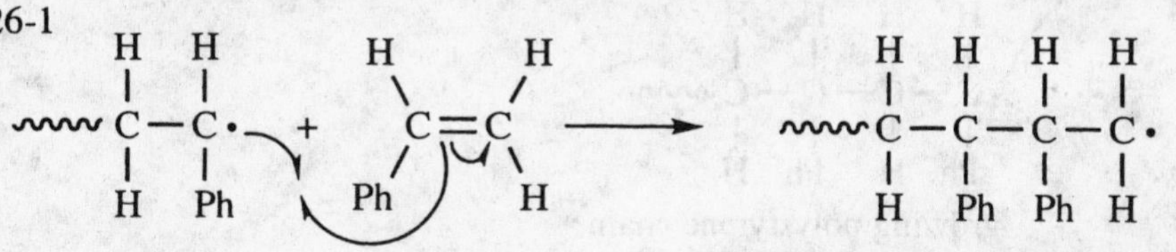

1° radical, and *not* resonance-stabilized—
this orientation is not observed

Orientation of addition always generates the more stable intermediate; the energy difference between a 1° radical (shown above) and a benzylic radical is huge. Moreover, this energy difference accumulates with each repetition of the propagation step. The phenyl substituents must necessarily be on alternating carbons because the orientation of attack is always the same—not a random process.

26-2

$$PhCOO-OOCPh \longrightarrow 2\ Ph\cdot + 2\ CO_2$$

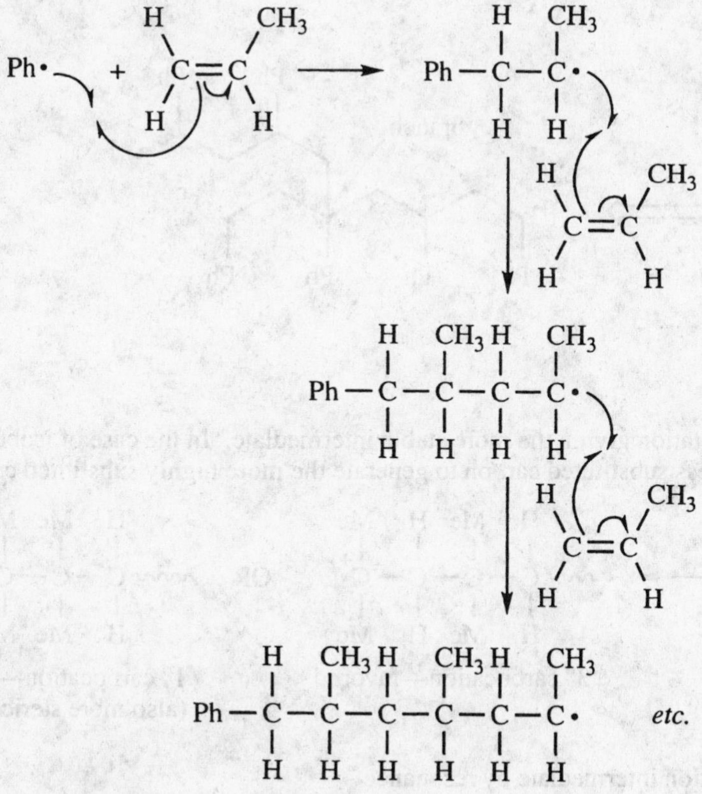

659

26-3 The benzylic hydrogen will be abstracted in preference to a 2° hydrogen because the benzylic radical is both 3° and resonance-stabilized, and the 2° radical is neither.

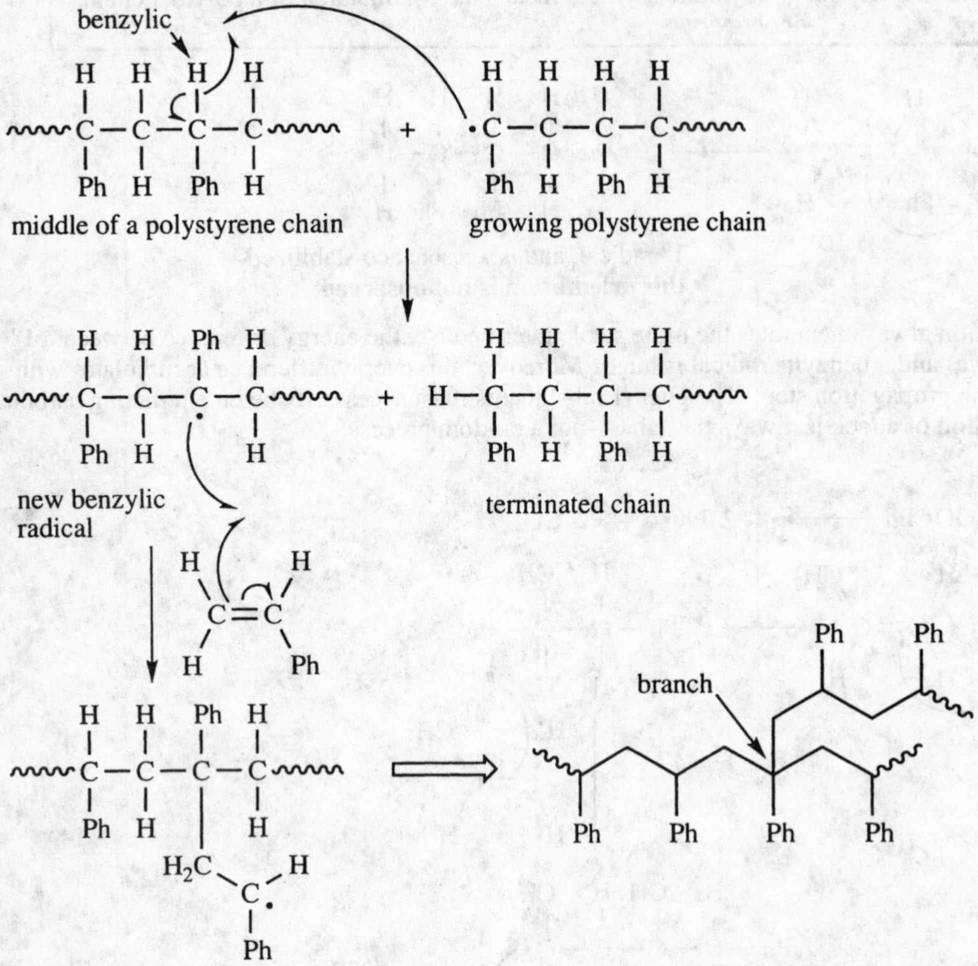

middle of a polystyrene chain   growing polystyrene chain

new benzylic radical   terminated chain

branch

26-4 Addition occurs with the orientation giving the more stable intermediate. In the case of isobutylene, the growing chain will bond at the less substituted carbon to generate the more highly substitited carbocation.

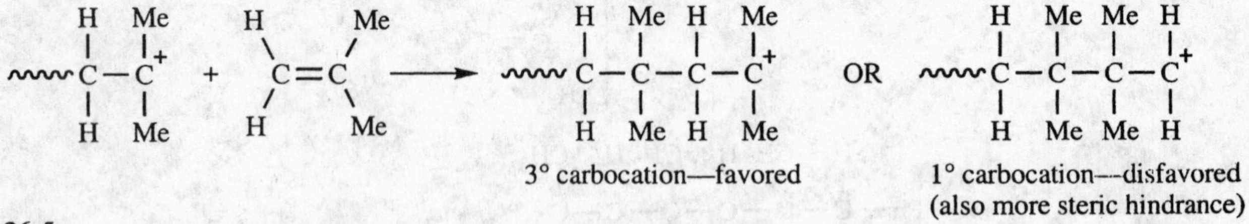

3° carbocation—favored

OR

1° carbocation—disfavored
(also more steric hindrance)

26-5

(a) chlorine can stabilize a carbocation intermediate by resonance

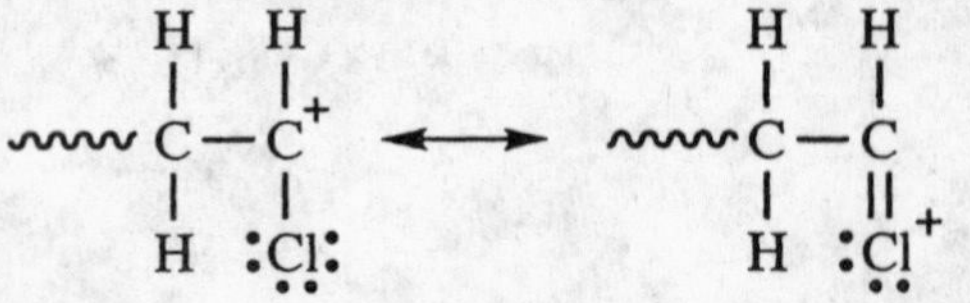

26-5 continued

(b) CH₃ can stabilize the carbocation intermediate by induction

$$\text{wwww}\overset{\overset{\displaystyle H}{|}}{\underset{\underset{\displaystyle H}{|}}{C}}-\overset{\overset{\displaystyle H}{|}}{\underset{\underset{\displaystyle CH_3}{|}}{C}}{}^{+}$$ 2° (not the best case imaginable, but still possible)

(c) terrible for cationic polymerization: both substituents are electron-withdrawing and would *destabilize* the carbocation intermediate

$$\text{wwww}\overset{\overset{\displaystyle H}{|}}{\underset{\underset{\displaystyle H}{|}}{C}}-\overset{\overset{\displaystyle COOCH_3}{|}}{\underset{\underset{\displaystyle C\equiv N}{|}}{C}}{}^{+}$$ *destabilized* carbocation

26-6   benzylic

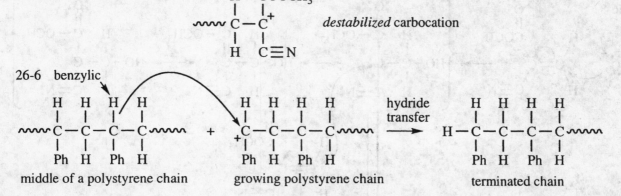

middle of a polystyrene chain   growing polystyrene chain   terminated chain

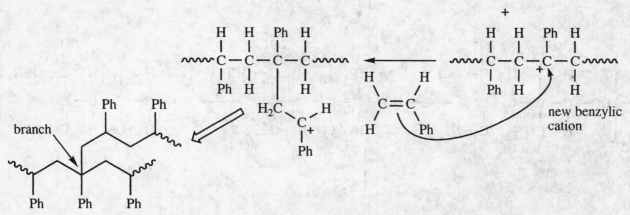

branch

Polystyrene is particularly susceptible to branching because the 3° benzylic cation produced by a hydride transfer is so stable. In poly(isobutylene), there is no hydrogen on the carbon with the stabilizing substituents; any hydride transfer would generate a 2° carbocation at the expense of a 3° carbocation at the end of a growing chain—this is an increase in energy and therefore unfavorable.

$$2°$$

$$\text{wwww}\overset{\overset{\displaystyle H}{|}}{\underset{\underset{\displaystyle H}{|}}{C}}-\overset{\overset{\displaystyle Me}{|}}{\underset{\underset{\displaystyle Me}{|}}{C}}-\overset{\overset{\displaystyle H}{|}}{\underset{\underset{\displaystyle H}{|}}{C}}-\overset{\overset{\displaystyle Me}{|}}{\underset{\underset{\displaystyle Me}{|}}{C}}\text{wwww}$$ + $$\overset{+}{\underset{3°}{}}\overset{\overset{\displaystyle Me}{|}}{\underset{\underset{\displaystyle Me}{|}}{C}}-\overset{\overset{\displaystyle H}{|}}{\underset{\underset{\displaystyle H}{|}}{C}}-\overset{\overset{\displaystyle Me}{|}}{\underset{\underset{\displaystyle Me}{|}}{C}}-\overset{\overset{\displaystyle H}{|}}{\underset{\underset{\displaystyle H}{|}}{C}}\text{wwww}$$ **no hydride transfer**

middle of a poly(isobutylene) chain   growing poly(isobutylene) chain

661

26-7

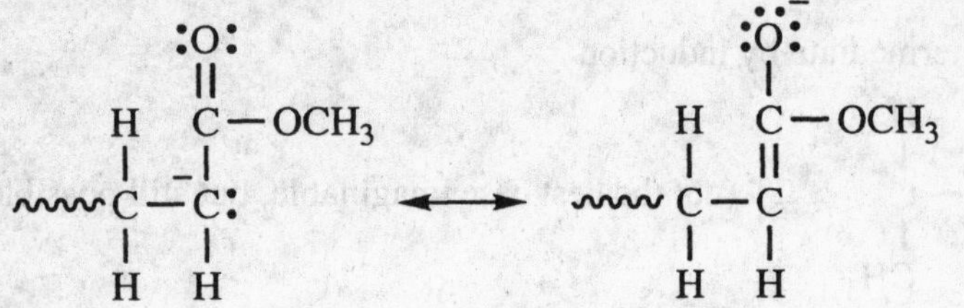

26-8

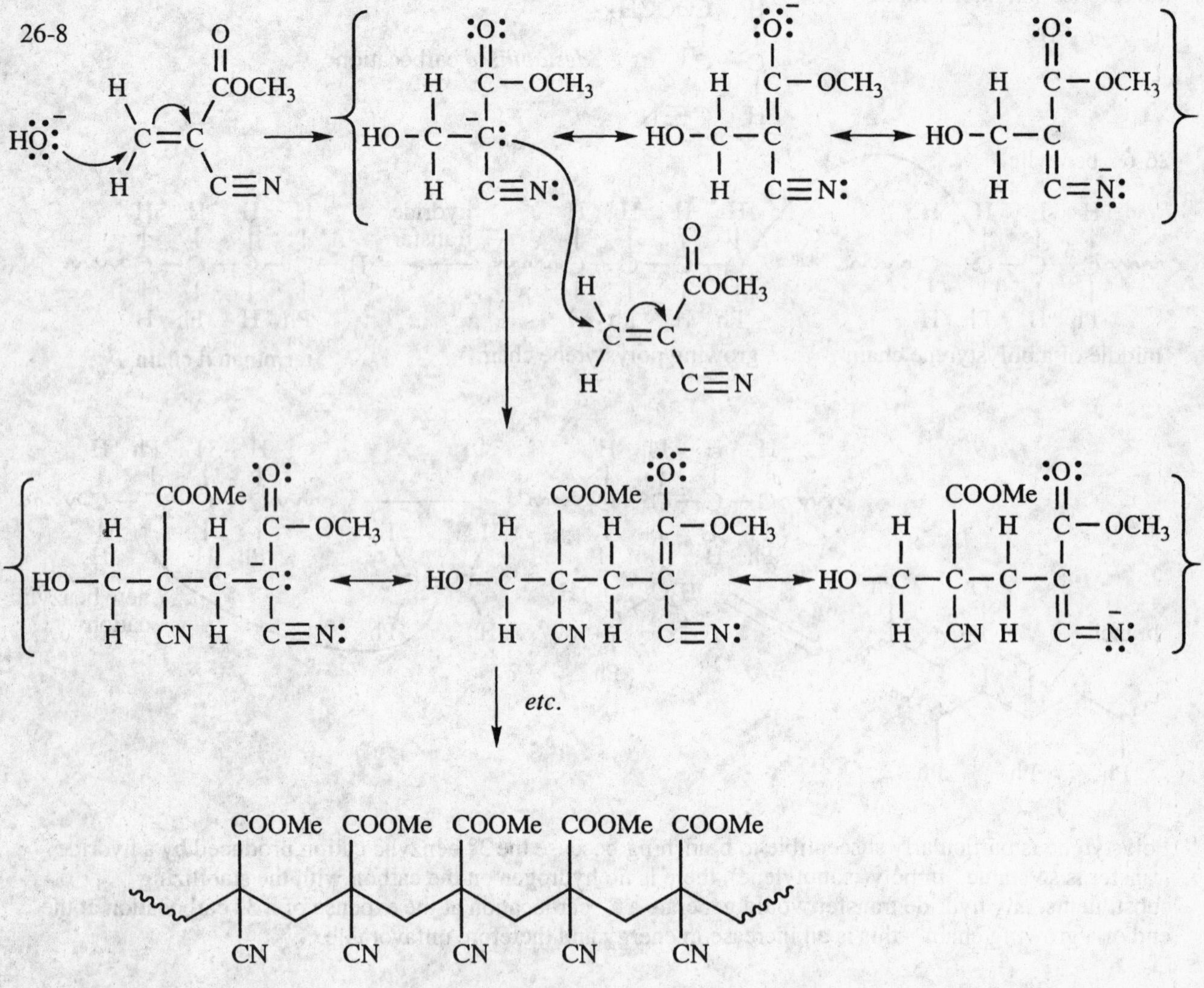

This polymerization goes so quickly because the anionic intermediate is highly resonance stabilized by the carbonyl and the cyano groups. A stable intermediate suggests a low activation energy which translates to a fast reaction.

26-9

(a)

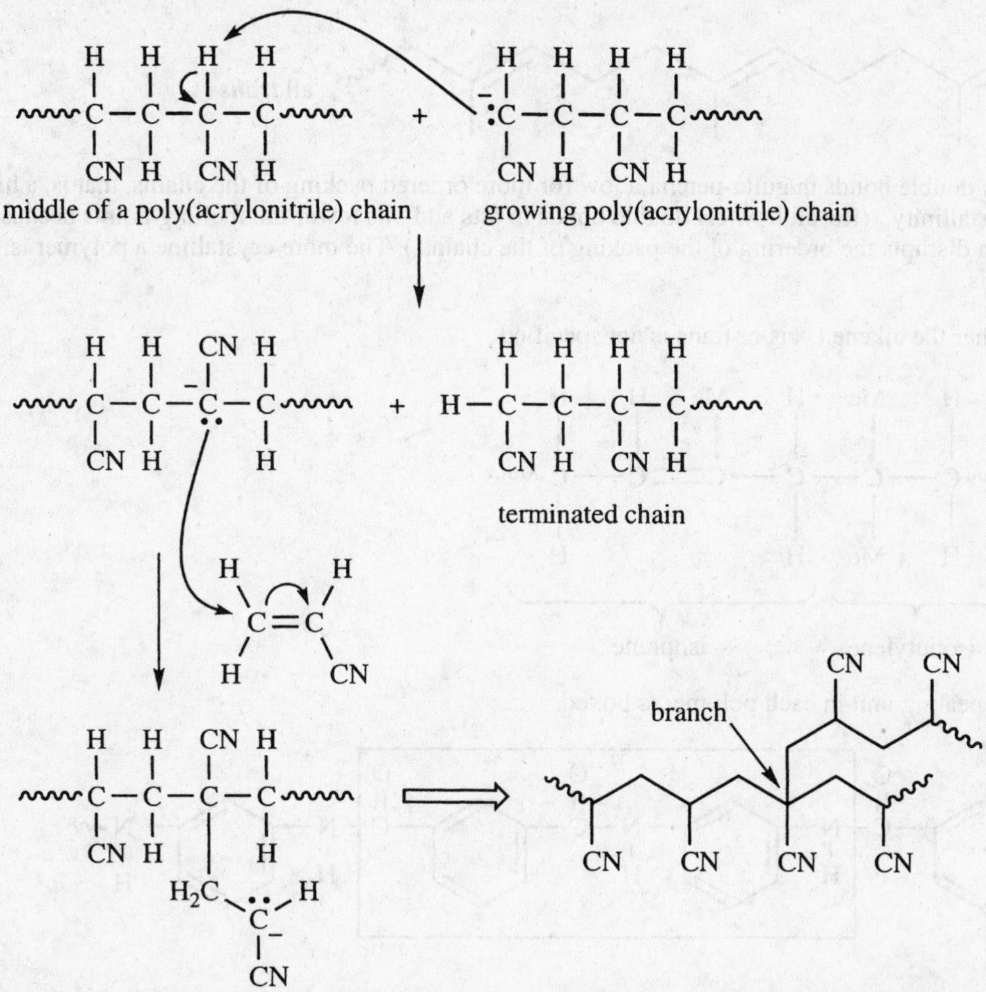

middle of a poly(acrylonitrile) chain   growing poly(acrylonitrile) chain

terminated chain

branch

(b) The chain-branching hydride transfer (from a cationic mechanism) or proton transfer (from an anionic mechanism) ends a less-highly-substituted end of a chain and generates an intermediate on a more-highly-substituted middle of a chain (a 3° carbon in these mechanisms). This stabilizes a carbocation, but greater substitution *destabilizes* a carbanion. Branching can and does happen in anionic mechanisms, but it is less likely than in cationic mechanisms.

26-10   isotactic poly(acrylonitrile)

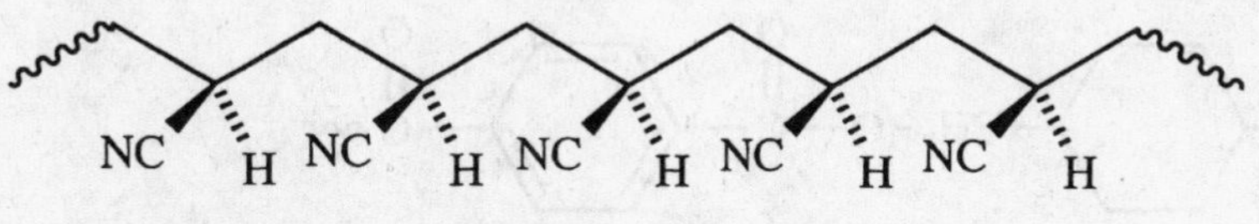

syndiotactic polystyrene

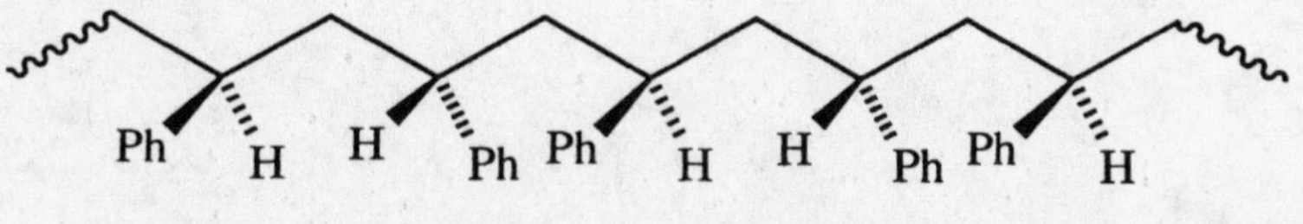

26-11

(a)

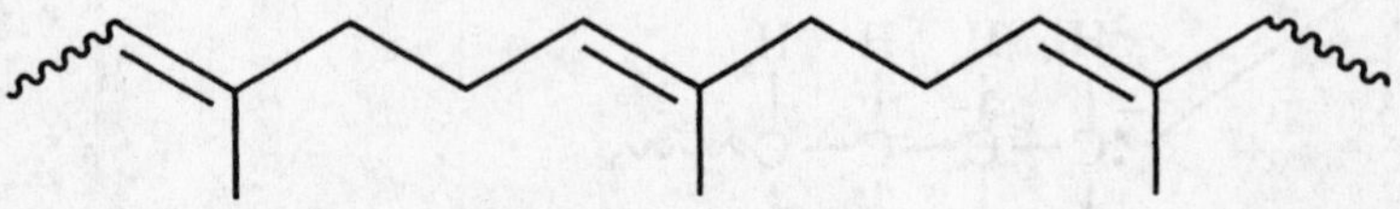

all *trans*

(b) The *trans* double bonds in gutta-percha allow for more ordered packing of the chains, that is, a higher degree of crystallinity. (Recall how *cis* double bonds in fats and oils lower the melting points because the *cis* orientation disrupts the ordering of the packing of the chains.) The more crystalline a polymer is, the less elastic it is.

26-12 Whether the alkene is cis or trans is not specified.

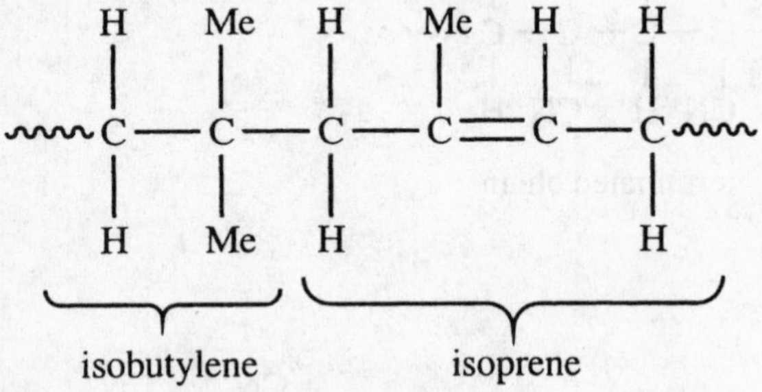

isobutylene   isoprene

26-13 The repeating unit in each polymer is boxed.

(a) Nomex®

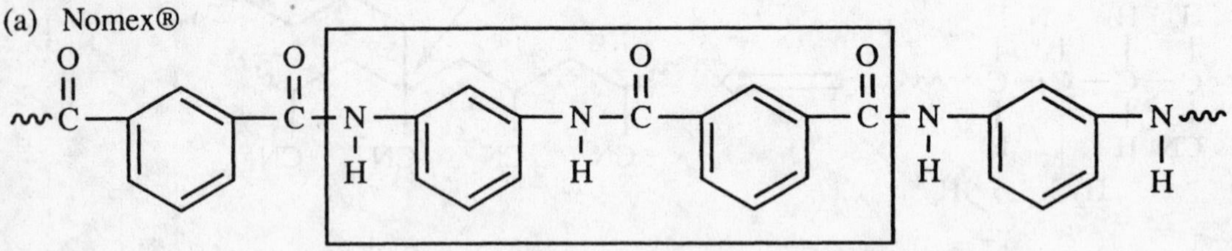

(b) Kevlar®

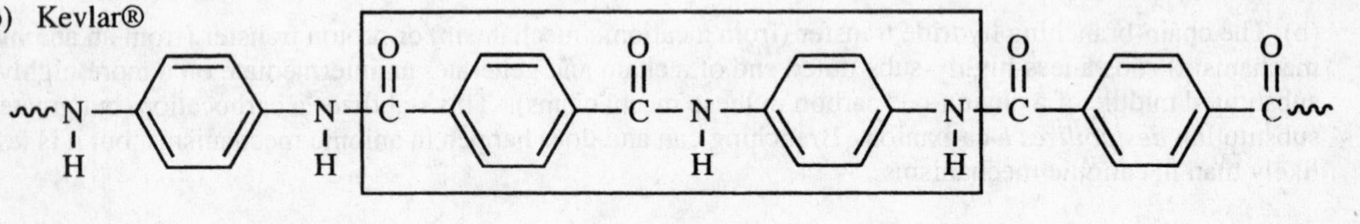

26-14 Kodel® polyester (only one repeating unit shown)

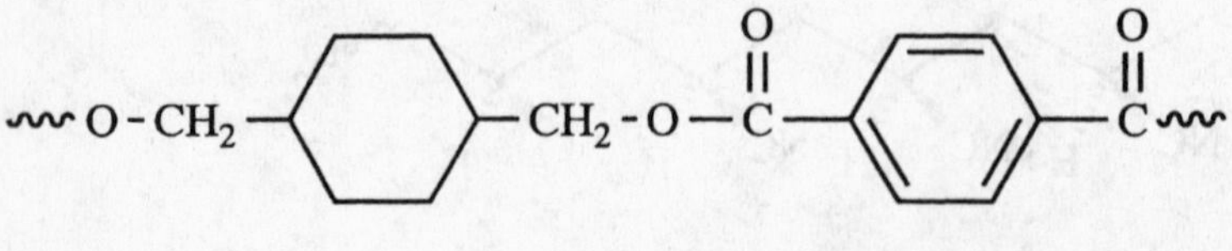

26-15

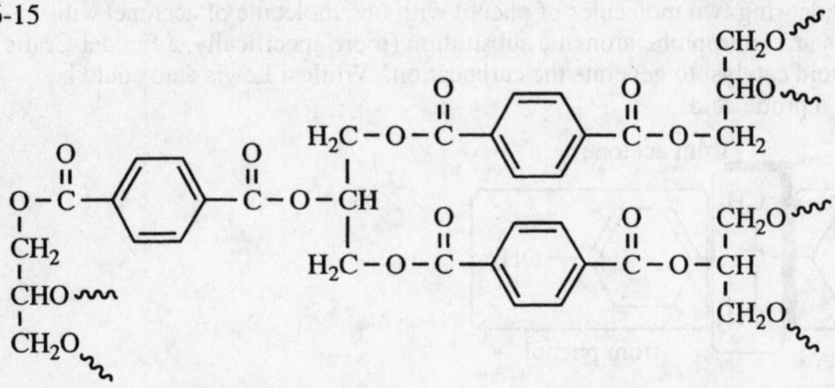

Glycerol is a trifunctional molecule, so not only does it grow in two directions to make a chain, it grows in three directions. All of its chains are cross-linked, forming a three-dimensional lattice with very little motion possible. The more cross-linked the polymer is, the more rigid it is.

26-16 For simplicity in this problem, bisphenol A will be abbreviated as a substituted phenol.

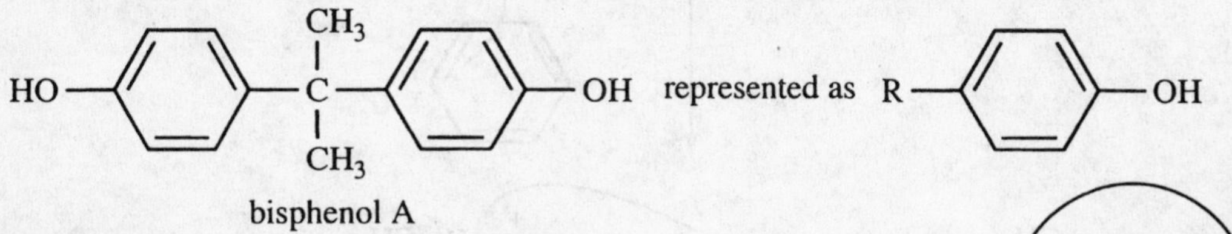

bisphenol A

mechanism

665

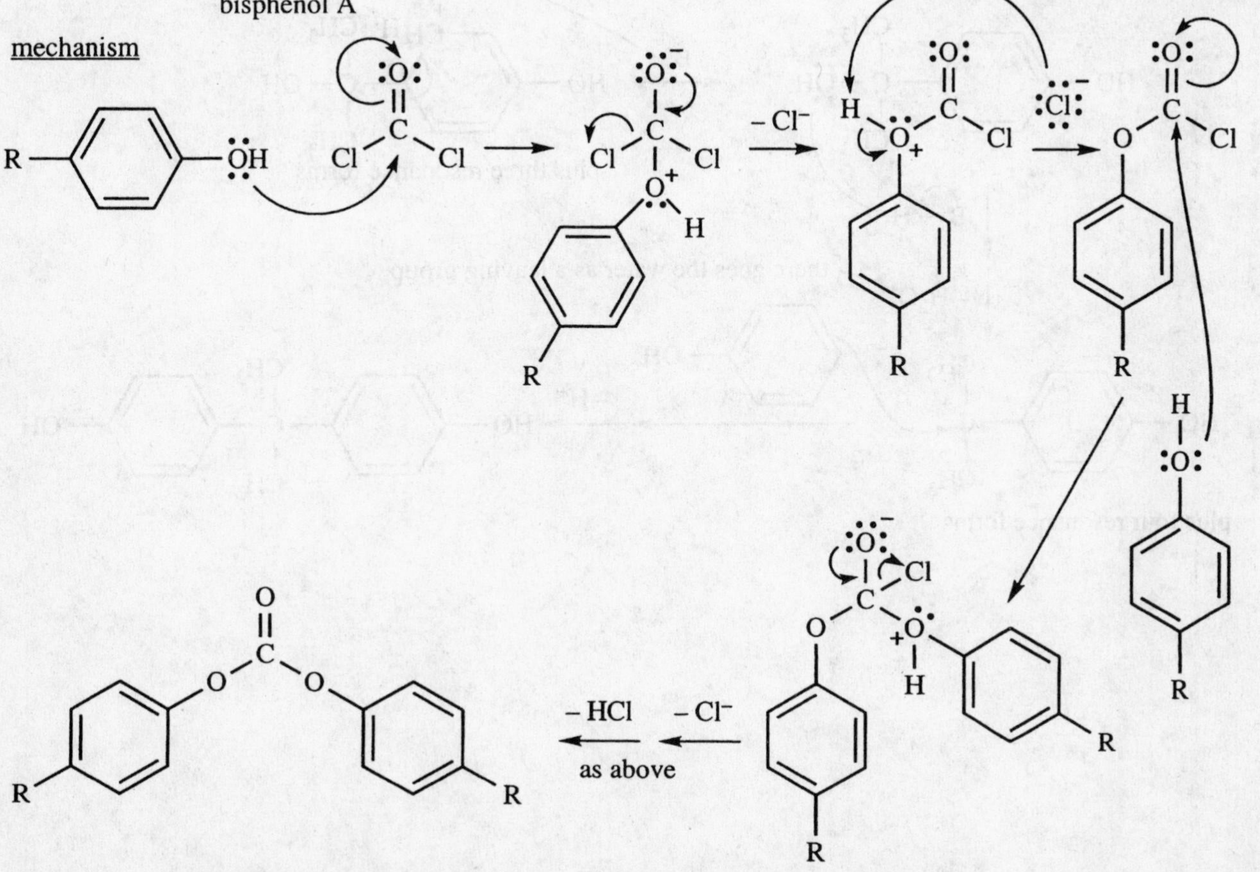

26-17 Bisphenol A is made by condensing two molecules of phenol with one molecule of acetone, with loss of a molecule of water. This is an electrophilic aromatic subsitution (more specifically, a Friedel-Crafts alkylation), and would require an acid catalyst to generate the carbocation. While a Lewis acid could be used, the mechanism below shows a protic acid.

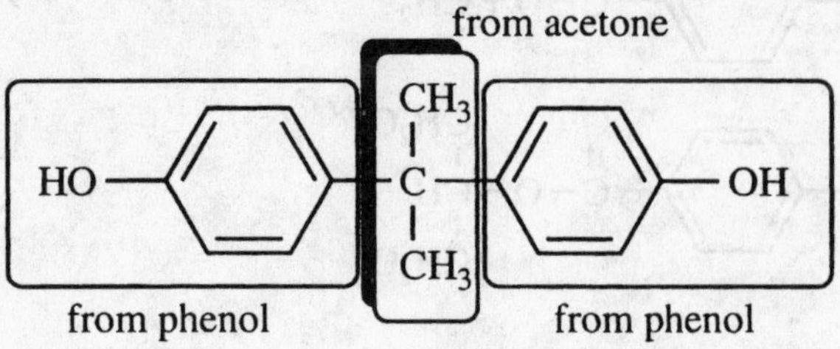

from acetone

from phenol   from phenol

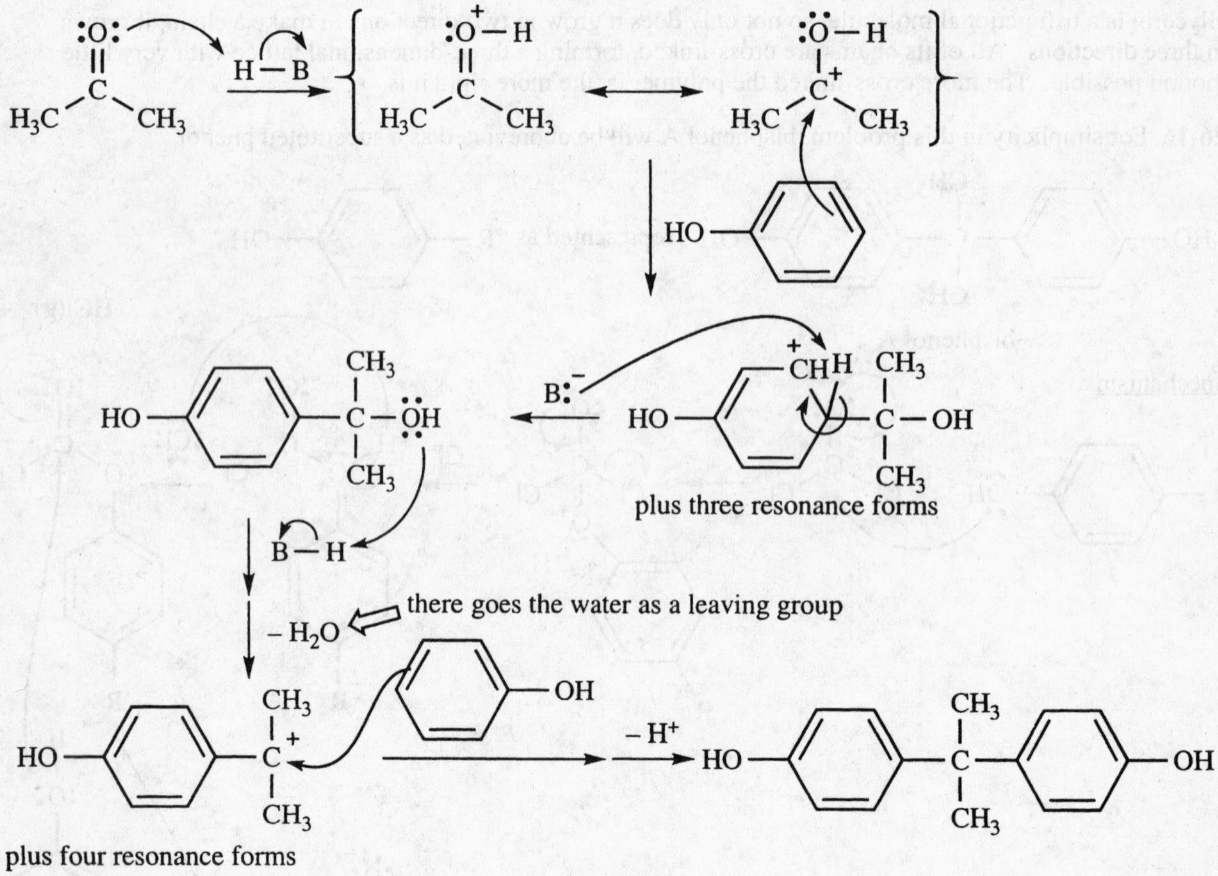

plus three resonance forms

there goes the water as a leaving group

plus four resonance forms

26-18

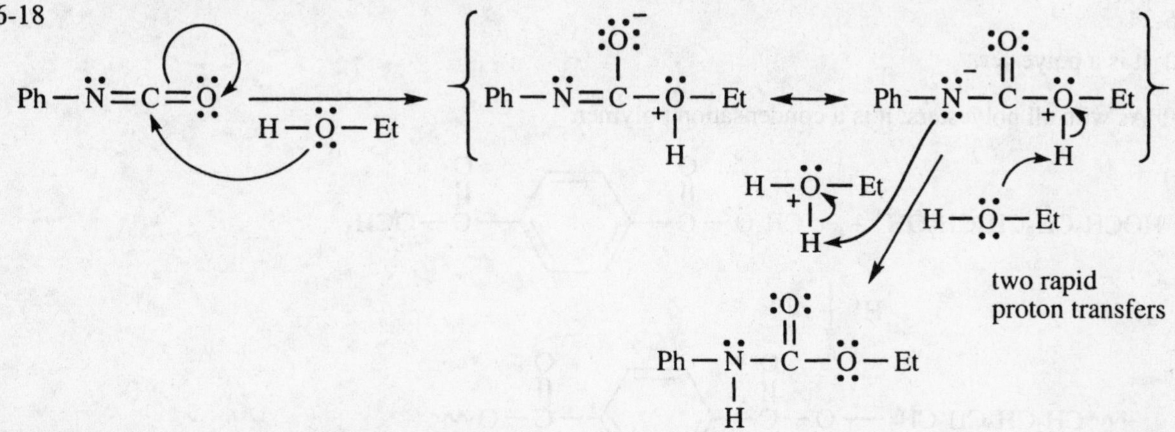

26-19 Glycerol is a trifunctional alcohol. It uses two of its OH groups in a growing chain. The third OH group cross-links with another chain. The more cross-linked a polymer, the more rigid it is.

26-20

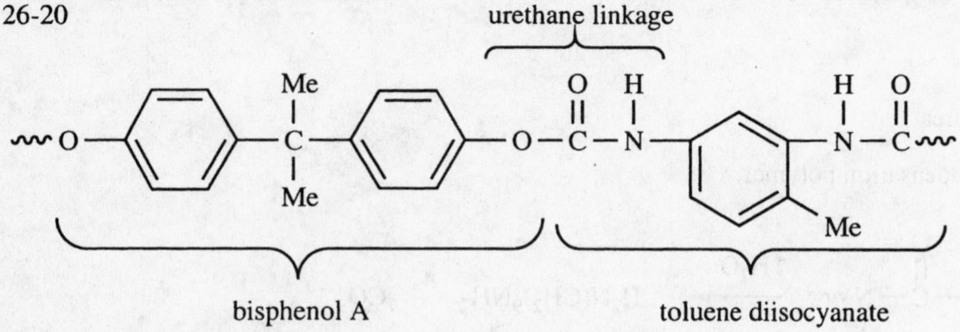

bisphenol A   toluene diisocyanate

26-21 Please refer to solution 1-20, page 12 of this Solutions Manual.

26-22

(a)

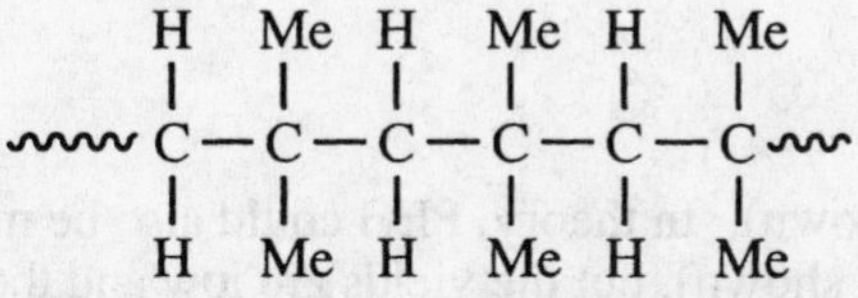

(b) Polyisobutylene is an addition polymer. No small molecule is lost, so this cannot be a condensation polymer.

(c) Either cationic polymerization or free-radical polymerization would be appropriate. The carbocation or free-radical intermediate would be 3° and therefore relatively stable. Anionic polymerization would be inappropriate as there is no electron-withdrawing group to stabilize the anion.

26-23

(a) It is a polyurethane.

(b) As with all polyurethanes, it is a condensation polymer.

(c)

$$\sim\sim CH_2CH_2CH_2 - N - C - O \sim\sim \xrightarrow{H_2O} HOCH_2CH_2CH_2NH_2 + CO_2$$

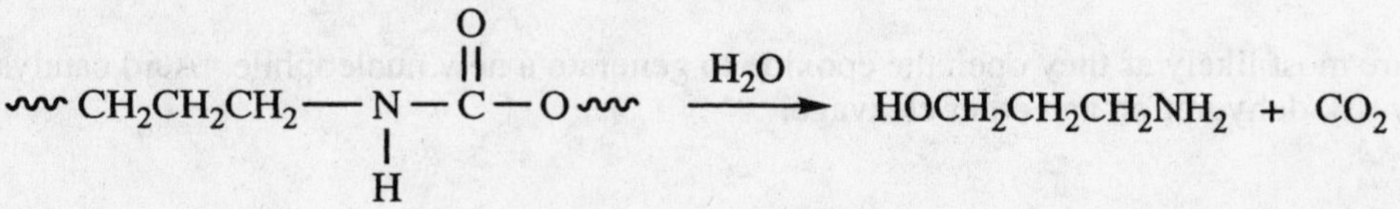

26-24

(a) It is a polyester.

(b) As with all polyesters, it is a condensation polymer.

(c)

$$HOCH_2CH_2CH_2CH_2OH \ + \ CH_3O-\overset{\overset{\displaystyle O}{\|}}{C}-\!\!\!\left\langle\!\!\bigcirc\!\!\right\rangle\!\!\!-\overset{\overset{\displaystyle O}{\|}}{C}-OCH_3$$

$$H^+ \downarrow \Delta$$

$$\sim\!\!\!\sim CH_2CH_2CH_2CH_2-O-\overset{\overset{\displaystyle O}{\|}}{C}-\!\!\!\left\langle\!\!\bigcirc\!\!\right\rangle\!\!\!-\overset{\overset{\displaystyle O}{\|}}{C}-O\!\sim\!\!\!\sim$$

$$+ \ CH_3OH$$

Using the dicarboxylic acid instead of the ester would produce water as the small neutral molecule lost in this condensation.

26-25

(a) Urylon® is a polyurea.

(b) A polyurea is a condensation polymer.

(c)

$$\sim\!\!\!\sim (CH_2)_9-\underset{\underset{\displaystyle H}{|}}{N}-\overset{\overset{\displaystyle O}{\|}}{C}-\underset{\underset{\displaystyle H}{|}}{N}\!\sim\!\!\!\sim \ \xrightarrow{H_2O} \ H_2N(CH_2)_9NH_2 \ + \ CO_2$$

26-26

(a) Polyethylene glycol, abbreviated PEG, is a polyether.

(b) PEG is usually made from ethylene oxide (first reaction shown). In theory, PEG could also be made by intermolecular dehydration of ethylene glycol (second reaction shown), but the yields are low and the chains are short.

$$n \ \triangle\!\!\!O \ + \ HO^- \ \longrightarrow \ HO\!\sim\!\!O\!\sim\!\!O\!\sim\!\!O\!\sim\!\!\!\sim$$

ethylene
oxide

$$HO\!\sim\!\!OH \ \xrightarrow[-\,H_2O]{\overset{\displaystyle H^+}{\Delta}} \ HO\!\sim\!\!O\!\sim\!\!O\!\sim\!\!O\!\sim\!\!\!\sim$$

ethylene
glycol

(c) Basic catalysts are most likely as they open the epoxide to generate a new nucleophile. Acid catalysts are possible but they risk dehydration and ether cleavage.

26-26 continued

(d) Mechanism of ethylene oxide polymerization (showing hydroxide as the base):

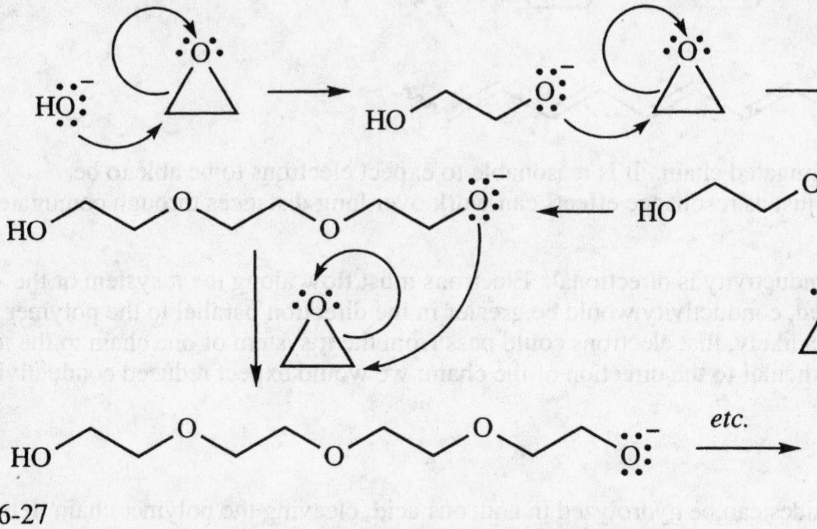

etc.

26-27

(a) Polychloroprene (Neoprene®) is an addition polymer.

(b) Polychloroprene comes from the diene, chloroprene, just as natural rubber comes from isoprene:

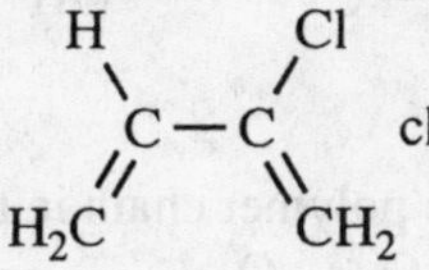

chloroprene

26-28

(a)

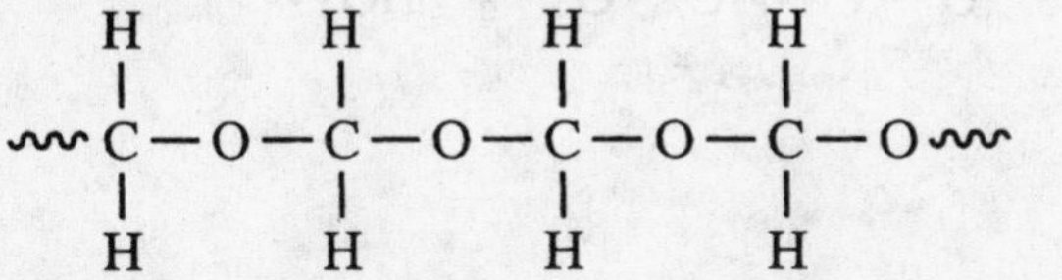

Delrin® (polyformaldehyde)

(b) All of these intermediates are resonance-stabilized.

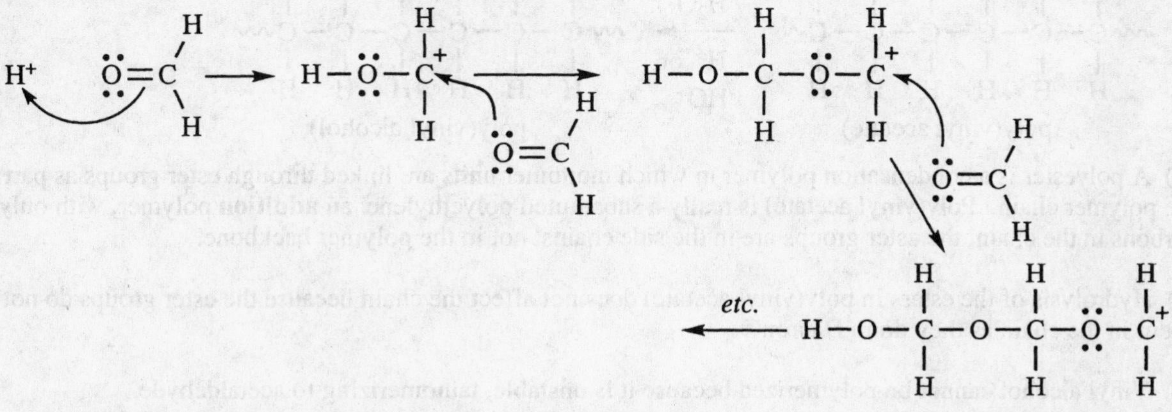

etc.

trimer

(c) Delrin® is an addition polymer; instead of adding across the double bond of an alkene, addition occurs across the double bond of a carbonyl group.

669

26-29

(a) *cis*

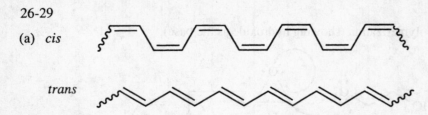

trans

(b) Each structure has a fully conjugated chain. It is reasonable to expect electrons to be able to be transferred through the π system, just as resonance effects can work over long distances through conjugated systems.

(c) It is not surprising that the conductivity is directional. Electrons must flow along the π system of the chain, so if the chains were aligned, conductivity would be greater in the direction parallel to the polymer chains. (It is possible, though less likely, that electrons could pass from the π system of one chain to the π system of another, that is, perpendicular to the direction of the chain; we would expect reduced conductivity in that direction.)

26-30

(a) A Nylon is a polyamide. Amides can be hydrolyzed in aqueous acid, cleaving the polymer chain in the process.

$$\text{mwN}-\underset{\underset{H}{|}}{\overset{\overset{O}{\|}}{C}}\text{mw}\overset{\overset{O}{\|}}{C}-\underset{\underset{H}{|}}{N}\text{mw} \xrightarrow{H_3O^+} \text{mw}\overset{+}{N}H_3 + HO-\overset{\overset{O}{\|}}{C}\text{mw}\overset{\overset{O}{\|}}{C}-OH + H_3\overset{+}{N}\text{mw}$$

(b) A polyester can be saponified in aqueous base, cleaving the polymer chain in the process.

$$\text{mw}O-\overset{\overset{O}{\|}}{C}\text{mw}\overset{\overset{O}{\|}}{C}-O\text{mw} \xrightarrow{NaOH} \text{mw}OH + {}^-O-\overset{\overset{O}{\|}}{C}\text{mw}\overset{\overset{O}{\|}}{C}-O^- + HO\text{mw}$$

26-31

(a)

$$\text{mwC}-\underset{H}{\overset{H}{|}}\text{C}-\underset{H}{\overset{O}{|}}\text{C}-\underset{H}{\overset{H}{|}}\text{C}-\underset{H}{\overset{O}{|}}\text{C}-\underset{H}{\overset{H}{|}}\text{C}-\underset{H}{\overset{O}{|}}\text{C}\text{mw} \xrightarrow[\substack{H^+ \text{ or} \\ HO^-}]{H_2O} \text{mwC}-\text{C}-\text{C}-\text{C}-\text{C}-\text{C}\text{mw}$$

poly(vinyl acetate)   poly(vinyl alcohol)

(b) A polyester is a condensation polymer in which monomer units are linked through ester groups as part of the polymer chain. Poly(vinyl acetate) is really a substituted polyethylene, an **addition** polymer, with only carbons in the chain; the ester groups are in the side chains, not in the polymer backbone.

(c) Hydrolysis of the esters in poly(vinyl acetate) does not affect the chain because the ester groups do not occur in the chain as they do in Dacron®.

(d) Vinyl alcohol cannot be polymerized because it is unstable, tautomerizing to acetaldehyde.

$$H_2C=\overset{\overset{OH}{|}}{CH} \underset{\longleftarrow}{\xrightarrow{\hspace{2cm}}} CH_3-\overset{\overset{O}{\|}}{CH}$$

26-32

(a)

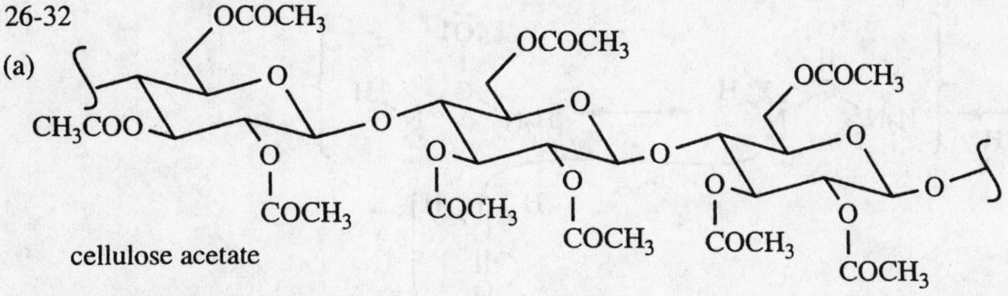
OCOCH3
CH3COO
O
COCH3

cellulose acetate

(b) Cellulose has three OH groups per glucose monomer, which form hydrogen bonds with other polar groups. Transforming these OH groups into acetates makes the polymer much less polar and therefore more soluble in organic solvents.

(c) The acetone dissolved the cellulose acetate in the fibers. As the acetone evaporated, the cellulose acetate remained but no longer had the fibrous, woven structure of cloth. It recrystallized as white fluff.

(d) Any article of clothing made from synthetic fibers is susceptible to the ravages of organic solvents. Solvent splashes leave dimples or blotches on Corfam shoes. (Yet, Corfam shoes could still provide protection for the toenail polish!)

26-33

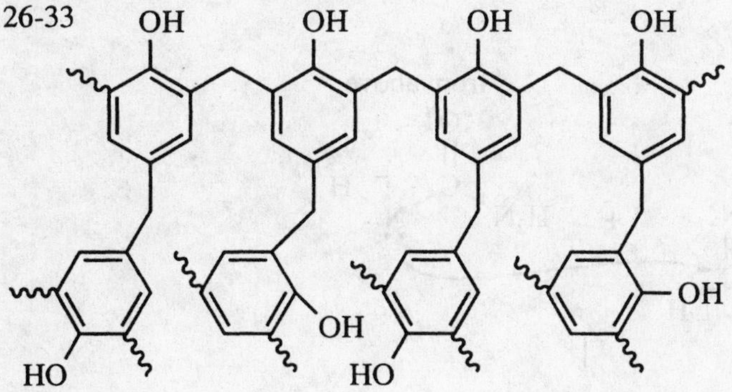
OH
HO

Bakelite is highly cross-linked through the ortho and para positions of phenol; each phenol can form a chain at two ring positions, then form a branch at the third position.

<u>mechanism</u>

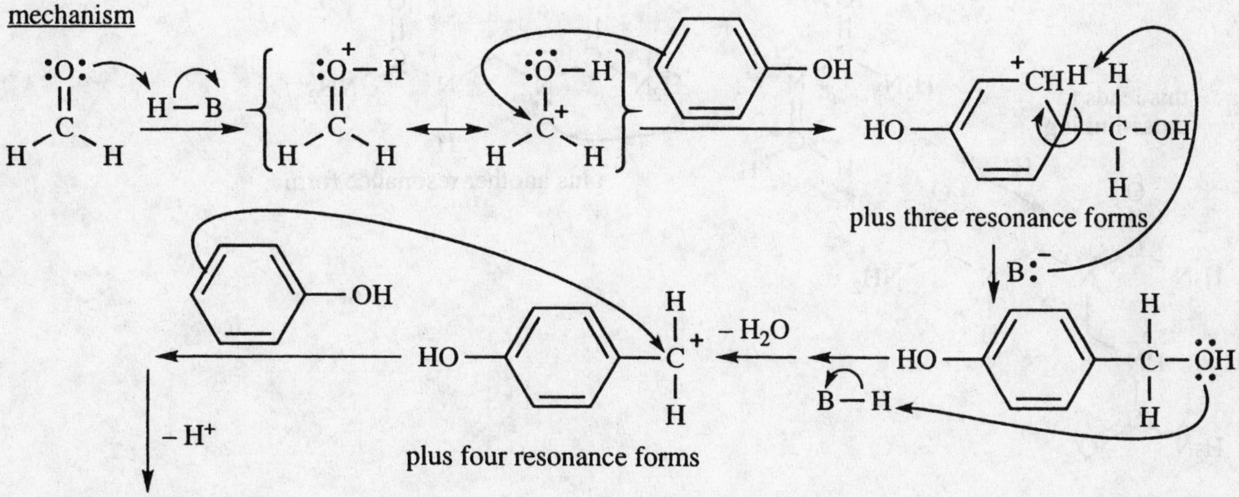
H—B
OH
HO
CH
C—OH
plus three resonance forms
B:
− H2O
B—H
− H+
plus four resonance forms

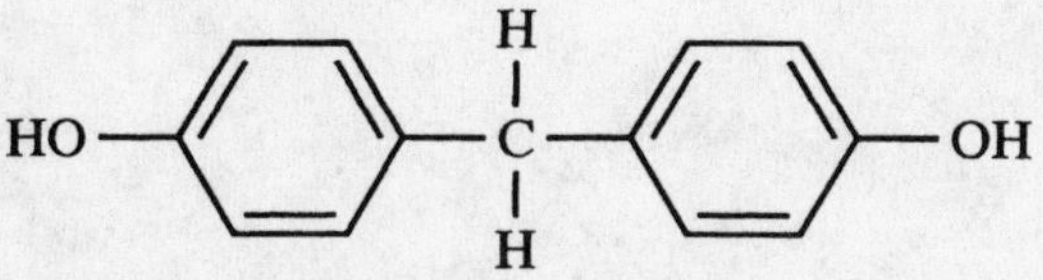
HO
OH

further coupling at ortho positions leads to cross-linked Bakelite

26-34

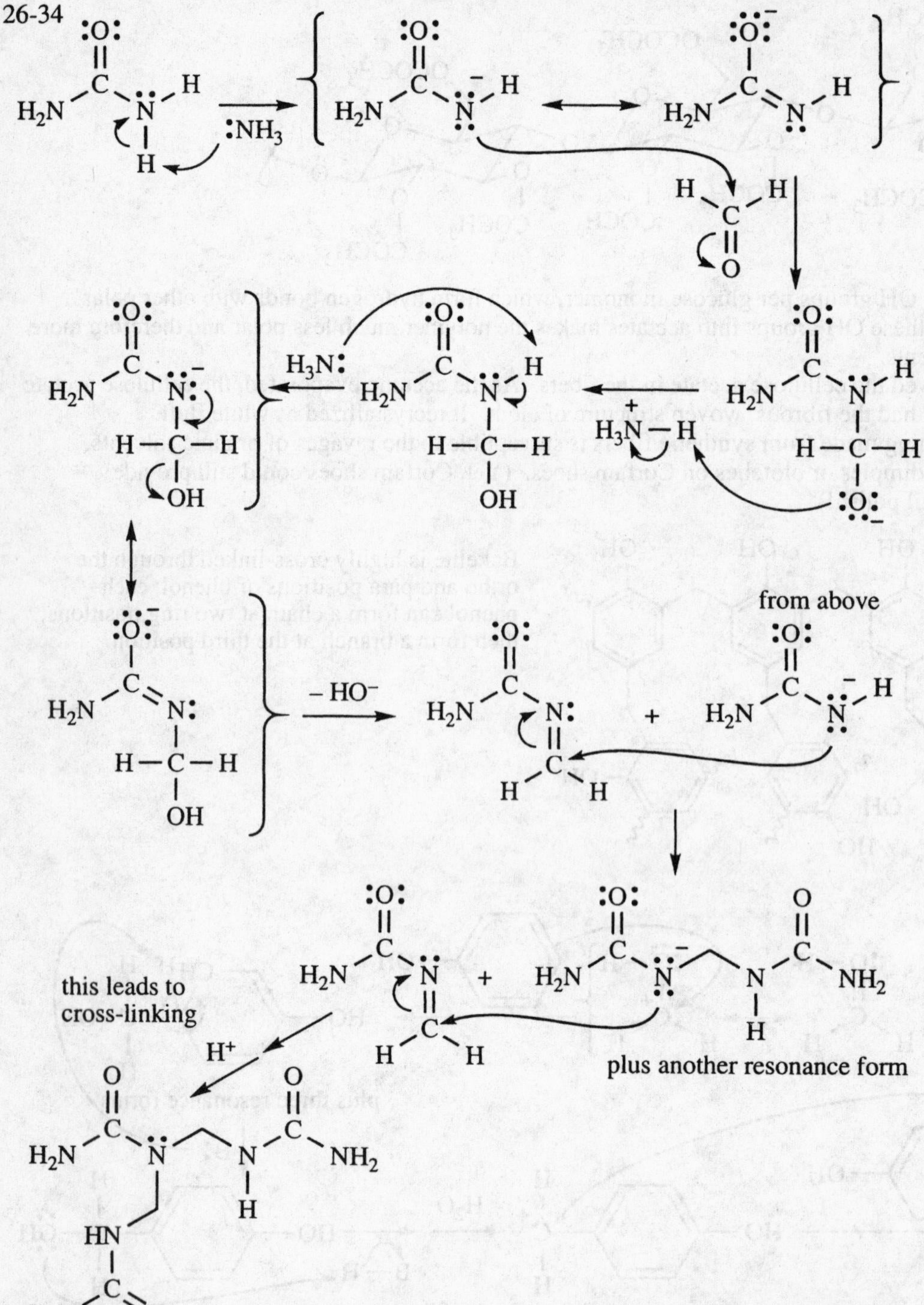

:NH3
H3N:
from above
– HO⁻
this leads to
cross-linking
H⁺
plus another resonance form

26-35

glycolic acid | lactic acid | glycolic acid | lactic acid | glycolic acid | lactic acid

26-36

cellulose = cotton

polypropylene

As we have seen repeatedly through this presentation of organic chemistry, physical and chemical behavior depend on *structure*. The structure of cotton, i.e. cellulose, has multiple oxygen atoms that form hydrogen bonds with water. When cotton gets wet, it holds onto the water tightly, as you have seen if you have put cotton clothes in a clothes dryer—it takes a long time to dry. Polypropylene is a hydrocarbon with no hydrogen bonding groups; the fiber feels dry because it cannot hold the water the way cotton can. Athletic garments are increasingly using polypropylene because they allow evaporation and cooling during periods of exertion; cotton is just the opposite.

26-37

(a) This addition polymer is called polyvinylidene chloride, trade name Saran®. It could be made by any of the three mechanism types: radical, cationic, or anionic.

monomer

(b) When the substituent is on every fourth carbon, and one double bond in the chain in every 4-carbon unit, the polymer must come from addition across a diene, probably under cationic conditions.

monomer

(c) This polyester is a condensation polymer of two monomers, a diol and a derivative of phthalic acid, either the anhydride, an ester, the acid chloride, or the acid itself. Heating the monomers will make the polymer; no catalyst is required if done at high temperature.

monomer

one of these is the other monomer

X = OH or Cl or OR

673

26-37 continued

(d) This polyamide (Nylon) is a condensation polymer made from two monomers, a diamine and a derivative of succinic acid, either the anhydride, an ester, the acid chloride, or the acid itself. Heating the monomers will make the polymer; no catalyst is required if done at high temperature.

H$_2$N ⌒⌒ NH$_2$
monomer

one of these is the other monomer

X = OH or Cl or OR

26-38

(a)

COOH + HO ⌒ OH $\xrightarrow{H^+}$ OCH$_2$CH$_2$OH
CH$_3$ CH$_3$ hydroxyethyl methacrylate

polymerize using radical or anionic conditions

polymer

(b) This polymer has a few properties that make it useful as the material in soft, extended-wear contact lenses. First, carboxylic acids usually are crystalline solids with high melting points, but esters and alcohols are low melting, often liquids, so the polymer with this ester is softer than the carboxylic acid or even the methyl ester. (The methyl ester, polymethyl methacrylate or Plexiglas, was the first material used in the original hard contact lenses.) Second, the ability of the free OH to form hydrogen bonds with water makes the contact lens more fluid and less irritating to the cornea. Third, a hidden advantage but very important for ocular health: the fluidity of the contact lens also permits oxygen to go through the lens. Because the cornea does not have a large blood flow, it needs to absorb oxygen from the air to maintain its health, and this enhanced gas permeability permits the contact lens to be worn for days at a time without compromising the health of the cornea. Thanks, polymers!

Note to the student: BON VOYAGE! I hope you have enjoyed your travels through organic chemistry.
               Jan Simek

Appendix 1—Summary of IUPAC Nomenclature of Organic Compounds

Introduction

The purpose of the IUPAC system of nomenclature is to establish an international standard of naming compounds to facilitate communication. The goal of the system is to give each structure a unique and unambiguous name, and to correlate each name with a unique and unambiguous structure.

I. Fundamental Principle

IUPAC nomenclature is based on naming a molecule's longest chain of carbons connected by single bonds, whether in a continuous chain or in a ring. All deviations, either multiple bonds or atoms other than carbon and hydrogen, are indicated by prefixes or suffixes according to a specific set of priorities.

II. Alkanes and Cycloalkanes (also called "aliphatic" compounds)

Alkanes are the family of saturated hydrocarbons, that is, molecules containing carbon and hydrogen connected by single bonds only. These molecules can be in continuous chains (called linear or acyclic), or in rings (called cyclic or alicyclic). The names of alkanes and cycloalkanes are the root names of organic compounds. Beginning with the five-carbon alkane, the number of carbons in the chain is indicated by the Greek or Latin prefix. Rings are designated by the prefix "cyclo". (In the geometrical symbols for rings, each apex represents a carbon with the number of hydrogens required to fill its valence.)

C_1	CH_4	methane		C_{12}	$CH_3[CH_2]_{10}CH_3$	dodecane
C_2	CH_3CH_3	ethane		C_{13}	$CH_3[CH_2]_{11}CH_3$	tridecane
C_3	$CH_3CH_2CH_3$	propane		C_{14}	$CH_3[CH_2]_{12}CH_3$	tetradecane
C_4	$CH_3[CH_2]_2CH_3$	butane		C_{20}	$CH_3[CH_2]_{18}CH_3$	icosane
C_5	$CH_3[CH_2]_3CH_3$	pentane		C_{21}	$CH_3[CH_2]_{19}CH_3$	henicosane
C_6	$CH_3[CH_2]_4CH_3$	hexane		C_{22}	$CH_3[CH_2]_{20}CH_3$	docosane
C_7	$CH_3[CH_2]_5CH_3$	heptane		C_{23}	$CH_3[CH_2]_{21}CH_3$	tricosane
C_8	$CH_3[CH_2]_6CH_3$	octane		C_{30}	$CH_3[CH_2]_{28}CH_3$	triacontane
C_9	$CH_3[CH_2]_7CH_3$	nonane		C_{31}	$CH_3[CH_2]_{29}CH_3$	hentriacontane
C_{10}	$CH_3[CH_2]_8CH_3$	decane		C_{40}	$CH_3[CH_2]_{38}CH_3$	tetracontane
C_{11}	$CH_3[CH_2]_9CH_3$	undecane		C_{50}	$CH_3[CH_2]_{48}CH_3$	pentacontane

cyclopropane   cyclobutane   cyclopentane

cyclohexane   cycloheptane   cyclooctane

> The IUPAC system of nomenclature is undergoing many changes, most notably in the placement of position numbers. The new system places the position number close to the functional group designation; however, you should be able to use and recognize names in either the old or the new style. Ask your instructor which system to use.

III. Nomenclature of Molecules Containing Substituents and Functional Groups

A. Priorities of Substituents and Functional Groups
LISTED HERE FROM HIGHEST TO LOWEST PRIORITY, except that the substituents within Group C have equivalent priority.

Group A—Functional Groups Named By Prefix Or Suffix

Functional Group	Structure	Prefix	Suffix
Carboxylic Acid	$R-\overset{\overset{O}{\|\|}}{C}-OH$	carboxy-	-oic acid (-carboxylic acid)
Aldehyde	$R-\overset{\overset{O}{\|\|}}{C}-H$	oxo- (formyl)	-al (carbaldehyde)
Ketone	$R-\overset{\overset{O}{\|\|}}{C}-R$	oxo-	-one
Alcohol	$R-O-H$	hydroxy-	-ol
Amine	$R-N\diagdown$	amino-	-amine

Group B—Functional Groups Named By Suffix Only

Functional Group	Structure	Prefix	Suffix
Alkene	$C=C$	--------	-ene
Alkyne	$-C\equiv C-$	--------	-yne

Group C—Substituent Groups Named By Prefix Only

Substituent	Structure	Prefix	Suffix
Alkyl (see next page)	$R-$	alkyl-	--------
Alkoxy	$R-O-$	alkoxy-	--------

(alkoxy groups take the name of the alkyl group, like methyl or ethyl, drop the "yl", and add "oxy"; CH_3O is methoxy; CH_3CH_2O is ethoxy)

Halogen			
	$F-$	fluoro-	--------
	$Cl-$	chloro-	--------
	$Br-$	bromo-	--------
	$I-$	iodo-	--------

Miscellaneous substituents and their prefixes

$-NO_2$	$-CH=CH_2$	$-CH_2CH=CH_2$	(phenyl ring)
nitro	vinyl	allyl	phenyl

Appendix 1, Summary of IUPAC Nomenclature, continued

<u>Common alkyl groups</u>—replace "ane" ending of alkane name with "yl". Alternate names for complex substituents are given in brackets.

—CH₃
methyl

—CH₂CH₃
ethyl

—CH₂CH₂CH₃
propyl (*n*-propyl)

—CH₂CH₂CH₂CH₃
butyl (*n*-butyl)

$$-CH\begin{array}{c} CH_3 \\ \\ CH_3 \end{array}$$
isopropyl
[1-methylethyl]

$$-CH_2-CH\begin{array}{c} CH_3 \\ \\ CH_3 \end{array}$$
isobutyl
[2-methylpropyl]

$$-CH\begin{array}{c} CH_3 \\ \\ CH_2CH_3 \end{array}$$
sec-butyl
[1-methylpropyl]

$$-\overset{CH_3}{\underset{CH_3}{\overset{|}{\underset{|}{C}}}}-CH_3$$
t-butyl or
tert-butyl
[1,1-dimethylethyl]

B. Naming Substituted Alkanes and Cycloalkanes—Group C Substituents Only

Organic compounds containing substituents from Group C are named following this sequence of steps, as indicated on the examples below:

• Step 1. Find the longest continuous carbon chain. Determine the root name for this parent chain. In cyclic compounds, the ring is usually considered the parent chain, unless it is attached to a longer chain of carbons; indicate a ring with the prefix "cyclo" before the root name. (When there are two longest chains of equal length, use the chain with the greater number of substituents.)

• Step 2. Number the chain in the direction such that the position number of the first substituent is the smaller number. If the first substituents have the same number, then number so that the second substituent has the smaller number, *etc*.

• Step 3. Determine the name and position number of each substituent. (A substituent on a nitrogen is designated with an "*N*" instead of a number; see Section **III.D.1.** below.)

• Step 4. Indicate the number of identical groups by the prefixes di, tri, tetra, *etc*.

• Step 5. Place the position numbers and names of the substituent groups, in alphabetical order, before the root name. In alphabetizing, ignore prefixes like *sec*-, *tert*-, di, tri, *etc*., but include iso and cyclo. Always include a position number for each substituent, regardless of redundancies.

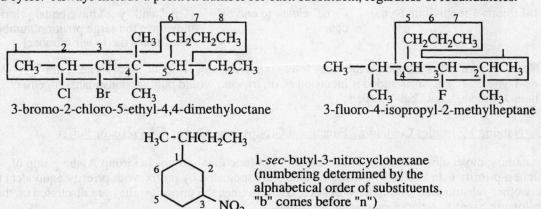

3-bromo-2-chloro-5-ethyl-4,4-dimethyloctane

3-fluoro-4-isopropyl-2-methylheptane

1-*sec*-butyl-3-nitrocyclohexane
(numbering determined by the alphabetical order of substituents, "b" comes before "n")

Appendix 1, Summary of IUPAC Nomenclature, continued

C. Naming Molecules Containing Functional Groups from Group B—Suffix Only

1. Alkenes—Follow the same steps as for alkanes, except:
a. Number the chain of carbons *that includes the C=C* so that the C=C has the lower position number, since it has a higher priority than any substituents;
b. Change "ane" to "ene" and assign a position number to the first carbon of the C=C; place the position number just before the name of functional group(s);
c. Designate geometrical isomers with a *cis,trans* or *E,Z* prefix.

4,4-difluoro-3-methylbut-1-ene

1,1-difluoro-2-methyl-buta-1,3-diene

5-methylcyclopenta-1,3-diene

Special case: When the chain cannot include an alkene, a substituent name is used. See Section V.A.2.a.

3-vinylcyclohex-1-ene

Numbering must be on EITHER a ring OR a chain, but not both.

2. Alkynes—Follow the same steps as for alkanes, except:
a. Number the chain of carbons *that includes the C≡C* so that the alkyne has the lower position number;
b. Change "ane" to "yne" and assign a position number to the first carbon of the C≡C; place the position number just before the name of functional group(s).
Note: The Group B functional groups (alkene and alkyne) are considered to have equal priority: in a molecule with both an ene and an yne, whichever is closer to the end of the chain determines the direction of numbering. In the case where each would have the same position number, the alkene takes the lower number. In the name, "ene" comes before "yne" because of alphabetization.

4,4-difluoro-3-methylbut-1-yne

pent-3-en-1-yne
("yne" closer to end of chain)

pent-1-en-4-yne
("ene" and "yne" have equal priority unless they have the same position number, when "ene" takes the lower number)

(Notes: 1. An "e" is dropped if the letter following it is a vowel: "pent-3-en-1-yne" , not "pent-3-ene-1-yne". 2. An "a" is added if inclusion of di, tri, *etc.*, would put two consonants together: "buta-1,3-diene", not "but-1,3-diene".)

D. Naming Molecules Containing Functional Groups from Group A—Prefix or Suffix

In naming molecules containing one or more of the functional groups in Group A, the group of highest priority is indicated by suffix; the others are indicated by prefix, with priority equivalent to any other substituents. The table in Section **III**.A. defines the priorities; they are discussed on the following pages in order of increasing priority.

678

Appendix 1, Summary of IUPAC Nomenclature, continued

Now that the functional groups and substituents from Groups A, B, and C have been described, a modified set of steps for naming organic compounds can be applied to all simple structures:

•Step 1. Find the highest priority functional group. Determine and name the longest continuous carbon chain that includes this group.
•Step 2. Number the chain so that the highest priority functional group is assigned the lower number. (The number "1" is often omitted when there is no confusion about where the group must be. Aldehydes and carboxylic acids must be at the first carbon of a chain, so a "1" is rarely used with those functional groups.)
•Step 3. If the carbon chain includes multiple bonds (Group B), replace "ane" with "ene" for an alkene or "yne" for an alkyne. Designate the position of the multiple bond with the number of the first carbon of the multiple bond.
•Step 4. If the molecule includes Group A functional groups, replace the last "e" with the suffix of the highest priority functional group, and include its position number just before the name of the highest priority functional group.
•Step 5. Indicate all Group C substituents, and Group A functional groups of lower priority, with a prefix. Place the prefixes, with appropriate position numbers, in alphabetical order before the root name.

1. Amines: prefix: amino-; suffix: -amine—substituents on nitrogen denoted by "N"

$CH_3CH_2CH_2 — NH_2$

propan-1-amine

3-methoxycyclohexan-1-amine
("1" is optional in this case)

N,N-diethylbut-3-en-2-amine

2. Alcohols: prefix: hydroxy-; suffix: -ol

$CH_3CH_2 — OH$

ethanol

but-3-en-2-ol

2-aminocyclobutan-1-ol
("1" is optional in this case)

3. Ketones: prefix: oxo-; suffix: -one (pronounced "own")

3-hydroxybutan-2-one

cyclohex-3-en-1-one
("1" is optional in this case)

4-(N,N-dimethylamino)pent-4-en-2-one

4. Aldehydes: prefix: oxo-, or formyl- (O=CH-); suffix: -al (abbreviation: —CHO)
An aldehyde can only be on carbon 1, so the "1" is generally omitted from the name.

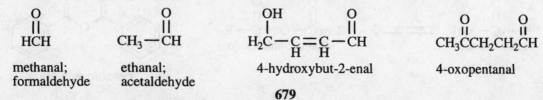

methanal;
formaldehyde

ethanal;
acetaldehyde

4-hydroxybut-2-enal

4-oxopentanal

679

Appendix 1, Summary of IUPAC Nomenclature, continued

Special case: When the chain cannot include the carbon of the aldehyde, the suffix "carbaldehyde" is used:

cyclohexanecarbaldehyde

5. Carboxylic Acids: prefix: carboxy-; suffix: -oic acid (abbreviation: —COOH)
A carboxylic acid can only be on carbon 1, so the "1" is generally omitted from the name.
(Note: Chemists traditionally use, and IUPAC accepts, the names "formic acid" and "acetic acid" in place of "methanoic acid" and "ethanoic acid".)

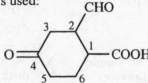

methanoic acid; formic acid

ethanoic acid; acetic acid

2-amino-3-phenylpropanoic acid

2,2-dimethyl-3,4-dioxobutanoic acid

Special case: When the chain numbering cannot include the carbon of the carboxylic acid, the suffix "carboxylic acid" is used:

2-formyl-4-oxocyclohexanecarboxylic acid ("formyl" is used to indicate an aldehyde as a substituent when its carbon cannot be in the chain numbering)

E. Naming Carboxylic Acid Derivatives
The six common groups derived from carboxylic acids are, in decreasing priority after carboxylic acids: salts, anhydrides, esters, acyl halides, amides, and nitriles.

1. Salts of Carboxylic Acids
Salts are named with cation first, followed by the anion name of the carboxylic acid, where "ic acid" is replaced by "ate" :

acetic acid	becomes	acetate
butanoic acid	becomes	butanoate
cyclohexanecarboxylic acid	becomes	cyclohexanecarboxylate

lithium 2-aminopropanoate

sodium chloroacetate

ammonium 2-methoxy-cyclobutanecarboxylate

2. Anhydrides: "oic acid" is replaced by "oic anhydride"

alkanoic acid

alkanoic anhydride

benzoic anhydride

Appendix 1, Summary of IUPAC Nomenclature, continued

3. Esters
Esters are named as "organic salts" that is, the alkyl name comes first, followed by the name of the carboxylate anion. (common abbreviation: —COOR)

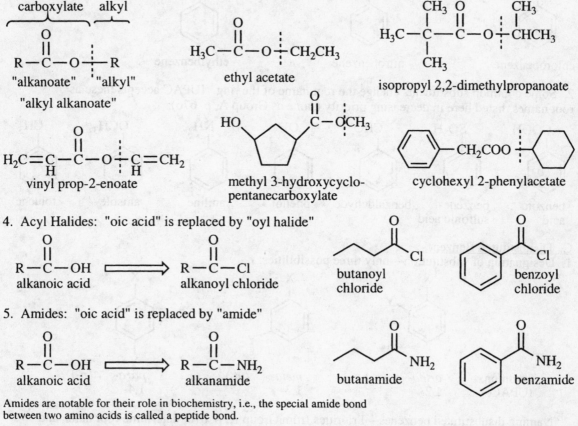

"alkanoate" "alkyl"

"alkyl alkanoate"

ethyl acetate

isopropyl 2,2-dimethylpropanoate

vinyl prop-2-enoate

methyl 3-hydroxycyclo-
pentanecarboxylate

cyclohexyl 2-phenylacetate

4. Acyl Halides: "oic acid" is replaced by "oyl halide"

alkanoic acid $\Longrightarrow$ alkanoyl chloride

butanoyl
chloride

benzoyl
chloride

5. Amides: "oic acid" is replaced by "amide"

alkanoic acid $\Longrightarrow$ alkanamide

butanamide

benzamide

Amides are notable for their role in biochemistry, i.e., the special amide bond
between two amino acids is called a peptide bond.

6. Nitriles: "oic acid" is replaced by "enitrile"

alkanoic acid $\Longrightarrow$ alkanenitrile

butanenitrile

benzonitrile

(common spelling differs
from IUPAC)

IV. Nomenclature of Aromatic Compounds
"Aromatic" compounds are those derived from benzene and similar ring systems. As with
aliphatic nomenclature described above, the process is: determining the root name of the parent
ring; determining priority, name, and position number of substituents; and assembling the name in
alphabetical order. *Functional group priorities are the same in aliphatic and aromatic
nomenclature.* See p. 676 for the list of priorities.

A. Common Parent Ring Systems

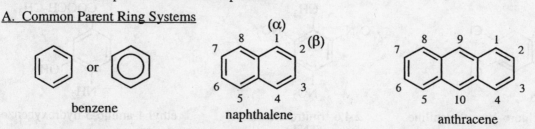

benzene

naphthalene

anthracene

Appendix 1, Summary of IUPAC Nomenclature, continued

B. Monosubstituted Benzenes

1. Most substituents keep their designation, followed by the word "benzene":

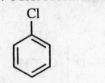

chlorobenzene

nitrobenzene

ethylbenzene

2. Some common substituents change the root name of the ring. IUPAC accepts these as root names, listed here in decreasing priority (same as Group A, p. 676):

benzoic   benzene-   benzaldehyde   phenol   aniline   anisole   toluene
acid   sulfonic acid

C. Disubstituted Benzenes

1. Designation of substitution—only three possibilities:

| common: | *ortho-* | *(o-)* | *meta-* | *(m-)* | *para-* | *(p-)* |
| IUPAC: | 1,2- | | 1,3- | | 1,4- | |

2. Naming disubstituted benzenes—Priorities from Group A, p. 676, determine root name and substituents

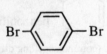   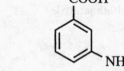   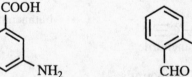

p-dibromobenzene   *m*-aminobenzoic acid   *o*-methoxybenzaldehyde   *m*-methylphenol
1,4-dibromobenzene   3-aminobenzoic acid   2-methoxybenzaldehyde   3-methylphenol

D. Polysubstituted Benzenes—must use numbers to indicate substituent position

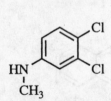

3,4-dichloro-*N*-methylaniline

2,4,6-trinitrotoluene
(TNT)

ethyl 4-amino-3-hydroxybenzoate

682

Appendix 1, Summary of IUPAC Nomenclature, continued

E. Aromatic Ketones
A special group of aromatic compounds are ketones where the carbonyl is attached to at least one benzene ring. Such compounds are named as "phenones", the prefix depending on the size and nature of the group on the other side of the carbonyl. These are the common examples:

acetophenone

propiophenone

butyrophenone

benzophenone

V. Nomenclature of Bicyclic Compounds

"Bicyclic" compounds are those that contain two rings. There are four possible arrangements of two rings that depend on how many atoms are shared by the two rings. The first arrangement in which the rings do not share any atoms does not use any special nomenclature, but the other types require a method to designate how the rings are put together. Once the ring system is named, then functional groups and substituents follow the standard rules described above.

Type 1. Two rings with no common atoms
These follow the standard rules of choosing one parent ring system and describing the other ring as a substituent.

ketone is the highest priority functional group, phenyl is substituent
⟹ 3-phenylcyclohexan-1-one   ("1" could be omitted here)

benzene is the parent ring system as it is larger than cyclopentane and it has three substituents
⟹ 1-cyclopentyl-2,3-dinitrobenzene

Type 2. Two rings with one common atom—spiro ring system
The ring system in spiro compounds is indicated by the word "spiro" (instead of "cyclo"), followed by brackets indicating how many atoms are contained in each path around the rings, ending with the alkane name describing how many carbons are in the ring systems including the spiro carbon. (If any atoms are not carbons, see section VI.) Numbering follows the smaller path first, passing through the sprio carbon and around the second ring.

spiro[3.4]octane

spiro[4.5]decane

Appendix 1, Summary of IUPAC Nomenclature, continued

Substituents and functional groups are indicated in the usual ways. Spiro ring systems are always numbered smaller before larger, and numbered in such a way as to give the highest priority functional group the lower position number.

spiro[3.4]oct-5-ene

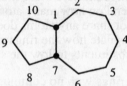

7,7-dimethylspiro[4.5]decan-2-one

Type 3. Two rings with two common atom—fused ring system

Two rings that share two common atoms are called fused rings. This ring system and the next type called bridged rings share the same designation of ring system. Each of the two common atoms is called a bridgehead atom, and there are three paths between the two bridgehead atoms. In contrast with naming the spiro rings, the *longer* path is counted first, then the shorter, then the shortest. In fused rings, the shortest path is always a zero, meaning zero atoms between the two bridgehead atoms. Numbering starts at a bridgehead, continues around the largest ring, through the other bridgehead and around the shorter ring. (In these structures, bridgeheads are marked with a dark circle for clarity.)

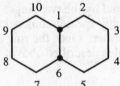

bicyclo[2.1.0]pentane
(path of 2 atoms and a
path of 1 atom)

bicyclo[4.4.0]decane
(path of 4 atoms in
each direction)

bicyclo[5.3.0]decane
(path of 5 atoms and
path of 3 atoms)

Substituents and functional groups are indicated in the usual ways. Fused rings systems are always numbered larger before smaller, and numbered in such a way as to give the highest priority functional group the lower position number.

Type 4. Two rings with more than two common atom—bridged ring system

Two rings that share more than two common atoms are called bridged rings. Bridged rings share the same designation of ring system as Type 3 in which there are three paths between the two bridgehead atoms. The longer path is counted first, then the medium, then the shortest. Numbering starts at a bridgehead, continues around the largest ring, through the other bridgehead and around the medium path, ending with the shortest path numbered from the original bridgehead atom. (In these structures, bridgeheads are marked with a dark circle for clarity.)

bicyclo[2.1.1]hexane
(paths of 2 atoms, 1
atom, and 1 atom)

bicyclo[2.2.2]octane
(three paths of 2 atoms)

bicyclo[3.2.1]octane
(paths of 3 atoms, 2 atoms,
and 1 atom)

Appendix 1, Summary of IUPAC Nomenclature, continued

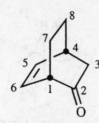

5,5-dibromo-
bicyclo[2.1.1]hexane

bicyclo[2.2.2]oct-5-en-2-one

8,8-dimethyl-
bicyclo[3.2.1]octan-1-ol

VI. Replacement Nomenclature of Heteroatoms

The term "heteroatom" applies to any atom other than carbon or hydrogen. It is common for heteroatoms to appear in locations that are inconvenient to name following basic rules, so a simple system called "replacement nomenclature" has been devised. The fundamental principle is to name a compound as if it contained only carbons in the skeleton, plus any functional groups or substituents, and then indicate which carbons are "replaced" by heteroatoms. The prefixes used to indicate these substitions are listed here *in decreasing priority and listed in this order in the name:*

Element	Prefix	Example
O	oxa	
S	thia	nonan-1-ol
N	aza	
P	phospha	
Si	sila	
B	bora	

2-thia-8-aza-4-sila-6-boranonan-1-ol

In the above example, note that the (imaginary) compound no longer has nine carbons, even though the name still includes "nonan". The heteroatoms have replaced carbons, but the compound is named as if it still had those carbons.

Where the replacement system is particularly useful is in polycyclic compounds. Shown below are three examples of commercially available and synthetically useful reagents that use this system.

parent hydrocarbon	reagent	abbreviation
bicyclo[2.2.2]octane	1,4-DiAzaBiCyclo[2.2.2]Octane (upper case added to explain abbreviation)	DABCO
bicyclo[5.4.0]undec-7-ene	1,8-DiazaBicyclo[5.4.0]Undec-7-ene (upper case added to explain abbreviation)	DBU

685

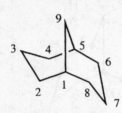

bicyclo[3.1.1]nonane

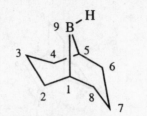

9-BoraBicyclo[3.1.1]Nonane
(upper case added to explain abbreviation)

9-BBN

VII. Designation of Stereochemistry; Cahn-Ingold-Prelog system

Is this alkene cis or trans?

F   I

Cl   Br

How can we distinguish this structure from its mirror image?

F

I ⋯⋯ Cl

Br

Compounds that exhibit stereoisomerism, whether geometric isomers around double bonds, substituent groups on rings, or molecules with asymmetric tetrahedral atoms (which are almost always carbons), require a system to designate relative and absolute orientation of the groups. The terms cis/trans, D/L in carbohydrates and amino acids, and *d/l* for optically active compounds, are limited and cannot be used generally, although each still is used in appropriate situations. For example, cis/trans still is used to indicate relative positions of substituents around a ring.

A system developed by chemists Cahn, Ingold, and Prelog, uses a series of steps to determine group priorities, and a definition of position based on the relative arrangement of the groups. In alkenes, the system is relatively simple:

high   high

low   low

this arrangement is defined as Z = zusammen, together, from both high priority groups on the same side of the C=C

high   low

low   high

this arrangement is defined as E = entgegen, opposite, from the two high priority groups on opposite sides of the C=C

As with alkenes, the orientation around an asymmetric carbon can be only one of two choices. In three dimensions, clockwise and counterclockwise are the only two directions that are definite, and even that description requires a fixed reference point. To designate configuration, the lowest (fourth) priority group is always placed farthest away from the viewer (indicated by a dashed line), and the group priorities will follow 1 to 2 to 3 in either a clockwise or a counterclockwise direction.

1 to 2 to 3 is clockwise, defined as the R = *rectus* configuration

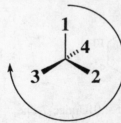

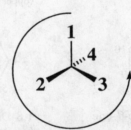

1 to 2 to 3 is counterclockwise, defined as the S = *sinister* configuration

The only step remaining is to determine the priority of groups, for which there is a carefully defined set of rules.

Rule 1. Consider the first atom of the group, the point of attachment. Atoms with higher atomic number receive higher priority. Heavier isotopes have higher priority than lighter isotopes.

$$I > Br > Cl > F \qquad\qquad O > N > C > H \qquad\qquad {}^{14}C > {}^{13}C > {}^{12}C$$

Rule 2. If the first atoms of two or more groups are the same, go out to the next atoms to break the tie. One high priority atom takes priority over any number of lower-priority atoms.

Rule 3. Treat multiple bonds as if they were all single bonds; one will be to the real atom, the others will be to imaginary atoms.

This is the hardest rule to put into practice. This example of an alkyne shows stepwise how to accomplish this. In the pictures, imaginary atoms (ones that did not start out in the structure, we had to invent them) are indicated by italics.

all atoms are real

started with 3 bonds to carbon, so have to replace them

halfway there—had to add two fake carbons

final—had to add four fake carbons

becomes

added a fake O to the real C, and a fake C to the real O

becomes

added two fake Ns to the C, and two Cs to the N

Examples applying the Cahn-Ingold-Prelog sytem

comparing F and Cl

comparing I and Br

E

E

Z

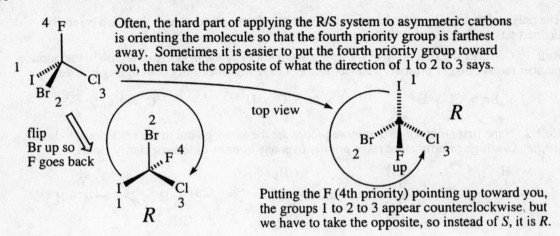

Often, the hard part of applying the R/S system to asymmetric carbons is orienting the molecule so that the fourth priority group is farthest away. Sometimes it is easier to put the fourth priority group toward you, then take the opposite of what the direction of 1 to 2 to 3 says.

top view

Putting the F (4th priority) pointing up toward you, the groups 1 to 2 to 3 appear counterclockwise, but we have to take the opposite, so instead of *S*, it is *R*.

Appendix 2: Summary of Acidity and Basicity

Imagine that you are at a family reunion where you can observe the competition for ice cream cones among your nieces and nephews. Pretty soon, you formulate a generalization: the older kids can hold onto their ice cream cones more strongly than the younger ones. Another way of saying it is that the older ones are less likely to give up their cones. You could even represent this information in a table showing a series of equilibria between the child with ice cream, and the free ice cream plus the hungry child. The differences in strength could also be quantitated: the larger the hunger factor, pK_H, the less likely the child will give up the ice cream.

Approximate pK_H Values of Children

pK_H			
12	12-year-old with ice cream	$\rightleftharpoons$	ice cream + hungry 12-year-old
10	10-year-old with ice cream	$\rightleftharpoons$	ice cream + hungry 10-year-old
8	8-year-old with ice cream	$\rightleftharpoons$	ice cream + hungry 8-year-old
6	6-year-old with ice cream	$\rightleftharpoons$	ice cream + hungry 6-year-old
4	4-year-old with ice cream	$\rightleftharpoons$	ice cream + hungry 4-year-old

What good is this table? It allows anyone to make predictions about what will happen when kids with ice cream are mixed with hungry kids.

$$\left.\begin{array}{c}\text{hungry 10-year-old} \\ + \\ \text{4-year-old with ice cream}\end{array}\right\} \overset{?}{\rightleftharpoons} \left\{\begin{array}{c}\text{10-year-old with ice cream} \\ + \\ \text{hungry 4-year-old}\end{array}\right.$$

If the hungry 10-year-old was left unattended in a room with the 4-year-old with ice cream, which side of this equilibrium would be favored when you came back in a few minutes: "reactants" or "products"? More likely than not, there would be an ice cream transfer in your absence; "products" would be favored. You could have predicted this from the table, and you could generalize: the hungry 10-year-old will be strong enough to rip the ice cream away from any kid lower on the table than the 10-year-old him/herself.

Let's predict the results of another equilibrium:

$$\left.\begin{array}{c}\text{hungry 6-year-old} \\ + \\ \text{12-year-old with ice cream}\end{array}\right\} \overset{?}{\rightleftharpoons} \left\{\begin{array}{c}\text{6-year-old with ice cream} \\ + \\ \text{hungry 12-year-old}\end{array}\right.$$

Is the hungry 6-year-old strong enough to pull the ice cream away from the 12-year-old? Not in most families. The table shows that the only chance for the hungry 6-year-old is to find a 4-year-old with ice cream. The 12-year-old with ice cream is pretty safe as long as the hunger table doesn't go to higher ages.

If you understand this analogy and can make predictions of ice cream transfer using the table, then you can understand how to predict the direction of equilibrium in acid-base reactions. Turn the page.

Appendix 2 continued, Summary of Acidity and Basicity

To set the stage ....

A few generalizations:

A) This Appendix deals with protic acids and bases, called Bronsted-Lowry acids. Similar statements can be made about Lewis acids but they are not the focus of this discussion.

B) Values of pK_a are measures of equilibrium constants, described further below. Values between 0 and 15.7 are measured by titration in water solution and are known accurately, to within 0.1 pK unit and sometimes better. Values outside of this range cannot be measured in water because of the leveling effect of water, and there is no universally accepted method for measuring these pK_a values. This lack of a single standard of measurement means that the values below 0 and above 15.7 should be considered relative, not absolute. If your instructor says the pK_a of methane is 46 and this book says it is 50, those should be considered the same value within experimental variation.

C) Acidity is a thermodynamic property, and the acid equilibrium constant, K_a, is a measure of the relative concentrations of species in the protonated and unprotonated form. As most organic acids are weak acids, meaning they are present mostly in the protonated form at equilibrium, the $K_a < 1$. Since the $pK_a = -\log K_a$, the pK_a values will be greater than 0, with the larger pK_a representing a weaker acid. If this is not clear, review text section 1-13.

D) Acids do not spontaneously spit out a proton! Despite our way of writing ionization equilibria as shown on the next page, acids do not give up a proton unless a base comes by to take the proton away. The reactions as drawn in the table should be considered half-reactions, just as the reactions in the electromotive series were half-reactions for balancing oxidation-reduction reactions in general chemistry.

I. Predicting equilibrium position

Look at the table on p. 691 and notice how it looks just like the "children-with-ice-cream" table. We can use this table to make predictions about equilibrium position in acid-base reactions just as we did for the children with ice cream.

1) *A base will deprotonate any acid stronger than its conjugate acid.* This is the most important principle of predicting acid-base reactions. On the table, this means that any base, hydroxide for example, can react with any acid more acidic than the conjugate acid of itself, water in our example. So hydroxide is a strong enough base to pull the proton from any of these: bicarbonate ion, a phenol, carbonic acid, a carboxylic acid, or a sulfonic acid. We can also predict that hydroxide is NOT a strong enough base to react with any acid above water on the table; for example, a mixture of hydroxide with an alkyne will favor the reactants at equilibrium, with only a small amount of products.

reactants favored

HO^- + $RC\equiv C-H$ $\rightleftharpoons$ H_2O + $RC\equiv C{:}^-$

pK_a 25 pK_a 15.7

weaker   *weaker*   *stronger*   *stronger*
base   *acid*   *acid*   *base*

2) Another way of predicting the position of an equilibrium is to assign "stronger" and "weaker" to the acid and base on each side of the equation, using the table to determine which is stronger and which is weaker. *Equilibrium will always favor the weaker acid and base.* This method will always give the same answer as the principle in #1 above.

To lead into the next section, look again at the table on p. 691 and notice two things: a) with only a couple of exceptions, all the acidic protons are on either oxygen or carbon; and b) generalizations can be made about the acidity of functional groups. Learning to correlate acidity with functional group is important in predicting reactivity of the functional group.

Approximate pKa Values of Organic Compounds

		pKa					
weaker acid	alkane	≈ 50	$R-\overset{\displaystyle |}{\underset{\displaystyle |}{C}}-H$	$\rightleftharpoons$	H^+	$+$ $R-\overset{\displaystyle |}{\underset{\displaystyle |}{C}}\!:^-$	*stronger base*

alkene ≈ 45 $=\overset{\diagup}{C}\underset{H}{\diagdown} \rightleftharpoons H^+ + =\overset{\diagup}{C}\!:^- $

amine 35-40 $-\ddot{N}-H \rightleftharpoons H^+ + -\ddot{N}\!:^-$

≈ 35 $H-H \rightleftharpoons H^+ + H\!:^-$

alkyne ≈ 25 $RC\equiv C-H \rightleftharpoons H^+ + RC\equiv C\!:^-$

ketone and ester 20-25 $-\overset{O}{\overset{\|}{C}}-\overset{|}{C}-H \rightleftharpoons H^+ + -\overset{O}{\overset{\|}{C}}-\overset{|}{C}\!:^-$

alcohol

≈ 18 $R-\overset{R}{\underset{R}{\overset{|}{\underset{|}{C}}}}-O-H \rightleftharpoons H^+ + R-\overset{R}{\underset{R}{\overset{|}{\underset{|}{C}}}}-\ddot{\ddot{O}}\!:^-$

≈ 17 $R-\overset{H}{\underset{R}{\overset{|}{\underset{|}{C}}}}-O-H \rightleftharpoons H^+ + R-\overset{H}{\underset{R}{\overset{|}{\underset{|}{C}}}}-\ddot{\ddot{O}}\!:^-$

↑ cannot be measured in water solution

≈ 16 $R-\overset{H}{\underset{H}{\overset{|}{\underset{|}{C}}}}-O-H \rightleftharpoons H^+ + R-\overset{H}{\underset{H}{\overset{|}{\underset{|}{C}}}}-\ddot{\ddot{O}}\!:^-$

- -

↓ measured in water solution

15.7 $H_2O \rightleftharpoons H^+ + HO^-$

10.3 $HCO_3^- \rightleftharpoons H^+ + CO_3^{2-}$

phenol ≈ 10 $Ar-O-H \rightleftharpoons H^+ + Ar-\ddot{\ddot{O}}\!:^-$

6.4 $H_2CO_3 \rightleftharpoons H^+ + HCO_3^-$

carboxylic acid 4-5 $R-\overset{O}{\overset{\|}{C}}-O-H \rightleftharpoons H^+ + R-\overset{O}{\overset{\|}{C}}-\ddot{\ddot{O}}\!:^-$

sulfonic acid < 0 $RSO_2-O-H \rightleftharpoons H^+ + RSO_2-\ddot{\ddot{O}}\!:^-$

stronger acid   *weaker base*

Appendix 2 continued, Summary of Acidity and Basicity

II. Correlation of Acidity with Functional Group

A. Oxygen Acids

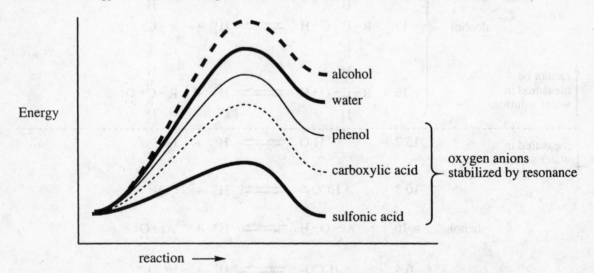

$$R-\overset{\overset{O}{\|}}{\underset{\underset{O}{\|}}{S}}-O-H \quad > \quad R-\overset{\overset{O}{\|}}{C}-O-H \quad > \quad \langle\!\!\bigcirc\!\!\rangle-O-H \quad > \quad HO-H \quad > \quad R-\overset{\overset{R}{|}}{\underset{\underset{R}{|}}{C}}-O-H$$

pK$_a$ < 0   pK$_a$ 4-5   pK$_a$ 10.0   pK$_a$ 15.7   pK$_a$ 16-19

sulfonic acid   carboxylic acid   phenol   water   alcohol

Sulfonic acids are the strongest of the oxygen acids but are not common in organic chemistry. Carboxylic acids, however, are everywhere and are considered the strongest of the common organic oxygen acids. (Note that "strong" and "weak" are relative terms: acetic acid, pK$_a$ 4.74, was a "weak" acid in general chemistry in comparison to sulfuric and hydrochloric acids, but acetic acid is a "strong" acid in organic chemistry relative to the other oxygen acids.) Phenols having OH groups on benzene or other aromatic rings are still stronger acids than water.

Why are phenols, carboxylic acids, and sulfonic acids stronger acids than water? Because their anions are stabilized by resonance. (Refer to text sections 1-13, 10-6, and 20-4, especially Figure 20-1.) Let's look at that statement in more detail.

Remember that acidity is a thermodynamic property; that is, acidity equilibrium depends on the difference in energy between the reactants and products. The more the anion is stabilized by resonance, the lower in energy it is, and the less positive the ΔG, as shown on the reaction energy diagram:

Why are alcohols weaker acids than water? There are two effects that contribute, both of which are consistent with the trend that 1° alcohols (pK$_a$ 16) are slightly stronger than 2° alcohols (pK$_a$ 17) which are slightly stronger than 3° alcohols (pK$_a$ 18). Alkyl groups are mildly electron-donating in their inductive effect (more about this later) and destabilize the anion, as shown in the energy diagram above. Second, the more crowded the anion is, the less it can be stabilized by hydrogen bonding with the solvent.

Appendix 2 continued, Summary of Acidity and Basicity

B. Carbon Acids

When we think of "acids", we do not usually think of protons on carbon, yet carbon acids and the carbanions that come from them are of tremendous importance in organic chemistry.

"Unstabilized" carbon acids are those that do not have any substituent to stabilize the anion. Alkanes, with only sp^3 carbons are the weakest acids with pK_a around 50. The vinyl carbon in a carbon-carbon double bond is sp^2 hybridized with the electrons of the anion slightly closer to the positive nucleus, leading to some stabilization of the anion. This type of stabilization is particularly important in alkynes with sp hybridized carbons.

$$R-\overset{|}{\underset{|}{C}}-H \rightleftharpoons R-\overset{|}{\underset{|}{C}}{:}^{-} \qquad =C\overset{\diagup}{\underset{H}{\diagdown}} \rightleftharpoons =C\overset{\diagup}{\underset{\cdot\cdot}{\diagdown}}{}^{-} \qquad RC\equiv C-H \rightleftharpoons RC\equiv C{:}^{-}$$

sp^3 hybridized sp^2 hybridized   sp hybridized
pK_a 50 pK_a 45 pK_a 25

C. Carbon Acids Alpha to Carbonyl (This topic is described in detail in text section 22-2B.)

Look at these huge differences in pK_a when the acidic group is next to a carbonyl.

$O-H \Rightarrow RCH_2O-H \qquad pK_a$ 16-18 $\qquad R-\overset{O}{\overset{||}{C}}-O-H \qquad pK_a$ 4-5

$N-H \Rightarrow RCH_2NH-H \qquad pK_a$ 35-40 $\qquad R-\overset{O}{\overset{||}{C}}-\overset{H}{\overset{|}{N}}-H \qquad pK_a$ 16

$C-H \Rightarrow RCH_2CH_2-H \qquad pK_a$ 50 $\qquad R-\overset{O}{\overset{||}{C}}-\overset{H_2}{\overset{|}{C}}-H \qquad pK_a$ 20

Hydrogens alpha to carbonyl are unusually acidic because of resonance stabilization of the anionic conjugate base.

D. Acidities of Acyl Functional Groups

In addition to the significant variation in the acidity of alpha hydrogens depending on which atom the H is bonded to, what is on the other side of the carbonyl also has a dramatic influence. In this case, the stabilization is more important on the starting material, not on the conjugate base. See the energy diagram on p. 694.

aldehyde	ketone	ester	amide
$H_2\overset{H}{\overset{\vert}{C}}-\overset{O}{\overset{\|}{C}}-H$	$H_2\overset{H}{\overset{\vert}{C}}-\overset{O}{\overset{\|}{C}}-CH_3$	$H_2\overset{H}{\overset{\vert}{C}}-\overset{O}{\overset{\|}{C}}-\overset{\cdot\cdot}{O}CH_3$	$H_2\overset{H}{\overset{\vert}{C}}-\overset{O}{\overset{\|}{C}}-\overset{\cdot\cdot}{N}(CH_3)_2$
pK_a 17	pK_a 20	pK_a 25	pK_a 30
no stabilization of starting material	mild stabilization of starting material by weak electron donation from CH_3	$H_2\overset{H}{\overset{\vert}{C}}-\overset{O^-}{\overset{\|}{C}}=\overset{+}{\underset{\cdot\cdot}{O}}CH_3$ significant resonance stabilization of starting material	$H_2\overset{H}{\overset{\vert}{C}}-\overset{O^-}{\overset{\|}{C}}=\overset{+}{N}(CH_3)_2$ strongest resonance stabilization of starting material

Appendix 2 continued, Summary of Acidity and Basicity

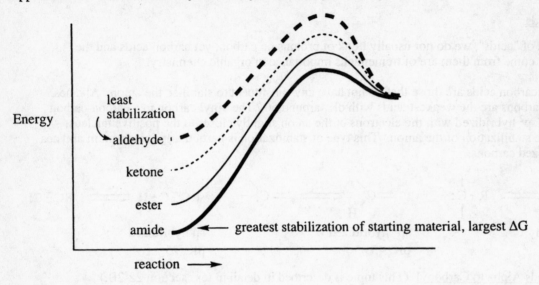

Energy

least
stabilization

↘aldehyde

ketone

ester

amide ← greatest stabilization of starting material, largest ΔG

reaction ⟶

E. Carbon Acids Between Two Carbonyls (This topic is described in detail in text section 22-15.)

While a hydrogen alpha to one carbonyl moves into the pK_a 20-25 range for ketones and esters respectively, a hydrogen between two carbonyls (cyano and nitro are similar to carbonyl electronically) is more acidic than water. The increased resonance stabilization of the conjugate base is largely responsible, but there are subtle variations depending on the type of functional group as noted at the bottom of p. 693. Look at the enormous influence of the nitro group.

pK_a 13.5 pK_a 11.2 pK_a 10.2 pK_a 9,0

pK_a 10.2 pK_a 5.8 pK_a 3.6

III. Correlation of Basicity with Functional Group

The bulk of this Appendix is on acidity because many more functional groups are acidic than are basic. Basically (oooh, sorry), only one functional group is basic: amines. There is variation among aliphatic, aromatic, and heteroaromatic amines; these are covered thoroughly in text sections 19-5 and 19-6. One point in the text, just before Table 19-3, deserves emphasis: for any conjugate acid-base pair:

$$pK_a + pK_b = 14$$

Appendix 2 continued, Summary of Acidity and Basicity

This simple algebraic relationship is very useful:

Sample problem. Is triethylamine (pK$_b$ 3.24) a strong enough base to deprotonate phenol (pK$_a$ 10.0)?

We need to calculate either the pK$_a$ of the conjugate acid of triethylamine or the pK$_b$ of the conjugate base of phenol to see which is stronger and weaker. Then we can say with certainty which side of the equilibrium will be favored.

pK$_a$ 10.0	pK$_b$ 3.24	pK$_b$ = **4.0** (14 – 10.0)	pK$_a$ = **10.76** (14 – 3.24)
stronger acid	*stronger base*	*weaker base*	*weaker acid*

Aha! Products are favored at equilibrium, so the correct answer to the question is "Yes, triethylamine is a strong enough base to deprotonate phenol."

Try this for fun: How weak must a base be before it does NOT deprotonate phenol? What algebraic rule can you formulate to predict whether any combination of acid and base will favor products or reactants?

IV. Substituent Effects on Acidity

So far, we have focused on acidities of different functional groups. Let's turn to more minor, more subtle, structural changes to see what effect substituents will have on the acidity of a group. Primarily, we imply *electronic* effects as opposed to *steric* effects, but this Appendix will conclude with a discussion of how steric and electronic effects can work together.

A. Classification of Substituents—Induction and Resonance

Substituent groups can exert an electronic effect on an acidic functional group in two different ways: through sigma bonds, where this is called an *inductive effect*, or through p orbitals and pi bonds which is called a *resonance effect*. Groups can also be electron-donating or electron-withdrawing by either of the mechanisms, so there are four possible categories for groups. Note that a group can appear in more than one category, even in conflicting groups!

a) Electron-donating by induction: only alkyl groups (abbreviated R) have electrons to share by induction;

b) Electron-withdrawing by induction: every group that has a more electronegative atom than carbon is in this category; some examples: F, Cl, Br, I, OH, OR, NH$_2$, NHR, NR$_2$, NO$_2$, C=O, CN, SO$_3$H, CX$_3$ where X is halogen;

c) Electron-donating by resonance: groups that have electron pairs to share: F, Cl, Br, I, OH, OR, NH$_2$, NHR, NR$_2$;

d) Electron-withdrawing by resonance: NO$_2$, C=O, CN, SO$_3$H.

Appendix 2 continued, Summary of Acidity and Basicity

B. Generalizations on Electronic Effects on Acidity (refer to text section 20-4B)

Electric charge is the key to understanding substituent effects. An acid is always more positive than its conjugate base; in other words, the conjugate base is always more negative than the acid. Electron-donating and electron-withdrawing groups will have opposite effects on the acid-base conjugate pair.

Electron-donating groups stabilize the more positive acid form and destabilize the more negative conjugate base. From the diagram, it is apparent that electron-donating groups widen the energy gap between reactants and products, making ΔG more positive, favoring reactants more than products. In essence, this weakens the acid strength.

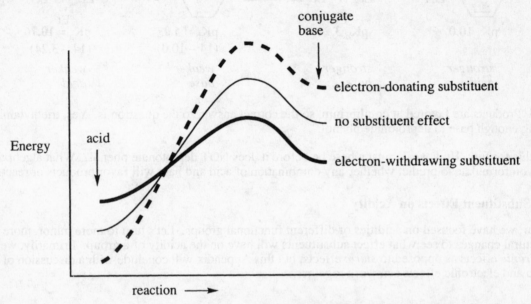

Electron-withdrawing groups destabilize the more positive acid form and stabilize the more negative conjugate base, narrowing the energy gap between reactants and products, making ΔG less positive. Products are increased in concentration at equilibrium which we define as a stronger acid.

Electron-withdrawing groups increase acid strength; electron-donating groups decrease acid strength.

C. Inductive Effects on Acidity

Text section 20-4B gives a thorough explanation of the inductive effect of electron-withdrawing groups on simple carboxylic acids, from which three generalizations arise:

A) Acidity increases with stronger electron-withdrawing groups. (See problem 20-33.)
B) Acidity increases with greater number of electron-withdrawing groups.
C) Acidity increases with closer proximity of the electron-withdrawing group to the acidic group.

We don't usually look to aromatic systems for examples of inductive effects, because the pi system of electrons is ripe for resonance effects. However, in analyzing the resonance forms of phenoxide on the next page, it becomes apparent that the negative charge is never distributed on the meta carbons. Meta substituents cannot exert any resonance stabilization or destabilization; at the meta position, substituents can exert only an inductive effect. The series of phenols demonstrates this phenomenon, consistent with aliphatic carboxylic acids.

Appendix 2 continued, Summary of Acidity and Basicity

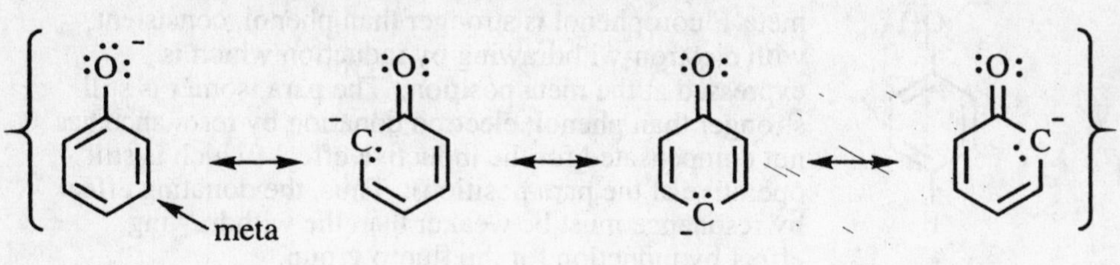

the delocalized negative charge does not increase electron density at meta position

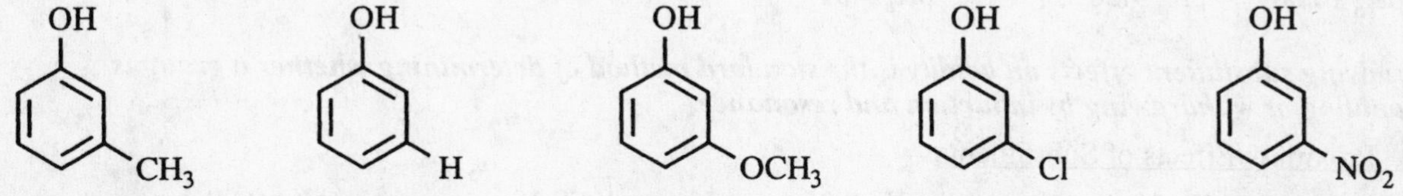

pK$_a$ 10.09
donating

pK$_a$ 10.00

pK$_a$ 9.65
withdrawing

pK$_a$ 9.02
withdrawing

pK$_a$ 8.39
withdrawing

D. Resonance Effects on Acidity—benzoic acids and phenols (review the solution to problem 20-45)

Resonance effects can be expressed with placement of substituents at ortho or para positions, but ortho has the complication of steric effects, so just para substitution is shown here.

Electron-withdrawing substituents increase the acidity of benzoic acids and phenols:

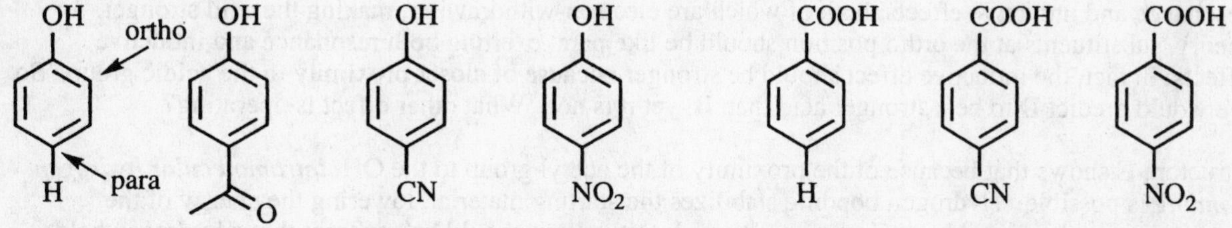

pK$_a$ 10.00

pK$_a$ 8.05

pK$_a$ 7.95

pK$_a$ 7.15

pK$_a$ 4.20

pK$_a$ 3.55

pK$_a$ 3.42

Electron-donating by resonance but electron-withdrawing by induction:

There is a group of substituents that donate by resonance but withdraw by induction: alkoxy groups and halogens are the most notable examples, and the acidity data provide an insight into which effect is stronger.

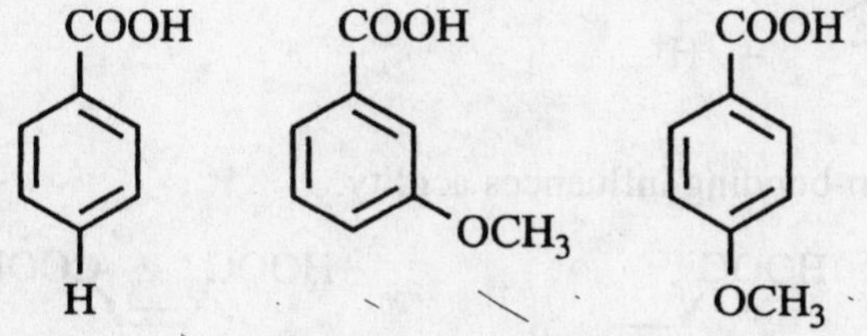

pK$_a$ 4.20

pK$_a$ 4.09

pK$_a$ 4.47

meta-Methoxybenzoic acid is stronger than benzoic acid, consistent with electron-withdrawing by induction which is expressed at the meta position. But the para isomer is *weaker* than benzoic acid; electron donation by resonance has not only compensated for the inductive effect (which is still operative at the para position) but has decreased the acidity even further. Thus, the donating effect by resonance must be stronger than the withdrawing effect by induction for the methoxy group.

See another example on the next page.

697

Appendix 2 continued, Summary of Acidity and Basicity

OH   OH   OH

H   F   F

pK_a 10.00 pK_a 9.28 pK_a 9.81

meta-Fluorophenol is stronger than phenol, consistent with electron-withdrawing by induction which is expressed at the meta position. The para isomer is still stronger than phenol; electron donation by resonance has not compensated for the inductive effect (which is still operative at the para position). Thus, the donating effect by resonance must be weaker than the withdrawing effect by induction for the fluoro group.

Studying substituent effects on acidity is the standard method of determining whether a group is donating or withdrawing by induction and resonance.

E. Proximity Effects of Substituents

pKa 10.0   pKa 8.05   pKa 9.19   pKa 9.90

OH   OH   OH   OH   O

A   **B**   **C**   **D**

Three effects influence the pK_a values of these substituted phenols. In **C**, the acetyl group at the meta position is electron-withdrawing by induction only. In **B**, the acetyl group at the para position exerts both resonance and inductive effects, both of which are electron-withdrawing, making the acid stronger. In theory, substituents at the ortho position should be like para, exerting both resonance and inductive effects; in fact, the inductive effect should be stronger because of closer proximity to the acidic group. So we would predict **D** to be a stronger acid than **B**, yet it is not. What other effect is operating?

Structure **E** shows that because of the proximity of the acetyl group to the OH, *intramolecular hydrogen bonding* is possible. Hydrogen bonding stabilizes the starting material, lowering the energy of the starting material and making ΔG more positive. Intuitively, it should be apparent that a hydrogen held between two oxygens will be more difficult to remove by a base. Also, after the proton has left as shown in structure **F**, the negative charge on the phenolic oxygen is close to the partial negative charge on the oxygen of the carbonyl, destabilizing product **F**, raising its energy, also making ΔG more positive. The proximity of the acetyl group influences both sides of the equation to make the acid weaker.

intramolecular
hydrogen bond

    H O^- $O^{\delta-}$
   O O

             $\rightleftharpoons$ + H$^+$

E   **F**

Here are two more examples where intramolecular hydrogen-bonding influences acidity.

HO—⬡—COOH   (OH)⬡—COOH   HOOC—CH=CH—COOH   HOOC—CH=CH—COOH

pK_1 4.6; pK_2 9.3 pK_1 2.75; pK_2 13.4 pK_1 3.02; pK_2 4.38 pK_1 1.94; pK_2 6.23

Appendix 2 continued, Summary of Acidity and Basicity

F. Steric Inhibition of Resonance

Another type of proximity effect arises when the placement of a substituent interferes with the orbital overlap required for resonance stabilization. This can be seen clearly in the acidity of substituted benzoic acids and in the basicity of substituted anilines.

Let's analyze this series of carboxylic acids.

COOH
H
pK_a 3.75

COOH
pK_a 4.20

COOH
CH_3
pK_a 4.34

COOH
CH_3
pK_a 4.27

COOH
pK_a 3.46

COOH
H_3C CH_3
pK_a 3.21

Formic acid, pK_a 3.75, serves as the reference carboxylic acid. Benzoic acid is weaker because the phenyl group is electron-donating by resonance, stabilizing the protonated form. Methyl substituents are known to be electron-donating by induction, strengthening the electron-donating effect, making the meta- and para-substituted acids weaker than benzoic acid.

Then come the anomalies. Alkyl groups are electron-donating by induction and should weaken the acids, but the ortho-t-butyl and the 2,6-dimethylbenzoic acids are not only stronger than benzoic acid, they are stronger than formic acid! Something has happened to turn the phenyl group into an electron-withdrawing group.

Phenyl is electron-donating by resonance but electron-withdrawing by induction, so what has happened is that the ortho substituents have forced the COOH out of the plane of the benzene ring so that there is no resonance overlap between the benzene ring and the COOH orbitals. The COOH "feels" the benzene ring as simply an inductive substituent. Resonance has been "inhibited" because of the steric effect of the substituent.

COOH group is not parallel with
the plane of the benzene ring—
no resonance interaction.

this three-dimensional view down the C-C bond between the COOH and the benzene ring shows that COOH is twisted out of the benzene plane

The same phenomenon is observed in substituted anilines. Anilines are usually much weaker bases than aliphatic amines because of resonance overlap of the nitrogen's lone pair of electrons with the pi system of benzene. When that resonance is disrupted, the aniline becomes closer in basicity to an aliphatic amine. (See problem 19-49(c).)

pK_b 8.94

$pK_b \approx$ 6-7
(estimated)

More examples on the next page.

699

Appendix 2 continued, Summary of Acidity and Basicity

Examples of steric inhibition of resonance:

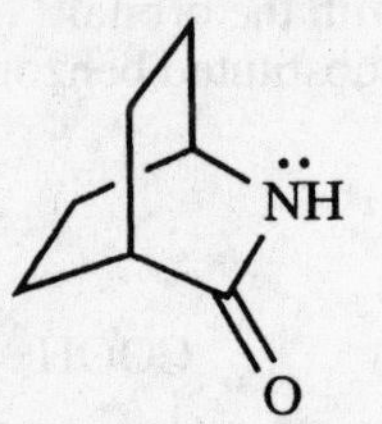

amide—not basic because of
resonance sharing of N lone
pair with carbonyl

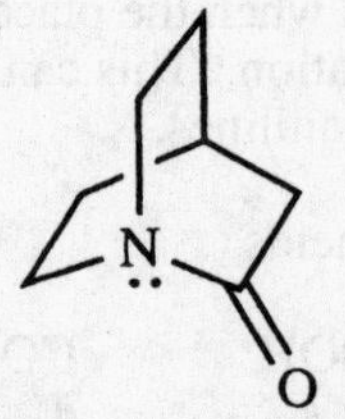

strong base similar to aliphatic amine;
geometry of bridged ring prevents
overlap of N lone pair with carbonyl

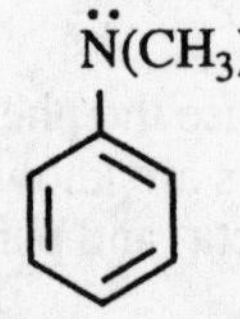

pK$_b$ 8.9

3° aromatic amine

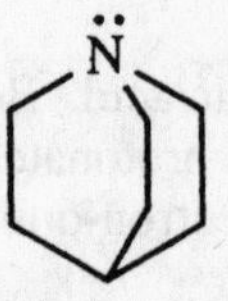

pK$_b$ 3.4

3° aliphatic amine

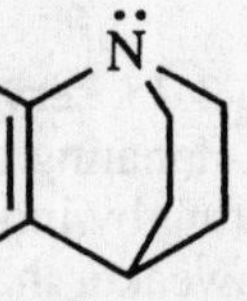

pK$_b$ 6.2

3° amine, and the N is
bonded to a benzene, but the
bridged ring system prevents
overlap of N lone pair with
benzene

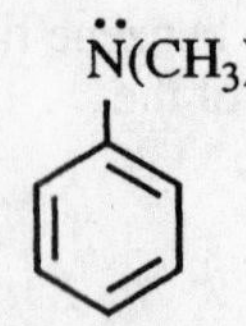

$N(CH_3)_2$

pK$_b$ 8.9

$(CH_2CH_3)_2N:$ $:N(CH_2CH_3)_2$
H_3CO OCH_3

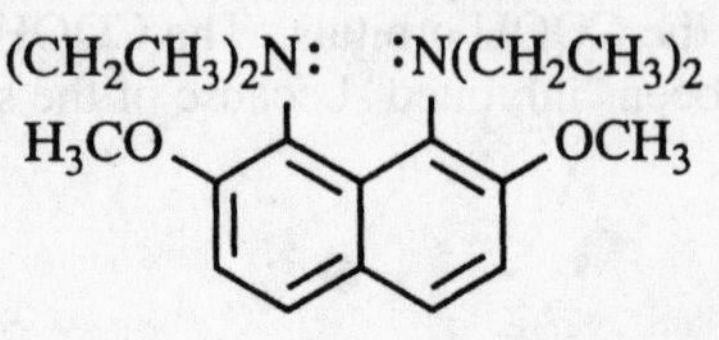

pK$_b$ − 2.3 (yes, negative!)

Not only is steric inhibition of
resonance important in this example,
but so is intramolecular hydrogen-
bonding in the protonated form. Draw
a picture.